Gardener's
Essential
Plant Guide

To my wife Lin for her love and support

Gardener's Essential Plant Guide

BRIAN DAVIS

Quantum
Books

A QUANTUM BOOK

This edition published by Bookmart Ltd,
Blaby Road, Wigston, Leicester, LE18 4SE

This edition printed 2007

ISBN: 978-1-84573-268-4

QUMTGE2

This book is published by
Quantum Publishing Ltd.
6 Blundell Street
London N7 9BH

Printed in Singapore by
Stamford Press (Pte) Ltd.

CONTENTS

Magnolia × *soulangiana* in flower

INTRODUCTION

The vast number of trees, shrubs, and climbing plants commercially available today offers a complexity of choice so wide and varied that choosing a plant for a particular position in the garden is a daunting task for even the most experienced gardener. Many publications fail to give the whole picture, often relying on information from established references that may not be up-to-date. With this in mind, I have compiled *The Gardener's Essential Plant Guide*, which incorporates my experience of running a large commercial nursery over the last quarter of a century and dealing with customers' questions and problems.

I have included over 4,000 varieties of commercially available trees, shrubs, and climbing plants. Every aspect of a plant's performance and needs that is important to the gardener is covered. These elements include height over five, ten, and twenty years, soil, sun/shade and cultivation requirements, history of origin, and so forth.

Unique to this publication are the additional entries of trees and shrubs that can successfully be grown against walls, fences, or other supports. It will be noted that a specific tree or shrub is often dealt with twice within the text. The entry in the "Dictionary of Shrubs and Climbing Plants" deals specifically with the use of the plant against a support. The environment in these areas is different to the open garden, and plants perform differently and require alternative cultivation techniques in this setting.

Whether browsing or using this book to landscape a garden, I hope that within its pages you will find the information and inspiration to carry out your task.

BRIAN DAVIS

HOW TO USE THIS BOOK

Botanical name
The heading for each entry consists of the accepted botanical or Latin name of an individual tree or shrub or of a plant genus at the time of publication. Where there is a synonym in current or recent use, this is bracketed following the main heading.

Common name
There are numerous common names of trees, shrubs, and climbing plants in use, often more than one for a single plant, and these may vary from region to region. Many are attractively descriptive and may help the gardener to identify a familiar tree, shrub, or plant, but they should not be relied upon when ordering from a nursery or mail-order source; only the universally accepted botanical name guarantees accurate identification.

Family name
This is not usually of practical importance to the gardener, but it indicates a family relationship between plants often of very different appearance and may help to identify the cultural requirements of a tree, shrub, or climbing plant, and some aspects of its performance.

Classification
A very wide range of plants can be considered for use as wall shrubs or climbing plants, and this entry records the classification under which the plant is normally found.

Deciduous, evergreen, or semievergreen
An evergreen retains leaves throughout the year; a deciduous plant sheds its leaves in autumn and winter. A tree, shrub, or climbing plant is described as semievergreen if it loses some but not all of its leaves during the colder months. Foliage retention is variable according to climate and conditions. In a mild winter, for example, a normally deciduous shrub may retain some of its foliage, in particularly harsh conditions an evergreen may shed a larger number of leaves than accounted for by normal aging and replacement of foliage. Details of all such variations are given as appropriate in the Varieties/Forms of interest section in each entry.

Special features
Every entry begins with a brief comment providing immediate reference to the most attractive features of the tree, shrub, or plant; its seasonal displays; and special requirements.

Origin
Knowledge of the native environment of a plant can help to explain the conditions required for successful cultivation. This section gives the country or region of natural origin, or indicates that the tree, shrub, or plant has been bred and propagated in cultivation in a garden or nursery environment.

Use
Any information relevant to selecting and planting the tree, shrub, or plant is given under this heading: details of shape, size, seasonal effect, suitable locations (including container planting), and special uses such as wall covering, screening, ground cover, or hedging.

Description
The following details are given under separate subheadings:
Flower Shape, size, color and fragrance, if any. The flowering season stated should be considered in relation to local conditions: for example, an early spring-flowering shrub will come into bloom sooner in a mild location than in an area subject to relatively harsh, extended winters.
Foliage Shape and arrangement on the stem, size, color, and texture, including seasonal varia-

tions such as a distinct change of color in autumn. The Botanical Glossary on page 387 includes illustrations of leaf shapes referred to within the dictionary entries.
Stem Color, texture, and growth habit, including special features such as climbing features, thorns, or attractive bark. The growth rate of the tree, shrub or plant is given in general terms of fast, medium, or slow: slow indicates an increase of no more than 6-8 in (15-20cm) each year; medium an average of 18-24 in (45-60cm) annually; fast 3 ft (1m) or more annually. A very fast growth rate can be up to 6 ft (2m) of additional growth in one year. Individual growth rates and annual increases, cannot be precisely predicted as performance depends on the location, type of soil, climatic conditions, and cultivation methods. The overall shape that the full-grown tree, shrub, or plant should achieve is also described.
Fruit Shape, size, color and specific uses, if any, including fleshy fruits, nuts, seedheads, and seed pods. It is noted whether the fruit is edible or, as in some cases, poisonous. Many of the plant varieties listed bear fruits that are edible as well as ornamental. Trees, shrubs, and climbing plants that are specially bred for their fruiting capabilities and that can be trained on to a support are included as well.

Hardiness
Hardiness is described in terms of the lowest temperature in the range at which the tree, shrub, or plant will be damaged. This allows for the considerable variations in weather conditions and outdoor temperatures that may be found even between areas that are geographically quite close. However, hardiness is also affected by soil type and moisture content, altitude, aspect and cultivation methods, and experience has shown that trees, shrubs, and plants can prove themselves hardy even in a location where they might have been expected to fail, or vice versa. The stated temperature is the most accurate guideline for choosing a tree, shrub or plant suited to your local climate, and details are given of susceptibility to other factors such as wind chill and unexpected late frosts.

Soil requirements
A number of trees, shrubs, and plants grow happily on any soil type, but for others the requirement for acid or alkaline soil is a vital factor in successful cultivation. Acid-loving plants cannot succeed on alkaline soil as they are unable to obtain the necessary nutrients: iron, magnesium, and nitrogen; lack of these elements causes a condition called chlorosis, signified by yellowing of the leaves, which can become terminal. Plants preferring alkaline soil are generally more tolerant of slightly acid types. Apart from these preferences, a tree, shrub, or plant's tolerance of other soil conditions such as drought or excess water is also described.

Sun/shade aspect
Too much or too little shade can cause actual damage to a tree or shrub, or may affect foliage color, or the rate or habit of growth. Shade conditions are indicated in terms of light, medium, or deep shade. Light shade refers to permanent but broken shade, such as underneath other trees or shrub, when some light is available for about two-thirds of the day. Medium shade refers to a deeper degree of shade cast by other plants, or on the sunless side of a building, but with some full light available during the day. Deep shade is that found under a heavy canopy of trees or in the shadow of a tall building with no direct sunlight. Some trees, shrubs, and plants prefer a position in full sun or will tolerate it, while others are likely to suffer sun scorch damage.
Information in this section also refers to the relationship between the planting position, and

movement of the sun, and exposure to atmospheric conditions, indicating whether the tree, shrub, or plant needs a warm, sunny, protected site or will tolerate a cool, more exposed position.

Pruning
Pruning advice describes the most suitable methods of pruning, the time of year when the plant should be trimmed or pruned, and advice on training to shape or to control overall growth. Further details of pruning are provided on page 379.

Training
A number of trees, shrubs, and climbing plants can or must be trained against walls, trellises, or other supports. This section outlines the various methods, techniques, and materials required for the specific plant.

Propagation and nursery production
This section describes the main propagation methods in relation to nursery production, the form in which the plant is found on sale, its general availability, and the best planting heights. The majority of these trees, shrubs, and plants can be obtained through local garden centers and nurseries: where specialist sources are recommended, seek advice from a national or local trade organization, specialist gardening publications, or garden societies. Many suppliers offer a mail-order facility that may be useful in obtaining less common varieties. Recommendation is given on purchasing bare-rooted, root-balled (balled and burlapped) or container-grown plants. These terms are fully explained in the Practical Glossary on page 383 and details of planting methods are shown on pages 376 and 378. The recommended planting heights are selected to encourage establishment and healthy growth; larger or smaller stock may be less successful.

Problems
Susceptibility to disease or damage and known problems of cultivation are described in this section, underlining any potential problems mentioned elsewhere in the text. Special points to note on purchase, planting, and cultivation are highlighted.

Varieties/Forms of interest
Related species, varieties, cultivars, and hybrids are listed under this heading, with full information on the characteristics and appearance of each form, and any variations on the details of soil conditions and cultivation requirements provided in the main entry. Plant names are clearly defined in bold type for easy identification of all the varieties. Descriptions are provided and cover such features as alternative flower colors, attractive leaf shapes and variegation, arching or pendulous stems, ornamental or edible fruits, variations of size and growth rate, special uses, and suitable planting sites.

Average height and spread
The heights and spreads given over five, ten, and twenty years are guidelines to the size of the plant and the appropriate planting area. The actual size of the tree, shrub, or plant will be influenced by local climate, soil conditions, and methods of cultivation.

Illustration
Every entry is illustrated with one or more color photographs showing the best feature of a particular species or variety, and with a line drawing showing the overall shape of the mature plant. Note that line illustrations are not scaled in proportion to each other and should be considered individually in relation to the average heights and spreads given.

Botanical name
A single subject or a group of related forms and varieties.

Family name
The main category of botanical classification.

Special features
Main attractions and cultivation details highlighted for quick reference.

Origin
The native environment of the plant or its origin in cultivation.

Description
Colour, size, shape, seasonal changes and special features of flowers, foliage, stems and fruits.

•ACER CAPPADOCICUM

KNOWN BY BOTANICAL NAME •

•*Aceraceae*

Deciduous •

• Attractive medium to large trees for spring and autumn foliage and colour.

• **Origin** From Caucasus, western Asia through to Himalayas.

Use As a large specimen tree for medium to large gardens. Ideal for estate and park planting and for avenue use when planted at 39ft (12m) apart. Good shade-producing tree.

• **Description** *Flower* Yellow-green, small to medium corymbs, up to 2½in (6cm) long, in late spring, early summer. *Foliage* Leaves palmate, 2-4in (5-10cm) long and wide, with 5-7 lobes. Lobes triangular with graceful, slender points. Glabrous green giving way to yellow in autumn. Some varieties have yellow or red new growth in spring. *Stem* Glabrous green when young, becoming dull grey-green. Upright at first, spreading with age, forming a round-topped tree. Medium rate of growth. *Fruit* Hanging keys or winged fruits 1-2in (3-5cm) wide produced in autumn. Light green when young, aging to yellow and finally brown.

Acer cappadocicum

• **Hardiness** Tolerates winter temperatures down to −13°F (−25°C).

Soil Requirements Any soil conditions. •

• **Sun/Shade aspect** Full sun to medium shade, preferring light shade.

Pruning None required. To reduce size prune only on wood up to three years old.

• **Propagation and nursery production** A. *cappadocicum* from seed, varieties by layers or grafting. Plant bare-rooted or container-grown. Best planting heights 3-10ft (1-3m). Must be sought from good general nurseries or specialist suppliers.

• **Problems** Often planted in areas where it exceeds the allocated space.

Varieties of interest A. c. *'Aureum'* New growth in spring has lime green to golden yellow foliage, contrasting well with darker green leaves on older wood. Spring colouring lost by midsummer, but leaves develop good yellow autumn colour. A. c. *'Rubrum'* New foliage growth in spring wine red, aging to purple-red, contrasting well with dark green leaves on older wood. Spring colouring lost by midsummer, but develops good yellow autumn colour.

• **Average height and spread**
Five years
20x10ft (6x3m)
Ten years
32x20ft (10x6m)
Twenty years
or at maturity
65x39ft (20x12m)

Common name
All variations in common usage are included.

Type
Evergreen, semi-evergreen or deciduous. (Leaves retained all year round, shed partially or fully.)

Use
Details of size, shape, seasonal effect, suitable location and special uses such as hedge-planting or screening.

Colour photograph
For every entry, shows the seasonal feature of primary attraction in a selected named form.

Soil requirements
Particular preferences or dislikes in soil types and conditions.

Pruning
Seasonal pruning requirements and advice on shaping and control.

Varieties of interest
Alphabetical listing of species, cultivars and related forms with description and cultivation notes.

Line drawing
For every entry, shows the growth pattern of trunk or main stems, branches and twigs making the overall shape of the tree or shrub.

Hardiness
Tolerance of outdoor temperature and winter conditions.

Sun/Shade aspect
Light requirements or shade protection according to a given scale from full sun to deep shade.

Propagation and nursery production
Details of the way the plant is grown and presented at purchase; availability; best planting heights.

Problems
Special cultivation requirements, susceptibility to disease or damage etc.

Average height and spread
The proportions of the tree or shrub at five, ten and twenty years after planting.

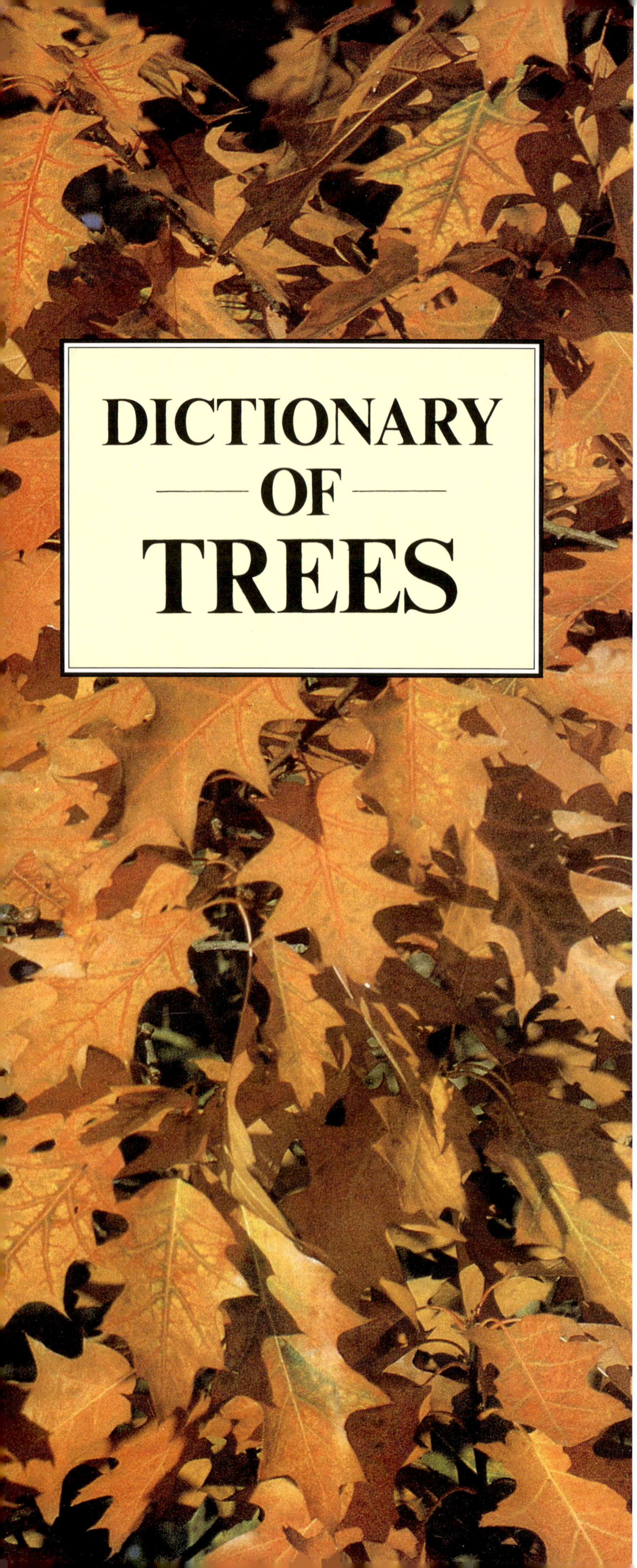

DICTIONARY —OF— TREES

ACACIA Hardy forms

WATTLE
Leguminosae
Evergreen
A most graceful and attractive tree for mild areas with no winter frost.

Origin From Australia.
Use As a quick growing specimen or screening tree. May be grown in a medium to large, frost-free conservatory or greenhouse.

Acacia dealbata in flower

Description *Flower* Yellow flowers in clusters or racemes, produced late winter and through spring. Normally fragrant, depending on variety. *Foliage* Attractive grey-green, cut-leaved foliage 6-10in (15-25cm) long and 6-8in (15-20cm) wide. *Stem* Blue-tinged grey-green, fast growing, upright, spreading with age. *Fruit* Insignificant.
Hardiness Minimum temperature 32°F (0°C); plant out of doors only in frost-free areas. Plants grown under cover may be moved out when all danger of frosts is past, restored to protection in early autumn.
Soil requirements Does best in light, well drained soil. Severe alkaline conditions will cause chlorosis.
Sun/Shade aspect Best in full sun; tolerates light shade.
Pruning None required, but can be cut back moderately hard and quickly rejuvenates.
Propagation and nursery production From seed or from semi-ripe cuttings taken in early spring. Purchase container-grown. Small plants establish better than mature specimens. Relatively easy to find in favourable locations. Best planting height 3-6ft (1-2m).
Problems Rarely survives in non mild areas.
Varieties of interest *A. armata* (Kangaroo Thorn) Profuse yellow flowers over the entire branch area in mid spring. Small prickles and small dissected leaves. Tall, large, bushy habit, reaching tree proportions in very favourable areas. *A. baileyana* (Cootamundra Wattle) Glaucous cut-leaved foliage and producing bright yellow racemes of flowers in late winter, early spring. Two-thirds average height and spread. *A. dealbata* (Silver Wattle) Fern-like, silver-green foliage, producing masses of yellow mimosa-like flowers. One of the hardiest forms. Tolerates 23°F (−5°C) if protected on a sheltered, sunny wall.
These forms are among the most tolerant. Others, equally attractive, require very mild conditions and may fail even in a conservatory if winter temperatures drop below 32°F (0°C).
Average height and spread
Five years
10x6ft (3x2m)
Ten years
20x13ft (6x4m)
Twenty years
or at maturity
32x20ft (10x6m)

ACER CAMPESTRE

HEDGE MAPLE, FIELD MAPLE
Aceraceae
Deciduous
A small, compact rural tree.

Origin From Europe and the British Isles.
Use As a small, single round-topped tree for winter bark and autumn colour. Also good for mass planting. Often grown as a round, short shrub or for hedge planting.
Description *Flower* Small, greenish, insignificant flowers produced only on mature wood. *Foliage* Trifoliate, 2-3in (5-8cm) long, light to mid-green foliage with spectacular yellow autumn colour, sometimes suffused with red. *Stem* Slow-growing to medium vigour, dependent on soil. Upright when young, quickly becoming very twiggy. Bark has corky consistency, especially on wood 2-3 years old or more. Grey-green when young, aging to grey-brown. Forms a round-topped shrub or small tree. *Fruit* Very mature trees may produce small grey to yellow-green, winged fruits.
Hardiness Tolerates −13°F (−25°C).
Soil requirements Any soil conditions. Tolerates a wide range of soil types.
Sun/Shade aspect Full sun to medium shade. May occasionally suffer mild wind or sun scorch to young foliage in late spring.
Pruning None required, but can be reduced in size and will rejuvenate.
Propagation and nursery production From seed. Plant bare-rooted or container-grown. Can be planted young and trained as a tree or bought as an established, single-stemmed specimen. Best planting heights 3-10ft (1-3m). If difficult to find in garden centres may be sought from general nurseries and forestry outlets.
Problems Less at home in urban areas than in rural locations.
Average height and spread
Five years
13x13ft (4x4m)
Ten years
16x16ft (5x5m)
Twenty years
or at maturity
20x20ft (6x6m)

ACER CAPPADOCICUM

KNOWN BY BOTANICAL NAME
Aceraceae
Deciduous
Attractive medium to large trees for spring and autumn foliage colour.

Origin From Caucasus, western Asia through to Himalayas.
Use As a large specimen tree for medium to large gardens. Ideal for estate and park planting and for avenue use when planted at 39ft (12m) apart. Good shade producing tree.

Acer cappadocicum **in autumn**

Description *Flower* Yellow-green, small to medium clusters, up to 2½in (6cm) long, in late spring, early summer. *Foliage* Leaves palmate, 2-4in (5-10cm) long and wide, with 5-7 lobes. Lobes triangular with graceful, slender points. Glabrous green giving way to yellow in autumn. Some varieties have yellow or red new growth in spring. *Stem* Glabrous green when young, becoming dull grey-green. Upright at first, spreading with age, forming a round-topped tree. Medium rate of growth. *Fruit* Hanging keys or winged fruits 1-2in (3-5cm) wide produced in autumn. Light green when young, aging to yellow and finally brown.
Hardiness Tolerates 4°F (−15°C).
Soil Requirements Any soil conditions.
Sun/Shade aspect Full sun to medium shade.
Pruning None required. To reduce size prune only on wood up to three years old.
Propagation and nursery production *A. cappadocicum* from seed, varieties by layers or grafting. Plant bare-rooted or container-grown. Best planting heights 3-10ft (1-3m). Must be sought from good general nurseries or specialist suppliers.
Problems Often planted in areas where it exceeds the allocated space.
Varieties of interest *A. c. 'Aureum'* New growth in spring has lime green to golden yellow foliage, contrasting well with darker green leaves on older wood. Spring colouring lost by midsummer, but leaves develop good yellow autumn colour. *A. c. 'Rubrum'* New foliage growth in spring wine red, aging to purple-red, contrasting well with dark green leaves on older wood. Spring colouring lost by midsummer, but develops good yellow autumn colour.
Average height and spread
Five years
20x10ft (6x3m)
Ten years
32x20ft (10x6m)
Twenty years
or at maturity
65x39ft (20x12m)

ACER GRISEUM

PAPERBARK MAPLE
Aceraceae
Deciduous
A delightful tree, but extremely slow-growing.

Origin Introduced from China into the UK in the very early 1900s, then spread worldwide.
Use As a small ornamental bark tree for winter effect, best planted in isolation to show off its full beauty. If grouped, should be well spaced.

Acer griseum − **winter bark**

Description *Flower* Small, sulphur yellow, hanging flowers in early spring. *Foliage* Olive green to grey-green, 2-3in (5-8cm) long and wide, trifoliate leaves of pendulous habit. Extremely good orange-brown autumn colouring. *Stem* Upright when young. Often loses terminal buds, but re-forms new leader shoots in spring. Branches usually borne quite low on the main stem, 3-4ft (1-1.2m) from ground level. Becomes twiggy and thin in mature growth. Its real beauty is produced on wood three years old or more, brown bark peeling away to show a golden brown underskin. *Fruit* Grey-green, aging to light brown,

Acer campestre **in autumn leaf**

hanging, twin-winged seeds, often produced in large numbers on the mature tree.
Hardiness Tolerates winter temperatures down to −13°F (−25°C), but tips of new shoots vulnerable to winter damage.
Soil Requirements Most soil conditions, but extremely alkaline soils may cause chlorosis.
Sun/Shade aspect Best in full sun, but tolerates light to moderate shade. Deep shade spoils the overall shape.
Pruning None required, although removal of the lower limbs where possible enhances the trunk effect.
Propagation and nursery production From seed, but very variable in its germination. Seed sown in sand-filled trays can take up to three years to germinate. Purchase container-grown up to 3ft (1m) in height. Larger plants may be found and these can be relatively safely lifted if the root ball is maintained, but will command a high price.
Problems Attains the stature of a small tree, but slow growth is characteristic. Plants 3ft (1m) are usually all that can be obtained: the tree takes time to establish itself, but is well worth waiting for.
Average height and spread
Five years
6x4ft (2x1.2m)
Ten years
13x6ft (4x2m)
Twenty years
or at maturity
20x10ft (6x3m)

ACER NEGUNDO

BOX MAPLE, ASH-LEAVED MAPLE
Aceraceae
Deciduous
An extremely useful selection of foliage-attractive Maples, ideal as small, ornamental, bushy trees. Hard pruning each spring greatly enhances the quality and variegation of the foliage.

Origin From North America.
Use As a bush or standard tree for all but the smallest gardens. Extremely attractive when planted 32ft (10m) apart in an avenue of one variety, or for an attractive hedge, bushes planted 4ft (1.2m) apart. Can be used in large containers if offered adequate annual feeding.
Description *Flower* Hanging, sulphur yellow, fluffy flowers in early spring. *Foliage* Light green, 6-8in (15-20cm) wide and long, pinnate leaves with 3-5 and sometimes up to 9 leaflets. Soft texture; slightly pendulous habit. Good yellow autumn colour. *Stem* Light to mid green, with grey-green texture when young. Upright at first, becoming dense with age.

Acer negundo 'Flamingo' in leaf

Liable to mechanical or wind breakage in exposed areas. Medium to fast rate of growth.
Fruit Hanging, winged fruits, grey-green when young, aging to light yellow-brown. On mature trees seed is plentiful.
Hardiness Tolerates 4°F (−15°C).
Soil Requirements Does well on all soil types, unless extremely dry or poor, when it survives but may not thrive. Severely alkaline soils may cause chlorosis.
Sun/Shade aspect Tolerates full sun or light to mid shade.

Acer negundo 'Elegans' in leaf

Pruning None required other than removal of dead or damaged wood, but in early years can be pollarded very hard in early spring to rejuvenate larger, more attractive foliage, especially in variegated forms, although not increasing overall height.
Propagation and nursery production *A. negundo* from seed, layering or root-stooling, but variegated forms grafted or budded except for a few found propagated by cuttings on their own roots. Can be purchased bare-rooted, container-grown or root-balled (balled-and-burlapped). For summer planting only container-grown or established containerized plants should be used. Best planting heights 3-10ft (1-3m).
Problems Any of the variegated forms may revert to producing all-green shoots, which should be removed immediately. Grafted plants may produce suckers in early years, which should be removed.
Varieties of interest *A. n. 'Auratum'* Golden yellow foliage from spring through summer. Susceptible to leaf scorch in very hot weather. Slightly smaller growing than the parent. *A. n. var. californicum* A green-leaved form producing pink, pendulous fruits. Originating in California, it is less hardy than most and scarce in cultivation outside its native environment. *A. n. 'Elegans'* syn. *A. n. 'Elegantissimum'* Bright, yellow-edged, variegated foliage. Slightly less than average height and spread. *A. n. 'Flamingo'* Pale to rosy pink variegated leaves at tips of all new growths from late spring through early summer and often into autumn. Mature leaves variegated white. A new variety responding very well to pollarding. *A. n. 'Variegatum'* syn. *A. n. 'Argenteovariegatum'* Broad, white leaf margins but very likely to revert to green. *A. n. var. violaceum* Young shoots purple to violet and covered with white bloom. Long, hanging, dark pink flower tassels in spring. Good autumn colours. May be scarce in cultivation. Slightly less hardy.
Average height and spread
Five years
13x10ft (4x3m)
Ten years
16x13ft (5x4m)
Twenty years
or at maturity
23x20ft (7x6m)

ACER PLATANOIDES

NORWAY MAPLE
Aceraceae
Deciduous
Large, quick-growing trees, requiring adequate space.

Origin From Europe and Caucasus.
Use For large gardens, estates or parks, and excellent for avenues when planted 39ft (12m) apart. An ideal screening or shade-producing tree.
Description *Flower* Small, green-yellow, erect clusters in mid to late spring. *Foliage* Five-lobed leaves 6-8in (15-20cm) wide and long. Lobes pointed and slightly toothed. Light green, soft-textured. Good yellow autumn colour. *Stem* Light green to grey-green. Upright when young, quickly spreading, forming a large, round-topped, fast-growing tree. *Fruit* Small, attractive hanging keys or winged fruits 1½-2½in (4-6cm) long, in early autumn.
Hardiness Tolerates winter temperatures below −13°F (−25°C).

Acer platanoides in autumn

Soil Requirements Any soil conditions; tolerates high alkalinity.
Sun/Shade aspect Full sun to light shade.
Pruning None required, but stems up to three years old can be reduced as necessary.
Propagation and nursery production *A. platanoides* from seed; named varieties by grafting or budding. Plant bare-rooted or container-grown. Can be planted at heights of 3-20ft (1-6m). Most forms readily available from garden centres and nurseries. Best planting height 10ft (3m).
Problems Often planted in an area where it quickly outgrows the allocated space.
Varieties of interest *A. p. 'Columnare'* Light green foliage. Very upright habit, forming a tall, cigar-shaped tree of slightly less than average height, ultimate spread 13-16ft (4-5m). *A. p. 'Crimson King'* syn. *A. p. 'Goldsworth Purple'* Dark purple foliage spring to autumn: red and orange autumn colours. Purple flowers and purple fruits. Two-thirds average height and spread. *A. p. 'Drummondii'* Attractive light green to grey-green foliage with outer-edge variegation of primrose yellow aging to white. Forms a round-topped tree of two-thirds average height and spread. *A. p. 'Erectum'* Light green foliage, green fruits and flowers. Upright habit. Good yellow autumn colours. Average height and one-third average spread. *A. p. 'Globosum'* Forms a complete and symmetrical globe-shaped tree. Good yellow autumn colours. Green flowers and green fruits. Half average height and spread. *A. p. 'Laciniatum'* Foliage light to mid green, deeply cut. Good yellow autumn colour. *A. p. 'Royal Red'* New foliage on new growth bright wine red. Purple-red

Acer platanoides 'Crimson King' in leaf

foliage on older wood, giving a sumptuous effect. Red and orange autumn colour. Fruits and flowers dark and bright red-purples. *A. p. 'Schwedleri'* New foliage on new growth red to purple-red, contrasting with dark green foliage on older wood. Good yellow to yellow-orange autumn colours. New stems purple-red. Red fruits and flowers.

Acer platanoides 'Drummondii' in leaf

Average height and spread
Five years
16x13ft (5x4m)
Ten years
32x26ft (10x8m)
Twenty years
or at maturity
64x52ft (20x16m)

ACER PSEUDOPLATANUS

PLANE-TREE MAPLE, SYCAMORE, SYCAMORE MAPLE
Aceraceae
Deciduous
Large specimen field trees with attractive spring and autumn foliage.

Origin From Europe.
Use As a screening tree or large individual specimen. Ideal for large avenues when planted at 39ft (12m) apart. Only medium to large estates, gardens or parks can accommodate this tree due to its size.

Description *Flower* Light green, scale-shaped flowers in hanging racemes, early to mid spring. *Foliage* Five-lobed, deeply veined, coarse-surfaced leaves, 4-7in (10-18cm) across. Lobes ovate with coarsely-toothed edges. Grey-green to mid green producing good yellow autumn colours. Some all yellow or variegated forms. *Stem* Grey-green to dark green, becoming green-brown. Upright and strong when young with limited branching, branching and spreading more with age. Forms large, quick-growing, round-topped tree of bold stature. *Fruit* Hanging racemes of winged fruits, 5in (12cm) long, or keys, light green, aging to yellow, finally brown, up to 2½in (6cm) long.
Hardiness Tolerates winter temperatures below −13°F (−25°C). Withstands high winds.
Soil Requirements Any soil type; tolerates high degree of alkalinity.
Sun/Shade aspect Full sun to light shade. Tolerates medium to deep shade but shape may become deformed.
Pruning None required, but branches up to 4 years old may be reduced and normally rejuvenate.
Propagation and nursery production *A. pseudoplatanus* from seed; named varieties grafted or budded. Plant bare-rooted or container-grown. Can be planted at heights of 3-20ft (1-6m). Best planting height 10ft (3m).
Problems Foliage develops a black spotted fungus from midsummer through to early autumn, but this only affects appearance. Sycamore seed is very viable and can become an invasive weed due to rapid germination.
Varieties of interest *A. p. 'Atropurpureum'* (Purple Sycamore, Purple Sycamore Maple) Strong purple-green shoots, supporting purple-green, large foliage with distinct purple undersides. Purple colouring deteriorates through late spring and summer. Good golden yellow autumn foliage. Two-thirds average height and spread. *A. p. 'Leopoldii'* Large, grey-green to mid green foliage, splashed with broad areas of yellow and gold variegation. Yellow autumn colours. Three-quarters average height and spread. *A. p. var. purpureum* Dark purple-green new foliage, purple undersides. Dark yellow autumn tints.

Half average height and spread. *A. p. 'Simon-Louis Freres'* New foliage growth in spring splashed with white and rose pink. Foliage on older wood splashed white, giving very attractive overall appearance. Two-thirds average height and spread. Difficult to find in production, but worth the search. *A. p. 'Worleei'* (Golden Sycamore, Golden Sycamore Maple) Spring foliage lime green to yellow, aging to golden yellow early summer, becoming less distinctive as summer progresses. Good yellow autumn colours. Three-quarters average height and spread.

Acer pseudoplatanus 'Simon-Louis Freres'

Average height and spread
Five years
16x13ft (5x4m)
Ten years
32x26ft (10x8m)
Twenty years
or at maturity
64x52ft (20x16m)

ACER PSEUDOPLATANUS 'BRILLIANTISSIMUM'

SHRIMP-LEAVED MAPLE
Aceraceae
Deciduous
An extremely attractive mop-headed tree with spring foliage colour.

Origin Of garden origin.
Use A small mop-headed tree useful for all sizes of garden. A satisfyingly round, architectural shape.
Description *Flower* Hanging clusters of green-red scale flowers in early to mid spring. *Foliage* Five-lobed, 4-7in (10-18cm) across. Lobes ovate with slightly toothed edges. New foliage deep shrimp pink, becoming flesh pink and finally green in summer. Good yellow autumn colours. *Stem* Light green, becoming green-brown. Stout, short, branching and slow-growing. *Fruit* Keys or winged fruits up to 1in (3cm) across. Light green, becoming green-brown.
Hardiness Tolerates winter temperatures down to −13°F (−25°C).
Soil Requirements Does well on all soil types, tolerating high alkalinity.
Sun/Shade aspect Best in full sun, but tolerates very light shade.
Pruning None required.
Propagation and nursery production Grafted on to stems of *A. pseudoplatanus*. Plant bare-rooted or container-grown. Plants budded either at ground level to form large bushes or at 5ft (1.5m) or 6ft (2m) height from ground-level for standard forms.
Problems Very slow-growing when compared with other *A. pseudoplatanus* forms, but useful for smaller gardens. Young trees often poorly shaped due to grafting and need at least three years to develop the mop-head form. Dead and damaged twigs may be attacked by coral spot. They should be re-

Acer pseudoplatanus 'Leopoldii' in spring

Acer pseudoplatanus 'Brilliantissimum'
in spring

moved, cuts treated with pruning compound.
Varieties of interest *A. p. 'Prinz Handjery'*
New foliage splashed shrimp pink with green
base colour, aging to green, with yellow
autumn colours. A large shrub or mop-
headed tree, reaching 13ft (4m) in height and
spread. Grafted on to stems of *A. pseudoplatanus*. Should be sought from specialist
nurseries.
Average height and spread
Five years
8x5ft (2.5x1.5m)
Ten years
12x8ft (3.5x2.5m)
Twenty years
or at maturity
14½x12ft (4.5x3.5m)

ACER RUBRUM

RED MAPLE
Aceraceae
Deciduous
Large trees with good autumn colour.

Origin From North America.
Use As a large tree for large gardens, estates
and parkland, either single or in groups. Can
be used for avenue planting at 39ft (12m)
apart.
Description *Flower* Hanging, thick clusters of
attractive red scale flowers in early to mid
spring. *Foliage* Leaves triangular, 3 or 5
lobed, 6-8in (15-20cm) long and wide, with
coarsely toothed edges. Dark green upper

surfaces, blue-white undersides. Brilliant
scarlet and yellow autumn colour in its native
environment, but somewhat unreliable in
other locations. *Stem* Relatively quick-
growing, upright when young, spreading with
age. Shoots grey-green to glabrous. *Fruit*
Keys or winged fruit, up to 1in (3cm) long,
dark or dull red.
Hardiness Tolerates −13°F (−25°C).
Soil Requirements Dislikes alkaline soils;
does well only on neutral to acid types.
Sun/Shade aspect Tolerates light shade but
prefers full sun.
Pruning None required, but obstructive lower
limbs can be removed. This makes a more
attractive trunk for the mature tree.
Propagation and nursery production *A. rubrum* from seed. Named varieties grafted on to
understock of *A. rubrum*. Can be planted
bare-rooted or container-grown. Planting
height can be 3-16ft (1-5m), best at 10ft (3m).
Problems Plants imported into regions other
than the native environment may make a
disappointing show of the autumn colour.
Varieties of interest *A. r. 'Scanlon'* Very scarce
but not impossible to find. Said to be more
reliable in its autumn colour. Slightly less than
average height and spread. *A. r. 'Schlesingeri'*
Also said to have more reliable autumn
colour and difficult but not impossible to find.
Slightly less than average height and spread.
Average height and spread
Five years
16x10ft (5x3m)
Ten years
32x20ft (10x6m)
Twenty years
or at maturity
64x48ft (20x15m)

ACER
Interesting foliage varieties

NAME DEPENDENT ON VARIETY
Aceraceae
Deciduous
Interesting leaf formations in trees for
larger gardens.

Origin Dependent on variety.
Use As ornamental trees with interesting
foliage, planted singly or in groups for
medium-size and larger gardens.
Description *Flower* Green or red scale-shaped
flowers in hanging clusters in early to mid
spring. Details dependent on variety. *Foliage*
Leaves light green, oblong to ovate or 3-5
lobed. Good autumn colour in some varieties.
Stem Grey-green. Upright when young, becoming branching in most cases. Medium to
fast rate of growth. *Fruit* Hanging small keys
or winged fruits in autumn. Green or red
depending on variety.

Hardiness Tolerates winter temperatures
down to −13°F (−25°C).
Soil Requirements Any soil conditions.
Sun/Shade aspect Full sun to light shade,
preferring full sun.
Pruning None required, but obstructive branches can be removed.
Propagation and nursery production Normally from seed, with some forms grafted. Most
forms must be sought from specialist nurseries. Plant bare-rooted. Best planting heights
3-10ft (1-3m).
Problems Often only young plants are
obtainable, requiring patience before the full
foliage effect appears.

Acer saccharinum in autumn

Varieties of interest *A. carpinifolium* (Hornbeam Maple) Shoots upright, glabrous to
grey-green. Foliage oblong, 3-4in (8-10cm)
long. Silky, hairy undersides when young and
double-toothed edges. Flowers green, hanging on slender stalks which are 1in (3cm) long,
produced in mid spring. Average height but
slightly less spread. From Japan. *A. crataegifolium* (Thorn-Leaved Maple) Young shoots
purple to glabrous green with white-striped
veins. Forms a round-topped tree. Leaves
ovate, 2-4in (5-10cm) long, with 3-5 lobes.
Some yellow autumn colour. Flower clusters
green-yellow, in late spring. Hanging keys or
winged fruits. Slightly less than average
height, but of equal spread. From Japan. *A.
nikoense* (Nikko Maple) Hairy, grey-green
shoots. Soft-textured trifoliate leaves on hairy
stalks, middle leaflet oval and 3-5in (8-12cm)
long, stalkless side leaflets small, ovate, with
shallow toothed edges. Rich red-orange in
colour in autumn. Yellow flowers hanging in
threes, produced in mid spring. Keys or
hanging winged fruits in autumn. One-third
more than average height and spread. From
Japan and China. *A. opalus* (Italian Maple)
Leaves 2-4in (5-10cm) wide, 5-lobed with
toothed edges, irregular, dark green to green-
blue upper surfaces, paler undersides. Some
yellow autumn colour. Yellow flowers in
short, hanging clusters in early spring, each
individual flower on a slender, glabrous,
green stalk. Fruits are yellow-green keys
produced in autumn. One-third more than
average height and spread. From Central and
Southern Europe. *A. saccharinum* syn. *A.
dasycarpum* (Silver Maple) Large, palmate
leaves, light grey-green when young, aging to
light or mid green, silver undersides. Good
yellow autumn colour. One-third more than
average height and spread, a quick-growing,
round-topped tree. *A. s. laciniatum* Deeply
lobed, palmate leaves wih broad lacerations.
Otherwise like the parent but of average
height and spread. *A. s. 'Lutescens'* New
spring foliage has lime green upper surfaces,
silver undersides, aging to pale green in
midsummer. Good autumn colours. Leaves
palmate and lobed. *A. s. 'Pyramidale'* Large,
palmate leaves, some dissection. Light green
to grey-green with silver undersides. Good
yellow autumn colour. Upright habit,

Acer rubrum in autumn

reaching one-third more than average height and 13ft (4m) in width over 25 years. *A. saccharum* (Sugar Maple) Glabrous green shoots. Leaves 4-6in (10-15cm) long, wih 3-5 lobes and coarsely toothed edges. Medium to dark green with grey undersides. Hanging, green-yellow flowers presented on thin thread-like stalks up to 3in (8cm) long. Glabrous keys or winged fruits up to 2in (5cm) long in early autumn. Highly valued as shade tree for handsome form, dense foliage and fine autumn colour. A large tree, ultimately reaching up to 80ft (25m) height and 64ft (20m) spread. Used for maple syrup and sugar production in north-eastern USA. *A. s. 'Bonfire'* Good red autumn colour. *A. s. 'Globosum'* Rounded form with fine yellow autumn colour. *A. s. 'Green Mountain'* A heat-tolerant variety with leathery, dark green foliage. *A. s. 'Monumentale'* A broad columnar form with yellow-orange autumn colour. These cultivars are widely available only in their native environment. *A. triflorum* Leaves trifoliate, up to 3½in (9cm) long, ovate to oblong with limited marginal toothing. Dark green upper surfaces, light downy undersides, especially when young. Good orange-yellow autumn colour. Greenish-yellow flowers produced in threes in early spring. Keys or winged fruit 1-1¼in (3-3.5cm) long, producing a hairy central nut. Slender, branching habit. From Manchuria and Korea.

Average height and spread
Five years
13x10ft (4x3m)
Ten years
16x13ft (5x4m)
Twenty years
or at maturity
23x20ft (7x6m)

ACER Snake bark varieties

SNAKE-BARKED MAPLES
Aceraceae
Deciduous
Good autumn foliage colour and attractive winter colour in trunks and stems.

Origin From China or Japan, depending on variety.
Use As individual specimen trees for winter stem effect or for group planting. Interesting grown on a trunk or as lower branching shrub. Suits all but the smallest gardens.
Description *Flower* Short racemes of green to yellow-green, hanging flowers in mid spring. *Foliage* Ovate, tooth-edged leaves 2-4in (5-10cm) long and wide, normally grey-green to mid-green, producing an excellent display of orange, yellow and bronze autumn colours. *Stem* Upright when young. Depending on variety, streaked with purple or green, and white veined on two-year-old or more wood, gaining attractiveness with increasing girth. Bright red or green buds enhance the winter effect. Forms a small tree with round or spreading habit. Medium rate of growth. *Fruit* Hanging bright red or red-purple, winged fruit in autumn, green in some varieties, aging to grey-brown.

Acer capillipes in autumn

Hardiness Tolerates 4°F (−15°C).
Soil Requirements Tolerates most soil conditions, but may show decreased growth on very alkaline or dry soil.
Sun/Shade aspect Full sun to light shade.
Pruning None required other than removal of damaged or badly crossing branches. Young shoots can be pruned to control overall size. Reducing size of mature tree is extremely difficult and not advised.
Propagation and nursery production From seed, layers or grafting, depending on variety. Obtain container-grown, bare-rooted or root-balled (balled-and-burlapped). Most forms readily available from general and specialist nurseries. Best planting heights 3-10ft (1-3m).
Problems Relatively slow to mature. Often offered at no more than 3ft (1m) in height, smaller than recognized size for ornamental trees. Full snake-bark effect is achieved after 5 years.

Acer hersii — winter branches

Varieties of interest *A. capillipes* Bark purple-red to coral-red when young, aging to purple with white veining. Good autumn colour. From Japan. *A. davidii* Glossy, purple-green bark with white stripes. Foliage dark green with very good autumn colours. Profuse autumn fruit production of green winged seeds, often tinged with red. Slightly less than average height and spread. From Central China. *A. d. 'Ernest Wilson'* Strong stem colour. Pink leaf petioles supporting ovate, pale green leaves. More compact than most and of slightly less than average height and spread. Somewhat difficult to find in commerical production. From the Yunnan Valley in China. *A. d. 'George Forrest'* A spreading variety with white-veined purple stems. Bright purple-red new growth in spring. Slightly less than average height but of equal spread. From the Yunnan in China. *A. grosseri* Seldom found in commercial production and may be sold as *A. grosseri var. hersii*. Light to mid green, ovate foliage changes to good autumn colour. Purple-green stems with white veining. Slightly less than average height and spread. From central China. *A. hersii* syn. *A. grosseri var. hersii* Green to grey-green, marbled bark. Ovate, mid-green foliage of good size, producing vivid autumn colours. Winged fruits produced in long racemes, light green, aging to yellow-green. From Central China. *A. pennsylvanicum* (Moosewood) Bright green young shoots, aging to grey-green with white-striped veining. Three-lobed foliage up to 6-8in (15-20cm) across; bright yellow autumn colour. Dislikes very alkaline soils. From eastern North America. *A. p. 'Erythrocladum'* Young shoots shrimp pink, aging to purple-green with white stripes. Growth production may be poor. A variey for use only as a large shrub. *A. rufinerve* Stems green-grey, almost glaucous. Mature wood has good white veining. Three-lobed leaves bright green with grey sheen, producing good red and yellow autumn colours. Two-thirds average height and spread. From Japan. *A. r. albolimbatum* New foliage in spring splashed irregularly with white

variegation, maintained to midsummer. Good autumn colour. Green stems with white veining. Otherwise similar to *A. rufinerve* but scarce in commercial production and must be sought from specialist nurseries.
Average height and spread
Five years
10x5ft (3x1.5m)
Ten years
16x10ft (5x3m)
Twenty years
or at maturity
23x16ft (7x5m)

AESCULUS × CARNEA

RED HORSE CHESTNUT
Hippocastanaceae
Deciduous
A truly magnificent tree when in full flower; it requires a good-sized planting space.

Origin Of garden origin, a cross between *A. hippocastanum* and *A. pavia*.
Use As a medium to large tree, useful for field planting, or in an avenue when planted 39ft (12m) apart. Can be used as an individual shade tree for the larger garden.
Description *Flower* Upright panicles of rose pink flowers, 8in (20cm) in length, late spring to early summer. *Foliage* Large compound palmate leaves, 5-12in (12-30cm) long, dark to olive green upper sides, dull silver to silver-grey undersides. Attractive yellow autumn colour. *Stem* Grey-green, smooth, strong, branching stems with green-brown terminal and axillary buds of sticky or resinous texture. *Fruit* Large grey-green fruits containing two brown seeds.
Hardiness Tolerates −13°F (−25°C).
Soil Requirements Most soil conditions, except waterlogged.
Sun/Shade aspect Best in full sun, but tolerates considerable shade.
Pruning None required but lower limbs may be pruned to restrict size.
Propagation and nursery production *A. × carnea* is grown from seed, *A. × c. 'Briotii'* budded or grafted. Plant bare-rooted or container-grown. Best planting heights 6-10ft (2-3m).
Problems Considerable height and spread, forming a dense canopy under which no other plant can grow.

Aesculus × carnea 'Briotii' in flower

Varieties of interest *A. × c. 'Briotii'* Very attractive large, dark red-pink flowers, slightly less height and spread than *A. × carnea*. *A. × c. 'Plantierensis'* Pink flowers late spring to early summer. Does not fruit.
Average height and spread
Five years
10x10ft (3x3m)
Ten years
16x16ft (5x5m)
Twenty years
or at maturity
39x39ft (12x12m)

15

AESCULUS FLAVA
(Aesculus octandra)

SWEET BUCKEYE, YELLOW BUCKEYE
Hippocastanaceae
Deciduous
An attractive, large, yellow-flowering tree suited to larger gardens.

Origin Native to North America.
Use As a specimen tree for large gardens, parks, estates or for avenue planting, when trees should be 32ft (10m) apart.
Description *Flower* Yellow to pale yellow florets, ¾in (2cm) long, formed in upright panicles 6-7in (15-18cm) long and 2-3in (5-8cm) wide in late spring and early summer.

Aesculus flava in flower

Foliage Compound palmate, five, sometimes seven, ovate, light to mid green leaves, slightly incurved, 3-7in (7-18cm) long and 1-3in (3-8cm) wide, with fine toothed edges. Downy undersides when young. Yellow autumn colours. *Stem* Young stems brown-green. Large, light green-brown winter buds. Medium rate of growth. Forms a medium to large, round-topped tree. *Fruit* Rounded fruit capsules 2-3in (5-8cm) long, light green aging to brown.
Hardiness Tolerates −13°F (−25°C).
Soil Requirements Most soil types, but distressed by extremely alkaline or dry conditions.
Sun/Shade aspect Best in full sun to light shade, but tolerates slightly deeper shade.
Pruning None required. Limbs causing obstruction can be removed.
Propagation and nursery production From seed. Not common in commercial production outside its native environment. Planting heights range from 3-16ft (1-5m), best at 6-10ft (2-3m). Plant bare-rooted or container-grown.
Problems Final size often underestimated. May suffer from coral spot; affected branches should be removed and cuts painted with pruning compound.
Average height and spread
Five years
16x10ft (5x3m)
Ten years
32x20ft (10x6m)
Twenty years
or at maturity
64x64ft (20x20m)

AESCULUS HIPPOCASTANUM

HORSE CHESTNUT, CONKER TREE
Hippocastanaceae
Deciduous
Large, statuesque, bold trees for spacious locations.

Origin From northern Greece and Albania.
Use As a large specimen tree for large gardens, parks, estates or for avenue planting at 39ft (12m) apart.
Description *Flower* White florets, 1in (3cm) long, red-patched at the base of each petal, produced in large conical panicles 8-12in (20-30cm) long late spring, early summer. Extremely bold and attractive. *Foliage* Mid to dark green compound palmate leaves, with 5-7 ovate, toothed leaflets 5-12in (12-30cm) long and 2-5in (5-12cm) wide. Some yellow autumn colour, but often dies untidily. *Stem* Fast-growing with strong, rigidly branching, grey-brown to grey-green shoots, upright when young, quickly spreading. Winter buds very sticky. Winter branches clearly show horseshoe-shaped scars at each leaf axil formed by the previous season's leaf form; hence the common name Horse Chestnut. *Fruit* Globe-shaped, 2-2½in (5-6cm) wide with distinct prickly covering containing one large, shiny brown seed, produced late summer, early autumn.
Hardiness Tolerates −13°F (−25°C).

Aesculus hippocastanum in flower

Soil Requirements Does well on all soil types; only dislikes extremely wet conditions.
Sun/Shade aspect Good from medium shade through to full sun.
Pruning None required, though obstructing branches may be removed. Will take severe pollarding, but remains misshapen for a number of years afterwards.
Propagation and nursery production From seed, or by grafting or budding for named varieties. Can be purchased and planted from 3-16ft (1-5m). Plant bare-rooted or container-grown. Trees relatively common in general nurseries; seek named varieties from specialist outlets.
Problems Ultimate size is often underestimated. Seeds readily, especially mature trees.

Once mature no underplanting is normally possible due to the heavy, dense leaf canopy. May suffer from coral spot. May lose large limbs in extreme gale-force winds.
Varieties of interest *A. h. 'Baumannii'* syn. *A. h. 'Flore Pleno'* A variety with large double white flowers which does not set seed.
Average height and spread
Five years
20x16ft (6x5m)
Ten years
39x32ft (12x10m)
Twenty years
or at maturity
80x48ft (25x15m)

AESCULUS INDICA

INDIAN CHESTNUT
Hippocastanaceae
Deciduous
An aristocrat among flowering trees for the larger garden, park or estate.

Origin From northern India.
Use As a large specimen tree or for avenue planting at 39ft (12m) apart. Suited to large gardens, parks or estates.
Description *Flower* Upright panicles of white florets blotched with yellow and red and tinged pale rose-pink, 8-12in (20-30cm) long and 4in (10cm) wide, produced early to mid summer; later-flowering than most other forms of *Aesculus*.
Foliage Leaves ovate to lanceolate, light green to green with orange hue, produced in sevens up to 12in (30cm) long and 4in (10cm) wide. Compound palmate form decreasing in size on more mature trees. Good salmon to orange-red autumn colours. *Stem* Grey-green to brown-green, branching, stout. Buds brown-grey, non-sticky, attractive in winter. Slow to medium growth rate. *Fruit* Spineless green-brown round fruits 2-2½in (5-6cm) across.
Hardiness Tolerates 14°F (−10°C).
Soil Requirements All soil conditions, except extremely wet.
Sun/Shade aspect Full sun to light shade.
Pruning None required, but branches causing obstruction can be reduced.
Propagation and nursery production From seed. Plant bare-rooted or container-grown. Normally 1½-6ft (50cm-2m) in height when

Aesculus indica in flower

purchased. Larger trees may be obtainable from specialist nurseries, but sometimes difficult to find due to a season's poor seed germination.
Problems Trees are purchased young, and take time to achieve full dimensions. May suffer from attacks of coral spot: remove affected branches and treat cuts with a pruning compound.
Average height and spread
Five years
13x10ft (4x3m)
Ten years
26x16ft (8x5m)
Twenty years
or at maturity
52x32ft (16x10m)

AESCULUS PAVIA

RED BUCKEYE
Hippocastanaceae
Deciduous
A very attractive small tree or large shrub with interesting flowers.

Origin From North America.
Use As a small, compact, flowering tree or large shrub for medium to large gardens.
Description *Flower* Bright red florets 1½in (4cm) long with the interesting characteristic of opening rarely, and then only slightly. Produced in upright panicles 2½-6in (6-15cm) long, in early summer. *Foliage* Five mid to dark green ovate leaflets, 2-5½in (5-13cm), downy undersides when young, forming a compound palmate leaf, irregular and doubly tooth-edged. Some yellow autumn colour. *Stem* Grey-green to grey-brown. Stout, short, branching. Slow-growing, forming a large round-topped shrub or small tree. *Fruit* Round to oval, smooth-skinned, green fruits, aging to brown-red.
Hardiness Tolerates −13°F (−25°C).
Soil Requirements Any soil conditions.
Sun/Shade aspect Full sun to light shade.
Pruning Rarely required, but limbs may be removed as necessary.
Propagation and nursery production From seed. Best purchased container-grown. Relatively difficult to find; must be sought from specialist nurseries. Usually available at no more than 1½ft (50cm) in height.
Problems May suffer from attacks of coral spot.
Average height and spread
Five years
5x5ft (1.5x1.5m)
Ten years
8x8ft (2.5x2.5m)
Twenty years
or at maturity
23x12ft (7x3.5m)

Aesculus pavia **in flower**

AILANTHUS ALTISSIMA
(Ailanthus glandulosa)

TREE OF HEAVEN
Simaroubaceae
Deciduous
A fast-maturing tree with interesting large foliage.

Origin From China.
Use As a lawn specimen for medium-sized or larger gardens.
Description *Flower* Green-white flowers with pungent odour, produced in terminal panicles mid to late summer. *Foliage* Leaves large, compound, pinnate, with 13-25 leaflets along a central stem which can be 1-2ft (30-60cm) long. Light green to grey-green; attractive brown-red when young. Some yellow autumn colour. *Stem* Grey-green to green-brown. Strong with very limited branching when young, more with age. Very quick-growing when young. Stems subject to wind damage, breaking under gale-force conditions. *Fruit* Oblong, green to green-red fruits sometimes produced in autumn following a hot dry summer or in a warm climate.

Ailanthus altissima **in spring**

Hardiness Tolerates 14°F (−10°C). Some terminal bud or stem damage experienced annually. Die-back is normal, and encourages branching habit.
Soil Requirements Most soil conditions, but moist, deep soil helps quick growth production.
Sun/Shade aspect Full sun to light shade.
Pruning None required, but remove any dead tips within reach.
Propagation and nursery production From seed, rooted cuttings or suckers. Can be purchased 3-10ft (1-3m) high, bare-rooted or container-grown. Relatively easy to find from general nurseries.
Problems Young trees rarely form any substantial branching head. Nurseries may offer a single shoot 6-10ft (2-3m) high with no side branches, but this rapidly develops once planted.
Varieties of interest *A. a.* '*Erythrocarpa*' Dark green leaves and red fruit. *A. a.* '*Pendulifolia*' Similar to the parent, but with longer leaves hanging down from stems. Both these forms are cultivars not always readily available.
Average height and spread
Five years
23x13ft (7x4m)
Ten years
46x26ft (14x8m)
Twenty years
or at maturity
69x39ft (21x12m)

ALBIZIA JULIBRISSIN

PINK MIMOSA, PINK SIRIS, SILK TREE
Leguminosae
Deciduous
Only for frost-free areas, but included here both for mild areas and for conservatory growing.

Origin From Iran to China and Taiwan.
Use As a small to medium tree of spreading habit for areas with very mild winter temperatures. Otherwise used as a large container-planted conservatory shrub for its attractive foliage.
Description *Flower* Terminal clusters of flowerheads on long stalks, producing a mop-like cluster of pink stamens up to 1in (3cm) across in mid to late summer. Dense all-over flowering makes a spectacular effect. *Foliage* Bipinnate leaves, 9-18in (22-45cm) long. Each of the 6-12 branches of the pinnate leaf carries 20-30 pairs of small oblong, grey-green leaflets. *Stem* Grey-green, upright when young, quickly spreading. Medium rate of growth, slower in container. *Fruit* Insignificant.
Hardiness Can only flourish in a completely frost-free area.
Soil Requirements Any soil types.
Sun/Shade aspect Full sun to very light shade.
Pruning None required; may even resent it.
Propagation and nursery production From seed. Purchase container-grown. Normally supplied 2-5ft (60cm-1.5m) in height. Relatively easy to find in suitable planting areas; otherwise only from specialist nurseries.
Problems Often chosen because seen growing well in warm climates, but not adaptable to harsher conditions.
Varieties of interest *A. j.* '*Rosea*' A smaller variety with bright pink flowers, reaching two-thirds average height and spread.
Average height and spread
Five years
10x32ft (3x10m)
Ten years
23x23ft (7x7m)
Twenty years
or at maturity
32x32ft (10x10m)
Heights and spreads will be greatly reduced when grown in containers.

Albizia julibrissin **in leaf**

ALNUS CORDATA

ITALIAN ALDER
Betulaceae
Deciduous
A useful tree for difficult, wet conditions.

Origin From Corsica and southern Italy.
Use As an individual specimen tree or for small group planting. Good avenue tree when planted 26ft (8m) apart.
Description *Flower* Male catkins, brown when young, forming in early winter and opening in early spring to yellow, 2-2½in (5-6cm) long, produced in threes and sixes. *Foliage* Leaves broad, ovate, deeply veined, finely toothed-edged, 2-3in (5-8cm) long. Dark green with shiny, slightly glabrous upper surfaces and light grey-green undersides. Some yellow autumn colour. *Stem* Dark green with brown tinge. Strong, upright, quick-growing, forming a pyramidal tree. *Fruit* Oval, upright, erect, dark brown cones, ½-1in (1-3cm) long, normally produced in threes in autumn, retained through winter into early spring.
Hardiness Tolerates winter temperatures below −13°F (−25°C).
Soil Requirements Does particularly well on moist soil; dislikes dry conditions.
Sun/Shade aspect Full sun to medium shade.
Pruning None required, but lower branches can be removed.
Propagation and nursery production From seed. Can be purchased from 1½ft (50cm) up to 13ft (4m) in height, bare-rooted or container-grown. Best planting heights 5½-6ft (1.8-2m). Relatively easy to find from general nurseries or from forestry outlets.
Problems May suffer stunted growth in very dry soil conditions.
Varieties of interest *A. rubra* Although not related to *A. cordata*, of similar appearance, reaching slightly less than average height, often with a pronounced pyramidal habit. New shoots red. Leaves more rounded with coarsely double-toothed edges. Catkins larger and longer. From western North America.
Average height and spread
Five years
20x8ft (6x2.5m)
Ten years
39x16ft (12x5m)
Twenty years
or at maturity
48x23ft (15x7m)

Alnus cordata in catkin

ALNUS GLUTINOSA

COMMON ALDER, BLACK ALDER
Betulaceae
Deciduous
Attractive, quick-growing trees for wet areas.

Origin From Europe, Asia Minor and North Africa.
Use As an individual specimen tree, especially in the case of ornamental varieties, or for group planting.

Alnus glutinosa 'Imperialis' in leaf

Description *Flower* Dark brown male catkins formed in early winter, opening to yellow late winter, early spring. Produced 3-5 per cluster and 2-4 in (5-10cm) long. *Foliage* Ovate to broad ovate leaves with coarsely toothed edges. Dark green shiny upper surface, dull undersides. Produced on leafstalks up to 1in (3cm) long. Good yellow autumn colour. *Stem* Light to mid green when young, aging to dark green, finally to green-purple. Upright pyramidal habit. Fast-growing. *Fruit* Small, oval, cone-like fruit ½in (1cm) long; produced in autumn, retained through winter into early spring.
Hardiness Tolerates winter temperatures below −13°F (−25°C).
Soil Requirements Especially tolerant of wet conditions; dislikes very dry soil.
Sun/Shade aspect Full sun to mid shade, with light shade for preference.
Pruning None required but lower branches can be removed.
Propagation and nursery production From seed or layering. Can be purchased from 1½ft (50cm) up to 13ft (4m). Plant bare-rooted or container-grown. Normally readily available from general nurseries or forestry outlets.
Problems None.
Varieties of interest *A. g. 'Aurea'* Bright golden yellow foliage, colour only good in spring. Reddish stems when young. Leaf petioles red. Somewhat susceptible to scorch in strong midday summer sun. Requires light dappled shade for best results. Two-thirds average height and spread. *A. g. 'Imperialis'* Deeply cut, almost lacerated foliage. Attractive light green. Difficult to find. Two-thirds average height and spread. *A. g. var. incisa* (Thorn-leaved Alder) Leaves smaller than average with a deeply toothed, lobed effect. Difficult to find.
Average height and spread
Five years
22x8ft (6x2.5m)
Ten years
33x16ft (10x5m)
Twenty years
or at maturity
40x23ft (12x7m)

ALNUS INCANA

GREY ALDER
Betulaceae
Deciduous
Attractive trees for difficult, wet areas.

Origin From Europe and North America.
Use As a freestanding tree for spring catkins and interesting fruit or for mass planting to achieve a screen or windbreak.
Description *Flower* Male catkins, 2-4in (5-10cm) produced in groups of three or four. Catkins first produced in autumn, brown and scaly, opening to yellow in early spring. *Foliage* Dark green upper surface with grey, downy underside. Ovate to oval, 2-4in (5-10cm) long, with tapered points and coarse tooth-edges. Some yellow autumn colour. *Stem* Grey-green when young, quickly becomes purple-green and finally brown-green. Upright, fast-growing, pyramidal habit. *Fruit* Small, brown, scale-like cones produced early autumn and maintained until following spring.
Hardiness Tolerates 4°F (−15°C).
Soil Requirements Prefers moist soils but tolerates a wide range of soil types.

Alnus incana 'Pendula' in leaf

Sun/Shade aspect Full sun to light shade, preferring light shade.
Pruning None required but lower branches can be removed.
Propagation and nursery production From seed or layering. Can be purchased from 1½ft (50cm) up to 13ft (4m). Plant bare-rooted or container-grown. Relatively easy to find in general nurseries.
Problems None.
Varieties of interest *A. i. 'Aurea'* Golden shoots, foliage and catkins when young. Colouring decreases towards late summer, but is followed by good autumn colour. Foliage susceptible to scorch in hot summers. Requires a moist position to do well. Half average height and spread. *A. i. 'Laciniata'* Light green leaves deeply cut and lacerated. Good catkins and fruit. *A. i. 'Pendula'* Weeping branches displaying dark green, silver-backed foliage. Rarely produces a trunk. Interesting catkins and fruit. Two-thirds average height and spread.
Average height and spread
Five years
20x8ft (6x2.5m)
Ten years
39x16ft (12x5m)
Twenty years
or at maturity
48x23ft (23x7m)

AMELANCHIER CANADENSIS (Amelanchier lamarckii)

SHADBLOW SERVICEBERRY, SNOWY MESPILUS
Rosaceae
Deciduous
An attractive small tree for all sizes of garden.

Origin From North America and Canada, and now naturalized throughout most of Europe. Commonly sold in Europe and the UK as *A. lamarckii*.
Use Grown as a standard tree to make a featured specimen for small, medium or large gardens.
Description *Flower* Racemes of white flowers produced in late spring, before or just after leaves appear. *Foliage* Leaves ovate, 2-3in (5-8cm) long, light green with slight orange-red veining and shading on some soil conditions. Brilliant orange-red autumn colours. *Stem* Strong, upright when young, becoming twiggy and arching with age. Often grown as a large shrub, but readily produces a single trunk to form a small round-headed tree. Medium rate of growth. *Fruit* Small, light red fruits produced in racemes in autumn. Degree of fruiting dependent on the heat and dryness of summer.
Hardiness Tolerates winter temperatures below −13°F (−25°C).
Soil Requirements Any soil type; tolerates alkalinity and acidity.
Sun/Shade aspect Best in full sun, but tolerates quite deep shade.
Pruning None required, but any lower shoots or branches may be removed.
Propagation and nursery production From layers or by removal of a sucker from the base. On bushy plants up to 2ft (60cm) a single stem may be encouraged to develop by pruning. Trees pretrained up to 6-10ft (2-3m) are available. Plant bare-rooted or container-grown. Relatively easy to find in production.
Problems When grown as a standard tree is sometimes grafted on to *Crataegus oxyacantha* or *Pyrus communis*. Both these forms can be poor-rooted and produce suckers which must be removed.
Varieties of interest *A. laevis* Oval foliage producing good autumn colours. Branches more open and flowers more widely spaced, but larger individual flowers per open raceme. From eastern North America.
Average height and spread
Five years
14½x10ft (4.5x3m)
Ten years
22x16ft (6.5x5m)
Twenty years
or at maturity
25x20ft (7.5x6m)

ARBUTUS

STRAWBERRY TREE
Ericaceae
Evergreen
Arbutus varieties do not truly reach tree proportions for a great number of years, but are worth waiting for.

Origin From south-eastern Europe and Asia Minor.
Use As a freestanding specimen shrub and eventually a small tree.

Arbutus andrachne — branches

Description *Flower* Medium-size cup-shaped flowers, pure white or tinged pink, hanging in small arching panicles, produced throughout spring and early summer. *Foliage* Leaves leathery, ovate to oval, 2-4in (5-10cm) long, dark green with paler undersides. Purple shading and purple veining. *Stem* Dark orange to orange-purple. Some varieties have a striking winter effect. Shrub-forming for the first 10 years of its existence, slowly evolving tree-like habit. Upright when young, spreading with age. *Fruit* Globe-shaped to round, strawberry-like fruits, ¼-½in (5mm-1cm) across, produced in autumn and maintained through winter, ripening in spring and often still showing with the next season's flowers.
Hardiness Minimum winter temperature 14°F (−10°C). Defoliation may occur in severe frost conditions, although the plant is rarely destroyed completely.
Soil Requirements Does well on all soil types except extremely alkaline conditions. Makes more growth when leafmould or peat added to soil. Young plant responds well to an annual mulch of organic matter 2-4in (5-10cm) deep and at least 3ft (1m) wide around the base.
Sun/Shade aspect Full sun to light shade, with light shade for preference.
Pruning None required.
Propagation and nursery production From seed or semi-ripe cuttings taken in early summer. Always purchase container-grown. Some varieties quite hard to obtain and must

be sought from specialist nurseries. Usually only to be purchased from 1-1½ft (30-50cm) in height.
Problems Slowness of growth is frequently underestimated.
Varieties of interest *A. andrachne* (Grecian Strawberry Tree) Dark red to cinnamon-coloured peeling bark, its main attraction in winter. Often difficult to find in commercial production. From south-eastern Europe and Asia Minor. *A. × andrachnoides* (Killarney Strawberry Tree) Attractive cinnamon-red winter branches. More lime-tolerant than most. Should be considered normally hardy. Winter and autumn flowering with fruits produced in spring. A cross between *A. andrachne* for its winter stems and *A. unedo* for its fruit. *A. menziesii* (The Madrona of California) Peeling smooth bark, revealing light red to terracotta underbark. Foliage oval, up to 6in (15cm) long and 2½in (6cm) wide. Fruits orange-red, globe-shaped and of good size, following large terminal pyramid panicles of white flowers in spring. Should be considered slightly tender in most areas with winter frost. *A. unedo* (Killarney Strawberry Tree) Abundant flowers produced in terminal panicles, white or pink shaded, late autumn to early winter, followed by bright red, granular-surfaced, globe-shaped, edible fruit. Fruits often used for the production of liqueur, especially in Portugal. Slow-growing for the first fifteen years of its life. Normally considered as a large shrub. Hardy but with some winter damage to foliage possible in severe wind-chill conditions. From southern Europe, south-western Ireland and Asia Minor. A popular and widely available variety. *A. u. 'Quercifolia'* Foliage dark green with purple undershading to veins, and dissected, similar to leaves of the common oak. Flowers and fruits identical to parent. Scarce but not impossible to find. *A. u. 'Rubra'* Dark, rich pink flowers. Good fruiting effects in hot summers. Slightly less vigorous and less hardy than parent. Difficult but not impossible to find. Two-thirds average height and spread.

Arbutus unedo in fruit and flower

Average height and spread
Five years
3x3ft (1x1m)
Ten years
6x6ft (2x2m)
Twenty years
or at maturity
16x16ft (5x5m)

BETULA PENDULA (Betula alba, Betula verrucosa)

SILVER, WHITE OR COMMON BIRCH
Betulaceae
Deciduous
Delightful small trees, suitable for most gardens.

Origin From Europe through to Asia Minor.
Use As individual specimens often planted in closely grouped formations, less than 3-5ft (1-1.5m) apart to give a traditional coppice effect. Good for planting as large-scale wind-break.
Description *Flower* Round to oval female catkins, up to 1in (3cm) long and scaly in

Amelanchier canadensis in flower

Betula pendula in autumn

texture, brown when juvenile in late autumn, growing and opening with yellow stamens in early spring. *Foliage* Broad, ovate, ½-2in (1-5cm) long. Light green in colour and with lobed edges. Good yellow autumn colour. *Stem* Dark purple-brown, pendulous twigs. Grey-brown young foliage becomes white-stemmed and branched with wood more than 2in (5cm) wide. Pyramidal habit, with branches pendulous towards outer ends. *Fruit* Small, hanging, scaly fruits in autumn.
Hardiness Tolerates winter temperatures below −13°F (−25°C).
Soil Requirements Tolerates a wide range of soil conditions but dislikes waterlogging.
Sun/Shade aspect Full sun to light shade, preferring full sun.
Pruning None required but removal of lower branches may enhance the appearance of the white trunk.
Propagation and nursery production From seed for basic form, or grafted for named forms. Can be purchased from 1½ft (50cm) up to 16ft (5m) depending on requirements. Best planting heights 5-6ft (1.5-2m). Plant bare-rooted or container-grown. Relatively easy to find.

Betula pendula 'Dalecarlica' in leaf

Problems Often required as a multi-stemmed tree, but rarely found in this form. Grouping achieves this effect. *Betulas* are relatively short-lived — 50 years being average maximum lifespan. Older trees can suddenly die for no apparent reason; this may be due to wood-boring insects.
Varieties of interest *B. p.* **'Dalecarlica'** (Swedish Birch) Light green, deeply lobed foliage, lobes lanceolate with indentations. Yellow autumn colour. Upright in habit. Horizontal branches and weeping twigs. Good white stems. *B. p.* **'Fastigiata'** Upright white stems, narrow columnar habit. Typical foliage, cat-

kins and fruit. *B. p.* **'Purpurea'** (Purple Leaf Birch) New foliage in spring bright purple, aging to purple-green, finally dark green. Purple catkins and fruit. Dark purple growth. Two-thirds average height and spread. Worthy of wider planting. *B. p.* **'Tristis'** Typical lobed foliage with good yellow autumn colour. Upright central trunk, pendulous side branches. Extremely attractive when mature. Good white stems. *B. p.* **'Youngii'** Pendulous, low, dome-shaped, spreading tree, no central leader. All branches and twigs weeping. Good white stems. Typical foliage with good yellow autumn colour. Two-thirds average height and spread.
Average height and spread
Five years
20x5ft (6x1.5m)
Ten years
32x10ft (10x3m)
Twenty years
or at maturity
39x16ft (12x5m)

BETULA
White-stemmed varieties

BIRCH
Betulaceae
Deciduous
Splendid tall trees with beautiful bark.

Origin From north-east Asia, the Himalayas, Japan and western China, through to Europe, depending on variety.
Use As individual trees or grouped, for display of autumn colour and winter bark.
Description *Flower* Female catkins, round to oval in shape, up to 4in (10cm) long and scaly in texture; small and brown when juvenile in late autumn, growing and opening with yellow stamens in early spring. *Foliage* Leaves 2-2½in (5-6cm) long; slender, pointed, ovate or oblong. Light green with grey sheen and prominent veins. Good yellow autumn colour. *Stem* Light grey-green when young, becoming grey-brown. Finely peeling, revealing white underbark of great winter attraction. Upright pyramidal trees of medium growth rate. *Fruit* Small, hanging, scaly fruits in autumn.
Hardiness Tolerant of winter temperatures below −13°F (−25°C).
Soil Requirements Tolerant of a wide range of soil conditions; dislikes waterlogging.
Sun/Shade aspect Full sun to light shade, preferring full sun.
Pruning No pruning other than removal of lower branches to enhance the appearance of the white trunk.
Propagation and nursery production From seed or grafts, depending on variety. Plant bare-rooted or container-grown. Trees can be purchased from 4-16ft (1.2-5m) in height. Some forms may be difficult to find and range of sizes limited; must be sought from specialist nurseries.
Problems Often required as a multi-stemmed tree, but this form is rarely found. Grouping is the way to achieve the effect. Trees are also relatively short-lived, 50 years being the average maximum lifespan. Older trees suddenly die for no apparent reason.
Varieties of interest *B. costata* Creamy-white bark with dark older bark retained, especially at leaf axils, giving a two-toned colour effect. Slightly more than average height and spread. Relatively easy to find, especially from specialist nurseries. *B. jacquemontii* Pure white stems showing well in winter light, with large areas of shaggy brown, peeling bark. Attractive autumn colours and large catkins. *B. nigra* (River Birch, Black Birch) Foliage diamond-shaped with soft green upper surfaces and glaucous undersides. Useful for planting in damp difficult areas. Extremely good on wet areas and damp ground. Native to central and eastern USA. *B. papyrifera* (Paper Birch, Canoe Birch) Large, ovate foliage, 1-4in (3-10cm) long. Bark peels in large complete sheets, revealing white underskin. Does well in moist conditions. From

North America. In its native environment may reach more than average height and spread. *B. platyphylla* Very large leaves with good autumn colour. Grey-white stems. From Manchuria and Korea. *B. populifolia* (Grey Birch) Ovate light green to grey-green foliage. Attractive autumn colours. Young bark grey-white. Two-thirds average height and spread. Originating in North America. *B. pubescens* (White Birch) Broad ovate leaves. Yellow catkins. White bark, revealed by peeling, dark, older bark. Difficult to find. *B. utilis* Foliage ovate, coarsely tooth-edged and 2-2½in (5-6cm) long. Female yellow catkins. Bark creamy white. New shoots red-brown. Slightly less hardy. From the Himalayas.

Betula jacquemontii − branches in winter

Average height and spread
Five years
20x5ft (6x1.5m)
Ten years
30x10ft (9x3m)
Twenty years
or at maturity
39x16ft (12x5m)

BETULA
Colored-stemmed varieties

BIRCH
Betulaceae
Deciduous
A fine addition to the garden for colourful winter stems: consider the location carefully to plant for best effect.

Origin From China, Korea, Japan, North America or Europe, depending on variety.
Use As individual specimen or planted in very closely grouped formations, less than 3-5ft (1-1.5m) apart to give a traditional coppice effect.
Description *Flower* Female catkins round to oval in shape, up to 1in (3cm) long and scaly-textured, small and brown when juvenile from late autumn, growing and opening with yellow stamens in early spring. *Foliage* Leaves ovate, 2-3in (5-8cm) long with toothed edges. Light grey-green when young, becoming light green with age. Good yellow autumn colour. *Stem* Light grey-green when young, aging to grey-brown; developing after 5-7 years various bark colours — orange-brown, yellow-orange, orange-grey or grey-green, depending on variety. Predominantly upright growth, medium vigour. *Fruit* Small, hanging, scaly fruits in autumn.
Hardiness Tolerant of winter temperatures below −15°F (−25°C).
Soil Requirements Tolerates a wide range of soil conditions.
Sun/Shade aspect Full sun to light shade, preferring full sun. Winter sunlight shows off coloured stems to best effect.

Pruning None required. Lower branches may be removed to enhance the appearance of trunk and stems.
Propagation and nursery production From grafting, layering or seed, dependent on variety. Plants can be obtained from 5-16ft (1.5-5m), depending on variety. Best planting heights 5-10ft (1.5-3m). Most forms must be sought from specialist nurseries, as not generally in regular production. Plant bare-rooted or container-grown.
Problems Rarely show full coloured-stem

Betula albo-sinensis var. septentrionalis

potential until 7-10 years after planting, but well worth waiting for.
Varieties of interest *B. albo-sinensis* Large attractive foliage with good autumn colour. Stems acquire an orange to orange-red peeling effect of good winter value. From China. *B. albo-sinensis var. septentrionalis* One of the finest of all orange-stemmed birches, with bark colours orange-brown through yellow-orange to orange-grey. Old leaves give good autumn colour. Relatively scarce in commercial production. From China. *B. ermanii* Bark orange brown, changing with age to creamy white. Large foliage, good autumn colour. Can be damaged by late spring frosts in frost-pocket areas. From Manchuria, Korea and Japan. *B. lenta* (Cherry Birch) Almost black bark, non-peeling. Large foliage, good yellow autumn colour. Bark sweet and aromatic to taste and smell. From North America. *B. lutea* (Yellow Birch) Bark yellow to yellow-grey and of flaky composition, with bitter taste but aromatic scent. Leaves ovate to oblong, 2-4in (5-10cm) long. Good yellow autumn colour. From North America. *B. maximowicziana* Bark pale orange or grey with brown shoots. Large leaves up to 6in (15cm) long and 4in (10cm) wide. Two types of catkins — male 5in (12cm) long and female 2½in (6cm) long. Both produced in groups of 2-4 in long racemes. Good yellow autumn colour. The largest-leaved birch. Slightly more than average height and spread. *B. medwediewii* Interesting large winter terminal buds, oval and of stiff habit. Foliage ovate to round, up to 5in (12cm) long with good autumn colours, but stem colour less interesting. From Corsica.
Average height and spread
Five years
20x5ft (6x1.5m)
Ten years
33x5ft (10x1.5m)
Twenty years
or at maturity
39x16ft (12x5m)

BUDDLEIA ALTERNIFOLIA

BUDDLEIA, BUTTERFLY BUSH
Loganiaceae
Deciduous
An interesting and attractive, small weeping tree which needs training while young.

Origin From China.
Use As a small-feature tree, planted singly or in pairs to emphasize a particular vista or walkway.
Description *Flower* Small bunches of very fragrant, small, trumpet-shaped, lilac flowers borne in early summer along graceful, arching branches. *Foliage* Grey-green, lanceolate,

Buddleia alternifolia **in flower**

small leaves, giving yellow autumn colour. *Stem* Grey-green to mid green, vigorous, long and upright. Often grown as a large shrub, attaining the height of a tree after a number of years. A single stem may be encouraged and the head allowed to grow to form a small mop-headed, weeping tree. *Fruit* Brown to grey-brown seedheads in autumn and winter.
Hardiness Tolerates 4°F (−15°C).
Soil Requirements Does best in a rich, deep soil. Tolerates a wide range of other soil types, but may not attain maximum beauty.
Sun/Shade aspect Best in full sun, but tolerates light, dappled shade.
Pruning None required, other than occasional removal of very old wood.
Propagation and nursery production From softwood cuttings taken in summer or hardwood cuttings in winter. Always purchase container-grown. A tree may be formed by planting and training a young shrub. Pre-trained standards can also be found. Best planting heights 2-3ft (60cm-1m).
Problems As a standard tree requires staking throughout its life to prevent physical damage by wind.
Average height and spread
Five years
12x12ft (3.5x3.5m)
Ten years
14½x14½ft (4.5x4.5m)
Twenty years
or at maturity
16x16ft (5x5m)

CARAGANA ARBORESCENS

PEA TREE, SIBERIAN PEASHRUB
Leguminosae
Deciduous
A group of trees offering large and small foliage forms, both being rewarding garden features.

Origin From Siberia and Manchuria.
Use As a small, low tree with attractive spring flowers, suitable for any size of garden.
Description *Flower* Small, yellow, pea-shaped

Caragana arborescens 'Pendula' **in flower**

flowers, borne in clusters of up to four on thin stalks mid to late spring. *Foliage* Oval to ovate, 1½-3in (4-8cm) long, produced in pairs and grouped 4-6 together; or narrow, thin, strap-like and lanceolate, very soft in texture, giving a cloud effect. Good yellow autumn colour. *Stem* Main stems spine-tipped, with secondary spines at leaf axils. Grey-green, fast-growing when young, slowing with age, forming a round-topped tree or large shrub. *Fruit* Small pods, 1½-2in (4-5cm) long, containing 4-6 seeds, produced in autumn.
Hardiness Tolerates winter temperatures below −13°F (−25°C).
Soil Requirements Any soil conditions; tolerates high alkalinity.

Caragana arborescens 'Lorbergii' **in flower**

Sun/Shade aspect Best in full sun, but tolerates light shade.
Pruning None required except for training. Mature tree may resent pruning.
Propagation and nursery production From seed for parent. Named varieties from grafting on to understock of *C.arborescens*. Tree can be purchased from 3ft (1m) to 10ft (3m). *C. arborescens* is more difficult to find than its varieties. Plant bare-rooted, root-balled (balled-and-burlapped) or container-grown.
Problems May be late to break leaf in spring and can appear to be dead, but grows quickly once started.
Varieties of interest *C. a. 'Lorbergii'* A gem among small trees, useful for large containers. Thin, strap-like leaves give a hazy appearance, grey-green when young becoming paler with age, with good yellow autumn colour. Yellow pea-flowers produced in each leaf axil. Young trees should be pruned hard, when rapid new growth emerges to make a good head shape. This process may be repe-

21

Caragana arborescens 'Walker' in flower

ated every 3-4 years to encourage new growth production. *C. a. 'Pendula'* A round-leaved variety, normally available on stems of 5ft (1.5m) or 8ft (2.5m), either size being suitable. Grey-green stems weep directly from grafted crown to the ground and spread outwards, forming a large canopy. When young, subject to suckering, which must be removed. Quickly attains its very gnarled, weeping effect, of architectural value. *C. a. 'Walker'* Soft, strap-like foliage with yellow flowers. A small weeping tree grafted on stems of 3ft (1m), 5ft (1.5m) or 6ft (2m) in height. Growth is completely pendulous, hugging central stem. Ideal for feature planting, patio work, large rock gardens or in containers. Tolerates high winds and drought. *C. frutex* Slow-growing, spineless. Grafted on to a 5ft (1.5m) stem to make a round, tight ball of light green foliage, covered with bright yellow pea-flowers in spring. *C. maximowicziana* Normally a shrub but can be found grafted on to stems of 5ft (1.5m), forming a large, round head. Heavily armed, spiny branches. Spines white in winter. Small, grey-green foliage; dark yellow flowers in spring. From China and Tibet. *C. pygmaea* Truly a shrub, but often found grafted on to 5ft (1.5m) stems, to produce grey-green foliage in a round mop-headed effect; yellow flowers in spring. Useful for containers, rock gardens, patios and feature planting.

Average height and spread
Five years
6x6ft (2x2m)
Ten years
13x13ft (4x4m)
Twenty years
or at maturity
30x20ft (9x6m)

CARPINUS BETULUS

COMMON HORNBEAM, EUROPEAN HORNBEAM
Carpinaceae
Deciduous
Interesting as ornamental trees, often neglected but worthy of wider planting.

Origin From Europe and Asia Minor.
Use As a small to medium tree for medium or large gardens; as individual specimens, heavily shaped and pruned features, large, tall hedges on stilts when trained horizontally, or trimmed trees.
Description *Flower* Hanging catkins 1½-3in (4-8cm) long, consisting of three-lobed bracts. Produced in spring. *Foliage* Oval to ovate, 2-4in (5-10cm) long. Light grey-green when young, aging to bright green, toothed edges and pronounced veins. Good yellow autumn colour. *Stem* Light grey-green when young, becoming grey-brown. Very twiggy, forming a round-topped tree. *Fruit* Hanging

racemes of flat, rounded seeds, light green to yellow and finally brown, retained well into winter.
Hardiness Tolerates −13°F (−25°C).
Soil Requirements Any soil conditions.
Sun/Shade aspect Full sun to medium shade, preferring full sun.
Pruning None required but can be cut, trimmed or shaped.
Propagation and nursery production From seed; named varieties grafted. Plant bare-rooted or container-grown. Trees from 3ft (1m) to 16ft (5m) can be obtained. Best planting heights 5-6ft (1.5-2m). Relatively easy to find in general nurseries, specific varieties in specialist nurseries.
Problems Can suffer attacks of coral spot, and sometimes greenfly, but this is seldom a lasting problem.
Varieties of interest *C. b. 'Columnaris'* Slender, upright pyramidal. Branches spiralling skywards, especially when young, becoming more open with age. Average height, but unlikely to reach more than 10ft (3m) width at maturity. Good for street planting or as an avenue feature when planted at 20ft (6m) apart. *C. b. 'Fastigiata'* syn. *C. b. 'Pyramidalis'* Narrow, upright branches to average height and reaching 13ft (4m) in width at maturity. Good for street planting or as an avenue feature when planted at 26ft (8m) apart. *C. b. 'Incisa'* Two-thirds average height and spread with very deeply cut, lacerated leaves. Relatively difficult to find in commercial production, but worth looking for. *C. b. 'Pendula'* Weeping main stem and branches make an attractive large-shrub effect. Central leader needs encouragement to gain height. *C. b. 'Purpurea'* Young growth purple-green, aging to dark green. Chiefly

Carpinus betulus in autumn

interesting in its spring effect. *C. caroliniana* (American Hornbeam) A tree reaching up to 39ft (12m) in height and spread, producing a fluted, grey-green trunk. Grey-green foliage oval to ovate and up to 4in (10cm) long, with double-toothed edges. Catkins 2-4in (5-10cm) long with three-lobed bracts in late spring. From eastern North America. Good autumn colours of orange and scarlet. Readily available in its native environment, less common elsewhere. *C. japonica* (Japanese Hornbeam) A tree of flat-topped, spreading habit, two-thirds average height and spread. Foliage 2-4in (5-10cm) long, grey-green with pronounced veins. Catkins 2in (5cm) long on slender, hanging stalks. An attractive form, somewhat difficult to find.
There are a number of other forms, but those listed above are best for garden use.
Average height and spread
Five years
13x6ft (4x2m)
Ten years
26x20ft (8x6m)
Twenty years
or at maturity
44x39ft (16x12m)

CARYA

HICKORY
Juglandaceae
Deciduous
A rare tree outside its native environment because of its cultivation requirements.

Origin From eastern North America.
Use As a large, freestanding, fruiting tree for large gardens, estates and parks.

Carya cordiformis in spring

Description *Flower* Unisexed flowers, male catkins produced in threes with 3-10 stamens and a limited number of female flowers produced in terminal spikes. *Foliage* Large pinnate leaves with 3-17 pairs of leaflets. Tooth-edged, light grey-green to dull green. Good yellow autumn colour. *Stem* Strong, unbranching, upright shoots. Tips often killed by winter, ensuring branching. Distinctive 'shaggy' bark. Upright when young, becoming spreading and bold with age, to form a large, round-topped tree. *Fruit* Globe-shaped to oblong, husky fruit. Splits into four parts, containing an edible nut.
Hardiness Tolerates 14°F (−10°C), but stem damage to previous year's young shoots can occur in harsh conditions.
Soil Requirements Most soil conditions; best on a rich, deep loam.
Sun/Shade aspect Best in full sun, but tolerates light shade.
Pruning None required other than removal of dead tips when young.
Propagation and nursery production From seed grown directly *in situ*. Resents any soil

and root disturbance. Young seedlings can be pot-grown and transplanted, but with limited success. The plant is scarce in gardens and parks outside its native habitats.

Problems Intolerant of soil or root damage.

Varieties of interest *C. aquatica* Interesting peeling bark, falling away in long scales. Good in damp conditions. Rarely found outside south-eastern USA. *C. cordiformis* syn. *C. amara* (Bitter Nut) Flaky bark. Attractive yellow buds in winter. Large leaves with good autumn colour. Thin-shelled, bitter-tasting nuts. From North America. *C. glabra* syn. *C. porcina* (Pig Nut) Round nuts, large foliage and good autumn colour. Average height and spread, but mature trees can sometimes well exceed this. Possibly one of the most successful outside its native environment. From North America. *C. ovalis* (Sweet Pig Nut) Fruits are sweet, edible nuts. Large leaves with good autumn colours. From eastern North America, where several named varieties may be offered for fruiting ability, but rarely seen elsewhere. *C. ovata* syn. *C. alba* (Shagbark Hickory) Shaggy bark which falls away in large chunks. Large leaves. Male catkins and small female flowers. White nuts, four-angled and very thin-shelled and sweet. From North America. *C. illinoinensis* Large leaves with good autumn colour, producing large quantities of pecan nuts. Large, having one-third more height and spread than average. From North America, and rarely succeeds outside its native environment. *C. tomentosa* (Mockernut) Interesting large winter buds. Elegant foliage and very good yellow autumn colours. Sweet, light-brown nuts. From eastern USA and rarely seen elsewhere.

Average height and spread
Five years
20x10ft (6x3m)
Ten years
39x26ft(12x8m)
Twenty years or at maturity
78x39ft (24x12m)

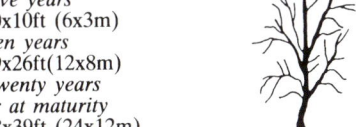

CASTANEA SATIVA

SPANISH CHESTNUT

Fagaceae
Deciduous
Large trees, well known for their fruit, but not always successful in production outside their native environments.

Origin From southern Europe, Asia Minor and North Africa.

Use As a large tree to provide shade for larger gardens, estates and parks. Can be coppiced for the production of strong staking or fencing poles. Also used in southern Europe for fruit production.

Description *Flower* Long, yellow catkins, up to 8in (20cm) in length, produced in late summer and originating in leaf axils of young shoots. Male and female catkins carried on the same tree. *Foliage* Oblong, 5-12in (12-30cm) long, coarsely tooth-edged, light green leaves, giving good yellow autumn colour. *Stem* Grey-green to grey-brown. Strong, quick-growing, upright, spreading with age. *Fruit* Brown chestnuts contained within a round, prickly-burred fruit case. Edible, and in some countries part of the staple diet and used as a commercial crop. Fruiting best in areas around the Mediterranean.

Hardiness Tolerates 4°F (−15°C).

Soil Requirements Any soil type; produces best fruit on rich, deep loam.

Sun/Shade aspect Full sun, but tolerates light shade.

Pruning None required; in fact may be spoilt by pruning.

Propagation and nursery production From seed or grafting for named or variegated varieties. *C. sativa* can normally be purchased from 3ft (1m) to 16ft (5m), with best planting heights 5½-10ft (1.8-3m). Named varieties normally offered as 3-5ft (1-1.5m) young trees. Plant bare-rooted or container-grown.

Castanea sativa **in fruit**

C. sativa readily available from general nurseries, but named varieties are extremely scarce and must be sought from specialist nurseries.

Problems Often planted for fruit production in locations too unlike its native environment to encourage successful fruiting.

Varieties of interest *C. s. 'Albomarginata'* Leaves bordered with bold white variegation. *C. s. 'Aureomarginata'* Leaves bordered with bold yellow variegation, fading to creamy yellow in summer. *C. s. 'Heterophylla'* Foliage light grey-green, linear and deeply cut with irregular lobed indentations.

Named fruiting forms may be available but are usually only prolifically fruit-bearing in southern Europe.

Average height and spread
Five years
23x10ft (7x3m)
Ten years
46x20ft (14x6m)
Twenty years or at maturity
60x39ft (18x12m)

CATALPA BIGNONIOIDES

INDIAN BEAN TREE, SOUTHERN CATALPA

Bignoniaceae
Deciduous
Attractive, large-leaved shade trees.

Origin From south-eastern USA.

Use As a large, green-leaved shade tree, good for use in medium-sized and larger gardens.

Description *Flower* Upright panicles 8-10in (20-25cm) long of white bell-shaped flowers with frilled edges, yellow markings and pur-

Catalpa bignonioides **in flower**

ple spotted throats, produced midsummer. *Foliage* Broad, ovate, large leaves, 6-10in (15-25cm) long and 3-8in (8-20cm) wide presented on long stalks, forming a large canopy of shade-giving foliage. Good yellow autumn colour. Foliage smells unpleasant when crushed. *Stem* Light grey-green, becoming green-brown. Strong and upright when young, limited branching, becoming slower and more spreading after the first five years. Timber susceptible to wind damage. Fast-growing and matures quickly to make an interesting round-topped tree. *Fruit* Long, narrow, green aging to black, slender pods, 8-15in (20-40cm) long, produced in early autumn and retained into early winter.

Hardiness Tolerates 14°F (−10°C), but stem damage can be caused by winter frosts, especially in the golden-leaved varieties. This may be an advantage in encouraging branching.

Soil Requirements Requires a deep, rich soil to do well. Shows signs of chlorosis on extremely thin alkaline soils.

Sun/Shade aspect Best in full sun; green-leaved varieties tolerate light shade.

Pruning None required, but tolerates cutting back to control or shape growth.

Catalpa bignonioides 'Aurea' **in leaf**

Propagation and nursery production From seed or layers for green-leaved varieties. Purple and golden-leaved varieties normally grafted onto understock of *C. bignonioides* or layered. Plant green and purple forms from 3ft (1m) to 13ft (4m), either bare-rooted or container-grown. Use container-grown plants for yellow-leaved forms. Due to quick growth rate and limited branching when young, a single stem of 6-10ft (2-3m) may be offered with little side branching, but once planted rapidly extends outwards.

Problems May be physically damaged by high winds or heavy snow; consider location when planting. Young trees rarely look attractive,

23

especially while in nursery production. **Varieties of interest** *C. b. 'Aurea'* Attractive, broad, large golden yellow leaves. A less hardy form, even slightly tender. In less mild areas should be considered as a large bush rather than a tree. One-third average height and spread, but may reach more in ideal conditions. Usually supplied at 2-2½ft (60-80cm). Trees grafted on to 5ft (1.5m) or 6ft (2m) stems can be purchased to give quick height. May be fan-trained on a large wall. *C. b. 'Variegata'* Attractive, large-leaved foliage, grey-green leaves margined with gold. Limited in commercial production, but not impossible to find. *C.* × *hybrida 'Purpurea'* New growth purple to purple-green, aging to dark green. White flowers. Two-thirds average height and spread. From central USA.

Average height and spread
Five years
16x16ft (5x5m)
Ten years
30x30ft (9x9m)
Twenty years or at maturity
39x39ft (12x12m)

CERCIDIPHYLLUM JAPONICUM

KNOWN BY BOTANICAL NAME
Cercidiphyllaceae
Deciduous
Attractive large multi-stemmed shrubs or small trees for autumn colour.

Origin From Japan.
Use As a large, multi-stemmed shrub or tree or as a small tree with a single trunk; in a favourable locality the improved growth rate forms a larger tree.
Description *Flower* Small, yellow flowers produced on short shoots at each leaf axil on 2-3 year-old wood. *Foliage* Attractive round to heart-shaped foliage, purple-red when young, becoming glaucous green with a purple tinge. Excellent autumn colours of rich red to yellow. *Stem* Stems purple-green when young, becoming purple-brown with age. Often multi-stemmed, requiring selection of a strong individual shoot to develop a standard tree, but forms an interesting, large, multi-stemmed clump. Slow to establish, but later increases growth production. Older trees spread to form a canopy. *Fruit* Insignificant.
Hardiness Tolerates 14°F (−10°C), but young foliage damaged by late spring frosts. Damage is rarely permanent due to a second leaf crop in early summer.
Soil Requirements Best on a neutral to acid soil, but tolerates limited alkalinity. The richer and deeper the soil, the more growth is achieved.

Cercidiphyllum japonicum in autumn

Sun/Shade aspect Full sun to light shade; best results probably in light shade.
Pruning None required other than the selection of a central stem for standard tree formation if preferred.
Propagation and nursery production From layers or seed, named and weeping varieties by grafting. Always purchase container-grown for best results. Plants normally available from 2ft (60cm) up to 5ft (1.5m), except weeping forms which are grafted on to 6-8ft (2-2.5m) stems. *C. japonicum* usually easy to find from general and specialist nurseries. Other forms require further search.
Problems Possible damage to early spring growth. May not achieve full height in northerly locations.
Varieties of interest *C. j. 'Pendulum'* Usually grafted on to tall stem. Weeping to ground-level in a wide-headed habit. *C. j. var. sinense* Almost identical to the parent, but not commonly available and must be sought from specialist nurseries. Originates from China. *C. magnificum* A form said to improve upon *C. japonicum* because of its larger, thicker leaves. Difficult to find, and its superior qualities are only a matter of preference.

Average height and spread
Five years
16x10ft (5x3m)
Ten years
26x20ft (8x6m)
Twenty years or at maturity
39x30ft (12x9m)

CERCIS SILIQUASTRUM

JUDAS TREE
Leguminosae
Deciduous
Very attractive large shrubs evolving into trees, requiring a great deal of time to reach any true stature.

Origin From southern Europe through to the Orient.
Use As freestanding large shrub or small tree. Can be successfully fan-trained on to a large wall.
Description *Flower* Numerous purple-rose flowers ½-1in (1-3cm) long, borne as leaves are produced, on both young and old branches in late spring, early summer. *Foliage* Deeply veined, broad kidney-shaped leaves, purple-green with a blue sheen. Good yellow autumn colours. *Stem* Dark brown, almost black stems. Very twiggy and branching. Good growth from base. Forms a large shrub, evolving into a multi-stemmed tree over 10 years. Moderately slow-growing. *Fruit* Pods, 3-4in (8-10cm) long, light grey-green, produced in autumn, aging to grey-brown and attractive in winter.
Hardiness Tolerates 14°F (−10°C).
Soil Requirements Does best on neutral to acid soil, but tolerates moderate alkalinity.
Sun/Shade aspect Full sun to medium shade, with light shade for preference.
Pruning None required, other than removal of unwanted limbs to achieve tree proportions. Pruning should not be undertaken until the tree is 5-10 years old.
Propagation and nursery production From seed or layers. Purchase container-grown or root-balled (balled-and-burlapped). Many are imported from Europe as root-balled plants in peat soil and may require careful weaning to become accustomed to normal soil. Established or trained trees rarely available and not recommended.
Problems This is not really a tree for a great number of years, especially in northerly climates, and for practical purposes should be considered as a large shrub.
Varieties of interest *C. canadensis* (Redbud) Large leaves and clusters of pale rose pink flowers in early summer. Less hardy than average. From central and eastern USA. Limited in production outside its native environment. *C. c. 'Forest Pansy'* Heart-shaped,

deep purple leaves maintained throughout summer, turning bright scarlet in autumn. Flowers inconspicuous. Two-thirds average height and spread, more shrub-forming than tree-shaped. Only obtainable from specialist nurseries. *C. chinensis* Flowers purple-pink. Slightly more than average height and spread. Rarely forms trunk or tree-like shape. Minimum winter temperature 23°F (−5°C). From China. Seldom seen in production. *C. occidentalis* Rose-coloured flowers on short stalks. Two-thirds average height and spread, and rarely growing to above 16ft (5m). From California. Minimum winter temperature 23°F (−5°C). Rarely grown outside its native environment. *C. racemosa* Flowers red-pink produced in racemes 4in (10cm) long in late spring. Less hardy. From China. Somewhat scarce in production. *C. siliquastrum 'Alba'* A pure white-flowering form from Europe and the Orient. Very scarce in commercial production in Europe, more readily available in North America.

Cercis siliquastrum in flower

Average height and spread
Five years
5x5ft (1.5x1.5m)
Ten years
10x10ft (3x3m)
Twenty years or at maturity
20x20ft (6x6m)
Continued growth reaches 39ft (12m) but extremely slowly, taking 50 years or more in most northerly locations.

CLADRASTIS LUTEA

YELLOW WOOD
Leguminosae
Deciduous
A gem among trees, for both flowers and autumn colour.

Origin Native to the USA.
Use As a freestanding large shrub or medium-sized tree given time.
Description *Flower* Hanging, fragrant white panicles, 8-15in (20-40cm) long and 4-6in (10-15cm) wide, in early summer. *Foliage* Light green, pinnate leaves with 7-9 small ovate to oval, broad leaflets on a yellow-green leaf-stalk 8-12in (20-30cm) long. Rich yellow autumn colour. *Stem* Yellow-green with bright yellow young terminal tips. Strong, upright when young, possibly multi-stemmed but should be encouraged to produce one central trunk with sideshoots removed. Forms a moderately wide-based, round-topped tree of good proportions. Medium rate of growth. *Fruit* Produces light green to yellow, long flat pods 2½-4in (6-10cm) in autumn.
Hardiness Tolerates 4°F (−15°C).
Soil Requirements Acid to neutral soil for best

Cladrastis lutea in spring

results. Resents alkalinity or waterlogging.
Sun/Shade aspect Full sun.
Pruning None required, except for training in early years.
Propagation and nursery production From seed or rooted cuttings, except for *C. sinensis*, which is grafted on to *C. lutea*. Purchase container-grown for best results. Must be sought from specialist nurseries. Best planting heights 3-6ft (1-2m).
Problems May be slow to form a tree, but in time becomes a truly spectacular sight.
Varieties of interest *C. sinensis* Fragrant flowers in upright panicles, white with some pink flushing, produced midsummer. Very scarce in commercial production and must be sought from very specialized sources. Of more than average height and spread. From China.
Average height and spread
Five years
10x3ft (3x1m)
Ten years
16x8ft (5x2.5m)
Twenty years
or at maturity
32x20ft (10x6m)

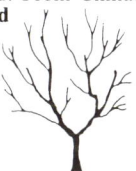

CORNUS NUTTALLII

PACIFIC DOGWOOD
Cornaceae
Deciduous
A large shrub which makes an attractive tree if trained.

Origin From western North America.
Use As a freestanding tree, or as background tree specimen in a large shrub border.

Cornus nuttallii in autumn

Description *Flower* Round, four-bract, pure white flowers aging to pink produced mid to late spring. *Foliage* Large, 6-8in (15-20cm) long, elliptic, mid to dark green with a grey sheen. Vivid orange-yellow autumn colours, with leaves colouring at an intermittent rate giving a fine contrast of green and orange. *Stem* Normally a large shrub, but with some training slowly forms a useful small tree of great beauty. *Fruit* Insignificant.
Hardiness Tolerates 4°F (−15°C).
Soil Requirements Best on neutral to acid soil, but tolerates some alkalinity.
Sun/Shade aspect Full sun to light shade.
Pruning None required, but needs initial training to take on the overall shape of a small tree.
Propagation and nursery production From layers or softwood cuttings taken in mid to late summer. Purchase container-grown. Not readily found in nursery production; may be difficult to find outside its native environment.
Problems Although listed here as a tree, is rarely sold as such, and usually cannot be obtained above 4ft (1.2m) in height.
Average height and spread
Five years
13x5ft (4x1.5m)
Ten years
20x8ft (6x2.5m)
Twenty years
or at maturity
32x16ft (10x5m)

CORYLUS COLURNA

TURKISH HAZEL
Corylaceae
Deciduous
An interesting bark and foliage tree, with a good architectural shape.

Origin From south-eastern Europe and western Asia
Use As a large specimen tree for large gardens, estates and parks.
Description *Flower* Yellow male catkins, 2-2½in (5-6cm) long produced mid to late spring. *Foliage* Deeply veined, oval to ovate, pointed leaves 2½-6in (6-15cm) long with coarsely toothed edges. Good yellow autumn colour. *Stem* Upright with scaling grey bark, an interesting winter feature. New shoots have downy, corrugated effect. Stems light green in colour, upright when young, rounded with age. Medium growth rate. *Fruit* Small nuts in clusters of 3-6 in mid autumn, downy, bristly outer shell, grey-green in colour, ripening to brown.
Hardiness Tolerates 4°F (−15°C).
Soil Requirements Most soil types, but dislikes extremely alkaline conditions.
Sun/Shade aspect Full sun to light shade, preferring light shade.

Corylus colurna − trunk

Pruning None required, but low branches can be removed if necessary.
Propagation and nursery production From seed. Plant bare-rooted or container-grown. Available at 3-10ft (1-3m), but range may be limited as the plant is not cultivated widely and must be sought from specialist nurseries.
Problems Late to break leaf in spring and may appear dead.
Average height and spread
Five years
13x5ft (4x1.5m)
Ten years
26x13ft (8x4m)
Twenty years
or at maturity
52x26ft (16x8m)

COTONEASTER
Small-leaved standards

KNOWN BY BOTANICAL NAME
Rosaceae
Deciduous, Semi-evergreen and Evergreen
Interesting 'toy' trees for tub and patio work.

Origin Dependent on variety.
Use As small, weeping, mop-headed trees for feature emphasis. Ideal for small gardens, or patio and tub planting.
Description *Flower* Small, white, four-petalled flowers with red calyx late spring to early summer, singly or in clusters. *Foliage* Small, ovate, dark green leaves. Size varies. *Stem* Due to grafting on tall understock, branches are pendulous or weeping. Normally red-green when young, aging to purple-brown. Most varieties extremely twiggy and branching. *Fruit* Red berries in autumn as main attraction, cascading down outer surface of weeping branches.
Hardiness Tolerates 4°F (−15°C).
Soil Requirements Any soil conditions.
Sun/Shade aspect Full sun to medium shade.
Pruning Previous season's growth should be halved in spring to encourage branching, and form a tight, shapely head.
Propagation and nursery production By grafting on to understock of *C. bullatus*. Best purchased container-grown but also available root-balled (balled-and-burlapped). Produced in limited numbers and must be sought from specialist nurseries.
Problems Likely to sucker. The relatively thin stems must be staked to prevent wind damage to the heavy heads.
Varieties of interest *C. 'Coral Beauty'* Semi-

Cotoneaster 'Coral Beauty' in fruit

evergreen. Round to ovate, grey-green leaves ½-1in (1-3cm) long, ovate leaves giving good orange autumn colour. Coral red berries follow white flowers in autumn. Slightly less than average height and spread. Of garden origin. *C. dammeri* Evergreen. Ovate, 1in (3cm) long, dark green leaves with grey-green undersides. Some leaves have a slight purple shading along the veins and outer edges. White flowers followed by red berries in autumn. From China. *C. horizontalis* (Fishbone or Herringbone Cotoneaster) Deciduous. Small, dark green ovate leaves ½-1in (1-3cm) long, good autumn colour. Very branching habit. Grey-green stems forming a herringbone pattern of branches. White flowers in spring, red berries in autumn. Fruits and flowers borne at each leaf terminal along mature branches. From western China. *C. 'Skogholm'* Long trailing branches, semi-evergreen to evergreen. Leaves small, ½-1in (1-3cm) lanceolate to oval, mid to dark green. White flowers followed by red berries in autumn.

Average height and spread
Five years
5x2½ft (1.5x0.75m)
Ten years
5x4ft (1.5x1.2m)
Twenty years
or at maturity
5x5ft (1.5x1.5m)

COTONEASTER
Large-leaved standards

KNOWN BY BOTANICAL NAME
Rosaceae
Evergreen or Semi-evergreen
Small useful, trees, but not without problems.

Origin Dependent on variety.
Use Ideal evergreen or semi-evergreen effect for small, medium or large gardens.
Description *Flower* Small, white, four-petalled flowers with red calyx, in clusters or singly, dependent on variety, late spring to early summer. *Foliage* Ovate, dark green foliage with silver undersides. *Stem* Normally red-green when young, aging to purple-brown. A single central stem should be encouraged, to form a small to medium height tree. Normally quick-growing. Rounded or slightly pendulous habit. *Fruit* Red or yellow berries, dependent on variety, cascading down the outer surface of the branches in autumn, the tree's main attraction.
Hardiness Tolerates 4°F (−15°C).
Soil Requirements Any soil conditions.
Sun/Shade aspect Full sun through to medium shade.
Pruning Little or none required, other than to

maintain a single central stem, especially when young.
Propagation and nursery production By grafting or budding on to a stem of *C. bullatus*. Purchase container-grown or root-balled (balled-and-burlapped). May be scarce in commercial production, as only small numbers trained each year so must be sought from specialist nurseries.
Problems In grafted plants, suckers of *C. bullatus* may appear at any height up the stem and should be removed as soon as seen. May suffer from attacks of fire blight; if this is confirmed the plant should be immediately burned. The spread of this disease may cause the plants to disappear from commercial production, especially in Europe.

Cotoneaster cornubia in fruit

Varieties of interest *C. 'Cornubia'* Broad, lanceolate leaves, 2½-4in (6-10cm) long. White flowers produced in large clusters, followed by clusters of bright red berries. Of garden origin. *C. 'Exburiensis'* Light to mid-green, broad, lanceolate leaves. White flowers followed by clusters of bright yellow berries. Raised in the Exbury Gardens, Hampshire, England. *C. franchetii* Dark, grey-green, 1-2in (3-5cm) long, ovate to round foliage borne on long arching branches. Stems dark brown, almost black, covered in individual white flowers in early summer, followed by dull red berries in autumn. From China. *C. 'Rothschildianus'* Large clusters of yellow berries set off against light green, broad, lanceolate leaves 2½-4in (6-10cm), making a truly spectacular sight in autumn. Raised in England. *C. salicifolius var. floccosus* Narrow, elliptic, shiny-surfaced, green leaves 2½-4in (6-10cm) long with white undersides on long, graceful, pendulous

branches, giving a drooping, fan-like appearance. Clusters of white flowers give way to clusters of tiny, dull red fruits. From China. *C. 'St. Monica'* Leaves 6-8in (15-20cm) long, lanceolate, dark green with a purple hue and grey undersides. Veins and leaf petioles purple-red. Purple to purple-brown, strong, upright branches. Large clusters of white flowers, followed by clusters of bright red fruits. Originated in a convent garden in Bristol, England.
Average height and spread
Five years
12x2½ft (3.5x0.75m)
Ten years
12x4ft (3.5x1.2m)
Twenty years
or at maturity
16x10ft (5x3m)

COTONEASTER
'HYBRIDUS PENDULUS'

WEEPING COTONEASTER
Rosaceae
Semi-evergreen
A useful, attractive weeping tree.

Origin Of garden origin.
Use As a small weeping tree for small, medium or large gardens.
Description *Flower* Numerous small white flowers produced in clusters in early summer. *Foliage* Ovate, 2-3in (5-8cm) long, dark green, with reddish veins on silver undersides. May be retained well into winter. *Stem* Slightly upright at first, quickly becoming weeping and finally spreading. Relatively quick-growing, the tree slows in vigour as it matures, making a round-topped, symmetrical effect. *Fruit* Clusters of red fruits in autumn along full length of mature stems, retained into winter.
Hardiness Tolerates 4°F (−15°C).

Cotoneaster 'Hybridus Pendulus' in fruit

Soil Requirements Any good soil, but shows some resistance on severe alkaline soils by stunted, poor growth.
Sun/Shade aspect Full sun to light shade, preferring full sun.
Pruning None required, but trimming back of previous season's growth in early spring improves shape of young trees.
Propagation and nursery production From cuttings with stems trained up to required height, or by grafting or budding on to stems of *C. bullatus*. Best purchased container-grown. Trees of 6-10ft (2-3m) usually available. For mail order, take care to state that a standard is required; the plant is also sold as a creeping ground-cover shrub before training has commenced. May be found from garden centres, general nurseries or specialist sources.
Problems Susceptible to stem canker, which can kill entire branches. On poor soils may well lose its vigour. Also susceptible to fire blight; if this disease is confirmed the tree must be completely destroyed by burning.
Average height and spread
Five years
8x4ft (2.5x1.2m)
Ten years
8x8ft (2.5x2.5m)
Twenty years
or at maturity
10x13ft (3x4m)

CRATAEGUS
Autumn foliage varieties

THORN
Rosaceae
Deciduous
Trees with pretty flowers and interesting autumn fruits, as well as fine foliage colours. Although deciduous, some varieties hold leaves well into autumn.

Origin Throughout the northern hemisphere.
Use A useful tree for medium-sized and larger gardens. Suitable for avenues at 32ft (10m) apart and street planting.
Description *Flower* Small white florets, ½in (1cm) across, produced in clusters 2-3in (5-8cm) wide in early summer, often with a musty scent enjoyed by bees. *Foliage* Ovate, 2-3in (5-8cm) long, light or mid green, sometimes glossy, depending on variety. Good yellow or orange autumn colour. *Stem* Medium rate of growth, forming a ball-shaped tree, very closely branched. Stems grey-green and attractive in winter. Most varieties have large thorns, normally curved and mahogany brown. *Fruit* Clusters of round, orange or crimson fruits produced in autumn; in some varieties very late to ripen.
Hardiness Tolerant of winter temperatures down to −13°F (−25°C).
Soil Requirements Most soil conditions, except very dry.
Sun/Shade aspect Full sun to medium shade, preferring light shade.
Pruning None required, but can be cut back relatively hard and any obstructing branches removed.
Propagation and nursery production From budding or grafting onto understock of *C. monogyna*. Plant container-grown or bare-rooted. Most varieties relatively easy to find from general nurseries. Plants 6-13ft (2-4m) can be obtained. Best planting heights 6-10ft (2-3m).
Problems Slow to establish and may need two full springs to recover from transplanting. Sharp thorns can be a hazard in close garden planting.
Varieties of interest *C. 'Autumn Glory'* Excellent yellow and orange autumn colour; white flowers and red berries. Of garden origin. *C. crus-galli* (Cockspur Thorn) A flat-topped tree of more spreading habit and slightly less height. Foliage ovate to narrowly ovate, up to 3in (8cm) long, with toothed edges. Good orange-yellow autumn colours. White flowers in May followed by large clusters of long-lasting red fruits. Large thorns up to 3in (8cm) long. From North America. *C. durobrivensis* Leaves ovate with good autumn colours. White flowers followed by shining crimson berries maintained well into winter. Two-thirds average height and spread and may also be grown as a large shrub. Limited in commercial production outside its native environ-

ment of North America. *C.* × *grignonensis* Leaves ovate, up to 2½in (6cm), long, very glossy upper surface, downy grey underside. Large clusters of white flowers followed by oval to globe-shaped red fruits in autumn. Two-thirds average height and spread. Originating in France. Not readily available, but not impossible to find. *C.* × *lavallei* syn. *C. carrierei* Ovate, dark glossy green foliage with paler undersides. Good autumn colour. White flowers in clusters with dominant anthers of red and yellow. Fruits orange-red, globe-shaped, maintained well into winter. Two-thirds average height and spread. *C. mollis* Ovate leaves up to 4in (10cm) long with double-toothed edges. Light, downy grey-green at first, aging to light green. Good autumn colours. Flowers white with yellow anthers followed by large, globe-shaped, red fruits with downy texture. From North America. *C. pedicellata* (Scarlet Haw) Foliage light to mid green, slightly glossy with tooth edges. Good yellow-orange autumn colour. Numerous short thorns. White flowers in early to mid spring, followed in autumn by bunches of scarlet fruits. Two-thirds average height and spread. From north-eastern USA. *C. phaenopyrum* (Washington Thorn) Lobed, sharply tooth-edged leaves up to 2½in (6cm) long and wide. Very shiny upper surfaces, duller undersides. Good autumn colours of scarlet and orange. Clusters of white flowers followed by scarlet fruits which stay on the tree well into winter. Has an almost triangular shape. Can be difficult to obtain. From North America. *C. pinnatifida* Interesting crimson fruits with small, dark, red-brown to black dots over their surface. Leaves light to mid green, slightly glossy with deeply cut lobes. Good orange-red autumn colour. Few thorns, in some cases not produced at all. Two-thirds average height and spread. From Northern China. Not readily available and must be sought from specialist nurseries. *C. prunifolia* Dark green, glossy foliage, round to ovate with slightly downy undersides. Round clusters of white flowers up to 2½in (6cm) across, on downy stalks, followed by rounded crimson fruits which are rarely maintained far into winter. Two-thirds average height and spread. Thought to be a cross between *C. macrantha* and *C. crus-galli*. *C. submollis* Similar to *C. mollis*, but not reaching such a great height. Plants may be intermixed in nursery production and difficult to differentiate; *C. submollis* has 10 stamens to the flower and *C. mollis* has 20.
Other varieties of good autumn-coloured *Crataegus* may be found, but the above are recommended for garden planting.

Average height and spread
Five years
13x6ft (4x2m)
Ten years
20x4ft (6x4m)
Twenty years
or at maturity
26x20ft (8x6m)

Crataegus prunifolia **in fruit**

CRATAEGUS
OXYACANTHA

THORN, MAY, HAWTHORN
Rosaceae
Deciduous
With the other cut-leaved species and varieties listed, a group of attractive and reliable trees for late spring flowering.

Origin From Europe.
Use Ideal small round-topped trees for all types of garden. Can be severely pruned to control or shape the growth. Also useful for street planting due to the neat, tight, compact habit.

Crataegus oxyacantha 'Paul's Scarlet'

Description *Flower* Clusters of white, pink or red flowers, single or double depending on variety. Flower clusters up to 2in (5cm) across, produced in late spring. Musty scent attractive to bees. *Foliage* Basically ovate, 2in (5cm) long, very deeply lobed with 3 or 5 indentations. Grey-green with some yellow autumn tints. *Stem* Light grey-green, becoming grey-brown. Strong, upright when young, quickly branching. Armed with small, extremely sharp spines up to ½in (1cm) long. Medium rate of growth, forming a round-topped tree which spreads with age. *Fruit* Small, dull red, round to oval fruits produced in autumn containing two stone seeds, a distinctive characteristic.
Hardiness Tolerant of −13°F (−25°C).
Soil Requirements Any soil conditions, but shows signs of distress on extremely dry areas, where growth may be stunted.
Sun/Shade aspect Tolerates full sun to medium shade, preferring light shade.
Pruning None required, but responds well to being cut back hard if necessary.
Propagation and nursery production From seed for the parent; all varieties budded or grafted. Plant bare-rooted or container-grown. Available from 3ft (1m) up to 13ft (4m). Best planting heights 5½-6ft (1.8-2m).
Problems The sharp spines can make cultivation difficult. Suckers of understock may appear and must be removed.
Varieties of interest *C. laciniata* syn. *C. orientalis* Attractive dark grey-green, cut-leaved foliage, light grey undersides. White flowers and large dull orange-red fruits. Two-thirds average height and spread. Somewhat scarce in production. From the Orient. *C. monogyna* (Hedgerow Thorn, Singleseed Hawthorn) Single white flowers. Red, round fruits containing one stone. Foliage dark green. Rarely offered as a tree, normally used as a hedgerow plant. From Europe. *C. 'Stricta'* White flowers followed by orange-red berries. Upright growth of average height, maximum spread of 4ft (1.2m). Useful as a street tree or for planting in confined spaces. *C.* × *mordenensis 'Toba'* Double, creamy white flowers in the spring, followed by red berries in autumn. *C.*

Crataegus oxyacantha 'Rosea Plena' in flower

oxyacantha *'Alba Plena'* Double white flowers in mid spring, followed by limited numbers of red berries in autumn. *C. o. 'Crimson Cloud'* Profuse single, dark pink to red flowers with yellow eyes in late spring. Good yellow-bronze-orange autumn colour. *C. o. 'Fastigiata'* Single white flowers, limited red berries. A narrow, columnar tree of average height and no more than 6ft (2m) spread. *C. o. 'Gireoudii'* New foliage on new growth mottled pink and white, aging to green. New foliage on old wood green. Bush-forming. Two-thirds average height and spread. A light annual pruning of outer extremities in early spring is recommended to encourage new variegated growth. Shy to flower, but can produce white, musty-scented flowers. Limited fruit production. Difficult to find and must be sought from specialist nurseries. *C. o. 'Paul's Scarlet'* syn. *C. o. 'Coccinea Plena'* Dark pink to red, double flowers produced late spring, early summer. Limited red berries in autumn. *C. o. 'Rosea Plena'* Double pink flowers. Produced late spring, early summer. Some limited berrying.

Average height and spread
Five years
13x4ft (4x1.2m)
Ten years
20x10ft (6x3m)
Twenty years
or at maturity
20x20ft (6x6m)

× CRATAEMESPILUS GRANDIFLORA

KNOWN BY BOTANICAL NAME
Rosaceae
Deciduous
An attractive tree, and something of a curiosity.

Origin Found in the wild in France, a natural cross between a *Crataegus* and a Medlar.
Use As a small, flowering specimen tree of botanical interest.
Description *Flower* Pure white flowers up to ½in (3cm) across, produced in twos or threes, from late spring to early summer. *Foliage* Leaves oval to ovate, 2-4in (5-10cm) long, lobed, light green with downy grey undersides. Some good orange to orange-yellow autumn colours. *Stem* Grey-green when young, branching from an early age, becoming purple-brown, froming a short-stemmed, round-topped tree *Fruit* Globular or oval, yellow-bronw, and edible.
Hardiness Tolerates 4°F (−15°C). Susceptible to mechanical wind damage, uprooting the relatively poor root system.
Soil Requirements Needs deep, rich loam to do well. Poor soils may lead to poor root

development and growth.
Sun/Shade aspect Best in full sun, but tolerates light shade.
Pruning None required; in fact, may resent it.
Propagation and nursery production From grafting or budding on to understock of *Crataegus monogyna*. Relatively scarce in production. Plants normally offered from 3-8ft (1-2.5m) but range of sizes may be limited. Plant bare-rooted or container-grown. May be offered as a multi-stemmed shrublet which will need some training to achieve true status.

× *Crataemespilus grandiflora* in fruit

Problems None, if good soil and conditions are provided, except risk of damage in high wind.
Average height and spread
Five years
13x4ft (4x1.2m)
Ten years
20x10ft (6x3m)
Twenty years
or at maturity
26x20ft (8x6m)

CYDONIA OBLONGA

QUINCE
Rosaceae
Deciduous
Beautiful spring-flowering trees, with autumn fruits both edible and attractive.

Origin Parent from northern Iran and Turkestan; many varieties of garden origin, particularly from France.
Use As freestanding ornamental and fruiting trees. Can be grown on a stem or as large bushes. May be fan-trained on all but the most exposed walls.
Description *Flower* Saucer-shaped, delicate mother-of-pearl to light rose-pink, produced in good numbers in mid to late spring. Slightly scented. *Foliage* Leaves ovate, tooth-edged, mid to dark green with silver undersides. Good yellow autumn colour. *Stem* Upright when young, branching with age. Attractive growth formation. Branches dark brown to purple-brown, forming a round-topped large shrub or small tree. Medium to fast rate of growth. *Fruit* Medium-sized, round or pear-shaped yellow fruits, depending on variety, abundantly produced in late summer, early autumn. Used to make quince jelly.

Cydonia oblonga 'Vranja' in fruit

Hardiness Tolerates −13°F (−25°C).
Soil Requirements Most soil conditions; only dislikes extremely dry or very waterlogged areas.
Sun/Shade aspect Best in full sun, to allow fruit to ripen.
Pruning None required, other than the removal of any crossing branches which may rub and form lesions. Centres may be thinned occasionally to speed up ripening of fruit. Pruning best carried out in winter.
Propagation and nursery production *C. oblonga* from seed; all named varieties by budding or grafting. Planted bare-rooted or container-grown. Must be sought from specialist nurseries. Best planting heights 5-8ft (1.5-2.5m).
Problems Fruiting can be a little erratic, especially after hard, late spring frosts at flowering time. Trees tend to look misshapen and irregular when young, but grow in well.
Varieties of interest *C. o. 'Meech's Prolific'* Round, squat, pear-shaped fruits produced in good quantities. Flowers not the best feature. *C. o. 'Portugal'* Small, pear-shaped fruits in profusion; a good culinary variety. Generous flower production. *C. o. 'Vranja'* Large, pear-shaped, yellow fruits. Large flowers, somewhat sparsely produced, but the tree is a fine ornamental form.
Average height and spread
Five years
10x8ft (3x2.5m)
Ten years
16x13ft (5x4m)
Twenty years
or at maturity
23x20ft (7x6m)

CYTISUS BATTANDIERI

MOROCCAN BROOM, PINEAPPLE BROOM
Leguminosae
Deciduous
An interesting tree for mild climates, in some areas considered semi-evergreen.

Origin From Morocco.
Use As a small, mop-headed, silver-leaved flowering tree.
Description *Flower* Long panicles of bright yellow, pineapple-scented flowers produced early to mid summer. *Foliage* Grey to silver-grey pinnate leaves. *Stem* Strong and upright when young, silver-grey with a downy covering. Turns grey-green to grey-brown with age. A single stem can be trained from a shrub or wall-type specimen to form a small standard mop-headed tree, but only in mild areas. *Fruit* May produce small green pea-pods, but of little attraction or reliability.
Hardiness Can be grown as a tree only in areas with temperatures no less than 23°F (−5°C) and well protected from high winds. Requires staking throughout its life.
Soil Requirements Most soil conditions, but shows signs of distress on very thin chalk soils.
Sun/Shade aspect Best in full sun.
Pruning None required; to confine growth young wood should be cut after flowering. Cutting old wood may lead to die-back, so is best planted with space to attain full maturity without pruning.

Cytisus battandieri in flower

Propagation and nursery production From grafting, cuttings or seed. Purchase container-grown. Young plants of 2ft (60cm) to 3ft (1m) have to be trained to standard shape. Rarely found in commercial production as a pre-trained standard.
Problems Only successful in mild areas.
Average height and spread
Five years
14½x6ft (4.5x2m)
Ten years
21x13ft (6.5x4m)
Twenty years
or at maturity
25x20ft (7.5x6m)

CYTISUS × PRAECOX and CYTISUS SCOPARIUS

BROOM, WARMINSTER BROOM
Leguminosae
Deciduous
Interesting, small, mop-headed 'toy' trees for spring flowering.

Origin Of garden origin; basic forms from Europe.
Use As small trees for patios or as highlighting features in a small or medium-sized garden.
Description *Flower* Scented pea-flowers, abundantly produced and numerous colours depending on variety. *Foliage* Small, lanceolate, grey-green leaves, very sparse. *Stem* C. × praecox and C. scoparius varieties can be grafted on to a short stem of Laburnum, to form small, mop-headed to slightly pendulous standards. Stems light green to grey-green. *Fruit* Small, grey-green pea-like pods may be produced, providing winter interest.

Hardiness Tolerates 14°F (−10°C).
Soil Requirements Best on a neutral to acid soil. Dislikes alkaline soils, which may cause chlorosis.
Sun/Shade aspect Full sun.
Pruning None required, but small sections of year-old growth may be removed after flowering to contain shape.
Propagation and nursery production By grafting on to stems of *Laburnum vulgaris*. Purchase container-grown or root-balled (balled-and-burlapped). Relatively scarce in production as only a small number of trees are propagated each year; must be sought from specialist nurseries. Two stem sizes usually available, 3ft (1m) and 5ft (1.5m).

Cytisus × praecox in flower

Problems A limited number of suckers may arise from the Laburnum understock, which should be removed when seen. Relatively short-lived plants; useful life 10-15 years.
Varieties of interest *C. × praecox 'Albus'* Pure white flowers. *C. × p. 'Allgold'* Golden yellow flowers. *C. scoparius 'Andreanus'* Crimson-red and chrome yellow flowers. *C. s. 'Burkwoodii'* Maroon and bright red flowers. *C. s. 'Windlesham Ruby'* Mahogany crimson flowers.
The above may be found grafted as mop-headed standards. Other suitable varieties may be available for garden planting.
Average height and spread
Five years
6x3ft (2x1m)
Ten years
6x3ft (2x1m)
Twenty years
or at maturity
6x3ft (2x1m)

DAVIDIA INVOLUCRATA

DOVE TREE, GHOST TREE, HANDKERCHIEF TREE
Davidiaceae
Deciduous
A beautiful flowering tree, requiring patience.

Origin From China.
Use As a large, spring-flowering tree for medium-sized and larger gardens.
Description *Flower* Small, central black flowers, flanked by two broad, ovate, pure white leaves 2½-6in (6-15cm) long, acting like flower petals to attract insects. Produced late spring, but tree will take up to 20 years from planting date to come into full flower. *Foliage* Large, deeply veined, tooth-edged leaves, mid green upper surfaces with white felted undersides. Broad to ovate, 2½-6in (6-15cm) long and 2-2½in (5-6cm) wide. Good yellow autumn colour. *Stem* Grey-green to purple-green. Strong upright shoots, over 6ft (2m) long in established plants. Branches in ascendant formation. Trees up to

20 years are pyramidal, eventually evolving a more rounded head. *Fruit* Small, round to pear-shaped fruits hanging on short stalks produced in good quantities in autumn. Green aging to green-brown.
Hardiness Tolerates 14°F (−10°C); some terminal bud damage may occur in winter, causing stem die-back.
Soil Requirements Does best on rich, deep loam which encourages quick growth formation. Tolerates moderate alkalinity.
Sun/Shade aspect Best in light shade when young, growing into full sun with time; but tolerates medium shade to full sun.
Pruning None required once established; between young shrublet and tree stages requires some training into single central stem. Winter die-back should be removed.
Propagation and nursery production From layering. Young shrublets normally available up to 18in (45cm), often multi-stemmed. After planting a number of strong new growths appear, normally in the second spring. The best and strongest shoot is selected and trained to tree shape. It is not advisable to buy plants of more than 3ft (1m), as they may become retarded or die back. Purchase root-balled (balled-and-burlapped) or container-grown. May have to be sought from specialist nurseries, although in good years for propagation the plants are generally available.

Davidia involucrata in flower

Problems It takes a long time to train a tree and bring it to flower, but the full effect is well worth waiting for.
Varieties of interest *D. vilmoriniana* In effect identical to *D. involucrata*, often sold as the same.
Average height and spread
Five years
6x4ft (2x1.2m)
Ten years
20x12ft (6x3.5m)
Twenty years
or at maturity
39x16ft (12x5m)
Continued growth can exceed 66x33ft (20x10m).

EUCALYPTUS

EUCALYPTUS, GUM TREE
Myrtaceae
Evergreen
Attractive blue-leaved and ornamental-stemmed trees, requiring mild conditions.

Origin From Australia and Tasmania.
Use As an ornamental foliage tree or large shrub for all but the smallest gardens. Very useful for flower arranging and floral decoration. Can be grown in large containers for conservatories or greenhouses, possibly moving outside in summer for patio display.

29

Eucalyptus gunnii in leaf

Description *Flower* White or cream tufted clusters in late autumn, early winter. *Foliage* The main attraction, ovate to broadly linear, depending on variety. Stiff leathery texture, blue-green to glaucous blue. Adult foliage often different from juvenile in shape and form. *Stem* Grey-green to blue-green, often with peeling bark revealing primrose-yellow underskin. Winter stems attractive. Often very quick growing. *Fruit* Rarely fruits outside its native environment.
Hardiness Minimum winter temperature 23°F (−5°C). Hard frost can cause severe damage.
Soil Requirements Light, well-drained soil for best results; tolerates a range of conditions.
Sun/Shade aspect Full sun to light shade, preferring full sun.
Pruning None required, but can be cut back hard and stooled to make a multi-stemmed large shrub or small tree. This can be done annually or biennially, and greatly enlarges foliage size; but decreases ultimate height, while increasing overall spread.
Propagation and nursery production From seed. Purchase container-grown. Plants best moved between 15in (40cm) and 4ft (1.2m). Larger trees for transplanting are not recommended. Hardiest varieties generally available. Outside favourable locations, more tender varieties must be sought from specialist nurseries.
Problems May look weak when young in nursery conditions, but develops quickly once planted out. Reacts very quickly to poor planting; requires good preparation for speedy results.
Varieties of interest *E. coccifera* Juvenile foliage glaucous blue, round to oval, up to 1½in (4cm) long. Adult foliage grey-green, narrow, oblong to lanceolate, up to 4in (10cm) long. Yellow flower umbels produced in clusters late autumn, early winter. May produce conical fruits. One-third average height and spread. From Tasmania. *E. dalrympleana* Attractive foliage and patchwork bark. Light brown bark with grey-white undercolour progressing in area as tree matures. Young foliage bronze; later grey-green, broadly lanceolate, up to 5in (12.5cm) long. *E. globulus* (Blue Gum) Juvenile foliage glaucous white, up to 6in (15cm) long and 2½in (6cm) wide. Adult foliage is dark, shiny green, lanceolate, up to 12in (30cm) long and 2in (5cm) wide. Tufted clusters of three individual white florets produced late autumn, early winter. From Tasmania. In its native environment can exceed 100ft (30m) but usually achieves around 48ft (15m) in height. Tender; for use in non-mild areas only as annual foliage bedding plant. *E. gunnii* (Cider Gum) Juvenile foliage orbicular, up to 2in (5cm) across, glaucous white, produced on short stalks. Adult foliage green, lanceolate, up to 4in (10cm) long. White flowers in threes produced late autumn to early winter. May reach more than average height and spread in mild areas. From Tasmania and Southern Australia. One of the hardiest of all Eucalyptus. *E. niphophila* (Snow Gum) Slow-growing with large, leathery, grey-green, ovate to lanceolate leaves 8in (20cm) long. Attractive trunk patched with green, grey and cream. Relatively hardy. From Australia. *E. parvifolia* Narrow, lanceolate, blue-green leaves, up to 6in (15cm) long. Tolerates more alkalinity than most. *E. pauciflora* syn. *E. coriacea* (Cabbage Gum) Juvenile foliage round to broad lanceolate, glaucous green, up to 8in (20cm) long and 6in (15cm) wide. Adult foliage produced on long stalks up to 8in (20cm) long and 1in (2.5cm) wide. White flowers in tufted clusters of 5-12. From Australia and Tasmania. Relatively hardy. *E. perriniana* Juvenile foliage small, silver-blue, ovate to round. Adult foliage lanceolate, glaucous blue. Stems white with dark purple-brown blotches. *E. pulverulenta* Attractive peeling bark. Juvenile foliage ovate to round, sometimes kidney-shaped, up to 2½in (6cm) long. Adult foliage similar, slightly larger. Tufted clusters of three white florets produced late autumn, early winter. Two-thirds average height and spread. From New South Wales.
There is an extremely large number of available Eucalyptus forms. Those listed are among the most hardy, but even so, careful selection is required to suit the plant to its particular location.
Average height and spread
Five years
16x6ft (5x2m)
Ten years
32x13ft (10x4m)
Twenty years
or at maturity
64x26ft (20x8m)

Eucalyptus niphophila – branches

EUODIA HUPEHENSIS
(Evodia hupehensis)

KNOWN BY BOTANICAL NAME
Rutaceae
Evergreen
An attractive but very slow-growing evergreen, flowering through late summer and early autumn.

Origin From North China and Korea.
Use As a freestanding, small to medium-sized tree for late summer flowering.
Description *Flower* Pungent smelling, white, panicle-shaped clusters of small flowers in late summer, early autumn. *Foliage* Compound pinnate leaves, mid green, glossy upper surfaces, dull grey undersides. Normally evergreen, but may be defoliated in severe winters. *Stem* Slow growing, light green. Attractive buds in winter. *Fruit* Clusters of red fruits in autumn, not always regularly produced.
Hardiness Tolerates 14°F (−10°C), but may defoliate in severe winters.
Soil Requirements Most soil conditions; tolerates high alkalinity but needs plenty of moisture on these soils to make any real growth.
Sun/Shade aspect Full sun to light shade.

Euodia hupehensis in flower

Pruning Rarely required, other than occasional shaping when young.
Propagation and nursery production From seed. Purchase container-grown. Extremely difficult to find and must be sought from specialist nurseries. Best planting heights 1½-3ft (50cm-1m). Rarely found larger.
Problems Although an evergreen, foliage attraction alone does not warrant mass planting. Its rate of growth and development compared with other trees is extremely slow.
Varieties of interest *E. daniellii* Flowers more clustered followed by fleshy black fruits. Similar growth to *E. hupehensis*. From China and Korea. Extremely difficult to find, but not impossible.
Average height and spread
Five years
5x3ft (1.5x1m)
Ten years
8x10ft (2.5x3m)
Twenty years
or at maturity
32x16ft (10x5m)

FAGUS SYLVATICA

COMMON BEECH, EUROPEAN BEECH
Fagaceae
Deciduous
Large, attractive trees with fine foliage, for medium to large gardens.

Origin From Europe, including Great Britain.
Use As a large standard tree for large gardens, estates, parks; for open landscape features or coppices. Often grouped to form a windbreak. Can be used for pleaching when trained horizontally. Useful avenue tree when planted at 39ft (12m) apart.
Description *Flower* Small, yellow-green to green-brown flowers in spring. *Foliage* Oval to ovate, 2-3in (5-8cm) long, slightly toothed edged and wavy. Bright green with a grey sheen, especially when young. Bronze-yellow

in autumn. Leaves retained in some forms well into winter. *Stem* Grey to grey-green when young, becoming green-brown and finally dark brown. Strong and upright when young, more branching with age, normally forming a round-topped tree quite quickly. *Fruit* Small, hanging husks protect the fruit, light green aging to brown and armed with small, soft spines.

Hardiness Tolerates −13°F (−25°C).

Soil Requirements Any soil type; capable of flourishing in alkaline conditions.

Sun/Shade aspect Full sun to light shade, preferring full sun.

Pruning None required, but takes extensive cutting back, and can be trained into shapes.

Propagation and nursery production From seed, or grafting for named varieties. Best planted container-grown, but also available bare-rooted. Best planting heights 3-8ft (1-2.5m): 5-5½ft (1.5-1.8m) is most suitable, but can be purchased from 15in (40cm) up to 20ft (6m). *F. sylvatica* usually available, but may have to be sought from forestry outlets. Named varieties mostly available from specialist nurseries.

Problems Mature trees may suffer from beech-bark fungus after 40 or more years. Often slow-growing when young; may be third spring before established and growing well. Mortality rate can be high for bare-rooted stock, especially in early years.

Varieties of interest *F. englerana* (Chinese Beech) Lanceolate, grey-blue young foliage up to 4in (10cm) long. Attractive fruits produced on long stalks. Graceful habit. Two-thirds average height and spread. Scarce but not impossible to find. *F. grandifolia* syn. *F. americana*, *F. ferruginea* (American Beech) Light green foliage 5in (12cm) long, ovate and coarsely toothed, glabrous upper surface, downy underside. Large fruit. Average height and spread, with suckering shoots. From North America, and rarely performs well elsewhere. *F. sylvatica 'Albovariegata'* Two-thirds average height and spread and slower-growing. Produces white variegated leaves which may scorch in strong summer sunlight. Scarce in commercial production. *F. s. 'Aurea Pendula'* An extremely rare weeping form which and must be sought from very specialized sources. New foliage golden yellow aging to yellow-green. Remains small for many years, ultimately reaching two-thirds average height and spread. *F. s. 'Cockleshell'* Interesting, very rounded foliage, bright green with bronze autumn colours. Upright in habit, reaching two-thirds average height and spread. Relatively scarce in commercial production, but not impossible to find. *F. s. 'Cristata'* (Cock's Comb Beech) Unusual foliage, leaves clustered, deeply lobed and curled. Scarce and difficult to find. Two-thirds average height and spread. *F. s. 'Dawyck'* syn. *F. s. 'Fastigiata'* (Dawyck Beech) Round to ovate, bright green glossy foliage. Narrow and

Fagus sylvatica var. heterophylla in leaf

upright. Fastigiate stems with loose cork-screwing effect. Yellow and bronze autumn tints. Reaches average height but maximum 13ft (4m) spread in mature specimens. Useful for street planting or for large-feature, upright trees. *F. s. 'Dawyck Gold'* Very scarce. Young foliage in spring golden yellow, aging to light yellow-green. Yellow-bronze autumn tints. Maximum height 30ft (10m), spread 13ft (4m). *F. s. 'Dawyck Purple'* Foliage purple-red when young, aging to purple in mid to late summer, coppery in winter. Leaves not held. Upright, pyramidal habit. One-third average height and reaching only 10ft (3m) in spread. *F. s. var. heterophylla* syn. *F. s. 'Asplenifolia'*, *F. s. 'Incisa'*, *F. s. 'Laciniata'* Foliage deeply cut and lobed to varying degrees, light to mid green, slightly glossy. Yellow to yellow-bronze autumn colour. Upright when young, broadly spreading with age. *F. s. 'Latifolia'* Large foliage, purple-red in spring, turning coppery in winter. *F. s. 'Luteovariegata'* Green foliage margined with yellow to gold in spring, variegation lessens as summer approaches. Two-thirds average height and spread. Relatively difficult to find in commercial production. *F. s. 'Pendula'* Leaves rounded and light glossy green. Requires training when young to obtain height. After 10-15 years of establishment makes strong, upright, ranging branches, which arch and become completely pendulous. Initially useful in most gardens, but in 30-50 years outgrows all but the largest sites. *F. s. var. purpurea* (Purple or Copper Beech) Rounded leaves, bright red-purple in spring, turning copper in winter but not retained. Slightly less than average height and spread. *F. s. 'Purpurea Pendula'* (Weeping Purple Beech) Upright branches quickly becoming pendulous.

Bright purple-red foliage in spring, aging to bronze-purple in winter. Small for many years, but ultimately grows very large. *F. s. 'Riversii'* (Rivers Purple) Large foliage, very dark purple in summer, turning bronze in winter. Leaves not retained. Slightly quicker-growing than most purple-leaved forms. *F. s. 'Roseomarginata'* Good size, ovate, purple foliage with pink or white leaf margins retained until midsummer, when it normally decreases or disappears. Two-thirds average height and spread, but still a large tree. *F. s. 'Rotundifolia'* Upright with very round, bright green foliage and coppery autumn tints. Average height and two-thirds average spread. *F. s. 'Tortuosa'* Branches twisted and contorted. Green foliage. One-third average height and spread. An interesting architectural tree but scarce in production. *F. s. 'Zlatia'* Bright golden yellow spring growth, aging to lime-green in summer. Two-thirds average height and spread. Spectacular effect, worthy of wider planting in medium to large gardens.

Fagus sylvatica var. purpurea in leaf

Average height and spread
Five years
10x10ft (3x3m)
Ten years
26x26ft (8x8m)
Twenty years
or at maturity
46x46ft (14x14m)
Continued growth over 50 years
reaches 82x52ft (25x16m)

Fagus sylvatica in autumn

Fagus sylvatica 'Roseomarginata' in leaf

FORSYTHIA
Trained as standard

GOLDEN BALL
Oleaceae
Deciduous
An interesting, small-feature, spring-flowering shrub trained as a standard tree.

Origin Native to Europe, but many forms of garden origin and a large number raised by Dr Karl Sax at the Arnold Arboretum, Massachusetts, USA.
Use As a small, mop-headed, spring-flowering tree, useful for small gardens, patio and tub work.
Description *Flower* Golden to lemon yellow, depending on variety, hanging, bell-shaped flowers covering the bare stems in great profusion early to mid spring. *Foliage* Ovate, 2-3in (5-8cm) long, tooth-edged foliage, light to mid green with yellow, purple or red-purple autumn colours. *Stem* Yellow to light brown-green twigs and stems with crusted bark. When used as a standard on a single stem needs staking throughout its life. Quick-growing. *Fruit* Small brown to grey-brown seedheads after flowering.
Hardiness Tolerates −13°F (−25°C).

Forsythia 'Spring Glory' in flower

Soil Requirements Any soil conditions.
Sun/Shade aspect Full sun to medium shade. Deep shade makes the tree open, lax and shy to flower.
Pruning One-third of the oldest flowering wood should be removed after flowering. Requires training to a central stem, supported by a cane and with side branches removed to encourage a mop-headed effect. When young can be trimmed to any shape, but this may reduce flowering.
Propagation and nursery production From cuttings. Plant bare-rooted or container-grown. Standard forms are not readily available; must be sought from specialist nurseries, or can be trained after planting as a shrub. Best planting heights 5½-6ft (1.8-2m).
Problems As a standard tree the stem is not entirely sound and will break in high winds; it requires good staking throughout its life.
Varieties of interest *F.* × *intermedia 'Spectabilis'* Profuse golden yellow flowers, relatively unfading. Of garden origin. *F. 'Lynwood'* syn. *F. intermedia, F. 'Lynwood Gold'* One of the best flowering forms with broad-petalled, large, clearly defined, rich yellow flowers. Of garden origin, found in Northern Ireland in the early 1900s. *F. 'Spring Glory'* Profuse bright yellow flowers in mid spring. Smooth, light brown stems. Probably the best variety for producing a small standard tree and may be found pretrained.
Average height and spread
Five years
13x5ft (4x1.5m)
Ten years
16x8ft (5x2.5m)
Twenty years
or at maturity
16x12ft (5x3.5m)

FRAXINUS EXCELSIOR

COMMON ASH, EUROPEAN ASH
Oleaceae
Deciduous
Large trees, including weeping forms attractive in all seasons; useful in medium-sized and larger gardens.

Origin Most of Europe.
Use As a large tree for large gardens, estates or parks. Can be used as an avenue tree when planted 39ft (12m) apart. Good for windbreaks.

Fraxinus excelsior 'Jaspidea'
in autumn

Description *Flower* Short racemes of green-yellow flowers in spring. *Foliage* Pinnate leaves, 10-12in (25-30cm) long, with 7-11 oblong to lanceolate leaflets, each up to 4in (10cm) long. *Stem* Grey-green, strong and upright. May exceed 10ft (3m) without branching when young, becoming branching with age. Predominant and attractive black buds in winter. Tree is upright for 20-25 years, then spreads and ultimately becomes pendulous. *Fruit* Hanging clusters of winged fruit 1in (3cm) long. Green when young, aging to green-yellow; still quite attractive in winter.

Hardiness Tolerates winter temperatures below −13°F (−25°C).
Soil Requirements Any soil type.
Sun/Shade aspect Full sun to medium shade, preferring full sun.
Pruning None required, but can safely be reduced in height.
Propagation and nursery production From seed for *F. excelsior* or named varieties from grafting. Weeping forms usually grafted on to a 6-10ft (2-3m) high stem of *F. excelsior*. Plant bare-rooted or container-grown. Normally available from general or specialist nurseries.
Problems Can quickly outgrow the space allowed. In many localities grows almost as a weed, since it seeds and germinates readily.
Varieties of interest *F. e. 'Aurea Pendula'* Golden yellow winter stems, punctuated by black side and terminal buds. Foliage light yellow in spring, aging to green-yellow in autumn. Reaches no more than 13ft (4m) in height, but spreads up to 26ft (8m) or more. A useful weeping tree in medium-sized or larger gardens. *F. e. 'Diversifolia'* syn. *F. e. 'Monophylla'* (One-leaved Ash) Light grey-green foliage, ovate to lanceolate. Sometimes divided into threes with toothed edges. White fluffy flower panicles in summer. Yellow autumn colours. Two-thirds average height and spread. *F. e. 'Jaspidea'* Upright growth. Stems yellow to golden yellow with distinct black buds in winter. New foliage light golden yellow. Yellow autumn tints. White upright panicles of fluffy flowers in summer and grey-brown fruits in autumn. *F. e. 'Pendula'* Initially a small weeping tree, reaching 20ft (6m) in height and spread. After 10-15 years often produces strong upright growth to ultimate height of 32ft (10m) or more before weeping habit increases spread to 32ft (10m). Upright shoot can be removed to maintain a relatively compact weeping form. *F. e. 'Westhof's Glory'* Large fluffy white flowers in late spring to early summer. Large green foliage turning to good yellow autumn colour. Relatively upright branches.
Average height and spread
Five years
20x6ft (6x2m)
Ten years
32x13ft (10x4m)
Twenty years
or at maturity
64x23ft (20x7m)
Continued growth over 50 years reaches 77x32ft (24x10m)

Fraxinus excelsior 'Pendula' in winter

FRAXINUS
Other varieties

ASH
Oleaceae
Deciduous
An interesting range of trees for medium to large gardens, with attractive foliage, flowers and fruits.

Origin Dependent on variety.
Use As shade trees for medium to large gardens upwards. Good for avenues when planted at 39ft (12m) apart. Ideal as wind-break.
Description *Flower* Normally upright, short racemes of fluffy white or cream flowers in late spring, early summer. *Foliage* Single, ovate or pinnate, according to variety. Up to 10in (25cm) long. Normally light green, but some variegated with white. *Stem* Grey-green with dominant black winter buds on most varieties. Normally upright and pyramidal, branching with age. Quick-growing. *Fruit* Hanging clusters of winged fruit 1in (3cm) long, green when young, aging to green-yellow. Retained well into winter as attractive grey-brown seed pods.
Hardiness Tolerates −13°F (−25°C).
Soil Requirements Any soil type.
Sun/Shade aspect Full sun to mid shade, preferring full sun.
Pruning None required, but can be reduced in height as necessary.
Propagation and nursery production By grafting or seed, depending on variety. Plant bare-rooted or container-grown. Trees available from 3-16ft (1-5m). Best planting heights 5½-10ft (1.8-3m). Most varieties listed here must be sought from specialist nurseries.
Problems May outgrow its allotted space.

Fraxinus ornus in flower

Varieties of interest *F. americana* (White Ash) Foliage pinnate, up to 15in (40cm) long, consisting of 7-9 lanceolate leaflets, each 4-6in (10-15cm) long. Light grey-green, giving good yellow autumn tints. Tufted, creamy white flowers followed by winged fruits up to 2in (5cm) long, grey-green aging to green-brown. From North America. *F. mariesii* Very grey, downy young wood, becoming grey-green with age. Pinnate foliage, up to 7in (18cm) long, with 3-5 oval or ovate leaves up to 3in (8cm) long. Flowers creamy white in loose open panicles up to 6in (15cm) long, produced in early summer, followed by fruits up to 1in (3cm) long, oblong, deep purple, held into winter. One-third average height and spread. Originating in China. *F. ornus* (Manna Ash) Pinnate foliage up to 8in (20cm) long with 5-7 leaflets each up to 2½in (6cm) long, grey-green with shallow toothed edges. Flowers scented, off-white, in panicles up to 4in (10cm) long, in late spring. Fruits narrow, oblong, up to 1in (3cm) long, produced in autumn and retained into winter. Two-thirds

average height and spread. From Southern Europe and Asia Minor. *F. oxycarpa 'Flame'* Pinnate foliage with 7-9 leaflets up to 10in (25cm) long. Creamy white flowers in short panicles. Narrow, light green, lanceolate fruits up to 1in (3cm) long. Vivid flame red autumn colours. Two-thirds average height and spread. Not readily available in commercial production, but can be found. Hybrid of American origin. *F. o. 'Raywood'* Similar variety in leaf form to *F. oxycarpa 'Flame'*, but with purple autumn colours. Half average height and spread. Of North American origin. *F. pennsylvanica* (Green Ash) Pinnate foliage up to 12in (30cm) long, with 7-9 oblong to lanceolate leaflets. Foliage light grey-green with yellow autumn colour. Fruits up to 2½in (6cm) long, green to yellow-green. Fast-growing, reaching two-thirds average height and spread. Originating in eastern North America, and rarely grown elsewhere. *F. p. 'Variegata'* Attractive white variegated foliage. Deserves wider planting, but extremely difficult to find.
Average height and spread
Five years
20x6ft (6x2m)
Ten years
32x13ft (10x4m)
Twenty years
or at maturity
64x23ft (20x7m)
Continued growth over 50 years reaches 84x32ft (26x10m)

GLEDITSIA
TRIACANTHOS

HONEYLOCUST
Leguminosae
Deciduous
Trees with a very wide range of habit, from tall specimens to attractive, compact ornamentals.

Origin From North America.
Use As a large, tall, shade tree for medium-sized or larger gardens. Green, golden and purple varieties as small, ornamental trees for any size of garden.
Description *Flower* Green-white male flowers formed in hanging racemes 2in (5cm) long. Female flowers limited and inconspicuous. Borne in midsummer. *Foliage* Pinnate or bipinnate, up to 8in (20cm) long with up to 32 leaflets 1in (3cm) long. Light green with a glossy sheen. Good yellow autumn colour. *Stem* Light grey-green, somewhat fragile. May suffer from wind damage. Spines from 2½in (6cm) to 12in (30cm) long, produced singly or up to 7 in a bunch. Forms upright tree, crown spreading with age. Mature bark grey and deeply channelled, of architectural interest. *Fruit* Pea-shaped, grey-green pods, up to 12-18in (30-45cm) long, sword-shaped and twisted, often retained well into winter.
Hardiness Tolerates 14°F (−10°C).

Gleditsia triacanthos in winter

Soil Requirements Best results on well-drained, deep, rich soil. Tolerates limited amount of alkalinity through to fully acid types.
Sun/Shade aspect Full sun to light shade.
Pruning None required, but shortening of previous season's growth in spring improves overall shape while young.

Gleditsia triacanthos 'Sunburst' in leaf

Propagation and nursery production From seed for the parent. Varieties grafted. Purchase container-grown 6-8ft (2-2.5m) high for best transplanting results. Should be sought from general or specialist nurseries, but some named varieties distributed through garden centres.
Problems *G. triacanthos* suffers from wind damage, losing quite large limbs, often without warning. Ornamental varieties are normally slower-growing but are small enough to resist wind damage.
Varieties of interest *G. t. 'Bujoti'* Pendulous branches covered in bright green, narrow, pinnate foliage. Good yellow autumn colour. White flowers. Reaches only one-third average height and spread. *G. t. 'Elegantissima'* Compact, only reaching one-third ultimate height and spread; can be considered as a large shrub. Interesting light green, fern-like foliage. Yellow autumn colour. *G. t. var. inermis* (Thornless Honeylocust) A small, round, mop-headed tree. Delicate light green pinnate leaves giving good autumn colour. Mature trees may produce white flowers. *G. t. 'Moraine'* Attractive light to mid green foliage. White flowers. Round habit. *G. t. 'Ruby Lace'* Foliage purple to purple-green in spring; new growth red-purple. Some yellow-bronze autumn colour. White flowers, these rarely occur. Spreading when young, eventually forming a round-topped tree. Reaches only one-third ultimate height and spread. *G. t. 'Skyline'* Attractive small light green foliage. White flowers. Upright pyramidal habit. Two-thirds average height and one-third average spread. Good for confined spaces. Difficult to find in commercial production. *G. t. 'Sunburst'* Beautiful bright golden yellow, pinnate foliage. Round-topped when young, becoming tall and upright with age. Yellow autumn colour. Two-thirds average height and spread. Responds well to pruning when young.
Average height and spread
Five years
13x6ft (4x2m)
Ten years
26x13ft (8x4m)
Twenty years
or at maturity
32x26ft (10x8m)
In favourable conditions continued growth reaches 45x32ft (14x10m)

GYMNOCLADUS DIOICUS

KENTUCKY COFFEE TREE
Leguminosae
Deciduous
An interesting large-leaved tree with seasonal foliage colour.

Origin From North America, where it was originally used by settlers as a substitute for coffee, hence the common name.
Use As a feature tree, for its very large ornamental leaves.
Description *Flower* Female flowers green-white in panicles 8-12in (20-30cm) long, male 2-4in (5-10cm) long, produced in early summer. Rarely flowers outside its native environment. *Foliage* Leaves up to 3ft (1m) long and 2ft (60cm) wide, produced in pinnate groups of 3 to 7 ovate leaflets. New spring foliage has pink colouring. Yellow autumn colour. *Stem* Grey-green bark. Moderately slow-growing, upright in habit forming narrow pyramid. *Fruit* Pod-shaped, dark green fruits. Rarely fruits outside native environment.

Gymnocladus dioicus in leaf

Hardiness Tolerates 4°F (−15°C).
Soil Requirements Best on neutral to acid, rich soil. Dislikes alkalinity.
Sun/Shade aspect Full sun to light shade, preferring light shade.
Pruning None required.
Propagation and nursery production From seed. Plant container-grown or root-balled (balled-and-burlapped). Plants over 3ft (1m) scarce in commercial production — may need to be sought from specialist nurseries.
Problems May be difficult to find. Slow-growing in most areas.
Average height and spread
Five years
10x5ft (3x1.5m)
Ten years
20x10ft (6x3m)
Twenty years
or at maturity
20x20ft (6x6m)
Continued growth over 50 years reaches 59x20ft (18x6m)

HYDRANGEA PANICULATA 'GRANDIFLORA'

KNOWN BY BOTANICAL NAME
Hydrangeaceae
Deciduous
Attractive, late-flowering shrubs trained as small 'toy' trees.

Origin From Japan, China and Taiwan.
Use As a small, mop-headed, flowering tree of interest for small garden and patio areas. Unsuitable for containers, gaining insufficient nutrients to produce new growth needed for flowering.
Description *Flower* Large panicles of sterile white bracts produced mid to late summer, retained through autumn fading to pink and finally brown in early winter. Interesting in the young, green form. *Foliage* Light green, broad, ovate leaves of good size. Good yellow

Hydrangea paniculata 'Grandiflora' in flower

autumn colour. *Stem* Light green to green-brown, aging to mahogany brown. Wood very brittle. *Fruit* None.
Hardiness Tolerates 4°F (−15°C).
Soil Requirements Does best on a rich, deep soil, but tolerates a wide range of conditions, alkaline or acid. Extreme alkalinity may cause distress.
Sun/Shade aspect Best in light shade, but tolerates full sun. Deep shade spoils overall shape.
Pruning All previous year's wood should be cut back to one or two buds in mid spring to encourage large flower production in the following autumn. Otherwise the tree becomes old, woody and brittle, with small flowers.
Propagation and nursery production From softwood cuttings grown on in nursery conditions to form a small mop-headed tree. May be found as a pretrained standard where an upright stem has been trained up to 5ft (1.5m) and then allowed to branch, but not always available in this form. It is not advisable to try training from a very young plant. Plant bare-rooted or container-grown. Where available, normally easy to find when in flower in autumn.
Problems Brittle and needs staking throughout its life, otherwise wind damage is likely.
Average height and spread
Five years
12x6ft (3.5x2m)
Ten years
14½x10ft (4.5x3m)
Twenty years
or at maturity
18x13ft (5.5x4m)

IDESIA POLYCARPA

IGIRI TREE
Flacourtiaceae
Deciduous
A foliage tree of general interest.

Origin From Japan and China.
Use Initially as a large-leaved tall shrub, ultimately becoming an interesting foliage tree.
Description *Flower* Large terminal panicles of male flowers in midsummer only on mature trees of 10-15 years or more. Female flowers insignificant. *Foliage* Large, ovate leaves, 10in (25cm) long and 12in (30cm) wide on long stalks. Light green with a grey sheen when young, quickly becoming dark green with glaucous grey undersides. Slightly curved, with veins reddening as summer progresses. Good yellow autumn colour. *Stem* Grey-green, branching, rigid, stout in appearance, quickly becoming old and woody-looking. Moderately slow-growing, forming a round-topped tree. *Fruit* Female

trees may produce pea-like, bright red berries in autumn.
Hardiness Tolerates 14°F (−10°C).
Soil Requirements Any soil type. Tolerates alkaline conditions if adequate depth of top-soil covers underlying chalk.
Sun/Shade aspect Best in light shade, but tolerates full sun.
Pruning None required.
Propagation and nursery production From seed or from layers. Purchase container-grown. Plants over 1½ft (50cm) are rarely found and must be sought from specialist nurseries.

Idesia polycarpa in leaf

Problems Slow rate of growth. Male plants are best for flower and females produce the seed, although nurseries do not segregate the sexes.
Average height and spread
Five years
10x5ft (3x1.5m)
Ten years
20x10ft (6x3m)
Twenty years
or at maturity
39x20ft (12x6m)

ILEX
Standard trees

HOLLY
Aquifoliaceae
Evergreen
Interesting small mop-headed evergreen trees.

Origin Dependent on variety; spread over a wide area of the northern hemisphere.
Use As freestanding, mop-headed trained trees for patio planting, large containers or for small to large gardens.
Description *Flower* Clusters of small white flowers with prominent stamens in summer. *Foliage* Ovate leaves, light, dark or medium green with silver and variegated forms, with or without spines, depending on variety. *Stem* Mid green to dark green, with purple hue.

Ilex aquifolium 'Argentea Marginata'

Upright when young, quickly branching. Standard forms have a pretrained central stem at 3ft (1m), 5ft (1.5m) or 6ft (2m), with tightly clustered branches forming a round head. Mature shrubs may after 20 or 30 years also form standard or multi-stemmed trees.
Fruit Red, orange or yellow berries, depending on variety.
Hardiness Tolerates −13°F (−25°C), but can suffer from wind-scorch, in severe winters. Heads become very heavy and need staking for a number of years following planting.
Soil Requirements Does well on all soil types, but best results on rich, moist, open soil.
Sun/Shade aspect Full sun to medium shade.
Pruning Small, individual suckers may arise from along the pretrained stem and should be removed with a sharp knife. Growth can be clipped into a round, square or pyramid shape, or left to develop freely.
Propagation and nursery production From softwood cuttings in early summer. Trees are trained up to a central stem and encouraged to grow from the desired height. Purchase container-grown. Supply is relatively limited; not always available and must be sought from specialist nurseries. Semi-trained trees may be offered up to 5ft (1.5m) where heads are not entirely formed: new shoots should be cut back by half when over 18in (45cm) long to encourage more branching and a thicker head.
Problems Not readily available.
Varieties of interest *I. × altaclarensis 'Camelliifolia'* Foliage purple-red when young, becoming large, dark green and shiny with age, very few spines. Purple stems. A female form bearing large clusters of red fruits. *I. × a. 'Hodginsii'* Leaves dark green and oval shaped with irregularly produced spines. A male form producing no berries but useful as a pollinator for female varieties. *I. × a. 'Lawsoniana'* Leaves edged dark green, bright yellow centres, completely spineless. A female form with orange-red berries. May revert to green, and any plain green shoots should be removed. *I. × a. 'Silver Sentinel'* Narrow, elliptic leaves, pale green to grey-green with creamy-white to yellow margins. Very few spines. A female form with orange-red berries. Quick-growing. *I. aquifolium 'Argentea Marginata'* (Broad-leaved Silver Holly) Spiny, dark green to grey-green leaves with white margins. Normally offered in its female form, but can also be found as a male plant. Ascertain the sex before purchasing. Berries orange to orange-red. *I. a. 'Argentea Pendula'* (Perry's Silver Weeping Holly) Dark to olive green foliage with white margins and some spines. Branches long and weeping. A female form with orange-red berries. *I. a. 'Bacciflava'* syn. *I. a. 'Fructuluteo'* (Yellow-fruited Holly) Foliage dark to mid green and spiny. A female form with bright yellow berries. *I. a. 'J. C. van Tol'* Foliage dark green, narrow, ovate and almost spineless. A female form producing large crops of red fruits. *I. a. 'Madame Briot'* Dark green leaves with dark yellow to gold margins. Purple stems. A female form with orange-red berries.
Average height and spread
Five years
12x5ft (3.5x1.5m)
Ten years
18x8ft (5.5x2.5m)
Twenty years
or at maturity
22x12ft (6.5x3.5m)

JUGLANS REGIA

ENGLISH WALNUT, PERSIAN WALNUT
Juglandaceae
Deciduous
Useful, quick-growing foliage tree, with the advantage of edible nuts.

Origin From southern and eastern Europe through to the Himalayas and China.
Use As a freestanding, fruiting tree for shade effect for medium to large gardens. Good for avenues when planted at 39ft (12m) apart.
Description *Flower* Of no interest. *Foliage* Large pinnate leaves up to 12in (30cm) long, with 7-9 leaflets. Attractive bronze-green when young, aging to grey-green. Some yellow autumn colour, but may lose leaves in late summer or early autumn in drought conditions. *Stem* Grey-green to grey. Stout, branching habit. Quick-growing when young, slowing after 20 years, forming a round-topped, large, slightly spreading tree. *Fruit* Round green seed cases in autumn. Fruits are edible walnuts.
Hardiness Tolerates 4°F (−15°C), but young spring growth may be slightly damaged by late frosts.
Soil Requirements Any soil conditions; tolerates alkaline or acid types.
Sun/Shade aspect Full sun to mid shade.
Pruning None required other than training in early years.
Propagation and nursery production From seed, except some named fruiting varieties which are grafted. Plant root-balled (balled-and-burlapped) or container-grown. Available from 3ft (1m) up to 10ft (3m), this being the safest planting range. Even with due care, establishment can be difficult.
Problems Often planted in an area where it is unable to attain its full growth. Can suffer wind damage.
Varieties of interest *J. nigra* (Black Walnut) Black stems with interesting winter effect. Dark foliage. Black walnuts have a distinctive flavour, used in baked goods rather than eaten fresh. Difficult to find in commercial production, and difficult to establish.
Average height and spread
Five years
13x5ft (4x1.5m)
Ten years
26x10ft (8x5m)
Twenty years
or at maturity
52x20ft (16x6m)
Continued growth over 50 years reaches 78x48ft (24x15m)

KALOPANAX PICTUS
(Acanthopanax ricinifolius)

KNOWN BY BOTANICAL NAME
Araliaceae
Deciduous
A tree with unusual spiny stems and interesting foliage.

Origin From Japan.
Use As an interesting feature. Relatively quick-growing and of medium height, suitable for medium-sized and larger gardens.

Kalopanax pictus in leaf

Description *Flower* Small clusters of white flowers in early autumn, medium to large, flattened heads, up to 2½in (6cm) across. *Foliage* Leaves with 5-7 lobes, up to 12in (30cm) across on young plants, becoming smaller on mature specimens. Good yellow autumn colour. *Stem* Upright, spreading slightly with age. Rigid, branching habit. Stems grey to grey-green with numerous stout prickles spaced intermittently over the entire stem surface. Medium rate of growth. *Fruit* Insignificant.
Hardiness Tolerates 14°F (−10°C).
Soil Requirements Any soil conditions except extremely alkaline or dry. Deep, rich soil encourages better foliage.
Sun/Shade aspect Full sun to medium shade, preferring light shade.

Juglans regia in leaf

Pruning None required.
Propagation and nursery production From seed, layers or root cuttings. Purchase root-balled (balled-and-burlapped) or container grown. Relatively difficult to find in commercial production; must be sought from specialist nurseries. Normally supplied 2ft (60cm) to 5ft (1.5m) as a single stout shoot. Larger plants are not recommended.
Problems Stem prickles are a cultivation hazard. Takes a number of years to reach recognized tree shape.
Varieties of interest *K. p. maximowiczii* Deeply cut leaves. This variety may be more readily available.
Average height and spread
Five years
10x4ft (3x1.2m)
Ten years
20x7ft (6x2.2m)
Twenty years
or at maturity
20x20ft (6x6m)

KOELREUTERIA PANICULATA

CHINESE RAIN TREE, GOLDEN RAIN TREE, PRIDE OF INDIA, CHINA TREE
Sapindaceae
Deciduous
Mid to late summer-flowering tree with interesting foliage and bark.

Origin From northern China.
Use As a small to medium height summer-flowering tree. Suitable for medium-sized or larger gardens as a solo specimen.

Koelreuteria paniculata
in flower

Description *Flower* Upright, open, terminal panicles of numerous yellow florets produced in mid to late summer. *Foliage* Light green with grey sheen. Pinnate, up to 12in (30cm) long, leaflets narrow with deeply toothed edges. Attractive light bronze when young. Good yellow autumn colour. *Stem* Light brown-green. Stiffly branched. Medium rate of growth, making a round-topped tree. *Fruit* Large, inflated, three-lobed yellow fruits in autumn.
Hardiness Tolerates 4°F (−15°C).
Soil Requirements Dislikes thin alkaline soils.
Sun/Shade aspect Best in full sun; dislikes shade.
Pruning None required.
Propagation and nursery production From seed or rooted cuttings. 'Fastigiata' is a grafted form. Plants can be transplanted root-balled (balled-and-burlapped), bare-rooted or container-grown. Plants normally supplied between 2ft (60cm) and 6ft (2m); larger plants usually scarce and not always a wise investment.
Problems Relatively short-growing and

straight-stemmed specimens are extremely scarce.
Varieties of interest *K. p. apiculata* Originating in China. Less available than *K. paniculata*. *K. p. 'Fastigiata'* More upright habit. Average height, but a maximum of 6ft (2m) spread. Relatively scarce in commercial production.
Average height and spread
Five years
16x6ft (5x2m)
Ten years
26x13ft (8x4m)
Twenty years
or at maturity
39x20ft (12x6m)

LABURNOCYTISUS ADAMII

PINK LABURNUM
Leguminosae
Deciduous
A very attractive tree to give interest and variety.

Origin A graft hybrid of *Laburnum anagyroides* and *Cytisus purpureus*, originated in the early 1800s.
Use As a single specimen given space to show its full beauty.
Description *Flower* Individual limbs bear either racemes of yellow laburnum flowers or shorter racemes of the pink *Cytisus purpureus* flower. Some limbs may even present both flowers, one type sparsely interspersed with the other. Small areas of true *Cytisus purpureus* may occur in small clusters at the ends of older branches. *Foliage* Light grey-green pinnate leaves, rather untidy. On mature trees tend to be yellow and sickly. Some yellow autumn colour. *Stem* Grey-green, rubbery texture. Upright when young, spreading with age. *Fruit* Small, poisonous pea-pods sometimes produced.
Hardiness Tolerates 4°F (−15°C).
Soil Requirements Does well on most soil conditions; tolerates heavy alkalinity or acid types. Dislikes wet conditions, when root damage may cause poor anchorage.
Sun/Shade aspect Best in full sun; tolerates light shade.
Pruning None required; may resent it.
Propagation and nursery production By grafting or budding on to *Laburnum vulgaris* understocks. Plant bare-rooted or container-grown. Best planting heights 5-10ft (1.5-3m).
Problems Root anchorage may be poor, easily dislodged in high winds. Needs staking for most of its life. Root system often feeble at time of purchase.

Average height and spread
Five years
13x5ft (4x1.5m)
Ten years
20x8ft (6x2.5m)
Twenty years
or at maturity
23x13ft (7x4m)

LABURNUM ALPINUM

ALPINE LABURNUM, SCOTCH LABURNUM
Leguminosae
Deciduous
Sometimes overlooked, but a very attractive tree, especially the dark green foliage. The weeping form can be used to great advantage as a specimen or feature plant in the smallest of gardens.

Origin From central and southern Europe.
Use As a small ornamental tree, flowering in late spring or early summer when many other flowering trees have finished. Also used as a bush or as a standard tree, and in its weeping form for tubs and containers.

Laburnum alpinum 'Pendulum' in flower

Description *Flower* Pendulous racemes of fragrant, golden yellow flowers in late spring or early summer. *Foliage* Olive green, trifoliate leaves, glossy upper surfaces and paler slightly hairy undersides. *Stem* Olive green upright stems when young. Spreading with age to form a dome-shaped large bush or standard tree. *Fruit* Poisonous, hanging, grey-green, small, pea-pod type fruits, often

Laburnocytisus adamii in flower

36

in great profusion, especially on mature plants.
Hardiness Tolerates −13°F (−25°C).
Soil Requirements Most soil types; likes very alkaline forms. Extremely wet conditions may lead to root-damage and unstable anchorage. Tub-grown plants need a large container and good potting soil.
Sun/Shade aspect Best in full sun, but tolerates quite deep shade.
Pruning None required; may even resent it.
Propagation and nursery production L. alpinum grown from seed; 'Pendulum' form is grafted. Can be purchased bare-rooted or container-grown.
Problems Susceptible to blackfly (aphid), a winter host to this pest. Poisonous pods are dangerous to children.
Varieties of interest *L. a. 'Pendulum'* A good weeping form, reaching only about 10ft (3m) in height unless grafted. Slow-growing and weeping to the ground, with a spread of 5ft (1.5m) in time. Good display of hanging yellow flowers late spring, early summer. Grafts on a 5ft (1.5m) or 8ft (2.5m) stem can look unsightly in early years.
Average height and spread
Five years
13x6ft (4x2m)
Ten years
13x10ft (4x3m)
*Twenty years
or at maturity*
23x13ft (7x4m)

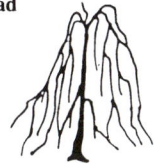

LABURNUM × WATERERI 'VOSSII'

GOLDEN CHAIN TREE, WATERER LABURNUM
Leguminosae
Deciduous
An interesting and well-known tree of great beauty for late spring and early summer flowering.

Origin Of garden origin.
Use As a small to medium-sized flowering tree suitable for medium to large gardens. Can be trained to cover archways and walkways, or as a wall climber for north or east walls.
Description *Flower* Long, hanging racemes of numerous deep yellow to golden yellow pea-flowers up to 12in (30cm) long, produced in late spring. *Foliage* Leaves trifoliate, each leaflet up to 3in (8cm) long. Grey-green to dark green, glossy upper surfaces and lighter, often hairy undersides. *Stem* Dark glossy green with a slight grey sheen. Strong, upright when young, branching and twiggy with age; finally a spreading tree of medium vigour. *Fruit* Grey-green, hanging pods containing

Laburnum × watereri 'Vossii' in flower

poisonous black, pea-shaped fruits. *L. × w. 'Vossii'* produces fewer seed-pods than other Laburnums.
Hardiness Tolerates −13°F (−25°C).
Soil Requirements Any soil conditions; tolerates high alkalinity.
Sun/Shade aspect Full sun to light shade, preferring full sun.
Pruning None required; may resent it.
Propagation and nursery production By grafting. Plant bare-rooted or container-grown. Stocked by most garden centres and nurseries. Normally 6-13ft (2-4m) in height at purchase; best planting heights 8-10ft (2.5-3m).
Problems All Laburnums have poor root systems and require permanent staking. Relatively short-lived trees, showing signs of distress after 40 or more years. Poisonous fruits can be dangerous to children.
Varieties of interest *L. anagyroides* Foliage smaller and grey-tinged. Average height but slightly more spreading in habit. Flowers earlier, smaller and paler yellow than those of 'Vossii'. *L. a. 'Aureum'* Light yellow to green-yellow foliage. Not a strong tree, can be disappointing. *L. a. 'Pendulum'* A weeping form with small yellow flowers, but less handsome than *L. alpinum 'Pendulum'*.
Average height and spread
Five years
13x6ft (4x2m)
Ten years
20x10ft (6x3m)
*Twenty years
or at maturity*
23x16ft (7x5m)

LAURUS NOBILIS

BAY LAUREL, BAY
Lauraceae
Evergreen
A small specimen tree that can be trained and shaped, suitable for outdoor planting in a mild climate or for container-growing indoors. Makes an attractive herb garden feature if kept trimmed.

Origin From southern Europe.
Use As a small, mop-headed tree for tubs and ornamental containers. Can be planted in open ground, but tends to grow very vigorously, becoming open and lax in habit. Suitable for small or large gardens or for indoor growing.
Description *Flower* Rarely flowers significantly when trained in tree form. *Foliage* Medium to large, elliptic. Dark green with glossy upper surfaces, dull, grey-green undersides. *Stem* Dark green. Very branching; can be trained to a tight round ball, square or pyramid. *Fruit* None produced when trained.

Laurus nobilis − **tub-grown**

Hardiness Minimum winter temperature 23°F (−5°C).
Soil Requirements A good quality potting medium for tub-growing. In the garden, does well on most soils.
Sun/Shade aspect Full sun to medium shade, preferring light shade.
Pruning Requires two annual clippings — in mid spring and in mid to late summer.
Propagation and nursery production From cuttings. Purchase container-grown. May need to be sought from specialist nurseries. Pretrained plants 3-10ft (1-3m) are normally available.
Problems Requires mild conditions.
Average height and spread
Five years
8x2½ft (2.5x0.75m)
Ten years
8x3ft (2.5x1m)
*Twenty years
or at maturity*
8x4ft (2.5x1.2m)

LIGUSTRUM LUCIDUM

WAX LEAF, GLOSSY, OR CHINESE PRIVET
Oleaceae
Evergreen
Attractive evergreens for sheltered areas.

Origin From China.
Use Suitable for medium-sized or larger gardens.
Description *Flower* Panicles of white, musty-scented flowers produced on mature wood late summer or early autumn. *Foliage* Pointed, ovate leaves, glossy green upper surfaces, duller undersides. Variegated forms. *Stem* Light green to green-purple. Single stem up to 5-6ft (1.5-2m) high encouraged from young, newly planted shrubs; allowed to branch to form small tree. Trees grafted at 3ft (1m) upwards on to stems of *L. vulgaris* may also be found. *Fruit* Clusters of dull blue to blue-black berries in autumn.
Hardiness Hardy.
Soil Requirements Most soil types; dislikes extremely dry, alkaline conditions or heavy waterlogging.
Sun/Shade aspect Full sun to medium shade. Tolerates deep shade, but may become slightly deformed.
Pruning None required unless training as standard. Pruning to shape is not recommended.
Propagation and nursery production From softwood cuttings or by grafting on a standard stem. Trained trees are relatively scarce in commercial production and must be sought from specialist nurseries. May be only 3ft (1m) high and needing further training after planting.
Problems The need to train after purchase can

Ligustrum lucidum in flower

deter inexperienced gardeners, but with a little care it is achieved relatively easily and quickly. Trees require lifelong staking as the evergreen head becomes very heavy, further strained by snow and high winds which can cause structural damage.
Varieties of interest *L. l. 'Aureum'* Bright yellow to golden spring foliage. Variegation declines as summer progresses, and may be completely lost. May be difficult to find. One-third less than average height and spread. *L. l. 'Excelsum Superbum'* Foliage margined and mottled with deep yellow or creamy white variegation. May be somewhat difficult to find. One-third less than average height and spread. *L. l. 'Tricolor'* A variegated form with leaf margins tinged pink when young, a pleasant addition to the colour. May be difficult to find. One-third less than average height and spread.
Average height and spread
Five years
6x3ft (2x1m)
Ten years
13x6ft (4x2m)
*Twenty years
or at maturity*
20x13ft (6x4m)

LIQUIDAMBAR STYRACIFLUA

SWEET GUM
Hamamelidaceae
Deciduous
Among the aristocrats of autumn-colour trees, suitable for most gardens.

Origin From eastern North America.
Use As a moderately compact, pyramidal tree, for good autumn colours and interesting architectural shape. Suitable for medium-sized and larger gardens.
Description *Flower* Small, inconspicuous, green-yellow flowers, without petals or in small catkins. Flowering variable and unreliable. *Foliage* Palmate leaves with 3 or 5 lobes, 3-6in (8-15cm) wide and 5-8in (12-20cm) long. Slightly glossy upper surfaces, downy undersides. Excellent orange-red autumn colour. *Stem* Upright pyramidal habit, branching regularly from low level. Medium rate of growth. Corky bark, grey-green. Can be trained with more main stem if required. *Fruit* Scarce and of no interest.
Hardiness Tolerates 4°F (−15°C).
Soil Requirements Does well on most soil conditions; tolerates high alkalinity. Best on moist, rich soil but dislikes waterlogging.
Sun/Shade aspect Full sun to light shade, preferring full sun. Deep shade decreases autumn colour-intensity and encourages more open habit.

Pruning None required.
Propagation and nursery production From seed, layering or grafting for named varieties. Planted root-balled (balled-and-burlapped) or container-grown. Plants normally available 2ft (60cm) to 8ft (2.5m), occasionally obtainable at 13ft (4m). Stocked in general nurseries, but named varieties may have to be sought from specialist sources.

Liquidambar styraciflua in autumn

Problems Slow to establish after transplanted; can take up to three years to produce good new growth. New leaf appears very late in spring, even in early summer.
Varieties of interest *L. formosana* Very similar to *L. styraciflua*, but the bark is not corky or winged. Slightly more tender. From central and southern China. *L. f. var. monticola* Slightly larger leaves than *L. formosana*, otherwise similar. *L. styraciflua 'Aurea'* Golden yellow splashed variegation in spring, decreasing as summer approaches. Two-thirds average height and spread. *L. s. 'Lane Roberts'* Non-corky bark. Dark wine red autumn colours. Not widely grown, but can be found. *L. s. 'Variegata'* White creamy margins on grey-green leaves. Good autumn colour. Two-thirds average height and spread. *L. s. 'Worplesdon'* Corky bark. Good autumn colours.
Average height and spread
Five years
13x6ft (4x2m)
Ten years
26x13ft (8x4m)
*Twenty years
or at maturity*
52x26ft (16x8m)

LIRIODENDRON TULIPIFERA

TULIP TREE, TULIP POPLAR
Magnoliaceae
Deciduous
Interesting large trees with attractive flowers, though these are not a feature of the young tree.

Origin From North America.
Use As a large specimen tree for estate and park planting.
Description *Flower* Fragrant green and yellow flowers borne singly in branch terminals in early summer; six upright petals and three spreading sepals with numerous stamens, resembling a tulip flower in shape and form. Takes up to 20 years to come into flower. *Foliage* Large grey-green, three-lobed foliage, the central lobe cut and with curved indentation. Leaves 2-8in (5-20cm) long. New foliage purple-green while in bud. Good yellow autumn colour. *Stem* Grey-green aging to dark green, finally green-brown. Strong, upright and fast-growing when young, slowing in growth with age, forming a broad-based, pyramidal tree. *Fruit* Flowers give way to tall dome-shaped calyx of little interest.
Hardiness Tolerates 14°F (−10°C); young spring foliage can be damaged by late frosts but rejuvenates.
Soil Requirements Most soil conditions. Produces more growth on a good, rich soil.
Sun/Shade aspect Full sun, but tolerates limited light shade, except golden variegated forms which must have full sun to maintain coloration.
Pruning None required, unless to form a single central trunk.
Propagation and nursery production From seed or layers. Variegated and upright forms grafted or budded. Plant root-balled (balled-and-burlapped) or container-grown. Normally available from 2ft (60cm) up to 10ft (3m); larger specimens occasionally available. Best planting heights up to 6ft (2m).

Liriodendron tulipifera in flower

Problems Needs thorough soil preparation. Can be difficult to establish; may sulk for a number of years before achieving good growth.
Varieties of interest *L. t. 'Aureomarginatum'* Bold yellow to green-yellow leaf margins, decreasing in colour as summer progresses but usually apparent throughout the season. Two-thirds average height and spread. Minimum winter temperature 23°F (−5°C), especially in early spring as new growth emerges.

Liriodendron tulipifera 'Aureomarginatum' **in leaf**

L. t. 'Fastigiatum' syn. *L. t. 'Pyramidale'* A thick, upright, columnar tree with good foliage colour and flowers. Average height but only one-third spread. Difficult to find in commercial production, but infrequently available.

Average height and spread
Five years
12x6ft (3.5x2m)
Ten years
23x13ft (7x4m)
Twenty years
or at maturity
46x26ft (14x8m)
Continued growth over 50 years reaches 100x33ft (30x10m)

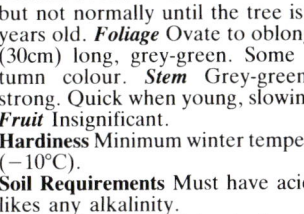

MAGNOLIA CAMPBELLII

GIANT HIMALAYAN PINK TULIP TREE
Magnoliaceae
Deciduous
A rarely seen, large-flowered tree Magnolia.

Origin From Nepal and surrounding regions.
Use As a large, freestanding tree for woodland areas, in association with Rhododendrons, Azaleas and other acid-loving plants.

Magnolia campbellii **in flower**

Description *Flower* Large, goblet-shaped, with pale pink inner colouring, deep rose pink outside. Produced late winter or early spring, but not normally until the tree is at least 20 years old. *Foliage* Ovate to oblong, up to 1ft (30cm) long, grey-green. Some yellow autumn colour. *Stem* Grey-green, upright, strong. Quick when young, slowing with age. *Fruit* Insignificant.
Hardiness Minimum winter temperature 14°F (−10°C).
Soil Requirements Must have acid soil. Dislikes any alkalinity.
Sun/Shade aspect Light shade when young, emerging into full sun with age. Must be planted away from early morning sun as frosted flowers are damaged if thawed too quickly.
Pruning None required except occasional removal of limbs or branches to encourage tree shape.
Propagation and nursery production From seed or cuttings. Sometimes grafted. Purchase container-grown or root-balled (balled-and-burlapped). Extremely difficult to find, especially in non-mild areas. Best planting heights 2ft (60cm) to 3ft (1m). Rarely offered above these sizes but may be found smaller. Not in general cultivation; must be sought from specialist nurseries.
Problems Garden potential limited because of tenderness and acid-soil requirement. Takes time to flower.

Varieties of interest *M. c. alba* A white-flowering form, more difficult to find than its parent. *M. c. 'Charles Raffill'* Rose pink buds opening to rosy purple outside, white with mauve margins inside. Becoming more readily available. *M. c. mollicomata* More hardy and normally flowers within 10-15 years of planting. Large pink to rose-purple, waterlily-type flowers. From Tibet and the Yunnan Valley.

Average height and spread
Five years
5x3ft (1.5x1m)
Ten years
13x6ft (4x2m)
Twenty years
or at maturity
26x13ft (8x4m)

MAGNOLIA GRANDIFLORA

EVERGREEN MAGNOLIA, SOUTHERN MAGNOLIA, BULL BAY
Magnoliaceae
Evergreen
Interesting evergreen trees for very mild areas.

Origin From south-eastern USA.
Use As a freestanding tree for gardens experiencing winter temperatures no lower than 23°F (−5°C). In wind-exposed areas, requires staking for most of its life.

Magnolia grandiflora **in flower**

Description *Flower* Creamy white, very fragrant flowers up to 10in (25cm) across, produced throughout summer and early autumn. Late flowers susceptible to autumn frost-damage. *Foliage* Elliptic leaves, 8-10in (20-25cm) long. Attractive bright green when young, maturing to dark green with glossy upper surfaces, and dull brown, felted undersides. *Stem* Grey-green, becoming green-brown. Normally upright with branches from ground level, but in mild areas can be free-standing with lower branches removed. Can also be trained as multi-stem. Takes a number of years to achieve tree stature. *Fruit* Green fruit pods 4in (10cm) long produced after flowering.
Hardiness Tolerates 14°F (−10°C).
Soil Requirements Most soil conditions. Needs 3ft (1m) of good topsoil over an underlying alkaline soil.
Sun/Shade aspect Full sun to mid shade.
Pruning None required, but lower branches can be removed. Can be trimmed to control size.
Propagation and nursery production From semi-ripe cuttings taken in early summer. Purchase container-grown. Normally supplied 2ft (60cm) to 3ft (1m), but occasionally available up to 8ft (2.5m). Smaller sizes

recommended. Available from general nurseries.
Problems Takes time to form an established tree. For growing only in very mild areas.
Varieties of interest *M. delavayi* Large grey-tinged dark green foliage with silver undersides. Creamy white to pale brown flowers in late summer and early autumn, fragrant, up to 6in (15cm) across, followed by upright seed heads. Tolerates most soils (even alkaline) if adequate topsoil is available. Difficult to find. *M. g. 'Exmouth'* Foliage dark green, polished upper surfaces, red or brown felted undersides. Large richly scented, creamy white flowers, produced early in its lifespan. Suitable for colder areas. *M. g. 'Ferruginea'* Large white scented flowers. Leaves elliptic, ovate, dark shiny green; pronounced brown felting on undersides. Upright. Tender; only for very mild areas. *M. g. 'Glenn St Mary'* Large, white flowers. Requires mild conditions. *M. g. 'Gloriosa'* Large, white flowers. Requires mild conditions. *M. g. 'Goliath'* Foliage dark glossy green to light green. Elliptic, slightly concave, with rounded leaf ends. Flowers globular, white and scented, produced 3-5 years after planting. *M. g. 'Majestic Beauty'* Large white flowers. Tender. *M. g. 'Maryland'* Large, white, fragrant flowers, slightly more open, produced two years after planting. Leaves elliptic, ovate, mid-green with shiny upper surfaces and grey-brown undersides.

Average height and spread
Five years
10x3ft (3x1m)
Ten years
13x6ft (4x2m)
Twenty years
or at maturity
26x13ft (8x4m)

MAGNOLIA KOBUS
Large star-flowering varieties

STAR MAGNOLIA
Magnoliaceae
Deciduous
Early spring-flowering trees or large shrubs, very beautiful in bloom.

Origin From Japan.
Use As a freestanding small tree, or for the back of a large shrub border if adequate space is allowed.
Description *Flower* White, multi-petalled, fragrant flowers produced in small numbers within 10-15 years of planting, soon increasing to a display of spectacular proportions. *Foliage* Mid to dark green small, elliptic leaves, some yellow autumn colour. *Stem* Dark green to green-brown. Strong, upright, branching and spreading with age. *Fruit* Small green capsules of little interest. **Hardiness** Tolerates −13°F (−25°C).
Soil Requirements Any soil type; tolerates moderate alkalinity.
Sun/Shade aspect Best in full sun, but tolerates light shade. To prevent frost damage in spring, plant away from early morning sun, to allow frosted flowers or flower buds to thaw out slowly without tissue damage.
Pruning None required, other than training as a single or multi-stemmed tree.
Propagation and nursery production From layers of semi-ripe cuttings taken in early summer. Purchase container-grown or root-balled (balled-and-burlapped). Found in garden centres and general nurseries. Obtainable from 2ft (60cm) to 3ft (1m). Larger plants occasionally available. *M. kobus* may also be found as a single-stemmed tree in specialist nurseries.
Problems Slow in coming to flower.
Varieties of interest *M. × loebneri* A cross between *M. kobus* and *M. stellata*. From an early age produces a profusion of multi-petalled, fragrant, white flowers in early to mid spring. Of garden origin. *M. × l. 'Leonard Messel'* Fragrant, multi-petalled flowers, deep pink in bud opening to lilac

pink. A cross between *M. kobus* and *M. stellata 'Rosea'*. Originating from the Nymans Gardens, Sussex, England. *M. × l. 'Merrill'* Large, white, fragrant flowers produced from an early age. Spectacular when mature. Raised in the Arnold Arboretum, Massachusetts, USA. *M. salicifolia* White, fragrant, narrow-petalled, star-shaped flowers in mid spring. Leaves, bark and wood are lemon-scented if bruised. From Japan.

Average height and spread
Five years
13x6ft (4x2m)
Ten years
20x13ft (6x4m)
Twenty years
or at maturity
30x26ft (9x8m)

MAGNOLIA × SOULANGIANA

TULIP MAGNOLIA, SAUCER MAGNOLIA
Magnoliaceae
Deciduous
Although not truly trees, these plants achieve tree-like proportions. The main form outclasses its own beautiful varieties and is highly recommended where a single specimen magnolia is required.

Origin Raised by Soulange-Bodin at Fromont, near Paris, France, in the early nineteenth century.
Use As a single or multi-stemmed small to medium feature tree, for medium-sized and larger gardens.
Description *Flower* Light pink with purple shading at petal bases and centres. Produced before leaves in early spring. *Foliage* Leaves light green to grey-green, elliptic, 4-5in (10-12cm) long. Yellow autumn colour. *Stem* Grey-green becoming grey-brown. Flower-buds large with hairy outer coating. Lower branches removed to form single or multi-stemmed tree after 15 or more years. *Fruit* May produce small green seed capsules.
Hardiness Tolerates 4°F (−15°C).
Soil Requirements Does well on heavy clay soils and most other types. Alkaline conditions will cause chlorosis.
Sun/Shade aspect Plant where there is no direct early morning sun in flowering time. Rapid thawing of flowers after late spring frosts causes browning.
Pruning Remove lower limbs to encourage single or multi-stemmed tree formation.
Propagation and nursery production From layers or semi-ripe cuttings taken in early summer. Purchase container-grown or root-balled (balled-and-burlapped). Plants normally supplied at recommended planting heights of 2ft (60cm) to 3ft (1m); plants

occasionally available up to 6ft (2m). Rarely found as pretrained standard trees.
Problems Slow to flower, taking 5 years or more to produce a really good display.
Varieties of interest *M. × soulangiana 'Alba Superba'* Large, scented, pure white, erect, tulip-shaped flowers, petals flushed purple at the base. Growth upright. Average height but slightly less spread. *M. × s. 'Alexandrina'* Vigorous and upright. Free-flowering with large white flowers, purple-flushed at the base. Sometimes difficult to obtain. *M. × s. 'Amabilis'* Tulip-shaped, ivory-white flowers with a light purple flush at the base. May be difficult to obtain. *M. × s. 'Brozzonii'* Large white flowers with purple shading at the base. Late-flowering. Not readily available at garden centres and nurseries; needs seeking out. *M. × s. 'Lennei'* Leaves broad, ovate, up to 10-12in (25-30cm) long. Flowers goblet-shaped and with fleshy petals, rose-purple outside, creamy white stained purple inside. Produced mid to late spring; in some seasons a second limited flowering in autumn. Thought to be of garden origin from Lombardy, Italy. *M. × s. 'Lennei Alba'* Ivory-white, goblet-shaped flowers, held upright along the branches. Average height but slightly more spreading. Not readily found in commercial production; needs seeking out. *M. × s. 'Picture'* Petals coloured purple outside, white inside. Flowers of erect habit. Flowering comes early in its lifespan. Leaves up to 10in (25cm) long, branches upright. Average height but with less spread. Not normally stocked by garden centres; must be sought from specialist nurseries. *M. × s. 'Rustica Rubra' ('Rubra')* Oval leaves 8in (20cm) long. Flowers cup-shaped and rosy red. A strong-growing variety, said to be a sport of *M.*

Magnolia × loebneri 'Merrill' in flower

Magnolia × soulangiana in flower

40

'Lennei' M. × s. 'Speciosa' White flowers with slight purple shading. Small leaves. Less than average height and spread. Not readily found in garden centres; must be sought from specialist outlets.

Average height and spread
Five years
13x6ft (4x2m)
Ten years
20x13ft (6x4m)
Twenty years or at maturity
32x26ft (10x8m)

MALUS
Weeping forms

WEEPING CRAB APPLE
Rosaceae
Deciduous
Useful weeping trees for small gardens or limited spaces.

Origin Of garden origin.
Use As small to medium-sized weeping trees for all but the smallest gardens.
Description *Flower* Wine red or pink-tinged white, depending on variety, 1½in (4cm) across, in clusters of 5-7, mid spring. *Foliage* Wine red, aging to greenish to green-red or light green, depending on variety. Ovate, up to 2in (5cm) long, with broad-toothed edges. Limited autumn colour. *Stem* Purple-green to dark purple or green-red, aging to mid green depending on variety. Arching, becoming pendulous, forming a spreading, weeping tree of irregular branch formation. Medium rate of growth. *Fruit* Wine red or bright red fruits, depending on variety, from late summer into early autumn.

Malus 'Red Jade' in fruit

Hardiness Tolerates 4°F (−15°C).
Soil Requirements Most soil types; dislikes extremely poor soil.
Sun/Shade aspect Coloured-leaved varieties must be in full sun; any degree of shade tends to lessen colouring. Green-leaved varieties tolerate light shade.
Pruning None required, but crossing branches of straggling limbs can be removed.
Propagation and nursery production From grafting or budding on to a wild crab understock. Plant bare-rooted or container-grown. Plants normally supplied from 4ft (1.2m) to 6ft (2m).
Problems Can suffer from apple scab and apple mildew.
Varieties of interest *M. 'Echtermeyer'* Purple-green foliage, flowers and fruit. Arching branches. *M. 'Red Jade'* Bright green foliage, yellow autumn colour. Pink-tinged white flowers and bright red fruits. Slightly less than average height and spread.
Average height and spread
Five years
6x5ft (2x1.5m)
Ten years
10x10ft (3x3m)
Twenty years or at maturity
10x20ft (3x6m)
Overall height will depend on initial height of plant at purchase.

MALUS
Fruiting varieties

FRUITING CRAB, CRAB APPLE
Rosaceae
Deciduous
Attractive, useful trees for blossom and fruit.

Origin Some varieties of natural origin, others of garden extraction.
Use As ornamental flowering trees with useful edible fruits for medium-sized or larger gardens.
Description *Flower* White, pink-tinged white or wine red, depending on variety. Flowers 1-1½in (3-4cm) across, produced singly or in multiple heads of 5-7 flowers in mid spring. *Foliage* Ovate, tooth-edged, 2in (5cm) long. Green or wine red, depending on variety. Some yellow autumn colour. *Stem* Light green to grey-green when young. Moderately upright, becoming green-brown, spreading and branching, forming a round-topped tree. Moderate rate of growth. *Fruit* Colours from yellow, orange-red, through to purple-red. Shaped like miniature apples, 1-2in (3-5cm) across, which they are. Produced in late summer and early autumn. All edible, and also used to make jelly.
Hardiness Tolerates 4°F (−15°C).
Soil Requirements Most soil conditions; dislikes extreme waterlogging.
Sun/Shade aspect Best in full sun to very light shade to enhance ripening of fruit. Purple-leaved varieties must have full sun to maintain leaf colour.
Pruning Dead, crossing branches should be removed. To encourage fruit production the centre should be kept open to allow for ripening by sun.
Propagation and nursery production Some species forms grown from seed, but plants mainly grafted or budded on wild apple understock. Planted bare-rooted or container-grown. Normally supplied at 5½ft (1.8m) to 10ft (3m) in height. Smaller or larger plants may be obtainable, but recommended planting size is 6-10ft (2-3m). Most varieties readily available from general or specialist nurseries, also sometimes from garden centres.
Problems Liable to fungus diseases such as apple scab and apple mildew, damaging both foliage and fruit. Some stem canker may occur; remove by pruning and treat cuts with pruning compound.
Varieties of interest *M. 'Dartmouth'* White flowers followed by sizeable red-purple fruits. Green foliage. *M. 'Dolgo'* A white flowering form with yellow fruits held well into autumn. Used as a universal pollinator in orchards or for garden-grown trees. *M. 'Golden Hornet'* White flowers, followed by good crop of

Malus 'John Downie' in fruit

bright yellow fruits which may remain on the tree well into winter. Green foliage. Can be used as a universal pollinator for garden or orchard apple trees. *M. 'John Downie'* Conical fruits, 1in (3cm) long, bright orange shaded scarlet. One of the best for making jelly. Green foliage. More susceptible to apple scab and mildew than most varieties. *M. 'Professor Sprenger'* Flowers pink in bud, opening to white. Good crop of amber fruits retained until midwinter. Green foliage. Half average height and spread. *M. 'Red Sentinel'* White flowers followed by deep red fruits maintained beyond midwinter. Green foliage. *M. × robusta* (Siberian Crab) Two forms available, both with pink-tinged white flowers. *M. × r. 'Red Siberian'* Large crop of red fruits. Green foliage. *M. × r. 'Yellow Siberian'* Yellow fruits. Green foliage. *M. sylvestris* (Common Crab Apple) Flowers white shaded with pink. Fruits yellow-green, sometimes flushed red, 1-1½in (3-4cm) wide. Not readily available; must be sought from specialist nurseries. The parent of many ornamental crabs and the garden apple. *M. 'Wintergold'* White flowers, pink in bud. Good crop of yellow fruit, retained into winter. Green foliage. *M. 'Wisley'* Strong-growing tree with limited purple-red to bronze-red flowers, with reddish shading and slight scent. Large purple-red fruits in autumn which although sparse are attractive for their size. Purple to purple-green foliage.
Average height and spread
Five years
13x5ft (4x1.5m)
Ten years
20x10ft (6x3m)
Twenty years or at maturity
26x20ft (8x6m)

Malus 'Golden Hornet' in fruit

MALUS
Purple foliage varieties

PURPLE-LEAVED CRAB APPLE
Rosaceae
Deciduous
Among the loveliest of foliage and flowering trees for spring.

Origin Of garden or nursery origin.
Use As freestanding, small to medium-sized trees for medium to large gardens.
Description *Flower* Wine red to purple-red flowers, up to 1in (3cm) across, in clusters of 5-7 produced in great profusion in mid spring. *Foliage* Ovate, sometimes toothed, purple-red to purple-bronze. *Stem* Purple-red to purple-green. Upright when young, quickly spreading and branching to form a round-topped tree. Moderate rate of growth. *Fruit* Small, wine red fruits in early autumn, sometimes inconspicuous against the purple foliage.

Malus 'Royalty' in flower

Hardiness Tolerates 4°F (−15°C).
Soil Requirements Does well on most soils; dislikes very poor or waterlogged types.
Sun/Shade aspect Full sun to very light shade. Deeper shade spoils foliage colour and shape of tree.
Pruning Any damaged or crossing branches should be removed in winter, also low or obstructing branches as necessary. Can be pruned back hard and rejuvenates over the next two or three seasons.
Propagation and nursery production From budding or grafting on to wild apple understock. Plant bare-rooted or container-grown. Can be purchased from 3ft (1m) up to 10ft (3m). Trees of 13-16ft (4-5m) occasionally available, but recommended planting heights 5½-8ft (1.8-2.5m).
Problems Can suffer severe attacks of apple mildew, lesser attacks of apple scab.
Varieties of interest *M. 'Eleyi'* Large red-purple flowers, followed by conical, purple-red fruits. Foliage dark red-purple, up to 4in (10cm) long. Initial growth somewhat weak. *M. 'Lemoinei'* An early nursery cross of merit. Purple foliage, crimson-purple flowers, bronze-purple fruits. *M. 'Liset'* Modern hybrid with good foliage and flowers, adequate dark red fruit. *M. 'Neville Copeman'* Foliage dull wine red, flowers pink-purple, fruits purple. Not as intensely coloured as some, but worth consideration. Not easy to find in commercial production. *M. 'Profusion'* An early nursery cross variety. Good purple-wine flowers, purple-red fruits and coppery-crimson spring foliage. Originally robust, but recently appears to be losing its overall vigour. *M. 'Red Glow'* Wine red flowers, large leaves and fruits. A good introduction, but not readily available. *M. 'Royalty'* Possibly the best purple-red leaf form. Large, disease-resistant, wine-coloured foliage and wine red

flowers followed by large purple-red fruits. Becoming more widely available.
Average height and spread
Five years
13x5ft (4x1.5m)
Ten years
20x10ft (6x3m)
Twenty years
or at maturity
26x20ft (8x6m)

MALUS
Green-leaved, flowering varieties

FLOWERING CRAB, CRAB, CRAB APPLE
Rosaceae
Deciduous
Very attractive and interesting, spring-flowering trees.

Origin Mostly of garden origin; a few direct species.
Use As medium-sized flowering trees for medium and large gardens. Best grown singly.
Description *Flower* White, pink-white or bi-coloured pink and white flowers 1½in (4cm) across, individually or in clusters of 5-7 flowers, producing a mass display. *Foliage* Green, ovate, 2in (5cm) long, tooth-edged, giving some yellow autumn colour. *Fruit* Normally green to yellow-green and of little attraction. *Stem* Purple-red to purple-green. Upright when young, quickly spreading and branching, forming a round-topped tree.
Hardiness Tolerates −13°F (−25°C).
Soil Requirements Most soil conditions; dislikes waterlogging.
Sun/Shade aspect Full sun to light shade, preferring full sun.
Pruning None required except removal of crossing and obstructing branches.
Propagation and nursery production From budding or grafting on to wild apple understock. Plant bare-rooted or container-grown. Can be purchased from 3ft (1m) up to 10ft (3m). Trees of 13-16ft (4-5m) occasionally available, but recommended planting heights 5½-8ft (1.8-2.5m).
Problems Can suffer from severe attacks of apple mildew and lesser attacks of apple scab.
Varieties of interest *M. baccata* White flowers up to 1½in (4cm) across in mid spring, followed by bright red, globe-shaped fruits. One-third more than average height and spread. From eastern Asia and north China. Normally sold in the form *M. baccata var. mandshurica*, which has slightly larger fruits. *M. Coronaria 'Charlottae'* Foliage ovate and coarsely toothed, up to 4in (10cm) long and 2in (5cm) wide. Semi-double, fragrant flow-

ers borne singly or in twos or threes up to 1½in (4cm) across; attractive mother-of-pearl to mid pink colouring. Fruits large, green-yellow, not conspicuous. *M. 'Evereste'* A dwarf, mass-flowering variety. Flowers pink-white. One-third average height and spread. Not readily available, but not impossible to find. *M. floribunda* A pendulous variety, branches on mature trees reaching to ground. Can also be grown as large shrub. Flowers rose red in bud, opening to pink, finally fading to white, produced in mid to late spring in great profusion. Foliage smaller than most, ovate and deeply toothed. *M. 'Hillieri'* Somewhat weak constitution but worth consideration. Flowers semi-double, 1½in (4cm) wide, crimson-red in bud, opening to bright pink. Slightly pendulous habit. Very thin wood. *M. hupehensis* Fragrant flowers soft pink in bud, opening to white. Fruits yellow with red tints. Two-thirds average height and spread. Somewhat upright in habit. From China and Japan. *M. 'Katherine'* Semi-double flowers, pink in bud, finally white. Bright red fruits with yellow flushing. Two-thirds average height and spread with a globular head. Not readily found in production, but worth some research. *M. 'Lady Northcliffe'* Carmine-red buds opening to white with blush shading. Fruits small, yellow and round. Two-thirds average height and spread. Not always available. *M. 'Magdeburgensis'* A tree similar to a cultivated apple. Flowers deep red in bud, opening to blush-pink, finally becoming white. Fruits light green to green-yellow and unimportant. Two-thirds average height and spread. Not readily available, but not impossible to find. *M. sargentii* Foliage oblong with three lobes, up to 2½in (6cm) long. Some yellow autumn colour. Flowers pure white with greenish centres in clusters of 5 and 6; petals overlap. Fruits bright red. Very floriferous. Shrubby and reaches only one-third average height and spread, possibly more when grown as a standard tree. May be best grown as large shrub, though good effect when trained into small tree. Originating in Japan. *M. × scheideckeri* Coarsely toothed, small, elliptic to ovate leaves, sometimes with 3-5 lobes. Flowers pink to deep rose in clusters of 3-6. Growth very slender. Fruits yellow and round. A shrubby variety reaching one-third average height and spread. *M. 'Snow Cloud'* A relatively new variety of upright habit, reaching average height but less than average spread. Profuse white double flowers, opening from pink buds, in mid spring. Fruits inconspicuous. Foliage dark green with autumn tints. *M. spectabilis* Grey-green foliage susceptible to apple scab. Flowers rosy red in bud, opening to pale blush pink, up to 2in (5cm) across and borne in clusters of 6-8 in early spring. Fruits yellow and globe-shaped. From China. *M. 'Strathmore'* Light green foliage. A profusion of pale pink flowers.

Malus floribunda in flower

Malus hupehensis in flower

Round-topped. *M. toringoides* Foliage ovate to lanceolate up to 3in(8cm) long. Deeply lobed new foliage; that produced on older wood is less indented. Pastel autumn colours. Flowers light pink in bud opening to creamy-white, produced in clusters of 6-8. Fruit globe-shaped, yellow with scarlet flushing. Two-thirds average height and spread with a graceful, flat-headed effect. From China. *M. transitoria* Small-lobed foliage with small pink-white flowers and rounded yellow fruits. Excellent autumn colour. Two-thirds average height and spread. From north-west China. *M. trilobata* Leaves maple-shaped, deeply lobed, three-sectioned, mid to dark green with good autumn colour. White flowers, followed by infrequently produced yellow fruits. Originating in Eastern Mediterranean and north-eastern Greece. Two-third average height and spread. Scarce in production and will have to be sought from specialist nurseries. *M. 'Van Eseltine'* Flowers rose-scarlet in bud, opening to shell pink, semi-double. Small yellow fruits. Two-thirds average height and spread, slightly columnar habit.

Average height and spread
Five years
13x5ft (4x1.5m)
Ten years
20x10ft (6x3m)
Twenty years
or at maturity
26x20ft (8x6m)

MALUS TSCHONOSKII

TSCHONOSKI CRAB APPLE
Rosaceae
Deciduous
One of the most spectacular autumn-tinted trees.

Origin From Japan.
Use As a moderately upright ornamental autumn-foliage tree.
Description *Flower* Rose-tinted at first, then white, 1½in (4cm) wide, produced in clusters of 4 or 6 in late spring. *Foliage* Ovate to broad, with slender point. 2-5in (5-12cm) long, grey felted undersides and grey-green uppers. Extremely good orange-red autumn colouring. *Stem* Grey-green to green-brown. Upright when young, spreading with age. Attractive red-black scaled buds in winter. Medium rate of growth. *Fruit* Small, brown-yellow fruits, globe-shaped and 1½in (4cm) wide, produced in autumn.
Hardiness Tolerates −13°F (−25°C).
Soil Requirements Most soil conditions, but shows signs of distress on extremely poor types.
Sun/Shade aspect Full sun to light shade. Requires a sunny position to show off the full potential of its autumn colour.
Pruning None required other than removal of badly crossing branches.

Propagation and nursery production By grafting or budding. Plant bare-rooted or container-grown. Plants obtainable from 3ft (1m) up to 10ft (3m). Plants up to 16ft (5m) occasionally available, but 5½-10ft (1.8-3m) recommended as the best planting heights. Readily available from general nurseries, and sometimes garden centres.

Malus tschonoskii in autumn

Problems Susceptible to apple mildew and apple scab. Also very susceptible to stem canker; complete removal of diseased branches is the only solution.
Average height and spread
Five years
13x3ft (4x1m)
Ten years
26x5½ft (8x1.8m)
Twenty years
or at maturity
39x13ft (12x4m)

MESPILUS GERMANICA

MEDLAR
Rosaceae
Deciduous
An interesting flowering and fruiting tree of architectural shape.

Origin From south-eastern Europe and Asia Minor.
Use As a medium to large ornamental tree with edible fruits.
Description *Flower* Single, 1-1½in (3-4cm) wide, white, 5-petalled, round flowers produced early summer. *Foliage* Elliptic to oblong, up to 5in (12cm) long. Grey-green with dull texture and downy undersides. Some yellow autumn colour. *Stem* Grey-green to grey-brown with some large, sparsely distributed spines. Branches upright when young, quickly arching to form a wide, round-topped tree. Ideally grown as a large shrub without central trunk. Roots can be unstable; without adequate staking uprooting can occur even to bush trees. *Fruit* Apple-shaped, edible, brown, 1½in (4cm) wide medlar fruits produced in autumn. Each fruit has open crown effect at top.
Hardiness Tolerates 14°F (−10°C).
Soil Requirements Most soil conditions; dislikes excessive alkaline types.
Sun/Shade aspect Best in full sun for ripening of fruit, but tolerates moderate shade.
Pruning None required other than removal of crossing branches.
Propagation and nursery production From layers, seed or grafting. Plant bare-rooted, container-grown or root-balled (balled-and-burlapped). Plants normally available from 4ft (1.2m) up to 10ft (3m), but plants of 6ft (2m) recommended for best results. Young trees always deformed and irregular in shape and likely to have sparse root system. Must be sought from specialist nurseries.

Mespilus germanica in fruit

Problems Poor root system means staking will be required for most of the tree's life.
Varieties of interest There are a number of named clones, but these are extremely scarce in commercial production. The basic form *M. germanica* often equals the named forms for fruit production.
Average height and spread
Five years
12x12ft (3.5x3.5m)
Ten years
14½x14½ft (4.5x4.5m)
Twenty years
or at maturity
20x20ft (6x6m)

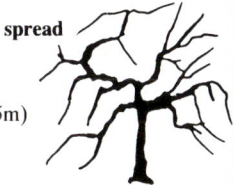

MORUS

MULBERRY
Moraceae
Deciduous
Attractive, interesting, ornamental fruiting trees.

Origin From China, North America and western Asia, depending on variety.
Use As an architectural form of fruiting tree ideal for medium-sized or larger gardens, except *Morus alba 'Pendula'*, which can be grown in a smaller area.

Morus nigra in fruit

Description *Flower* Very short, male or female catkins; produced in early spring, of little interest. *Foliage* Ovate, lobed or un-lobed, grey-green leaves. Individual branches may carry both leaf shapes. Good yellow autumn colour. The lighter green leaves of *Morus alba* are used to feed silkworms. *Stem* Grey-green, corky, almost rubbery texture.

Forms a spreading, short-trunk tree which may need to be staked for most of its life to prevent wind damage. *Fruit* Blackberry-shaped, dark red to black clusters of fruit. Edible and very juicy.
Hardiness Tolerates 14°F (−10°C).
Soil Requirements Best results on rich, moist, deep soil. Must be well drained.
Sun/Shade aspect Best in full sun to allow ripening of fruit, but tolerates light shade.
Pruning None required; may resent it.
Propagation and nursery production From hardwood cuttings taken in early winter. Purchase container-grown. Plants normally available from 2ft (60cm) up to 7ft (2.2m), occasionally as small standard trees up to 10ft (3m).
Problems Roots very fleshy and often poorly anchored. Cropping is extremely heavy in some localities and fruits fall when ripe.
Varieties of interest *M. alba* (White Mulberry) Quick-growing. Light green-grey stems with large, ovate, light green foliage. Black fruits in autumn. One-third more than average height and spread, and slightly more tender. From China. *M. a. 'Laciniata'* Leaves deeply cut and toothed. Good autumn colour. Fruits as parent. *M. a. 'Pendula'* Stout, stiff, pendulous branches, grey-green stems, light green foliage. Architecturally attractive. Black fruits. Worth considering as a weeping tree for any garden. *M. nigra* (Common or Black Mulberry) Good fruiting ability in all areas. Of reliable hardiness. From western Asia. *M. rubra* (Red Mulberry) Fruits red to dark purple, and sweet. May exceed average height and spread by one-third or more. From North America, and not always successful as a garden tree elsewhere.
Average height and spread
Five years
12x4ft (3.5x1.2m)
Ten years
16x8ft (5x2.5m)
Twenty years
or at maturity
26x16ft (8x5m)
Continued spreading over 35 years may be in excess of 32ft (10m)

NOTHOFAGUS

SOUTHERN BEECH
Fagaceae
Deciduous and Evergreen
Interesting field trees, though not very hardy.

Origin From South America or New Zealand.
Use In mild areas, a quick-growing field tree of substantial height and spread, also good for windbreak. Used in timber production. Of use in large gardens, estates and parks.
Description *Flower* Small, inconspicuous flowers, either male or female, presented singly or in threes. *Foliage* Ovate leaves up to 1½in (4cm) long, deciduous or evergreen depending on variety. Deciduous give good yellow autumn colour. *Stem* Grey-green to grey-brown. Fast-growing. Predominantly upright, more spreading after 20-30 years, forming a pyramidal tree. *Fruit* Small-husk, dark brown fruits produced in autumn containing three small inedible nuts.
Hardiness Minimum winter temperature 14°F (−10°C).
Soil Requirements Does best on well-drained, loamy soil. Some varieties dislike alkaline conditions.
Sun/Shade aspect Full sun to light shade, preferring full sun.
Pruning None required, but lower limbs can be removed to increase trunk size.
Propagation and nursery production From seed or layers. *N. dombeyi* and *N. procera* from cuttings. Best purchased container-grown. Plants from 2ft (60cm) to 10ft (3m) normally available; larger plants occasionally offered. Best planting heights 2ft (60cm) to 10ft (3m). Not always readily available; must be sought from specialist nurseries.

Nothofagus obliqua in autumn

Problems Hardiness in some areas is suspect, and no firm ruling is possible.
Varieties of interest *N. antartica* Deciduous. Small dark green leaves with silver reverse, some yellow autumn colour. Dark brown shoots and very twiggy branches. From Chile. *N. betuloides* Evergreen foliage. Scarce outside its native environment. Slightly tender to tender. Not recommended for planting in areas with winter temperatures below 23°F (−5°C). From Chile. *N. dombeyi* Normally evergreen, but can become deciduous in hard winters. Foliage dark green with grey undersides and toothed edges. From Chile. *N. fusca* An evergreen tree which may become deciduous or semi-evergreen in areas with winter temperatures below 23°F (−5°C). New foliage bright red, aging to mid to dark green. Slender, upright habit. Quick-growing. From New Zealand. *N. obliqua* (Roble Beech) Deciduous foliage giving good autumn colour. Fast-growing. From Chile. *N. procera* Deciduous foliage with good autumn colour. Quick-growing. Propagates from cuttings. Slightly less than average height and spread. Originating in Chile.
These are the hardiest of *Nothofagus* varieties, and most suitable for garden planting.
Average height and spread
Five years
13x6ft (4x2m)
Ten years
26x16ft (8x5m)
Twenty years
or at maturity
64x32ft (20x10m)

NYSSA SYLVATICA

NYSSA, BLACK GUM, SOUR GUM, TUPELO
Nyssaceae
Deciduous
Extremely attractive, ornamental, autumn-colour large shrubs or trees.

Origin From eastern North America.
Use As a freestanding, large shrub or tree to produce good autumn colour.
Description *Flower* Racemes of green-yellow florets 6-10in (15-25cm) long, on slender, arching stalks produced mid to late summer. *Foliage* Ovate to oval, up to 5in (12cm) long and 3in (8cm) wide. Glossy surface, duller undersides. Brilliant red-yellow autumn colours, the main attribute. *Stem* Grey-green, becoming green-brown. Always branching, forming a large shrub which evolves into a tree. Moderately quick-growing when young, slowing with age. *Fruit* Oval, blue-black fruits formed late summer to early autumn.
Hardiness Tolerates 14°F (−10°C).
Soil Requirements Requires acid soil. Dislikes alkalinity.
Sun/Shade aspect Best in full sun, but toler-

ates light shade. Positioning for best effect under autumn sunlight.
Pruning None required, but can be slowly trained into small tree by pruning.
Propagation and nursery production From seed or layers. Purchase container-grown at heights between 12in (30cm) and 3ft (1m). Larger trees are normally unsuccessful when transplanted. Not always readily available; must be sought from specialist nurseries.
Problems Its requirement for acid soil is often overlooked.
Varieties of interest *N. sinensis* Leaves ovate and oval, pointed and tooth-edged, soft-textured. Good autumn colours. Slightly less than average height and spread and less hardy. Requires very acid soil.

Nyssa sylvatica in autumn

Average height and spread
Five years
6x6ft (2x2m)
Ten years
13x13ft (4x4m)
Twenty years
or at maturity
32x26ft (10x8m)

OSTRYA CARPINIFOLIA

EUROPEAN HOP HORNBEAM
Carpinaceae
Deciduous
An attractive large shrub or tree with yellow autumn foliage.

Origin From central southern Europe through to Asia Minor.
Use As a large shrub and ultimately a tree, for autumn colouring. Good for single and mass planting.
Description *Flower* Clusters of small, inconspicuous, green-yellow flowers. *Foliage* Ovate to oval, 2in (5cm) long, with tapering points and double-toothed edges. Young leaves grey-green with downy texture, aging to light to medium green. Good yellow autumn colour. *Stem* Grey-green becoming green-brown. Intially forming a branching shrub, later becoming a multi-stemmed, medium-sized tree. Relatively slow-growing in all but ideal conditions. *Fruit* Hop-like, hanging in clusters, grey-green to green-yellow.
Hardiness Tolerates 4°F (−15°C).
Soil Requirements All soil conditions, but shows signs of distress on extremely dry soil.
Sun/Shade aspect Full sun to medium shade.
Pruning None required
Propagation and nursery production From seed. Purchase container-grown. Plants relatively difficult to obtain, and must be sought from specialist nurseries. Normally offered from 3-5ft (1-1.8m).
Problems None.
Varieties of interest *O. japonica* Leaves grey-blue, long, slender and pointed. Attractive

autumn colouring. Nuts coloured glabrous blue. From Japan and China. Reaches more than average height and spread in its native environment. *O. virginiana* (Ironwood, American Hop Hornbeam) Stems very dark. Foliage oval or rounded with tapering point. Good autumn colouring. Spindle-shaped, blue nutlets in autumn. From eastern North America. Not in general garden use outside its native environment.

Ostrya carpinifolia **in leaf**

Average height and spread
Five years
6x5ft (2x1.5m)
Ten years
16x10ft (5x3m)
Twenty years
or at maturity
32x20ft (10x6m)
Continued growth can reach more than 39x26ft (12x8m)

OXYDENDRUM ARBOREUM

SORREL TREE, SOURWOOD TREE
Ericaceae
Deciduous
An attractive foliage shrub or small tree for acid soils, giving good autumn colours.

Origin From eastern and south-eastern USA.
Use As a large shrub or small tree, for autumn colour, for solo or mass planting.
Description *Flower* Drooping, white, terminal panicles, 6-10in (15-25cm) long, opening in late summer. *Foliage* Oblong to oval, slightly lanceolate foliage 4-6in (10-15cm) long, mid green giving way to brilliant red and orange in autumn. *Stem* Bright green with red hue, becoming green-brown, finally brown. Shrub-forming. Moderately slow-growing, becoming a tree only after 10-15 years. Upright when young, spreading with age. *Fruit* Insignificant.
Hardiness Tolerates 14°F (−10°C).
Soil Requirements Must have acid soil and

good drainage. Dislikes any degree of alkalinity.
Sun/Shade aspect Tolerates light shade, but best in full sun. Position for best effect under autumn sunlight.
Pruning None required.
Propagation and nursery production From seed, cuttings or layers. Purchase container-grown or root-balled (balled-and-burlapped). Relatively difficult to find in nursery production and must be sought from specialist sources. Normally offered from 15in (40cm) to 2½ft (80cm). Larger plants may be available but are not recommended.
Problems Needs some attention to soil requirements. Slow rate of growth.
Average height and spread
Five years
4x3ft (1.2x1m)
Ten years
10x6ft (3x2m)
Twenty years
or at maturity
20x13ft (6x4m)
Continued growth may reach 65ft (20m) in height.

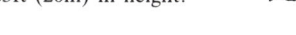

PARROTIA PERSICA

PERSIAN PARROTIA
Hamamelidaceae
Deciduous
An aristocrat among autumn-foliage large shrubs and trees.

Origin From northern Persia to the Caucasus.
Use As a large multi-stemmed shrub or tree for autumn colour and winter flowering.
Description *Flower* Short clusters of red anthers, surrounded by dark brown or black hairy bracts, an attractive winter feature. Abundant flowers only on trees of more than 10-15 years. *Foliage* Ovate, up to 5in (12cm) long, with toothed edges. Dark green, turning to vivid orange, yellow and red in autumn. *Stem* Grey-green to grey-brown. Spreading and branching. Rarely produces a significant trunk. For 10-15 years may be considered a large shrub, then developing into a multi-stemmed tree. *Fruit* Small, inconspicuous nuts in autumn.
Hardiness Tolerates −13°F (−25°C).
Soil Requirements Requires neutral to acid soil for best results. Tolerates limited alkalinity, but this inhibits overall performance.
Sun/Shade aspect Full sun.
Pruning None required other than to keep within bounds.
Propagation and nursery production From seed or layers. Purchase root-balled (balled-and-burlapped) or container-grown. Normally supplied from 8in (20cm) to 3ft (1m), but can be infrequently obtained from 6-10ft

(2-3m). Readily available from general nurseries or specialist outlets.
Problems Generally considered a tree, but appears as a large shrub for the first 10-15 years.

Parrotia persica **in autumn**

Varieties of interest *P. p. 'Pendula'* Good autumn colouring and more weeping branches than the parent. Scarce in production, but not impossible to find. *Parrotiopsis jacquemontiana* Previously known as *Parrotia jacquemontiana* and closely related to *P. persica*. Flowers consist of 4-6 creamy-white bracts, produced in late spring. Foliage ovate, light green, with good yellow autumn colour. Requires acid to neutral soil. From the Himalayas. Difficult to find and must be sought from specialist nurseries.
Average height and spread
Five years
6x6ft (2x2m)
Ten years
13x13ft (4x4m)
Twenty years
or at maturity
32x32ft (10x10m)

PAULOWNIA TOMENTOSA (Paulownia imperialis)

FOXGLOVE TREE, ROYAL PAULOWNIA, PRINCESS TREE, EMPRESS TREE
Scrophulariaceae
Deciduous
A truly interesting, large-flowering, spring tree for milder areas.

Origin From China.
Use As a moderately large shrub or multi-stemmed tree for medium and large gardens. Best freestanding.
Description *Flower* Wide, pyramid-shaped panicles 8-12in (20-30cm) long, upright in habit, carrying blue to purple, 2in (5cm) long, 5-lobed, foxglove-shaped flowers in late spring. Flowers not normally borne until 5-10 years after planting. *Foliage* Ovate to round, 3-5 lobed leaves on long stalks. Each leaf on established trees 8-10in (20-25cm) wide and long. Soft grey-green, grey woolly undersides. Shallow-toothed edges. *Stem* Hollow, fast-growing, upright, grey-green stems. Terminal bud dies in winter causing secondary branching, to form a short trunk and multi-stemmed round head, spreading with age. *Fruit* Pointed oval capsules up to 2in (5cm) long with winged seeds of little interest.
Hardiness Tolerates 4°F (−15°C), but some terminal stem damage may occur in areas with winter temperatures below 23°F (−5°C). This natural effect allows the tree to branch more freely from buds below the damage. Mild weather encourages early growth in spring, but this is vulnerable

Oxydendrum arboreum **in autumn**

Paulownia tomentosa in flower

to damage from late spring frosts.
Soil Requirements Requires deep, rich soil for adequate new growth. Tolerates limited alkalinity.
Sun/Shade aspect Best in full sun. Dislikes shade.
Pruning None required other than removal of dead shoots.
Propagation and nursery production From seed or self-perpetuated root suckers. Purchase root-balled (balled-and-burlapped) or container-grown. Plants normally available no more than 3ft (1m) in height and may deteriorate during winter. Aim to establish a rooted crown providing strong new shoots and select the strongest to form the eventual tree. Not readily available; must be sought from specialist nurseries.
Problems No cultivation problems, but cannot be bought as an established tree.
Average height and spread
Five years
16x5ft (5x1.5m)
Ten years
23x13ft (7x4m)
Twenty years
or at maturity
47x23ft (14x7m)

PHELLODENDRON AMURENSE

AMUR CORK TREE
Rutaceae
Deciduous
An interesting medium to large tree rarely seen in gardens.

Origin From Manchuria.
Use As a medium-sized, quick-growing tree with pleasant foliage and late spring flowers, for medium or large gardens.
Description *Flower* Upright panicles of green--yellow flowers 3-4in (8-10cm) long, produced late spring or early summer. *Foliage* Bright green pinnate leaves, up to 15in

Phellodendron amurense in leaf

(40cm) long with 5-11 leaflets, each 4in (10cm) long. Some yellow autumn tints. Used as feed in butterfly farms. *Stem* Stout, rigid, upright habit, spreading with age. Stems dark green with purple hue. Terminal buds and stems may die over winter, with no lasting ill effect. Quick-growing. *Fruit* Small fruits, ½in (1cm) wide, in late summer, early autumn.
Hardiness Tolerates 14°F (−10°C).
Soil Requirements Requires deep, rich loam to do well. Tolerates moderate alkalinity.
Sun/Shade aspect Best in full sun; tolerates light shade.
Pruning None required.
Propagation and nursery production From seed or softwood cuttings, taken in midsummer. Purchase root-balled (balled-and-burlapped) or container-grown. Plants supplied between 10in (25cm) and 10ft (3m), but supply is erratic, offering little choice. Not common in commercial production, but available from specialist nurseries.
Problems Early spring growth produced in mild weather may be damaged by late frosts.
Varieties of interest There is a number of other varieties and forms, but the above species is recommended for garden planting.
Average height and spread
Five years
13x6ft (4x2m)
Ten years
26x13ft (8x4m)
Twenty years
or at maturity
39x20ft (12x6m)

PHOTINIA VILLOSA

ORIENTAL PHOTINIA
Rosaceae
Deciduous
A gem among autumn-flowering small trees, for neutral or acid soils.

Origin From western China.
Use As a large shrub or small tree featured for its autumn colouring.
Description *Flower* Clusters of small white flowers in late spring, each up to 2in (5cm) across. *Foliage* Leaves up to 3½in (8cm) long, ovate with toothed edges. Dark green upper surfaces with blue sheen, grey, downy undersides. Extremely vivid orange-yellow autumn colours. *Stem* Light grey-brown when young, darkening with age. Twiggy, not trunk-forming. Initially a small shrub, evolving to a small tree after 20 years. Moderately upright habit. *Fruit* Small, bright red, oval fruits in late summer or early autumn.
Hardiness Tolerates 4°F (−15°C).
Soil Requirements Requires a neutral to acid soil for best results. Dislikes alkalinity.

Sun/Shade aspect Full sun; select position to show off autumn colours.
Pruning None required.
Propagation and nursery production From seed, cuttings or layers. Purchase container-grown. Normally available from 8in (20cm) up to 4ft (1.2m). Must be sought from specialist nurseries.
Problems None except small at purchase and not fast-growing.
Average height and spread
Five years
6x3ft (2x1m)
Ten years
12x6ft (3.5x2m)
Twenty years
or at maturity
16x13ft (5x4m)

PLATANUS × HISPANICA (Platanus × acerifolia)

LONDON PLANE TREE
Platanaceae
Deciduous
Useful large trees for very large gardens.

Origin Previously there were thought to be two basic forms of this plant — *P.* × *hispanica*, originating from Spain, and *P.* × *acerifolia* said to have been raised in Oxford, England, but these are now considered to be the same and may be found under either name.
Use As a freestanding large tree, or grouped for large gardens, estates, parks and for avenues when planted 39ft (12m) apart. Excellent for street planting in cities — seems more at home in urban than in country environment.
Description *Flower* Flowers borne in globe-shaped heads, grey-green to yellow-green, produced late spring or early summer. *Foliage* Palmate 5-lobed leaves, up to 12in (30cm) across. Light grey-green, giving way to yellow autumn colour. *Stem* Grey-green when young. Vigorous, slowing with age. Peeling, mottled bark effect, with creamy, light brown patchwork display in trees over 10 years old. Quickly forms an upright tree which spreads with age. *Fruit* Conspicuous hanging brown to grey-brown balls in autumn, retained well into winter or to following spring.
Hardiness Tolerates 4°F (−15°C).
Soil Requirements Most soil conditions — dislikes extremely alkaline types
Sun/Shade aspect Full sun to light shade, preferring full sun.
Pruning None required, but lower limbs can be pruned to remove obstruction and enhance trunk appearance.
Propagation and nursery production From seed or layers. Plant container-grown or bare-rooted. Best planting heights 6-10ft (2-3m).

Photinia villosa in autumn

Platanus × *orientalis* in leaf

Populus × *candicans* 'Aurora' in leaf

Varieties of interest *P.* × *hispanica* '*Suttrieri*' Palmate foliage splashed boldly with white variegation maintained into midsummer. Extremely scarce in production and must be sought from specialist nurseries. Rarely found and the usually at only 1½-2½ft (50-80cm) in height. *P.* × *orientalis* Deeply cut and lobed green leaves, each leaflet with 5-7 lobes. Grey-green undersides. Flowers and fruit similar to *P.* × *hispanica*. Average height, but more spreading habit. Susceptible to stem canker. *P. orientalis var. digitata* Distinctive deeply cut leaves, good yellow autumn colour. Broadly spreading habit.

Average height and spread
Five years
13x6ft (4x2m)
Ten years
32x13ft (10x4m)
Twenty years
or at maturity
63x20ft (20x6m)

POPULUS ALBA

WHITE POPLAR, SILVER-LEAVED POPLAR

Salicaceae

Deciduous

Interesting silver-grey foliage. Ultimately a large tree but controllable when young.

Origin From Europe and central Asia.
Use As a large specimen tree for medium-sized and larger gardens; or as a windbreak, especially when pollarded. Can be grown as a large shrub if cut back every 2-3 years.
Description *Flower* Small, pale yellow catkins in mid spring. *Foliage* Broad, ovate wavy edged leaves, grey upper surfaces and white undersides, aging to green-grey and silver. Good yellow autumn colour. Foliage on young trees 3-5 lobed, up to 5in (12cm) long; older foliage up to 2in (5cm) long. *Stem* Fast-growing, upright, limited branching. Covered in white woolly bloom. Becomes more round-headed with age. *Fruit* Insignificant.
Hardiness Tolerates −13°F (−25°C).
Soil Requirements Requires moist, rich soil, acid or alkaline, for best results. Tolerates moderate waterlogging. Shows distress on very dry, poor soils.
Sun/Shade aspect Full sun, to enhance grey-white leaf colouring.
Pruning None required but may be pollarded hard in early spring to increase stem and foliage colour and reduce overall size.
Propagation and nursery production From hardwood cuttings. Plant bare-rooted or container-grown. Plants normally available from 2ft (60cm) up to 16ft (5m). Best planting sizes for most situations from 5-8ft (1.5-2.5m). Readily available from general nurseries and forestry outlets.
Problems Young containerized and nursery plants always look poorly shaped; require open-ground conditions to produce best re-

sults. Older trees may experience stem canker, but this usually causes isolated damage. Roots are spreading; not recommended for planting near buildings or drains.
Varieties of interest *P.a.* '*Pyramidalis*' A narrow, upright, silver-leaved, fastigiate tree, useful for street or avenue planting at 16ft (5m) apart or as individual garden specimen. Average height, but never more than 10ft (3m) in width. Previously called *P.* '*Bolleana*'. *P. a.* '*Richardii*' Simultaneous silver and yellow leaves throughout summer. Good grey stems in winter. Bushy, round habit. Two-thirds average height and spread. Relatively scarce, but not impossible to find.

Populus alba in leaf

Average height and spread
Five years
16x13ft (5x4m)
Ten years
32x20ft (10x6m)
Twenty years
or at maturity
66x33ft (20x10m)
Continued growth over 50 years reaches 66x46ft (20x14m)

POPULUS
Balsam-scented varieties

BALSAM POPLAR

Salicaceae

Deciduous

Fast-growing, leafy screening trees, requiring space to mature.

Origin Dependent on variety
Use As quick-growing large trees; as individual specimens, for use as coppice or windbreaks, or for screening industrial buildings.
Description *Flower* Yellow male catkins up to 4in (10cm) long borne in mid to late spring. *Foliage* Broadly ovate, with pointed ends. Grey-green when young, and giving off pleasant balsam scent in spring and early summer. Aging to bright green. Good yellow autumn colour. *Stem* Ribbed, green-brown to brown, strong, upright stems capable of 6-10ft (2-3m) of growth in a single year. Normally forms a moderately upright tree for many years, then spreading with age. Very quick-growing but has a relatively short lifespan of 40-50 years. *Fruit* Insignificant.
Hardiness Tolerates −13°F (−25°C).
Soil Requirements All soil types; tolerates extremely wet conditions.
Sun/Shade aspect Full sun to medium shade.
Pruning Responds to hard cutting back, producing stronger, larger leaves, but can be left untrimmed. Cutting back new growth every 3-4 years encourages branching and a more luxuriant leaf canopy.
Propagation and nursery production From hardwood cuttings. Plant bare-rooted or container-grown. Available from 2ft (60cm)

up to 16ft (5m) or more. Planting heights of 8-12ft (2.5-3.5m) usually most rewarding. Readily available from forestry outlets, sometimes from general nurseries. Some forms must be sought from specialist nurseries.
Problems Root spread can interfere with nearby buildings or drains. May suffer from attacks of stem canker, but not seriously.
Varieties of interest *P. balsamifera* (Balsam Poplar) Attractive, bright green, scented foliage in spring. Round-headed. Fast-growing.From North America. *P.* × *candicans* (Balsam Poplar of Gilead) Attractive, light green foliage. Good balsam scent. Fine autumn colours. Strong, quick growth. *P.* × *c.* '*Aurora*' New foliage on new growth splashed white and pink, an interesting contrast to green background leaves. Balsam scent in spring. Annual pollarding or triennial hard pruning recommended to increase intensity of foliage colour and size. *P. trichocarpa* (Black Cottonwood) Fast-growing, upright. Useful fast-screening trees when planted 10-13ft (3-4m) apart. Quickly attains a thick screen of winter twigs or summer foliage. Responds well to regular pruning back in early spring, enhancing foliage and scent for coming season. Good yellow autumn colour. Leaf fungus may occur in wet seasons, disfiguring but not of lasting damage. *P. t.* '*Fritzi Pauley*' Large leaves, good scent and good growth.
Average height and spread
Five years
16x13ft (5x4m)
Ten years
32x20ft (10x6m)
Twenty years
or at maturity
65x32ft (20x10m)
Continued growth reaches up to 85ft (26m) in height.

Populus trichocarpa in leaf

Populus lasiocarpa in leaf

POPULUS LASIOCARPA

KNOWN BY BOTANICAL NAME

Salicaceae

Deciduous

A shapely tree of architectural proportions, worthy of wider planting.

Origin From China.
Use As an interesting structural and foliage tree for medium-sized or larger gardens, leaf size being a particular curiosity.
Description *Flower* Pale yellow male catkins 4in (10cm) long, in early spring. *Foliage* Leaves 10in (25cm) long and up to 8in (20cm) wide, or larger in favourable conditions. Ovate, with pointed ends and incurved bases. Light green to grey-green upper surfaces, grey-green downy undersides. Red mid-rib and leaf stalk. Yellow autumn tints. *Stem* Upright, with distinct lateral branches forming a narrow-based pyramid. Stems grey-green to grey-brown with terminal buds of winter interest. Moderately slow-growing. *Fruit* Insignificant.
Hardiness Tolerates 14°F (−10°C), but does best in sheltered positions.
Soil Requirements Tolerates most soil conditions, acid or alkaline, except very dry or extremely waterlogged.
Sun/Shade aspect Tolerates shade, but best in full sun to light shade.
Pruning None required other than initial training of central trunk.
Propagation and nursery production From cuttings or grafting. Purchase container-grown. A limited number available grafted on a stem, although these do not ultimately make the best trees. Best planting heights 2ft (60cm) up to 3ft (1m). Must be sought from specialist nurseries.
Problems Mature trees can develop stem canker, sometimes serious. Watch for early signs of infestation, remove damaged wood immediately and provide adequate feeding to overcome lost growth.
Average height and spread
Five years
6x3ft (2x1m)
Ten years
13x6ft (4x2m)
Twenty years
or at maturity
26x13ft (8x4m)
Continued growth over 50 years reaches 32x26ft (10x8m)

POPULUS NIGRA

BLACK POPLAR

Salicaceae

Deciduous

A particularly fine estate or parkland tree.

Origin From Europe and western Asia.
Use Not suitable for garden planting due to enormous size. Used as an estate tree or for parks. Good for avenues when planted at 39ft (12m) apart. Often seen in large plantations for commercial timber production.

Populus nigra − trunk

Description *Flower* Male catkins up to 3in (8cm) long with attractive red anthers, produced mid spring. *Foliage* Ovate or triangular, up to 4in (10cm) long, grey-green. Highly attractive in early spring when first opening. Good yellow autumn colour. *Stem* Strong and very vigorous. Trunk may have girth of up to 6ft (2m) after 30 or 40 years. Bark grey-green to grey-brown, rugged texture. Buds dull brown with slight resinous coating. *Fruit* Insignificant.
Hardiness Tolerates winter temperatures below −13°F (−25°C).
Soil Requirements Any soil conditions, but best growth on deep, moist, rich soil.

Sun/Shade aspect Full sun to light shade. Tolerates shade; especially when group planted.
Pruning All side branches can be removed to create a main trunk of 39ft (12m) or more.
Propagation and nursery production From hardwood cuttings. Plant bare-rooted. Normally available at 5-8ft (1.5-2.5m). Larger plants may be available, but are not recommended. Usually obtainable from forestry outlets.
Problems None.
Varieties of interest *P. n. var. betulifolia* (Manchester Poplar) Closer foliage and branched effect. A large tree of estate and parkland dimensions. *P. 'Robusta'* Foliage red-green when young, up to 5in (12cm) long, triangular to ovate, markedly pointed. Used predominantly in large plantations for timber-production.
Average height and spread
Five years
16x13ft (5x4m)
Ten years
32x20ft (10x6m)
Twenty years
or at maturity
66x33ft (20x10m)
Continued growth over 50 years reaches 79x48ft (25x15m)

POPULUS NIGRA 'ITALICA'

LOMBARDY OR ITALIAN POPLAR

Salicaceae

Deciduous

The famous Lombardy Poplar with its characteristic columnar shape.

Origin From Italy.
Use As a tall, upright pillar for medium to large gardens. Good for avenue use planted at a minimum of 32ft (10m) apart to achieve the best appearance. A useful screening tree for both private and industrial complexes, if adequate space is available.

Populus nigra 'Italica'

48

Description *Flower* Male catkins with attractive red anthers, up to 3in (8cm) long, in mid spring. *Foliage* Ovate or triangular, up to 4in (10cm) long, grey-green and giving good yellow autumn colour. Very attractive in early spring when first opening. *Stem* Grey-green to grey-brown. Upright and fastigiate, forming a tall, symmetrical, pencil-like pillar. Medium growth rate. *Fruit* Insignificant.
Hardiness Tolerates 4°F (−15°C).
Soil Requirements Most soil conditions, but poor growth when waterlogged or extremely dry.
Sun/Shade aspect Full sun to very light shade. Deeper shade spoils overall shape.
Pruning None required.
Propagation and nursery production From hardwood cuttings. Plant bare-rooted or container-grown. Normally offered from 5-12ft (1.5-3.5m). Larger plants occasionally available. Best planting heights 5-8ft (1.5-2.5m). Relatively easy to find from general nurseries, and sometimes from specialist sources.
Problems Can suffer from Poplar stem canker, but this is rarely a major concern. Mature tree annually loses a number of the softer, thinner branches.
Average height and spread
Five years
16x3ft (5x1m)
Ten years
32x6ft (10x2m)
Twenty years
or at maturity
64x13ft (20x4m)
Continued growth over 50 years reaches 82x19ft (25x5m)

POPULUS SEROTINA 'AUREA'

GOLDEN ITALIAN POPLAR
Salicaceae
Deciduous
A beautiful golden-leaved tree, requiring considerable space.

Origin From southern Europe.
Use As a very large specimen tree or pollarded for its intense golden foliage.

Populus serotina 'Aurea' in leaf

Description *Flower* Male catkins with prominant red anthers, up to 4in (10cm) long, produced on mature trees in spring. *Foliage* Triangular to ovate leaves, 3in (8cm) long, rich yellow aging through summer to yellow-green, finally producing golden yellow autumn tints. Foliage may be larger on young or hard-pruned trees. *Stem* Grey-green, strong, upright, spreading to make a tree of substantial proportions. Eventually round-topped. Medium to quick growth when young, slowing with age. *Fruit* Insignificant.
Hardiness Tolerates 4°F (−15°C).

Soil Requirements Most soil conditions; dislikes extreme dryness or waterlogging.
Sun/Shade aspect Must have full sun or leaf colour is lost.
Pruning Can be pollarded hard every three or five years, a method which keeps it within bounds for medium-sized gardens and increases leaf size and colour.
Propagation and nursery production From hardwood cuttings. Plant bare-rooted or container-grown. Normally provided at best planting size from 5ft (1.5m) to 8ft (2.5m) in height. Larger trees may be available, but are not recommended.
Problems Young nursery trees have a fairly weak appearance, and it takes two or more seasons for the tree to become fully established and show its full potential.
Average height and spread
Five years
16x13ft (5x4m)
Ten years
32x20ft (10x6m)
Twenty years
or at maturity
65x32ft (20x10m)
Continued growth over 50 years reaches 82x48ft (25x15m)

POPULUS TREMULA

EUROPEAN ASPEN, ASPEN POPLAR
Salicaceae
Deciduous
Very attractive foliage trees providing a sense of movement in a garden plan.

Origin From Europe, North Africa and Asia Minor.
Use As an individual specimen for its shimmering leaves. For screening when planted in a coppice or line. Good windbreak. For medium or large gardens.
Description *Flower* Male catkins up to 4in (10cm) long, in late winter. *Foliage* Triangular, ovate, sometimes rounded or broad ovate up to 4in (10cm) long. Grey-green with notable white undersides. Foliage moves freely in any wind or breeze, displaying white undersides to fine effect. *Stem* Grey-green. Upright central trunk with horizontal side limbs slightly ascending. Medium rate of growth. *Fruit* Insignificant
Hardiness Tolerates 4°F (−15°C).
Soil Requirements Most soil conditions, but dislikes extremely wet or dry locations.
Sun/Shade aspect Best in full sun to show off the shimmering foliage.
Pruning None required, but young trees benefit from shortening of side growths for the first two or three years after planting to give thick stem growth and covering.
Propagation and nursery production From hardwood cuttings. Plant bare-rooted or

container-grown. Plants normally obtainable from 3-13ft (1-4m) but best planting heights 5-8ft (1.5-2.5m). Must be sought from specialist nurseries.
Problems None.
Varieties of interest *P. t. 'Erecta'* Very upright, spiralling, pyramidal habit. Useful as a spot or feature tree. Same leaf characteristics as the parent. *P. t. 'Pendula'* A broader weeping form with long arching branches reaching to ground level. Requires training and staking to obtain height. Good architectural shape in summer and winter. Two-thirds average height and spread.
Average height and spread
Five years
6x3ft (2x1m)
Ten years
13ftx6ft (4x2m)
Twenty years
or at maturity
26x13ft (8x4m)
Continued growth over 50 years reaches 32x20ft (10x6m)

PRUNUS
Almond trees

ALMOND
Rosaceae
Deciduous
Among the earliest to flower of spring-blossoming trees.

Origin From the Mediterranean to central Asia.
Use As small to medium flowering trees with the potential of a fruit crop on *P. dulcis*. Best planted singly to show off blossom effect.

Prunus × amygdalo-persica in flower

Description *Flower* White or rose-pink. Single or double, depending on variety. Flowers up to 2in (5cm) across, usually borne on bare branches in late winter to early spring. *Foliage* Ovate to lanceolate, up to 6in (15cm) long, with finely toothed edges. Mid green to grey-green with red hue, particularly along veins and leaf stalks. Prominent glands at leaf bases. Limited autumn colour. *Stem* Red-

Populus tremula in leaf

green with slight grey sheen. Moderately vigorous when young, becoming branching and twiggy with age. Initially an upright tree, but quickly spreading. *Fruit* Elliptic, velvety, green with red hue. Splits and opens when ripe to show the smooth stone with pitted marks used as almond nut. *P. dulcis* is the true almond fruiting form.
Hardiness Tolerates 14°F (−10°C).
Soil Requirements Does best on well-drained soil.
Sun/Shade aspects Best in full sun to assist fruit-ripening. This is variable outside mild locations.
Pruning None required; resents it and bleeds a gummy resin from stems and trunk.
Propagation and nursery production From grafting or budding on to understocks of *P. amygdalus* or *P. persica*. Plant bare-rooted or container-grown. Normally offered from 5½ft (1.8m) to 9ft (2.8m). Larger trees not recommended. Available from general nurseries or specialist sources.
Problems Very susceptible to the virus diseases peach-leaf curl or peach-leaf blister, both extremely difficult to control. May also suffer from peach aphids. If soil preparation is inadequate the tree reacts by not producing new growth.

Prunus dulcis **in fruit**

Varieties of interest *P. × amygdalo-persica* syn. *P. amygdalo* 'Pollardii' (Flowering Almond) A cross between a peach and an almond. Now presented by nurseries and garden centres as *P. amygdalo* 'Pollardii'. Fragrant, bright pink flowers in early spring. Fruits not edible. Originating in Australia. *P. dulcis* syn *P. amygdalus*, *P. communis* Pink flowers. Producing edible almond nuts and grown commercially. *P. triloba* Rose-pink double flowers. Non-fruiting. Obtainable grown on a 5-6ft (1.5-2m) standard stem to make a mop-headed tree. Or can be grown as large shrub.
Average height and spread
Five years
13x5ft (4x1.5m)
Ten years
20x10ft (6x3m)
Twenty years
or at maturity
23x16ft (7x5m)

PRUNUS AVIUM

GEAN, MAZZARD, WILD CHERRY, SWEET CHERRY
Rosaceae
Deciduous
A spectacular spring-flowering tree.

Origin From Europe, where it is a wild tree.
Use As a large, freestanding or group-planted tree for windbreaks, screening and other general use in medium-sized and larger gardens. Used as a nursery understock on which most ornamental cherries are grown by budding or by grafting.
Description *Flower* Single white flowers, up to 1½in (4cm) across, borne close to the stem in late spring. *Foliage* Ovate to oval, up to 5in (12cm) long. Light green with a grey sheen. Good yellow autumn colour. *Stem* Dark brown to mahogany brown. Upright, strong-growing when young and quick to form a pyramidal tree with gnarled effect. *Fruit* Small, black to red-black round fruits, late summer to early autumn.

Prunus avium 'Flore Pleno' **in flower**

Hardiness Tolerates −13°F (−25°C).
Soil Requirements Most soil conditions. Dislikes waterlogging.
Sun/Shade aspects Full sun to light shade, preferring light shade.
Pruning None required.
Propagation and nursery production From seed. Planted bare-rooted or container-grown. Normally offered from 3-20ft (1-6m). Best planting sizes 5½-8ft (1.8-2.5m). May have to be sought from forestry outlets.
Problems May suffer mechanical root damage. In severe winters bark may crack and bleed a resinous substance, usually without lasting harm. Often seen suckering on weeping cherries and from surface root growth: suckers should be removed. Lifespan of *P. avium* is approximately 50 years, after which time its overall vigour rapidly decreases.
Varieties of interest *P. a.* 'Flore Pleno' Large, pure white, long lasting double flowers, each with over 30 petals, hanging in long, bunched clusters. For medium-sized and larger gardens. *P. a.* 'Pendula' Weeping or drooping branches covered in single white flowers. Of interest, but rarely seen and not for general garden use.
Average height and spread
Five years
16x10ft (5x3m)
Ten years
32x20ft (10x6m)
Twenty years
or at maturity
48x39ft (15x12m)

PRUNUS
Early-flowering Cherries

FLOWERING CHERRY
Rosaceae
Deciduous
Heralds of the spring. Extremely fine flowering trees for early blossom effect.

Origin Of garden origin and hybrid crosses.
Use As small to medium early-flowering trees for any size of garden.
Description *Flower* Single flowers, white, shell pink, mid pink or purple-pink dependent on variety, produced on bare branches in great profusion from late winter to early spring, or spring-flowering. *Foliage* Ovate, up to 4in (10cm) long, with toothed edges. Light to mid green with yellow, often spectacular, autumn colour. *Stem* Thin growth,

very branching habit, forming either a round-topped and spreading or upright tree, depending on variety. Wood grey to grey-green. Medium rate of growth. *Fruit* Insignificant.
Hardiness Tolerates 4°F (−15°C).
Soil Requirements Most soil conditions; dislikes extremely dry or poor soils.
Sun/Shade aspect Will tolerate full sun to mid shade, preferring light shade.
Pruning None required other than the removal of any crossing or crowded branches. Prune while tree is dormant and treat cuts quickly with pruning compound to prevent fungus disease.
Propagation and nursery production From budding or grafting on to understocks of *P. avium*. Planted bare-rooted or container-grown. Normally sold at 5½-10ft (1.8-3m). Most varieties readily available from general nurseries, but a few must be sought from specialist outlets.

Prunus 'Okame' **in flower**

Problems Flower buds may be damaged by birds, in quiet areas, sometimes severely. Young trees can look weak at purchase due to thin growth, but improve after planting.
Varieties of interest *P.* 'Accolade' A cross between *P. sargentii* and *P. subhirtella*. Large clusters of rich pink flowers up to 8in (20cm) across, individual flowers up to 1½in (4cm) across. Good autumn colour. Average height with slightly more spread, branches pendulous in habit on mature trees. *P.* 'Hally Jolivette' A cross between *P. subhirtella* and *P. yedoensis × subhirtella*. Thin stems produce clusters of small semi-double, pale pearly pink flowers through early spring. Predominantly used as a large shrub, later reaching the proportions of a small tree. Raised in the Arnold Arboretum, Massachusetts, USA. *P. × hillieri* 'Spire' A cross between *P. incisa* and *P. sargentii*. Soft pink flowers and good autumn colours. Useful for all sizes of garden. Upright to conical, reaching average height, but basal spread only 10ft (3m). Raised in the Hilliers Nurseries in Winchester, England. *P. incisa* (Fuji Cherry) Foliage small, light green with good autumn colours. Flowers pink in bud, opening to white. Occasionally produces small purple black fruits. Round-topped habit. Best grown as a large shrub, but can be found as a small standard tree. Slightly less than average height and spread. *P. i.* 'February Pink' Flowers slightly earlier than *P. incisa* and with more pink colouring. *P.* 'Kursar' A cross between *P. campanula* and *P. kurilensis*. Profuse, deep pink flowers in mid spring. Young foliage red-bronze; good yellow autumn tints. Forms a small, round-topped tree with interesting dark purple stems. Not readily available, but may be found in specialist nurseries. *P.* 'Moerheimii' syn. *P. incisa* 'Moerheimii' Flowers pink in bud, opening to white with blush pink shading. Round-topped, weeping branches, forming a dome-shaped effect of two-thirds average height and spread. May be difficult to find,

Prunus × yedoensis **in flower**

and must be sought from specialist nurseries. **P. 'Okame'** A cross between *P. campanulata* and *P. incisa*. A delightful round-topped small tree, early-flowering and reaching two-thirds average height and spread. Profuse crimson-rose flowers in early spring. Good yellow autumn tints. **P. 'Pandora'** Shell pink flowers up to 1in (3cm) across in early spring. Young foliage bronze-red in spring with good yellow autumn colours. Extremely reliable. **P. 'Umineko'** A cross between *P. incisa* and *P. speciosa*. White flowers in mid spring. Good autumn colour. Upright, of average height and two-thirds average spread. **P. × yedoensis** Spectacular masses of almond-scented, blush-white flowers in early spring. Branches slightly pendulous. From Japan, and apparently of garden origin.

Average height and spread
Five years
12x8ft (3.5x2.5m)
Ten years
23x12ft (7x3.5m)
Twenty years
or at maturity
30x23ft (9x7m)

PRUNUS
Weeping Cherries

WEEPING CHERRY, JAPANESE CHERRY
Rosaceae
Deciduous
Truly aristocratic weeping trees, of exceptional spring beauty.

Origin From Japan. Mostly of garden or nursery origin.
Use As ornamental weeping trees, ideal for small, medium or large gardens. Best planted singly and isolated.
Description *Flower* Single or double, white or rose pink depending on variety, produced mid to late spring, often in profusion *Foliage* Ovate, 5in (12cm) long, slightly tooth-edged. Light to mid green with some yellow autumn colour. *Stem* Weeping, pendulous stems to ground level, forming a circular, cascading fountain. Wood grey-brown and of winter interest. Medium rate of growth. *Fruit* Insignificant.
Hardiness Tolerates 4°F (−15°C).
Soil Requirements Most soil conditions, but resents poor soil.
Sun/Shade aspect Best in full sun. Tolerates limited light shade.
Pruning None required, but spread can be reduced by removal of outer weeping branches. Best carried out while tree is still dormant and wounds treated with pruning compound to prevent fungus disease.
Propagation and nursery production From grafting or budding on to understocks of *P. avium*. Plant bare-rooted or container-

grown. Many varieties readily available, but some must be sought from specialist nurseries, including all forms of *P. subhirtella*. Normally supplied 5-8ft (1.5-2.5m) in height. Select a suitable shape and structure, allowing for increasing height once planted.
Problems Often planted in areas too small to accommodate it; may outgrow its location quite quickly. Must not be allowed to develop suckering growths, which may appear on main stem below grafted or budded point.

Prunus 'Kiku-shidare Sakura' **in flower**

Varieties of interest *P. 'Hilling's Weeping'* Thin, slender, hanging branches, completely covered with a profusion of small white flowers in mid spring. Slightly less than average height and spread. *P. 'Kiku-shidare Sakura'* (Flowering Cherry) Large, double, pink flowers produced in mid to late spring, on the full length of the pendulous branches. Sometimes referred to as Cheal's Weeping Cherry. *P. subhirtella 'Pendula'* Thin, willow-like branches forming a low, spreading crown, with a profusion of small, off-white flowers in early to mid spring. One of the earliest to flower. *P. s. 'Pendula Rosea'* A rose pink flowering form, blossoming profusely in early spring. Height dependent on grafting. Average spread. *P. s. 'Pendula Rubra'* Dark pink to purple-pink flowers, very profuse in early spring. Height dependent on grafting.

Prunus × yedoensis 'Shidare Yoshino'

Average spread. *P. × yedoensis 'Ivensii'* Thin, hanging branches abundantly festooned with pure white flowers in mid spring. Extremely difficult to find in commercial production, and needs extensive search. *P. × yedoensis 'Shidare Yoshino'* syn. *P. × y. purpendens* (Yoshino Weeping Cherry of Japan) Vigorous, quick-growing, pendulous branches covered in mid spring with pink-budded flowers aging to pure white usually produced in short racemes. Slightly fragrant. Good yellow autumn colour. Readily available from specialist and general nurseries, but not normally stocked by garden centres.

Average height and spread
Five years
10x10ft (3x3m)
Ten years
13x16ft (4x5m)
Twenty years
or at maturity
16x26ft (5x8m)

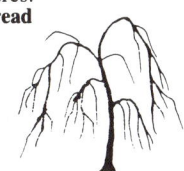

PRUNUS PADUS

EUROPEAN BIRD CHERRY
Rosaceae
Deciduous
Spring-flowering trees of great beauty.

Origin From northern Europe.
Use As a freestanding ornamental tree for all but the smallest gardens. Good singly or in groups. Can be used for windbreaks with spectacular spring display.

Prunus padus var. colorata **in flower**

Description *Flower* Fragrant white racemes, either drooping or spreading, up to 6in (15cm) long and ½in (1cm) wide, in late spring or early summer when many other spring-flowering trees have finished. *Foliage* Dark green upper surfaces, duller hairy undersides. Oval, up to 5in (12cm) long with finely toothed edges. Some good yellow autumn tints. *Stem* Strong, upright and pyramidal when young, spreading and branching with age. Red-green to green-brown with attractive winter colour. Medium rate of growth. *Fruit* Small round black cherries with a bitter taste, produced irregularly on flowered stalks in late summer and autumn.
Hardiness Tolerates −13°F (−25°C).
Soil Requirements Any soil conditions.
Sun/Shade aspect Full sun to light shade.
Pruning None required.
Propagation and nursery production From seed for *P. padus*, or named varieties by grafting on to understocks of *P. padus*. Can also be budded on to *P. avium*. Plant bare-rooted or container-grown. Normally offered 5½-10ft (1.8-3m) in height; larger specimens occasionally available, but smaller trees best for planting.
Problems None.
Varieties of interest *P.p. 'Albertii'* Shorter, fuller racemes of white flowers. Average

Prunus padus 'Watereri' in flower

Prunus serrula – trunk in winter

spread but less height, more shrubby habit. *P. p. var. colorata* Short, somewhat irregular racemes of flowers, deep purple-pink in bud, paler when open. New foliage purple to purple-green. A beautiful sight from a distance, but loses effect at close quarters. *P. p. 'Purple Queen'* Foliage deep purple when young, aging to purple-green. Racemes of dark purple-pink flowers, good contrast with foliage. *P. p. 'Watereri'* Racemes up to 8in (20cm) long, white, fragrant; both pendulous and horizontal in formation. Good dark green foliage. Interesting winter purple stems.

Average height and spread
Five years
13x6ft (4x2m)
Ten years
26x13ft (8x4m)
Twenty years
or at maturity
39x20ft (12x6m)

PRUNUS SARGENTII

SARGENT CHERRY
Rosaceae
Deciduous
Beautiful spring-flowering and autumn-foliage trees, requiring some space to develop.

Origin Of garden origin.
Use As a large spreading tree for spring flowers and with fine autumn foliage colouring. Planted singly or grouped for good effect.
Description *Flower* Single, pale pink flowers borne profusely on the bare branches in early

Prunus sargentii in autumn

spring. *Foliage* Ovate to oval, up to 4in (10cm) long. Light green when young, aging to mid or dark green. Brilliant autumn colours of yellow, orange and flame. *Stem* Grey-green. Quick-growing and upright at first, rapidly spreading to form a wide, flat-topped tree, often large in relation to other cherries. *Fruit* May produce small, oval crimson fruits up to ½in (1cm) long.
Hardiness Tolerates −13°F (−25°C).
Soil Requirements Most soil conditions, but resents poor soils.
Sun/Shade aspect Tolerates quite deep shade, but prefers full sun or light shade.
Pruning None required.
Propagation and nursery production From budding or grafting. Plant bare-rooted or container-grown. Normally supplied at 5½-10ft (1.8-3m).
Problems None.
Varieties of interest *P. s. 'Rancho'* Slightly earlier flowering with larger flowers, but having no true advantage over its parent.
Average height and spread
Five years
12x8ft (3.5x2.5m)
Ten years
20x16ft (6x5m)
Twenty years
or at maturity
30x32ft (9x10m)

PRUNUS SERRULA
(Prunus serrula tibetica)

PEELING BARK CHERRY
Rosaceae
Deciduous
Highly attractive winter trunk and branches.

Origin From western China.
Use As a small garden tree suitable for medium-sized and larger gardens to show off attractive coloured trunk and branches in winter.
Description *Flower* Small, uninteresting white flowers in mid spring. *Foliage* Ovate to lanceolate, leaves 2in (5cm) long, slightly tooth-edged. Light green, becoming darker green. *Stem* Main trunk branches show mahogany glossy surface interspersed with brown bark. Glossed area increases with age. A fine sight seen under strong winter sun. *Fruit* Small, round, black cherries.
Hardiness Tolerates −13°F (−25°C).
Soil Requirements Any soil type.
Sun/Shade aspect Full sun to light shade. Best seen in a position receiving maximum light in winter.
Pruning None required other than removal of lower branches to create maximum possible length of trunk.

Propagation and nursery production From grafting or budding. Occasionally from seed. Best container-grown but can be used bare-rooted. Trees normally available 5-10ft (1.5-3m) from specialist nurseries. Not normally stocked by garden centres.
Problems Young trees have only thin stems or trunks and do not show full potential for 5-10 years.
Average height and spread
Five years
13x6ft (4x2m)
Ten years
16x10ft (5x3m)
Twenty years
or at maturity
23x13ft (7x4m)

PRUNUS SERRULATA 'AMANOGAWA'

UPRIGHT CHERRY, LOMBARDY CHERRY
Rosaceae
Deciduous
A well-known, useful, narrow, upright flowering tree.

Origin From Japan.
Use As a tall sentinel, singly or in pairs, or making lines or avenues when planted at 16ft (5m) apart. Extremely useful for accentuating a particular area.

Prunus serrulata 'Amanogawa' in autumn

Description *Flower* Large clusters of semi-double, pale pink aging to off-white, slightly scented flowers abundant on the vertical branches in mid to late spring. *Foliage* Ovate to broadly lanceolate, up to 8in (20cm) long, with toothed edges. Good yellow, orange and flame autumn colours. *Stem* Grey-green to grey-brown. Upright, fastigiate habit, forming a tall, narrow pillar. *Fruit* Insignificant.
Hardiness Tolerates 4°F (−15°C).
Soil Requirements Does well on most soils, but shows signs of distress on very poor conditions.
Sun/Shade aspect Best in full sun or very light shade. Deeper shade spoils the overall shape.
Pruning None required, but protruding

branches can be removed, the cuts dressed with pruning compound to prevent fungus infection.

Propagation and nursery production From grafting or budding on to understocks of *P. avium*. Normally supplied bare-rooted or container-grown, from 3-10ft (1-3m). Larger trees occasionally available, but ideal planting heights are 3-6ft (1-2m). Readily available from most garden centres and general nurseries.

Problems Heavy snowfalls can open up the centre of the tree and cause branches to splay out sideways. A binding of soft string or hosepipe, which does not cut into the upright stems, holds the shape together.

Varieties of interest There are no other suitable varieties of this form. The early-flowering variety *P. × hillieri 'Spire'* is another good upright cherry.

Average height and spread
Five years
10x3ft (3x1m)
Ten years
16x5ft (5x1.5m)
Twenty years
or at maturity
20x7ft (6x2.2m)

PRUNUS SERRULATA
Large-flowering Japanese Cherries

JAPANESE FLOWERING CHERRY
Rosaceae
Deciduous
Very popular, spring-flowering trees, widely planted, with a large range of planting potential.

Origin From Japan.
Use As medium to large flowering trees for spring blossom. Can also be grown as large bushes if suitable plants can be obtained for planting. Well presented in open grass areas as tall shrubs or short trees.
Description *Flower* A profusion of dark pink, white or cream, double or single flowers depending on variety, in late spring. *Foliage* Ovate to oval leaves, 3in (8cm) long with toothed edges. Light green to mid green. Yellow or orange autumn colour depending on variety. *Stem* Grey-green to grey-brown. Upright and round-topped or spreading, depending on variety. *Fruit* Insignificant.
Hardiness Tolerates temperatures down to −13°F (−25°C).
Soil Requirements Most soil conditions, but shows signs of distress on extremely poor soils.
Sun/Shade aspect Full sun to light shade.
Pruning None required other than to confine growth as necessary. Crossing branches or limbs can be removed in winter and the cuts treated with pruning compound to prevent fungus diseases.
Propagation and nursery production From budding or grafting. Normally supplied bare-rooted or container-grown from 5-10ft (1.5-3m). Larger trees occasionally available, but often slow to establish and superseded by younger trees. Most varieties readily available from garden centres and general nurseries.
Problems Some varieties, particularly *P. 'Kanzan'* can suffer from silver-leaf virus. There is no cure and affected trees must be burned to prevent further contamination.
Varieties of interest *Prunus 'Asano'* syn. *P. serrulata var. geraldiniae* Large clusters of double deep pink flowers in mid spring. Leaves light green-bronze when young, an attractive contrast with the flowers. Two-thirds average height and spread. Not readily available. *P. 'Fudanzakura'* syn. *P. serrulata var. fudanzakura* Single white flowers, pink in bud, from late winter to early spring during mild periods. Young leaves coppery-red to red-brown in spring. Difficult to find in commercial production. *P. 'Hisakura'* Large,

Prunus 'Kanzan' in flower

double or semi-double, pale pink flowers. Young leaves brown-bronze in colour. A very old variety, often confused with *P. 'Kanzan'*. *P. 'Hokusai'* Clusters of semi-double, pale pink flowers in mid spring. A round-topped tree with wide-spreading branches. *P. 'Ichiyo'* Double, shell-pink flowers with frilled edges in mid spring. Young foliage bronze-green. Upright habit. Relatively easy to find from specialist nurseries. *P. 'Imose'* Flowers double, mauve-pink, produced in hanging, loose clusters in mid spring. Young foliage copper-coloured, becoming bright green with good yellow autumn colour. Irregularly produces small, round, black fruits. Difficult to find. *P. 'Kanzan'* Very large clusters of double, purple-pink flowers produced in great profusion in mid spring. Young growth copper-red to red-brown, becoming dark green. Good autumn colours. Often confused with *P. 'Hisakura'* but they are distinct varieties, though similar in effect. *P. 'Ojochin'* Single flowers up to 2in (5cm) across in mid spring, pale pink in bud opening to pink-white in long hanging clusters of 7 or 8 florets. Young foliage bronze-brown becoming leathery and dark green with age. *P. 'Pink Perfection'* Pale pink flowers in mid to late spring. Rounded, fairly open habit. Young leaves bronze. Good yellow autumn tints. *P. 'Shimidsu Sakura'* syn. *P. serrulata var. longipes* Flowers double, pink in bud opening to pure white and hanging in clusters along the undersides of all branches in mid to late spring. Young foliage bright green, attractive. Good yellow autumn colours. Wide and spreading, forming a broad, flattened crown. *P. 'Shirofugen'* Flowers purple-pink in bud, aging through light pink to white. Foliage coppery when young, becoming light green. Good autumn tints. Forms a

Prunus 'Tai Haku' in flower

Prunus 'Ukon' in flower

round-topped tree initially, spreading with age. *P. 'Shirotae'* syn. *P. 'Mount Fuji'* Single or semi-double flowers up to 2in (5cm) across, pure white and fragrant. Young foliage light green; leaves have a distinctive fringed edge. Wide-spreading habit. Horizontal branches dipping sometimes to the ground. *P. 'Shosar'* Single, clear pink flowers in early to mid spring. Good orange-yellow autumn colours. Upright, pyramidal habit. *P. 'Tai Haku'* (Great White Cherry) Single, pure white flowers up to 2in (5cm) across, produced in great profusion in mid-spring. Young foliage copper-red. Spreading branches. Can be grown as large shrub or tree, and extremely attractive in both forms. *P. 'Taoyama Zakura'* Fragrant, semi-double, pale pink flowers in mid spring whitening with age. Good red-brown to copper-coloured young foliage. Low-growing with spreading habit. Relatively difficult to find in commercial production. *P. 'Ukon'* (Green Cherry) Hanging clusters of semi-double, pale green to green-yellow or cream flowers in mid spring, best seen against a background of clear blue sky. Interesting autumn foliage colours of red, yellow and orange.

Average height and spread
Five years
12x8ft (3.5x2.5m)
Ten years
23x16ft (7x5m)
Twenty years
or at maturity
25x25ft (8x8m)

Prunus serrulata var. pubescens **in autumn**

PRUNUS SERRULATA VAR. PUBESCENS

KOREAN HILL CHERRY, ORIENTAL CHERRY
Rosaceae
Deciduous
Interesting dark autumn colour to foliage.

Origin From China, Korea and Japan.
Use For the unusual autumn foliage colouring in medium-sized and larger gardens.
Description *Flower* Small, single, pink flowers, produced in profusion in early spring. *Foliage* Ovate, broad, up to 4in (10cm) long on younger wood, smaller on old. Dark green giving way to beautiful deep purple or purple-red autumn colour. *Stem* Dark brown to mahogany brown. Twisted formation, rarely producing straight shoots. Forms a spreading domed tree with horizontal emphasis. Slow to medium growth rate. *Fruit* May produce tiny, uninteresting wine red fruits.
Hardiness Tolerates 4°F (−15°C).
Soil Requirements Most soil conditions, but poor soil produces poor growth.
Sun/Shade aspect Full sun to light shade. Sun enhances autumn colour; ideally seen against a blue sky.
Pruning None required.
Propagation and nursery production From budding or grafting on to understocks of *P. avium*. Plant bare-rooted or container-grown. Normally offered at 5-10ft (1.5-3m). Normally stocked by general and specialist nurseries.
Problems A twisted appearance is a natural characteristic of the tree's growth.
Average height and spread
Five years
8x6ft (2.5x2m)
Ten years
16x13ft (5x4m)
Twenty years
or at maturity
26x30ft (8x9m)

PRUNUS SUBHIRTELLA 'AUTUMNALIS'

WINTER-FLOWERING OR AUTUMN-FLOWERING CHERRY, HIGAN CHERRY
Rosaceae
Deciduous
Useful winter-flowering trees when mild weather allows.

Origin From Japan.
Use For autumn, winter or early spring flowering in larger gardens. May also be considered as an attractive large shrub.
Description *Flower* Small white flowers, emerging on the bare branches in any mild weather period from late autumn to early spring. Flowering can be extensively delayed in cold winters, encouraging more vigorous spring blossoming. *Foliage* Small, ovate, up to 2in (5cm) long. Light green to grey-green.

Good yellow autumn colours. *Stem* Wispy, branching, thin, grey-brown to grey-green, forming a round-topped, slightly spreading tree, or large shrub if grown without a central trunk. *Fruit* Insignificant.
Hardiness Tolerates 4°F (−15°C).
Soil Requirements Most soils; may show distress on extremely poor types.
Sun/Shade aspect Tolerates light shade, but best seen in full light against a dark background.
Pruning None required.
Propagation and nursery production From budding or grafting on to understock of *P. avium*. Plant bare-rooted or container-grown. Offered in bush form without central trunk, or with stem of 5-6ft (1.5-2m) approximately. These are best planting heights, although larger trees may be available.

Prunus subhirtella 'Autumnalis Rosea'

Problems Severe winter weather may kill some of the flowers, especially if they open before a cold spell. In wet, mild winters white varieties may be spotted pink by excessive rain. Young plants in nurseries rarely look substantial; trees take three or more seasons to establish and gain size.
Varieties of interest *P. s. 'Autumnalis Rosea'* Rosy pink flowers. *P. s. 'Fukubana'* Best grown as a large shrub, but in time can be encouraged to form a small tree. Produces semi-double, rose madder flowers in early spring.
Average height and spread
Five years
5x6ft (1.5x2m)
Ten years
16x13ft (5x4m)
Twenty years
or at maturity
24x24ft (7x7m)

PRUNUS
Peach trees

FLOWERING OR ORNAMENTAL PEACH
Rosaceae
Deciduous
Unfortunately, these trees are susceptible to diseases which at present have little chance of cure. They are difficult to grow, but despite the problems are attractive for early spring blossom.

Origin From China.
Use As a small or medium, early spring-flowering tree.
Description *Flower* Single or semi-double, white, dark pink or purple-pink dependent on variety, produced on bare branches in early spring. *Foliage* Ovate to broadly lanceolate, up to 4in (10cm) long. Some yellow autumn colour, but not reliable. *Stem* Green to dark green with purple shading, becoming purple-brown. Strong, upright when young, quickly branching. Susceptible to winter die-back. Moderately upright at first, but tending to break down in vigour, becoming gnarled and somewhat pendulous. *Fruit* Small peaches form but rarely ripen and are not edible.

Prunus persica 'Klara Meyer' **in flower**

Hardiness Tolerates 14°F (−10°C), but some winter stem damage possible.
Soil Requirements Requires a rich, deep, well drained soil for good results.
Sun/Shade aspect Full sun.
Pruning Lateral growths can be pruned back following flowering; this controls peach-leaf curl to a certain extent by producing strong new growth.
Propagation and nursery production From grafting or budding onto understocks of *P. persica*. Normally offered bare-rooted or container-grown from 3-8ft (1-2.5m). Best planting heights are 5½-8ft (1.8-2.5m). Relatively scarce in production due to the problems of peach-leaf curl.
Problems May suffer winter frost damage. *P. persica* and its forms highly susceptible to peach-leaf curl, an airborne fungus disease extremely difficult to control. Pruning and good growth conditions are the best solution.
Varieties of interest *P. armeniaca* (Apricot) White or pink-shaded single flowers in mid to late spring. Sometimes produces yellow-red tinted fruits. In favourable mild areas can be grown for fruiting value. From China. There are also many named clone varieties useful for their fruiting ability in mild, sunny areas. *P. davidiana* Single white flowers in late winter. Dark green, ovate foliage up to 5in (12cm) long. Good yellow autumn colour. Upright habit. Smooth brown bark. Rarely seen in commercial production, but should be more widely grown. Less susceptible to fungus problems. *P. d. 'Alba'* A pure white form. Difficult to obtain. *P. d. 'Rubra'* A purple-pink form. Difficult to obtain. *P. persica*

'Crimson Cascade' Double crimson flowers and weeping branches. **P. p. 'Iceberg'** Semi-double, pure white flowers. **P. p. 'Klara Mayer'** Double peach-pink flowers. One of the best flowering peaches and most resistant to peach-leaf curl.

Average height and spread
Five years
6x3ft (2x1m)
Ten years
13x10ft (4x3m)
Twenty years
or at maturity
20x13ft (6x4m)

PRUNUS
Plum trees

PURPLE-LEAVED PLUM

Rosaceae
Deciduous
Beautiful purple-leaved trees ideal for gardens of any size.

Origin From western Asia and Caucasus.
Use As a small purple-leaved tree of good proportions. Can be planted in a line to form a large windbreak or screening. Good for any size of garden, due to its ability to be trimmed and trained.
Description *Flower* Masses of single or double flowers, white or pink, depending on variety, produced very early in spring on bare branches. *Foliage* Ovate, 2in (5cm) long, dark purple to purple-black, soft-textured compared with other *Prunus*. Some flame-red autumn colour. *Stem* Dark purple. Strong and upright when young, more branching and slower-growing with age, forming a round-topped tree of medium vigour. *Fruit* Small, round, dark purple to purple-black fruits in autumn.
Hardiness Tolerates −13°F (−25°C).
Soil Requirements Any soil type, except extremely dry or poor conditions.
Sun/Shade aspect Full sun to very light shade. Deeper shade causes foliage colour to turn green.
Pruning None required, but can be clipped extensively from an early age to form a tight, formal shape.
Propagation and nursery production From hardwood cuttings taken in winter. Some varieties grafted. Available bare-rooted or container-grown. Normally offered as bushes up to 3ft (1m) or as standard trees from 5-10ft (1.5-3m). Larger trees may be available, but up to 10ft (3m) is best planting height. Relatively easy to find in general nurseries or garden centres.
Problems Young trees in nursery production often look short and stunted, requiring open ground to produce good growth and foliage.
Varieties of interest *P. americana* (American Red Plum) White flowers up to 1in (3cm) across, produced in clusters of two or five.

Prunus cerasifera 'Nigra' in leaf

Yellow fruits becoming very red as autumn progresses. Widely grown in the USA and Canada, not common in Europe. **P. × blireana** Double pink flowers up to 1in (3cm) across, very early in spring just before leaves emerge. Attractive purple foliage. **P. cerasifera** White flowers. Green, ovate leaves. Red or yellow round fruits in autumn. Best as large bush. **P. c. 'Atropurpurea'** syn. **P. c. 'Pissardii'** Flowers pink in bud, opening to white, and borne in good numbers before leaves appear. Purple fruits, unreliably produced. Useful as a screening tree or single specimen. **P.c. 'Nigra'** Flowers pink, single, small, borne early in spring before leaves emerge. Dark foliage and purple stems. **P. c. 'Rosea'** Salmon pink flowers, losing colour with age, boldly produced on purple stems. Rare in cultivation, but worthy of some research. **P. c. 'Trailblazer'** Masses of grey to grey-white flowers. Large purple-green leaves. Edible plums up to 1in (3cm) long in autumn. Crops can be heavy, giving the tree a pendulous habit of growth. **P. cerasus 'Rhexii'** Attractive double white flowers hanging in clusters in late spring. Foliage ovate, mid to dark green. Fruits red, aging to black, with an acid taste. May be difficult to find. **P. 'Cistena'** Not truly a tree, but shrub plants can be purchased grafted or budded on to 5ft (1.5m) high stems to form small mop-headed trees. White flowers produced on purple stems before leaves appear. Dark wine red foliage. Good autumn colour. Easily controlled by clipping. **P. spinosa** (Blackthorn) Masses of white flowers from early to mid spring. Dark green, elliptic leaves. Branches very twiggy with spines. Fruits produced as sloes. From Europe and North Africa. **P. s. 'Purpurea'** Rich purple leaves and white flowers. Forms a neat, compact large bush or small tree. Difficult to find; must be sought from specialist

Prunus cerasifera 'Trailblazer' in fruit

nurseries. Slightly less than average height and spread.
Other forms of *P. cerasifera* are available but the above are most suitable for garden planting.

Average height and spread
Five years
10x5ft (3x1.5m)
Ten years
20x10ft (6x3m)
Twenty years
or at maturity
26x16ft (8x5m)

PTELEA TRIFOLIATA

HOP TREE

Rutaceae
Deciduous
Interesting multi-stemmed trees for flowers, fruits and autumn colour.

Origin From eastern North America.
Use As a large multi-stemmed shrub, evolving into a tree with time, and of many attractions. Plant singly or in groups.
Description *Flower* Clusters up to 3in (8cm) across of dull, green-white florets, each up to ½in (1cm) across, in early summer. *Foliage* Trifoliate, light green or golden leaves on long stalks; leaflets 3in (8cm) long, lanceolate to ovate with finely toothed edges. Good yellow autumn colour. *Stem* Grey-green, upright. Predominantly a large shrub, evolving into a multi-stemmed tree. Moderately quick-growing. *Fruit* Clusters of single-seeded, round, winged fruits of great beauty.
Hardiness Tolerates 4°F (−15°C).

Prunus × blireana in flower

Ptelea trifoliata in fruit

Soil Requirements Most soil types; finer fruit and foliage produced on deep, rich soils.
Sun/Shade aspect Full sun to mid shade, preferring light shade.
Pruning None required other than occasional removal of lower limbs to encourage a tree shape.
Propagation and nursery production From layers. Available bare-rooted or container-grown; normally offered bare-rooted or container-grown as multi-stemmed shrubs of 3-5ft (1-1.5m). Found in good general nurseries or specialist outlets.
Problems Fruits seeding on the tree can become unsightly in their brown, decaying state.
Varieties of interest *P. t. 'Aurea'* Requires full sun to show off golden foliage. Flowers and fruit identical to the parent. Slightly less height and vigour, but equal hardiness.
Average height and spread
Five years
6x6ft (2x2m)
Ten years
13x13ft (4x4m)
Twenty years
or at maturity
20x20ft (6x6m)

PTEROCARYA FRAXINIFOLIA

WING NUT
Juglandaceae
Deciduous
Useful large trees with interesting hanging flowers, autumn fruits and glossy foliage.

Origin From Iran.
Use As a feature tree of good architectural shape with flowers, foliage and fruits of interest. Can be planted as windbreak, but mass planting restricts the shape.

Pterocarya fraxinifolia in flower

Description *Flower* Long, hanging racemes, up to 20in (50cm) long, of round, yellow to yellow-green, scaly florets, catkin-like and hanging directly downwards. *Foliage* Compound, pinnate, up to 18in (45cm) long with 11-27 oblong leaflets, each up to 5in (12cm) long. Dark glossy green upper surfaces, duller glabrous undersides. Good yellow autumn colour. *Stem* Strong, upright stems. Rapid growth slowing with maturity. Purple to purple-green winter colour. Forms a round-topped, slightly spreading tree. *Fruit* Hanging strings 9in (23cm) long of round, green fruits up to 1in (3cm) long, produced in autumn. Of interest and attraction.
Hardiness Tolerates 14°F (−10°C). May suffer from die-back on young stem tips in extremely cold areas.
Soil Requirements Moist, slightly heavy loam; needs good moisture supplies to develop leaf size.
Sun/Shade aspect Full sun to light shade. Gains a better shape in an open locality.
Pruning None required, but can be trained occasionally.
Propagation and nursery production From seed, layers and sometimes suckers. Normally available bare-rooted or container-grown from 3-6ft (1-2m) in height, but must be sought from specialist nurseries.
Problems Winter die-back occurs but is not a major problem.
Varieties of interest *P. × rehderana* Rarely seen; almost identical to *P. fraxinifolia*. *P. stenoptera* Similar to *P. fraxinifolia*, but slightly smaller and narrower, and with shorter catkins.
Average height and spread
Five years
13x10ft (4x3m)
Ten years
26x20ft (8x6m)
Twenty years
or at maturity
52x39ft (16x12m)
Continued growth over 50 years reaches 65x52ft (18x16m)

PYRUS

ORNAMENTAL PEAR
Rosaceae
Deciduous
A range of trees with varying characteristics of merit for garden planting.

Origin From Europe through to Asia and Japan.
Use As a freestanding ornamental tree useful for any size of garden.
Description *Flower* Clusters up to 3in (8cm) across, composed of 5-7 white, single, cup-shaped florets, each up to 1½in (4cm) across, produced in mid spring. *Foliage* Ovate to linear, sometimes round. Green or silver-grey, depending on variety. Good autumn colours. *Stem* Grey-green to grey-brown. Upright, spreading or pendulous, depending on variety. Medium rate of growth in all cases. *Fruit* Small, oval or rounded pears up to 2in (5cm) long, in autumn. Fruits not edible.
Hardiness Tolerates −13°F (−25°C).
Soil Requirements Any soil type, but shows distress in extremely poor soils.
Sun/Shade aspect Green forms tolerate moderate shade. Silver-leaved varieties prefer full sun.
Pruning None required, but dead or crossing branches should be removed. May resent any other pruning.

Pyrus calleryana 'Chanticleer' in flower

Propagation and nursery production From grafting or budding on to understocks of pear, quince or *P. communis*. Available bare-rooted or container-grown. Normally offered from 4-10ft (1.2-3m) in height; occasionally available up to 13ft (4m). Best planting heights 5-8ft (1.5-2.5m). Most forms easily found in garden centres or general nurseries; some must be sought from specialist nurseries.
Problems All varieties are poor-rooted, especially when young. Extra peat or other organic material should be added to the planting hole to encourage establishment of young roots. Also needs extra watering in spring, and staking.
Varieties of interest *P. amygdaliformis* Foliage grey-green and ovate, white undersides. White flowers and insignificant fruits. Forms a round tree, becoming slightly pendulous with age. Rarely available. *P. calleryana 'Chanticleer'* Dark glossy green leaves maintained to late autumn and early winter, then turning orange-red. Narrow, columnar habit. Useful for all planting areas. *P. communis* (Wild Pear) Attractive white flowers in spring. Light green, ovate foliage with good autumn colour. Useful for windbreaks and exposed areas. Limited in commercial production; must be sought from specialist nurseries. *P. c. 'Beech Hill'* Dark green, round leaves; interesting leathery texture,

Pyrus salicifolia 'Pendula' **in leaf**

Quercus rubra **in autumn**

glossy surface and wavy edges. Good autumn colours. Upright pyramidal habit. Useful for all gardens. *P. elaeagrifolia* Silver-grey ovate, sometimes lanceolate, foliage. Interesting white flowers. Slightly more tender than most but above average height. *P. nivalis* Attractive ovate, white to silver-grey foliage. Graceful white flowers. Small, globed, yellow-green fruits. Two-thirds average height and spread. Must be sought from specialist nurseries. *P. salicifolia 'Pendula'* (Weeping Willow-leaved Pear) Narrow, lanceolate leaves up to 2in (5cm) long. A round, mop-headed pendulous tree. Not usually as large as other forms when supplied. Attractive lawn specimen.
Average height and spread
Five years
10x6ft (3x2m)
Ten years
16x13ft (5x4m)
Twenty years
or at maturity
26x20ft (8x6m)

QUERCUS
Autumn foliage varieties

SCARLET OAK
Fagaceae
Deciduous
Large trees with very good autumn colours for large sites.

Origin From North America.
Use As large specimen trees, singly or grouped, for large gardens, estates and parks Ideal for avenues when planted 39ft (12m) apart.
Description *Flower* Hanging male catkins and single or multiple spikes of female flowers in early spring. Light green, but of little attraction. *Foliage* Ovate leaves up to 6in (15cm) long, with 7-9 lobed indentations. Light green with grey sheen. Flame red or orange autumn colours, which in mild conditions are maintained well into winter. *Stem* Grey-green, moderately vigorous. Upright when young, spreading with age. *Fruit* Small, round acorns produced in autumn.
Hardiness Tolerates −13°F (−25°C).
Soil Requirements Most soil conditions; acid types may lead to better growth and autumn colour.
Sun/Shade aspect Full sun to light shade; better autumn colours in full sun.
Pruning None required but lower limbs may be removed to emphasise trunk.
Propagation and nursery production From seed or grafting depending on variety. Normally supplied bare-rooted or container-grown from 3-10ft (1-3m). Larger trees occasionally available, but not recommended. Best planting heights 5-8ft (1.5-1.8m). Not usually stocked by garden centres; most

varieties must be sought from general or specialist nurseries.
Problems Oaks are notoriously slow to establish and may take three springs to settle down completely.
Varieties of interest *Q. coccinea* (Scarlet Oak) Average height and spread with trunk up to 10ft (3m) in circumference. Good scarlet autumn colour. Scarce in production. *Q. c. 'Splendens'* Spectacular scarlet-red autumn colours. A form raised by grafting, not readily available but can be found in specialist nurseries at no more than 6ft (2m) in height. *Q. palustris* (Pin Oak) Branches slender, slightly pendulous with a graceful habit. Shiny green foliage up to 4in (10cm) long with 5-7 lobes. Good autumn colour. Buds glabrous in winter. Slightly more than average height but with less spread. *Q. petraea* (Durmast Oak) Large leaves with hairy underside midribs. Good autumn orange-red colour. Acorns are produced without stalks. *Q. rubra* syn. *Q. borealis maxima* (Red Oak) Foliage up to 10in (25cm) long in ideal conditions, ovate with 3-5 pointed lobes. Autumn colour brown-red with dull texture. Fast-growing, reaching one-third greater height and spread than average. From North America. *Q. rubra 'Aurea'* Soft yellow growth in spring, turning green in summer. Limited yellow-orange autumn colour. Requires light dappled shade; tends to scorch in full sun. One-third average height and spread. Not readily available, but can be found in specialist nurseries.
Average height and spread
10x8ft (3x2.5m)
Ten years
23x16ft (7x5m)
Twenty years
or at maturity
39x26ft (12x8m)
Continued growth over 50 years reaches 78x39ft (24x12m)

QUERCUS ROBUR

COMMON OAK, ENGLISH OAK
Fagaceae
Deciduous
True giants requiring generous space.

Origin From Europe through Asia.
Use As a large specimen tree for large gardens, parks, estates, farmland.
Description *Flower* Male catkins and female flowers, single or in multiples on short spikes, light green, in early spring. *Foliage* Ovate to oblong leaves up to 4in (10cm) long, with 3-6 lobes along each side and two smaller lobes at the base. Light green with a glossy sheen when young. Good yellow autumn colours. *Stem* Grey-green to green-brown. Upright when young, quickly branching to form a large, wide, spreading tree with thick trunk,

or a shorter, almost shrubby effect. Moderately quick-growing when young, slowing with age. *Fruit* Acorns up to 1in (3cm) long in autumn; heavy production.
Hardiness Tolerates winter temperatures below −13°F (−25°C).
Soil Requirements Any soil conditions.
Sun/Shade aspect Full sun to light shade, but young trees tolerate deeper shade.
Pruning None required unless to form a single central stem when young.

Quercus robur **in fruit**

Quercus robur − **trunk**

Propagation and nursery production From seed; grafting for named forms. Available bare-rooted or container-grown. Best planting heights 5-8ft (1.5-1.8m). Not normally stocked by garden centres or general nurseries; can be sought from specialist nurseries or forestry outlets.
Problems *Q. robur* is extremely difficult to establish in early years and can sulk, showing no signs of new growth for up to two years. Extra organic compost or peat should be added when planting; adequate water-supply

Quercus frainetto in leaf

is especially vital in the first spring following planting.
Varieties of interest *Q. frainetto* (Hungarian Oak) Leaves ovate, often as much as 8in (20cm) in length, with deeply lobed edges. Yellow autumn colours. Interesting bark. Good on alkaline soils. From south-east Europe. *Q. robur 'Concordia'* New spring foliage yellow, more golden as autumn approaches. Susceptible to sun-scorch. Weak, slow growth. Reaches only 16ft (5m) in height and 10ft (3m) in width over 30-40 years. Very difficult to find; only available from very specialised nurseries. *Q. r. 'Fastigiata'* (Cyprus Oak) Green to grey-brown stems, forming a columnar habit. Only 10ft (3m) ultimate width, with 39ft (12m) height. Available from general and specialist nurseries. *Q. r. 'Pendula'* (Weeping Oak) Pendulous, weeping to ground level, forming a tall, wide-spreading tree requiring a great deal of space.
Average height and spread
Five years
10x8ft (3x2.5m)
Ten years
23x16ft (7x5m)
Twenty years
or at maturity
39x26ft (12x8m)
Height and spread variable according to climate, altitude and seed source.

QUERCUS
Evergreen varieties

EVERGREEN OAKS
Fagaceae
Evergreen
Useful evergreen tress, but requiring space and time to develop.

Origin From the Mediterranean area.
Use As a slow-growing, ultimately large, evergreen tree for medium-sized and larger gardens. Interesting avenue trees planted at 32ft (10m) apart for estates and parks. Good for seaside planting, windbreaks and tall hedges, especially *Q. ilex*.
Description *Flower* Male hanging catkins; light green female flowers, produced singly or in multiples on short spikes, both in early spring. Flowers conspicuous, due to contrast with dark evergreen foliage. *Foliage* Ovate to lanceolate, up to 3in (8cm) long. Undersides glabrous grey, upper surfaces dark green with glossy, leathery texture. Wavy margins common on foliage of older trees. *Stem* Grey-green to grey-brown. Moderately slow, forming a round-topped tree. *Fruit* Small acorns, up to 1in (3cm) long, orbicular to oval, in autumn.
Hardiness Tolerates temperatures down to

14°F (−10°C) but suffers defoliation in wind chill factors below 10°F (−12°C) especially when young.
Soil Requirements Most soil conditions, only shows distress on very dry types.
Sun/Shade aspect Best in full sun to light shade.
Pruning None required, but can be reduced in size by cutting back young wood only. If required as a hedge, trimming should start at early age on young wood only.
Propagation and nursery production From seed or softwood cuttings; some forms grafted. Purchase container-grown. Normally supplied from 15in (40cm) up to 6ft (2m). Larger trees occasionally available, but smaller sizes more reliable for establishment. Must be sought from general and specialist nurseries. Rarely stocked in garden centres.

Quercus × *turneri* in leaf

Problems All oaks are difficult to establish, slow to develop in the 2-3 years following planting. Extra organic compost or peat should be added to the soil before planting and adequate moisture supplied in spring.
Varieties of interest *Q. ilex* (Holm Oak) Good, dark green, glossy foliage. A widely planted evergreen form. Slow-growing. Forms a round-topped tree. Good for seaside conditions. *Q.* × *kewensis* Oblong to ovate, triangular-lobed leaves up to 3in (8cm) long. Dull green upper leaf surfaces, glossy undersides, with pronounced vein network. Raised in Kew Gardens, London in the early 1900s. Two-thirds average height and spread. Best grown as a single specimen tree or in a widely-spaced group. *Q. suber* (Cork Oak) Very thick, corky-textured bark, grey-green, aging to black. Foliage ovate with toothed edges and grey-green, felted undersides.

Small acorns in autumn. Two-thirds average height and spread. From southern Europe and North Africa. Used for the production of cork in Mediterranean areas. *Q.* × *turneri* (Turner's Oak) Semi-evergreen, often losing all leaves in mid to late winter. Foliage up to 4in (10cm) long, oblong to ovate with 4-6 lobes at each edge. Good acorn production. Two-thirds average height and spread. Medium rate of growth. Raised in Essex, England, in the mid 1700s. Leaf shapes may very considerably. Not readily available; must be sought from specialist nurseries.
Average height and spread
Five years
10x8ft (3x2.5m)
Ten years
23x16ft (7x5m)
Twenty years
or at maturity
39x26ft (12x8m)
Heights variable, depending on altitude and seed-source.

QUERCUS
Other interesting varieties

OAK
Fagaceae
Deciduous
Interesting trees for larger gardens and arboretums.

Origin From Asia and Mediterranean areas.
Use As large, freestanding specimen trees, for group planting or avenues.
Description *Flower* Male catkins and light green female flowers, singly or in multiples on short spikes, both in early spring. *Foliage* Usually ovate and lobed; varietal features as mentioned below. *Stem* Grey-green to grey-brown. Slow to medium rate of growth. Upright when young, becoming more rounded with age. *Fruit* Acorns, light green to grey-green, in autumn.
Hardiness Tolerates −13°F (−25°C).
Soil Requirements Most soil conditions.
Sun/Shade aspect Best in light shade to full sun, but may tolerate deeper degrees of shade with care.

Quercus phellos in leaf

Pruning None required other than encouragement of strong, central trunk.
Propagation and nursery production From seed; some varieties grafted. Available bare-rooted or container-grown. Normally supplied from 3-10ft (1-3m); larger trees occasionally available. Planting heights of 5-8ft (1.5-2.5m) recommended for best establishment. All forms must be sought from specialist nurseries or forestry outlets.
Problems All Oaks are difficult to establish and slow to develop for 2-3 years after planting. Extra organic compost or peat

should be added to the soil before planting and adequate moisture supplied in the spring following planting.

Varieties of interest *Q. canariensis* (Algerian Oak) Attractive light green foliage. A scarce variety, difficult to find. Good on extremely alkaline soils. For estates or arboretums only. From North America and Port of Spain. *Q. castaneifolia* (Spanish Chestnut-leaved Oak) Foliage resembles chestnut leaves, up to 7in (17cm) long, with dark green upper surfaces and slightly glaucous undersides. Good acorn production. One-third more than average height and spread. For estates and arboretums only. From the Caucasus and through Iran. *Q. cerris* (Turkey Oak) Dark green foliage up to 5in (12cm) long, oval to oblong with lobed or deeply toothed edges. Dark green in colour. Good yellow autumn colours. Large-trunked and one-third more than average height and spread; requires a lot of space. From southern Europe and Asia. Readily available from general and specialist nurseries. *Q. c. 'Variegata'* White variegated foliage. Reaches only one-third average height and spread. Extremely difficult to find in commercial production, but worth seeking as a collector's item. *Q. phellos* (Willow Oak) Foliage narrow and linear, up to 2in (5cm) long. Light green with a soft, thin texture and attractive yellow autumn colours. Small acorns. A graceful tree of two-thirds average height and spread. For medium to large gardens. From North America.

Average height and spread
Five years
10x8ft (3x2.5m)
Ten years
23x16ft (7x5m)
Twenty years
or at maturity
39x26ft (12x8m)
Continued growth over 50 years reaches 80x59ft (25x18m)

RHUS

SUMAC, SUMACH
Anacardiaceae
Deciduous
Small trees for autumn colours.

Origin From eastern North America.
Use As a freestanding tree or large shrub, planted singly or for inclusion in large shrub borders; or, if pruned hard, for mass planting.
Description *Flower* Flowers produced on wood two years old or more. Green panicles on male forms and dark pink on female forms; either or both forms may appear on a single plant. *Foliage* Large, pinnate leaves up to 18in (40cm) long and 6in (15cm) wide. Light to mid green, with edges of leaflets deeply toothed. Some lacerated and deeply cut. Very

Rhus typhina in autumn

good rich autumn colours of yellow, red and orange. *Stem* Smooth or covered with red-brown hairs. Shoots upright when young, becoming branching, giving the effect of deer antlers. Forming a large, round-topped shrub which ultimately appears as a small, single or multi-stemmed 'tree'. Fast growing when young, slowing with age. *Fruit* Hairy plumes of scarlet fruits, produced in autumn and may be maintained into winter, eventually turning brown, on female plants only. Nurseries normally offer the female form, as fruits are a main characteristic of the plant.
Hardiness Tolerates −13°F (−25°C); in very severe conditions some stem damage may occur.
Soil Requirements Any soil type.
Sun/Shade aspect Full sun to light shade.
Pruning None required when grown as a tree. For maximum foliage effect, plants treated as shrubs can be cut to ground level and will rejuvenate with foliage three times the size of that of unpruned plants. Unfortunately will not flower when grown under this practice, and reach less ultimate height.
Propagation and nursery production From root suckers freely produced. Planted bare-rooted or container-grown. *R. typhina* forms are relatively easy to obtain from most nursery outlets, but *R. glabra* forms may be more difficult to find. Best planting heights 2½-6ft (60cm-2m)
Problems Can be invasive, producing root suckers some distance from the parent plant. Young nursery plants always look gaunt and unattractive, but performance improves after planting out.
Varieties of interest *R. glabra* (Smooth Sumac) Slightly more than average spread. Smooth stems. Good autumn colour and good fruits in autumn. From eastern North America. *R. g. 'Laciniata'* Fern-like foliage, very deeply cut. Good orange, yellow and red autumn colours. May revert to non-lacerated form. *R. potaninii* Bright glossy green, pinnate foliage. Very shy to flower but may produce greenish white followed by red fruits. Difficult to find; try a specialist nursery. One-third more than average height and spread. From China. *R. trichocarpa* A large shrub or small tree with good orange-red autumn colour. Smooth stemmed. Yellow fruits. From China, Japan and Korea. *R. typhina* (Staghorn Sumac) Good autumn colours. Winter fruits in conical clusters, crimson aging to brown. Very free with root suckers. *R. t. 'Laciniata'* Very fine, deeply cut, pinnate foliage of fern-like appearance. Attractive shades of pastel orange and yellow in autumn.
Average height and spread
Five years
6x6ft (2x2m)
Ten years
13x13ft (4x4m)
Twenty years
or at maturity
20x20ft (6x6m)

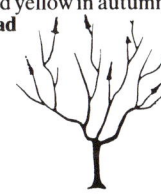

ROBINIA HISPIDA

ROSE ACACIA, PINK ACACIA, BRISTLY LOCUST
Leguminosae
Deciduous
Extremely attractive flowers in late spring and early summer when other flowering trees are finished. The tree is susceptible to wind damage, but it is worth trying to find a suitable location.

Origin From south-eastern USA.
Use As a shrub or shrubby tree, although for exposed locations it is best grown as a large fan-shaped shrub or tree against a sheltered wall.
Description *Flower* Racemes of deep rose laburnum-type flowers 1-1½in (3-4cm) long, borne in late spring, early summer. *Foliage* Pinnate leaves, 6in (15cm) long, grey-green to light green with good yellow autumn tints. *Stem* Brittle, grey-brown to light brown. Moderately upright in first year, later becoming very twiggy. Young shoots are attractively brown and bristly. *Fruit* May produce small pea-shaped pods in hot, dry summers.

Robinia hispida 'Macrophylla' in flower

Hardiness Tolerates 4°F (−15°C), but may be damaged by strong winds or heavy snow breaking the main limbs.
Soil Requirements Most soil conditions, but resents heavy waterlogging.
Sun/Shade aspect Tolerates light shade, but prefers warm, sunny position.
Pruning None required, but can be reduced in size by cutting back the new young shoots.
Propagation and nursery production *R. hispida* from seed. The form 'Macrophylla' is grafted or layered. Best purchased container-grown, but also available bare-rooted. Growth habit slightly irregular when young, so nursery plants do not show to advantage.
Problems Vulnerable to wind damage.
Varieties of interest *R. h. 'Macrophylla'* Larger flowers similar to those of *Wisteria*.
Average height and spread
Five years
6x6ft (2x2m)
Ten years
13x13ft (4x4m)
Twenty years
or at maturity
20x20ft (6x6m)

ROBINIA
Pink-flowering forms

FALSE ACACIA
Leguminosae
Deciduous
Attractive late spring-flowering trees, not well known and deserving more attention.

Origin South-western USA; some named varieties from France.
Use As large shrubs or small trees for flowering display. Ideal for all sizes of garden. Can be fan-trained on to a large wall if required, a procedure which shows off the flowers to best advantage.
Description *Flower* Clusters of pea-flowers, up to 3in (8cm) long on wood two years old or more, in early summer. *Foliage* Pinnate leaves, up to 6in (15cm) long with 9-11 oblong

Robinia kelseyi in flower

or ovate leaflets each 2in (5cm) long. Light grey-green with yellow autumn colours. *Stem* Light grey-green to grey-brown with small prickles. Upright when young, spreading and branching with age. Branches and twigs appear dead in winter, but produce leaves from apparently budless stems in late spring. Grown as large shrubs or as single-stemmed trees. *Fruit* Small, grey-green, bristly pea-pods, up to 4in (10cm) long, in late summer and early autumn.
Hardiness Tolerates 14°F (−10°C) but stems may suffer some tip damage in severe winters.
Soil Requirements Most soil conditions; particularly tolerant of alkaline types. Resents waterlogging.
Sun/Shade aspect Full sun to very light shade.
Pruning None required. Young shoots can be shortened in early spring to encourage strong regrowth but this curtails flowering.
Propagation and nursery production From seed or grafting. Purchase container-grown. Normally available from 5-8ft (1.5-2.5m). Best planting heights 5-6ft (1.5-2m). Must be sought from general or specialist nurseries; most varieties not offered by garden centres.
Problems Notorious for poor establishment; container-grown trees provide best results. Branches may be damaged by high winds and need shelter.
Varieties of interest *R.* × *ambigua* Light pink flowers. Pinnate leaves with 13-21 light grey-green leaflets. Must be sought from specialist nurseries. *R.* 'Casque Rouge' Rose-pink to pink-red flowers. An interesting variety from France. Difficult to find. *R. fertilis* 'Monument' Possibly best grown as a large suckering shrub, but can be encouraged to produce a single stem. Rosy red flowers. Half average height and spread. From south-eastern USA. *R.* × *hillieri* Slightly fragrant lilac-pink flowers. Originally raised in Hillier's Nurseries, Hampshire, England. *R. kelseyi* Flowers bright purple-pink. Attractive pale grey-green foliage with 9-11 leaflets. From south-eastern USA. Must be sought from specialist nurseries. *R. luxurians* Rose pink flowers. Leaves up to 12in (30cm) long, pinnate and with 15-25 oval bright green, leaflets. Slightly more than average height and spread. Not readily available; must be sought from specialist nurseries. From south-western USA.

Average height and spread
Five years
10x6ft (3x2m)
Ten years
20x13ft (6x4m)
Twenty years
or at maturity
39x20ft (12x6m)

ROBINIA PSEUDOACACIA
White-flowering varieties

ACACIA, BLACK LOCUST, FALSE ACACIA
Leguminosae
Deciduous
A stately tree, with several varieties of good shape and form.

Origin Native to the USA.
Use As a medium to large tree for medium-sized and larger gardens. Some named varieties suitable for smaller gardens.
Description *Flower* Racemes of fragrant white flowers hanging in clusters up to 7in (17cm) long in early summer; florets have blotched yellow bases. Size can vary with age of tree and location. *Foliage* Pinnate, up to 10in (25cm) long, with 11-23 oval to ovate leaflets. Foliage light grey-green with good yellow autumn colour. *Stem* Grey-green to grey-brown and covered in thorns. Upright, quick-growing when young, slowing and branching with maturity. Appears completely dead in winter, but breaks leaf from almost budless stems. Produces suckers at ground level, often far from the central stem. *Fruit* Small, grey-green pea-pods up to 4in (10cm) long, in autumn.
Hardiness Tolerates 14°F (−10°C). Some stem-tip damage may occur in severe winters. High winds can cause physical damage to branches.
Soil Requirements Most soil conditions; dislikes waterlogging. Good on dry sandy soils.
Sun/Shade aspect Full sun to medium shade, preferring full sun.
Pruning None required, but may be reduced in size when young.
Propagation and nursery production From seed. Named varieties from grafting on to understocks of *R. pseudoacacia*. Best purchased container-grown or root-balled (balled-and-burlapped). Difficult to establish bare-rooted. Plants offered from 5-10ft (1.5-3m). Larger trees may be available but rarely transplant well. Stocked by general nurseries and specialist outlets.

Robinia pseudoacacia 'Bessoniana' in flower

Problems Subject to wind damage or mechanical breakage. Sometimes difficult to establish; extra organic or peat composts should be added to soil, and adequate watering supplied in first spring.
Varieties of interest *R. p.* 'Bessoniana' White flowers in early summer. The best white-flowering form for small gardens. Two-thirds average height and spread. *R. p.* 'Inermis' (Mop-head Acacia, Thornless Black Locust) Rarely flowers. Forms a tight, mop-headed tree 13x13ft (4x4m). Very slow-growing. Useful for all sizes of gardens for its interesting mature shape. Must be sought from specialist nurseries. *R. p.* 'Pyramidalis' syn. *R. p.* 'Fastigiata' Rarely flowers. Narrow, upright growth, slightly twisting branches. Forms an upright pillar of average height and maximum 6ft (2m) width. *R. p.* 'Tortuosa' Interesting contorted stems and bright green foliage, yellow autumn colour. Rarely flowers. Good architectural winter shape. Normally supplied at 3-5ft (1-1.5m). Useful for small gardens grown on a short trunk or as a bush. Two-thirds average height and spread.

Average height and spread
Five years
10x5ft (3x2.5m)
Ten years
26x16ft (8x5m)
Twenty years
or at maturity
32x32ft (10x10m)
Continued growth over 50 years reaches 56x36ft (17x11m)

ROBINIA PSEUDOACACIA
'FRISIA'

GOLDEN ACACIA
Leguminosae
Deciduous
An outstanding golden-foliaged tree.

Origin Of garden origin, from Europe.
Use As a freestanding tree or large shrub for medium-sized and larger gardens. Can be fan-trained on a large wall.

Robinia pseudoacacia 'Frisia' in leaf

Description *Flower* Short racemes of white, pea-flowers, produced only on very mature trees in midsummer. *Foliage* Pinnate with 7-9 leaflets 6in (15cm) long. Bright yellow to yellow-green in spring, lightening in early summer. Turns deeper bright yellow in late summer to early autumn. *Stem* Brown to grey-brown. Strong shoots with definite red prickles on new growth. Wood may appear dead in winter but quickly grows away in late spring or early summer from apparently budless stems. An upright tree when young, becoming basal-spreading with age or can be grown as a large trunkless shrub. *Fruit* Insignificant.
Hardiness Tolerates 14°F (−10°C). Some stem-tip damage may occur in severe winters.
Soil Requirements Most soil conditions; growth is limited on very alkaline, permanently wet or poor soils. Grows best on moist, rich, loamy soil.
Sun/Shade aspect Full sun or very light shade. Deeper shade causes foliage to turn green.
Pruning None necessary, but shortening back side branches when young, on standards or bushes, encourages thicker, more attractive foliage and also controls size. Trees not fully established can benefit from the treatment, making new growth, and hence new roots.
Propagation and nursery production From grafting on to *R. pseudoacacia*. Purchase

container-grown. Best planting heights 6-10ft (2-3m). Normally a single stem with very limited lateral branches is needed for a standard form. Single stem container-grown plants of 3ft (1m) can be used for fan-trained or bushy trees. Standard form readily available from most garden centres. Shorter bushes or trainable stock should be sought from general or specialist nurseries.
Problems Branches may be brittle and easily damaged by severe weather conditions. Late to break leaf, often bare until early summer, but grows rapidly once started. Notoriously difficult to establish unless container-grown.
Average height and spread
Five years
10x6ft (3x2m)
Ten years
20x13ft (6x4m)
Twenty years
or at maturity
39x20ft (12x6m)

SALIX ALBA

WHITE WILLOW
Salicaceae
Deciduous
A fast-maturing, medium-sized tree with graceful silver stems and leaves.

Origin Throughout the northern hemisphere.
Use As small, individual trees; for group planting or for screening. Can also be grown as shrubs.
Description *Flower* White to yellow-white catkins produced on bare stems in early to mid spring. Size and shape vary according to variety. *Foliage* Narrow, ovate to lanceolate, 3-6in (8-15cm) long, silver-grey or green with grey sheen, depending on variety. *Stem* Red, orange or silver, depending on variety. Strong, upright when young, branching with age. *Fruit* Insignificant.
Hardiness Tolerates winter temperatures down to −13°F (−25°C).
Soil Requirements Any soil type or condition. Extremely tolerant of waterlogging.
Sun/Shade aspect Full sun to medium shade. Full sun shows off winter stem colour.
Pruning Can be left unpruned but many coloured-stem varieties give a better display if pruned back very hard, pollarding to a crown each year in mid to late spring. Otherwise may be trimmed biennially.
Propagation and nursery production From hardwood cuttings taken in early to mid winter. Available bare-rooted or container-grown. Young plants may have thin habit and poor root systems, but once planted out rapidly produce large trees. Most varieties relatively easy to find; some must be sought from specialist nurseries or forestry outlets.
Problems Speed of growth and ultimate size often underestimated; plants may quickly

Salix alba 'Sericea' in leaf

outgrow the area allowed. Some varieties suffer from stem canker and atttacks of willow mildew and scab, needing hard cutting back of diseased wood.
Varieties of interest *S. a. 'Chermesina'* syn. *S. a. 'Britzensis'* (Red-stemmed Willow) Bright orange-scarlet winter stems. Very responsive to annual or biennial pruning. From Europe, northern Asia and North Africa. *S. a. 'Liempde'* Silver-white foliage and silver stems. From Europe. Not readily available; must be sought from specialist nurseries. *S. a. 'Sericea'* syn. *S. a. 'Argentea'* (Silver Willow) Attractive lanceolate, grey-green foliage with silver hue, excellently seen against a dark or blue sky. Stems grey-green, slightly open in habit. Can be retained as a large shrub with annual or biennial pruning. From Europe, northern Asia and North Africa. *S. a. 'Vitellina'* (Golden-stemmed Willow, Yellow-stemmed Willow) Bright to dark yellow, strong, upright shoots for winter effect. A male form producing small yellow catkins on bare stems of unpruned trees. Responds well to annual or biennial pruning.

Salix alba 'Vitellina' − stems in winter

Average height and spread
Five years
23x8ft (7x2.5m)
Ten years
26x12ft (8x3.5m)
Twenty years
or at maturity
30x18ft (9x5.5m)

SALIX DAPHNOIDES

VIOLET WILLOW
Salicaceae
Deciduous
Useful trees with attractive winter stems.

Origin Throughout the northern hemisphere.
Use As an interesting, ornamental tree for winter stems and attractive silver catkins in early spring.
Description *Flower* Silver-white catkins yellowing with age, produced on bare stems in early spring. *Foliage* Narrow, ovate to lanceolate 3-6in (8-15cm) long. Purple-green with grey sheen. Some yellow autumn colour. *Stem* Dark purple to violet shoots, covered with white bloom. Upright, strong, quick-growing when young, spreading with age. *Fruit* Insignificant.
Hardiness Tolerates winter temperatures down to −13°F (−25°C).
Soil Requirements Most soil types and conditions. Extremely tolerant of waterlogging.
Sun/Shade aspect Full sun to medium shade; full sun shows off coloured stems.
Pruning Can be left unpruned but gives a better display of coloured stems if pruned back very hard annually or biennially.
Propagation and nursery production From hardwood cuttings taken in early to mid winter. Available bare-rooted or container-grown as standard trees 3-10ft (1-3m) in height. Easy to find although some varieties must be sought from specialist nurseries and forestry outlets.
Problems Often planted in an area which restricts ultimate size.
Varieties of interest *S. acutifolia 'Blue Streak'* Branches arching, thin, graceful, carrying silver catkins in mid to late spring. Whiter,

Salix acutifolia 'Pendulifolia' in winter

more downy covering on purple stems. *S. a. 'Pendulifolia'* Weeping branches with thin, downy white covering on purple stems. Narrow lanceolate leaves. *S. daphnoides 'Aglaia'* White stems and large catkins. Somewhat difficult to find in commercial production; can be sought from specialist nurseries.
Average height and spread
Five years
20x5ft (6x1.5m)
Ten years
23x8ft (7x2.5m)
Twenty years
or at maturity
26x14½ft (8x4.5m)

SALIX MATSUDANA 'TORTUOSA'

CONTORTED WILLOW
Salicaceae
Deciduous
A very attractive tree for winter stems, good material for flower arranging.

Origin From China and Korea.
Use As an individual small tree of interest for winter stem formation. Suited to water features.

Salix matsudana 'Tortuosa' − branches

Description *Flower* Small, inconspicuous catkins produced in early spring. *Foliage* Lanceolate, 3-6in (8-15cm) long, grey-green. Good yellow autumn colour. *Stem* Grey-green, twisted and contorted. Slow to establish, then produces good rate of growth, but slows again with age. Forms a narrow-based pyramid for the first 15 years, then becomes spreading with gnarled trunk. *Fruit* Insignificant.
Hardiness Tolerates 4°F (−15°C), some die-back on stem tips in winter. Large branches may be damaged but once cut back quickly generate new growth.
Soil Requirements Any soil type.
Sun/Shade aspect Full sun or very light shade.
Pruning None required, but can be severely reduced as necessary and quickly will rejuvenate

Propagation and nursery production From hardwood cuttings taken in late autumn or early winter. Available bare-rooted or container-grown. Normally offered from 3-10ft (1-3m). Best planting sizes 3-6ft (1-2m). Readily available from general nurseries. Not normally stocked by garden centres.
Problems Can suffer stem canker, especially on mature trees. Full size and shape often underestimated.
Average height and spread
Five years
6x3ft (2x1m)
Ten years
20x10ft (6x3m)
Twenty years
or at maturity
32x23ft (10x7m)

SALIX
Other standard forms

WILLOWS
Salicaceae
Deciduous
Useful large shrubs or trees for wet difficult soils. Large-growing and needing space.

Origin Throughout the northern hemisphere.
Use As individual specimen trees or for windbreaks when mass planted.
Description *Flower* White to yellow-white catkins produced on bare stems in early to mid spring. Size and shape variable, dependent on variety. *Foliage* Elliptic to ovate 3-6in (8-15cm) long. Grey-green or green; some varieties give good autumn colour. *Stem* Green or grey-green. Upright and vigorous, rapidly reaching ultimate height, branching with age. *Fruit* Insignificant.
Hardiness Tolerates winter temperatures down to −13°F (−25°C).

Salix caprea in catkin

Soil Requirements Any soil type or condition. Extremely tolerant of waterlogging.
Sun/Shade aspect Full sun to medium shade.
Pruning Can be left unpruned to form a large tree, or pollarded back to central crown to induce new annual growth with good foliage and catkin production, though this curtails overall height.
Propagation and nursery production From hardwood cuttings taken in early to mid winter. Available bare-rooted or container-grown. Young plants appear thin in habit with poor root systems, but once planted out rapidly produce large trees. Normally supplied from 3-10ft (1-3m) in height. Larger trees may be available but are not recommended. Young trees of 3ft (1m) can easily be trained into standard forms over one or two seasons. Most varieties relatively easy to find; some must be sought from specialist nurseries or forestry outlets.

Problems Trees are often planted with insufficient allowance for ultimate height and spread.
Varieties of interest *S. aegyptiaca* syn. *S. medemii* (Egyptian Musk Willow) Attractive open habit. Stems grey to grey-green. Foliage ovate with toothed edges, grey-green and covered in silky hairs. Good catkin display on bare stems in early spring. From North America. *S.* '*Caerulea*' (Cricket-bat Willow) Foliage lanceolate, grey-green to sea-green with silver undersides. Upright branches forming a pyramidal or conical tree relatively quickly. Used commercially; grown for timber used to make cricket bats. Rarely used for garden planting. *S. caprea* (Goat Willow, Great Willow, Pussy Willow) Strong, upright, slightly spreading light green to grey-green stems. Long, round, ovate leaves with grey undersides. Good yellow autumn colour. A male form with large yellow catkins on bare stems mid to late spring. Should be pruned annually or biennially to improve catkin production. From Europe and western Asia. *S.* × *smithiana* Bright green stems. Large yellow catkins on bare stems in spring. Large, lanceolate, tooth-edged, bright green foliage with slightly hairy undersides. From the UK. Stems respond well to annual or biennial pruning. *S.* '*The Hague*' Branches grey-green, strong, upright, becoming spreading. Large silver catkins on bare stems in early spring. Foliage large, ovate, tooth-edged, grey-green. Best left unpruned for catkin production. Of garden origin.
Average height and spread
Five years
23x8ft (7x2.5m)
Ten years
26x12ft (8x3.5m)
Twenty years
or at maturity
30x18ft (9x5.5m)

SALIX
Weeping varieties

WEEPING WILLOW
Salicaceae
Deciduous
A range of large and small weeping trees. Large trees require special siting and appreciation of overall size.

Origin From Europe through to central Asia.
Use As a freestanding, large or small weeping tree for medium-sized and larger gardens. Associates well with a large water feature.
Description *Flower* Catkins up to 2in (5cm) long produced in spring, light grey-green becoming yellow when ripe. *Foliage* Narrow, lanceolate leaves up to 4in (10cm) long, or

ovate up to 3in (8cm) long, dependent on variety. *Stem* Graceful, thin, arching branches. Quick to mature, becoming slower after twenty years. Forms a wide-based, high, weeping tree or small feature tree, dependent on variety. *Fruit* Insignificant.
Hardiness Tolerates 4°F (−15°C).
Soil Requirements Requires good, rich, deep, moist soil to do well. Poor soils will lead to fungus disease.
Sun/Shade aspect Best in full sun. Shade deforms the ultimate shape.
Pruning None required, but can be reduced in size dramatically as necessary and quickly achieves new growth. Responds well to triennial hard pruning in spring, making up to 10ft (3m) or more new growth by midsummer, with improvement in overall health.

Salix × *chrysocoma in spring*

Propagation and nursery production From hardwood cuttings, some varieties grafted. Normally supplied from 5-10ft (1.5-3m). Larger trees are available, but not recommended, as young trees grow so rapidly.
Problems *S.* × *chrysocoma* suffers badly from willow scab and willow mildew; can become completely defoliated and deformed. Control is very difficult; severe spring pruning is recommended in the year following attack to remove all damaged branches and encourage new growth. Triennial hard pruning should keep the disease within bounds.
Varieties of interest *S. babylonica* The true Weeping Willow, with green stems and narrow, lanceolate leaves. A tree of great proportions. Not readily available; must be sought from specialist nurseries. *S. caprea* '*Pendula*' (Kilmarnock Willow) Stout, weeping branches, forming a neat compact head cascading to ground level. Reaches only 6-10ft (2-3m),

Salix caprea 'Pendula' in catkin

depending on initial height of stem. Final spread up to 10ft (3m) after many years. Stems purple-green to dark green. Silver-white catkins yellowing with age. Useful for all sizes of garden, particularly in association with a small pool feature. Widely available from garden centres and nurseries. **S. × chrysocoma** (Golden Weeping Willow) Attractive golden yellow foliage, new foliage green-yellow. Very attractive at bud-break in spring. Susceptible to willow scab and willow mildew. Requires careful siting in all but the largest gardens. Readily available from nurseries and garden centres. **S. matsudana 'Pendula'** Stems yellow with a downy grey covering. Leaves lanceolate with slender points and slightly toothed edges, giving yellow autumn colour. Stout, weeping branches, forming a large tree two-thirds average height and spread. Originating in China and Korea. Not readily available from nurseries; must be sought from specialist outlets. **S. purpurea 'Pendula'** A small, weeping tree reaching no more than 10ft (3m) in height and spread. Purple to purple-green, thin, graceful stems. Foliage narrow and lanceolate, grey-purple to purple-green. Yellow autumn colour. Ideal for small water features.

Average height and spread
Five years
16x16ft (5x5m)
Ten years
32x32ft (10x10m)
Twenty years
or at maturity
64x64ft (20x20m)
Continued growth over 40 years reaches 80x80ft (25x25m)

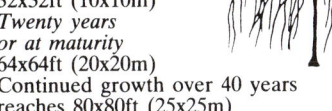

SASSAFRAS ALBIDUM

KNOWN BY BOTANICAL NAME
Lauraceae
Deciduous
A beautiful foliage tree but with problems of establishment outside its native environment.

Origin From North America.
Use As a freestanding tree with attractive foliage, especially in autumn.
Description *Flower* Racemes up to 2in (5cm) long of green-yellow flowers in late spring, of little attraction. *Foliage* Produces three leaf shapes, all ovate and up to 6in (15cm) long — without lobes, lobed on one side, lobed on both sides. Glaucous grey undersides, upper surfaces grey-green with red hue. Good scarlet and orange autumn colours. *Stem* Green with purple shading. Upright, slow rate of growth. *Fruit* Round to oval, dark blue fruits in autumn.
Hardiness Minimum winter temperatures 4°F (−15°C).
Soil Requirements Best on neutral to acid soil; tolerates limited alkalinity if soil is deep and rich in organic compost.
Sun/Shade aspect Full sun to very light shade.
Pruning None required.
Propagation and nursery production From seed or root suckers. Purchase container-grown. Trees normally offered 2ft (60cm) to 3ft (1m) in height. Not readily available; must be sought from specialist nurseries.
Problems Can be difficult to establish. Relatively scarce in production outside its native environment and not widely planted.

Average height and spread
Five years
6x3ft (2x1m)
Ten years
13x6ft (4x2m)
Twenty years
or at maturity
32x10ft (10x3m)
Ultimate height and spread
48x32ft (15x10m)

Sophora japonica in leaf

SOPHORA JAPONICA

JAPANESE PAGODA TREE, SCHOLAR TREE
Leguminosae
Deciduous
An attractive foliage tree worthy of wider garden planting.

Origin From China and Korea.
Use As a medium to large tree for medium-sized and larger gardens. Good effect when fan-trained on to a wall.
Description *Flower* Creamy white flower panicles up to 10in (25cm) long in late summer, early autumn, in the terminals of each branch. *Foliage* Pinnate leaves 10in (25cm) long with 9-17 oval to ovate 2in (5cm) long leaflets. Grey-green when young, dark to mid green with age. *Stem* Dark green, glossy. Strong, upright and quick-growing when young, slowing and branching with age. Initially an upright tree, aging to a round-headed effect. Mature trees have interesting corrugated bark. *Fruit* Grey-green pea-shaped pods up to 3in (8cm) long in late summer, early autumn.
Hardiness Tolerates winter 14°F (−10°C).
Soil Requirements Any soil conditions.
Sun/Shade aspect Best in full sun, but tolerates very light shade.
Pruning None required.
Propagation and nursery production From seed. Purchase container-grown. Normally offered from 4-10ft (1.2-3m). Larger trees occasionally available, but smaller trees more readily established. Available from general and specialist nurseries.
Problems *S. japonica* has an extremely poor root system and container-grown plants do

Sassafras albidum in leaf

Sophora japonica in flower

much better than bare-rooted trees, which must be given peat or organic compost planting and careful watering in the first season. **Varieties of interest** *S. j. 'Pendula'* Usually grafted on a stem 8-10ft (2.5-3m) or more, to form a cascading pattern of stiff branches with attractive pinnate leaves. May be considered a small weeping tree, reaching 16ft (5m) height and 26ft (8m) spread after many years. Other forms of *Sophora* are less hardy or only successful in their native environment.

Average height and spread
Five years
13x6ft (4x2m)
Ten years
16x13ft (8x4m)
Twenty years
or at maturity
64x26ft (20x8m)

SORBUS ARIA

WHITEBEAM, MOUNTAIN ASH
Rosaceae
Deciduous
Beautiful silver-foliaged tree with white spring flowers and orange autumn berries.

Origin From Europe.
Use As freestanding trees of great beauty, or for mass planting for screening. Suitable for medium-sized and larger gardens. Can be trimmed tight to form archways or pillars, showing off new foliage each spring. May be fan-trained on large walls but training decreases flowering and fruit production. Good avenue when planted 32ft (12m) apart.
Description *Flower* Clusters of white fluffy flowers, 4in (10cm) across, in late spring. *Foliage* Ovate to oval, up to 4in (10cm) long with downy white undersides and dark green to grey-green upper surfaces. Good yellow autumn colour. *Stem* Grey-green to grey-brown, with downy covering. Strong, upright and quick-growing when young, slowing and branching with age, often slightly pendulous at branch tips. Forms an upright pyramid. *Fruit* Clusters of scarlet-red, oval to globe-shaped fruits 4in (10m) across, in autumn.
Hardiness Tolerates temperatures down to −13°F (−25°C).
Soil Requirements Any soil conditions; particularly tolerant of severe alkalinity.
Sun/Shade aspect Best in full sun to maintain colouring.
Pruning None required, but young growth can be cut back hard to keep within bounds or to train.

Sorbus aria 'Lutescens' in leaf

Sorbus 'Mitchellii' in autumn

Propagation and nursery production Parent plant from seed; named varieties grafted or budded on to understocks of *S. aria*. Available bare-rooted or container-grown. Normally offered from 3-10ft (1-3m). Larger trees occasionally available, but less good for planting and establishment.
Problems None.
Varieties of interest *S. a. 'Chrysophylla'* Leaves predominantly yellow through summer with some silver undersides, turning buttercup yellow in autumn. White flowers. Orange-red fruits. *S. a. 'Lutescens'* Best of all silver Whitebeams. Bright silver-white foliage, white flowers and orange berries. Good for training. *S. a. 'Decaisneana'* syn. *S. a. 'Majestica'* Large, round to ovate foliage 6in (15cm) long, grey upper surfaces with white, down undersides. Orange fruits in autumn. *S. bristoliensis* Small, ovate, silver leaves. Little clusters of white flowers and orange-red berries. Not readily available, but worth researching for a planting collection or arboretum. Two-thirds average height and spread. Originating in the wild in the Bristol valley, UK. *S. folgneri* Slender growths with lanceolate to narrowly ovate, finely tapering leaves up to 4in (10cm) long. White felted undersides and dark grey upper surfaces. Good yellow autumn colour. Clusters of white flowers 4in (10cm) across, followed by red, oval fruits. Half average height and spread. Not easy to find, and will have to be sought from specialist nurseries. From China. *S. hybrida 'Gibbsii'* Dark grey leaves with silver undersides, lobed with 3-5 indentations. Large clusters of white flowers up to 5in (12cm) across, followed by orange-red berries in autumn. Two-thirds average height and spread. Not readily available; must be sought from specialist nurseries. *S. intermedia* A late form of *S. aria*. Leaves lobed, up to 4in (10cm) long, with steel-grey upper surfaces and white undersides. White flowers and orange-red fruits. *S. latifolia* (Service Tree of Fontainebleau) Large, silver-grey foliage with downy undersides, round to ovate, 4in (10cm) long or more, jagged toothed edges. White flower clusters up to 3in (8cm) across, followed by globe-shaped, brown-red fruits. From Europe. *S. 'Mitchellii'* Among the largest of Whitebeam foliage forms. Leaves up to 8in (20cm) long and 6in (15cm) wide. Grey-green to silver-grey upper surfaces with white, downy, felted undersides. Good yellow autumn colour. White flower clusters up to 4in (10cm) across, followed by red fruits in autumn, more sparse than in most varieties. Stems brown, solid, stout, forming a perfect pyramid. Two-thirds average height and spread. Not readily available, but found in specialist nurseries.

Average height and spread
Five years
10x6ft (3x2m)
Ten years
20x13ft (6x4m)
Twenty years
or at maturity
39x26ft (12x8m)

SORBUS AUCUPARIA

ROWAN, EUROPEAN MOUNTAIN ASH
Rosaceae
Deciduous
Well-known, attractive flowering and fruiting small trees.

Origin From Europe.
Use As small, attractive, ornamental trees suitable for all gardens, freestanding or grouped.
Description *Flower* Clusters of fluffy white flowers up to 5in (12cm) across produced in late spring, early summer. *Foliage* Pinnate, up to 9in (23cm) long with 11-15 leaflets, lanceolate with sharply toothed edges. Dark green with a slight grey sheen. Good orange-red or yellow autumn colours. *Stem* Green to green-brown, becoming grey-brown. Strong, upright when young, quickly spreading and eventually very branching. Mature wood has rubbery consistency. *Fruit* Clusters of red fruits in autumn, enjoyed by birds.
Hardiness Tolerates winter temperatures down to −13°F (−25°C).
Soil Requirements Most soil types, including very alkaline conditions.
Sun/Shade aspect Best in full sun, but tolerates light shade.
Pruning None required.
Propagation and nursery production Parent plant from seed; named forms by grafting or budding on to understocks of *S. aucuparia*. Available bare-rooted or container-grown. Normally supplied from 3-10ft (1-3m); larger plants occasionally available but best planting heights 5-8ft (1.5-2.5m). Found in general nurseries and garden centres.
Problems Birds take fruits readily in autumn. *S. aucuparia* variable in size, flowering and fruiting ability; unless mass planting is intended named forms are more reliable.
Varieties of interest *S. a. 'Asplenifolia'* (Cut-leaved Mountain Ash) Attractive deeply toothed and dissected foliage, light green to grey-green. White flowers and orange-red fruits. *S. a. 'Beissneri'* (Orange-stemmed Mountain Ash) Winter stems orange to

Sorbus aucuparia in fruit

flowers up to 5in (12cm) across in late spring, early summer. *Foliage* Pinnate, up to 9in (23cm) long with 9-11 leaflets. Grey-green with good yellow autumn colour. *Stem* Dark brown to brown-grey. Slow-growing short-branched and bushy nature. Forms a short round-topped tree or large shrub. *Fruit* Hanging clusters of pearl-white fruits in late summer or early autumn, maintained well into winter. Fruit stems red.
Hardiness Tolerates 4°F (−15°C).
Soil Requirements Any soil types.
Sun/Shade aspect Full sun to very light shade; fruits show well in autumn sunlight.
Pruning None required.
Propagation and nursery production From seed, grafting or budding. Available bare-rooted or container-grown. Normally offered from 3-6ft (1-2m). Relatively difficult to find; must be sought from specialist nurseries.
Problems Moderate height and vigour should be taken into account when positioning, not overestimating space needed to accommodate ultimate height and spread.

Average height and spread
Five years
6x6ft (2x2m)
Ten years
10x10ft (3x3m)
Twenty years
or at maturity
13x13ft (4x4m)

orange-red, a real attraction if planted in a bright position. About two-thirds average height and spread. *S. a. 'Fastigiata'* (Upright Mountain Ash) Green foliage, white flowers and dark red berries. Upright branches, forming a tall pillar. Susceptible to stem canker and of weaker constitution than its parent. *S. a. 'Pendula'* (Weeping Mountain Ash) White flowers and red berries. A strong, ranging, weeping tree reaching 10-13ft (3-4m) height and 23-26ft (7-8m) spread. May suffer from stem canker. *S. a. 'Sheerwater Seedling'* Good, large foliage, white flowers and red fruits in autumn. A good selected form of upright growth. *S. a. 'Xanthocarpa'* (Yellow-berried Mountain Ash) Yellow berries in autumn. Light green foliage. Strong-growing and attractive. *S. commixta* Large flowers, large red fruits, strong growth and good foliage. Uniform habit. A good selected form.

Average height and spread
Five years
8x5ft (2.5x1.5m)
Ten years
16x8ft (5x2.5m)
Twenty years
or at maturity
32x16ft (10x5m)
Continued growth over 40 years reaches 50x30ft (15x9m)

SORBUS CASHMIRIANA

KASHMIR MOUNTAIN ASH
Rosaceae
Deciduous
One of the best autumn-fruiting trees.

Origin From Kashmir.
Use As a small tree or large shrub with attractive white autumn fruits planted singly or grouped.
Description *Flower* Hanging clusters of white

SORBUS DOMESTICA

SERVICE TREE
Rosaceae
Deciduous
A curiosity of garden interest for its fruit, but with no outstanding beauty.

Origin From Europe.
Use As an interesting fruiting tree suitable for medium-sized and larger gardens.
Description *Flower* Panicles of white flowers, up to 4in (10cm) long, in late spring, early summer. *Foliage* Pinnate foliage, up to 9in (23cm) long with 13-21 grey-green oblong leaflets, narrow and up to 2in (5cm) long. Some yellow autumn colour. *Stem* Grey-green when young, becoming grey-brown. Upright at first, spreading with age. *Fruit* Small, hanging, pear-shaped or apple-shaped fruits up to 1in (3cm) long, green to green-brown, sometimes red on sun side. Edible, but must be bletted.
Hardiness Tolerates 4°F (−15°C).
Soil Requirements Any soil conditions.
Sun/Shade aspect Full sun to light shade.
Pruning None required.
Propagation and nursery production Available bare-rooted or container-grown, but scarce in nursery production, and normally

Sorbus aucuparia 'Beissneri' − branches

Sorbus cashmiriana in fruit

Sorbus domestica in fruit

offered only up to 3ft (1m). Needs extensive search for suitable planting stock.
Problems At maturity produces abundant fruits which drop and create a nuisance underfoot.
Varieties of interest *S. torminalis* (Wild Service Tree) Leaves 5in (12cm) long, pinnate, with 3-5 pairs of ovate, deeply cut leaflets. Glossy green when young, aging to greygreen. White flowers in early summer, followed by oval, speckled brown fruits. Good red or golden yellow autumn colour. Half average height and spread.
Average height and spread
Five years
8x5ft (2.5x1.5m)
Ten years
16x8ft (5x2.5m)
Twenty years
or at maturity
32x16ft (10x5m)
Continued growth over 40 years reaches 50x30ft (15x9m)

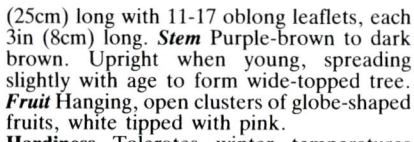

SORBUS HUPEHENSIS

KNOWN BY BOTANICAL NAME
Rosaceae
Deciduous
A spectacular sight when fruiting in autumn.

Origin From China.
Use As a freestanding tree or particularly impressive for group planting.
Description *Flower* White flower clusters up to 4in (10cm) across in early summer. *Foliage* Grey-green with purple hue, up to 10in (25cm) long with 11-17 oblong leaflets, each 3in (8cm) long. *Stem* Purple-brown to dark brown. Upright when young, spreading slightly with age to form wide-topped tree. *Fruit* Hanging, open clusters of globe-shaped fruits, white tipped with pink.
Hardiness Tolerates winter temperatures down to −13°F (−25°C).
Soil Requirements Most soil conditions, but dislikes extremely alkaline types.
Sun/Shade aspect Full sun to light shade. Fruits show well in autumn sun.
Pruning None required.
Propagation and nursery production From grafting or budding on to understocks of *S. aucuparia.* Best purchased container-grown, but also available bare-rooted. Normally supplied at 5-10ft (1.5-3m). Larger trees occasionally available, but 5½-10ft (1.8-3m) best for establishment.
Problems None.
Varieties of interest *S. h. obtusa* Deep pink fruits in autumn, otherwise similar to the parent.
Average height and spread
Five years
8x5ft (2.5x1.5m)
Ten years
16x8ft (5x2.5m)
Twenty years
or at maturity
32x16ft (10x5m)

SORBUS 'JOSEPH ROCK'

ROCK'S VARIETY
Rosaceae
Deciduous
On the right soil, one of the finest yellow-fruiting autumn trees.

Origin Of garden origin.
Use As a freestanding specimen tree planted singly or grouped; or sited in a large shrub border as an autumn attraction.
Description *Flower* White flower clusters in mid spring to early summer. *Foliage* Pinnate leaves, up to 6in (15cm) long with 9-11 leaflets. Some yellow-orange autumn tints. Mature foliage can be a little disappointing in both summer and autumn. *Stem* Grey-green, becoming grey-brown. Pyramidal, becoming ascending at maturity, forming a tight, neat small tree of medium vigour. *Fruit* Hanging clusters of yellow to yellow-orange fruits, profuse and well displayed in autumn.
Hardiness Tolerates 4°F (−15°C).
Soil Requirements Tolerates most soil conditions, but requires rich, deep loam to show best foliage and fruiting ability. Rapidly becomes old and stunted on poor soils.
Sun/Shade aspect Full sun. Autumn and winter sun show fruits to best advantage.
Pruning None required.
Propagation and nursery production From grafting or budding on to *S. aucuparia.* Normally available bare-rooted or container-grown from 5½-10ft (1.8-3m). Larger trees may be available, but smaller more likely to succeed. Readily found in garden centres and general nurseries.
Problems Cannot do well on poor soil conditions. Often sells out due to its wide popularity; early ordering from nurseries may be advisable.
Average height and spread
Five years
8x5ft (2.5x1.5m)
Ten years
16x8ft (5x2.5m)
Twenty years
or at maturity
32x16ft (10x5m)

Sorbus hupehensis in fruit

Sorbus 'Joseph Rock' in fruit

SORBUS
Autumn foliage varieties

KNOWN BY BOTANICAL NAME
Rosaceae
Deciduous
Among the finest trees for autumn foliage colour.

Origin From Europe.
Use Freestanding or group planted to show off autumn tints.
Description *Flower* Clusters of fluffy white flowers, up to 5in (12cm) across in late spring, early summer. *Foliage* Pinnate, up to 9in (23cm) long with 11-15 lanceolate, sharply tooth-edged leaflets. Dark green with slight grey sheen. Excellent plum, orange and red autumn colours. *Stem* Green to green-brown, becoming grey-brown. Strong, upright when young, quickly becomes spreading and eventually very branching. *Fruit* Clusters of orange-red fruits in autumn, enjoyed by birds.

Sorbus 'Embley' in autumn

Hardiness Tolerates −13°F (−25°C).
Soil Requirements Any soil type, including alkaline conditions.
Sun/Shade aspect Best in full sun, but tolerates light shade.
Pruning None required.
Propagation and nursery production By grafting or budding. Available bare-rooted or container-grown. Normally supplied from 3-10ft (1-3m), larger plants occasionally available. Heights of 5-8ft (1.5-2.5m) recommended as best for establishment. Readily found in general and specialist nurseries.
Problems None.
Varieties of interest *S. 'Embley'* syn. *S. discol-*

or Spectacular dark plum red autumn tints with some orange and yellow. Orange-red fruits not plentiful. *S. sargentiana* Spectacular orange to orange-red autumn colours. Large leaves. Stems light cream-green with red buds and thicker than most forms. Red to orange fruits. Normally available at only 3-6ft (1-2m) in height, upright when young becoming rounder with age.

Average height and spread

Five years
8x5ft (2.5x1.5m)
Ten years
16x8ft (5x2.5m)
Twenty years
or at maturity
32x16ft (10x5m)

SORBUS
Good fruiting forms

KNOWN BY BOTANICAL NAME
Rosaceae
Deciduous
Useful autumn-fruiting trees of varied shape and growth rate.

Origin From Europe through central Asia and China, depending on variety.
Use As small ornamental trees for all but the smallest gardens, planted singly or grouped.
Description *Flower* Clusters of white, fluffy flowers, up to 4in (10cm) across, in early spring. *Foliage* Pinnate leaves, up to 8in (20cm) long. Light to mid-green, slightly grey leaf stem. Good yellow to orange autumn colours. *Stem* Green-brown to brown. Medium or fast rate of growth, dependent on variety, forming rounded or upright trees. *Fruit* Red or yellow in clusters, dependent on variety, in late summer, early autumn.
Hardiness Tolerates winter temperatures down to −13°F (−25°C).
Soil Requirements Any soil conditions.
Sun/Shade aspect Full sun or very light shade. Autumn fruit and foliage colour shows well in full sun.
Pruning None required.
Propagation and nursery production From grafting or budding. Normally supplied bare-rooted or container-grown from 5-10ft (1.5-1.8m). Larger trees occasionally available but smaller plants generally more successful.
Problems Availability is a problem as many of the large range of varieties are only in limited commercial production.
Varieties of interest *S. americana* (American Mountain Ash) Leaves 10in (25cm) long with 11-17 leaflets each up to 4in (10cm) long. Dark glossy green with glabrous undersides. Flowers creamy white in late spring, early

Sorbus vilmorinii in fruit

summer. Large, globe-shaped, red fruits. Good autumn colour. Interesting sticky buds in winter. Forms a round-topped tree of medium growth rate. From eastern USA.: Available from specialist nurseries. *S. 'Apricot Queen'* Apricot yellow fruits. *S. decora* Very similar to *S. americana* and may be offered as an alternative. Differs in having 9-11 broader leaflets. *S. esserteauana* Pinnate leaves up to 12in (30cm) long with 11-13 leaflets, each 4in (10cm) long. Oblong to lanceolate with coarsely toothed edges. Dark green with green-white undersides, giving good autumn red tints. Large flower panicles up to 6in (15cm) across, in late spring. Red to rich scarlet fruits in autumn. Strong and vigorous. Slightly more than average height. Pyramidal habit. From China. Available from specialist nurseries. *S. e. 'Flava'* Identical to *S. esserteauana* except having yellow fruits. Available from specialist nurseries. *S. 'Golden Wonder'* Golden yellow fruits in autumn. *S. 'Kirsten Pink'* Pink fruits. Small foliage. *S. matsumurana* (Japanese Mountain Ash) Leaves pinnate, up to 9in (23cm) long with 9-13 leaflets, each 3in (8cm) long. Foliage grey-green. Good autumn colours. White flowers, followed by sparse clusters of orange-red berries in autumn. Two-thirds average height and spread. Slightly spreading with some irregular growth. From Japan. Available from specialist nurseries. *S. pohuashanensis* Leaves up to 6in (15cm) long with 11-15 leaflets. Good autumn colour. White flowers and globe-shaped, orange-red fruits. Two-thirds average height and spread. From North China. *S. 'McClaren D84'* syn. *S. poterifolia* Leaves up to 8in (20cm) long with 15-19 leaflets. Fruits rose-pink, globe-shaped, presented in large, open bunches in

Sorbus sargentiana in fruit

Sorbus matsumurana in autumn

autumn. Half average height and spread. Upright habit. From China. Rather scarce in commercial production. *S. prattii* Leaves pinnate with up to 29 coarsely toothed leaflets. Clusters of pearly-white fruits hanging along undersides of branches in autumn. Thin, graceful branches. Half average height and spread. Limited in commercial production. From western China. *S. scalaris* Leaves 8in (20cm) long, with up to 37 leaflets, each narrow and oblong, very dark green upper surfaces and grey felted undersides, turning rich red in autumn. Flowers off-white up to 5in (12cm) across, in late spring to early summer. Bunches of bright red, glossy, globe-shaped fruits in autumn. Two-thirds average height and spread. From China. Not readily available in commercial production. *S.* × *thuringiaca* 'Fastigiata' Stout, strong, dark green-brown, upright shoots. Large dark green, pinnate foliage. White flowers and red fruits in autumn. Narrow and fastigiate. Average height and only 6ft (2m) spread. *S. vilmorinii* Small purple-green foliage with grey sheen. White flowers, mauve berries. Two forms exist; one bushy, reaching half average height and spread, the true variety, the other more upright and strong-growing.

Average height and spread
Five years
8x5ft (2.5x1.5m)
Ten years
16x8ft (5x2.5m)
Twenty years
or at maturity
32x16ft (10x5m)

STUARTIA PSEUDOCAMELLIA (Stewartia pseudocamellia)

KNOWN BY BOTANICAL NAME
Theaceae
Deciduous
An acid-loving large shrub or small tree, with good flowers and beautiful autumn colours.

Origin From Japan.
Use As a large, freestanding shrub or small tree for gardens, or as a small arboretum specimen.
Description *Flower* Cup-shaped, pure white flowers up to 2in (5cm) across, in mid to late summer. Petals incurved with jagged margins and very silky texture. Stamens white with pronounced orange-yellow anthers. Trees take 5-10 years to come into full flowering. *Foliage* Ovate, slightly tooth-edged, 4in (10cm) long. Grey-green giving good orange-red autumn colours. *Stem* Young shoots grey-green. Quick-growing once established. Stems aging to brown-grey, often with red-orange sheen in spring. Forms a slow-growing large pyramidal shrub, ultimately tree-forming. *Fruit* Insignificant.
Hardiness Tolerates 14°F (−10°C).
Soil Requirements Requires an acid soil, moist, rich and deep for good growth. Dislikes any alkalinity.
Sun/Shade aspect Best in light shade. Tolerates full sun but may suffer leaf scorch in summer.
Pruning Rarely required.
Propagation and nursery production From seed or layers. Purchase container-grown. Normally supplied between 2ft (60cm) and 3ft (1m), the best heights for planting. Must be sought from specialist nurseries.
Problems Must have acid soil. Vulnerable to sun scorch.
Varieties of interest *S. koreana* Broader foliage than *S. pseudocamellia* and larger flowers. Difficult to obtain; must be sought from specialist nurseries. From Korea. *S. malacodendron* Foliage up to 4in (10cm) long, ovate or oval, with good autumn colours. Pure white flowers 3in (8cm) across, violet anthers and white stamens. One-third average height and spread. Normally considered a

Stuartia pseudocamellia in flower

---

large shrub. From south-eastern USA and extremely rare outside its native environment. *S. sinensis* Interesting peeling, dull, orange-brown bark over a grey undersurface. Foliage oval to oblong, up to 4in (10cm) long, slightly tooth-edged, bright green with good orange-yellow autumn colours. White fragrant flowers up to 2in (5cm) across, with yellow anthers. Two-thirds average height and spread. From China. Difficult to obtain. Other *Stuartia* varieties are offered, but those listed are best for general garden use.

Average height and spread
Five years
5x3ft (1.5x1m)
Ten years
13x6ft (4x2m)
Twenty years
or at maturity
32x16ft (10x5m)

STYRAX JAPONICA

JAPANESE SNOWBELL
Styracaceae
Deciduous
A splendid small tree, requiring extreme patience if it is to reach its true potential.

Origin From Japan.
Use As a small, individual freestanding tree.
Description *Flower* Hanging, white, five-petalled flowers presented on graceful stalks along the undersides of all branches in early summer. *Foliage* Oval, 3in (8cm) long, light green with grey sheen. Yellow autumn colours. *Stem* Grey to grey-green. Slow-growing for some years after planting, then rapidly gaining height but slowing again with maturity. Upright, with horizontal branches at regular intervals, forming a narrow, pyramidal tree. *Fruit* Small, oval, dark green, hanging fruits along undersides of branches in early autumn.
Hardiness Tolerates 14°F (−10°C).
Soil Requirements Does best on an acid soil with a light, open texture but tolerates limited alkalinity.

Styrax japonica in flower

Sun/Shade aspect Best in light shade, but tolerates full sun.
Pruning None required.
Propagation and nursery production From seed or layers. Purchase container-grown or root-balled (balled-and-burlapped). Not readily available, and must be sought from specialist nurseries. Normally supplied at no more than 15in (40cm) to 3ft (1m) in height, or occasionally up to 5ft (1.5m).
Problems Slow-growing and needing time to show off its full potential.
Varieties of interest *S. hemsleyana* A good alternative to *S. japonica* for planting. Flowers slightly larger, but not to advantage. Foliage ovate, up to 5in (12cm) long. Slightly more than average height. Relatively quick-growing. *S. obassia* Foliage round to orbicular, up to 8in (20cm) long. Hanging, open clusters of white flowers. Extremely difficult to find. Originating in Japan.
Other *Styrax* forms may be seen occasionally but may not be hardy outside their native environment.

Average height and spread
Five years
5x3ft (1.5x1m)
Ten years
10x6m (3x2m)
Twenty years
or at maturity
20x13ft (6x4m)

SYRINGA Cultivars

LILAC, STANDARD LILAC
Oleaceae
Deciduous
Attractive spring-flowering 'toy' trees, trained from the so-called Standard Lilac commonly grown as a shrub.

Origin Of garden origin; many raised by Victor Lemoine and his son, Émile, in their nursery in Nancy, France. Varieties may show only slight differences of size or colour.
Use As small specimen trees for spot planting and to emphasize garden features.
Description *Flower* Single or double florets in large, fragrant panicles, late spring or early summer. Colour range blue to lilac, pink to red or purple; also white, yellow, and some bicoloured forms. Colour, size and flowering time dependent on variety. *Foliage* Medium-sized, ovate leaves, dark green to mid green. *Stem* Grey-green to grey-brown. Stout shoots with pronounced buds, yellow or red-purple in winter, dependent on variety. Standard forms pretrained from ground level; or by budding or grafting at 5-8ft (1.5-2.5m) on single stems. Round, single-stemmed, mop-headed or slightly pyramidal trees.

Syringa 'Monique Lemoine' in flower

Fruit Grey-brown seedheads of some winter attraction.

Hardiness Tolerates winter temperatures down to −13°F (−25°C).

Soil Requirements Most soil conditions, but may show signs of chlorosis on severe alkaline types.

Sun/Shade aspect Full sun to medium shade, preferring full sun.

Pruning Very little required but removal of seedheads in winter encourages flower production. Can be drastically cut back to control vigour but this leads to reduced flowering in the next 2-3 years.

Propagation and nursery production From budding and grafting using *S. vulgaris* or the common Privet, *Ligustrum vulgare* or *L. ovalifolium* as understock. Can also be raised from semi-ripe cuttings taken early to mid summer. Single, individual shoots are trained from ground level; budding or grafting takes place at 3-5ft (1-1.5m) from ground level on pretrained understock. Available bare-rooted or container-grown. Standard forms grown as mop-headed trees are relatively scarce in production.

Problems Top growth of standard forms somewhat weak and very susceptible to wind damage; requires good staking.

Varieties of interest *S. 'Charles Joly'* Double, dark red-purple flowers in large panicles produced late spring to early summer. *S. 'Katherine Havemeyer'* Large, broad, bold lavender-blue flowers. Strong, rounded, bright green foliage. *S. 'Monique Lemoine'* syn. *S. 'Madame Lemoine'* Large, pure white double flowers produced in mid spring. *S. 'Souvenir de Louis Spaeth'* Single wine red flowers in broad trusses produced in mid to late spring. Strong-growing.

Many varieties can be grown as standards. Those listed are most suitable for garden plants and available commercially.

Average height and spread
Five years
7x3ft (2.2x1m)
Ten years
12x5ft (3.5x1.5m)
Twenty years
or at maturity
16x8ft (5x2.5m)

TILIA

LIME, LINDEN
Tiliaceae
Deciduous
Large trees with good foliage and winter stems. Some varieties can be confined by pruning.

Origin Most from Europe and North America; some varieties from Iran.

Use As large specimen trees for large gardens or estates. As trained trees for smaller gardens, either pleached as hedges or as heavily pollarded specimens. Useful for streets and avenues when planted at 39ft (12m) apart, with limited tolerance of pollution.

Description *Flower* Dull-white to yellow-green, up to 1in (3cm), each with five sepals and five petals, produced on short stalks and hanging usually in groups of three but can be up to 40. Interesting massed effect but not of individual beauty. *Foliage* Round to ovate young foliage 4-7in (10-17cm) long and wide. Smaller leaves produced on old wood. Light green with pronounced veining and grey undersides. Good yellow autumn colour. *Stem* Grey-green to grey-brown. Predominantly upright, strong, vigorous. Rarely branching when young but becoming very twiggy with age. Winter stems grey-green, yellow, red or orange-red dependent on variety. *Fruit* Small nut-like fruits in autumn; of little attraction.

Hardiness Tolerates winter temperatures down to −13°F (−25°C).

Soil Requirements Tolerates most soils, but shows poor spring foliage growth or early autumn leaf drop on starved, difficult soil conditions.

Sun/Shade aspect Full sun to light shade.

Pruning Can be cut hard; either as pleached trees trained horizontally to form an aerial hedge or by annual or triennial pollarding back to a central crown. Mature trees unpruned for 10-20 years can have main limbs cut back hard to central crown, rejuvenating over 5-10 years. May be left unpruned but hard pruning enhances coloured stems. Weeping forms normally left to grow unpruned.

Propagation and nursery production From seed, layers, budding or grafting depending on variety. Available bare-rooted or container-grown. Normally offered from 5½-10ft (1.8-3m). Larger trees also available, up to 20-23ft (6-7m). Most forms readily available from general nurseries.

Problems Leaves subject to sooty mould, black fungus growth which is sticky and disfiguring. Control is difficult in such large trees and to some extent the problem is inevitable. Some forms of *Tilia* give off a poisonous nectar which affects honey-bees, especially *T. petiolaris*, and should not be planted near to commercial bee-keeping properties.

Varieties of interest *T. americana* (American Lime or Linden, Basswood) Grey-green shoots. Round to ovate leaves, up to 8in (20cm) long. Dark, dull green upper surfaces with contrasting bright green undersides. Good yellow autumn colour. From North America. *T. a. 'Erecta'* A narrowed, upright form of *T. americana*. *T. cordata* (Small-leaved Lime, Littleleaf Linden) Round leaves 3in (8cm) across, mid to dark green upper surfaces and grey undersides. Hanging clusters of predominantly yellow-white, fragrant flowers. Round fruits with felted texture in winter. Slightly less than average height and spread. From Europe, including the UK. *T. c. 'Erecta'* An upright form of *T. cordata*. *T. c. 'Gold Spire'* Bright green foliage, bright yellow in autumn. Narrow, upright habit. Good street or avenue potential. *T. × euchlora* (Yellow-twigged Lime, Crimean Linden) Leaves ovate to round, up to 4in (10cm) long. Glossy green upper surfaces, greyer undersides. Yellow and white fragrant flowers in midsummer, followed by oval, pointed, downy fruits. Yellow winter stems. Responds well to pleaching or pollarding. Two-thirds average height and spread. *T. × europaea* (Common Lime, European Linden) Leaves broad, ovate, up to 4in (10cm) long. Mid green with slight red hue along veins. Yellow-white, fragrant flowers. *T. henryana* Large, broad, ovate leaves, over 6in (15cm) wide and long. New foliage on new growth may be larger, making a spectacular display. Half average height and spread. Scarce in production. Originating in China. May be slightly tender. *T. petiolaris* (Weeping Silver Lime, Pendent Silver Linden) Silver-grey stems and shoots. Leaves ovate, up to 4in (10cm) long, grey-green upper surfaces, white to silver, felted undersides. Flowers white, very fragrant but dangerous to bees, produced mid to late summer. Fruits rounded, grooved and warted. Average height but up to 64ft (20m) spread. From south-eastern Europe, but true origin not known. *T. platyphyllos* Leaves ovate to round, up to 5in (12cm) long. Mid green with some red veining. Winter shoots red. Responds well to pollarding or pleaching. From Europe. *T. p. 'Erecta'* A more upright variety of *T. platyphyllos*. *T. p. 'Rubra'* Winter twigs dark red to bright red after pollarding or pleaching in previous year. *T. tomentosa* (Silver Lime or Linden) Leaves ovate to round, grey-green upper surfaces, white felted undersides giving a shimmery appearance in light wind. Dull white, hanging flowers in mid to late summer. Distinct, broad, pyramidal habit. From south-eastern Europe.

Average height and spread
Five years
13x6ft (4x2m)
Ten years
26x13ft (8x4m)
Twenty years
or at maturity
53x26ft (16x8m)
Continued growth over 50 years reaches 79x32ft (24x10m)

Tilia cordata **in flower**

ULMUS

ELM
Ulmaceae
Deciduous
Picturesque upright and weeping trees, recently much depleted by Dutch Elm disease and less commonly planted for this reason.

Origin Mostly from Europe.
Use As large specimen field trees for estates, parks and farmland. Several varieties suitable for smaller gardens.
Description *Flower* Tiny, uninteresting, light green bisexual flowers in early spring. *Foliage* Oval, 4in (10cm) long, coarse-textured on both surfaces. Light green to mid green. Grey when young. Yellow autumn colour. *Stem* Grey-green aging to grey-brown. Old stems often heavily barked with coarse, corky texture. Predominantly upright, spreading with age. Also weeping forms. Basal suckers appear on some varieties, often to the dimensions of small or medium trees. *Fruit* Scaly, light grey-green fruits in autumn.
Hardiness Tolerates winter temperatures down to −13°F (−25°C).
Soil Requirements Any soil conditions.
Sun/Shade aspect Full sun to light shade.
Pruning None required.

Ulmus × hollandica 'Wredei' in leaf

Propagation and nursery production From seed, suckers or grafting. Available bare-rooted or container-grown, but extremely difficult to find in Europe, UK and USA, due to the spread of Dutch Elm disease, which has eliminated many varieties. Plants from 5½-10ft (1.8-3m) normally transplant successfully when young.
Problems All varieties are susceptible to Dutch Elm disease which has devastated Elm populations in the past 30 years. Few are now offered for planting. Some said to be resistant to the disease are not truly immune, or are not the true Field Elm.
Varieties of interest *U. angustifolia var. cornubiensis* (Cornish Elm) Small, oval to round foliage. Good yellow autumn colour. Originating in the south-west of England and in Brittany, France. *U. 'Camperdownii'* (Weeping Elm) Closed canopy of growth. One-quarter average height and spread. Previously thought to be immune to Dutch Elm disease, but trees are now becoming infected. *U. carpinifolia* (Smooth-leaved Elm) Smooth-textured, slightly grey foliage. Grows suckering basal shoots. From Europe, including the UK, through to western Asia. *U. glabra* (Witch or Scotch Elm) Large oval foliage up to 7in (17cm) across. Light green with grey undersides. Slightly spreading and arching habit. *U. g. 'Lutescens'* Yellow leaves in spring, turning green in autumn. Appears less susceptible to Dutch Elm disease, but this is not guaranteed. *U. × hollandica 'Wredei'* syn.

U. × h. 'Wredei Aurea' A tight, upright golden pillar. Reaching only one-quarter of average height and no more than one-tenth average width. Thought to be resistant to Dutch Elm disease, but seen recently to be succumbing. *U. procera* (English Elm) Small, broad, ovate, light green leaves with rough texture. At one time seen throughout UK and France, now almost completely lost to Dutch Elm disease.
Average height and spread
Five years
10x6ft (3x2m)
Ten years
20x13ft (6x4m)
Twenty years
or at maturity
37x26ft (12x8m)
Continued growth over 50 years reaches 64x33ft (20x10m)

VIBURNUM
Standard trees

SPRING-FLOWERING VIBURNUM, STANDARD VIBURNUM
Caprifoliaceae
Deciduous
Useful specially trained 'toy' standard trees, but difficult to find.

Origin Basic spring-flowering form from Korea; most varieties of garden origin.
Use As small, mop-headed trees ideal for emphasizing particular garden features. Can be grown in large containers in good potting medium with area and depth not less than 2½ft (80cm).
Description *Flower* Medium to large rounded clusters of tubular florets in white or pink, most highly scented. *Foliage* Ovate, 4in (10cm) long, grey-green. Some limited autumn display. *Stem* Upright, in spring-flowering forms covered with greyish scale. Branching and dome-forming with age. Grown from ground level with a single stem or budded and grafted at a height of 5ft (1.5m), forming a small mop-headed tree. *Fruit* Insignificant.
Hardiness Tolerates 14°F (−10°C).
Soil Requirements Any soil conditions.
Sun/Shade aspect Full sun to medium shade, preferring light shade.
Pruning None required other than the removal of suckering growths which may appear from below soil level.
Propagation and nursery production Spring-flowering varieties by budding or grafting on to a stem of *V. lantana*. Purchase container-grown. Not readily available; must be sought from specialist nurseries.

Problems Needs good staking throughout life, or wind damage breaks off the heavy heads very easily.
Varieties of interest *V. × carlcephalum* Large, ovate to round, grey-green leaves with some good autumn colours. Clusters of large, white, tubular flowers 4-5in (10-12cm) across, pink in bud and very fragrant. *V. carlesii* Ovate to round, downy grey to grey-green leaves with grey felted undersides; good range of red autumn colours. Clusters of pure white, tubular flowers in mid to late spring, pink in bud and strongly fragrant. Some small, black, clustered fruits, relatively insignificant. *V. × juddii* Grey-green, ovate foliage with good autumn colours. Clusters of pink, tinted, tubular flowers in mid to late spring at the terminals of branching stems. Open habit when young, becoming denser with age. From USA. *V. opulus 'Sterile'* Round, snowball flowers with ray florets, produced in early summer. Light green, lobed foliage with yellow autumn colour. Grown on a central stem to produce a head at 5ft (1.5m) above ground level.
Average height and spread
Five years
8x3ft (2.5x1m)
Ten years
10x5ft (3x1.5m)
Twenty years
or at maturity
12x6ft (3.5x2m)

WEIGELA 'BRISTOL RUBY'

KNOWN BY BOTANICAL NAME
Caprifoliaceae
Deciduous
An interesting, useful 'toy' or mop-headed tree, though not always available trained in this form.

Origin Basic form from Japan, Korea and north China; 'Bristol Ruby' of garden origin.
Use As a freestanding, small mop-headed tree, grown singly or in pairs to emphasize a particular garden feature. Can be grown in large containers but needs good feeding.
Description *Flower* Dark red, funnel-shaped flowers in late spring to early summer and produced intermittently throughout spring, summer and autumn. *Foliage* Dark green ovate foliage, 4in (10cm) long. Yellow autumn colour. *Stem* Grey-green to grey-brown. Upright, spreading with age. A single central stem is encouraged to form a mop-headed structure. *Fruit* Dark brown to mid brown seedheads, of some attraction in winter.
Hardiness Tolerates winter temperatures down to −13°F (−25°C), but susceptible to wind damage throughout its life.

Viburnum opulus 'Sterile' in flower

Weigela 'Bristol Ruby' in flower

Soil Requirements Any soil type.
Sun/Shade aspect Full sun to medium shade, preferring full sun.
Pruning Remove old wood as much as possible after flowering to encourage rejuvenation and good reproduction of flowering wood. Prune annually from two years after planting.
Propagation and nursery production From semi-ripe cuttings; individual single stem encouraged to grow upright to form standard tree shape. Not readily found in this form.
Problems The heavy head on a single stem is vulnerable to wind damage.
Varieties of interest Any vigorous forms of *Weigela* can be considered for training as standard.
Average height and spread
Five years
8x4ft (2.5x1.2m)
Ten years
10x5½ft (3x1.8m)
Twenty years
or at maturity
10x7ft (3x2.2m)

WISTERIA
Standard trees

WISTERIA, STANDARD WISTERIA
Leguminosae
Deciduous
Interesting training of a normally climbing plant to form a spreading, weeping tree.

Origin Mostly from China.
Use As a spreading, weeping, mop-headed, small tree. Best grown as a single lawn specimen in medium-sized or larger gardens.
Description *Flower* Racemes of blue, white or pink flowers, depending on variety, up to 12in (30cm) long, produced in late spring and cascading in a long waterfall effect. *Foliage* Pinnate and consisting of 11 oval to oblong leaflets, each leaf up to 12in (30cm) long, light green in colour with some yellow autumn tints. *Stem* Single, upright stem encourage. Grey-green quickly becoming grey-brown, old and gnarled. Requires support throughout its life. After 10-20 years forms a widely spread, arched and weeping effect. *Fruit* Grey-green pea-pods in autumn.
Hardiness Tolerates 4°F (−15°C).
Soil Requirements Most soil conditions, but dislikes extreme alkalinity.
Sun/Shade aspect Best in light dappled shade, but tolerates full sun.
Pruning Tendrils from new summer growth cut back hard in early autumn to two buds to encourage central crown of mainly flowering wood. Mop-headed effect formed by shortening tendrils by one-third.
Propagation and nursery production From

grafting of named varieties on to *W. sinensis*, then encouraging a central, single trunk and short cut-back head. Normally supplied from 6-8ft (2-2.5m). Not readily available and must be sought from specialist nurseries. Garden training is by selecting a single stem and encouraging it as sole wood producer by staking and removing all unrequired tendrils. Head may be trained on to wire umbrella support.
Problems Trees may take between 5-10 years to come into flower. Can be somewhat unruly and misshapen, and require extensive annual pruning.
Varieties of interest *W. sinensis* Blue, hanging flowers up to 12in (30cm) long. Only use grafted plants from a good known source: seed-raised trees are unreliable. *W. s. 'Alba'* White racemes of flowers up to 10in (25cm) long. *W. s. 'Rosea'* Pink to rose pink, hanging flower racemes 10in (25cm) long.
Any *Wisteria* variety can be trained in this way. The above are selected from the range of flower colours.
Average height and spread
Five years
10x6ft (3x2m)
Ten years
13x13ft (4x4m)
Twenty years
or at maturity
16x20ft (5x6m)

Wisteria sinensis in flower

ZELKOVA CARPINIFOLIA

ELM ZELKOVA
Ulmaceae
Deciduous
An interesting tree with attractive foliage.

Origin From the Caucasus.
Use As a freestanding, large specimen tree for fine shape, foliage texture and autumn tints. Suitable for medium-sized or larger gardens. Makes good specimen field tree, but somewhat slow.
Description *Flower* Insignificant. *Foliage* Ovate to round, 4in (10cm) long, deeply veined and heavily tooth-edged. Light green with a slight grey sheen. Good autumn colours of red to purple-red becoming orange-red to yellow. *Stem* Grey-green to grey-brown. Upright when young, quickly spreading to become twiggy and branching. Forms a large, handsome tree, but relatively slowly. *Fruit* Small round, hanging fruits, of little interest.
Hardiness Tolerates winter temperatures down to −13°F (−25°C).
Soil Requirements Does best on a well-drained soil; tolerates high alkalinity or acidity.

Zelkova serrata in autumn

Sun/Shade aspect Full sun to light shade, preferring full sun.
Pruning None required.
Propagation and nursery production From seed or layers. Available bare-rooted or container-grown. Normally supplied at 3-5½ft (1-1.8m). Larger trees occasionally available: trees up to 10ft (3m) usually transplant safely. Must be sought from specialist nurseries.
Problems Slow growth rate often underestimated. Young trees appear weak and do not establish and grow away with any vigour until 3-5 years from planting.
Varieties of interest *Z. serrata* Relatively long, ovate leaves. Slightly less hardy. In some areas difficult to find.
Average height and spread
Five years
8x3ft (2.5x1m)
Ten years
16x10ft (5x3m)
Twenty years
or at maturity
32x20ft (10x6m)
Continued growth over 50 years reaches 64x32ft (20x10m)

DICTIONARY
— OF —
SHRUBS

ABELIA × GRANDIFLORA

ABELIA, GLASSY ABELIA
Caprifoliaceae
Semi-evergreen
An attractive range of varieties, suited to most temperate areas; makes a good effect when fan-trained against a sheltered, sunny wall in colder areas.

Origin Unknown. The parent plant was named after Dr Clarke Abel (1780-1826).
Use As medium height, late-flowering shrub; planted 2½ft (80cm) apart makes an informal hedge.
Description *Flower* Pink and white, small to medium-sized, hanging, bell-shaped flowers borne in small clusters on wood 2-3 years old or more, from late summer to early autumn. *Foliage* Leaves 1-2½in (3-6cm) long, ovate, pointed, olive-green with reddish shading. May fall from late autumn, or be retained into early spring following mild winter. *Stem* Light green with brownish shading, upright when young, becoming more branching in second and third year, and darker brown. New shoots grow from base up through the older framework, to form a rounded shape. Medium growth rate. *Fruit* Small, translucent seeds in clusters.

Abelia × grandiflora in flower

Hardiness Tolerates 14°F (−10°C).
Soil Requirements Any soil, but new growth less vigorous on dry soils. On extremely alkaline soils, chlorosis may occur.
Sun/Shade aspect Prefers full sun, tolerates light shade.
Pruning On established shrubs, remove one-third of the oldest shoots each year in early spring to encourage new flowering shoots.
Propagation and nursery production From semi-ripe cuttings taken in late summer. Always purchase container-grown. Found in garden centres and general nurseries. Best planting heights 15in-2½ft (40-80cm).
Problems May appear weak when purchased but once planted grows quickly. Wood can be extremely brittle so handle carefully.
Varieties of interest *A. chinensis* Taller growing than *A. × grandiflora* with fragrant white, rose-tinted flowers, early to mid summer. A more tender variety from China. *A. 'Edward Goucher'* Produces purple-pink trumpet flowers freely; foliage grey-green. Height and hardiness as *A. × grandiflora*. *A. floribunda* Cherry-red flowers 2in (5cm) long, borne in early summer. Slightly tender in most areas so fan-train on sun-facing wall. Tall growing. Of Mexican origin. *A. × grandiflora 'Francis Mason'* Golden variegated leaf form, a shorter shrub and slightly more tender than its parent. Recent introduction, bears pale pink to white flowers throughout late summer and early autumn. *A. schumannii* A less vigorous and more tender form than *A. × grandiflora*, bearing lilac-pink flowers throughout sum-

Abelia × grandiflora 'Francis Mason' in leaf

mer. *A. triflora* Scented pink-tinged white flowers borne in threes. Eventually reaching 6ft (2m) or more. Minimum winter temperature 23°F (−5°C).

Average height and spread
Five years
4x3ft (1.2x1m)
Ten years
5x5ft (1.5x1.5m)
Twenty years
or at maturity
5x6ft (1.5x2m)

ABELIOPHYLLUM DISTICHUM

KOREAN ABELIALEAF, WHITE FORSYTHIA
Oleaceae
Deciduous
A much overlooked, undemanding flowering shrub.

Origin From Korea.
Use As a small, fragrant, early spring flowering shrub.
Description *Flower* Tiny white, sweetly scented flowers with mauve-tinged basal shading when in bud, borne late winter to early spring. *Foliage* Leaves ovate, 1-3in (3-8cm) long, light green, often sparsely presented. *Stem* Young, upright shoots, aging to twiggy, arching effect. Forms upright, slightly spreading shrub. Slow to medium growth rate. *Fruit* Insignificant.
Hardiness Tolerates 4°F (−15°C).

Abeliophyllum distichum in flower

Soil Requirements Does well on all soils, except very alkaline.
Sun/Shade aspect Full sun or light to medium shade.
Pruning None required.
Propagation and nursery production From semi-ripe cuttings taken in midsummer. Purchase container-grown; may be difficult to find. Best planting heights 8in-1½ft (20-50cm).
Problems Looks weak and uninteresting in a container — hence its neglect.

Average height and spread
Five years
3x3ft (1x1m)
Ten years
4x3ft (1.2x1m)
Twenty years
or at maturity
4x3ft (1.2x1m)

ABUTILON MEGAPOTAMICUM

TRAILING ABUTILON
Malvaceae
Deciduous
An interesting range of tender flowering shrubs for mild areas.

Origin From Brazil.
Use As a freestanding shrub in very mild areas, as a wall shrub in colder regions, or in a conservatory.
Description *Flower* Medium-sized, hanging, bell-shaped flowers with yellow petals, red calyx and purple anthers, borne in late summer and early autumn. *Foliage* Leaves ovate, 2-4in (5-10cm) long, with toothed edges. Olive green with purple-red shading and veins. *Stem* Upright, slightly arching, requiring support such as a wall or post. Medium growth rate. *Fruit* Insignificant.
Hardiness Minimum temperature range between 32°F (0°C) and 23°F (−5°C). May die back in winter but normally renews from ground level in late spring.
Soil Requirements Any soil, but does best on moist, rich types.
Sun/Shade aspect Full sun to light shade.
Pruning Remove one-third of old flowering wood on established shrubs, early to mid spring.
Propagation and nursery production From softwood cuttings taken in early summer. Purchase container-grown; available from specialist nurseries. Best planting heights 2-3ft (60cm-1m).
Problems A shrub of weak constitution, and not hardy.
Varieties of interest *A. megapotamicum 'Kentish Belle'* Larger, orange-yellow flowers in midsummer. *A. megapotamicum 'Variegatum'* Orange-yellow flowers and yellow variegated leaves. *A. milleri* Orange petals and crimson stamens. Very large dark green leaves. All listed varieties are tender, minimum temperature 32°F (0°C). In very mild areas a number of other varieties may be found. *A. megapotamicum* itself is the hardiest.

Average height and spread
Five years
5x5ft (1.5x1.5m)
Ten years
6x6ft (2x2m)
Twenty years
or at maturity
6x6ft (2x2m)

ACANTHOPANAX SIEBOLDIANUS (Acanthopanax pentaphyllus)

FIVELEAF ARALIA
Araliaceae
Deciduous
A shrub rarely seen but worth searching for.

Origin From China.
Use As medium height foliage shrub.
Description *Flower* Rarely flowers. *Foliage* Leaves three-parted, compound palmate, 1½-3½in (4-9cm) long, light green with small curved thorn at base. Good yellow autumn colour. *Stem* Grey-green, upright, becoming branching with age, forming a ball-shaped shrub. Medium growth rate. *Fruit* Rarely fruits.
Hardiness Tolerates 23°F (−5°C).
Soil Requirements Most soils, but resents severe alkalinity or dryness.
Sun/Shade aspect Full sun to light shade.
Pruning None required although may be cut back to encourage new foliage.

Acanthopanax sieboldianus 'Variegatus'

Propagation and nursery production By division or softwood cuttings taken in summer. Plant from open ground, or buy container-grown. Best planting heights 1½-3ft (50cm-1m).
Problems None.
Varieties of interest *A. s. 'Variegatus'* Attractive creamy white variegated foliage. Requires full sun to maintain variegation.

Average height and spread
Five years
5x5ft (1.5x1.5m)
Ten years
6x6ft (2x2m)
Twenty years
or at maturity
8x8ft (2.5x2.5m)

Abutilon megapotamicum in flower

ACER GINNALA

AMEUR MAPLE
Aceraceae
Deciduous
Listed here because it is best grown as a flowering shrub, the Ameur Maple may be trained on a single stem as a small tree.

Origin From China, Manchuria and Japan.
Use As a single large shrub or small tree for mass planting for autumn effect.
Description *Flower* Small, tufted, yellow, hanging flowers in spring. *Foliage* Leaves three-lobed, medium to small, 3½in (9cm) long and 2½in (6cm) wide, light green with brilliant orange and crimson autumn colour. *Stem* Light green shaded bright brown. Upright when young, becoming more spreading and twiggy to form round-topped shrub or small tree, of medium growth rate. *Fruit* Insignificant.
Hardiness Tolerates winter temperatures down to −13°F (−25°C).
Soil Requirements Any soil; tolerates high degrees of alkalinity.
Sun/Shade aspect Prefers full sun, tolerates light shade.
Pruning None required, but size can be reduced by removing a limb occasionally.

Acer ginnala in autumn

Propagation and nursery production From seed. Plant bare-rooted or container-grown. Purchase at 3-6ft (1-2m) from specialist nurseries.
Problems None.
Average height and spread

Five years
10x10ft (3x3m)
Ten years
16x16ft (5x5m)
Twenty years
or at maturity
26x26ft (8x8m)

ACER JAPONICUM

JAPANESE MAPLE, FULL-MOON MAPLE
Aceraceae
Deciduous
A large, flowering shrub with vivid autumn colour.

Origin From Japan.
Use As single, handsome shrub for autumn effect.
Description *Flower* Small, hanging, clusters of dark red flowers appearing in early spring with the young leaves. *Foliage* Leaves five-fingered, deeply lobed, 3-5in (8-12cm) long and wide, light green, sometimes red-veined. *Stem* Light green to green-brown. Upright when young, later spreading and twiggy; moderately fast growth rate. *Fruit* Small,

Acer japonicum 'Aconitifolium' in spring

hanging, winged fruits, light grey-green, aging to green-brown.
Hardiness Tolerates −13°F (−25°C).
Soil Requirements Prefers neutral to acid soil, but tolerates alkalinity given enough moisture during growing period. Dryness will stunt growth and may cause chlorosis.
Sun/Shade aspect Best in light shade; tolerates full sun.
Pruning None required. May be cut back but apply pruning compound to all cuts to prevent fungus disease.
Propagation and nursery production From seed. Named varieties grafted on to understock of *A. japonicum*. Purchase container-grown or root-balled (balled-and-burlapped). Relatively easy to find. Best planting heights 2-4ft (60cm-1.2m).
Problems In some areas liable to coral spot, a fungus disease attacking damaged or cut wood; prune away affected wood and paint cuts with pruning compound.
Varieties of interest *A. j.* 'Aconitifolium' syn. *A. j.* 'Laciniatum', *A. j.* 'Filicifolium'. Very dissected foliage, light grey-green; yellow, orange and red autumn colour. Flowers hanging in red droplets in spring; one of the best forms. *A. j.* 'Vitifolium' Fan-shaped leaves, purple and plum-coloured in autumn. Fairly hard to find.
Average height and spread
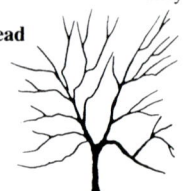
Five years
4x6ft (1.2x2m)
Ten years
6x10ft (2x3m)
Twenty years
or at maturity
10x13ft (3x4m)

ACER JAPONICUM 'AUREUM'

GOLDEN-LEAVED JAPANESE MAPLE
Aceraceae
Deciduous
A very attractive shrub providing colour from early spring to autumn.

Origin From Japan.
Use For large rock gardens or other feature areas; good standing alone near water if light shade provided.
Description *Flower* Small, hanging clusters of red-purple flowers in early spring. *Foliage* Leaves 3-5in (8-12cm) long and wide. Lemon-yellow in spring, golden yellow in summer, vivid orange-red in autumn. *Stem* Upright when young, branching and twiggy with age, forming upright shrub. Slower growth rate than other Japanese Maples. *Fruit* Small, red-tinged grey-green fruits in some summers.
Hardiness Tolerates winter temperatures down to 4°F (−15°C).
Soil Requirements Prefers neutral to acid soil,

but tolerates alkalinity given adequate moisture.
Sun/Shade aspect Best in very light shade. Full sun may lead to scorching of foliage in mid-summer.
Pruning None required.
Propagation and nursery production From grafting on to *A. japonicum*. Purchase container-grown or root-balled (balled-and-burlapped). Found in garden centres and specialist nurseries. Best planting heights 1-2ft (30-60cm).
Problems Slow-growing, compared to related Maples. May be liable to coral spot on diseased or damaged wood.

Acer japonicum 'Aureum' in spring

Average height and spread
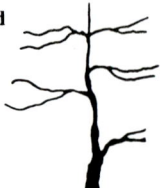
Five years
4x3ft (1.2x1m)
Ten years
6x6ft (2x2m)
Twenty years
or at maturity
10x8ft (3x2.5m)

ACER PALMATUM

JAPANESE MAPLE
Aceraceae
Deciduous
A magnificent range of garden plants; protection from strong midday summer sun and spring frosts is essential.

Origin From Japan, with many varieties of garden origin.
Use As ornamental shrubs, either singly or grouped. In time can form small multi-stemmed trees. Well placed near water where shade is adequate.
Description *Flower* Small, hanging, purple-red clusters of flowers in spring. *Foliage* Leaves palmate, 2-4in (5-10cm) long and wide. Grey-green to mid green, veins becoming red. Extremely fine shades of red in autumn. Many varieties with purple, red or yellow foliage in spring; some also variegated white. *Stem* Purple or green-brown. Upright when young, becoming very twiggy and forming slightly upright to spreading plant in later years. Moderate growth rate. *Fruit* Small, brown winged fruits in some summers.
Hardiness Tolerates minimum temperature 4°F (−15°C) but late spring frosts may damage the very soft foliage. New regrowth may occur.
Soil Requirements Prefers neutral to acid soil; tolerates alkalinity given adequate moisture.
Sun/Shade aspect Plant out of midday summer sun.
Pruning None required. Lower branches may be removed to improve shape.
Propagation and nursery production *A. palmatum* from seed. All cultivars by grafting or softwood cuttings taken in summer. Purchase

Acer palmatum 'Atropurpureum' in summer

container-grown or root-balled (balled-and-burlapped). General range available from most garden centres or nurseries but some varieties harder to find. Best planting heights 15in (40cm) to 6ft (2m).
Problems Soft foliage vulnerable to late winter frosts, and to sun scorch in summer; needs a protected location. In some areas may suffer from coral spot; diseased wood should be removed and cuts painted with a pruning compound.
Varieties of interest *A. p. 'Atropurpureum'* Deep purple leaves in spring and summer, turning brilliant reds and scarlets in autumn. In time forming a large shrub and even a small, multi-stemmed tree, although slightly smaller than *A. palmatum*. *A. p. 'Aureum'* Very soft yellow to golden yellow foliage. Slightly less vigorous than the parent. Rarely found but obtainable. *A. p. 'Bloodgood'* One of the finest purple-leaved varieties. Blood red new growth, aging to deep, almost black-purple in summer; brilliant red in autumn. Growth more upright. *A. p. 'Chitoseyama'* Bronze green foliage, turning bright red in autumn. Forms a tighter mound than *A. palmatum* and is slightly less vigorous. *A. p. 'Reticulatum'* syn. *A. p. 'Flavescens'* Soft pale yellow to pale green foliage, conspicuously veined darker green. A strong grower. *A. p. 'Shinonome'* Bright red new growth in spring, aging to purple-red through summer and with good autumn colour. More vigorous than *A. palmatum*. *A. p. 'Shishio'* New foliage in spring vivid crimson-red, aging to pink-tinged green. Good orange-red autumn colours. Requires light shade to prevent leaf scorch. Must be sought from specialist nurseries.
Average height and spread
Five years
6x6ft (2x2m)
Ten years
12x12ft (3.5x3.5m)
Twenty years
or at maturity
16x16ft (5x5m)

ACER PALMATUM 'DISSECTUM'

CUT-LEAF JAPANESE MAPLE
Aceraceae
Deciduous
These handsome architectural foliage shrubs must have adequate protection from late spring frosts and strong sunlight.

Origin From Japan.
Use As a feature shrub for a very large rock garden or placed as focal point in a group of shrubs. Ideal in tubs at least 2½ft (80cm) in diameter, placed in light shade.

Description *Flower* Small, red, hanging flower clusters, early spring. *Foliage* Very dissected palmate leaves, 2-4in (5-10cm) long and wide, light green in spring, turning red-green in summer, brilliant orange-flame in autumn. Purple-leaved forms also with good autumn colour. *Stem* Light green with silver markings. Arching to form dome-shaped, symmetrical shrub. Some purple-leaved varieties are slow-growing. *Fruit* Small to medium, hanging, light green or purple-red winged fruits on mature shrubs.
Hardiness Tolerates a minimum temperature of 4°F (−15°C), but unprotected the very soft foliage can be damaged by late spring frosts.
Soil Requirements Prefers neutral to acid soil; tolerates alkalinity given adequate moisture.
Sun/Shade aspect Best in light dappled shade. Strong summer midday sun will scorch foliage and cause serious damage.
Pruning None required; will grow completely symmetrically on its own, even if damaged when young.
Propagation and nursery production From grafting on to *A. palmatum*. Purchase container-grown or root-balled (balled-and-burlapped). Most varieties easy to find but some will need a search. Best planting heights 12in-2½ft (30-80cm).
Problems Young shrubs are thin-looking until 2-3 years after planting, but gradually acquire a symmetrical shape. Liable to minor attacks of coral spot; remove by pruning and paint the cut with pruning compound. May suffer from aphid attacks.
Varieties of interest *A. p. 'Dissectum Atropurpureum'* An attractive variety with mid purple foliage and good autumn colours. *A. p. 'Dissectum Crimson Queen'* Deep purple-red foliage and good autumn colours. Very dis-

Acer palmatum 'Dissectum' in autumn

sected leaves. Slightly less than average height and spread. *A. p. 'Dissectum Filigree'* Light green foliage, deeply divided, with silver bar down upper surface of each vein. Light grey-green arching stems. Hard to find. *A. p. 'Dissectum Filigree Lace'* Purple to purple-red foliage, deeply divided. Highly susceptible to wind and sun scorch. Slightly less vigorous. Good autumn colour. *A. p. 'Dissectum Garnet'* A stronger growing variety with very good dark purple dissected leaves. Good autumn colours. *A. p. 'Dissectum Inabashidare'* Good purple foliage and good autumn colours. Very graceful, arching habit. Quicker and stronger growing than most. *A. p. 'Dissectum Nigrum'* One of the darkest of all the purple-leaved varieties. Slightly less vigorous than *A. p. 'Dissectum'*. *A. p. 'Dissectum Ornatum'* Bronze foliage in spring and summer. Good autumn colours. *A. p. 'Dissectum Rubrifolium'* Bright red leaves. Good autumn colours. *A. p. 'Dissectum Variegatum'* A rare variety with white and green variegated leaves. Light shade and protection from frosts are essential. Slightly less than average height and spread.

Acer palmatum 'Dissectum Garnet' in summer

Average height and spread
Five years
2½x3ft (80cmx1m)
Ten years
4x5ft (1.2x1.5m)
Twenty years
or at maturity
5x6ft (1.5x2m)

ACER PALMATUM VAR. HEPTALOBUM

SEVEN-LOBED LEAVED JAPANESE MAPLE
Aceraceae
Deciduous
More robust Maples than other *palmatum* forms, but summer foliage colour is less intense.

Origin From Japan.
Use As a large shrub, ideal for woodland or near water, for autumn colour.
Description *Flower* Small, hanging clusters of red flowers in spring. *Foliage* Leaves light to olive green, seven-lobed, palmate, 3-5in (8-12cm) long and wide, turning orange and red in autumn. Some varieties purple leaved in spring and summer. *Stem* Upright when young, spreading with age. Light to mid green with orange shading, forming a rounded shrub. Medium growth rate. *Fruit* Small, red-green aging to grey-brown winged fruits.
Hardiness Tolerates winter temperatures down to 4°F (−15°C).
Soil Requirements Does well on both neutral and acid soils; tolerates alkalinity given adequate moisture.
Sun/Shade aspect Best in light dappled shade,

Acer palmatum var. heptalobum 'Osakazuki' in autumn

but tolerates more sun than *A. palmatum* or *A. p. 'Dissectum'* and their varieties.
Pruning None required. If branches removed paint cuts with pruning compound.
Propagation and nursery production From grafting on to *A. palmatum*. Purchase container-grown or root-balled (balled-and-burlapped). Obtain from specialist nurseries. Best planting heights 2-4ft (60cm-1.2m).
Problems Few, although leaves may be damaged by frost, scorched by sun or lacerated by hailstones or gales.
Varieties of interest *A. p. h. 'Elegans Purpureum'* Dark bronze-crimson to purple foliage, turning orange-red in autumn. *A. p. h. 'Lutescens'* Light yellow to mid yellow foliage. Very scarce. *A. p. h. 'Osakazuki'* Olive green foliage with purple shading; one of the best of all Maples for brilliant red autumn colour. *A. p. h. 'Rubrum'* Blood red leaves in spring, turning to green-red through summer; good autumn colour.
Average height and spread
Five years
6x6ft (2x2m)
Ten years
10x10ft (3x3m)
Twenty years
or at maturity
13x13ft (4x4m)

ACER PALMATUM 'LINEARILOBUM'

FIVE-FINGERED JAPANESE MAPLE

Aceraceae
Deciduous
A very attractive foliage form well worth adding to any collection of Japanese Maples.

Origin From Japan.
Use As a single shrub of interesting leaf formation.
Description *Flower* Small, hanging clusters of red flowers may be produced on mature plants. *Foliage* Leaves formed of five, narrow lobes 3-5in (8-12cm) long, resembling fingers. Good orange autumn colour. *Stem* Upright when young, becoming arching to form dome-shaped plant. Medium growth rate. *Fruit* Small, hanging clusters of green winged fruits, orange to orange-brown in autumn.
Hardiness Tolerates winter minimum of 4°F (−15°C).
Soil Requirements Prefers neutral to acid soil; tolerates alkalinity given adequate moisture.
Sun/Shade aspect Plant out of midday summer sun to prevent leaf scorch.
Pruning None required.
Propagation and nursery production From grafting on to *A. palmatum*. Purchase container-grown or root-balled (balled-and-

burlapped). Not easy to find; obtainable from specialist nurseries. Best planting heights 1½-2½ft (50-80cm).
Problems None.
Varieties of interest *A. p. 'Atropurpureum'* Purple-leaved variety with the same leaf formation and good autumn colours. Slightly less vigorous.

Acer palmatum 'Linearilobum' in autumn

Average height and spread
Five years
5x5ft (1.5x1.5m)
Ten years
6x6ft (2x2m)
Twenty years
or at maturity
8x8ft (2.5x2.5m)

ACER PALMATUM 'SENKAKI'

CORAL-BARKED MAPLE

Aceraceae
Deciduous
The stem of this shrub, bright coral in winter, is its chief virtue, most attractive given the right position.

Origin From Japan.
Use As single upright shrub for winter effect.
Description *Flower* Small, hanging, red flowers, mainly in early spring. *Foliage* Leaves small to medium, 2-3in (5-8cm) long and wide, palmate, tooth-edged. Light salmon-green when young, aging to light green to light orange. Good yellow autumn colour.

Stem Upright when young, becoming more twiggy. Stems vivid coral-red throughout winter. Medium growth rate. *Fruit* Small, winged fruits on mature shrubs.
Hardiness Tolerates winter minimum of 4°F (−15°C).
Soil Requirements Prefers neutral to acid soil; tolerates alkalinity given adequate moisture.
Sun/Shade aspect Plant out of midday summer sun to prevent leaf scorch.
Pruning None required.

Acer palmatum 'Senkaki' in winter

Propagation and nursery production By grafting on to *A. palmatum*. Plant container-grown or root-balled (balled-and-burlapped). Obtainable from specialist nurseries. Best planting heights 15in-3ft (40cm-1m).
Problems Liable to coral spot; remove diseased wood and treat affected area with pruning compound.
Average height and spread
Five years
5x2½ft (1.5mx80cm)
Ten years
6x4ft (2x1.2m)
Twenty years
or at maturity
13x6ft (4x2m)

ACER PALMATUM 'VERSICOLOR'

VARIEGATED JAPANESE MAPLE

Aceraceae
Deciduous
The white and pink variegation on the pale green leaves of this shrub is of particular interest.

Origin From Japan.
Use As an unusual variegated Japanese Maple in a group of Maples or other shrubs.
Description *Flower* Mature plants may produce small, hanging clusters of red flowers in autumn. *Foliage* Leaves palmate, 2-3in (5-8cm) long and wide, with five lobes. Light green with white and pink-splashed variegation. Some yellow autumn colour. *Stem* Thin, purple-green. Upright when young, becoming more twiggy to form a moderately upright, goblet-shaped shrub. Slow to medium growth rate. *Fruit* Mature shrubs may in some summers produce small winged light green fruits.
Hardiness Minimum winter temperature 23°F (−5°C).
Soil Requirements Prefers neutral to acid soil; tolerates high alkalinity given adequate moisture.
Sun/Shade aspect Plant out of summer sun to prevent leaf scorch.
Pruning None required but remove winter die-back as soon as obvious in spring.

Propagation and nursery production From grafting on to *A. palmatum*. Purchase container-grown or root-balled (balled-and-burlapped). Hard to find, but obtainable from specialist nurseries. Best planting heights 15in-3ft (40cm-1m).
Problems Liable to severe die-back from winter frost damage. Foliage may begin to revert to green and any signs of this should be removed.
Varieties of interest *A. p. 'Butterfly'* A very attractive variety, the centre of each leaf being striped white and pink. Very susceptible to leaf scorch and late spring frosts and protection must be given. Approximately two-thirds average height. *A. p. 'Pink Edge'* Palmate leaves splashed pink and white on light green base. Less vigorous and attaining two-thirds height height and spread. *A. p. 'Ukigumo'* More variegated colour, but reversion can be a problem. Scarce.

Acer palmatum 'Ukigumo' in leaf

Average height and spread
Five years
4x3ft (1.2x1m)
Ten years
5½x5ft (1.8x1.5m)
Twenty years
or at maturity
6x5½ft (2x1.8m)

AESCULUS PARVIFLORA

BUCKEYE, BOTTLEBRUSH BUCKEYE
Hippocastanaceae
Deciduous
A handsome flowering and foliage shrub of architectural shape.

Origin From south-eastern USA.
Use For feature planting, as a large specimen for a shrub border or as a single isolated specimen.
Description *Flower* Flowers in upright, loose spikes, creamy white to white, 8-12in (20-30cm) long. *Foliage* Leaves palmate, olive green with red shading and veins, dull silver undersides. Good yellow autumn colour. *Stem* After 10 years or more forms a two-tiered effect. The knotted, slightly bent upright stems form a canopy at 10-13ft (3-4m) bearing the leaves, surmounted by the flower panicles. Below this canopy, short shoots, 1½-2½ft (50-80cm) long, grow and put out leaves one-third larger than those of the upper adult foliage. These shoots mostly die after a season or two, but are rapidly replaced from ground level, thus achieving the two-tiered effect. If left unattended may grow into a wide-based, suckering thicket. Medium growth rate. *Fruit* Small, grey-brown fruits very erratically produced.
Hardiness Tolerates temperatures down to −13°F (−25°C).
Soil Requirements Any soil, although severe

Aesculus parviflora in flower

alkalinity will induce chlorosis. Extremely dry soils may slow down growth and impair performance.
Sun/Shade aspect Thrives in quite deep shade or full sun, provided moisture is adequate.
Pruning None required. Remove any dead shoots or limbs at maturity.
Propagation and nursery production Propagated by layers or stooling. Purchase container-grown or root-balled (balled-and-burlapped). Not always easy to find. Best planting heights 15in-3ft (40cm-1m).
Problems None, except that it is slow to show its true beauty.
Average height and spread
Five years
5x5ft (1.5x1.5m)
Ten years
10x10ft (3x3m)
Twenty years
or at maturity
13x13ft (4x4m)

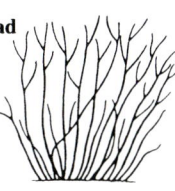

AMELANCHIER LAMARCKII
(Amelanchier laevis)

ALLEGHENY SERVICEBERRY, SNOWY MESPILUS, JUNE BERRY, SERVICE BUSH
Aceraceae
Deciduous
An extremely attractive spring-flowering shrub, worth a place in any medium-sized to large garden.

Origin From North America and Canada; now naturalized throughout most of Europe.
Use As a large shrub standing alone or for shrub borders; it can also be purchased trained as a tree.
Description *Flower* Loose, open racemes of white flowers produced profusely, late spring,

before or just after leaves appear. *Foliage* Leaves ovate, 1½-3in (4-8cm) long, light green, sometimes with slight orange-red veining and shading; good autumn colour. *Stem* Strong, upright when young, becoming twiggy and arching, produced from ground level annually as suckers and quickly forming a large round-topped shrub. *Fruit* Small, light red fruits produced on old flowering racemes in autumn, especially after a good summer.
Hardiness Tolerates winter temperatures below −13°F (−25°C).
Soil Requirements Any soil; tolerates high alkalinity and acidity.
Sun/Shade aspect Prefers full sun, but tolerates shade well.
Pruning None required. May be cut back hard if desired.
Propagation and nursery production From layers or removal of a sucker from base of shrub. Plant bare-rooted or container-grown. Easy to find under one of its various names. Best planting heights 2-4ft (60cm-1.2m).
Problems None.
Varieties of interest *A. 'Ballerina'* A new, compact-growing form with large, white flowers.
Average height and spread
Five years
10x10ft (3x3m)
Ten years
16x16ft (5x5m)
Twenty years
or at maturity
20x20ft (6x6m)

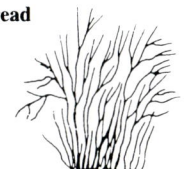

AMORPHA FRUTICOSA

FALSE INDIGO, INDIGO BUSH AMORPHA, BASTARD INDIGO
Leguminosae
Deciduous
A rarely-seen, summer-flowering shrub for the large garden.

Origin From southern USA.
Use For a sunny shrub border where there is room for its ultimate height and spread.
Description *Flower* Racemes of purple-blue flowers with yellow centres, in summer. *Foliage* Grey to grey-green, pinnate leaves 1-2in (3-5cm) long, giving yellow autumn colour. *Stem* Grey-green, upright when young, arching and becoming twiggy to form rounded shape. Medium growth rate. *Fruit* Insignificant.
Hardiness Tolerates 14°F (−10°C).
Soil Requirements Tolerates any soil, but the richer and damper the soil, the more growth and hence more flowers.
Sun/Shade aspect Full sun; dislikes any shade.
Pruning Prune back hard in spring to encourage foliage and flowers.

Amelanchier lamarckii in autumn

Amorpha fruticosa **in flower**

Propagation and nursery production From softwood cuttings or seed. Purchase container-grown. May need to be sought from specialist nurseries except in its native environment. Best planting heights 1-2½ft (30-80cm).
Problems Rather untidy habit of growth.
Varieties of interest *A. canescens* (Lead Plant) A smaller shrub with violet flowers and small acacia-type foliage, 2-3in (5-8cm) long. From eastern North America.
Average height and spread
Five years
5x5ft (1.5x1.5m)
Ten years
8x8ft (2.5x2.5m)
*Twenty years
or at maturity*
10x10ft (3x3m)

ANDROMEDA POLIFOLIA

BOG ROSEMARY
Ericaceae
Evergreen
An unusual flowering plant for moist, acid soils; now the only plant in this genus which formerly included *Leucothoe*, *Oxydendrum* and *Pieris*.

Origin From Europe, northern Asia and North America.
Use For use in an acid garden in association with heathers and other ericaceous plants such as rhododendrons, azaleas and mountain laurels.
Description *Flower* Clusters of soft pink to purple-pink hanging bell-shaped flowers, mid to late spring. *Foliage* Leaves slender, narrow, glaucous green, 1-1½in (3-4cm) long, with blue, hairless upper sides and white downy undersides. *Stem* Slender, arching branches, becoming twiggy and mat-forming with age. Slow-growing. *Fruit* Insignificant.
Hardiness Minimum winter temperature 23°F (-5°C).
Soil Requirements Must have moist, rich acid soil. Resents any alkalinity.
Sun/Shade aspect Prefers very light shade; tolerates full sun if moisture adequate.
Pruning None required.
Propagation and nursery production From softwood cuttings taken in late summer. Purchase container-grown or root-balled (balled-and-burlapped). Obtainable from specialist growers. Best planting heights 8-15in (20-40cm).
Problems None.
Varieties of interest *A. glaucophylla* Leaves have very downy undersides; pitcher-shaped flowers produced slightly later than those of *A. polifolia*. *A. p. 'Compacta'* A dwarfer, more compact form.

Andromeda polifolia 'Compacta' **in flower**

Average height and spread
Five years
2x2ft (60x60cm)
Ten years
3x4ft (1x1.2m)
*Twenty years
or at maturity*
3x5ft (1x1.5m)

ARALIA ELATA

JAPANESE ANGELICA TREE, JAPANESE ARALIA
Araliaceae
Deciduous
A truly handsome plant for foliage effect.

Origin From Japan.
Use As a large architectural plant to stand on its own.
Description *Flower* Very large upright panicles, up to 2ft (60cm), of white flowers with branching side panicles, borne in central terminal of each mature branch, early autumn. *Foliage* Leaves olive green to grey-green, very large, pinnate, borne towards ends of upright stems. Leaves can be as much as 2-4ft (60cm-1.2m) long and 3ft (1m) wide on good soil conditions. *Stem* Strong, upright stems lightly branched, grey to grey-green, covered in short, hairy spines. Of winter architectural value. Fast-growing when young, slows with age. *Fruit* Insignificant.
Hardiness Tolerates winter temperatures down to 4°F (-15°C).
Soil Requirements Any soil, although may show signs of chlorosis on alkaline soils. Dry, poor soils may hinder development.
Sun/Shade aspect Prefers full sun, tolerates very light shade.
Pruning None required.
Propagation and nursery production From seed. May produce root suckers that can be transplanted. Purchase container-grown or root-balled (balled-and-burlapped). Variegated forms may be more difficult to find than green-leaved types. Best planting heights 15in-3ft (40cm-1m).
Problems None.
Varieties of interest *A. e. 'Aureovariegata'* A very attractive variety with golden variegated leaves, slightly more tender than the parent. Difficult and expensive to obtain. *A. e. 'Variegata'* Leaves margined and blotched with yellow-white variegation. Slightly less vigorous than the green form, but makes a spectacular display if well placed. Slightly more tender than the parent and much more difficult to find.
Average height and spread
Five years
6x5ft (2x1.5m)
Ten years
10x8ft (3x2.5m)
*Twenty years
or at maturity*
12x10ft (3.5x3m)

Aralia elata **in leaf and flower**

ARCTOSTAPHYLOS UVA-URSI

RED BEARBERRY, KINNIKINICK
Ericaceae
Evergreen
An interesting ground cover subject if the right soil conditions are available.

Origin Widespread as an alpine shrub throughout the cooler areas of the northern hemisphere.
Use As a low, creeping shrub for ground cover together with other ericaceous plants, such as rhododendrons and azaleas.
Description *Flower* Small, white, tinged pink, hanging bell-shaped flowers, late spring, early summer. *Foliage* Leaves small, ovate, ½-1in (1-3cm) long, olive green with some red markings to the veins. *Stem* Short, twiggy, forming a flat, creeping plant. Slow growth rate. *Fruit* Small, round, slightly glossy, red fruits in autumn.
Hardiness Tolerates 4°F (−15°C).
Soil Requirements Requires acid soil. Will not tolerate chalky, alkaline types.
Sun/Shade aspect Needs full sun; resents shade.
Pruning None required.

Arctostaphylos uva-ursi in flower

Propagation and nursery production From softwood cuttings taken in summer or from self-rooted layers lifted from parent plant. Purchase container-grown or root-balled (balled-and-burlapped). Fairly hard to find, but obtainable from specialist nurseries. Best planting heights 4-8in (10-20cm).
Problems Soil type very limited, restricting areas in which shrub may be planted.
Average height and spread
Five years
1½x2ft (50x60cm)
Ten years
1½x3ft (50cmx1m)
Twenty years
or at maturity
1½x4ft (50cmx1.2m)

ARONIA ARBUTIFOLIA

RED CHOKEBERRY
Rosaceae
Deciduous
A group of acid-loving plants well worth growing for their rich autumn foliage and edible fruits.

Origin From eastern North America.
Use As a flowering shrub with good autumn colour for planting among other ericaceous plants, such as rhododendrons and azaleas.
Description *Flower* Medium-sized, single white flowers borne in late spring. *Foliage*

Aronia prunifolia in autumn

Leaves ovate, 1½-3in (3-8cm) long, olive green with red to purple shading, brilliantly coloured in autumn. *Stem* Light to mid green shaded red. Upright, becoming twiggy at the ends with age, to form an upright, spreading, suckering shrub of medium vigour. *Fruit* Red, edible fruits in autumn.
Hardiness Tolerates winter temperatures down to −13°F (−25°C).
Soil Requirements Acid soil, dislikes any alkalinity.
Sun/Shade aspect Thrives in light shade to full sun, but tolerates quite heavy shade.
Pruning None required. May be pruned to reduce size.
Propagation and nursery production From rooted suckers or layers. Purchase container-grown. Available from specialist nurseries. Best planting heights 2-3ft (60cm-1m).
Problems May become a little invasive.
Varieties of interest *A. 'Erecta'* A more upright variety, otherwise similar to the parent. *A. melanocarpa* (Black Chokeberry) White flowers, good autumn colour and black berries. From eastern North America. *A. m. 'Brilliant'* Exceptionally fine autumn colouring. Of garden origin. *A. prunifolia* (Purple Chokeberry) A cross between *A. arbutifolia* and *A. melanocarpa* with purple black fruits. Good orange-red autumn colour. From south-eastern USA.
Average height and spread
Five years
5x4ft (1.5x1.2m)
Ten years
6x5½ft (2x1.8m)
Twenty years
or at maturity
6x8ft (2x2.5m)

ARTEMISIA ABROTANUM

SOUTHERNWOOD, LAD'S LOVE
Compositae
Deciduous
A very attractive, aromatic foliage shrub for general use in the garden.

Origin From southern Europe.
Use As an aromatic grey shrub for low borders and herb gardens.
Description *Flower* Silver-yellow flowers, early summer. *Foliage* Leaves grey-green, very sweetly aromatic, finely divided, 1-2½in (3-6cm) long and ½in (1cm) wide. *Stem* Grey-green, upright, with foliage closely bunched up entire stem. Rarely branching, new growth from ground level each spring. Fast growth rate. *Fruit* Insignificant.
Hardiness Winter minimum 14°F (−10°C).
Soil Requirements Any soil, but prefers well-drained area.
Sun/Shade aspect Best in full sun or very light shade. Deep shade leads to open shape.

Pruning When established prune back hard each spring to encourage new growth and improve overall appearance; also benefits from regular light trimming.
Propagation and nursery production From softwood cuttings taken mid to late summer. Always purchase container-grown; easy to find in nurseries and garden centres. Best planting heights 15in-2½ft (40-80cm).
Problems Can become unattractively woody and spreading if not cut back annually or triennially.
Varieties of interest *A. arborescens* A shrubby variety with a mass of filigree silver leaves. Slightly less vigorous than *A. abrotanum*.

Artemisia abrotanum in leaf

Average height and spread
Five years
2½x2½ft (80cmx80cm)
Ten years
2½x4ft (80cmx1.2m)
Twenty years
or at maturity
2½x4ft (80cmx1.2m)

ATRIPLEX HALIMUS

TREE PURSLANE, SALT BUSH
Chenopodiaceae
Semi-evergreen
A silver-leaved shrub often overlooked, but worth inclusion in a collection of attractive foliage shrubs.

Origin Native to southern Europe where it grows wild on cliffs and sandy areas.
Use To form a background in a silver ornamental border or as contrast with purple-leaved shrubs.

Atriplex halimus in leaf

Description *Flower* Small, sulphur yellow flowers, late spring to early summer. *Foliage* Ovate, ½-2½in (1-6cm) long, silver-grey leaves producing very attractive foliage effect. Some yellowing of older foliage when it dies contrasts well with young grey foliage. *Stem* Grey to grey-green, upright, becoming twiggy and arching to form a loose, straggly, moderately upright shrub of medium growth rate. *Fruit* Insignificant.

Hardiness Minimum temperature range 14°F (−10°C) to 4°F (−15°C).

Soil Requirements Prefers light, well-drained soil; tolerates other types.

Sun/Shade aspect Best in full sun.

Pruning None required and may be resented.

Propagation and nursery production From softwood cuttings taken in late summer. Purchase container-grown. Relatively easy to find from nurseries and larger garden centres. Best planting heights 15in-2½ft (40-80cm).

Problems May become a little straggly in the wrong soil conditions, and unable to produce new silver foliage.

Average height and spread
Five years
4x4ft (1.2x1.2m)
Ten years
4x5½ft (1.2x1.8m)
Twenty years
or at maturity
4x5½ft (1.2x1.8m)

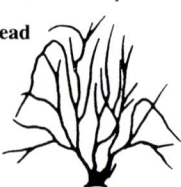

AUCUBA JAPONICA

JAPANESE AUCUBA, SPOTTED LAUREL, HIMALAYAN LAUREL, JAPANESE LAUREL
Cornaceae
Evergreen
A truly stately evergreen, especially the new introductions becoming available, excellent for general garden planting.

Origin From Japan.

Use As a large evergreen for difficult, shady areas, used massed or singly.

Description *Flower* Monoecious. Sulphur-yellow panicles produced in late spring. Each variety carries either male or female flowers, which are similar in appearance; for fruiting, plants of both sexes are needed. *Foliage* Leaves dark, glossy green, lanceolate 3-8in (8-20cm) long and 1-3in (3-8cm) wide. *Stem* Bright green and glossy. Strong, upright and branching, forming a round-topped shrub. Medium growth rate. *Fruit* On female flowering forms, bright red, round fruits in racemes appear from autumn and remain through winter and possibly into spring. Produced only if male plant grows nearby.

Hardiness Withstands −13°F (−25°C), but some damage may be caused by wind chill.

Soil Requirements Tolerates almost any soil, including dry and alkaline.

Sun/Shade aspect Dislikes full sun; tolerates very deep shade.

Pruning None required, but may be cut back hard to control size.

Propagation and nursery production From softwood cuttings taken in mid to late sum-

mer. Purchase container-grown. Easy to find; many new variegated forms becoming available.

Problems None, apart from wind chill hazard.

Varieties of interest *A. j. 'Crotonifolia'* A variety with spotted and blotched golden leaves. Slightly less vigorous than *A. japonica*. Male. *A. j. 'Mr Goldstrike'* A new golden variegated variety with red berries. Female. *A. j. 'Picturata'* Dark green leaves boldly splashed chrome yellow, slightly less vigorous than *A. japonica*. Male. *A. j. 'Salicifolia'* A green-leaved form with very narrow, toothed-edged, dark green foliage. Freely fruiting but not easy to find. Female. *A. j. 'Variegata'* Leaves splashed golden and yellow well over half their area. One of the most variegated forms. Female. *A. j. 'Variegata Gold Dust'* A very good form with golden variegated foliage and red berries in autumn. Female.

Average height and spread
Five years
4x4ft (1.2x1.2m)
Ten years
5½x5½ft (1.8x1.8m)
Twenty years
or at maturity
13x13ft (4x4m)

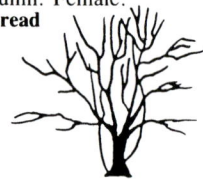

AZARA DENTATA

KNOWN BY BOTANICAL NAME
Flacourtiaceae
Evergreen
These scented flowering evergreens are on the tender side, requiring protection in winter.

Origin From Chile.

Use As a freestanding, evergreen shrub in mild areas, or wall shrub in colder climates. Can be grown in tubs in medium-sized conservatories or greenhouses.

Description *Flower* Clusters of fragrant yellow flowers, early spring, borne in profusion. *Foliage* Leaves ovate or oblong, 1-1½in (1-4cm), bright green to dark glossy green with felted undersides. *Stem* Light green to mid green. Upright when young, becoming more twiggy with age and spreading. Moderate rate of growth. *Fruit* Insignificant.

Hardiness Winter minimum 23°F (−5°C).

Soil Requirements Does well on most soils but dislikes excessive alkalinity and waterlogging.

Sun/Shade aspect Tolerates full sun to mid shade.

Pruning None required.

Propagation and nursery production From softwood cuttings taken in midsummer. Purchase container-grown. Obtainable from specialist nurseries. Best planting heights 2-3ft (60cm-1m).

Problems None.

Varieties of interest *A. lanceolata* Narrow, lanceolate leaves and mustard yellow flowers in early summer which are as fragrant as those of *A. dentata*. Possibly more tender. *A. microphylla* In mild areas forms a small tree up to 23ft (7m) in height and 20ft (6m) in spread after 20 years. Small, round or ovate leaves, ½-1in (1-3cm) long, giving a fan effect. Yel-

Azara dentata in flower

low flowers, vanilla scented, produced on the undersides of the twigs in early spring. One of the hardiest forms. *A. m. 'Variegata'* A form of *A. microphylla* with cream variegated leaves. Slower growing than its parent and more tender. *A. serrata* Often confused with *A. dentata*, producing similar scented flowers under the edges of each leaf. Leaves more serrated. In hot climates, or in hot summers, small white berries may be produced. One of the hardier forms.

Average height and spread
Five years
5x5ft (1.5x1.5m)
Ten years
8x8ft (2.5x2.5m)
Twenty years
or at maturity
12x12ft (3.5x3.5m)

BALLOTA PSEUDODICTAMNUS

KNOWN BY BOTANICAL NAME
Labiatae
Evergreen
A very attractive edging or filling shrub, a fine feature for a white or grey planting scheme.

Origin From the Mediterranean area, through Greece and Crete.

Use As a low-growing shrub for summer foliage effect. Can be used as a low, informal hedge planted at 2ft (60cm) apart, as an edging for rose borders, or as a bedding plant for hanging baskets and tub displays.

Description *Flower* Whorls of lilac pink flowers, midsummer. *Foliage* Leaves round, 1-1½in (3-4cm) long and wide, silver-grey to grey-white, the stem being covered in a white down. *Stem* Soft, upright, grey, down-covered stems, forming a neat, low, bun-shaped plant. Fast growth rate in spring. *Fruit* Insignificant.

Hardiness Minimum winter temperature 32°F (0°C). Natural die-back occurs in winter, and if the plant is frosted this may be terminal.

Soil Requirements Requires a good, rich, light soil for maximum new foliage and growth. Dislikes waterlogged or very dry soils.

Sun/Shade aspect Full sun. Any shade will detract from the grey foliage and hinder growth.

Pruning Prune surviving plants back hard in mid spring, just as growth starts, to encourage a compact shape. Otherwise will become very straggly, open and deformed.

Propagation and nursery production From soft, semi-ripe cuttings taken in midsummer and over-wintered under protection. A cutting planted out in spring rapidly increases in size and produces usable plant within first few weeks. This ensures that replacements are

Aucuba japonica 'Crotonifolia' in winter

Ballota pseudodictamnus in leaf

available in case the established plant dies. Purchase container-grown and plant out directly. Best planting heights 4-12in (10-30cm).

Problems As this is a tender sub-shrub, producing no true woody structure, in winter it may die back to ground level and will not survive in colder areas. In mild areas it may survive unharmed.

Average height and spread
Five years
1½ftx8in-2ft (50x20-60cm)
Ten years
1½ftx8in-2ft (50x20-60cm)
Twenty years
or at maturity
1½ftx8in-3ft (50x20cm-1m)

BERBERIS AGGREGATA

Origin From western China.
Use As an individual shrub or effective for mass planting. If planted at 2½ft (80cm) apart in a single line makes an informal, impenetrable hedge.
Description *Flower* Small, yellow, hanging, semi-double to double flowers, late spring, early summer. *Foliage* Leaves small, ½-¾in (1-2cm) long, ovate, light green, turning good orange-red in autumn. *Stem* Upright when young, becomes very twiggy and dense with

sharp spines at each leaf joint. Moderate growth rate. *Fruit* Pale deep orange clusters of succulent small fruits produced on wood 2 years old; one of the main attractions.
Hardiness Tolerates winter temperatures down to −13°F (−25°C).
Soil Requirements Tolerates almost any, including alkaline and acid, wet and dry.
Sun/Shade aspect Equally tolerant of full sun and fairly deep shade.
Pruning Remove one-third of wood 2 years old or more annually after fruiting, late autumn or winter. Handle carefully, the thorns are sharp.
Propagation and nursery production From semi-ripe cuttings taken in early to midsummer. Seed will germinate but resulting plants can be variable in performance. Plant bare-rooted or container-grown. From general nurseries and sometimes garden centres. Best planting heights 15in-2½ft (40-80cm).
Problems Spiny stems make handling difficult.
Varieties of interest *B. 'Buccaneer'* Orange-red fruit of good size and uniformity. Light to mid green foliage, good autumn colour. Upright habit. *B. 'Pirate King'* Dull red to purple-red, oval berries with grey surface covering. Mid green foliage, giving orange-red autumn colour. Stems red-brown. Upright in habit to average height with slightly less spread. May be hard to find.
Average height and spread

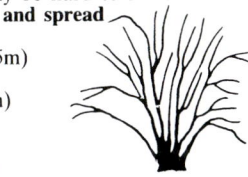

Five years
5½x5ft (1.8x1.5m)
Ten years
6½x6ft (2.2x2m)
Twenty years
or at maturity
10x10ft (3x3m)

Berberis aggregata in fruit

BERBERIS DARWINII

Origin From Chile and Argentina. First discovered by the naturalist Charles Darwin, from whom it takes its name, on the voyage of the 'Beagle' in 1835.
Use As a spring-flowering evergreen for shrub borders or for mass planting. Plant 2½ft (80cm) apart in single line for informal hedge.
Description *Flower* Bright orange, hanging, double, cup-shaped flowers borne in clusters on wood 2 years old or more in great profusion in early to mid spring. *Foliage* Leaves ovate, ½-1in (1-3cm) long, with five-pointed edges, olive green upper surfaces, silver undersides. Although evergreen some of the leaves of very old branches may turn red in autumn when old leaves are discarded. *Stem* Light to mid green, aging to green-brown. Strong, upright when young, becoming spreading and twiggy with age. Small spines, flexible when young, at most leaf and stem axils, which become woody with age. Medium growth rate. *Fruit* Clusters of blue-black hanging fruits on mature wood in autumn.
Hardiness Unlikely to survive temperatures below 14°F (−10°C). Young new growth may be damaged by late spring frosts, but without lasting effect.
Soil Requirements Does well in most, but dislikes thin, chalk soils or very dry areas.
Sun/Shade aspect Prefers light dappled shade, tolerates sun or shade.

Berberis darwinii in flower

Pruning On mature shrubs remove some wood 3-4 years old to ground level each spring to encourage rejuvenation. After hard spring pruning flowering will be reduced during the 2-3 years it takes to rejuvenate.
Propagation and nursery production From cuttings taken in midsummer. Buy plants raised from cuttings as seed-raised stock can be very variable. Plant container-grown. Available from garden centres and general nurseries. Best planting heights 15in-2½ft (40-80cm).
Problems Whole mature plants, or sections may die very suddenly in summer, perhaps due to damage to roots by drying out or waterlogging but the reason is not fully understood.
Average height and spread
Five years
5x5ft (1.5x1.5m)
Ten years
7x6ft (2.2x2m)
Twenty years
or at maturity
10x10ft (3x3m)

BERBERIS DICTYOPHYLLA
(Berberis dictyophylla 'Albicaulis')

WHITE-STEMMED BARBERRY
Berberidaceae
Deciduous
A rarely seen form of *Berberis*, the white stems making an effective winter display.

Origin From western China.
Use As a medium height shrub for winter stem display.
Description *Flower* Semi-double, yellow, hanging, cup-shaped flowers borne singly along stem on wood 2 years old or more. *Foliage* Leaves grey-white to grey-green with red-tinged veining, lanceolate, ½-1in (1-3cm) long with white underside. *Stem* Upright red stems with a white bloom covering, the most attractive part of the plant. Small, pliable thorns at leaf axils. Slow to medium growth rate. *Fruit* Small, red fruits covered in grey-white bloom, sparsely produced in autumn.
Hardiness Winter minimum 4°F (−15°C).
Soil Requirements Good, deep, rich soil to encourage annual production of new, strong, white winter stems.

Berberis dictyophylla in leaf

Sun/Shade aspect Full sun to light shade. Needs sun or good light to show its winter display to full advantage.
Pruning Rarely required, but remove very old wood to ground level occasionally to aid rejuvenation of stems.
Propagation and nursery production Very difficult to propagate and hard to obtain; raised from cuttings taken during summer months. Best planting heights 8in-2ft (20-60cm).
Problems None, except its scarcity and its rather slow growth habit.
Varieties of interest *B. temolaica* A stronger growing form displaying the same white bloom on its young upright stems. On older stems the bloom turns purple-brown in winter. Egg-shaped berries, covered with a grey bloom, in autumn. Remove one-third of the oldest wood each spring to encourage new white growth. From south-eastern Tibet.
Average height and spread
Five years
3x2½ft (1mx80cm)
Ten years
6x4ft (2x1.2m)
Twenty years
or at maturity
10x5ft (3x1.5m)

BERBERIS JULIANAE
Tall-growing evergreen forms

WINTERGREEN BARBERRY
Berberidaceae
Evergreen
A handsome evergreen worthy of wider planting.

Origin From China.
Use As a big, attractive, evergeen shrub for large group planting, screening or mass planting. Plant 3ft (1m) apart in single line for a tall evergreen hedge or screen.
Description *Flower* Double, cup-shaped, lemon-yellow, hanging flowers borne singly or in clusters, late spring to early summer. *Foliage* Leaves olive green, lanceolate, 3-4in (8-10cm) long and 1in (3cm) wide, with spiny edges. Although evergreen, discards some older leaves in autumn and these may turn vivid orange or orange-red in autumn, contrasting well with younger olive green foliage. *Stem* Strong, moderately vigorous and upright when young, becoming slower, arching and twiggy with age. Light green, aging to brown-green. Very long, dark green, sharp thorns borne triply at each leaf axil. *Fruit* Hanging, cylindrical, blue to blue-black with downy texture in profusion on wood 2 years old in autumn.
Hardiness Withstands temperatures down to −13°F (−25°C), although late spring frosts can damage new shoots; but with no lasting effect.
Soil Requirements Most soils, but resents extreme dryness.
Sun/Shade aspect Full sun to fairly deep shade.
Pruning None required. May be reduced in size and will regenerate the following spring.
Propagation and nursery production From cuttings taken in summer or from seed. If possible buy stock raised from cuttings from a known good parent form rather than seed-produced plants. Plant container-grown. Best planting heights 16in-2½ft (40-80cm). Generally available.
Problems Sharp thorns make close cultivation difficult. Can be slow to establish itself.
Varieties of interest *B. atrocarpa* Similar to *B. julianae*, more vigorous though slightly less hardy, and with longer, lighter green, lanceolate leaves, very long, pale green, sharp thorns, yellow flowers and blue fruit. Height and spread at maturity up to 3ft (1m) more. Fast-growing when young, slowing with age. *B. gagnepainii* Leaves holly-like, 2-3in (5-8cm) long, light olive to grey-green, attractively arranged on stem. Flowers and fruit similar to *B. julianae*. May produce underground suckers from base and in some soil conditions may even become invasive. Good

as evergreen screen. Medium growth rate. *B. knightii* Leaves deep olive green with slightly more spiny leaf-edges. Similar flowers and fruit. Very good as tall hedge if planted 3ft (1m) apart. May not be easy to find. Medium to fast growth rate. *B. sargentiana* Tough, leathery, oblong leaves, 2-3in (5-8cm) long with pronounced net-like veining. Yellow flowers and blue-black berries. Height and spread 6ft (2m). Slow to medium growth rate. From western China. *B. 'Walitch Purple'* A lower growing variety up to 5ft (1.5m) in height with a spread of 5ft (1.5m). Leaves holly-like, 1-1½in (3-4cm) long and 1½-2in (4-5cm) wide, olive green tinted with red. Yellow flowers and blue berries. Plant 2½ft (80cm) apart in single line to make attractive, impenetrable hedge. Slow to medium growth rate.
Average height and spread
Five years
6x6ft (2x2m)
Ten years
10x10ft (3x3m)
Twenty years
or at maturity
13x13ft (4x4m)

BERBERIS
Low-growing evergreen forms

KNOWN BY BOTANICAL NAME
Berberidaceae
Evergreen
Useful low evergreens for individual planting or for ground cover.

Origin From western China or of garden origin.
Use As low, evergreen shrubs. May 'be planted 2ft (60cm) apart in a single line to make a good, low hedge.
Description *Flower* Small, double, individual yellow flowers, hanging mainly from undersides of arching branches, mid to late spring. *Foliage* Leaves dark green, glossy, spiny, ovate, ½in-1in (1-3cm) long, silver undersides. Some good red autumn colour when old leaves are discarded. *Stem* Upright, light green, spiny when young, becoming arching and dense, to form a round, low mound. Slow growth rate. *Fruit* Small, oblong, black fruits in autumn.
Hardiness Survives winter temperatures down to −13°F (−25°C).
Soil Requirements Any soil conditions.
Sun/Shade aspect Full sun to deep shade.
Pruning None required. May be reduced in size.
Propagation and nursery production From softwood cuttings taken in midsummer. Plant

Berberis julianae in flower

82

container-grown. Best planting heights 8-15in (20-40cm). Generally available.
Problems Thorns may make cultivation of other plants nearby difficult. Takes several years to form an attractive bush of any size and patience is needed for good final results.
Varieties of interest *B.* × *bristolensis* Larger leaves, 1½-2½in (4-6cm) long. Good for a low hedge and may be clipped for this pur-

Berberis verruculosa **in flower**

pose. Faster growing than average, with slightly more height and spread. Of garden origin. *B. buxifolia 'Nana'* Semi-evergreen. Small, ovate, light green leaves, turning plum colour in autumn. The small, double, bright yellow flowers are borne singly at each leaf joint, followed by purple-blue grape-shaped berries in autumn. Forms a rounded shrub, height 1½ft (50cm) spread 6ft (2m) at maturity. Slow growth rate. *B. calliantha* Leaves much larger, 1½-2½in (4-6cm) long, holly-shaped but retaining silver undersides. New growth red. Flowers borne singly or in pairs on undersides of branches; black to blue-black fruits in autumn. Slow growth rate. From south-eastern Tibet. *B. candidula* (Paleleaf Barberry) A very compact form with light green foliage and forming a dome-shaped shrub. Leaves almost lanceolate with silvery-white undersides, bright yellow flowers borne singly, shy to fruit. Height up to 3ft (1m), spread 4ft (1.2m). Slow growth rate. From western China. *B. c. 'Amstelveen'* Small, vigorous habit of growth. Slightly arching branches. Good for mass planting or as ground cover. Moderate growth rate. Of European origin. *B. c. 'Telstar'* Very bright green foliage with tight, compact habit. Good for mass planting and ground cover. Moderate growth rate. *B. 'Chenaultii'* (Chenault Barberry) Ovate or lanceolate leaves, dull green above, with grey undersides and small spines around outer edges. Yellow double flowers hang from undersides of branches; blue-black berries in dry summers. Height and spread about 5ft (1.5m) at maturity. Moderate growth rate. *B. panlanensis* One of the best evergreen forms. Bright green, linear leaves with small spines, forming a neat, tight, bun-shaped shrub. Small yellow flowers and black berries. Plant 2ft (60cm) apart in straight line for low, informal hedge. Slow growth rate. From western China. *B. 'Park-juweel'* A cross between *B. gagnepainii* and *B. 'Chenaultii'*. Taller growing, with thorny stems, although ovate leaves are almost spineless. Some good autumn colour which often stays into early spring. Medium growth rate. Of garden origin. *B. verruculosa* Leaves dark, shiny upper surface with silver undersides. Semi-double, yellow flowers and black fruit. Arching habit. Slow growing. From western China.
Average height and spread
Five years
2x2½ft (60x80cm)
Ten years
3x5ft (1x1.5m)
Twenty years
or at maturity
4x6ft (1.2x2m)

BERBERIS LINEARIFOLIA

KNOWN BY BOTANICAL NAME
Berberidaceae
Evergreen
A spectacular shrub if planted in a slightly shaded position.

Origin From Argentina and Chile.
Use As an evergreen, spring-flowering shrub. Ideal for the woodland border or shady shrub border.
Description *Flower* Flowers bright orange, double, cup-shaped, hanging singly or sometimes in clusters, borne in profusion in mid to late spring. *Foliage* Leaves linear, 1½in (4cm) long, light green to mid green, with silver undersides. *Stem* Upright or laterally spreading new growth. Bright green, becoming twiggy, but still spreading with age. Moderately long, sparsely presented spines at some leaf axils. Medium growth rate. *Fruit* Ovate, hanging, blue-black fruits borne on mature wood, but shy to fruit.
Hardiness Tolerates 4°F (−15°C), but leaves may be damaged by severe wind chill.

Berberis linearifolia **in flower**

Soil Requirements Does well on most, but unhappy on very thin or dry soils.
Sun/Shade aspect Tolerates full sun, prefers medium to light shade.
Pruning None required. May safely be cut back moderately hard if desired.
Propagation and nursery production From cuttings taken in early spring to midsummer or grating on to *B. thunbergii var. atropurpurea*. Purchase container-grown or root-balled (balled-and-burlapped). Generally

Berberis × *lologensis* **in flower**

available. Best planting heights 15in-2½ft (40-80cm).
Problems Roots may be damaged by drought or excessive moisture, causing shrub to die. Can be very slow to establish. Prepare soil by adding liberal quantities of sedge peat or other organic compost to give plant best chance of success. Suckers on grafted plants must be removed.
Varieties of interest *B. l. 'Orange King'* Larger, more intensely orange flowers, otherwise similar to the parent.
Average height and spread
Five years
4x5½ft (1.2x1.8m)
Ten years
6x8ft (2x2.5m)
Twenty years
or at maturity
10x10ft (3x3m)

BERBERIS × LOLOGENSIS

KNOWN BY BOTANICAL NAME
Berberidaceae
Evergreen
Although often temperamental in establishing itself, this impressive shrub is well worth trying in the medium-sized to large garden.

Origin Of garden origin; a cross between *B. darwinii* and *B. linearifolia*.
Use As an upright shrub of spectacular effect for the shrub border or as a specimen plant.
Description *Flower* Comparatively large, cup-shaped, orange to bright orange flowers borne in great profusion up entire length of wood 2 years old or more in late spring. *Foliage* Leaves holly-like, tri-pointed ovate, 1in (2cm) long and wide, dark olive green with silver reverse. *Stem* Upright, strong, light green when young, aging to green-brown. Maintains an upright, stout column. Slow to medium growth rate. *Fruit* Attractive, ovate to round, light blue to blue-black fruit in some profusion in autumn.
Hardiness Tolerates 14°F (−10°C), although extreme wind chill may severely damage foliage.
Soil Requirements Does well on most, but unhappy on thin, chalk soils or in very dry areas.
Sun/Shade aspect Prefers light shade but tolerates full sun on moist soils.
Pruning None required.
Propagation and nursery production From cuttings taken in summer, but normally grafted on to *B. thunbergii* or *B. t. var. atropurpurea*, which may give some weak suckering growth. Purchase container-grown or root-balled (balled-and-burlapped).

Generally available. Best planting heights 15in-2½ft (40-80cm).
Problems Can be very slow to establish. Prepare soil by adding liberal quantities of sedge peat or other organic compost to give plant best chance of success.
Varieties of interest *B. × lologensis 'Gertrude Hardyzerii'* Very good dark orange flowers. Of European origin. *B. × lologensis 'Apricot Queen'* Apricot-orange flowers.
Average height and spread
Five years
3x2½ft (1mx80cm)
Ten years
6x4ft (2x1.2m)
Twenty years
or at maturity
10x5ft (3x1.5m)

BERBERIS × STENOPHYLLA

KNOWN BY BOTANICAL NAME
Berberidaceae
Evergreen
A useful range of evergreen, spring-flowering shrubs, offering a wide variation in height and habit.

Origin Of garden origin, with *B. darwinii* in its parentage.
Use As a single, spring-flowering shrub for the shrub border, for mass planting or, if planted 3ft (1m) apart in a single line, for an informal hedge.

Berberis × stenophylla in flower

Description *Flower* Yellow, double, cup-shaped, flowers hanging singly and in threes along arching branches, early to late spring. *Foliage* Leaves narrow, lanceolate, convexed, curving, ½-1in (1-3cm) long, olive green upper surfaces, silver-white undersides. *Stem* Light to bright green when young, aging to dark green and brown. Single or triple thorns at each leaf axil. Upright, becoming spreading and gracefully arching, forming large mound of equal height and spread. Some growth can occur at ground level, with shoots arising from underground stolons. Medium to fast growth rate. *Fruit* Small, oblong, blue to blue-black fruit borne freely along branches in some hot summers.
Hardiness Tolerates 4°F (−15°C) but winter wind chill may damage foliage.
Soil Requirements Does well on most, but unhappy on very dry or thin chalk soils.
Sun/Shade aspect Full sun to heavy deep shade.
Pruning If cut to ground level will rejuvenate in 2-3 years. May be clipped very hard when used as hedge.
Propagation and nursery production From cuttings taken in summer. Purchase container-grown; widely available. Best planting heights 1-2½ft (30-80cm).

Problems May be slow to establish.
Varieties of interest *B. × stenophylla 'Autumnalis'* Produces a second crop of flowers in autumn. A slower growing variety, attaining two-thirds the height and spread of its parent. *B. × s. 'Claret Cascade'* Arching purple-red branches carry dark, claret red flower buds which open to dark orange in mid spring. Foliage dark green with silver underside. Of two-thirds average height and spread. *B. × s. 'Corallina Compacta'* Flower buds bright coral red, yellow on opening. Small foliage, ½in (1cm) long. A very slow dwarf form, rarely exceeding 12-15in (30-40cm) in height and spread. *B. × s. 'Crawley Beauty'* Apricot-yellow flowers in spring; small leaves ½-¾in (1-2cm) long. Makes a low hedge if planted 1½ft (50cm) apart in a single line. A low, slower growing variety reaching 2¼ft (70cm). *B. × s. 'Irwinii'* Flowers deep yellow, borne in profusion along arching stems. Even smaller leaves, narrow, lanceolate, ½in (1cm) long, olive green. When leaves die in autumn, some contrasting red-orange tints seen. Reaching only 2½ft (80cm) in height and 3ft (1m) in width. Slow growth rate. *B. × s. 'Pink Pearl'* An interesting variety of doubtful vigour. The flowers may be creamy-yellow, orange-pink or bi-coloured on different shoots on the same plant. The leaves on new growth are pink or creamy striped in some good seasons; this striping can be very variable. Medium to slow growth rate, reaching 3ft (1m) height and spread.
Average height and spread
Five years
5x5ft (1.5x1.5m)
Ten years
6x6ft (2x2m)
Twenty years
or at maturity
10x10ft (3x3m)

BERBERIS THUNBERGII

COMMON BARBERRY, JAPANESE BARBERRY
Berberidaceae
Deciduous
A group of attractive foliage shrubs of very varied height and spread.

Origin From Japan.
Use As freestanding single shrubs or for mass planting for autumn effect. If planted 2ft (60cm) apart in a single line makes a good informal hedge.
Description *Flower* Small, hanging, cup-shaped flowers, white tinged pink with red calyx, produced in early to mid spring. *Foliage* Leaves elliptic, ½-1in (1-3cm) long, light green in spring, aging to red-orange in autumn. Also purple, yellow and variegated leaved forms. *Stem* Upright, thorny, becoming more bushy and spreading with age. Medium growth rate. *Fruit* Small, oblong,

Berberis thunbergii 'Aurea' in leaf

---

bright red, glossy, hanging fruits on mature stems in autumn.
Hardiness Tolerates winter temperatures down to −13°F (−25°C).
Soil Requirements Does well on most soils but dislikes dryness.
Sun/Shade aspect Full sun to light shade. Purple and variegated leaved forms need full sun or leaf colour will be lost.
Pruning Remove occasional stem 3-4 years old to ground level to encourage rejuvenation of new growth and better leaf colour.
Propagation and nursery production From seed, but may vary in appearance and overall performance. Named varieties from semi-ripe cuttings taken in summer. Plant bare-rooted or container-grown. Widely available. Best planting heights 1½-2½ft (50-80cm).
Problems If pruning neglected can become very old and woody after 10-15 years. Even so, cutting back hard will rejuvenate it after 2-3 years.
Varieties of interest *B. sieboldii* From Japan. A small compact shrub with many suckering shoots being produced from ground level. Racemes of pale yellow flowers precede oval leaves producing round orange berries with a shiny texture. Good autumn foliage colour. Height and spread 3ft (1m). Slow growth rate. *B. thunbergii var. atropurpurea* Red-purple leaves through spring and summer, even more intense orange-red in autumn. Mature plants fruit freely. Reaches slightly more in height and spread than *B. thunbergii* itself. If planted 2½ft (80cm) apart in single line makes an attractive hedge. *B. t. a. 'Nana'* (Little Favourite) Dwarf form. Intense purple foliage through spring and summer with good autumn colour and small red fruit. Must be planted in sun. Height 1½ft (50cm), spread 2ft (60cm). Slow growth rate. Plant 1½ft (50cm) apart in a single line for a low hedge. *B. t. 'Aurea'* A golden-leaved form, slightly more rounded leaves, lime green to pale yellow in early spring, aging to deeper yellow through summer, then becoming red-margined as autumn approaches, culminating in orange-flame. To get the best results this variety needs light dappled shade to protect it from strong midday sun, and so avoid leaf scorch. Height 3ft (1m), spread of 4ft (1.2m). Slow to medium growth rate. *B. t. 'Bagatelle'* Similar to *B. t. a. 'Nana'*, but more compact and new foliage is strong red-purple. *B. t. 'Erecta'* An upright growing form. Round to ovate, light to mid green leaves in spring producing in autumn one of the most spectacular foliage displays of any small shrub. On mature shrubs small, white, cup-shaped flowers with red calyxes, glossy red cylindrical fruits in autumn. Height 4ft (1.2m), spread 2-2½ft (60-80cm). Slow to medium growth rate. *B. t. 'Golden Ring'* Leaves round, purple in spring, developing intense gold margin in warm, dry summer and in autumn turning red-flame. Flowers white with red calyx, later red fruits on mature wood. Plant in full sun to maintain purple colouring. Height and spread 5½ft (1.8m). Medium growth rate. *B. t. 'Green Carpet'* Leaves round and ovate, very light green, orange-red in autumn. Forms a spreading shrub up to 5ft (1.5m) in spread, height 3ft (1m). *B. t. 'Green Ornament'* Dark green leaves, upright growth. Height 5ft (1.5m), spread of 3ft (1m). Good autumn colour. Medium growth rate. *B. t. 'Halmond's Pillar'* Dark wine red foliage in spring maintained through summer; good autumn colour. Flowers white with red calyx, fruits red. Upright growth, height 4ft (1.2m) and spread 2½ft (80cm). *B. t. 'Harlequin'* Leaves small, round to ovate, new leaves on new growth being mottled cream and pink, against purple base. Mottling increases in intensity as early spring growth matures and by midsummer makes a very attractive display. Height 4ft (1.2m) at maturity, forming a tight, compact shrub. Medium growth rate. *B. t. 'Kelleriis'* Light to mid green foliage, splashed with white variegation on all new growth. Good red-orange autumn colour. Flowers white with red calyx, followed by red berries. Bushy growth, reaching height and spread of 5ft (1.5m). Slow to medium rate of growth. *B. t.*

Berberis thunbergii 'Rose Glow' in leaf

'Kobold' A dwarf form. Leaves light green, tightly compacted, round and ovate in spring, aging to darker green through summer, very good orange in autumn. Small, white, hanging, cup-shaped flowers with red calyxes, developing into small, red, shiny, cylindrical fruits. A very useful dwarf shrub for rock gardens or, for a very low hedge, planted 1½ft (50cm) apart. Height 1½ft (50cm), spread 2½-3ft (80cm-1m). Slow growth rate. **B. t. 'Lombarts Purple'** Upright habit. Bright purple foliage, aging to duller purple in summer. Good orange-red autumn colour. Small, hanging, cup-shaped white flowers with red calyxes followed by red fruits. Of two-thirds average height and spread. Medium growth rate. **B. t. × ottawensis 'Purpurea'** One of the tallest purple-leaved related forms of *B. thunbergii*. Leaves comparatively large, purple, ovate, with extremely good autumn colour. Good racemes of golden yellow flowers in spring, mainly on old stems, culminating in hanging clusters of red fruits. Plant singly, or 3ft (1m) apart in single line for an informal hedge, or use with other white or golden variegated shrubs. Height 12ft (3.5m), spread 13ft (4m). Medium growth rate. **B. t. 'Red Chief'** Foliage almost lanceolate when young, very vivid purple-red, aging to darker purple at maturity and on old wood. Stems also bright red to red-purple, forming a truly spectacular display in early spring to midsummer. Small, hanging, cup-shaped white flowers with red calyxes in spring, followed by bright red cylindrical fruits in autumn. Good orange-red to orange-purple autumn colour with stems remaining purple throughout the winter. Plant 2½ft (80cm) apart in a single line for an informal hedge which may be clipped, though at cost of new red-stemmed foliage. Height 6ft (2m), spread 8ft (2.5m). Medium growth rate. **B. t. 'Red Pillar'** An upright form, forming a moderately dense pillar. Leaves red to red-purple, flowers small, white, hanging, cup-shaped, with red calyxes, followed by bright red, cylindrical berries in autumn. Good orange-red autumn colour. Ideal for the larger rock garden, as a feature at the end of a herbaceous border or corner of a group planting, or for emphasizing a gateway. Height 3-4ft (1-1.2m), spread 1½-2ft (50-60cm). Medium to slow rate of growth. **B. t. 'Rose Glow'** An attractive variety with purple foliage in spring, which later becomes variegated pink and white, the variegation being produced by new growth in late spring to early summer and maintained into autumn, when some autumn colour may be produced. Flowers small, white, cup-shaped; mature into red berries in autumn. If planted 2½ft (80cm) apart in a single line makes a medium-sized, informal hedge. Contrasts well with other white or golden variegated foliage shrubs for the shrub border, or ideal for mass planting. Height 8ft (2.5m), spread 10ft (3m) at maturity. Medium

growth rate. **B. vulgaris 'Atropurpurea'** (Purple-leaved Barberry) Not a *B. thunbergii* variety, but very closely related in form and performance. A purple-leaved form of *B. vulgaris*, the Common Barberry, producing ovate foliage, giving good autumn colour. Possibly *B. vulgaris*, which produces green foliage, is more widely seen. Both forms bear yellow flowers and red fruit. Height and spread 10ft (3m).

Average height and spread
Five years
5½x5ft (1.8x1.5m)
Ten years
8x8ft (2.5x2.5m)
*Twenty years
or at maturity*
9x10ft (2.8x3m)

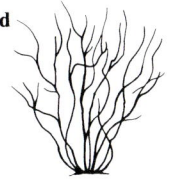

BERBERIS WILSONIAE

WILSON'S BARBERRY
Berberidaceae
Deciduous
A very attractive form of *Berberis*, its pink and orange berries deserving a place in any garden if a suitable planting area is available.

Origin From western China. Found by Ernest Wilson, the plant collector, who named it in honour of his wife.
Use As interesting small to medium height shrub for autumn foliage and fruiting effect. Extremely good for mass planting and if planted at 2½ft (80cm) apart in a single line makes attractive informal hedge.

Berberis wilsoniae in fruit

Description *Flower* Small, double, hanging, cup-shaped, pale yellow flowers hanging from undersides of arching branches in early to late spring. *Foliage* Leaves small, ovate, light green to slightly grey-green in spring and summer, with light orange-salmon autumn tints. *Stem* Light green to green-brown. Upright when young, arching with age, with spines at each leaf joint. Slow to medium rate of growth. *Fruit* Round, salmon to light orange, almost translucent fruit in profusion along wood 2-3 years old or more in autumn, combining with the salmon-tinted foliage to give a spectacular display.
Hardiness Tolerates winter temperatures down to −13°F (−25°C).
Soil Requirements Any soil type.
Sun/Shade aspect Full sun to medium shade.
Pruning None required. Remove one or more very mature shoots annually after 4 or 5 years to encourage new growth and maintain health. May be cut back hard if required.
Propagation and nursery production From cuttings taken in late spring to early summer. Obtain stock propagated from cuttings rather than from seed, as seed-raised plants can be very variable. Plant bare-rooted or container-grown. Generally available. Best planting heights 15in-2½ft (40cm-80cm).
Problems The prickles can make cultivation around this plant a little uncomfortable; use gloves.
Varieties of interest B. rubrostilla Upright stems, red-brown in colour. Slightly larger grey-green leaves. Yellow hanging flowers in spring, followed by long, ovate, red to red-orange fruits. Good autumn colours. Height 5ft (1.5m), spread 4ft (1.2m). **B. jamesiana** Leaves 1-2½in (3-6cm) long, reticulate, dull green. Yellow flowers in racemes in spring, followed by pale pink to white fruits. Scarce. Height 8ft (2.5m), spread 10ft (3m). **B. koreana** Leaves large, light green, ovate, 1-1½in (3-4cm) long. Large yellow flowers in short racemes in mid spring with good autumn tints. Ovate, orange-red fruits in autumn.
Average height and spread
Five years
3x4ft (1x1.2m)
Ten years
4x6ft (1.2x2m)
*Twenty years
or at maturity*
4x8ft (1.2x2.5m)

BETULA NANA

DWARF BIRCH, ROCK BIRCH
Betulaceae
Deciduous
A delightful miniature version of the beautiful Silver Birch, but unfortunately the bark is not silvered to any degree.

Origin A native shrub of the most northern temperate regions.
Use As a small, low, interesting plant for rock gardens where it can be used as a focal point at the edge, or combined with dwarf heathers and conifers. Can also be grown in troughs or tubs
Description *Flower* May produce small, stunted, yellow, male catkins in spring. *Foliage* Leaves tooth-edged, small, round to ovate, ½in (1cm) long, giving good yellow autumn colours. *Stem* Upright when young, becoming

Betula nana in catkin

arching. Light green at first, aging to grey-brown. Forms a low, bun-shaped plant. Moderately slow growth rate. *Fruit* Small, hanging, birch-type seed spirals on very mature plants in autumn.
Hardiness Tolerates winter temperatures down to −13°F (−25°C).
Soil Requirements Any soil type.
Sun/Shade aspect Full sun to light shade.
Pruning None required.
Propagation and nursery production From seed or layers. Purchase container-grown. Best planting heights 1-2ft (30-60cm).
Problems None.
Average height and spread
Five years
3x3ft (1x1m)
Ten years
4½x4½ft (1.4x1.4m)
Twenty years
or at maturity
5½x5½ft (1.8x1.8m)

BUDDLEIA ALTERNIFOLIA

FOUNTAIN BUDDLEIA, ALTERNATE-LEAF BUTTERFLY BUSH
Loganiaceae
Deciduous
A truly beautiful shrub if given enough space.

Origin From China.
Use As large, late summer to early autumn-flowering, graceful, arching shrub. Possibly best on its own, but can be used in mixed border if space allows. If a central single shoot is encouraged, can become small weeping 'tree'. Can also be fan-trained on a large sun-facing wall, particularly suitable for the variety *B. alternifolia 'Argentea'*.

Buddleia alternifolia in flower

Description *Flower* Small bunches of very fragrant, lilac-coloured, small, trumpet-shaped flowers borne along graceful, arching branches, early summer. *Foliage* Leaves grey-green, lanceolate, 1½-4in (4-10cm) long, giving yellow autumn colour. *Stem* Grey-green to mid green, vigorous, long, upright, becoming arching branches, forming a wide canopy. *Fruit* Brown to grey-brown seedheads in autumn and winter.
Hardiness Tolerates 4°F (−15°C).
Soil Requirements Prefers good, rich, deep soil, although tolerates other soil types.
Sun/Shade aspect Best in full sun; tolerates slight dappled shade.
Pruning Thin out one-third of old wood after flowering.
Propagation and nursery production From softwood cuttings in summer or hardwood cuttings in winter. Purchase container-grown. Best planting heights 15in-3ft (40cm-1m).

Problems When offered for sale it resembles an old, woody shrub. Once planted out, however, it grows quickly and often fills a larger space than anticipated.
Varieties of interest *B. alternifolia 'Argentea'* Slightly more tender and lower growing, with attractive silver foliage and slightly paler blue flowers. Not always easy to find but worth searching out for a sheltered site. Best protected by a sunny, sheltered wall.
Average height and spread
Five years
6x6ft (2x2m)
Ten years
10x10ft (3x3m)
Twenty years
or at maturity
13x13ft (4x4m)

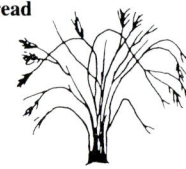

BUDDLEIA CRISPA

KNOWN BY BOTANICAL NAME
Loganiaceae
Deciduous
An attractive late-flowering shrub, worth pursuing.

Origin From northern India.
Use As medium height, late summer to early autumn-flowering shrub for sheltered, protected areas.

Buddleia crispa in flower

Description *Flower* Fragrant, round panicles of tubular lilac flowers with orange throats, in late summer. *Foliage* Leaves light green to grey-green, broadly lanceolate, 2-5in (5-12cm) long, with deeply toothed edges. *Stem* Upright when young, spreading with age. Young stems covered with dense white felt, into late summer. Forms a round shrub. *Fruit* Light brown seedheads, attractive in winter.
Hardiness Minimum winter temperature 23°F (−5°C).
Soil Requirements Prefers rich, deep soil, tolerates alkalinity and acidity.
Sun/Shade aspect Full sun to very light shade.

Pruning Remove one-third of old wood on established plants to ground level each spring to encourage growth of young flowering wood.
Propagation and nursery production From softwood cuttings in summer or hardwood cuttings in winter. Purchase container-grown. Best planting heights 1-2ft (30-60cm).
Problems Not easily found.
Varieties of interest *B. colvilei* Long racemes of deep rose, tubular flowers, borne at the ends of the branches. Foliage ovate, grey-green with some yellow autumn colour. Slightly more tender than *B. crispa*. Difficult to find; must be sought from specialist nurseries. *B. c. 'Kewensis'* An attractive form with dark red flowers. Difficult to find.
Average height and spread
Five years
6x6ft (2x2m)
Ten years
10x10ft (3x3m)
Twenty years
or at maturity
10x10ft (3x3m)

BUDDLEIA DAVIDII

BUTTERFLY BUSH, SUMMER LILAC
Loganiaceae
Deciduous
A beautiful shrub, called Butterfly Bush because of its attraction of butterflies.

Origin From central and western China. Modern forms of garden origin.
Use As large, late summer to early autumn-flowering shrub. Can be trained into small mop-headed tree or fan-trained on a sunless wall.

Buddleia davidii 'Black Knight' in flower

Buddleia davidii 'Fascinating' in flower

Buddleia davidii 'Harlequin' in flower

Buddleia fallowiana 'Alba' in flower

Description *Flower* Fragrant, tubular, pale to mid blue flowers borne in racemes in mid to late summer. Named varieties offer a range of colours from white to almost black. *Foliage* Leaves broad, lanceolate, 4-12in (10-30cm) long, dark to grey-green upper surfaces, light grey to silver undersides; some yellow autumn colour. *Stem* Light to dark green. Strong, upright, becoming slightly spreading, forming a tall shrub. *Fruit* Racemes of light to mid brown seedheads in winter.
Hardiness Tolerates 14°F (−10°C).
Soil Requirements Does best on good, rich, deep soil; tolerates a wide range, both acid and alkaline.
Sun/Shade aspect Full sun to light shade.
Pruning Previous year's wood must be cut back hard in early spring to within 4in (10cm) of its origin. This increases flower size by at least half as much again and extends life of shrub.

Buddleia davidii 'White Profusion' in flower

Propagation and nursery production From softwood cuttings in summer or hardwood cuttings in winter. Purchase container-grown. Best planting heights 1½-3ft (50cm-1m).
Problems Will be short-lived, with small flowers, unless pruned. May look old and woody when purchased, but grows quickly once planted.
Varieties of interest *B. davidii 'African Queen'* Dark violet flowers, a strong growing variety. *B. d. 'Black Knight'* Dark purple, almost black flowers, with orange eye in centre. Strong to medium growth rate. *B. d. 'Border Beauty'* Crimson red flowers. A lower grow-

ing variety, two-thirds average height and spread. *B. d. 'Darkness'* Deep blue to purple-blue flowers. Wide spreading, arching shrub. *B. d. 'Empire Blue'* Violet-blue racemes of tubular flowers with orange eye. Strong, upright stems. *B. d. 'Fascinating'* Extra-large racemes of tubular, lilac-pink flowers. Strong growing. *B. d. 'Fortune'* Long, round racemes of soft lilac flowers with orange eyes. *B. d. 'Harlequin'* Flowers rich purple with broad, lanceolate, creamy white variegated leaves. Two-thirds average height and spread. *B. d. 'Ile de France'* Rich violet flowers. Strong growing. *B. d. 'Nanho Alba'* White flowers, slender habit, narrow foliage. Half average height. *B. d. 'Nanho Blue'* Deep blue flowers in short racemes. Half average height with more spreading, graceful, arching, branches. *B. d. 'Nanho Purple'* Purple-blue flowers, low growing. Half average height. *B. d. 'Opera'* Deep purple-red flowers borne on a strong growing shrub. *B. d. 'Orchid Beauty'* Pure mauve flowers. Two-thirds average height. *B. d. 'Peace'* Racemes of white flowers in tubular florets; orange eye in throat. Arching branches. Two-thirds average height. *B. d. 'Purple Prince'* Very large racemes of purple-red flowers. A very strong growing variety with upright branches. *B. d. 'Royal Red'* Rich purple-red racemes of flowers borne on long graceful, arching branches. Two-thirds average height. *B. d. 'Royal Red Variegata'* A white variegated form of the above. *B. d. 'White Bouquet'* Pure white, very fragrant flowers; florets may have yellow eyes. *B. d. 'White Profusion'* Pure white flowers with yellow eyes. Two-thirds average height.
Average height and spread
Five years
10x10ft (3x3m)
Ten years
13x13ft (4x4m)
Twenty years
or at maturity
13x13ft (4x4m)

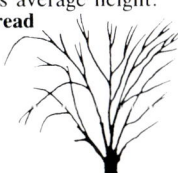

Origin From China.
Use As a medium height shrub for sheltered gardens. Can be grown freestanding or on sun-facing wall.
Description *Flower* Pure white, fragrant flowers in short to medium racemes; late summer. *Foliage* Leaves, broad, lanceolate, 8-12in (20-30cm) long, grey to white, downy. Very attractive. *Stem* White, downy, upright when young, becoming spreading with age, forming a medium height, round shrub. *Fruit* Brown to light brown seedheads in winter.

Hardiness Winter minimum 23°F (−5°C).
Soil Requirements Thrives on rich deep soil, either alkaline or acid.
Sun/Shade aspect Best in full sun.
Pruning Cut back hard in early spring to encourage new silver-white foliage and increase flower size.
Propagation and nursery production From softwood cuttings taken in summer or hardwood cuttings in winter. Purchase container-grown. In less temperate areas keep young plants under cover in winter to plant next spring in case parent does not survive. May not be easy to find. Best planting heights 15in-2½ft (40-80cm).
Average height and spread
Five years
5x5ft (1.5x1.5m)
Ten years
6x6ft (2x2m)
Twenty years
or at maturity
8x8ft (2.5x2.5m)

Origin From Chile and Peru.
Use As a large, early summer-flowering shrub for border planting and screening.
Description *Flower* Globes of bright yellow to yellow-orange, tubular flowers borne in loose panicles, midsummer. *Foliage* Leaves broad, lanceolate, 4-8in (10-20cm) long, dark green, upper surface with rough, deep-veined texture, silver underside. Some yellow autumn

Buddleia globosa in flower

colour. *Stem* Strong, light green to dark green, aging to brown. Upright, forming a large spreading, round-topped shrub. *Fruit* Insignificant.
Hardiness Tolerates 4°F (−15°C).
Soil Requirements Best on a rich, deep soil. Extreme alkalinity may lead to chlorosis.
Sun/Shade aspect Prefers full sun, tolerates light shade.
Pruning Remove one-third of the oldest wood from mature specimens annually after flowering to rejuvenate. If plant has been allowed to become very old and woody, can be cut back hard and will regenerate in 2-3 years.
Problems None.
Average height and spread
Five years
6x6ft (2x2m)
Ten years
10x10ft (3x3m)
Twenty years
or at maturity
13x13ft (4x4m)

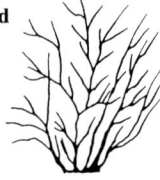

BUDDLEIA 'LOCHINCH'
(Buddleia × fallowiana 'Lochinch')

KNOWN BY BOTANICAL NAME
Loganiaceae
Deciduous
A very attractive, late summer-flowering shrub, the foliage contrasting well with the lavender-blue flowers.

Origin Of garden origin. A cross between *B. fallowiana* and *B. davidii*.
Use As a large flowering shrub for a shrub border, requiring a little protection. Can be fan-trained on sunny wall.
Description *Flower* Lavender-blue, scented flowers in long, broad racemes, each tubular floret having a large orange eye. *Foliage* Leaves 8-10in (20-25cm) long, broadly lanceolate, grey-green upper surface, silver-grey underside. *Stem* Grey-green, strong, upright, branching, forming a large, spreading-topped shrub. *Fruit* Light brown seedheads, autumn and winter.
Hardiness Tolerates 4°F (−15°C).
Soil Requirements Does best on rich, deep soil, but tolerates most types.
Sun/Shade aspect Needs full sun or foliage colour deteriorates.
Pruning Responds well to very hard pruning in spring, or can be left unpruned to grow as a very large shrub which can be cut back hard if it becomes out of hand; will then rejuvenate itself.
Propagation and nursery production From semi-ripe cuttings taken in summer or hardwood cuttings in winter. Purchase container-

grown. Best planting heights 1½-3ft (50cm-1m).
Problems None, but may appear unattractive when purchased.
Average height and spread
Five years
8x8ft (2.5x2.5m)
Ten years
12x12ft (3.5x3.5m)
Twenty years
or at maturity
13x13ft (4x4m)

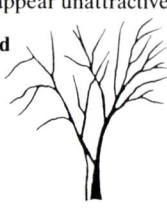

BUDDLEIA × WEYERIANA

KNOWN BY BOTANICAL NAME
Loganiaceae
Deciduous
An interesting flowering Buddleia with scented flowers and grey foliage.

Origin Of garden origin. A cross between *B. davidii* and *B. globosa*.
Use As a large, late summer to early autumn flowering shrub. Can be fan-trained on a wall.

Buddleia × weyeriana **in flower**

Description *Flower* Racemes of ball-shaped clusters of orange-yellow flowers with orange eyes; pleasant scent. *Foliage* Leaves broad, lanceolate, 4-8in (10-20cm) long, grey to grey-green. *Stem* Grey-green, upright, spreading with age, forming a large, round-topped shrub. *Fruit* Insignificant.
Hardiness Tolerates winter temperatures down to 4°F (−15°C).
Soil Requirements Prefers rich, deep soil but does well on any type.

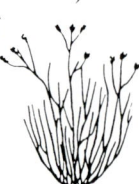

Buddleia 'Lochinch' **in flower**

Sun/Shade aspect Best in full sun or very light shade.
Pruning None required but can be reduced each spring to 4in (10cm) from point of origin of last year's shoots.
Propagation and nursery production From semi-ripe cuttings taken in summer, or hardwood cuttings in winter. Purchase container-grown. May be difficult to find. Best planting heights 15in-2½ft (40-80cm).
Problems Foliage abnormally susceptible to damage by hailstones in late spring, but this is a temporary problem.
Varieties of interest *B. × w.* 'Golden Glow' Golden-yellow flowers. *B. × w.* 'Sun Gold' Orange-yellow flowers.
Average height and spread
Five years
6x6ft (2x2m)
Ten years
8x8ft (2.5x2.5m)
Twenty years
or at maturity
10x10ft (3x3m)

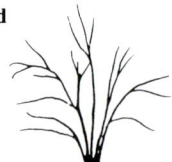

BUPLEURUM FRUTICOSUM

KNOWN BY BOTANICAL NAME
Umbelliferae
Evergreen
A distinctively coloured flowering evergreen.

Origin From southern Europe.
Use As a medium-sized, evergreen shrub. Best freestanding or trained on a sunless wall.
Description *Flower* Ball-shaped clusters of green-cream to yellow-green flowers from midsummer to early autumn. *Foliage* Leaves elliptic, ½-2in (1-5cm) long, dark, glossy, grey-green with silver undersides. *Stem* Light green to dark olive green, forming a rounded shrub, somewhat loose in habit. Slow growth rate. *Fruit* Brown seedheads, interesting in winter.

Bupleurum fruticosum **in flower**

Hardiness Established shrubs withstand winter temperatures down to 4°F (−15°C), but young plants are less hardy. Good in exposed coastal sites.
Soil Requirements Any soil conditions.
Sun/Shade aspect Best in full sun; tolerates shade well but becomes looser in habit in deep shade.
Pruning None required. May be trimmed or cut back to maintain shape.
Propagation and nursery production From seed or from softwood cuttings taken in summer. Purchase container-grown. Best planting heights 8in-1½ft (20-50cm).
Problems Not easy to find.
Average height and spread
Five years
3x3ft (1x1m)
Ten years
5x5ft (1.5x1.5m)
Twenty years
or at maturity
6x6ft (2x2m)

BUXUS SEMPERVIRENS

COMMON BOX, BOXWOOD
Buxaceae
Evergreen
Used in the right position for hedging or formal display, or with the variegated or golden forms for winter effect, an extremely useful range of evergreen shrubs.

Buxus sempervirens 'Elegantissima' in leaf

Origin Possibly from North Africa and western Asia; now distributed throughout southern Europe and in some parts of the British Isles grows in the wild.
Use As an evergreen for use in its own right or for topiary. Plant 1½ft (50cm) apart for an attractive hedge, reaching, in time, 6ft (2m) or even more. Extremely good in tubs and containers if good quality potting medium is used.
Description *Flower* Fluffy, sulphur yellow, flowers borne in small clusters at each leaf axil on mature wood. *Foliage* Leaves round to ovate, 1-1½in (3-4cm) long, small, dark green, slightly glossy on upper side, light grey on underside, borne in profusion closely along stem. *Stem* Grey-green and upright when young, aging to green-brown to light brown. Becoming twiggy and spreading with age, forming a round shrub. Moderately slow to medium growth rate. Unless trimmed, matures into small tree after 60-70 years. *Fruit* Small blue-black fruits on mature shrubs.
Hardiness Withstands winter temperatures down to −13°F (−25°C).
Soil Requirements The better the soil, the more growth, but tolerates any type.
Sun/Shade aspect Tolerates full sun to deep shade, although variegated and golden-leaved forms have less attractive colouring in shade.
Pruning None required but can be trimmed to almost any desired shape for topiary purposes. Large, mature or large, deformed shrubs can be drastically reduced in height and spread, rejuvenating over 2-3 years.
Propagation and nursery production Propagated from semi-ripe cuttings taken in early to mid summer from new growth. Purchase root-balled (balled-and-burlapped) or container-grown. Best planting heights 8in-2ft (20-60cm).
Problems Once established takes many plant nutrients from soil and its invasive roots can be detrimental to other plants nearby. Rather slow to develop unaided on poor soil. Plants sold are normally small and patience is required to achieve the full effect.
Varieties of interest *B. s. 'Aurea Pendula'* syn. *B. s. 'Aurea Maculata Pendula'* Creamy yellow, mottled, dark green foliage presented on arching to weeping shoots. May be hard to

find. Slow to mature but achieves a fine architectural shape. *B. s. 'Aureovariegata'* syn. *B. s. 'Aurea Maculata'* Attractive green leaf with gold mottling. New growth, when produced in late spring or summer, is gold to buttercup yellow. Not good for trimming because of its lax growth but makes an attractive round mound with time, up to 6ft (2m) in height and spread. *B. s. 'Elegantissima'* Leaves lanceolate to round, attractive creamy white variegation. Upright in habit, but much dwarfer and slower than the parent. Eventually produces a round shrub, height 3-5ft (1-1.5m). Useful as a low hedge if planted 15in (40cm) apart. *B. s. 'Gold Tip'* Leaves lanceolate to round, light to dark grey-green, gold-tipped. Same growth rate as *B. sempervirens* with which it is often confused. Not easy to find or clearly defined as a type in its own right. *B. s. 'Handsworthensis'* syn. *B. s. 'Handsworthii'* Leaves round or oblong and much larger than the parent. Attractive in its own right and one of the best of all the hedging forms. Possibly taller, more upright than *B. sempervirens*. Not always easy to find. *B. s. 'Latifolia Maculata'* syn. *B. s. 'Japonica Aurea'* Dark olive green leaves blotched dull yellow. Young spring growth bright yellow to golden yellow. Makes a good, dense low hedge. A slow-growing form, reaching 3-5ft (1-1.5m) in height with slightly more spread at maturity. Not always easy to find. *B. s. 'Rotundifolia'* Similar to *B. s. 'Handsworthensis'* although the leaves are more round. Not easy to find but worth searching for. *B. s. 'Suffruticosa'* (Edging Box) Leaves smaller than *B. sempervirens*, more ovate and brighter green. A very low, dwarf, slow-growing form used for edging flower beds and paths, producing a very

formal effect. Plant 4-6in (10-15cm) apart to make a small, trimmed, square edging to a border. Plants can be purchased bare-rooted or container-grown but are unattractive at this stage, needing time to mature into a plant or hedge of any interest. Not always easy to find in sufficient quantity. If planning a formal edging, make sure first that enough stock is available.

Average height and spread
Five years
3x3ft (1x1m)
Ten years
6x6ft (2x2m)
Twenty years
or at maturity
13x13ft (4x4m)

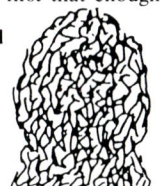

CALLICARPA BODINIERI VAR. GIRALDII
(Callicarpa giraldiana)

BEAUTY BERRY
Verbenaceae
Deciduous
An unusual, moderately large-growing shrub, particularly attractive for its violet berries.

Origin From east to west China.
Use As a late-flowering, autumn-fruiting shrub which when well established produces a spectacular display in groups or in mixed shrub borders.

Callicarpa bodinieri 'Profusion' in fruit

Description *Flower* Small, lilac-pink flowers produced in clusters at leaf axils on wood 2 years old or more, late summer. *Foliage* Leaves light grey-green turning purple in autumn with attractive purple leaf stalks. Lanceolate to elliptic with purple ends. *Stem* Light grey, aging to grey-green, upright when young, and becoming more branching with age. Forming a spreading, round, open shrub. *Fruit* Large clusters of violet fruits follow late summer flowers, often in some profusion, on mature shrubs. Although the shrub is dioecious it seems to fruit better if planted in pairs.
Hardiness Tolerates 4°F (−15°C).
Soil Requirements Any soil except extremely alkaline, which will lead to chlorosis.
Sun/Shade aspect Full sun to light or medium shade.
Pruning None required. Large shoots or limbs may safely be reduced if desired.
Propagation and nursery production From softwood cuttings taken in midsummer. Purchase container-grown. Not always easily obtainable, but may be found in specialist nurseries. Best planting heights 1½-3ft (50cm-1m).
Problems Planted individually it may not fruit well and requires a partner for best results. May be distressed by extremely dry summers.

Buxus sempervirens in leaf

Varieties of interest *C. b. 'Profusion'* A decidedly improved form, with larger fruits, more reliably produced and possibly today replacing *C. b. var. giraldii* itself. *C. dichotoma* syn. *C. purpurea*, *C. koreana* (Purple Beautyberry) Leaves more ovate, flowers earlier than the form and produces dark lilac fruits but not in such quantities as *C. b. giraldii*. A much shorter-growing variety, height and spread 5ft (1.5m). From China, Korea and northern Taiwan. *C. japonica 'Leucocarpa'* (Japanese Beautyberry) An attractive light green-leaved variety with yellow autumn colours. Clusters of white flowers in late summer, followed by bunches of white berries. If not found in garden centres and nurseries, worth a search among specialist sources.

Average height and spread
Five years
6x6ft (2x2m)
Ten years
10x10ft (3x3m)
Twenty years
or at maturity
13x13ft (4x4m)

AUSTRALIAN BOTTLE BRUSH
Myrtaceae
Evergreen
If the right conditions can be offered, an extremely useful, attractive plant.

Callistemon citrinus 'Splendens' in flower

Origin From Australia and New Zealand.
Use As a large, summer-flowering shrub for mild districts. Useful as a large wall shrub, or for growing in big containers, where must be planted in lime-free potting medium, or for a large conservatory where adequate protection can be provided in winter. Very good in coastal areas.
Description *Flower* Tufted, brush-like spikes of red flowers, very dense in formation, mid to late summer. *Foliage* Leaves narrow, lanceolate, 1-1½in (3-4cm) long, light green, often with red-orange shading or pronounced coloured veins. If crushed, release an aromatic, lemon scent. *Stem* Light green to grey-green, aging to grey-brown. Upright when young, becoming more arching to form a graceful, wide-spreading shrub. Medium growth rate, slowing with age. *Fruit* May produce interesting tufted, light brown seedheads.
Hardiness Tolerates 23°F (−5°C).
Soil Requirements Good, rich, acid soil; dislikes any alkalinity.
Sun/Shade aspect Full sun; does not tolerate any shade.
Pruning None required. Remove an old shoot occasionally to rejuvenate from the base.
Propagation and nursery production From seed or softwood cuttings taken in late spring

or early summer. Purchase container-grown. Not always easy to find. Best planting heights 15in-2½ft (40-80cm).
Problems Not to be grown in alkaline soils or in locations with winter conditions well below freezing.
Varieties of interest *C. c. 'Splendens'* More brilliant flowers. Slightly less height than *C. citrinus* but possibly more hardy.
There are a number of other varieties and forms of *Callistemon* only suitable for very mild conditions. Seek local advice on planting very tender types.

Average height and spread
Five years
6x5ft (2x1.5m)
Ten years
10x8ft (3x2.5m)
Twenty years
or at maturity
10x10ft (3x3m)

SCOTCH HEATHER, LING
Ericaceae
Evergreen
A very attractive ground cover plant, requiring acid soil.

Origin Widely spread throughout Europe to Asia; found in its wild form in large tracts of open land and heathland.
Use As a summer-flowering, low carpeting, ground-covering shrub.
Description *Flower* Numerous bell-shaped, medium to small flowers, purple-pink flowers, also white, pink or purple, dependent on variety, in mid summer to early autumn. *Foliage* Leaves small, numerous, lanceolate, ½in (1cm) long, ranging from dark green to light green dependent on variety or to yellow, gold and purple. Some varieties produce darker purple, orange or gold autumn and winter tints. *Stem* Short, dense, upright when young, spreading with age. Green or brown shoots, forming low, spreading mound of growth. Medium growth rate. *Fruit* Light grey-green seedheads, attractive in winter.
Hardiness Tolerates 14°F (−10°C); harsh temperatures can damage or even destroy the shrubs.
Soil Requirements Must be grown in acid soils; dislikes any alkalinity.
Sun/Shade aspect Prefers full sun, will tolerate light shade.
Pruning Trim lightly each spring with clippers or hedging shears to encourage new clean foliage and profuse summer flowers.
Propagation and nursery production From very small cuttings taken early summer to midsummer. Purchase container-grown. Normally easy to obtain, especially from nurser-

Calluna vulgaris 'Kinlochruel' in flower

ies in areas having predominantly acid soil. Best planting heights 3-4in (8-10cm), if offered in 2¾in (7cm) pots, or 4-6in (10-15cm).
Problems Susceptible to extremes of cold or drought. Annual mulching with wet sedge peat in autumn gives some protection.
Varieties of interest *C. v. alba* (White Heather) Single flowers in midsummer. This variety, particularly associated with Scotland, is widely spread throughout temperate areas. *C. v. 'Alba Plena'* syn. *C. v. 'Alba Flore Pleno'* Double white flowers in midsummer. *C. v. 'Aurea'* Single purple flowers in midsummer. Foliage gold, turning red-bronze in winter. Slightly less vigorous than the parent. *C. v. 'County Wicklow'* Double, shell pink flowers in late summer on a round, dwarf, spreading shrub. *C. v. 'Cuprea'* Flowers single, pale mauve in midsummer. Young shoots golden in summer, turning bronze in autumn and winter. *C. v. 'Golden Feather'* Flowers single, light pink in midsummer, golden feathery foliage changing in autumn to orange. One of the most attractive forms. *C. v. 'Gold Haze'* Single, white flowers in midsummer. Golden foliage with darker autumn colours. *C. v. 'Goldsworth Crimson'* Single, deep crimson flowers in midsummer. More upright than most, height 2½ft (80cm). *C. v. 'H. E. Beale'* Double, bright rose-pink flowers in long racemes in mid to late summer. Dark green foliage. Good for cutting. Strong-growing, height 2ft (60cm). *C. v. 'Joan Sparkes'* Double, mauve flowers in midsummer. Good golden foliage. Low-growing. *C. v. 'Joy Vanstone'* Light purple, single flowers in midsummer. Golden foliage, aging to rich orange in winter. Height 1½ft (50cm). *C. v. 'Kinlochruel'* Double, white flowers in mid to

Calluna vulgaris 'County Wicklow'

late summer, set off by deep green foliage. Low growing, height 10in (25cm). *C. v. 'Mullion'* Deep mauve, single flowers, very densely displayed in long racemes in midsummer. Very low, creeping variety, height 6-8in (15-20cm). *C. v. 'Orange Queen'* Single, purple-pink flowers in mid to late summer. New spring foliage golden, turning to deep orange in autumn. *C. v. 'Peter Sparkes'* Double, pink flowers in late summer to early autumn. Grey-green foliage. Height 1½ft (50cm). *C. v. 'Red Haze'* Single, purple-red flowers with rich yellow foliage. Early flowering. Height 15in (40cm). *C. v. 'Robert Chapman'* Deep pink flowers in mid to late summer, yellow-orange foliage. *C. v. 'Silver Night'* Single, purple-pink flowers in late summer, early autumn; silver-green foliage. Height 12in (30cm). *C. v. 'Silver Queen'* Single, pale mauve flowers in midsummer. Silver foliage. Height 2ft (60cm). *C. v. 'Spring Torch'* Light purple flowers in late summer, early autumn. Red foliage. *C. v. 'Spitfire'* Single, pink flowers in midsummer. Golden foliage, turning bronze-red in autumn and winter. Height 8-12in (20-30cm). *C. v. 'Sunset'* Single, pink flowers in midsummer.

Calluna vulgaris 'Tib' in flower

Variegated yellow foliage, turning gold and orange. Height 8-12in (20-30cm). *C. v. 'Tib'* Double rose red flowers borne profusely in early summer. Dark green foliage, erect habit. Height 1-2ft (30-60cm). *C. v. 'Tricolorifolia'* Single, pink flowers in midsummer. Young spring foliage bronze-red, aging in autumn to deep green. Height 2ft (60cm). *C. v. 'Winter Chocolate'* Single purple-pink flowers in late summer to early autumn. Cream spring foliage. Height 15in (40cm).
The above is a selection of the most useful varieties of this wide range. New varieties are being introduced all the time and may supersede some of these.
Average height and spread
Five years
15inx2½ft (40x80cm)
Ten years
15inx3ft (40cmx1m)
Twenty years
or at maturity
15inx3ft (40cmx1m)

CALYCANTHUS

ALLSPICE, PALE SWEETSHRUB
Calycanthaceae
Deciduous
An unusual shrub with star-shaped flowers that deserves to be seen more often.

Origin From North America.
Use As a large, summer-flowering, scented shrub with interestingly shaped flowers.
Description *Flower* Crimson-brown, star-shaped, slightly scented flowers, medium to

Calycanthus fertilis in flower

large, borne throughout summer on mature wood 2 years old or more. *Foliage* Leaves large, ovate, lanceolate, 3-6in (8-15cm) long, light to mid green, turning yellow in autumn. Aromatic perfume. *Stem* Grey-green to green-brown, upright and branching, forming a round clump. Medium growth rate. *Fruit* Insignificant.
Hardiness Tolerates a winter minimum range from 4°F (−15°C) down to −13°F (−25°C), but some branches may die back in winter.
Soil Requirements Does best on deep, rich soil; dislikes extreme alkalinity.
Sun/Shade aspect Prefers light shade, tolerates full sun.
Pruning None required. May be reduced in size but new growth will not flower for several years.
Propagation and nursery production From softwood cuttings or layers. Purchase container-grown; may be difficult to obtain. Best planting heights 15in-2½ft (40-80cm).
Problems Suffers from die-back in very cold conditions.
Varieties of interest *C. fertilis* syn. *C. glaucus* Good-sized flowers; possibly the hardiest of the forms. From northern USA. *C. floridus* (Carolina Allspice) A relatively rare form outside its native environment, often confused with *C. fertilis*, but can be distinguished by the downy texture on underside of its leaves. *C. occidentalis* (Californian Allspice or Sweetshrub) More tender and with larger leaves and foliage than *C. fertilis* but roughly same height.
Average height and spread
Five years
3x3ft (1x1m)
Ten years
6x6ft (2x2m)
Twenty years
or at maturity
10x10ft (3x3m)

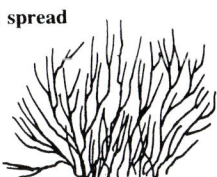

CAMELLIA JAPONICA

KNOWN BY BOTANICAL NAME
Theaceae
Evergreen
Used in the right place, an extremely beautiful plant.

Origin From China and Japan; most varieties now of garden or nursery origin.
Use As an evergreen shrub for acid soils. Very good for large tubs or containers. Can be fan-trained on a sheltered shady wall to good effect.
Description *Flower* Large, cup-shaped, flowers in a wide range of colours; may be single, semi-double, anemone or peony-shaped, loose double or tight double, depending on variety. Size ranges from small to very large.

Foliage Dark, glossy-green upper surfaces, with grey-green undersides. Ovate to oblong, 3-4in (8-10cm) long and 1½in (4cm) wide. *Stem* Bright to dark green, upright, forming a stiff, solid shrub; a few varieties are more laxly presented. Slow to medium growth rate. *Fruit* Insignificant.
Hardiness Tolerates 14°F (−10°C), but may shed leaves in harsh conditions, occasionally causing plant to fail.
Soil Requirements Must have acid soil; dislikes any alkalinity.
Sun/Shade aspect Prefers light to mid shade; dislikes full sun.
Pruning None required. May be cut back to keep within bounds. Young plants may be improved by removing one-third of current season's growth, after flowering, for first 2-3 years.

Camellia × williamsii 'Donation' in flower

Propagation and nursery production From cuttings in early to mid summer. Purchase container-grown. A limited number of varieties can be found in garden centres; less common varieties must be sought from specialist nurseries. Planting heights 1½-6ft (50cm-2m), ideally 2-2½ft (60-80cm).
Problems Often planted on alkaline soils, where it fails, or in full sun, which it dislikes. Flowers can be damaged by frost in exposed areas.
Varieties of interest *Camellia 'Cornish Snow'* Single, small white flowers; a very attractive small-leaved variety. *C. japonica 'Adolphe Audusson'* Semi-double, blood-red flowers. *C. j. 'Apollo'* Semi-double, rose red flowers, sometimes with white blotches. *C. j. 'Arejishi'* Rose red, peony-shaped flowers. *C. j. 'Betty Sheffield Supreme'* Semi-double, white, peony-shaped flowers with rose pink or red

edges to each petal. *C. j. 'Contessa Lavinia Maggi'* Double, white or pale pink flowers with cerise stripes. *C. j. 'Elegans'* Peach pink, large flowers. Anemone flower formation. *C. j. 'Madame Victor de Bisschop'* Semi-double, white flowers. *C. j. 'Mars'* Red, semi-double flowers. *C. j. 'Mathotiana Alba'* Double, white flowers of great beauty. *C. j. 'Mathotiana Rosea'* A double pink form. *C. j. 'Mercury'* Deep crimson flowers, semi-double in form. *C. j. 'Nagasaki'* Semi-double, rose pink flowers with white stripes. *C. j. 'Tricolor'* Semi-double white flowers with carmine or pink stripe. *C. × 'Mary Christian'* Single, clear pink flowers. Tall-growing. *C. × williamsii 'Donation'* Clear pink, semi-double flowers. Possibly the best known Camellia. Height 8ft (2.5m).
The above are just a selected few of the many hundreds of varieties available.

Average height and spread
Five years
3x3ft (1x1m)
Ten years
6x6ft (2x2m)
Twenty years
or at maturity
10x10ft (3x3m)

CARAGANA PYGMAEA

DWARF SALT TREE, PYGMY PEASHRUB
Leguminosae
Deciduous
A group of low shrubs, more interesting than spectacular.

Origin From China and Siberia.
Use As a low shrub for dry or very alkaline conditions, can be grown in tubs and containers for which good quality potting medium must be used. May also be obtainable as small weeping standard when grafted on to a 4ft (1.2m) stem of *C. arborescens*.
Description *Flower* Small, individual, orange-yellow, hanging pea-flowers, borne on undersides of branches in late spring, early summer.

Caragana pygmaea in leaf and flower

Foliage Leaves short, narrow, lanceolate, grey-green, ½in (1cm) long, giving some autumn colour. *Stem* Upright, somewhat wispy when young and slightly arching; thickening with age but maintaining somewhat lax formation. Light grey-green to green-brown in colour. Slow growth rate. *Fruit* Grey-green pea-pods in late summer and early autumn.
Hardiness Tolerates winter temperatures down to −13°F (−25°C).
Soil Requirements Does well on any soil.
Sun/Shade aspect Full sun or light shade.
Pruning None required, in fact may resent it.
Propagation and nursery production From cuttings taken in mid to late summer, or from layers. Purchase container-grown or bare-rooted. May not be easy to find; must be

sought from specialist nurseries. Best planting heights 1½-2½ft (50-80cm)
Problems Must be grown in open soil to show to best advantage.
Varieties of interest *C. arborescens 'Nana'* An attractive, small plant for rock gardens or as a special feature amongst dwarf conifers. May not be easy to find. Height 6in (15cm). *C. frutex glabosa* A glabrous blue to blue-green foliage shrub of neat shape. Best planting heights 4-8in (10-20cm). Eventual height 2ft (60cm). Very slow growth rate. From eastern Europe and central Asia.

Average height and spread
Five years
2x2ft (60x60cm)
Ten years
3x3ft (1x1m)
Twenty years
or at maturity
5x4ft (1.5x1.2m)

CARPENTERIA CALIFORNICA

KNOWN BY BOTANICAL NAME
Philadelphiaceae
Evergreen
A magnificent flowering shrub, well worth the trouble of finding a favourable planting site.

Origin From California.
Use As an evergreen summer-flowering shrub for mild areas; can be grown fan-trained on a sunny wall in less mild regions.
Description *Flower* Medium to large, pure white, saucer-shaped flowers with yellow anthers, borne in midsummer on mature wood. *Foliage* Leaves light to bright green, broad, lanceolate, 2-4in (5-10cm) long. *Stem* Light to dark green, upright at first, slightly spreading with age, forming a mound-shaped shrub. Medium growth rate. *Fruit* Insignificant.
Hardiness Reacts violently to temperatures below 23°F (−5°C), but normally regenerates from ground level.
Soil Requirements Deep, rich soil; tolerates both acidity and alkalinity.
Sun/Shade aspect Must be in full sun and in cooler areas requires the protection of a sun-facing wall.
Pruning May be cut back hard and will regenerate, but takes a year or two to come into full flower. Remove one-third of oldest wood each spring to maintain health.
Propagation and nursery production From seed or from softwood cuttings taken in mid to late summer. Purchase container-grown. May be difficult to find and should be sought from specialist nurseries. Best planting heights 8in-2ft (20-60cm).

Problems As a young plant, the shrub appears weak, but develops well after planting.
Varieties of interest *C. c. 'Ladham's Variety'* Said to be more free-flowering than the parent, with larger flowers. Difficult to find.

Average height and spread
Five years
3x3ft (1x1m)
Ten years
5x5ft (1.5x1.5m)
Twenty years
or at maturity
5x5ft (1.5x1.5m)

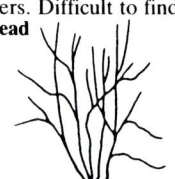

CARYOPTERIS INCANA × MONGOLICA 'ARTHUR SIMMONDS' (Caryopteris × clandonensis)

BLUE SPIRAEA, BLUEBEARD, BLUE-MIST SHRUB
Verbenaceae
Deciduous
A very useful late-flowering hybrid shrub with attractive grey foliage. Formerly *C. × clandonensis* and still sometimes referred to by its old name.

Origin From Japan, Korea, China and Taiwan; the form is thought to be of garden origin.
Use As a small, low, late summer to early autumn-flowering shrub, for individual or mass planting. Planted 1½ft (50cm) apart makes a very pretty low hedge.
Description *Flower* Tufts of pale blue flowers presented profusely in clusters at each leaf axil of the top 4in (10cm) of each new upright shoot, late summer to early autumn. *Foliage* Leaves grey-green, aromatic, lanceolate, 2-3in (5-8cm) long, with slightly toothed edges. *Stem* New grey-green to grey, upright stems, becoming slightly spreading, produced each spring from crown. Fast to medium growth rate. *Fruit* Small grey to grey-brown seed-heads in autumn.
Hardiness Minimum winter temperature 23°F (−5°C).
Soil Requirements Any soil suitable, especially alkaline.
Sun/Shade aspect Needs full sun to produce good shape and to flower well.
Pruning All shoots must be cut back almost to ground level in mid to late spring.
Propagation and nursery production From semi-ripe cuttings taken early to mid summer. Purchase container-grown. This particular variety should be fairly easy to find under either of its names. Best planting heights 8in-2ft (20-60cm).
Problems Often very late to start growth from ground level each spring; young plants

Carpenteria californica in flower

Caryopteris incana × mongolica 'Arthur Simmonds' in flower

offered for sale at this season may have dead stems and look rather unattractive.
Varieties of interest *C. incana* syn. *C. tangutica, C. mastacanthus* (Common Bluebeard) Violet-blue flowers, one of the main sources of the modern hybrids. Slightly less vigorous and more tender than the form. *C. i. × m. 'Ferndown'* Dark violet to blue-violet flowers; an extremely good form. *C. i. × m. 'Heavenly Blue'* Mid blue flowers and more compact growth. *C. i. × m. 'Kew Blue'* Flowers clear blue, grey-green foliage.
Average height and spread
Five years
2x2½ft (60x80cm)
Ten years
2¼x2½ft (70x80cm)
Twenty years
or at maturity
2½x2½ft (80x80cm)

Cassinia fulvida in flower

CASSINIA FULVIDA
(Diplopappus chrysophyllus)

GOLDEN HEATHER
Compositae
Evergreen
An acid-loving small shrub worthy of a place in gardens with suitable soil.

Origin From New Zealand.
Use As a foliage plant for acid gardens. Can be planted with heathers as taller focal point.
Description *Flower* Small, white, daisy-shaped flowers borne at ends of each erect stem in midsummer. *Foliage* Leaves very small, ½in (1cm) long, round to ovate, almost covering the stems. Light golden yellow when

young, aging to dark gold in autumn. *Stem* Light green to gold. Upright, becoming very dense and thick, forming upright shrub initially but becoming more lax and spreading. New stems and foliage are sticky to touch in mid to late spring. Slow growth rate. *Fruit* Small, fluffy, grey seedheads in autumn.
Hardiness Minimum winter temperature 14°F (−10°C), but may prove to be slightly tender.
Soil Requirements Dry, acid soil; dislikes any alkalinity.
Sun/Shade aspect Must be in full sun; any shade decreases golden colouring and shrub becomes very open and lax.
Pruning None required although one-third of shoots 2-3 years old may be reduced to ground level to rejuvenate plant and maintain shape. Any straggly branches may be cut right back to point of origin.
Propagation and nursery production From semi-ripe cuttings taken in midsummer or from self-produced layers. Purchase container-grown. Not easy to find. Best planting heights 1-2ft (30-60cm).
Problems Can become a little untidy looking. May show die-back of an individual branch or section.
Average height and spread
Five years
3x3ft (1x1m)
Ten years
5x5ft (1.5x1.5m)
Twenty years
or at maturity
5x5ft (1.5x1.5m)

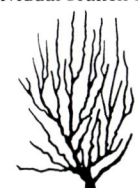

CEANOTHUS
Deciduous forms

CALIFORNIA LILAC
Rhamnaceae
Deciduous
A very attractive range of autumn-flowering shrubs, worthy of most gardens.

Origin From California; but many varieties of garden origin.
Use As a freestanding, autumn-flowering shrub, in a shrub border, or may be fan-trained on a sun-facing wall for protection in less temperate areas. Plant 3ft (1m) apart in single line for an informal hedge of spectacular beauty when in flower.
Description *Flower* Flowers in panicles up to 3-4in (8-10cm) long, various shades of blue, pink or white, in late summer and in some varieties held into early autumn. *Foliage* Leaves medium to large, ovate, 3-5in (8-12cm) long, tooth-edged, light green to olive green, some varieties having pink to pink-red leaf stalks. Some yellow autumn colour. *Stem*

Upright, strong new growth produced each spring with flowers borne at the tips. Light to mid green, aging to dark brown in autumn, forming a bun-shaped shrub. Fast to medium growth rate. *Fruit* Insignificant.
Hardiness Tolerates 14°F (−10°C).
Soil Requirements Good, rich, deep soil; tolerates poorer types if given adequate feeding but liable to chlorosis when severe alkalinity present.
Sun/Shade aspect Best in full sun to very light shade.
Pruning Prune back hard in spring to 4in (10cm) from point of origin of previous year's growth.
Propagation and nursery production From softwood cuttings in summer or by hardwood cuttings in late autumn to early winter. Purchase container-grown; most varieties easy to find, especially at flowering times. Best planting heights 15in-3ft (40cm-1m).

Ceanothus 'Marie Simon' in flower

Problems If insufficiently pruned becomes weak and performs insipidly. Shoots can be broken by strong winds occasionally.
Varieties of interest *C. 'Gloire de Versailles'* A highly popular variety with large panicles of well-spaced, powder blue flowers, midsummer to late summer. *C. 'Henri Defosse'* Beautiful large panicles of deepest blue to purple-blue flowers, late summer. Dark green, medium-sized, elliptic leaves. *C. 'Marie Simon'* Rose pink flowers borne in good-sized panicles, midsummer to late summer. Mid green, oblong leaves with purple-red main veins and leaf stalks. Possibly more tender than average. One-third average height and spread. *C. 'Perle Rose'* Flowers carmine, borne in good-sized panicles, midsummer to late summer. Foliage smaller than

Ceanothus 'Gloire de Versailles' in flower

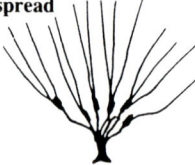

C. 'Marie Simon' and with red to purple-red veins. May show die-back in winter. May be slower to establish than other varieties.

Average height and spread
Five years
5x5ft (1.5x1.5m)
Ten years
6x6ft (2x2m)
Twenty years
or at maturity
8x8ft (2.5x2.5m)

CEANOTHUS
Evergreen forms

CALIFORNIA LILAC
Rhamnaceae
Evergreen
Evergreen *Ceanothus*, if grown in a mild area or protected from winter winds and cold, can give a very spectacular effect in spring and summer.

Ceanothus thyrsiflorus var. repens in flower

Origin From southern USA. Many varieties of garden origin.

Use As a fan-trained wall shrub, or freestanding if area is mild enough or adequate protection provided. In mild districts may be grown as a small tree, attaining this form over 10-15 years.

Description *Flower* Various shades of blue, or white, tufted flowers borne freely in mid to late spring; some varieties early or late summer and even autumn. *Foliage* Leaves mostly ovate, ½-1½in (1-4cm) long, light to dark green; in a few varieties broad to narrow lanceolate. All with shiny upper surface and dull grey underside. In some varieties leaves have pronounced tooth edge, others convex, inturned shapes. *Stem* Light green to grey-green. Upright when young, becoming very twiggy to form a round shrub. After 3 years or more may become lax in habit. Medium to fast growth rate. *Fruit* Insignificant.

Hardiness Minimum winter temperature 14°F (−10°C). Foliage very susceptible to scorch by extremely cold winds.

Soil Requirements Good, deep, rich soil; tolerates both acidity and mild alkalinity. Thin chalk or limestone soils will induce severe chlorosis.

Sun/Shade aspect Prefers full sun, tolerates light to medium shade.

Pruning None required, but removing a few shoots 3-4 years old annually after flowering will encourage new growth. Treat severe winter damage by cutting back completely to non-damaged wood to regenerate new growth from below, or just on, soil level.

Propagation and nursery production From semi-ripe cuttings taken in midsummer. Purchase container-grown. Garden centres and nurseries in many areas stock a representative range of varieties; otherwise specific varieties may be sought from specialist nurseries.

Problems Leaves liable to scorching by cold winds. Will not attain full height and spread in unsuitable areas and likely to experience chlorosis on unsuitable soils.

Varieties of interest *C. americanus* (New Jersey Tea) Panicles of white flowers in early to mid summer. Dark green ovate leaves. A slightly tender variety reaching two-thirds average height and spread. From eastern and central USA. *C. arboreus* (Tree Ceanothus) Deep, vivid blue flowers in panicles borne in spring; large, ovate, dark green leaves. Slightly more tender than the average, and attains one-third more height and spread. *C. a. 'Trewithen Blue'* Called after its place of origin, Trewithen Gardens near Truro in Cornwall, England. Flowers slightly scented and deeper blue than *C. arboreus*. *C. 'A. T. Johnson'* Mid to pale blue panicles of flowers, late spring; some early autumn flowering. A light green, large-leaved variety. Very vigorous in habit, in some situations exceeding average heights. *C. 'Autumnal Blue'* Good-sized panicles of dark blue flowers, late summer and autumn. One of the hardiest varieties. *C. 'Burkwoodii'* Rich blue flowers borne mainly late spring and early summer, with good displays intermittently until autumn. Slightly more tender and slightly less height and spread than the average. *C. 'Cascade'* Powder blue flowers in open panicles, spring. Foliage light green and more lanceolate than normal. Branches more lax and open, forming attractive, almost pendulous habit. *C. 'Delight'* Deep blue flowers, produced in panicles 3-4in (8-10cm) long in mid to late spring. Leaves broad, lanceolate, mid green. Said to be one of the hardiest varieties. *C. dentatus* (Santa Barbara Ceanothus) Bright blue flowers, late spring, small, tooth-edged dark green leaves. *C. 'Dignity'* Dark blue flower panicles and dark green foliage. Normally flowers in spring, sometimes intermittently in autumn. *C. 'Edinburgh'* syn. *C. 'Edinensis'* Mid blue panicles of flowers, spring; broad, olive green leaves. Less than average hardiness. *C. impressus* Deep blue flowers, small, but borne in great profusion. Distinctive foliage effect, with small, almost curled, dark green leaves, veins being very deeply impressed within the surface. New shoots red to purple-red in colour. One of the hardiest of the *Ceanothus* varieties. *C. i. 'Puget Blue'* Deeper blue flowers and larger foliage. Possibly less hardy than its parent. *C. 'Italian Skies'* Mid to soft sky blue panicles of flowers, borne in trusses on branching stems in spring. Medium-sized, round to ovate, light green leaves. Less hardy than average. *C. × lobbianus 'Russellianus'* Bright blue flowers, freely borne in mid to late spring. Less hardy than average. *C. prostratus* (Squaw Carpet) Bright blue flowers borne freely in spring on this creeping, spreading plant with small, dark green to light green, broad to lanceolate leaves. Forms a mat 1½ft (50cm) in height and 5ft (1.5m) in area after 10-20 years. *C. rigidus* Very dark blue flowers in small, short, tufted panicles profusely borne mid to late spring. Interesting foliage, very dark olive green, small and crinkled. Hard to find. *C. 'Southmead'* Sky blue flowers in late spring and early summer. A very dense-growing shrub, with light green, broad, lanceolate leaves. Slightly less hardy than average. *C. thyrsiflorus* An abundance of medium-sized, well-spaced, mid blue flower panicles in spring and early summer. Dark green leaves. One of the hardiest varieties. *C. t. 'Blue Mound'* Covered in short panicles of deep blue flowers, late spring and early summer. Dark green leaves. Height 5ft (1.5m), spread 6ft (2m). *C. t. var. repens* (Creeping Blue Blossom) Rich blue flowers in abundance, mid spring. Good-sized, dark green, tooth-edged foliage. A low, spreading shrub attaining height 3ft (1m) and spread 6ft (2m) over 5 years, up to 6ft (2m) by 10ft (3m) over 10 years. *C. 'Topaz'* Large, well-spaced panicles of indigo blue flowers, mid to late summer. Large, round or ovate, mid green leaves. In cold climates should be considered semi-evergreen or even deciduous. *C. × veitchianus* Deep blue flowers, late spring and early summer. Medium-sized, dark green, broadly lanceolate leaves. Taller than most varieties and said to be one of the hardiest. *C. 'Yankee Point'* Panicles of light blue flowers in mid spring. Light to mid green, medium-sized narrow, ovate leaves. Compact habit.

Average height and spread
Five years
6x6ft (2x2m)
Ten years
10x10ft (3x3m)
Twenty years
or at maturity
13x13ft (4x4m)

Ceanothus impressus 'Puget Blue' in flower

CERATOSTIGMA WILLMOTTIANUM

SHRUBBY PLUMBAGO
Plumbaginaceae
Deciduous
This shrub is one of the jewels of autumn with its blue flowers and yellow foliage colour.

Origin From western China.
Use As a low, late summer to early autumn-flowering shrub for herbaceous or perennial borders, or for a grouped planting of shrubs.
Description *Flower* Five-petalled, saucer-shaped flowers of deepest blue with yellow stamens, borne singly or in clusters, opening intermittently through late summer to early autumn frosts. *Foliage* Leaves ovate, 2in (5cm) long, dark green with purple-red veining and shading. Good yellow autumn colour. *Stem* Upright, light green new shoots produced each spring, aging to purple-green and dying completely in winter. Fast growth rate. *Fruit* Small, tufted clusters of grey to grey-brown seedheads.
Hardiness Tolerates 14°F (−10°C), especially if the base of the plant is protected with dried twigs and leaves.
Soil Requirements Deep, rich soil required to achieve the best effect of flowers and foliage. Tolerates both dry and moist areas.

Ceratostigma willmottianum **in flower**

Sun/Shade aspect Best in full sun. Dislikes shade.
Pruning All stems should be cut back hard to ground level annually in early to mid spring.
Propagation and nursery production From softwood cuttings taken in midsummer. Purchase container-grown. Generally available, especially in late summer when in flower. Best planting heights 1-1½ft (30-50cm).
Problems Because of the plant's growth cycle, stock often looks weak and insipid in winter and early spring but once planted out grows rapidly the following spring.
Varieties of interest *C. griffithii* Large, deep purple-blue flowers. A variety with broader, oblong foliage, with attractive orange-red shading and good autumn colour. Slightly less hardy and lower growing with less spread.
Average height and spread
Five years
2x2ft (60x60cm)
Ten years
3x3ft (1x1m)
Twenty years
or at maturity
3x3ft (1x1m)

CHAENOMELES JAPONICA
and similar varieties

ORNAMENTAL QUINCE, JAPANESE FLOWERING QUINCE, JAPONICA
Rosaceae
Deciduous
A fine flowering shrub, equally well-presented as a freestanding or wall-trained specimen.

Origin From China, most varieties being of garden origin.
Use As a single, freestanding specimen, in a shrub border, or as a wall shrub for cool, relatively exposed sites. Plant 3ft (1m) apart for a large, informal, open-growing hedge.
Description *Flower* Single flowers shaped like apple blossom borne in profusion on wood 2 years old or more, early to mid spring. Colours range through white, pink, apricot, flame, orange and red, dependent on variety. *Foliage* Leaves elliptic, medium-sized, 3-4in (8-10cm) long, light to dark green. Some autumn colour. *Stem* Upright when young and light green-brown, becoming dark brown, more twiggy and producing isolated large rigid thorns. Forming a round-topped shrub or fan-shaped wall shrub. Medium growth rate. *Fruit* Large, pear-shaped fruits follow the flowers, ripening to an attractive bright yellow.
Hardiness Tolerates −13°F (−25°C).
Soil Requirements Does well on any soil but liable to chlorosis in very alkaline areas.
Sun/Shade aspect Does well in full sun to heavy shade.
Pruning Remove all previous season's growth after flowering if not required for training or shaping.
Propagation and nursery production From semi-ripe cuttings taken in midsummer. Purchase container-grown. Best planting heights 15in-2½ft (40-80cm). Wide range of varieties generally available; individual varieties from specialist nurseries.
Problems Intermittently produces very sharp thorns. May suffer fungus disease; prune out affected wood.
Varieties of interest *C. j. var. alpina* Orange-red flowers borne freely, late spring. A little shy to fruit due to its less vigorous habit. A low, dwarf, spreading variety, height at maturity 3ft (1m) with spread up to 6ft (2m). *C. speciosa 'Atrococcinea'* Large, deep crimson flowers. *C. s. 'Brilliant'* Large brilliant red to clear scarlet flowers. *C. s. 'Cardinalis'* Crimson-scarlet flowers. Average height but more than average spread. *C. s. 'Geisha Girl'* Very attractive deep apricot flowers. Later flowering. *C. s. 'Moerloosii'* (Apple Blossom) Apple blossom pink and white flowers, more sparsely produced than some varieties. *C. s. 'Nivalis'* A pure white-flowered variety,

Chaenomeles speciosa 'Nivalis' **in flower**

Chaenomeles × superba 'Crimson and Gold'

green-white on first opening. Fewer flowers than average, but growth more vigorous. *C. s. 'Simonii'* Deep blood red flowers freely produced. Low-growing, height 3ft (1m), spread 6ft (2m). *C. s. 'Snow'* Snow white flowers. A good variety. *C. s. 'Umbilicata'* Deep pink flowers, larger than most. *C × superba 'Choshan'* Semi-double, peach-apricot flowers. Low growing, height 3ft (1m), spread 5ft (1.5m). Not easy to find. *C. × s. 'Coral Sea'* Coral pink, good-sized flowers and good fruits. *C. × s. 'Crimson and Gold'* Bright red flowers with pronounced golden anthers, good fruit production. *C. × s. 'Elly Mossel'* Large, bright scarlet flowers, good fruit. *C. × s. 'Fire Dance'* Rich orange-scarlet flowers. *C. × s. 'Knap Hill Scarlet'* Smaller, brilliant orange-scarlet flowers, freely borne. Height slightly less than average. *C. × s. 'Hollandia'* A very good scarlet-red flowering variety with good fruits. *C. × s. 'Nicoline'* Large red flowers; average height but with more spread. *C. × s. 'Pink Lady'* Good deep pink flowers and good fruits. *C. × s. 'Rowallane'* Brilliant crimson flowers and small fruits; a lower, spreading variety.
There are many varieties of *C. speciosa* and *C. × superba*. Those listed are a good representative selection.
Average height and spread
Five years
6x6ft (2x2m)
Ten years
10x10ft (3x3m)
Twenty years
or at maturity
10x10ft (3x3m)

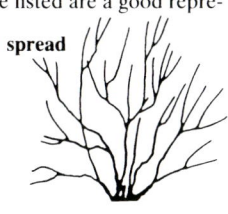

CHIMONANTHUS PRAECOX
(Chimonanthus fragrans)

FRAGRANT WINTERSWEET
Calycanthaceae
Deciduous
An interesting scented, flowering shrub for winter display.

Origin From China.
Use Freestanding or on a sunny wall, where it will produce more flowers.
Description *Flower* Lemon yellow, waxy, hanging, bell-shaped flowers with purple anthers, frost-hardy, borne on mature branches 3-4 years old, late winter. *Foliage* Leaves light green to yellow-green, medium-sized, elliptic, 3-7in (8-17cm) long; some yellow autumn colour. *Stem* Light green, strong, upright when young, aging into brittle, twiggy branches. If grown as a wall shrub, stems ripen more readily and flower earlier. Forms an upright, tall shrub, or if fan-trained a spreading wall shrub. Medium growth rate, slowing with age. *Fruit* Insignificant.

95

Hardiness Tolerates 14°F (−10°C), but expect some die-back on growth produced in late autumn.
Soil Requirements Does well on most soils, especially alkaline.
Sun/Shade aspect Full sun ripens the wood best for flowering; tolerates light shade.
Propagation and nursery production From semi-ripe cuttings taken in midsummer. Purchase container-grown. Available from specialist nurseries. Best planting heights 15in-2½ft (40-80cm).
Problems Can be slow to flower, especially in years when wood is slow to ripen. Looks unattractive in a container, but grows rapidly once planted out.

Chimonanthus praecox in flower

Varieties of interest *C. p. 'Grandiflorus'* Deeper yellow flowers with an interesting red stain in the throat. More difficult to find. *C. p. 'Luteus'* Clear bright yellow with even more waxy texture, the bell-shaped flowers are more open than in the parent. Not easy to find but worth pursuing.
Average height and spread
Five years
4x3ft (1.2x1m)
Ten years
6x5ft (2x1.5m)
Twenty years
or at maturity
8x8ft (2.5x2.5m)

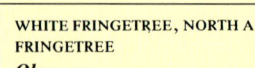

CHIONANTHUS VIRGINICUS

WHITE FRINGETREE, NORTH AMERICAN FRINGETREE
Oleaceae
Deciduous
An intriguing summer-flowering shrub which deserves wider planting.

Origin From eastern North America.
Use As a large freestanding shrub or in a fair-sized shrub border for interesting midsummer flowering.
Description *Flower* Slightly fragrant, white, 4-5 narrow strap-shaped petals being produced on mature wood 3 years old or more, early to mid summer. *Foliage* Leaves lanceolate, 3-8in (8-20cm) long, broad, olive green, giving some yellow autumn colour. *Stem* Upright and strong when young, becoming more spreading with age, forming domeshaped shrub. Medium to slow growth rate. *Fruit* In its native environment may produce small damson-like fruits.
Hardiness Tolerates 14°F (−10°C).
Soil Requirements Does well in any soil.
Sun/Shade aspect Best in full sun.
Pruning None required but any obstructive branches may be removed.
Propagation and nursery production From layering or from seed. Purchase container-grown. Readily available in its native environ-

Chionanthus virginicus in flower

ment but not easily found elsewhere. Best planting heights 15in-2½ft (40-80cm).
Problems Extreme cold leads to die-back of previous season's growth.
Varieties of interest *C. retusus* (Chinese Fringetree) A tender variety requiring temperatures no lower than 23°F (−5°C). More likely to produce damson-like fruits in Mediterranean areas. From China.
Average height and spread
Five years
5x5ft (1.5x1.5m)
Ten years
8x8ft (2.5x2.5m)
Twenty years
or at maturity
12x12ft (3.5x3.5m)

CHOISYA TERNATA

MEXICAN ORANGE BLOSSOM
Rutaceae
Evergreen
A very attractive, scented, late spring to early summer-flowering shrub.

Origin From Mexico.
Use As an evergreen shrub for summer flowering, standing on its own or in a mixed border. Plant 2½ft (80cm) apart in single line for an informal hedge.
Description *Flower* Fragrant, single, white, orange-scented flowers, borne in flat-topped clusters, late spring to early summer. *Foliage* Leaves glossy, mid to dark green, trifoliate, 3-6in (8-15cm) long, which when crushed give off aromatic scent. *Stem* Light to bright green,

glossy, upright, becoming spreading and twiggy with age, forming broad-based, dome-shaped shrub. Medium growth rate when young or pruned back, slowing with age. *Fruit* Insignificant.
Hardiness Tolerates 14°F (−10°C). Leaf damage can occur in lower temperatures or in severe wind chill. In some winters, may die back to ground level but can rejuvenate itself in following spring.
Soil Requirements Does well on most, although very severe alkaline soils may lead to chlorosis.
Sun/Shade aspect Equally good in full sun or deep shade.
Pruning Two methods of pruning are advocated. Cut back to within 1½ft (50cm) of ground level after 3-4 years, so that it can rejuvenate itself, and repeat process every third or fourth year following. This keeps the foliage glossy and encourages flowering. Otherwise, on mature shrubs, remove one-third of the oldest wood to ground level after flowering to encourage rejuvenation from centre and base.
Propagation and nursery production From softwood cuttings taken in summer. Purchase container-grown. Plants are always relatively small when purchased, but quickly mature when planted out. Best planting heights 1-2ft (30-60cm). Availability variable; if not in garden centres and nurseries, must be sought from specialist sources.
Problems If pruning is neglected plant becomes old, woody and unproductive.
Varieties of interest *C. ternata 'Sundance'* Yellow-green in spring, quickly becoming golden yellow which persists through winter. Slightly more tender. Two-thirds average height and spread.
Average height and spread
Five years
3x4ft (1x1.2m)
Ten years
6x5½ft (2x1.8m)
Twenty years
or at maturity
6x6ft (2x2m)

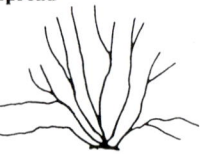

CISTUS

ROCK ROSE
Cistaceae
Evergreen
A tender shrub worth experimenting with for the beauty it can give in its summer display of flowers.

Origin From southern Europe and North Africa.
Use As a low shrub, very freely flowering in summer used singly or in mixed borders. Can be fan-trained on sunny or sheltered wall.
Description *Flower* Single, flat flowers lasting

Choisya ternata in flower

Cistus × cyprius in flower

only for one day but followed by new buds opening in rapid succession, early to late summer. White, or white with brown or purple spots, through shades of pink to dark purple-pink. *Foliage* Leaves ovate or lanceolate, medium-sized, 1-2in (3-5cm) long, light green or grey; glossy surfaced in the green-leaved varieties, grey-leaved having a grey down or bloom. When young, leaf buds and new growth are often sticky. *Stem* Light green or grey-green depending on variety. Some varieties grow upright, others of spreading habit. Medium growth rate. *Fruit* Insignificant.
Hardiness Minimum winter temperature 23°F (−5°C).

Cistus × purpureus in flower

Soil Requirements Does well on all soil types, tolerating quite high alkalinity.
Sun/Shade aspect Best in full sun for free flowering.
Pruning Trim back lightly each year to increase flowering.
Propagation and nursery production From softwood cuttings taken early to mid summer. Purchase container-grown. A wide range of varieties normally obtainable in mild areas; otherwise must be sought from specialist sources. Best planting heights 8in-2½ft (20-80cm).
Problems Container-grown plants often look leggy and unsightly, but are the best choice because the wood is hard. Plants looking soft and fleshy at purchase have been grown under glass and are less reliable, but for early to mid summer planting shrubs of any age are suitable.
Varieties of interest *C. × aiguilari* Large white flowers throughout early and mid summer. Grey-green foliage. *C. × a. 'Maculatus'*

Large, flat flowers with central ring of crimson blotches on each of the five petals. Dark green, glossy foliage, 2-4in (5-10cm) long. Upright; height and spread over 3ft (1m) in favourable conditions. *C. × corbariensis* Small, pure white flowers, opening from crimson-tinted buds. Small, light green to grey-green leaves. Low-growing; height 2ft (60cm), spread 3ft (1m). Hardier than most. *C. × cyprius* Large white flowers with crimson blotches on each of the five petals. Shiny green foliage, 4-6in (10-15cm) long. Upright and tall-growing to height and spread of 6ft (2m). Said to be one of the hardiest. *C. × c. 'Albiflorus'* Pure white flowers. Otherwise similar to *C. × cyprius*. *C. 'Greyswood Pink'* A profusion of pink flowers. Downy grey to grey-green foliage. *C. ladanifer* (Gum Cistus) White flowers up to 4in (10cm) in width with chocolate stain at base of each crimped petal. Leaves narrow, lanceolate, 3-4in (8-10cm) long, bright to mid green with glossy surface. Strong-growing, height up to 6ft (2m) in favourable conditions. *C. l. 'Albiflorus'* A pure white form of the above. *C. laurifolius* Flat, white flowers with yellow marking in centre. Very leathery, dark blue-green leaves, 4-5in (10-12cm) long. A more hardy variety, height up to 6ft (2m). *C. × lusitanicus* Large, white flowers with crimson basal blotches, leaves green. Height 2½-3ft (80cm-1m). *C. × lusitanicus 'Decumbens'* Covered in small, white flowers, again with crimson basal blotches. A wide-growing variety, height 2ft (60cm), spread up to 4ft (1.2m). *C. 'Peggy Sannons'* Flowers a soft shade of pink which contrasts well with foliage of grey-green with a downy texture, which is repeated on the upright stems. Height 4ft (1.2m). *C. populifolius* White flowers with yellow staining at the

Cistus 'Silver Pink' in flower

base of each petal. Rounded, poplar-shaped, hairy leaves produced on an erect, upright shrub. One of the hardiest. Height 6ft (2m). *C. × pulverulentus* syn. *C. albidus × crispus, C. 'Warley Rose'* Cerise flowers, foliage sage green and waxy-textured. A low spreading shrub, height 2ft (60cm), spread 3ft (1m). *C. × purpureus* Flowers rosy crimson, each with a chocolate basal blotch. Narrow, green to grey-green foliage moderately sparsely produced on strong, upright stems. Height 5ft (1.5m), spread 3ft (1m). *C. salvifolius* White flowers with a yellow basal stain to each petal, sage green leaves forming a foil. Height 1½ft (50cm), spread 2½ft (80cm). *C. 'Silver Pink'* Silver-pink flowers, borne unusually on long clusters, very attractive grey foliage. Height 2ft (60cm), spread 2½ft (80cm). *C. × skanbergii* Clear pink flowers and small grey leaves make this one of the most attractive of all low-growing *Cistus* varieties. Height 2ft (60cm), spread 2½ft (80cm). *C. 'Sunset'* Deep cerise-pink flowers, foliage grey. Upright, height 2-2½ft (60-80cm), spread 2ft (60cm).
Average height and spread
Five years
2x2ft (60x60cm)
Ten years
2½x2½ft (80x80cm)
Twenty years
or at maturity
3x3ft (1x1m)

CLERODENDRUM BUNGEI

KNOWN BY BOTANICAL NAME

Verbenaceae
Deciduous
An attractive, autumn-flowering shrub with good foliage as long as the correct position can be found.

Origin From China.
Use As a foliage plant with late autumn flowers, best planted in a position where it can be restrained from sending out underground shoots.

Clerodendrum bungei in flower

Description *Flower* Large clusters of purple-red flowers with a pungent odour produced in the terminals of each shoot, late summer to early autumn. *Foliage* Leaves large, 4-8in (10-20cm) long and 4-6in (10-15cm) wide, heart-shaped, deep purple when young, aging to purple-green through summer. When crushed, leaves give off a pungent odour. Foliage destroyed by the first winter frosts. *Stem* Upright, green with purple shading, ending in flower bud, dying away in winter and normally regenerating in spring. Can become invasive, spreading underground by root suckers which often emerge some distance from the shrub. Fast growth rate in spring and early summer. *Fruit* Insignificant.

Hardiness This is a sub-shrub and winter damage to shoots above ground level must be expected; the whole plant dies to ground level and is rejuvenated in following spring. Roots may also be damaged where winter temperatures of less than 4°F (−15°C) are experienced.
Soil Requirements Any, including moderately dry soil.
Sun/Shade aspect Best in light shade to full sun to maintain leaf colour.
Pruning Prune all shoots that have survived winter to ground level in early spring to encourage new shoots with large foliage and flowering heads.
Propagation and nursery production From rooted suckers lifted when seen. Purchase container-grown. In autumn and winter normally available as rooted base showing little or no signs of shoots.
Problems An invasive shrub which needs control except in very large gardens.
Average height and spread
Five years
2½x5ft (80cmx1.5m)
Ten years
3x6ft (1x2m)
Twenty years
or at maturity
3x8ft (1x2.5m)

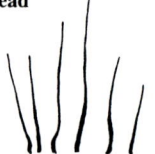

CLERODENDRUM TRICHOTOMUM

KNOWN BY BOTANICAL NAME
Verbenaceae
Deciduous
If planted in the right position to form a specimen, this is a truly magnificent sight when in flower or fruit.

Origin From China and Japan.
Use As a freestanding specimen shrub.
Description *Flower* Open clusters of fragrant, white, star-shaped flowers with pronounced maroon calyxes at the base, late summer through to early autumn. *Foliage* Leaves purple-green, oblong, 4-9in (10-23cm) long and 2-5in (5-12cm) wide. Some yellow autumn colour. *Stem* Upright, becoming spreading, to make a wide-headed, canopy-like small 'tree', with little or no lower growth. Fast growth rate when young, slowing with age. *Fruit* Bright blue berries follow flowering while the maroon calyxes to each flower persist, contrasting very attractively.
Hardiness Tolerates 14°F (−10°C). Young plants may suffer from winter die-back so that the overall height is curtailed, arresting the shrub's formation of a canopy shape.
Soil Requirements Does well on most, although severely alkaline soils may lead to chlorosis.
Sun/Shade aspect Prefers medium to light

Clerodendrum trichotomum in flower

shade, tolerates full sun if moisture is adequate.
Pruning None required, except for removal of any winter-damaged growth.
Propagation and nursery production From seed or from rooted suckers. Best purchased container-grown. Difficult to find.
Problems Apart from some stem damage in winter, the only other problem is that it does take time to reach its full potential.
Varieties of interest *C. t. var. fargesii* Smoother leaves and stems, and said to fruit with greater freedom.
Average height and spread
Five years
5x5ft (1.5x1.5m)
Ten years
8x8ft (2.5x2.5m)
Twenty years
or at maturity
12x12ft (3.5x3.5m)

CLETHRA ALNIFOLIA

SUMMERSWEET CLETHRA, SWEET PEPPER BUSH, WHITE ALDER
Clethraceae
Deciduous
Used in a wild or woodland garden, a very attractive autumn-flowering shrub.

Origin From eastern North America.
Use As a freestanding shrub for wild or woodland gardens.

Clethra alnifolia in flower

Description *Flower* Fluffy panicles, 2-4in (5-10cm) long, scented, creamy-white to white or pink, dependent on variety, in early to late autumn. *Foliage* Leaves broad, lanceolate, 1½-4in (4-10cm) long and up to 2in (5cm) wide, light green with slightly toothed edges, giving good yellow autumn colour. *Stem* Upright when young, light green to grey, aging to grey-brown, becoming more twiggy and forming a wide-based, suckering clump. Slow to medium growth rate. *Fruit* Insignificant.
Hardiness Tolerates 4°F (−15°C).
Soil Requirements Acid, dislikes any alkalinity.
Sun/Shade aspect Prefers light shade; tolerates full sun or moderate shade.
Pruning None required.
Propagation and nursery production From softwood cuttings taken in midsummer, from seed or from layers. Purchase container-grown; may be available from specialist sources if not found in nurseries or garden centres.
Problems None, apart from its need for a lime-free soil.
Varieties of interest *C. a. 'Paniculata'* A vigorous and superior variety with large panicles of flowers. *C. a. 'Rosea'* A beautiful variety, the buds and flowers being tinged pink and the leaves glossy. *C. a. 'Pink Spire'*

An improved European form with pinker, more spiky panicles of flowers. *C. acuminata* (White Alder) A variety with racemes of fragrant cream flowers and good yellow autumn colour. From south-eastern USA and not readily available outside its native environment. *C. tomentosa* A later-flowering variety similar to *C. alnifolia*, with greyer foliage. Originating in south-eastern USA.
Average height and spread
Five years
3x3ft (1x1m)
Ten years
4x6ft (1.2x2m)
Twenty years
or at maturity
7x10ft (2.2x3m)

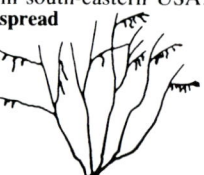

COLLETIA CRUCIATA

ANCHOR PLANT
Rhamnaceae
Evergreen
An unusual shrub without foliage but viciously armed; a cactus-like feature which must be carefully sited.

Origin From Brazil and Uruguay.
Use Planted on its own, as an interesting shrub of pure architectural value. Ideal for a very large rock garden or to surmount a planting of heathers or other low-growing shrubs.
Description *Flower* Small, white, fragrant, pitcher-shaped, borne in profusion on mature shrubs midsummer to early autumn. *Foliage* No foliage. *Stem* Light grey-green stems, surmounted by branches which take on a flat, triangular shape, each triangulation topped by a very sharp spike, forming a round-topped shrub somewhat resembling a cactus. Slow-growing. *Fruit* Insignificant.
Hardiness Tolerates 14°F (−10°C), but susceptible to frost when young and needing protection in these early stages.

Colletia cruciata

Soil Requirements Any, including dry to very dry soils.
Sun/Shade aspect Full sun to light shade.
Pruning None required.
Propagation and nursery production From seed. Purchase container-grown. Not easy to find. Best planting heights 8-15in (20-40cm).
Problems Its spines are vicious, so handle with due care.
Varieties of interest *C. armata* Flowers vanilla-scented, borne profusely late summer to early autumn. Shoots form rounded spines, each tipped with a single thorn. From Chile. *C. a. 'Rosea'* A stronger-growing variety, height and spread 8ft (2.5m). A very scarce form with flowers pink in bud, opening to pink-tinged white.

Colletia armata **in flower**

Average height and spread
Five years
2x2½ft (60x80cm)
Ten years
3x4ft (1x1.2m)
Twenty years
or at maturity
4x5½ft (1.2x1.8m)

COLUTEA ARBORESCENS

BLADDER SENNA
Leguminosae
Deciduous
A shrub to give an attractive display of summer flowers and autumn fruit, well worth a place in any medium-sized shrub border.

Origin From southern Europe and the Mediterranean.
Use Either freestanding or in a group planting. Valued both for its flowers and interesting seed pods.
Description *Flower* Yellow pea-flowers borne from midsummer to late autumn. *Foliage* Leaves light green to grey-green, pinnate, 2-4in (5-10cm) long, producing yellow autumn colour. *Stem* Light green to grey-green, aging to grey-brown. Upright, becoming twiggy and spreading with age, forming a rounded shrub, or a canopy shape with a bare base. *Fruit* Almost translucent grey to grey-green, large seed pods; if a pod is squeezed, it will explode and release a black seed. Produced early to late autumn, together with some late flowers.
Hardiness Tolerates 4°F (−15°C).

Colutea arborescens **in flower and fruit**

Soil Requirements Any soil, tolerates high alkalinity.
Sun/Shade aspect Best in full sun, tolerates light shade.
Pruning If it becomes invasive and too large, cut back hard and remove some shoots from the outer edges.
Propagation and nursery production From seed. Purchase container-grown. Not normally seen in garden centres; must be sought from a good nursery or specialist source. Best planting heights 2-3ft (60cm-1m).
Problems May have a poor root system; a short stake will prevent damage by wind rocking.
Varieties of interest *C. a. 'Copper Beauty'* Dark orange to copper-red flowers, producing pods and good autumn colour. *C. × media* Bronze and yellow flowers, again with pods. Foliage slightly more grey-green than that of *C. arborescens*. Slightly less vigorous height and spread. Of garden origin. *C. orientalis* Copper flowers and blue, glaucous foliage. Forms a more rounded shape, with less height. A variety from the Caucasian region.

Average height and spread
Five years
6x6ft (2x2m)
Ten years
8x8ft (2.5x2.5m)
Twenty years
or at maturity
10x10ft (3x3m)

CONVOLVULUS CNEORUM

KNOWN BY BOTANICAL NAME
Convolvulaceae
Evergreen
An attractive silver-foliaged, white-flowering, low shrub.

Origin From south-eastern Europe.
Use For a small border, large rock garden or grouped with other silver-leaved plants. Effective in a large container; use good potting medium.

Convolvulus cneorum **in flower**

Description *Flower* Green-white to pure white, medium-sized, funnel-shaped, emerging from pale pink buds, late spring to midsummer. *Foliage* Leaves lanceolate, 1-1½in (3-4cm) long, silver-grey; best in spring and summer, although retained through winter. Severe cold may damage, although not totally destroy foliage. *Stem* Short, grey-green to silver-grey, upright. Forms a low, dome-shaped shrub. Medium growth rate. *Fruit* Insignificant.
Hardiness Minimum winter temperature 14°F (−10°C).
Soil Requirements Dry, well-drained soil; tolerates some alkalinity.

Sun/Shade aspect Must be in full sun. Any shade will spoil foliage colour.
Pruning A light trim in early spring encourages new flowering growth and good foliage.
Propagation and nursery production From softwood cuttings taken early summer. Purchase container-grown; in areas which experience cold winters, spring or early summer planting is best. Best planting heights 8-15in (20-40cm).
Problems Young plants may look small and distorted when purchased in early spring, but once planted out grow quickly.
Average height and spread
Five years
1½x2ft (50x60cm)
Ten years
1½x2½ft (50x80cm)
Twenty years
or at maturity
1½x2½ft (50x80cm)

CORDYLINE AUSTRALIS

CABBAGE TREE OF NEW ZEALAND
Agavaceae
Evergreen
A shrub with tropical foliage effect, useful for gardens in mild areas.

Origin From New Zealand.
Use As a feature shrub or, in time, a small tree, to give a Mediterranean sub-tropical appearance. Does well in large containers, if good potting medium is used.
Description *Flower* Small, fragrant, terminal panicles, creamy white, produced on mature plants more than 5 years old. *Foliage* Leaves 3ft (1m) long or more, narrow, sword-like, produced in rosettes at the end of each branch. *Stem* Short, squat when young, becoming elongated with age as produces new terminal leaves and discards lower, aged leaves. Brown-grey with some bark peeling. At first a low shrub, in 30 years or more a small tree. *Fruit* Insignificant.
Hardiness Winter minimum 23°F (−5°C). Good in coastal areas.
Soil Requirements Well-drained dry soil.
Sun/Shade aspect Best in full sun to very light shade.
Pruning None required, but lower leaves that have died may be removed.
Propagation and nursery production From seed. Purchase container-grown. Fairly easy to find in temperate areas. In colder climates less easily obtainable and will need to be protected under glass in winter. Best planting heights 15in-6ft (40cm-2m).
Problems Often purchased in a favourable area and transported to a colder location, where it soon succumbs to cold winters.
Varieties of interest *C. a. 'Atropurpurea'* A variety with purple leaves, slightly more tender than the parent.
Average height and spread
Five years
3x3ft (1x1m)
Ten years
6x6ft (2x2m)
Twenty years
or at maturity
13x10ft (4x3m)
Height and spread may be greater in very mild areas.

Cordyline australis

99

CORNUS ALBA

RED-BARKED DOGWOOD, TATARIAN
DOGWOOD
Cornaceae
Deciduous
A very useful range of plants with year-round attractions.

Origin From Siberia to Manchuria and through North Korea.
Use For large mass planting or as a winter attraction for the shrub border. Variegated forms blend well with other shrubs of purple or red summer foliage. If planted 3ft (1m) apart all varieties make an attractive hedge which may be clipped into a formal shape, although foliage formation may be a little lax and open.
Description *Flower* White, borne in clusters, late spring, early summer on mature wood more than 2 years old. *Foliage* Leaves medium-sized, ovate, 2-5in (5-12cm) long, with slightly lacerated edges. Dark to olive green with red veins; undersides slightly silver. Attractive red to plum-red autumn colour. *Stem* Its main feature. Almost upright when young, strong and quick-growing. Smooth, glossy, red to bright red stems in winter, becoming green-red to green-brown in summer. Stems are the main feature. At maturity (or without hard pruning in spring) becoming a duller red in winter, slow in growth, more twiggy and branching. On mature unpruned plants, the branches may become arching. *Fruit* May produce clusters of white to blue-white, soft, translucent fruits in early autumn.

Cornus alba 'Elegantissima' in leaf

Hardiness Tolerates winter temperatures below −13°F (−25°C).
Soil Requirements Does well on all soils. Tolerates extreme waterlogging and the moisture produces stronger shoots, but performance also good on dry soils.
Sun/Shade aspect Full sun to medium shade. Deep shade will lead to some deterioration of overall shape.
Pruning Can be cut back very hard in early spring to generate strong, well-coloured winter stems and, in variegated forms, a larger variegated leaf. Unfortunately, pruning back hard as recommended eliminates the flowering and fruiting displays.
Propagation and nursery production From hardwood cuttings, layering or self-layered off-shoots. Plant bare-rooted or container-grown. Best planting heights 2-3ft (60cm-1m).
Problems Often planted in areas for which it is too large, or is not sufficiently pruned back and therefore gets out of hand.
Varieties of interest *C. alba 'Aurea'* Yellow to yellow-green leaves which are somewhat thinner and softer than those of parent; may suffer leaf scorch in strong sunlight, so needs

Cornus alba 'Sibirica' in winter

light shade. A slightly less vigorous form which may be hard to find. *C. a. 'Elegantissima'* syn. *C. a. 'Sibirica Variegata'* Leaves have broad, white margins with mottling of the green inner section. One of the most popular forms. *C. a. 'Gouchaltii'* An extremely attractive golden variegated form, giving way in some areas in favour of *C. a. 'Spaethii'*. *C. a. 'Kesselringii'* An attractive, almost black-stemmed variety with dark purple-green foliage. Very effective used with red-stemmed forms and with the yellow-stemmed *C. stolonifera 'Flaviramea'*. Slightly less vigorous than the parent. *C. a. 'Sibirica'* syn. *C. a. 'Atrosanguinea'* (Westonbirt Dogwood) New yellow foliage is very soft and easily damaged, both by strong sunlight and late spring frosts. Extremely attractive if grown in a moist, fertile planting area. The common name refers to the Westonbirt Arboretum in Gloucestershire, England. *C. a. 'Spaethii'* A good form for its yellow-margined, mid green leaves; not as vigorous as the parent, but forming a bushy yellow clump. Equally good if unpruned, although leaves will be smaller. The Long Ashton form, named after the Long Ashton Horticultural Experimental Station in the UK, where it was produced, will become the accepted form of this variety in commercial horticulture. *C. a. 'Variegata'* Often confused with *C. a. 'Elegantissima'*, but with strong white bands of variegation around dark green centres. *C. amomum* Attractive, ovate, dark green foliage with good autumn colour. White flower clusters. Green-red stems. *C. baileyi* Red stems. Mid green ovate foliage with good orange-red autumn colours. White flower clusters. *C. sanguinea* (Common Dogwood) Attractive red-green winter stems. Mid green, slightly glossy, ovate

Cornus alba 'Spaethii' in leaf

foliage with orange-red autumn colours. White flowers give way to red clusters of berries.
Average height and spread
Five years
8x10ft (2.5x3m)
Ten years
8x13ft (2.5x4m)
Twenty years
or at maturity
10x13ft (3x4m)

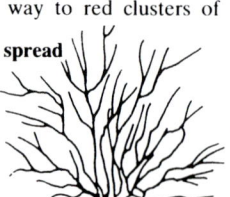

CORNUS ALTERNIFOLIA

PAGODA CORNUS, PAGODA DOGWOOD,
PAGODA TREE
Cornaceae
Deciduous
Cornus alternifolia is interesting in its own right, but the variety 'Argentea' has exceptional foliage colour.

Origin From eastern North America.
Use As a freestanding feature shrub displaying the pagoda effect, or, in time, as a small tree.

Cornus alternifolia 'Argentea' in leaf

Description *Flower* White, in flat clusters, borne on short stalks above the flat, tiered branches in mid spring. *Foliage* Leaves small, ovate, 1-1½in (3-4cm) long, light green to mid green, produced on short, upright twigs from main horizontal branches. Good autumn colour. *Stem* Horizontal branches protrude from central upright stem, producing a tiered pagoda effect. Small branchlets standing upright from stem support foliage and

flowers. Light green when young, aging to red-green, finally to brown-green. Medium growth rate. *Fruit* In very hot, dry summers may produce small red fruits, aging to black.
Hardiness Tolerates 14°F (−10°C).
Soil Requirements Any, tolerates both alkalinity and acidity, and a certain amount of excess moisture.
Sun/Shade aspect Best in full sun or light shade. Deep shade causes elongation of stems, spoiling overall shape.
Pruning None required, but advisable to remove any intermediate stems or twigs which may occur between the tiers.
Propagation and nursery production From seed or softwood cuttings taken midsummer. Purchase container-grown. May need some searching for in specialist nurseries. Best planting heights 1½-3ft (50cm-1m).
Problems May produce more than one central leading stem and this should be corrected by pruning.
Varieties of interest *C. a. 'Argentea'* White-silver, variegated foliage of great beauty. Good red-edged autumn colour. Reaching less than two-thirds the height and spread of its parent. Hard to find and, because it has to be grafted, fetches a high nursery price. Best planting heights 15in-2½ft (40-80cm). Slower growth rate.
Average height and spread
Five years
5x3ft (1.5x1m)
Ten years
10x6ft (3x2m)
Twenty years
or at maturity
16x12ft (5x3.5m)

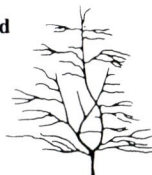

CORNUS CANADENSIS

BUNCH BERRY, CREEPING DOGWOOD
Cornaceae
Semi-evergreen
A very attractive ground cover flowering plant useful in shady sites; classified as a sub-shrub.

Origin From USA and Canada.
Use As ground cover for woodland or shady areas.
Description *Flower* Four-petalled, white bracts produced in summer above a low carpet of foliage. *Foliage* Leaves ovate, 1in (3cm) long, olive green with some red shading, giving good autumn colour. Some foliage retained through winter, although dying away in early spring. *Stem* Slow-growing, light green to red-green suckering shoots produced each spring, mostly dying away in winter, but some being retained through winter, supporting the semi-evergreen foliage. *Fruit* Red, strawberry shaped, from early to late autumn onwards.

Cornus canadensis in flower

Cornus controversa 'Variegata' in leaf

Hardiness Tolerates 4°F (−15°C).
Soil Requirements A good, sandy, open, leafy soil, enabling it to send runners through underground effortlessly.
Sun/Shade aspect Light shade to medium shade. Dislikes full sun.
Pruning None required.
Propagation and nursery production From division or self-produced layers. Purchase container-grown. Not easy to find. Best planting heights 4-8in (10-20cm).
Problems Suitable soil essential for good results.
Average height and spread
Five years
8inx3ft (20cmx1m)
Ten years
8inx6ft (20cmx2m)
Twenty years
or at maturity
8inx10ft (20cmx3m)

CORNUS CONTROVERSA

WEDDING CAKE TREE
Cornaceae
Deciduous
A truly choice plant in both its green and variegated forms, with an intriguing growth pattern.

Origin From Japan, China and Taiwan.
Use As a large, tiered shrub of architectural shape. Often considered to be a tree and may achieve tree-like proportions.
Description *Flower* Umbels of white flowers, late spring to early summer, standing upright on short stalks above the tiered branches. *Foliage* Leaves medium-sized, 3-5in (8-12cm) long, ovate, light green with a brown-red hue, colouring depending partly on vigour of shrub — less vigorous plants have more colour, and vice versa. Very good bright orange-scarlet autumn colour. Variegated-leaved forms even more spectacular. *Stem* Initially twiggy, spreading, tiered formation. In second and third year stem forms strong upward growth 2-3ft (60cm-1m) long from below first tiered branches; in third and fourth year produces a second tier of foliage at same time as lower tier expands and extends with twiggy growth. In fifth and sixth years a new upward strong shoot forms; process continues until shrub reaches ultimate height, giving a tiered 'wedding cake' effect. *Fruit* Small, ovate, white to white-blue, occasionally produced in hot summers.
Hardiness Tolerates winter temperatures ranging from 4°F (−15°C) down to −13°F (−25°C) although new upward growth and emerging young foliage can be damaged by extreme conditions and late spring frosts.
Soil Requirements Does well on all, but matures more quickly on deep, rich soils. Extreme alkalinity may lead to chlorosis.
Sun/Shade aspect Best in full sun to light shade. Deep shade causes shrub to lose its interesting structural shape.
Pruning None required in early years. After 15-20 years infilling twigs between each tier

may be pruned to retain the architectural structure.
Propagation and nursery production Green-leaved form from seed or semi-ripe cuttings. Not always easy to find.
Problems None.
Varieties of interest *C. c. 'Variegata'* An extremely beautiful white variegated form with wide clear margins around the slightly convexed leaves. Propagated by grafting on various understocks. Hard to find and costly.
Average height and spread
Five years
6x6ft (2x2m)
Ten years
10x10ft (3x3m)
Twenty years
or at maturity
23x20ft (7x6m)

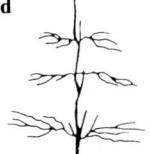

CORNUS FLORIDA
(Benthamidia florida)

NORTH AMERICAN FLOWERING DOGWOOD
Cornaceae
Deciduous
Planted in woodland, or with other ericaceous plants — azaleas and rhododendrons, heathers — a truly magnificent flowering shrub.

Origin From eastern USA.
Use As a woodland shrub or for a large shrub border. In ideal conditions, as in its native environment, may be grown as a small tree.
Description *Flower* Each head with four medium-sized white bracts resembling petals, late spring to early summer, bracts being slightly twisted and curled. Some pink flowering forms. *Foliage* Leaves oblong, 2-3in (5-8cm) long, purple-green, giving good red-orange autumn colour. *Stem* Upright when young, becoming branching with age. Medium to fast growth rate when young, slowing with age. *Fruit* Insignificant.
Hardiness Tolerates 14°F (−10°C).
Soil Requirements Prefers neutral to acid soil; dislikes any alkalinity.
Sun/Shade aspect Prefers full sun, but tolerates light shade.
Pruning None required.
Propagation and nursery production From softwood cuttings taken in midsummer or from layers. Purchase container-grown. Readily available in its native environment; scarce elsewhere. Best planting heights 2-2½ft (60-80cm).
Problems None, given suitable conditions.
Varieties of interest *C. f. 'Apple Blossom'* Apple blossom pink bracts. Good autumn colour. *C. f. 'Cherokee Chief'* Flower bracts deep rose red. Good autumn colour. *C. f. 'Cherokee Princess'* Large, round, white flower bracts. Good autumn colour. *C. f. 'Pendula'* White flower bracts; good autumn colour. An interesting weeping form, of less than average height, but with more spread. *C. f. 'Rainbow'* White flower bracts. Golden-

Cornus florida 'Rainbow' in leaf

margined dark green leaves, in autumn turning plum-purple. Slightly less hardy and less than average height and spread. *C. f. 'Rubra'* One of the most sought-after varieties, with rosy pink flower bracts in spring. Leaves red when young. Slightly less hardy and less than average height and spread. *C. f. 'Spring Song'* Deep rose-red flower bracts. *C. f. 'Tricolor'* White flower bracts in spring. Foliage green with white irregular margin, flushed rose-pink, turning purple in winter with rose-red edges. Less than average hardiness, height and spread. *C. f. 'White Cloud'* Large white flower bracts freely borne. Foliage bronze-green.

Cornus florida 'Rubra' in flower

Average height and spread
Five years
6x6ft (2x2m)
Ten years
13x13ft (4x4m)
Twenty years
or at maturity
16x16ft (5x5m)
In favourable conditions, can reach twice these dimensions.

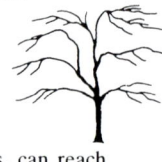

CORNUS KOUSA

CHINESE DOGWOOD, JAPANESE DOGWOOD, KOUSA DOGWOOD
Cornaceae
Deciduous
A truly magnificent late spring-flowering shrub with good autumn colour.

Origin From Japan and Korea.
Use As a freestanding shrub for a woodland garden; also well seen when planted with azaleas and rhododendrons. Can be used in a large shrub border but requires space to mature. After 20 or 30 years becomes a small tree.
Description *Flower* Four large creamy white bracts resembling petals, aging to pink-tinged white in late spring, early summer. *Foliage* Leaves elliptic, slightly curled, 2-3in (5-8cm) long, olive green with some purple shading, giving exceptional orange-red autumn colour.

Foliage in autumn retained well and is not usually damaged by wind. *Stem* Upright, light green to grey-green, becoming grey-brown and branching with age. Forms upright, slightly spreading shrub in early years, becoming more spreading with age. Slow growth rate at first, then medium, finally becoming less vigorous. *Fruit* Dull red, trawberry-like fruits on mature shrubs.
Hardiness Tolerates 4°F (−15°C).
Soil Requirements Neutral to acid; in moist conditions and light shade, however, may be happy on slightly alkaline soil.
Sun/Shade aspect Prefers light shade; tolerates medium shade to full sun.
Pruning None required.
Propagation and nursery production From layers or softwood cuttings taken in midsummer. Purchase container-grown. Moderately easy to find. Best planting heights 2-3ft (60cm-1m).
Problems Slow to establish, taking 3-4 years to flower really well.
Varieties of interest *C. capitata* syn. *Benthamia fragifera* Grey-green foliage. Sulphur-yellow bracts, followed in early autumn by strawberry-shaped, orange-red fruits. Winter minimum 23°F (−5°C). *C. kousa chinensis* A Chinese variety which produces larger flowering bracts than its parent. *C. k. 'Gold Spot'* A variety with white flower bracts and with the green foliage mottled with golden variegation. Hard to find. *C. k. 'Milky Way'* An American variety of garden origin with larger white flower bracts. *C. 'Norman Haddon'* Light grey-green foliage with good autumn colour. Pink bracts in late spring. Fruits insignificant. Likely to shed leaves completely in winter.

Cornus kousa in autumn

Average height and spread
Five years
4x3ft (1.2x1m)
Ten years
8x6ft (2.5x2m)
Twenty years
or at maturity
12x13ft (3.5x4m)
In favourable conditions can reach 30ft (10m) height and spread.

CORNUS MAS

CORNELIAN CHERRY, CORNEL
Cornaceae
Deciduous
Cornus mas and *C. m. 'Variegata'* cannot be recommended too highly as shrubs for the larger garden, providing winter flowers and autumn fruits.

Origin From central and southern Europe.
Use As a large winter-flowering shrub, used singly or in mixed plantings, and for its autumn colour. Can also be grown to form, after 20-30 years, a small tree.
Description *Flower* Tufted, small, yellow flowers produced profusely on bare stems, late winter to early spring. *Foliage* Leaves ovate, 1½-2in (4-5cm) long, mid green to grey-green, giving yellow autumn colour. *Stem* Light green to green-brown. Upright when young, becoming branching with age to form a round, slightly spreading, tall shrub. Medium growth rate when young, slowing with age. *Fruit* Red, cherry-shaped, edible fruits produced in autumn, suitable for jam-making.

Cornus kousa in flower

Cornus mas in flower

Hardiness Tolerates 4°F (−15°C).
Soil Requirements Does well on any soil.
Sun/Shade aspect Full sun to light shade. Deep shade leads to unsightly shape.
Pruning None required, but can be pruned to keep it within bounds.
Propagation and nursery production From layers or softwood cuttings taken in midsummer. Can also be grown from seed. Plant container-grown or root-balled (balled-and-burlapped). Best planting heights 2-2½ft (60-80cm).

Cornus mas 'Variegata' in leaf

Problems None, provided it is planted in a site where it can achieve full growth potential.
Varieties of interest *C. m. 'Aurea'* Yellow foliage in spring, aging to lime green in summer and autumn. A less hardy variety of less than average height and spread. Not easy to find. *C. m. 'Elegantissima'* syn. *C. m. 'Tricolor'* Silver variegated foliage, which has a pink flush at maturity. Slightly less hardy and slower growing, reaching one-third the height and spread of its parent. *C. m. 'Variegata'* A white variegated shrub or finally a small tree, more free-fruiting than the green-leaved forms. Becoming more readily available and well worth a place in the garden.
Average height and spread
Five years
6x6ft (2x2m)
Ten years
13x13ft (4x4m)
Twenty years
or at maturity
16x20ft (5x6m)
In favourable conditions can reach 25x20ft (8x6m).

CORNUS NUTTALLII

PACIFIC DOGWOOD, PACIFIC CORNEL
Cornaceae
Deciduous
A fine specimen among large, early summer-flowering shrubs.

Origin From western North America.
Use For woodland areas, where may develop into small tree, or for shrub borders, provided adequate space is allowed for final dimensions.
Description *Flower* Large, round, four-bract flowers, pure white aging to pink, produced mid to late spring. *Foliage* Leaves large, 6-8in (15-20cm) long, elliptic, mid to dark green with grey sheen, turning vivid orange-yellow in autumn. Colouring takes place intermittently, giving effective contrast of green leaves and autumn colour. *Stem* Upright when young, purple-brown, eventually spreading at top to tree proportions. *Fruit* Insignificant.
Hardiness Tolerates 4°F (−15°C).
Soil Requirements Prefers neutral to acid soil; tolerates some alkalinity.

Cornus nuttallii in flower

Sun/Shade aspect Full sun to light shade.
Pruning None required, but lower or obstructing limbs may be removed in winter.
Propagation and nursery production From layers or softwood cuttings, mid to late summer. Purchase container-grown. Readily found in its native environment; less common elsewhere. Best planting heights 2-6ft (60cm-2m).
Problems Appears one-sided, long and leggy when purchased, but once planted quickly acquires a good, rounded shape.
Varieties of interest *C. 'Eddie's White Wonder'* Large, white, four-bract flowers. Large, ovate, dark green foliage. Upright habit. *C. nuttallii 'Gold Spot'* A form with gold-splashed leaves. *C. n. 'North Star'* Large, narrow white flower bracts, new shoots purple. May be difficult to find.
Average height and spread
Five years
6x5ft (2x1.5m)
Ten years
13x8ft (4x2.5m)
Twenty years
or at maturity
26x16ft (8x5m)

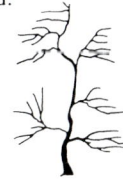

CORNUS STOLONIFERA 'FLAVIRAMEA' (Cornus sericea 'Flaviramea')

YELLOW-STEMMED DOGWOOD, GOLDEN-TWIG DOGWOOD
Cornaceae
Deciduous
The bright yellow stems of this shrub form an effective winter display.

Origin From North America.
Use As a winter shrub with attractive stems; good used with red-stemmed or black-stemmed *Cornus alba* varieties.
Description *Flower* Clusters of small white flowers produced only on wood 2 or 3 years

old, early to mid spring. *Foliage* Leaves light to mid green, medium-sized, 2-5in (5-12cm) long, elliptic, giving good yellow autumn colour. *Stem* Yellow-green, strong, upright, becoming twiggy with age, turning yellow or golden-yellow, autumn and winter; its main attraction. *Fruit* Mature shrubs may produce clusters of white to blue-white fruits in autumn.
Hardiness Tolerates winter temperatures below −13°F (−25°C).
Soil Requirements Any soil; tolerates almost waterlogged conditions.
Sun/Shade aspect Best in full sun where yellow stems show to advantage in winter.

Cornus stolonifera 'Flaviramea' in winter

Pruning For best yellow stems, prune to ground level in early spring, to induce rejuvenation through spring and summer. Unpruned shrubs lose some stem colour, but do flower and fruit; pruned shrubs do not.
Propagation and nursery production From hardwood cuttings, taken late autumn to early winter, or from softwood cuttings taken in summer. Plant bare-rooted or container-grown. Best planting heights 2-3ft (60cm-1m). Fairly easy to find.
Problems Vigorous, and often insufficient is space allowed for its mature dimensions.
Varieties of interest *C. s. 'Kelsey's Dwarf'* A totally different form, with light green, 2in (5cm) long, ovate leaves giving good autumn colour. Very dense stem formation, orange to orange-red in winter. Low-growing, reaching only 3ft (1m) in height and spread.
Average height and spread
Five years
6x6ft (2x2m)
Ten years
8x10ft (2.5x3m)
Twenty years
or at maturity
8x12ft (2.5x3.5m)

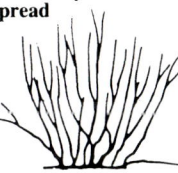

COROKIA COTONEASTER

WIRE-NETTING BUSH
Cornaceae
Evergreen
Very good for rock gardens and special garden features.

Origin From New Zealand.
Use As low, spreading shrubs for large rock gardens. Can be grown in containers to good effect.
Description *Flower* Small, single, bright yellow, star-like flowers, late spring to early summer. *Foliage* Leaves very small, ovate, ½-¾in (1-2cm) long, dark green to purple-green. Some yellow autumn colour. *Stem* Contorted, very twiggy branchlets forming an intricate tracery, making small, dome-shaped, slightly spreading shrub. Very slow growth rate. Stems purple-green to brown-

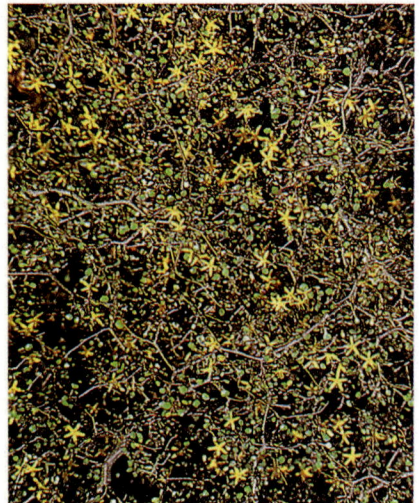
Corokia cotoneaster in flower

purple. *Fruit* In hot summers produces small, round, orange fruits.
Hardiness Minimum temperature range 23°F (−5°C) down to 14°F (−10°C).
Soil Requirements Prefers well-drained soil.
Sun/Shade aspect Best in full sun or very light shade.
Pruning None required.
Propagation and nursery production From softwood cuttings taken in early summer. Purchase container-grown. Not easy to find. Best planting heights 4-12in (10-30cm).
Problems Extremely slow-growing. Young plants often look weak and insipid before being planted out, but gradually acquire an interesting, contorted form.
Varieties of interest *C. × virgata* Small yellow star-shaped flowers. Freely produces small bright orange fruits. Oblong to lanceolate, grey-green leaves with white undersides. Best planting heights 15in-2ft (40-60cm). A larger, more upright, slightly tender form, height 3ft (1m), spread 2½ft (80cm).
Average height and spread
Five years
1½x2½ft (50x80cm)
Ten years
2x3ft (60cmx1m)
Twenty years
or at maturity
2x4ft (60cmx1.2m)

Origin From southern Europe.
Use As a long-flowering, freestanding shrub, or fan-trained against a sunny, sheltered wall.
Description *Flower* Small to medium, yellow pea-flowers, very freely produced, mid to late spring, with further intermittent flowering through summer and early autumn. *Foliage* Leaves round to ovate, ½in (1cm) long, grey-green, giving some yellow autumn colour on older leaves. *Stem* Grey-green. Upright, becoming branching, forming a loose, dome-shaped shrub or fan-shaped wall specimen. *Fruit* Small, grey-green pea pods.
Hardiness Minimum winter temperature 23°F (−5°C).
Soil Requirements Any soil, including very alkaline.
Sun/Shade aspect Full sun to very light shade.
Pruning Remove some branches 3-4 years old to ground level occasionally to help annual rejuvenation.
Propagation and nursery production From softwood cuttings taken midsummer. Purchase container-grown. Often looks weak when purchased, but once planted out quickly

establishes itself. Best planting heights 2-2½ft (60-80cm).
Problems May suffer from blackfly; not a major problem.
Varieties of interest *C. emerus* Dark yellow pea-flowers produced mainly in spring, but continuing intermittently into summer and early autumn. A slightly more hardy form. *C. glauca 'Variegata'* Creamy-white variegated foliage, with yellow flowers produced throughout summer, but with main display in spring. Best planting heights 15in-2ft (40-60cm). More tender than the parent, with slightly less height and spread.

Coronilla emerus in flower

Average height and spread
Five years
5x5ft (1.5x1.5m)
Ten years
6x6ft (2x2m)
Twenty years
or at maturity
8x8ft (2.5x2.5m)

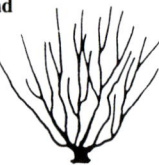

Origin From South America, the Pampas Plains.
Use Singly, as a feature plant in a large lawn or at the end of a shrub border. Very impressive mass planted as group of five or more individual clumps.

Cortaderia selloana 'Gold Band'

Description *Flower* Large silver-white plumes up to 1½ft (50cm) long or more, early to late autumn. *Foliage* Dark to olive green, lanceolate, 3-6ft (1-2m) long grass spears with very sharp serrated edges. Fast spring foliage growth, slowing in summer. *Stem* Foliage is borne from central clump and plumes presented on tall, olive green, cane-like structures up to 12ft (3.5m) high. Forms a narrow base with spreading top and plumes presented in V-shape formation. Fast to establish and mature. *Fruit* Plumes age into green-white seedheads, which persist into autumn and winter.

Cortaderia selloana 'Rendatieri'

Hardiness Tolerates 4°F (−15°C).
Soil Requirements Any, but good rich soil produces best plumes.
Sun/Shade aspect Best in full sun; tolerates shade but plumes may be smaller.
Pruning Remove dead flower canes.
Propagation and nursery production From seed or by division. Purchase container-grown. Easy to find. Best planting heights 15in-2½ft (40-80cm).
Problems Leaf edges are extremely sharp.
Varieties of interest *C. s. 'Gold Band'* Much

Cortaderia selloana

smaller, less freely produced flower plumes, each leaf having bright gold edge. A very attractive variety, especially if used near water. Two-thirds average height and spread. *C. s. 'Pumila'* Large white plumes, dwarf form, reaching a height of 6ft (2m) and spread of 5ft (1.5m) after 10 years. *C. s. 'Rendatieri'* Large, open, feathery, hanging plumes tinged pink. A slightly larger and more spreading form. *C. s. 'Silver Cornet'* Creamy-white flower plumes, lanceolate leaves, silver-white late summer and early autumn. Slightly less than average height and spread. *C. s. 'Sunningdale Silver'* Very large, silvery to pure white, fluffy, open plumes.

Average height and spread
Five years
12x3ft (3.5x1m)
Ten years
12x6ft (3.5x2m)
Twenty years
or at maturity
12x10ft (3.5x3m)

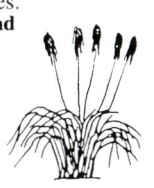

CORYLOPSIS SPICATA

COWSLIP BUSH, WINTER HAZEL, SPIKE WINTERHAZEL
Hamamelidaceae
Deciduous
Given time to mature, an extremely attractive group of ornamental shrubs.

Origin From Japan.
Use As a freestanding shrub or in a large border or group planting. Ideal for woodland areas.
Description *Flower* Racemes of small to medium-sized, fragrant, hanging, pale yellow, cowslip-shaped flowers, very early in spring. *Foliage* Leaves round to ovate, 2-4in (5-10cm) long, light green to grey-green when young, aging to light olive green, veins and certain areas sometimes with orange-red shading. Undersides covered in soft, downy, glaucous coating. *Stem* Upright, becoming spreading and branching to a wide, round-topped shrub. Grey-brown when young, aging to darker grey-brown. Medium growth rate. *Fruit* Insignificant.

Corylopsis spicata **in flower**

Hardiness Tolerates 4°F (−15°C).
Soil Requirements Best on neutral to acid soil, dislikes alkalinity.
Sun/Shade aspect Prefers light to medium shade, tolerates full sun if moisture adequate.
Pruning None required.
Propagation and nursery production From layers or softwood cuttings taken in midsummer. Purchase container-grown or root-balled (balled-and-burlapped). Best planting heights 2-3ft (60cm-1m).
Problems Can be slow to establish.
Varieties of interest *C. glabrescens* Larger leaves and flowers on longer, thinner trusses,

followed by glabrous fruits. From Japan. A more spreading form. *C. pauciflora* Primrose yellow flowers, cowslip-scented, borne on short, three-flowered racemes in some profusion, mid spring. Small, glabrous fruits follow. Leaves ovate, 1-1½in (3-4cm) long, with toothed edges, giving good autumn colour. A very much smaller variety, rarely reaching 6ft (2m) in height and spread, and a much thinner, denser, branched form. Slightly less hardy than *C. spicata*. From Japan and Taiwan. Best planting heights 1-2ft (30-60cm). *C. veitchiana* Primrose yellow flowers in long racemes, each flower having very attractive brick red anthers; new foliage in spring is purple-red. A taller, more erect form. From western China. *C. willmottiae* Flowers soft yellow, cowslip-shaped and scented. Ovate leaves, often red-purple when young with purple veins on undersides. From western China. *C. w. 'Spring Purple'* Similar to its parent, but both new and mature stems rich purple, making a fine effect in winter and early spring.

Average height and spread
Five years
5x5ft (1.5x1.5m)
Ten years
8x8ft (2.5x2.5m)
Twenty years
or at maturity
12x12ft (3.5x3.5m)

CORYLUS AVELLANA and CORYLUS MAXIMA

HAZEL, FILBERT, COBNUT
Corylaceae
Deciduous
Useful foliage shrubs, some with edible nuts.

Origin From Europe, western Asia and North Africa.
Use Ideal for a quick-growing screen, or to form attractive hedging. Golden and purple-leaved varieties good in large shrub borders.
Description *Flower* Male flowers produced as 2-2½in (5-6cm) long yellow catkins, late winter to early spring. Female flowers, small, red and tufted, produced inconspicuously at the same time. *Foliage* Leaves large, round, light to mid green, 2-4in (5-10cm) long, giving yellow autumn colour. *Stem* Light grey-green when young, aging to grey-green to grey-brown. Fast, strong-growing, upright, becoming branching and twiggy. Forming an upright shrub for the first 10-15 years, then spreading. *Fruit* Edible hazelnuts, red-brown with an attractive grey-green downy husk, early to late autumn.
Hardiness Tolerates winter temperatures down to −13°F (−25°C).

Corylus avellana 'Contorta' **in winter**

Soil Requirements Any soil.
Sun/Shade aspect Full sun to moderately deep shade.
Pruning None required, but can be coppiced after 5-6 years by removing all wood to ground level to encourage regeneration of clean young wood, particularly important with golden and purple foliage varieties. The branching, twiggy tops of each pruned stem can be used to support perennial plants or peas.
Propagation and nursery production From layers or rooted shoots taken from the plant base. Plant bare-rooted or container-grown. Relatively easy to find in nurseries, but not always in garden centres. Best planting heights 2-3ft (60cm-1m).
Problems May look a little uninteresting when young. Squirrels may take nuts before they can be harvested.
Varieties of interest *C. a. 'Aurea'* (Golden-leaved Hazel) Round foliage, lime green at first, quickly aging to soft yellow, then darker yellow as autumn approaches. Unless pruned as described above every 2-3 years can become weak and uninteresting in foliage size and colour. Slightly less height and spread than the parent. *C. a. 'Contorta'* (Corkscrew Hazel, Harry Lauder's Walking Stick, Contorted Hazel) Branches twisted and contorted, round light green foliage also contorted. Good yellow catkins in winter, but rarely followed by nuts. Ideal as a winter stem feature providing an unusual element for flower arranging. Over 25 years reaches up to 10ft (3m) in height and 13ft (4m) in spread. Best planting heights 15in-3ft (40cm-1m). *C. a. 'Heterophylla'* syn. *C. a. 'Laciniata'*, *C. a. 'Quercifolia'* (Cut-leaved Hazel) Attractive, deeply-lobed leaves, grey-green in spring

Corylus avellana 'Aurea' **in leaf**

Small yellow catkins rarely followed by nuts. Of slightly less vigour than the parent, and less than average height and spread. *C. maxima* (Cobnut) Leaves large, 4-6in (10-15cm) long, light to mid green, rounded heart-shaped, giving good yellow autumn colour. The edible cob nuts are large and longer than those of *C. avellana* and protected by large, grey-green husks. Reaching up to 16ft (5m) in height and spread. *C. m. 'Purpurea'* syn. *C. m. 'Atropurpurea'* (Purple-leaf Filbert, Purple Giant Filbert) Leaves large, 4-5in (10-12cm) long, round, deep purple and very attractive. Purple catkins, 2½in (6cm) long, produced early to late spring, followed by large edible nuts contained within red-purple husk. Young stems and shoots are upright and purple. Benefits from being cut down to ground level at least once in every 4-5 years, when it rapidly produces new growth.

Corylus maxima 'Purpurea' in leaf

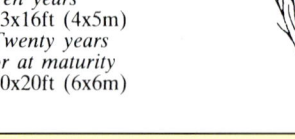

Average height and spread
Five years
10x10ft (3x3m)
Ten years
13x16ft (4x5m)
Twenty years
or at maturity
20x20ft (6x6m)

CORYNABUTILON VITIFOLIUM
(Abutilon vitifolium)

FLOWERING MAPLE, VINE-LEAVED ABUTILON
Malvaceae
Deciduous
A tall, impressive plant, but susceptible to winter cold. In very mild areas it may be considered semi-evergreen. Recently reclassified as *Corynabutilon vitifolium*, the shrub is often still available under the name *Abutilon vitifolium*.

Origin From Chile.
Use As a tall flowering shrub, or small tree, for mild areas; grown on a wall or in a conservatory for colder regions.
Description *Flower* Large, saucer-shaped, deep to pale mauve flowers, borne freely late spring through early summer. *Foliage* Leaves vine-shaped, 4-6in (10-15cm) long with a grey, downy covering. Some yellow autumn colour. *Stem* Upright, grey-green with a grey downy covering. Fast to medium growth rate. *Fruit* Insignificant.
Hardiness Minimum temperature range 32°F (0°C) down to 23°F (−5°C). Best protected by a wall in colder areas.
Soil Requirements Most soils, although thin alkaline soils will cause chlorosis.
Sun/Shade aspect Full sun. Resents any shade.
Pruning None required, although individual branches or limbs may be removed.

Corynabutilon vitifolium in flower

Propagation and nursery production The basic form from seed; named varieties may be produced from softwood cuttings taken mid to late summer. Often difficult to find. Purchase container-grown. Best planting heights 15in-3ft (40cm-1m).
Problems Stems at risk to severe temperature changes in winter and to wind chill factor; if stems die off regeneration from the base is possible.
Varieties of interest *C. suntense* Slightly later-flowering, with mauve flowers. More tender than the parent. *C. vitifolium album* A white-flowering form of the Vine-leaved Abutilon. *C. v. 'Veronica Tennant'* Produces masses of large mauve flowers, deeper in colour than those of the parent.

Average height and spread
Five years
13x6ft (4x2m)
Ten years
16x10ft (5x3m)
Twenty years
or at maturity
20x13ft (6x4m)

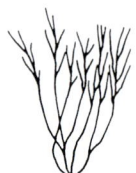

COTINUS COGGYGRIA

SMOKE TREE, SMOKE BUSH, BURNING BUSH, CHITAM WOOD, VENETIAN SUMACH
Anacardiaceae
Deciduous
A shrub for summer and autumn attraction, producing fine foliage colours, profuse flowers and good structural shape.

Origin From central and southern Europe.
Use As a foliage shrub for autumn colour and flowers, either on its own or in a large shrub border.
Description *Flower* Large, open, pale pink inflorescences resembling plumes 6-8in (15-20cm) long borne profusely on all wood 3 years old or more in summer, persisting into early and late autumn, turning smoky grey. *Foliage* Leaves ovate to oblong, 1½-2in (4-5cm) long, grey-green when young opening to mid green, vivid orange-yellow in autumn, purple-leaved forms turning scarlet-red. *Stem* Light green, becoming streaked with orange or red shading, finally grey-brown. Fast growing and upright when young, becoming slower and very branching and twiggy to form a round-topped, spreading shrub. *Fruit* Inflorescences change to seedheads.
Hardiness Tolerates 4°F (−15°C). Some winter die-back may occur at tips of new growth.
Soil Requirements Prefers rich, deep soil, but tolerates any.
Sun/Shade aspect Green-leaved varieties tolerate very light shade to full sun. Purple-

Cotinus coggygria 'Foliis Purpureis' in flower

leaved varieties must have full sun, otherwise they turn green.
Pruning Established plants can be cut to ground level each spring; this produces strong new growth, up to 6ft (2m) in one season with large foliage at the expense of flowers. Or leave completely unpruned to achieve good flowering after 3 or 4 years. Otherwise, it is acceptable to remove mature shoots 1-3 years old each spring, so inducing some foliage rejuvenation and improved flowering.
Propagation and nursery production From layers. Purchase container-grown. Relatively easy to find in nursery production; some

Cotinus coggygria 'Royal Purple' in leaf

Cotinus obovatus **in flower**

varieties will be found in garden centres. Best planting heights 15in-2½ft (40-80cm).
Problems Some purple-leaved varieties susceptible to mildew. Slow to establish, taking 2-3 years after planting to gain full stature.
Varieties of interest *C. c. 'Flame'* One of the best varieties for autumn colour. Pink flowers and bright red-orange foliage in autumn. Hard to find and often confused with *C. coggygria* itself. *C. c. 'Foliis Purpureis'* Pink inflorescence, young foliage rich plum-purple, aging to lighter red to purple-red, late summer. Good autumn colours. *C. c. 'Notcutt's Variety'* Pink to purple-pink inflorescence with good red-purple autumn colours. Very deep purple leaves, slightly larger than those of its parent. Of slightly less height and spread. *C. c. 'Royal Purple'* Purple-pink inflorescence, purple-wine foliage, almost translucent when seen with sunlight behind it, becoming duller purple towards autumn, finally red. Slightly smaller than *C. coggygria*. Of garden origin. *C. c. 'Rubrifolius'* Pink inflorescence, deep wine red leaves when young, translucent in sunlight, becoming red-green towards autumn. *C. obovatus* syn. *C. o. americanus* (Chitam Wood) Light pink inflorescence. Round, ovate, light to mid green foliage with some orange shading, brilliant orange-red in autumn. Leaves much larger than *C. coggygria*. Forming a large, spreading shrub. From south-eastern USA.
Average height and spread
Five years
5x5½ft (1.5x1.8m)
Ten years
10x10ft (3x3m)
Twenty years or at maturity
20x20ft (6x6m)

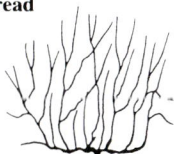

COTONEASTER
Low, spreading evergreen forms

KNOWN BY BOTANICAL NAME
Rosaceae
Evergreen
Useful flowering and fruiting evergreens for ground cover effect.

Origin Mainly native to China, but many varieties now of garden or nursery origin.
Use As low shrub edging on a shrub border; for ground cover, either singly or grouped; effective in large tubs.
Description *Flower* Small or tiny, white, four-petalled flowers, with red calyxes, borne singly or in clusters, dependent on variety, late spring to early summer. *Foliage* Leaves ½-¾in (1-2cm) long, lanceolate to ovate and

in some varieties round. Dark shiny green with grey undersides, veins and stalks often red-shaded. *Stem* Green to dark brown, spreading, forming a low cover or creeping carpet. Slow to medium growth rate. *Fruit* Round to ovate, glossy fruits, normally red, some purple-red dependent on variety.
Hardiness Tolerant of winter temperatures down to −13°F (−25°C).
Soil Requirements Tolerant of any soil.
Sun/Shade aspect Good in full sun to medium shade.
Pruning None required other than keeping within bounds.
Propagation and nursery production From semi-ripe cuttings taken early to mid summer. Purchase container-grown. Best planting heights 8-15in (20-40cm). There is a wide range of varieties, most of which are easy to find.
Problems Relatively small when purchased and rather slow-growing.
Varieties of interest *C. adpressus var. praecox* Small, round, semi-evergreen, ovate leaves, giving some contrasting autumn red and orange colour from old leaves which are discarded. Orange-red, good-sized fruits in autumn. Height 1½ft (50cm), spread 4ft (1.2m) after 10 years. Also known sometimes as *C. 'Nan-shan'*. From western China. *C. buxifolius* A very dense dwarf evergreen variety with white flowers and grey to dull green round foliage. Oval to round red fruits in autumn. Ideal for rock gardens. From south-western India. *C. congestus* A very thick, creeping form with white flowers and blue-green foliage, forming a series of mounds and valleys. Good red round fruits. From the Himalayas. *C. c. var. procumbens* A very slow-growing, creeping hummock, with

Cotoneaster microphyllus **in fruit**

rather sparsely produced red fruit in autumn. Height 4in (10cm), spread 1½ft (50cm). *C. dammeri* A wide-spreading, low variety with long ranging shoots. White flowers, bright red fruits in autumn. Leaves ovate, dark green above with grey-green underside, some having slight purple shading on veins and outer edges. Quite free in habit and has self-rooting tips. Good for ground cover on banks and beneath other larger shrubs. From China. *C. d. var. radicans* A vigorous variety with ovate leaves and flowers borne closely together, often in pairs. Red fruits in autumn. Slightly less than average height and spread. *C. 'Donards Gem'* Small, round, ovate leaves, semi-evergreen, borne on a low, round bush. Good red autumn fruits. *C. microphyllus* Round, dark grey-green glossy leaves, ½in (1cm) long, very thickly produced along very twiggy branches. Round, dark scarlet fruits, autumn. Ideal for covering banks or draping over walls. Very hardy. Slower growth than average. Forms a low clump, spreading with age. From the Himalayas and south-western China. *C. m. var. cochleatus* Slower than its parent and lower growing. Dull grey-green, ovate leaves and dull grey-red, round fruits, autumn. Of less height and vigour but of equal hardiness. From western China, south-eastern Tibet through eastern Nepal. *C. m. var. thymifolius* Very small, narrow, ovate, shiny, dark green leaves produced very thickly and close to stems and branches. Purple-red berries in autumn. Forms a round mound. Ideal for rock gardens and tubs. Not easy to find. *C. salicifolius 'Gnome'* Small, elliptic to round, dark grey-green leaves; small red fruits. Forms low mound. Of garden origin. *C. s. 'Parkteppich'* (Park Carpet) Leaves evergreen, narrow to medium width, lanceolate, dark green upper surface with grey underside, purple markings on veins. Round, red fruit, borne singly and in clusters. Forms a wide-spreading ground cover. Of garden origin. *C. s. 'Repens'* Leaves broad, lanceolate, dark to grey-green, evergreen; small red fruits in autumn. Very prostrate, height reaching no more than 8in (20cm), but spreading up to 6ft (2m). Of garden origin. *C. 'Skogholme'* Leaves small, lanceolate to oval, mid green to dark green. Large, oval to round, coral red fruits in autumn. From western China.
Average height and spread
Five years
1½x4ft (50cmx1.2m)
Ten years
2x6ft (60cmx2m)
Twenty years or at maturity
2x10ft (60cmx3m)

COTONEASTER
Medium height, spreading evergreen forms

KNOWN BY BOTANICAL NAME
Rosaceae
Evergreen
Useful medium height flowering and fruiting shrubs.

Origin Mostly native to China or of garden origin.
Use For ground cover of attractive foliage with autumn and winter fruits, for mass planting or in very large containers. Can also be fan-trained on a wall.
Description *Flower* Small, white, four-petalled flowers with small red calyxes, borne singly or in clusters, late spring to early summer, dependent on variety. *Foliage* Leaves small, 1-1½in (3-4cm) long, round to broad ovate, dark green or grey-green, dependent on variety. *Stem* Upright when young, becoming spreading with age, forming a broad, spreading mound. Medium growth rate. *Fruit* Round to ovate, glossy, normally red, but some purple-red varieties.

Cotoneaster conspicuus 'Decorus' in fruit

Cotoneaster horizontalis 'Variegatus' in leaf

Hardiness Tolerant of winter temperatures down to −13°F (−25°C).
Soil Requirements Any soil conditions.
Sun/Shade aspect Full sun to medium shade.
Pruning None required but may be cut back to keep in bounds.
Propagation and nursery production From semi-ripe cuttings taken in midsummer. Purchase container-grown. Most varieties relatively easy to find. Best planting heights 2-2½ft (60-80cm).
Problems May suffer from fire blight.
Varieties of interest *C. conspicuus* Arching branches with small, ovate, grey to grey-green foliage, forming a wide spreading mound, covered in small white flowers, late spring to early summer. Bright red fruits freely borne. From south-eastern Tibet. *C. c. 'Decorus'* Equal to the parent in flowers and fruit, lower growing, height up to 4ft (1.2m) at maturity, spread 8ft (2.5m). *C. 'Hybridus Pendulus'* Often seen as a small weeping tree, but also less commonly grown as low, spreading shrub. Round, medium-sized, dark green, ovate foliage with purple-red leaf stalks and veining. Flowers borne in small to medium-sized clusters, early summer, followed by round, bright red fruits, autumn. Of garden origin and varied parentage. *C. salicifolius 'Autumn Fire'* Narrow, ovate, dark green leaves, white flowers followed by orange-red fruit borne in clusters. Some foliage dies in winter, turning scarlet and enhancing the overall effect of fruit and foliage colour.
Average height and spread
Five years
3x5ft (1x1.5m)
Ten years
5x8ft (1.5x2.5m)
Twenty years
or at maturity
5x12ft (1.5x3.5m)

COTONEASTER
Medium height, spreading deciduous forms

KNOWN BY BOTANICAL NAME
Rosaceae
Deciduous
An excellent range of foliage, flowering and fruiting shrubs.

Origin Mostly native to western China or of garden origin.
Use As deciduous shrubs of autumn colour interest. Can be fan-trained on a shady, exposed wall and will attain more overall height.
Description *Flower* Masses of small white, four-petalled, cup-shaped flowers, each with a red calyx, late spring to early summer. *Foliage* Leaves round, ovate, ½-1in (1-3cm) long, dark green to grey-green, depending on

variety. Good autumn colours. *Stem* Upright when young, becoming very branching, twiggy and spreading with age. Soft grey-green when young, aging to grey-brown. Medium growth rate. *Fruit* Round, small, red, glossy fruits in autumn.
Hardiness Tolerant of winter temperatures down to −13°F (−25°C).
Soil Requirements Any soil conditions.
Sun/Shade aspect Full sun to medium shade.
Pruning None required but can be reduced in size in early spring.
Propagation and nursery production From semi-ripe cuttings taken in midsummer. Purchase container-grown for best results; bare-rooted plants also available. Best planting heights 2-4ft (60cm-1.2m).
Problems Relatively slow growing.

Cotoneaster horizontalis in fruit

Varieties of interest *C. 'Coral Beauty'* Small, round, grey-green, ovate leaves giving good orange autumn colour. White flowers followed by coral red fruit, autumn. Slightly less than average height and spread. Of garden origin. *C. horizontalis* (Fishbone or Herringbone Cotoneaster) Grey-green stems forming a herringbone pattern, the branches spreading out and becoming fan-shaped and covered with small, round, green to grey-green leaves, turning bright red in autumn. Single white flowers followed by round, red fruits borne at each leaf axil along mature branches. Slightly less than average height and spread. *C. h. 'Major'* Very similar in growth habit to its parent, but slightly more vigorous, with larger, rounder leaves and larger, but fewer fruit. *C. h. 'Robusta'* A larger, quicker, stronger-growing variety in which herringbone pattern is more widely spaced. Leaves larger, rounder, almost cup-shaped. Good red, round autumn fruits. One-third more than average height and spread. Not always

easy to find. *C. h. 'Variegatus'* Herringbone growth pattern. Light grey-green foliage, round, with creamy white margins and lined with red round the outer edge. White flowers and red fruit, less glossy and less freely produced. Two-thirds average height and spread. Of garden origin.
Average height and spread
Five years
2x3ft (60cmx1m)
Ten years
3x6ft (1x2m)
Twenty years
or at maturity
3x10ft (1x3m)

COTONEASTER
Tall evergreen forms

KNOWN BY BOTANICAL NAME
Rosaceae
Evergreen
A very useful group of large evergreen flowering and fruiting shrubs.

Origin Mostly native to western China or of garden origin.
Use As tall, evergreen shrubs for flower and fruit effect. Ideal for screening; can be trained as small trees.
Description *Flower* White flowers with red calyxes borne either in clusters or singly, dependent on variety, in early summer. *Foliage* Ovate, lanceolate to round, 2-4in (5-10cm) long, dark to mid or grey-green, dependent on variety. *Stem* Strong, upright, green to green-red or grey-green, becoming branching with age. Fast growth rate. *Fruit* Yellow, red or orange fruits, dependent on variety, produced in clusters in early autumn.
Hardiness Tolerant of winter temperatures down to −13°F (−25°C).
Soil Requirements Most soil conditions, but may show distress on very thin alkaline types.
Sun/Shade aspect Full sun to medium shade.
Pruning None required but may be reduced in size; varieties used for hedging may be clipped hard.
Propagation and nursery production From semi-ripe cuttings taken early summer or by grafting in winter. Purchase container-grown. Most varieties fairly easy to find in nurseries and some in garden centres. Best planting heights 2-4ft (60cm-1.2m).
Problems Certain varieties are susceptible to the fungus disease fire blight, while others are liable to stem canker. In areas where the fire blight problem cannot be controlled, some named varieties are likely to be withdrawn from commercial production. Stem canker can be dealt with by cutting back affected shoots and treating with pruning compound.
Varieties of interest *C. 'Cornubia'* Broad, lanceolate leaves. Large clusters of white

Cotoneaster 'Cornubia' in fruit

flowers followed by bright red fruit. Of garden origin. **C. 'Exburiensis'** Light to mid green, broad, lanceolate leaves. White flowers followed by clusters of bright yellow fruit. Two-thirds average height and spread. Raised in the Exbury Gardens, Hampshire, England. **C. franchetii** Dark grey green, 1-1½in (3-4cm) long, ovate to round foliage borne on long arching branches. Stems dark brown to almost black, covered in single white flowers, early summer in followed by dull red fruit, autumn. May be grown as a small, single-stemmed tree. Two-thirds average height and spread. From China. **C. franchetii var. sternianus** syn. **C. wardii** Round, silver-grey to grey-green foliage. White flowers followed by good orange fruit. Two-thirds average height and spread. Can be trained as a small, single-stemmed tree. From southern Tibet and northern Burma. **C. frigidus** Leaves 4-6in (10-15cm) long, broad, elliptic, dark green. Clusters of white flowers followed by crimson-red fruits, persisting well into winter. Can be trained as a small single-stemmed tree. Thought to originate from the Himalayas. **C. henryanus** Leaves long, lanceolate, dark green, slightly corrugated, with duller grey undersides. Clusters of white flowers, borne on brown-red, strong, upright shoots, followed by red fruit. Can be trained as a small, single-stemmed tree. **C. 'Inchmery'** Foliage light green, slightly broader, lanceolate. Clusters of white flowers, followed by coral yellow fruit in autumn. Can be trained as a small, single-stemmed tree. Two-thirds average height and spread. Raised in the Inchmery Gardens, Hampshire, England. **C. lacteus** Elliptic, dark green foliage, white

clusters of flowers followed by dark red round fruits borne profusely and held sometimes until they rot. Good for hedging if planted 3ft (1m) apart or as large screen when planted in single line 6ft (2m) apart. Can also be trained as a small, single-stemmed tree. From China. **C. pannosus** Rounded, sage-green leaves set off small clusters of green-red fruits, becoming deep red through winter. Two-thirds average height and spread. From western China. **C. 'Rothschildianus'** Large clusters of yellow fruit set off against light green, broad, lanceolate leaves, 4-6in (10-15cm) long, spectacular in autumn. Can be trained as a small single-stemmed tree. Raised at Exbury Gardens, Hampshire, England. **C. salicifolius var. floccosus** Long, graceful, pendulous branches with narrow, elliptic, shiny-surfaced leaves, green with white undersides, giving a drooping, fan-like appearance. Clusters of white flowers give way to small, dull, red fruits. Very good for using on exposed, shady wall as climber. Can also be trained as a small tree on a single stem. Two-thirds average height and spread. From China. **C. 'St. Monica'** Leaves very long, lanceolate, up to 6-8in (15-20cm) long, dark purple-green with grey undersides, veins and leaf petioles purple-red. Large clusters of white flowers give way to bright red fruits, borne on purple to purple-brown, strong, upright branches. Not easy to find but one of the most stately varieties. Originated in a convent garden in Bristol, England. **C. × watereri** Foliage semi-evergreen, elliptic, mid to dark green. White flowers and red fruit. Can be grown as an individual shrub or on single stem as a tree. Ideal for screening. The name *C. × watereri* covers a large number of

very similar seedling hybrids, crosses between *C. frigidus*, *C. henryanus* and *C. salicifolius*.

Average height and spread
Five years
10x10ft (3x3m)
Ten years
16x16ft (5x5m)
*Twenty years
or at maturity*
23x23ft (7x7m)

COTONEASTER
Tall deciduous forms

KNOWN BY BOTANICAL NAME

Rosaceae
Deciduous
A group of shrubs often overlooked but which can give good autumn foliage and fruit effects.

Origin Mostly from China and the Himalayas, or of garden origin.
Use As a freestanding, fruiting shrub or for larger shrub borders. Some varieties ideal for hedging.
Description *Flower* White, four-petalled flowers with prominent red calyxes, borne either singly or in clusters, late spring to early summer. *Foliage* Basically elliptic, 1-1½in (3-4cm) long, normally green to grey-green with good autumn colours. *Stem* Strong, quick-growing, upright, becoming spreading with age. *Fruit* Red fruits, borne singly or in clusters, depending on variety.
Hardiness Tolerant of winter temperatures down to −13°F (−25°C).
Soil Requirements Any soil.
Sun/Shade aspect Tolerates full sun to medium shade.

Cotoneaster bullatus in fruit

Pruning None required, but can be reduced in size. Varieties used for hedging may be hard clipped.
Propagation and nursery production From softwood cuttings, grafting or seed, depending on variety. Purchase container-grown. Best planting heights 2-4ft (60cm-1.2m).
Problems On grafted varieties understock of *C. bullatus* can sucker and given time kill the intended variety; watch out for suckering growth and remove immediately.
Varieties of interest C. bullatus Foliage large, round to elliptic, dark green with corrugated surfaces, giving good autumn colour. Clusters of white flowers give way to dark red fruit, enhancing autumn colour of foliage. Often used as grafting understock for many of the large-leaved, evergreen forms. From western China. **C. distichus** syn. **C. rotundifolius** Foliage very distinct, round, ¾-1in (2-3cm) long and wide, glossy green with good red autumn colour. Elliptic fruits borne singly in each leaf axil, being held well into early spring. Two-thirds average height and spread. Often found in commercial production under the incorrect name of *C. rotundifolius*. From the Himalayas and south-western China. **C. divaricatus** Very good orange-red autumn foliage, contrasting with some later held green leaves. Fruit dark red and glossy. Useful on its own or if planted 2½ft (80cm) apart makes semi-formal hedge. From western China. **C. 'Firebird'** Foliage dark grey-green, 3-4in (8-10cm) long, elliptic, giving

Cotoneaster 'Rothschildianus' in fruit

good autumn colour. Large, orange-red fruits borne in thick clusters in autumn. Possibly more spreading than average and very hardy. Not easy to find, but obtainable. A hybrid of *C. bullatus*, probably being a cross between this and *C. franchetii*. *C. simonsii* Semi-evergreen, round to elliptic leaves, borne on upright stems covered in late spring with single, or small clusters, of white flowers in each leaf axil, culminating in red fruits. If planted 2½ft (80cm) apart in single line, makes semi-formal hedge. Very good for propagation from seed. From Assam.

Average height and spread
Five years
6x6ft (2x2m)
Ten years
10x10ft (3x3m)
Twenty years
or at maturity
13x13ft (4x4m)

CRATAEGUS MONOGYNA 'COMPACTA'
(Crataegus 'Inermis Compacta')

DWARF THORN, SINGLESEED HAWTHORN
Rosaceae
Deciduous
A true novelty, worth inclusion in the larger rock garden or among a collection of heathers.

Origin From Europe.
Use As an unusual shrub for a large rock garden or other position where it will not be overpowered by surrounding plants.
Description *Flower* May produce clusters of white flowers sparsely, late spring. *Foliage* Leaves light green to grey-green, deeply lobed, elliptic, 1½-2in (4-5cm) long, characteristic of the common Hawthorn, May or Quickthorn. Yellow autumn colour. *Stem* Very dwarf, slow-growing, stout, stiff, thornless branches. Light grey-green when young, aging to grey-brown. *Fruit* May produce round, orange-red haws in autumn.
Hardiness Tolerates winter temperatures down to −13°F (−25°C).
Soil Requirements Any soil.
Sun/Shade aspect Full sun to light shade.
Pruning None required.
Propagation and nursery production From grafting on to *C. monogyna*. Purchase container-grown; very scarce but not impossible to find. Best planting heights 8in-1½ft (20-50cm).
Problems None.
Average height and spread
Five years
2x2ft (60x60cm)
Ten years
3x3ft (1x1m)
Twenty years
or at maturity
4x4ft (1.2x1.2m)

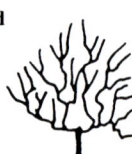

Crataegus monogyna 'Compacta'

CRINODENDRON HOOKERANUM
(Tricuspidaria lanceolata)

LANTERN TREE
Elaeocarpaceae
Evergreen
A magnificent shrub if it can be given the correct soil and climatic conditions.

Origin From Chile.
Use As a specimen shrub for acid gardens. Good in association with dwarf rhododendrons, azaleas and heathers.
Description *Flower* Crimson-red, lantern-shaped flowers hanging from long stalks along underside of branches, on mature wood only, late spring to early summer. *Foliage* Leaves dark green, lanceolate, 1½-2in (4-5cm) long, with silver undersides. *Stem* Dark green, upright. Slow to medium growth rate. *Fruit* Rarely fruits.

Crinodendron hookeranum in flower

Hardiness Minimum winter temperature 23°F (−5°C).
Soil Requirements Acid soil; dislikes any alkalinity.
Sun/Shade aspect Full sun to light shade.
Pruning None required.
Propagation and nursery production From softwood cuttings taken in early summer. Purchase container-grown; rather hard to find. Best planting heights 15in-2½ft (40-80cm).
Problems Slow to establish and to achieve full beauty.
Varieties of interest *C. patagua* syn. *Tricuspidaria dependens* White, bell-shaped flowers, late summer. A stronger growing shrub, reaching greater height and spread. Tender,

---

best protected by a wall in cold areas. Very hard to find.
Average height and spread
Five years
3x2½ft (1mx80cm)
Ten years
6x3ft (2x1m)
Twenty years
or at maturity
10x5ft (3x1.5m)

CYTISUS BATTANDIERI

MOROCCAN BROOM, PINEAPPLE BROOM
Leguminosae
Deciduous
An elegant summer-flowering shrub, which in mild areas may be semi-evergreen.

Origin From Morocco.
Use As a freestanding shrub for mild areas or as a wall shrub for colder localities. In mild areas, can be grown with single stem as small tree.
Description *Flower* Long, upright panicles of bright yellow, pineapple-scented flowers, early to mid summer. *Foliage* Leaves grey to silver-grey, pinnate up to 4in (10cm) long and wide, comprising three leaflets, 1½in (4cm) long. Silver-grey when young, aging to grey-green to grey-brown. Strong, upright, becoming branching and spreading with age. Fast-growing when young, slowing with age. *Fruit* Insignificant.

Cytisus battandieri in flower

Hardiness Tolerates winter temperatures in the range 23°F (−5°C) down to 14°F (−10°C).
Soil Requirements Does well on most soils but unhappy on very thin chalk types.
Sun/Shade aspect Prefers full sun.
Pruning None required; best planted where it can attain full maturity without pruning. If size must be restricted cut young wood after flowering. Cutting into old wood may lead to die-back.
Propagation and nursery production From seed, although good flowering forms are best grafted on to *Laburnum vulgaris*, as seedling stock can be very variable. May also be found produced from semi-ripe cuttings taken mid-summer. May be available in nursery production; not common in garden centres. Best planting heights 2-4ft (60cm-1.2m).
Problems None, if given space to attain its full height and spread.
Average height and spread
Five years
10x6ft (3x2m)
Ten years
16x13ft (5x4m)
Twenty years
or at maturity
20x20ft (6x6m)

CYTISUS
Low-growing forms

SPREADING BROOM
Leguminosae
Deciduous
Beautiful spring-flowering shrubs with spreading growth that can be used as ground cover or trained over a wall.

Origin Mostly of garden origin or from southern Europe.
Use As low spreading carpet for spring flowering. Some varieties ideal for cascading over walls. Effective used in very large containers.

Cytisus × kewensis in flower

Description *Flower* Normally pale to golden yellow pea-flowers but colour dependent on variety. Flowers borne in profusion, mid spring. *Foliage* Leaves very sparsely produced, small, lanceolate, ½-¾in (1-2cm) long, grey-green. *Stem* Light grey-green aging to dark green. Very low-growing, spreading, creeping, carpet-forming shrub. Slow to medium growth rate. *Fruit* Insignificant.
Hardiness Tolerates 14°F (−10°C).
Soil Requirements Tolerates a wide range of conditions, particularly happy on alkaline soils, provided depth of soil is not less than 12in (30cm) above limestone base.
Sun/Shade aspect Full sun to very light shade.
Pruning Very little required; current season's growth may benefit from being shortened, but never cut into older wood.
Propagation and nursery production From softwood cuttings in taken early summer. Purchase container-grown. Availability varies. Best planting heights 4-12in (10-30cm).
Problems Some die-back of stems and yellowing may be seen in very wet summers or extremely cold winters.

Cytisus purpureus in flower

Varieties of interest *C. × beanii* Golden yellow flowers, early to mid spring, forming a very dense carpet. A mound-forming shrub of medium growth rate. Of garden origin. *C. decumbens* Bright yellow flowers, mid spring. Forms a dense, flat carpet. An extremely prostrate, slow-growing form, reaching average spread but no more than 8in (20cm) in height. *C. × kewensis* A low creeping shrub sending out long, searching, thin branches, covered in a mass of creamy yellow flowers along the upper surfaces in mid spring. Raised in Kew Gardens, England, in about 1891. *C. procumbens* A low, slow-growing shrub, covered with bright yellow flowers in mid to late spring. From south-eastern Europe. *C. purpureus* (Purple Broom) A low shrub of medium growth rate, covered in lilac-pink pea-flowers, mid spring. Obtainable from specialist nurseries. From south-eastern Europe.
Average height and spread
Five years
1½x2½ft (50x80cm)
Ten years
1½x4ft (50cmx1.2m)
Twenty years
or at maturity
1½x5ft (50cmx1.5m)

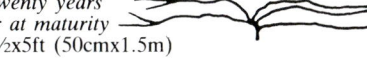

CYTISUS MULTIFLORUS
(Cytisus albus)

WHITE SPANISH BROOM
Leguminosae
Deciduous
A beautiful medium height, late spring, early summer, white-flowering shrub.

Origin From Spain, Portugal and north-western Africa.
Use As a medium to tall flowering shrub, best for solo planting but can be sited in medium-sized to large shrub borders.

Cytisus multiflorus in flower

Description *Flower* Small, pure white, pea-flowers borne over entire branch length, late spring to early summer. *Foliage* Leaves sparsely produced, small, grey-green, lanceolate, ½-¾in (1-2cm) long, borne mainly at lower end of stems. *Stem* Upright becoming spreading, with graceful habit. Fast growth rate, slowing with age. *Fruit* Small, grey-green, pea-pods.
Hardiness Tolerates 14°F (−10°C).
Soil Requirements Neutral to acid; dislikes alkalinity but may tolerate it provided adequate depth of topsoil available.
Sun/Shade aspect Full sun.
Pruning Best left unpruned to reach full size but if desired current season's growth may be halved after flowering. Cutting older wood may lead to total die-back.

Propagation and nursery production From semi-ripe cuttings taken in early summer. Purchase container-grown. Not always easy to find in garden centres, but may be found in nurseries or from specialist sources. Best planting heights 2-3ft (60cm-1m).
Problems Rather short-lived. The shrub has a very poor root system and may need staking and tying for support.
Average height and spread
Five years
6x5ft (2x1.5m)
Ten years
10x8ft (3x2.5m)
Twenty years
or at maturity
10x8ft (3x2.5m)

CYTISUS × PRAECOX

WARMINSTER BROOM
Leguminosae
Deciduous
Beautiful spring-flowering shrubs worthy of any garden.

Origin Of garden origin. Originally found in Wiltshire, England.
Use In shrub borders as a useful, low to medium height shrub for single or mass planting. Good in tubs of adequate size. Sometimes supplied grafted on to 4ft (1.2m) high stem of *Laburnum vulgaris* to make small to medium height weeping standard.
Description *Flower* Small pea-flowers, white, cream, yellow, pink or red, according to variety, freely borne, late spring. *Foliage* Leaves lanceolate, grey-green, ½-¾in (1-2cm) long, sparsely produced, chiefly towards base of branches. *Stem* Grey-green to light green, upright when young, becoming arching to form a very attractive shape. Medium to fast growth rate. *Fruit* Insignificant.
Hardiness Tolerates 14°F (−10°C).
Soil Requirements Does well on most soils but extreme alkalinity may lead to severe chlorosis.
Sun/Shade aspect Full sun.
Pruning None required, but small sections of year-old growth may be removed after flowering.
Propagation and nursery production From softwood cuttings taken in early summer. Purchase container-grown. *C. × praecox* and *C. × p. 'Allgold'* are fairly easy to find, but other varieties less common. Best planting heights 15in-2ft (40-60cm).
Problems Rather short-lived and should be replaced after 10-15 years. Has a poor root system and mature plants may need staking in exposed areas.
Varieties of interest *C. × p. 'Albus'* Pure white flowers. Vigorous and of slightly more height and spread than average. *C. × p. 'Allgold'*

Cytisus × praecox 'Albus' in flower

Golden yellow flowers. *C. × p. 'Buttercup'* Buttercup yellow flowers. *C. 'Hollandia'* Purple-red flowers. *C. 'Zeelandia'* Pink and mauve bicoloured flowers. One-third more than average height and spread.

Average height and spread
Five years
2½x3ft (80cmx1m)
Ten years
3x5ft (1x1.5m)
Twenty years
or at maturity
4x6ft (1.2x2m)

CYTISUS SCOPARIUS

BROOM
Leguminosae
Deciduous
Very attractive spring-flowering shrubs, but relatively short-lived.

Origin Basic form from Europe, but mostly of garden origin.
Use As freestanding flowering shrubs for larger shrub borders, or for single or mass planting.

Cystisus scoparius 'Andreanus' in flower

Description *Flower* Scented pea-flowers, pink, red, amber, yellow, bronze, or bicoloured, late spring, early summer. *Foliage* Leaves sparsely produced, small, lanceolate, ½-¾in (1-2cm) long, grey-green, borne chiefly on low branches. *Stem* Long, angular, upright, becoming spreading; light green to grey-green. Fast-growing when young, slowing with age. *Fruit* Small, grey-green pods may be produced in winter.
Hardiness Tolerates 14°F (−10°C).
Soil Requirements Prefers neutral to acid soil, dislikes alkalinity but tolerates it provided depth of topsoil is adequate.
Sun/Shade aspect Best in full sun.
Pruning Best unpruned so as to reach full size, but if desired current season's growth can be halved after flowering. Cutting into older wood may lead to total die-back.
Propagation and nursery production From softwood cuttings taken in early summer. Purchase container-grown. Limited range of varieties at nurseries and garden centres; some varieties may have to be sought from specialist sources. Best planting heights 2-2½ft (60-80cm).
Problems Brooms are all relatively short-lived and after 10-12 years should be replaced. Varieties of *C. scoparius* root poorly; taller plants may need staking.
Varieties of interest *C. nigricans* Yellow pea-flowers produced in terminal racemes in mid to late summer. Of slightly more than average height and spread. Scarce in production. From central and south-eastern Europe through to central Russia. *C. scoparius 'Andreanus'* Bicoloured, crimson-red and

Cytisus scoparius 'Cornish Cream' in flower

chrome yellow flowers. *C. s. 'Cornish Cream'* Bicoloured cream-yellow flowers borne on strong, upright branches. *C. s. 'Fulgens'* Bicoloured amber and rich crimson flowers carried on stiff, strong branches. *C. s. 'Red Favourite'* Deep red flowers, upright stems. **Hybrids:** *C. 'Burkwoodii'* Maroon and bright red flowers. Upright when young, becoming arching with age. *C. 'C.E. Pearson'* Rose pink, yellow and red flowers. *C. 'Criterion'* Bicoloured, brown-purple flowers. Strong-growing. Slightly more than average height and spread. *C. 'Daisy Hill'* Bicoloured crimson and cream flowers, graceful arching stems. Above average spread. *C. 'Donard Seedling'* Bicoloured mauve, red and pink flowers on strong, stout branches. Above average height and spread. *C. 'Dorothy Walpole'* Bicoloured crimson-pink flowers. Strong growth. *C. 'Eastern Queen'* Bicoloured amber and crimson flowers. Strong growth. *C. 'Golden Cascade'* Large, golden yellow flowers. Arched, weeping stems. *C. 'Goldfinch'* Golden yellow flowers. Upright. One-third less than average height and spread. *C. 'Killiney Salmon'* Red-salmon flowers. Graceful habit. Slightly more than average height and spread. *C. 'Lord Lambourne'* Bicoloured cream and maroon flowers. *C. 'Minstead'* Flowers white flushed lilac with darker outer wings. Thin stems and graceful, lax habit. One-third more than average height and spread. *C. 'Moonlight'* Delicate sulphur yellow flowers. *C. 'Windlesham Ruby'* Mahogany-crimson flowers. Arching branches.

Average height and spread
Five years
6x5ft (2x1.5m)
Ten years
10x6ft (3x2m)
Twenty years
or at maturity
10x8ft (3x2.5m)

Cytisus 'Killiney Salmon' in flower

DABOECIA CANTABRICA
(Menziesia polifolia)

CONNEMARA HEATH, ST. DABOEC'S HEATH, IRISH HEATH
Ericaceae
Evergreen
Used on the right soil, a very attractive low carpeting shrub.

Origin From western Europe, Ireland in particular.
Use For low ground cover used with dwarf rhododendrons or conifers, or on its own as mass ground cover for lightly shaded areas, or for planting on large rock gardens.
Description *Flower* Small, hanging, pitcher-shaped flowers, rose-purple or white, dependent on variety, borne on upright flower spikes early summer through to late autumn. *Foliage* Leaves very small, lanceolate, ½in (1cm) long, dark purple-green with glossy upper surfaces and duller silver undersides. *Stem* Upright, short, dark purple-green to purple-brown, forming a low, spreading carpet. Medium to slow growth rate. *Fruit* May produce small, brown seedheads of some winter attraction.

Daboecia cantabrica in flower

Hardiness Tolerates 14°F (−10°C), but foliage may be damaged by severe wind chill.
Soil Requirements Neutral to acid; dislikes any alkalinity.
Sun/Shade aspect Prefers light shade; tolerates full sun provided moisture is adequate.
Pruning Trim lightly with shears, early spring.
Propagation and nursery production From softwood cuttings taken early summer. Purchase container-grown. May be found in nurseries in areas of acid soil or from specialist sources. Best planting heights 4-8in (10-20cm).
Problems None.
Varieties of interest *D. c. 'Alba'* White, hanging, pitcher-shaped flowers. Lime green foliage. *D. c. 'Atropurpurea'* Rose-purple flowers. Darker foliage than the parent. *D. c. 'Bicolor'* White and rose-purple striped flowers, often on same branchlet. *D. c. 'Praegerae'* Narrow, pitcher-shaped rich purple flowers. A low-growing variety, only two-thirds average height and spread.

Average height and spread
Five years
8inx2ft (20x60cm)
Ten years
8inx3ft (20cmx1m)
Twenty years
or at maturity
8inx3ft (20cmx1m)

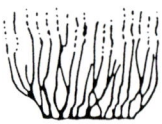

DANAE RACEMOSA
(Ruscus racemosus)

ALEXANDRIAN LAUREL
Liliaceae
Evergreen
A useful evergreen plant for very shady areas.

Origin From Asia Minor through to Iran.
Use As a low, shade-loving shrub. Popular with flower arrangers.
Description *Flower* Small, inconspicuous, sulphur-yellow flowers in early summer. *Foliage* Pointed, lanceolate, 1½-4in (4-10cm) long, dark green and shiny. Not true leaves, but flattened stems of leaf-like appearance. *Stem* Dark green, shiny, upright, flattened, and resembling leaves, forming a low suckering clump. Moderately slow growth rate. *Fruit* Round, orange-red fruits in autumn, following hot summers.
Hardiness Tolerates 14°F (−10°C).

Danae racemosa in leaf

Soil Requirements Any soil conditions, including relatively dry areas.
Sun/Shade aspect Not only tolerates but prefers deep shade.
Pruning None required, but one-third of oldest wood may be removed occasionally in spring to rejuvenate.
Propagation and nursery production By division. Purchase container-grown or as clumps from open ground. Not readily available. Best planting heights 8in-1ft (20-30cm).
Problems None.
Average height and spread
Five years
2x2½ft (60x80cm)
Ten years
2½x4ft (80cmx1.2m)
Twenty years
or at maturity
3x5½ft (1x1.8m)

DAPHNE BLAGAYANA

KNOWN BY BOTANICAL NAME
Thymelaeaceae
Deciduous
One of the most beautiful scented shrubs, but difficult to obtain and grow well.

Origin From the mountain forests of south-eastern Europe.
Use As a scented, low shrub for woodland areas.
Description *Flower* Clusters of creamy white, richly scented flowers, mid to late spring. *Foliage* Leaves light grey-green, ovate to round, 1½-2½in (4-6cm) long. *Stem* Light green to grey-green, rambling shoots, forming a creeping, open, mat. Slow growth rate. *Fruit* Round, white-blue fruits, midsummer.

Daphne blagayana in flower

Hardiness Tolerates minimum winter temperature range between 23°F (−5°C) and 14°F (−10°C).
Soil Requirements Very specific; open, deep, slightly acid, humus-rich soil, as in a woodland environment.
Sun/Shade aspect Light to medium shade, resents sunlight.
Pruning None required.
Propagation and nursery production From semi-ripe cuttings taken in early summer. Purchase container-grown. Not easy to find. Best planting heights 8-15in (20-40cm).
Problems Its very specific requirements must be met to achieve success.
Average height and spread
Five years
1x2ft (30x60cm)
Ten years
1½x2½ft (50x80cm)
Twenty years
or at maturity
2x2½ft (60x80cm)

DAPHNE × BURKWOODII

BURKWOOD'S DAPHNE
Thymelaeaceae
Deciduous
An attractive and very sweetly scented, early summer-flowering shrub.

Origin Introduced by the celebrated nurseryman Burkwood of Surrey, England in the early 1930s.
Use As a flowering, scented shrub used on its own or as focal point for other low underplanting.

Daphne × burkwoodii in flower

Description *Flower* Clusters of pale pink, sweetly scented flowers borne at terminals of each branch, late spring to early summer. *Foliage* Leaves narrow, round-ended, lanceolate, 1-1½in (3-4cm) long, grey-green, profusely borne in rosettes around stems. Some yellow autumn colour. *Stem* Grey-green, upright, becoming branching at upper ends, forming a wide-topped, narrow-based shrub. *Fruit* May produce small orange-red fruits in hot summers.
Hardiness Tolerates winter temperatures in the range 23°F (−5°C) down to 14°F (−10°C).
Soil Requirements Prefers neutral to acid soil, but tolerates some alkalinity provided adequate soil moisture and good topsoil structure is available.
Sun/Shade aspect Prefers light shade, tolerates medium shade to full sun.
Pruning None required.
Propagation and nursery production From softwood cuttings taken in early summer. Purchase container-grown. Fairly easy to find in nurseries, less easily in garden centres. Best planting heights 15in-2½ft (40-80cm).
Problems Virus prone, but this does not seem to cause trouble with this particular form. May look small and weak at purchase, but grows quickly when planted.
Varieties of interest *D. × b.* 'Somerset' Larger flowers and perhaps better constitution than *D. × burkwoodii*. Said by some authorities to be actually the same shrub. *D. × b.* 'Somerset Gold Edge' Flowers slightly smaller, but equally scented; leaves gold-margined. Slightly less height and spread. Relatively hard to find.
Average height and spread
Five years
2x1½ft (60x50cm)
Ten years
2½x2¼ft (80x70cm)
Twenty years
or at maturity
2½x2½ft (80x80cm)

DAPHNE CNEORUM

GARLAND FLOWER, ROSE DAPHNE
Thymelaeaceae
Evergreen
A beautiful small, scented shrub, worthy of any small garden or small planting feature.

Origin From central and southern Europe.
Use As a low-growing, spreading hummock for small borders or large rock gardens. Can be grown as a dwarf shrub in large containers and troughs.
Description *Flower* Fragrant, rose pink flowers borne in clusters, mid to late spring. *Foliage* Leaves narrow, lanceolate, ½-1in

(1-3cm) long, small, slightly blue-grey with yellow autumn colour. *Stem* Short branching, forming a low spreading mound. Slow growth rate. *Fruit* Brown-yellow, round fruits in some hot summers.
Hardiness Tolerates 23°F (−5°C).
Soil Requirements Prefers neutral to acid soil, but tolerates some alkalinity provided depth of topsoil over underlying chalk or limestone is adequate.
Sun/Shade aspect Prefers light shade, tolerates full sun.
Pruning None required.

Daphne cneorum in flower

Propagation and nursery production From softwood cuttings taken early summer. Purchase container-grown. May be difficult to find. Best planting heights 4-8in (10-20cm).
Problems Difficult and slow to establish, taking 2-3 years.
Varieties of interest *D. c. 'Alba'* Extremely rare, pure white-flowering form, with *D. c. 'Pygmaea'* in its parentage. Two-thirds average height and spread and slightly more tender. *D. c. 'Eximia'* Flowers rose pink, crimson when in bud. A variety with larger flowers and leaves. Of average spread but less height. *D. c. 'Pygmaea'* A very rare, pink, free-flowering, scented variety, of very prostrate habit. *D. c. 'Variegata'* Lighter pink flowers, scented. A moderately strong form with creamy leaf margins.
Average height and spread
Five years
8inx1½ft (20x50cm)
Ten years
1x2½ft (30x80cm)
Twenty years
or at maturity
1¼x3ft (40cmx1m)

DAPHNE COLLINA

Origin From southern Italy.
Use As a low, scented shrub to plant among alpines, heathers or dwarf conifers. Ideal for medium-sized to large rock gardens or can be grown in a large tub or trough.
Description *Flower* Fragrant clusters of rose pink flowers at terminals of each leaf shoot, mid to late spring. *Foliage* Leaves ovate to lanceolate, ½-1in (1-3cm) long, light green to grey-green, borne close to stems. *Stem* Short, stout, branching, forming a neat round mound. Slow growth rate. *Fruit* Light red and round, but rarely produced.
Hardiness Tolerates 23°F (−5°C).
Soil Requirements Prefers acid to neutral soil, but tolerates some alkalinity provided topsoil is adequate.

Daphne collina in flower

Sun/Shade aspect Prefers light shade, tolerates full sun.
Pruning None required.
Propagation and nursery production From semi-ripe cuttings taken in early summer. Purchase container-grown. Rather hard to find. Best planting heights 4-8in (10-20cm).
Problems None.
Varieties of interest *D. c. var. neapolitana* Rose pink, fragrant flowers in clusters, mid spring to early summer. A slightly less tall variety with grey-green to light green foliage.
Average height and spread
Five years
8inx15in (20x40cm)
Ten years
15inx2ft (40x60cm)
Twenty years
or at maturity
2x3ft (60cmx1m)

DAPHNE LAUREOLA

Origin From southern and western Europe, including the southern region of the British Isles.
Use As a low, evergreen, winter-flowering shrub for woodland and shady areas.
Description *Flower* Thick, round clusters of fragrant yellow, trumpet-shaped flowers, borne late winter to early spring. *Foliage* Leaves leathery textured, broad, lanceolate,

Daphne laureola in flower

2-4in (5-10cm) long, mid to dark green with shiny upper surfaces and duller grey-green undersides. *Stem* Upright, slightly lax, light to mid green, becoming branching to form dome-shaped shrub of slightly more width than height. *Fruit* Clusters of small, black fruits follow flowering.
Hardiness Tolerates 14°F (−10°C).
Soil Requirements Thrives in deep, rich, leafy loam; tolerates some alkalinity.
Sun/Shade aspect Prefers mid to light shade, dislikes any sun.
Pruning None required. Safe to remove any outer branch which has become too lax.
Propagation and nursery production From semi-ripe cuttings taken midsummer or from seed. Purchase container-grown. Not easy to find, even within its native territory. Best planting heights 8-12in (20-30cm).
Problems Only thrives under very specific soil and shade conditions.
Varieties of interest *D. l. var. philippi* Smaller flowers than parent and ovate leaves. Two-thirds average height. More readily available. From the Pyrenees.
Average height and spread
Five years
2x2¼ft (60x70cm)
Ten years
2½x2¾ft (80x90cm)
Twenty years
or at maturity
3x4ft (1x1.2m)

DAPHNE MEZEREUM

Origin From Europe, including British Isles, through Asia Minor and Siberia.
Use As a low, winter-flowering shrub for borders, large rock gardens or the edges of a shrub border.
Description *Flower* Small, purple-red, or white, dependent on variety, scented, trumpet-shaped flowers borne thickly along the entire length of upright shoots, late winter to early spring. *Foliage* Leaves small, round-ended, lanceolate, 1½-2in (4-5cm) long, grey-green with some yellow autumn colour. *Stem* Light grey to grey-green, upright and of rubbery texture, forming a goblet-shaped shrub. Slow to medium growth rate. *Fruit* Round, yellow aging to red, poisonous fruits borne more or less freely, depending on dryness of season. Seed often germinates into young plants at base of parent plant.
Hardiness Tolerates 14°F (−10°C).
Soil Requirements Prefers rich, deep, leafy

Daphne mezereum 'Alba' in flower

loam; tolerates moderately high alkalinity but dislikes waterlogging.
Sun/Shade aspect Prefers light shade, tolerates full sun.
Pruning None required.
Propagation and nursery production From seed or layers. Plants produced from layers from a known parent can have better flower production. Purchase container-grown or root-balled (balled-and-burlapped). Normally available when in flower from both nurseries and garden centres. Best planting heights 15in-2½ft (40-80cm).
Problems Genetic virus can lead to sudden death of mature specimens, as with most Daphnes. The virus can cause the foliage to look diseased, though without affecting flowering or fruiting.

Daphne mezereum in flower

Varieties of interest *D. bholua* Flowers red-mauve in bud, opening to white with red-mauve reverses to each petal. Sweetly scented and borne in terminal clusters in midwinter. Extremely scarce in production. *D. genkwa* Fragrant lilac-blue flowers produced on branches which are leafless in early spring. From China and Taiwan. Extremely difficult to find; must be sought from specialist nurseries. *D. mezereum 'Alba'* White to green-white scented flowers, late winter to early spring. A slightly less vigorous form and more difficult to find. *D. m. 'Alba Bowle's Hybrid'* A hybrid propagated from layers, producing slightly larger green-white flowers than *D. m. 'Alba'*. Stock always small when supplied, and usually with single stem. *D. m. 'Grandiflora'* Larger flowers, rich purple, often appearing mid to late autumn. Very scarce. *D. m. 'Rubrum'* A European variety with very good large, purple-red flowers.
Average height and spread .
Five years
2x1½ft (60x50cm)
Ten years
2½x2¼ft (80x70cm)
Twenty years
or at maturity
2½x3ft (80cmx1m)

DAPHNE ODORA

FRAGRANT DAPHNE, WINTER DAPHNE
Thymelaeaceae
Evergreen
One of the most fragrant flowering shrubs for late winter and early spring.

Origin From China and Japan.
Use As freestanding, early spring-flowering, scented shrub.
Description *Flower* Clusters of small to medium-sized, very scented flowers, rose pink or white, dependent on variety, in late winter to early spring. *Foliage* Leaves broad, lanceolate, 1-3in (2-8cm) long, grey-green to mid green with greyer undersides. Some yellow variegated edges. *Stem* Fairly stout, grey-green, upright shoots, becoming more spreading and branching to form a wide-spreading low mound. Stems of mature plants often very weak. Slow to medium growth rate. *Fruit* Small, round, yellow fruits sometimes produced in very hot, dry summers.
Hardiness Minimum winter temperature 23°F (−5°C). May need protection in winter.
Soil Requirements Prefers rich, deep loam, tolerates moderately high alkalinity.
Sun/Shade aspect Prefers light shade, dislikes full sun.
Pruning None required.
Propagation and nursery production From softwood cuttings taken early to mid summer. Purchase container-grown. *D. odora* itself may be rather hard to find. Best planting heights 8-15in (20-40cm).
Problems Can become extremely lax and woody; rejuvenation by pruning is rarely successful. Old plants may deteriorate in health and appearance.
Varieties of interest *D. o. 'Alba'* A pure white-flowering form. Scarce in commercial production. *D. o. 'Aureomarginata'* Clusters of very scented, purple-pink, trumpet-shaped flowers. Broad, lanceolate leaves edged with prominent gold bands, late spring through to summer, decreasing in intensity as autumn and winter approach. Slightly less than average height and spread, but more hardy. Fairly easy to find.
Average height and spread
Five years
1½x2½ft (50x80cm)
Ten years
2¼x3ft (70cmx1m)
Twenty years
or at maturity
5½x3ft (1.8x1m)

Daphne odora 'Aureomarginata' in flower

DAPHNE TANGUTICA

KNOWN BY BOTANICAL NAME
Thymelaeaceae
Evergreen
This is a shrub of somewhat lax habit by comparison with other Daphnes, but truly warrants a featured place in the garden.

Origin From China.
Use As a low-growing, interesting dwarf shrub for small rock gardens or container gardens, or combined with other low plants and shrubs.
Description *Flower* Fragrant flowers in small clusters, white tinged purple on the inside, rose purple outside, formed at the terminals of each branch, mid to late spring. *Foliage* Leaves narrow, ovate, 1-3in (3-8cm) long, olive green with silver undersides. *Stem* Mid to dark green. Stout, branching, upright becoming spreading, forming a broad, open mound. Slow growth rate. *Fruit* Rarely fruits.
Hardiness Tolerates 14°F (−10°C).

Daphne tangutica in flower

Soil Requirements Prefers rich, deep, woodland type, but is fairly amenable to other soils.
Sun/Shade aspect Best in light shade. Tolerates medium shade and full sun as long as adequate moisture available.
Pruning None required.
Propagation and nursery production From semi-ripe cuttings taken in early summer. Purchase container-grown. Normally available from specialist nurseries. Best planting heights 8-12in (20-30cm).
Problems May start to drop its leaves in extremely wet, cold winters. If this occurs lighten the surrounding soil by working in sedge peat or sharp sand. Guard against further waterlogging by providing some sort of temporary canopy.
Varieties of interest *D. pontica* Bright green, glossy-surfaced, ovate leaves. Yellow-green flowers mid to late spring, blue-black fruits in autumn. Average height but slightly more spread. Scarce in production. From Asia Minor. *D. retusa* Scented, deep rose purple flower clusters, late spring to early summer. Available from specialist nurseries. Very slow-growing, only reaching two-thirds average height and spread. From western China.
Average height and spread
Five years
4inx1ft (10x30cm)
Ten years
1¼x1¼ft (40x40cm)
Twenty years
or at maturity
2½x2½ft (80x80cm)

DECAISNEA FARGESII

KNOWN BY BOTANICAL NAME
Lardizabalaceae
Deciduous
Interesting green flowers which produce turquoise blue seed pods after a good summer.

Origin From China.
Use A tall, unusual upright shrub for growing either as a specimen or on a wall, where it seems to produce better flowers and fruit.
Description *Flower* Open racemes, up to 12in (30cm) long, of lime green to yellow-green tubular flowers, produced at ends of upright shoots, late spring to early summer. *Foliage* Large, pinnate leaves, 2-3ft (60cm-1m) long, with each leaflet 3-6in ((8-15cm) long. Light grey-green with yellow autumn colour. *Stem* Upright, becoming leaning with age, forming an upright central clump with some spreading side growths. Medium to fast growth rate. *Fruit* Metallic blue bean pods up to 15in (40cm) long in autumn following dry hot summers.

Decaisnea fargesii in flower

Hardiness Tolerates 14°F (−10°C).
Soil Requirements Does well on most soils but very moist, well-drained soil produces better growth and therefore more flowers and, in favourable years, fruit.
Sun/Shade aspect Prefers light shade, but tolerates full sun provided adequate moisture available.
Pruning None required. Weak, unattractive stems may be removed.
Propagation and nursery production From seed. Purchase container-grown or root-balled (balled-and-burlapped). Available from specialist nurseries. Best planting heights 15in-3ft (40cm-1m).
Problems Can take several years to produce fruit.

Average height and spread
Five years
6x6ft (2x2m)
Ten years
10x13ft (3x4m)
*Twenty years
or at maturity*
13x13ft (4x4m)
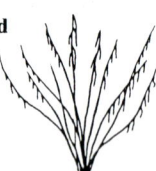

DESFONTAINEA SPINOSA

KNOWN BY BOTANICAL NAME
Potaliaceae
Evergreen
A truly spectacular midsummer-flowering shrub, given moisture, shade and protection.

Origin From Chile and Peru.
Use As an evergreen, flowering shrub to be grown on its own for best effect.
Description *Flower* Scarlet, 1½in (4cm) long, tubular flowers, yellow shaded in mouth, midsummer. *Foliage* Leaves oval, 1-2in (3-5cm) long, dark green, glossy, holly-like, with

Desfontainea spinosa in flower

silver undersides. *Stem* Upright, moderately stout, dark green, forming a clump of upright flowering shoots. Slow growth rate. *Fruit* Insignificant.
Hardiness Minimum winter temperature range 23°F (−5°C) down to 14°F (−10°C).
Soil Requirements Acid soil, dislikes any alkalinity or waterlogging. Thrives on soil high in leaf mould as in woodland situation.
Sun/Shade aspect Prefers light to medium shade, unsuited to full sun.
Pruning None required.
Propagation and nursery production From semi-ripe cuttings taken midsummer or from rooted suckers from outer edges of mature clumps. Purchase container-grown. Hard to find; may be available from specialist nurseries. Best planting heights 8-15in (20-40cm).
Problems Often takes a number of years to settle down.
Varieties of interest *D. s. 'Harold Comber'* A variety collected in the wild, with varying flower colour, from vermilion to orient red. Very scarce.

Average height and spread
Five years
1½x2ft (50x60cm)
Ten years
2¼x3ft (70cmx1m)
*Twenty years
or at maturity*
5½x4ft (1.8x1.2m)

DEUTZIA

KNOWN BY BOTANICAL NAME
Philadelphaceae
Deciduous
A range of very free-flowering early summer shrubs of widely varying sizes and shapes.

Origin Mainly from China, but many varieties of garden or nursery origin.
Use For shrub borders or group plantings, or as a freestanding specimen.
Description *Flower* Small, short panicles of bell-shaped flowers in shades of white or pink through to purple-pink, depending on variety; late spring to early summer. *Foliage* Leaves small to medium, ovate, 1½-4in (4-10cm) long, olive green, giving some yellow or purple autumn colour depending on variety. *Stem* Mostly upright, strong, light grey to grey-green, becoming very twiggy and branching with age; some varieties more arching and pendulous. Medium growth rate. *Fruit* Grey-brown seedheads in autumn, retained into winter.
Hardiness Tolerates −13°F (−25°C).
Soil Requirements Prefers moist, good soil, tolerates high alkalinity. If grown on dry soils must be watered, particularly during droughts.

Sun/Shade aspect Full sun to light shade.
Pruning Remove one-third of oldest flowering wood after flowering.
Propagation and nursery production From semi-ripe cuttings taken midsummer or from hardwood cuttings taken late autumn and winter. Purchase container-grown or bare-rooted. Most varieties fairly easy to find. Best planting heights 2-2½ft (60-80cm).
Problems May not survive drought.

Deutzia gracilis in flower

Varieties of interest *D. chunii* Single flowers, pink outside, white within, with yellow anthers, produced in 2in (5cm) panicles. Very scarce. *D. 'Contraste'* Panicles of star-shaped, semi-double flowers, soft lilac-pink on outer side, rich purple inner side. Reaches one-third average height and spread. An outstanding variety. Of garden origin. *D. corymbosa* Open, short racemes of single white flowers in early summer. Half average height and spread. Best planting heights 15in-2ft (40-60cm). *D. × elegantissima* Rose pink, fragrant paniculated clusters of flowers, early to mid summer. Half average height and spread with much thinner, more twiggy growth. *D. × elegantissima 'Rosealind'* Single, deep carmine-pink flowers in short panicles, borne freely on thin, wispy, arching branches, early summer. Half average height and spread. Best planting heights 15in-1½ft (40-50cm). *D. gracilis* (Slender Deutzia, Japanese Snow Flower) Pure white single flowers. From Japan. *D. × kalmiiflora* Very short racemes or singly produced flowers, pink to flushed carmine, borne on arching branches, late spring. Good plum-purple autumn colours. One-third average height and spread. Of garden origin. Best planting heights 15in-2ft (40-60cm). *D. longifolia 'Veitchii'* Clusters

Deutzia × kalmiiflora in flower

of single, lilac-pink flowers, early summer. Foliage long, more lanceolate and lighter green than most Deutzias. Of garden origin. *D. 'Magicien'* Large, mauve-pink, single flowers with white edges and purple reverses. Of garden origin. *D. × magnifica* Double pure white flowers, early summer. A strong-growing variety. Of garden origin. *D. monbeigii* Single, white, star-like flowers profusely produced in late summer. Smaller leaves with white undersides and slightly arching branches. One-third average height and spread. From China. *D. 'Mont Rose'* Single to semi-double, rose pink flowers with darker highlights, produced in large paniculated clusters, early summer. Of garden origin. *D. pulchra* Racemes of pure white flowers in long, hanging, lily-of-the-valley type formation. Two-thirds average height and spread and very hardy. From Taiwan. *D. × rosea 'Carminea'* Flowers rose-carmine with paler shading. Two-thirds average height and spread. Of garden origin. *D. scabra* (Fuzzy Deutzia) Single white flowers in panicle-shaped clusters, early summer. Very upright habit. From Japan. *D. s. 'Candidissima'* Pure white, double flowers, early summer. Upright. Of garden origin. *D. s. 'Plena'* Double, rose purple flowers with white shading. Of garden origin. *D. s. 'Pride of Rochester'* Double white flowers produced in early summer. Upright. Of garden origin. *D. setchuenensis* Clusters of small white, star-like flowers profusely borne, mid to late summer. Two-thirds average height and spread. Slightly less hardy. From China.

Average height and spread
Five years
6x3ft (2x1m)
Ten years
10x6ft (3x2m)
Twenty years
or at maturity
13x10ft (4x3m)

Deutzia 'Mont Rose' in flower

DIERVILLA

DIERVILLA, BUSH-HONEYSUCKLE
Caprifoliaceae
Deciduous
Attractive, medium height, summer-flowering shrubs, with often underestimated autumn colour.

Origin Most varieties from North America.
Use Freestanding or for shrub borders. Ideal for mass planting.
Description *Flower* Short, yellow, tubular flowers, each with two distinct lips, borne early to late summer. *Foliage* Leaves ovate, 2-6in (5-15cm) long, light to mid green with some yellow or purple-red autumn colours, depending on variety. *Stem* Light green to mid green shaded red. Upright when young, becoming spreading with age. Medium to fast growth rate, depending on variety. *Fruit* Insignificant.

Diervilla sessilifolia in flower

Hardiness Minimum winter temperature −13°F (−25°C).
Soil Requirements Does well on most soils but dislikes very dry areas.
Sun/Shade aspect Full sun to very light shade.
Pruning Remove one-third of oldest flowering wood after flowering.
Propagation and nursery production From semi-ripe cuttings taken midsummer. Purchase container-grown or bare-rooted. Readily available in its native environment but scarce in commercial production elsewhere. Best planting heights 15in-2½ft (40-80cm).
Problems None.
Varieties of interest *D. lonicera* (Dwarf Bush-

honeysuckle) Pale yellow flowers of honeysuckle shape borne on small suckering shrub. Foliage giving good autumn colours in dry, cold positions. *D. rivularis* (Georgia Bush-honeysuckle) Lemon yellow flowers, midsummer. Foliage giving good autumn colours. From south-eastern USA. *D. sessilifolia* (Southern Bush-honeysuckle) Sulphur yellow flowers borne in short panicles, early to late summer. Very good red and purple autumn foliage. Two-thirds average height and spread. From south-eastern USA. *D. × splendens* Sulphur yellow flowers, early to mid summer. Good autumn foliage colours. One-third more than average height. Of garden origin.

Average height and spread
Five years
5x5ft (1.5x1.5m)
Ten years
8x8ft (2.5x2.5m)
Twenty years
or at maturity
12x12ft (3.5x3.5m)

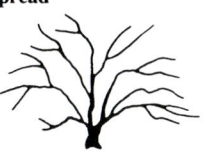

DIPELTA FLORIBUNDA

KNOWN BY BOTANICAL NAME
Caprifoliaceae
Deciduous
A very beautiful flowering shrub which, once its reluctance to root has been overcome by advances in propagation techniques, will become one of the most desirable spring-flowering garden shrubs.

Origin From central and western China.
Use As a flowering shrub, for planting singly or in a shrub border.
Description *Flower* Large, scented, bell-shaped flowers, pink flushed yellow at throat, late spring. *Foliage* Leaves elliptic, 4-7in (10-17cm) long, slightly pendulously presented. Light green, with good yellow autumn colour. *Stem* Strong, upright, light grey to grey-brown, forming with time an upright, spreading-topped shrub. Medium to fast growth rate. *Fruit* Attractive, round to oval, flat discs, ½in (1cm) across, light green aging to pink-green. From midsummer onwards.
Hardiness Tolerates 14°F (−10°C).
Soil Requirements Does well on most soils, but dislikes extreme alkalinity.
Sun/Shade aspect Prefers light shade, tolerates medium shade to full sun.
Pruning Remove one-third of old flowering wood after flowering.
Propagation and nursery production From semi-ripe cuttings taken early to mid summer. Very difficult to root. Purchase container-grown. Very hard to find. Best planting heights 15in-2ft (40-60cm).
Problems May be attacked by aphids.
Varieties of interest *D. ventricosa* Deep rose pink outer petals with paler inner surfaces and orange throats. Three-quarters average

Dipelta floribunda in flower

height, average spread. Scarce. **D. yunnanensis** Cream tinted rose pink, trumpet-shaped flowers with orange throats.
Average height and spread
Five years
6x3ft (2x1m)
Ten years
10x6ft (3x2m)
*Twenty years
or at maturity*
13x6ft (4x2m)

DISANTHUS CERCIDIFOLIUS

KNOWN BY BOTANICAL NAME
Hamamelidaceae
Deciduous
A beautiful autumn foliage shrub which needs an acid soil.

Origin From south-eastern China and Japan.
Use As a freestanding shrub or planted with dwarf heathers and rhododendron hybrids.
Description *Flower* Insignificant, tiny, purple flowers in early autumn. *Foliage* Leaves large, round to slightly heart-shaped, deeply veined. Dark green to olive green in summer, giving very fine crimson and claret red autumn colours. *Foliage* Light green when young, becoming spreading and branching with age, making a large, round, dense shrub. Slow to medium growth rate. *Fruit* Insignificant.

Disanthus cercidifolius **in autumn**

Hardiness Tolerates 14°F (−10°C).
Soil Requirements Acid soil; dislikes any alkalinity. Ideal in humus-rich soil.
Sun/Shade aspect Prefers very light shade, but needs afternoon sun in autumn to bring out full beauty of autumn foliage.
Pruning None required.
Propagation and nursery production From layers taken from parent plants. Purchase container-grown. Not easy to find and may take extensive searching in specialist nurseries. Best planting heights 15in-2ft (40-60cm).
Problems Moderately slow to establish.
Average height and spread
Five years
3x3ft (1x1m)
Ten years
6x6ft (2x2m)
*Twenty years
or at maturity*
10x10ft (3x3m)

DISTYLIUM RACEMOSUM

KNOWN BY BOTANICAL NAME
Hamamelidaceae
Evergreen
A curious shrub with interesting flowers in early spring.

Origin From Japan and southern Taiwan.
Use As a small to medium height, evergreen shrub of curiosity value.
Description *Flower* No petals; flowers consist of numerous red stamens produced from the stems, early to mid spring. *Foliage* Leaves narrow, ovate, 1-1½in (3-4cm) long, dark green with lighter underside. Glossy and

Distylium racemosum **in flower**

leathery textured. *Stem* Grey-green to brown-green, slowly forming a round-topped, slightly spreading shrub. Moderately slow growth rate. *Fruit* Insignificant.
Hardiness Tolerates 14°F (−10°C).
Soil Requirements Neutral to acid, but tolerates very slight alkalinity.
Sun/Shade aspect Prefers light dappled shade, but tolerates full sun if adequate moisture is available.
Pruning None required.
Propagation and nursery production From seed or softwood cuttings taken midsummer. Purchase container-grown. Scarce in commercial production. Best planting heights 1-1½ft (30-50cm).
Problems Very slow to mature.
Average height and spread
Five years
3x3ft (1x1m)
Ten years
6x6ft (2x2m)
*Twenty years
or at maturity*
13x13ft (4x4m)

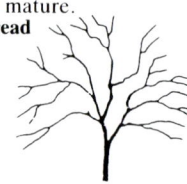

DORYCNIUM HIRSUTUM

KNOWN BY BOTANICAL NAME
Leguminosae
Deciduous or Semi-evergreen
A pleasing, low, spreading, grey and silver-leaved shrub covered all over with silvery hairs.

Origin From the Mediterranean area and southern Portugal.
Use As a shrub with interesting foliage, flowers and fruit for shrub borders, particularly good for grey borders.

Dorycnium hirsutum **in flower**

Description *Flower* Pink-tinged white pea-flowers, borne in terminals of current season's growth, late summer and early autumn. *Foliage* Leaves grey to silver-grey, fine, ovate, ¾-1in (2-3cm) long, on short stalks. *Stem* Grey-green; upright when young, becoming spreading. Dies down to ground level each winter. *Fruit* Small, brown to red-tinged pea-pods in autumn; offset by silvery hairs covering whole shrub.
Hardiness Tolerates 14°F (−10°C). As a sub-shrub, growth die-back to the plant base is normal in winter, although this may not occur in mild conditions.
Soil Requirements Open, well-drained soil; tolerates both acidity and alkalinity.
Sun/Shade aspect Full sun.
Pruning None required if winter die-back occurs. If not, clean off old shoots in early spring.
Propagation and nursery production From softwood cuttings taken midsummer and overwintered under protection for spring planting. Purchase container-grown; may have to be sought from specialist nurseries. Best planting heights 8-12in (20-30cm).
Problems None.
Average height and spread
Five years
1½x3ft (50cmx1m)
Ten years
2¼x4ft (70cmx1.2m)
*Twenty years
or at maturity*
2¼x4ft (70cmx1.2m)

DRIMYS WINTERI

WINTER'S BARK
Winteraceae
Evergreen
An interesting evergreen useful for variation of foliage in a group planting of ericaceous subjects, such as rhododendrons, azaleas and witch hazel.

Origin From South America.
Use As a woodland shrub, in a shrub border or group planting. Effective as a wall shrub.

Drimys winteri **in leaf**

Description *Flower* White to ivory white loose clusters of jasmine-scented flowers, mid spring. *Foliage* Leaves large, 5-10in (12-25cm) long, ovate, grey-green with glaucous undersides and leathery texture. *Stem* Upright, light to bright green to grey-green, becoming branching and spreading with age. *Fruit* Insignificant.
Hardiness Minimum winter temperature 23°F (−5°C); foliage resents severe wind chill.
Soil Requirements Acid, preferably open, woodland, leafy soil; dislikes any alkalinity.
Sun/Shade aspect Best in light or dappled shade.
Pruning None required, but any arching, spreading branches may be removed to confine the shrub to the planting area.

Propagation and nursery production From layers or semi-ripe cuttings taken late summer. Purchase container-grown or root-balled (balled-and-burlapped). Rather hard to find; must be sought from specialist nurseries. Best planting heights 2-2½ft (60-80cm).
Problems Wind chill may cause extensive defoliation and some die-back.
Average height and spread
Five years
6x6ft (2x2m)
Ten years
10x10ft (3x3m)
Twenty years
or at maturity
13x13ft (4x4m)

Origin Varieties from Europe and from North America.
Use As attractive foliage shrubs, planted singly or in groups.
Description *Flower* Small, inconspicuous, scented, sulphur yellow flowers, mid to late spring. *Foliage* Leaves broad, lanceolate, 2-4in (5-10cm) long, grey-green to silver with yellow autumn colour. *Stem* Grey-green, upright, becoming arching, forming a large, round-topped shrub. Medium to fast growth rate. *Fruit* Small, oval, ½in (1cm) long, amber fruits with silvery texture.

Elaeagnus commutata in leaf

Hardiness Tolerates 14°F (−10°C).
Soil Requirements Does well on almost all soils, tolerating quite dry areas, but unhappy on very alkaline types.
Sun/Shade aspect Full sun.
Pruning None required but can be cut back and will rejuvenate.
Propagation and nursery production From softwood cuttings taken early summer. Purchase container-grown. Easy to find. Best planting heights 1½-4ft (50cm-1.2m).
Problems Susceptible to drought or waterlogging. Any root disturbance may lead to some of the roots dying, followed by foliage failing, even to ground level. Normally regenerates but benefits from any efforts to correct soil problem.
Varieties of interest *E. angustifolia* (Oleaster, Russian Olive) Lanceolate leaves up to 3in (8cm) long, bright silver in spring aging to silver-grey with yellow autumn colour. May produce oval, dull orange fruits on mature shrub. Spiny, becoming a large shrub or eventually a small tree. In its native environment may reach one-third more than average

height and spread. From southern Europe. *E. commutata* syn. *E. argentea* (Silver Berry) Very good silver foliage with yellow fragrant flowers, mid spring. Silver, egg-shaped fruits. A variety producing underground stoloniferous shoots which arise from ground level. From North America. *E. multiflora* (Cherry Eleagnus) Semi-evergreen if winter conditions are harsh. Ovate to lanceolate leaves, mid green with grey sheen and silver undersides. Yellow-white scented flowers, mid to late spring, followed in good summers by orange cherry-like fruits. Slightly less than average height, average spread. *E. umbellata* (Autumn Olive) Light green, ovate to lanceolate leaves, with silvery undersides. Stems thorny. Flowers creamy white, funnel-shaped, late spring to early summer. Red fruit in autumn. Somewhat loose and open in habit; one-third more than average height.
Average height and spread
Five years
4x4ft (1.2x1.2m)
Ten years
6x6ft (2x2m)
Twenty years
or at maturity
10x10ft (3x3m)

Origin From Japan.
Use As quick-growing evergreens used alone or as foil for other shrubs in shrub borders, or as a hedge, if planted 2½ft (80cm) apart in single line. Can also be trimmed and trained as a small mop-headed tree or fan-trained on a wall.
Description *Flower* Small, inconspicuous, strongly scented, sulphur yellow flowers, early summer to late autumn. *Foliage* Leaves ovate, 3-6in (8-15cm) long, grey-green, with some gold-variegated varieties. *Stem* Upright when young, becoming very twiggy and dense with age, forming an upright shrub or round bush, depending on variety. Normally grey-green when young, aging to grey-brown. Medium to fast growth rate. *Fruit* Insignificant.
Hardiness Tolerates 4°F (−15°C).
Soil Requirements Tolerates most soils, although unhappy on extremely alkaline or dry soils.
Sun/Shade aspect Tolerates full sun to deep shade although shrub may be more lax and less well-shaped in deep shade.
Pruning Can be trimmed to form hedge or to control size, or can be left unpruned. Cut long new shoots of young plants back hard by

Elaeagnus × *ebbingei* in leaf

Elaeagnus × *ebbingei* 'Gilt Edge' in leaf

two-thirds in spring to encourage a bushy habit.
Propagation and nursery production From semi-ripe cuttings taken early to mid summer. Purchase container-grown; most forms fairly easy to find. Best planting heights 1½-4ft (50cm-1.2m).
Problems Roots very susceptible to any dramatic change in soil conditions such as drought or waterlogging, which may cause whole areas of root to die, resulting in death of limb or even whole side of shrub directly related to damaged root area. Cultivation damage can also lead to this demise. If gold variegated varieties start to revert, producing all green foliage, remove this at once.
Varieties of interest *E.* × *ebbingei 'Gilt Edge'* Attractive gold-margined leaves. Severe wind chill and extreme waterlogging may damage foliage. Slightly tender; minimum winter temperature 23°F (−5°C). Two-thirds average height and spread. Of garden origin. *E.* × *e. 'Limelight'* Centres of leaves irregularly splashed gold to pale gold, variegation best in spring and summer declining slightly in autumn. Upright habit. *E. macrophylla* Slightly broader, larger, grey-green leaves with silver undersides. Good flower production. Less hardy than *E.* × *ebbingei* and rather hard to find in colder areas.
Average height and spread
Five years
4x4ft (1.2x1.2m)
Ten years
6x6ft (2x2m)
Twenty years
or at maturity
10x10ft (3x3m)

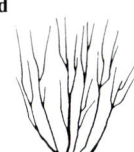

Origin From Japan; most variegated forms of garden origin.
Use Freestanding to show off its winter foliage or in a mixed shrub border. Plant 2½ft (80cm) apart for semi-formal or informal hedge. Or train as a clipped, single-stemmed, mop-headed tree.
Description *Flower* Small, inconspicuous, fragrant, sulphur yellow or silver-white flowers, mid to late autumn. *Foliage* Leaves 2-3in (5-8cm) long, ovate with pointed ends, dark olive green with glossy upper surfaces and dull undersides. Variegated forms, grey-brown in spring, take until early summer to turn variegated. *Stem* Grey-green, aging to brown-green. Upright when young, quickly becoming spreading, branching and twiggy, forming a round, spreading shrub. Medium growth rate. *Fruit* Insignificant.

Elaeagnus pungens 'Maculata' in leaf

Hardiness Tolerates winter temperature range 4°F (−15°C) down to −13°F (−25°C).
Soil Requirements Tolerates most soils but unhappy on extremely alkaline or dry conditions.
Sun/Shade aspect Tolerates full sun to deep shade, but may be lax and poorly shaped in deep shade.
Pruning Can be trimmed to form hedge or to control size. Can also be left unpruned but young plants benefit from having previous season's shoots cut back by half in spring to induce more bushy habit.
Propagation and nursery production By grafting in winter or semi-ripe cuttings taken late spring, early summer. Purchase container-grown or root-balled (balled-and-burlapped). Best planting heights 1½-4ft (50cm-1.2m).
Problems May suffer from aphid attack. Shoots of variegated forms showing reversion to green should be cut out at once. On grafted shrubs, rip out any suckers of understock.

Elaeagnus pungens 'Variegata' in leaf

Varieties of interest *E. p. 'Dicksonii'* Leaves narrow, almost holly-like with bright gold margins. Two-thirds average height and spread. Slightly less hardy than *E. pungens*. Of garden origin. *E. p. 'Fredericia'* Light green, cream-splashed, narrow, very pointed leaves. Very branching habit. Slightly less than average height and spread. *E. p. 'Gold Rim'* A European variety having round leaves with gold margins. Two-thirds average height and spread. *E. p. 'Maculata'* Dark leaves with central gold patches. Reaching just less than average height and spread. Very susceptible to green reversion; remove any reverted shoots at once or golden colouring will be destroyed. Of garden origin and very widely grown. *E. p. 'Variegata'* Foliage margined creamy white; some sulphur yellow fragrant flowers on very old plants. Prune back old

limbs occasionally to preserve foliage size and quality. Two-thirds average height and spread. Of garden origin.
Average height and spread
Five years
4x4ft (1.2x1.2m)
Ten years
6x6ft (2x2m)
Twenty years
or at maturity
12x13ft (3.5x4m)

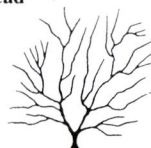

ELSHOLTZIA STAUNTONII

Labiatae
Deciduous
A little known, but interesting autumn-flowering shrub.

Origin From northern China.
Use As an unusual autumn-flowering shrub for the front of shrub borders or for a specific position.
Description *Flower* Panicles of bright lilac-purple to lilac-pink flowers profusely borne in upright formation from autumn to early winter. *Foliage* Leaves lanceolate 4-6in (10-15cm) long, with toothed edges, light green to grey-green. When crushed give off a pleasant minty odour. Some yellow autumn colour. *Stem* Grey-green, upright when young, becoming spreading with age. Dies back in winter in a sub-shrub habit, new growth appearing from rooted base in mid to late spring. Medium growth rate. *Fruit* Insignificant.
Hardiness Minimum winter temperature 23°F (−5°C).
Soil Requirements Does well on most soils but unhappy on thin alkaline types.
Sun/Shade aspect Prefers full sun, dislikes shade.
Pruning In early spring remove any shoots which have not died down to ground level in winter.
Propagation and nursery production From layers or softwood cuttings taken midsummer. Purchase container-grown. Rather hard to find. Best planting heights 2-2½ft (60-80cm).
Problems Young plants when purchased often look very weak.
Average height and spread
Five years
3x3ft (1x1m)
Ten years
5x5½ft (1.5x1.8m)
Twenty years
or at maturity
5x7ft (1.5x2.2m)

EMBOTHRIUM COCCINEUM

Proteaceae
Semi-evergreen
A spectacularly beautiful shrub of very precise planting requirements.

Origin From Chile.
Use As a late spring-flowering specimen shrub for woodland areas.
Description *Flower* Orange-scarlet, consisting of a cluster of strap-like petals, late spring to early summer. *Foliage* Leaves slightly broad, lanceolate, 4-7in (10-17cm) long, grey-green with some red shading to veins. Some autumn colour when older leaves die. *Stem* Very upright, becoming branching but still upright with age. Grey-green. Medium to fast growth rate. *Fruit* Insignificant.

Embothrium coccineum 'Longifolium'

Hardiness Tolerates winter temperatures down to 14°F (−10°C).
Soil Requirements Deep, rich, peat soil with adequate moisture, but well drained. Dislikes any alkalinity.
Sun/Shade aspect Best in a lightly shaded woodland clearing.
Pruning None required. May be reduced in height, rejuvenating itself, but may be slow to flower fully after pruning.
Propagation and nursery production From layers or seed. Purchase container-grown. Obtainable from specialist nurseries. Best planting heights 15in-2½ft (40-80cm).
Problems Its precise soil and shade requirements must be met.

Elsholtzia stauntonii in flower

Varieties of interest *E. c. var. lanceolatum*
Flowers scarlet and closely set on stems.
Narrow, lanceolate leaves. Possibly more
hardy. Sometimes referred to as *E. lanceolatum 'Norquinco Valley'*. *E. c. 'Longifolium'*
Long narrow leaf formation. Possibly evergreen and the most hardy variety, but also the
most difficult to find.
Average height and spread
Five years
6x3ft (2x1m)
Ten years
13x6ft (4x2m)
Twenty years
or at maturity
26x13ft (8x4m)

ENKIANTHUS CAMPANULATUS

REDVIEW ENKIANTHUS
Ericaceae
Deciduous
An impressive flowering and foliage
shrub, grown in the correct environment.

Origin From Japan.
Use Freestanding or mass-planted for spring
flowers, useful for cutting, and spectacular
autumn colour.
Description *Flower* Small to medium, hanging, sulphur yellow to bronze, cup-shaped,
long lasting flowers freely borne in small
clusters, mid to late spring. *Foliage* Leaves
round to ovate, 1-1½in (3-4cm) long, light
green to grey-green, attractively presented.
Very good red and yellow autumn colours.
Stem Light green to green-brown, upright
when young, becoming very branching and
twiggy to form a rounded, spreading shrub at
maturity. Slow to medium growth rate. *Fruit*
Insignificant.
Hardiness Tolerates 4°F (−15°C).

Enkianthus campanulatus in flower

Soil Requirements Acid soil, dislikes any
alkalinity. The richer, deeper and more open
the soil, the better results.
Sun/Shade aspect Light shade, performs best
in woodland situation.
Pruning None required but to encourage very
old plants to rejuvenate remove one or more
old shoots each year to ground level.
Propagation and nursery production From
layers. Purchase container-grown. Moderately easy to find from specialist nurseries. Best
planting heights 1½-3ft (50cm-1m).
Problems Often planted in unfavourable conditions where it rarely shows its true beauty.
Varieties of interest *E. c. 'Alba'* A variety with
white flowers, rather scarce. *E. cernus var.
rubens* Deep red fringed flowers. Very good
autumn colour. Two-thirds average height
and spread and slightly less hardy. From
Japan. *E. chinensis* Large yellow-red flowers
with darker veins through each petal. Foliage

larger than parent and presented on red leaf
stalks. Very good autumn colour. Extremely
scarce. In favourable conditions may reach
height of over 20ft (6m) and resemble small
tree. From western China and parts of Burma. *E. perulatus* Hanging, urn-shaped white
flowers, mid spring. Good scarlet autumn
colour. Very scarce. A much slower growing
variety reaching two-thirds average height
and spread. From Japan.
Average height and spread
Five years
5x3ft (1.5x1m)
Ten years
8x6ft (2.5x2m)
Twenty years
or at maturity
10x10ft (3x3m)

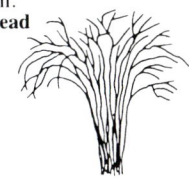

ERICA ARBOREA

TREE HEATH, BRIAR
Ericaceae
Evergreen
A stately, spring-flowering heath for acid
soils.

Origin From the Mediterranean area.
Use As a tall subject for gardens with acid
soils, to form a large, massed effect.
Description *Flower* White, sweetly scented,
globe-shaped flowers borne freely, mid to late
spring. *Foliage* Leaves short, ⅛in (3mm)
long, dark grey-green with glossy surface,
densely presented on branches. *Stem* Upright, possibly spreading on outer edges of
shrub at maturity. Silver-grey, but inconspicuous due to close grouping of foliage on
stems. Slow to medium growth rate. *Fruit*
Grey to grey-white seedheads retained into
winter.
Hardiness Minimum winter temperature 4°F
(−15°C).
Soil Requirements Acid soil, dislikes any
alkalinity.
Sun/Shade aspect Full sun to very light shade.
Pruning None required.
Propagation and nursery production From
semi-ripe cuttings taken early to mid summer.
Purchase container-grown. Difficult to find
outside acid-soil areas. Best planting heights
8-15in (20-40cm).
Problems May damaged or broken by wind or
snow and needs protection.
Varieties of interest *E. a. 'Alpina'* White
flowers. One third average height but of
average spread, said to be hardier than the
form. From central southern Europe. *E.
australis* Whorls of 3-4 pink flowers, mid
spring. Height and spread up to 6ft (2m). *E. a.
'Mr. Robert'* White flowers. Upright habit.
Less hardy than *E. arborea*. *E. a. 'Riverslea'*

Rosy purple flowers. Less hardy than *E.
arborea*.
Average height and spread
Five years
3x3ft (1x1m)
Ten years
6x10ft (2x3m)
Twenty years
or at maturity
13x16ft (4x5m)

ERICA CILIARIS

DORSET HEATH
Ericaceae
Evergreen
An attractive heath for acid soils.

Origin From southern England and western
Ireland, many varieties being of garden
origin.
Use As a low, summer-flowering, carpeting
shrub for use with rhododendrons, azaleas
and other ericaceous shrubs.

Erica ciliaris 'Corfe Castle' in flower

Description *Flower* Flowers 1-5in (3-12cm)
long, produced in threes at ends of branches,
in shades of pink and white from midsummer
through to mid autumn. *Foliage* Leaves
ovate, narrow, ⅛ (3mm) long, dark grey-
green to light green. *Stem* Upright, covered in
grey down. *Fruit* Grey-brown seedheads retained into autumn.
Hardiness Minimum winter temperature 14°F
(−10°C).
Soil Requirements Acid soil, resents any
alkalinity.

Erica arborea in flower

Sun/Shade aspect Full sun to very light shade.
Pruning Trim lightly after flowering, or very early spring, to prevent open, lax habit.
Propagation and nursery production From softwood cuttings taken midsummer. Purchase container-grown. Best planting heights 4-6in (10-15cm). If hard to find in non-acid areas obtain from specialist nursery.
Problems Will not succeed on alkaline soils.
Varieties of interest *E. c. 'Corfe Castle'* Rich pink flowers. *E. c. 'David Maclintock'* Bi-coloured pink and white flowers. Slightly more than average height. *E. c. 'Globosa'* Pale pink flowers. Round, shrub-forming. *E. c. 'Maweana'* Large purple-pink flowers. *E. c. 'Mrs. C.H. Gill'* Cherry-red flowers. *E. c. 'Multiflorus'* Pale rose pink flowers. Less hardy. *E. c. 'Stoborough'* Pure white flowers. Slightly taller.
Average height and spread
Five years
1x1ft (30x30cm)
Ten years
1x1½ft (30x50cm)
Twenty years
or at maturity
1x2½ft (30x80cm)

ERICA CINEREA

BELL HEATHER, BELL FLOWER HEATHER
Ericaceae
Evergreen
Interesting additions to the heather collection for acid-soil areas.

Origin From western Europe, including the British Isles.
Use For larger rock gardens, for banks where it can be mass planted, or grown with rhododendrons, azaleas and conifers as low spreading ground cover.
Description *Flower* Small, hanging, bell-shaped white flowers, midsummer to early autumn. Also crimson through pink and purple, depending on variety. *Foliage* Leaves thin, lanceolate, ¼-½in (5mm-1cm) long, light green in spring, aging to dark green and, in some varieties, purple in autumn. *Stem* Wiry, wispy stems moderately upright when young, becoming spreading with age to form spreading, loose mound. Grey-green to grey-brown. Fast to medium growth rate. *Fruit* Small brown seedheads may be retained into winter.
Hardiness Tolerates 4°F (−15°C).
Soil Requirements Neutral to acid; dislikes alkalinity.
Sun/Shade aspect Full sun to light shade.
Pruning Trim lightly after flowering.
Propagation and nursery production From softwood cuttings taken midsummer. Purchase container-grown. Normal outlets offer a very wide range, but particular varieties may

have to be sought from specialist nurseries. Best planting heights 4-6in (10-15cm).
Problems Very susceptible to failure from drought. Protect with annual mulch of sedge peat in early spring. Severe wind chill or prolonged frozen soil may also lead to untimely death.
Varieties of interest *E. c. 'Alba Minor'* White flowers. Dwarf. *E. c. 'Atropurpurea'* Deep pink flowers. *E. c. 'C. D. Eason'* Dark red-pink flowers. *E. c. 'Cindy'* Good purple flowers. *E. c. 'Contrast'* Deep purple flowers. Bronze leaves. *E. c. 'Elan Valley'* Lilac and white flowers. *E. c. 'Foxhollow'* Mahogany-maroon flowers. *E. c. 'Golden Drop'* Pink flowers. Orange leaves, coppery red in autumn. *E. c. 'Knaphill Pink'* Deep pink flowers. Olive foliage. *E. c. 'Lavender Lady'* Pink and red flowers. *E. c. 'Pentreath'* Rich red-purple flowers. *E. c. 'Pink Ice'* Deep purple flowers. *E. c. 'P.S. Patrick'* Purple flowers. *E. c. 'Purple Beauty'* Rich purple flowers. *E. c. 'Pygmea'* Pink flowers. Only 6in (15cm) in height. *E. c. 'Rosabelle'* Salmon flowers. Dwarf. *E. c. 'Rozanne Waterer'* Maroon flowers. Purple leaves. *E. c. 'Stephen Davis'* Scarlet flowers. *E. c. 'Vivienne Patricia'* Soft purple flowers.
Average height and spread
Five years
1x1½ft (30x50cm)
Ten years
1x1½ft (30x50cm)
Twenty years
or at maturity
1x1½ft (30x50cm)

ERICA × DARLEYENSIS

KNOWN BY BOTANICAL NAME
Ericaceae
Evergreen
Good winter-flowering heathers for both acid and neutral soils.

Origin This hybrid is named after Darley Dale, Derbyshire, England, where it originated.
Use As a ground cover shrub used with rhododendrons, azaleas and conifers, or for mass planting on banks. A little vigorous for use on all but large rock gardens.
Description *Flower* Colours range through white to pink to dark pink, depending on variety, early to late spring. *Foliage* Leaves thin, narrow, lanceolate, ¼-½in (5mm-1cm) long, light green in spring, aging to dark green and in some varieties dark olive green in autumn. *Stem* Moderately loose, open formation, making a good ground-covering shrub. Fast to medium growth rate. *Fruit* May retain brown seedheads into autumn.
Hardiness Tolerates 4°F (−15°C).

Erica × darleyensis 'Ghost Hills' in flower

Soil Requirements Neutral to acid, dislikes high alkalinity.
Sun/Shade aspect Full sun to light shade.
Pruning Trim lightly after flowering to induce new growth.
Propagation and nursery production From very small softwood cuttings taken early summer. Plant container-grown. Wide range of varieties normally obtainable but particular varieties may have to be sought from specialist nurseries. Best planting heights 4-6in (10-15cm).
Problems Very susceptible to drought; protect with annual mulch of sedge peat, early spring. Severe wind chill or prolonged frozen soil will also lead to untimely death.

Erica × darleyensis 'Silberschmelze' in flower

Varieties of interest *E. × d. 'Arthur Johnson'* Deep purple-pink flowers. *E. × d. 'Darley Dale'* Pale purple flowers. *E. × d. 'Furzey'* Rose pink flowers. Slightly shorter than average. *E. × d. 'Ghost Hills'* Rose pink flowers. *E. × d. 'George Rendell'* Rose pink flowers. *E. × d. 'Jack H. Brummage'* Pink flowers. Golden foliage. *E. × d. 'Silberschmelze'* syn. *E. × d. 'Alba'* (Molten Silver, Silver Beads) A pure white flowering variety of spectacular effect. Flowers sweetly scented.
Average height and spread
Five years
1½x2ft (50x60cm)
Ten years
1½x2ft (50x60cm)
Twenty years
or at maturity
1½x2ft (50x60cm)

Erica cinerea 'Purple Beauty' in flower

ERICA ERIGENA
(Erica mediterranea)

MEDITERRANEAN HEATH
Ericaceae
Evergreen
A heather flowering in late winter and early spring, requiring an acid soil.

Origin From southern France, Spain and Northern Ireland. The Irish form is often referred to as *E. hibernica*, but to all intents and purposes it is the same as the Mediterranean form.
Use For mass planting on banks or larger rock gardens with rhododendrons and other ericaceous plants. Good for underplanting.
Description *Flower* Small, narrow, hanging, fragrant, bell-shaped flowers in a range of colours from white to shades of pink, dependent on variety, winter to late spring. *Foliage* Leaves lanceolate, ¼-½in (5mm-1cm) long, light green becoming darker green, then purple-green in autumn and winter. *Stem* Inconspicuous as covered by foliage; forms carpet with slight mounding effect. Fast to medium growth rate. *Fruit* Brown seedheads of some winter attraction.

Erica erigena 'W. T. Rackliff' in flower

Hardiness Tolerates 4°F (−15°C).
Soil Requirements Does best in deep, moist, acid soil; cannot tolerate alkalinity.
Sun/Shade aspect Full sun to light shade. Medium to deep shade will deform shape and lead to fewer flowers.
Pruning Trim after flowering to encourage new flowering shoots.
Propagation and nursery production From softwood cuttings taken midsummer. Purchase container-grown. In some areas readily available but certain varieties may have to be sought from specialist nurseries. Best planting heights 4-6in (10-15cm).
Problems Very susceptible to drought; protect with annual mulch of sedge peat, early spring. Severe wind chill or prolonged frozen soil may also lead to untimely death.
Varieties of interest *E. e. 'Alba'* Pure white flowers. *E. e. 'Brightness'* Rose pink flowers. Dark foliage. *E. e. 'Inch Salmon'* Buds salmon pink, opening to clear pink. Early flowering. *E. e. 'Superba'* Rose pink. Up to 4ft (1.2m) in height and spread. *E. e. 'W.T. Rackliff'* White flowers. Dark green foliage.
Average height and spread
Five years
2x2½ft (60x80cm)
Ten years
2x2½ft (60x80cm)
*Twenty years
or at maturity*
2x2½ft (60x80cm)

ERICA HERBACEA
(Erica carnea)

HEATH, HEATHER, WINTER-FLOWERING HEATHER
Ericaceae
Evergreen
Extremely useful, winter-flowering, carpeting shrubs of wide popularity.

Origin From the Alps and central Europe. Many varieties of garden origin.
Use For mass planting on banks or large rock gardens with rhododendrons and other ericaceous plants. Good for underplanting.
Description *Flower* Flowers small, bell-shaped, ranging from white, through pink to purple, borne singly in profusion along branches, midwinter to early spring. *Foliage* Leaves narrow, lanceolate, ½in (1cm) long, very thickly clustered along stems. Light to dark purple-green, depending on variety. Colour may vary from dark in winter to much lighter in spring, again depending on variety, with some golden yellow forms. *Stem* Grey-green, almost unseen, due to close foliage formation. Upright when young, quickly becoming spreading and mat-forming. Fast to medium growth rate. *Fruit* Brown seedheads follow flowers.
Hardiness Tolerates 4°F (−15°C), but may be scorched by severe wind chill.
Soil Requirements Prefers neutral to acid soil, tolerates mild alkalinity and said to be somewhat lime tolerant.
Sun/Shade aspect Prefers very light shade, tolerates full sun. Becomes very open, leggy and shy to flower in deep shade.
Pruning Trim lightly after flowering, to encourage new flowering shoots for following winter and early spring.
Propagation and nursery production From small softwood cuttings taken early to mid summer. Purchase container-grown. Many garden centres and nurseries carry very wide range of varieties; some may have to be sought from specialist nurseries. Best planting heights 4-6in (10-15cm).
Problems Very susceptible to failure from drought; protect with mulch of sedge peat, early spring. Severe wind chill or prolonged frozen soil may lead to untimely death.
Varieties of interest *E. h. 'Aurea'* Pink flowers. Golden foliage. *E. h. 'Ann Sparks'* Carmine flowers. Orange and yellow foliage. *E. h. 'December Red'* Deep pink flowers in early spring. Dark green foliage. *E. h. 'James Backhouse'* Pale pink flowers. *E. h. 'Myretown Ruby'* Deep purple-pink flowers. *E. h. 'Pink Spangles'* Rosy pink flowers. Good grower. *E. h. 'Queen Mary'* Lilac pink flowers. Early flowering. *E. h. 'Ruby Glow'* Ruby pink flowers. *E. h. 'Springwood Pink'* Pink flowers, spreading. *E. h. 'Springwood White'*

Erica herbacea 'Springwood White' in flower

White flowers. Spreading. *E. h. 'Startler'* Deep pink flowers. *E. h. 'Vivellii'* Deep pink flowers. Dark foliage. *E. h. 'Winter Beauty'* Dark pink flowers. *E. h. 'Westwood Yellow'* Pink flowers, yellow foliage.
Average height and spread
Five years
1½x2½ft (50-80cm)
Ten years
1½x2½ft (50-80cm)
*Twenty years
or at maturity*
1½x2½ft (50-80cm)

ERICA TERMINALIS
(Erica stricta)

CORSICAN HEATH
Ericaceae
Evergreen
An interesting heather of some stature.

Origin From western Mediterranean regions and also naturalized in Northern Ireland.
Use As an upright feature shrub, either singly or in groups. Useful to raise the height of a heather garden.
Description *Flower* Small, pink, bell-shaped, hanging flowers produced at end of each upright branch, late summer. *Foliage* Leaves small, lanceolate, ¼-½in (5mm-1cm) long, light green, aging to darker green in winter; closely grouped along stems. *Stem* Upright, erect branches forming pillar effect. Fast to medium growth rate for ultimate size. *Fruit* Brown seedheads retained into winter.
Hardiness Tolerates 4°F (−15°C).

Erica herbacea 'December Red' in flower

Soil Requirements Does well on both acid and alkaline soils, but distressed by extreme alkalinity.
Sun/Shade aspect Full sun to light shade. Dislikes deep shade.
Pruning Trim lightly after flowering to encourage new flowering shoots.
Propagation and nursery production From softwood cuttings taken midsummer. Purchase container-grown. Many garden centres and nurseries offer wide range of varieties, but some may have to be sought from specialist nurseries.

Erica terminalis in flower

Problems Very susceptible to drought; protect with annual mulch of sedge peat, early spring. Severe wind chill or prolonged frozen soil may also lead to untimely death.
Varieties of interest *E. t. 'Thelma Woolner'* Deep pink flowers. Shorter habit.
Average height and spread
Five years
3x2½ft (1mx80cm)
Ten years
3x2½ft (1mx80cm)
Twenty years
or at maturity
3x2½ft (1mx80cm)

ERICA TETRALIX

CROSS-LEAVED HEATH, BOG HEATHER
Ericaceae
Evergreen
Not easy to find, but worth pursuing if the right acid soil conditions can be given.

Origin From northern and western Europe, including the British Isles.
Use For planting with rhododendrons and other ericaceous shrubs. Good for underplanting.
Description *Flower* Terminal heads of small, narrow, bell-shaped flowers, closely branched, pure white to shades of pink, early summer through to early autumn. *Foliage* Leaves lanceolate, narrow, ¼-½in (5mm-1cm) long, grey-green, densely gathered in cross formation on upright stems, hence its common name. *Stem* Upright, becoming slightly spreading with age, grey-green, heavily covered with foliage. Fast to medium growth rate for ultimate size. *Fruit* Brown to mahogany-brown seedheads retained well into winter.
Hardiness Tolerates 14°F (−10°C).
Soil Requirements Acid soil, dislikes any alkalinity.
Sun/Shade aspect Full sun to very light shade.
Pruning Trim lightly each spring to produce new growth.
Propagation and nursery production From softwood cuttings taken midsummer. Purchase container-grown. Best planting heights 4-6in (10-15cm). Many garden centres and nurseries offer a wide range but specific

Erica tetralix 'Alba Mollis' in flower

varieties may have to be sought from specialist sources.
Problems Very susceptible to drought; protect with annual mulch of sedge peat in spring. Prolonged frozen soil or wind chill may lead to untimely death.
Varieties of interest *E. t. 'Alba Mollis'* White flowers, freely borne, offset by grey foliage. *E. t. 'Alba Praecox'* White flowers, grey foliage. Early flowering. *E. t. 'Con Underwood'* Crimson flowers. Dwarf. *E. t. 'L.E. Underwood'* Pale pink flowers, darker in bud, borne on silver-grey foliage. Mound forming. *E. t. 'Moonstone Pink'* Pink flowers. *E. t. 'Melbury White'* Large white flowers, silver foliage. *E. t. 'Pink Glow'* Bright pink flowers, grey-green foliage. *E. t. 'Pink Star'* Pink flowers, facing upwards. *E. t. 'Rosea'* Rose pink flowers and grey foliage. *E. × watsonii 'Dawn'* A cross between *E. ciliaris* and *E. tetralix.* Soft rose pink flowers. *E. × williamsii* Another cross between these species. Pale pink flowers and pale yellow foliage.
Average height and spread
Five years
1¼x2ft (40x60cm)
Ten years
1¼x2ft (40x60cm)
Twenty years
or at maturity
1¼x2ft (40x60cm)

ERICA VAGANS

CORNISH HEATH
Ericaceae
Evergreen
An infrequently used group but worthy of the right planting situation.

Origin From south-western Europe and the British Isles.
Use As low spreading ground cover to use with rhododendrons and azaleas.
Description *Flower* Sprays of small, hanging, tubular flowers, white, pink or dark pink depending on variety, midsummer to early autumn. *Foliage* Leaves lanceolate, ¼-½in (5mm-1cm) long, light to dark green, silver reverses. *Stem* Long, arching, creeping branches forming a spreading, low, dwarf shrub. Medium to fast growth rate for ultimate size. *Fruit* Brown seedheads retained into winter.
Hardiness Tolerates 4°F (−15°C).
Soil Requirements Acid, dislikes any alkalinity. Best in moorland or woodland situations.
Sun/Shade aspect Full sun to light shade; dislikes deep shade.
Pruning Trim lightly each spring to encourage new growth.
Propagation and nursery production From softwood cuttings taken midsummer. Purchase container-grown. Best planting heights 4-6in (10-15cm). Many nurseries and garden centres offer a wide range but specific

varieties may have to be sought from specialist sources.
Problems Very susceptible to drought; protect with an annual mulch of sedge peat, early spring. Prolonged exposure to frozen soil or wind chill may also lead to death.

Erica vagans 'Lyonesse' in flower

Varieties of interest *E. v. 'Alba'* Pure white flowers. Height 2ft (60cm). *E. v. 'Brick Glow'* Rose to bright rose flowers. Bright green foliage. *E. v. 'Dianna Hookstone'* Deep pink flowers. Early. *E. v. 'Lyonesse'* Brown marking on pure white flower petals. *E. v. 'Mrs. D.F. Maxwell'* Soft coral pink flowers. Height 1¼ft (40cm), spread 2ft (60cm). *E. v. 'Pyrenees Pink'* Shell pink flowers borne on low shrub. Height 1¼ft (40cm), 2ft (60cm) spread. *E. v. 'St. Keverne'* Lilac-pink flowers. Height 1½ft (50cm). *E. v. 'Valerie Proudley'* White flowers. Pale gold foliage. Low growing.
Average height and spread
Five years
2½x2½ft (80x80cm)
Ten years
2½x2½ft (80x80cm)
Twenty years
or at maturity
2½x2½ft (80x80cm)

ERYTHRINA CRISTA-GALLI

CORAL TREE, COXCOMB
Leguminosae
Deciduous
A truly magnificent shrub for a sun-facing wall.

Origin From Brazil.
Use As a medium-sized, summer-flowering sub-shrub for a sunny wall.
Description *Flower* Racemes of deep scarlet pea-flowers, waxy-textured, borne in large clusters at ends of shoots, mid to late summer. *Foliage* Leaves medium-sized, trifoliate, 4-5in (10-12cm) long, leaflets ovate. Grey-green, giving some yellow autumn colour. *Stem* Strong, upright, becoming arching with weight of flowers, grey to grey-green. Small sharp spines at each leaf axil. Dies to ground level each winter, regenerating following spring. Fast growing, annually produced stems. *Fruit* Insignificant.
Hardiness Minimum winter temperature 23°F (−5°C).
Soil Requirements Good, light, open soil.
Sun/Shade aspect Plant facing full sun, in a protected position against a wall.
Pruning None required, but in early spring remove any old shoots retained through winter.
Propagation and nursery production Produced by root division and from seed. Purchase container-grown without stems while dor-

Erythrina crista-galli in flower

mant, if available, or as plants up to 1¼-1½ft (40-50cm). Extremely scarce; should be sought from specialist nurseries.
Problems Difficult to establish, taking some years to flower.
Average height and spread
Five years
5x5ft (1.5x1.5m)
Ten years
6x8ft (2x2.5m)
Twenty years
or at maturity
6x10ft (2x3m)

ESCALLONIA

KNOWN BY BOTANICAL NAME
Escalloniaceae
Evergreen
Handsome early summer-flowering shrubs with an attractive range of flower colours.

Origin From South America. Nearly all varieties offered are of garden or nursery origin.
Use As a freestanding evergreen or for large shrub borders. If planted 2½ft (80cm) apart in single row makes good semi-formal evergreen hedge.
Description *Flower* Single or in short racemes, bell-shaped flowers, various shades of pink to pink-red; some white forms. Late spring through to early summer, with intermittent flowering through late summer and early autumn. *Foliage* Leaves ovate, 1-1½in (3-4cm) long, with indented edges. Light to dark green glossy upper surfaces, grey undersides. Size of foliage varies according to variety. *Stem* Upright when young, becoming arching according to variety, and branching with age. Grey-green. Fast growth rate when young, slowing with age. *Fruit* Insignificant.

Hardiness Tolerates 14°F (−10°C).
Soil Requirements Tolerates most soils but liable to chlorosis on extremely alkaline types.
Sun/Shade aspect Full sun through to medium shade.
Pruning Remove one-third of old flowering wood after main flowering period. If cut to ground level will rejuvenate after second or third spring.
Propagation and nursery production From softwood cuttings taken midsummer. Purchase container-grown. Most varieties fairly easy to find. Best planting heights 2-2½ft (60-80cm).
Problems Susceptible to severe wind chill and may lose leaves completely. Normally rejuvenates from ground level, but may take 2-3 years to reach previous height.

Escallonia 'Donard Seedling' in flower

Varieties of interest *E. 'Apple Blossom'* Flowers apple blossom pink, large and normally borne singly. Slightly upright. *E. 'C.F. Ball'* Rich red flowers and medium-sized foliage. Slightly arching. *E. 'Crimson Spire'* Crimson flowers and medium-sized foliage. Upright, good for hedging. Slightly less than average spread. *E. 'Donard Beauty'* Large rose carmine flowers and large green foliage. Slightly arching. Takes its name from the Slieve Donard Nurseries in Northern Ireland, where several of the varieties listed below were raised. *E. 'Donard Brilliance'* Large rose red flowers and large foliage. Upright, becoming spreading with age. *E. 'Donard Gem'* Pale pink flowers, slightly scented. *E. 'Donard Radiance'* Medium-sized, rich pink flowers and large foliage. Slightly arching. *E. 'Donard Seedling'* Medium-sized flowers, pink in bud, opening to white tinted with rose. Large

Escallonia 'Iveyi' in flower

leaves. Slightly arching. *E. 'Donard Star'* Large deep rosy pink flowers. Large foliage. Upright, becoming spreading. *E. 'Edinensis'* Medium-sized, carmine-pink flowers. Large foliage. Arching. *E. 'Gwendolyn Anley'* Small flowers, pink in bud, opening to paler pink. Small foliage. Arching, spreading and very twiggy. Two-thirds average height and spread. Best planting heights 12-15in (30-40cm). *E. 'Ingramii'* Rose pink flowers, medium size. Large foliage. Arching. Height and spread 6ft (2m). *E. 'Iveyi'* Large pure white flowers. Large dark green, shiny foliage. Upright, becoming slightly spreading with age. *E. 'Langleyensis'* Small, bright carmine-rose flowers in profusion along arching branches. Small leaved. *E. macrantha* Medium-sized rose carmine flowers, large, scented, dark green foliage. Upright and branching. Good in coastal areas. Less than average hardiness. Height and spread slightly larger than average in favourable areas. *E. 'Peach Blossom'* Good-sized clear pink flowers. Large foliage. Arching. *E. rubra 'Woodside'* syn. *'Pygmaea'* Small, dark pink to red flowers. Small cut foliage, light green, not glossy. Height 1½ft (50cm), spread 2ft (80cm). Best planting heights 12-15in (30-40cm). *E. 'Slieve Donard'* Small pale pink flowers in profusion, borne on long, arching branches. Slightly less than average height and spread.
Average height and spread
Five years
6x6ft (2x2m)
Ten years
10x13ft (3x4m)
Twenty years
or at maturity
10x13ft (3x4m)

EUCRYPHIA

BRUSH BUSH
Eucryphiaceae
Evergreen or Deciduous
A magnificent specimen shrub, if a suitable site can be found.

Origin Many forms of garden origin and garden crosses.
Use As a freestanding specimen shrub flowering late summer, often attaining the height of small tree.
Description *Flower* White, single saucer-shaped flowers with pronounced stamens, freely borne. Size depends on variety, from numerous small flowers to flowers up to 2in (5cm) across, late summer to early autumn. *Foliage* Mostly evergreen. Ovate, 1½-3in (4-8cm) long, some with indented outer edges. Glossy upper surfaces, grey undersides. *Stem* Upright, becoming branching with age, but maintaining upright effect.

Escallonia 'Crimson Spire' in flower

Eucryphia × nymansensis 'Nymansay' in flower

Grey-green aging to green-brown. Medium growth rate. *Fruit* Insignificant.
Hardiness Tolerates winter temperature range of 23°F (−5°C) down to 14°F (−10°C).
Soil Requirements Neutral to acid soil, tolerating low alkalinity provided soil is rich, deep and moist.
Sun/Shade aspect Prefers light shade or woodland clearing, but as long as roots are shaded, tops may be in full sun.
Pruning None required. Individual limbs can be cut back in late spring, encouraging new growth.
Propagation and nursery production From semi-ripe cuttings taken late summer. Purchase container-grown. Some varieties may have to be sought from specialist nurseries. Best planting heights 15in-3ft (40cm-1m).
Problems In severe cold, especially with high wind chill, foliage can be damaged and branches denuded; normally new growth occurs following late spring.
Varieties of interest *E. cordifolia* syn. *E. Ulmo* White flowers and heart-shaped evergreen leaves. Possibly tolerates more alkalinity than most. Slightly less hardy than average and slightly less in height and spread. *E. glutinosa* White flowers 2½-2¾in (6-7cm), mid to late summer. Foliage deciduous, pinnate, grey-green, giving good autumn colour. A very upright form best grown in woodland area. From Chile. *E. × intermedia 'Rostrevor'* Produces very good display of fragrant white flowers from 1-2in (3-5cm) across, with yellow centres. Branches more pendulous and graceful than in most varieties. Fast growth rate. *E. lucida* Fragrant, white, hanging flowers 2in (5cm) across, mid to late summer. Very thick, dark grey-green evergreen foliage, oblong with glaucous underside. Slightly less hardy. From Tasmania. *E. milliganii* Flowers 1-1½in (3-4cm) across, white and cup-shaped, often produced freely on young shrub. Small dark grey-green leaves open from sticky buds. Half average height and spread. From Tasmania. *E. × nymansensis 'Nymansay'* The most popular of all the varieties. Pure white flowers, 2½in (6cm) across, late summer to early autumn. Interesting dark green evergreen foliage. Raised in the Nymans Gardens, Sussex, England.
Average height and spread
Five years
6x3ft (2x1m)
Ten years
13x6ft (4x2m)
Twenty years
or at maturity
26x13ft (8x4m)

EUONYMUS ALATUS

WINGED EUONYMUS
Celastraceae
Deciduous
One of the finest of all autumn-colour shrubs,

Origin From China and Japan.
Use As a specimen feature shrub to show off its autumn colours and interesting winged stems in winter.

Euonymus alatus in autumn

Description *Flower* Small, green-yellow, four-petalled flowers, mid spring. *Foliage* Leaves ovate, 1-3in (3-7cm) long, light grey-green, turning vivid scarlet, autumn. *Stem* Short, branching with four corky wings at edges of squarish stems; grey-green to grey-brown. Forms a round, structurally pleasing shrub. Slow to medium growth rate. *Fruit* May produce small green-tinged pink hanging fruits in autumn.
Hardiness Tolerates winter temperatures down to −13°F (−25°C).
Soil Requirements Does well on all soil types, although the richer and moister the soil, the more winged bark appears on stems.
Sun/Shade aspect Best in full sun to light shade. Deeper shade will spoil shape, making it more open and lax, and reduce autumn colour.

Pruning None required.
Propagation and nursery production From softwood cuttings. Purchase container-grown. Fairly easy to find in nurseries, less often in garden centres. Best planting heights 15in-2½ft (40-80cm).
Problems Relatively slow rate of growth can be discouraging to the gardener.
Varieties of interest *E. a. 'Compacta'* As its name implies, a more compact variety of close habit. Not always easy to obtain. Best planting heights 1-1½ft (30-50cm). *E. phellomanus* Winged stems in winter, pink fruits and some leaf colour in autumn. Best planting heights 1½-4ft (50cm-1.2m). A stronger-growing variety with up to two-thirds greater height and spread.
Average height and spread
Five years
3x3ft (1x1m)
Ten years
5x6ft (1.5x2m)
Twenty years
or at maturity
6x13ft (2x4m)

EUONYMUS EUROPAEUS

SPINDLE, COMMON SPINDLE, EUROPEAN EUONYMUS
Celastraceae
Deciduous
An excellent shrub for autumn display, but requiring a broad planting area to accommodate eventual height and spread.

Origin From Europe.
Use On its own as a fruiting shrub giving good autumn colour. Also good for mixed borders or for mass planting.
Description *Flower* Small, hanging, clusters of inconspicuous green to green-yellow flowers, late spring to early summer. *Foliage* Leaves narrow, ovate, 1-3½in (3-9cm) long, light grey-green, giving good red and scarlet autumn colour. *Stem* Square, upright, strong, becoming branching and spreading with age. Light green to grey-green, aging to green-brown. Fast growth rate. *Fruit* Interesting hanging, pink to pink-red fruits which open to display pendulous red capsules in autumn.
Hardiness Tolerates winter temperatures below −13°F (−25°C).
Soil Requirements Prefers alkaline, but tolerates all soils.
Sun/Shade aspect Full sun to light shade. Tolerates deeper shade but becomes lax.
Pruning None required, but may be reduced in height and spread.
Propagation and nursery production From seed or softwood cuttings. Purchase container-grown. All varieties easy to find. Best planting heights 1½-4ft (50cm-1.2m).
Problems Can be a host plant for aphids,

Euonymus europaeus 'Red Cascade' in fruit

which do not damage the shrub but can spread to vegetables grown nearby.

Varieties of interest *E. e. 'Albus'* Pure white fruits retained into winter. Slightly less than average height and spread. *E. e. 'Red Cascade'* An improved modern variety of garden origin with very good rosy red fruits and good strong autumn colours. Slightly less than average height and spread.

Average height and spread
Five years
6x6ft (2x2m)
Ten years
12x13ft (3.5x4m)
Twenty years
or at maturity
13x20ft (4x6m)

EUONYMUS FORTUNEI
(Euonymus radicans)

WINTERCREEPER EUONYMUS
Celastraceae
Evergreen
Very useful ground-covering or climbing foliage shrubs for shady, exposed areas.

Origin From China; most modern varieties of garden origin.
Use As a low edging shrub for shrub borders or for ground cover; ideal for banks and large areas or can be fan-trained on a shady, exposed wall. Also found as small mopheaded standard.

Euonymus fortunei 'Emerald Gaiety' **in leaf**

Description *Flower* Small, rather inconspicuous, green flowerheads. *Foliage* Leaves small, ovate, ¾-1in (2-3cm) long, ranging from green, through to golden and silver variegated, depending on variety; some changing to pink in winter, again according to variety. *Stem* Spreading, creeping branches, forming a flat, ground-covering shrub, or tall, wide-spreading wall shrub. Grey-green to green-brown. Slow to medium growth rate. *Fruit* Sometimes produces small, round, red fruits; rarely seen in variegated forms.
Hardiness Tolerates winter temperatures down to −13°F (−25°C), but may lose some leaves in very cold condition.
Soil Requirements Accepts wide range of soil conditions; once established, will tolerate quite dry areas.
Sun/Shade aspect Full sun to fairly deep shade, although some golden varieties prefer full sun to medium shade.
Pruning None required, although old shoots may be removed occasionally to help rejuvenation.
Propagation and nursery production From softwood cuttings. Purchase container-grown. Most varieties easy to find. Best planting heights 8in-1½ft (20-50cm).
Problems Container-grown shrubs sometimes

Euonymus fortunei 'Emerald 'N' Gold' **in leaf**

appear weak when purchased, but once planted grow quickly.
Varieties of interest *E. f. var. 'Coloratus'* Green leaves, turning purple in winter, becoming green again in spring. Best effect is achieved if grown on dry or starved soils. One of the quickest growing varieties. *E. f. 'Emerald Gaiety'* Round grey-green leaves with white margins, turning pink in very cold winter weather. *E. f. 'Emerald'N'Gold'* Round to ovate, dark grey-green leaves with gold margins. Variegation turns pink-red in very cold conditions. One of the most popular of low, spreading varieties. *E. f. 'Gold Spot'* Dark green foliage with bright gold splashes. Branches more upright and stronger than most. Very attractive in winter and early spring. *E. f. 'Kewensis'* Slender, small, dainty, almost round, mid green leaves. Very prostrate with slender, spreading branches, forming a small hummock. A scarce variety, good for rock gardens. *E. f. var. radicans* Ovate to elliptic, dark green leaves with toothed edges and leathery texture. A variety commonly used for ground cover. From Japan, China and Korea. *E. f. 'Silver Queen'* Elliptic leaves, creamy yellow in spring, becoming green and with creamy white margin through summer. Two-thirds average height and spread. *E. f. 'Sunshine'* Bold gold edges, grey-green leaves. Slightly quicker growth rate than most variegated forms. *E. f. 'Variegatus'* A small-leaved variety with grey-green ovate foliage, margined white, tinged pink in cold winter weather. Good as climbing form and ground cover. *E. f. vegetus* Leaves ovate to round, very thick, dull green. Very low, spreading form, good fruiting, bearing round, red fruits in autumn.

Average height and spread
Five years
2x3ft (60cmx1m)
Ten years
2x6ft (60cmx2m)
Twenty years
or at maturity
2x10ft (60cmx3m)

EUONYMUS JAPONICUS

JAPANESE EUONYMUS
Celastraceae
Evergreen
A useful range of evergreens including green and variegated forms, most of which should be treated as slightly tender.

Origin From Japan, with many forms of garden origin.
Use As an evergreen shrub used on its own, for large shrub borders or as wall shrubs. If planted at 2½ft (80cm) apart makes a good evergreen hedge, especially in coastal areas.

Description *Flower* Clusters of green-yellow flowers, late spring and early summer. *Foliage* Leaves medium-sized, elliptic, up to 2½in (6cm) long, shiny upper surfaces, dull, matt undersides. All-green, or some silver or gold variegation, depending on variety. *Stem* Upright, becoming branching with age, after 10 years or more becoming spreading. Forms round-topped, predominantly upright shrub. Light to dark green. Slow growth rate in cold areas, faster in warmer regions. *Fruit* Pinkish capsules with orange, hanging seeds, borne if the shrub is planted in an ideal situation.
Hardiness Green forms tolerate 4°F (−15°C). Variegated forms minimum winter temperature 23°F (−5°C).
Soil Requirements Any soil conditions, including sandy, coastal locations.
Sun/Shade aspect Full sun to moderately deep shade but variegated forms may resent very deep shade and lose some variegation.

Euonymus japonicus 'Aureopictus' **in leaf**

Pruning None required, but may be clipped and trimmed.
Propagation and nursery production From softwood cuttings taken in early summer. Purchase container-grown. Most forms easy to find. Best planting heights 15in-3ft (40cm-1m).
Problems Very severe wind chill may damage foliage, but shrub normally regenerates following spring. Some variegated forms, especially golden varieties, may revert to green, but this can be controlled by pruning.
Varieties of interest *E. j. 'Albomarginatus'* Leaves 2-2½in (5-6cm) long, mid to dark green with white outer margins. Slightly less than average height and spread. *E. j. 'Aureopictus'* Dark green foliage with bold gold centre. *E. j. 'Macrophyllus'* Elliptic green

leaves broader and larger than those of *E. japonicus*. *E. j.* **'Macrophyllus Albus'** Large, broad leaves with white margins. Two-thirds average height and spread. *E. j.* **'Microphyllus'** syn. *E.* **'Myrtifolius'** A very small, green-leaved variety, very dense and compact. Height and spread 3ft (1m). *E. j.* **'Microphyllus Pulchellus'** syn. *E.* **'Microphyllus Aureus'** Very small, with golden variegated small leaves. Height and spread 3ft (1m). *E. j.* **'Microphyllus Variegatus'** Very small, box-like leaves with white outer margins. Height and spread 3ft (1m). Heavy layers of snow can damage foliage and should be removed. *E. j.* **'Ovatus Aureus'** syn. *E.* **'Aureovariegatus'** Leaves larger, ovate, with creamy yellow margins. Should always be planted in full sun to maintain variegation. Two-thirds average height and spread. *E. j. robusta* A green-leaved variety with good fruiting. Strong-growing. *E. kiautschovicus* syn. *E. patens* (Spreading Euonymus) Light green foliage, small yellow-green flowers, followed by pink autumn fruits. Low-growing, very good for ground cover in polluted city areas. *E. k.* **'Manhattan'** More compact in habit. Extremely good for street planting.

Average height and spread
Five years
5x5ft (1.5x1.5m)
Ten years
8x8ft (2.5x2.5m)
Twenty years or at maturity
12x12ft (3.5x3.5m)

EUONYMUS SACHALINENSIS (Euonymus planipes)

KNOWN BY BOTANICAL NAME
Celastraceae
Deciduous
A large, spreading shrub, requiring adequate space, but worthy of display for its autumn fruits.

Origin From north-eastern Asia.
Use For a large shrub border or as an individual specimen.
Description *Flower* Small clusters of yellow-green flowers, spring. *Foliage* Leaves large, elliptic, 3-5in (8-12cm) long, dark green tinged red, late summer; good red and orange autumn colour. More sparsely presented than other *Euonymus* foliage. *Stem* Long, upright, becoming ranging and spreading to form open, loose, spreading shrub. Light to mid green when young, aging to green-red. Smooth-barked and producing maroon-red, long, pointed, slightly sticky, winter buds. Medium to fast growth rate. *Fruit* Large red seed capsules with pink to red hanging fruit borne regularly along undersides of mature,

Euonymus sachalinensis **in fruit**

arching branches, suspended on 2in (5cm) red stalks.
Hardiness Tolerates winter temperatures down to −13°F (−25°C).
Soil Requirements Any soil; tolerates high alkalinity.
Sun/Shade aspect Prefers full sun to light shade, tolerates medium shade. Deeper shade tends to spoil shape of shrub.
Pruning None required, but individual limbs may be reduced in size.
Propagation and nursery production From layers or softwood cuttings taken in early summer. Purchase container-grown or root-balled (balled-and-burlapped). May need to be sought from specialist nurseries. Best planting heights 15in-3ft (40cm-1m).
Problems None.

Average height and spread
Five years
6x6ft (2x2m)
Ten years
13x13ft (4x4m)
Twenty years or at maturity
16x16ft (5x5m)

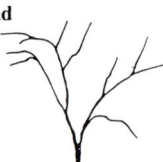

EXOCHORDA × MACRANTHA

KNOWN BY BOTANICAL NAME
Rosaceae
Deciduous
A beautiful group of shrubs, with *Exochorda × macrantha 'The Bride'* offering an exceptional garden variety.

Origin Basic forms from China; others of garden origin.
Use As a late spring-flowering shrub worth a featured position.
Description *Flower* Racemes of single, white, bold, saucer-shaped flowers arching outwards and downwards from main stems, late spring.

Exochorda × macrantha 'The Bride' **in flower**

Foliage Leaves medium-sized, lanceolate, 1½-3in (4-8cm) long, grey-green, giving some yellow autumn colour. *Stem* Upright when young, quickly arching to form a weeping shrub wider than its height. Grey to grey-green. *Fruit* Sometimes produces small, red, round fruits.
Hardiness Tolerates 4°F (−15°C).
Soil Requirements Does well on most soil conditions, but extreme alkalinity will lead to chlorosis which, if not fatal to the shrub, will certainly cut it back.
Sun/Shade aspect Prefers full sun to light shade; may become leggy and spreading in medium shade.
Pruning Remove one-third of old flowering wood after flowering to encourage new growth for flowering in subsequent seasons.
Propagation and nursery production From

layers or softwood cuttings taken in early summer. Purchase container-grown or root-balled (balled-and-burlapped). Fairly easy to find.
Problems None, apart from its dislike of extremely alkaline soils.
Varieties of interest *E. giraldii* Flowers freely produced on an arching, branched shrub. Originating in north-western China. *E. g. wilsonii* Possibly the largest-flowered variety, each flower 2in (5cm) across. Originating in central China. *E. × macrantha 'The Bride'* The most floriferous variety. Good-sized flowers, interesting arching habit, and tolerates higher alkalinity than its parent. Slightly less than average height. *E. racemosa* A very spreading shrub covered in flowers in early summer. This variety dislikes alkaline soils more than most. From China.

Average height and spread
Five years
5x5½ft (1.5x1.8m)
Ten years
6x8ft (2x2.5m)
Twenty years or at maturity
8x10ft (2.5x3m)

FABIANA IMBRICATA

KNOWN BY BOTANICAL NAME
Solanaceae
Evergreen
A very useful, slightly tender, flowering shrub for acid soils.

Origin From Chile.
Use As a medium height, spreading shrub for gardens with acid soils.
Description *Flower* White tubular flowers thickly covering the branches, making entire branch appear to be a flower spike; late spring. *Foliage* Leaves small, ½in (1cm) long, lanceolate, bright green to grey-green, bunched closely on stems. *Stem* Upright at first, quickly becoming arching, forming a wide spreading shrub. Bright green to light grey-green. Slow growth rate. *Fruit* Insignificant.
Hardiness Minimum winter temperature 23°F (−5°C).
Soil Requirements Acid soil, dislikes any alkalinity.
Sun/Shade aspect Prefers full sun or very light shade. Resents deeper shade.
Pruning None required other than occasional cutting back of any overlong branches.
Propagation and nursery production From softwood cuttings taken in early summer. Purchase container-grown. Rather difficult to find.
Problems Its slight tenderness and need for acid soil restricts its use.
Varieties of interest *F. i.* **'Prostrata'** Small pale mauve-tinted tubular flowers freely borne, late spring to early summer. A low-growing

Fabiana imbricata **in flower**

form achieving a dense, round mound, half average height and spread. Useful for tops of walls and large rock gardens. *F. i. var. violacea* Covered with a mass of lavender-blue flowers, late spring to early summer.

Average height and spread
Five years
2½x5ft (80cmx1.5m)
Ten years
4x8ft (1.2x2.5m)
Twenty years or at maturity
4x12ft (1.2x3.5m)

× FATSHEDERA LIZEI

ARALIA IVY
Araliaceae
Evergreen
Very good for large-scale ground cover in shade, as long as adequate moisture is available.

Origin Of garden origin, said to be a cross between *Hedera helix* and *Fatsia japonica* '*Moseri*'.
Use As a moderately spreading evergreen for shady areas. Can be fan-trained on a shaded, exposed wall, where it will grow up to 13ft (4m), or grown over another shrub or low tree stump.

× *Fatshedera lizei* in leaf

Description *Flower* Clusters of round, green flowers on 2-2½in (5-6cm) stalks, borne almost throughout year, either as buds, flowers or dead flowers, but only on mature branches 3-4 years old. *Foliage* Leaves large, palmate, 4-10in (10-25cm) wide, leathery textured with shiny upper surfaces and duller undersides. *Stem* Rambling to form a wide-spreading carpet, or climbing shrub. Light to dark green. Fast growth rate in favourable conditions. *Fruit* Clusters of black fruits follow flowers.
Hardiness Tolerates 14°F (−10°C).
Soil Requirements Prefers rich, moist soils but does well on any.
Sun/Shade aspect Very good in deep shade, but tolerates medium to light shade. Unhappy in full sun, unless adequate moisture is available.
Pruning None required, although very old shoots can be pruned back hard.
Propagation and nursery production From softwood cuttings taken late spring to early summer. Purchase container-grown. Easy to find. Best planting heights 15in-2½ft (40-80cm).
Problems When purchased in container looks weak and floppy but grows quickly once planted out.
Varieties of interest × *F. l.* '*Variegata*' Grey-green leaves with creamy white margins. Rather hard to find. More tender than the parent.

Average height and spread
Five years
3x6ft (1x2m)
Ten years
4x13ft (1.2x4m)
Twenty years or at maturity
5x20ft (1.5x6m)

FATSIA JAPONICA

CASTOR OIL PLANT, JAPANESE FATSIA
Araliaceae
Evergreen
One of the best of all shade-loving, large shrubs for quick covering effect.

Origin From Japan.
Use As a freestanding evergreen for shady areas or as a wall shrub if carefully trained. Very good for maritime areas.
Description *Flower* Clusters of silver-green, opening to milk white, flowers in spring, which are maintained over a long period, produced from round clusters of buds developed in autumn. *Foliage* Leaves large to very large, dark to mid green, 2½-6½in (6-16cm) wide and 9in-1½ft (23-50cm) long, palmate, with 7, 9 or 11 leaflets. Glossy upper surface and paler underside. *Stem* Light to mid green. Strong, stout, upright, forming a tall, rigid structure. Fast growth rate in the right conditions. *Fruit* Large clusters of black fruits follow flowers. If removed this will increase leaf size.
Hardiness Tolerates 4°F (−10°C). Exceptionally severe wind chill will cause loss of leaves, but the shrub normally recovers.
Soil Requirements Does well on most types, but rich, deep, moist soil produces largest leaves.
Sun/Shade aspect Best in deep to medium shade; dislikes full sun.
Pruning None required but can be reduced in size and will rejuvenate itself from below pruning cut.
Propagation and nursery production From semi-ripe cuttings taken in early summer. Purchase container-grown. Easy to find. Best planting heights 1½-3ft (50cm-1m).
Problems When purchased, may look weak and sickly but once planted out soon grows away.
Varieties of interest *F. j.* '*Variegata*' A variety with white to creamy white variegation to lobes and tips of foliage. Slightly less than average height and spread. Tender; will not withstand temperatures below freezing.
Average height and spread
Five years
6x6ft (2x2m)
Ten years
10x10ft (3x3m)
Twenty years or at maturity
13x13ft (4x4m)

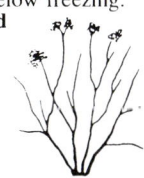

FEIJOA SELLOWIANA

GUAVA, PINEAPPLE GUAVA
Myrtaceae
Evergreen
An interesting, attractive, foliage evergreen, even without the summer flowers, and worthy of a sunny, sheltered planting position.

Origin From Brazil and Uruguay.
Use In mild areas as a freestanding large shrub or as a wall shrub in colder areas.
Description *Flower* Fleshy, crimson and white petals, with central bunch of long crimson stamens, but shy to flower in most areas. Flowers edible. *Foliage* Leaves round to ovate 1½-3in (4-8cm) long, grey-green with white felted undersides and grey leaf stalks. Some older leaves dying to give contrasting autumn colour. *Stem* Grey-green when young, aging to grey-brown, forming round, branching shrub. Slow to medium growth rate. *Fruit* Egg-shaped, good-sized yellow fruits, edible, with aromatic flavour. Rarely fruits outside its native environment.

Feijoa sellowiana in leaf

Hardiness Minimum winter temperature 23°F (−5°C).
Soil Requirements Does well on most soils, disliking only extreme alkalinity.
Sun/Shade aspect Prefers light shade, tolerates full sun.
Pruning None required.
Propagation and nursery production From softwood cuttings. Purchase container-grown. Difficult to find but not impossible.
Problems None.
Varieties of interest *F. s.* '*David*' A large-foliage variety from New Zealand. *F. s.* '*Variegata*' Foliage variegated creamy white.

Fatsia japonica in fruit

Two-thirds average height and spread. More tender than the parent.

Average height and spread
Five years
3x3ft (1x1m)
Ten years
6x6ft (2x2m)
Twenty years
or at maturity
13x13ft (4x4m)

FICUS CARICA

COMMON FIG, FIG
Moraceae
Deciduous
A very decorative fruiting shrub, and there is nothing better on a late summer or early autumn afternoon than picking the ripe figs from your own plants.

Origin From western Asia.
Use As a freestanding large shrub or as a wall-grown specimen for its edible figs and ornamental foliage. Can be containerized but produces less height and spread.
Description *Flower* Pale green, insignificant.

Ficus carica **in fruit**

Foliage Leaves large, palmate, 8-10in (20-25cm) long and wide, mid green to grey-green, abundantly produced, giving good yellow autumn colour. *Stem* Light green when young, becoming spreading and arching with time, needing support. *Fruit* Large, green-skinned, edible fruits becoming wrinkled and purple-black when ripe.
Hardiness Tolerates 14°F (−10°C), but may be damaged by severe wind chill.
Soil Requirements Does well on most soil types. For best fruiting roots must be restricted within a 6ft (2m) box sunk in soil, or the shrub will grow too vigorously and not fruit adequately.
Sun/Shade aspect Full sun to light shade needed to ripen fruit.
Pruning None required other than keeping within bounds.
Propagation and nursery production From softwood cuttings taken late summer. Purchase container-grown. Easy to find. Best planting heights 1½-2½ft (50-80cm).
Problems Can become invasive unless root development restrained.
Varieties of interest *F. c. 'Brown Turkey'* Large, brown to purple-skinned pear-shaped fruits. Flesh creamy white with a red tinge. *F. c. 'Brunswick'* Yellow-skinned, pear-shaped fruits, flesh white flushed red. Hard to find, but not impossible. *F. c. 'White Marseilles'* White-skinned, round fruits, flesh almost translucent, very sweet and juicy. Very scarce.

Average height and spread
Five years
10x6ft (3x2m)
Ten years
13x13ft (4x4m)
Twenty years
or at maturity
13x20ft (4x6m)

Forsythia 'Lynwood' **in flower**

FORSYTHIA

FORSYTHIA, GOLDEN BALL
Oleaceae
Deciduous
Forsythias are very widely planted, but some of the more unusual varieties, especially the early-flowering forms, deserve more attention.

Origin Native to Europe, but with many forms of garden origin.
Use As a freestanding, spring-flowering shrub or for large shrub borders. Plant 2½-3ft (80cm-1m) apart in single line for informal flowering hedge. Can be trimmed formally but will produce fewer flowers. May be trained as small tree on 3-6ft (1-2m) high stem. Very good for mass planting and makes interesting windbreak.
Description *Flower* Lemon yellow to golden yellow, depending on variety, hanging, bell-shaped flowers profusely covering bare stems, early to mid spring. *Foliage* Leaves ovate, 1½-2½in (4-6cm) long, tooth-edged. Light to mid green, many varieties giving purple to red-purple autumn colours. *Stem* Green to grey-green, or purple-green to purple-brown, sometimes yellow, depending on variety. Upright when young, arching and spreading with age, forming a round-topped shrub. Dwarf, low varieties have slow growth rate, taller forms fast. *Fruit* Insignificant.
Hardiness Tolerates winter temperatures down to −13°F (−25°C).
Soil Requirements Does well on any soil.
Sun/Shade aspect Full sun to medium shade. In deep shade becomes very open, lax and shy to flower.
Pruning Remove one-third of oldest flowering wood to ground level, or to lowest young, strong shoot, after flowering. Old, neglected shrubs will rejuvenate if pruned to ground level but will take 2-3 springs to flower fully.
Propagation and nursery production From semi-ripe cuttings taken midsummer or hardwood cuttings taken in winter. Plant bare-rooted or container-grown. Most varieties easy to find; some must be sought from specialist sources. Best planting heights 1-3ft (30cm-1m).
Problems None.
Varieties of interest *F. 'Arnold Dwarf'* Yellow-green flowers intermittently produced. A dwarf variety, very compact. Height 3ft (1m), spread 5-6ft (1.5-2m). Raised in the Arnold Arboretum, Massachusetts, USA. *F. 'Arnold Giant'* Very large, rich, golden yellow flowers, hanging on short shoots. Strong growing. Raised in the Arnold Arboretum. *F. 'Beatrix Farrand'* Canary yellow, hanging flowers with purple throat markings. Good autumn colour. Raised in the Arnold Arboretum. *F. giraldiana* Pale yellow branches, produced somewhat sparsely in early spring, on long, arching, purple-brown branches. One of the earliest to flower. More lax and open than average. From north-western China. *F. × intermedia* (Border Forsythia) Flowers very closely presented on upright branches. Good golden yellow, fading to white-yellow. *F. × i. 'Spectabilis'* Very good golden yellow flowers, relatively unfading, freely produced. Of garden origin. *F. 'Karl Sax'* Deep canary yellow flowers. A very strong-growing American variety. *F. 'Lynwood'* Often called *F. intermedia*, 'Lynwood' or 'Lynwood Gold'. One of the best flowering forms with broad petalled, large, clearly defined, rich yellow flowers. Of garden origin. *F. ovata* (Korean Lilac, Early Forsythia) Flowers yellow to amber-yellow. Foliage ovate with good yellow autumn colour. A dwarf variety, height 5ft (1.5m) after some considerable time. Forms very dense, thick shrub with attractive grey to grey-green stems. *F. ovata 'Tetragold'* Very well formed, large, rounded, bell-shaped, golden yellow flowers. Slightly earlier in flowering than most varieties. A strong-growing form, height and spread 6ft (2m). *F. 'Robusta'* Covered with deep yellow flowers, early spring. Often confused with *F. ovata* and *F. ovata 'Tetragold'*, but much stronger-growing. Height and spread 10ft (3m). *F. 'Spring Glory'* Abundant, bright yellow flowers, mid spring. Stems smooth textured and light brown in colour. Slightly less than average height and spread. This particular variety is best for producing a small standard tree. *F. suspensa* (Climbing Forsythia, Weeping Forsythia) Flowers narrower and more slender than most, light yellow, early spring. Stems attractive, grey-green. Long, arching branches, height up to 10ft (3m), spread 13-16ft (4-5m). *F. s. var. atrocaulis* Pale lemon yellow flowers produced on black-purple stems, arching in habit, early spring. One of the earliest to flower. More open than most varieties in final habit. *F. s. 'Nymans'* Flowers primrose-yellow, of good size and borne in an open semi-pendulous shrub. Early flowering. A variety from the Nymans Garden, England. *F. 'Tremonia'* An interesting variety with good golden yellow flowers borne freely in spring, and giving a parsley-leaved appearance. Good autumn colours. Half average height and spread. *F. viridissima* (Greenstem Forsythia) Pale yellow flowers, stems grey-green and very square. A medium-sized variety, height 10ft (3m). From China. *F. v. 'Bronxensis'* Small lemon yellow flowers, somewhat sparsely produced, sometimes even failing to flower one year. A very dwarf, twiggy, dense shrub. Of garden origin.

Average height and spread
Five years
8x5ft (2.5x1.5m)
Ten years
12x8ft (3.5x2.5m)
Twenty years
or at maturity
13x12ft (4x3.5m)

FOTHERGILLA

KNOWN BY BOTANICAL NAME
Hamamelidaceae
Deciduous
If given the correct planting situation, extremely interesting flowering shrubs with excellent autumn colour.

Origin From south-eastern USA.
Use As an early spring-flowering shrub, but particularly for autumn colour. Ideal for woodland planting.
Description *Flower* No actual petals; clusters of fragrant white stamens producing small to medium, rounded inflorescence mid to late spring, before leaves appear. *Foliage* Leaves ovate, long, 1-2½in (3-6cm) across, grey-green, often with grey, glaucous undersides. Extremely good yellow, orange and flame autumn colours. *Stem* Short, very branching and forming ball-shaped shrub, spreading after 15-20 years. Grey to grey-brown. Often produces suckers from ground level, but not invasively and they should not be removed. Slow to medium growth rate. *Fruit* Insignificant.
Hardiness Tolerates winter temperature range 4°F (−15°C) down to −13°F (−25°C).

Fothergilla major in leaf

Soil Requirements Acid soil, dislikes any alkalinity. Best in moist, rich, deep woodland type.
Sun/Shade aspect Prefers light shade, deep shade spoiling overall shape. Dislikes full sun but for best effect of foliage colour, needs position where sun strikes it in autumn.
Pruning None required.
Propagation and nursery production From layers or seed. Purchase container-grown or

Fothergilla monticola in flower

root-balled (balled-and-burlapped). Best planting heights 1-2½ft (30-80cm). Readily available in its native environment; more difficult to find elsewhere.
Problems Slow to establish; takes some time to develop its full display.
Varieties of interest *F. gardenii* (Dwarf Fothergilla) Good spring flowering and fine autumn colour. Somewhat erect habit, low-growing. Height and spread 3ft (1m). *F. major* (Large Fothergilla) Good spring flower production. Foliage round, very glaucous, extremely good autumn colours. Slower growing than average. From the Allegheny Mountains of USA. *F. monticola* (Alabama Fothergilla) Good spring flower production and very good autumn colours. Slow growth rate. From south-eastern USA.
Average height and spread
Five years
3x3ft (1x1m)
Ten years
6x6ft (2x2m)
Twenty years
or at maturity
13x13ft (4x4m)

FREMONTODENDRON CALIFORNICUM

FREMONTIA
Sterculiaceae
Evergreen
One of the most spectacular shrubs for planting on a wall.

Origin From California and Arizona, USA.
Use As a freestanding shrub for very mild areas or as wall shrub for colder areas where may reach 26-30ft (8-9m) if required.

Fremontodendron californicum 'California Glory' in flower

Description *Flower* No actual petals; large, saucer-shaped yellow calyx with golden yellow stigma and stamens protruding from centre, profusely borne, late spring through to late autumn. *Foliage* Leaves heart-shaped, 3-7 lobed, 2-4in (5-10cm) long, grey-green. *Stem* Upright when young, becoming more spreading to form a tall, upright shrub in very mild areas, or a fan-shaped shrub on a sheltered, sun-facing or lightly shaded wall. Grey-green, down-covered. Fast growth rate. *Fruit* Insignificant.
Hardiness Tolerates 14°F (−10°C), although often wrongly considered more tender.
Soil Requirements Any soil, tolerating high alkalinity.
Sun/Shade aspect Prefers full sun, tolerates very light shade.
Pruning None required but if current season's growth is cut back in early spring, this will increase flowering and control size.
Propagation and nursery production From softwood cuttings taken early spring, or from seed. Purchase container-grown. May not be

easy to find outside its native environment. Best planting heights 15in-3ft (40cm-1m).
Problems When young, susceptible to severe winter weather, but after reaching 5ft (1.5m) rarely affected by cold. The dusty down from the stems can be very painful if it gets into the eyes or on the skin.
Varieties of interest *F. c. 'California Glory'* A variety with larger, lemon yellow to yellow flowers. From Mexico. *F. c. mexicanum* Flowers golden yellow but more star-like in appearance with narrow sepals; five-lobed leaves. Slightly more tender. Very hard to find outside its native environment. From California and Mexico.
Average height and spread
Five years
10x10ft (3x3m)
Ten years
16x16ft (5x5m)
Twenty years
or at maturity
20x20ft (6x6m)

FUCHSIA
Hardy varieties

FUCHSIA, HARDY FUCHSIA, SHRUBBY FUCHSIA
Onagraceae
Deciduous
A range of exotic-looking, summer-flowering, small, low shrubs.

Origin Mostly from South America, but widely distributed.
Use As a freestanding, summer-flowering shrub, on its own or in a large shrub border. If planted 2½ft (80cm) apart all varieties make an informal hedge. If grown in tubs and containers size must exceed 2½ft (80cm) in width and 2ft (60cm) in depth.
Description *Flower* Small to medium-sized, hanging flowers, in most hardy forms having red sepals. Flowering late spring through to early autumn. *Foliage* Leaves ovate, 2-3in (5-8cm) long, light to mid green tinged purple. Some golden and silver variegated forms. *Stem* Upright when young, becoming arching; light green to purple-green. Normally dies to ground level in winter and rejuvenates following spring. Fast growth rate in spring. *Fruit* Some varieties produce purple-black or red edible fruits in autumn.
Hardiness Minimum winter temperature 14°F (−10°C). Protect roots in winter except in very mild areas.
Soil Requirements Any well-drained soil; tolerates both alkalinity and acidity.
Sun/Shade aspect Prefers full sun, tolerates light to mid shade.
Pruning Reduce to ground level in early spring to encourage new strong-growing shoots, although in areas where temperature does not fall below 23°F (−5°C) in winter may

Fuchsia 'Golden Treasure' in flower and leaf

Fuchsia 'Madame Cornelissen' in flower

be left unpruned to form large shrub up to 6ft (2m) height and spread.

Propagation and nursery production From softwood cuttings. Purchase container-grown. Most varieties fairly easy to find, late spring and early summer. Best planting height 8in-2ft (20-60cm).

Problems Although called hardy, fuchsias do not withstand very cold conditions and may suffer damage in temperatures below 23°F (−5°C), and even in relatively mild winters benefit from root protection.

Varieties of interest There are a large number of varieties, many considered hardy in one locality and not in another. The following varieties are those most worth planting for winter temperature range 23°F (−5°C) to 14°F (−10°C). *F. 'Alice Hoffman'* Small flowers with scarlet calyx and white petals, purple tinged leaves. Two-thirds average height. *F. 'Chillerton Beauty'* Good-sized flowers, white calyx shaded deep rose with violet to soft violet petals. *F. 'Eva Boerg'* Cream sepals, magenta petals. Very attractive but slightly less hardy than most. Two-thirds average height and spread. *F. 'Golden Treasure'* Small red and purple flowers, golden foliage. *F. 'Lady Thumb'* Red and white flowers, freely borne, small leaves. Compact, one-third average height and spread. *F. 'Madame Cornelissen'* Good-sized flowers, scarlet sepals, white petals. Very attractive. *F. magellanica 'Alba'* Flowers white, tinged soft pink, hanging, narrow, borne profusely. Foliage light green, small, some yellow autumn colour. Branches upright to arching, forming an attractive shaped shrub. One of the hardiest varieties. *F. m. 'Pumila'* Red and violet-blue flowers, narrow leaves. A very dwarf, bunshaped shrub. Height and spread 1½ft

Fuchsia magellanica 'Versicolor' in flower

(50cm). *F. m. 'Riccartonii'* Slender, narrow flowers of scarlet and violet. One of the hardiest and most common forms. Good for hedging. *F. m. 'Variegata'* syn. *F. gracilis 'Variegata'* Good scarlet and violet flowers. Leaves margined white with grey-green centres. Arching branches. Very attractive as mass planting. *F. m. 'Versicolor'* syn. *F. m. 'Tricolor'* Scarlet and violet, narrow, hanging flowers. Silver-grey foliage with white and pink outer margin. Very attractive for mass planting or singly. *F. 'Margaret'* Large semi-double flowers, scarlet and violet. Leaves large, green-purple, round to ovate. Strong, upright branches. *F. 'Mrs. Popple'* A good flowering form. Crimson and violet flowers with protruding crimson stamens and style. Round to ovate leaves, dark green tinged purple. *F. procumbens* A small, trailing variety with yellow and violet flowers and red stamens. Very small foliage on arching branches. Large, magenta, edible fruits freely produced. Slightly more tender than most and difficult to find. *F. 'Tom Thumb'* Scarlet and violet flowers. Very dwarf and compact, height and spread 1½ft (50cm).

Average height and spread
Five years
3x3ft (1x1m)
Ten years
4x5ft (1.2x1.5m)
Twenty years
or at maturity
4x5ft (1.2x1.5m)

Fuchsia magellanica 'Alba' in flower

GARRYA ELLIPTICA

TASSEL BUSH, SILK TASSEL TREE
Garryaceae
Evergreen
It is hard to imagine a finer sight than *Garrya elliptica* fully hung with catkins in late spring.

Origin From California and Oregon, USA.
Use As a freestanding, large shrub in areas where temperature does not fall below 23°F (−5°C) in winter, or as a wall shrub for any except extremely exposed, shady walls in colder areas. Good in maritime areas.
Description *Flower* Male plants have long, grey-green, hanging catkins. Female plants have insignificant flowers and lacking catkins are not normally considered worth growing. *Foliage* Leaves broadly ovate, 1½-3in (4-8cm) long, dark green, glossy upper surfaces, glaucous undersides and of leathery texture. *Stem* Upright, becoming branching and spreading with age. Grey-green to green-brown. Medium growth rate. *Fruit* Small, round, purple-brown fruits in long clusters produced by female plants when there is a male plant nearby.
Hardiness Tolerates 14°F (−10°C).
Soil Requirements Prefers fertile soil but does well on any.
Sun/Shade aspect Full sun to medium shade. In deep shade becomes long, open and straggly and catkins diminish.
Pruning None required but can be reduced by removing major limbs in late winter or early spring and will rejuvenate in succeeding spring.
Propagation and nursery production From semi-ripe cuttings taken early to mid summer. Purchase container-grown. Availability varies. Best planting heights 1½-3ft (50cm-1m).
Problems Rather untidy habit of growth. Black spots often appear on old foliage just before it drops in late winter and early spring. In cold, unproductive springs there may be a delay between fall of old evergreen foliage and growth of new leaves, leaving the plant temporarily defoliated. This can also be apparent when new plants are planted, with a gap arising in mid spring.

Garrya elliptica 'James Roof' in catkin

Varieties of interest *G. elliptica 'James Roof'* Good, bright, leathery leaves, with catkins twice the length of *G. elliptica* and thicker in texture. Strong-growing. *G. fremontii* Similar to its parent but with more ovate, twisted leaves and short catkins. From western USA.

Average height and spread
Five years
8x5ft (2.5x1.5m)
Ten years
12x8ft (3.5x2.5m)
Twenty years
or at maturity
14½x12ft (4.5x3.5m)

× GAULNETTYA

KNOWN BY BOTANICAL NAME

Ericaceae
Evergreen
A low, dense, evergreen, early summer-flowering shrub with red fruits in autumn.

Origin A hybrid of garden origin, a cross between *Gaultheria 'Shallon'* and *Pernettya mucronata*.
Use As ground cover for summer and autumn interest.
Description *Flower* Short, urn-shaped white flowers, borne on short sprays in leaf axils, late spring to early summer. *Foliage* Broad, ovate, 1-2in (3-5cm) long, dark green and leathery-textured. *Stem* Branching, becoming very spreading, light green, producing new shoots from below ground, forming wide spreading, carpet-like shrub. Slow growth rate, except when young. *Fruit* Red fruits freely borne in autumn.

× *Gaulnettya wisleyensis 'Ruby'* in fruit

Hardiness Tolerates 14°F (−10°C).
Soil Requirements Acid soil; dislikes any alkalinity.
Sun/Shade aspect Prefers medium shade, but tolerates full sun.
Pruning None required; straggly shrubs can be cut to ground level in early spring and will quickly rejuvenate.
Propagation and nursery production From suckers or from cuttings taken late summer. Purchase container-grown. Best planting heights 8-15in (20-40cm). Fairly easy to find in acid soil areas.
Problems Can become invasive.
Varieties of interest × *G. wisleyensis 'Ruby'* White flowers borne late spring, early summer, followed by deep ruby red fruit, autumn. × *G. w. 'Wisley Pearl'* White flowers in late spring, early summer followed by blood-red fruits in autumn.
Average height and spread
Five years
1½x3ft (50cmx1m)
Ten years
1½x6ft (50cmx2m)
Twenty years
or at maturity
1½x10ft (50cmx3m)

GAULTHERIA

BOX BERRY, CANDA TREE, CHECKERBERRY, CREEPING WINTERGREEN, TEA BERRY, WINTERGREEN

Ericaceae
Evergreen
When used as ground cover with rhododendrons and azaleas, makes a fine carpet.

Origin Mostly from North and Central America, some varieties from Japan and New Zealand, and from various other parts of the world.
Use As ground cover, forming a spreading carpet with flowers and fruit, for acid soil areas.
Description *Flower* Small, white, urn-shaped flowers borne late spring to early summer.

Gaultheria shallon in flower

Foliage Round to ovate, 1½-4in (4-10cm) long, light green when young, aging to darker green, with purple shading. *Stem* Branching, becoming very spreading to form a wide carpet. Light green. Produces new shoots from below ground. Slow growth rate, except when young. *Fruit* Round, white or blue through purple to bright red, depending on variety.
Hardiness Tolerates 14°F (−10°C).
Soil Requirements Acid soil, dislikes any alkalinity.
Sun/Shade aspect Prefers medium to light shade but good in full sun.
Pruning None required but if it becomes straggly with age, cut to ground level in early spring, and it will quickly rejuvenate.
Propagation and nursery production From suckers. Purchase container-grown for best planting results. Fairly easy to find in acid soil areas. Best planting heights 8-15cm (20-40cm).
Problems Can become invasive.
Varieties of interest *G. procumbens* (Checkerberry, Wintergreen) Small white flowers, followed by red fruits in winter, leaves dark green. Forms a creeping, evergreen carpet. One-third less than average height and spread. From North America. *G. shallon* Good-sized pinky-white flowers, dark purple fruits in large quantities. Broad, leathery green leaves with some red-purple shading. A strong-growing variety making good ground cover. From western North America.
Average height and spread
Five years
1½x3ft (50cmx1m)
Ten years
1½x6ft (50cmx2m)
Twenty years
or at maturity
1½x10ft (50cmx3m)

GENISTA AETNENSIS

MOUNT ETNA BROOM

Leguminosae
Deciduous
A graceful, spreading, large, early summer-flowering shrub.

Origin From Sardinia and Sicily.
Use As a freestanding, very graceful shrub or small tree.
Description *Flower* Masses of single, golden yellow pea-flowers, hanging from pendulous branches, early summer. *Foliage* Sparsely produced, ½in (1cm) long, linear, grey-green. *Stem* Upright when young, quickly becoming pendulous and graceful to form a wide, weeping, large shrub of almost tree size. Grey-green to grey-brown. Medium to fast growth rate when young, slowing with age. *Fruit* Insignificant.

Hardiness Tolerates 14°F (−10°C).
Soil Requirements Does well on most soils; dislikes waterlogged areas. On very light, sandy soils will need staking firmly as root anchorage may become very poor.
Sun/Shade aspect Prefers full sun, dislikes shade.
Pruning Best left unpruned as dislikes pruning immensely. Only when it is very young can this be attempted, or on very young wood.
Propagation and nursery production From softwood cuttings or from seed. Purchase container-grown. Not always easy to find. Best planting heights 1½-2½ft (50-80cm).
Problems Space must be allowed for the ultimate size of its long, arching, spreading branches as pruning is not advised.

Genista aetnensis in flower

Varieties of interest *G. tenera 'Golden Showers'* syn. *G. cinerea 'Golden Showers'* Golden yellow flowers abundantly borne on long, arching branches in early summer.
Average height and spread
Five years
8x8ft (2.5x2.5m)
Ten years
12x12ft (3.5x3.5m)
Twenty years
or at maturity
14½x14½ft (4.5x4.5m)

GENISTA HISPANICA

SPANISH GORSE

Leguminosae
Deciduous
An attractive, low, mounded shrub with a number of uses in the garden.

Origin From south-western Europe, predominantly Spain.
Use As a low, freestanding shrub or for a large shrub border. If planted 2ft (60cm) apart makes spring-flowering low hedge.

Genista hispanica in flower

Description *Flower* Light yellow, single pea-flowers freely borne on leading 2-4in (5-10cm) of upright shoots, mid spring to late summer. *Foliage* Leaves narrow, lanceolate ½in (1cm) long, grey-green, very sparsely produced and insignificant. *Stem* Upright, spiny, becoming slightly spreading to form a round, prickly mound. Grey-green. Slow growth rate. *Fruit* Insignificant.
Hardiness Tolerates winter temperatures down to −13°F (−25°C).
Soil Requirements Does well on most soils, but very severe alkalinity may lead to chlorosis of stems. Dislikes extremely waterlogged areas.
Sun/Shade aspect Prefers full sun. In shade will develop into an open, coarse shrub with poor flowering.
Pruning None required but will take hard cutting back to remove old, woody stems and will slowly rejuvenate.
Propagation and nursery production From softwood cuttings taken in early summer. Purchase container-grown. Availability varies. Best planting heights 8-15in (20-40cm).
Problems Its slightly prickly nature can make cultivation difficult.
Average height and spread
Five years
2x2½ft (60x80cm)
Ten years
2x3ft (60cmx1m)
Twenty years
or at maturity
3x5ft (1x1.5m)

GENISTA LYDIA

KNOWN BY BOTANICAL NAME

Leguminosae
Deciduous
One of the most interesting shrubs, especially when planted to cascade down a wall or bank, where flowers are shown to best effect.

Origin From the eastern Balkans.
Use For mass planting on banks, for individual or group planting in large shrub borders. Ideal for spreading and cascading over walls. Can be planted in large containers of not less than 2½ft (80cm) diameter and 2ft (60cm) depth.
Description *Flower* Single, golden yellow pea-flowers entirely covering branches, late spring and early summer. *Foliage* Leaves small, ½in (1cm) long, lanceolate, grey-green, sparsely produced and insignificant. *Stem* Arching branches, forming a spreading, arching mound, bright green to grey-green. Slow to medium growth rate. *Fruit* Insignificant.
Hardiness Tolerates 14°F (−10°C).

Soil Requirements Almost any soil, tolerating both acidity and alkalinity. Dislikes waterlogging.
Sun/Shade aspect Prefers full sun, dislikes any shade.
Pruning None required.
Propagation and nursery production From softwood cuttings. Purchase container-grown. Availability varies. Best planting heights 8in-1½ft (20-50cm).
Problems Often looks weak when purchased but once planted quickly improves shape.
Varieties of interest *G. pilosa* Numerous yellow pea-flowers, midsummer. Not a direct variety of *G. lydia*, but has similar growth habit, forming a very low, spreading carpet, only 12in (30cm) in height with very slow spread of up to 6ft (2m).
Average height and spread
Five years
1½x4ft (50cmx1.2m)
Ten years
2x6ft (60cmx2m)
Twenty years
or at maturity
3x10ft (1x3m)

Genista lydia in flower

GENISTA TINCTORIA 'ROYAL GOLD'

DYER'S GREENWOOD, COMMON WOADWAXEN

Leguminosae
Deciduous
An extremely useful, summer-flowering shrub, with golden flowers and yellow autumn foliage tints.

Origin Of garden origin.
Use As a freestanding, summer-flowering shrub. Can be used as informal hedge planted 2ft (60cm) apart in single line.
Description *Flower* Single, golden yellow pea-flowers borne profusely throughout foliage,

midsummer. *Foliage* Leaves small, linear, 1in (3cm) long, light grey-green giving some yellow autumn tints. *Stem* Very upright, becoming slightly twiggy with age and forming an upright, round-topped shrub. Grey-green. Medium growth rate. *Fruit* Insignificant.
Hardiness Tolerates minimum winter temperature range from 14°F (−10°C) down to 4°F (−15°C).
Soil Requirements Any soil, including extremely alkaline.
Sun/Shade aspect Prefers full sun. Any shade spoils shape and reduces flowering ability.
Pruning None required, in fact resents it.

Genista tinctoria 'Royal Gold' in flower

Propagation and nursery production From softwood cuttings taken early in summer. Purchase container-grown. Availability varies. Best planting heights 15in-2ft (40-60cm).
Problems Very susceptible to blackfly.
Average height and spread
Five years
2½x3ft (80cmx1m)
Ten years
3x5ft (1x1.5m)
Twenty years
or at maturity
4x5ft (1.2x1.5m)

GRISELINIA LITTORALIS

KNOWN BY BOTANICAL NAME

Cornaceae
Evergreen
A useful evergreen for mild locations or, given adequate protection, for colder regions.

Origin From New Zealand.
Use As a freestanding shrub or for large shrub borders. If planted 3ft (1m) apart makes a fine evergreen hedge. In coastal locations or under mild conditions, will reach dimensions of small tree.
Description *Flower* Inconspicuous; some plants carry all male flowers, others all female. *Foliage* Leaves broad, ovate, 1-3in (3-8cm) long, bright green, slightly curled, leathery-textured, densely produced, on attractive green leaf stalks. *Stem* Upright, becoming branching with time. Bright green to dark green. Medium to fast growth rate. *Fruit* Insignificant.
Hardiness Minimum winter temperature 23°F (−5°C).
Soil Requirements Almost any soil, but may show signs of chlorosis on extreme alkalinity.
Sun/Shade aspect Full sun to light shade.
Pruning None required but can be cut to ground level if it becomes old and woody, and will rejuvenate. Can be trimmed very hard into any desired shape.
Propagation and nursery production From softwood cuttings taken in early summer. Purchase container-grown. Not always easy

Griselinia littoralis in leaf

to find; most readily available in mild or coastal locations. Best planting heights 15in-3ft (40cm-1m).
Problems The shrub must be considered tender or slightly tender. Some winter die-back will occur, but this is not necessarily terminal.
Varieties of interest *G. l. 'Dixon's Gold'* A very bold variety with leaves marked creamy white. Originating from *G. l. 'Variegata'* and found in a Jersey garden. *G. l. 'Variegata'* Foliage variegated white. A tender variety; winter minimum 32°F (0°C). Most likely to be found in mild coastal areas.
Average height and spread
Five years
5x5ft (1.5x1.5m)
Ten years
8x8ft (2.5x2.5m)
Twenty years
or at maturity
16x16ft (5x5m)

HALESIA

MOUNTAIN SILVERBELL, MOUNTAIN
SNOWBALL TREE, CAROLINA SILVERBELL
Styracaceae
Deciduous
A very attractive large specimen shrub or small tree for the larger garden.

Origin From south-eastern USA.
Use As a freestanding flowering shrub.
Description *Flower* Small to medium-sized, hanging, nodding, bell-shaped flowers borne in clusters of 3 or 5 along underside of each branch, mid to late spring. *Foliage* Medium-

Halesia monticola in flower

sized, broad, ovate, 2-5in (5-12cm) long, light grey-green, giving good yellow autumn colour. *Stem* Upright when young, becoming branching and twiggy with age, forming a large, round-topped shrub. After 20 years may form a small tree. Grey to grey-green. Medium growth rate. *Fruit* Small, green, winged fruits in autumn.
Hardiness Tolerates 14°F (−10°C).
Soil Requirements Does well on most types, best on well-drained soil.
Sun/Shade aspect Prefers full sun to light shade.
Pruning None required.
Propagation and nursery production From layers of semi-ripe cuttings taken in early summer. Purchase container-grown. Must be sought from specialist nurseries. Best planting heights 1½-2½ft (50-80cm).
Problems Often weak looking when purchased. Begins to achieve full potential only after 3-5 years.
Varieties of interest *H. carolina* syn. *H. tetraptera* (Carolina Silverbell) White, bell-shaped, nodding flowers grouped along branches, early to mid spring. Fruits are four-winged and pear-shaped. *H. monticola* (Mountain silverbell) A variety with larger flowers and fruit, fruit clusters up to 2in (5cm) in length. From the mountain regions of south-eastern USA. *H. m. vestita* A large-flowering form with flower clusters up to 1-1½in (3-4cm) across, sometimes pink-tinged. When mature, the leaves are downy with a glabrous covering.
Average height and spread
Five years
5x6ft (1.5x2m)
Ten years
10x13ft (3x4m)
Twenty years
or at maturity
20x20ft (6x6m)

× HALIMIOCISTUS

KNOWN BY BOTANICAL NAME
Cistaceae
Evergreen
A group of somewhat neglected shrubs of truly spectacular flowering ability.

Origin Throughout central and southern Europe, but most forms of garden origin.
Use As small shrubs for large rock gardens and small groups on the edge of a large shrub border. Plant in single row 1½ft (50cm) apart for a low, informal hedge. Spectacular when mass planted.
Description *Flower* Small, round, white flowers, 1½-2in (4-5cm) across, opening in succession in sunny weather and each lasting only one day. Flowering often stops temporarily when weather is dull and cloudy. *Foliage*

Leaves broad, lanceolate, 1-1½in (3-4cm) long, dark green to grey-green. Often resinous and sticky to touch. *Stem* Slightly upright when young, quickly becoming spreading and very branching to make low, spreading hummock. Mid green to grey-green. Medium growth rate when young, slower with age. *Fruit* Insignificant.
Hardiness Minimum winter temperature 23°F (−5°C).
Soil Requirements Does well on most types, but alkaline soils will cause severe chlorosis.
Sun/Shade aspect Prefers full sun, dislikes any shade.
Pruning Trim lightly in early spring, to encourage new flowering growth.
Propagation and nursery production From semi-ripe cuttings taken in midsummer. Purchase container-grown. Availability varies. Best planting heights 8-15in (20-40cm).
Problems None, except that young plants are unattractively small when offered.

× *Halimiocistus ingwersenii* in flower

Varieties of interest × *H. ingwersenii* Pure white flowers borne throughout summer and early autumn. Forms dwarf, spreading shrub. Discovered in Portugal and believed to be a cross. × *H. sahucii* Very free flowering. Pure white flowers, late spring to early autumn. Dark green, linear leaves, slightly resinous. Discovered in the south of France. × *H. wintonensis* Pearly white flowers 2in (5cm) wide, pencilled zones of crimson-maroon, yellow stained at base of each petal, late spring through early summer. Foliage grey to grey-green on a low, hummock-like shrub. Originating in Hillier's Nursery, Winchester, England.
Average height and spread
Five years
1¼x1¼ft (40x70cm)
Ten years
1½x3ft (50cmx1m)
Twenty years
or at maturity
2x4ft (60cmx1.2m)

HALIMIUM

KNOWN BY BOTANICAL NAME
Cistaceae
Evergreen
A small group of attractive shrubs from the Mediterranean region, worthy of a place in a temperate garden but needing some protection.

Origin From Portugal and Spain.
Use As a small shrub for large rock gardens or in small groups at the edges of large shrub borders. In very mild areas may be planted in single row 1½ft (50cm) apart as low, informal hedge.
Description *Flower* Small, flat, round, saucer-shaped flowers, yellow to golden yellow,

135

Halimium ocymoides
in flower

some brown or dark purple markings according to variety. Flowers produced daily from late spring through summer, intermittently in early autumn. *Foliage* Ovate, ½-1½in (2-4cm) long, grey to grey-green, slightly resinous. *Stem* Slightly upright when young, becoming spreading and branching with age, forming low, round hummock. Grey to grey-green. Medium growth rate when young, slowing with age. *Fruit* Insignificant.
Hardiness Tolerates minimum winter temperature 23°F (−5°C).
Soil Requirements Needs well-drained soil, tolerates both alkalinity and acidity.
Sun/Shade aspect Full sun, dislikes any shade.
Pruning Trim lightly in mid spring to encourage new flowering growth.
Propagation and nursery production From semi-ripe cuttings taken in midsummer. Purchase container-grown. Must be sought from specialist nurseries. Best planting heights 6-12in (15-30cm).
Problems None, except that young plants are unattractively small when offered.
Varieties of interest *H. lasianthum* syn. *Cistus formosus* Golden yellow flowers in spring, each petal blotched dark purple at base. Grey-green foliage. *H. ocymoides* syn. *Helianthemum algarvense* Bright yellow flowers in summer, brown to dark brown markings at petal bases. Foliage grey to grey-green. *H. umbellatum* White flowers, early summer. More erect than most. Originating in Mediterranean region.
There are a number of other varieties, but those mentioned are among the most reliable.
Average height and spread
Five years
1¼x2¼ft (40x70cm)
Ten years
1½x3ft (50cmx1m)
Twenty years
or at maturity
2x4ft (60cmx1.2m)

HAMAMELIS

Origin Most forms originating from stock from Japan; most varieties now of garden origin.
Use As a freestanding winter-flowering shrub.
Description *Flower* Clusters of strap-like petals, ranging from lemon yellow through gold, brown, orange to dark red according to variety, borne winter through early spring, withstanding harsh weather conditions undamaged. Some varieties fragrant. Exact

flowering time depends on temperature; the first semi-mild spell following the shortest day of the year often triggers first flower bud to open. *Foliage* Leaves ovate, 3-5in (8-12cm) long, with good orange-yellow autumn colours. *Stem* Upright when young, becoming spreading and branching with age, to form open, loose, spreading shrub. Grey to grey-green. Medium growth rate. *Fruit* Insignificant.
Hardiness Tolerates −13°F (−25°C).
Soil Requirements Neutral to acid; moderate to heavy alkalinity will rapidly cause signs of chlorosis. Mulch annually with peat or garden compost to depth of 2in (5cm) over 1-2sq yds (1-2sq m) to maintain health and encourage flowering.
Sun/Shade aspect Prefers full sun to light shade; too much shade will spoil shape and reduce flowers. Best planted where winter sunlight can enhance flowering effect.
Pruning None required.
Propagation and nursery production *H. virginiana* grown from seed, and is the understock on to which all other varieties are grafted. Purchase container-grown or root-balled (balled-and-burlapped). Best planting heights 2-4ft (60cm-1.2m).
Problems Young grafted plants often fail so it is advisable to purchase plants 4 years old or more, although these are expensive.

Hamamelis mollis in flower

Varieties of interest *H. × intermedia 'Diane'* Flowers 1-1½in (3-4cm) across, rich copper-red. Good autumn foliage colours. Slightly less than average height and spread. *H. × 'Jelena'* Flowers 1½in (4cm) across, bright copper-orange. *H. × 'Ruby Glow'* Flowers 1in (3cm) across, copper-red. Narrow foliage with good autumn colour. *H. × 'Westersteide'* Flowers 1in (3cm) across, clear yellow and freely borne, produced later than most. Small to medium-sized foliage. Good autumn colour. *H. japonica 'Zuccariniana'* (Japanese Witch Hazel) Grey, curled flower buds open to release pale yellow, lemon-scented flowers in early spring. Flowers less than 1in (3cm) across but borne profusely. *H. mollis* (Chinese Witch Hazel) Pure golden yellow, very fragrant flowers 1½in (4cm) across, late winter. Large oval to round leaves with good autumn colour. *H. m. 'Brevipetala'* Scented, bronze-yellow flowers, 1½in (4cm) across, borne on strong upright branches, more vigorous than most. Broad ovate leaves with good autumn colour. *H. m. 'Pallida'* Silver-yellow, sweetly scented flowers 1½in (4cm) across. Broad ovate fleaves with good autumn colour. Spreading habit. One of the most beautiful forms. *H. virginiana* (Common Witch Hazel) Slightly scented, golden yellow flowers, 1in (3cm) across. Good golden yellow autumn colour. Apart from its use as understock, this is the basic form from which the essence known as witch hazel is distilled.

Average height and spread
Five years
5x6ft (1.5x2m)
Ten years
10x13ft (3x4m)
Twenty years
or at maturity
16x20ft (5x6m)

HEBE
Low-growing varieties

Origin Mostly from New Zealand; many varieties of garden origin.
Use As a low, evergreen, summer-flowering shrub for shrub borders, for low hedge if planted 1¼ft (40cm) apart, or for containers more than 2ft (60cm) wide and 1¼ft (40cm) deep.
Description *Flower* Small, short, upright racemes of flowers, white or shades of blue, depending on variety, borne late spring through summer and sporadically in early autumn. *Foliage* Leaves round, narrow or ovate, ½-1in (2-3cm) long, green or variegated depending on variety. *Stem* Short, becoming branching, forming a low, round hummock or spreading carpet, depending on variety. Dark green to grey-green. Medium growth rate when young, slowing with age. *Fruit* Flowers give way to brown seed capsules.
Hardiness Minimum winter temperature 23°F (−5°C).
Soil Requirements Well-drained, open soil, either alkaline or acid.
Sun/Shade aspect Prefers full sun, tolerates light shade.
Pruning Trim over lightly each spring to encourage new growth.
Propagation and nursery production From softwood cuttings taken early summer. Purchase container-grown. Availibility varies. Best planting heights 6-12in (15-30cm).
Problems Hebes often look sickly and stunted when purchased, but once planted out grow quickly.
Varieties of interest *H. albicans* Small white flowers from early summer. Round to ovate, glaucous leaves. *H. a. 'Pewter Dome'* White flowers, grey-green foliage, forming dome-shaped shrub. *H. a. 'Red Edge'* White flowers. Grey foliage edged with red. *H. 'Autumn Glory'* Deep purple blue flowers in medium-sized racemes, continuously from early summer through to late autumn. Dark purple-

Hebe 'Autumn Glory' in flower

Hebe pinguifolia 'Pagei' in flower

green, round foliage. **H. 'Carl Teschner'** Small racemes of violet blue flowers, early summer. A very low-growing shrub with small, narrow, lanceolate leaves, forming a flat carpet. **H. glaucophylla 'Variegata'** syn. **H. darwiniana variegata** Small pale blue spikes of flowers. Grey-green leaves with silver margins, somewhat sparsely distributed on lax, open growth. **H. macrantha** Short racemes of pure white flowers, early summer. Stems surrounded by leathery, bright green foliage. More tender than most and good in coastal areas. **H. pinguifolia 'Pagei'** Small white flowers, mainly early summer, but some produced later, surmounting carpet of glaucous grey-green foliage. **H. p. 'Quick Silver'** White flowers, silver grey foliage. Open spreading habit, carpet forming. **H. rakaiensis** syn. **H. subalpina** Very bright green, ovate foliage. Sometimes, but not consistently, covered with racemes of white flowers with blue central markings. Forms a symmetrical, dome-shaped shrub. Very good for low hedge.

Average height and spread
Five years
1x2½ft (30x80cm)
Ten years
1½x3ft (50cmx1m)
Twenty years
or at maturity
1½x3ft (50cmx1m)

HEBE
Tall-growing varieties

VERONICA
Scrophulariaceae
Evergreen
Although not hardy, among the most spectacular of all summer-flowering shrubs and worth replacing if killed by cold conditions.

Origin Most forms from New Zealand or of garden origin.
Use As a quick-growing, summer-flowering shrub for mass planting or for borders. Can be planted 2½ft (80cm) apart in straight line to make a useful, informal hedge. Some forms good in containers not less than 2½ft (80cm) in diameter and 2ft (60cm) in depth if good potting medium used.
Description *Flower* Racemes of white, dark red, blue or purple flowers, depending on variety. Flowers in spikes 2-6in (5-15cm) long. *Foliage* Leaves broad, round, ovate or lanceolate, 1-4in (3-10cm) long, depending on variety. Light to mid to dark green, purple-green or silver, or silver variegated. *Stem* Upright becoming spreading with age. Light to mid green, sometimes with purple shading. Fast growth rate. *Fruit* Small, brown seed clusters follow flowers, attractive in winter.

Hardiness Minimum winter temperature 23°F (−5°C). Large-leaved varieties tend to be less hardy than small-leaved.
Soil Requirements Prefers light, open soil, tolerating alkalinity and acidity.
Sun/Shade aspect Prefers full sun. Any shade will spoil shape, and decrease flowering.
Pruning Trim over annually in early spring. Cut back hard almost to ground level every 3-4 years to induce new strong healthy shoots.
Propagation and nursery production From semi-ripe cuttings taken in summer. Purchase container-grown. Most readily available when in flower. Best planting heights 15in-2½ft (40-80cm).
Problems Stock available for purchase between midsummer and early spring often looks unattractively woody but once planted grows quickly. Younger flowering plants should only be planted late spring or early summer to avoid risk of succumbing to winter frosts.
Varieties of interest *H. 'Alicia Amherst'* Racemes of purple-blue flowers, early to mid summer, offset by long, mid green, lanceolate leaves. Slightly more tender than average. *H. 'Amy'* Deep purple flowers, purple-green foliage. Slightly more tender. *H. × andersonii* Lavender-blue flowers, early to mid summer, broad, round, dark to mid green foliage. Slightly less hardy than most. *H. × a. 'Variegata'* Lavender-blue flowers surmounting broad, round, silver-margined grey-green foliage. Slightly more tender. *H. anomala* Small, short racemes of inconspicuous white flowers surmounting very small, round, bright green leaves, similar to Boxwood. One of the most hardy. *H. 'Bowles Hybrid'* Pale lavender racemes of flowers, early to mid summer, surmounting narrow, lanceolate, purple-green leaves. One of the most hardy. *H. brachysiphon* syn. *H. traversii* Racemes of short, white flowers, early summer, surmounting round, bright green foliage. One of the most hardy. *H. × franciscana 'Blue Gem'* syn. *Elliptica 'Blue Gem'* Mid length to very long racemes of pale violet-blue flowers, borne throughout summer and early autumn; long, broad, lanceolate, mid green leaves. *H. × f. 'Variegata'* syn. *H. Elliptica* Long pale violet-purple flowers borne throughout summer and early autumn. Long, broad, lanceolate leaves with creamy-white margins. *H. 'Great Orme'* Pink racemes of good-sized flowers, produced profusely throughout summer. Originated on the Great Orme peninsula, north Wales, where it still grows. Foliage broadly lanceolate and light green. *H. hulkeana* Panicles of lavender-blue flowers, open and long. Round, curved edged foliage, purple-green. Very scarce and more tender than most. *H. 'La Seduisante'* Long deep magenta-purple flowers, early summer through to early autumn, surmounting light green, broad, lanceolate foliage. *H. 'Marjorie'* Mid length pale violet flowers, leaves

Hebe salicifolia in flower

mid green, short, round to ovate. One of the hardiest. *H. 'Mauvena'* Short, mid blue flowers. Light to mid green foliage. *H. 'Midsummer Beauty'* Long, pale to mid purple-blue flowers produced over long period. Mid to dark green foliage with red underside. *H. 'Mrs. E. Tennant'* Light violet-blue, large flowers. Mid to dark green foliage. *H. 'Mrs. Winder'* Short bright blue-mauve flowers. Leaves bronze-purple on chestnut coloured stems. *H. 'Purple Queen'* Purple to deep purple, long flowers. Purple-green to purple leaves. More tender than most. *H. salicifolia* Numerous short racemes of white flowers in midsummer. Long, narrow, lanceolate, light green leaves. *H. s. 'Variegata'* Light blue flowers in summer, surmounting long, narrow, lanceolate leaves, edged white. Weak in constitution and less than average height and spread. Tender and does not tolerate freezing conditions. *H. 'Simon Delaux'* Rich crimson-red flowers in large racemes throughout summer. Foliage round, light green to mid green. *H. 'Waikiki'* Short racemes of deep blue flowers, summer. Bronze-tinted foliage and bronze stems, of average height, but with more than average spread and relatively hardy.

Average height and spread
Five years
3x3ft (1x1m)
Ten years
4x6ft (1.2x2m)
Twenty years
or at maturity
4x10ft (1.2x3m)

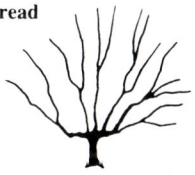

Hebe 'Midsummer Beauty' in flower

HEBE CUPRESSOIDES

KNOWN BY BOTANICAL NAME
Scrophulariaceae
Evergreen
An interestingly shaped shrub, grown for its foliage effect.

Origin From New Zealand.
Use As a freestanding feature shrub for large rock gardens or standing alone, like a sentinel.
Description *Flower* Small, pale blue flowers, early to mid summer, sometimes unreliably produced. *Foliage* Leaves very small, ovate, grey-green, very closely set on stems, giving effect similar to conifer foliage. *Stem* Upright, round to slightly pointed, forming tightly bunched clumps. Slow growth rate. *Fruit* Insignificant.
Hardiness Tolerates 4°F (−15°C).
Soil Requirements Good, rich, open, well-drained soil; tolerates alkalinity and acidity.
Sun/Shade aspect Prefers full sun. Resents any shade, leading to loss of characteristic shape.
Pruning None required. Individual limbs may be shortened but it will take some time to regain shape.

Hebe cupressoides **in flower**

Propagation and nursery production From softwood to semi-ripe cuttings taken early summer. Purchase container-grown. Availability varies. Best planting heights 6-12in (15-30cm).
Problems Young plants in containers often look weak and insipid when purchased, but once planted out grow quickly.
Varieties of interest *H. c. 'Boughton Dome'* A very dwarf, silvery green variety, making a tight, bun-shaped shrub up to 2ft (60cm) in height, 3ft (1m) in spread. Slightly less hardy than its parent.
Average height and spread
Five years
2½x2½ft (80x80cm)
Ten years
3x3ft (1x1m)
Twenty years
or at maturity
4x4ft (1.2x1.2m)

HEBE OCHRACEA

CORD-BRANCHED VERONICA
Scrophulariaceae
Evergreen
Interesting for its cord-like foliage, a characteristic not shared by any other shrub.

Origin From New Zealand.
Use As a curious shrub of distinctive growth habit. Can be grown in containers not less than 2½ft (80cm) in diameter and 2ft (60cm) in depth. Use good potting medium and feed to encourage growth.

Hebe armstrongii **in flower**

Description *Flower* Small, white, single flowers borne some years along cord-like foliage, in many years producing no flowers. *Foliage* Leaves tiny, round, ovate, packed close against stems, old gold with predominantly yellow tinged tip, giving effect of golden whip-cord. *Stem* Long, arching, spreading and covered completely in foliage. Slow growth rate. *Fruit* None.
Hardiness Minimum winter temperature 23°F (−5°C).
Soil Requirements Good, rich soil, deep and well-drained.
Sun/Shade aspect Full sun. Any shade spoils overall shape and effect.
Pruning None required.
Propagation and nursery production From semi-ripe cuttings taken in early summer. Purchase container-grown. May be available from nurseries, not often in garden centres. Best planting heights 6-12in (15-30cm).
Problems Stem very susceptible to waterlogging; large areas of growth will die if shrub is too wet, too dry or hungry for nutrients, so these conditions must be guarded against.
Varieties of interest *H. armstrongii* Olive green foliage. Often confused with *H. ochracea* at point of sale. *H. ochracea 'James Stirling'* A variety with green-gold, whip-cord branches and stronger, more vigorous growth. Neater habit of growth.
Average height and spread
Five years
2x2½ft (60x80cm)
Ten years
2½x3ft (80cmx1m)
Twenty years
or at maturity
3x4ft (1x1.2m)

HEDERA
Ground cover varieties

IVY
Araliaceae
Evergreen
As large-scale evergreen ground cover for all areas.

Origin From various areas of both the northern and southern hemispheres.
Use As low, spreading ground cover planted 3ft (1m) apart, for shady areas.
Description *Flower* Round heads of green to lime green flowers, borne mainly early to late spring, either in bud or open throughout the year. *Foliage* Leaves small to large, diamond-shaped, 1½-5in (4-12cm) long with glossy upper surfaces and duller undersides, ranging from dark through to mid green, golden, silver or variegated according to form. *Stem* Sprawling, spreading, rooting at leaf axils, forming a dense, thick mat-forming cover.

Rarely reaches above 1ft (30cm) but if spreads to upright object, such as tree or post, regains climbing aptitude. *Fruit* Round clusters of black, poisonous fruits, autumn and winter.
Hardiness Tolerates 4°F (−15°C).
Soil Requirements Any soil, including very alkaline and very acid. Tolerates very dry or wet areas, although may lose some variegation on moist soils. In very dry areas should be watered to help it establish itself.
Sun/Shade aspect Prefers medium shade, good from deep shade through to full sun.
Pruning None required, except to keep it within bounds.
Propagation and nursery production From semi-ripe cuttings or rooted natural layers. Purchase container-grown. Best planting heights 8in-2½ft (20-80cm). Readily available. May be sold climbing on a cane, which should be removed when planting as ground cover.

Hedera canariensis 'Gloire de Marengo'

Problems Can become invasive.
Varieties of interest *H. canariensis 'Gloire de Marengo'* Dark green leaves up to 5in (12cm) long and wide. Silvery grey outer margins, becoming white. Good for ground cover in dark areas, although may lose some variegation. *H. colchica* (Persian Ivy) Ovate to elliptic leaves 6-8in (15-20 cm) long and wide, dark green and leathery textured; leaves may be larger on healthy, established new growth. *H. c. 'Dentata Variegata'* Bright green to green-grey leaves with yellow to creamy yellow outer margins when young. Becoming creamy white with age. *H. c. 'Sulphur Heart'* (Paddy's Pride) Large, heart-shaped, ovate leaves. Grey-green, aging to dark green, centre splashed yellow, aging to pale lime

green, late spring and summer. Leaves occasionally completely yellow. *H. helix 'Buttercup'* Rich golden yellow in spring, aging to yellow-green, finally pale green in autumn, but regaining the yellow in early winter. Small leaves, up to 1-1½in (3-4cm) long and wide. May appear green when purchased at certain times of year. *H. h. 'Cristata'* Attractively shaped foliage, very twisted at edges, round, pale green, aging to purple-brown to purple in winter, regaining light green following spring. *H. h. 'Glacier'* Almost diamond-shaped foliage, up to about 1¼in (3cm) long and wide, grey-green, becoming silver-grey with narrow white margins. *H. h. 'Gold Heart'* syn. *H. h. 'Jubilee'* Foliage of same size as *'Glacier'* with distinct diamond-shape. Dark glossy green with central yellow variegation. *H. h. 'Hibernica'* (Irish Ivy) Five-lobed leaves, dark glossy green, 2¾-6in (7-15cm) long and wide. Strong-growing, one of the best for ground cover. Extremely hardy, and tolerating maximum shade.

There are a number of attractive ornamental varieties which can be used for ground cover and in fact any variety will perform this duty.

Average height and spread
Five years
1x13ft (30cmx4m)
Ten years
1x16ft (30cmx5m)
Twenty years or at maturity
1x20ft (30cmx6m)

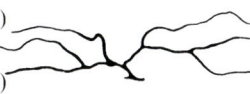

HEDERA
Shrub-forming varieties

SHRUBBY IVIES
Araliaceae
Evergreen
Attractive and useful, upright evergreen shrubs.

Origin From various areas of both the northern and southern hemispheres.
Use As slow-growing, upright, shrub-forming evergreens.
Description *Flower* Heads of green to lime green flowers, borne mainly early to late spring but either in bud or open throughout year. *Foliage* Leaves small to large, diamond-shaped, 1½-5in (4-12cm) long, with glossy upper surface, duller underside, ranging from dark through mid green to golden or silver variegated forms. *Stem* Densely foliaged, upright stems. Some varieties with slightly looser habit. Slow growth rate. *Fruit* Round clusters of black poisonous fruits, autumn and winter.
Hardiness Tolerates 4°F (−15°C).
Soil Requirements Any soil, including very alkaline and very acid. Tolerates both dry and wet areas, although forms more growth on moister soils. In very dry areas should be

Hedera colchica 'Arborescens'

watered to help it establish itself.
Sun/Shade aspect Prefers medium shade, good from deep shade through to full sun.
Pruning None required.
Propagation and nursery production From semi-ripe cuttings or rooted natural layers. Purchase container-grown. Best planting heights 4-15in (10-40cm). May be difficult to find.
Problems Can be slow to attain any substantial size.
Varieties of interest *H. colchica 'Arborescens'* (Shrubby Persian Ivy) Foliage light green, oblong to ovate. Up to 3in (8cm) in width and length. Forming a dense, mound-shaped, evergreen clump. Good for flower and fruit production. *H. helix 'Arborescens'* (Shrubby Common Ivy) Dark green, broad, spreading mounds of foliage with flowers and fruit. *H. h. 'Conglomerata'* (Shrubby Common Ivy) Small foliage, lobed, with wavy edges no more than 1-1½in (3-4cm) in width. Dark green upper surface, silvery underside. Forms either small low spreading hummock or upright column, depending on training. Very slow growth rate. *H. h. 'Erecta'* Narrow shaped leaves, dark upper surface, duller undersides. Very slow-growing upright shoots. Small enough to be used in rock gardens and confined areas.
Average height and spread
Five years
2x2ft (60x60cm)
Ten years
3x2½ft (1mx80cm)
Twenty years or at maturity
4x3ft (1.2x1m)

HEDYSARUM MULTIJUGUM

KNOWN BY BOTANICAL NAME
Leguminosae
Deciduous
An attractive flowering shrub in which rose purple flowers are enhanced by the soft green foliage.

Origin From Mongolia.
Use As a freestanding, summer-flowering shrub or as a fan-trained wall shrub, particularly in less favourable areas.

Hedysarum multijugum in flower

Description *Flower* Racemes of openly spaced, small, rose purple pea-flowers, borne mainly during mid to late summer but produced intermittently throughout summer and early autumn. *Foliage* Leaves pinnate, 4-6in (10-15cm) long, with narrow leaflets, grey-green to sea-green, some yellow autumn colour. *Stem* Upright at first, becoming arching with age. Grey-green to green. Growth rate fast when new, slowing with age. *Fruit* Insignificant.

Hardiness Tolerates 14°F (−10°C), but stems may be killed back in winter; rejuvenation from base usually occurs in the following spring.
Soil Requirements Prefers light, open soil, dislikes extreme waterlogging but tolerates both alkalinity and acidity.
Sun/Shade aspect Full sun to very light shade.
Pruning None required but one-third or more of oldest wood may be removed in early spring, to encourage new shoots for flowering late summer.
Propagation and nursery production From softwood cuttings taken in early summer. Purchase container-grown from specialist nurseries. Best planting heights 1-2ft (30-60cm).
Problems Rather slow to establish. In areas where its hardiness may be suspect best not planted before late spring.
Average height and spread
Five years
4x5ft (1.2x1.5m)
Ten years
6x8ft (2x2.5m)
Twenty years or at maturity
10x12ft (3x3.5m)

HELIANTHEMUM

ROCK ROSE, SUN ROSE
Cistaceae
Evergreen
Although sometimes regarded as alpine plants, these are strictly small shrubs and should be treated accordingly.

Origin From central and southern Europe, mostly in mountain areas. Many varieties of garden origin.
Use As a low, ground-covering, carpeting, flowering shrub, especially effective cascading down short walls, in tubs and containers of adequate size, in which it may also be grown as an alpine shrub, and in large rock gardens.

Helianthemum nummularium 'Cerise Queen'

Description *Flower* Small, single or double, according to variety, flat saucer-shaped flowers borne profusely, early summer to mid summer, intermittently to mid autumn. In shades varying from white, through yellow, pink, orange and red, depending on variety. *Foliage* Leaves ovate, 1-2in (3-5cm) long, dark glossy green or grey-green, according to variety. *Stem* Slightly upright when young, quickly becoming spreading and arching, forming a low, spreading shrub. Grey to green-grey. Fast growth rate when young, slowing with age. *Fruit* Small, grey-brown seedheads, attractive in late autumn and winter sunlight.
Hardiness Tolerates 14°F (−10°C).
Soil Requirements Well-drained, open soil essential, any waterlogging causing shrub to fail. Tolerates acidity and mild alkalinity but

Helianthemum nummularium 'Yellow Queen'

distressed by extreme alkalinity.
Sun/Shade aspect Prefers full sun, tolerates very light shade. Deeper shade rapidly spoils compact habit and reduces flowering.
Pruning Trim lightly with hedging shears, early to mid spring, to induce new healthy flowering shoots in summer.
Propagation and nursery production From semi-ripe cuttings taken in early summer. Purchase container-grown. Best planting heights 4-6in (10-15cm).
Problems Plants are always rather small when sold, but grow rapidly once planted out.
Varieties of interest *H. alpestre* syn. *H. serpyllifolium* Bright yellow flowers and grey-green foliage. From the mountains of southern and central Europe. *H. lunulatum* Yellow, single flowers with orange-stained throats. Very low and compact. Originating in Italy. *H. nummularium 'Afflick'* Deep orange single flowers with darker centres. Green foliage. *H. n. 'Amy Baring'* Deep yellow to buttercup yellow single flowers. Good compact form with green foliage. *H. n. 'Ben Hope'* Carmine single flowers with deep orange centres. Light grey-green foliage. *H. n. 'Ben Ledi'* Bright, deep rose to pink single flowers. Glossy, dark green foliage. *H. n. 'Ben More'* Orange single flowers with darker centres. Dark green foliage. *H. n. 'Ben Nevis'* Yellow single flowers with bronze centres. Green foliage. *H. n. 'Cerise Queen'* Rosy-red double flowers. Green foliage. *H. n. 'Cherry Pink'* Deep pink single flowers. Grey foliage. *H. n. 'Fireball'* syn. *H. n. 'Mrs. Earl'* Deep orange, double flowers. Green foliage. *H. n. 'Fire Dragon'* syn. *H. n. 'Mrs Clay'* Flame-orange single flowers. Neat green foliage. *H. n. 'Henfield Brilliant'* Deep orange single flowers. Green foliage. *H. n. 'Jubilee'* Primrose yellow, double flowers. Green foliage. *H. n. 'Lemon Queen'* Lemon yellow single flowers. Green foliage. *H. n. 'Old Gold'* Golden yellow single flowers. Green foliage. Reaching only 2in (5cm) and forming low, covering mat. *H. n. 'Raspberry Ripple'* Rose pink single flowers with white tips. Grey-green foliage. *H. n. 'Rhodanthe Carneum'* syn. *H. n. 'Wisley Pink'* Pale pink single flowers with orange centres. Grey-green foliage. *H. n. 'Rose Perfection'* Soft rose pink single flowers. Grey-green foliage. *H. n. 'Salmon Bee'* Salmon-pink single flowers. Grey foliage. *H. n. 'St. John's College'* Bright golden yellow single flowers. Green foliage. *H. n. 'Sudbury Gem'* Deep rose pink single flowers with flame red centres. Green foliage. *H. n. 'The Bride'* White single flowers with yellow blotches. Green foliage. *H. n. 'Wisley Primrose'* Primrose yellow single flowers. Grey foliage. *H. n. 'Wisley Yellow'* Yellow single flowers. Grey foliage. *H. n. 'Wisley White'* White single flowers. Grey foliage. *H. n. 'Yellow Queen'* Bright lemon yellow single flowers.
Average height and spread
Five years
4inx2¼ft (10x70cm)
Ten years
8inx3ft (20cmx1m)
Twenty years
or at maturity
1x3ft (30cmx1m)

HELICHRYSUM

KNOWN BY BOTANICAL NAME
Compositae
Evergreen
Useful in the range of grey foliage shrubs, if its tenderness is taken into account and it is not relied upon as a permanent feature.

Origin Some varieties from southern Europe, others from South Africa and New Zealand.
Use On its own as a spot plant to highlight bedding or as a grey-leaved shrub for front edging of borders. Good in containers, *H. petiolatum* in particular for hanging baskets.
Description *Flower* Clusters of lemon to golden yellow flowers, depending on variety, of less interest than foliage. One variety, *H. splendidum*, is an everlasting flower, retaining its flowerheads well into winter. *Foliage* Leaves, the main attraction, lanceolate to ovate, ½-1½in (2-4cm) long, silver-grey, often giving off aromatic scent when crushed. *Stem* Upright at first, some laxer growth late summer, early autumn, forming medium-height, round-topped, spreading shrub. Grey-green to silver-grey. Fast growth rate when young, slowing with age. *Fruit* Grey-brown seedheads have some attraction.
Hardiness Minimum winter temperature 23°F (−5°C).
Soil Requirements Well-drained soil, dislikes waterlogging.
Sun/Shade aspect Prefers full sun, dislikes any shade.
Pruning Annual light to medium trimming in late spring will encourage new growth. Can be reduced to ground level in mid to late spring, and normally will rejuvenate.
Propagation and nursery production From semi-ripe cuttings taken late spring, early summer. Purchase container-grown. Advisable to propagate some stock each summer to carry through under protection for replanting following spring in case the parent plant succumbs to winter. Best planting heights 3-8in (8-20cm).
Problems Plants may become old and woody, in which case reduce to ground level and they will rapidly rejuvenate.
Varieties of interest *H. petiolatum* Good-sized, round to ovate, very silvery foliage with grey, woolly covering, produced on long, trailing stems also covered with white woolly down; yellow flowers. Good for hanging baskets and large tubs, or as summer bedding in very mild areas where temperature can be relied upon to stay above freezing point. Of average height with possibly one-third more spread. From South Africa. Many new cultivars of *H. petiolatum* also available, with varying types of silver and yellow-silver foliage. *H. plicatum* Long, lanceolate, silver-grey foliage, giving yellow autumn tints. Grey, upright stems with yellow, terminal

Helichrysum splendidum **in flower**

clusters of flowers, mid to late summer. From south-eastern Europe. *H. serotinum* syn. *H. angustifolia* (Curry plant) Thick, closely produced, upright stems, with thin narrow leaves sage green to silver-grey; when crushed giving off a strong curry-like scent. One of the hardiest forms. From southern Europe. *H. splendidum* (Everlasting Immortelles) Small, ovate, grey-green leaves, symmetrically displayed around upright, grey-green stems, forming a globe shape up to 3ft (1m) in height and spread in favourable areas. Yellow flowers retained dried on the plant until midwinter. From South Africa.
Average height and spread
Five years
2x2½ft (60x80cm)
Ten years
2x2½ft (60x80cm)
Twenty years
or at maturity
2x2½ft (60x80cm)

HIBISCUS SYRIACUS

TREE HOLLYHOCK, FLOWER OF AN HOUR, ROSE OF SHARON, SHRUB ALETHEA
Malvaceae
Deciduous
A very stately, attractive, late summer-flowering shrub.

Origin From eastern Asia.
Use As late summer, early autumn-flowering shrub of great distinction, for planting on its own to show off its shape.

Hibiscus syriacus 'Blue Bird' **in flower**

Description *Flower* Large, trumpet-shaped flowers, single or double according to variety, ranging through white, pink, blue, red and combinations of these colours, borne late summer to mid autumn; each lasting only one day and succeeded by others the next. *Foliage* Leaves dentated, ovate, 1½-2½in (4-6cm) long, often not produced until early summer, light grey to mid green, giving some yellow autumn colour. *Stem* Upright, forming a tall, upright, globe-shaped shrub. Grey to grey-green, of some winter attraction. Slow growth rate. *Fruit* Insignificant.
Hardiness Tolerates 4°F (−15°C), but flower buds may be damaged by severe late spring frosts in some localities.
Soil Requirements Open, well-drained soil; does well on slightly alkaline through acid conditions.
Sun/Shade aspect Prefers very light shade, tolerates full sun.
Pruning None required. On very mature shrubs individual branches may be reduced to control size.
Propagation and nursery production From semi-ripe cuttings taken in midsummer, or from layers. Some named varieties grafted on to seedling-produced understocks. Purchase container-grown or root-balled (balled-and-burlapped). Specimens 3-4 years old, although more costly, are best to buy because younger plants often fail. Most varieties must be sought from specialist nurseries, but some stocked by garden centres, especially when in

Hibiscus syriacus 'Woodbridge' in flower

flower. Best planting heights 15in-2½ft (40-80cm).
Problems Often thought to be dead in spring, due to very late leaf formation.
Varieties of interest *H. s. 'Admiral Dewey'* Double, pure white flowers. *H. s. 'Ardens'* syn. *H. 'Caeruleus Plenus'* Very large, double flowers; pale rose-purple petals with maroon blotch at base. *H. s. 'Blue Bird'* Clear violet-blue single flowers with darker central shading. *H. s. 'Duc de Brabant'* Double, dark red flowers. Late-flowering. *H. s. 'Hamabo'* Large, single, pale rose flowers with darker red central blotches. Not to be confused with a distinct form called *H. hamabo* which is not hardy. *H. s. 'Jeanne d'Arc'* Semi-double, clear white flowers. *H. s. 'Lady Stanley'* syn. *H. s. 'Elegantissimus'* White double flowers with pink shading and red central blotch. Early flowering. *H. s. 'Leopoldi'* Semi-double pink and white flowers with red speckles. Rather hard to find. *H. s. 'Mauve Queen'* Single, mauve-blue flowers. *H. s. 'Monstrosus'* Single, white flowers with maroon blotch in centre. *H. s. 'Pink Giant'* Large, single, pink flowers. *H. s. 'Red Heart'* Single, pure white flowers with prominent red blotch in centre of petals. *H. s. 'Rubus'* Single, red flowers. A compact growing variety. *H. s. 'Russian Violet'* Single, deep violet-pink flowers. Good for very cold areas. *H. s. 'Speciosus'* Double, white flowers with red centres. *H. s. 'Violet Clair Double'* syn. *H. s. 'Puniceus Plenus', H. s. 'Roseus Plenus'* Large, double, purple-blue flowers. *H. s. 'William E. Smith'* Large, single, pure white flowers. *H. s. 'Woodbridge'* Large, single, bright pink-red to bright red flowers.
Average height and spread
Five years
3x3ft (1x1m)
Ten years
5x5ft (1.5x1.5m)
*Twenty years
or at maturity*
6x6ft (2x2m)

HIPPOPHAE RHAMNOIDES

SEA BUCKTHORN
Elaeagnaceae
Deciduous
A very useful, quick-growing screening plant which should be more often considered for windbreaks and coastal protection, although it is invasive.

Origin From Europe through temperate Asia.
Use As a silver-leaved shrub on its own or for mass planting. Especially good for windbreaks in seaside areas. May also be trained as small, single or multi-stemmed tree.

Description *Flower* Small, inconspicuous sulphur yellow flowers, male or female, singly sexed on individual plants. *Foliage* Leaves lanceolate to narrowly elliptic, 1-2½in (3-6cm) long, silver grey to grey-green, giving some good yellow autumn colour. *Stem* Strong, upright, quickly becoming spreading, forming a spreading, large, dense thicket grey to grey-green, becoming grey-brown. Branches armed with short, sharp spines. Fast growth rate when young, slowing with age. *Fruit* Female shrubs may produce round, orange-yellow berries in profusion; if fruit is desired they must be planted in groups including male shrubs.

Hippophae rhamnoides in fruit

Hardiness Tolerates 4°F (−15°C).
Soil Requirements Prefers light, sandy soils, tolerating wide range of conditions, both alkaline and acid.
Sun/Shade aspect Full sun or very light shade to maintain good silver foliage. Good in coastal areas.
Pruning None required but if cut to ground level will regenerate following spring, although female shrubs may not fruit for a season or two.
Propagation and nursery production From seed. Regrettably, to date, nurseries do not sex shrubs and propagate from cuttings vegetatively, so when purchasing stock one cannot be sure of obtaining a female or male shrub. Plant container-grown for best results; also available bare-rooted. Fairly easy to find. Best planting heights 1½-3ft (50cm-1m).
Problems May become invasive where it is extremely happy, both by germination of seeds and by far-ranging underground suckers. Planted in wild, unspoiled areas it may take over and become a pest.

Average height and spread
Five years
6x6ft (2x2m)
Ten years
13x13ft (4x4m)
*Twenty years
or at maturity*
18x18ft (5.5x5.5m)

HOHERIA

KNOWN BY BOTANICAL NAME
Malvaceae
Deciduous
Interesting, very free-flowering, handsome shrubs, but not without problems.

Origin From New Zealand.
Use As a tall, midsummer-flowering shrub, requiring some protection in winter. Good grown on a sunny wall, which provides shelter.
Description *Flower* Clusters of round, white, saucer-shaped, fragrant flowers, midsummer. *Foliage* Leaves ovate, 2-4½in (5-11cm) long, sometimes toothed and lobed, especially when young. Grey to grey-green, giving some yellow autumn colour. *Stem* Upright, becoming branching and spreading with age, forming a tall, round-topped shrub. Grey to grey-green, aging to grey-brown. Medium growth rate. *Fruit* Insignificant.
Hardiness Minimum winter temperature 23°F (−5°C).
Soil Requirements Prefers deep, rich, light soil, acid to neutral, but tolerates some alkalinity.
Sun/Shade aspect Prefers light shade, tolerates full sun.
Pruning None required but any over-sized branches may be pruned back.
Propagation and nursery production From softwood cuttings or layers. Purchase container-grown. All varieties must be sought from specialist nurseries. Best planting heights 1¼-3ft (40cm-1m).
Problems Appears quite tough, so its tenderness is not always appreciated. Stems can be damaged quite severely in winter.
Varieties of interest *H. glabrata* syn. *Gaya lyallii, Plagianthus lyallii* White flowers, very heavily produced in clusters on large shrub which in very mild areas may form a small tree. Reputed to be more hardy than most. *H. lyallii* syn. *Gaya lyallii, Gaya lyallii ribifolia, Plagianthus lyallii* Clusters of white flowers borne in profusion, midsummer. Foliage more glabrous grey than most. Somewhat variable in growth production, forming a large shrub or small tree. The most widely available variety. *H. populnea* Pure, white flowers, 1in (3cm) across, in large clusters, flowering later than most varieties. Large, broad, ovate leaves very similar to poplar

Hoheria lyallii in flower

leaves. *H. p. 'Foliis Purpureis'* White flowers. Leaves plum-coloured on undersides. Very scarce. *H. p. 'Variegata'* White flowers. Leaves yellow-green, aging to white, with deep green margins. Very scarce. *H. sexstylosa* Pure white flowers, narrow mature leaves, borne on large upright shrub or small tree of variable growth pattern. Said by some authorities to be more hardy than most, although it seems wisest to treat all varieties as tender.

There are a number of other varieties, most of which are extremely tender or even more scarce in production.

Average height and spread
Five years
6x6ft (2x2m)
Ten years
12x12ft (3.5x3.5m)
Twenty years or at maturity
14½x14½ft (4.5x4.5m)

HOLODISCUS DISCOLOR (Spiraea discolor)

OCEAN SPRAY
Rosaceae
Deciduous
An attractive, summer-flowering shrub which may cause some difficulty in establishing, but once established, quickly becomes a large, clump-forming shrub.

Origin From North America.
Use As a freestanding, flowering shrub or for mass planting to create a thicket effect.
Description *Flower* Large, feathery panicles of creamy white flowers, midsummer. *Foliage* Leaves pinnate 2-4in (5-10cm) long, grey-green with grey-white undersides, giving some yellow autumn colour. *Stem* Upright, becoming arching, green-brown. Normally all stems are formed from a basal crown or from suckers. *Fruit* Flowers give way to brown seedheads, maintained into late autumn, early winter.
Hardiness Tolerates −13°F (−25°C), but susceptible to frost while young and unestablished.
Soil Requirements Does well on most soils, both alkaline and acid, and tolerates quite moist conditions. Good on heavy clay, once established.
Sun/Shade aspect Prefers light to medium shade; tolerates wide range of sun and shade.
Pruning None required except on very old shoots which are best cut to ground to induce new basal growth.
Propagation and nursery production From semi-ripe cuttings taken in summer, or more often by removal of rooted suckers from

parent plant. Purchase container-grown. If hard to find should be sought from specialist nurseries. Best planting heights 1¼-2½ft (40-80cm).
Problems Not easy to raise as young plant but once established is very easy to grow.

Average height and spread
Five years
6x6ft (2x2m)
Ten years
10x12ft (3x3.5m)
Twenty years or at maturity
10x20ft (3x6m)

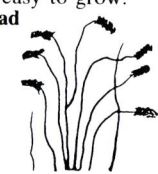

HYDRANGEA ARBORESCENS

SMOOTH HYDRANGEA, TREE HYDRANGEA
Hydrangeaceae
Deciduous
An interesting, little known group of Hydrangeas well worth planting.

Origin From eastern USA.
Use As a flowering shrub, often grown alongside perennials and also suitable for woodland areas. Can be grown in large containers not less than 2½ft (80cm) across and 2ft (60cm) deep. Use good potting medium and feed during growing season.
Description *Flower* Large, round clusters, up to 6in (15cm) across with many creamy white ray florets, green at first, from summer through to early autumn. Dead flowerheads retained into winter. *Foliage* Leaves large, 3-7in (8-18cm) long, elliptic, with grey-green upper surfaces and glabrous grey undersides. *Stem* Upright when young, becoming spreading, forming a tall, dome-shaped shrub, somewhat open and lax. Bright green to grey-green. Medium growth rate. *Fruit* Insignificant.
Hardiness Tolerates 4°F (−15°C).
Soil Requirements Prefers good loam soil but does well on most soil types, both acid and alkaline.
Sun/Shade aspect Performs best in medium to light shade; tolerates deep shade to full sun.
Pruning None required, but any sickly, late-produced branches may be removed. Flowerheads should be left to drop off, as new flower bud for following year is borne in terminal of stem, just behind old flower bud.
Propagation and nursery production From layers or semi-ripe cuttings taken early summer. Purchase container-grown. Fairly easy to find in nurseries. Best planting heights 1¼-2½ft (40-80cm).
Problems When purchased as young plant tends to look sickly but when planted out quickly shows its full potential.
Varieties of interest *H. a. 'Annabelle'* Good,

Hydrangea arborescens 'Annabelle' in flower

tight, round, very large flowers, up to 10in (25cm) across. Reaches two-thirds average height and spread; the best form for most gardens. Of garden origin. *H. a. 'Grandiflora'* A very attractive form with large flowers up to 8in (20cm) across. Possibly more lax in habit of growth.

Average height and spread
Five years
3x5ft (1x1.5m)
Ten years
6x8ft (2x2.5m)
Twenty years or at maturity
10x12ft (3x3.5m)

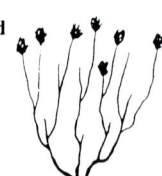

HYDRANGEA ASPERA

ROUGH-LEAVED HYDRANGEA
Hydrangeaceae
Deciduous
One of the aristocrats of the Hydrangea group and a truly delightful sight when in full flower in late summer.

Origin From western and central China, Taiwan and the Himalayas.
Use As a freestanding shrub, either singly or mass planted, especially in shady, woodland areas. Can be fan-trained on a cool, exposed wall where it looks particularly well.
Description *Flower* Large, porcelain blue clusters with ring of lilac-pink or white ray florets on outer edges; midsummer, maintained into early autumn. *Foliage* Leaves large to very large, 4-9in (10-22cm) long and 3-6in (8-15cm) wide, ovate, purple-green, down-covered with silver undersides; variable in length and width. *Stem* Cream and brown to cream and purple-brown stems.

Holodiscus discolor in flower

Hydrangea aspera in flower

Upright when young, producing attractive coloured bark and becoming branching, gnarled and spreading to form a dome-shaped shrub. Stem formation and colour of interest in winter. Medium growth rate. *Fruit* None, but may retain old flowerheads with ray florets turning brown.
Hardiness Tolerates 14°F (−10°C).
Soil Requirements Neutral to acid soil, tolerates some alkalinity.
Sun/Shade aspect Medium to light shade. Dislikes full sun and deep shade may spoil shape.
Pruning None required, and in fact may resent it.
Propagation and nursery production From semi-ripe cuttings taken in early summer. Purchase container-grown. Rather hard to find, must be sought from specialist nurseries. Best planting heights 1¼-2½ft (40-80cm).
Problems Often looks uninteresting when purchased but grows rapidly once planted out.
Average height and spread
Five years
5x5ft (1.5x1.5m)
Ten years
8x8ft (2.5x2.5m)
Twenty years
or at maturity
12x12ft (3.5x3.5m)

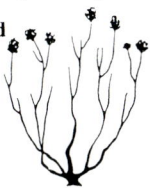

HYDRANGEA INVOLUCRATA

KNOWN BY BOTANICAL NAME
Hydrangeaceae
Deciduous
A delightful little gem of a plant, not easy to obtain but worth seeking out.

Origin From Japan and Formosa.
Use As a low, summer-flowering shrub. Can be used in large containers 2½ft (80cm) across and 2ft (60cm) deep, if a good potting medium is used and adequate feeding is given in the growing season.

Hydrangea involucrata **in flower**

Description *Flower* Bell-shaped flowers, blue to rosy lilac, depending on soil type, with white or pale blue outer ray florets sometimes tinted pink. Actual flower colour determined by acidity or alkalinity of soil; more acid soils give rise to paler colour. Flowering summer to early autumn, flowerheads maintained well into winter. *Foliage* Leaves elliptic, 4-9in (10-22cm) long, light to mid green. Some yellow autumn colour. *Stem* Grey to grey-green, becoming creamy brown. Short and branching, gradually forming a dome-shaped, spreading shrub. Medium growth rate. *Fruit* None.
Hardiness Tolerates 4°F (−15°C) but new spring growth may be damaged if late spring frosts occur as foliage is opening. Normally

regenerates new growth.
Soil Requirements Grows well on most soil conditions, but shows some distress on extremely alkaline types.
Sun/Shade aspect Light shade. Dislikes full sun or very deep shade.
Pruning None required.
Propagation and nursery production From softwood cuttings. Purchase container-grown. Must be sought from specialist nurseries. Best planting heights 1¼-2½ft (40-60cm).
Problems Very difficult to find. Very slow growing and difficult to find good propagating wood, hence its scarcity.
Varieties of interest *H. i. 'Hortensis'* Double, white to creamy white florets, aging to rose-tinted and at stages having apricot colouration. Very difficult to find. Of garden origin.
Average height and spread
Five years
2x2ft (60x60cm)
Ten years
3x3ft (1x1m)
Twenty years
or at maturity
4x4ft (1.2x1.2m)

HYDRANGEA MACROPHYLLA
Hortensia varieties

BIGLEAF HYDRANGEA, MOP-HEADED HYDRANGEA
Hydrangeaceae
Deciduous
One of the more spectacular summer-flowering shrubs, worthy of any garden, although there is some colour variation and a fairly precise planting position is required.

Origin Basic forms from Japan; most varieties now offered are of garden or nursery origin.
Use As a freestanding, flowering shrub, either singly, massed or in groups. Can be used in large containers not less than 2½ft (80cm) across and 2ft (60cm) deep but must be given adequate in-season feeding.
Description *Flower* Large, round, sterile heads, ranging through white, pink, red, blue or combinations of these, aging to darker, more metallic shades from midsummer to well into winter, when flowers turn bronze-brown. Exact shade depends both on variety and on soil in which it is grown. Many forms on acid soil will be blue, on neutral soil red and on alkaline soil pink. Plants may be shy to flower unless planted as indicated by soil, sun and shade requirements. *Foliage* Leaves ovate, 4-8in (10-20cm) long and 6-8in (15-20cm) wide, light to mid green, sometimes dark green on rich moist soil. Yellow autumn

colouring. *Stem* Creamy brown mature stems, upright, branching and forming a round, tall, mound-shaped shrub. Medium to fast growth rate. *Fruit* None.
Hardiness Tolerates 4°F (−15°C), although late spring frosts may damage new foliage.
Soil Requirements Deep, moist soil. Any drying out will be shown by poor growth and flagging foliage. Water freely in time of drought.
Sun/Shade aspect Light shade. Dislikes both full sun and deep shade, which spoils shape and reduces flowering.
Pruning None required. Oversized shrubs may be cut to ground level, but will take 3-4 seasons to come back to full flowering. On mature plants thin one-third of weakest growth annually, in very cold areas not before early spring.

Hydrangea macrophylla 'Miss Belgium'

Propagation and nursery production From softwood cuttings. Purchase container-grown. Most common forms fairly easy to find in garden centres but some forms may need searching for in specialist nurseries. Best planting heights 1½-2½ft (50-80cm).
Problems Light shade and good, rich, moist soil are essential requirements for good flowering.
Varieties of interest Because the flower colour varies according to alkaline or acid conditions, the first colour given below is that for alkaline soil, the second that which can be expected on acid soil. *H. m. 'Altona'* Cherry pink or mid blue. *H. m. 'Ami Pasquier'* Crimson or purple-red. *H. m. 'Amethyst'* Pink or purple. *H. m. 'Ayesha'* Pink or grey-lilac. *H. m. 'Blue Prince'* Rose red or cornflower blue. *H. m. 'Deutschland'* Pale pink or pink-blue. *H. m. 'Europa'* Pale pink-blue or clear

Hydrangea macrophylla 'Generale Vicomtesse de Vibraye' **in flower**

blue. *H. m. 'Generale Vicomtesse de Vibraye'* Pink or clear blue. *H. m. 'Hamburg'* Pink-rose or purple-rose. *H. m. 'Harry's Red'* Red or deep red. *H. m. 'Holstein'* Pink or sky blue. *H. m. 'King George'* Pale pink or dark, rich, rose pink. *H. m. 'Klus Supreme'* Deep pink or dark blue. *H. m. 'Madame Emile Mouilliere'* White with some pink shading or pure white. *H. m. 'Maréchal Foch'* Deep pink or deep blue. *H. m. 'Miss Belgium'* Red to rose red. *H. m. 'Niedersachsen'* Pale pink or pale blue. *H. m. 'President Doumer'* Purple-red or blue. *H. m. 'Sister Teresa'* Pure white on both soil types.

Average height and spread
Five years
3x3ft (1x1m)
Ten years
6x6ft (2x2m)
Twenty years
or at maturity
10x10ft (3x3m)

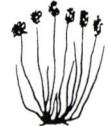

HYDRANGEA MACROPHYLLA
Lacecap varieties

LACECAP HYDRANGEA
Hydrangeaceae
Deciduous
Given the right soil conditions and required shade, these later summer-flowering shrubs provide a most effective display.

Origin From Japan, but most varieties of garden or nursery origin.
Use As an individual shrub, for mass planting, good for shady areas. Or for planting in tubs not less than 2½ft (80cm) across and 2ft (60cm) deep. Use good potting medium and feed adequately in season. Can be forced for the purpose of making a houseplant and planted out after flowering.

Hydrangea macrophylla 'Blue Wave'

Description *Flower* Round central clusters of flowers, white, through pink to blue surrounded by bold ray florets, late summer and early autumn. *Foliage* Leaves large, ovate, 4-8in (10-20cm) long and 4-6in (10-15cm) wide, light to mid green, giving some plum autumn colour. *Stem* Upright when young, becoming spreading with age, forming a wide, spreading, dome-shaped shrub. Grey to creamy brown. Medium to fast growth rate. *Fruit* None. Dead brown flowerheads may be maintained well into autumn.
Hardiness Tolerates 4°F (−15°C); late spring frosts may damage newly opened foliage, but this readily rejuvenates.
Soil Requirements Prefers rich, deep acid soil; tolerates wide range but dislikes very dry soils. Blue flower colour is turned pink by alkaline soils. White varieties maintain colour on both types of soil.
Sun/Shade aspect Light shade. Dislikes both

Hydrangea macrophylla 'White Wave'

full sun and deep shade. Lack of light spoils shape and dryness leads to very poor results.
Pruning Thin lightly, removing older, weaker shoots each winter, or in spring in colder areas. Avoid cutting back hard as this stops the following year's flowering.
Propagation and nursery production From cuttings. Purchase container-grown. All varieties fairly easy to find, especially when in flower. Best planting heights 1½-2½ft (50-80cm).
Problems Requirements of light shade and good, rich, moist soil must be met for full flowering.
Varieties of interest *H. m. 'Blue Wave'* Central, fertile, tufted flowers with numerous outer, large ray florets. Pink on alkaline soils, blue on acid, almost gentian blue in very acid conditions. *H. m. 'Lanarth White'* Central fertile flowers pink on alkaline soils, blue on acid, with outer ray florets pure white. Slightly more compact than most. *H. m. 'Lilacina'* Central flowers amethyst blue with pink ray florets on alkaline or blue on acid soils. May be difficult to find. *H. m. 'Maculata'* syn. *H. m. 'Variegata'* Flowers white to pink-white in colour. May be blue-white on acid soils. Leaves green to grey-green with creamy white margins. *H. m. 'Mariesii'* Central fertile flowers rosy pink to blue-pink on acid soils, with outer ray florets varying shades of blue, or pink on alkaline soils. *H. m. 'Sea Foam'* Blue, fertile central flowers surrounded by white ray florets. Very good in coastal areas but not fully hardy inland. Not always easy to find. *H. m. 'Tricolor'* Flowers pale pink on alkaline soil or pure white on acid, slightly smaller than those of most varieties and less reliably produced. Good, mid green foliage with grey or pale yellow outer variegation.

H. m. 'Veitchii' Central fertile flowers pink on alkaline soils and blue on acid, having outer white florets which fade to pink. One of the hardiest varieties. *H. m. 'White Wave'* syn. *H. m. 'Mariesii Alba'*, *H. m. 'Mariesii Grandiflora'* Central fertile flowers pink to lilac-blue on acid soils, large white outer ray florets. Dark green, large, healthy foliage. A very good growing form.

Average height and spread
Five years
3x4ft (1x1.2m)
Ten years
5½x6ft (1.8x2m)
Twenty years
or at maturity
5½x12ft (1.8x3.5m)

HYDRANGEA PANICULATA

PANICK HYDRANGEA
Hydrangeaceae
Deciduous
Spectacular, late summer-flowering shrubs, requiring specific pruning to do well.

Origin From Japan, China and Taiwan.
Use As a freestanding shrub planted singly or in a shrub border, for mass planting or fan-trained on a wall. Can be trained or purchased as small, short standard, but needs protected garden to prevent wind damage.
Description *Flower* Large panicles of sterile white bracts, mid to late summer, fading to pink, finally to brown. *Foliage* Leaves broad, ovate, 3-6in (8-15cm) long and 1½-3in (4-8cm) wide, light green, giving yellow autumn colour. *Stem* Upright and brittle. Light green to green-brown, becoming mahogany-brown with age. Fast growth rate. *Fruit* None, but dead flowerheads maintained into early winter.
Hardiness Tolerates 4°F (−15°C).
Soil Requirements Prefers rich, deep soil; tolerates wide range including alkaline or acid but dislikes extreme alkalinity.
Sun/Shade aspect Prefers light shade, tolerates full sun. Deep shade spoils shape.
Pruning Ensure new growth and improved flowering by cutting back hard each spring, reducing all previous year's wood back to two buds from point of origin. Otherwise the shrub becomes old, woody and very brittle, with small flowers.
Propagation and nursery production From softwood cuttings taken in early summer. Purchase container-grown or root-balled (balled-and-burlapped). Easy to find, especially when in flower. Best planting heights 15in-3ft (40cm-1m).

Hydrangea paniculata 'Grandiflora' in flower

Problems As the shrub is very brittle take care to avoid damaging it.
Varieties of interest *H. p. 'Grandiflora'* (Pee Gee Hydrangea) Very large white panicles of flowers, fading to pink. The most common form and the one most grown as a standard. *H. p. 'Praecox'* Earlier, white flowers, although slightly less profuse, larger foliage and stronger stems. *H. p. 'Tardiva'* A later-flowering form, with slightly more open flowers and a wider base to the panicles.
Average height and spread
Five years
6x6ft (2x2m)
Ten years
10x10ft (3x3m)
Twenty years
or at maturity
13x13ft (4x4m)

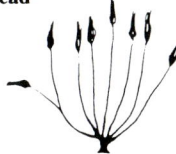

HYDRANGEA PETIOLARIS

CLIMBING HYDRANGEA
Hydrangeaceae
Deciduous
A good ground cover shrub which, though deciduous, is very twiggy and mat-forming, excellent for dry, difficult soils due to its wide, spreading habit.

Origin From Japan and South Korea.
Use Either as a ground-covering, low shrub for banks and large areas or as a climber for exposed walls or on trees, where it tolerates deep shade and is extremely hardy.

Hydrangea petiolaris in flower

Description *Flower* Flat inner heads of small, white, fertile flowers, surrounded by ring of white ray florets, slightly open in habit. Aging to pink, finally to brown and retained well into winter. *Foliage* Leaves pointed, ovate, 2-4½in (5-11cm) long, bright green with good yellow autumn colour. *Stem* Light green, quickly aging to mahogany-brown in winter, with some peeling bark revealing paler browny-cream underskin. Buds pointed and attractive. Forms a twiggy, very wide-spreading, ground-hugging carpet. Slow growth rate when young, becoming faster once established. *Fruit* None, but retains the dead flowerheads into winter.
Hardiness Tolerates winter temperatures down to −13°F (−25°C).
Soil Requirements Does well on all soils including acid and alkaline.
Sun/Shade aspect Full sun through to medium shade.
Pruning None required other than to control size.
Propagation and nursery production From semi-ripe cuttings taken in summer or from layers produced by established plants. Purchase container-grown. Easy to find in nurseries where it may be supplied tied to an upright cane; the shrub rapidly forms a ground-

hugging carpet once cane is removed. Best planting heights 8in-2ft (20-60cm).
Problems Can be very brittle and easily damaged during mechanical cultivation.
Average height and spread
Five years
2x6ft (60cmx2m)
Ten years
2½x13ft (80cmx4m)
Twenty years
or at maturity
3x20ft (1x6m)

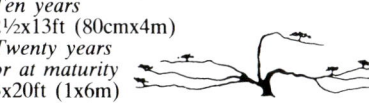

HYDRANGEA QUERCIFOLIA

OAK LEAF HYDRANGEA
Hydrangeaceae
Deciduous
A very good autumn-flowering shrub of interesting shape, with foliage giving very good autumn colours.

Origin From south-eastern USA.
Use As a medium-sized, autumn-flowering shrub with extremely good autumn foliage colours, freestanding or for mass planting. Can be trained on an exposed or shady wall where it will attain 3ft (1m) more than average height and possibly an extra 5ft (1.5m) in width over 5 years.
Description *Flower* Panicles of white florets produced from midsummer, maintaining their white texture well into early autumn when they begin to turn pink and finally, in winter, brown. *Foliage* Leaves round to ovate, 4-8in (10-20cm) long and 2-4in (5-10cm) wide, with large pointed lobes similar in shape to oak leaves. Dark green, deeply veined, turning vivid orange-red in autumn. *Stem* Upright when young, becoming lax and ranging. Brown, peeling bark. Very brittle in constitution. Medium growth rate. *Fruit* None. Dead flowerheads maintained into winter.
Hardiness Tolerates 14°F (−10°C).
Soil Requirements Does well on most soils, both alkaline and acid, tolerating equally considerable dryness or moisture.
Sun/Shade aspect Full sun to deep shade.
Pruning None required, but may be reduced in size and quickly rejuvenates.
Propagation and nursery production From semi-ripe cuttings taken midsummer. Purchase container-grown. Easy to find in nurseries, less common in garden centres. Best planting heights 1¼ft-2ft (40-60cm).

Hydrangea quercifolia in flower

Problems The wood is brittle and young plants when purchased look extremely mis-shapen, but once planted they quickly form good, round shrubs.
Average height and spread
Five years
3x4ft (1x1.2m)
Ten years
5x6ft (1.5x2m)
Twenty years
or at maturity
6x8ft (2x2.5m)

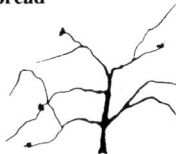

HYDRANGEA SARGENTIANA

SARGENT'S HYDRANGEA
Hydrangeaceae
Deciduous
A handsome structural shrub, worthy of a place in a sheltered position.

Origin From China.
Use As a freestanding shrub in its own right or as a focal point among lower ground-covering shrubs, or as a large wall shrub.
Description *Flower* Large, flat, clusters of blue central fertile flowers, surrounded by white ray florets, slightly pink on alkaline soils; mid to late summer. *Foliage* Leaves large to very large, pointed, ovate, 6-10in (15-25cm) long and 3-7in (8-18cm) wide, and velvety textured. Red-green, giving some yellow autumn colour. *Stem* Strong, upright branches. Young shoots covered with a thick, hairy covering. Black terminal winter buds. Medium growth rate, slowing with age. *Fruit* None. Retains sterile florets into winter.

Hydrangea sargentiana in flower

Hardiness Tolerates 14°F (−10°C), but needs protection from wind chill; even so, some terminal die-back may occur.
Soil Requirements Prefers moist, rich soil. Tolerates mildly alkaline or acid conditions.
Sun/Shade aspect Prefers medium or light shade, dislikes full sun.
Pruning None required, but any winter-damaged shoots should be removed.
Propagation and nursery production From cuttings or layers taken early summer. Purchase container-grown or root-balled (balled-and-burlapped). Rather hard to find. Best planting heights 1¼-2½ft (40-80cm).
Problems Appearance of young shrub is unattractive, consisting of a root system and single short stem less than 1½ft (50cm) high.
Average height and spread
Five years
4x4ft (1.2x1.2m)
Ten years
6x6ft (2x2m)
Twenty years
or at maturity
10x10ft (3x3m)

HYDRANGEA SERRATA

KNOWN BY BOTANICAL NAME
Hydrangeaceae
Deciduous
Given the right acid soil conditions, *H. serrata* varieties form an extremely attractive group of shrubs for late summer flowering.

Origin From Japan and Korea.
Use As an individual freestanding shrub for autumn flowers and foliage, or for a spectacular massed display.
Description *Flower* Central fertile flowers in flat clusters, white, pink or blue, depending

on variety and alkaline or acid content of soil; outer ray florets blue, deepening as autumn approaches. *Foliage* Leaves medium-sized, elliptic, 2-4in (5-10cm) long, purple to olive green, giving way to good red, orange and bronze autumn colours. *Stem* Upright, becoming slightly branching with age, forming an upright, round-topped shrub. Light green when young, becoming brown to grey-brown. Medium growth rate. *Fruit* None. Retains dead flowerheads into winter.
Hardiness Tolerates 14°F (−10°C).
Soil Requirements Prefers acid soils, which produce good flower colours. Tolerates some alkalinity but flower colour will deteriorate.
Sun/Shade aspect Best in light shade. Dislikes full sun and deep shade equally.
Pruning Remove old or very weak wood each autumn after flowering or in spring in colder areas.
Propagation and nursery production From softwood cuttings. Purchase container-grown. Usually available from good general nurseries, although some varieties less easy to find. Best planting heights 1¼-2½ft (40-80cm).

Hydrangea serrata 'Preziosa' in flower

Problems Plants often look gaunt and uninteresting while in containers, but grow quickly once planted out.
Varieties of interest *H. pia* Small clusters of pink flowers, aging to red-pink. Ovate foliage 4-6in (10-15cm) long and 2-3in (5-8cm) wide, slightly red-tinged with some autumn colour. Two-thirds 'average height and spread. *H. serrata 'Bluebird'* syn. *H. s. var. acuminata, H. acuminata 'Bluebird'* Central, blue, fertile flowers on acid soils, becoming more pink on alkaline. Forming dome-shaped clusters, surrounded by large ray florets, which will be red-purple on alkaline soils, deep blue on acid types. *H. s. 'Grayswood'* Central, blue, fertile flowers on acid soils, pink on alkaline, with outer ring of ray florets, firstly white, changing to rose and finally to deep crimson on acid soils, much paler on alkaline. *H. s. 'Preziosa'* More rounded heads, with rose pink florets which turn red-purple in autumn. Stems upright and purple-red. Foliage purple-tinged at first, aging to dark green and giving good purple autumn colour. *H. s. 'Rosalba'* Ray florets white at first, turning crimson later; best colours on acid soils. A small shrub, reaching one-third average height and spread.
Average height and spread
Five years
2x2½ft (60x80cm)
Ten years
2½x3ft (80cmx1m)
Twenty years
or at maturity
3x4ft (1x1.2m)

HYDRANGEA VILLOSA
(Hydrangea rosthornii)

KNOWN BY BOTANICAL NAME
Hydrangeaceae
Deciduous
One of the supreme Hydrangea varieties worthy of any garden able to give it the precise conditions it requires. By some authorities considered to be a form of *H. aspera*, although quite different in foliage formation.

Origin From western China.
Use As an individual shrub on its own, or among other shrubs in large plantations; ideal for woodland situations. Can be used in tubs not less than 2½ft (80cm) in diameter and 2ft (60cm) in depth. Use good lime-free potting medium and feed adequately during the growing season.
Description *Flower* Round, flat, lilac-blue clusters with toothed outer marginal sepals, late summer to early autumn. May be more pink on alkaline soils. *Foliage* Leaves narrow, elliptic, 4-9in (10-23cm) long, purple-green, velvet textured. Some autumn colours. *Stem* Numerous upright stems, surmounted by slight branching, each branch producing a terminal flower. Grey-green when young, aging to creamy brown, attractive in winter. Fast growth rate when young, becoming slower with age. *Fruit* None. Flowers retained into winter as brown flowerheads.
Hardiness Tolerates 14°F (−10°C). Foliage may be destroyed by late spring frosts, but normally recovers.
Soil Requirements Prefers acid soil, but tolerates considerable alkalinity. Does best on rich, deep, moist woodland leaf mould.
Sun/Shade aspect Light dappled shade. Dislikes full sun. Very useful in exposed, shady positions if wind protection is provided.
Pruning Remove one-third of oldest stems on established plants in early spring to encourage good basal rejuvenation and maintain foliage and flower size.
Propagation and nursery production From softwood cuttings taken in early summer, or from micropropagated nursery stock. Purchase container-grown or root-balled (balled-and-burlapped). Best planting heights 1¼-2ft (40-60cm).
Problems Stock appears short and deformed when purchased.
Average height and spread
Five years
3x3ft (1x1m)
Ten years
6x6ft (2x2m)
Twenty years
or at maturity
10x10ft (3x3m)

Hydrangea villosa in flower

HYPERICUM CALYCINUM

HYPERICUM, AARONSBEARD, ST JOHN'S WORT
Guttiferae
Evergreen
A shrub often given a bad name because of its invasive habit, but which, planted in the right position, is a true friend for use as ground cover in any large garden.

Origin From south-eastern Europe and Asia Minor.
Use As a ground-covering shrub for large areas. Ideal for consolidation of banks and planting in inhospitable areas, but needs adequate soil preparation before planting and must be weeded and kept cultivated in first and second springs after planting. Plant 1½ft (50cm) apart to achieve best ground cover.
Description *Flower* Large, conspicuous, golden yellow flowers with pronounced central yellow anthers, borne mainly early summer, but continuing to late autumn. *Foliage* Leaves large, ovate, 3-4in (8-10cm) long, semi-evergreen to evergreen, depending on winter conditions. Dark green with purple hue, some yellow autumn colour contrasting with green foliage. *Stem* Short, squat, upright flower-bearing shoots producing underground stolon suckers which can become invasive. Stems below ground white, thick and fleshy; above ground, light green when young, becoming brown with age. Medium to fast growth rate at ground level. *Fruit* Rare, but may produce small black insignificant fruits.

Hypericum calycinum in flower

Hardiness Tolerates winter temperatures down to −13°F (−25°C), although some foliage damage and die-back may occur in very severe winters; rapidly rejuvenates in the following spring.
Soil Requirements Any soil, both alkaline and acid. Will tolerate very dry conditions once established, but keep watered at early stage.
Sun/Shade aspect Does well in full sun through to deep shade. Strongly recommended for ground cover in deeply shaded areas.
Pruning If cut to ground level early each spring, rapidly reproduces itself with better foliage and flowers.
Propagation and nursery production By softwood cuttings or rooted suckers. Best planted container-grown. Easy to find. Best planting heights 4-8in (10-20cm).
Problems Can become invasive and this must be borne in mind when selecting site. Spread is determined by production of underground suckers and could well exceed that shown.
Average height and spread
Five years
1x3ft (30cmx1m)
Ten years
1x4ft (30cmx1.2m)
Twenty years
or at maturity
1x6ft (30cmx2m)

HYPERICUM
Flowering forms

HYPERICUM, ROSE OF SHARON, ST JOHN'S WORT
Guttiferae
Deciduous and Semi-evergreen
One of the brightest of summer-flowering shrubs and worthy of any garden.

Origin Most varieties of garden origin.
Use As an individual summer-flowering shrub of small to medium height or for mass planting. If planted 2½ft (80cm) apart in a single line, all flowering forms make attractive low, informal hedges.
Description *Flower* Good-sized, yellow, cup-shaped flowers borne profusely, early summer to late autumn. *Foliage* Leaves elliptic, 1-3in (3-8cm) long and 1in (3cm) wide, mid green, glossy and sometimes shaded red. Retained well into late autumn, even early winter. Some yellow autumn colour. *Stem* Upright, branching at extremities, mahogany-brown. Fast growth rate in spring when young, becoming slower with age. *Fruit* May produce small, red-black fruits, but less interesting than in those recognized for fruiting performance.
Hardiness Tolerates 4°F (−15°C).
Soil Requirements Does well on most soils, both acid and alkaline, but unhappy on very dry areas.
Sun/Shade aspect Prefers light shade, tolerates full sun and medium shade.
Pruning Remove one-third of oldest shoots in very early spring, to induce new basal growth. If this method is undertaken the shrub will reach average height and spread and flowering size will be good. Otherwise, prune back to ground level in early spring, to induce vigorous new flowering shoots for the following summer. Flower size is increased, but the shrub will reach only two-thirds average height and spread with this treatment.
Propagation and nursery production From semi-ripe cuttings taken in early summer. Plant bare-rooted or container-grown. Best planting heights 1-2ft (30-60cm). Availibility varies.
Problems Fairly trouble-free. Unlike *H. calycinum* these flowering forms, of which a selection is given below, are not invasive.
Varieties of interest *H. beanii 'Gold Cup'* Flowers cup-shaped, up to 2½in (6cm) across and deep golden yellow. Lanceolate foliage interestingly arranged on stems and in distinct opposing rows along arching branches. Hard but not impossible to find. *H. coris* Golden yellow flowers borne in panicles some 5in (12cm) long at ends of branches, mostly during the summer. Forms bold, floriferous, small clump ideal for dry walls or rock gardens. A very low-growing variety,

reaching 1ft (30cm) in height; semi-evergreen to evergreen. Should be only lightly pruned in spring. From central and southern Europe. *H. forrestii* syn. *H. patulum forrestii, H. p. henryi* Large, saucer-shaped, golden yellow flowers borne early summer through to late autumn. One of the largest flowering forms. Slightly tender. From south-western China, Assam and Burma. *H. 'Hidcote'* syn. *H. patulum 'Hidcote', H. p. 'Hidcote Gold'* Golden yellow, saucer-shaped flowers of good size, especially if pruned back hard, early summer through to late autumn. Semi-evergreen. Generally considered the best variety for flowering. *H. × moseranum* Flowers golden yellow with red shading and very prominent red anthers, early summer through to late autumn. Very attractive, large, ovate foliage, green with grey sheen. Sometimes slightly tender, but normally tolerates 14°F (−10°C). Reacts to extreme weather conditions very unfavourably, but is one of the most attractive of flowering varieties. Two-thirds average height and spread. *H. prolificum* Bright yellow flowers, very freely borne, with short petals and long central anthers formed in terminal clusters, summer to autumn. Very densely branched, forming a round, bushy shrub reaching two-thirds average height and spread. Native to eastern and central USA. *H. 'Rowallane'* Large, rich golden yellow, bowl-shaped flowers, 2-2¾in (5-7cm) wide, with almost waxy texture. Semi-evergreen form with long ranging branches which must be cut to ground level annually for best flowering. In favourable areas reaches slightly more than average height and spread. Slightly tender.

Average height and spread
Five years
2x2ft (60x60cm)
Ten years
3x3ft (1x1m)
Twenty years
or at maturity
4x4ft (1.2x1.2m)

HYPERICUM
Fruiting forms

HYPERICUM, ST JOHN'S WORT
Guttiferae
Deciduous
Interesting additions to the garden range of *Hypericum* varieties.

Origin Most varieties from western and southern Europe, some from China.
Use As freestanding shrubs, or for mass planting, or in medium-sized to large shrub borders.
Description *Flower* Yellow, cup-shaped flowers of various sizes, depending on variety,

borne mainly midsummer but also early summer, through to late autumn. *Foliage* Leaves ovate, 1-3in (3-8cm) long and ½-2in (1-5cm) wide, light green to yellow green when young, aging to purple. Some autumn colour. *Stem* Upright stems, branching at extremities, mahogany-brown. Medium growth rate. *Fruit* Clusters of red or black capsules, dependent on variety, late summer and early autumn.
Hardiness Tolerant of winter temperatures down to −13°F (−25°C).
Soil Requirements Any soil, including acid and alkaline, but dislikes very dry conditions.
Sun/Shade aspect Full sun to light shade.
Pruning Thin out, removing one-third of oldest branches very early in spring. This encourages strong spring and summer growth, which can give a good secondary crop of fruit.
Propagation and nursery production From softwood cuttings taken early summer, or hardwood cuttings taken late autumn, early winter. Purchase bare-rooted or container-grown. Most varieties available from good nurseries, smaller choice in garden centres. Best planting heights 1¼-2½ft (40-80cm).
Problems Most fruiting forms of *Hypericum* suffer badly from the fungus disease rust. Shrubs very severely affected should be cut to ground level in early spring. Also susceptible to leaf mildew, a green mould appearing late summer and early autumn.

Hypericum persistens 'Elstead' in flower

Varieties of interest *H. androsaemum* (Tutsan Hypericum) Small, yellow flowers with prominent central anthers, surmounted in autumn by red fruits very erectly presented, which finally turn black. Foliage very susceptible to mildew and slightly to rust. Two-thirds average height and spread. From western and southern Europe. *H. × inodorum* syn. *H. androsaemum × hircinum, H. elatum* Pale yellow flowers, summer and early autumn, surmounted from late summer by white, upright, long fruits which finally turn red. Leaves round, ovate when young, very oblong in maturity, light green at first, aging to purple. This form can be very variable and foliage very susceptible to both rust and mildew. Originating from France and Madeira. *H. kouytchense* syn. *H. penduliflorum, H. patulum grandiflorum, H. patulum 'Sungold'* Flowers up to 2in (5cm) across, good golden yellow with long stamens, midsummer to early autumn. Foliage, some evergreen, ovate. Bright red fruits of stork's head shape borne on round, compact shrub. Very scarce. One-third average height and spread. *H. persistens 'Elstead'* syn. *H. elatum 'Elstead'* Flowers yellow and small, followed by fruits which are long, upright and finally ripening to salmon red. Sparser leaves than the rest and more green and orange-red, especially in autumn. Less susceptible to mildew, but prone to attacks of rust. A very good form and the one most commonly available. Of garden origin.

Hypericum 'Hidcote' in flower

Average height and spread
Five years
2½x2½ft (80x80cm)
Ten years
3x3ft (1x1m)
Twenty years
or at maturity
4x4ft (1.2x1.2m)

HYPERICUM
Variegated and golden foliage forms

HYPERICUM, ST JOHN'S WORT
Guttiferae
Deciduous
The variegated forms of *Hypericum* are less vigorous than the flowering forms and the golden varieties in particular are susceptible to sun scorch; they are nevertheless well worth planting, even if it means trial and error to achieve good results.

Origin Most varieties of garden origin.
Use As a low-growing shrub for edging a shrub border or for mass planting, provided the position is in light shade.
Description *Flower* Normally small, yellow, cup-shaped flowers, early summer through to early autumn. *Foliage* Leaves lanceolate or elliptic, depending on variety, 1-3in (3-8cm) long, either silver or golden variegated. *Stem* Thin, arching, multi-branched, forming low clump of growth. Medium growth rate. *Fruit* May produce small, red to red-black fruits, but inferior to those of recognized fruiting forms of *Hypericum*.

Hypericum × moseranum 'Tricolor' in flower

Hardiness Tolerates 14°F (−10°C).
Soil Requirements Does well on most, including alkaline or acid types, but dislikes extremely dry areas.
Sun/Shade aspect Light shade. Medium to deep shade will decrease amount of variegation and full sun will scorch.
Pruning For best foliage results should be pruned either annually or biennially down to ground level in early spring to induce new growth, but all forms are rather weak in constitution and need some feeding each spring to encourage adequate growth production.
Propagation and nursery production From softwood cuttings. Purchase container-grown. Availibility varies. Best planting heights 1-1½ft (30-50cm).
Problems Constitution can be a little weak. Golden varieties liable to leaf scorch in strong summer sunlight.
Varieties of interest *H. × moseranum 'Tricolor'* Narrow, elliptic leaves striped gold, with some red to red-orange shading. Shrub surmounted by small, but interesting, golden

yellow flowers. Possibly the best variegated form. *H. 'Patulum Variegata'* Foliage long, narrow, almost lanceolate, streaked with white variegation against grey-green background. Small yellow flowers and sometimes small, inconspicuous red fruits. Of garden origin. *H. 'Summer Gold'* Lime green to pale yellow spring foliage, aging to golden yellow. Susceptible to late spring frost damage and leaf scorch from strong sunlight. A recent variety and possibly the best pure golden foliage form. *H. 'Ysella'* Foliage very soft lime green all over in spring. Rounded, ovate leaves. Some small pale yellow flowers. Very susceptible to late spring frosts and liable to leaf scorch from strong summer midday sun. Also susceptible to attacks of rust.
Average height and spread
Five years
1½x1½ft (50x50cm)
Ten years
2¼x2¼ft (70x70cm)
Twenty years
or at maturity
3x3ft (1x1m)

ILEX × ALTACLARENSIS

ALTACLAR HOLLY, HIGHCLERE HOLLY
Aquifoliaceae
Evergreen
A beautiful range of evergreen shrubs with female plants producing interesting berries and foliage.

Origin Basic form raised in Highclere, Berkshire, England as a cross between *I. aquifolium* and *I. perado*; from this came all the varieties.
Use As a freestanding shrub, or for the back of a large shrub border. Ideal for mass planting as screen. If planted 3ft (1m) apart in straight line forms a dense evergreen hedge up to 13ft (4m) or more in height. Can be grown in tubs not less than 2½ft (80cm) in diameter and 2ft (60cm) in depth, but will not attain its full height. May also be trained on a single stem to form a mop-headed standard.
Description *Flower* Small clusters of small white flowers with prominent stamens, late spring, early summer. *Foliage* Leaves ovate, 2-4in (5-10cm) long and 2-3in (5-8cm) wide, some spines at outer lobed edges. Glossy, waxy upper surfaces, dull undersides. Mainly green, but with silver and golden variegated forms. *Stem* Upright, forming a pyramidal shrub or small tree. Light to mid green, glossy. After 15 years a low basal skirt of foliage may be formed. Medium growth rate. *Fruit* Round berries, green at first, becoming bright glossy red, borne on female plants in autumn and maintained into winter. Male shrubs do not fruit. Male and female specimens must be planted together to achieve fruits on the female.

Ilex × altaclarensis 'Golden King'

Hardiness Tolerates winter temperatures down to −13°F (−25°C).
Soil Requirements Prefers rich, moist, open soil but does well on most.
Sun/Shade aspect Full sun to medium shade. In deep shade will lose its compact pyramidal shape, become sprawling and less likely to fruit.
Pruning None required, but can be reduced in size, or trimmed in early spring.
Propagation and nursery production From semi-ripe cuttings taken in early summer. Purchase container-grown or root-balled (balled-and-burlapped). Availability varies. Best planting heights 1½-4ft (50cm-1.2m).
Problems Although less spiny than most hollies, cultivation at ground level can be uncomfortable. Young plants when purchased often look uninteresting, but shape improves once planted out. Newly planted shrubs may lose leaves entirely in early spring following planting, due to dropping of old foliage and production of new.
Varieties of interest *I. × a. 'Atkinsonii'* Large, dark green, glossy foliage with spiny edges. A male form producing no fruits, but often used as a pollinator for other forms. *I. × a. 'Camelliifolia'* Large, camellia-like foliage, red-purple when young, becoming dark green and shiny with age. A female form with large clusters of dark red fruits and purple stems. Forms a good pyramidal shrub. *I. × a. 'Golden King'* Almost round, shiny, spineless leaves, green with bright yellow to gold margins. A female form, producing red to orange-red fruit. One of the finest of all golden variegated hollies. Slightly less than average height. *I. × a. 'Hodginsii'* Dark green, oval leaves with some spines produced irregularly. A male form producing no fruit, but useful as a pollinator. *I. × a. 'Lawsoniana'* Spineless leaves, dark green with bright yellow centres. A female form with orange-red fruit. Reversion may be a problem and any completely green shoots should be removed at once. *I. × a. 'Silver Sentinel'* Creamy white to yellow margins around pale green to grey-green, narrow, elliptic leaves. Very few spines. A female form with orange-red fruit. Quick-growing and upright in habit. Slightly less than average spread.
Average height and spread
Five years
6x5ft (2x1.5m)
Ten years
13x8ft (4x2.5m)
Twenty years
or at maturity
16x12ft (5x3.5m)

ILEX AQUIFOLIUM

COMMON HOLLY, ENGLISH HOLLY
Aquifoliaceae
Evergreen
An extremely interesting and useful range of very hardy evergreen shrubs.

Origin From Europe and western Asia.
Use As a freestanding shrub or for the back of a large shrub border. Ideal for mass planting as screens. If planted 3ft (1m) apart in straight line, forms an evergreen hedge up to 13ft (4m) in height. Can be grown in tubs not less than 2½ft (80cm) in diameter and 2ft (60cm) in depth, but will not attain its full height. Can also be trained on a single stem to form a mop-headed standard.
Description *Flower* Small clusters of small white flowers with prominent stamens, late spring to early summer. *Foliage* Leaves ovate, lobed, 1-3in (3-8cm) long and 1-2in (3-5cm) wide, very spiny, with spines produced on edges of leaves. Glossy green, golden or silver variegated, depending on variety. *Stem* Upright, becoming spreading with age, forming a wide, pyramidal, (moderately open shrub. Mid to dark green, glossy. Often forms a single stem, eventually becoming small tree. Fast to medium growth rate when young, slowing with age. *Fruit* Female forms, if pollinated by a male form nearby, produce

Ilex aquifolium 'Pyramidalis' in fruit

red, round fruits in autumn, maintaining them into winter.
Hardiness Tolerates winter temperatures down to −13°F (−25°C).
Soil Requirements Does well on any soil.
Sun/Shade aspect Full sun to medium shade. In deep shade will lose its compact pyramidal shape, become sprawling and be less likely to fruit.
Pruning None required, but can be reduced in height and spread, or closely trimmed, in early spring.
Propagation and nursery production From semi-ripe cuttings taken in early summer. Purchase container-grown or root-balled (balled-and-burlapped). Most varieties easy to find in nurseries and some offered through garden centres. Best planting heights 1½-4ft (50cm-1.2m).
Problems Spiny leaves make cultivation at ground level difficult. Newly planted shrubs may lose leaves entirely in early spring following transplanting due to dropping of old foliage and production of new. Young plants when purchased often look uninteresting, but quickly improve shape once planted out.
Varieties of interest *I. a. 'Argentea Marginata'* (Broad-leaved Silver Holly) White margins on dark green to grey-green spiny leaves. Normally offered in female form, but can also be found as a male plant so care should be taken to ascertain sex before purchasing. Fruit orange to orange-red in female form, retained into midwinter, but fruit production requires a nearby male shrub for pollination. *I. a. 'Argentea Pendula'* (Perry's Silver Weeping Holly) Long, weeping branches forming somewhat squat shrub with orange-red fruits on female form. Foliage dark to olive green with white margins. *I. a. 'Bacciflava'* syn. *I. a. 'Fructuluteo'* (Yellow-fruited Holly) Female form with bright dark green to mid green spiny leaves and good bright yellow fruits maintained into winter. *I. a. 'Golden Milkboy'* syn. *I. a. 'Aurea Mediopicta Latifolia'* Leaves flat in presentation, green with large golden centre variegation and spiny edges. May revert and all-green shoots should be removed when seen. Male form, no fruits. Very attractive foliage. *I. a. 'Golden Queen'* syn. *I. a. 'Aurea Regina'* Very dark green leaves with broad, bright golden yellow margins. Male form, no fruits. Slightly less than average height and spread. *I. a. 'Golden van Tol'* Foliage less spiny than most forms, more ovate to lanceolate, and with very good golden margins. Female, producing red fruits maintained into winter. *I. a. 'Handsworth New Silver'* Narrow, ovate leaves with creamy white margins around deep green to grey-green leaves. Female, producing orange to orange-red fruits. *I. a. 'J.C. van Tol'* Narrow, ovate, dark green foliage, almost spineless. Female, producing large regular crops of red fruits. *I. a. 'Madame Briot'* Glossy, dark yellow to gold margins around

dark green leaves, presented on purple stems. Female, producing orange-red fruits. *I. a. 'Myrtifolia Aureomaculata'* Smaller foliage than most, narrow and ovate, very dark green with central golden variegation. Male form, no fruit. Slightly less than average height and spread. *I. a. 'Pendula'* Female form, good red fruits presented on long, weeping shoots, covered with dark green, spiny leaves, making a very thick, round mound. Slightly less than average height, but more spread. *I. a. 'Pyramidalis'* Female form, produces clusters of red fruits. Foliage narrow to ovate, dark green and glossy. Conical, upright in shape, becoming wider with age. Often confused with '*J.C. van Tol*'. *I. a. 'Silver Milkboy'* Flattish, spiny leaves, dark green with central creamy white variegation. Some reversion may occur and all-green shoots should be removed when seen. Male form, no fruits. Slightly less than average height and spread. *I. a. 'Silver Queen'* Young shoots purple to black in colour. Creamy white margins to dark green leaves. Male form, no fruits.
Average height and spread
Five years
6x5ft (2x1.5m)
Ten years
13x8ft (4x2.5m)
Twenty years
or at maturity
20x12ft (6x3.5m)

ILEX AQUIFOLIUM 'FEROX'

HEDGEHOG HOLLY
Aquifoliaceae
Evergreen
An interesting novelty with unusual foliage attraction.

Origin A sport from *I. aquifolium*.
Use As a featured novelty evergreen or variegated evergreen shrub.
Description *Flower* Small clusters of small white male flowers with prominent stamens, late spring to early summer. Flowering may be intermittent. *Foliage* Leaves lobed, twisted, curled, ovate, 2-3in (5-8cm) long and 1-2in (3-5cm) wide. Dark green, also silver and gold variegated, depending on variety, covered with spines not only on outer edges but also emerging from upper and lower leaf surfaces, giving a hedgehog-like appearance. *Stem* Light green, short branches, forming a round shrub which is subject to die-back, often becoming irregularly shaped after 15-20 years. Slow growth rate. *Fruit* None, as it is a male form.
Hardiness Tolerates winter temperatures down to −13°F (−25°C).
Soil Requirements Does well on any soil.

Sun/Shade aspect Full sun to medium shade. In deep shade loses its overall shape, and becomes sprawling and less likely to flower.
Pruning None required.
Propagation and nursery production From softwood cuttings. Purchase container-grown. Must be sought from specialist nurseries. Best planting heights 1¼-2½ft (40-80cm).
Problems Handling is extremely difficult owing to dense mass of spines. Slow-growing, needs time and patience.

Ilex aquifolium 'Ferox Argentea'

Varieties of interest *I. a. 'Ferox Argentea'* (Silver Hedgehog Holly) Leaves margined broadly in white. Slightly less than average height and spread. *I. a. 'Ferox Aurea'* (Golden Hedgehog Holly) Golden variegated. Hard to find. Slightly less than average height and spread.
Average height and spread
Five years
2½x2½ft (80x80cm)
Ten years
4x4ft (1.2x1.2m)
Twenty years
or at maturity
6x6ft (2x2m)

ILEX CRENATA

BOX-LEAVED HOLLY, JAPANESE HOLLY
Aquifoliaceae
Evergreen
A very attractive group of low evergreens, especially when used in rock gardens and other areas where they can be shown individually to best effect.

Origin From Japan.
Use As an individual low shrub, ideal as a specimen plant for large rock gardens or to emphasize a particular garden feature. If planted 1½ft (50cm) in a single line can be clipped into low, informal hedge.
Description *Flower* Small, inconspicuous white flowers with small petals and prominent stamens, late spring to early summer. *Foliage* Small, round to ovate, ½-1½in (1-4cm) long, grey-green to dark green, some golden varieties. *Stem* Short, very branching, forming a low hummock, and eventually a spreading bowl shape. Grey-green to dark green. Slow growth rate. *Fruit* All listed forms female. Small individual black fruits in late autumn and early winter.
Hardiness Tolerates 4°F (−15°C).
Soil Requirements Does well on most, but unhappy on very dry soils. Golden-leaved forms require good rich soil to encourage new growth and ensure production of golden foliage.
Sun/Shade aspect Green forms good from full sun through to mid shade. Golden variegated forms must have light dappled shade, or leaf scorch will occur. In deep shade golden colour decreases.

Ilex crenata 'Golden Gem'

Pruning None required.
Propagation and nursery production From semi-ripe cuttings taken in midsummer. Purchase container-grown or root-balled (balled-and-burlapped). May have to be sought from specialist nurseries. Best planting heights 1¼-2ft (40-60cm).
Problems *I. c. 'Golden Gem'*, is extremely susceptible to leaf scorch from strong summer sun, so must be planted in light shade. This and other forms are rather small when sold and need time to achieve their true potential.
Varieties of interest *I. c. 'Aureovariegata'* syn. *I. c. 'Variegata'* Pale green to grey-green, ovate leaves, new spring growth having yellow splashes. This variegation is normally held through to early summer. *I. c. 'Convexa'* Small, glossy, dark, rounded foliage, very densely clothing branches, giving the effect of Box *(Buxus sempervirens)*. Black, shiny fruits. Slightly less than average height and spread. *I. c. 'Golden Gem'* Some three-quarters of new spring foliage golden variegated. Variegation decreases as winter approaches and in the second season eventually turns green, but is surmounted by new crop of golden foliage each spring. Fruiting limited, but unfortunately any amount of fruit production can indicate that the shrub is in some distress. Slightly less than average height and spread. *I. c. 'Helleri'* Small, green, round to ovate leaves and black fruits in winter. Forms flat hummock-shaped shrub, ideal for rock gardens. Half average height and spread. *I. c. 'Mariesii'* Although this variety is indeed a holly, it looks so like Box *(Buxus sempervirens)* that it is often mistaken for it. Small, round, dark green leaves, very closely bunched in small, twiggy groups. A female form, with black fruit, often used in bonsai culture and equally good for small rock gardens or among an alpine collection. *I. c. 'Stokes Variety'* Small, grey-green leaves, black fruit. Forms low, tight mound. Useful for large rock gardens.
Average height and spread
Five years
1¼x2½ft (40x80cm)
Ten years
2x3ft (60cmx1m)
Twenty years
or at maturity
2½x5ft (80cmx1.5m)

ILEX × MESERVEAE

BLUE HOLLY, MERSERVE HOLLY
Aquifoliaceae
Evergreen
A colourful winter foliage shrub, with good red berries on female forms.

Origin From the USA.
Use As an interesting foliage and fruiting shrub for winter effect, planted singly or in a shrub border. If planted at 3ft (1m) apart in a single line makes an attractive hedging plant. Can be trained on a single stem to form a mop-headed standard.

Description *Flower* Small clusters of white flowers with small petals and pronounced stamens, late spring to mid summer. *Foliage* Leaves ovate, 1½-3in (4-8cm) long and 1-2in (2-5cm) wide, lobed with short spines at end of each pointed lobe. Blue-green to purple-green in winter. Attractive and different from other holly varieties. *Stem* Strong upright shoots, forming an upright pyramidal shape. Purple to purple-green. Medium to fast growth rate. *Fruit* Large clusters of very deep red, large, round berries in autumn and early winter. Both male and female forms must be planted to obtain good fruiting performance.
Hardiness Tolerates 14°F (−10°C). May suffer complete defoliation in severe winters, but normally refoliates the following spring.
Soil Requirements Most soils, but shows distress in waterlogged conditions.
Sun/Shade aspect Full sun to mid shade. Deep shade spoils overall shape and can decrease fruiting ability.
Pruning None required but can be trimmed as necessary.

Ilex × meserveae

Propagation and nursery production From semi-ripe cuttings taken in early summer. Purchase container-grown or root-balled (balled-and-burlapped). Normally easy to find in garden centres, especially in winter. Best planting heights 1½-4ft (50cm-1.2m).
Varieties of interest *I. × m. 'Blue Angel'* Female. Foliage in spring very blue-green, turning purple in winter, degree of purple depending on coldness of the winter. White flowers; large, dark red fruits in autumn. Average height, but possibly greater spread. *I. × m. 'Blue Prince'* Male. No berries, but used as a pollinator for female forms. Deep blue-green leaves. Slightly less than average height, but equal spread. *I. × m. 'Blue Princess'* Female. Red fruits and good blue-green foliage.
Average height and spread
Five years
5½x3ft (1.8x1m)
Ten years
10x6ft (3x2m)
Twenty years
or at maturity
20x12ft (6x4m)

ILEX PERNYI

PERNY HOLLY
Aquifoliaceae
Evergreen
An interesting evergreen foliage shrub.

Origin From central and western China.
Use As an individual shrub for foliage interest.
Description *Flower* Female, white flowers with small petals and prominent stamens, produced late spring to early summer. *Foliage* Leaves triangular, 1-2in (3-5cm) long and 1in (3cm) wide. Dark olive green with lighter undersides. Few spines. *Stem* Short branched, forming a round shrub. Dark grey-green. Slow to medium growth rate. *Fruit* Small, sparsely produced, bright red fruits, autumn and early winter.
Hardiness Tolerates 14°F (−10°C).
Soil Requirements Does well on most soils but the richer the soil, the more growth.
Sun/Shade aspect Prefers medium to light shade. In deep shade may become open and lax; in full sun growth rate may be slightly decreased.

Ilex pernyi

Pruning None required, but may be trimmed.
Propagation and nursery production From semi-ripe cuttings taken in early summer. Purchase container-grown. May have to be sought from specialist nurseries. Best planting heights 1¼-2½ft (40-80cm).
Problems Young plants when purchased tend to look misshapen and uninteresting, and take time to mature.
Varieties of interest *I. p. var. veitchii* syn. *I. bioritsensis* Very similar to parent form but with larger leaves. Female form producing small red fruits; the form most widely available. From western China and Taiwan.
Average height and spread
Five years
3x3ft (1x1m)
Ten years
6x6ft (2x2m)
Twenty years
or at maturity
10x10ft (3x3m)

ILEX VERTICILLATA

WINTERBERRY
Aquifoliaceae
Deciduous
An interesting shrub, being deciduous and quite unlike its close relations.

Origin From eastern North America.
Use As a feature shrub of particular botanical interest, with good spring and autumn foliage and bright red berries.

Ilex verticillata

Description *Flower* Small, white, female flowers with small petals and prominant stamens, in late spring to early summer. *Foliage* Leaves ovate 1-1½in (3-4cm) long and ½-1in (1-3cm) wide. Purple shaded with dark green, especially when opening in spring. Yellow autumn colour. *Stem* Moderately upright, purple-green, forming a large rounded shrub. Medium growth rate, slowing with age. *Fruit* Bright red fruits, held well into winter.
Hardiness Tolerates 14°F (−10°C).
Soil Requirements Must be grown on acid soils; dislikes any alkalinity.
Sun/Shade aspect Full sun to medium shade.
Pruning None required, but can be trimmed as necessary.
Propagation and nursery production From seed or from semi-ripe cuttings taken early summer. Purchase container-grown. Availability varies. Best planting heights 1½-3ft (50-1m).
Problems None.
Average height and spread
Five years
6x6ft (2x2m)
Ten years
13x10ft (4x3m)
Twenty years
or at maturity
20x13ft (6x4m)

INDIGOFERA HETERANTHA
(Indigofera gerardiana)

INDIGO BUSH
Leguminosae
Deciduous
An attractive, mid to late summer-flowering shrub useful for its ability to adapt to dry conditions.

Origin From the north-western Himalayas.
Use As a midsummer, early autumn-flowering shrub for dry, sunny areas. Very good as a fan-trained wall shrub, especially in colder areas. In this situation it will attain one-third more height and spread than if freestanding.
Description *Flower* Moderately open racemes of purple-pink pea-flowers in midsummer through to early autumn. *Foliage* Leaves pinnate, 2-4in (5-10cm) long with 13-21 leaflets up to ½in (1cm) long. Grey-green, giving some good yellow autumn colour. *Stem* Long, arching, becoming twiggy and shrub-forming in second year in milder areas. Grey-green. Fast growth rate when young, slowing with age. *Fruit* May produce small, grey-green pea-pods of some winter interest.
Hardiness Tolerates 14°F (−10°C). Stems may die back to ground level in severe winters, but normally rejuvenate in spring.

Soil Requirements Does well on all soil conditions, especially dry areas.
Sun/Shade aspect Full sun to very light shade.
Pruning If not destroyed by winter cold, reduce long arching stems by two-thirds or more in spring to encourage new growth. Leave unpruned in mild areas to encourage a large freestanding shrub.
Propagation and nursery production From softwood cuttings taken in late spring, early summer. Purchase container-grown. May need to be sought from specialist nurseries. Best planting heights 1¼-2½ft (40-80cm).
Problems Often very late to produce new growth, which may not appear until early summer. May look weak and insipid when purchased.

Indigofera heterantha **in flower**

Varieties of interest *I. potaninii* Longer racemes of pink flowers, up to 4-5in (10-12cm) long, late summer to early autumn. In areas with quite severe winter conditions, best grown on a sunny, sheltered wall. Slightly less than average height and spread. From north-western China.
Average height and spread
Five years
4x5ft (1.2x1.5m)
Ten years
5½x8ft (1.8x2.5m)
Twenty years
or at maturity
6x12ft (2x3.5m)

ITEA ILICIFOLIA

HOLLY-LEAF SWEETSPIRE
Iteaceae
Evergreen
One of the most spectacular late summer-flowering shrubs, but needing protection against cold conditions.

Origin From central China.
Use As a freestanding shrub for mild areas, flowering late summer to autumn. In colder areas may be used as a wall shrub in sheltered, lightly shaded or sunny sites, or may tolerate more exposed or shaded positions provided adequate protection from prevailing wind is available.
Description *Flower* Racemes up to 1¼ft (40cm) long of fragrant green to green-white flowers, aging to yellow-green, reminiscent of large catkins, late summer. *Foliage* Leaves lobed, dark green, glossy, ovate, 2-5in (5-12cm) long, similar to those of Holly, with purple hue underlying base colour and silver undersides. *Stem* Strong and upright when young, becoming weeping with age, forming a round, drooping effect when freestanding or a cascading, weeping shrub when grown against a wall or fence. Light green to purple-green. Fast growth rate when young, slowing with age. *Fruit* Insignificant.
Hardiness Tolerates winter temperature

Itea ilicifolia **in flower**

range from 23°F (−5°C) down to 14°F (−10°C). Leaves easily damaged by severe wind chill.
Soil Requirements Does well on any soil.
Sun/Shade aspect Prefers light shade, tolerates full sun to medium shade.
Pruning None required.
Propagation and nursery production From semi-ripe cuttings taken in early summer. Purchase container-grown. Obtainable from specialist nurseries. Best planting heights 1¼-2ft (40-60cm).
Problems Gives appearance of poor rooting and irregular shape when young. Requires open ground to perform at its best.
Average height and spread
Five years
3x3ft (1x1m)
Ten years
6x6ft (2x2m)
Twenty years
or at maturity
10x10ft (3x3m)

ITEA VIRGINICA

VIRGINIA SWEETSPIRE
Iteaceae
Deciduous
An interesting, attractive, small, summer-flowering shrub.

Origin From eastern North America.
Use As a summer-flowering shrub, grown on its own or in large shrub borders.
Description *Flower* Round, upright, fragrant, creamy white racemes, midsummer. *Foliage* Leaves ovate, light green, 1½-4in (4-10cm) long, with toothed edges, giving good autumn colour in the right environment. *Stem* Up-

Itea virginica **in flower**

151

right, forming an upright shrub. Light grey-green. Medium to fast growth rate. *Fruit* Insignificant.
Hardiness Tolerates 4°F (−15°C).
Soil Requirements Does well on almost all soils except for extremely alkaline where it may show distress.
Sun/Shade aspect Full sun to light shade.
Pruning None required.
Propagation and nursery production From seed or semi-ripe cuttings taken in early summer. Purchase container-grown or bare-rooted. Available from nurseries in its native environment; more difficult to find elsewhere. Best planting heights 2-4ft (60cm-1.2m).
Problems As a young plant, rarely gives any indication of the performance to come in later years.
Average height and spread
Five years
6x3ft (2x1m)
Ten years
13x10ft (4x3m)
Twenty years
or at maturity
20x16ft (6x5m)

JASMINUM
Shrub-forming varieties

SHRUBBY JASMINE, JESSAMINE
Oleaceae
Deciduous or Evergreen
Very useful winter and summer-flowering shrubs and very diverse in their range of performance.

Origin From China and the Himalayas, depending on variety.
Use As a freestanding shrub for either summer or winter flowering, depending on variety. *J. humile revolutum* and *J. nudiflorum* do well as wall shrubs, the latter also as a conservatory plant.
Description *Flower* Small, short, yellow, trumpet-shaped flowers borne either in winter or summer, depending on variety. *Foliage* Leaves trifoliate, bright green, up to ¾in (2cm) long, evergreen or deciduous according to variety, giving some yellow autumn colour. *Stem* Upright or more lax in growth, depending on variety, bright green in winter. Medium growth rate, slowing with age. *Fruit* Insignificant.
Hardiness Tolerates 4°F (−15°C).
Soil Requirements Any soil, often tolerating extremely poor conditions.
Sun/Shade aspect Tolerates full sun through to medium shade and even deep shade in some circumstances.
Pruning Remove one-third of old flowering

Jasminum nudiflorum in flower

shoots to ground level on mature established shrubs after flowering.
Propagation and nursery production From semi-ripe cuttings taken in early summer. Purchase container-grown. Most forms easy to find. Best planting heights 1¼-2½ft (40-80cm).
Problems None.
Varieties of interest *J. humile revolutum* Deep yellow clusters of flowers in midsummer. Leaves bright green, evergreen, ovate to round, presented in groups of 5-7 leaflets. Some good yellow autumn colour on older leaves. From China. *J. nudiflorum* (Winter Jasmine) Bright yellow flowers covering bare, leafless, long, dark green, arching branches, in autumn, winter and early spring. Extremely good wall shrub or for covering banks. Very hardy. From China. *J. parkeri* Tiny yellow flowers in summer on dwarf, prostrate, spreading shrub, forming a low mound of grey-green stems covered with small, pinnate leaves. Very good for rock gardens and large tub planting. May be hard to find. Best planting heights 4-8in (10-20cm). From the western Himalayas.
Average height and spread
Five years
5x5ft (1.5x1.5m)
Ten years
6x6ft (2x2m)
Twenty years
or at maturity
10x10ft (3x3m)

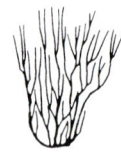

KALMIA LATIFOLIA

CALICO BUSH, SPOON WOOD, MOUNTAIN LAUREL
Ericaceae
Evergreen
Equal in flowering performance to any rhododendron and worthy of inclusion in acid-soil gardens.

Origin From eastern North America.
Use As an interesting flowering shrub to grow with rhododendrons, requiring similar conditions. Also good in large containers of 2½ft (80cm) diameter and 2ft (60cm) depth, using an acid potting medium.
Description *Flower* Large clusters of bright pink, saucer-shaped flowers, attractive in bud, mid spring. Flowers may not be produced in any quantity until 3-4 years after planting. *Foliage* Leaves large, elliptic, 4-6in (10-15cm) long and 1½-2in (4-5cm) wide, alternately presented. Dark green with purple veins. *Stem* Upright when young, quickly becoming spreading and ranging with age. Dark green to olive green. Slow to medium growth rate. *Fruit* Insignificant.
Hardiness Tolerates 14°F (−10°C).

Kalmia latifolia in flower

Soil Requirements Acid soil, resents any alkalinity. Deep, moist, rich soil gives best results.
Sun/Shade aspect Prefers full sun, provided soil moisture is adequate.
Pruning None required but individual limbs can be removed from overgrown, aged plants which will take 2-3 years before resuming flowering.
Propagation and nursery production From layers. Also sometimes from softwood cuttings taken in summer. Purchase container-grown or root-balled (balled-and-burlapped). Fairly easy to find in nurseries, especially when in flower. Best planting heights 1½-2½ft (50-80cm).
Problems Can be damaged in extreme wind chill conditions but normally rejuvenates. Patience required for its slowness in coming to flower.
Varieties of interest *K. l. 'Clementine Churchill'* Deep red flowers. Foliage narrower and slightly more red-purple. Possibly a little more tender than average. Very scarce. *K. l. 'Myrtifolia'* Smaller pink flowers, and much smaller leaves. More tender; for areas with winter temperatures not below 23°F (−5°C). A more compact form, reaching only one-third average height.
Average height and spread
Five years
5x5ft (1.5x1.5m)
Ten years
6x6ft (2x2m)
Twenty years
or at maturity
10x10ft (3x3m)

KALMIA
Low-growing varieties

KNOWN BY BOTANICAL NAME
Ericaceae
Evergreen
A range of plants requiring specialized soil conditions, but producing interesting displays.

Origin From northern and north-eastern USA.
Use As a low-growing mound of grey-purple foliage and spring flowers for acid gardens.
Description *Flower* Saucer-shaped to bell-shaped flowers, red-pink to purple-pink, according to variety, mid to late spring. *Foliage* Leaves narrow, lanceolate to ovate, 1-1½in (3-4cm) long, dark green tinged purple-grey with red to purple veins. *Stem* Light green to green-brown, becoming spreading and carpet forming, to make a small mound of growth. Slow growth rate. *Fruit* Insignificant.
Hardiness Tolerant of 14°F (−10°C).
Soil Requirements Must have deep, rich, moist, acid soil.

Sun/Shade aspect Full sun to very light shade. Any deeper degree of shade will decrease growth and flowering performance.
Pruning None required but may be reduced in size.
Propagation and nursery production From layers or softwood cuttings taken in early summer. Purchase container-grown or root-balled (balled-and-burlapped). If difficult to find, best sought from nurseries specializing in acid-loving plants. Best planting heights 8in-2ft (20-60cm).
Problems Shrubs tend to be small, wispy and uninteresting when purchased.

Kalmia polifolia in flower

Varieties of interest *K. angustifolia* (Sheep Laurel) Rosy red, saucer-shaped flowers in early summer. Narrow, green-red foliage. *K. a. 'Rubra'* Flowers deep rosy red, often maintained over a long period. Deep green foliage. *K. polifolia* (Bog Myrtle) Flowers rose-purple borne in terminal clusters, mid spring. Small, narrow, green leaves with glaucous blue undersides, borne on long, wiry stems. Very good for planting in wet, boggy areas. Never reaches more than 1½ft (50cm) in height and 3ft (1m) in spread.
Average height and spread
Five years
2x2ft (60cmx60cm)
Ten years
2½x3ft (80cmx1m)
Twenty years
or at maturity
3x4ft (1x1.2m)

Origin From China and Japan.
Use As a spring-flowering shrub for a shrub border or for mass planting on banks and difficult areas. Can be grown on a wall of any aspect.
Description *Flower* Yellow, buttercup-shaped flowers, single or double, dependent on variety, mid to late spring. *Foliage* Leaves elliptic, 1½-3in (4-8cm) long, bright green, giving some yellow autumn colour. Also silver and white variegated forms. *Stem* Strong, upright, spreading. Lateral underground shoots are produced which emerge some distance from parent to make a wide, upright thicket. Good winter stem attraction if pruned as suggested. *Fruit* Insignificant.
Hardiness Tolerates winter temperatures down to −13°F (−25°C).

Kerria japonica 'Pleniflora' in flower

Soil Requirements Any soil type and condition, often tolerating extremely poor areas.
Sun/Shade aspect Full sun to medium shade.
Pruning Cut back one-third to half of oldest flowering wood to ground level, after flowering.
Propagation and nursery production From rooted underground suckers or from hardwood cuttings taken in winter, or from semi-ripe cuttings taken in early summer. Purchase container-grown. All forms fairly easy to find. Best planting heights 2-3ft (60cm-1m).
Problems Needs open ground to perform to best advantage.
Varieties of interest *K. j. 'Pleniflora'* syn. *K. j. 'Flore Pleno'* Double, golden yellow flowers on an upright, vigorous shrub. Good green stems in winter. *K. j. 'Splendens'* A single-flowered variety with larger buttercup yellow flowers than most varieties. *K. j. 'Variegata'* syn. *K. j. 'Picta'* Masses of single yellow buttercup-shaped flowers, mid to late spring. Foliage ovate, creamy-white variegated on bright green wispy, wiry stems. Does not become invasive. Best planting heights 8-15in (20-40cm). A variety reaching only 3ft (1m) in height and spread.
Average height and spread
Five years
10x5ft (3x1.5m)
Ten years
10x8ft (3x2.5m)
Twenty years
or at maturity
10x12ft (3x3.5m)

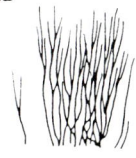

Origin From western China.
Use As a medium-sized to large shrub to stand on its own forming an attractive symmetrical clump or for a shrub border. Can be used as an informal hedge planted 3ft (1m) apart.
Description *Flower* Bell-shaped, soft pink with yellow throat, hanging in small clusters along wood 3 years old, late spring, early summer and midsummer. *Foliage* Leaves medium to small, ovate, 1-1½in (3-4cm) long, slightly tooth-edged, light olive green to grey-green with red shading and silver undersides. Yellow autumn colour. *Stem* Young shoots light green to green-brown. Strong and upright being produced mainly from ground level. Growth more than 2 years old becomes slightly arched, spreading and twiggy. Fast to medium growth rate. *Fruit* Small, grey-brown, slightly translucent seeds.

Hardiness Tolerates low winter temperatures in the range 4°F (−15°C) down to −13°F (−25°C).
Soil Requirements Any soil, no preferences.
Sun/Shade aspect Prefers full sun, tolerates light shade.
Pruning Remove one-third of oldest flowering wood by cutting to ground level after the flowering period.
Propagation and nursery production From soft to semi-ripe cuttings taken midsummer, or from hardwood cuttings taken in winter. Plant bare-rooted or container-grown. Best planting heights 1½-2½ft (50-80cm).
Problems When purchased container-grown presents very fragile, weak appearance but rapidly becomes robust once planted out.

Kolkwitzia amabilis in flower

Varieties of interest *K. a. 'Pink Cloud'* A cultivar of garden origin, with large, strong pink flowers, possibly better for garden planting than the parent.
Average height and spread
Five years
5x5ft (1.5x1.5m)
Ten years
8x8ft (2.5x2.5m)
Twenty years
or at maturity
10x10ft (3x3m)

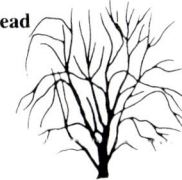

Origin From the Mediterranean region.
Use As a freestanding, large evergreen for mild locations. Can be clipped to a pyramid, ball or short pompom top standard. Does well in containers not less than 2½ft (80cm) in

diameter and 2ft (60cm) in depth; use good quality potting medium and feed adequately. Often grown as a very useful herb for the kitchen.

Description *Flower* Small tufts of almost petalless flowers with very prominent stamens, yellow-green. Flowers are either male or female, the latter being more commonly propagated. *Foliage* Leaves medium to large, elliptic, 2-4in (5-10cm) long, dark green, glossy, upper surfaces with dull grey-green undersides. *Stem* Upright, becoming branching, forming a pyramidal shape. Can be trained to a single stem. Mid green, sometimes tinged purple. Medium to fast growth rate, slowing with age. *Fruit* Small, shiny, black fruits sometimes produced on mature shrubs growing in mild areas. Fruit not always an asset as it can mean that the shrub is maturing or even in ill health.

Hardiness Minimum winter temperature 23°F (−5°C). Very susceptible to leaf scorch in cold winter winds.

Laurus nobilis 'Aurea' in leaf

Soil Requirements Prefers light open soil, but tolerates most types, including considerable alkalinity and acidity.

Sun/Shade aspect Full sun to light shade; tolerates deep shade.

Pruning Can be clipped hard, as required. Reduced to ground level, it rejuvenates but takes some years to reach former height.

Propagation and nursery production From semi-ripe cuttings taken midsummer. Purchase container-grown. If not found in garden centres and nurseries, may be sought from suppliers of culinary herbs. Best planting heights 1½-6ft (50cm-2m).

Problems Expensive if bought already trained to a pyramid shape.

Varieties of interest *L. n. 'Aurea'* An attractive variety with golden yellow evergreen leaves. Slightly less than average height and spread, and slightly more tender.

Average height and spread
Five years
5x3ft (1.5x1m)
Ten years
8x6ft (2.5x2m)
Twenty years
or at maturity
16x13ft (5x4m)

LAVANDULA

LAVENDER
Labiatae
Evergreen
Lavenders are among the most useful of all summer-flowering low shrubs, with the bonus of the very pleasant lavender scent.

Origin From Europe.
Use As low shrubs for edging in shrub borders, planted singly or in groups. Dwarf varieties are useful for underplanting, for example in rose beds. Larger types may be planted 15in (40cm) apart in a single line to make a low informal hedge.

Lavandula angustifolia 'Hidcote' in flower

Description *Flower* Borne in upright spikes, produced above foliage, in varying shades of blue, pink or white, depending on variety, mid to late summer. Very sweetly scented with typical lavender perfume. Attractive and fragrant when dried. *Foliage* Sparsely produced, narrow, lanceolate, ½-1in (2-3cm) long, grey-green to silver-green. *Stem* Short, limited branching, upright stems, surmounted by tall, bare, flowering stems. Grey to silver-grey. Fast growth rate, slowing with age. *Fruit* Large seedheads retained into winter.

Hardiness Tolerates 14°F (−10°C).
Soil Requirements Light, well-drained soil; dislikes waterlogging.
Sun/Shade aspect Full sun, dislikes any shade.
Pruning A light to medium trimming each spring, just as new growth starts, encourages rejuvenation of new foliage and increases the number of flower spikes. In addition, the removal of dying flower spikes is advised. Old plants can be cut back hard in spring, and will normally rejuvenate.

Propagation and nursery production From semi-ripe cuttings taken in early summer. Purchase container-grown or bare-rooted. Some forms easy to find, although specific varieties may have to be sought from specialist sources. Best planting heights 4-15in (10-40cm).

Problems Young plants when purchased in late spring or early summer look very good, but as autumn approaches can become old and woody; however once planted they quickly produce new growth in following spring.

Varieties of interest *L. angustifolia* syn. *L. officinalis*, *L. spica* (Old English Lavender) Mid blue flower spikes, produced on long slender stems, good silver-grey foliage. *L. a. 'Alba'* A white form of Old English Lavender. Off-white flowers on long stems, produced mid to late summer. Rather hard to find. *L. a. 'Folgate'* Lavender-blue flowers, late summer, narrow, silvery grey-green leaves. A good, compact form. *L. a. 'Grappenhall'* Lavender-blue flowers in midsummer, grey-green leaves. *L. a. 'Hidcote'* Violet-blue, very thick spikes of scented flowers in midsummer, surmounting grey-green foliage. Possibly one of the best varieties, good for hedging and underplanting. Compact in growth, rarely reaching more than 1½ft (50cm) in height and spread. *L. a. 'Loddon Pink'* Blue to pink-blue spikes of flowers on long stems in midsummer, surmounting grey-green foliage. *L. a. 'Munstead'* Lavender-blue flowers in midsummer. Good for hedging or underplanting. Height 1½ft (50cm). *L. a. 'Nana Alba'* A white form, reaching no more than 1ft (30cm) in height. Scarce and hard to find. *L. a. 'Rosea'* Blue-pink to pink flowers, in midsummer. A good, compact plant. *L. a. 'Twicket Purple'* Lavender-blue to purple flowers, in midsummer. A tight-growing form with grey-green foliage. *L. a. 'Vera'* (Dutch Lavender) Lavender-blue flowers in midsummer. Foliage grey-green and broader than most. Good, strong-growing variety, slightly more than average height and spread.

Average height and spread
Five years
1½x2ft (50x60cm)
Ten years
2½x2½ft (80x80cm)
Twenty years
or at maturity
3x3ft (1x1m)

LAVANDULA STOECHAS

FRENCH LAVENDER, SPANISH LAVENDER
Labiatae
Evergreen
An interesting, fragrant, midsummer-flowering shrub, but rather difficult to find.

Origin From the Mediterranean area.
Use As a small shrub for edging a shrub border, or for group plantings. Good in large rock gardens. Plant in threes and fives for best effect.
Description *Flower* Dense terminal heads of scented, dark purple flowers, of an interesting shape, borne mid to late summer. *Foliage* Leaves narrow, lanceolate, ½-1in (1-3cm) long, grey-green, aromatic. *Stem* Short, branching, spreading habit, forming a spreading mound with flowers borne on long, tall, upright flower stems. Medium growth rate when young, very slow with age, except for the flower spikes. *Fruit* Insignificant, but

Lavandula stoechas in flower

154

dead flowerheads are maintained well into late autumn and early winter.
Hardiness Tolerates 14°F (−10°C).
Soil Requirements Prefers light open soil, dislikes waterlogging.
Sun/Shade aspect Full sun, dislikes any shade.
Pruning A light trimming in early spring encourages rejuvenation and increases the number of flower spikes.
Propagation and nursery production From semi-ripe cuttings taken in midsummer. Purchase container-grown. Easy to find in specialist nurseries, seldom elsewhere. Best planting heights 4-12in (10-30cm).
Problems Often overlooked when offered for sale, as it does not present itself well in containers, although once planted achieves rapid spring growth.
Average height and spread
Five years
1x2ft (30x60cm)
Ten years
1½x3ft (50cmx1m)
Twenty years
or at maturity
1½x3ft (50cmx1m)

LAVATERA OLBIA

TREE MALLOW, TREE LAVATERA
Malvaceae
Deciduous
A late-flowering shrub with beautiful pink flowers, requiring specific pruning, but a fine addition to the garden.

Origin From southern France.
Use As a late-flowering summer shrub, either on its own or in a large shrub border.

Lavatera olbia in flower

Description *Flower* Single, large, saucer-shaped flowers, pink to silver-pink, produced from early summer through until autumn frosts. Flowers open progressively from mid-way down flowering shoot, last for a short period and then perish as the next 2-3 flowers open, continuing in this way to end of flowering spike. *Foliage* Leaves broad, elliptic, five-lobed, 3-4in (8-10cm) long and wide, light green to grey-green. *Stem* Strong, upright, grey-green shoots produced from a central crown each spring. Very fast annual growth of new shoots. *Fruit* Insignificant.
Hardiness Minimum winter temperature 23°F (−5°C).
Soil Requirements Light, open, dry soil. Requires adequate feeding to encourage production of long flower shoots.
Sun/Shade aspect Full sun. Tolerates very light shade.
Pruning All previous year's shoots should be cut right back annually in early spring. Hard pruning encourages flowering.
Propagation and nursery production From hardwood cuttings taken in winter or from

semi-ripe cuttings in early summer. Purchase container-grown. Normally found in specialist nurseries. Best planting heights 15in-2ft (40-60cm).
Problems Can be late to break into growth and may show little life until early summer, but grows rapidly once dormancy is broken. Plants in containers tend to look sickly and unsightly. Flower size and colour is dependent on maximum sunlight — the duller the summer, the less the flower size and colour intensity.
Average height and spread
Five years
8x6ft (2.5x2m)
Ten years
10x10ft (3x3m)
Twenty years
or at maturity
10x16ft (3x5m)

LEPTOSPERMUM

MANUKA, TEA TREE, NEW ZEALAND TEA TREE
Myrtaceae
Evergreen
With the correct soil and adequate protection, an extremely attractive summer-flowering shrub.

Origin From New Zealand; many forms now of garden origin.
Use As a summer-flowering shrub in open borders for very mild areas or as a wall shrub in harsher climates, on sunny or lightly shaded, sheltered walls. Very good for growing in large containers 2½ft (80cm) in diameter and 2ft (60cm) in depth, using a good quality potting medium, and can be used as patio shrubs; but need protection under glass in areas experiencing winter temperatures below 23°F (−5°C).
Description *Flower* Masses of long-lasting small, saucer-shaped flowers ranging from white through pink to red, depending on variety, borne profusely along upright stems and branches, early to mid summer. *Foliage* Leaves small, narrow, round-ended lanceolate, ½-¾in (1-2cm) long, normally purple-green. *Stem* Purple to grey-purple. Long, upright shoots with a feathery, branching framework, forming an upright, pyramidal to cigar-shaped shrub; some spreading forms. Medium to slow growth rate. *Fruit* Insignificant.
Hardiness Tolerates 23°F (−5°C).
Soil Requirements Moist, rich, neutral to acid soil; dislikes any alkalinity.
Sun/Shade aspect Full sun, tolerates light shade.
Pruning None required. Branches can be shortened and will rejuvenate.

Propagation and nursery production From semi-ripe cuttings taken early to mid summer. Purchase container-grown. Normally found in nurseries in coastal regions and garden centres in mild, acid areas. Some varieties may be hard to find. Best planting heights 1¼-2½ft (40-80cm).
Problems Must have a favourable climate and the right soil type to survive.
Varieties of interest *L. cunninghamii* White flowers in midsummer, silver-grey leaves and red stems. One of the hardiest forms. From Australia. *L. scoparium 'Chapmanii'* Bright pink flowers contrasting well with brown-green foliage. *L. s. 'Decumbens'* Covered with pale pink flowers, early summer. A low, semi-spreading form, reaching one-third average height and spread. *L. s. 'Jubilee'* Double pink flowers, becoming scarlet-red. Foliage red-purple. Long flowering. *L. s. 'Nichollsii'* Carmine-red flowers against dark purple-bronze foliage. *L. s. 'Red Damask'* Double red flowers, very long lasting. Purple-bronze foliage. *L. s. 'Snow Flurry'* Pure white double flowers. *L. s. 'Spectro Color'* White outer petals, red centres to each flower, produced very freely in late spring, early summer. Foliage light green with purple margins, veins and stems.
Average height and spread
Five years
6x3ft (2x1m)
Ten years
10x6ft (3x2m)
Twenty years
or at maturity
13x10ft (4x3m)

LESPEDEZA THUNBERGII
(Lespedeza sieboldii)

THUNBERG LESPEDEZA, BUSH CLOVER
Leguminosae
Deciduous
An interesting autumn-flowering shrub, although its growth habit is sometimes ungainly.

Origin From China and Japan.
Use As a late summer, early autumn-flowering shrub, planted singly or in a shrub border.
Description *Flower* Good-sized, open, hanging racemes of small, rose purple pea-flowers. *Foliage* Leaves trifoliate, round-ended, 1-1½in (3-4cm) long, silver-grey giving some yellow autumn colour. Develops late in season, growth sometimes not appearing until early summer. *Stem* Long, ranging branches, annually produced, forming loose, wide-spreading, weeping shrub, often arching down to the ground. Fast annual growth rate.
Fruit Insignificant

Leptospermum scoparium 'Jubilee' in flower

Hardiness Minimum winter temperature 23°F (−5°C).
Soil Requirements Rich, open, well-drained soil. Tolerates acidity and alkalinity.
Sun/Shade aspect Prefers full sun, tolerates very light shade. Any deeper shade spoils shape of shrub and severely reduces flowers.
Pruning Cut to ground level every spring it will produce long, ranging flowering shoots.
Propagation and nursery production From semi-ripe cuttings taken in early summer or hardwood cuttings taken in early autumn. Purchase container-grown. Must be sought from specialist nurseries. Best planting heights 1-2ft (30-60cm).

Lespedeza thunbergii **in flower**

Problems The weeping, ranging habit of this shrub should be borne in mind before planting it in a particular position.
Varieties of interest *L. bicolor* Short racemes of purple-pink flowers, mid to late summer. More upright in habit. *L. thunbergii 'Alba Flora'* A white flowering form. Very scarce.
Average height and spread
Five years
4x6ft (1.2x2m)
Ten years
4x6ft (1.2x2m)
Twenty years
or at maturity
4x6ft (1.2x2m)

Origin From south-eastern USA.
Use As a medium-sized shrub for underplanting in shady, acid-soil areas.
Description *Flower* Hanging racemes of white pitcher-shaped flowers, mid to late spring. *Foliage* Leaves broad, lanceolate, 2-4in (5-10cm) long, green, leathery-textured, with underlying purple tinge. Becoming rich red or bronze-purple in autumn and winter. *Stem* Upright when young, quickly becoming arching and spreading, forming a dense canopy of growth. Some underground suckering of new growth. Light green with purple shading. Medium growth rate slowing with age. *Fruit* Insignificant.
Hardiness Tolerates 14°F (−10°C).
Soil Requirements Rich, deep, acid to neutral soil. Dislikes any alkalinity.
Sun/Shade aspect Prefers light shade but tolerates quite deep shade; dislikes full sun.
Pruning Remove one-third of oldest wood to ground level each spring on established plants to encourage growth of new, clean, attractive foliage.
Propagation and nursery production From softwood cuttings taken in early summer; or from layers. Purchase container-grown or root-balled (balled-and-burlapped). Fairly easy to find in nurseries in acid soil areas. Best planting heights 1¼-2½ft (40-80cm).
Problems None, if planted on acid soil.
Varieties of interest *L. f. 'Rainbow'* syn. *L. f.*

Leucothoe fontanesiana 'Rainbow'

'Multicolor' A variety with white flowers and attractive creamy yellow and pink variegated foliage. For best results it should be planted in medium to light shade; dense shade may spoil foliage colour. *L. f. 'Rollissonii'* Similar to the parent, but with narrower leaves. White pitcher-shaped flowers and good winter foliage colour.
Average height and spread
Five years
3x3ft (1x1m)
Ten years
5x6ft (1.5x2m)
Twenty years
or at maturity
5x10ft (1.5x3m)

Origin From the Himalayas.
Use As flowering shrub on its own or in a large shrub border. Good for mass planting, especially for winter effect. Combines well with water features.
Description *Flower* Long, hanging, broad, thick, heavy racemes of white flowers surmounted by purple-red bracts, late summer. *Foliage* Leaves pointed, elliptic, 2-7in (5-

18cm) long, dark green to olive green, with some red shading, giving some yellow autumn colour. *Stem* Hollow, strong, upright shoots, annually produced, becoming slightly spreading with age and forming an upright, spreading-topped shrub. Bright green stems, good for winter display. Fast annual growth rate. *Fruit* Flowers succeeded by broad, thick racemes of purple fruit, surmounted by purple bracts and produced in late autumn.
Hardiness Tolerates 14°F (−10°C).
Soil Requirements Almost any soil but may show signs of distress and chlorosis on extremely alkaline areas, due to lack of moisture rather than alkaline content.
Sun/Shade aspect Full sun to medium shade.
Pruning Reduce completely to ground level in early to mid spring, to encourage rejuvenation.
Propagation and nursery production From seed. Purchase container-grown. Availability varies. Best planting heights 1½-2½ft (50-80cm).
Problems Fruits attract insects. In ideal conditions the shrub may seed itself so freely as to become a nuisance.
Average height and spread
Five years
8x8ft (2.5x2.5m)
Ten years
8x12ft (2.5x3.5m)
Twenty years
or at maturity
8x12ft (2.5x3.5m)

Origin From Japan.
Use As novelty shrub for large rock gardens, as an individual specimen, or in a good-sized container, using good potting medium.
Description *Flower* Short, triangular panicles of white, musty-scented flowers sometimes produced by very mature shrubs in midsummer. *Foliage* Leaves almost round, very dark green, glossy upper surfaces, grey-green, dull undersides, very leathery, thick texture and closely bunched on stems. *Stem* Short, gnarled branch system, grey-green, very slow growth rate. *Fruit* Insignificant.
Hardiness Tolerates 4°F (−15°C).
Soil Requirements Most soils, including acid

Leycesteria formosa **in flower**

Ligustrum japonicum 'Rotundifolium' in leaf

and alkaline but dislikes extremely dry or extremely wet types.

Sun/Shade aspect Prefers light shade, tolerates full sun to medium shade.

Pruning None required.

Propagation and nursery production From semi-ripe cuttings, early to mid summer. Sometimes grafted on *L. vulgare*. Purchase container-grown. Fairly difficult to find. Best planting heights 1-1½ft (30-50cm).

Problems Very slow growth rate and takes 2-5 years from planting for a plant of any size to be formed.

Average height and spread

Five years
1½x1ft (50x30cm)
Ten years
2½x1½ft (80x50cm)
Twenty years
or at maturity
3x3ft (1x1m)

LIGUSTRUM LUCIDUM

GLOSSY PRIVET, CHINESE PRIVET, WAXLEAF PRIVET

Oleaceae

Evergreen

A useful group of evergreen shrubs with some striking variegated varieties, not so widely planted as they deserve.

Origin From China.

Use As a tall-growing evergreen shrub with white summer flowers. Can be trimmed to a single stem, becoming in time a conical evergreen tree.

Description *Flower* Good-sized panicles of musty scented flowers on mature wood, late summer. *Foliage* Pointed, ovate leaves, 3-6in

Ligustrum lucidum 'Tricolor' in leaf

(8-15cm) long, glossy green upper surfaces, duller undersides; some white, silver, and gold variegated forms. *Stem* Strong, vigorous, upright, becoming branching and spreading with age, light green to green-purple, normally producing large, dome-shaped shrub, or small to medium-sized tree of attractive stem formation. Medium to fast growth rate. *Fruit* Clusters of dull blue to blue-black fruit in late summer, early autumn.

Hardiness Tolerates 4°F (−15°C).

Soil Requirements Any soil, but dislikes extremely dry, alkaline conditions or extremely waterlogged areas.

Sun/Shade aspect Full sun to medium shade. Tolerates deep shade, but may become slightly deformed.

Pruning None required. Can be cut back hard and will quickly rejuvenate, even from very old stems. Can be clipped for hedging.

Propagation and nursery production From semi-ripe cuttings taken mid to late summer, or hardwood cuttings taken late autumn to early winter. Purchase container-grown or root-balled (balled-and-burlapped). Fairly easy to find. Best planting heights 1½-2½ft (50-80cm).

Problems The green form is vigorous and may outgrow the area allowed for it. Golden-leaved varieties may suffer sun scorch in strong sunlight.

Varieties of interest *L. l. 'Aureum'* Bright yellow to golden spring foliage, this colouring declining as summer progresses. Two-thirds average height and spread and less hardy than the parent. May be hard to find. *L. l. 'Excelsum Superbum'* Foliage margined and mottled with deep yellow or creamy white. Can be trained as small tree but slow to achieve this and seems to be less hardy in this form. Two-thirds average height and spread. May be hard to find. *L. l. 'Tricolor'* White to cream variegation, leaf margins tinged pink when young, making a very pleasant colour combination. Two-thirds average height and spread. Slightly less hardy than the parent. May be hard to find.

Average height and spread

Five years
6x6ft (2x2m)
Ten years
13x13ft (4x4m)
Twenty years
or at maturity
26x26ft (8x8m)

LIGUSTRUM OVALIFOLIUM

OVAL-LEAVED PRIVET, HEDGING PRIVET, CALIFORNIA PRIVET

Oleaceae

Deciduous to Semi-evergreen

An extremely useful ally for the purpose of hedging or screening, with golden and silver variegated forms which produce an abundance of material for flower arranging.

Origin From Japan.

Use As hedging; if planted in single line 1½ft (50cm) apart makes a very compact, solid, semi-evergreen hedge. Can also be used as tall-growing windbreak.

Description *Flower* Short, upright panicles of off-white flowers with a musty scent, borne only on untrimmed shrubs, in midsummer. *Foliage* Leaves ovate, pointed, 1-1½in (3-4cm) long, mid to dark green with glossy upper surfaces and lighter, duller green undersides. Often maintained to midwinter in mild conditions, but cannot be relied upon as a full evergreen. *Stem* Strong, green-grey to green-brown, upright, vigorous, becoming very branching with age, forming a round-topped, dome-shaped shrub. Fast growth rate. *Fruit* Small clusters of dull black berries produced only on mature, untrimmed shrubs.

Hardiness Tolerates −13°F (−25°C), but loses more leaves the colder the winter conditions.

Soil Requirements Most soils, but distressed by extremely dry or very thin, alkaline types.

Sun/Shade aspect Tolerates full sun to very deep shade, but heavy shade leads to very open and lax habit.

Pruning May be reduced to ground level and will rejuvenate quite quickly, or may be trimmed back hard; both operations decrease flowering on mature plants.

Propagation and nursery production From semi-ripe cuttings taken in late spring or early summer, or hardwood cuttings taken in winter. Plant bare-rooted or container-grown. Availability varies. Best planting heights 1¼-3ft (40cm-1m).

Ligustrum ovalifolium 'Argenteum' in leaf

Problems Privet roots are extremely invasive and draw all plant nutrients out of surrounding soil up to 6-10ft (2-3m) from base of shrub. Consequently it is difficult for other plants to grow in its immediate vicinity. Golden-leaved varieties may suffer sun scorch in strong sunlight.

Varieties of interest *L. o. 'Argenteum'* Leaves grey-green with creamy white margins. Mature, unpruned shrubs may produce off-white flowers not particularly visible against foliage. Current season's growth should be cut by half or more in early spring, to encourage prolific regeneration of creamy white foliage. Two-thirds average height and spread. *L. o. 'Aureum'* (Golden Privet) Rich golden yellow leaves with green centres. Shrubs make a spectacular hedge if planted 18in (50cm) apart in single line. Also good left freestanding to grow to ultimate size. If required for foliage effect, current season's growth should be cut by half or more in early spring to encourage regeneration of golden foliage. Two-thirds average height and spread.

Average height and spread

Five years
10x10ft (3x3m)
Ten years
13x13ft (4x4m)
Twenty years
or at maturity
20x20ft (6x6m)

LIGUSTRUM QUIHOUI

KNOWN BY BOTANICAL NAME

Oleaceae

Deciduous

An aristocrat among ornamental flowering privets, making an extremely fine display in late summer.

Origin From China.

Use As a late summer-flowering shrub for shrub borders or as an individual or group specimen.

Description *Flower* Large, 10-12in (25-30cm) long, graceful white to creamy white flower panicles produced profusely, mid to late

summer. *Foliage* Leaves narrow, ovate, 1-1½in (3-4cm) long, somewhat sparsely produced on branches. Grey to olive green giving some yellow autumn colour. *Stem* Upright at first, quickly becoming ranging and arching, producing an informal, open and somewhat lax effect. Light grey-green when young, becoming dark green to green-brown with age. Medium growth rate, slowing with age. *Fruit* Clusters of blue to blue-black fruits, early to late autumn.
Hardiness Tolerates 4°F (−15°C).
Soil Requirements Most soils, but may be distressed by very alkaline or waterlogged conditions.
Sun/Shade aspect Full sun to medium shade.
Pruning None required. May be cut back hard and will rejuvenate.

Ligustrum quihoui in flower

Propagation and nursery production From semi-ripe cuttings taken midsummer or from hardwood cuttings in early winter. Purchase container-grown. Rather hard to find, should be sought from specialist nurseries. Best planting heights 2-2½ft (60-80cm).
Problems The young plant gives no indication of its true flowering potential.
Average height and spread
Five years
4x4ft (1.2x1.2m)
Ten years
5½x5½ft (1.8x1.8m)
Twenty years
or at maturity
13x13ft (4x4m)

LIGUSTRUM SINENSE

CHINESE PRIVET
Oleaceae
Deciduous
A handsome shrub, with an attractive variegated form providing useful material for flower arranging.

Origin From China.
Use As a tall, flowering evergreen for windbreaks or for large shrub borders.
Description *Flower* Good-sized panicles of white flowers freely produced in midsummer. *Foliage* Leaves narrow, pointed, ovate, 1-3in (3-8cm) long, mid green, giving some limited autumn colour. *Stem* Strong and upright when young, becoming branching with age, forming a high, wide-spreading clump. Light grey-green to green-brown. Medium growth rate. *Fruit* Clusters of black to blue-black fruits follow flowers in early autumn, and are often retained into winter.
Hardiness Tolerates winter temperatures down to −13°F (−25°C).
Soil Requirements Most soils but dislikes extremely dry, alkaline or waterlogged conditions.
Sun/Shade aspect Full sun to medium shade. In deep shade may become open, lax and shy to flower.

Ligustrum sinense in flower

Pruning None required. May be trimmed very hard and will regenerate rapidly. Can even be cut to ground level.
Propagation and nursery production From semi-ripe cuttings taken in midsummer or hardwood cuttings in late winter. Purchase container-grown or bare-rooted. Variegated forms fairly easy to find, green form less often seen. Best planting heights 2-2½ft (60-80cm).
Problems The true beauty of this privet is not always appreciated, because it is often kept trimmed and not allowed to reach full stature.
Varieties of interest *L. s. 'Variegatum'* Narrow, ovate, grey-green leaves, with white to creamy white margins; white flowers. Slightly less than average height and spread. Current season's growth may be cut by half in early spring, to encourage new, prolific growth of variegated leaves.
Average height and spread
Five years
6x10ft (2x3m)
Ten years
13x20ft (4x6m)
Twenty years
or at maturity
20x26ft (6x8m)

LIGUSTRUM 'VICARYI'

GOLDEN VICARY PRIVET, GOLDEN PRIVET
Oleaceae
Deciduous to semi-evergreen
An attractive golden-leaved privet not as invasive as some types, nor subject to sun scorch as are other golden varieties.

Origin Of garden origin, a cross between *L. ovalifolium 'Aureum'* and *L. vulgare*.
Use As a golden foliage shrub, especially useful for flower arranging. Will make a moderately formal small hedge if planted in a single line 1½ft (50cm) apart.
Description *Flower* Small to medium, closely bunched panicles of creamy white flowers produced on mature shrubs, early to midsummer. *Foliage* Leaves elliptic, 1-2in (3-5cm) long, delicately suffused lime green, spring, aging to golden yellow in summer. *Stem* Upright, becoming slightly spreading, but maintaining its dome formation. Grey-green becoming green-brown. Medium growth rate when young, becoming slower. *Fruit* Small clusters of black fruit on mature, unpruned shrubs.
Hardiness Tolerates winter temperatures down to −13°F (−25°C).
Soil Requirements Most soils but dislikes extremely dry, alkaline or very waterlogged conditions.
Sun/Shade aspect Full sun to light shade. Deeper shade lessens intensity of golden variegation.

Pruning For best foliage effect, current season's growth should be cut by at least half in early spring, to encourage prolific regeneration of golden foliage.
Propagation and nursery production From semi-ripe cuttings taken early to mid summer or from hardwood cuttings in late winter to early spring. Purchase container-grown. Obtainable from specialist nurseries. Best planting heights 1¼-2½ft (40-80cm).

Ligustrum 'Vicaryi' in leaf

Problems None.
Average height and spread
Five years
3x5ft (1x1.5m)
Ten years
5x10ft (1.5x3m)
Twenty years
or at maturity
10x16ft (3x5m)

LIGUSTRUM VULGARE

COMMON PRIVET, EUROPEAN PRIVET
Oleaceae
Deciduous
A pleasant flowering shrub which makes good windbreaks or screening.

Origin From Europe.
Use As a large summer-flowering shrub of use only for very large shrub borders or as screening or windbreaks, especially in coastal areas.
Description *Flower* Short, wide-based racemes of off-white flowers in mid to late summer. *Foliage* Narrow, ovate, pointed leaves, 1-1½in (3-4cm) long, dark green.

Ligustrum vulgare in flower

Stem Somewhat lax and open compared to other forms, producing round, loose shrub or, with time, small, multi-stemmed tree. Grey to grey-green, becoming grey-brown. Fast growth rate. *Fruit* Small clusters of black fruits in autumn, which may be retained into winter.
Hardiness Tolerates winter temperatures below −13°F (−25°C).
Soil Requirements Any soil conditions.
Sun/Shade aspect Full sun to deep shade.
Pruning None required. May be trimmed and cut as hedging. Large, old shrubs can be cut back hard to ground level and will rejuvenate.
Propagation and nursery production From semi-ripe cuttings taken in midsummer or hardwood cuttings in late winter to early spring. Plant bare-rooted or container-grown. Best obtained from nurseries specializing in large-scale forestry production. Best planting heights 1¼-3ft (40cm-1m).
Problems A little uninteresting when young and does not display its potential until mature.
Average height and spread
Five years
10x10ft (3x3m)
Ten years
13x13ft (4x4m)
Twenty years
or at maturity
16x16ft (5x5m)

LIPPIA CITRIODORA (Aloysia triphylla)

LEMON VERBENA, SHRUBBY VERBENA, LEMON PLANT
Verbenaceae
Deciduous
A very attractive shrub with aromatic, lemon-scented foliage and stems.

Origin From Chile.
Use As a moderately low shrub for a sunny borders. Can be fan-trained on sunny or lightly shaded wall with good effect. Can be trained after planting as a short standard mop-head tree but in this form needs over-wintering under cover. Suitable for containers and tubs 2½ft (80cm) across and 2ft (60cm) deep; use a good potting medium.
Description *Flower* Short to medium-sized, open terminal panicles of small pale blue-mauve florets in midsummer. *Foliage* Leaves lanceolate, 3-4in (8-10cm) long, grey-green to sea green, giving off lemony, aromatic scent when crushed. *Stem* In cooler areas dies back completely in winter, in milder areas is maintained as a loose, open shrub. Grey-green, lemon-scented. *Fruit* Insignificant.
Hardiness Minimum winter temperature 23°F

(−5°C). In cold areas may need protection of a wall.
Soil Requirements Open, well-drained, warm soil.
Sun/Shade aspect Full sun to very light shade; requires a warm situation for best results.
Pruning Either cut to ground level or remove one-third of oldest wood each spring to encourage new strong shoots with good aromatic foliage.

Lippia citriodora in flower

Propagation and nursery production From semi-ripe cuttings taken in midsummer for overwintering under cover and planting out in spring, or earlier if planting area is completely frost-free. Purchase container-grown. Best planting heights 1¼-2ft (40-60cm).
Problems Often a poor looking specimen when purchased container grown, but grows rapidly once planted out.
Average height and spread
Five years
2½x2½ft (80x80xm)
Ten years
3x3ft (1x1m)
Twenty years
or at maturity
4x4ft (1.2x1.2m)

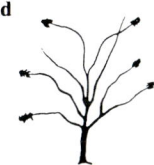

LOMATIA FERRUGINEA

KNOWN BY BOTANICAL NAME
Proteaceae
Evergreen
A rarely-seen shrub, but if the correct protection, climatic conditions and acid soil are provided, it can make an impressive subject.

Origin From Chile.
Use As a specimen foliage shrub with other acid-loving shrubs such as rhododendrons.
Description *Flower* Short racemes of small buff and scarlet flowers in midsummer. *Foliage* Leaves fern-like, very divided, 8in

(20cm) long and 4in (10cm) wide, deep green. *Stem* Upright, erect stems with red-brown velvety texture, forming upright shrub with occasional spreading branch. Medium growth rate. *Fruit* Insignificant.
Hardiness Minimum winter temperature 23°F (−5°C).
Soil Requirements Acid soil, dislikes any alkalinity.
Sun/Shade aspect Full sun to light shade.
Pruning None required.
Propagation and nursery production From softwood cutings taken in late summer, or from seed. Purchase container-grown. Rather hard to find. Best planting heights 1¼-2ft (40-60cm).
Problems Often looks very poor and weak in containers, but on correct acid soil quickly presents itself attractively.
Varieties of interest *L. myricoides* syn. *L. longifolia* Long, narrow, light grey-green leaves up to 3in (8cm) long and ½in (1cm) wide. Long, strap-like white flowers, very fragrant, borne in late summer. *L. tinctoria* A variety with pinnate to double pinnate, very narrow, long leaflets, light to mid green. Flowers sulphur yellow in bud, becoming creamy white, produced in long racemes at the terminals of each shoot. Half average height and spread. Less hardy than *L. ferruginea*. From Tasmania.

Lomatia myricoides in flower

Average height and spread
Five years
6x6ft (2x2m)
Ten years
10x10ft (3x3m)
Twenty years
or at maturity
13x13ft (4x4m)
In mild, favourable conditions may reach 26ft (8m) height and spread.

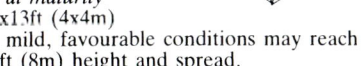

LONICERA FRAGRANTISSIMA

HONEYSUCKLE, WINTER HONEYSUCKLE, SHRUBBY HONEYSUCKLE
Caprifoliaceae
Semi-evergreen
A delightful scented, winter-flowering shrub.

Origin From China.
Use As a winter-flowering shrub for a large shrub border. Often grown as a wall shrub where it tolerates exposed conditions and may grow rather taller than it would in an open situation.
Description *Flower* Small, sweetly scented, creamy white flowers, produced on almost leafless branches in mild spells, late autumn through to mid spring, standing up well to light to moderate frosts. *Foliage* Leaves ovate, 1-2½in (3-6cm) long, dark green ting-

Lonicera fragrantissima **in winter**

ed purple with lighter undersides. Some autumn colour. *Stem* Upright when young, becoming branching and spreading with age, forming a round, mound-shaped shrub. Light grey-green when young, becoming darker green with some purple veining. Medium growth rate when young, becoming slower with age. *Fruit* May produce bunches of red fruits in early to mid spring.
Hardiness Tolerates winter temperatures down to −13°F (−25°C).
Soil Requirements Any soil, tolerating quite dry areas.
Sun/Shade aspect Light to medium shade, but tolerates full sun.
Pruning None required, but may be reduced in size.
Propagation and nursery production From semi-ripe cuttings taken in early summer. Purchase container-grown. Fairly easy to find. Best planting heights 1¼-2½ft (40-80cm).
Problems Young plants give no indication of their true potential.
Varieties of interest *L. × purpusii* White fragrant flowers, early winter to early spring. Slightly more vigorous than *L. fragrantissima* and with larger leaves but fewer flowers. *L. × p. 'Spring Purple'* White fragrant flowers in winter. New foliage purple-green and stems purple when young. *L. standishii* Fragrant white flowers in winter, produced on stems covered in fine downy hair. Red fruits in early summer.

Average height and spread
Five years
4x6ft (1.2x2m)
Ten years
6x10ft (2x3m)
Twenty years
or at maturity
10x13ft (3x4m)

LONICERA INVOLUCRATA
(Lonicera ledebourii)

TWIN BERRY
Caprifoliaceae
Deciduous
A good, large-growing flowering and fruiting shrub.

Origin From western North America.
Use For a large shrub border; also good for mass planting and suited to coastal or industrial areas.
Description *Flower* Attractive yellow flowers, each with two bright red bracts, early summer. *Foliage* Leaves good-sized, ovate, 2-4in (5-10cm) long, dark green tinged purple. Some yellow autumn colour. *Stem* Strong, upright when young, becoming branching and spreading with age. Light green aging to

grey-brown. Fast growth rate when young, slowing with age. *Fruit* Shiny black fruits, each with the two red flower bracts intact, produced in early autumn and often retained into early winter.
Hardiness Tolerates winter temperatures down to −13°F (−25°C).
Soil Requirements Any soil, extremely adaptable.
Sun/Shade aspect Full sun to medium shade. In deep shade will be lax and likely to produce few flowers.
Pruning Remove one-third of old flowering shoots to ground level in early spring, to encourage good new growth, which will increase flowers and fruit.
Propagation and nursery production From semi-ripe cuttings taken in midsummer. Purchase container-grown, bare-rooted or root-balled (balled-and-burlapped). Easier to find in general nurseries than in garden centres. Best planting heights 2-2½ft (60-80cm).
Problems It is a large shrub and to do well must be grown as such. Attempts to curtail it reduce flower and fruit displays.

Lonicera involucrata **in flower**

Varieties of interest *L. maackii* (Amur honeysuckle) Ovate, mid green foliage with some yellow autumn colouring. Leaves 2-2½in (5-6cm) long and 1in (3cm) wide. Flowers white, fragrant, yellowing with age, produced in midsummer, followed by long-lasting dark red berries in late summer, early autumn. One-third greater than average height and spread.

Average height and spread
Five years
5x6ft (1.5x2m)
Ten years
7x10ft (2.2x3m)
Twenty years
or at maturity
10x13ft (3x4m)

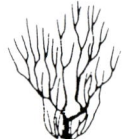

LONICERA NITIDA

BOXLEAF HONEYSUCKLE, POOR MAN'S BOX
Caprifoliaceae
Evergreen
A useful hedging or large-scale ground cover shrub, the golden-leaved varieties having a decorative effect.

Origin From western China.
Use As a medium-sized evergreen shrub. Good for ground cover or for mass planting on banks. Planted 1ft (30cm) apart in a single line forms a good, low hedge, but if allowed to grow above 3ft (1m) can become very lax.
Description *Flower* Inconspicuous, small, scented, sulphur yellow flowers. *Foliage* Leaves small, ovate, ¼-½in (5mm-1cm) long, mid green, shiny, with silver undersides, freely produced tightly along branches and shoots. Slight purple tinge when young. *Stem*

Upright and branching when young, becoming very full and twiggy with age; mature plants may become lax and be damaged by heavy snow. Purple-green when young, becoming dark green with purple shading. Fast growth rate. *Fruit* Insignificant.
Hardiness Tolerates 14°F (−10°C).
Soil Requirements Most soils but dislikes extreme dryness or waterlogging.
Sun/Shade aspect Full sun to mid shade. In deep shade may become very lax and open, and golden-leaved varieties will lose their colour.

Lonicera nitida 'Baggesen's Gold' **in flower**

Pruning If left freestanding, remove one-third of old wood each spring to rejuvenate. May be cut to ground level and will regrow, and may also be trimmed for hedging.
Propagation and nursery production From semi-ripe cuttings taken in late summer or hardwood cuttings in midwinter. Plant bare-rooted or container-grown. Best planting heights 1-2ft (30-60cm).
Problems Becomes woody and uninteresting as it ages. Hard pruning corrects this.
Varieties of interest *L. n. 'Baggesen's Gold'* Yellow foliage, often turning more golden in winter when in full sun; in light shade turns yellow-green in autumn. Dislikes dry or waterlogged soils and shows ill effects immediately by losing foliage. A useful evergreen for a shrub border, hedging, training or growing in a tub. Two-thirds average height and spread. *L. n. 'Yunnan'* syn. *L. yunnanensis* A green-leaved variety with larger foliage, more freely flowering and fruiting. Upright, typically offered by nurseries for hedging.

Average height and spread
Five years
3x6ft (1x2m)
Ten years
5x12ft (1.5x3.5m)
Twenty years
or at maturity
5½x14½ft (1.8x4.5m)

LONICERA PILEATA

PRIVET HONEYSUCKLE
Caprifoliaceae
Semi-evergreen
A very useful shrub for large-scale ground cover.

Origin From China.
Use As a large, spreading, low shrub for large shrub borders or mass planted for ground cover on banks or areas of poor soil.
Description *Flower* Small, scented, sulphur yellow flowers in summer. *Foliage* Leaves small, round to ovate, ½-1½in (1-4cm) long light, bright green, borne profusely, making a dense covering. Some yellow autumn colour. *Stem* Upright at first, quickly becoming arching and spreading with roots forming on

undersides of branches wherever they touch ground, from which new upright shoots form. Bright green when young, soon turning grey-green to grey-brown. Medium to fast growth rate. Forms large, spreading carpet. *Fruit* Small, ovate, violet-purple fruits formed on undersides of branches. Excessive fruit production may indicate distress.
Hardiness Tolerates winter temperatures down to −13°F (−25°C).
Soil Requirements Most soils, except extremely alkaline conditions which will lead to chlorosis.
Sun/Shade aspect Full sun to fairly deep shade. In deeper shade may become more lax and open.
Pruning None required. May be reduced in size as necessary.
Propagation and nursery production From semi-ripe cuttings taken in early summer or hardwood cuttings in winter. Also from self-layered plants. Purchase container-grown. Fairly easy to find. Best planting heights 8-12in (20-30cm) with a spread of 1-2ft (30-60cm).

Lonicera pileata in leaf

Problems The vigour of this shrub is often underestimated and it is planted in area which will not accommodate maximum spread. Young plants may lose all foliage after planting, but regrowth occurs in the following spring.
Average height and spread
Five years
1¼x4ft (40cmx1.2m)
Ten years
2½x6ft (80cmx2m)
Twenty years or at maturity
4x13ft (1.2x4m)

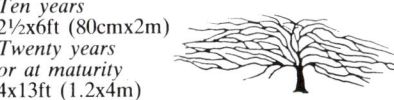

LONICERA SYRINGANTHA

LILAC-SCENTED SHRUBBY HONEYSUCKLE
Caprifoliaceae
Deciduous
A shrub in which sparseness of flowers is compensated by their beautiful fragrance.

Origin From China and Tibet.
Use As a large, spring-flowering shrub with attractive foliage.
Description *Flower* Small, lilac, very fragrant flowers borne sparsely in small clusters along the branches in late spring to early summer. *Foliage* Leaves ovate, ½-1in (1-3cm) long, grey to grey-green, with some yellow autumn colours. *Stem* Upright when young, quickly becoming very branching and twiggy, forming a large, arching shrub. Medium to fast growth rate. *Fruit* Small, round, red fruits in mid to late summer.
Hardiness Tolerates winter temperatures down to −13°F (−25°C).

Lonicera syringantha in flower

Soil Requirements Most soils, but dislikes extremely alkaline types.
Sun/Shade aspect Full sun to light shade. Any deeper shade will disfigure the shrub.
Pruning Remove one-third of oldest flowering wood to ground level after flowering to induce rejuvenation.
Propagation and nursery production From semi-ripe cuttings taken in early summer. Purchase container-grown. More likely to be found in nurseries than garden centres. Best planting heights 1¼-2½ft (40-80cm).
Problems Overall size is often underestimated, so the shrub is planted in areas unable to accommodate full spread.
Average height and spread
Five years
3x6ft (1x2m)
Ten years
6x10ft (2x3m)
Twenty years or at maturity
10x13ft (3x4m)

LONICERA TATARICA

TATARIAN HONEYSUCKLE
Caprifoliaceae
Deciduous
An interesting late spring, early summer-flowering shrub, provided mildew attacks are controlled.

Origin From USSR.
Use In a large shrub border or for mass planting.
Description *Flower* Small, pink flowers profusely borne in mid spring to early summer. White and red flowering varieties also available. *Foliage* Elliptic, grey-green, somewhat sparsely produced. *Stem* Upright, becoming branching with age, grey-green. Medium growth rate when young, becoming slower with age. *Fruit* Small, red fruits, late summer, early autumn.
Hardiness Tolerates winter temperatures down to −13°F (−25°C).
Soil Requirements Any soil.
Sun/Shade aspect Prefers light shade, tolerates full sun to medium shade.
Pruning Remove one-third of old wood after flowering period to induce new growth.
Propagation and nursery production From semi-ripe cuttings taken in early summer. Purchase container-grown. Fairly easy to find in nurseries and some garden centres, especially when in flower. Best planting heights 1¼-2½ft (40-80cm).
Problems Suffers badly from attacks of mildew which can lead to complete defoliation.
Varieties of interest *L. t. 'Alba'* White flowers and red fruits. *L. t. 'Hack's Red'* Deep rose pink flowers and red fruits.

Lonicera tatarica in flower

Average height and spread
Five years
10x10ft (3x3m)
Ten years
10x13ft (3x4m)
Twenty years or at maturity
10x20ft (3x6m)

LYCIUM BARBARUM

DUKE OF ARGYLL'S TEA TREE
Solanaceae
Deciduous
A very useful shrub for coastal areas, standing up to salt-laden winds.

Origin From China, but naturalized throughout Europe and western Asia.
Use In areas where a single shrub is required for covering a large expanse. Possibly too invasive for general planting. Often used as windbreak for coastal sites.
Description *Flower* Small, purple, trumpet-shaped flowers in clusters 1¾-3in (4-8cm) in diameter, produced from each leaf axil from early to mid summer. *Foliage* Leaves narrow, linear, ½-1in (1-3cm) long, sea green, giving some yellow autumn colour. *Stem* Upright when young, quickly becoming arching, forming a spreading, arching shrub. Grey to grey-green with sparsely distributed small spines. Medium to fast growth rate. *Fruit* Small, egg-shaped, orange-red fruits in autumn.
Hardiness Tolerates 4°F (−15°C).
Soil Requirements Light, open, sandy soil.

161

Dislikes waterlogging.
Sun/Shade aspect Full sun, dislikes shade.
Pruning Reduce one-third of oldest shoots on mature shrubs to ground level in early spring, to encourage growth of new shoots for following year.
Propagation and nursery production From seed or semi-ripe cuttings taken in early summer. Purchase container-grown. Availability varies but may be found in coastal areas. Best planting heights 1¼-2ft (40-60cm).
Problems Can become a little invasive if happily situated, but if not, may be difficult to establish.

Lycium chilense **in fruit**

Varieties of interest *L. chilense* Spreading branches covered with yellow, white and purple funnel-shaped flowers, mid to late summer, followed by red, egg-shaped fruits. A spineless variety more open and of slightly less than average height and spread. Less hardy than *L. barbarum*. From Chile.
Average height and spread
Five years
6x6ft (2x2m)
Ten years
6x10ft (2x3m)
Twenty years or at maturity
10x16ft (3x5m)

MAGNOLIA GRANDIFLORA

EVERGREEN MAGNOLIA, LAUREL MAGNOLIA, SOUTHERN MAGNOLIA, BULL BAY

Magnoliaceae
Evergreen
When grown in the correct position, given space and time to mature, this is a truly spectacular evergreen flowering shrub.

Origin From south-eastern USA.
Use As a freestanding shrub, or small tree in frost-free areas. In harsher environments can be grown on a sunny or moderately sunny wall, if this can accommodate its ultimate size.
Description *Flower* Large, up to 10in (25cm) across, creamy white, very fragrant flowers, produced late spring or throughout summer and early autumn. Late flowers susceptible to early autumn frost damage. *Foliage* Leaves elliptic, large, 8-10in (20-25cm) long, attractive bright green when young, maturing to dark green with glossy upper surfaces and duller, brown, felted undersides. *Stem* Strong, upright, becoming branching with age, forming a tall shrub, or small single-stemmed tree in frost-free areas. Mid to dark green. Medium rate of growth. *Fruit* Green fruit pods, 4in (10cm) long, following flowering.
Hardiness Tolerates 14°F (−10°C).
Soil Requirements Does well on most soils, including alkaline as long as there is 6ft (2m) of good topsoil above underlying alkaline soil.

Magnolia grandiflora 'Exmouth' **in flower**

Sun/Shade aspect Full sun to mid shade.
Pruning None required but can be trimmed to control size.
Propagation and nursery production From semi-ripe cuttings taken in early summer. Purchase container-grown. One or other of the varieties likely to be found in most general nurseries and some garden centres. Best planting heights 2-4ft (60cm-1.2m).
Problems May take 5-6 years to come into flower.
Varieties of interest *M. g.* 'Exmouth' Large, creamy white flowers, richly scented, produced at a relatively early age in its life span. Dark green, polished, narrowed foliage with red or brown felted undersides. A useful variety for the colder winter temperature ranges. *M. g.* 'Ferruginea' Large, white, scented flowers. Leaves elliptic to ovate, dark shiny green, brown and heavily felted undersides. Upright and of average height but slightly less spread. More tender than other varieties and should be planted only in very mild areas. *M. g.* 'Goliath' Flowers globe-shaped, white and scented, produced after 3-5 years of planting. Foliage elliptic, slightly concave with rounded ends. Dark glossy green to light green, grey-green undersides. Slightly less than average spread. Not suited to temperatures below 23°F (−5°C). *M. g.* 'Maryland' Large, white, fragrant flowers of more open shape, said to be produced two years after planting. Foliage elliptic to ovate; mid green, shiny upper surfaces and grey-brown undersides. Slightly less than average height and spread.
Average height and spread
Five years
5½x3ft (1.8x1m)
Ten years
10x6ft (3x2m)
Twenty years or at maturity
26x13ft (8x4m)

MAGNOLIA
Large-growing, star-flowered varieties

STAR MAGNOLIA
Magnoliaceae
Deciduous
Large, early spring-flowering shrubs for very attractive featured flower display.

Origin From Japan.
Use As a large freestanding shrub, or small tree. A good feature at the back of a large shrub border if adequate space is available.
Description *Flower* Multi-petalled, star-shaped, white or pink, dependent on variety, fragrant flowers produced in numbers up to

15-20 years after planting, after which flowering increases to give glorious display in mid to late spring. *Foliage* Leaves small, elliptic, 2½-4in (6-10cm) long, light to mid green. Some yellow autumn colour. *Stem* Strong, upright, becoming branching with age, eventually forming a very dense, branching twiggy framework. Dark green to green-brown. Medium growth rate. *Fruit* Small green fruit capsules in late summer.
Hardiness Tolerates winter temperatures down to −13°F (−25°C).
Soil Requirements Most soil types, tolerates alkaline conditions provided depth of topsoil is adequate.
Sun/Shade aspect Prefers full sun, tolerates light shade. As a precaution against frost damage in spring, plant shrub in an area where it does not get early morning sun, so that flowers thaw out slowly, thus limiting tissue damage.

Magnolia × loebneri 'Leonard Messel'

Pruning None required. If desired, can be pruned as a single-stem tree with good results.
Propagation and nursery production From layers or semi-ripe cuttings taken in early summer. Purchase container-grown or root-balled (balled-and-burlapped). Best planting heights 1½-3ft (60cm-1m).
Problems Some varieties, such as *M. kobus*, are slow to come into flower and can take as long as 15 years to produce a full display.
Varieties of interest *M. kobus* (Northern Japanese Magnolia, Kobus Magnolia) White, fragrant flowers produced only after 10-15 years from date of planting. *M. × loebneri* A cross between *M. kobus* and *M. stellata*, which from an early age produces a profusion of multi-petalled, fragrant, white flowers in early to mid spring. Does well on all soil

Magnolia × loebneri 'Merrill' in flower

types, including alkaline. Reaches two-thirds average height and spread. Of garden origin. **M. × l. 'Leonard Messel'** Fragrant, multi-petalled flowers are deep pink in bud, opening to lilac-pink. Said to be a cross between *M. kobus* and *M. stellata 'Rosea'*. Of garden origin, from Nymans Gardens, Sussex, England. **M. × l. 'Merrill'** Large, white, star-shaped, fragrant flowers produced from an early age on a shrub two-thirds average height and spread. Raised in the Arnold Arboretum, Massachusetts, USA. **M. salicifolia** White, fragrant, star-shaped flowers with 6 narrow petals in mid spring. Slightly more than average height but slightly less spread, forming either a large shrub or small tree. Leaves, bark and wood are lemon-scented if bruised. From Japan.

Average height and spread
Five years
6x6ft (2x2m)
Ten years
13x13ft (4x4m)
Twenty years
or at maturity
23x26ft (7x8m)

MAGNOLIA LILIIFLORA
(Magnolia quinquepeta)

LILY MAGNOLIA, LILY-FLOWERED MAGNOLIA
Magnoliaceae
Deciduous
A useful Magnolia coming into bloom in late spring, so unlikely to be at risk of frost damage.

Origin From Japan.
Use As a freestanding shrub on its own or for a large shrub border.
Description *Flower* Buds resemble slender tulips. Open flowers purple outside, creamy white within, late spring to early summer. Occasionally flowers appear later in summer. *Foliage* Leaves elliptic, 4-7in (10-18cm) long, dark glossy green, forming good foil for flowers. *Stem* Upright, becoming spreading and forming a round-shaped shrub. Brown-purple to green-purple, grey shaded with age. Slow to medium rate of growth. *Fruit* Insignificant.
Hardiness Tolerates 14°F (−10°C), although some tip damage to shoots may occur in winter.
Soil Requirements Neutral to acid soils, dislikes alkalinity.
Sun/Shade aspect Prefers very light shade, tolerates medium shade to full sun.
Pruning None required but can be reduced in size in early winter.
Propagation and nursery production From layers or semi-ripe cuttings taken in early summer. Purchase container-grown or root-

balled (balled-and-burlapped). May only be found in specialist nurseries. Best planting heights 1½-3ft (50cm-1m).
Problems Young shrub may look weak and take a number of years to gain a substantial framework.

Magnolia liliiflora 'Nigra' in flower

Varieties of interest M. l. 'Nigra' Buds resembling slender tulips, gradually opening to a reflexed star shape. Flowers deep purple outside, creamy white stained purple inside, late spring through to early summer. From Japan.
Average height and spread
Five years
3x3ft (1x1m)
Ten years
6x6ft (2x2m)
Twenty years
or at maturity
13x13ft (4x4m)

MAGNOLIA ×
SOULANGIANA

SAUCER MAGNOLIA, TULIP MAGNOLIA
Magnoliaceae
Deciduous
Very popular flowering shrubs, which must be given adequate space to develop. None of the beautiful varieties surpasses the splendour of *M. × soulangiana* itself.

Origin Raised by M. Soulange-Bodin at Fromont, near Paris, France in the early 19th century.
Use As a freestanding shrub or eventually a

small tree or for a large shrub border, where adequate space available. All varieties do well as wall-trained shrubs.
Description *Flower* Light pink with purple shading in centre and at base of each petal; flowers produced before leaves in early spring. Flower buds large with hairy outer coat. Some secondary flowering in early summer. *Foliage* Leaves elliptic to ovate, 3-6in (8-15cm) long, light green to grey-green. Some yellow autumn colour. *Stem* Upright, strong when young and light grey-green. In maturity branches become very short, twiggy and almost rubbery in texture, forming a basal skirt. Can be trained as single or multi-stemmed standard trees in favourable areas. Medium rate of growth. *Fruit* Insignificant.
Hardiness Tolerates 4°F (−15°C).
Soil Requirements Does well on heavy clay soils and other types, except extremely alkaline areas which will lead to chlorosis.
Sun/Shade aspect Must be planted away from early morning sun. Otherwise flowers frozen by late spring frosts thaw too quickly and cell damage causes browning.
Pruning None required, but best to remove any small crossing branches in winter to prevent rubbing.
Propagation and nursery production From layers or semi-ripe cuttings taken in early summer. Purchase container-grown or root-balled (balled-and-burlapped). *M. × soulangiana* easy to find but some varieties best sought from specialist nurseries. Best planting heights 2-4ft (60cm-1.2m).
Problems Can take up to 5 years or more to flower well.
Varieties of interest M. × s. 'Alba Superba' syn. **M. × s. 'Alba'** Large, scented, pure white, erect, tulip-shaped flowers, flushed purple at petal bases. Growth upright and strong, but forms slightly less spread than the parent. **M. × s. 'Alexandrina'** Large, upright, white flowers with purple-flushed petal bases. A good, vigorous, upright, free-flowering variety, sometimes difficult to obtain. **M. × s. 'Amabilis'** Ivory white, tulip-shaped flowers, flushed light purple at bases of inner petals. Upright habit. May have to be obtained from specialist nurseries. **M. × s. 'Brozzonii'** Large, longer than average, white flowers with purple shading at base. Not always available in garden centres and nurseries. **M. × s. 'Lennei'** Flowers goblet-shaped with fleshy petals rose purple outside, creamy white stained purple on inner sides, in mid to late spring; in some seasons, repeated limited flowering in autumn. Broad, ovate leaves, up to 25-30cm (10-12in) long. **M. × s. 'Lennei Alba'** Ivory white, extremely beautiful goblet-shaped flowers presenting themelves upright along branches. May need to be obtained from specialist nurseries. Slightly more than average spread. **M. × s. 'Picture'** Purple outer colouring to petals, white inner sides. Flowers erectly borne, often appearing early in the shrub's lifespan. Leaves up to 10in (25cm) long. Somewhat upright branches, reaching

Magnolia × soulangiana 'Alba Superba'

less than average spread. Best sought from specialist nurseries. *M. × s. 'Speciosa'* White flowers with very little purple shading, leaves smaller than average. Slightly less than average height and spread. Best sought from specialist nurseries.

Average height and spread
Five years
6x6ft (2x2m)
Ten years
13x13ft (4x4m)
Twenty years
or at maturity
26x26ft (8x8m)

MAGNOLIA STELLATA

STAR MAGNOLIA, STAR-FLOWERED MAGNOLIA
Magnoliaceae
Deciduous
A well-loved, exceptionally beautiful early spring-flowering garden shrub.

Origin From Japan.
Use As a feature shrub in its own right, or for a large shrub border.
Description *Flower* Slightly scented, white, narrow, strap-like, multi-petalled, star-shaped flowers, 2-2½in (5-6cm) across, borne in early spring before leaves. Usually flowers within two years of planting. *Foliage* Leaves medium-sized, elliptic, 2-4in (5-10cm) long, light green, giving some yellow autumn colouring. *Stem* Upright when young, quickly branching and spreading, to form a dome-shaped shrub slightly wider than it is high. Grey-green. Slow growth rate. *Fruit* Insignificant.
Hardiness Tolerates 4°F (−15°C).
Soil Requirements Any soil type except extremely alkaline.
Sun/Shade aspect Full sun to light shade. As a precaution against frost damage in spring, plant in a position where the shrub will not get early morning sun, so allowing frozen flowers to thaw out slowly, and incur less tissue damage and browning.
Pruning None required.
Propagation and nursery production From layers or cuttings taken in early summer. Purchase container-grown or root-balled (balled-and-burlapped). Normally available from garden centres and nurseries, especially in spring when in flower. Best planting heights 1¼-2½ft (40-80cm).
Problems Young plants when purchased may look small and misshapen due to their slow growth, and can seem expensive, but the investment is well worthwhile.
Varieties of interest *M. s. 'King Rose'* A variety with good pink flowers. *M. s. 'Rosea'* Star-shaped flowers deep pink in bud, opening to flushed pink. *M. s. 'Royal Star'* Slightly larger white flowers with numerous petals making a very full star shape. *M. s. 'Rubra'*

Magnolia stellata **in flower**

MAGNOLIA
Summer-flowering varieties

KNOWN BY BOTANICAL NAME
Magnoliaceae
Deciduous
Summer-flowering Magnolias are among the most magnificent of summer shrubs.

Flowers multi-petalled and purple-pink, deeper colouring while in bud. Rather scarce. *M. s. 'Water Lily'* Larger flowers with more petals. Extremely attractive, but usually a little weaker in constitution than the parent.

Average height and spread
Five years
2½x3ft (80cmx1m)
Ten years
5x6ft (1.5x2m)
Twenty years
or at maturity
8x12ft (2.5x3.5m)

Origin From western China.
Use As a freestanding long-flowering specimen shrub grown on its own to show off the additional beauty of its stem formation.
Description *Flower* Initially, hanging egg-shaped buds opening into white, fragrant, cup-shaped flowers. As buds open, flowers turn outwards. Flowering mainly in late spring and early summer, then intermittently until late summer. *Foliage* Leaves elliptic, 4-5in (10-12cm) long, light grey-green giving some autumn colour. Liable to be blackened by hard early autumn frosts. *Stem* Strong, forming a goblet-shaped shrub. Grey-green. Medium growth rate. *Fruit* Crimson-red fruit clusters in late summer, early autumn.
Hardiness Tolerates 14°F (−10°C).
Soil Requirements Does well on most soils, only disliking extreme alkalinity; specific requirements of certain varieties are described below.
Sun/Shade aspect Full sun through to medium shade.
Pruning None required. May be reduced in size with care; best done in late autumn or early winter.
Propagation and nursery production From layers. Purchase container-grown or root balled (balled-and-burlapped). Fairly easy to find in nurseries, less often stocked by garden centres. Best planting heights 1½-4ft (50cm-1.2m).
Problems Often planted in areas where it is unable to reach its full size and potential.
Varieties of interest *M. denudata* syn. *M. conspicua* Cup-shaped, pure white, fragrant flowers, with broad, thick, fleshy petals, produced mid spring to early summer. Foliage ovate with rounded ends, 3-6in (8-15cm) long, mid to grey-green. Upright habit, becoming rounded with time, reaching the dimensions of a small tree. *M. × highdownensis* White fragrant, hanging flowers with purple central cone 4in (10cm) across. Good on very chalky soils. Of slightly less than average height and spread. Rather hard to find except in specialist nurseries. Found in Highdown Gardens, Sussex, England and possibly a clone of *M. wilsonii*. *M. hypoleuca* syn. *M. obovata* White to creamy white, fragrant flowers up to 6in (15cm) across with central crimson stamens, early summer, followed by interesting fruit clusters in late summer, early autumn. Obovate leaves, up to 6in (15cm) long, grey-green. Dislikes alkaline soils, best on neutral to acid types. A large shrub or small tree. From Japan . *M. sieboldii* syn. *M. parviflora* Fragrant, white, cup-shaped flowers, 3-3½in (8-9cm) across, with rose pink to crimson central stamens. Stems upright whn young, spreading with age. Foliage grey-green wih some yellow autumn colour. Two-thirds average height and spread. *M. sinensis* White, lemon-scented, hanging flowers with red central cones, 4-6in (10-15cm) across, early summer. Foliage grey-green, 5in (12cm) long, obovate. Slightly less than average height and slightly more than average spread.

Magnolia sinensis **in flower**

From western China. *M. virginiana* (Swamp Bay, Sweetbay Magnolia) Creamy white, fragrant flowers, 2in (5cm) across, globular and slightly hanging, early to late summer. Ovate to elliptic, semi-evergreen leaves with glossy upper surfaces and blue-white undersides. Can be trained as a small tree. Tolerates some alkalinity as long as adequate depth of good topsoil available. Rather hard to find. From eastern USA. *M. × watsonii* Strongly scented, creamy white, saucer-shaped flowers facing upwards, with rosy crimson anthers and pink sepals, 5½-6in (13-15cm) across, early to midsummer. Grey-green, oval, leathery leaves. May become small tree in favourable areas. Will tolerate some alkalinity provided adequate depth of good topsoil available. Extremely difficult to find, although not impossible. From Japan. *M. wilsonii* Hanging, white, saucer-shaped, slightly scented flowers with crimson stamens, late spring, early summer. Foliage elliptic to lanceolate, 4-5in (10-12cm) long, grey-green. A wide, spreading shrub. Will tolerate a limited amount of alkalinity, provided adequate depth of topsoil available. Scarce in production. From western China.
Average height and spread
Five years
5x6ft (1.5x2m)
Ten years
10x13ft (3x4m)
Twenty years
or at maturity
26x20ft (8x6m)

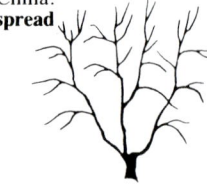

MAHOBERBERIS
AQUISARGENTII

KNOWN BY BOTANICAL NAME
Berberidaceae
Evergreen
A useful spring-flowering evergreen.

Origin *Mahoberberis* itself is a cross between *Mahonia* and *Berberis*; this variety, raised in a garden in Sweden, is thought to be a cross between *M. aquifolium* and *B. sargentiana*.
Use As a small to medium height evergreen for mass planting or shrub borders or, if planted 2½ft (80cm) apart, for an informal hedge.
Description *Flower* Terminal clusters of soft yellow flowers in spring. *Foliage* Leaves vary, being either up to 8in (20cm) long, narrow, elliptic, lanceolate, with spines and toothed edges, or shorter, ovate, lanceolate, with margins covered in 1in (3cm) long sharp spines. Dark green with shiny upper surface and pale, duller undersides. Some flame-red autumn colour on older leaves. *Stem* Upright, with small spined leaflets at each leaf axil. Bright to dark green. Medium rate of growth. *Fruit* Clusters of black fruit in late summer, early autumn.

Hardiness Tolerates 14°F (−10°C).
Soil Requirements Prefers leafy, open, rich soil where good foliage production will be achieved. Tolerates very wide range, but may do less well on other types.
Sun/Shade aspect Prefers light shade, good from full sun to medium shade.
Pruning Every 3-4 years, remove one-third of oldest wood to ground level after flowering to encourage rejuvenation.
Propagation and nursery production From semi-ripe cuttings taken in early summer. Purchase container-grown. May have to be sought from specialist nurseries. Best planting heights 1¼-2½ft (40-80cm).

Mahoberberis aquisargentii **in flower**

Problems Sometimes extremely prone to it own particular form of mildew which is difficult to control.
Average height and spread
Five years
2½x2½ft (80x80cm)
Ten years
3x3ft (1x1m)
Twenty years
or at maturity
4x4ft (1.2x1.2m)

MAHONIA AQUIFOLIUM

OREGON GRAPE
Berberidaceae
Evergreen
A very useful ground-covering shrub, with shrub-forming varieties of interesting colouring.

Origin From western USA.
Use As a ground cover plant, good for mass planting, and extremely useful for retaining banks and difficult ground due to underground shoots which consolidate the planting area. Shrubby varieties for single or mass planting.
Description *Flower* Yellow, upright, thick racemes of musty-scented flowers borne at the end of each branch in early spring. *Foliage* Leaves pinnate with pointed tips to each leaflet, up to 6-8in (15-20cm) long with 3-7 leaflets each 3in (8cm) long. Dark green tinged red, silver-grey undersides, turning purple-red in autumn and winter, regaining green-purple in the succeeding spring. *Stem* Having some spined leafy stem axils. Light green. Underground shoots emerging some distance from parent to form a spreading, flat-topped, possibly invasive carpet. Some varieties shrub-forming. Medium growth rate, slowing with age. *Fruit* Attractive blue-black fruits in autumn.
Hardiness Tolerates winter temperatures down to −13°F (−25°C).
Soil Requirements Any soil, but distressed by extreme dryness.
Sun/Shade aspect Deep shade through to full sun, provided adequate moisture available.

Mahonia aquifolium **in flower**

Pruning Every 3-4 years reduce old woody growth to ground level in early spring; shrub quickly rejuvenates with improved health.
Propagation and nursery production From rooted shoots, or semi-ripe cuttings taken in early summer, or from seed. Plant bare-rooted, root-balled (balled-and-burlapped) or container-grown. Best planting heights 8in-2ft (20-60cm).
Problems Unless pruned as recommended, can look tired and shabby within 4-5 years.
Varieties of interest *M. a. 'Apollo'* Bright yellow upright racemes of flowers in early spring. Good bright red-green foliage, turning purple-green in winter. A particularly good shrub-forming variety which does not sucker or become invasive. Reaches 5ft (1.5m) in height and 6ft (2m) in spread. *M. a. 'Atropurpurea'* Yellow flowers. Foliage purple-red to bronze-red in winter, maintaining colour throughout year. A good ground cover variety. *M. a. 'Moseri'* Bronze-red to apricot-red new growth requiring light dappled shade to protect colour. Not invasive, in fact rather hard to establish. Difficult to find but recommended as a novelty for a collection of Mahonias. Shrub-forming, reaching 5ft (1.5m) in height and 6ft (2m) in spread. *M. a. 'Orange Flame'* Yellow upright flowers, bright orange new growth. Requires a very specific degree of light dappled shade: if too much shade orange colouring will not appear; if unshaded, new growth will be sun-scorched. Shrub-forming. Not reliable, but a good item for a collection.
Average height and spread
Five years
2½x3ft (80cmx1m)
Ten years
3x6ft (1x2m)
Twenty years
or at maturity
3x10ft (1x3m)

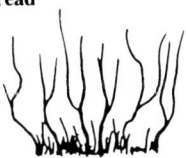

MAHONIA 'CHARITY'

KNOWN BY BOTANICAL NAME
Berberidaceae
Evergreen
Among the aristocrats of tall-growing, winter-flowering shrubs.

Origin Of garden origin. Raised in the Savill Gardens at Windsor, England.
Use As freestanding large shrubs in their own right or in large shrub borders.
Description *Flower* Circular groups of fragrant, bright yellow racemes, up to 8-10in (20-25cm) long in late autumn through to midwinter. Flowers borne at terminals of each upright shoot, at first upright, but becoming weeping with age. *Foliage* Leaves large, 1½ft (50cm) long, bright green, pinnate, with 10-20 pairs of leaflets, each up to 10in (25cm) long with points at lobes of leaflets. *Stem* Very upright and unbranching

unless correctly pruned. Light grey-green with spined small leaflets at each leaf axil.
Fruit Racemes of blue-black fruits with white downy covering follow flowers in mid to late winter.
Hardiness Tolerates −13°F (−25°C). Young foliage may suffer leaf scorch in severe winters, but normally rejuvenates following spring. Some damage may be caused by late spring frosts.
Soil Requirements Prefers open, leafy, moist, peaty soil. Tolerates extremely wide range of types but may show distress on very dry, alkaline conditions.
Sun/Shade aspect Full sun to very light shade. In deeper shade the shrub becomes open and lax.
Pruning Very important that young plants have terminal clusters of foliage removed after flowering to encourage branching. Flowering shoots should be cut annually just below terminal bud after flowering and fruiting for same purpose; otherwise growth becomes very tall, upright and somewhat ungainly.
Propagation and nursery production From semi-ripe cuttings taken in early summer. Very slow to root. Purchase container-grown or root-balled (balled-and-burlapped). Relatively easy to find. Best planting heights 1¼-3ft (40cm-1m).
Problems None if pruning advice followed.

Mahonia 'Charity' **in flower**

Varieties of interest *M. 'Buckland'* Good, thick racemes of yellow flowers, late autumn to early winter, followed by blue-black fruits. Flowers are initially upright in habit, but become lax with age. Foliage pinnate, mid green and 10in (25cm) long, but with smaller leaflets than *M. 'Charity'*. Upright form, slightly more than average spread. Obtainable from specialist nurseries. *M. 'Lionel Fortescue'* Upright racemes of yellow scented flowers with up to 15-20 racemes in each terminal cluster. Florets thickly borne along flowering racemes. Good large foliage, up to 12in (30cm) long and pinnate. One-third more than average spread. Obtainable from specialist nurseries. *M. lomariifolia* Flowers deep yellow, produced from terminal clusters of up to 20 flower spikes, racemes retaining an upright habit, 6-10in (15-25cm) long and carrying numerous small florets. Foliage 10in (25cm) long, pinnate, dark to olive green with moderately small leaflets. Minimum winter temperature 23°F (−5°C). Reaches one-third more than average height. *M. 'Winter Sun'* Upright racemes of scented yellow flowers, winter, on good rounded shrub. Dark green, pinnate leaves. Average height, spread 10ft (3m) over 20 years. Usually found in specialist nurseries.
Average height and spread
Five years
5x3ft (1.5x1m)
Ten years
10x5½ft (3x1.8m)
Twenty years
or at maturity
16x8ft (5x2.5m)

MAHONIA JAPONICA

KNOWN BY BOTANICAL NAME
Berberidaceae
Evergreen
A delightful feature for any average-sized garden for its scented flowers in winter.

Origin From Japan.
Use As a freestanding shrub in its own right, for mass planting or for large shrub borders.
Description *Flower* Terminal racemes of yellow to lemon yellow, very fragrant flowers, at first upright, but becoming weeping with age. *Foliage* Leaves large, broad, pinnate, with 5-9 pairs of diamond-shaped leaflets up to 8-10in (20-25cm) long. Dark green on acid soil, red-green on alkaline, with soft spines at end of each lobe of leaf axil. *Stem* Upright when young, with spiny leaflets at each leaf axil, becoming spreading with age, forming a round-topped, slightly spreading shrub. Medium growth rate. *Fruit* Racemes of blue-black fruits follow flowers in late winter and early spring.
Hardiness Tolerates winter temperatures down to −13°F (−25°C).

Mahonia japonica **in flower**

Soil Requirements Any soil but foliage colour is affected by alkalinity and acidity.
Sun/Shade aspect Prefers medium or light shade, tolerates full sun.
Pruning All flowering terminals should be pruned back each year after fruiting to encourage branching and increase flowering the following year. Any lax shoots which come out above general shape should be cut back.
Propagation and nursery production From semi-ripe cuttings taken in early summer. Purchase container-grown. Easy to find. Best planting heights 1¼-3ft (40cm-1m).
Problems None.
Varieties of interest *M. bealei* Differs from *M. japonica* by having shorter, stouter, more erect flowers and shorter, rounder and slightly concave, red-green, pinnate leaves. Has generally been superseded by *M. japonica* for garden use, but may be offered as an alternative.
Average height and spread
Five years
5x5½ft (1.5x1.8m)
Ten years
8x9ft (2.5x2.8m)
Twenty years
or at maturity
12x10ft (3.5x3m)

MAHONIA PINNATA
(Mahonia fascicularis)

KNOWN BY BOTANICAL NAME
Berberidaceae
Evergreen
An interesting foliage shrub with good flowers in late winter.

Origin From California and southern USA.
Use As a medium height shrub for large shrub borders. Good for mass planting.
Description *Flower* Clusters of 4-5 thick, upright racemes of lemon or pale yellow flowers borne in late winter along the branches in leaf

Mahonia pinnata **in flower**

axils. *Foliage* Leaves sea green, 4-6in (10-15cm) long, pinnate, with 3-7 leaflets protected by numerous prickles at edges of pointed, lobed leaflets. *Stem* Upright when young, branching with age, forming a bulbous-based, pointed, upright shrub. Slow to medium growth rate. *Fruit* May produce small, blue-black fruits in some locations, but not always reliable.
Hardiness Tolerates 14°F (−10°C).
Soil Requirements Any soil except extremely dry conditions.
Sun/Shade aspect Prefers light shade, tolerates medium shade to full sun.
Pruning Cut back one or more mature branches occasionally to ground level after flowering to encourage rejuvenation.
Propagation and nursery production From semi-ripe cuttings taken in early autumn. May also be grown from seed, but very variable. Purchase container-grown. Easy to find in general nurseries. Best planting heights 1¼-2ft (40-60cm).
Problems May become temporarily defoliated in extremely cold or wet conditions.
Average height and spread
Five years
3x3ft (1x1m)
Ten years
5x5ft (1.5x1.5m)
Twenty years
or at maturity
6x6ft (2x2m)

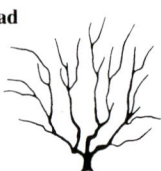

MAHONIA 'UNDULATA'
(Mahonia aquifolium undulata)

KNOWN BY BOTANICAL NAME
Berberidaceae
Evergreen
A good early spring, evergreen, flowering shrub as long as suitable conditions can be given.

Origin A garden hybrid.
Use As a medium-sized shrub for shrub borders, or for mass planting.
Description *Flower* Short, stout, upright racemes of yellow, candle-shaped flowers borne in clusters of 3-4 at leaf axils on mature wood in late winter and early spring. *Foliage* Leaves 4-6in (10-15cm) long, pinnate, with 5-7 leaflets, spiked, lobed and slightly concaved. Olive green tinged red, with silver undersides to leaflets. Some purple-red autumn colour as older leaves die and are replaced by new, often extending well into winter, even early spring. *Stem* Upright, strong when young, aging to lax, spreading, woody framework. Dark green to green-brown. Small, spiked leaflets at each leaf axil. If pruned as suggested, maintains good shape, if not can be ungainly and somewhat unin-

teresting. Slow to medium growth rate. *Fruit* Clusters of blue to silver-blue fruits sometimes produced in early autumn, following dry, late springs and early summers.
Hardiness Tolerates winter temperatures down to −13°F (−25°C), but may suffer defoliation in extreme wind chill conditions.
Soil Requirements Prefers deep, rich soil; tolerates very wide range, but distressed by very poor, dry types.
Sun/Shade aspect Best in light shade, tolerates full sun to medium shade if soil conditions are good.
Pruning Every 4-5 years, on mature, established shrubs, remove one-third of oldest shoots in late spring to encourage rejuvenation. Other shoots should be shortened by a quarter for same purpose.
Propagation and nursery production From semi-ripe cuttings. Purchase container-grown or root-balled (balled-and-burlapped). Best planting heights 1¼-2ft (40-60cm).

Mahonia 'Undulata' **in flower**

Problems Poor or dry soil conditions will lead to general deterioration of overall appearance. Pruning and feeding usually correct this within two seasons.
Average height and spread
Five years
3x3ft (1x1m)
Ten years
6x6ft (2x2m)
Twenty years
or at maturity
13x13ft (4x4m)

MELIANTHUS MAJOR

HONEY BUSH
Melianthaceae
Evergreen
A useful evergreen shrub, but suitable only for very mild areas or sheltered gardens.

Origin From South Africa.
Use As a wall shrub for a warm, sunny, sheltered wall, where it may survive even in an area with hard winter frosts. In colder areas, for conservatory and summer patio use.
Description *Flower* Crimson to tawny-crimson, tubular flowers in erect terminal racemes up to 6in (15cm) long, borne on mature shrubs throughout summer, but only in very favourable areas. *Foliage* Leaves pinnate, up to 1-1¼ft (30-40cm) long, with 9-11 leaflets 3-4in (8-10cm) long. Deeply toothed edges. Sea green to glaucous blue. *Stem* Hollow, upright stems, grey-green. In cold areas may die back to ground level in winter but shows a fast rate of growth in following spring. *Fruit* Insignificant.
Hardiness Ideally, temperatures must go no lower than 32°F (0°C).

Melianthus major in leaf

Microglossa albescens in flower

Soil Requirements Rich, moist, deep soil to produce good foliage.
Sun/Shade aspect Full sun to very light shade, otherwise leaf colour and shrub shape is spoiled.
Pruning None required.
Propagation and nursery production From root division or from seed. Purchase container-grown. Very hard to find outside favourable mild areas and even in these locations available only from specialist nurseries. Best planting heights 4in-2ft (10-60cm).
Problems Should not be bought where available in a favourable area for transplanting in a cooler climate, as it will not adapt.
Average height and spread
Five years
3x3ft (1x1m)
Ten years
6x6ft (2x2m)
Twenty years
or at maturity
10x10ft (3x3m)

MICROGLOSSA ALBESCENS (Aster albescens)

SHRUBBY ASTER
Compositae
Deciduous
An interesting low-growing sub-shrub for those who wish to collect unusual plants.

Origin From the Himalayas and China.
Use As a low-growing sub-shrub for edging large shrub borders, or for use in small mixed groups of similar heights.
Description *Flower* Terminal clusters of pale lilac-blue, daisy-like flowers in midsummer. *Foliage* Leaves broad, lanceolate, 2-5in (5-12cm) long, tooth-edged, grey-green to dark green. *Stem* Upright, dying quickly to ground level in winter. Purple to purple-green. Fast growth rate in late spring, slowing in summer. *Fruit* May produce soft, grey-brown seedheads.
Hardiness Stems die completely to ground level in winter. The rooted plant base will not survive temperatures much below 23°F (−5°C).
Soil Requirements Well-drained, rich soil. Dislikes waterlogging.
Sun/Shade aspect Full sun to very light shade.
Pruning Winter will probably kill all top growth, but if not, remove wood to ground level in early spring.
Propagation and nursery production From semi-ripe cuttings taken in early summer. Purchase container-grown from specialist nurseries. Plant in spring to establish during following summer. Best planting heights 4-8in (10-20cm).

Problems Resents wet, cold conditions and may perish over winter. Best propagated so young plants may be kept under cover through winter for planting in spring, to replace established plants which do not survive.
Average height and spread
Five years
2x3ft (60cmx1m)
Ten years
3x3ft (1x1m)
Twenty years
or at maturity
3x3ft (1x1m)

MYRICA

BAYBERRY, CALIFORNIAN BAYBERRY, SWEET GALE, BOG MYRTLE, WAX MYRTLE
Myriaceae
Evergreen and Deciduous
Extremely useful shrubs for boggy, acid ground where many other plants will not grow.

Origin Most varieties from North America.
Use As an interesting feature for moist, acid soil areas.
Description *Flower* Small, brown to yellow-brown, male and female catkins, early spring. *Foliage* Leaves oblong to lanceolate, with serrated edges, 1-1½in (3-4cm) long, aromatic. Light green with some purple shading in winter. *Stem* Upright, becoming branching, forming large, dome-shaped shrub or small, short, multi-stemmed tree, dependent on variety. Brown-red. Slow to medium growth rate, but faster in ideal conditions. *Fruit* Dark

Myrica gale in catkin

purple clusters of small, blackberry-type fruit, retained well into midwinter.
Hardiness Tolerates 14°F (−10°C)
Soil Requirements Moist, acid soil, dislikes any alkalinity.
Sun/Shade aspect Full sun to very light shade.
Pruning None required, in fact may resent it.
Propagation and nursery production From semi-ripe cuttings taken in early spring, or from seed. Purchase container-grown. Best planting heights 1¼-2ft (40-60cm).
Problems Very specific soil requirement must be met for good results.
Varieties of interest *M. californica* (Californian Bayberry) May become a small tree, up to 26ft (8m) in height and spread. Interesting catkins and dark purple fruits. Native to western USA. Evergreen. *M. cerifera* (Wax Myrtle) Reaching tree proportions in ideal conditions with a height and spread of 32ft (10m) and more. Narrow, obovate leaves. Wax covering of fruits used to produce fragrant candles. Native to eastern USA and Caribbean Islands. Evergreen. *M. gale* (Sweet Gale, Bog Myrtle) Male and female catkins, produced on separate shrubs in early spring, are about 2in (5cm) long, golden brown and sparkling. Round or ovate, dark green to green-red aromatic foliage. Requires a very boggy, acid soil, often where no other plants will grow. From Europe, north-eastern Asia and North America. Deciduous. *M. pennsylvanica* (Bayberry) Reaches half as much again in height and spread. Foliage oblong to ovate and aromatic, giving some good autumn colour. Small, white to grey-white fruits of winter interest. Requires dry, arid area, with acid soil. Very good for maritime areas. Native to eastern North America. Deciduous.
Average height and spread
Five years
3x3ft (1x1m)
Ten years
6x6ft (2x2m)
Twenty years
or at maturity
13x13ft (4x4m)

MYRTUS

MYRTLE
Myrtaceae
Evergreen
An attractive flowering evergreen shrub for mild locations.

Origin From southern Europe and the Mediterranean.
Use As a freestanding, evergreen shrub, or for colder sites as a wall shrub. Can be grown in containers 2½ft (80cm) in diameter and 2ft (60cm) in depth; use good potting medium. Achieves less size when grown this way.

167

Descrption *Flower* Small, tufted, white flowers in mid to late summer. *Foliage* Leaves round to ovate, 1-2in (3-5cm) long, dark green with duller undersides. *Stem* Upright, becoming branching with time. Purple green to dark green. Fast growth rate when young, slowing with age. *Fruit* Small, purple-black fruits in autumn.
Hardiness Minimum winter temperature 23°F (−5°C).
Soil Requirements Does best on well-drained soil, tolerating both alkalinity and acidity.
Sun/Shade aspect Full sun to mid shade.
Pruning None required but can be cut back hard or trimmed.
Propagation and nursery production From semi-ripe cuttings taken in early summer. Purchase container-grown. *M. communis* easy to find, but other varieties should be obtained from specialist nurseries. Best planting heights 1¼-2ft (40-60cm).
Problems Sometimes thought to be hardier than it actually is.

Myrtus communis in flower

Varieties of interest *M. apiculata* syn. *M. luma* Tufted white flowers borne singly at each leaf axil cover the entire shrub during late summer and early autumn, followed by red to black fruits which are edible and sweet. Foliage oval, dark, dull green, ending in an abrupt point. Bark cinnamon-coloured and peels off in patches, showing creamy under surface. From Chile. In mild sheltered locations reaches size of small tree. *M. communis* (Common Myrtle) White, round, tufted flowers, mid to late summer, with purple-black fruits in autumn. Leaves aromatic and very densely cropped. From southern Europe through Mediterranean region. *M. c. tarentina* White tufted flowers in summer with white berries in autumn. Leaves narrower than those of *M. communis*. From the Mediterranean region. *M. c. 'Variegata'* A form with broad creamy white margins to the foliage. A tender variety which will not withstand temperatures below freezing.
Average height and spread
Five years
3x3ft (1x1m)
Ten years
6x6ft (2x2m)
Twenty years
or at maturity
13x10ft (4x3m)

NANDINA DOMESTICA

NANDINA, SACRED BAMBOO, HEAVENLY BAMBOO, CHINESE SACRED BAMBOO
Berberidaceae
Evergreen
A useful shrub for the small garden, with good colouring in the evergreen foliage and attractive summer flowers.

Origin From Japan, China and India.
Use As an isolated, low shrub of interest. Best planted as a single feature with other low-growing shrubs.
Description *Flower* Terminal panicles presented on long stems above main leaf canopy, midsummer. *Foliage* Leaves pinnate, 1-1½ft (30-50cm) long with lanceolate to ovate leaf-

Nandina domestica 'Fire Power' in autumn

lets 1½-4in (4-10cm) long. Attractively red-shaded in spring, orange-red in autumn. *Stem* Long, strong, upright, green with red shading, forming a typical bamboo clump effect, with an upright, domed final stage. Slow to medium growth rate. *Fruit* Insignificant.
Hardiness Minimum winter temperature 14°F (−10°C).
Soil Requirements Prefers light open soil but does well on all types.
Sun/Shade aspect Prefers light dappled shade, tolerates medium shade to full sun.
Pruning None required, but old flowering spikes may be removed and one-third of upright shoots occasionally cut back to induce new growth from base level.
Propagation and nursery production By division. Purchase container-grown. Available from garden centres and nurseries. Best planting heights 1-2ft (30-60cm).
Problems In severe winters may die back to ground level but seems to rejuvenate, even if somewhat late in the season.
Varieties of interest *N. d. 'Fire Power'* Extremely good, large, orange-red foliage in spring. Light green summer growth and autumn reds and scarlets. From New Zealand. *N. d. 'Nana Purpurea'* Large, more bronze-red foliage in both spring and autumn. A little shy to flower. A lower-growing variety reaching two-thirds height and spread.
Average height and spread
Five years
1½x2½ft (50-80cm)
Ten years
2½x3ft (80cm-1m)
Twenty years
or at maturity
3x4ft (1x1.2m)

NEILLIA THIBETICA
(Neillia longiracemosa)

KNOWN BY BOTANICAL NAME
Rosaceae
Deciduous
An underrated, attractive flowering shrub worthy of wider planting.

Origin From western China.
Use As a freestanding, early summer-flowering shrub, for a large or medium-sized border, or for mass planting.
Description *Flower* Attractive terminal racemes of pink tubular flowers in late spring, early summer. *Foliage* Leaves ovate with narrow points, 1½-4in (4-10cm) long, and some leaves 3-lobed with narrow points. *Stem* Upright, downy, branching at top. Always produced from ground level. Becoming twiggy and branching with time, forming upright, round-topped shrub. Medium growth rate. *Fruit* Insignificant.

Hardiness Tolerates 14°F (−10°C).
Soil Requirements Any soil, but dislikes extremely wet conditions.
Sun/Shade aspect Full sun through to light shade.
Pruning Reduce one-third of oldest flowering wood to ground level after flowering to induce growth of new spring flowering shoots in succeeding year.
Propagation and nursery production From semi-ripe cuttings taken in early summer. Purchase container-grown. May have to be sought from specialist nurseries. Best planting heights 1¼-2ft (40-80cm).

Neillia thibetica in flower

Problems Container-grown plant at purchase gives no indication of its coming beauty as a flowering shrub.
Average height and spread
Five years
6x6ft (2x2m)
Ten years
6x10ft (2x3m)
Twenty years
or at maturity
6x13ft (2x4m)

NERIUM OLEANDER

OLEANDER
Apocynaceae
Evergreen
Although extremely tender, this shrub is worth of mention, especially for conservatory use.

Origin From the Mediterranean region.
Use In frost-free gardens as an interesting flowering evergreen shrub; in areas where frost is expected should be grown in containers not less than 2ft (60cm) in diameter and 1½ft (50cm) in depth; can be moved outdoors

in summer and kept under glass in winter, or grown under glass at all times.
Description *Flower* Medium-sized, flat, 5-petalled flowers ranging from white, through yellow, buff and various shades of pink, midsummer to mid autumn. *Foliage* Leaves lanceolate, 3-6in (8-15cm) long, borne in clusters. Grey-green to sea green with some silver and gold variegated forms. **Stem** Upright stems, becoming branching at extremities, forming a narrow based, wide-headed shrub. Grey-green to sea green. Medium growth rate. *Fruit* Insignificant.
Hardiness Minimum temperature 32°F (0°C).
Soil Requirements Light open soil, tolerating both alkalinity and acidity.
Sun/Shade aspect Full sun to light shade.
Pruning None required. Individual branches may be shortened and will rebranch.

Nerium oleander **in flower**

Propagation and nursery production From softwood cuttings taken in early spring. Purchase container-grown Best planting heights 1¼-2ft (40-60cm).
Problems Only available in favourable areas.
Average height and spread
Five years
3x3ft (1x1m)
Ten years
6x6ft (2x2m)
Twenty years
or at maturity
13x13ft (4x4m)

OLEARIA

DAISY BUSH, TREE DAISY, TREE ASTER, AUSTRALIAN DAISY BUSH
Compositae
Evergreen
Evergreen shrubs for late spring and summer flowering in mild areas.

Origin From Australia and New Zealand.
Use As a freestanding shrub planted singly, for inclusion in medium-sized or large shrub borders or, in less favourable areas, as a wall shrub for a sheltered, sunny aspect.
Description *Flower* Clusters of daisy-like flowers, white, creamy white, pink or blue, depending on variety, produced through mid-summer. *Foliage* Sea green to grey-green, normally ovate but with some variation of shape and size according to variety. *Stem* Upright, becoming branching with age. Grey-green. Slow to medium growth rate. *Fruit* Small brown-grey seedheads maintained into winter.
Hardiness Minimum winter temperature 23°F (−5°C).
Soil Requirements Best results on well-drained, open soil. Neutral to acid types preferred but a wider range tolerated if moisture is adequate.
Sun/Shade aspect Full sun; dislikes shade.
Pruning Remove one-third of old flowering

Olearia phlogopappa subrepanda **in flower**

wood after flowering to encourage rejuvenation.
Propagation and Nursery Production From semi-ripe cuttings or softwood cuttings taken early to mid summer. Purchased container-grown. May be available in favourable locations, especially mild coastal areas. Best planting heights 14in-2½ft (40-80cm).
Problems Often bought where available in mild areas for planting out elsewhere, but will not survive a cold climate.
Varieties of interest *O.* × *haastii* Masses of fragrant white daisy-like flowers, mid to late summer. Grey-green to sea green, ovate leaves, ½-1in (2-3cm) long, white undersides and smooth grey upper surface. In mild coastal areas may be considered as good hedging shrub. One of the hardiest varieties. *O. macrodonta* (New Zealand Holly) White clusters of daisy flowers in mid summer. Foliage holly-like, dark grey-green with pointed lobes and silver undersides 2-5in (5-12cm) long. *O. mollis* Large flowerheads of daisy-like white flowers in late spring. A variety reaching only half average height and spread. Wavy-edged, light grey to silvery grey, toothed leaves up to 1½-2in (4-5cm) long. *O. nummulariifolia* Small, fragrant, white, daisy flowers, produced singly in axils of each leaf joint in midsummer. Round, thickly textured, light green to yellow-green leaves ½in (2cm) long and thickly presented on stems. A variety reaching two-thirds average height and spread. From New Zealand. *O. phlogopappa* syn. *O. gunniana* (Tasmanian Daisy Bush) White flower panicles 1in (3cm) across, early spring, surmounting medium-sized, aromatic, ovate, tooth-edged, 1¾in (4cm) long leaves, thickly crowded on upright stems. From Tasmania and south-eastern Australia. *O. p. subrepanda* syn. *O. subrepanda* White daisy flowers in late spring. Narrow, ovate, tooth-edged, grey, hairy-covered leaves. Two-thirds average height and spread. *O.* × *scilloniensis* Masses of white daisy-like flowers in early summer completely smother entire shrub. Narrow, ovate foliage, ½-¾in (1-2cm) long, grey-green, toothed-edged with silver undersides. From the Scilly Islands, UK. *O. stellulata* '*Splendens*' Blue, white or pink daisy-like flowers, depending on form. Narrow, ovate to lanceolate, ½-¾in (1-2cm) long, grey-green foliage with bold toothed edges, somewhat susceptible to mildew. Reaching 3ft (1m) in height and 2½ft (80cm) in width after 10 years. Treat as tender and in colder areas plant as a wall shrub. Take cuttings to overwinter under cover as guarantee of future planting.
Average height and spread
Five years
3x3ft (1x1m)
Ten years
6x6ft (2x2m)
Twenty years
or at maturity
10x10ft (3x3m)

OSMANTHUS

KNOWN BY BOTANICAL NAME.
Oleaceae
Evergreen
A useful group of evergreen shrubs with scented flowers.

Origin Most varieties from western China or Japan.
Use As a freestanding shrub or for medium-sized to large shrub borders. Can be used for hedging planted 2ft (60cm) apart in a single line.
Description *Flower* Small, white or cream, trumpet-shaped flowers, sweetly scented, late autumn or early spring, depending on variety. *Foliage* Leaves elliptic, ovate or oblong, ½-1½in (1-4cm) long, some larger according to variety, some with toothed, spiny edges. *Stem* Upright, becoming branching. Dark green to grey-green. Slow to medium growth rate. *Fruit* Broad seedheads may be formed after flowering.
Hardiness Tolerates 14°F (−10°C) but some varieties more tender, minimum 23°F (−5°C).
Soil Requirements Most soils, but resents waterlogging.
Sun/Shade aspect Prefers light dappled shade. Tolerates full sun through to medium shade.
Pruning None required. May be cut and trimmed after flowering to keep within bounds.
Propagation and nursery production From semi-ripe cuttings taken in early summer. Purchase container-grown. Most varieties fairly easy to find. Best planting heights 1¼-2½ft (40-80cm).
Problems Liable to leaf scorch in cold winters.
Varieties of interest *O. americanus* (Devil Wood) A variety reaching one-third more than average height and spread, even becoming small tree. White fragrant flowers borne in short panicles in axils of each leaf joint, spring. Dark blue fruits, late summer. Slightly tender. Outside its native environment, needs warm, sheltered position, or can be grown in conservatory. From south-eastern USA. *O. armatus* White, sweetly scented flowers, autumn. Elliptic to oblong, 3-6in (8-15cm) long, lanceolate leaves, thick and rigid in texture and often up to 7-8in (18-20cm) long. Edges of leaves have a series of hooked spiny teeth. From western China. *O.* '*Burkwoodii*' Round to ovate leaves, 1½-2in (4-5cm) long. Dark shiny upper surface, leathery textured, silvery undersides with toothed edges. Clusters of tubular white, very fragrant flowers, early to mid spring. Will tolerate shallow chalk soils. A nursery hybrid formerly known as *Osmarea* '*Burkwoodii*'. *O. delavayi* Dark grey-green foliage. Reaching two-thirds average height and spread. Covered with a good display of very sweetly scented white flowers, early to

Osmanthus delavayi in flower

mid spring. Foliage may be damaged by severe wind chill. *O. heterophyllus* syn. *O. aquifolium, O. ilicifolius* (Holly Osmanthus, false Holly) Small to medium holly-shaped leaves, dark green, shiny upper surface, paler undersides, having spines and toothed edges, 1½-2in (4-5cm) long. White, scented, tubular flowers in small clusters, autumn. Useful for hedging. From Japan. *O. h. 'Aureomarginatus'* Holly-like leaves with golden yellow to deep yellow variegated margins. White flowers. Two-thirds average height and spread. *O. h. 'Gulftide'* Leaves twisted and lobed with strong spines. White scented flowers. Very dense habit of growth. Slightly less than average height and spread. *O. h. 'Purpureus'* Young growth holly-shaped, purple, later becoming very dark green tinged purple. White flowers in autumn. Two-thirds average height and spread. *O. h. 'Rotundifolius'* Almost round, tooth-edged, leathery, dark green foliage on dark, green-black stems. White trumpet flowers, scented, freely borne in spring. Very scarce but not impossible to find. *O. h. 'Variegatus'* Holly-like leaves, grey-green with creamy white bold bordered edges. Shy to flower. Grows slowly to two-thirds average height and spread.
Average height and spread
Five years
6x6ft (2x2m)
Ten years
10x10ft (3x3m)
Twenty years
or at maturity
13x13ft (4x4m)

Osmanthus heterophyllus 'Variegatus' in leaf

170

OSMARONIA CERASIFORMIS (Nuttallia cerasiformis)

OSO BERRY
Rosaceae
Deciduous
An interesting late spring-flowering shrub, if somewhat difficult to obtain.

Origin From California.
Use As a spring-flowering shrub for individual planting or for medium-sized to large shrub borders.
Description *Flower* Short, hanging racemes of fragrant, creamy white, currant-like flowers in late winter, early spring. *Foliage* Leaves oblong to lanceolate, 3in (8cm) long, toothed-edged, sea green. *Stem* Upright with some branching on top two-thirds of stem. Often suckering from base to form a wide, upright thicket. Medium growth rate when young, slowing with age. *Fruit* Small, plum-like, brown ripening to purple fruits in autumn on female shrubs if male specimen is planted nearby.
Hardiness Tolerates 14°F(−10°C).
Soil Requirements Light, open soil, tolerating all but extreme alkalinity.
Sun/Shade aspect Full sun to light shade.
Pruning Remove one-third of old flowering wood each year after flowering to encourage new growth.

Osmaronia cerasiformis in flower

Propagation and nursery production From removal of rooted suckers. Purchase container-grown. Difficult but not impossible to find. Best planting heights 1¼-2½ft (40-80cm).
Problems None.
Average height and spread
Five years
3x3ft (1x1m)
Ten years
6x6ft (2x2m)
Twenty years
or at maturity
6x6ft (2x2m)

OZOTHAMNUS ROSMARINIFOLIUS

KNOWN BY BOTANICAL NAME
Compositae
Evergreen
A shrub of interesting architectural shape, with a good contrast of dark foliage and white flowers.

Origin From Tasmania and south-eastern Australia.
Use As an upright, short-growing shrub, either on its own or in a medium-sized shrub border.
Description *Flower* Small, white, scented flowers opening from red buds, mid to late summer. *Foliage* Leaves small, lanceolate,

Ozothamnus rosmarinifolius in flower

¼-½in (1-2cm) long, dark green, curled with silver backing, densely formed on upright stems. *Stem* Very upright, covered with a column of foliage growth similar to that of some conifers. Slow to medium growth rate. *Fruit* Grey-brown seedheads have some autumn and winter attraction.
Hardiness Minimum winter temperature 23°F (−5°C).
Soil Requirements Well-drained, open soil.
Sun/Shade aspect Prefers light dappled shade to full sun. Deeper shade spoils overall effect.
Pruning None required but remove one or more shoots on mature shrubs each year to encourage basal rejuvenation.
Propagation and nursery production From semi-ripe cuttings taken in early summer. Purchase container-grown. Availability varies. Best planting heights 1¼-2ft (40-60cm).
Problems None, if not subjected to severe cold in winter.
Average height and spread
Five years
2½x2ft (80x60cm)
Ten years
4x4ft (1.2x1.2m)
Twenty years
or at maturity
4x4ft (1.2x1.2m)

PACHYSANDRA

JAPANESE PACHYSANDRA, JAPANESE SPURGE, MOUNTAIN SPURGE
Buxaceae
Evergreen
A very useful ground-cover plant, as long as the correct acid soil conditions can be supplied.

Origin From Japan.
Use As a low-growing carpeting shrub for acid soils. Good for mass planting.
Description *Flower* Spikes of green-white flowers, aging to white, produced at terminals of previous year's shoots, early to mid spring. *Foliage* Leaves diamond-shaped with toothed edges, 1-3in (3-8cm) long, glossy green, produced at upper ends of upright shoots, giving some yellow colour interest when leaves are dying and being replaced. Medium rate of spread on suitable soil. *Stem* Suckering to form a low-growing carpet, light grey-green. *Fruit* Insignificant.
Hardiness Tolerates 14°F (−10°C).
Soil Requirements Moist, open, rich, leafy, acid to neutral soil.
Sun/Shade aspect Tolerates full sun through to deep shade provided adequate moisture is available.
Pruning If untidy or damaged, hard trimming in early spring leads to good new growth.
Propagation and nursery production From division, or semi-ripe cuttings taken in early summer. Purchase container-grown. Fairly

easy to find, especially in acid soil areas. Best planting heights 8-12in (20-30cm).
Problems When purchased young plants always look sickly, but once planted out in appropriate soil they quickly establish and send out underground suckers.
Varieties of interest *P. terminalis 'Variegata'* Good flowers but of slightly less than average height and spread. Each leaf is boldly edged with white. A slightly tender variety, winter minimum 23°F (−5°C).

Pachysandra terminalis 'Variegata' in leaf

Average height and spread
Five years
8inx3ft (20cmx1m)
Ten years
1x5ft (30cmx1.5m)
Twenty years
or at maturity
1x6ft (30cmx2m)

PAEONIA LUTEA VAR. LUDLOWII

YELLOW TREE PEONY
Paeoniaceae
Deciduous
A handsome peony with yellow summer flowers.

Origin The basic form is from the Yunnan in China and the form *P. l. ludlowii* was collected in Tibet.
Use As a freestanding shrub on its own, or in medium-sized to large shrub borders.
Description *Flower* Medium-sized, 2-2½in (5-6cm) across, golden yellow, saucer-shaped flowers, produced in terminal clusters in early summer. *Foliage* Leaves large, lobed and cut, pinnate, up to 8-10in (20-25cm) across, each leaflet 2-4in (5-10cm) long. Light green to mid green, giving some yellow autumn colour. *Stem* Stout, upright, grey-green shoots with leaves borne at end, forming round, slightly spreading, large shrub. Maintained throughout winter. Medium to tast growth rate. *Fruit* Green fruit pods in late summer, early autumn which, when opened, contain 3-5 large, round fruits.
Hardiness Tolerates 14°F (−10°C).
Soil Requirements Most soils, showing chlorosis only on extreme alkalinity. Dislikes waterlogging.
Sun/Shade aspect Full sun to light shade.
Pruning Beginning 3-5 years after planting, remove one-third of oldest flowering shoots each year in early spring, to encourage new growth.
Propagation and nursery production From seed. Purchase container-grown. Moderately easy to find. Best planting heights 1-2ft (30-60cm).
Problems Young plants may be offered as a single shoot.
Varieties of interest *P. delavayi* Flowers blood

Paeonia lutea var. ludlowii in flower

red with golden stamens, supported on mid to dark green foliage. Shrubby, branching effect to one-third average height and spread. Rather hard to find from non-specialist sources. *P. suffruticosa* (Japanese Tree Peony, Mountain Peony) Large, double, cup-shaped flowers up to 6in (15cm) across, produced in midsummer in a range of colours from white through flesh pink, rose pink, red to dark red, or pale yellow to yellow, depending on variety. Foliage pinnate, 10in (25cm) long with leaflets 2-4in (5-10cm) long and two lobed, basal leaflets. Grey-green tinged with red, especially along veins. Upright in habit, spreading to form a goblet-shaped shrub with time. Rigid when young, drooping slightly with age. Young foliage may be damaged by spring frosts. Half average height and spread.

Average height and spread
Five years
5x6ft (1.5-2m)
Ten years
8x13ft (2.5x4m)
Twenty years
or at maturity
12x16ft (3.5x5m)

PALIURUS SPINA-CHRISTI

CHRIST'S THORN, JERUSALEM THORN
Rhamnaceae
Deciduous
An interesting shrub for specimen planting, traditionally reputed to have been used in Christ's Crown of Thorns.

Origin From southern Europe to the Himalayas and through to northern China.

Paliurus spina-christi in flower

Use As an individual specimen shrub.
Description *Flower* Small, open racemes of yellow to green-yellow flowers, late summer. *Foliage* Leaves ovate, ¾-1½in (2-4cm) long, slightly tooth-edged, light yellow-green, giving good yellow autumn colour. *Stem* Upright when young, quickly becoming branching, twiggy and spreading. Mid green to green-yellow. Armed with numerous pairs of thorns of unequal length. Slow to medium growth rate. *Fruit* Rounded, curiously shaped, green-yellow fruits in autumn and early winter.
Hardiness Tolerates 14°F (−10°C).
Soil Requirements Most soils, provided adequate feeding is supplied.
Sun/Shade aspect Full sun to very light shade. Dislikes deeper shade.
Pruning None required, but may be reduced in size as necessary.
Propagation and nursery production From semi-ripe cuttings taken in late spring, early summer, or from layers. Difficult to propagate. Purchase container-grown. Rather hard to find. Best planting heights 8in-1½ft (20-50cm).
Problems A little slow to develop into an interesting plant.

Average height and spread
Five years
3x3ft (1x1m)
Ten years
6x6ft (2x2m)
Twenty years
or at maturity
13x13ft (4x4m)

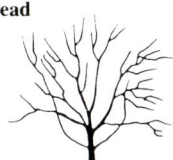

PARAHEBE CATARRACTAE
(Veronica catarractae)

KNOWN BY BOTANICAL NAME
Scrophulariaceae
Evergreen
An interesting, small evergreen shrub for very mild areas.

Origin From New Zealand.
Use As low, spreading ground cover, or an individual mound for larger rock gardens.
Description *Flower* Small flowers, white to rose purple with crimson central rings, produced in erect racemes in late summer. Most nurseries provide a form with blue flowers, but this is very variable. *Foliage* Leaves dark green, ¾-1in (1-3cm) long, ovate to lanceolate with coarse serrated edges. *Stem* Greenish-brown. Spreading, arching, forming a low mound. Slow to medium growth rate. *Fruit* Insignificant.
Hardiness Minimum temperature 32°F (0°C).
Soil Requirements Does best on moist, open,

Parahebe catarractae **in flower**

acid soil, but tolerates quite high degrees of alkalinity.
Sun/Shade aspect Full sun; dislikes shade.
Pruning None required, but old branches may occasionally be reduced in length to encourage rejuvenation.
Propagation and nursery production From soft to semi-ripe cuttings taken in early summer. Purchase container-grown. Difficult to find in all but very mild areas. Best planting heights 4-8in (10-20cm).
Problems None.
Average height and spread
Five years
1½x1½ft (50x50cm)
Ten years
1½x2½ft (50-80cm)
Twenty years
or at maturity
1½x3ft (50cmx1m)

PERNETTYA MUCRONATA

KNOWN BY BOTANICAL NAME
Ericaceae
Evergreen
Useful, fruiting, evergreen shrubs for acid soils.

Origin From Chile and Argentina. Many varieties of garden origin.
Use For mass planting in small or large groups.
Description *Flower* Small clusters of hanging, cup-shaped, small, white flowers in early summer. Male and female forms. *Foliage* Leaves small, tooth-edged, ovate to oblong, ½-¾in (1-2cm) long, glossy, very dark green,

Pernettya mucronata 'Alba' **in fruit**

with red veins and leaf stalks, closely presented along stems. *Stem* Upright when young, becoming branching and spreading with age, bright red. Slow to medium growth rate. *Fruit* Round, shiny fruits, ranging from white, through to pink, to red, purple and shades in between, dependent on variety, borne freely and maintained well into winter. Best fruits produced when more than one shrub is planted. *P. mucronata* has purple fruit.
Hardiness Tolerates 4°F (−15°C), but may lose some leaves in severe wind chill.
Soil Requirements Neutral to acid soil, dislikes any alkalinity.
Sun/Shade aspect Full sun to mid shade.
Pruning On mature plants, cut back rambling shoots more than one year old to ground level in early spring.
Propagation and nursery production From semi-ripe cuttings taken in early summer. Also from seed, but produces plants with variable size and colour of fruits. Purchase container-grown. Availability varies. Best planting heights 8in-2ft (20-60cm).

Pernettya mucronata 'Bell's Seedling'

Problems The need for acid soil is often overlooked; alkalinity causes actual distress.
Varieties of interest *P. m. 'Alba'* Medium-sized white fruits, pink-tipped and slightly pink shaded. *P. m. 'Bell's Seedling'* Very good fruiting variety with large, dark red berries. *P. m. 'Davis's Hybrids'* Fruits of various colours produced slightly later than average on healthy shrub. *P. m. 'Lilacina'* Red-lilac fruits. Slightly shorter than average. *P. m. 'Mulberry Wine'* Magenta fruits aging to purple. Stems greener than parent. *P. m. 'Pink Pearl'* Large, lilac-pink fruits. *P. m. 'Sea Shell'* Large, soft shell pink fruits, ripening to rose pink. *P. m. 'White Pearl'* Large white fruits.
Average height and spread
Five years
1½x2½ft (50x80cm)
Ten years
2½x4ft (80cmx1.2m)
Twenty years
or at maturity
2½x4ft (80cmx1.2m)

PEROVSKIA ATRIPLICIFOLIA

RUSSIAN SAGE
Labiatae
Deciduous
An attractive, grey-leaved shrub flowering profusely in late summer.

Origin From Afghanistan through the western Himalayas to Tibet.
Use As a grey-leaved sub-shrub for bedding out on its own, for mass planting, or for edging small to medium shrub borders.

Can also be treated as a herbaceous plant.
Description *Flower* Terminal panicles up to 1-1¼ft (30-40cm) long, lavender-blue in late summer, contrasting well with grey foliage and stems. *Foliage* Leaves rhomboid to ovate, deeply toothed, 1-2½in (3-6cm) long, aromatic, grey, giving some yellow autumn colour. *Stem* Upright, grey, terminating in flower spikes. Somewhat fragile and needs light support. Fast annual spring growth. *Fruit* Grey seedheads.
Hardiness Minimum winter temperature 14°F (−10°C).
Soil Requirements Well-drained, open soil.
Sun/Shade aspect Full sun.

Perovskia atriplicifolia 'Blue Spire' **in flower**

Pruning Reduce to ground level each spring to induce new, attractive shoots and late summer flowers.
Propagation and nursery production From semi-ripe cuttings taken in late spring. Purchase container-grown. Availability varies. Best planting heights 4in-1½ft (10-50cm).
Problems Can be late to come into leaf in spring and may appear dead even into early summer.
Varieties of interest *P. a. 'Blue Spire'* Large panicles of lavender-blue flowers in late summer. Deeply cut grey-green leaves.
Average height and spread
Five years
2½x3ft (80cmx1m)
Ten years
2½x3ft (80cmx1m)
Twenty years
or at maturity
2½x3ft (80cmx1m)
Heights include length of flower spikes

PHILADELPHUS
Low-growing varieties

MOCK ORANGE
Philadelphaceae
Deciduous
Useful low-growing, scented summer-flowering shrubs. These are often mistakenly given the common name Syringa.

Origin Most varieties of garden origin.
Use For planting singly, for medium to large shrub borders, or for mass planting.
Description *Flower* White, often fragrant flowers, ¾in (2cm) wide, single or double, depending on variety, borne in midsummer. *Foliage* Leaves ovate and slightly tooth-edged, 1½-4in (4-10cm) long, light to mid green, giving good yellow autumn colour. Some varieties with variegation. *Stem* Upright, forming an upright, round-topped shrub. Grey-green. Medium to fast growth rate. *Fruit* Insignificant.
Hardiness Tolerates winter temperatures down to −13°F (−25°C).

Soil Requirements Any soil including both alkaline and acid types.
Sun/Shade aspect Full sun through to medium shade.
Pruning With established plants 3 years old or more remove one-third of oldest flowering shoots after flowering.
Propagation and nursery production From semi-ripe cuttings taken in early summer or hardwood cuttings in winter. Plant bare-rooted or container-grown. Best planting heights 8in-1½ft (20-50cm).
Problems Commonly harbours blackfly which can transfer to other garden crops.

Philadelphus coronarius 'Variegatus' **in leaf**

Varieties of interest *P. 'Boule d'Argent'* Double white, scented flowers, compact habit. Hard to find but worth the attempt. *P. coronarius 'Variegatus'* Single to semi-double, white, scented flowers, early summer. Grey-green leaves with bold, creamy white margins. Light dappled shade essential; strong sunlight scorches and deforms the shrub. *P. 'Manteau d'Hermine'* Double white to creamy white, very fragrant flowers. One of the most popular of low-growing forms. *P. microphyllus* Single, very small, white, scented flowers on very twiggy growth. *P. 'Silver Showers'* Masses of white, scented double flowers on upright, graceful stems.
Average height and spread
Five years
2½x2½ft (80x80cm)
Ten years
3x3ft (1x1m)
Twenty years
or at maturity
4x4ft (1.2x1.2m)

PHILADELPHUS
Medium height varieties

MOCK ORANGE
Philadelphaceae
Deciduous
A useful group of summer-flowering, fragrant shrubs.

Origin Most varieties of garden origin.
Use For planting singly, or in medium to large shrub borders, or for mass planting.
Description *Flower* White, often fragrant flowers, ¾-1½in (2-4cm) wide, single or double, depending on variety, some with purple throat markings, borne in midsummer. *Foliage* Leaves ovate, 1½-4in (4-10cm) long, slightly tooth-edged, light to mid green with good yellow autumn colour. Some varieties variegated. *Stem* Upright, forming an upright, round-topped shrub and grey-green. Medium to fast growth rate. *Fruit* Insignificant.
Hardiness Tolerates winter temperatures down to −13°F (−25°C).

Philadelphus 'Belle Etoile' **in flower**

Soil Requirements Most soils, including alkaline and acid types.
Sun/Shade aspect Full sun through to medium shade.
Pruning On established plants 3 years old or more, remove one third of older flowering wood to ground level after flowering.
Propagation and nursery production From semi-ripe cuttings taken in early summer or from hardwood cuttings in winter. Purchase container-grown or bare-rooted. Best planting heights 1¼-2½ft (40-80cm).

Problems Harbours blackfly which can transfer to other garden crops.
Varieties of interest *P. 'Avalanche'* Small, pure white, single, scented flowers borne in profusion. An upright-growing shrub with smaller than average foliage. *P. 'Belle Etoile'* White, single, fragrant flowers with maroon central blotches. *P. 'Erectus'* Pure white, single, scented flowers, very free flowering. Distinct, upright branches. *P. 'Galahad'* Single, scented, white flowers. Stems mahogany-brown in winter. Rather hard to find. *P. 'Innocence'* Single, white, fragrant flowers, foliage sometimes having creamy white variegation. *P. 'Sybille'* Almost square, purple-stained, single, white flowers, orange-scented, on arching branches.
Average height and spread
Five years
3x3ft (1x1m)
Ten years
5x5ft (1.5x1.5m)
Twenty years
or at maturity
5x6ft (1.5x2m)

PHILADELPHUS
Tall-growing varieties

MOCK ORANGE
Philadelphaceae
Deciduous
A useful group of large, fragrant, summer-flowering shrubs.

Origin Most varieties of garden origin.
Use As large shrubs for planting singly, or in medium to large shrub borders, or for mass planting.
Description *Flower* Single or double, depending on variety, very fragrant, ¾-1½in (2-4cm) wide, white, or white with purple throat markings, borne in clusters early to midsummer. *Foliage* Leaves ovate 1½-4in (4-10cm) long, slightly tooth-edged, light to mid green, good yellow autumn colour. Some varieties with golden foliage. *Stem* Upright, forming an upright, round-topped shrub. Grey-green. Fast growth rate. *Fruit* Insignificant.
Hardiness Tolerates winter temperatures down to −13°F (−25°C).
Soil Requirements Most soils, including alkaline and acid types.
Sun/Shade aspect Full sun through to medium shade.
Propagation and nursery production From semi-ripe cuttings taken in early summer or from hardwood cuttings in winter. Plant bare-rooted or container-grown. Best planting heights 1¼-3ft (40cm-1m).
Problems Harbours blackfly which can transfer to other garden crops.

Philadelphus 'Manteau d'Hermine' **in flower**

Philadelphus coronarius 'Aureus' **in flower**

Philadelphus 'Virginal' in flower

Varieties of interest *P. 'Beauclerk'* Large, somewhat square, fragrant, single, white flowers, with pink centres. *P. 'Burfordensis'* Large, fragrant, semi-double, white flowers with yellow stamens. More upright in growth than most. *P. 'Burkwoodii'* Very fragrant, single, cup-shaped white flowers with pink-stained centres. *P. coronarius* Single, white to creamy white, scented flowers. A large shrub, upright becoming spreading. The form from which many garden hybrids derived. *P. c. 'Aureus'* Single, white, scented flowers, somewhat obscured by lime green to golden yellow foliage, which is susceptible to late frost damage and strong summer sun scorching. Plant in light dappled shade; in deeper shade foliage turns green. *P. 'Minnesota Snowflake'* Large, double, fragrant, 'white flowers, weighing down branches. May reach small tree proportions in time. *P. 'Norma'* Single, large, white flowers, less fragrant than most, on long arching branches. *P. 'Virginal'* Fragrant, pure white, double flowers up to 1½in (4cm) across, covering upright branches in midsummer. The most popular variety. It is worth looking out for a very rare variegated form. *P. 'Voie Lactée'* Strong-growing. Broad-petalled, pure white, single, scented flowers up to 2in (5cm) across.

Average height and spread
Five years
5x3ft (1.5x1m)
Ten years
8x6ft (2.5x2m)
Twenty years
or at maturity
13x13ft (4x4m)

PHILLYREA

JASMINE BOX
Oleaceae
Evergreen
Evergreen shrubs with small but fragrant spring flowers

Origin From Africa, southern Europe and the Mediterranean regions.
Use As useful, compact, medium-sized, early summer-flowering evergreens for planting singly or in a shrub border. Plant 2½ft (80cm) apart in single line for an informal hedge.
Description *Flower* Small, tubular, fragrant flowers borne in clusters on mature shrubs, late spring, early summer. *Foliage* Leaves ovate to lanceolate, small or medium-sized, 2-5in (5-12cm) long and ½-1¾in (1-4.5cm) wide, leathery textured. Dark glossy green upper surfaces, duller glabrous undersides. *Stem* Branching from an early age, to form a round, medium to large, dome-shaped shrub. Light olive green to green-brown. Medium growth rate. *Fruit* Small, oval, purple to purple-black fruits in autumn.

Hardiness Tolerates 4°F (−15°C).
Soil Requirements Almost any soil, but distressed by extremely alkaline or very wet conditions.
Sun/Shade aspect Prefers light shade, tolerates full sun to medium shade.
Pruning None required. May be clipped and shaped.
Propagation and nursery production From semi-ripe cuttings taken in early summer. Purchase container-grown. Normally available from specialist nurseries but becoming more widely stocked. Best planting heights 1¼-2½ft (40-80cm)
Problems Young plants when purchased often look misshapen and require time to develop.

Phillyrea decora in leaf

Varieties of interest *P. angustifolia* Flowers creamy yellow, fragrant, produced in clusters at leaf axils in late spring, early summer on mature shrubs. Foliage narrow, ovate, 1½-2in (4-5cm) long, mid grey-green foliage, lighter green when young. Very good for coastal planting. *P. decora* syn. *P. medwediewii* Small clusters of fragrant white flowers, in mid to late spring. Foliage large, ovate to lanceolate, leathery texture, dark glossy green upper surfaces and paler glabrous undersides. The best variety for general garden use. Now classified as *Osmarea decora* by some authorities. *P. latifolia* Flowers white, some fragrance, but less interesting than most, borne in late spring, sometimes followed by small, blue-black fruits. Small, glossy green leaves presented opposite each other on arching stems. Slightly less than average height.

Average height and spread
Five years
4x4ft (1.2x1.2m)
Ten years
6x6ft (2x2m)
Twenty years
or at maturity
10x10ft (3x3m)

PHLOMIS

KNOWN BY BOTANICAL NAME
Labiatae
Evergreen
Attractive, silver-leaved shrubs for summer flowering.

Origin From the Mediterranean region.
Use As low to medium height, grey-foliaged subjects for shrub borders or for mass planting.

Phlomis fruticosa in flower

Description *Flower* Yellow or pink whorls of interesting shaped flowers, colours depending on variety, late summer. *Foliage* Leaves ovate or lanceolate, 1-2in (3-5cm) long with grey-green, downy surface, giving some yellow autumn colour. Some varieties with more lanceolate, metallic blue new growth. *Stem* Upright when young, grey-green and downy, becoming open and straggly with age. Old growth of less interest. Fast growth rate in spring, slowing with season. *Fruit* Insignificant.
Hardiness Minimum winter temperature 14°F (−10°C).
Soil Requirements Dry, open soil, dislikes any waterlogging.
Sun/Shade aspect Full sun, dislikes any shade.
Pruning As best foliage and flowers borne on new wood, trim by one-third in late spring to induce new growth.
Propagation and nursery production From softwood cuttings taken in mid to late summer. Purchase container-grown. *P. fruticosa* normally stocked by garden centres and general nurseries; other varieties less readily available. Best planting heights 8in-1½ft (20-50cm).

Phlomis italica in flower

Problems May die off in a severe winter so rooted cuttings should be overwintered under protection, to insure replacements for spring.
Varieties of interest *P. chrysophylla* Golden yellow flowers, early to midsummer, soft grey-green foliage, with new shoots sage green. From the Lebanon. *P. fruticosa* (Jerusalem Sage) Yellow whorls of bright yellow flowers, mid to late summer. Foliage grey-green, downy covering. From the Mediterranean region. The variety normally obtainable. *P. italica* Pale lilac-pink terminal spikes of flowers, mid to late summer. Leaves and stems covered with white hairs. Slightly more tender and slightly more upright than average.
Average height and spread
Five years
3x3ft (1x1m)
Ten years
3x4ft (1x1.2m)
Twenty years
or at maturity
3x4ft (1x1.2m)

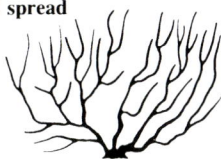

PHORMIUM

NEW ZEALAND FLAX
Agavaceae
Evergreen
Architecturally interesting foliage shrubs of tropical effect.

Origin From New Zealand.
Use As an individual feature, ideal for large tubs not less than 2½ft (80cm) across and 2ft (60cm) deep, using good potting medium. Good for large herbaceous borders or standing alone for sub-tropical foliage effect. Useful for coastal or industrial areas, accepting high degrees of pollution.
Description *Flower* Interestingly shaped upright panicles of bronze-red flowers, produced on shrubs 3-4 years old or more, midsummer. Full flower spike can grow to 13ft (4m) in height. *Foliage* Leaves upright, 8-9ft (2.5-2.8m) long and 4-5in (10-12cm) wide, leathery textured, lanceolate, emerging from basal rosette. *Stem* None. Leaves originate from small ground crown which regenerates from inside with new foliage, discarding some older outer leaves each year. Medium to fast growth rate. *Fruit* Upright flower spikes age to interestingly shaped dark brown-red seedheads.
Hardiness Minimum winter temperature 14°F (−10°C).
Soil Requirements Well-drained, open soil, dislikes any waterlogging.
Sun/Shade aspects Full sun, dislikes any shade.
Pruning None required, but dead or damaged foliage can be removed to ground level

Phormium tenax 'Purpureum'

Phormium tenax 'Sundowner'

Propagation and nursery production By division through removal of self-generated side plants. Plant container-grown. Availability varies. Best planting heights 1¼-3ft (40cm-1m).
Problems Can take 2-3 years to look effective, longer still before it flowers fully.
Varieties of interest *P. cookianum 'Cream Delight'* syn. *P. colensoi 'Cream Delight'* Foliage less rigid than most, olive-green with a creamy central band. Flower panicles reaching 3ft (1m). Two-thirds average height and spread. *P. tenax* (New Zealand Flax) Rigid clumps of upright, lanceolate leaves, leathery in texture, grey to grey-green with red-tinged veins and outer edges, especially towards autumn and winter. Bronze-red flower panicles in summer. *P. t. 'Bronze Baby'* Foliage bronze-purple, fewer flowers. Useful for tubs and where lower planting is required. Reaches one-third average height and spread but more lax in habit. *P. t. 'Dazzler'* Upright leaves, becoming pendulous, red-brown with carmine-red bands. Very showy. Less hardy than average and two-thirds average height and spread. *P. t. 'Marie Sunrise'* Red-purple spears with rose pink and bronze veining. Slightly lax. Two-thirds average height and spread. *P. t. 'Purpureum'* Strong, upright, broad, lanceolate, bronze-purple foliage. Good flowers. *P. t. 'Sundowner'* Foliage striped with cream outer band, light green centre, and purple midrif band. Excellent multi-coloured effect. More tender than most. *P. t. 'Tricolor'* Leaves green, striped with white, with red margins. *P. t. 'Variegatum'* Leaves having creamy white margins on bright green base colour, with some red veining. *P. t. 'Yellow Wave'* Golden yellow leaves with green edges. Slightly more lax and

more tender than the parent and reaching two-thirds average height and spread.
Average height and spread
Five years
3x3ft (1x1m)
Ten years
5x6ft (1.5x2m)
Twenty years
or at maturity
5x7ft (1.5x2.2m)

PHOTINIA

KNOWN BY BOTANICAL NAME
Rosaceae
Evergreen
A group of plants making the most spectacular foliage display in winter and early spring.

Origin From New Zealand and Australia.
Use As a winter foliage shrub for larger shrub borders. Planted 3ft (1m) apart makes a tall, informal hedge, especially in mild and coastal areas. Good fan-trained as a wall shrub where cutting and retaining induces plentiful red winter foliage.

Photinia × fraseri 'Red Robin'

Description *Flower* Clusters of white flowers in late spring, but outside its native environment rarely flowers except in warm areas. *Foliage* Leaves large, ovate to lanceolate, 4-8in (10-20cm) long and 1½-3½in (4-9cm) wide, often tooth-edged foliage. Dark glossy green when mature. Young growth, produced from late autumn, starts as dark red or brilliant red and increases in intensity through winter into early spring, becoming bronze during mid spring to early summer. *Stem* Upright, becoming branching with age, forming an upright, tall, dense shrub. Dark green tinged red. Medium growth rate. *Fruit* Rarely fruits outside its native environment or very hot climates, where it produces clusters of red or orange-red fruits.
Hardiness Tolerates 4°F (−15°C), although young foliage may become slightly misshapen or split in severe cold.
Soil Requirements Does well on most soils, only disliking extreme alkalinity.
Sun/Shade aspect Full sun to light shade; tolerates mild shade.
Pruning None required, but can be cut back moderately hard.
Propagation and nursery production From semi-ripe cuttings taken in early summer. Purchase container-grown. Most varieties easy to find. Best planting heights 1¼-3ft (40cm-1m).
Problems Generally purchased as a single shoot and requires 1-3 years to flourish following planting.
Varieties of interest *P. × fraseri 'Birmingham'* Very dark coppery red to dark red new growth. Slightly broader foliage. Raised in North America. *P. × f. 'Red Robin'* syn. *P.*

glabra 'Red Robin' One of the most spectacular winter foliage shrubs. New growth brilliant red in late autumn, held through winter and spring, aging to bronze in late spring and summer. From New Zealand. *P. glabra 'Rubens'* Foliage slightly smaller and broader than parent. May also be more likely to flower, with white clusters of flowers, followed by red fruits. Young growth brilliant red in winter and spring. Slightly more tender. From Japan. *P. serrulata* (Chinese Hawthorn) Upright coppery red foliage from mid winter through spring and maintained over long period. Sometimes bears clusters of white flowers in mid to late spring, followed by red fruits in early autumn.

Average height and spread
Five years
5x6ft (1.5x2m)
Ten years
6x10ft (2x3m)
Twenty years
or at maturity
13x13ft (4x4m)

PHYGELIUS

CAPE FIGWORT
Scrophulariaceae
Deciduous
Interesting, late summer-flowering sub-shrubs.

Origin From South Africa
Use As a low shrub for small or medium shrub borders, or grouped in large borders; also for herbaceous borders for late flowering effect, or as wall shrubs reaching up to 6ft (2m).
Description *Flower* Yellow or orange, depending on variety, tubular flowers, up to 1½in (4cm) long, hanging from upright panicles, late summer through to early autumn. *Foliage* Leaves ovate to lanceolate, 2-3in (5-8cm) long, tooth-edged, dark glossy green upper surfaces, grey undersides. Some yellow autumn colour. *Stem* Upright shoots, annually produced, dark green, normally dying down or being damaged in winter; if grown against a wall achieves more woody structure. Fast annual spring and summer growth rate. *Fruit* Insignificant.
Hardiness Minimum winter temperature 23°F (−5°C); in need of winter protection.
Soil Requirements Does well on most soils.
Sun/Shade aspect Prefers full sun, tolerates light shade.
Pruning Shrubs grown in open borders should be pruned to ground level in early spring, to encourage new strong growths and autumn flowers. Wall-grown shrubs should be lightly trimmed and new growths tied in as they appear.
Propagation and nursery production From seed or semi-ripe cuttings taken in midsummer. Purchase container-grown. In cold areas young plants may be overwintered under

Phygelius capensis 'Coccineus' **in flower**

176

---

cover in case replacements are needed. Fairly easy to find; yellow-flowering forms less often available. If purchased when in flower best kept under cover until following spring. Best planting heights 8-15in (20-40cm).
Problems As sub-shrubs, these plants die back or become very woody in winter and require pruning to ground level each spring.
Varieties of interest *P. aequalis 'Yellow Trumpet'* Large, yellow, tubular flowers, up to 1½in (4cm) long, hanging from upright panicles, 1-1½ft (30-50cm) early autumn. Light green foliage. May not withstand any frost. *P. capensis 'Coccineus'* Crimson-scarlet, hanging, tubular flowers, 1in (3cm) long, reminiscent of fuchsias, produced from tall, terminal, panicles, 1ft (30cm) long in late summer, early autumn. Good as wall climber in mild areas.

Average height and spread
Five years
3x3ft (1x1m)
Ten years
6x6ft (2x2m)
Twenty years
or at maturity
6x6ft (2x2m)
Heights include flower spike

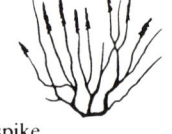

PHYSOCARPUS OPULIFOLIUS 'LUTEUS'

NINEBARK, EASTERN NINEBARK
Rosaceae
Deciduous
Useful golden foliage shrubs, as long as due care is taken to meet their specific cultural needs.

Origin Parent plant from eastern North America, with varieties of garden origin.
Use As medium to tall foliage shrubs, the golden colour good in association with purple-leaved shrubs. Ideal for large to medium shrub borders or for mass planting.

Physocarpus opulifolius 'Luteus' **in leaf**

Description *Flower* Clusters of off-white flowers, early to mid summer. *Foliage* Leaves three-lobed 1½-2in (4-5cm) wide and long, tooth-edged, clear yellow or golden yellow, depending on variety. *Stem* Mahogany-brown new growth, becoming brown-grey. If pruned as advised, giving good winter interest. Medium growth rate. *Fruit* Brown seedheads maintained into winter.
Hardiness Tolerates winter temperatures down to −13°F (−25°C).
Soil Requirements Most soils but shows signs of chlorosis in severely alkaline conditions.
Sun/Shade aspect Very light dappled shade, deeper shade causes foliage to turn green. Strong summer sun may scorch golden foliage.
Pruning Remove one-third of oldest flowering wood on established shrubs after flower-

---

ing to encourage rejuvenation for coming year and to maintain a good supply of mahogany-brown winter stems.
Propagation and nursery production From semi-ripe cuttings taken in early summer or hardwood cuttings in winter. Purchase container-grown. Easy to find, especially in summer. Best planting heights 1¼-2½ft (40-80cm).
Problems Leaf scorch can occur in bright sunlight.
Varieties of interest *P. o. 'Dart's Gold'* Produces more plentiful, bright golden foliage. Two-thirds average height and spread.
Average height and spread
Five years
5x5ft (1.5x1.5m)
Ten years
8x8ft (2.5x2.5m)
Twenty years
or at maturity
8x8ft (2.5x2.5m)

PIERIS
(Andromeda)

PIERIS, LILY OF THE VALLEY SHRUB, ANDROMEDA
Ericaceae
Evergreen
Beautiful foliage and flowering shrubs for acid soils.

Origin Forms ranging over a wide area of the world, and many of garden origin.
Use Ideal for woodland gardens, with rhododendrons and azaleas. Can be grown in tubs not less than 2½ft (80cm) in diameter and 2ft (60cm) in depth, using lime-free potting medium. Variegated forms good for large, protected, slightly shaded rock gardens.
Description *Flower* White or pink, depending on variety, hanging or upright broad panicles of pitcher-shaped flowers, mid to late spring. *Foliage* Leaves broad, lanceolate 1¼-3¼in (3.5-7.5cm) long, tooth-edged foliage, red-orange in mid to late spring, aging to bronze and finally light to mid green. Some white-variegated forms. *Stem* Upright when young, becoming spreading with age, forming a dome-shaped shrub. Light green to grey-green. Slow to medium growth rate. *Fruit* Insignificant.

Pieris japonica **in flower**

Hardiness Tolerates 14°F (−10°C), but new foliage is often severely damaged by late spring frosts. Requires protection from cold winds.
Soil Requirements Acid to neutral soil, dislikes any alkalinity.
Sun/Shade aspect Prefers light to medium shade, dislikes full sun.
Pruning None required but can be reduced in size following flowering.
Propagation and nursery production From

Pieris japonica 'Fire Crest' in spring

semi-ripe cuttings taken in early summer or from layers. Purchase container-grown or root-balled (balled-and-burlapped). Easy to find. Best planting heights 1-2½ft (30-80cm).
Problems Spring foliage, its main attribute, is easily damaged, so take particular care when choosing planting position.
Varieties of interest *P. floribunda* White flowers produced in erect terminal panicles, early to mid spring. Some new red foliage but not among the most interesting. Two-thirds average height and spread and slow-growing. One of the hardiest forms. From south-eastern USA. *P. formosa* Large, leathery textured leaves with fine toothed margins. Coppery red new growth, aging to dark glossy green, showing off white flower panicles produced in clusters in mid spring. More tender than most and possibly exceeding average height and spread. From the eastern Himalayas. *P. f. forrestii* Young growth bright red and of good size. Flowers white, slightly scented, formed in long conical panicles, borne later than most, mid to late spring. From south-western China through north-eastern Burma. *P. f. 'Wakehurst'* Large, vivid red young foliage. White, waxy, hanging panicles of flowers in midsummer. More susceptible to frost damage than most. Slightly less than average height and spread. *P. forrestii 'Forest Flame'* Comparatively long, large, brilliant red new growth in spring, changing from red through pink to creamy white and finally green. Large terminal hanging panicles of flowers, mid spring. One of the hardiest varieties. *P. japonica* Glossy green foliage, longer and broader than most, coppery red when young. Flowers borne in drooping panicles with waxy textured white florets, mid to late spring. A variety reaching two-thirds average height and spread. From Japan. *P. j. 'Christmas Cheer'* Flowers flushed pink with dark carmine-rose shading at tip, giving a bicoloured effect. Borne even on young plants and often appearing in mid to late winter. Some bronze-red new foliage growth. *P. j. 'Fire Crest'* Attractive red new foliage with racemes of waxy, white, scented flowers. *P. j. 'Pink Delight'* Flowers pink in bud, opening to white. Bronze spring foliage, turning green. Of slightly less than average height and spread. *P. j. 'Scarlet O'Hara'* Young spring foliage red beoming green. Creamy white, speckled red flowers. *Pieris j. 'Variegata'* Leaves grey-green with creamy white variegation and flush of pink when young. Some flowers, but grown for its variegated foliage. *P. taiwanensis* Small foliage, dark green to olive green with some bronze new growth. Upright panicles of white flowers in mid spring. Very low growing and reaches only one-third average height and spread. From Taiwan.
Average height and spread
Five years
2½-3ft (80cmx1m)
Ten years
4x6ft (1.2x2m)
Twenty years
or at maturity
6x13ft (2x4m)

PIPTANTHUS LABURNIFOLIUS (Piptanthus nepalensis)

EVERGREEN LABURNUM
Leguminosae
Evergreen
An interesting evergreen shrub, particularly useful as a wall specimen.

Origin From the Himalayas.
Use As a freestanding shrub or grown against a wall, where it may reach more height and spread than average.

Piptanthus laburnifolius in flower

Description *Flower* Bright yellow pea-flowers in short racemes, mid to late spring. *Foliage* Leaves trifoliate, 3-6in (8-15cm) long, with dark glossy green upper surfaces and duller undersides. Old leaves drop in autumn giving some good autumn colour. *Stem* Upright, with some branching, forming a narrow-based, wide-headed shrub. Dark glossy green. Medium to fast growth rate. *Fruit* Medium to long pea-pods, grey to grey-brown in autumn and winter.
Hardiness Tolerates 14°F (−10°C), but may be defoliated in severe winters, with some stem die-back.
Soil Requirements Prefers well-drained soil but accepts most, tolerating high alkalinity.
Sun/Shade aspect Full sun or shade, but may be slightly taller in shade.
Pruning Each year shorten back one or two stems on established shrubs after flowering, to rejuvenate good wood.
Propagation and nursery production From seed. Purchase container-grown. Easy to find in some areas, otherwise obtainable from specialist nurseries. Best planting heights 1¼-2½ft (40-80cm).
Problems Apart from possible winter die-

back, occasional shoots can die from drought, waterlogging, or mechanical damage to roots.
Average height and spread
Five years
6x5ft (2x1.5m)
Ten years
10x8ft (3x2.5m)
Twenty years
or at maturity
13x12ft (4x3.5m)

PITTOSPORUM

PITTOSPORUM, KOHUHA
Pittosporaceae
Evergreen
Useful shrubs for mild areas or grown in containers as patio shrubs needing winter protection.

Origin From New Zealand.
Use In mild areas as a freestanding shrub or in a medium to large shrub border. If planted 3ft (1m) apart makes semi-formal hedge. In less mild areas, best grown in containers not less than 2½ft (80cm) in diameter and 2ft (60cm) in depth. May be grown as conservatory plants. Use good potting medium and provide winter protection. May also be used as a wall shrub on warm, sunny, very sheltered sites. Extremely good material for flower arranging.
Description *Flower* Small, unusual, chocolate brown to purple, honey-scented flowers in late spring, only on very mature shrubs in warm, favourable areas. *Foliage* Leaves ovate to round, 1-2½in (3-6cm) long, glossy, thick textured, in varying shades of olive green, purple, silver, gold, or variegated. *Stem* Upright stems, becoming branching and twiggy with age, forming an upright, pyramidal shrub. Dark brown to almost black. Medium growth rate. *Fruit* Insignificant.
Hardiness Minimum winter temperature 23°F (−5°C).
Soil Requirements Well-drained soil.
Sun/Shade aspect Full sun or very light shade.
Pruning None required but can be trimmed quite harshly; if cut to ground level will rejuvenate.
Propagation and nursery production From semi-ripe cuttings taken in early summer. Purchase container-grown. Easy to find, especially in mild coastal areas. Best planting heights 1-3ft (30cm-1m).
Problems Often seen and purchased in mild coastal areas and taken inland for planting in the open, where it rarely succeeds.
Varieties of interest *P. tenuifolium* Attractive grey-green foliage, leaves slightly twisted with good black twigs. Flowers well in mild areas. One of the hardiest. *P. t. 'Garnettii'* Round, white-variegated leaves, flushed pink in winter. One of the hardiest of variegated forms. Slightly less than average height and

Pittosporum tenuifolium 'Silver Queen'

177

spread. Of garden origin. *P. t. 'Irene Paterson'* Marbled, creamy white foliage. Interesting leaf display, forming slightly rounder shrub than most. Half average height and spread. *P. t. 'James Stirling'* Small, silver-green, rounded or oval leaves, very densely presented on black-purple twigs. Tender and will not withstand frost. Slightly less than average height and spread. *P. t. 'Purpureum'* Young pale green leaves turn to deep bronze-purple, aging to purple-green, giving an outer canopy of good purple foliage. Two-thirds average height and spread. *P. t. 'Silver Queen'* Narrow, ovate leaves, pointed and tightly bunched on black stems, forming neat, conical shape. Foliage variegated silver-grey, black twigs. Two-thirds average height and spread. *P. t. 'Tresederi'* Amber leaves, mottled gold when young. *P. t. 'Warnham Gold'* Yellow to yellow-green young foliage, aging to golden yellow. Half average height and spread. Of garden origin. *P. tobira* (Japanese Pittosporum, Tobiri Pittosporum) Large, long, obovate, bright glossy green leaves, presented in whirls, bearing orange-blossom scented, creamy coloured flowers in midsummer. Good in mild coastal areas; in less favourable conditions use as a wall shrub or conservatory specimen. In very mild areas, useful for hedging planted 2½ft (80cm) apart in single line. Two-thirds average height with one-third more spread, making a round-topped shrub. From China, Taiwan and Japan, now frequently seen in southern Europe. *P. t. 'Variegatum'* Variegation irregular, with creamy white margins and central patches. Needs favourable, mild location. In colder areas use as a conservatory shrub.

Average height and spread
Five years
5x3ft (1.5x1m)
Ten years
10x6ft (3x2m)
Twenty years
or at maturity
20x10ft (6x3m)

PONCIRUS TRIFOLIATA (Citrus trifoliata)

HARDY ORANGE, JAPANESE BITTER ORANGE, TRIFOLIATE ORANGE
Rutaceae
Deciduous
An interesting, architecturally shaped shrub with attractive flowers and fruit. By some authorities now referred to as *Aegle sepiaria*.

Origin From northern China
Use As a freestanding shrub for architectural shape and interesting flowers and fruits.
Description *Flower* Very sweetly scented, large, 1½-2in (4-5cm) across, white, orange-blossom type flowers in mid to late spring.

Poncirus trifoliata in flower

Foliage Leaves somewhat sparsely presented, obovate, 1-2½in (3-6cm) long, light, bright green, giving some yellow autumn colour. *Stem* Branching, stout, armed with spines 1in (3cm) long, grey to bright green. Forming a ball-shaped shrub of architectural interest. Slow growth rate. *Fruit* Green, ripening to yellow, globular, miniature orange fruits 1½-1¾in (4-4.5cm) across.
Hardiness Tolerates 14°F (−10°C).
Soil Requirements Well-drained, open, light soil, tolerates alkaline and acid types.
Sun/Shade aspect Full sun to light shade.
Pruning None required.
Propagation and nursery production From seed. Purchase container-grown. Obtainable from specialist nurseries. Best planting heights 8in-1½ft (20-50cm).
Problems Relatively short-lived; can deteriorate after 25 years. Dislikes soil cultivation close to roots.

Average height and spread
Five years
3x3ft (1x1m)
Ten years
6x6ft (2x2m)
Twenty years
or at maturity
10x10ft (3x3m)

POTENTILLA

SHRUBBY CINQUEFOIL, BUTTERCUP SHRUB, FIVE FINGER
Rosaceae
Deciduous
Gems of summer-flowering shrubs, offering a wide choice of flower and foliage colour and different heights.

Origin *P. arbuscula* and *P. dahurica* are from the Himalayas, northern China and Siberia. *P. fruticosa* is native to a wide area of the northern hemisphere, but most varieties now of garden origin.
Use Grouped in a border of any size or planted singly. If planted 2ft (60cm) apart makes an informal, low hedge.
Description *Flower* Small, single saucer-shaped flowers, up to 1in (3cm) across, in colours dependent on variety, ranging from white through primrose, yellow, pink, orange and red, borne mainly in midsummer, but from early summer through to early autumn. *Foliage* Leaves cut and lobed, pinnate, with 3, 5 or 7 leaflets, linear to oblong, ½-1in (1-3cm) long, overlapping. Light green to sage green to silver green, depending on variety. Some yellow autumn colour. *Stem* Grey to grey-brown, appearing dead when bare, but quickly producing foliage in spring. Some varieties grow directly upright to form round-topped shrubs. Others are mound-forming, a few

---

almost prostrate. *Fruit* Brown-grey seed-heads formed after flowering.
Hardiness Tolerant of winter temperatures down to −13°F (−25°C).
Soil Requirements Wide range of soils, only distressed by extremely dry, wet or very alkaline conditions.
Sun/Shade aspect Prefers full sun, tolerates mid shade.
Pruning Each year after planting remove one-third of growth, and occasionally cut back older stems to ground level to induce maximum rejuvenation. Very old established, neglected shrubs can either be cut extremely hard, when they will produce new growth, or thinned over a period of years to bring them back into full production.

Potentilla 'Daydawn' in flower

Propagation and nursery production From softwood or semi-ripe cuttings taken in spring or early summer. Purchase container-grown. Most varieties fairly easy to find. Best planting heights 8in-2ft (20-60cm).
Problems Can appear to have died in winter. If not pruned annually becomes woody.
Varieties of interest *P. arbuscula* Good yellow flowers over a long period, contrasting with grey-green foliage. Forming a bushy mound with arching branches. Slightly less than average height and spread. *P. a. 'Beesii'* syn. *P. fruticosa 'Beesii'*, *P. a. 'Nana Argentea'* Bright golden yellow flowers, contrasting well with bright silvery-grey foliage. Forms a flattish mound. Reaches one-third average height and spread. *P. dahurica 'Abbotswood'* Pure white flowers, grey-green foliage. Mound-forming. Reaching two-thirds average height and spread. *P. d. 'Manchu'* syn. *P. mandshurica fruticosa* White flowers, grey-green foliage. Very low and carpet-forming,

Potentilla dahurica 'Abbotswood' in flower

Potentilla 'Elizabeth' in flower

reaching only 8in (20cm) in height with spread of 2ft (60cm). *P. d. 'Mount Everest'* syn. *P. fruticosa 'Mount Everest'* Pure white, large flowers. Strong-growing, making a round shrub. *P. d. var. veitchii* syn. *P. fruticosa veitchii* Pure white flowers, grey-green foliage. Upright, forming spreading up shape, two-thirds average height with slightly more spread. From western and central China. *P. 'Dart's Golddigger'* Covered in buttercup yellow flowers, against grey-green foliage. Round and bushy. Two-thirds average height and spread. *P. 'Daydawn'* Peach pink to cream flowers, which maintain their colour in strong sunlight. Bushy habit. Reaching two-thirds average height and spread. *P. 'Elizabeth'* syn. *P. fruticosa 'Elizabeth'* Covered in large, canary yellow flowers against grey-green foliage background. Bushy habit. Often confused with and sold as *P. arbuscula*. *P. fruticosa* Smaller, yellow flowers covering an upright shrub with much smaller, divided leaves, light green to mid green. Upright habit. *P. 'Jackman's Variety'* Bright golden yellow flowers. Strong, upright-growing shrub with green foliage. One of the best forms for hedging. *P. 'Katherine Dykes'* Covered in primrose yellow flowers, with grey-green foliage. Bushy habit. A strong-growing variety, reaching slightly more than average height and spread. *P. 'Longacre'* Good-sized sulphur yellow flowers, green foliage. Bushy habit. Reaching two-thirds average height and spread. *P. 'Maanelys'* syn. *P. 'Moonlight'* Covered in soft yellow to primrose yellow flowers throughout summer. Grey-green foliage. Upright habit. *P. parvifolia* Covered in small golden yellow flowers. Mid green to dark green foliage. Tight, compact, upright habit, reaching slightly less than average height and equal spread. *P. p. 'Buttercup'* syn. *P. fruticosa 'Buttercup'* Small, deep golden yellow flowers, green foliage. Bushy habit. *P. p. 'Klondike'* syn. *P. fruticosa 'Klondike'* One of the best golden yellow forms. Small green foliage. Upright to bushy in habit. Slightly less than average height and spread. *P. 'Primrose Beauty'* syn. *P. fruticosa 'Primrose Beauty'* Primrose yellow flowers with dark yellow centres. Grey to grey-green foliage. Can be somewhat untidy with its large brown seedheads. Bushy habit. *P. 'Princess'* Flowers softest rose pink. Green foliage. A small shrub of bushy habit, reaching one-third average height and equal spread. *P. 'Red Ace'* Red to orange-red flowers against a finely cut green leaf background. Needs light shade to maintain good red colour. Bushy habit. Reaching two-thirds average height and spread. *P. 'Royal Flush'* Deep rose pink flowers set against largish green leaves. Somewhat susceptible to die-back, especially in wet conditions. May be unreliable. Bushy habit. *P. 'Sanved'* syn. *P. sandudana* Good-sized white flowers, bright green leaves on

upright, bushy shrub. *P. 'Sunset'* Orange flowers flecked with brick red. Green foliage. Best in light shade for good colour effect. Bushy habit. *P. 'Tangerine'* Pale copper yellow to tangerine orange-yellow flowers. Requires light shade to maintain flower colour. Bushy habit. Two-thirds average height and spread. *P. 'Tilford Cream'* Creamy white flowers of good size, borne in profusion, with slightly grey-green foliage. Bushy habit. Reaches two-thirds average height and spread. *P. 'Vilmoriniana'* syn. *P. fruticosa 'Vilmoriniana'* Very pretty primrose yellow flowers, attractive silver foliage. Probably found only in specialist nurseries. Can be difficult to establish. Upright, reaching slightly more than average height.

Average height and spread
Five years
2½x2½ft (80x80cm)
Ten years
4x4ft (1.2x1.2m)
Twenty years
or at maturity
4x4ft (1.2x1.2m)

PRUNUS × CISTENA

PURPLE-LEAF SAND CHERRY, DWARF CRIMSON CHERRY
Rosaceae
Deciduous
A useful foliage shrub with year-round attractions.

Origin Of garden origin.
Use As medium height, purple-leaved shrub with early spring flowers and interesting winter wood. For use in small to medium shrub borders or grouped in a larger border. Ideal for mass planting. If planted 2ft (60cm) apart in single line makes low hedge. Can be trained as small standard tree or clipped to various shapes. Good in containers more than 2½ft (80cm) in diameter and 2ft (60cm) in depth; use good potting medium.
Description *Flower* Small, flat, off-white flowers borne on bare stems, early spring. *Foliage* Leaves ovate, 2in (5cm) long, purple-red from mid spring onwards. Purple retained well into late summer, when it becomes reddish orange-purple, giving good autumn flame colour. *Stem* Upright, becoming twiggy with age, forming goblet-shaped shrub. Dark purple-red. Fast growth rate when young, slowing with age. *Fruit* Small, purple-black fruits in autumn.
Hardiness Tolerates 4°F (−15°C).
Soil Requirements Any soil.
Sun/Shade aspect Full sun to very light shade, to maintain good foliage colour. Too deep shade turns the purple foliage to green.
Pruning None required, but can be cut or trimmed fairly hard.

Propagation and nursery production From hardwood cuttings taken early winter. Purchase container-grown or bare-rooted. Availability varies. Best planting heights 1-2½ft (30-80cm).
Problems Young plants can often look a little thin and weak, but grow rapidly once planted.

Prunus × cistena **in summer**

Average height and spread
Five years
3x2½ft (1mx80cm)
Ten years
5x4ft (1.5x1.2m)
Twenty years
or at maturity
6x5½ft (2x1.8m)

PRUNUS GLANDULOSA

CHINESE BUSH CHERRY, DWARF FLOWERING ALMOND
Rosaceae
Deciduous
Useful spring-flowering shrub of great flowering ability.

Origin From central and northern China.
Use As a small, individual shrub for large rock gardens or specialist planting, or for small or medium-sized shrub borders.
Description *Flower* Double, many-petalled, small pink or white flowers, depending on variety, produced profusely along entire length of upright shoots, mid to late spring, before foliage is formed. *Foliage* Leaves small, ovate ¾-1¼in (2-3.5cm) long, slightly tooth-edged, light grey-green. Some autumn colour. *Stem* Upright, thin, wispy stems, forming a goblet-shaped shrub. Brown to

Prunus glandulosa 'Albiplena' **in flower**

mahogany-brown. Slow growth rate. *Fruit* Insignificant.
Hardiness Best planted in a sheltered position with winter temperatures no lower than 23°F (−5°C).
Soil Requirements Well-drained soil; shows signs of chlorosis on alkaline soils.
Sun/Shade aspect Prefers full sun to light shade.
Pruning Upright shoots should be cut hard following flowering will rejuvenate quickly flowering profusely in the following spring.
Propagation and nursery production From grafting or budding. Purchase container-grown. Not always readily available in garden centres but should be found in general or specialist nurseries. Best planting heights 1-2ft (30-60cm).
Problems Susceptible to a form of peach-leaf curl mildew. Pruning as recommended helps to control these problems.
Varieties of interest *P. g. 'Albiplena'* A variety with double white flowers profusely produced along stems. More resistant to disease and fairly robust in its habit of growth. *P. g. 'Sinensis'* Bright pink, double flowers. Not a robust variety.

Average height and spread
Five years
2½x2ft (80x60cm)
Ten years
4x2½ft (1.2mx80cm)
Twenty years or at maturity
5½x3ft (1.8x1m)

PRUNUS LAUROCERASUS

ENGLISH LAUREL, CHERRY LAUREL
Rosaceae
Evergreen
A group of shrubs deserving more attention than they usually get, as flowering evergreens with wide application for difficult planting areas.

Origin From eastern Europe and Asia Minor but now extensively planted throughout northern Europe.
Use As large, tall, wide-spreading evergreens for screening. Can be trained as small trees, either single-stemmed or multi-stemmed. If planted 3ft (1m) apart in single line makes very large, impenetrable hedge. Some good ground cover varieties.
Description *Flower* Terminal and axiliary racemes, 2-5in (5-12cm) long, of small white flowers, closely bunched. Flowers borne only on older, mature wood in mid spring and if shrub has been cut back or trimmed it will not flower. *Foliage* Leaves large, up to 6-8in (15-20cm) long, ovate, slightly pointed. Glossy bright green upper surfaces, becoming dark olive green with age, with duller undersides. Leathery texture. Slightly incurved.

Prunus laurocerasus 'Otto Luyken' in **flower**

Prunus laurocerasus 'Zabeliana' in leaf

Some varieties with narrow lanceolate leaves. *Stem* Strong, upright shoots. Bright green, aging to brown, becoming branching with age, forming a large, round-topped shrub. Some goblet-shaped or spreading forms. Medium to fast growth rate. *Fruit* Racemes of red fruits, aging to black and retained into early winter, resembling cherries.
Hardiness Tolerates winter temperatures down to −13°F (−25°C), but in severe winters may be defoliated.
Soil Requirements Most soils, only disliking conditions of extreme dryness or extreme alkalinity, where signs of distress occur.
Sun/Shade aspect Full sun through to heavy shade, where it may be more open and lax.
Pruning None required but can be pruned very hard to control size. If overgrown and neglected, can be reduced to ground level and will rejuvenate very quickly. Hard pruning should be carried out in late winter to early spring.
Propagation and nursery production From semi-ripe cuttings taken early summer. Plant bare-rooted or container-grown. Easy to find. Best planting heights 1¼-3ft (40cm-1m).
Problems Often planted in areas where it does not have space to achieve its full potential; the merits of certain flowering and fruiting forms are insufficiently appreciated.
Varieties of interest *P. l. 'Camelliifolia'* Leaves somewhat twisted, very dark green and camellia-like. Good flower production. May have to be sought from specialist nurseries. *P. l. 'Magnoliifolia'* A very attractive evergreen with leaves 10in (25cm) long and up to 4in (10cm) wide. Good white racemes of flower and berries. Obtainable from specialist nurseries. *P. l. 'Otto Luyken'* Comparatively small, narrow, pointed ovate, shiny dark green leaves. Forms a goblet-shaped shrub. Extremely useful for mass planting, as a low hedge, or for large containers. Reaching only 5ft (1.5m) in height and spread. *P. l. 'Rotundifolia'* Large, broad, round-ended, light green leaves aging to mid green. Most useful variety for hedging. *P. l. 'Schipkaensis'* Narrow, dark green leaves, good flowers. Useful tall ground cover. Reaches height of 10ft (3m) with spread of 16ft (5m). *P. l. 'Variegata'* Primrose yellow to creamy white mottled variegation to foliage. Forms a slightly goblet-shaped, short-growing shrub, reaching up to 6ft (2m) in height and spread. *P. l. 'Zabeliana'* A low-growing, spreading form. Bright green, narrow leaves, making excellent ground cover for open positions, large banks, or under trees. Can be planted with some success in containers 2½ft (80cm) in diameter and 2ft (60cm) in depth. Spreading branches, reaching out to 13-16ft (4-5m) with a height of 3-6ft (1-2m).

Average height and spread
Five years
10x10ft (3x3m)
Ten years
20x20ft (6x6m)
Twenty years or at maturity
26x26ft (8x8m)

PRUNUS LUSITANICA

PORTUGAL LAUREL
Rosaceae
Evergreen
A useful evergreen shrub for difficult sites, often underestimated.

Origin From Spain and Portugal.
Use As a freestanding shrub, or ultimately a small multi-stemmed tree, or as background planting for medium to large borders. Ideal for mass planting as windbreak or screens, especially in coastal areas. Can be trained as a small, mop-headed, trimmed standard and used where Bay Laurel *(Laurus nobilis)* is too tender. If planted 2½ft (80cm) apart, makes a large hedge.

Prunus lusitanica in flower

Description *Flower* Racemes, 6-7in (15-18cm) long, of white flowers with a hawthorn scent, borne on mature wood in early summer. *Foliage* Leaves ovate, 2½-5in (6-12cm) long, dark green with red veins and red leafstalks. Glossy upper surfaces, duller undersides. *Stem* Upright when young, red to olive green, becoming branching with age and possibly forming, after 30 years or more, a small multi-stemmed tree. Medium to fast growth rate. *Fruit* Racemes of red fruits in autumn, aging to purple to purple-black.
Hardiness Tolerates winter temperatures down to −13°F (−25°C).
Soil Requirements Any soil.
Sun/Shade aspect Prefers medium to light shade, tolerates full sun to very deep shade.
Pruning None required but can be cut back hard and will rejuvenate fairly quickly.
Propagation and nursery production From semi-ripe cuttings taken in early summer. Purchase container-grown or root-balled (balled-and-burlapped). Easy to find in

Prunus lusitanica 'Variegata' in leaf

general nurseries, less often in garden centres. Best planting heights 1-2½ft (30-80cm). **Problems** If left to grow freely, its ultimate height and size is often underestimated. On very old shrubs branches may die back for no apparent reason, but if cut out hard and low down, they generally rejuvenate quite quickly.
Varieties of interest *P. l. var. azorica* Larger leaves than the parent and slightly curled. Bright green with red tints and red leafstalks. Stems reddish and more open in habit. Obtainable from specialist nurseries. Two-thirds average height and spread. *P. l. 'Variegata'* A variegated form, with grey-green, glossy foliage with white marginal variegation, sometimes pink-tinged in winter. Variegation is held even in medium to deep shade, although in very deep shade the shrub may become very open and lax. Should be considered slightly less hardy. Reaches only half average height and spread.
Average height and spread
Five years
3x3ft (1x1m)
Ten years
8x8ft (2.5x2.5m)
Twenty years
or at maturity
16x16ft (5x5m)

PRUNUS MUME

JAPANESE APRICOT
Rosaceae
Deciduous
Beautiful shrubs, although difficult to obtain.

Origin From China and Korea. Long cultivated in Japan, hence its common name.
Use As a very attractive large shrub on its

Prunus mume 'O-moi-no-wac' in flower

own or for medium to large shrub borders.
Description *Flower* Double or semi-double, scented, white or pink, dependent on variety, produced on bare stems very early in spring. *Foliage* Leaves ovate, 1¼-2in (3.5-5cm) long, slightly tooth-edged, light to mid green. Some yellow autumn colour. *Stem* Upright when young, moderately strong, becoming spreading with age, forming a wide, large, spreading shrub, which may in time become small multi-stemmed tree. Medium to fast growth rate. *Fruit* Insignificant.
Hardiness Tolerates 14°F (−10°C).
Soil Requirements Any soil.
Sun/Shade aspect Full sun to light shade.
Pruning None required.
Propagation and nursery production From grafting on to seedlings of *P. amygdalus*. Best purchased container-grown. Very hard to find. Best planting heights 8in-2½ft (20-80cm).
Problems May suffer from peach leaf curl and peach mildew.
Varieties of interest *P. m. 'Beni-shi-don'* Very fragrant, cup-shaped, rose madder flowers. *P. m. 'O-moi-no-wac'* Fragrant, cup-shaped, semi-double, white flowers in early spring.
Average height and spread
Five years
6x6ft (2x2m)
Ten years
13x13ft (4x4m)
Twenty years
or at maturity
20x20ft (6x6m)

PRUNUS TENELLA

DWARF RUSSIAN ALMOND
Rosaceae
Deciduous
A beautiful early spring-flowering shrub with some autumn colouring.

Origin From south-eastern Europe, western Asia through to eastern Siberia.
Use As an attractive early spring-flowering shrub with some autumn colouring.
Description *Flower* Small, round, pink flowers profusely produced on upright stems in early spring. *Foliage* Leaves lanceolate to ovate, 1½-3½in (4-9cm) long, slightly tooth-edged, grey-green, giving some orange-yellow autumn colour. *Stem* Upright, long, non-branching stems; producing branches and twigs only at upper limits. Light brown to mahogany-brown. Forms an upright shrub. Slow to medium growth rate. *Fruit* Insignificant.
Hardiness Tolerates 4°F (−15°C).
Soil Requirements Any soil except extremely alkaline types.
Sun/Shade aspect Full sun to light shade. Dislikes deep shade.

Pruning None required and in fact may resent it.
Propagation and nursery production From grafting or from budding using *P. amygdalus* as understock. Purchase container-grown or root-balled (balled-and-burlapped). Availability varies. Best planting heights 1-2ft (30-60cm).
Problems Despite its common name it can reach quite large proportions and this must be allowed for.
Varieties of interest *P. t. 'Fire Hill'* Flowers brilliant rose red, profusely produced. Very good autumn colour. Forms a suckering upright thicket. Difficult to find. Two-thirds average height and spread.
Average height and spread
Five years
5x3ft (1.5x1m)
Ten years
5½x4ft (1.8x1.2m)
Twenty years
or at maturity
6x5ft (2x1.5m)

PRUNUS TRILOBA
(Prunus triloba multiplex)

KNOWN BY BOTANICAL NAME
Rosaceae
Deciduous
An attractive, spring-flowering, large shrub or small mop-headed tree.

Origin From China, of garden origin.
Use As a freestanding, large flowering shrub of attractive shape and performance. Good as background in a large shrub border. Can be trained as an attractive small mop-headed tree.
Description *Flower* Double, rosette-shaped, peach-pink flowers of good size borne on entire length of bare stems in mid spring. *Foliage* Leaves ovate, 1½-2½in (4-6cm) long, tooth-edged, mid green giving some yellow autumn colour. *Stem* Upright when young, becoming spreading and pendulous with age, forming a round-topped, large, mound-shaped shrub. Mahogany-brown. Medium to fast growth rate. *Fruit* Insignificant.
Hardiness Tolerates 4°F (−15°C).
Soil Requirements Any soil.
Sun/Shade aspect Full sun to light shade.
Pruning Cut back one-third of old flowering shoots moderately hard after flowering to induce new flowering wood for subsequent years.
Propagation and nursery production From semi-ripe cuttings taken in early summer, or from budding or grafting. Standards normally budded or grafted on to stems of *P. avium*. Purchase root-balled (balled-and-burlapped)

Prunus tenella 'Fire Hill' in flower

Prunus triloba **in flower**

or container-grown. Best planting heights 1¼-2½ft (40-80cm).
Problems Young plants do not show their full potential.

Average height and spread
Five years
4x4ft (1.2x1.2m)
Ten years
5½x5½ft (1.8x1.8m)
Twenty years
or at maturity
8x8ft (2.5x2.5m)

PSEUDOSASA JAPONICA
(Arundinaria japonica)

BAMBOO, JAPANESE BAMBOO
Gramineae
Evergreen
Useful foliage plants for screening or ground cover, depending on variety. From recent reclassification the shrub is now correctly named *Pseudosasa japonica*, but is still often sold as *Arundinaria japonica*.

Origin From Japan.
Use As tall screening clumps, particularly near water. Lower growing varieties as ground cover.
Description *Flower* Large, open, fluffy plumes of sulphur yellow flowers may be produced in very hot summers; the plants often die after flowering. *Foliage* Leaves 7-8in (18-20cm) long and 1-2in (3-5cm) wide, glossy, dark green, with grey undersides, borne singly from each of the nodes on the upper section of each cane. *Stem* Upright, bright green, aging to yellow green. Shoots bran-

Pseudosasa murieliae **in leaf**

ching upwards from base, producing useful canes. Fast growth rate once established.
Fruit Insignificant.
Hardiness Tolerates temperatures down to 4°F (−15°C).
Soil Requirements Thrives on a good, moist but well-drained soil. Resents very dry areas and severe alkalinity.
Sun/Shade aspect Light to quite deep shade.
Pruning None required.
Propagation and nursery production By division of the mature clumps. Best purchased container-grown but may be planted bare-rooted as large clump. Available from garden centres and nurseries. Best planting heights 2-3ft (60cm-1m) for tall varieties and 1-2ft (30-60cm) for shorter forms.
Problems Spasmodic flowering often followed by severe deterioration for a number of years. Extremely severe cold or drought can spoil appearance. Spreading roots can become invasive.

Pseudosasa viridistriata **in leaf**

Varieties of interest *P. humilis* Relatively fast-growing with narrow, light green, lanceolate leaves, 2-7in (5-18cm) long, forming a low thicket of canes, reaching a height at maturity of 6ft (2m). Very good for ground cover or for large unsightly banks or waste places. *P. murieliae* Same size as *P. japonica* but with narrower, lighter green leaves, 2½-4½in (6-11cm) long, forming very thick, dense growth, with numerous canes; thinner than *P. japonica* and slightly more branching at the ends. Very useful for screening and forming barriers. Fast growth rate. *P. nitida* Dark purple stems, otherwise similar to *P. murieliae*. *P. pumila* A dwarf species of very neat habit, reaching 3ft (1m); bright, glossy green foliage, 2-6in (5-15cm) long. Well worth a space in the larger garden. Slow growth rate. *P. pygmaea* A very dwarf variety, reaching no more than 10in (25cm) high with leaves 4-5in (10-12cm) long and 1in (3cm) wide, with slightly glaucous covering. Growth rate slow at first. *P. variegata* A low-growing, spreading variety, reaching a height of 3-5ft (1-1.5m) forming a dense carpet. Leaves mid green striped white, 5-7in (12-18cm) long. Use for mass planting, larger rock gardens and for tubs; especially good near water. Medium growth rate. *P. viridistriata* syn. *P. auricoma* Varies from 3-6ft (1-2m) in height and carpet forming. Very attractive golden yellow striped foliage, 4-6in (10-15cm) long. For tubs and larger rock gardens. Medium growth rate.
This listing represents a selection of bamboos of proven garden merit from a large number available.

Average height and spread
Five years
6x3ft (2x1m)
Ten years
8x6ft (2.5x2m)
Twenty years
or at maturity
10x10ft (3x3m)

PYRACANTHA

PYRACANTHA, FIRETHORN
Rosaceae
Evergreen
Fruiting shrubs often considered only for wall-training and overlooked for planting as freestanding features.

Origin Most forms originating in China; many varieties of garden origin.
Use As a large, fruiting shrub for a medium to large shrub border. If planted 2½ft (80cm) apart in single line makes a useful informal hedge up to 6ft (2m). Can be fan-trained or shaped to any requirement for wall covering. Exceptionally good on cold, very exposed or shady walls.
Description *Flower* Good-sized clusters of white flowers with musty scent in early summer. Loved by bees. *Foliage* Leaves ovate, ¾-1¾in (2-4.5cm) long, glossy, light to mid green; some grey and variegated leaved varieties. *Stem* Sharp spines at most leaf axils. Red-brown, aging to dark green to brown. Forming an upright, wide-topped shrub or a flat fan shape for walls. Some attractive pendulous forms. Fast growth rate. *Fruit* Clusters of round, good sized fruits, yellow, orange or red, depending on variety, early autumn, maintained well into winter.
Hardiness Tolerates winter temperatures down to −13°F (−25°C), although severe wind chill may damage some leaves and tips of younger growth. Damage is usually corrected in following spring growth.
Soil Requirements Most soils, but distressed by extreme alkalinity.
Sun/Shade aspect Full sun to deep shade, but produces fewer fruit in deeper shade, and may also become more open.
Pruning None required. Can be drastically reduced in size, although following year's flowering and fruiting will be poorer. Pruning is best carried out late winter to early spring, to avoid attacks of fungus disease. Can be trimmed and shaped as desired.
Propagation and nursery production From semi-ripe cuttings taken in early summer. Propagates readily from seed, but seed-raised plants variable in fruit quality although useful for hedging or for large-scale planting. Easy to find a good range of named varieties. Best planting heights 1¼-3ft (40cm-1m).
Problems Subject to the airborne fungus disease fire blight, in summer months, completely destroying all foliage. There is no cure or prevention. All affected or suspect shrubs should be burnt immediately.
Varieties of interest *P. 'Alexander Pendula'* A variety with interesting, long, weeping branches, forming a somewhat unruly shrub. Fruit coral red. *P. angustifolia* Grey-green, long, narrow, oblong leaves with grey, felted undersides. Fruit orange-yellow, may be maintained into winter. *P. atalantioides* syn. *P. gibbsii* (Gibbs Firethorn) Strong, quick-growing variety with good-sized, oval, dark, glossy green leaves. Scarlet fruits produced in autumn. Good for difficult shady walls, but may be withdrawn from production due to fire blight. *P. a. 'Aurea'* syn. *P. 'Flava'* A yellow-fruited form of the above, which may also be withdrawn for the same reason. *P. coccinea 'Lalandei'* A somewhat upright but wide form with small, cut leaves. Orange-red fruits in autumn, may be maintained into winter. *P. c. 'Sparklers'* Rather small, narrow, ovate foliage variegated with white margins. Orange-red fruits. Reaching two-thirds average height and spread. *P. 'Golden Charmer'* Good dark foliage. Large clusters of golden-yellow fruits. *P. 'Harlequin'* Grey-green foliage with silver-white edges, sometimes pink-tinged, especially in cold weather. Small orange-red fruits. Two-thirds average height and spread. *P. 'Mojave'* Good, large, mid to dark green foliage. Large orange-red fruits in autumn, may be maintained into winter. Very winter hardy and resistant to wind scorch. *P. 'Orange Charmer'* Dark green foliage. Fruits large, deep orange in colour. *P.*

Pyracantha 'Orange Charmer' in flower

'Orange Glow' Interesting dark green foliage with dark purple-black stems. Good clusters of orange-red fruits very freely produced in autumn. Smaller than normal. *P. 'Red Cushion'* Light green foliage, aging to grey-green. White flowers, followed by red fruits in autumn. Low-growing, reaching only 3ft (1m) in height, 6ft (2m) spread. *P. rogersiana* Very dark, oblong to lanceolate leaves. Especially recommended for shaded walls, giving a display of orange-red fruits in autumn. *P. r. 'Flava'* Small foliage, clusters of bright yellow fruits. Good for shaded walls. Somewhat weeping habit. *P. 'Shawnee'* Dense foliage and numerous spines. Fruit yellow to light orange, colouring early in season and maintained well into winter. Raised in the National Arboretum, USA. Said to be resistant to fire blight and scab fungus. *P. 'Soleil d'Or'* Good mid to light green foliage. Very large clusters of deep yellow fruits. *P. 'Teton'* Good small dark green foliage. Clusters of orange-yellow fruits. Upright growth.

Average height and spread

Five years
6x4ft (2x1.2m)
Ten years
12x6ft (3.5x2m)
Twenty years
or at maturity
13x10ft (4x3m)

RHAMNUS

BUCKTHORN
Rhamnaceae
Evergreen and Deciduous
Evergreen forms are of extreme garden interest in sheltered areas. Deciduous forms are useful for mass planting and as windbreaks.

Origin From Europe and the Mediterranean. Use As large fruiting shrubs. Deciduous forms often used as large windbreaks, evergreen variegated forms as interesting specimen shrubs in mild areas and for fan-training on walls.
Description *Flower* Inconspicuous, small pale cream flowers, early summer. *Foliage* Leaves ovate, 1-2½in (3-6cm) long, green or white variegated, depending on variety. Deciduous forms give good yellow autumn colour. *Stem* Upright, becoming very branching with age. Some forms with spines, forming dense thickets. Dark purple to purple-brown. Fast growth rate. *Fruit* Red or black fruits depending on variety, in autumn. Spectacular in warm locations.
Hardiness Minimum winter temperature 4°F (−15°C).
Soil Requirements Most soil types, but dislikes extremely wet conditions.
Sun/Shade aspect Medium to full shade.

Pruning None required and is best left free growing, but evergreen forms may be trimmed.
Propagation and nursery production Variegated forms from semi-ripe cuttings taken in summer. Deciduous forms normally grown from seed. Purchase container-grown, or deciduous forms bare-rooted. Evergreen variegated forms fairly easy to find. Deciduous forms may have to be sought from forestry outlets, as seldom stocked elsewhere. Best planting heights 1¼-3ft (40cm-1m).
Problems None.

Rhamnus alaterna 'Argenteovariegata' in leaf

Varieties of interest *R. alaterna 'Argenteovariegata'* Evergreen. Grey-green leaves with irregular white to creamy white margins. Often produces good crop of red fruits in warm locations. Attractive, upright, pyramidal habit. Should be considered slightly tender; minimum winter temperature 23°F (−5°C). *R. cathartica* (Common Buckthorn of Europe) Deciduous. Foliage grey-green, attractive and spiny. Produces shiny black fruits in autumn. A variety normally used for hedging and other rural purposes but also useful for large-scale planting to form screen for wind protection. *R. frangula* (Alder Buckthorn, Black Dogwood) Deciduous. Ovate leaves with good yellow autumn colour, whitish flowers, producing red fruits aging to black in autumn. Good for large-scale windbreaks, especially in moist soil conditions. Not suitable for general garden use. A large shrub of above average height and spread, possibly forming a small, multi-stemmed tree with age.

Average height and spread

Five years
6x3ft (2x1m)
Ten years
12x6ft (3.5x2m)
Twenty years
or at maturity
16x13ft (5x4m)

RHODODENDRON
Large-flowering hybrids

KNOWN BY BOTANICAL NAME
Ericaceae
Evergreen
Well-loved and very handsome spring-flowering shrubs.

Origin Crosses from parent plants gathered from throughout the northern hemisphere, many hybridized and of garden origin.
Use As a freestanding shrub on its own or for mass planting; extremely good for accentuating a landscape design. Can be grown in containers for some years provided they are not less than 2½ft (80cm) in diameter and 2ft (60cm) in depth; use good lime-free potting medium.
Description *Flower* Large trusses of bell-shaped flowers up to 6in (15cm) across the truss and 7-8in (18-20cm) long, in colours ranging from white, through pink, to blue and red, depending on variety. Some flowers multicoloured or with petal markings. Mid spring to early summer. *Foliage* Leaves ovate, medium to large, from 4-8in (10-20cm) depending on variety, glossy dark green upper surface with duller, grey-green undersides. Leathery textured. *Stem* Upright, strong shoots when young, grey-green to dark green aging to grey-brown. Becoming branching and in some cases arching, forming a round-topped shrub with base slightly wider than top. Medium growth rate. *Fruit* May produce brown-grey seedheads which should be removed.
Hardiness Tolerates 4°F (−15°C).
Soil Requirements Neutral to acid soil. Dislikes any alkalinity.
Sun/Shade aspect Prefers light shade, but tolerates full sun through to deep shade, if, adequate moisture is available.
Pruning None required but large, mature shrubs can be reduced in size after flowering by cutting back even to very old wood.
Propagation and nursery production From grafting. Purchase root-balled (balled-and-burlapped) or container-grown. Fairly easy to find, although some varieties, especially when in flower, will be found only in specialist nurseries. Best planting heights 1¼-2½ft (40-80cm). Shrubs up to 6ft (2m) may be moved if required.
Problems Acid soil is essential and it is useless to attempt to grow rhododendrons on alkaline soils. Flower buds may be attacked by insects which are extremely difficult to control; remove affected buds to reduce infestation in future years.
Varieties of interest *R. 'Bagshot Ruby'* Large trusses, dense in formation, consisting of ruby red wide-mouthed funnel-shaped flowers, flowering late spring. *R. 'Betty Wormald'* Deep crimson in bud. Deep rose pink, funnel-shaped flowers with a wide, wavy-edged

Rhododendron 'Britannia' in flower

183

Rhododendron 'Pink Pearl' in flower

mouth and some maroon to black-crimson spots on inner sides. Large to very large trusses. Late spring. **R. 'Blue Peter'** Flower trusses conical and tightly formed, borne in late spring. Funnel-shaped flowers have frilled edges and are cobalt blue, aging to white in throat with ring of maroon spots. Strong-growing. **R. 'Britannia'** Flowers gloxinia-shaped, scarlet-crimson and in tight trusses, borne in late spring. Forms a round shrub, which in time also forms a low ground skirt. Slightly less than average height. **R. 'Countess of Athlone'** Wide, funnel-shaped, wavy-edged flowers in late spring. Buds purple, opening to mauve with yellow to green-yellow basal markings in cone-shaped trusses. **R. 'Countess of Derby'** Buds pink, opening to pink but intensifying with age. Flowers wide, funnel-shaped, with red-brown spots and streaks inside, borne in tight cone-shaped trusses in late spring. **R. 'Cynthia'** Strong-growing variety and quick to form a dome-shaped shrub. Rose-crimson, cone-shaped large flowers, with black-crimson markings, borne in late spring. **R. 'Earl of Donoughmore'** Good-sized trusses of bright red flowers with orange glow in late spring. **R. 'Fastuosum Flore Pleno'** Somewhat lax trusses of rich mauve funnel-shaped flowers with ring of brown-crimson markings on inner side and wavy edges, borne in late spring. Forming a dome-shaped shrub of good proportions. **R. 'General Eisenhower'** Large trusses of carmine-red flowers in late spring. A well-shaped shrub. **R. 'Goldsworth Orange'** Good-sized trusses of orange to pale orange flowers, each tinged with apricot pink in early summer. A somewhat low, spreading bush, reaching two-thirds average height and average spread. **R. 'Goldsworth Yellow'** Round trusses of funnel-shaped flowers, pink in bud and primrose yellow with brown markings on inner surface when open in late spring. Good foliage on dome-shaped, spreading shrub. **R. 'Gomer Waterer'** Round trusses of good-sized funnel-shaped flowers, divided at the mouth; white with pale mauve flush towards outer edges and mustard coloured blotches at base of petals. Late spring to early summer. Good-sized shrub with attractive leathery, oval to oblong foliage. **R. 'Kluis Sensation'** Large flowers, bright scarlet with darker red spots on outer edges, in late spring. **R. 'Kluis Triumph'** Deep red flowers in good-sized trusses borne on large shrub in late spring. **R. 'Lord Roberts'** Round trusses of funnel-shaped flowers, deep crimson with black markings on inner side, late spring to early summer. Upright habit. **R. 'Madame de Bruin'** Cone-shaped trusses of cerise-red flowers in late spring. Good foliage on strong-growing shrub. **R. 'Moser's Maroon'** Maroon-red flowers with darker inner markings in each truss, in late spring to early summer. New foliage copper. Strong-growing, upright habit. **R. 'Mrs. G.W. Leak'** Lax, conical trusses of wide, funnel-shaped flowers, light

rosy pink mottling, becoming darker towards base of tubes. Black-brown and crimson markings. Late spring. **R. 'Old Port'** Thick trusses of wide, funnel-shaped flowers, plum-coloured with black-crimson markings, in late spring to early summer. Strong-growing, large leaves. **R. 'Pink Pearl'** Large cone-shaped trusses of wide-mouthed, funnel-shaped flowers in late spring. Rose pink buds, opening to lilac-pink flowers. Outer margins becoming white with age. Large, strong-growing, upright shrub. One of the most widely planted varieties. **R. ponticum** Mauve to lilac-pink, tubular flowers in good-sized, slightly open trusses, borne in late spring. A large, round shrub, extremely useful for windbreaks and hedges. Can be invasive in woodland areas. **R. p. 'Variegatum'** Purple flower trusses in late spring. Foliage grey-green with white variegation. Reaches two-thirds average height and spread. **R. 'Purple Splendour'** Good-shaped trusses of funnel-shaped, wide-mouthed flowers in late spring to early summer. Rich royal purple-blue with black embossed markings on purple-brown background. Strong-growing, upright branches. **R. 'Sappho'** Cone-shaped trusses of wide-mouthed funnel-shaped flowers in late spring. Buds mauve, opening to pure white, with rich purple overlaid with black blotch on inner side. Open, round shrub. **R. 'Susan'** Large trusses of blue-mauve flowers with darker outer edges and purple spots within, in mid to late spring. A strong-growing shrub. **R. 'Unique'** Flower trusses dome-shaped, creamy white with pinkish shading and crimson spots within, in late spring. Interesting small foliage on dome-shaped shrub, reaching two-thirds average height and spread.

Average height and spread

Five years
4x5ft (1.2x1.5m)
Ten years
6x8ft (2x2.5m)
Twenty years or at maturity
10x13ft (3x4m)

RHODODENDRON
Dwarf hybrids

DWARF RHODODENDRON
Ericaceae
Evergreen
Among the most beautiful spring-flowering shrubs, but requiring specific soil conditions.

Origin Varieties listed are mostly of garden origin.
Use As low, dwarf, spring-flowering shrubs for use with heathers and dwarf conifers. Ideal for growing in tubs not less than 2½ft (80cm) in diameter and 2ft (60cm) in depth;

use good lime-free potting medium. Very good for medium to large rock gardens and for mass planting.
Description *Flower* Borne in small trusses in a wide range of colours from blue, through pink, purple, red and white from early, mid or late spring, depending on variety. *Foliage* Leaves generally small, round to ovate, ½-1¼in (1-3.5cm) long, dark green with glossy upper surfaces and dull undersides. *Stem* Very short, branching habit, forming a dome-shaped, low, spreading shrub. Slow growth rate. *Fruit* Insignificant.
Hardiness Tolerates 4°F (−15°C).
Soil Requirements Neutral to acid soils. Dislikes any alkalinity or waterlogging.
Sun/Shade aspect Prefers light shade, tolerates from full sun to mid shade.
Pruning None required.
Propagation and nursery production From semi-ripe cuttings taken early to mid summer. Purchase container-grown or root-balled (balled-and-burlapped). Most varieties fairly easy to find, especially when in flower. Best planting heights 8in-2ft (20-60cm).
Problems Buds susceptible to insect attacks; remove damaged buds. Severe wind chill may damage foliage and buds.

Rhododendron 'Blue Diamond' in flower

Varieties of interest R. 'Bluebird' Violet-blue trusses borne in mid spring on dwarf, round, tight, small-leaved shrub. **R. 'Blue Diamond'** Intense lavender-blue flowers borne in mid spring in small terminal clusters on slow-growing, tight shrub. **R. 'Blue Tit'** Funnel-shaped lavender-blue flowers in small terminal clusters, colour deepening with age. Borne in mid spring. Dense mould-shaped shrub. **R. 'Bow Bells'** Flowers wide and bell-shaped, deep cerise in bud, opening to soft pink, richer pink on outer side. Trusses loosely presented in mid to late spring on compact

Rhododendron 'Elizabeth' in flower

Rhododendron 'Praecox' in flower

shrub. Young foliage copper. *R. 'Bric-a-Brac'* Pure white, wide open flowers, 2½in (6cm) across, with chocolate anthers, borne early to mid spring. *R. 'Carmen'* Large, bell-shaped flowers, 1¾-2in (4.5-5cm) across, dark crimson with pale pink throats in mid spring. Larger leaves than most dwarf forms, up to 2-2½in (5-6cm) long, ovate, covering a dwarf, prostrate shrub. *R. 'Elisabeth Hobbie'* Trusses of 5-10, almost translucent, bell-shaped, scarlet red flowers in mid spring. *R. 'Elizabeth'* Flowers trumpet-shaped, 2½-2¾in (6-7cm) across, rich dark red, borne in mid to late spring. Slightly taller than most varieties, with spreading habit. *R. 'Humming Bird'* Hanging flowers, scarlet-red with scarlet inner shading, borne in early-spring. Dome-shaped, dwarf shrub of compact habit. *R. 'Moonstone'* Rose-crimson buds, opening to cream and pale primrose bell-shaped flowers in mid to late spring. Low, dome-shaped shrub. *R. 'Pink Drift'* Flowers lavender-rose in small clusters, produced in late spring. Foliage small, grey-green and aromatic. Good for rock gardens and other areas where a low, very compact, shrub is required. *R. 'Scarlet Wonder'* Trumpet-shaped flowers, ruby red with frilly margins borne in late spring. A dwarf shrub, forming a tight mound of close foliage. *R. 'Yellow Hammer'* Bright yellow flowers, produced in pairs at terminal and axillary buds in mid spring, occasionally also in autumn. Upright, narrow habit.

Average height and spread
Five years
2x2ft (60x60cm)
Ten years
2½x3ft (80cmx1m)
Twenty years
or at maturity
3x5ft (1x1.5m)

RHODODENDRON
Low-growing species

DWARF RHODODENDRON
Ericaceae
Evergreen
A varied range of spring-flowering shrubs for use on acid soils.

Origin Mostly from northern hemisphere, with a wide range of individual locations.
Use As dwarf, spring-flowering shrubs for use with heathers and dwarf conifers, particularly for medium to large rock gardens. Can be grown in tubs not less than 2½ft (80cm) across and 2ft (60cm) deep; use good, lime-free potting medium.
Description *Flower* Small trusses of flowers in colours ranging through yellow, blue, purple, pink, red and white. From early to late spring. *Foliage* Leaves round to ovate, ¼-1¾in

(5mm-2.5cm) long, dark green or grey, dependent on variety, with glossy upper surfaces and duller undersides. *Stem* Very short, branching habit, forming a dome-shaped, low, spreading shrub. Slow growth rate. *Fruit* Insignificant.
Hardiness Tolerates 4°F (−15°C).
Soil Requirements Neutral to acid soils, dislikes any alkalinity and waterlogging.
Sun/Shade aspect Prefers light shade, tolerates full sun to mid shade.
Pruning Generally requires no pruning. Slow-growing.

Rhododendron racemosum in flower

Propagation and nursery production From semi-ripe cuttings taken early to mid summer. Purchase root-balled (balled-and-burlapped) or container-grown. Most varieties fairly easy to find. Best planting heights 8in-2ft (20-60cm).
Problems May be subject to insect attacks; remove damaged buds. Flower buds and

foliage may be damaged by very severe wind chill.
Varieties of interest *R. ferrugineum* (Alpen Rose of Switzerland) Trusses of small, tubular, rose-crimson flowers in early summer. Red underside to foliage. Flat, dome-shaped shrub of spreading habit. *R. hirsutum* Tubular, rose-pink flowers in clusters in early summer, stems and leaves fringed with bristles. Dwarf, compact, many branched small alpine shrub, two-thirds mature height and spread. *R. impeditum* Purple-blue, funnel-shaped flowers in mid to late spring, produced on low, very small-leaved mound of scaly branches. Dwarf alpine reaching one-third average height and spread. *R. moupinense* Sweet-scented, pink to deep rose, funnel-shaped flowers in late winter, early spring; foliage ovate to elliptic with scaly undersides. Small shrub with bristly branches. May need some protection from east winds while in flower. *R. obtusum 'Amoenum Coccineum'* Carmine-rose flowers, foliage glossy green, oval. Low-growing, thickly branched, spreading shrub, branches being covered in hairs. Semi-evergreen in some situations. *R. pemakoense* Flowers funnel-shaped, lilac-pink to purple, profusely borne in early to mid spring, needing some protection from frost when in bud and flower. A variety reaching one-third mature height and spread, suckering as it goes and producing small, very low-spreading carpet. *R. 'Praecox'* Funnel-shaped flowers in open clusters, purple-crimson in bud, opening to rosy purple, produced late winter to early spring. Foliage can be slightly deciduous, yellow older leaves contrasting with dark glossy green new leaves; aromatic when crushed. *R. racemosum* Funnel-shaped flowers, pale to bright pink, produced in axillary buds, forming racemes along branches. Foliage oblong to elliptic, leathery-textured with blue glaucous undersides. In favourable conditions may exceed mature height and spread. *R. saluenense* Clusters of rose-purple to purple-crimson, funnel-shaped flowers in mid to late spring. A mat-forming shrub with very thick, grey-green, aromatic foliage, ovate to elliptic. Reaching only one-third mature height and spread. *R. williamsianum* Shell pink, bell-shaped flowers in early spring, round, heart-shaped leaves. Can exceed mature height and spread. *R. yakushimanum* Trusses of large, bell-shaped flowers borne in late spring, rose-pink in bud, opening to apple blossom pink, eventually aging to white. Dark, glossy green foliage, curving at edges with brown-blue undersides.

Average height and spread
Five years
2x2ft (60x60cm)
Ten years
2½x3ft (80cmx1m)
Twenty years
or at maturity
3x5ft (1x1.5m)

Rhododendron yakushimanum in flower

RHODODENDRON
Dwarf, evergreen Azaleas

DWARF AZALEA
Ericaceae
Evergreen to Semi-evergreen
Well-known, very floriferous, spring-flowering shrubs.

Origin Mostly of garden origin, from various parts of Europe and Japan.
Use As low, mounded, spring-flowering shrubs used beside water features, in woodland gardens and in medium to large rock gardens. Can be grown in tubs not less than 1½ft (50cm) in diameter and depth; use good quality, lime-free potting medium.
Description *Flower* Bell-shaped, borne profusely at each leaf axil in colours ranging from white through primrose to yellow, gold, apricot, red, purple with some variations of one or more colours; mid to late spring. *Foliage* Leaves broad, ovate, 1-1½in (3-4cm) long, dark glossy green upper surfaces, duller underside; may be semi-evergreen in severe wind chill. Good orange-red given by older leaves discarded in autumn contrasts with remaining younger foliage. *Stem* Upright when young, very quickly becoming branching, forming a spreading, mounded shrub of neat effect. Medium growth rate when young, slowing with age. *Fruit* Insignificant.

Azalea (Kurume hybrid) 'Hinomayo' in flower

Hardiness Minimum winter temperature 4°F (−15°C).
Soil Requirements Acid to neutral soil, dislikes any alkalinity. Best on peat or leaf mould type soil, although tolerates fairly wide range of acid contions.
Sun/Shade aspect Prefers light shade, tolerates full sun to medium shade.
Pruning None required, in fact may resent it. Severely winter-damaged wood may be reduced to ground level and may rejuvenate.
Propagation and nursery production From semi-ripe cuttings taken late spring to early summer. Plant root-balled (balled-and-burlapped) or container-grown. Best planting heights 4in-1½ft (10-50cm). Availability varies. May be found in areas of acid or neutral soil, particularly during flowering season.
Problems Often purchased on impulse and soil requirements not fully considered. The plant may survive 2-3 years on unsuitable soil, but cannot root properly.
Varieties of interest *Kurume types:* 'Addy Wery' Deep vermilion-red. 'Blaauw's Pink' Salmon-pink with light shading. 'Hatsugirl' Bright crimson-purple, dwarf. 'Hinodegirl' Bright crimson. 'Hinomayo' Clear pink. 'Imashojo' (Christmas Cheer) Bright red. 'Kureno-yuki' White, dwarf. 'Rosebud' Rose pink, low-growing. *Vuyk hybrids:* 'Beethoven' Purple. 'Blue Danube' Violet. 'Palestrina' Pure white. 'Vuyks Rosy Red' Satin rose.

Azalea (Vuyk hybrid) 'Palestrina' in flower

'Vuyks Scarlet' Bright crimson. *Dwarf hybrid crosses:* 'Favorite' Deep rose-pink. 'John Cairns' Dark orange-red. 'Leo' Orange. 'Mother's Day' Rose red. 'Naomi' Salmon-pink. 'Orange Beauty' Salmon-orange. 'Silvester' Rosy red.
There is an increasing number of dwarf Azaleas. This personal selection is given as a guide, but many other modern varieties and some older forms are well worth planting, given the right type of soil.

Average height and spread
Five years
2x2ft (60x60cm)
Ten years
2½x3ft (80cmx1m)
Twenty years
or at maturity
3x4ft (1x1.2m)

RHODODENDRON
Tall-growing Azaleas

AZALEA
Ericaceae
Deciduous
Among the finest and most well-known of all spring-flowering shrubs for acid soils.

Origin Most varieties of garden origin.
Use Singly, or for mass planting. Extremely attractive in woodland areas and in water features. Can be used in tubs 2½ft (80cm) across and 1½ft (50cm) deep; use lime-free potting medium.
Description *Flower* Large, open, trumpet-shaped, borne in clusters in colours ranging

from white through primrose to yellow and gold; through pink to apricot to orange-flame, dark red; combinations of various colours also available. Mid to late spring. *Foliage* Leaves ovate, 1-3in (3-8cm) long, sometimes larger on new growth. Light green, sometimes grey-green, occasionally with downy covering, giving very good yellow, red and orange autumn colours. *Stem* Strong, upright, light brown to yellow-brown when young, becoming branching and twiggy, forming upright shrub at first but becoming more spreading with age. Fast growth rate when young, becoming very slow with age. *Fruit* Brown seedheads follow flowers.
Hardiness Tolerates 4°F (−15°C).
Soil Requirements Neutral to acid soils, preferring peat or leaf mould type. Dislikes any degree of alkalinity and resents waterlogging.
Sun/Shade aspect Prefers light to medium shade, tolerates degrees either side, but growth may be impaired.
Pruning None required but extremely old shrubs can be cut back hard after flowering and will rejuvenate; fewer flowers produced for 1-3 years following pruning.
Propagation and nursery production From grafting, semi-ripe cuttings taken in midsummer, or layers. Plant root-balled (balled-and-burlapped) or container grown. Best planting heights 1¼-4ft (40cm-1.2m), ideally 2-2½ft (60-80cm). Wide range of varieties available in acid areas, especially when in flower.
Problems Often purchased on impulse and soil requirements not taken into consideration, so the shrub fails to root properly. Old plants may attract harmless grey-green lichen covering on older stems. Flower buds may be damaged by severe wind chill. Buds may be attacked by bud-boring azalea insect; if so remove damaged buds as soon as seen.
Varieties of interest *Knaphill hybrids:* 'Balzac' Orange. 'Berryrose' Rose-pink with yellow flush. 'Brazil' Tangerine. 'Cecile' Salmon-pink. 'Fireball' Deep red. 'Gallipoli' Pale tangerine, flushed pink with yellow. 'Gibraltar' Flame orange with yellow flush. 'Homebush' Carmine-pink. 'Hotspur' Orange-red. 'Hugh Wormald' Deep golden yellow. 'Klondyke' Orange in bud, opening to yellow. 'Persil' White with orange flush. 'Satan' Geranium red. 'Strawberry Ice' Light pink with gold flare. 'Tunis' Deep crimson with orange flare. *Mollis hybrids:* 'Apple Blossom' Apple blossom pink. 'Chevalier de Reali' Pale yellow with orange flare. 'Christopher Wren' Yellow, orange spotting. 'Dr. M. Oosthoek' Orange-red. 'Golden Sunlight' Golden yellow. 'Kosters Brilliant Red' Orange-red. *Ghent hybrids:* Azalea coccinea speciosa Orange red. 'Daviesii' White with yellow flare. Fragrant. 'Nancy Waterer' Golden yellow. 'Narcissiflorum' Pale yellow. Fragrant. 'Pallas' Orange-red with yellow flare. Azalea pontica (Rhododendron luteum) Yellow, scented. Good autumn colour.

Azalea (Mollis hybrid) 'Apple Blossom' in flower

Azalea (Knaphill hybrid) 'Gibraltar' in flower

There is an extremely wide range of deciduous, tall-growing Azaleas; this is a personal selection of worthwhile varieties generally available.

Average height and spread
Five years
3x3ft (1x1m)
Ten years
5x5ft (1.5x1.5m)
Twenty years
or at maturity
7x9ft (2.3x2.8m)

RHODOTYPOS SCANDENS

WHITE JEW'S MALLOW, BLACK JETBEAD

Rosaceae
Deciduous
A rarely seen but useful late spring-flowering shrub, though somewhat irregular in performance.

Origin From China, Korea and Japan.
Use As an interesting, late spring to early summer-flowering shrub for shrub borders or for mass planting.

Rhodotypos scandens in flower

Description *Flower* White flowers, 1-1½in (3-4cm) across, produced intermittently, late spring to early summer. *Foliage* Leaves ovate, 1-1¾in (3-4.5cm) long, slightly tooth-edged, light green, *Stem* Upright, branching at upper ends, light green. Some winter interest. Medium growth rate, slowing with age. *Fruit* Round, shiny black fruits in late summer, early autumn.
Hardiness Tolerates winter temperatures

down to −13°F (−25°C).
Soil Requirements Prefers light, open soil, tolerates wide range, both acid and alkaline.
Sun/Shade aspect Prefers full sun, tolerates light shade.
Pruning Remove one-third of old flowering wood after flowering to encourage rejuvenation.
Propagation and nursery production From semi-ripe cuttings taken in early summer or rooted shoots removed in winter. Purchase container-grown. May need to be obtained from specialist nurseries. Best planting heights 2-2½ft (60-80cm).
Problems None.
Average height and spread
Five years
3x2½ft (1mx80cm)
Ten years
6x4ft (2x1.2m)
Twenty years
or at maturity
6x6ft (2x2m)

RIBES

FLOWERING CURRANT

Grossulariaceae
Deciduous and Evergreen
A wide group with varying characteristics from which useful garden plants can be found.

Origin Throughout the northern hemisphere.
Use As a spring-flowering shrub for medium to large shrub borders, for mass planting or as a single specimen plant. Some varieties good fan-trained on walls.
Description *Flower* Small, hanging racemes, mainly in shades of pink, some forms yellow and green-yellow, depending on variety, mid to late spring. *Foliage* Leaves three-lobed, 1½-2in (4-5cm) long and wide, light green with some golden-leaved forms. Some brown-yellow autumn colour on deciduous forms. *Stem* Strong, upright, branching, forming a round-topped shrub. Light green to grey-green. Medium growth rate. *Fruit* Black fruits, some edible.
Hardiness Tolerates winter temperatures down to −13°F (−25°C).
Soil Requirements Most soils, only disliking extremely wet and extremely dry conditions.
Sun/Shade aspect Full sun to light shade.
Pruning Remove one-third of old wood on mature shrubs after flowering. Evergreen varieties require no pruning.
Propagation and nursery production From semi-ripe cuttings taken in early summer or, with some varieties, from hardwood cuttings in winter. Plant bare-rooted or container-grown. Most varieties easy to find, especially when in flower. Evergreen forms may need to

Ribes laurifolium in flower

be sought from specialist nurseries. Best planting heights 1¼-3ft (40cm-1m).
Problems Mature plants of *R. sanguineum* and its forms often die in summer, usually because of damage by severe waterlogging, excessive drought, or close cultivation killing roots close to surface. Some *Ribes* varieties must not be planted because they transmit damaging plant diseases; those listed below are acceptable for garden use.
Varieties of interest *R. alpinum* (Alpine Currant) Green-yellow, small, hanging flowers in early spring. Red fruits in late summer, early autumn. Upright stems forming a tight, dense, thicket. Very good for deep shade areas. Half average height and spread. From northern and central Europe. ***R. americanum*** (American Blackcurrant) Long, funnel-shaped, yellow flowers, good crimson-yellow autumn colour. Resembles a fruiting blackcurrant. Reaches two-thirds average height and spread. From eastern North America. ***R. × gordonianum*** Racemes of flowers bronzed-red on outside, with yellow inner shading. Rarely seen form, will need searching for. Of garden origin. ***R. henryi*** Flowers green-yellow, produced on new growth in late winter, early spring. A low, spreading, evergreen forming flat carpet. Rather hard to find. More tender than most, minimum winter temperature 23°F (−5°C). From central China. ***R. laurifolium*** Racemes of green-white flowers in late winter, early spring. Fruits red, becoming black with age. Elliptic, glabrous leaves, leathery in texture. Evergreen, dwarf shrub, slightly more tender and should be given protection of wall in areas where temperature falls below 23°F (−5°C) in winter. From western China. ***R. odoratum*** (Buffalo Currant, Clove Currant) Golden yellow clove-scented flowers in early spring. Berries black, very sweet and edible. Light green to

Ribes sanguineum 'Brocklebankii' in flower

Ribes sanguineum 'Pulborough Scarlet' in flower

green-blue lobed foliage with good orange-salmon autumn colour, upright stems. **R. sanguineum** (Flowering Currant) Racemes of mid pink flowers, followed by black fruits with blue-tinged bloom. From western North America. **R. s. 'Album'** Flowers pink-white, hanging in racemes. **R. s. 'Atrorubens'** Flowers blood red. Slightly darker foliage. **R. s. 'Brocklebankii'** Pale pink flowers and lobed golden yellow foliage. Needs to be in very light shade to prevent foliage from scorching in hot summer sun. Half average height and spread. **R. s. 'King Edward VII'** Intense crimson flowers. Very popular variety. Two-thirds average height and spread. **R. s. 'Pulborough Scarlet'** Deep red racemes, with good foliage. Upright habit. One of the best forms. **R. s. 'Splendens'** Rosy crimson flowers, larger than most and on long racemes. Good foliage. **R. s. 'Tydeman's White'** A very good white flowered form with racemes of good size. Two-thirds average height and spread. **R. speciosum** An interesting form with hanging,

Ribes speciosum in flower

fuchsia-like red flowers, mid to late spring. Foliage shaped like gooseberry leaves. In areas where temperature falls below 23°F (−5°C) it needs to be protected by a sunny wall. From California. **R. viburnifolium** Flowers terracotta red in small upright racemes and produced in mid spring, followed by red fruits in autumn. Round, shiny green foliage, giving off a turpentine-like scent when crushed. Tender, but worth planting where adequate protection can be given from any degree of frost. An evergreen form from California.

Average height and spread
Five years
5x3ft (1.5x1m)
Ten years
6x6ft (2x2m)
Twenty years or at maturity
8x10ft (2.5x3m)

ROMNEYA COULTERI

TREE POPPY, CALIFORNIA TREE POPPY
Papaveraceae
Deciduous
One of the aristocrats of mid to late summer-flowering sub-shrubs, with truly magnificent flowers.

Origin From California.
Use As a clump-forming sub-shrub, producing attractive foliage and flowers. Best grown alone, allowing room to develop fully.
Description *Flower* Large, 4-6in (10-15cm) across, fragrant, white, slightly waxy petals, surmounted in centre by golden yellow stamens, resembling large poppy, midsummer to mid autumn. *Foliage* Leaves 1¼-3in (3.5-8cm) wide and long, deeply lobed and cut, grey to grey-green with some yellow autumn colour. *Stem* Upright and glaucous blue, dying to ground level each winter. Fast annual spring growth rate. *Fruit* Insignificant.
Hardiness Once established, tolerates 4°F (−15°C).
Soil Requirements Moderately dry, light open soil.
Sun/Shade aspect Full sun.
Pruning Any shoots which do not die back in winter should be reduced to ground level in early spring.
Propagation and nursery production From root cuttings taken in mid to late summer. Purchase container-grown. Hard to find; normally offered in autumn and winter potted as root clumps.
Problems Can be difficult to establish, but once established can be invasive and difficult

to control, and must be given a spacious site.
Varieties of interest R. × hybrida 'White Cloud' Said to be stronger growing and larger flowering than the basic form. Of American origin.
Average height and spread
Five years
3x3ft (1x1m)
Ten years
3x6ft (1x2m)
Twenty years or at maturity
3x10ft (1x3m)

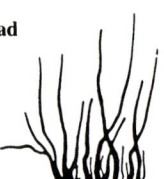

ROSMARINUS

ROSEMARY
Labiatae
Evergreen
A well-known range of aromatic herbs for culinary and ornamental use.

Origin From southern Europe and Asia Minor.
Use As a low to medium height, scented foliage shrub for use in shrub borders, or in herb gardens for culinary purposes. If planted 2ft (60cm) apart makes an informal, low hedge.
Description *Flower* Small, blue flowers of varying shades produced in leaf axils, often in small clusters, on branches 2 years old or more. Flowers late spring to early summer, and intermittently later in summer and sometimes in autumn. *Foliage* Leaves narrow, lanceolate, ¼-1¼in (5mm-3.5cm) long, grey to grey-green with white underside. Some golden variegated forms. Aromatic when crushed. *Stem* Upright, becoming arching and spreading with age, grey-green. Medium growth rate when young, slowing with age. *Fruit* Insignificant.
Hardiness Tolerates 14°F (−10°C).
Soil Requirements Light open soil, dislikes waterlogging and liable to chlorosis on extreme alkalinity.
Sun/Shade aspect Full sun, but tolerates light shade.
Pruning Remove one-third of old wood each spring to ground level to keep the shrub under control and healthy.
Propagation and nursery production From semi-ripe cuttings taken late spring to early summer. Purchase container-grown. Most varieties easy to find. Best planting heights 8in-2ft (20-60cm).
Problems Can become very woody, but pruning should control this.
Varieties of interest R. angustifolia (Narrow-leaved Rosemary) Very narrow, lanceolate foliage and deepest blue flowers. Not easy to find. **R. a. 'Corsican Blue'** Narrow foliage, bright blue flowers. Difficult but not impossible to find. **R. lavandulaceus** syn. **R. officinalis**

Romneya coulteri in flower

Rosmarinus officinalis in flower

Rubus tridel in flower

prostratus A flat, prostrate, low, carpeting form, reaching 1½ft (50cm) in height with a spread of 3-5ft (1-1.5m). Attractive blue flowers well presented on a low-growing shrub. More tender than most and should be given protection in areas experiencing temperatures below 23°F (−5°C). *R. officinalis* (Common Rosemary) Grey-green, aromatic foliage with white undersides. Pale to mid blue flowers produced in axillary clusters along wood 2 years old or more. One of the best forms for hedging and for culinary use. *R. o. 'Albus'* A variety with white flowers. More tender than most. Not easy to find. *R. o. 'Aurea variegata'* Foliage intermittently splashed pale gold. Light blue flowers. *R. o. 'Benenden Blue'* Bright blue flowers and dark green, narrow foliage. Reaches two-thirds average height and spread. *R. o. 'Fastigiatus'* syn. *R. o. 'Pyramidalis'* Mid blue flowers. An erect-growing form, forming an upright pillar. Often sold as *'Miss Jessop's Upright'*. *R. o. 'Roseus'* Lilac-pink flowers, more tender than average. Reaches two-thirds average height and spread. *R. o. 'Severn Sea'* Brilliant blue flowers. A low-growing variety, reaching one-third average height and spread. *R. o. 'Tuscan Blue'* Good bright blue flowers, broader leaves than most. One-third average height and spread.

Average height and spread
Five years
2½x3ft (80cmx1m)
Ten years
4x5ft (1.2x1.5m)
Twenty years
or at maturity
4x5ft (1.2x1.5m)

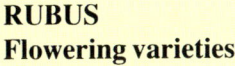

RUBUS
Flowering varieties

FLOWERING BRAMBLE
Rosaceae
Deciduous
Medium height shrubs with attractive flowers, performing well in shade.

Origin From North America, or of garden origin.
Use As medium height, flowering shrubs, useful for their tolerance of high degrees of shade. Arching habit attractive in association with water features.
Description *Flower* Single, white or pink to pinkish-purple flowers, borne singly or in small clusters in mid to late summer. *Foliage* Trifoliate, 1¼-5in (3.5-12cm) wide and long, depending on variety. Light to mid green, with good yellow autumn colour. *Stem* Upright, becoming spreading. Light orange-brown to red-brown, attractive in winter. Medium to fast growth rate. *Fruit* Red aging

to black, bramble-type fruits, but unreliably produced.
Hardiness Tolerates winter temperatures down to −13°F (−25°C).
Soil Requirements Most soils suitable. May show distress in very dry areas.
Sun/Shade aspect Prefers medium shade, tolerates full sun to very deep shade.
Pruning Remove one-third of oldest flowering wood to ground level after flowering.
Propagation and nursery production From semi-ripe cuttings taken in early summer; some varieties from root suckers or by division. Purchase container-grown. Best planting heights 1¼-2½ft (40-80cm).
Problems None.
Varieties of interest *R. odoratus* Clusters of purple-rose, fragrant flowers 1¾-2in (4.5-5cm) across, midsummer to early autumn. Leaves large, palmate, velvety, mid grey-green giving good autumn colour. Thornless brown stems with peeling bark. An upright variety. *R. spectabilis* (Salmon Berry) Single, bright magenta-rose, fragrant flowers in mid to late spring. Trifoliate leaves, 4in (10cm) ling and 3in (8cm) wide. Strong, suckering growth with upright, finely prickly stems. From eastern North America. *R. tridel* Large, single, glossy-surfaced white flowers, 2in (5cm) across, with pronounced golden yellow stamens. Fruits very rarely produced. Brown shoots long, upright, arching at the ends, with peeling bark. *R. tridel 'Benenden'* An improved form, said to have larger flowers.

Average height and spread
Five years
3x3ft (1x1m)
Ten years
6x6ft (2x2m)
Twenty years
or at maturity
10x10ft (3x3m)

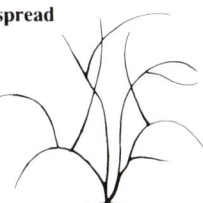

RUBUS
Fruiting varieties

FRUITING BRAMBLE, BLACKBERRY
Rosaceae
Deciduous
Spreading, somewhat unruly, medium height shrubs with good fruit production.

Origin From USA, China, Japan and Korea.
Use As rambling, medium height shrubs for fruit production and some varieties for attractive flowers or ornamental stems, good for shady sites.
Description *Flower* Small clusters of single or double flowers, pink or white depending on variety, in early summer. *Foliage* Leaves trifoliate, 1¼-1¾in (3.5-4.5cm) wide and long. Dark green to red-green with silver undersides, sometimes felted. *Stem* Long, ranging, arching stems, purple to purple-green or orange-red, depending on variety. Sharp, sometimes long thorns. Medium to

Rubus odoratus in flower

189

fast growth rate. *Fruit* Berries bright orange-red or black, depending on variety, and edible, produced in early autumn.
Hardiness Tolerant of winter temperatures down to −13°F (−25°C).
Soil Requirements Most soils suitable.
Sun/Shade aspect Prefers medium shade, tolerates full sun to deep shade.
Pruning Cut all stems to ground level after fruiting.
Propagation and nursery production From semi-ripe cuttings taken in early summer; some varieties from root suckers or by division. Purchase container-grown. Best planting heights 1¼-2½ft (40-80cm).
Problems Arching-stemmed varieties may become invasive.

Rubus phoenicolasius in fruit

Varieties of interest *R. deliciosus* Leaves 3-5 lobed, green, carried on arching branches. White, single flowers, 2in (5cm) across, in late spring, early summer. Fruits purple when mature, but this occurs only in a favourable environment. Two thirds average height and spread. From the Rocky Mountains, Colorado, USA. *R. fruticosus* (Bramble, Blackberry) Small to large flowers, light pink to dark pink, followed by production of the typical, edible blackberries. Fruiting hybrid forms best for garden use. *R. illecebrosus* (Strawberry-raspberry) White flowers produced singly or in small clusters at ends of branches. Flowers 1¼-1¾in (3.5-4.5cm) across, midsummer. Fruits red, large, edible, sweet, but with little substance. Small, erect stems with pinnate leaves produced annually. Reaches only 1½ft (50cm) in height and 3ft (1m) in spread. *R. phoenicolasius* (Wineberry) Small, pale-pink flowers in bristly clusters, midsummer, giving way to bright orange-red, very sweet, edible fruits. Trifoliate leaves, downy, coarsely toothed edged with white felted undersides. Very red bristly stems. Possibly best fan-trained on a wall, but can be used freestanding. Remove old fruit canes after fruiting. From Japan and Korea. *R. ulmifolius* 'Bellidiflorus' Long, arching branches, upright at first, producing panicles of double pink flowers from midsummer to early autumn. Can be fan-trained on a wall and tolerates an exposed site.
Average height and spread
Five years
3x3ft (1x1m)
Ten years
6x6ft (2x2m)
Twenty years
or at maturity
10x10ft (3x3m)

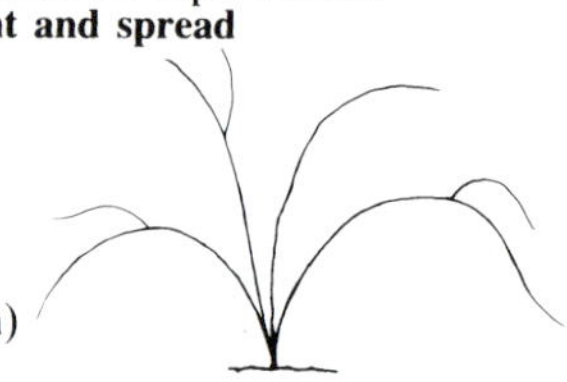

190

RUBUS
Ground cover varieties

GROUND-COVERING BRAMBLE
Rosaceae
Evergreen
Low-growing, carpeting shrubs for ground cover, ideal for shady areas.

Origin From China.
Use As ground cover for large areas or confined spaces, depending on variety, all tolerating deep shade.
Description *Flower* Single, white flowers produced in early summer, of little interest. *Foliage* Leaves trifoliate, dark green with silver undersides. Some red autumn colour. *Stem* Arching or very ground-hugging, depending on variety. Orange-red to orange-brown, covered with soft brown to grey brown hairy thorns. Tips and buds root on contact with soil. Slow to medium growth rate. *Fruit* Red, aging to black, edible bramble-type fruits, sparsely presented.
Hardiness Tolerant of winter temperatures down to −13°F (−25°C).
Soil Requirements Most soils suitable.
Sun/Shade aspect Prefers medium shade, tolerates full sun to very deep shade.
Pruning Varieties with arching stems respond to being cut back hard annually in early spring. Ground-hugging varieties require no pruning.
Propagation and nursery production From semi-ripe cuttings taken in early summer, or from self perpetuating root suckers. Purchase container-grown. Best planting heights 8-18in (20-50cm).
Problems Arching-stemmed varieties may become invasive.
Varieties of interest *R. calycinoides* syn. *R. fockeanus* A very low-spreading, slow-growing variety, forming flat ground cover. Five-lobed leaves, 1-1¼in (3-3.5cm) long and wide, glossy green with grey felted undersides. White flowers borne singly or in short clusters in midsummer, often hidden below foliage. Good ground cover for rock gardens. Tolerates quite deep shade. Reaches 4½ft (1.3m) in height with 3ft (1m) in spread. From Taiwan. *R. microphyllus* 'Variegatus' A low, suckering shrub, producing upright, slightly branched, prickly shoots. Foliage three-lobed, 1¾-2¾in (4.5-7cm) across, mottled green with creamy pink markings. Attractive ground cover effect. *R. tricolor* A creeping form with deeply veined leaves 2¾-4in (7-10cm) long, glossy dark green tinged red, with white felted undersides. White flowers, 1-1½in (3-4cm) across, produced thinly in leaf axils in late summer, occasionally followed by bright red edible fruits. Very good ground cover for shady areas, even under a dense tree

canopy, where should be planted 2ft (60cm) apart. Reaches only 1½ft (50cm) in height but spreads 10-13ft (3-4m), rooting as it goes. From western China.
Average height and spread
Five years
1½x5ft (50cmx1.5m)
Ten years
1½x8ft (50cmx2.5m)
Twenty years
or at maturity
1½x10ft (50cmx3m)

RUBUS
White-stemmed forms

WHITE-STEMMED BRAMBLES
Rosaceae
Deciduous
Medium height shrubs, grown for their very attractive winter stems and silver-grey foliage.

Origin From the Himalayas, China and North America.
Use As medium height shrubs, best planted singly or grouped in locations where they are able to attract winter light to show the white stems to best advantage.
Description *Flower* Single flowers, white or pink depending on variety, in late summer to early autumn. *Foliage* Leaves trifoliate, 1½-2in (4-5cm) long and wide. Grey-green, covered in silvery white bloom. Some varieties lacerated. *Stems* Upright, annually produced. Purple with a white bloom, hairy or with solid thorns. Medium to fast growth rate. *Fruit* Rarely fruits, but may produce small red-black bramble-type fruits.
Hardiness Tolerates winter temperatures down to −13°F (−25°C).
Soil Requirements Most soils suitable.
Sun/Shade aspect Tolerates full sun to very deep shade, but a position in winter sun shows the full beauty of the stems.
Pruning Remove all growth to ground level in spring to induce new stems for following winter.
Propagation and nursery production From semi-ripe cuttings taken in early summer or from root suckers. Purchase container-grown. Best planting heights 1½-2ft (40-60cm).
Problems May become invasive in ideal conditions.
Varieties of interest *R. biflorus* Stems white, sometimes brilliant white, upright when young, becoming spreading with age, covered with soft hairy prickles. Leaves 5-lobed and white felted. Small white flowers produced in small clusters at ends of branches, forming yellow edible fruits in autumn. From the Himalayas. *R. cockburnianus* syn. *R. giral-*

Rubus tricolor in fruit

Rubus cockburnianus — winter stems

dianus Strong, vivid white stems with purple underlay. Fern-like leaves with 7 to 9 leaflets, dark green-grey upper surfaces, white undersides. Small, single, insignificant rose-purple flowers, followed by black fruits coated with blue bloom. Strong-growing in favourable conditions, reaching 13ft (4m) height and spread. From northern and central China. *R. thibetanus* Purple-brown stems covered with white bloom. Foliage pinnate, fern-like, bright grey-white and covered in silky hairs. Small purple flowers. Fruits red to black. Good for winter effect. Upright, becoming very spreading with age. From western China.

Average height and spread
Five years
3x3ft (1x1m)
Ten years
6x6ft (2x2m)
Twenty years
or at maturity
10x10ft (3x3m)

RUSCUS ACULEATUS

BUTCHER'S BROOM, BOX HOLLY
Liliaceae
Evergreen
Interesting, dark, fruiting evergreens for ground cover in shady positions.

Origin From southern Europe.
Use As ground-covering shrubs for deeply shaded areas.
Description *Flower* Very small flowers borne on edges of flattened stems. *Foliage* None. *Stem* Unusual formation described as 'cladode'. Stems have flattened branches up to ½in (1cm) wide and 2in (5cm) long, with spined tips. Very dark green. Stems also produced by young green suckers on outer edges of parent clump. Slow growth rate. *Fruit* Bright, scarlet-red, cherry-like fruits in autumn, maintained into winter. Fruiting unpredictable as there are male, female and hermaphrodite (dual-sexed) forms and females need a male plant nearby in order to produce good fruit.
Hardiness Tolerates winter temperatures down to 13°F (−25°C).
Soil Requirements Most soils, including very dry areas.
Sun/Shade aspect Prefers deep to medium shade, dislikes sun.
Pruning None required.
Propagation and nursery production From division of self-produced suckering growth. Purchase container-grown. Hard to find, especially the hermaphrodite form. Best planting heights 4-15in (10-40cm).
Problems Some winter damage may be caused to flattened branchlets but if old damaged

shoots are removed shrub will rejuvenate. **Varieties of interest** *R. hypoglossum* An interesting form with numerous diamond-shaped, glossy green, leafy stems arising from central clump. Flowers emerge from centre of upper surface of cladode leaf. Female plants produce large red, cherry-like fruits in autumn. Very hard to find but worth the effort.

Ruscus aculeatus

Average height and spread
Five years
1½x2ft (50x60cm)
Ten years
3x3ft (1x1m)
Twenty years
or at maturity
3x4ft (1x1.2m)

RUTA GRAVEOLENS

RUE, HERB OF GRACE
Rutaceae
Evergreen
Useful low shrubs with attractive foliage and flowers for front edging of shrub borders.

Origin From southern Europe.
Use As a blue-grey evergreen for foreground planting in small to large shrub borders. If planted 2ft (60cm) apart makes low, informal hedge. Often planted in herb gardens. Grows well in large tubs, with good potting medium.
Description *Flower* Clusters of small mustard yellow flowers borne on short stalks above foliage, mid to late summer. *Foliage* Leaves grey-green to glaucous green or creamy white variegated, pinnate, oval, 3-5in (8-12cm)

long, very divided, giving fern-like appearance. Strong, pungent, aromatic scent. Contact with crushed stems can blister skin. *Stem* Upright, becoming slightly spreading with age, grey-green. Medium to fast growth rate in spring. *Fruit* Yellow-green seedheads follow flowers.
Hardiness Tolerates 14°F (−10°C), but evergreen foliage may be damaged in winter.
Soil Requirements Prefers light, well-drained soil, but tolerates wide variation.
Sun/Shade aspect Full sun to very light shade.
Pruning Lightly trim over the whole shrub in early spring, to remove any damaged foliage, and to encourage new spring growth. May look a little bare during early spring as a result.
Propagation and nursery production From semi-ripe cuttings taken in early summer. Purchase container-grown. Blue-leaved forms fairly easy to find, variegated forms obtainable from specialist nurseries. Take cuttings in late autumn and overwinter them under cover to ensure a succession of plantings if parent does not survive. Best planting heights 4-12in (10-30cm).
Problems When purchased young plants may look small and unsightly but once planted out grow rapidly, producing good foliage.

Ruta graveolens 'Jackman's Blue' in flower

Varieties of interest *R. g. 'Jackman's Blue'* The best glaucous blue foliage variety, of good, bushy, compact habit. Interesting mustard-coloured flower clusters. *R. g. 'Variegata'* Grey-green leaves, variegated creamy white. Slightly smaller and more upright than the parent, and slightly more tender.

Average height and spread
Five years
2x2½ft (60x80cm)
Ten years
2½x3ft (80cmx1m)
Twenty years
or at maturity
3x3ft (1x1m)

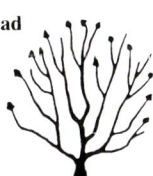

SALIX
Short and low-growing varieties

WILLOW
Salicaceae
Deciduous
A wide range of shrubs of varying interest, suitable for rock gardens and for use by small pools.

Origin Various locations throughout northern hemisphere.
Use As a low-growing shrub, for attractive stems, foliage or catkins. Very small varieties useful for rock gardens and water features.

191

Some forms can also be purchased as small, mop-head standards.

Description *Flower* White to yellow-white or even black catkins produced on bare stems in mid spring. Shape and size of catkins is variable, depending on specific variety. *Foliage* Leaves elliptic to ovate, ½-2in (1-5cm) long, size depending on variety. Grey, purple-grey, grey-green or green, some giving good autumn colour. *Stem* Clump or mat-forming, colours ranging from purple, through grey, to grey-green, depending on variety. Slow to medium growth rate. *Fruit* Insignificant.

Hardiness Tolerates winter temperatures down to −13°F (−25°C).

Soil Requirements Any soil, extremely good on wet and waterlogged areas.

Sun/Shade aspect Full sun to medium shade. Coloured-stemmed varieties best in full sun to show off full winter display.

Pruning None required, but may be cut back.

Salix alba integra maculata in leaf

Propagation and nursery production From semi-ripe cuttings taken in early summer or hardwood cuttings in winter, depending on variety. Purchase container-grown. Most forms easy to find, although some varieties may need a search. Best planting heights 8in-1½ft (20-50cm).

Problems Although the shrubs are small and twiggy, the eventual overall height and width is often underestimated.

Varieties of interest *S. alba integra maculata* syn. *S. 'Hakuro Nishikii'*, *'S. Fuiri-Koriyangi'* Young foliage completely pink, aging to cream and finally to light green with white mottling. Thin, light green stems, forming a bun-shaped shrub. Trim back new growth by two-thirds annually in early spring, to enhance foliage production. Requires protec-

Salix lanata in catkin

tion from strong midday sun, with very light dappled shade being required for best effect. Of garden origin, from Japan. *S. bockii* Very slender, numerous red twigs along spreading branches. Stems covered with grey down when young. Round, mid green foliage very thickly clustered along the stems and twigs. Small, grey catkins appear, very unusually, late summer to early autumn. Very scarce. From western China. *S. × boydii* A very low form, upright in habit, quickly becoming old in appearance. Leaves round, mid green with grey down when young. Rarely produces catkins. A variety so small it is ideal for troughs or small pockets in rock gardens. Extremely difficult to find, but not impossible. Reaches only one-third average height and spread. From Scotland. *S. hastata 'Wehrhahnii'* Large, silver, male catkins appear on the red-purple upright stems in profusion before round to ovate, dark green leaves are formed. Yellow autumn colour. Very quickly looks old and woody. May be found trained as small, short standard tree. Of garden origin. *S. helvetica* Yellow-green, short-branched, twiggy growth covered with pale yellow narrow catkins at same time as small, lanceolate, silver-grey leaves with white undersides appear, covering branches. A useful low, very slow-growing shrub, especially when used in connection with water features. May be found trained as small standard tree. Two-thirds average height and spread. From the European Alps. *S. lanata* (Woolly Willow) Ovate to round, silver-grey, down-covered leaves borne on short, grey-green, twiggy, goblet-shaped shrub. Small yellow-grey catkins produced with leaves, late spring. A useful

Salix melanostachys in catkin

low-growing shrub for large rock gardens and for pool effect. Foliage may be damaged by slugs or heavy rain; stems may suffer die-back in winter. Very slow-growing. Two-thirds average height and spread. From northern Europe and northern Asia. *S. l. 'Stuartii'* Slightly more dwarf variety than its parent. Yellow-green shoots with orange buds and silver-white to silver-grey downy leaves with yellow catkins. Very slow-growing. Two-thirds average height and spread. *S. lapponum* (Lapland Willow, Downy Willow) Dense branchwork of grey-green stems covered in spring with ovate, grey, downy leaves, preceded by silky, grey to yellow catkins. Very slow-growing. Two-thirds average height and spread. From the mountains of Europe and Siberia. *S. melanostachys* syn. *S. 'Kurome'* (Black Catkin Willow) Purple-brown to green-brown twigs, forming a round shrub. Dark green, ovate, tooth-edged leaves with blue undersides. Black male catkins with brick red anthers, produced before leaves. A most attractive form, grown for its catkins. *S. moupinensis* Short, glossy, purple-red, attractive shoots. Foliage ovate, dark grey-green with glabrous undersides, dark green-grey upper. Silvery yellow catkins in spring. Extremely slow in growth. Often confused with *S. fargesii*, but shorter-growing. May reach slightly more than average height and spread. From China. *S. purpurea 'Gracilis'* syn. *S. p. 'Nana'* Very wispy, slender branches, forming goblet-shaped shrub with small to very small, grey-purple to dark green, narrow, lanceolate leaves. Small, slender, purple-silver catkins, early spring. Very attractive, and useful for planting near water. *S. repens argentea* Long, grey-green, semi-prostrate branches spreading up to 6-10ft (2-3m), co-

Salix hastata 'Wehrhahnii' in catkin

vered in ovate, silver-grey catkins standing upright on upper surfaces of branches, before leaves form. Useful for coastal areas and sandy soils. In some nurseries may be found growing as a small weeping or standard tree, although very scarce in this form.

Average height and spread
Five years
2x3ft (60cmx1m)
Ten years
3x5ft (1x1.5m)
Twenty years
or at maturity
5x6ft (1.5x2m)

SALIX
Medium height varieties

WILLOW
Salicaceae
Deciduous
Useful foliage, stem or catkin effects, depending on variety, some shrubs showing fine architectural habit.

Origin From throughout the northern hemisphere.
Use As individual shrubs for stem, foliage or catkin attraction depending on variety, or for large-scale planting to provide windbreaks and screens. Very good beside water features such as pools or streams.
Description *Flower* Yellow-white to white catkins produced on bare stems, mid spring, shape and size depending on variety. *Foliage* Leaves elliptic to ovate, 1½-3½in (4-8cm) long. Grey, purple-grey, grey-green or green, depending on variety, some varieties giving good autumn colour. *Stem* Upright, vigorous, either glossy surfaced or covered in a bloom. Green to grey-green, silver-grey, gold, yellow or orange-red, depending on variety, making very good winter display. Some varieties with attractive coloured bud formation. Very fast growth rate in most cases. *Fruit* Insignificant.
Hardiness Tolerant of winter temperatures down to −13°F (−25°C).
Soil Requirements Any soil. Extremely good in wet and waterlogged areas.
Sun/Shade aspect Full sun to medium shade. Varieties with coloured stems are best in full sun to show off their full display in winter.
Pruning Many varieties should be pruned back hard in early spring, either annually or biennially to encourage new stem growth and larger foliage. Certain varieties require other pruning methods as described below.
Propagation and nursery production From hardwood cuttings taken early to mid winter. Plant container-grown or bare-rooted. Most varieties fairly easy to find, although some may have to be sought from specialist nurseries or forestry outlets. Best planting heights 1¼-4ft (40cm 1.2m).

Salix irrorata — winter stems

Problems The ultimate size and rapidity of growth is often underestimated. Some varieties may suffer from stem canker and attacks by willow mildew and scab. Cutting back of diseased wood will help to eliminate these fungus diseases.
Varieties of interest *S. cinerea* (Grey Sallow) Long, broad, ovate leaves, grey-green with more intense grey on undersides, and red veins. Good display of yellow catkins on bare grey to grey-green stems. From western Asia through Tunisia and Europe. *S. elaeagnos* syn. *S. incana, S. rosmarinifolia* (Hoary Willow) Stems grey-green, becoming dark red-brown with grey covering. Wispy, upright, forming goblet shape. Narrow, slender catkins, produced with foliage, the main attraction, in early spring. Leaves very similar to those of Rosemary, being narrow, lanceolate and grey to grey-green. Responds to hard pruning every 5-6 years to rejuvenate the overall appearance. Reaches ultimate height and spread over longer than average period. From central southern Europe and Asia Minor. *S. exigua* (Coyote Willow) Long, upright, slender branches forming goblet shape, covered in long, narrow, lanceolate, very slightly tooth-edged, silver-grey foliage. Pale yellow catkins produced with foliage, early spring. Prune hard every 3-4 years to enlarge foliage and maintain good, upright-growing shrub. One of the most statuesque of all Willows, attaining best height and spread in its native environment. Plant individually to achieve best shape. From western USA and northern Mexico. *S. fargesii* Attractive, short, very slow-growing, red-brown, glossy, polished stems with dark red winter buds,

eventually forming large goblet-shaped shrub. Foliage large, elliptic to oblong, up to 7in (18cm) long. Dark glossy green with purple veining. Yellow hanging catkins up to 4-6in (10-15cm) long, produced after leaves, late spring to early summer. No pruning required. Very handsome in habit, although slow to form large shrub. From central China. *S. gracilistyla* (Rosegold Pussy Willow) Stems grey-green to grey-purple, stout, upright at first, becoming spreading with age. Foliage silvery grey when young, becoming green with age. Good yellow autumn colour. Very attractive yellow catkins, early spring, before foliage. Best hard pruned every 4-5 years. From Japan, Korea and Manchuria. *S. Hakuro'* Grey-green, slender, upright stems. Young foliage variegated creamy white, aging to grey-green. Biennial hard pruning recommended to encourage variegated growth. Obtain from specialist nurseries. Of garden origin, from Japan. *S. 'Harlequin'* Large lanceolate leaves, with cream and pink mottled variegation of new foliage growth in early summer. Of garden origin. *S. humilis* (Prairie Willow) Leaves ovate to lanceolate, dark green upper surfaces, blue glaucous undersides. Yellow catkins, stamens on female forms having red markings, red anthers on male, produced before leaves. From eastern North America. Reaches more than average ultimate height and spread in native environment. *S. irrorata* Very attractive purple, upright stems, covered in white bloom coating, originating from central stool. Buds having very attractive red and black scales, becoming more marked through late winter and early spring. Small, yellow, male catkins, red anthers. Foliage lanceolate to oblong, glossy, dark green with glaucous

Salix exigua in leaf

Salix magnifica in catkin

Salix sachalinensis 'Sekka' in catkin

undersides. Somewhat pendulous. Prune biennially to encourage new growth for winter attraction. One of the best of all grey-stemmed Willows. From south-western USA. *S. magnifica* (Magnolia-leaved Willow) Large, ovate, completely uncharacteristic, magnolia-shaped leaves, up to 8in (20cm) long and 5in (12cm) wide. Upright yellow catkins, 6-8in (15-20cm) long, produced in late spring with foliage. No pruning required. Slow-growing, making 1½-2½ft (50-80cm) of growth annually. From western China. *S. phylicifolia* (Tea-leaf Willow) Bright green to yellow-green stems, very branching in habit, forming wide goblet-shaped shrub. Round to ovate, green foliage with blue undersides, giving very good yellow autumn colour, possibly its main attribute. Yellow catkins produced with leaves in spring. Slow-growing, taking longer than average to reach ultimate height and spread. No pruning required. From northern Europe. *S. purpurea* (Purple Osier Willow) Purple stems aging to purple-green. Narrow, oblong foliage, green to dull green upper surfaces, paler glaucous undersides. Narrow, slender catkins produced in spring before foliage. Thin, upright, becoming arching and graceful in habit. Possibly best pruned hard every 4-5 years to improve overall apearance. From Europe and central Asia. *S. p. 'Eugenei'* Attractive, bright grey-green stems, veined dull red, forming conical habit. Narrow, lanceolate, grey-green hanging foliage, which later gives dull green autumn colour, preceded in early spring by very attractive grey-pink male catkins, its main attraction. *S. sachalinensis 'Sekka'* syn. *S. 'Setsuka'* Stems purple to purple-green, some fastigiate, curling at ends, suggesting large, flat, hockey sticks; much used in flower arranging. Silver catkins produced both on normal and flat stems in early spring. Foliage lanceolate, grey-green to purple-green, making a large, wide, rambling shrub. Responds well to triennial hard pruning to encourage production of more large, flat, stems. Large shrub, often underestimated for its ultimate size, spreading out to 26ft (8m) and achieving average height. *S. × tsugaluensis 'Ginme'* A spreading variety with large silver catkins on bare stems, early spring. Foliage oblong, bright green with paler undersides and orange-tinted when young. Good yellow autumn colour. From Japan.

Average height and spread
Five years
6x6ft (2x2m)
Ten years
10x10ft (3x3m)
Twenty years
or at maturity
13x13ft (4x4m)

SALIX
Tall-growing varieties

WILLOW
Salicaceae
Deciduous
Large screening shrubs with attractive foliage and stems requiring adequate space.

Origin From throughout the northern hemisphere.
Use As large, tall shrubs for difficult wet soils. Very good planted with water features, for windbreaks, or as individual shrubs to show off winter stems and summer foliage colour. Most forms can be trained as small single-stemmed trees.
Description *Flower* White to yellow-white catkins of variable size and shape, according to variety, produced on bare stems early to mid spring. *Foliage* Leaves elliptic to ovate, 1½-6in (4-15cm) long, size varying according to variety. Foliage grey, purple-grey, grey-green or green, some varieties giving good autumn colour. *Stem* Upright, vigorous, either glossy surfaced or covered in white-grey bloom. Ranging from green to grey-green, silver grey, purple, gold, yellow or orange-red, depending on variety; some having very attractive coloured bud formation. Fast growth rate. *Fruit* Insignificant.

Hardiness Tolerant of winter temperatures down to −13°F (−25°C).
Soil Requirements Any soil, extremely good in waterlogged areas.
Sun/Shade aspect Full sun to medium shade. Varieties with coloured stems best in full sun.
Pruning None required, but many coloured-stem varieties give better display if pruned back very hard in mid to late spring, annually or biennially.
Propagation and nursery production From hardwood cuttings taken early to mid winter. Plant bare-rooted or container-grown. In both cases young plants may look thin in habit, with poor root systems, but once planted out rapidly produce large shrubs. Most varieties relatively easy to find, although some may have to be sought from specialist nurseries or forestry outlets. Best planting heights 2-6ft (60cm-2m).
Problems Rapidness of growth and ultimate size often underestimated; plants can quickly outgrow the allotted space. Some varieties may suffer from stem canker and attacks by willow mildew and scab; cutting back diseased wood helps to eliminate these fungus diseases.
Varieties of interest *S. acutifolia 'Blue Streak'* Attractive black to purple stems covered in a pronounced white bloom, giving good winter effects and graceful arching habit. Foliage pointed lanceolate, purple-green with grey to white bloom. Pale yellow catkins in spring. Prune hard in mid to late spring. Can be grown as a small tree. Parent from USSR and eastern Asia. *S. a. 'Pendulifolia'* Slender, purple-black stems with grey to white bloom. Long, lanceolate, hanging foliage, grey-green tinged purple. Catkins white, aging to pale yellow, medium-sized and narrow, produced on bare stems. Prune hard every 2-3 years to encourage new growth and larger foliage. Can be grown as small tree. *S. adenophylla* (Furry Willow) Stems grey to grey-green. Foliage ovate with toothed egdes, covered in layer of silky grey-green hairs. Good yellow catkin display on bare stems, early spring. Attractive open habit. Can be grown as small tree. From north and north-eastern USA. *S. aegyptiaca* syn. *S. medemii* (Egyptian Musk Willow) Young branches and twigs coated in grey bloom. Leaves lanceolate, grey-green surface, grey undersides in youth. Good large male yellow catkins, mid to late spring, on bare stems. Can be grown as small tree. From southern USSR and the mountains of Asia. *S. alba* (White Willow) Foliage lanceolate, covered in grey or silver-grey silky hairs. Branches forming conical shape, often arching at ends. Small yellow male catkins on bare stems, mid spring. Can be grown as fast-growing tree or pruned to be maintained as large shrub. All forms very good in moist sandy soils and good for maritime areas. From Europe and northern Asia through North Africa. *S. a. 'Chermesina'* syn. *S. a. 'Britzensis'* (Scarlet Willow) A variety grown

Salix alba 'Sericea' in leaf

for its bright orange-scarlet winter stems. Prune annually or biennially. From Europe, northern Asia and North Africa. *S. a. 'Sericea'* syn. *S. a. 'Argentea'* (Silver Willow) A variety often seen as a small tree but with annual or biennial pruning can be retained as a large shrub. Lanceolate, grey-green foliage tinged silver, extremely attractive when seen against grey or blue sky. Stems grey-green, slightly open in habit. Good effect whether or not pruned. From Europe, northern Asia and North Africa. *S. a. 'Vitellina'* (Golden Willow) Very bright yellow to dark yellow, strong, upright shoots, effective in winter. Male form, producing small yellow catkins on bare stems in late spring, on unpruned shrubs. Prune annually or biennially. Can be grown as a small tree. *S. caprea* (Goat Willow, Great

Salix caprea in catkin

Sallow, Palm Willow, Pussy Willow, Sallow) Strong, upright, slightly spreading, light green to grey-green stems. Long, round to ovate leaves with grey undersides, giving good autumn colour. Male form producing large yellow catkins on bare stems, mid to late spring. Prune annually or biennially to improve catkin production. Can be trained as small tree. From Europe and western Asia. *S. daphnoides* (Violet Willow) Strong, upright, becoming slightly spreading. Dark purple-violet shoots covered in distinctive white bloom. Large white, aging to yellow, male catkins in early spring, on bare stems. Long, narrow, tooth-edged purple-grey to green foliage with grey undersides. Extremely good winter stem display, especially when shrubs pruned annually or biennially. Can be grown either as large shrub or small tree. From northern Europe, central Asia and the Himalayas. *S. d. 'Aglaia'* Even whiter stems and larger catkins than *S. daphnoides*. May have to be sought from specialist nurseries. *S. × smithiana* Bright green stems, with large yellow catkins on bare stems in spring. Large, lanceolate, tooth-edged, bright green foliage with slightly hairy undersides. Stems respond well to annual or biennial pruning. Can be grown as small tree. From the British Isles. *S. 'The Hague'* Branches grey-green, strong, upright, becoming spreading. Interesting red or black, shielded buds in winter. Large silver catkins on bare stems in early spring. Foliage large, ovate, tooth-edged, grey-green. Best left unpruned for catkin production. Of garden origin.

Average height and spread
Five years
13x5ft (4x1.5m)
Ten years
16x8ft (5x2.5m)
Twenty years
or at maturity
20x14½ft (6x4.5m)

SALVIA OFFICINALIS

COMMON SAGE
Labiatae
Evergreen
Useful foliage shrubs for culinary uses or flower arranging, some with brightly coloured flowers.

Origin From southern Europe.
Use As a culinary herb for herb gardens or as a low shrub for front edging of shrub borders. If planted 2ft (60cm) apart in single line makes small, low, informal dividing hedge.
Description *Flower* Blue-purple flower spikes in summer; some varieties with red flowers. *Foliage* Leaves ovate, 1-1½in (3-4cm) long, grey-green, with a strong aromatic scent. Some varieties with purple, golden, silver or pink variegated foliage. *Stem* Soft, upright, becoming spreading and woody with age, grey-green. Fast growth rate in spring. *Fruit* Insignificant.
Hardiness Tolerates 14°F (−10°C), except variegated forms winter minimum of 23°F (−5°C).
Soil Requirements Prefers rich, light, warm soil but tolerates wide range, both acid and alkaline.
Sun/Shade aspect Full sun for good foliage colour.
Pruning In spring, reduce last season's growth by half, to keep the shrub young and healthy.
Propagation and nursery production From softwood cuttings taken in early summer. Purchase container-grown. Take cuttings late autumn and overwinter under cover to insure replacements in case parent plant succumbs in winter. Best planting heights 4in-1ft (10-30cm).
Problems Unpruned shrubs readily become woody and uninteresting.
Varieties of interest *S. microphylla* syn. *S. grahamii* Bright red flowers, aging to deeper red, up to 1¼in (3.5cm) long, midsummer to late autumn. Tender in all areas where temperature falls below 32°F (0°C). Reaches one-third more than average height and spread. From Mexico. *S. officinalis 'Icterina'* syn. *S. officinalis aureovariegatus* Golden variegated foliage. Mid blue flowers in midsummer to autumn. Possibly the hardiest of all. *S. o. 'Purpurascens'* (Purple-leaf Sage) Panicles of blue flowers in spring. Large, long, ovate leaves, purple when young, becoming grey-purple and finally green-purple. Can look somewhat shabby in winter. From southern Europe. *S. o. 'Tricolor'* (Tricoloured Sage) An interesting variety with basic purple foliage mottled white and pink. Mid blue flowers in midsummer to autumn. Rather shabby in winter.

Salvia officinalis 'Icterina' in leaf

Average height and spread
Five years
2½x3ft (80cmx1m)
Ten years
3x4ft (1x1.2m)
Twenty years
or at maturity
3x4ft (1x1.2m)

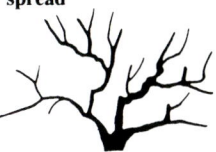

SAMBUCUS

ELDER
Caprifoliaceae
Deciduous
Among the most handsome of foliage shrubs if pruned as recommended.

Origin From throughout the northern hemisphere.
Use As a freestanding shrub for flowering or foliage effect, depending on variety, or for mass planting. Very good as single specimens for wild or woodland gardens.
Description *Flower* Convex clusters of tufted flowers or single tufts produced on unpruned wood one year old or more. *Foliage* Leaves large, pinnate, 3-6in (8-15cm) long and wide, normally green to grey-green, with good autumn colour. White-variegated, purple and gold-leaved forms, some deeply dissected. *Stem* Upright with limited branching, becoming spreading in time and forming a dome-shaped tall shrub. Grey to grey-brown. Fast growth rate when young. *Fruit* Clusters of red, purple or black fruits depending on variety, only on unpruned shrubs.

Salvia officinalis 'Purpurascens' in flower

Sambucus nigra 'Albovariegata' in leaf

Hardiness Tolerates 14°F (−10°C). Some winter die-back may occur but shrub normally recovers.

Soil Requirements Any soil, tolerating wet conditions.

Sun/Shade aspect Prefers very light shade, accepts full sun to medium shade.

Pruning Most varieties need one-third of oldest wood removed to ground level each spring. With golden-leaved or variegated varieties, prune all growth to ground level in early spring, to encourage new growth and new, good-coloured foliage. This will reduce the overall height and spread.

Propagation and nursery production From seed for some varieties, from layering or grafting for specific types. Plant bare-rooted, root-balled (balled-and-burlapped) or container-grown. Most varieties fairly easy to find, but one or two varieties may need a search. Best planting heights 2-3ft (60cm-1m), except in early to mid spring when it is better to use a shrub that has been cut back hard.

Problems Without the recommended pruning, plants rapidly become woody and uninteresting.

Varieties of interest *S. canadensis* (American Elderberry) Leaves with 5-11 leaflets to each pinnate leaf. White clusters of flowers on convexed flower heads up to 8in (20cm) across, mid to late summer, followed by clusters of black fruits. A strong-growing shrub for which thinning technique of pruning recommended. From south-eastern Canada and eastern USA. *S. c. 'Maxima'* Very large leaves up to 16in (45cm) long. Rose-purple flower stalks, surmounted by purple flowers up to 1ft (30cm) across. Requires very hard pruning each spring to encourage new shoots with large leaves and good flower clusters. *S. nigra* (Common Elder, Bour Tree) Green, pinnate leaves with 5-7 leaflets. White flowers with musky scent, early to mid summer, giving way to large bunches of shiny black fruits, used for wine making. Very good on extremely alkaline conditions. From Europe, Africa and western Asia. *S. n. 'Albovariegata'* Pinnate leaves with 5-7 leaflets. Grey-green with creamy white margins. Must be hard pruned for good foliage effect in coming spring and summer. *S. n. 'Aurea'* (Golden Elder) Foliage golden yellow, becoming more gold with age. Must be pruned hard each spring to encourage better foliage production; alternatively, remove one-third of oldest wood if flowers and fruit desired. *S. n. 'Laciniata'* (Fern-leaved Elder, Parsley-leaved Elder) Foliage pinnate and finely divided giving fern-like appearance. Remove one-third of oldest wood each year to encourage production of good fern-like leaves. Shrub deteriorates if pruning neglected. *S. n. 'Pulverulenta'* Green pinnate leaves striped and mottled white. Responds well to hard pruning in spring. Scarce. *S. n. 'Purpurea'* Young leaves purple-green, aging to green-purple. Responds equally well to hard pruning in spring, or to thinning. *S. racemosa 'Plumosa Aurea'* (Red-berried Elder) One of the best of all golden foliage shrubs if pruned hard each spring. If left unpruned, deeply cut, divided foliage becomes small, but rich yellow flowers are produced, followed by red berries. Choice is therefore either to cut back hard each spring for foliage, or to leave alone

or just lightly thin, to achieve yellow flowers and red berries but poorer foliage display. Liable to scorching in very hot summers. From Europe and western Asia. *S. r. 'Sutherland'* Very finely divided, golden yellow foliage. Should be cut back hard each spring to induce good new golden foliage. Of garden origin.

Sambucus racemosa 'Plumosa Aurea'

Average height and spread

Five years
10x6ft (3x2m)
Ten years
13x12ft (4x3.5m)
Twenty years
or at maturity
20x16ft (6x5m)

SANTOLINA

COTTON LAVENDER, LAVENDER COTTON
Compositae
Evergreen
Attractive foliage shrubs for front of borders, edging and mass planting.

Origin From the Mediterranean area.

Use As a low-growing, compact, silver or green foliage shrub for edges of shrub borders, for large rock gardens of for use among bedding plants. If planted 1¼ft (40cm) apart in a single line makes an informal low hedge.

Description *Flower* Small round flowerheads, primrose to lemon to golden yellow, depending on variety, produced on thin stalks above main foliage in midsummer. *Foliage* Leaves very small, linear, finely divided, 1-1½in (3-4cm) long, densely packed, green or grey to silver-grey. *Stem* Very short, twiggy stems, forming a low-spreading, bun-shaped shrub. Green or grey-green. Fast spring growth rate. *Fruit* Insignificant.

Hardiness Minimum winter temperature 14°F (−10°C); dislikes wet, cold conditions.

Soil Requirements Well-drained, open soil; tolerates alkalinity and acidity.

Sun/Shade aspect Full sun.

Pruning Once signs of spring growth are seen, trim lightly to encourage production of new foliage.

Propagation and nursery production From softwood cuttings taken in early summer, or semi-ripe cuttings taken in late summer for overwintering under cover. Purchase container-grown. All forms fairly easy to find. Best planting heights 4in-1ft (10-30cm).

Problems Some forms of grey-leaved types are extremely confused in production and it is difficult to be certain of obtaining a particular variety required.

Varieties of interest *S. chamaecyparissus* syn. *S. incana* Low mound of silver foliage, surmounted by bright yellow flowers in mid-summer. From southern France and the Pyrenees. *S. c. corsica* syn. *S. c. 'Nana'* A

Sambucus nigra 'Purpurea' in leaf

Santolina chamaecyparissus in leaf

much more compact variety with grey-silver foliage. Small lemon yellow flowers in summer. From Corsica and Sardinia. **S. neapolitana** A low shrub, more upright in habit, with very feathery divided leaves. Attractive, graceful habit. Bright lemon yellow flowers in midsummer. From north-western and central Italy. **S. n. 'Sulphurea'** Grey-green foliage with pale primrose yellow flowers. Slightly upright in habit. **S. virens** syn. **S. viridis** A low mound-forming shrub with green to bright green thread-like foliage, surmounted by bright lemon yellow flowers in midsummer.

Average height and spread
Five years
1½x2¼ft (50x70cm)
Ten years
1½x3ft (50cmx1m)
Twenty years
or at maturity
1½x3ft (50cmx1m)

SARCOCOCCA

SARCOCOCCA, CHRISTMAS BOX, SWEET BOX
Buxaceae
Evergreen
A delightful feature for any garden; shrubs with highly fragrant winter flowers.

Origin From western and central China.
Use As a low, winter-flowering, evergreen shrub for planting on its own, in mass plantings or at edge of shrub borders. If planted 1ft (30cm) apart makes small, low hedge.

Description *Flower* Small, very fragrant, white, male flowers with small, insignificant female flowers produced in same cluster, early winter. *Foliage* Leaves narrow, ovate, 1-2½in (3-6cm) long, dark green or purple-green, depending on variety. Shiny upper surfaces, duller undersides. *Stem* Dark glossy green or grey-green tinged red or purple, depending on variety. Forming slightly upright to spreading shrub. Slow growth rate. *Fruit* Black or red fruits, depending on variety.
Hardiness Tolerates 4°F (−15°C).
Soil Requirements Most soils, tolerating alkalinity and acidity, but needs adequate plant food to do well. Ideal for leafy woodland type soil. Dislikes poor, unfed soil.
Sun/Shade aspect Full sun to medium shade. May tolerate deep shade, but become more open and lax.
Pruning None required.
Propagation and nursery production From semi-ripe cuttings taken in early summer or from rooted suckering growth at sides. Purchase container-grown or root-balled (balled-and-burlapped). Best planting heights 8in-1½ft (20-50cm).
Problems Young plants often very small, but grow quickly in good fertile soil.
Varieties of interest *S. confusa* Foliage very dark glossy green, ovate, pointed and slightly lobed. White fragrant flowers in winter, followed by shiny black fruits. *S. hookerana var. digyna* Green-purple, narrow, lanceolate leaves, slightly pendulous. Scented pink-white flowers followed by black fruits. Upright, slender branches, purple-green in colour. Freely suckering from base but not

invasive. Best on leafy, woodland, open soil. From western China. **S. h. d. 'Purple Stem'** Foliage, flowers and fruit identical to those of its parent. Stems, leaf stalks and veining within leaves flushed deep purple and attractive in winter. **S. humilis** Leaves elliptic, shiny, deep green with duller undersides. White male flowers have pink anthers and are scented in winter, followed by black fruit. A very suckering form, though not invasive. Reaches only half average height and spread. From western China. **S. ruscifolia** Dark green, leathery, thick, shiny, ovate leaves produced on short, mounded, low-growing shrub. Red fruits. From central China. **S. r. chinensis** The form most commonly offered. Larger leaves than *S. ruscifolia* itself, although slightly narrower. Often confused with *S. confusa*, but differs in that it has dark red fruits following white, scented winter flowers. From central and western China.

Average height and spread
Five years
10inx1ft (25x30cm)
Ten years
3x1¼ft (1mx40cm)
Twenty years
or at maturity
5x2½ft (1.5mx80cm)

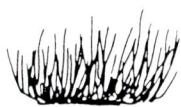

SENECIO

SHRUBBY RAGWORT
Compositae
Evergreen
Attractive, silver-leaved shrubs with white or yellow daisy-like summer flowers.

Origin From New Zealand.
Use As a low, grey foliage shrub for front of shrub borders. Effective for mass planting. Planted 2ft (40cm) apart makes low informal hedge.
Description *Flower* Clusters of yellow, daisy-shaped flowers in early summer, maintained over long period. *Foliage* Leaves ovate, 1-2½in (3-6cm) long, sometimes cut-leaved, silver-grey. *Stem* Upright, becoming branching and spreading with age. Grey. Slow to medium growth rate. *Fruit* Insignificant.
Hardiness Tolerates 14°F (−10°C).
Soil Requirements Well-drained, open soil, dislikes waterlogging.
Sun/Shade aspect Full sun to very light shade.
Pruning Trim lightly each spring to keep shrub healthy. Once growth starts, old woody shrubs can be cut back hard in early spring, and will rejuvenate from ground level.
Propagation and nursery production From semi-ripe cuttings taken early to mid summer. Purchase container-grown. Best planting heights 1-1½ft (30-50cm).
Problems Can become woody, but an annual trimming as suggested should prevent this.
Varieties of interest *S. cineraria* (Dusty Miller, Silver Groundsel) Very white, felted, cut-leaved foliage, surmounted by yellow daisy-shaped flowers in summer. Many named varieties offered as annuals for bedding rather than as permanent shrubs, but in areas where temperature does not fall below 23°F (−5°C) may be considered hardy. *S. compactus* Small, ovate, wavy-edged grey-green leaves with white undersides. Young shoots and flower stalks are also white-felted. Bright yellow daisy-shaped flowers produced early to mid summer. Dislikes waterlogging. Low-growing variety. Tender and should not be subjected to frost. *S. 'Dunedin Sunshine'* syn. *S. laxifolius* Ovate, silver-grey foliage surmounted by daisy-shaped flowers in terminal panicles. Can become woody without adequate pruning. In addition to trimming, one-third of oldest wood should be cut back to ground level in spring to induce new growth. *S. greyi* Very downy, soft grey foliage. Yellow flowers in summer. Tender and should not be subjected to forst. Scarce. *S. hectori* Leaves grey-green with white undersides and toothed edges. Semi-evergreen. Clusters of white

197

Senecio 'Dunedin Sunshine' in flower

flowers, mid to late summer. Needs protection, minimum winter temperature 23°F (−5°C). May be hard to find. *S. leucostachys* Very lax growth covered with very finely divided silver-white pinnate leaves, covered with good-sized white, daisy-shaped flowers throughout summer. From Patagonia. *S. monroi* Oblong or oval leaves, dark steel grey with wavy curling edges. Flower spikes and young shoots grey to white felted. Terminal clusters of yellow daisy-shaped flowers. Reaches two-thirds average height and spread. *S. reinoldii* syn. *S. rotundifolia* Large, round, leathery, dark grey-green, glossy foliage. Terminal round, open clusters of yellow daisy flowers in summer and early autumn. Minimum winter temperature 23°F (−5°C). In favourable conditions reaches 6ft (2m) height and spread.

Average height and spread
Five years
2½x2½ft (80x80cm)
Ten years
3x3ft (1x1m)
Twenty years
or at maturity
3x5ft (1x1.5m)

SKIMMIA

KNOWN BY BOTANICAL NAME
Rutaceae
Evergreen
On the right soil, a very good evergreen, autumn-fruiting shrub.

Origin From Japan and China.
Use As interesting evergreen flowering and fruiting shrubs for woodland gardens or for small to medium shrub borders. Can be grown in tubs not less than 2ft (60cm) in diameter and 17in (45cm) in depth; use lime-free potting medium.
Description *Flower* Terminal panicles, 3-5in (8-12cm) tall, of white, fragrant flowers, late spring. Shrub can be either male or female and in some cases both sexes on same shrub. To produce fruit, ratio of male to female can be any number of female forms to one male, provided grown within 10-12 sq yds (12-14 sq m) of each other. *Foliage* Leaves large, ovate to elliptic, 3-4½in (8-11cm) long, leathery, dark green with purple veins, and silver undersides. *Stem* Upright, becoming spreading and branching with age, forming dome-shaped shrub. Light green, aging to dark green, sometimes tinged purple. Slow to medium growth rate, depending on soil. *Fruit* Female forms produce bunches of glossy red or white fruits, depending on variety, when male planted nearby, in autumn.
Hardiness Tolerates 4°F (−15°C).
Soil Requirements Acid to neutral soil. Dislikes any alkalinity, very quickly showing

chlorosis; also dislikes waterlogging or extremely dry conditions.
Sun/Shade aspect Full sun to medium shade. Tolerates deep shade but may become more open, lax and bearing fewer flowers.
Pruning None required. Branches may be cut back and will rejuvenate from base.
Propagation and nursery production From semi-ripe cuttings taken midsummer, or from rooted suckers. Can also be grown from seed, but sex unpredictable. Purchase container-grown. Availability varies. Best planting heights 1-2ft (30-60cm).
Problems Often planted in non-acid soils where quickly deteriorates.
Varieties of interest *S. japonica* (Japanese Skimmia) Often raised from seed and therefore variable as to sex and habit with either obovate or elliptic leaves. *S. j. 'Foremanii'* syn. *S. j. 'Fisheri', S. j. 'Veitchii'* Female form, strong-growing, broad, ovate, light to mid green foliage with slightly purple veining. White clusters of flowers give way to bunches of brilliant red fruits in autumn. *S. j. 'Fragrans'* Male form producing panicles of scented white flowers resembling those of lily of the valley and used as pollinator for female forms. *S. j. 'Fructu-albo'* Female form producing white fruits. Foliage much paler green and less luxuriant than most. Often poor-growing. *S. j. 'Nymans'* Female, with longer, broader leaves than most and good red fruits. *S. j. 'Rogersii'* Female, with good red fruits. Dwarfer and slower-growing than most, becoming very dense and compact. *S. j. 'Rubella'* A male form with dark purple-green, ovate to lanceolate leaves. Flowers very deep red in bud, in large panicles, opening to white

flowers with yellow anthers, early spring. Very good when used as pollinator for female varieties. *S. laureola* Lanceolate to obovate leaves, dark green in colour. Leaves when crushed give off pungent scent. Green-yellow fragrant flowers give way to bright red fruits on female forms. Very low-growing, dense shrub. From the Himalayas.
Average height and spread
Five years
1¼x1¼ft (40x40cm)
Ten years
2x2ft (60x60cm)
Twenty years
or at maturity
3x3ft (1x1m)

SOPHORA TETRAPTERA

NEW ZEALAND KOWHAI
Leguminosae
Evergreen
An extremely interesting large shrub or small tree for foliage and flowers, provided that adequate protection or a very mild location is available.

Origin From New Zealand.
Use In areas where no winter frost is experienced, can be grown as a freestanding shrub or even a small tree. In less favourable areas, may require shelter of sunny or moderately sunny wall. Good as conservatory plants.
Description *Flower* Small, yellow pea-flowers, 1½-2in (4-5cm) long, in hanging, drooping clusters, late spring. *Foliage* Leaves small, ovate to oblong to elliptic, 1½-4½in (4-11cm) long, in formations of 20-40, light

Sophora tetraptera in leaf

Skimmia japonica 'Foremanii' in fruit

198

green to green-brown. *Stem* Upright, becoming attractively branched, slightly drooping. Light green-brown to brown. Slow to medium growth rate. *Fruit* Unusual seed pods of bearded appearance, each having 4 broad wings, late summer to early autumn.
Hardiness Requires shelter of wall in areas where temperature falls below 32°F (−0°C).
Soil Requirements Light, open, well-drained soil.
Sun/Shade aspect Full sun.
Pruning None required.
Propagation and nursery production From seed or semi-ripe cuttings taken in early summer. Purchase container-grown. May need to be obtained from specialist nurseries. Best planting heights 1¼-2½ft (40-80cm).
Problems Often seen in mild areas and purchased for transplanting to colder regions where it rarely succeeds.
Varieties of interest *S. microphylla* A very similar form to *S. tetraptera*, but with smaller leaflets and flowers. Juvenile foliage may be dense and wiry.
Average height and spread
Five years
3x3ft (1x1m)
Ten years
6x6ft (2x2m)
Twenty years
or at maturity
10x10ft (3x3m)

SORBARIA

TREE SPIRAEA
Rosaceae
Deciduous
Elegant, statuesque, large summer-flowering shrubs for the larger garden.

Origin From Afghanistan through to central and western China and Japan.
Use As individual shrubs, for mass planting, or for back of medium to large shrub borders.
Description *Flower* Large terminal panicles of creamy white flowers up to 8in (20cm) or more long, produced midsummer through early autumn. *Foliage* Leaves pinnate, 8-15in (20-40cm) long, with 13-19 ovate leaflets 3-4in (8-10cm) long. Light green with grey-silver undersides, giving yellow autumn colour. *Stem* Tall, unbranching, thicket-forming, giving good winter attraction. Light green, becoming red-brown. Fast growth rate when young or pruned, slowing with age. *Fruit* Brown flowerheads often maintained into early winter.
Hardiness Tolerant of winter temperatures down to −13°F (−25°C).
Soil Requirements Prefers rich, deep soil, moist and well-drained, but tolerates wide range of conditions.

Sorbaria arborea **in flower**

Sun/Shade aspect Full sun to light shade. Deeper shade spoils overall shape.
Pruning Remove one-third of oldest stems each spring to ground level to keep the shrubs rejuvenated.
Propagation and nursery production From semi-ripe cuttings or from division of rooted base. Purchase container-grown. Obtain from specialist nurseries. Best planting heights 2-3ft (60cm-1m).
Problems In very favourable areas may become invasive.
Varieties of interest *S. aitchisonii* Glabrous leaves having 11-20 sharply toothed, tapered leaflets. Large conical terminal panicles of creamy white flowers, late summer. From Afghanistan and Kashmir. *S. arborea* Leaves with 13-16 slender pointed leaflets with downy undersides. Conical terminal panicles of white flowers at end of current season's growth, mid to late summer. Elegant, large strong-growing specimen shrub with upright branches. *S. sorbifolia* Leaves composed of 13-25 tooth-edged glabrous leaflets. Smaller, narrow flowers in erect panicles, mid to late summer. Reaches two-thirds average height and spread. Forms a suckering, sometimes invasive thicket. From northern Asia.
Average height and spread
Five years
6x6ft (2x2m)
Ten years
10x13ft (3x4m)
Twenty years
or at maturity
13x20ft (4x6m)

SORBUS REDUCTA

CREEPING MOUNTAIN ASH
Rosaceae
Deciduous
An unusual, small suckering shrub with flowers, berries and fine autumn colours.

Origin From northern Burma and western China.
Use As a low thicket-forming shrub ideal for large rock gardens or open areas. Should stand alone for full effect.
Description *Flower* Clusters of white flowers 3in (8cm) across, late spring to early summer. *Foliage* Leaves 3-4in (8-10cm) long, on red petioles, composed of 9-15 sharply serrated, dark, shiny green leaflets, turning bronze to orange-red in autumn. *Stem* Short, upright, suckering, forming a thicket. Green to green-brown. Slow growth rate, but sending out faster shoots underground to emerge some distance away. *Fruit* Small, globular, white flushed rose pink fruits in autumn.
Hardiness Tolerates −13°F (−25°C).
Soil Requirements Any soil.
Sun/Shade aspect Full sun or light shade.

Pruning None required.
Propagation and nursery production From rooted suckers or seed. Purchase container-grown. Rather scarce, obtain from specialist nurseries. Best planting heights 3-15in (7-40cm).
Problems May suffer from leaf mildew in wet summers.

Sorbus reducta **in autumn**

Average height and spread
Five years
8inx1¼ft (20x40cm)
Ten years
1x3ft (30cmx1m)
Twenty years
or at maturity
2½x6ft (80cmx2m)

SPARTIUM JUNCEUM

SPANISH BROOM
Leguminosae
Deciduous
An attractive midsummer yellow-flowering shrub.

Origin From Mediterranean area and south-eastern Europe.
Use As a freestanding shrub or for medium to large shrub borders. May be trained single-stemmed as small tree. Very good in coastal areas
Description *Flower* Small, open racemes of yellow pea-flowers in mid to late summer, lasting over long period. *Foliage* Leaves very sparse, narrow, lanceolate, round-ended, ½-¾in (1-2cm) long, and of little significance. *Stem* Long, reed-like stems, forming an upright shrub, becoming more spreading with age. Dark green. Fast growth rate when young, slowing with age. *Fruit* Small, silver-green to grey-green pea-pods following flowers.
Hardiness Tolerates 4°F (−15°C).
Soil Requirements Most soils, particularly good in alkaline areas. Dislikes waterlogging.
Sun/Shade aspect Full sun through to light shade.
Pruning Resents pruning. Any reduction in size may be attained only by pruning wood one year old or less.
Propagation and nursery production From semi-ripe cuttings taken in early summer, or from seed. Purchase container-grown. Normally stocked by nurseries. Best planting heights 2-3ft (60cm-1m).
Problems Can become very drawn and leggy in sheltered gardens.

199

Spartium junceum in flower

Average height and spread
Five years
6x3ft (2x1m)
Ten years
10x6ft (3x2m)
Twenty years
or at maturity
13x10ft (4x3m)

SPIRAEA
Low-growing varieties

KNOWN BY BOTANICAL NAME

Rosaceae
Deciduous
Attractive, small, low shrubs for late spring and early summer flowers.

Origin From the Himalayas, through China and Japan.
Use For front of large shrub borders, for mass planting, or for low hedge if planted 1¼ft (40cm) apart.
Description *Flower* White, pink or pink-red clusters of flowers, depending on variety, in midsummer. *Foliage* Leaves ovate, 1-3in (3-8cm) long, often tooth-edged, mainly light to dark green; some golden-leaved forms. Some good autumn colour in all varieties. *Stem* Short, very branching and twiggy, forming round-topped mound. Fast growth rate when young, slowing with age. *Fruit* Insignificant, but brown seedheads of winter interest.
Hardiness Tolerant of winter temperatures down to −13°F (−25°C).

Spiraea × bumalda 'Gold Flame'

Soil Requirements Most soils, disliking only extremely alkaline or dry types.
Sun/Shade aspect Prefers full sun, tolerates light to medium shade.
Pruning According to variety. Most of those listed in this section merely need thinning by removing one-third of oldest flowering wood in early spring, to encourage rejuvenation. Some other varieties should be cut to ground level, as recommended below.
Propagation and nursery production From semi-ripe cuttings taken early to mid summer.

Purchase container-grown. Most varieties easy to find, especially when in flower. Best planting heights 1¼-2ft (40-60cm).
Problems No real problems if pruned correctly; otherwise can become woody.
Varieties of interest *S. albiflora* syn. *S. japonica alba* Light green lanceolate foliage with coarsely-toothed edges and glaucous green undersides. Clusters of fluffy-textured white flowers borne at ends of branches, midsummer. Thin wood by one-third in spring. From Japan. *S. × bumalda* '*Anthony Waterer*' New foliage occasionally appears pink and cream variegated, but not as a general rule. Leaves are ovate, tooth-edged, dark green with a reddish hue along the veins. Bright clusters of dark pink-red flowers are produced in early to mid summer. Should be cut to ground level early each spring to encourage good foliage and maximum flowering performance; if just thinned it will achieve more height and spread but have smaller flowers. *S. × b.* '*Gold Flame*' New foliage in spring orange-apricot, becoming orange-red and finally gold. May be scorched in strong sunlight. Dark pink-red flowers in early to mid summer. Should be reduced to ground level in early spring to encourage rejuvenation of attractive growth and large flowers. *S. japonica* '*Bullata*' syn. *S. crispifolia, S. bullata* Very small, broad, ovate leaves, deeply veined, dark grey-green with lighter undersides. Flowers small, dark crimson, produced in flat clusters in midsummer and presented well above the dark foliage. No pruning required. Reaches two-thirds average height and spread. *S. j. var. fortunei* syn. *S. j. wulfenii* Foliage incised and tooth-edged purple-green-red with glabrous undersides. Very attractive. Rich dark pink flowers in midsummer. Branches may be thinned or pruned back hard. From Europe and central China. *S. j.* '*Golden Princess*' Small mounds of golden yellow foliage, susceptible to scorching by strong summer sun, early to mid summer. Thin wood lightly in spring. *S. j.* '*Little Princess*' A dwarf, low, spreading carpet of light green, tooth-edged foliage, surmounted by rose-crimson flowers in midsummer. Thin wood lightly in spring. *S.* '*Shirobana*' Interesting light green to mid green, tooth-edged leaves. Flowers, either pink, all white, or half pink, half white, make an interesting flowering combination in early summer. May be thinned or cut back hard in spring. Of garden origin.

Spiraea 'Shirobana' in flower

Spiraea × bumalda 'Anthony Waterer' in flower

Average height and spread
Five years
1¼x1½ft (40x50cm)
Ten years
2x2¼ft (60x70cm)
Twenty years
or at maturity
2x2¼ft (60x70cm)

SPIRAEA
Tall-growing varieties

KNOWN BY BOTANICAL NAME
Rosaceae
Deciduous
Extremely fine shrubs for spring and summer-flowering, worthy of a place in any medium-sized to large garden.

Spiraea douglasii in flower

Origin Mostly from the Himalayas through China and Asia. Varieties of garden origin.
Use As flowering shrubs for individual planting, for shrub borders, or for mass planting. Some varieties good for thicket formation. Can be used as informal hedge when planted 2ft (60cm) apart in single line.
Description *Flower* Either clusters of small, white flowers, or erect panicles of purple rose pink, in spring, early summer or midsummer, depending on variety. *Foliage* Leaves small to medium, ovate, 1-3in (3-8cm) long; some varieties with toothed edges. Light green, all giving some degree of autumn colour. *Stem* Normally upright, becoming very branching with age. Light grey-green, green or light brown, depending on variety. Slow growth rate if unpruned, fast when young and after pruning. *Fruit* May produce grey to grey-brown seedheads.
Hardiness Tolerant of winter temperatures down to −13°F (−25°C).
Soil Requirements Most soils, although may show signs of chlorosis on extremely alkaline types.
Sun/Shade aspect Full sun to medium shade; the more sun, the better the overall appearance.
Pruning Spring-flowering varieties should be pruned relatively hard following flowering with up to one-third of oldest growth removed to ground level. Some summer-flowering forms require cutting completely to ground level in early spring to encourage rejuvenation. Other forms require no pruning whatsoever. Specific pruning recommendations are given for each of the forms listed below.
Propagation and nursery production From semi-ripe cuttings taken in early summer, some varieties from hardwood cuttings in winter. Purchase container-grown. All varieties fairly easy to find. Best planting heights 1½-2½ft (40-80cm).
Problems Some stronger thicket-producing forms can quickly outgrow desired area and even become invasive.
Varieties of interest *S. arcuata* Angular shaped stems, upright when young, becoming arching with age. Small, light to mid green ovate leaves. Small thick clusters of white flowers, covering shrub mid to late spring. Remove one-third of old flowering wood after flowering. From the Himalayas. *S. × arguta* (Bridal Wreath, Foam of May, Garland Wreath) Long, upright, wispy branches with numerous twiggy formations. Small, narrow, grey-green, oval leaves with slightly toothed edges and blue, glabrous undersides. Masses of small clusters of white flowers, entirely covering branches in mid to late spring. Some good yellow autumn colour. Remove one-third of old flowering wood after flowering. *S. × billiardii 'Triumphans'* syn. *S. menziesii 'Triumphana'* (Billiard Spiraea) Upright large panicles of purple-rose flowers borne on dense, thicket-forming shrub in midsummer. Responds well either to no pruning, or to very hard pruning, which will improve flower size. Dislikes extremely alkaline conditions. *S. douglasii* A suckering, upright thicket-forming shrub with reddish shoots. Flowers purple-rose in terminal panicles, early to mid summer. Foliage oblong and coarsely-toothed with grey felted undersides. Some good yellow autumn colour. Dislikes extremely alkaline soils. Can either be left unpruned to form large thicket, or cut to ground level in very early spring and allowed to rejuvenate, causing later flowering. From north-western USA. *S. menziesii* A suckering, thicket-forming shrub with small pyramid-shaped terminal panicles of purple-rose flowers in mid to late summer. Can either be reduced to ground level in early spring, for best flowering performance, or left unpruned to form wide, dense thicket. *S. nipponica var. tosaensis* syn. *S. 'Snowmound'* (Snowmound Nippon Spiraea) Arching branches form a mound-shaped shrub covered in white flowers in early summer. Remove one-third of oldest flowering wood in spring. *S. prunifolia* (Bridal Wreath Spiraea) Long arching branches with ovate, finely tooth-edged leaves, light to mid green, giving brilliant orange-red autumn colour. Flowers white, double, button-like, in stalkless clusters along branches, mid to late spring. Remove one-third of old flowering wood after flowering. *S. thunbergii* (Thunberg Spiraea) Long, graceful, arching, twiggy branches forming spreading shrub. Foliage narrow, ovate, sharply toothed, 1¼-1½in (3.5-4cm) long, grey-green to blue-green. Good yellow autumn colour. White flowers in clusters along entire length of branches, late to mid spring. One of the earliest flowering forms. Remove one-third of old flowering wood after flowering. *S. × vanhouttei* (Vanhoutte Spiraea) A good plum-coloured autumn foliage shrub. Leaves rhomboidal and coarsely toothed, single or sometimes 3-5 lobed. Dense umbels of white flowers in early summer. Remove one-third of old flowering wood after flowering. Of garden origin.
Average height and spread
Five years
3x3ft (1x1m)
Ten years
5x5ft (1.5x1.5m)
Twenty years
or at maturity
6x6ft (2x2m)

STACHYURUS PRAECOX

KNOWN BY BOTANICAL NAME
Stachyuraceae
Deciduous
A winter-flowering shrub which deserves to be more widely planted.

Origin From Japan.
Use As a freestanding, very freely flowering winter shrub, or for medium to large shrub borders. Ideal for feature planting, for example, a pair framing a central path or vista.
Description *Flower* Numerous racemes, 1½-2½in (4-6cm) long, rigid and hanging, borne along length of branches consisting of over 20 pale yellow, cup-shaped flowers, from early to mid winter through to early spring. Exact flowering time is dependent on winter temperatures. Flowers produced on wood 2 years old or more. Frost resistant. *Foliage* Leaves large, elliptic to broad elliptic, pointed, 3-6in (8-15cm) long and 2-3in (5-8cm) wide. Dark green tinged purple, with purple veining. Some yellow autumn colour. *Stem* Strong, upright, arching when young, becoming more branching in subsequent years. Purple-green. Forming a large, round, dome-shaped shrub. Medium to fast growth rate. *Fruit* Insignificant.
Hardiness Tolerates 4°F (−15°C).
Soil Requirements Most soils, accepting wide range.

Spiraea × arguta in flower

201

Sun/Shade aspect Full sun to mid shade.
Pruning On established shrubs 5 years old or more, remove one-third of old flowering wood to ground level following flowering, to encourage rejuvenation.
Propagation and nursery production From layers or semi-ripe cuttings taken early to mid summer. Purchase container-grown. Fairly easily found in nurseries. Best planting heights 2-4ft (60-1.2m).
Problems Young plants often appear one-sided, with one or more strong shoots, but grow quickly once planted.

Stachyurus praecox in flower

Varieties of interest *S. chinensis* Foliage slightly narrower and possibly duller green than that of *S. praecox*. Racemes shorter, but comprising more numerous small florets. Flowering dependent on weather, but can be slightly earlier. *S. c. 'Magpie'* White to creamy white margin variegation on leaves. Often splashed with pale green mottling and tinged rose pink. Yellow flowers. Extremely scarce.
Average height and spread
Five years
6x6ft (2x2m)
Ten years
10x10ft (3x3m)
Twenty years or at maturity
13x13ft (4x4m)

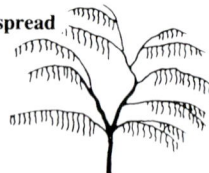

STAPHYLEA COLCHICA

BLADDERNUT, COLCHIS BLADDERNUT
Staphyleaceae
Deciduous
One of the aristocrats of flowering shrubs, especially when seen against a bright blue sky. Unfortunately, not easy to find.

Origin From southern Caucasus.
Use As a single feature shrub or for a woodland garden for its flowers and fruit.
Description *Flower* White, fragrant, upright or hanging panicles, over 5in (12cm) long, in mid spring. *Foliage* Leaves ovate to oblong, composed of 3-5 leaflets up to 4in (10cm) long. Dull grey-green upper surfaces, glossy undersides. Some yellow autumn colour. *Stem* Upright, with limited branching. Light green to grey-green, becoming grey-brown. Forming an upright, tall shrub, even small tree after 15-20 years. Fast growth rate when young, slowing with age. *Fruit* Conspicuous, translucent, bladder-shaped capsules, 3-4in (8-10cm) long, in late summer, early autumn.
Hardiness Tolerates 14°F (−10°C), but may suffer some damage to stems in severe winters, even in sheltered locations.
Soil Requirements Best on neutral to acid fertile soil, but tolerates moderate alkalinity.
Sun/Shade aspect Full sun or light shade.

Staphylea colchica in fruit

Pruning None required except to remove any damaged winter stems, cutting well back below damaged area.
Propagation and nursery production From seed or layers. Purchase container-grown. Rather hard to find. Best planting heights 2-2½ft (60-80cm).
Problems Susceptibility to severe winters can be a problem if the shrub is to establish and show its full potential. Must be provided with the right type of soil.
Varieties of interest *S. holocarpa* Flowers pink in bud, opening to white in short hanging panicles, mid to late spring. Foliage trifoliate, oblong to lanceolate leaflets, light· grey-green. From central China. Difficult to find. *S. h. 'Rosea'* Flowers pink to soft pink, hanging in clusters, mid to late spring. New foliage trifoliate, bronze. Difficult to find, but not impossible. *S. pinnata* (Antoney Nut) White flowers in thin, hanging panicles, late spring to early summer. Foliage pinnate, grey-green, 5-7 leaflets. Fairly easy to find. From central Europe.
Average height and spread
Five years
5x3ft (1.5x1m)
Ten years
8x5ft (2.5x1.5m)
Twenty years or at maturity
16x8ft (5x2.5m)

STEPHANANDRA INCISA

CUTLEAF STEPHANANDRA
Rosaceae
Deciduous
Versatile low shrubs which deserve to be more widely planted.

Origin From Japan and Korea.
Use As a low front shrub for any size of shrub border. Ideal for ground cover and bank stabilization when mass planted. Makes useful, informal, dividing hedge when planted in a single line 1¼ft (40cm) apart. Can be grown in tubs of not less than 2ft (60cm) in diameter and 1¼ft (40cm) in depth; use good quality potting medium.
Description *Flower* Small panicles of green-white flowers profusely borne on wood 2 years old or more, in early to mid summer. *Foliage* Leaves ovate, with cut, toothed edges, lobed, 1-2¾in (3-7cm) long, giving good pale orange and yellow autumn colours. *Stem* Upright when young, bright brown, quickly becoming arching; very twiggy with age, forming wide, low, dome-shaped shrub of winter colour attraction. Medium growth rate, slowing with age. *Fruit* Insignificant.
Hardiness Tolerates winter temperatures down to −13°F (−25°C).

Stephanandra incisa in flower

Soil Requirements Most soil types, except very dry.

Pruning Remove one-third of old flowering wood after flowering to encourage new shoots. Untidy old shrubs may be cut back very hard to ground level and will rejuvenate.

Propagation and nursery production From semi-ripe cuttings taken in early summer or from hardwood cuttings in winter. Purchase container-grown. Fairly easy to find, especially in nurseries. Best planting heights 8-15in (20-40cm).

Problems None.

Varieties of interest *S. i. 'Crispa'* A variety with leaves more crinkled and cut, forming a very low mound. Excellent for ground cover, especially on difficult soils. Reaching two-thirds average height and spread.

Average height and spread
Five years
1½x2ft (50x60cm)
Ten years
2½x2½ft (80x80cm)
Twenty years
or at maturity
3x5ft (1x1.5m)

STEPHANANDRA TANAKAE

TANAKA STEPHANANDRA
Rosaceae
Deciduous
A shrub offering outstanding winter stem display, worthy of any garden.

Origin From Japan.
Use As a freestanding single shrub, or grouped for winter stem effect.

Stephanandra tanakae in winter

Description *Flower* Racemes of green-white flowers, late spring to early summer. *Foliage* Leaves triangular or ovate up to 5in (12cm) long, 3-5 lobed and deeply toothed. Some yellow autumn colour. *Stem* Strong, upright branches arching at ends. Deep mahogany-brown making attractive winter display. Fast growth rate when pruned correctly. *Fruit* Insignificant.

Hardiness Tolerates winter temperatures down to −13°F (−25°C).

Soil Requirements Needs rich fertile soil for new stem production, but does well on most.

Sun/Shade aspect Full sun to medium shade; plant where winter sunlight catches the stems, for best effect.

Pruning Remove one-third to half of oldest wood each spring to ground level to induce strong new growths for stem colour display in following winter.

Propagation and nursery production From semi-ripe cuttings taken in early summer or hardwood cuttings in mid to late winter. Purchase container-grown. May need to be sought from specialist nurseries. Best plant-

ing heights 2-2½ft (60-80cm).
Problems Of little interest in summer. However, provided pruned as recommended, comes into its own in winter.

Average height and spread
Five years
5x5ft (1.5x1.5m)
Ten years
6x6ft (2x2m)
Twenty years
or at maturity
10x10ft (3x3m)

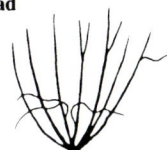

STRANVAESIA DAVIDIANA

CHINESE STRANVAESIA
Rosaceae
Semi-evergreen
An interesting, semi-evergreen, fruiting shrub.

Origin From western China.
Use As a background shrub for a medium to large shrub border, for individual planting as specimen in open area or woodland garden, or for mass planting. Makes large, informal hedge or screen if planted 3ft (1m) apart in single line.

Description *Flower* Clusters of white to off-white flowers borne on wood 2 years old or more, in early summer. *Foliage* Leaves lanceolate to broad lanceolate, 2-4in (5-10cm) long, leathery-textured, mid green with duller, lighter undersides. Some bright red autumn colour, contrasting with remaining semi-evergreen green leaves; extent of evergreen depending upon severity of winter. Better foliage on young shoots, decreasing in interest and colour intensity on older wood. *Stem* Upright, vigorous stems, dark green tinged red. Becoming spreading and branching with age. Fast growth rate when young, slowing with age. *Fruit* Dull red-orange fruits, resistant to birds, produced in clusters, late summer through early autumn.

Hardiness Tolerates 4°F (−15°C), but some stem damage may be caused in very severe winters.

Soil Requirements Does well on most, but performs less well on very poor soils.

Sun/Shade aspect Full sun to medium shade.

Pruning None required for its own sake but from time to time reduce young shoots not more than 4 years old to ground level in early spring, to induce new, upright, very leafy shoots, improve overall appearance and still maintain older fruiting branches.

Propagation and nursery production From semi-ripe cuttings taken in early summer or by layering. Purchase container-grown. Some nurseries offer plants from seed, but resulting shrubs can be extremely variable. Availability

varies. Best planting heights 2-3ft (60cm-1m).

Problems This shrub and its forms are susceptible to the fungus disease fire blight. Its future is uncertain, since production may be discontinued on this account. Affected shrubs should be destroyed immediately by burning. Also liable to stem canker fungus especially on older stems, with older branches being strangled by the collar-like effect of the canker.

Varieties of interest *S. d. 'Fructuluteo'* Yellow fruits in autumn, but somewhat sparser than the parent. Two-thirds average height and spread. *S. d. 'Prostrata'* Prostrate habit, good foliage and moderately good fruit production. One-third average height and equal spread. *S. d. var. undulata* Slightly later flowering than the form, but otherwise very similar. Two-thirds average height and equal spread.

Average height and spread
Five years
5x5ft (1.5x1.5m)
Ten years
8x8ft (2.5x2.5m)
Twenty years
or at maturity
16x16ft (5x5m)

SYMPHORICARPOS

SNOWBERRY, CORAL BERRY, INDIAN CURRANT
Caprifoliaceae
Deciduous
Useful shrubs with various applications, though suckering forms need careful control.

Origin From eastern USA or of garden origin.
Use Suckering forms ideal for ground cover, bank retention and large-scale planting. Also good planted singly in medium to large shrub borders. Some forms useful for informal hedging when planted 2ft (60cm) apart in a single line.

Description *Flower* Small, inconspicuous, pink, bell-shaped flowers, late spring, early summer. *Foliage* Leaves ovate to round, ½-2in (1-5cm) long, dull green tinged grey. Some silver and gold variegated forms. Moderate autumn colour. *Stem* Upright, very twiggy and branching. Light brown to green-brown, becoming darker brown with age. Some forms thicket-forming with suckering shoots, sometimes invasive. Other forms simply shrub-forming. Fast growth rate when young, slowing with age. *Fruit* Round, marble-like fruits, white through shades of pink; depending on variety.

Hardiness Tolerant of winter temperatures down to −13°F (−25°C).

Soil Requirements Any soil.

Sun/Shade aspect Full sun through to deep

Stranvaesia davidiana in fruit

Symphoricarpos × doorenbosii 'Mother of Pearl' in flower

shade. Variegated foliage forms must have full sun.

Pruning Remove one-third of oldest wood on established shrubs in early spring, to encourage new shoots. Very old shrubs can be pruned back hard in spring and will rejuvenate.

Propagation and nursery production From semi-ripe cuttings taken in early summer. Some forms raised from seed. Purchase container-grown. Most forms easy to find. Best planting heights 1¼-2½ft (40-80cm).

Problems In confined areas suckering forms can be a nuisance as they are invasive.

Varieties of interest *S. albus* (Snowberry) Upright, slender shoots from suckering base. Foliage oval to ovate, light green with downy undersides. Marble-like, round, pure white fruits up to ¼in (5mm) across produced freely in late summer, early autumn and retained well into winter. From eastern North America. *S. × chenaultii 'Hancock'* Bright green, small, ovate foliage. Interesting small pink bell-shaped flowers, replaced in autumn by small purple-pink to purple-red fruits, with pink-white unexposed sides. Extremely good ground cover. Reaches only a quarter of average height and up to two-thirds average spread. *S. × doorenbosii* Upright branches, becoming twiggy with age. Slightly larger flowers, fruits tinged pink. Raised in Holland. *S. × d. 'Erect'* A more upright-growing form with rose-lilac fruits. Useful for hedging. *S. × d. 'Magic Berry'* Good production of rose pink fruits. Reaches two-thirds average height and equal spread. *S. × d. 'Mother of Pearl'* Good production of white, flushed pink fruits. Reaches half average height and two-thirds average spread. *S. × d. 'White Hedge'* Upright-growing, closely knit branches, covered in white clusters of fruits in autumn. Good for hedging. *S. orbiculatus* syn. *S. vulgaris* (Indian Currant, Coral Berry) Small, round, ovate foliage very freely produced with light green upper surfaces and blue-green undersides. Purple-rose, small, round to oval fruits produced in clusters, early autumn. From eastern USA. *S. o. 'Albovariegatus'* An interesting variety with white-margined, variegated leaves. Limited fruiting. May revert to green form and any reverted shoots should be removed once seen. Reaches two-thirds average height and equal spread. *S. o. 'Variegatus'* Small round to ovate foliage, light grey-green with yellow margins, making useful golden variegated shrub. Rarely fruits. May revert back to green form and any reverted shoots should be removed as soon as seen. Reaches two-thirds average height and spread. *S. rivularis* (Snowberry) A variety with glabrous blue stems. Elliptic, light green leaves up to 2½in (6cm) long and normally lobed. A suckering shrub with white, marble-like fruits freely produced in autumn. May become invasive. From western North America.

Average height and spread
Five years
5x3ft (1.5x1m)
Ten years
8x6ft (2.5x2m)
*Twenty years
or at maturity*
8x13ft (2.5x4m)

Symphoricarpos orbiculatus 'Variegatus' in leaf

SYMPLOCOS PANICULATA

SAPPHINEBERRY, ASIATIC SWEETLEAF
Symplocaceae
Deciduous
An interesting fruiting shrub for wild or woodland gardens.

Origin From the Himalayas, China, Taiwan and Japan.
Use As a large feature shrub to show off its fruit, or in a group planting.
Description *Flower* Fragrant, small white flowers in panicles, late spring to early summer. For best pollination and fertilization of flowers, plant two or more shrubs together. *Foliage* Leaves mainly ovate, though somewhat variable in shape, ¾-1¾in (2-4.5cm) long, light green, with some yellow autumn colour. *Stem* Very dense and branching, grey-green. After many years may resemble small tree. Fast growth rate when young, slowing with age. *Fruit* Ultramarine blue fruits, perhaps its main attraction, produced in autumn and carried into winter. Best fruiting follows hot, dry summer.

Symplocos paniculata in fruit

Hardiness Tolerates 4°F (−15°C).
Soil Requirements Acid to neutral, rich deep fertile soil produces best results; tolerates slight alkalinity.
Sun/Shade aspect Prefers light dappled shade, tolerates full sun.
Pruning None required.
Propagation and nursery production From seed or layers. Purchase container-grown, from specialist nurseries. Best planting heights 1¼-2ft (40-60cm).
Problems Young plants take a number of years to fruit fully.
Average height and spread
Five years
5x3ft (1.5x1m)
Ten years
10x6ft (3x2m)
*Twenty years
or at maturity*
16x13ft (5x4m)

SYRINGA AMURENSIS

AMUR LILAC
Oleaceae
Deciduous
An interesting Lilac suitable for planting in larger gardens.

Origin From Manchuria to Korea.
Use As a freestanding specimen, either on its own or in groups.
Description *Flower* Large open panicles of creamy white flowers, very late spring or early summer. *Foliage* Leaves pointed, ovate,

Syringa amurensis in flower

2-5in (5-12cm) long and 1½-2½in (4-6cm) wide, light green, giving some yellow autumn colour. *Stem* Upright. On mature branches, older bark peels showing chestnut brown young bark with clearly defined horizontal pores. Forms attractive, upright, elegant, graceful shrub. Fast growth rate. *Fruit* Brown-grey seedheads in autumn, maintained into winter.
Hardiness Tolerates 14°F (−10°C), although late spring frosts may damage new growth.
Soil Requirements Most soils, but distressed by extremely alkaline types.
Sun/Shade aspect Full sun to light shade.
Pruning Mature shrubs benefit from occasional thinning of older branches. Remove seedheads to encourage increased formation of flower buds for following year.
Propagation and nursery production From layers or grafting. Purchase container-grown. Rather hard to find and may need a search. Best planting heights 2-2½ft (60-80cm).
Problems Spring frost damage to new foliage can be severe.
Average height and spread
Five years
4x3ft (1.2x1m)
Ten years
9x6ft (2.8x2m)
Twenty years
or at maturity
13x10ft (4x3m)

SYRINGA × CHINENSIS

CHINESE LILAC, ROUEN LILAC
Oleaceae
Deciduous
A charming Lilac but not widely available.

Origin Raised in the Botanic Gardens at Rouen in northern France in the late 1700s.
Use For individual planting in wild or woodland gardens.
Description *Flower* Slightly drooping panicles of very fragrant flowers, soft lavender-blue or lilac-red, depending on variety, late spring. *Foliage* Leaves large, 2-3in (5-8cm) long, formed of pinnate to ovate leaflets, light to mid green, giving some autumn colour. *Stem* Graceful, wispy branches, forming a thick, bushy, round shrub. Light grey-green. Medium growth rate. *Fruit* Grey-brown seedheads may be retained into winter.
Hardiness Tolerates 4°F (−15°C).
Soil Requirements Most soils, but distressed by extremely alkaline types.

Sun/Shade aspect Full sun through to light shade.
Pruning On established plants more than 4 years old, remove a quarter of oldest wood after flowering to induce new growth and larger flowering display. Old flowerheads should also be removed.
Propagation and nursery production From layers, semi-ripe cuttings taken in early summer, or from grafting. Purchase container-grown or root-balled (balled-and-burlapped). Difficult to find and must be sought from specialist nurseries. Best planting heights 1¼-2ft (40-60cm).
Problems Can be lax in habit and grow larger than expected.
Varieties of interest *S. × c. 'Saugeana'* syn. *S. × c. 'Rubra'* A variety of rounded bushy habit with lilac-red, scented flowers. Not easy to find.

Syringa × chinensis in flower

Average height and spread
Five years
3x3ft (1x1m)
Ten years
6x6ft (2x2m)
Twenty years
or at maturity
10x10ft (3x3m)

SYRINGA × HYACINTHIFLORA

LILAC
Oleaceae
Deciduous
A range of Syringa varieties very attractive in their loosely flowing flower display.

Origin Originally raised in France in the late 1800s. Modern varieties of North American origin.
Use As a freestanding spring-flowering shrub or for background planting in medium to large shrub borders. If planted 3ft (1m) apart in single line makes informal hedge or screen.
Description *Flower* Panicles of typical lilac tubular flowers produced in slightly open formation in colours ranging from cyclamen-purple, through blue to pink, depending on variety, mid spring. Not very scented. *Foliage* Leaves large, ovate, pointed, 6-8in (15-20cm) long and 3-4in (8-10cm) wide, mid green. Yellow autumn colour. *Stem* Upright, becoming branching; light grey-green. Fast growth rate when young, slowing with age. *Fruit* Old seedheads may be retained attractively into winter, but the shrub benefits if these are removed as soon as possible.

Syringa × hyacinthiflora 'Blue Hyacinth'

Hardiness Tolerates winter temperatures down to −13°F (−25°C).
Soil Requirements Most soils, but dislikes extremely alkaline types.
Sun/Shade aspect Full sun to light shade. In deeper shade will be more open and less likely to flower.
Pruning None required. Cutting back hard prevents flowering for 2-3 years. Remove old seedheads directly after flowering.
Propagation and nursery production From budding or grafting. Purchase container-grown, bare-rooted or root-balled (balled-and-burlapped). Best planting heights 2-3ft (60cm 1m).
Problems Often a delay of up to 3 years after planting before shrub comes into full flower.
Varieties of interest *S. × h. 'Alice Eastwood'* Claret-purple buds open to display double cyclamen-purple flowers. *S. × h. 'Blue Hyacinth'* A single variety with mauve to pale blue flowers. *S. × h. 'Clarke's Giant'* Rosy mauve in bud, opening to large single florets in panicles up to 1ft (30cm) long. *S. × h. 'Esther Staley'* Red in bud, opening to single pink flowers very freely borne.
Average height and spread
Five years
5x3ft (1.5x1m)
Ten years
8x6ft (2.5x2m)
Twenty years
or at maturity
12x9ft (3.5x2.8m)

Syringa × *josiflexa 'Bellicent'* in flower

SYRINGA × JOSIFLEXA 'BELLICENT'

CANADIAN LILAC
Oleaceae
Deciduous
Given space, a true treasure of a flowering shrub for the larger garden.

Origin Raised in Ottawa, Canada in the early 1900s.
Use As a freestanding, specimen shrub, or as background planting for large shrub borders.
Description *Flower* Large, fragrant panicles, up to 10in (25cm) long and plume-like, of clear rose pink, open flowers; slightly arching in display. *Foliage* Leaves ovate, 4-6in (10-15cm) long and 2-3in (5-8cm) wide. Very dark green and attractive, giving some yellow autumn colour. *Stem* Upright when young, spreading with age and forming a wide-topped, narrow-based shrub. Stout, upright branches infrequently producing side shoots at low levels, branching more towards ends. Grey-green to grey-brown. Fast growth rate when young, slowing with age. *Fruit* Old seedheads, grey-brown, may be retained well into winter if not removed as advised.
Hardiness Tolerates winter temperatures down to −13°F (−25°C).
Soil Requirements Most soils but distressed by chlorosis on alkaline types.
Sun/Shade aspect Full sun to light shade.
Pruning Remove one-third of oldest shoots biennially to induce rejuvenation on established shrubs more than 5 years old. Seedheads should also be removed as soon as seen.
Propagation and nursery production By budding, or from semi-ripe cuttings taken in early summer. Purchase container-grown, bare-rooted or root-balled (balled-and-burlapped). Best planting heights 2-3ft (60cm-1m).
Problems Young plants can look stunted and lopsided, but grow rapidly once planted. Often planted in areas which restrict full growth potential.
Average height and spread
Five years
4x3ft (1.2x1m)
Ten years
10x6ft (3x2m)
Twenty years
or at maturity
16x13ft (5x4m)

SYRINGA JOSIKAEA

HUNGARIAN LILAC
Oleaceae
Deciduous
A large shrub requiring space, but of interest for the larger garden.

Origin From central and eastern Europe.
Use As a freestanding shrub, singly or in groups. Can be used at the back of large shrub borders.
Description *Flower* Upright panicles, 6-8in (15-20cm) long, of fragrant, deep violet-mauve flowers, early summer. *Foliage* Leaves long, broad, lanceolate, dark green, 1-2in (3-5cm) long, heavily ribbed with indented veins, silver-grey undersides. *Stem* Upright, rigidly branching, forming a round, branching shrub. Light grey-brown. Medium growth rate, slowing with age. *Fruit* Old seedheads, grey-brown, may be retained into winter but should not be allowed to remain.
Hardiness Tolerates winter temperatures down to −13°F (−25°C).
Soil Requirements Does well on all soil conditions, even tolerating severely alkaline conditions.

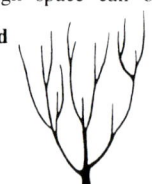

Sun/Shade aspect Prefers full sun, accepts medium shade.
Pruning Remove one-third of old flowering shoots after flowering on shrubs established more than 4-5 years, to induce new growth. Seedheads should also be removed to encourage next year's flowering.
Propagation and nursery production From semi-ripe cuttings taken in early summer. Purchase container-grown, bare-rooted or root-balled (balled-and-burlapped). May be difficult to find, and must be sought from specialist nurseries. Best planting heights 2-2½ft (60-80cm).
Problems None, if enough space can be allowed.
Average height and spread
Five years
3x3ft (1x1m)
Ten years
6x6ft (2x2m)
Twenty years
or at maturity
16x16ft (5x5m)

SYRINGA MEYERI VAR. PALIBIN (Syringa palibiniana, Syringa velutina)

KOREAN LILAC, MEYER LILAC
Oleaceae
Deciduous
A lovely feature for the rock garden or other small planting areas.

Origin From Korea.
Use As a small, neatly shaped shrub for small to medium rock gardens, tubs or large troughs. May be planted in groups.
Description *Flower* Small panicles of pale lilac to lilac-pink scented flowers produced even on very young shrubs, late spring to early summer. *Foliage* Leaves small, ovate to rhomboidal or sometimes rounded, 1¾-3½in (4.5-9cm) long, dark green with velvety texture. Some yellow autumn colour. *Stem* Very branching, forming a round, tight, twiggy shrub. Grey-green. Very slow growth rate, producing no more than 4-6in (10-15cm) growth each year. *Fruit* Insignificant.
Hardiness Tolerates 4°F (−15°C).
Soil Requirements Any soil.
Sun/Shade aspect Prefers light shade; accepts full sun to medium shade.
Pruning None required.
Propagation and nursery production From semi-ripe cuttings taken in early summer, from layers or by grafting. Purchase container-grown. Easy to find. Best planting heights 8-15in (20-40cm).

Syringa josikaea in flower

Syringa meyeri var. palibin in flower

Problems Takes a number of years to form a shrub of any appreciable size.

Average height and spread
Five years
1ftx10in (30x25cm)
Ten years
2¼x1½ft (70x50cm)
Twenty years
or at maturity
5x4ft (1.5x1.2m)

SYRINGA MICROPHYLLA 'SUPERBA'

DAPHNE LILAC, LITTLELEAF LILAC
Oleaceae
Deciduous
A useful flowering shrub, giving two displays of scented flowers, in spring and again in autumn.

Origin Originally from northern and western China, but the form 'Superba' is of garden origin.
Use As a freestanding small shrub, for medium to large shrub borders. May be found grown as small mop-headed standard on 5ft (1.2m) stem.
Description *Flower* Small panicles of fragrant, rosy pink flowers in mid to late spring, often with some autumn display. *Foliage* Leaves small, ovate, pointed, up to 2in (5cm) long. *Stem* Thin, wispy, light brown to grey-brown, forming round-topped shrub of slightly less width than height. Slow growth rate. *Fruit* Insignificant.
Hardiness Tolerates winter temperatures down to −13°F (−25°C).

Syringa microphylla 'Superba' in flower

Soil Requirements Does well on all but severely alkaline soils.
Sun/Shade aspect Full sun to light shade.
Pruning Remove one-third of old flowering wood in late summer after first flowering to induce new growth. Shrubs more than 5 years old can be pruned in this way every 2 or 3 years.
Propagation and nursery production From semi-ripe cuttings taken in early summer, or from budding or grafting on to *Ligustrum vulgaris*. Plant bare-rooted or container-grown. Availability varies. Best planting heights 1-2ft (30-60cm).
Problems Young plants often look small and insipid and require a year or more to develop.
Average height and spread
Five years
2x1½ft (60x50cm)
Ten years
2½x2½ft (80x80cm)
Twenty years
or at maturity
4x3ft (1.2x1m)

SYRINGA × PERSICA

PERSIAN LILAC
Oleaceae
Deciduous
Small-flowering, fragrant Lilacs of attractive shape.

Origin Said to have been cultivated in England in the mid 1600s.
Use As a freestanding shrub of interesting shape, or for medium to large shrub borders. Can be fan-trained on any except extremely exposed, cold walls.
Description *Flower* Fragrant small panicles of mauve flowers produced late spring, early summer. *Foliage* Leaves lanceolate, 1¼-2½in (3.5-6cm) long, dark green, giving some autumn colour. *Stem* Upright when young, spreading with age, forming a round, goblet-shaped shrub. Dark green tinged purple. Medium growth rate. *Fruit* Insignificant.
Hardiness Tolerates 4°F (−15°C).
Soil Requirements Most soils, although extremely alkaline types may cause chlorosis.
Sun/Shade aspect Full sun to very light shade.
Pruning None required.
Propagation and nursery production From budding, layers or semi-ripe cuttings taken in early summer. Purchase container-grown. Fairly easy to find. Best planting heights 1¼-2½ft (40-80cm).
Problems Young plants may be shy to flower and can take 2-3 years to establish.
Varieties of interest *S. × p. 'Alba'* A variety with off-white, scented flower panicles. Interesting but not showy. *S. × p. 'Laciniata'* syn. *S. × p. afghanica* Very attractive, very finely divided leaves unlike those of any other *Syringa*, no more than 1¼in (3.5cm) long. Flowers slightly fragrant, lilac-pink but often sparsely produced. Two-thirds average height and spread.
Average height and spread
Five years
3x3ft (1x1m)
Ten years
5½x5½ft (1.8x1.8m)
Twenty years
or at maturity
8x8ft (2.5x2.5m)

SYRINGA × PRESTONIAE

CANADIAN LILAC
Oleaceae
Deciduous
Elegant large-flowering shrubs which need space.

Origin A range of hybrids produced by crossing *S. reflexa* and *S. villosa*, called after Isabella Preston, who first raised them in the Division of Horticulture, Ottawa, Canada, in the early 1920s.
Use As a freestanding large shrub, or group of shrubs, or as background planting in large shrub borders. If planted 6ft (2m) apart in a single line makes a large, wide, screening hedge.
Description *Flower* Large, graceful, elegant panicles of scented pink to red-pink or purple-pink flowers, depending on variety, produced late spring, early summer. All varieties single-flowered. *Foliage* Leaves long, lanceolate,

Syringa × persica in flower

Syringa × prestoniae 'Elinor' in flower

2-6in (5-15cm) long and 2-3in (5-8cm) wide, dark green with prominent veins. **Stem** Upright when young, strong-growing, forming narrow-based, wide-topped, large shrub. Grey-green to grey-brown. Fast growth rate when young, becoming slower with age. **Fruit** Grey-brown seedheads may be retained into winter.

Hardiness Tolerates winter temperatures down to −13°F (−25°C).

Soil Requirements Does well on most, but alkaline types may produce narrow, yellow, chlorotic leaf margins. On very alkaline soils this may become acute and cause distress.

Sun/Shade aspect Full sun to medium shade.

Pruning On established plants more than 5 years old remove one-third of oldest flowering wood to ground level after flowering to encourage rejuvenation from base. Otherwise the shrub can become very large and woody and will produce fewer flowers.

Propagation and nursery production From semi-ripe cuttings taken in early summer, from budding, grafting or from layers. Purchase container-grown for best results, but may also be planted bare-rooted or root-balled (balled-and-burlapped). Some varieties rather scarce. Best planting heights 2-3ft (60cm-1m).

Problems Often underestimated for its overall size and in too small an area may become dominating.

Varieties of interest *S. × p. 'Audrey'* Deep pink flowers, early summer. May not be easy to find. *S. × p. 'Desdemona'* Pink panicles of flowers, late spring. Not easy to find. *S. × p. 'Elinor'* Pale lavender flowers purple-red in bud, borne in upright panicles, late spring, early summer. Given space, a very useful flowering shrub. The most widely available variety. *S. × p. 'Isabella'* Mallow purple flowers in erect panicles, late spring, early summer. May need some searching to obtain.

Average height and spread
Five years
6x6ft (2x2m)
Ten years
13x13ft (4x4m)
Twenty years
or at maturity
20x20ft (6x6m)

SYRINGA REFLEXA

KNOWN BY BOTANICAL NAME
Oleaceae
Deciduous
An aristocrat among Lilacs, but very demanding of space.

Origin From central China.
Use As a large individual flowering shrub, for planting in groups of three or five, or as a background shrub in large shrub borders. If planted 10ft (3m) apart makes wide, formal screen.

Description *Flower* Panicles at first upright, becoming drooping, up to 6-8in (15-20cm) long, purple-pink outside, pale colour within, late spring, early summer. *Foliage* Leaves large, oval, 3-8in (8-20cm) long and 2-3in (5-8cm) wide, dark green with rough texture to upper and lower surfaces. *Stem* Strong, stout, becoming branching and forming round-topped, large, robust shrub. Grey-green to grey-brown. Medium growth rate, slowing with age. *Fruit* Brown-green seedheads retained into winter.

Hardiness Tolerates winter temperatures down to −13°F (−25°C).

Soil Requirements Most soils; may show signs of distress on very alkaline types.

Sun/Shade aspect Full sun to medium shade.

Pruning On established shrubs more than 4-5 years old, remove one-third of old flowering wood after flowering to encourage rejuvenation from base.

Propagation and nursery production From semi-ripe cuttings taken in early summer, by division of side-rooted suckering growth, by budding or by grafting. Purchase container-grown or root-balled (balled-and-burlapped). Not always easy to find. Best planting heights 2-2½ft (60-80cm).

Problems Often planted in areas unable to accommodate its overall size.

Average height and spread
Five years
6x6ft (2x2m)
Ten years
13x13ft (4x4m)
Twenty years
or at maturity
16x16ft (5x5m)

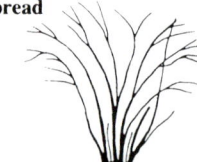

SYRINGA SWEGINZOWII

KNOWN BY BOTANICAL NAME
Oleaceae
Deciduous
A very elegantly shaped Lilac with attractive, scented flowers.

Origin From western China.
Use As freestanding shrubs of elegant shape or for background planting in larger shrub borders.

Syringa sweginzowii in flower

Description *Flower* Flesh pink, fragrant flowers, presented in long, loose panicles, late spring, early summer. *Foliage* Leaves pointed, ovate, 2-4in (5-10cm) long, mid to dark green, with some yellow autumn colour. *Stem* Upright, becoming arching, forming graceful, arching, wide-topped, narrow-based shrub. Green-grey, becoming green-brown. Medium growth rate. *Fruit* Insignificant.

Hardiness Tolerates winter temperatures down to −13°F (−25°C).

Soil Requirements Most soils, except extremely alkaline types, which cause chlorosis.

Sun/Shade aspect Full sun to medium shade.

Pruning Remove one-third of oldest flowering shoots after flowering to induce new shoots from ground level.

Propagation and nursery production From semi-ripe cuttings taken in early summer, from layers or by budding or grafting. Best planted container-grown, but may be planted bare-rooted or root-balled (balled-and-burlapped). Best planting heights 2-2½ft (60-80cm).

Problems Often planted in areas too small for its final overall dimensions.

Syringa reflexa in flower

Average height and spread
Five years
6x6ft (2x2m)
Ten years
10x10ft (3x3m)
*Twenty years
or at maturity*
16x16ft (5x5m)

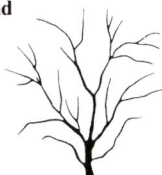

SYRINGA VULGARIS

COMMON LILAC

Oleaceae
Deciduous
Widely planted, useful, free-growing shrubs for large areas.

Origin From eastern Europe.
Use As a large screening shrub or ultimately a small tree.
Description *Flower* Lilac-blue, very fragrant flowers in upright pyramid-shaped panicles, late spring. Some purple and white flowering varieties. *Foliage* Leaves ovate, 2-6in (5-15cm) long and 1-2in (3-5cm) wide. Light to mid green, glossy upper surfaces, dull grey undersides. Some yellow autumn colour. *Stem* Upright when young, becoming branching and spreading. Grey-green, becoming grey-brown. Forms large, slightly pointed, broad shrub, which can in time become a small, multi-stemmed tree. Fast growth rate when young, slowing with age. *Fruit* Insignificant.
Hardiness Tolerates winter temperatures down to −13°F (−25°C).
Soil Requirements Any soil.
Sun/Shade aspect Prefers full sun, tolerates light or medium shade.
Pruning None required. Best left to grow large but old stems and unwanted suckers may be removed occasionally. If cut back hard will take 2-3 years to come back into flower. Seedheads should be removed as soon as possible to induce better flowers in following year.
Propagation and nursery production From suckers or layers. Plant bare-rooted or container-grown. Although extremely common may be difficult to find in commercial production. Best planting heights 2-3ft (60cm-1m).
Problems Can be invasive.
Varieties of interest *S. v. 'Alba'* A white-flowering form. *S. v. 'Rubra'* Wine red flowers.
Average height and spread
Five years
6x6ft (2x2m)
Ten years
13x13ft (4x4m)
*Twenty years
or at maturity*
20x20ft (6x6m)

Syringa 'Primrose' in flower

SYRINGA
Cultivars

LILAC

Oleaceae
Deciduous
One of the most spectacular groups of large scented shrubs for late spring or early summer-flowering.

Origin All varieties of garden origin; many were raised by Victor Lemoine and his son Emile in their nurseries in Nancy, France at the turn of the century. Alice Harding, author of the definitive book on this genus, was also responsible for a large number of varieties in production today.
Use As freestanding, flowering shrubs, either singly or in groups. Ideal for mass or specialist planting, also as background planting for larger shrub borders. Can be planted 3ft (1m) apart in single line to make an attractive, informal low hedge or screen. Also sometimes obtainable as small mop-headed standard on stem 5-6ft (1.5-2m) high.
Description *Flower* Large single or double florets in large fragrant panicles, in colours ranging from blue to lilac, through pink to red or purple, to white or yellow, with some bicoloured forms. Late spring, early summer. Colour, size of shrub, degree of fragrance and flowering time dependent on variety. *Foliage* Leaves dark green to mid green, medium-sized, ovate, 2-6in (5-15cm) long and 1¼-2½in (3.5-6cm) wide. *Stem* Stout shoots, grey-green to grey-brown with pronounced buds, either green-yellow or red-purple, depending on variety, in winter. Forms an upright, conical shrub; may form a small multi-stemmed tree after 10-15 years. Fast growth rate when young, slowing with age. *Fruit* Grey-brown seedheads maintained into winter.
Hardiness Tolerant of winter temperatures down to −13°F (−25°C).
Soil Requirements Most soils, but may show signs of chlorosis on severely alkaline types.
Sun/Shade aspect Prefers full sun, tolerates up to medium shade.
Pruning Requires very little pruning, but remove dead seedheads in winter to increase flowering. Old stems and unwanted suckers can be removed as necessary. Cutting back hard reduces flowering in next 2-3 years.
Propagation and nursery production From budding or grafting using either *S. vulgaris*, *Ligustrum vulgare* or *L. ovalifolium* as understocks. Can also be raised from semi-ripe cuttings taken early to mid summer. Best purchased container-grown, but can also be planted bare-rooted or root-balled (balled-and-burlapped). Availability varies. Best planting heights 2-3ft (60cm-1m).
Problems Often planted in areas where full growth potential is restricted. Young plants may take up to 3-5 years after planting to come into full flower.
Varieties of interest *Single-flowered varieties:* *S. 'Congo'* Large panicles of rich, lilac-red flowers in mid to late spring, becoming paler with age. Must be sought from specialist nurseries. *S. 'Etna'* Flowers rich purple to claret red, late spring to early summer, aging to lilac-pink. Must be sought from specialist nurseries. *S. 'Firmament'* Flowers mauve in bud, opening to clear lilac-blue, mid spring. A very free-flowering variety. *S. 'Lavaliensis'* Very attractive pale pink flowers. Smaller than average, bright green foliage. Slightly less than average height and spread. Difficult to find. *S. 'Marechal Foch'* Panicles of bright crimson-rose flowers in mid spring. Must be sought from specialist nurseries. *S. 'Massena'* Large, broad panicles of deep red-purple flowers in late spring, early summer. Must be sought from specialist nurseries. *S. 'Maud Nottcutt'* Very stately panicles of white flowers in mid spring. One of the best single-flowering varieties, good for flower-arranging material. *S. 'Night'* Very dark purple flowers in mid spring. May be hard to find. *S. 'Primrose'* Pale primrose yellow to yellow-white flowers in mid spring. A dense, compact shrub. *S. 'Reamur'* Large panicles of deep red-purple flowers, shaded with violet, in early summer. Must be sought from specialist nurseries. *S. 'Sensation'* Large panicles of purple to purple-red, white-edged florets in mid to late spring. Flower variegation may be lost in some seasons. A shrub of loose, open habit. Must be sought from specialist nurseries. *S. 'Souvenir de Louis Spaeth'* Large, broad trusses of wine red flowers in mid to late

Syringa vulgaris 'Rubra' in flower

Syringa 'Souvenir de Louis Spaeth' in flower

spring. A strong-growing shrub of spreading habit. *S. 'Vestale'* Large, loose panicles of pure white flowers, in mid to late spring. Light green foliage. Compact habit. *Double-flowering varieties:* *S. 'Belle de Nancy'* Lilac-pink flowers opening from purple-red buds in mid spring. *S. 'Charles Joly'* A very dark red-purple form. Large flower panicles produced in late spring, early summer. *S. 'Edward J. Gardner'* Large semi-double, light pink flowers, in mid spring to late spring. *S. 'Kathleen Havermeyer'* Wide, tight flower panicles, purple-lavender, becoming paler lilac-pink, in late spring, early summer. *S. 'Madame A. Buchner'* Semi-double, mauve-shaded rose-pink florets in feathery flowerheads. Very good scent. Tall, open growth habit. *S. 'Madame Lemoine* Large white flower panicles in mid spring. A large, wide, round shrub, one of the most popular forms. *S. 'Michel Buchner'* Pale rose-lilac flowers in dense panicles, very fragrant, in late spring. A large, upright shrub. *S. 'Mrs. Edward Harding* Semi-double, purple-red, fragrant flowers in mid to late spring. Tall-growing, open habit. *S. 'Paul Thirlon'* Deep rose-red flowers in late spring, early summer. May be difficult to find; should be sought from specialist nurseries. *S. 'Souvenir d'Alice Harding'* Large panicles of soft white flowers in late spring, early summer.

Average height and spread
Five years
4x3ft (1.2x1m)
Ten years
8x5ft (2.5x1.5m)
Twenty years
or at maturity
16x10ft (5x3m)

TAMARIX

TAMARISK
Tamaricaceae
Deciduous
Useful early or late-flowering shrubs for sheltered areas and coastal regions.

Origin From south-eastern to central Europe, through to western Asia.
Use As a single shrub or grouped. Can be used as mid to background planting for larger shrub borders. If planted 3ft (1m) apart in single line makes a useful hedge or screen, which may be clipped. Good in coastal areas.
Description *Flower* Long, graceful, plume-like racemes of small pink flowers borne at terminals of branches, some forms flowering in spring on current season's wood, others on previous season's growth. *Foliage* Leaves, very small, ovate, scale-like, thickly borne on branches, light green or glaucous blue, colour depending on variety. Some yellow autumn colour. *Stem* Strong, upright when young, quickly becoming branching and arching, forming spreading, delicately shaped shrub or, in very favourable areas, small tree. Green to purple-green or purple, dependent on variety. Medium to fast growth rate. *Fruit* Insignificant.
Hardiness Tolerates 14°F (−10°C).
Soil Requirements Prefers light, well-drained soil, dislikes any waterlogging.
Sun/Shade aspect Full sun or very light shade.
Pruning Shrubs left unpruned may attain very interesting shape but sometimes become un-

gainly. It is a matter of choice whether a neat or a natural, straggly effect is required ultimately. If pruning varieties flowering on current season's growth, cut back hard in early to mid spring. For varieties flowering on previous season's growth, cut back hard following flowering.
Propagation and nursery production From semi-ripe cuttings. Purchase container-grown. Easy to find, especially in coastal areas.
Problems Often bought in favourable locations and moved into a colder climate, where often it does not succeed.
Varieties of interest *T. parviflora* (Small-flowered Tamarix) Flowers deep pink, produced in mid to late spring on previous season's growth. Stems purple to purple-brown. Light green foliage. *T. ramosissima* syn. *T. pentandra* (Five-stamen Tamarix) Flowers rose pink, borne on current season's growth in late summer, early autumn. Glaucous green foliage, red-brown stems. *T. r. 'Rubra'* Deep pink flowers, purple in bud, produced on current season's growth in late summer, early autumn. Glaucous foliage, red-brown stems. *T. ramossima 'Pink Cascade'* Racemes of pink flowers produced in early to late summer on current season's growth. Glabrous green foliage, red-brown branches. *T. tetandra* Pink flowers produced in graceful racemes in late spring, early summer on previous season's wood. Dark branches, green foliage.
Average height and spread
Five years
5x5ft (1.5x1.5m)
Ten years
8x8ft (2.5x2.5m)
Twenty years
or at maturity
12x12ft (3.5x3.5m)

TEUCRIUM FRUTICANS

SHRUBBY GERMANDER
Labiatae
Evergreen
A useful silver-grey shrub requiring some winter protection.

Origin From southern Europe to North Africa.
Use For medium to large shrub borders or for planting with other silver-leaved subjects. Can be fan-trained on sunny or moderately sunny, sheltered wall. If planted 2½ft (80cm) apart makes an informal hedge in favourable areas.
Description *Flower* Small, light blue flowers in small terminal racemes, throughout summer. *Foliage* Leaves ovate, ½-1½in (1-4cm) long, light grey-green upper surfaces, white under-

Syringa 'Madame Lemoine' in flower

Tamarix parviflora in flower

Teucrium fruticans in leaf

sides. Some limited yellow autumn colour. *Stem* Upright stems, becoming branching, forming spreading, arching, dome-shaped shrub. Grey to silver-grey. Fast growth rate in spring, slowing with age. *Fruit* Insignificant.
Hardiness Minimum winter temperature 23°F (−5°C).
Soil Requirements Well-drained soil, dislikes any waterlogging.
Sun/Shade aspect Full sun, in sheltered position.
Pruning Previous season's growth should be shortened by two-thirds to three-quarters in mid spring, to encourage new foliage and flowering.
Propagation and nursery production From semi-ripe cuttings. Purchase container-grown. Generally available in mild districts, less easy to find elsewhere.
Problems Susceptibility to cold and wet conditions is not always appreciated.
Varieties of interest *T. chamaedrys* (Wall Germander) Dark, steel grey foliage and dark, red-mauve flowers in mid to late summer. Upright habit, reaching only one-quarter average height and spread.
Average height and spread
Five years
3x3ft (1x1m)
Ten years
5x5ft (1.5x1.5m)
Twenty years
or at maturity
6x6ft (2x2m)

TRACHYCARPUS FORTUNEI

CHUSAN PALM
Palmaceae
Evergreen
An interesting structural shrub, as long as adequate winter protection can be provided.

Origin From central China.
Use In very mild areas only, as a single or grouped, tall, tree-like feature shrub to give Mediterranean effect. Grows well in a large container but requires winter protection.
Description *Flower* Small terminal panicles, arching in habit, of yellow flowers borne in large numbers on mature shrubs, early summer. *Foliage* Leaves large, palm-like, fan-shaped, up to 3ft (1m) wide and 2ft (60cm) long, borne on strong, dark green leafstalks in clusters at summit of tall trunk, retained for many years. *Stem* Central, single trunk growing with formation of terminal leaf cluster, dark green, covered with furry grey hairs. Growth rate depends on planting location; the milder the location, the more growth. *Fruit* Round, marble-like fruits, blue-black on mature, shrubs in autumn.
Hardiness Minimum winter temperature 23°F (−5°C).
Soil Requirements Well-drained soil, dislikes waterlogging.

Sun/Shade aspect Full sun to very light shade.
Pruning None required except occasional removal of any very old or damaged leaves.
Propagation and nursery production From seed or division. Purchase container-grown. Difficult to find outside mild temperate locations.
Problems Often purchased in suitable growing areas and moved to colder sites, where it rapidly succumbs to winter frosts.
Average height and spread
Five years
3x3ft (1x1m)
Ten years
10x6ft (3x2m)
Twenty years
or at maturity
32x10ft (10x3m)

Height depends on location. Those shown are for ideal conditions.

ULEX EUROPAEUS

FURZE, GORSE, WHIN
Leguminosae
Deciduous
Not truly a garden shrub, but suitable for large-scale ground cover.

Origin From western Europe including the UK.
Use Ideal for covering large areas on poor soil, including banks and disused areas.

Ulex europaeus 'Plenus' in flower

Description *Flower* Lemon yellow through yellow, to gold and yellow pea-flowers, mainly late spring, early summer but produced over very wide period, sometimes throughout year. *Foliage* None. *Stem* Very twiggy and branching, covered in impenetrable sharp spines. Dark green, becoming grey-brown. Fast growth rate when young, becoming slow with age. *Fruit* Insignificant.
Hardiness Tolerates 4°F (−15°C), but may be damaged by severe wind chill in cold winters.
Soil Requirements Best on neutral to acid soil; tolerates mild alkalinity and poor soils.
Sun/Shade aspect Best in full sun or very light shade. Dislikes any deeper shade intensely.
Pruning None required, but can be cut to ground level and will rejuvenate very rapidly.
Propagation and nursery production From seed or semi-ripe cuttings taken early summer. Purchase container-grown. Will not tolerate being moved bare-rooted. May be difficult to find, and must be sought from specialist nurseries.
Problems Likely to suffer from severe drought. Spines make stems very difficult to handle.
Varieties of interest *U. e.* 'Plenus' Lemon yellow, double flowers. Two-thirds average height and spread.
Average height and spread
Five years
2½x3ft (80cmx1m)
Ten years
5x6ft (1.5x2m)
Twenty years
or at maturity
8x13ft (2.5x4m)

Trachycarpus fortunei in leaf

VACCINIUM

BLUEBERRY, WHORTLEBERRY
Ericaceae
Deciduous and Evergreen
A range of shrubs requiring extremely acid, moist soil conditions, but producing interesting flowers, fruit and autumn colours.

Origin Mostly from the USA.
Use As an ornamental shrub with good autumn colours, ideal used with rhododendrons and azaleas, or as a fruiting shrub. Tall-growing or carpeting forms can be used for large-scale planting.
Description *Flower* Small, round or bell-shaped, hanging flowers, normally white, often with red tints, early spring or summer, depending on variety. *Foliage* Leaves lanceolate or ovate, ½-1½in (1-4cm) long, depending on variety. Grey-green, all having good autumn colour. *Stem* Normally upright, and spreading, but with some forms low and carpeting, depending on variety. Grey-green to light green tinged purple. Slow to fast growth rate, depending on variety. *Fruit* Normally purple to purple-red with glabrous texture. Mostly edible.
Hardiness Tolerates 14°F (−10°C).
Soil Requirements Acid to very acid soil, tolerating highly acid conditions where very few or no other plants will grow.
Sun/Shade aspect Full sun to very light shade.
Pruning None required except for occasional thinning.
Propagation and nursery production From seed, semi-ripe cuttings or layers. Purchase container-grown, bare-rooted or root-balled (balled-and-burlapped). Best planting heights 8in-2½ft (20-80cm).
Problems The need for a truly acid soil is often underestimated.
Varieties of interest *V. angustifolium* (Low-bush Blueberry) A low-growing variety consisting of wiry stems covered in lanceolate leaves with bristle-toothed surfaces, giving good autumn colours. Fruits blue-black, very sweet and edible. Slow-growing, reaches only 1½ft (50cm) in height and 3ft (1m) in spread. *V. arboreum* (Farkleberry) Ovate foliage, often semi-evergreen, 1¾-2½in (4.5-6cm) long, leathery in texture with blue down. Very good autumn colours. Flowers white and bell-shaped, produced either singly or in small, short racemes. Fruits black, not edible. Medium growth rate. From south-eastern USA. *V. arctostaphylos* (Caucasian Whortleberry) Shoots red when young. Leaves up to 4-5½in (10-13cm) long, narrow, elliptic to ovate with finely toothed edges, giving purple-red autumn colour, often maintained well into winter. Flowers bell-shaped, crimson, tinted white, waxy texture, produced in

prominent racemes in summer, and often into autumn. Black, shiny fruits. Medium growth rate. Two-thirds average height and spread. From the Caucasus. *V. corymbosum* (Swamp Blueberry, High-bush Blueberry) Leaves ovate to lanceolate, up to 3in (8cm) long, bright green, turning vivid scarlet and bronze in autumn. Flowers pale pink or white, in clusters in late spring. Good-sized fruits, black with blue bloom, sweet and edible. Medium growth rate. Two-thirds average height and spread. From eastern USA where it is grown as a commercial fruit crop. *V. hirsutum* (Hairy Huckleberry) A low, suckering shrub with hairy stems, narrow, ovate to elliptic leaves, grey-green, giving some good autumn colour. Flowers round, white, tinged pink, produced in small racemes, late spring. Blue-black round fruits covered with thin hairs, sweet and edible. Slow-growing, reaching only a fifth of average height and a quarter of average spread. From south-eastern USA. *V. myrsinites* (Evergreen Blueberry) Foliage oval, finely tooth-edged, grey-green to dark green, giving some autumn colour. White flowers tinged rose pink in bud, borne in small clusters at leaf axils, mid to late spring. Fruits blue-black, edible. Slow growing, reaching only a fifth of average height and a quarter of average spread. From south-eastern USA. *V. myrtillus* (Bilberry, Whortleberry, Whinberry, Blueberry) Small, suckering patch of bright green stems, leaves ovate, finely toothed, with some autumn colour. Flowers globe-shaped, green-pink, either singly or in pairs in leaf axils, mid spring through to early summer. Black, edible, bloom-covered fruits in early summer. Slow-growing, reaching only one-fifth average height and a quarter of average spread. A native of Europe and the Caucausus. *V. ovatum* (Cranberry) An evergreen form with ovate to oblong, leathery-textured leaves, bright copper-red when young, becoming dark polished green with age. Bell-shaped flowers, white or pink, freely produced in short racemes in late spring, early summer. Fruits red, ripening to black. Medium growth rate, reaching two-thirds average height and spread. From western North America. *V. oxycoccos* Small, hanging, pink, bell-shaped flowers with recurved petals and yellow beak-like stamens, produced on thread-like stems, late spring, early summer. Edible round, red fruits. A very low, slow-growing, spreading shrub, reaching no more than 1ft (30cm) in height and 3ft (1m) in spread. From North America through northern hemisphere to Japan. *V. parvifolium* (Red Bilberry) Globular pink flowers borne individually in each leaf axil, late spring and early summer, followed by bright red, edible fruits. Medium growth rate. Reaching two-thirds average height and one-third average spread. From western USA and north into Alaska. *V. uliginosum* (Bog Whortleberry) A low,

spreading variety forming a round, mound-shaped shrub. Foliage ovate, blue-green. Flowers pale pink, oval, either borne singly or in clusters in leaf axils, late spring, early summer, followed by globular black fruits with bloomy coating, which are edible and sweet. Reaching only 1-1¼ft (12-15cm) in height with a spread of 2-2¼ft (60-70cm). Slow to medium growth rate. Native throughout the northern hemisphere. *V. virgatum* Long, arching branches, foliage ovate to lanceolate, grey-green, giving good rich autumn red and orange. White flowers tinged pink borne in small clusters in leaf axils, late spring, early summer. Black fruits. Medium to slow growth rate. From eastern North America. *V. vitis-idaea* (Cowberry) A low, creeping shrub. Foliage ovate and glossy dark green with greyer, pale undersides. Flowers white, tinged pink, bell-shaped, produced in short terminal racemes, mid to late summer. Edible globe-shaped red fruits. Slow-growing, reaching no more than 1ft (30cm) in height with a spread of 2-2½ft (60-80cm). Originating from northern regions of North America.
Average height and spread
Five years
3x3ft (1x1m)
Ten years
6x6ft (2x2m)
Twenty years
or at maturity
13x13ft (4x4m)

VIBURNUM
Early-flowering varieties

KNOWN BY BOTANICAL NAME
Caprifoliaceae
Deciduous
A range of extremely useful, normally large, scented flowering shrubs.

Origin Forms originating from China; many of garden origin.
Use As freestanding shrubs on their own, for winter flowering effect. Ideal for woodland and wild gardens and as mid or background planting for medium to large shrub borders.
Description *Flower* Clusters of pink or white, small, tubular, fragrant flowers produced on bare stems, early to mid spring. Also some intermittent flowering from late autumn through winter during periods of mild weather. *Foliage* Leaves lanceolate to ovate, ½-4in (1-10cm) long, dark green with red tinge, intensifying towards late summer, early autumn. Some autumn colour. *Stem* Strong upright growth, becoming branching and twiggy with age, forming upright shrub of somewhat stout habit. Purple-green when young, becoming mahogany-brown with age. Medium growth rate. *Fruit* Insignificant.
Hardiness Tolerant of −13°F (−25°C), but any flowers produced in winter may be damaged by sudden temperature changes.
Soil Requirements Any soil, including alkaline.
Sun/Shade aspect Full sun to medium shade.
Pruning After 4-5 years from planting, remove one-third of oldest flowering growth, early to mid spring, thereafter pruning every 2 or 3 years, depending on shrub's vigour.
Propagation and nursery production From semi-ripe cuttings taken in early summer. Purchase container-grown, bare-rooted or root-balled (balled-and-burlapped). Availability varies. Best planting heights 2-3ft (60cm-1m).
Problems Generally exceeds the allotted space after 10-15 years.
Varieties of interest *V. × bodnantense* A cross between *V. farreri* and *V. grandiflorum*. Strong, upright habit, covered with clusters of scented, rose-tinted flowers. Raised in the Bodnant Gardens, North Wales, UK. *V. × b. 'Dawn'* A clone, vigorous in habit, with larger foliage and larger, pink scented flowers. A decided improvement. *V. × b. 'Deben'* Flowers pink in bud, opening to white. Sweetly

Vaccinium corymbosum **in fruit**

Viburnum × bodnantense in flower

scented. Flowers may be damaged by severe frosts or excessive wetness. Raised in the Notcutt Nurseries, East Anglia, UK. *V. farreri* syn. *V. fragrans* A spreading variety, especially in early years, forming a rounded shrub. Flowers pink, opening to white, scented, tubular, in clusters at end of main shoots or in lateral clusters. Foliage more oval than those of most forms, with toothed edges, bronze when young, aging to dark green. Some autumn colours. From northern China. *V. f. candidissimum* Slightly less freely flowering, producing terminal and lateral clusters of white, tubular, scented flowers. Foliage light green. *V. f. 'Nanum'* syn. *V. f. 'Compactum'* A very dwarf, compact variety. Shy to flower, but infrequently produces small isolated clusters of pink, tubular-scented flowers. Small, bronze-green, ovate young foliage, aging to darker green. Some autumn bronze-orange. Up to 1½ft (50cm) in height and spread. Of garden origin. *V. foetens* syn. *V. coreana* Gaunt, stout stems, mahogany-brown tinged purple, forming large, goblet-shaped shrub. Clusters of good-sized, tubular, white, very fragrant flowers, susceptible to frost damage. Large, ovate, dark green leaves, tinged purple. Scarce; must be obtained from specialist nurseries. Very slow-growing, eventually reaching up to 6ft (2m) in height and spread.

Average height and spread
Five years
5x3ft (1.5x1m)
Ten years
8x6ft (2.5x2m)
Twenty years
or at maturity
12x13ft (3.5x4m)

VIBURNUM
Large-leaved varieties

KNOWN BY BOTANICAL NAME
Caprifoliaceae
Evergreen
Large, aristocratic flowering shrubs for screening and bold effect.

Origin From China.
Use Large, evergreen shrubs used individually or for mass planting. Ideal for screening, or as background planting for medium to large shrub borders.
Description *Flower* White to dull white clusters of small, short, tubular flowers, many clusters up to 4-6in (10-15cm) across, produced mid spring through early summer, according to variety. *Foliage* Leaves large, ovate, 2-6in (5-15cm) long, dark to mid green upper surface. Some smooth, some glossy, others felted, often with grey undersides. *Stem* Normally strong, upright, becoming branching, smooth or felty-textured, depend-

ing on variety, forming large, wide, round-topped shrub. *Fruit* Blue or black clusters of round or oval fruits in autumn.
Hardiness Tolerant of winter temperatures down to −13°F (−25°C).
Soil Requirements Any soil.
Sun/Shade aspect Prefers light shade, tolerates full sun to medium shade.
Pruning May be left unpruned. Alternatively, remove one-third of oldest wood on shrubs 5 years old or more to encourage rejuvenation.
Propagation and nursery production From semi-ripe cuttings taken in early summer. Purchase container-grown or root-balled (balled-and-burlapped). All forms fairly easy to find.
Problems Space must be allowed for eventual size of mature shrubs.
Varieties of interest *V. buddleifolium* An almost evergreen shrub, which occasionally defoliates in hard winters. White clusters of flowers up to 2¾in (6.5cm) across in early summer, followed by clusters of red fruits which turn black. Oblong to lanceolate leaves, pale green, velvety-textured and with grey felted undersides. Two-thirds average height and spread. From central China. *V. cinnamomifolium* Small clusters of dull white flowers up to 4-6in (10-15cm) across in early summer; dark, glossy, leathery-textured leaves. Clusters of egg-shaped, blue-black, shiny fruits in autumn. Good in light to medium shade. May be slightly less hardy and more difficult to find. Two-thirds average height and spread. From China. *V. henryi* A variety of open, upright habit. Pyramid-shaped panicles of yellow white flowers, early

to late summer. Elliptic, glossy green, leathery-textured leaves on stiff, upright red-green branches. Bright red fruits, later turning black. Obtain from specialist nurseries. From central China. *V. × hillieri* syn. *V. × hillieri 'Winton'* Semi-evergreen to evergreen, depending on winter severity. Creamy white panicles of flowers, early summer, red fruits, later turning black. Foliage narrow, oval, dark to mid green tinged copper, turning bronze-red in winter. From China. *V. japonicum* (Japanese Viburnum) White fragrant flowers in rounded trusses produced in early summer, only on mature shrubs, followed by red fruits. Leathery-textured foliage, up to 6in (15cm) long and up to 4in (10cm) wide, glossy dark green with paler undersides. Stout leaf stalks, orange-green. Two-thirds average height and spread. From Japan. *V. 'Pragense'* Clusters of pink buds opening to creamy white, produced at terminals of branches in late spring. Corrugated, elliptic leaves 2-3in (5-8cm) long, dark green with white felted undersides. Extremely hardy. Obtain from specialist nursery. Two-thirds average height and spread. Raised in Prague, Czechoslovakia. *V. × rhytidophylloides* (Lantanaphyllum Viburnum) Clusters of yellow-white flower buds produced in autumn and carried through winter before opening in late spring. Foliage elliptic to ovate or oblong to ovate, rough-textured, often hanging limp in cold weather, giving drooping, almost shabby effect. Some orange-brown autumn colour. Very good for screen planting. Of garden origin. *V. rhytidophyllum* (Leatherleaf Virburnum) A strong-growing variety, with clusters of creamy yellow flowers in late spring, formed in previous autumn and carried through win-

Viburnum rhytidophyllum in flower

Viburnum henryi in flower

213

ter as felted, closed buds. Leaves large, up to 8in (20cm) long, elliptic or oblong, with corrugated surface; dark glossy green, undersides grey and covered with matted hairs underneath. Oval red fruits, finally turning black, produced in autumn and maintained well into early winter. For good fruiting effect, a number of shrubs should be planted fairly closely together. Relatively easy to find. From central and western China. *V. r.* 'Roseum' Flowers tinted rose pink, otherwise identical to its parent. *V. r.* 'Variegatum' A rarely seen variety with white-splashed leaves, otherwise identical to its parent. *V. utile* (Service Viburnum) Clusters of white, scented flowers, late spring, followed by blue-black fruits in autumn. Sparsely produced branches. Ovate to oblong, glossy dark green foliage with white undersides. Obtain from specialist nurseries. Two-thirds average height and spread. From central China.

Average height and spread
Five years
5x5ft (1.5x1.5m)
Ten years
10x10ft (3x3m)
Twenty years
or at maturity
16x16ft (5x5m)

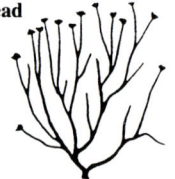

VIBURNUM
Best fruiting varieties

KNOWN BY BOTANICAL NAME
Caprifoliaceae
Deciduous
Useful large shrubs for fruiting effect.

Origin From China.
Use As large fruiting shrubs, either standing alone or grouped, or for background planting in medium to large shrub borders.
Description *Flower* Clusters of small, white flowers, mid spring to early summer. *Foliage* Leaves ovate, 2-4in (5-10cm) long, mid to dark green, often with good autumn colour. *Stem* Strong, upright shoots, becoming branching and spreading with age, normally dark green or purple-green, forming a large, wide-topped, narrow-based shrub. Fast growth rate when young, slowing with age. *Fruit* In bunches, either red or black, depending on variety, produced profusely in autumn.
Hardiness Tolerant of winter temperatures down to −13°F (−25°C).
Soil Requirements Any soil.
Sun/Shade aspect Prefers light shade, tolerates full sun to medium shade.
Pruning On established plants more than 5 years old, remove one-third of oldest wood every 2-3 years to encourage new basal growth and keep the shrub healthy.

Propagation and nursery production From semi-ripe cuttings taken in early summer or from seed. Purchase container-grown. Most forms will need to be obtained from specialist nurseries.
Problems Shrubs take 5 years or more to reach maturity and produce good fruiting displays.
Varieties of interest *V. betulifolium* Clusters of white flowers, early summer, set off by ovate to rhomboidal, coarsely toothed, dark to olive green foliage. Red fruits, hanging in large bunches from long, arching branches, maintained well into winter. It is often recommended that several shrubs should be planted nearby to encourage fertilization, but it has been found that single, isolated plants will fruit well at maturity. May have to be obtained from specialist nurseries. From central China. *V. hupehense* White clusters of flowers, late spring, early summer, give way to interesting egg-shaped, orange-yellow fruits in autumn, aging to red in early winter. Ovate, coarsely toothed leaves, light to mid green, giving good autumn colour. Obtain from specialist nurseries. From central China. *V. lantana* (Wayfaring Tree) Creamy white clusters of flowers 3-4in (8-10cm) across, late spring, early summer, giving way to clusters of oblong fruits, red at first, aging to black. Grey-green, felted, upright leaves, broad and ovate, 6in (15cm) long, velvety-textured on undersides; turning dark crimson in autumn. Fairly easy to find. Used as understock for all Viburnums that are grafted. From central and southern Europe, spreading through northern Asia to Asia Minor and North Africa. *V. l.* 'Variegatum' syn. *V. l.* 'Auratum' Young foliage golden yellow to yellow-green. Colour not retained well and turns light green in mid to late summer, although fruiting is very good. *V. lentago* (Sheepberry, Nannyberry) Upright growth, strong in stature. Flowers produced in terminal clusters, creamy white, in late spring, early summer, giving way to clusters of blue-black, bloom-covered, damson-shaped fruits. Ovate leaves, dark shiny green tinged purple or red. Good rich autumn colour. From eastern North America. *V. l.* 'Pink Beauty' A variety with pink flowers, otherwise identical to its parent. *V. prunifolium* (Black Haw) Upright branches with horizontal side branches. Clusters of white flowers give way to large blue-black, bloom-covered, edible fruits. Foliage bright green and shiny-textured, ovate to obovate, giving very good autumn colour. Hard to find outside its native environment. From eastern North America. *V. sargentii* White flowers with purple anthers, early summer, give way to large clusters of translucent red fruits lasting well into winter. Strong, grey-green stems support maple-shaped leaves with good autumn colour. Stems develop a cork-like bark. From north-eastern Asia.

Average height and spread
Five years
4x3ft (1.2x1m)
Ten years
7x6ft (2.2x2m)
Twenty years
or at maturity
13x13ft (4x4m)

VIBURNUM
Spring-flowering, scented forms

KNOWN BY BOTANICAL NAME
Caprifoliaceae
Deciduous and Semi-evergreen
Some of the most highly scented of spring-flowering shrubs.

Origin Basic forms from Korea, with many garden hybrids.
Use As fragrant spring-flowering, medium-sized shrubs for planting individually, in groups, or in medium to large shrub borders. Many forms do well in containers over 2ft (60cm) in diameter and 1¼ft (40cm) in depth; use good quality potting medium. Occasionally obtainable as short, mop-headed standards and can be used as feature planting.
Description *Flower* Medium to large, round, full clusters, consisting of many tubular flowers, varying shades of pink to white in early to late spring. All highly scented. *Foliage* Leaves ovate, medium-sized, 2-4in (5-10cm) long, grey-green, some yellow autumn display. *Stem* Upright, covered with grey scale when young, becoming branching to form dome-shaped shrub. Medium growth rate. *Fruit* May produce blue-black fruits in autumn.
Hardiness Tolerant of 4°F (−15°C).
Soil Requirements Most soils, disliking only very dry or very wet types.
Sun/Shade aspect Prefers light shade, accepts full sun to medium shade.
Pruning None required, but remove any suckering growths appearing below graft or soil level.
Propagation and nursery production Normally from grafting on to an understock of *V. lantana.* Some varieties from semi-ripe cuttings taken in early summer. Purchase container-grown or root-balled (balled-and-burlapped). Best planting heights 2-3ft (60cm-1m). Most varieties fairly easy to find, especially when in flower. Particular varieties and standard forms may have to be obtained from specialist nurseries.
Problems All forms, particularly *V. carlesii* and its varieties, suffer from aphid attack. Root systems of all forms are very fibrous and surface-rooting and react badly, sometimes succumbing completely, to damage caused by cultivation drought or waterlogging.
Varieties of interest *V. bitchiuense* (Bitchiu Viburnum) Clusters of pink, scented flowers, mid to late spring. Foliage ovate to elliptic, dark metallic green. Open habit. From Japan. *V. × burkwoodii* (Burkwood Viburnum) Clusters of pink buds open into fragrant, white tubular flowers, early to mid spring. followed by clusters of blue-black fruits. Semi-evergreen ovate foliage with dark green, shiny surface. As leaves die off in autumn they turn scarlet, red and orange, contrasting with remaining dark green foliage. Forms large round-topped shrub or can be fan-trained as a wall shrub. Reaches one-third more than average height and spread. *V. × b.* 'Anne Russell' Semi-evergreen. Large clusters of pale pink, fragrant flowers in mid spring, dark pink in bud. *V. × b.* 'Chenaultii' Semi-evergreen. Flowers similar to *V. × burkwoodii,* but does not reach same overall proportions. Not easy to find. Two-thirds average height and spread. *V. × b.* 'Fulbrook' Large white flowers, pink in bud and sweetly scented. *V. × b.* 'Park Farm Hybrid' A form with larger, more vigorous habit of growth. Flowers, mid spring, slightly larger than the form. Good glossy green foliage. *V. × carl-*

Viburnum betulifolium in fruit

Viburnum × burkwoodii in flower

cephalum A deciduous variety producing large, white, tubular florets, pink in bud, very fragrant. Complete clusters are 4-5in (10-12cm) across and extremely attractive. Large, ovate to round, grey-green foliage, may produce good autumn colours. *V. carlesii* (Koreanspice Viburnum) Clusters of pure white, tubular flowers, opening from pink buds, with strong scent, in mid to late spring. Ovate to round, downy, grey to grey-green leaves with grey felted undersides, producing good red-orange autumn colouring. Some forms of *V. carlesii* are weak in constitution and named varieties may be more successful. *V. c. 'Aurora'* Red flower buds, opening to fragrant pink tubular flowers produced in clusters, mid to late spring. Good ovate grey-green foliage. Good constitution. *V. c. 'Charis'* Good, vigorous growth. Flowers red in bud, opening to pink and finally fading to white. Very good scent. Foliage clean and grey-green. May be difficult to find, but not impossible. *V. c. 'Diana'* A good clone of compact habit. Flowers pink, red in bud. Good fragrance. May be difficult to find but not impossible. *V. × juddii* (Judd Viburnum) Clusters of scented, pink-tinted tubular flowers, produced in terminals of branching stem, mid to late spring. Grey-green ovate foliage with some autumn colour. Open in habit when young, becoming denser with age. Two-thirds average height and spread.

Average height and spread
Five years
3x3ft (1x1m)
Ten years
5x5ft (1.5x1.5m)
Twenty years
or at maturity
6x6ft (2x2m)

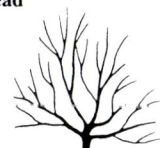

Viburnum × carlcephalum in flower

VIBURNUM
Interesting foliage varieties

KNOWN BY BOTANICAL NAME
Caprifoliaceae
Deciduous
Excellent foliage shrubs, but somewhat scarce and difficult to find.

Origin From China and Japan.
Use As foliage shrubs for woodland or wild gardens.
Description *Flower* Clusters of white flowers, early to mid spring. *Foliage* Normally good autumn colours. Some varieties with interesting waxy surface. *Stem* Strong-growing, upright green stems. Some varieties with suckering habit. *Fruit* Red or black, depending on variety, in clusters in autumn.
Hardiness Tolerant of 14°F (10°C).
Soil Requirements Any soil.
Sun/Shade aspect Prefers light shade, but tolerates full sun to medium shade.
Pruning On shrubs more than 5 years old, remove one-third of oldest wood to ground level every 2-3 years, to induce new growth from ground level and prevent shrub from becoming woody.
Propagation and nursery production From layers or semi-ripe cuttings taken in early summer. Purchase container-grown. Rather hard to find, but not impossible. Best planting heights 1-2ft (30-60cm).
Problems None.
Varieties of interest *V. acerifolium* (Dock-

mackie) Maple-shaped, three-lobed leaves, light to mid green, coarsely tooth-edged with undersides covered with black dots. Good rich dark crimson autumn colours. Terminal clusters of white flowers, early summer, followed by oval red fruits which age to purple-black. Reaches two-thirds average height and spread. From eastern North America. *V. alnifolium* (Hobble Bush) Large ovate foliage, very strongly veined, up to 4-8in (10-20cm) across. Foliage has downy upper surface when young, with more dense down on undersides, turning deep claret red in autumn. Large, lacecap white flowers produced late spring, early summer, followed by red fruits, turning to black-purple in autumn. Prefers shady position. Suckering from ground level forming a wide, tall, spreading shrub. *V. cylindricum* An evergreen variety with large, hanging, narrow or ovate, sometimes broad ovate, dull green leaves, with paler undersides and waxy upper surfaces. Tubular white flowers with lilac stamens, followed by black, egg-shaped fruits. Minimum winter temperature 23°F (−5°C). Of slightly more than average height and spread. From the Himalayas and western China. *V. furcatum* Broad, ovate, grey-green foliage, giving good orange-brown autumn colours. Clusters of white flowers with white marginal ray florets. Red fruits aging to black, in autumn. Upright habit.

Average height and spread
Five years
3x3ft (1x1m)
Ten years
6x6ft (2x2m)
Twenty years
or at maturity
10x10ft (3x3m)

VIBURNUM DAVIDII

DAVID'S VIBURNUM
Caprifoliaceae
Evergreen
A very useful, interesting, evergreen, female fruiting shrub.

Origin From western China.
Use As a low, spreading, evergreen shrub with interesting fruiting effect, for planting in small groups including both male and female forms, or in shrub borders.
Description *Flower* Clusters of white, short, tubular flowers in late spring, early summer. Individual shrub will bear either male or female flowers. *Foliage* Leaves large, ovate, 4-6in (10-15cm) long and 1½-2in (4-5cm) wide, dark glossy green, leathery texture with duller undersides. *Stem* Short, twiggy, relatively slow-growing, forming a low dome-shaped shrub. *Fruit* Female plants produce

Viburnum alnifolium in autumn

Viburnum davidii in fruit

clusters of very attractive turquoise-blue fruits, providing a male form planted close by. Not always easy to obtain sexed plants and so be sure of obtaining fruit.
Hardiness Tolerant of 13°F (−25°C), although foliage may show signs of blackening if temperature falls below 14°F (−10°C).
Soil Requirements Good, fertile soil for best results but does well on most soil conditions, only disliking extreme dryness or waterlogging, where growth will be inhibited, or root damage will be caused.
Sun/Shade aspect Prefers light shade. Good in full sun to medium shade.
Pruning None required but can be reduced in size.
Propagation and nursery production From semi-ripe cuttings taken in early summer. Purchase container-grown and sexed if possible. Relatively easy to find, but not always as sexed form. Best planting heights 8in-1½ft (20-50cm).
Problems Can suffer from attacks of leaf mildew in very wet humid summers. Takes a number of years from planting to develop to its full potential.
Average height and spread
Five years
2½x2½ft (80x80cm)
Ten years
4x4ft (1.2x1.2m)
Twenty years
or at maturity
5x6ft (1.5x2m)

VIBURNUM OPULUS

GUELDER ROSE, WATER ELDER, EUROPEAN CRANBERRYBUSH VIBURNUM

Caprifoliaceae
Deciduous
Useful fruiting, autumn display shrubs with flat lacecap or snowball-shaped flowers.

Origin From Europe, northern and western Asia and North Africa.
Use For mass planting in difficult areas, or planted individually or in medium to large shrub borders. Ideal specimens for wild gardens. Some varieties occasionally obtainable as short mop-headed standards.
Description *Flower* Two types, depending on variety. Either white, flat lacecap flowers with central small fertile flowers surrounded by ring or ray florets; or a globular snowball-like cluster of florets. The flat, lacecap forms are fertile whereas the round snowball forms are not. *Foliage* Leaves light green, 2-5in (5-12cm) long, often with orange tinge or shading in late summer. Five-lobed, maple-shaped leaves, giving good autumn colour. *Stem* Upright when young, becoming spreading and branching with age. Grey to grey-

green. *Fruit* Lacecap, fertile forms produce clusters of red and yellow round fruits in autumn, depending on variety. Snowball-flowering varieties do not fruit.
Hardiness Tolerates winter temperatures down to −13°F (−25°C).
Soil Requirements Any soil, tolerating alkalinity and acidity, dry or even boggy or waterlogged areas.
Sun/Shade aspect Full sun, but tolerates up to medium shade.
Pruning Can be left unpruned to form in-

Viburnum opulus 'Nottcutt's Variety' in fruit

teresting clump. Alternatively, on shrubs more than 5 years old remove one-third of oldest wood to ground level in spring. Repeat every 2-3 years to induce new growth from ground level and to prevent shrub from becoming woody.
Propagation and nursery production From semi-ripe cuttings taken in early summer or hardwood cuttings taken in winter. Purchase container-grown, bare-rooted or root-balled (balled-and-burlapped). Most forms fairly easy to find. Best planting heights 2-3ft (60cm-1m).
Problems Can outgrow its desired area unless adequate space is allowed.

Viburnum opulus 'Sterile' in flower

Varieties of interest *V. o. 'Aureum'* White, lacecap flowers with some red fruits in autumn. New spring growth bright yellow, aging to yellow-green. Unfortunately will scorch in full sun and therefore requires very light shade protection. Not readily available, but not impossible to find. One-third average height and spread. *V. o. 'Compactum'* White, lacecap flowers in early summer; red fruits in autumn. A very short-growing variety reaching no more than 4ft (1.2m) in height and 5ft (1.5m) in spread. *V. o. 'Fructuluteo'* White, lacecap flowers, early summer. Lemon yellow fruits, tinged pink and aging to chrome yellow, retaining faint shading on pink. Obtain from specialist nurseries. *V. o. 'Nanum'* A lacecap form very shy to flower, but producing good autumn leaf colour. Very dwarf, tight habit only reaching up to 2½ft (80cm) in height and spread. Obtain from specialist nurseries. *V. o. 'Nottcutt's Variety'* A very good selected form with large white lacecap flowers, followed by bunches of succulent red fruits in autumn. *V. o. 'Xanth-*

Viburnum opulus 'Xanthocarpum' in fruit

ocarpum' White lacecap flowers, pure golden yellow fruit, becoming darker and attaining translucent appearance when ripe. Spectacular especially when grown in tandem with *V. o. 'Nottcutt's Variety'*. *V. o. 'Sterile'* syn. *V. o. 'Roseum'* (Snowball Shrub) Gobular, creamy white, snowball-type flowerheads. Non-fruiting.

Average height and spread
Five years
5x5ft (1.5x1.5m)
Ten years
10x10ft (3x3m)
Twenty years
or at maturity
14½x14½ft (4.5x4.5m)

VIBURNUM PLICATUM

DOUBLEFILE VIBURNUM, JAPANESE SNOWBALL, LACECAP VIBURNUM
Caprifoliaceae
Deciduous
Extremely attractive snowball or lacecap-flowering spring shrubs in a wide range of sizes.

Origin From Japan and China.
Use For small, medium or large shrub borders, according to variety. For mass planting or as individual specimens.
Description *Flower* Either globular snowball heads of white florets, or flat, lacecap, hydrangea-like flowers with central fertile small flowers, surrounded by white ray florets. Both types produced late spring, early summer. *Foliage* Leaves 2-4in (5-10cm) long, ovate, with pleated effect and pronounced channelling along veins, light to mid green. Some good autumn colour, particularly after a dry summer or in very dry soils. *Stem* Upright when young, quickly becoming spreading to form dome-shaped, wide shrub. Light green to green-grey. Can spread to extent of producing flat tiers of growth, a habit common to all but more pronounced in some varieties. Medium growth rate. *Fruit* Clusters of oval, red to red-orange fruits in autumn, on mature shrubs more than 5 years old on relatively dry to average soil. Fruiting can be erratic.
Hardiness Tolerates 4°F (−15°C).
Soil Requirements Tolerates wide range of soils but extremely waterlogged or dry areas will damage its fine-textured root system.
Sun/Shade aspect Prefers light shade, tolerates from full sun to medium shade.
Pruning None required.
Propagation and nursery production From semi-ripe cuttings taken in early summer or from layers. Purchase container-grown or root-balled (balled-and-burlapped). Most varieties fairly easy to find: Best planting heights 1¼-3ft (40cm-1m).
Problems Very susceptible to root damage, from cultivation such as hoeing, or from

***Viburnum plicatum 'Mariesii'* in flower**

***Viburnum plicatum 'Pink Beauty'* in flower**

***Viburnum plicatum 'Watanabe'* in flower**

extreme drought or waterlogging. If roots are damaged, a section of top growth will die back. In severe cases this can destroy the shrub completely.
Varieties of interest *Lacecap flowering varieties: V. plicatum 'Cascade'* Large, white flat flowers. Fertile inner small tufted florets, surrounded by a ring of bold white ray florets. Red fruits after hot summer. Large, ovate, pointed foliage, giving good autumn colour, on branches arching to give semi-weeping effect. Less likely to die back than other varieties. Two-thirds average height and spread. *V. p. 'Lanarth'* White, large, flat, lacecap flowers produced in defined tiers along horizontal branches. Large, ovate, mid to dark green foliage, giving good autumn colour. Sometimes listed as shorter growing shrub; initially when this variety was catalogued it was confused with *V. p. 'Mariesii'*, 'Lanarth' being described as the lower of the two varieties, but the reverse is the case. *V. p. 'Mariesii'* White lacecap flowers borne on very horizontal, tiered branches. Good fruiting with fine autumn colour. Reaching half average height and spread. *V. p. 'Pink Beauty'* Ray florets of white lacecap flowers, aging to attractive pink. Small to medium, ovate, dark to olive green foliage. Good autumn colour. Two-thirds average height and spread. *V. p. 'Rowallane'* White lacecap flowers in late spring. Small, ovate, tooth-edged foliage, mid to dark green, giving some autumn tints. Closely tiered branch effect. Extremely reliable fruiting form. May be more difficult to find than most. Two-thirds average height and spread. *V. p. var. tomentosum* Flowers 2-4in (5-10cm) across, creamy white and surrounded by white ray florets, giving way to red, oval fruits, which age eventually to black.

Bright green, pleated, ovate foliage. Good autumn colour. Not easy to find. From Japan, China and Taiwan. *V. p. 'Watanabe'* syn. *V. semperflorens* Very compact, producing good-sized white lacecap flowers in early summer through to mid autumn. Some orange-red fruits, late summer through to autumn. Reaching only two-thirds average height and spread. *Snowball flowering varieties: V. macrocephalum* Semi-evergreen. White, round, large, globular heads of sterile flowers, 2¾-6in (7-15cm) across, late spring. Medium-sized, ovate foliage, light green, up to 2-4in (5-10cm) long. Somewhat less hardy than most, requiring a favourable sunny position on a wall in areas where temperature falls below 23°F (−5°C). Extremely difficult to obtain, but not impossible. *V. plicatum 'Grandiflorum'* (Japanese Snowball) Large, round to globular, sterile heads of white florets. New flowerheads in spring attractive green, aging to white, taking on pink margin and finally overall pink towards end of flowering period in midsummer. Foliage large, ovate, mid green tinged purple. Good autumn colour. Must have good moist soil and light dappled shade to do well; resents full sun. Two-thirds average height and spread.

***Viburnum plicatum 'Grandiflorum'* in flower**

Average height and spread
Five years
8x6ft (2.5x2m)
Ten years
12x10ft (3.5x3m)
Twenty years
or at maturity
16x16ft (5x5m)

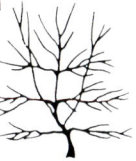

Viburnum tinus in fruit and flower

VIBURNUM TINUS

LAURUSTINUS
Caprifoliaceae
Evergreen
An attractive evergreen shrub with interesting winter flowers.

Origin From the Mediterranean regions of south-eastern Europe.
Use For mass planting or screening effect, or for mid to background planting in medium to large shrub borders. If planted 2ft (60cm) apart in single line makes useful formal or informal hedge, the latter producing more flowers because unclipped.
Description *Flower* Small to medium-sized clusters of white tubular flowers often pink in bud, late autumn through until late spring, resistant to frost. Opening to peak performance in relatively mild spells. *Foliage* Leaves broad, ovate, 1½-4in (4-10cm) long, evergreen, dark green with lighter silver undersides. Susceptible to wind scorch in severe cold. *Stem* Upright when young, becoming spreading and branching with age, forming a round-topped, broad-based shrub. Dark green to green-brown. Fast growth rate when young, slowing with age. *Fruit* Clusters of oval black fruits in autumn.
Hardiness Tolerates 14°F (−10°C), but may suffer from wind scorch in severe winters.
Soil Requirements Most soils; tolerates both alkalinity and acidity but resents extremely dry areas, and waterlogged conditions.
Sun/Shade aspect Prefers light shade, good from full sun to medium shade.
Pruning Can be reduced drastically and will rejuvenate relatively quickly, flowering from early stage. Alternatively, on shrubs more

than 5 years old, remove one-third of oldest wood to ground level in early spring, to encourage rejuvenation from base. May also be trimmed hard if used as hedging.
Propagation and nursery production From semi-ripe cuttings taken in early summer. Purchase container-grown. Not always easy to find. Best planting heights 1-2½ft (30-80cm).
Problems Suffers from severe winter cold, especially wind chill. Can appear old and straggly if pruning is neglected.
Varieties of interest *V. t. 'Eve Price'* Flowers carmine-red in bud, opening to white with pink shading. Foliage smaller than parent. Two-thirds average height and spread. *V. t. 'French White'* A good, strong growing variety, producing large white flowers. *V. t. 'Gwenllian'* Flowers deep pink in bud, opening to white with pink tinge. Small leaves, compact habit, reaching only two-thirds average height and spread. *V. t. lucidum* Flowers large, white, in early to late spring. Good, vigorous growing form with larger glossy leaves than parent. Slightly tender; minimum winter temperature 23°F (−5°C). *V. t. 'Purpureum'* White flowers. New growth tinged purple, older foliage very dark green. Minimum winter temperature 23°F (−5°C). Two-thirds average height and spread. *V. t. 'Variegatum'* White flowers. Attractive white to creamy white variegated foliage. Two-thirds average height and spread. Minimum winter temperature 23°F (−5°C).

Average height and spread
Five years
3x3ft (1x1m)
Ten years
6x6ft (2x2m)
Twenty years
or at maturity
12x12ft (3.5x3.5m)

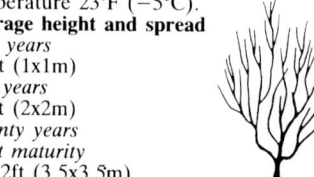

Viburnum tinus 'Variegatum' in flower

VINCA MAJOR

GREATER PERIWINKLE, LARGE PERIWINKLE
Apocynaceae
Evergreen
Used in the right position, very useful ground-covering shrubs.

Origin From central and south-eastern Europe and North Africa.
Use As bold, broad, ground-covering carpet. Variegated forms can be trained 3-5ft (1-1.5m) up walls or cascading down to 6ft (2m). Extremely good for retention of steep banks.
Description *Flower* Bright blue, star-shaped, 1in (3cm) across, borne in leaf axils of wood one year old or more, mid spring through to midsummer. *Foliage* Leaves ovate, dark green, glossy, 1-2¾in (3-7cm) long, with dull undersides. Also some white or golden variegated forms. *Stem* Long, ranging, light green shoots, arching and rooting at contact with soil, forming thick, ground-covering, suckering mass. *Fruit* Insignificant.
Hardiness Tolerates winter temperatures down to −13°F (−25°C).
Soil Requirements Any soil except very heavy clay where unable either to spread main roots or to root readily from tips.
Sun/Shade aspect Prefers light shade, good from full sun to shade.

Vinca major 'Variegata' in leaf

Pruning Cut back to ground level in early spring, to induce good evergreen foliage and large, numerous flowers. Unpruned shrubs can become very shabby and untidy.
Propagation and nursery production From softwood cuttings taken in early summer or from self-rooted shoots which are readily obtainable. Very easy to find. Best planting heights 4-12in (10-30cm).
Problems Can become invasive in confined areas, and may even become a nuisance.
Varieties of interest *V. difformis* Not a form of *V. major* but included here because of its close resemblance. Narrow, lanceolate to ovate, pointed, foliage, 2-3½in (5-9cm) long. Long-stalked single flowers produced in leaf axils, pale lilac-blue with 5 long tapering petals, autumn and early winter. *V. m. 'Maculata'* A rarely seen variety with central splash of green-yellow on each young leaf, splash deteriorating as leaf ages, being produced only in open, sunny positions. *V. m. 'Variegata'* Large, creamy white margined foliage. Pale blue flowers. Identical to its parent in flowering and spread. Very useful ground cover.
Average height and spread
Five years
1¼x3ft (40cmx1m)
Ten years
1¼x3ft (40cmx1m)
Twenty years
or at maturity
1¼x3ft (40cmx1m)

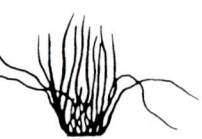

Vitex agnus-castus **in flower**

VINCA MINOR

COMMON PERIWINKLE, LESSER PERIWINKLE
Apocynaceae
Evergreen
Useful carpeting ground cover.

Origin From Europe and western Asia.
Use As a very low, small, ground-covering, carpeting shrub, easily contained.
Description *Flower* Single, small, purple, blue or white, depending on variety, star-shaped flowers, 1in (3cm) across, borne in leaf axils on short, upright, flowering shoots from early summer. Some purple and some white forms. Double-flowering varieties also obtainable. *Foliage* Leaves small, ovate to elliptic, sometimes lanceolate, 1¼in (3.5cm) long, dark green or white or golden variegated, depending on variety. *Stem* Arching, ground-covering, hugging branches which root readily at tips and then continue onwards producing new plants as they go. Mid green. *Fruit* Insignificant.
Hardiness Tolerates winter temperatures down to −13°F (−25°C).
Soil Requirements Most soils, disliking only extremely heavy clay types.
Sun/Shade aspect Prefers light shade, accepts full sun to deep shade.
Pruning Trim lightly in early spring to induce good new foliage and good flowering.
Propagation and nursery production From semi-ripe cuttings taken in early summer or from rooted suckers. Purchase container-grown or bare-rooted. Best planting heights 4-8in (10-20cm) or less.
Problems It can take up to 2 years after planting to form a good, carpeting mat.
Varieties of interest *V. m. 'Alba'* Single, white flowers and light green foliage. *V. m. 'Atropurpurea'* Single, deep plum-purple flowers. Green foliage. *V. m. 'Aureovariegata'* Single, mid blue flowers. Foliage blotched yellow. Best in light shade. *V. m. 'Aureovariegata Alba'* Single white flowers. Gold and cream variegated foliage. *V. m. 'Aureo Flore Pleno'* Double, sky blue flowers. Dark green foliage. *V. m. 'Bowles' Variety'* Single, azure blue flowers, slightly larger than normal. Dark green foliage. *V. m. 'Gertrude Jekyll'* Single, white flowers. Light green foliage. *V. m. 'Multiplex'* Double plum-purple flowers. Narrow, almost lanceolate leaves, dark green. *V. m. 'Variegata'* Single white flowers. Leaves variegated creamy white.
Average height and spread
Five years
6inx1¼ft (15x40cm)
Ten years
6inx2ft (15x60cm)
Twenty years
or at maturity
6inx2½ft (15x80cm)

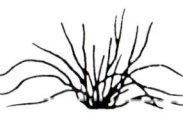

Vinca minor 'Aureovariegata' **in leaf**

VITEX AGNUS-CASTUS

CHASTE TREE
Verbenaceae
Deciduous
An attractive shrub, not often seen but useful for its autumn flowers.

Origin From the Mediterranean area and central Asia.
Use As a freestanding, interesting, autumn-flowering shrub or as a wall-grown, fan-trained specimen.
Description *Flower* Fragrant, violet, slender racemes produced at ends of current season's growth, early to mid autumn. *Foliage* Leaves compound, consisting of 5-7 ovate to lanceolate leaflets, 2-6in (5-15cm) long, on short stalks borne in pairs, along stems and shoots. *Stem* Graceful, grey, downy shoots, arching to form a round, slightly spreading shrub. Medium growth rate. *Fruit* Insignificant.
Hardiness Minimum winter temperature 14°F (−10°C).
Soil Requirements Well-drained soil, dislikes any degree of waterlogging.
Sun/Shade aspect Full sun, to ripen current season's growth and encourage flowering.
Pruning None required.
Propagation and nursery production From seed or semi-ripe cuttings taken in early summer. Purchase container-grown. Obtain from specialist nurseries. Best planting heights 1¼-2ft (40-60cm).
Problems Likely to succumb in wet, cold winters; hardier in dry, cold conditions.
Average height and spread
Five years
3x3ft (1x1m)
Ten years
5½x6ft (1.8x2m)
Twenty years
or at maturity
8x12ft (2.5x3.5m)

WEIGELA

KNOWN BY BOTANICAL NAME
Caprifoliaceae
Deciduous
Shrubs for late spring and early summer-flowering.

Origin Most forms from Japan, Korea, North China and Manchuria, but many cultivars and hybrid varieties are of garden origin.
Use As a freestanding shrub on its own or as mid to background shrub for shrub borders. If planted 2½ft (80cm) apart in single line, makes flowering, informal hedge. Can be fan-trained, especially the variegated forms, for use on cold, exposed or shady walls. Small

mop-headed standards on 5-5½ft (1.5-1.8m) high stems are obtainable, mainly in the red-flowering forms.
Description *Flower* Funnel-shaped, good-sized flowers in varying shades from yellow, through white, to pink and red. Flowers produced on wood 2 years old or more, late spring through early summer, possibly with intermittent flowering through late summer and early autumn. *Foliage* Leaves ovate, 1½-5in (4-12cm) long, dark to mid green with some light green varieties and golden and silver variegated. Some yellow autumn colour. *Stem* Upright, becoming spreading with age. Grey-green to grey-brown. Medium growth rate. *Fruit* Seedheads dark to mid brown, of some attraction in winter.
Hardiness Tolerates −13°F (−25°C).
Soil Requirements Any soil.
Sun/Shade aspect Prefers full sun, tolerates light to medium shade.
Pruning From two years after planting, remove one-third of old flowering wood annually after flowering to encourage rejuvenation and good production of flowering wood.

Weigela florida 'Albovariegata' **in leaf**

Propagation and nursery production From semi-ripe cuttings taken in early summer or hardwood cuttings in winter. Purchase container-grown. Most varieties easy to find, but some may have to be obtained from specialist nurseries. Best planting heights 1¼-2½ft (40-80cm).
Problems If unpruned can become too woody and flowers will diminish in size and number. Large, established shrubs can be cut to ground level and will regenerate, but will take two years to come into flower. *Weigela* was once classified with the closely related *Diervilla* but in recent years these shrubs have been classified separately.

Varieties of interest *W. florida 'Albovariegata'* Attractive creamy white edges to ovate leaves. Pale to mid pink flowers produced profusely on stems 2 or 3 years old. Useful as wall climber for cold, exposed walls. Two-thirds average height and spread. *W. f. 'Aureovariegata'* A variety with yellow variegation, often producing pink to red tinged leaves, particularly during autumn. Pink flowers profusely produced in late spring to early summer. Useful as wall shrub for cold, exposed walls. *W. f. 'Foliis Purpureis'* Attractive purple-flushed leaves produced in spring, which age and become duller as summer

Weigela 'Looymansii Aurea' in flower

Weigela florida 'Foliis Purpureis' in flower

Weigela 'Bristol Ruby' in flower

progresses. Purple-pink flowers in late spring, early summer. Not widely enough planted. One-third average height and spread. *W. middendorffiana* Arching branches with attractive grey-green winter wood. Flowers bell-shaped, sulphur yellow with dark orange markings on lower lobes, mid to late spring. Ovate, light green foliage with some yellow autumn colour. Prefers light shade, although not fussy. An all-round, attractive variety, reaching two-thirds average height and spread. From Japan, Northern China and Manchuria. *W. praecox 'Variegata'* A variety with ovate to obovate, creamy white variegated foliage, rigid and deeply veined. Flowers honey-scented, rose pink with yellow markings in throat, late spring, early summer. Obtain from specialist nurseries. From Japan, Korea and Manchuria. *Weigela cultivars and hybrids:* *'Abel Carrière'* Large, trumpet-shaped, rose-carmine flowers with gold markings in throat, opening from purple-carmine buds. Good, bold, green foliage. *'Avalanche'* A good, strong-growing, white flowering variety. May have to be obtained from specialist nurseries. *'Boskoop Glory'* Large, trumpet-shaped, salmon-pink flowers. A beautiful form. Two-thirds average height and spread. *'Bristol Ruby'* Possibly the most popular of all flowering forms. Ruby red flowers profusely borne on upright, strong shrub in late spring, early summer. *'Candida'* Pure white flowers with slightly green shading. Light green foliage on arching stems. Two-thirds average height and spread. *'Eva Rathke'* Bright crimson-red flowers with yellow anthers, produced over long period from late spring until late summer. *'Looymansii Aurea'*

Flowers pale pink, contrasting with foliage which is light golden yellow in spring aging to lime-yellow in autumn. Must be in light shade or it scorches. Obtain from specialist nurseries. Often looks weak when young. One-third average height and spread. *'Mont Blanc'* Fragrant white flowers, strong-growing. Obtain from specialist nurseries. *'Newport Red'* Good, dark red flowers. Two-thirds average height and spread. *'Stelzneri'* Good mid pink flowers borne in profusion. Interesting upright growth. Not widely available, but not impossible to find. *'Styriaca'* Carmine-red flowers produced in good quantities in late spring, early summer. Strong, very old-fashioned variety.

Average height and spread
Five years
4x4ft (1.2x1.2m)
Ten years
5½x5½ft (1.8x1.8m)
Twenty years or at maturity
7x7ft (2.2x2.2m)

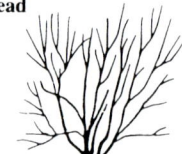

XANTHOCERAS SORBIFOLIUM

YELLOWHORN
Sapindaceae
Deciduous
An interesting if somewhat rare spring-flowering shrub with attractive foliage.

Origin From northern China.
Use As an isolated large, specimen shrub, and eventually a small tree.

Weigela 'Candida' in flower

Xanthoceras sorbifolium in leaf

Description *Flower* White flowers with carmine eyes, over 1in (3cm) wide, presented in upright panicles on previous year's wood, late spring. *Foliage* Leaves pinnate, consisting of 9-17 lanceolate, sharply toothed leaflets 8-10in (20-25cm) long, giving some yellow autumn colour. *Stem* Upright, becoming spreading, even to size of small tree. Greygreen. *Fruit* Shaped like a child's top, three-valved, walnut-like seed pods, containing large numbers of chestnut-like seeds.
Hardiness Tolerates 14°F (−10°C).
Soil Requirements Most soils, tolerating extreme alkalinity.
Sun/Shade aspect Full sun to light shade.
Pruning None required except to remove damaged or crossing stems in early spring.
Propagation and nursery production From seed. Purchase container-grown. Very hard to find, must be sought from specialist nurseries.
Problems Slightly slow to develop and plants are usually small when purchased.
Average height and spread
Five years
5x5ft (1.5x1.5m)
Ten years
8x8ft (2.5x2.5m)
Twenty years
or at maturity
13x13ft (4x4m)

YUCCA

Liliaceae
Evergreen
Very useful structural shrubs with elegant midsummer flowers.

Origin From central America, Mexico and southern USA.
Use As a feature shrub for a tropical effect. Can be grown in large containers, but this greatly reduces overall size and flower production.
Description *Flower* Large, creamy white flowers 1½-3in (4-8cm) long, produced closely on long, upright, conical panicles 3-6ft (1-2m) high, midsummer to early autumn. *Foliage* Leaves narrow, lanceolate, 6-18in (15-50cm) long, grey-green. Often toothed with pointed ends. Some variegated forms. *Stem* No true stems. Leaves originate from a basal rosette but after many years this may extend up to 3ft (1m) high and can become multi-trunked. Leaves have fast growth rate; trunk forms slowly. *Fruit* Seedheads retained into winter, of some attraction.
Hardiness Tolerates 14°F (−10°C).
Soil Requirements Well-drained, dry soil.
Sun/Shade aspect Full sun, resents shade.
Pruning Rarely needs pruning, but any large, dead leaves should be removed.
Propagation and nursery production From seed or division. Purchase container-grown. Most varieties easy to find, but some scarcer forms obtainable from specialist nurseries. Best planting heights 1-2½ft (30-80cm).
Problems Yuccas take 2-4 years from planting to produce useful growth.
Varieties of interest *Y. brevifolia* (Joshua Tree) A variety reaching size of small tree in favourable conditions where winter temperatures do not fall below 23°F (−5°C). Forms central, upright trunk, with some branching. Long, narrow, recurved green leaves with finely toothed margins. Dense panicles of creamy or green-white flowers in late summer. Outside its native environment may be hard to find. From south-eastern USA. *Y. filamentosa* (Adam's Needle Yucca) In place of trunk produces large clumps of lanceolate, mid grey foliage. Edges of leaves covered with small, curled, white threads. Flowers 2-2¾in (5-6.5cm) long, borne in upright panicles 3-6ft (1-2m) tall, mid to late summer. One of the most popular varieties. From south-eastern USA. *Y. f. 'Variegata'* Leaf margins striped yellow, white flowers. Slightly less vigorous than the parent. *Y. flaccida* A stem-

Yucca gloriosa **in flower**

less variety, forming tufts of long, green to glaucous green, lanceolate leaves with white curly threads along edges. Flowers 2-2½in (5-6cm) long, in panicles from 1½-3ft (50cm-1m) in height, in mid to late summer. From south-eastern USA. *Y. f. 'Ivory'* A very good flowering form. Flowers creamy white, stained green, in large panicles. Foliage identical to that of parent. *Y. florida 'Variegata'* Broad, long leaves striped creamy yellow, aging to creamy-white. Some flower production. *Y. gloriosa* (Spanish Dagger, Moundlily Yucca) A trunk-forming species with stems 3-6ft (1-2m) high with limited branching. Stiff, stout leaves with dangerous spined tips, green to glaucous green, 1-2ft (30-60cm) long and up to 4in (10cm) wide, presented in terminal clusters. Flowers creamy white, sometimes tinged red on outside, in upright, conical panicles up to 6ft (2m) high, from midsummer to early autumn. From south-eastern USA. *Y. recurvifolia* Recurving leaves, up to 3ft (1m) tall, glaucous when young, becoming green with age. Creamy white, upright panicles of flowers 2-3ft (60cm-1m) high, in late summer. From south-eastern USA. *Y. r. 'Variegata'* Foliage with pale green central band of variegation against grey-green outer colouring. *Y. whipplei* A stemless form,

making a round, globe-shaped clump. Narrow spine tips, stout, grey-green leaves. Fragrant flowers green-white tinged with purple, formed in large panicles up to 10ft (3m) tall, late spring through early summer. Slightly tender, winter minimum 23°F (−5°C), and needs very well-drained soil. From California.
Average height and spread
Five years
3x3ft (1x1m)
Ten years
6x6ft (2x2m)
Twenty years
or at maturity
13x13ft (4x4m)

ZENOBIA PULVERULENTA

Ericaceae
Deciduous
Not well known, but a useful foliage and flowering shrub for acid soils.

Origin From eastern USA.
Use As a low mound-forming shrub to grow among rhododendrons and azaleas in woodland sites.
Description *Flower* White, bell-shaped, aniseed-scented flowers, similar to those of lily of the valley but larger, produced in pendulous axil clusters in early to mid summer. *Foliage* Oblong to ovate, 1½-2½in (4-6cm) long, lightly toothed and covered with glaucous bloom, fading with age. Good autumn colour. *Stem* Upright at first, becoming arching, forming a round, slightly spreading shrub. Grey-green tinged with red, orange-red autumn colour. Medium growth rate. *Fruit* Insignificant.
Hardiness Tolerates 14°F (−10°C).
Soil Requirements Acid to neutral soil, resents any alkalinity.
Sun/Shade aspect Light to medium shade. Resents full sun.
Pruning None required.
Propagation and nursery production From semi-ripe cuttings taken in early summer or from layers. Purchase container-grown. Obtain from specialist nurseries. Best planting heights 1-2ft (30-60cm).
Problems None, if correct soil conditions are provided.
Average height and spread
Five years
2x2ft (60x60cm)
Ten years
3x3ft (1x1m)
Twenty years
or at maturity
4x4ft (1.2x1.2m)

Zenobia pulverulenta **in autumn**

DICTIONARY
— OF —
SHRUBS
AND
CLIMBING
PLANTS

ACTINIDIA ARGUTA
(*A. polygama*)

KNOWN BY BOTANICAL NAME

Actinidiaceae   *Woody Climber*
Deciduous

An attractive foliaged climber, at its best when showing its autumn colour.

Origin From Japan and Korea.
Use As a free-growing climbing plant for covering large areas. Produces fruit in favourable areas. Good on walls, fences, pergolas or over buildings; can be allowed to ramble through large shrubs and trees to provide interest.
Description *Flower* $\frac{3}{4}$ in (19 mm) across, white with dark purple anthers, fragrant; borne in pairs. Normally only produced in warm climates. *Foliage* Oval, pointed, with toothed edges, up to 5 in (12 cm) long and 2 in (5 cm) wide. Light green turning a good yellow in autumn. *Stem* Grey-green when young, becoming light brown; slightly downy texture. Twining and twisting. Attractive in winter. Medium to fast growing. *Fruit* Oblong, green-yellow, edible. Only produced in warm climates.
Hardiness Tolerates a minimum winter temperature of 13°F (−10°C).
Soil requirements Requires a moist, rich soil, tolerating acid, neutral and moderately alkaline types.
Sun/Shade aspect All but the most exposed aspect. Light shade to full sun.
Pruning Not normally required but can be contained by removal of any offending lateral shoots. It quickly rejuvenates itself with vigorous new growth.
Training Allow to ramble through trees or large shrubs. Provide wires or other large-scale support systems. It twines and does not normally require tying in.
Propagation and nursery production From seed or semi-ripe cuttings. Requires some heat to encourage rooting. Always purchase container grown, in early autumn to early summer, from specialist nurseries. Best planting height 18 in (45 cm) up to 4 ft (1.2 cm).
Problems Can be shy to flower and therefore shy to fruit in all but the warmest areas, although it is worth growing for the autumn foliage effect.

Actinidia arguta **in leaf**

Similar forms of interest *A. a.* 'Cordifolia' Narrow foliage, scarce. *A. a.* 'Aureo-Variegata' Golden-variegated foliage. Scarce.
Average height and spread
Five years
12 × 12 ft (3.7 × 3.7 m)
Ten years
20 × 20 ft (6 × 6 m)
Twenty years
30 × 30 ft + (9 × 9 m +)
Protrudes up to 2 ft (60 cm) from support.

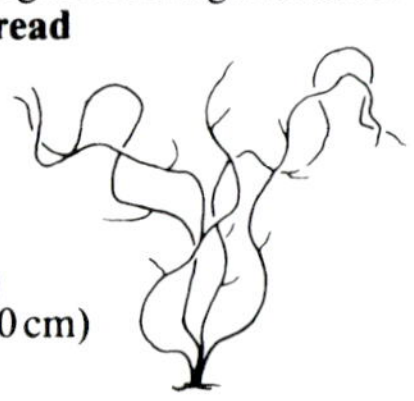

ACTINIDIA CHINENSIS

CHINESE GOOSEBERRY, KIWI FRUIT
Actinidiaceae   *Woody Fruiting Climber*
Deciduous
Extremely ornamental foliage adorning a vigorous grower which can produce edible fruit in hot summers.

Origin From China.
Use As a fast-growing climber for walls, fences, or through trees and large shrubs.
Description *Flower* Creamy white becoming buff yellow, 1½ in (4 cm) wide, five-petalled, incurving cup shaped, in early to mid summer. Male or female on different plants both needed for pollination. *Foliage* Large, almost round, heavily-veined, 5–8 in (12–20 cm) across. Downy undersides. Light green when young becoming more brown/green with age, good yellow/light orange autumn colour. *Stem* Mid green when young becoming light brown. Vigorous, twisting yet not clinging, wide ranging habit. Medium to fast growing. *Fruit* Small, hairy, oblong, round-ended. Up to 2 in (5 cm) long with gooseberry flavour, not always reliable in all but hottest areas.
Hardiness Tolerates a minimum winter temperature of 14°F (−10°C). Some damage to tips of growth may be caused in spring by frost but normally to no great harm.
Soil requirements A deep, well-fed, light soil for best results although it is tolerant to a wide range except extremely waterlogged.
Sun/Shade aspect All but the most exposed aspects. Full sun to very light shade.
Pruning Train shoots to cover required area, prune back all surplus shoots either after fruiting or in late summer to two buds from the point of origin.
Training Tie young shoots of newly planted plants to wires on walls and fences; they normally become self-twining and supporting. In trees and large shrubs, clings by twining.
Propagation and nursery production From semi-ripe cuttings taken in summer. Purchase container grown from good garden centres and specialist nurseries. Best planting height 1½–3 ft (45–91 cm).
Problems Often planted in areas too small to accommodate it. Can be shy to fruit. Male and female plants may be difficult to find.
Similar forms of interest The following commercial varieties may be available: *A. c.* 'Atlas' A good male form for pollinating other varieties; free flowering. *A. c.* 'Heywood' Good female form, heavy cropping on warm south walls, will require a male variety for pollination. *A. c.* 'Tomurii' Male, free flowering, disease resistant.
Average height and spread
Five years
10 × 10 ft (3 × 3 m)
Ten years
20 × 20 ft (6 × 6 m)
Twenty years
39 × 39 ft (12 × 12 m)
Protrudes up to 3 ft (91 cm) from support.

Actinidia chinensis **in fruit**

ACTINIDIA KOLOMIKTA

KNOWN BY BOTANICAL NAME
Actinidiaceae   *Woody Climber*
Deciduous
One of the most attractive of all foliage climbers but does require a warm situation.

Origin From China and Japan.
Use As an attractive foliaged climber for sunny positions both on walls and fences.
Description *Flower* Unattractive white fragrant flowers with yellow anthers ½ in (1 cm) wide; borne in groups of one to three in early summer. *Foliage* 3–6 in (7.5–15 cm) long 2–4 in (5–10 cm) wide oblong ovate leaves with pronounced veins and toothed edges; tips start white and age to pink, contrasting with dark green remainder, some yellow autumn colour. *Stem* Slender, deep mahogany brown. Not normally self clinging. Medium to fast growing. *Fruit* Oval, yellow, 1 in (2.5 cm) long; sweet and edible but not normally used for culinary or dessert purposes.
Hardiness Tolerates a minimum winter temperature of 14°F (−10°C).
Soil requirements Tolerates both alkaline and acid conditions but may produce more growth on neutral to acid types. Well drained and well fed soil is advised.
Sun/Shade aspect Requires some shelter from exposed aspects. Tolerates light shade but prefers full sun.
Pruning Normally requires none other than cutting in early spring to keep within bounds.
Training Tie main vines to wires against walls and fences as required.
Propagation and nursery production From semi-ripe cuttings. Always purchase container grown. Best planting height 1½–3 ft (45–91 cm). Generally only found in good garden centres and specialist nurseries.
Problems Can, on very dry soils, lose its leaves prematurely in late summer. Attractive to cats which claw the vines, causing damage.
Similar forms of interest None.
Average height and spread
Five years
5 × 5 ft (1.5 × 1.5 m)
Ten years
13 × 13 ft (4 × 4 m)
Twenty years
20 × 20 ft (6 × 6 m)
Protrudes up to 3 ft (91 cm) from support.

Actinidia kolomikta **in leaf**

AKEBIA QUINATA

KNOWN BY BOTANICAL NAME
Lardizabalaceae   *Woody Climber*
Deciduous to evergreen
An attractive climbing shrub which if allowed to ramble gives an interesting display of unusual shaped fruit.

Origin From China, Japan and Korea.
Use For growing up through other shrubs or small trees or against walls and fences.
Description *Flower* Pendent racemes 3–5 in (7.5–10 cm) long of male flowers up to ¼ in (5 mm) wide, pale purple in colour. Fragrant. Chocolate-purple female flowers, usually in pairs and 1–1¼ in (2.5–3 cm) wide, are produced in April. *Foliage* Five leaflets carried on a single stalk up to 3–5 in (7.5–12 cm) long; each leaflet oblong to oval in shape, 1½–3 in (4–7.5 cm) long with short 1½ in (4 cm) stalk; light to mid green giving good yellow autumn colour. *Stem* Light green to grey green, loosely twining, wiry in nature. Medium to fast growing. *Fruit* Attractive sausage-shaped grey/violet fruit, 2½–3½ in (6–9 cm) long, splitting lengthwise when ripe. Produced in early autumn.
Hardiness Tolerates a minimum winter temperature of 4°F (−15°C).
Soil requirements Tolerates most soil conditions except waterlogged. Good on alkaline types.
Sun/Shade aspect Needs some protection in exposed aspects. From light shade to full sun, but needs protection from strong, midday summer sun.
Pruning Allow to grow free; every five to six years lightly trim in early spring with hedging shears.
Training Leave to ramble over wires on walls and fences, or over shrubs and trees.
Propagation and nursery production From

Akebia quinata **in flower**

semi-ripe cuttings requiring basal heat to root in midsummer or by layering shoots. Always purchase container grown. Best planting height 1½–3 ft (45–91 cm). Not always readily available – may need searching for from specialist nurseries.
Problems A little unruly in its habit. Flowers and fruit may be hidden both by its own foliage and that of the host it is climbing in.
Similar forms of interest *A. lobata* (syn. *A. trifoliata*) Very scarce in production, flowers smaller, fruits possibly larger and pale violet in colour.
Average height and spread
Five years
6 × 6 ft (1.8 × 1.8 m)
Ten years
18 × 18 ft (5.5 × 5.5 m)
Twenty years
20 × 30 ft (6 × 9 m)
Protrudes up to 2 ft (60 cm) from support.

AMPELOPSIS BREVIPENDUNCULATA

KNOWN BY BOTANICAL NAME

Vitaceae   *Woody Climber*
Deciduous

An interesting vine with good autumn foliage colour.

Origin From North East Asia.
Use Attractive autumn foliage climber to cover walls, fences and pergolas; when used on the latter makes a good shade cover.
Description *Flower* Small, light green, uninteresting. *Foliage* Three- or five-lobed broadly ovate leaves, up to 6 in (12 cm) long. Coarse texture. Downy undersides with pronounced veins often purple red in colour. Good yellow/orange autumn colour. *Stem* Light to green/brown, becoming darker, twining in habit, may be self clinging on old brick walls. Medium to fast growing. *Fruit* Bright blue, grape-like in shape, ¼–½ in (5 mm–1 cm) wide. May require warm summers to fruit well.
Hardiness Tolerates a minimum winter temperature of 14°F (−10°C).
Soil requirements Dislikes extremely wet, dry or poor conditions. Does well on both acid or alkaline soil types.
Sun/Shade aspect Does well in all aspects. Light shade to full sun.
Pruning Not normally required other than that needed for shaping, although in confined spaces can be cut hard back in spring without ill effect.
Training Tie young shoots to wires or wall fixings; normally becomes self-entwining and clinging on walls and fences or over pergolas.
Propagation and nursery production From semi-ripe cuttings. Always purchase container grown. Best planting height 1½–3 ft (45–91 cm). Will need searching for from specialist nurseries.
Problems Can become invasive in good conditions. In too deep shade can become open and lax in habit. In wet autumns may fail to produce good autumn colour.
Forms of interest See further entry.
Average height and spread
Five years
5 × 5 ft (1.5 × 1.5 m)
Ten years
13 × 13 ft (4 × 4 m)
Twenty years
20 × 20 ft (6 × 6 m)
Protrudes up to 2 ft (60 cm) from support.

Ampelopsis brevipendunculata **in fruit**

AMPELOPSIS BREVIPENDUNCULATA 'ELEGANS' (*A. heterophylla* 'Elegans')

KNOWN BY BOTANICAL NAME

Vitaceae   *Woody Climber*
Deciduous

Although requiring more attention than many climbers the display of pink shoots and foliage makes it worth the effort.

Origin From North East Asia.
Use As an attractive coloured foliage climber for sheltered walls and fences or under protection in greenhouses or conservatories.
Description *Flower* Small clusters of creamy white inconspicuous flowers in mid to late spring. May in hot summers produce ⅜ in (2 mm) wide clear blue flowers with black spots. *Foliage* Hand-shaped leaves with some lobed indentations on outer edges varying in size from 2–4 in (5–10 cm) long; some with toothed edges. Grey/green undersides, upper surface white to pink with green variegation. *Stem* Attractive, pink to red when young becoming green to green/brown with age, not self-clinging but twining, interlacing itself around a support. Slow to medium growth rate. *Fruit* None of interest.
Hardiness Tolerates a minimum winter temperature of 23°F (−5°C).
Soil requirements Moderately alkaline to acid, requiring a high degree of organic content with good moisture retaining qualities.
Sun/Shade aspect Requires a sheltered aspect. Light shade for preference, will tolerate full sun if adequate moisture is available; if not, scorching may be a problem.
Pruning Prune back all side shoots produced last year to within two buds of origin except shoots which are required for training the main framework, so encouraging a high production of good new pink foliage.
Training Allow to ramble over wires and secure as required, or allow to scramble through an uninteresting shrub both in the open or under protection.
Propagation and nursery production From semi-ripe cuttings taken in early to mid summer. Always purchase container grown. Not readily available and may have to be searched for. Best planting height 1–2 ft (30–60 cm).
Problems Foliage scorching may be caused by late frosts and strong midday summer sun. Can be attacked by mildew in mid to late summer.
Similar forms of interest None.
Average height and spread
Five years
5 × 5 ft (1.5 × 1.5 m)
Ten years
10 × 10 ft (3 × 3 m)
Twenty years
15 × 12 ft (4.6 × 3.7 m)
Protrudes up to 2 ft (60 cm) from support.

ARISTOLOCHIA MACROPHYLLA (*A. sipho, A. durior*)

DUTCHMAN'S PIPE

Aristolochiaceae   *Woody Climber*
Deciduous

Of all climbing plants this has the most interesting foliage and flowers, given the right conditions.

Origin From Eastern North America.
Use As a climber for sheltered walls, fences, and pillars.
Description *Flower* 1–1½ in (2.5–4 cm) long yellow/green siphon-shaped flowers with open mouth effect at top, coloured

Aristolochia macrophylla **in leaf**

purple/brown around edges; produced in pairs at leaf axles and carried on tall shoots in early summer. *Foliage* Large, kidney- or heart-shaped, sometimes blunt, sometimes pointed; 4–10 in (10–25 cm) long and wide with downy undersides, light green upper surface; presented on stalks 1–3 in (2.5–7.5 cm) long, yellow autumn colour. Very attractive for shape and size. *Stems* Long, light grey/green turning green/brown twining stems, not self supporting. Medium to fast rate of growth. *Fruit* None of interest.
Hardiness Tolerates a minimum winter temperature of 4°F (−15°C).
Soil requirements Moderately alkaline to acid, requires a good rich organic content to maintain adequate moisture to support the large leaf structure.
Sun/Shade aspect Requires a sheltered aspect. Best in very light shade but will tolerate degrees either side.
Pruning Shorten back previous season's growth in early spring to encourage good production of new foliage and flowers.
Training Tie to wires in a fan shape to show foliage off to best effect.
Propagation and nursery production Soft semi-ripe cuttings taken in late spring/early summer; requires some additional bottom heat to assist rooting. Purchase container grown or root-balled (balled-and-burlapped) from specialist nurseries or better garden centres. Best planting height 1–3 ft (30–91 cm).
Problems May be difficult to find. Can take two years to establish before really good new growth is seen, in which time foliage will be small.
Similar forms of interest None.
Average height and spread
Five years
6 × 4 ft (1.8 × 1.2 m)
Ten years
10 × 10 ft (3 × 3 m)
Twenty years
20 × 20 ft (6 × 6 m)
Protrudes up to 2 ft (60 cm) from support.

AZARA DENTATA

KNOWN BY BOTANICAL NAME

Flacourtiaceae   *Wall Shrub*
Evergreen

These scented flowering evergreens are on the tender side, requiring the protection of a wall in winter.

Origin From Chile.
Use As a fan-trained shrub for walls and fences in sheltered areas or in conservatories and greenhouses.
Description *Flower* Clusters of fragrant yellow flowers in spring, borne in profusion. *Foliage* Leaves ovate or oblong, 1–1½ in (2.5–4 cm) long, bright green to glossy dark green with felted undersides. *Stem* Light

Azara dentata in flower

green to mid green. Upright when young, becoming more twiggy and spreading with age. Moderate rate of growth. *Fruit* Insignificant.

Hardiness Tolerates a minimum winter temperature of 23°F (−5°C).

Soil requirements Does well on most soils but dislikes excessive alkalinity and waterlogging.

Sun/Shade aspect Very sheltered aspect. Tolerates full sun to mid shade.

Pruning None required.

Training Requires wires or individual anchor points to secure and encourage the fan-trained shape.

Propagation and nursery production From softwood cuttings taken in mid summer. Best planting height 2–3 ft (60–91 cm). Always purchase container grown. Obtainable from specialist nurseries.

Problems None, apart from its lack of hardiness.

Similar forms of interest *A. lanceolata* Narrow, lanceolate leaves and mustard yellow flowers in early summer which are as fragrant as those of *A. dentata*. *A. serrata* Often confused with *A. dentata*, producing similar scented flowers under the edges of each leaf. Leaves more serrated. In hot climates, or in hot summers, small white berries may be produced. One of the hardier forms.

Average height and spread
Five years
5 × 5 ft (1.5 × 1.5 m)
Ten years
8 × 8 ft (2.4 × 2.4 m)
Twenty years
12 × 12 ft (3.7 × 3.7 m)
Protrudes up to 4 ft (1.2 m) from support.

AZARA MICROPHYLLA

KNOWN BY BOTANICAL NAME
Flacourtiaceae   *Wall Shrub*
Evergreen

A shrub offering attractive evergreen foliage and formation in a sheltered position.

Origin From Chile.

Use As a freestanding or fan-trained shrub for large walls and fences.

Description *Flower* Numerous very small, vanilla-scented, yellow to yellow/green flowers carried in clusters at leaf joints between late winter and early spring. Flowering can be very variable in performance. *Foliage* Very attractive small oval leaflets, 1 in (2.5 cm) long, round-ended, toothed edges, dark shiny green, carried uniformly along branches in interesting formation. *Stem* Light green to dark green, becoming grey/green. Upright, slow to

medium growth rate. Responds well to fan-training. *Fruit* None of interest.

Hardiness Tolerates a minimum winter temperature of 14°F (−10°C). Late spring frost may damage new growth.

Soil requirements Tolerates a wide range of soil conditions, only disliking extremely wet or dry types.

Sun/Shade aspect Best in light to medium shade but will tolerate full sun if required as long as adequate moisture is available.

Pruning Not normally required but can have individual limbs removed in spring if necessary for training.

Training Tie to wires or individual anchor points in a fan shape or allow to grow free-standing.

Propagation and nursery production From semi-ripe cuttings taken in early summer. Should always be purchased container grown; may have to be sought from specialist nurseries. Best planting height 6 in–3 ft (15–91 cm).

Problems Can reach the dimensions of a small tree given time and this should be allowed for in initial planting.

Similar forms of interest *A. m.* 'Variegata' Edges of leaves creamy yellow. An interesting plant less hardy than its parent.

Average height and spread
Five years
5 × 3 ft (1.5 × 91 cm)
Ten years
8 × 5 ft (2.4 × 1.5 m)
Twenty years
16 × 10 ft (4.9 × 3 m)
Protrudes up to 3 ft (91 cm) from support if fan-trained, 7 ft (2.1 m) untrained.

Azara microphylla

BERBERIDOPSIS CORALLINA

KNOWN BY BOTANICAL NAME
Flacourtiaceae   *Wall Shrub*
Evergreen

Technically a low, sprawling shrub but when used as a climber shows off its flowers to the best advantage.

Origin From Chile.

Use As a small climbing shrub for sheltered walls and fences.

Description *Flower* Crimson, ¼ in (4 mm) long, globe-shaped, hanging in racemes and contrasting well with foliage. Mid to late summer. *Foliage* Evergreen, oblong, up to 3 in (7.5 cm) in length, 1½ in (4 cm) wide. Tooth-edged, mid green with some orange/red shading towards autumn. *Stem* Not self-clinging. Light green to green/brown, sprawling and spreading. Slow to medium rate of growth. *Fruit* May produce ¼ in (5 mm) round, red berries following hot summers, in late summer/early autumn.

Hardiness Tolerates a minimum winter temperature of 23°F (−5°C) but requires protection from cold winter winds.

Soil requirements Neutral to acid, may tolerate very limited amounts of alkalinity. High degree of organic material required in soil to retain moisture for good growth.

Berberidopsis corallina in flower

Sun/Shade aspect Very sheltered aspect. Best in light shade but will tolerate degrees either side. Does well under the protection of greenhouses or conservatories.

Pruning Not normally required.

Training Allow to ramble through wires or other support. Individual branches may be supported and tied.

Propagation and nursery production Sow seed in mid to late spring. Take semi-ripe cuttings in mid summer or layer plants in autumn. Always purchase container grown, best size 6–24 in (15–60 cm). May be difficult to find, only available from specialist nurseries.

Problems Its hardiness is suspect and it may be difficult to obtain but it is worth the effort.

Similar forms of interest None.

Average height and spread
Five years
5 × 3 ft (1.5 m × 91 cm)
Ten years
10 × 6 ft (3 × 1.8 m)
Twenty years
15 × 9 ft (4.6 × 2.7 m)
Protrudes up to 12 in (30 cm) from support.

225

BIGNONIA CAPREOLATA
(*Doxantha capreolata*)

KNOWN BY BOTANICAL NAME

Bignoniaceae　　*Tender Woody Climber*
Evergreen

An extremely spectacular flowering climber but requiring greenhouse or conservatory protection in all but the warmest of locations.

Origin From the southern states of the USA.
Use As a spreading climbing plant for very sheltered walls or fences in the open or for scrambling over wires or greenhouse or conservatory roofs.
Description *Flower* Long tubular flowers up to 2 in (5 cm) long, yellow/red in colour, produced on stalks in clusters of two to five in mid spring through to late summer, depending on planting location. *Foliage* Oblong leaflets make up a branching leaf presented at the end of long stalks, light green in colour. Yellow autumn colour. *Stem* Light green long tendrils, twisting but not self clinging. Medium to fast growing. *Fruit* Narrow capsules with leathery appearance, light grey/green in colour.
Hardiness Tolerates a minimum winter temperature of 32°F (0°C).
Soil requirements Light sandy soil although must have moisture retention. Neutral to acid.
Sun/Shade aspect Very sheltered aspect or under protection of greenhouse or conservatory. Light shade to full sun. Will tolerate deeper degree of shade but may be shy to flower.

Bignonia capreolata **in flower**

Pruning Trim in spring to keep in desired area.
Training Tie when young then allow to ramble over wires or other framework.
Propagation and nursery production Short cuttings taken early to mid summer and inserted in a sand rooting medium with some bottom heat. Take care not to overwater in early stages. Should always be purchased container grown; best planting height 1–3 ft (30–91 cm). Not readily available – will have to be sought from specialist nurseries.
Problems Its hardiness is often overstated and availability may be difficult.
Similar forms of interest None.
Average height and spread
Five years
6 × 6 ft (1.8 × 1.8 m)
Ten years
12 × 12 ft (3.7 × 3.7 m)
Twenty years
24 × 18 ft (7.3 × 5.5 m)
Protrudes up to 3 ft (91 cm) from support.

BILLARDIER'A LONGIFLORA

KNOWN BY BOTANICAL NAME

Pittosporaceae　　*Tender Woody Climber*
Evergreen

An attractive evergreen climber requiring a protected situation to withstand winters.

Origin From Tasmania.
Use As a climber for sheltered walls and fences outside or for use under protection in greenhouse or conservatory in exposed, cold areas.

Billardier'a longiflora **in flower**

Description *Flower* Yellow/green turning purple, borne singly over the total area of climber in mid summer. *Stem* Light green turning finally to green/brown, twining not self clinging. Medium rate of growth. *Fruit* Attractive and interesting oval-shaped, blue, 1 in (2.5 cm) long fruits in mid autumn. *Foliage* Hanging, narrow, lance-shaped light green leaves, 1¾ in (4.5 cm) long and ⅜ in (1 cm) wide, leathery texture; may be sparsely presented.
Hardiness Tolerates a minimum winter temperature of 32°F (0°C).

Soil requirements Neutral to acid although may tolerate small degrees of alkalinity. Requires a high organic content for best results.
Sun/Shade aspect Requires a very sheltered aspect. Prefers light shade but will tolerate degrees either side.
Pruning Trim lightly in spring.
Training Allow to grow over wires or up some type of framework.
Propagation and nursery production Semi-ripe cuttings taken in mid summer inserted into a sand/peat mixture with some heat to roots, or raise from seed. Should always be purchased container grown, will need some searching for from specialist nurseries; best planting height 1–2 ft (30–60 cm).
Problems Not fully hardy.
Similar forms of interest None.
Average height and spread
Five years
5 × 5 ft (1.5 × 1.5 m)
Ten years
10 × 10 ft (3 × 3 m)
Twenty years
15 × 15 ft (4.6 × 4.6 m)
Protrudes up to 18 in (45 cm) from support.

BOUGAINVILLEA SPECTABILIS

KNOWN BY BOTANICAL NAME

Nyctaginaceae　　*Tender Greenhouse Climber*

Deciduous

Although tender in all but the mildest of areas, bougainvillea is included in this publication for its use as a climber for conservatories and large greenhouses.

Origin From Brazil.
Use As a climber for conservatories and greenhouses, planted in large containers or in greenhouse borders.
Description *Flower* Tubular flowers surrounded by large magenta bracts, up to 1½ in (4 cm) wide and long, carried in panicles 9–12 in (23–30 cm) long. *Foliage* Pointed, oval, grey/green to dull green, 1½ in (4 cm) long by ¾ in (2 cm) wide. Normally leathery in texture. *Stem* Angular, branching, grey/green, stiff, vigorous. Medium to fast growth rate. *Fruit* Of no interest.
Hardiness Tolerates a minimum winter temperature of 50°F (10°C).
Soil requirements If grown in large containers a good quality potting compost should be used. If grown in soil, the latter should be lightened with the addition of 25 per cent sand and 25 per cent sedge peat.

Bougainvillea spectabilis **in flower**

Sun/Shade aspect Must be in a fully protected aspect. Best in full sun but will tolerate light shade.

Pruning Prune all previous season's shoots, other than those needed to form a structure, back to 1 in (2.5 cm) from the base annually in early spring.

Training Tie to wires or individual anchor points.

Propagation and nursery production From soft to semi-ripe cuttings up to 3 in (7.5 cm) long, taken in spring, preferably with a small portion of old wood attached. Root in a pot containing very sandy soil in a protected frame with bottom heat. May be grown from seed but not easy. Always purchase container grown; will have to be sought from specialist growers, florists and houseplant suppliers. Best planting height 6 in–2 ft (15–60 cm).

Problems Foliage may be attacked by insects such as red spider or whitefly. Roots often attacked by mealy bug. Proprietary controls should be used. Keep ventilation as open as possible, particularly in winter, but do not allow temperature to drop below 50°F (10°C).

Similar forms of interest *B. s.* 'Lady Wilson' Cerise flowers. *B. s. lateritia* Brick-red bracts. *B. s. lindleyana* 'Mrs Louise Wathen' (syn. *B. s. l.* 'Orange King') Cinnamon-coloured bracts. *B. s.* 'Mrs Butt' Bright rose bracts. All varieties are difficult to obtain outside very temperate areas.

Average height and spread
Five years
6 × 6 ft (1.8 × 1.8 m)
Ten years
12 × 12 ft (3.7 × 3.7 m)
Twenty years
24 × 24 ft (7.3 × 7.3 m)
Protrudes up to 3 ft (91 cm) from support.

BUDDLEIA ALTERNIFOLIA

FOUNTAIN BUDDLEIA, ALTERNATE-LEAF BUTTERFLY BUSH

Loganiaceae   *Wall Shrub*
Deciduous

A truly beautiful wall shrub, given enough space.

Origin From China.

Use As a large late summer to early autumn flowering, graceful, arching wall shrub. Wall use is particularly suitable for the variety *B. a.* 'Argentea'.

Description *Flower* Small bunches of very fragrant, lilac-coloured, small, trumpet-shaped flowers borne along graceful, arching branches in early summer. *Foliage* Leaves grey/green, lanceolate, 1½–4 in (4–10 cm) long, giving yellow autumn colour. *Stem* Grey-green to mid green, vigorous, long, upright, becoming arching. Fast growing. *Fruit* Brown to grey/brown seedheads in autumn and winter.

Hardiness Tolerates a minimum winter temperature of 4°F (−15°C).

Soil requirements Prefers good, rich, deep soil, although tolerates other soil types.

Sun/Shade aspect Tolerates all but the most severe of aspects. Best in full sun, tolerates slight dappled shade.

Pruning Thin out one third of growth after flowering on established shrubs.

Training Requires wires or individual anchor points to secure and encourage the fan-trained shape.

Propagation and nursery production From softwood cuttings in summer or hardwood cuttings in winter. Always purchase container grown; best planting height 15–36 in (39–91 cm). Available in garden centres and nurseries.

Problems When offered for sale it resembles an old, woody shrub. Once planted out, however, it grows quickly and often fills a larger space than anticipated.

Buddleia alternifolia **in flower**

Similar forms of interest *B. a.* 'Argentea' Slightly more tender and lower growing, with attractive silver foliage and slightly paler blue flowers. Not always easy to find but worth searching out for a sheltered site. Best protected by a sunny, sheltered wall.

Average height and spread
Five years
8 × 8 ft (2.4 × 2.4 m)
Ten years
12 × 12 ft (3.7 × 3.7 m)
Twenty years
15 × 15 ft (4.6 × 4.6 m)
Protrudes up to 6 ft (1.8 m) from support.

BUDDLEIA (Tender Forms)

KNOWN BY BOTANICAL NAME

Loganiaceae   *Wall Shrub*
Deciduous or evergreen

There are a number of varieties of buddleia that can only be grown in favourable areas with the protection of a wall or fence.

Origin Various.

Use As large fan-trained or freestanding shrubs for walls and fences.

Description *Flower* Racemes 3–12 in (7.5–30 cm) long in a range of colours through white, pink, blue and orange in mid summer to late autumn. *Foliage* Lanceolate, 4–8 in (10–20 cm) long, 1–3 in (2.5–7.5 cm) wide, light green or grey/green, often downy depending on variety. *Stem* Grey/green, often downy when young, becoming grey/brown. Upright. Fast rate of growth. *Fruit* Most forms produce small brown seedheads which have some limited winter attraction.

Hardiness Tolerates a minimum winter temperature of 14°F (−10°C) but only with the protection of a large wall or fence.

Soil requirements Tolerates a wide soil range, only disliking extremely dry conditions. Requires a high degree of organic material and plant nutrient for best results.

Sun/Shade aspect Very sheltered aspect. Full sun for preference but tolerates light shade.

Pruning Prune back all previous season's growth, other than that required for training the main framework, to within 2 in (5 cm) of its origin in early to mid spring.

Training Tie to wires or individual anchor points in a fan shape or allow to grow free-standing.

Propagation and nursery production From semi-ripe cuttings taken in early to mid summer or hardwood cuttings taken in winter. Always purchase container grown; most varieties will have to be sought from specialist nurseries. Best planting height 6 in–3 ft (15–91 cm).

Problems Often exceeds the allotted area. May suffer some winter die-back but normally regenerates.

Forms of interest *B. auriculata* Lax growth carries fragrant creamy white flowers with yellow throats in panicles up to 2 in (5 cm) across and 8 in (20 cm) long. Lanceolate foliage 4 in (10 cm) long, 1 in (2.5 cm) wide, white felted underside, grey/green upper side in late summer to late autumn. From South Africa. *B. caryopteridifolia* Leaves oval to lanceolate, tooth-edged, up to 6 in (15 cm) long, 1 in (2.5 cm) wide. Attractive grey/green. Stems covered with white woolly down. Fragrant lavender-blue flowers in panicles up to 3 in (7.5 cm) long in late spring/early summer. From China. *B. colvillei* Long racemes of deep rose, tubular flowers, borne at the ends of the branches. Foliage ovate, grey/green with some yellow autumn colour. Difficult to find and must be sought from specialist nurseries. *B. c.* 'Kewensis' An attractive form with dark red flowers. Difficult to find. *B. crispa* Fragrant, tubular lilac flowers, grey/green foliage. *B. fallowiana* 'Alba' Panicles of pure white flowers up to 15 in (38 cm) long in early to mid

Buddleia fallowiana 'Alba' **in flower**

summer. Attractive grey/green foliage, 10 in (25 cm) long and up to 4 in (10 cm) wide. From China. *B. madagascariensis* Bold orange flowers, very prolific. Tender even in mild areas against walls but if circumstances suit, well worth the effort. *B. salviifolia* Semi-ever-green lanceolate foliage up to 4 in (10 cm) long and 2 in (5 cm) wide, grey/green. Panicles of white to pale lilac flowers with orange markings in each tubular floret in mid to late summer. From South Africa.

Average height and spread
Five years
6 × 6 ft (1.8 × 1.8 m)
Ten years
12 × 12 ft (3.7 × 3.7 m)
Twenty years
18 × 18 ft (5.5 × 5.5 m)
Protrudes up to 4 ft (1.2 m) from support if fan-trained, 8 ft (2.4 m) untrained.

BUPLEURUM FRUTICOSUM

KNOWN BY BOTANICAL NAME
Umbelliferae   *Wall Shrub*
Evergreen

Distinctively coloured flowers on an evergreen shrub.

Origin From southern Europe.
Use As a medium sized, evergreen wall shrub for shady walls.
Description *Flower* Ball-shaped clusters of green/cream to yellow/green flowers from mid summer to early autumn. *Foliage* Elliptic, ½–2in (1–5cm) long, dark, glossy, grey/green with silver undersides. *Stem* Light green to dark olive-green, forming a rounded shrub, somewhat loose in habit. Medium to slow growth rate. *Fruit* Brown seedheads, interesting in winter.
Hardiness Established shrubs withstand winter temperatures down to 4°F (−15°C), but young plants are less hardy. Good in exposed coastal sites.
Soil requirements Any soil conditions.
Sun/Shade aspect All but the most exposed walls. Best in full sun. Tolerates light shade but becomes looser in habit in deep shade.
Pruning None required. May be trimmed or cut back to maintain shape.
Training Requires wires or individual anchor points to secure and encourage the fan-trained shape.

Bupleurum fruticosum in flower

Propagation and nursery production From seed or from softwood cuttings taken in summer. Best planting height 8–18 in (20–45 cm). Should always be purchased container grown.
Problems Not easy to find.
Similar forms of interest. None.
Average height and spread
Five years
3 × 3 ft (91 × 91 m)
Ten years
6 × 6 ft (1.8 × 1.8 m)
Twenty years
8 × 8 ft (2.4 × 2.4 m)
Protrudes up to 3 ft (91 cm) from support.

CALLISTEMON CITRINUS

AUSTRALIAN BOTTLE BRUSH
Myrtaceae   *Wall Shrub*
Evergreen

If the right conditions can be offered, an extremely useful, attractive shrub.

Origin From Australia and Tasmania.
Use As a large, summer flowering wall shrub for mild districts. Ideal in large conservatories and greenhouses.

Callistemon citrinus in flower

Description *Flower* Tufted, brush-like spikes of red flowers, very dense in formation, mid to late summer. *Foliage* Narrow, lanceolate leaves, 1–1½ in (2.5–4 cm) long, light green, often with red/orange shading or pronounced coloured veins. They release an aromatic lemon scent when bruised. *Stem* Light green to grey/green, ageing to grey/brown. Upright when young, becoming more spreading with age. Medium rate of growth, slowing with age. *Fruit* May produce interesting, tufted, light brown seedheads.
Hardiness Tolerates a minimum winter temperature of 23°F (−5°C).
Soil requirements Good, rich, acid soil. Dislikes any alkalinity. If planted in a container in conservatories and greenhouses, the diameter of the container must be at least 21 in (55 cm) and a lime-free potting compost must be used.
Sun/Shade aspect Requires a very protected aspect in full sun. Does not tolerate any shade.
Pruning None required. Remove an old shoot occasionally to rejuvenate from the base.
Training Requires wires or individual anchor points to secure and encourage the fan-trained shape.
Propagation and nursery production From seed or softwood cuttings taken in late spring or early summer. Best planting height 15–30 in (38–76 cm). Purchase container grown. Not always easy to find.
Problems Not to be grown in alkaline soils or in locations with winter conditions well below freezing.

Similar forms of interest *C. c. 'Splendens'* More brilliant flowers. Slightly less height but possibly more hardy.
Average height and spread
Five years
6 × 6 ft (1.8 × 1.8 m)
Ten years
10 × 10 ft (3 × 3 m)
Twenty years
12 × 12 ft (3.7 × 3.7 m)
Protrudes up to 3 ft (91 cm) from support.

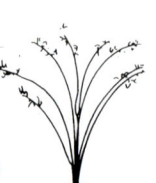

CAMELLIA JAPONICA

KNOWN BY BOTANICAL NAME
Theaceae   *Wall Shrub*
Evergreen

Although not normally considered as a wall shrub, in the right conditions can become an extremely attractive fan-trained specimen.

Origin From China, Korea and Japan.
Use As a fan-trained shrub for shady, protected walls and fences. Can be used to good effect in a large conservatory as a wall specimen.
Description *Flower* Large, cup-shaped flowers in a wide range of colours, may be single, semi-double, anemone or peony-shaped, loose double or tight double, depending on variety. Size ranges from small to very large. *Foliage* Dark, glossy-green upper surfaces, with grey/green undersides. Ovate to oblong, 3–4 in (7.5–10 cm) long and 1½ in (4 cm) wide. *Stem* Bright to dark green. Upright. Forming a stiff, solid shrub that can be fan-trained. A few varieties are more laxly presented. Slow to medium rate of growth. *Fruit* Insignificant.
Hardiness Tolerates a minimum winter temperature of 14°F (−10°C), but may shed leaves in harsh conditions, occasionally causing plant to fail.
Soil requirements Must have an acid soil: dislikes any alkalinity. If in a conservatory or greenhouse a container of not less than 21 in (55 cm) must be used, with a lime-free compost.
Sun/Shade aspect A sheltered sunless aspect. Prefers light to mid shade, dislikes full sun.
Pruning None required. May be cut back to keep within bounds. Young plants may be improved by removing one third of previous season's growth, after flowering, for first 2–3 years.
Training Requires wires or individual anchor points to secure and encourage the fan-trained shape.

Camellia japonica 'Adolphe Audusson' in flower

Propagation and nursery production From cuttings in early to mid summer. Best planting height 1½–6 ft (45 cm–1.8 m), ideally 2–2½ ft (60–76 cm). Purchase container grown. A limited number of varieties can be found in garden centres, less common varieties must be sought from specialist nurseries.

Problems Often planted on alkaline soils, where it fails, or in full sun, which it dislikes. Flowers can be damaged by frost in exposed areas.

Similar forms of interest *C. j.* 'Cornish Snow' Single, small white flowers, a very attractive small-leaved variety. *C. j.* 'Adolphe Audusson' Semi-double, blood-red flowers. *C. j.* 'Apollo' Semi-double rose-red flowers, sometimes with white blotches. *C. j.* 'Arejishi' Rose-red, peony-shaped flowers. *C. j.* 'Betty Sheffield Supreme' Semi-double, white, peony-shaped flowers with rose pink or red edges to each petal. *C. j.* 'Contessa Lavinia Maggi' Double white or pale pink flowers

Camellia japonica 'Elegans'

with cerise stripes. *C. j.* 'Elegans' Peach pink, large flowers. Anemone flower formation. *C. j.* 'Madame Victor de Bisschop' Semi-double, white flowers. *C. j.* 'Mars' Red, semi-double in form. *C. j.* 'Mathotiana Alba' Double white flowers of great beauty. *C. j.* 'Mathotiana Rosea' A double pink form. *C. j.* 'Mercury' Deep crimson flowers, semi-double in form. *C. j.* 'Nagasaki' Semi-double, rose pink flowers with white stripes. *C. j.* 'Tricolor' Semi-double white flowers with carmine or pink stripe. *C.* × 'Mary Christian' Single, clear pink flowers. Tall growing. *C.* × *williamsii* 'Donation' Clear pink, semi-double flowers. Possibly the best known camellia. Height 8 ft (2.5 m).

The above are just a selected few of the many hundreds of varieties available.

Average height and spread
Five years
3 × 3 ft (91 × 91 cm)
Ten years
6 × 6 ft (1.8 × 1.8 m)
Twenty years
10 × 10 ft (3 × 3 m)
Protrudes up to 3 ft (91 cm) from support.

CAMPSIS GRANDIFLORA (*C. chinesis*, *Bignonia grandiflora*, *Tecoma grandiflora*)

TRUMPET CREEPER

Bignoniaceae   *Woody Climber*
Deciduous

A spectacular flowering climber but requiring considerable space and a warm wall location for maximum flower production.

Origin From Eastern Asia, China and Japan.
Use As a large rambling climbing vine for sunny locations on walls and fences. Also good for covering large pergolas, gazebos etc. and for climbing large, high-canopied trees where adequate sunshine is available.

Campsis grandiflora in flower

Description *Flower* Panicles of six to 12 deep orange/red, wide mouthed, trumpet-shaped flowers up to 3 in (7.5 cm) long. The mouth of each flower is attractively divided into five segments, resembling five lips. Late summer/early autumn. *Foliage* Seven to nine light green oval leaflets up to 3 in (7.5 cm) long with coarse-toothed edges make up a pinnate shaped leaf. Good yellow autumn colour. *Stem* Light green when young, becoming yellow/brown, finally brown, twining, not self-clinging. Fast rate of growth. *Fruit* None of interest.

Hardiness Tolerates a minimum winter temperature of 14°F (−10°C).

Soil requirements Moderately alkaline to acid. Requires a deep, well fed, well drained soil with a high organic content for best results.

Sun/Shade aspect Sheltered aspect. Must be in full sun to ripen previous season's growth and encourage production of subsequent flowers.

Pruning Long, unwanted shoots produced in late spring and summer can be cut hard back after leaf fall in autumn, except those which are required for further training for the vine shape.

Training Tie to wires or trellis on walls and fences. When grown up trees, will require some early support by tying in.

Propagation and nursery production By use

---

of root cuttings, removal of self-rooted suckers, hardwood cuttings taken in winter or seed collected following a hot summer. Best size to purchase 18 in–4 ft (45 cm–1.2 m). Should always be container grown. Will have to be sought from good general nurseries or specialist growers.

Problems Not quick to cover a required area. Can be shy to come into flower and may require several years before it produces any type of good display. Late to break leaf in spring – often as late as the end of early summer – especially in the year following planting.

Forms of interest See further entry.

Average height and spread
Five years
8 × 8 ft (2.4 × 2.4 m)
Ten years
16 × 16 ft (5 × 5 m)
Twenty years
32 × 32 ft (9.7 × 9.7 m)
Protrudes up to 3 ft (91 cm) from support.

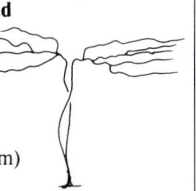

CAMPSIS RADICANS

TRUMPET CREEPER

Bignoniaceae   *Woody Climber*
Deciduous

This group of campis can be more relied upon to flower than *C. grandiflora* but still may take some time to settle and become established.

Origin From south-eastern USA and of garden origin.
Use For sunny walls and fences, for covering large pergolas, gazebos or other similar structures and for climbing large, high-canopied trees where adequate sunshine is available.
Description *Flowers* Four to twelve trumpet-shaped flowers 2–3 in (5–7.5 cm) long and up to 1½ in (4 cm) wide at the mouth presented in clusters at the ends of shoots in late summer to early autumn. The mouth of the flower has an interesting five-lobed effect, giving a lip-like appearance. Orange/red in colour, some red and yellow varieties. *Foliage* Up to 11 almost oval leaflets up to 4 in (10 cm) long with toothed edges form pinnate leaves 8–12 in (20–30 cm) long, light to mid green in colour. Good yellow autumn colour. *Stem*

Campsis × *tagliabuana* 'Madam Gallen' in flower

Campsis radicans in flower

Light grey/green when young ageing to creamy brown. Twining with retaining tendrils at leaf joints. Fast rate of growth. *Fruit* None of interest.
Hardiness Tolerates a minimum winter temperature of 4°F (−15°C).
Soil requirements Moderately alkaline to acid; requires a deep, well fed, well drained soil with a high organic content for best results.
Sun/Shade aspect Must be in full sun to ripen previous season's growth and encourage production of subsequent flowers.
Pruning Long, unwanted shoots produced in late spring and summer can be cut hard back after leaf fall in autumn except those that are required for further training.
Training Support with trellis, individual anchor points or wires; there is a certain amount of self-clinging by tendrils, but the weight of branches normally calls for secondary securing.
Propagation and nursery production By use of root cuttings, removal of self-rooted suckers, hardwood cuttings taken in winter or seed collected following a hot summer. Best size to purchase 1½–4 ft (45 cm–1.2 m). Should always be purchased container grown. Available from good garden centres, nurseries and specialist growers.
Problems Can be slow to come into flower and sometimes late to break leaf in spring, often as late as early summer. Often planted in areas where it is unable to fulfil its full potential because of lack of space or an unsuitable aspect.
Similar forms of interest *C. r.* 'Flava' A variety with all yellow flowers. *C. r.* 'Atropurpurea' Deep scarlet flowers; not readily available. *C. r.* 'Yellow Trumpet' Good bold yellow flowers. *C.* × *tagliabuana* 'Madam Gallen' A good free flowering variety with large salmon red flowers and good foliage.
Average height and spread
Five years
10 × 10 ft (3 × 3 m)
Ten years
20 × 20 ft (6 × 6 m)
Twenty years
30 × 30 ft (9 × 9 m)
Protrudes up to 3 ft (91 cm) from support.

CARAGANA ARBORESCENS 'LORBERGII'

SALT TREE
Leguminosae **Hardy Tree**
Deciduous
The attractive feathery foliage is shown off well when fan-trained.

Origin From Siberia and Manchuria.
Use As a small fan-trained tree or shrub for large walls.
Description *Flower* Small, yellow, pea-shaped ¼ in (5 mm) long flowers borne in clusters of up to four on thin stalks, mid to late spring. *Foliage* Very thin, wispy, feathery, light grey/green leaves up to 2 in (5 cm) long. Yellow autumn colour. *Stem* Grey/green with

---

attractive pronounced buds on branches. Moderately fast growing, slowing with age. *Fruit* Small pods, 1½–2 in (4–5 cm) long, containing four to six seeds, produced in autumn.
Hardiness Tolerates a minimum winter temperature of 0°F (−18°C).
Soil requirements Any soil conditions; tolerates high alkalinity.
Sun/Shade aspect Tolerates all aspects; very wind resistant. Best in full sun, but tolerates light shade.
Pruning Prune young bush trees hard in spring following planting. Select and train resulting five to seven shoots and tie into a fan-trained shape. In subsequent years, remove all side growths back to 2 in (5 cm) from their origin after flowering and maintain main branches in fan shape.
Training Requires tying to wires or individual anchor points.
Propagation and nursery production From grafting on to understock of *C. arborescens*. Ensure that purchased plants are bush and grafted or budded at ground level. Plant bare-rooted, root-balled (balled-and-burlapped) or container grown; best planting height 3 ft (91 cm).

Caragana arborescens 'Lorbergii' in flower

Problems May be late to break leaf in spring and can appear to be dead, but grows quickly once established.
Similar forms of interest None.
Average height and spread
Five years
6 × 6 ft (1.8 × 1.8 m)
Ten years
13 × 13 ft (4 × 4 m)
Twenty years
18 × 18 ft (5.5 × 5.5 m)
Protrudes up to 3 ft (91 cm) from support.

CARPENTERIA CALIFORNICA

KNOWN BY BOTANICAL NAME
Philadelphiaceae **Wall Shrub**
Evergreen
A magnificent flowering shrub, well worth the trouble of finding a favourable wall planting site.

Origin From California.
Use As an evergreen, summer-flowering shrub for mild areas. Ideal for fan-training on a sunny wall, particularly in mild regions.
Description *Flower* 2–3 in (5–7.5 cm) wide, pure white, saucer-shaped flowers with yellow anthers, borne in mid summer on mature wood. *Foliage* Leaves light to bright green, broad, lanceolate, 2–4 in (5–10 cm) long. *Stem* Light to dark green, upright at first, slightly spreading with age, forming a good fan-shape with training. Medium rate of growth. *Fruit* Small brown seedheads give limited winter attraction.
Hardiness Reacts badly to temperatures below 23°F (−5°C) but normally rejuvenates from ground level.
Soil requirements Deep, rich soil. Tolerates both acidity and alkalinity.
Sun/Shade aspect A sheltered aspect in full sun.
Pruning Remove one third of oldest wood each spring to maintain health. May be cut back hard and will rejuvenate, but can take up to two years to flower again.
Training Requires wires or individual anchor points to secure and encourage the fan-trained shape.
Propagation and nursery production From seed or from softwood cuttings taken in mid to late summer. Best planting height 8–24 in (20–60 cm). Purchase container grown. May be difficult to find and should be sought from specialist nurseries.
Problems When young the shrub appears weak, but it develops well after planting.
Similar forms of interest *C. c.* 'Ladham's Variety' Said to be more free-flowering than the parent, with larger flowers. Difficult to find.
Average height and spread
Five years
5 × 5 ft (1.5 × 1.5 m)
Ten years
7 × 7 ft (2.1 × 2.1 m)
Twenty years
9 × 9 ft (2.7 × 2.7 m)
Protrudes up to 4 ft (1.2 m) from support.

Carpenteria californica in flower

Catalpa bignonioides in flower

CATALPA BIGNONIOIDES

INDIAN BEAN TREE, SOUTHERN CATALPA

Bignoniaceae　　　　　　　　　**Hardy Tree**
Deciduous

An attractive, large-leaved tree that can be adapted for fan-training against a wall, where it can gain some additional protection.

Origin From south-eastern USA.
Use As a large, fan-trained tree for walls.
Description *Flower* Upright panicles 8–10 in (20–25 cm) long of white, bell-shaped flowers with frilled edges, yellow markings and purple spotted throats, produced mid summer. *Foliage* Broadly ovate leaves, 6–10 in (15–25 cm) long and 3–8 in (7.5–20 cm) wide, presented on long stalks. Good yellow autumn colour. Foliage smells unpleasant

Catalpa bignonioides 'Aurea' in leaf

when crushed. *Stem* Light grey/green, becoming green/brown. Strong and upright. Pruning increases branching. Medium to fast growth rate, becoming slower and more spreading after the first five years. *Fruit* Long, narrow, green, ageing to black, slender pods, 8–15 in (20–38 cm) long, produced in early autumn and retained into early winter.
Hardiness Tolerates a minimum winter temperature of 4°F (−15°C), but stem damage can be caused by winter frosts, especially in the golden-leaved varieties; however, this may in fact be an advantage in encouraging the plant to branch.

Soil requirements Requires a deep, rich soil to do well. Shows signs of chlorosis on extremely thin alkaline soils.
Sun/Shade aspect Requires a moderately sheltered aspect. Golden-leaved varieties scorch in full sun.
Pruning Prune young trees hard in spring following planting. Select and train resulting five to seven shoots and tie into a fan-trained shape. In subsequent years, remove all side growths back to two points from their origin and maintain main branches in fan shape.
Training Will require tying to wires or individual anchor points.
Propagation and nursery production From seed or layers for green-leaved varieties. Purple and golden-leaved varieties normally grafted onto understock of *C. bignonioides* or layered. Purchase bare-rooted or container grown; best planting height 3–6 ft (91 cm–1.8 m). Select only young trees that have not formed any main branches. For *C. b.* 'Aurea', buy trees that have been grafted or budded at ground level.
Problems Flowering may be decreased by fan-training but leaves will increase in size, particularly on golden-leaved varieties. May be damaged by high winds or heavy snow; consider location when planting. Young trees rarely look attractive, especially while in nursery production.
Similar forms of interest *C. b.* 'Aurea' Attractive, broad, large, golden-yellow leaves. A less hardy form, even slightly tender. One-third average height and spread, but may reach more in ideal conditions. *C. b.* 'Variegata' Attractive large-leaved foliage, grey/green leaves margined with gold. Limited in commercial production but not impossible to find. *C. × hybrida* 'Purpurea'

Catalpa × hybrida 'Purpurea' in leaf

New growth purple to purple/green, ageing to dark green. White flowers. Two-thirds average height and spread. From central USA.
Average height and spread
Five years
16 × 16 ft (4.9 × 4.9 m)
Ten years
30 × 30 ft (9 × 9 m)
Twenty years
39 × 39 ft (12 × 12 m)
Protrudes up to 5 ft (1.5 m) from support.

CEANOTHUS
(Evergreen Forms)

CALIFORNIA LILAC

Rhamnaceae　　　　　　　　　**Wall Shrub**
Evergreen

If grown in a mild area or protected from winter winds and cold, can give a very spectacular effect in spring and summer when grown as a fan-trained wall shrub.

Origin From southern states of the USA. Many varieties of garden origin.
Use As a fan-trained wall shrub for walls and fences.
Description *Flower* Various shades of blue flowers, some tufted, borne in panicles or umbels in mid to late spring, some varieties early or late summer and even autumn. *Foliage* Leaves mostly ovate, ½–1½ in (1–4 cm) long, light to dark green, in a few varieties broad to narrow lanceolate. All with shiny upper surfaces and dull grey undersides. In some varieties leaves have pronounced tooth edge, others convex, inturned shapes. *Stem* Light green to grey/green. Upright when young, becoming very twiggy. Medium rate of growth. *Fruit* Insignificant.

Ceanothus arboreus 'Trewithen Blue' in flower

Hardiness Tolerates a minimum winter temperature of 14°F (−10°C). Foliage very susceptible to scorch by cold winter winds.
Soil requirements Good, deep, rich soil. Tolerates both acidity and mild alkalinity. Thin chalk or limestone soils will induce severe chlorosis.
Sun/Shade aspect Requires a sheltered aspect; prefers full sun, tolerates light to medium shade.
Pruning Prune shoots by one third on 3–4 year old shrubs, annually after flowering. This will encourage new growth. Treat severe winter damage by cutting back into non-damaged wood.
Training Requires wires or individual anchor points to secure and encourage the fan-trained shape.
Propagation and nursery production From semi-ripe cuttings taken in mid summer. Always purchase container grown. Garden centres and nurseries generally stock a representative range of varieties; specific varieties may be sought from specialist nurseries.

Problems Leaves liable to scorching by cold winds. Will not attain full height and spread in unsuitable areas and likely to experience chlorosis on unsuitable soils.
Forms of interest *C. americanus* (New Jersey Tea) Panicles of white flowers in early to mid summer; dark green ovate leaves. A slightly tender variety reaching two thirds average height and spread. From eastern and central USA. *C. arboreus* (Tree Ceanothus) Deep, vivid blue flowers in panicles borne in spring. Large, ovate, dark green leaves. Slightly more tender than the average and attains one third more height and spread. *C. a.* **'Trewithen Blue'** Flowers slightly scented and deeper blue than *C. arboreus*. *C.* **'A. T. Johnson'** Mid to pale blue panicles of flowers, late spring, some early autumn flowering. A light green, large-leaved variety. Very vigorous in habit, in some situations exceeding average height and spread. *C.* **'Autumnal Blue'** Good-sized panicles of dark blue flowers, late summer and autumn. One of the hardiest varieties. *C.* **'Blue Cushion'** Very deep blue flowers, spreading but close-growing. *C.* **'Burkwoodii'** Rich blue flowers borne mainly late spring and early summer, with good displays intermittently until autumn. Slightly more tender and slightly less height and spread than the average. *C.* **'Cascade'** Powder blue

Ceanothus dentatus in flower

Ceanothus **'Cascade'** in flower

flowers in open panicles in spring. Foliage light green and more lanceolate than normal. Branches more lax and open, forming attractive, almost pendulous habit. *C.* **'Concha'** Bright blue summer flowers. Scarce. *C.* **'Delight'** Deep blue flowers, produced in panicles 3–4 in (7.5–10 cm) long in mid to late spring. Leaves broad, lanceolate and green. Said to be one of the hardiest varieties. *C. dentatus* (Santa Barbara Ceanothus) Bright blue flowers in late spring, small, tooth-edged dark green leaves. *C.* **'Dignity'** Dark blue flower panicles and dark green foliage. Normally flowers in spring, sometimes intermittently in autumn. *C. divergens* Deep blue flowers, spreading habit. *C.* **'Edinburgh'** (syn. *C.* **'Edinensis'**) Mid blue panicles of flowers in spring. Broad, olive-green leaves. Less than average hardiness. *C.* **'Emily Brown'** Fluffy violet/blue flowers in early summer. Low-growing. May be more tender. *C.* **'Floribundus'** Large clusters of mid blue flowers in late spring. *C.* **'Hurricane Point'** Cornflower-blue flowers late spring/early summer. Good foliage. Low growing. *C. impressus* Deep blue flowers, small, but borne in great profusion. Distinctive foliage effect, with small, curled, dark green leaves, veins being very deeply impressed within the surface. New shoots red to purple/red in colour. One of the hardiest of the ceanothus varieties. *C. i.* **'Puget Blue'** Deeper blue flowers and larger foliage. Possibly less hardy

than its parent. *C.* **'Indigo'** Indigo blue flowers in early summer. *C.* **'Italian Skies'** Mid to soft sky-blue panicles of flowers, borne in trusses on branching stems in spring. Medium-sized, round to ovate light green leaves. Less hardy than average. *C.* **'Joyce Coulter'** Deep blue flowers. May be scarce; obtain from specialist nursery. *C.* **'Julia Phelps'** Deep cobalt-blue flowers and deep green leaves. *C.* × *lobbianus* **'Russellianus'** Bright blue flowers, freely borne in mid to late spring. Less hardy than average. *C. pappillosus roweanus* Dark blue flowers in late spring; sticky leaves. Tender. *C. prostratus* (Squaw Carpet) Bright blue flowers borne freely in spring on this creeping, spreading plant with small, dark green to light green, broad to lanceolate leaves. *C.* **'Ray Hartman'** Large deep blue flowers in mid spring to late summer. Hardy. *C. rigidus* Very dark blue flowers in small, short tufted panicles profusely borne mid to late spring. Interesting foliage, very dark olive green, small and crinkled. Hard to find. *C.* **'Snow Flurries'** Snow-white flowers. Less hardy than average. *C.* **'Southmead'** Sky blue flowers in late spring and early summer. A very dense-growing shrub, with light green, broad, lanceolate leaves. Slightly less hardy than average. *C. thyrsiflorus* An abundance of medium-sized, well-spaced, mid blue flower panicles in spring and early summer. Dark green leaves. One of the hardiest varieties. *C. t.* **'Blue Mound'** Covered in short panicles of deep blue flowers, late spring and early summer. Dark green leaves. *C. t.* **'Repens'** (Creeping Blue Blossom) Rich blue flowers in abun-

Ceanothus **'Topaz'** in flower

dance in mid spring. Good-sized, dark green, tooth-edged foliage. *C. t. repens* **'Gnome'** Light blue flowers in spring, deep green leaves. Low habit. *C.* **'Topaz'** Large, well-spaced panicles of indigo blue flowers, mid to late summer. Large, round or ovate, mid green leaves. In cold climates should be considered semi-evergreen or even deciduous. *C.* × *veitchianus* Deep blue flowers, late spring and early summer. Medium-sized, dark green, broadly lanceolate leaves. Taller than average varieties and said to be one of the hardiest. *C.* **'Yankee Point'** Panicles of light blue flowers in mid spring. Light to mid green, medium-sized, narrow, ovate leaves. Compact habit.

Average height and spread
Five years
8 × 8 ft (2.5 × 2.5 m)
Ten years
12 × 12 ft (3.5 × 3.5 m)
Twenty years
18 × 18 ft (5.5 × 5.5 m)
Protrudes up to 5 ft (1.5 m)
from support.

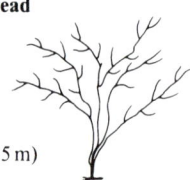

CEANOTHUS (Deciduous Forms)

CALIFORNIA LILAC

Rhamnaceae   **Wall Shrub**
Deciduous
A very attractive range of late summer/early autumn shrubs, ideal for fan-training on walls.

Origin From California but many varieties of garden origin.
Use As a fan-trained wall shrub for walls and fences.
Description *Flower* Flowers in panicles up to 3–4 in (7.5–10 cm) long, various shades of blue, pink or white, in late summer and in some varieties held into early autumn. *Foliage* Leaves medium to large, ovate, 3–5 in (7.5–12 cm) long, tooth-edged, light green to olive-green, some varieties having pink to pink/red leaf stalks. Some yellow autumn colour. *Stem* Upright, strong new growth produced each spring with flowers borne at the tips. Light to mid-green, ageing to dark brown in autumn. Fast to medium rate of growth.
Hardiness Tolerates a minimum winter temperature of 14°F (−10°C).
Soil requirements Good, rich, deep soil, tolerates poorer types if given adequate feeding but liable to chlorosis when severe alkalinity is present.
Sun/Shade aspect Tolerates all but the most exposed of aspects. Best in full sun to very light shade.

232

Pruning Prune back hard in spring all previous season's growth to 4 in (10 cm) from point of origin except shoots needed to form a fan shape.

Training Requires wires or individual anchor points to secure and encourage the fan-trained shape.

Propagation and nursery production From softwood cuttings in summer or by hardwood cuttings in late autumn to early winter. Best planting height 15 in–3 ft (38–91 cm). Purchase container grown. Most varieties easy to find, especially at flowering time.

Problems If insufficiently pruned becomes weak and performs insipidly. Shoots can occasionally be broken by strong winds.

Forms of interest *C.* **'Gloire de Versailles'** A popular variety with large panicles of well-spaced, powder blue flowers, mid to late summer. *C.* **'Henri Defosse'** Beautiful large panicles of deepest blue to purple/blue flowers, late summer. Dark green, medium

Ceanothus 'Gloire de Versailles' in flower

sized, elliptic leaves. *C.* **'Marie Simon'** Pale pink flowers borne in good-sized panicles, mid to late summer. Mid green oblong leaves with purple/red main veins and leaf stalks. Possibly more tender than average. *C.* **'Perle Rose'** Flowers carmine, borne in good-sized panicles, mid to late summer. Foliage smaller than average and with red to purple/red veins. May suffer die-back in winter. May be slower to establish than other varieties.

Average height and spread
Five years
5 × 5 ft (1.5 × 1.5 m)
Ten years
6 × 6 ft (1.8 × 1.8 m)
Twenty years
8 × 8 ft (2.4 × 2.4 m)
Protrudes up to 5 ft (1.5 m) from support.

Origin From North East Asia.

Use For growing over large buildings such as garages and sheds, through large established trees and shrubs or over large constructions such as pergolas.

Description *Flower* Small green flowers carried in clusters of up to four in early summer, of little interest. Flowers may be of single sex. *Foliage* Oval, up to 5 in (12 cm) long, with points; carried on short stalks up to 1 in (2.5 cm) long; light to mid green, very good yellow autumn colouring. *Stem* Twisting, twining, not self-clinging; light grey/green when young becoming light creamy brown with age. Some limited winter attraction in good light. Very fast growing. *Fruit* The main attraction. Capsules, bright yellow in colour when ripe, open to reveal a scarlet-coated seed within. Carried in large numbers on mature climbers. The hermaphrodite-flowered form is self fertile and bears fruit without a pollinator; otherwise male and female plants will be necessary.

Hardiness Tolerates a minimum winter temperature of 4°F (−15°C.)

Celastrus orbiculatus in fruit

Soil requirements Does well on all soil types, both alkaline and acid, with no particular preference except for adequate root run.

Sun/Shade aspect Full sun to medium shade with no particular preference.

Pruning Not normally considered practical as it covers an extremely large area but can be reduced in size if required after fruiting.

Training Leave to ramble through whatever type of construction or tree it is to cover. Self supporting by twining effect but not self-clinging.

Propagation and nursery production From seed collected in autumn and sown directly or layer lower shoots. Best planting height 18 in–4 ft (45 cm–1.2 m). Purchase container grown; available from good garden centres and general nurseries.

Problems Its overall size is often underestimated and it must be allowed to achieve this size to produce good displays of fruit. Some all male forms may exist when propagated from seed, but most plants produced in commercial horticulture are of the hermaphrodite form so the problem of also finding space for a female plant normally does not arise.

Similar forms of interest None.

Average height and spread
Five years
10 × 10 ft (3 × 3 m)
Ten years
20 × 20 ft (6 × 6 m)
Twenty years
30 × 30 ft (9 × 9 cm)
Protrudes up to 4 ft (1.2 m) from support.

Origin From southern Europe through to the Orient.

Use As a fan-trained or freestanding large shrub for walls.

Description *Flower* Numerous purple/rose pea-shaped flowers ½–1 in (1–2.5 cm) long, borne as leaves are produced, on both young and old branches in late spring, early summer. *Foliage* Deeply veined, broad, heart-shaped leaves, purple/green with a blue sheen. Good yellow autumn colours. *Stem* Dark brown to almost black stems. Very twiggy and branching. Good growth from base. Forms a large fan-trained shrub. Moderately slow growing. *Fruit* Pods, 3–4 in (7.5–10 cm) long, light grey/green, follow the flowers, ageing to grey/brown in autumn and retained in winter.

Hardiness Tolerates a minimum winter temperature of 4°F (−15°C).

Soil requirements Does best on neutral to acid soil types, but will tolerate moderate alkalinity.

Sun/Shade aspect Tolerates a moderately exposed aspect. Full sun to medium shade, with light shade for preference.

Pruning Cut young plants hard back in mid spring in second year of planting, which will induce strong shoots. Select five to seven to form a fan-trained shrub. Continue to tie in these shoots through their life. Any large protruding branches can be removed in winter.

Training Will require tying to wires or individual anchor points for fan-training, or allow to grow freestanding.

Propagation and nursery production From seed or layers. Purchase container grown or root-balled (balled-and-burlapped). Some of the latter are grown in peat soil and may require careful weaning to become accustomed to normal soil. Plant young plants from 18 in (45 cm) to 4 ft (1.2 m) tall; established or trained trees are rarely available and not recommended.

Ceanothus 'Marie Simon' in flower

Cercis siliquastrum in flower

Problems This shrub takes a number of years to reach any significant size. Some forms are very scarce in production and are difficult to find.

Similar forms of interest *C. canadensis* (Redbud) Large leaves and clusters of pale rose pink flowers in early summer. Less hardy than average. From central and eastern USA; limited in production outside its native environment. *C. c.* **'Forest Pansy'** Heart-shaped deep purple leaves maintained throughout summer, turning bright scarlet in autumn. Flowers inconspicuous. Two thirds average height and spread. Only obtainable from specialist nurseries. *C. chinensis* Flowers purple/pink. Slightly more than average height and spread. From China. Seldom seen in production. *C. occidentalis* Rose-coloured flowers on short stalks. Two thirds average height and spread. From California. Rarely found outside its native environment. *C. racemosa* Flowers red/pink, produced in racemes 4 in (10 cm) long in late spring. Less hardy. From China. Somewhat scarce in production. *C. siliquastrum* **'Alba'** A pure white-flowering form from Europe and the Orient. A plant that is very scarce in commercial production in Europe but is more readily available in North America.

Average height and spread
Five years
5 × 5 ft (1.5 × 1.5 m)
Ten years
10 × 10 ft (3 × 3 m)
Twenty years
20 × 20 ft (6 × 6 m)
Eventually reaches 39 ft (12 m) but extremely slowly, taking 50 years or more in most northerly locations. Protrudes up to 4 ft (1.2 m) from support if fan-trained, 12 ft (3.5 m) untrained.

CESTRUM ELEGANS
(*C. purpureum*)

KNOWN BY BOTANICAL NAME
Solanaceae   *Tender Wall Shrub*
Evergreen
A tender shrub for mild areas and benefiting from the protection of a wall.

Origin From Mexico.
Use As a wall shrub in mild areas or for conservatories and greenhouses in large containers or in a border.
Description *Flower* Pendent panicles, up to 8 in (20 cm) long and 4 in (10 cm) wide at base, red/purple to pink, each floret up to 1 in (2.5 cm) long and $\frac{3}{4}$ in (2 cm) wide, with pointed lobes at the ends. Flowers throughout summer. *Foliage* Lanceolate to oblong, up to 5 in (12 cm) long with a downy covering. Grey/green. *Stem* Grey/green, upright becoming arching at ends with weight of flowers. Medium growth rate. *Fruit* Rarely fruits, but if it does will produce globe-shaped, red/purple berries, $\frac{1}{2}$ in (1 cm) wide.
Hardiness Tolerates a minimum winter temperature of 12°F (−5°C) in very sheltered locations. Dislikes wind chill conditions.
Soil requirements If grown in containers, use a good quality potting compost. If grown in the soil provide a well drained, average garden soil, either alkaline or acid.
Sun/Shade aspect Requires a very sheltered aspect or the protection of a greenhouse or conservatory. Full sun to very light shade.
Pruning None, but individual limbs can be removed if required.
Training Freestanding. May need tying to anchor points for support with age.

Cestrum elegans in flower

Propagation and nursery production From cuttings taken from side shoots up to 3 in (7.5 cm) long in mid summer. Will require bottom heat of 65–70°F (18–21°C) to encourage rooting. Always purchase container grown, will have to be sought from specialist nurseries. Best planting height 6 in–2 ft (15–60 cm) tall.
Problems Often not winter hardy. Difficult to find.
Similar forms of interest *C. roseum* Attractive rose/pink flowers.
Average height and spread
Five years
3 × 3 ft (91 × 91 cm)
Ten years
5 × 5 ft (1.5 × 1.5 m)
Twenty years
7 × 7 ft (2.1 × 2.1 m)
Protrudes up to 5 ft (1.5 m) from support.

CHAENOMELES

ORNAMENTAL QUINCE, JAPANESE
FLOWERING QUINCE, JAPONICA
Rosaceae   *Wall Shrub*
Deciduous
A fine flowering shrub, very attractive when pruned correctly as a fan-trained wall shrub.

Origin From China, most varieties being of garden origin.
Use As a fan-trained shrub for walls and fences.
Description *Flower* Single flowers shaped and sized like apple blossom borne in profusion on wood two years old or more, early to mid spring. Colours range through white, pink, apricot, flame, orange and red, depending on variety. *Foliage* Leaves elliptic, medium-sized, 3–4 in (7.5–10 cm) long, light to dark green. Some yellow autumn colour. *Stem* Upright when young and light green/brown, becoming dark brown, more twiggy and producing isolated large rigid thorns. Medium rate of growth. *Fruit* Large, pear-shaped fruits follow the flowers, ripening to an attractive bright yellow.
Hardiness Tolerates a minimum winter temperature of 0°F (−18°C)

Chaenomeles speciosa 'Moerloosii' in fruit

Soil requirements Does well on any soil but liable to chlorosis in very alkaline areas.
Sun/Shade aspect Tolerates all aspects. Does well in full sun to deep shade.
Pruning Apart from growth required for fan shape, remove all previous season's growth back to two buds in spring before flowering, making sure that flowering buds are not removed.
Training Requires wires or individual anchor points to secure and encourage the fan-trained shape.
Propagation and nursery production From semi-ripe cuttings taken in mid summer. Best planting height 15–30 in (38–76 cm). Always purchase container grown. Wide range of forms generally available, or obtain from specialist nurseries.
Problems Intermittently produces very sharp thorns. May suffer fungus disease such as canker; prune out affected wood.
Forms of interest *C. japonica* Orange/red flowers. *C. j.* **'Alpina'** Orange/red flowers borne freely, late spring. A little shy to fruit due to its less vigorous habit. *C. j.* **'Issai Red'** Small red flowers in abundance. *C. j.* **'Issai White'** Many small white flowers. *C. speciosa* **'Atrococcinea'** Large, deep crimson flowers. *C. s.* **'Brilliant'** Large brilliant red to clear scarlet flowers. *C. s.* **'Cardinalis'** Crimson-scarlet flowers. *C. s.* **'Geisha Girl'** Very attractive deep apricot flowers. Later flowering. *C. s.* **'Moerloosii'** (Apple Blossom) Pink and white flowers, more sparsely produced than some forms. *C. s.* **'Nivalis'** A pure white-

Chaenomeles speciosa 'Moerloosii' in flower

flowered variety, green/white on first opening. Fewer flowers than average, but growth more vigorous. *C. s.* **'Rosea Plena'** Double rich rose-pink flowers. *C. s.* **'Simonii'** Deep blood red flowers freely produced. Low growing. *C. s.* **'Snow'** Snow-white flowers. A good variety. *C. s.* **'Umbilicata'** Deep pink flowers, larger than most. *C. s.* **'Eximia'** Upright deep brick-red flowers. *C. s.* **'Verbooms Vermillion'** Upright growing, bright red flowers. *C. × superba* **'Aurora'** Peach/pink flowers, unusual. *C. × sup.* **'Ballerina'** Large deep red flowers. *C. × sup.* **'Boule de Feu'** Vermillion flowers, strong-growing. Good yellow fruit. *C. × sup.* **'Choshan'** Semi-double, peach/apricot flowers; low-growing. Easy to find *C. × sup.* **'Coral Sea'** Coral-pink good sized flowers and good fruits. *C. × sup.* **'Crimson and Gold'** Bright red flowers with pronounced golden anthers. Good fruit production. *C. × sup.* **'Elly Mossel'** Large, bright scarlet flowers, good fruit. *C. × sup.* **'Ernest Finken'** Upright brilliant red flowers. *C. × sup.* **'Etna'** Rich vermillion flowers, good colour. *C. × sup.* **'Fascination'** Vivid orange flowers. *C. × sup.* **'Fire Dance'** Rich, orange/scarlet flowers, good fruit. *C. × sup.* **'Hever Castle'** Shrimp pink flowers. *C. × sup.* **'Knap Hill Scarlet'** Smaller, brilliant orange/scarlet flowers, freely borne. Height slightly less than average. *C. × sup.* **'Hollandia'** An excellent scarlet/red flowering variety with good fruits. *C. × sup.* **'Nicoline'** Large red flowers, average height but with more spread. *C. × sup.* **'Pink Lady'** Good deep pink flowers and good fruits. *C. × sup.* **'Port Elliot'** Large red flowers, good growth.

Chaenomeles speciosa 'Simonii' in flower

Chaenomeles × superba 'Crimson and Gold' in flower

C. × sup **'Ohld'** Large red flowers of good stature. *C. × sup.* **'Rowallane'** Brilliant crimson flowers and small fruits. *C. × sup.* **'Taxus Scarlet'** Scarlet red flowers. *C. × sup.* **'Vermillion'** Vermillion red flowers. *C. × sup.* **'Vesuvius'** Scarlet red flowers.
There are many varieties of *C. speciosa* and *C. × superba*. Those listed are a good representative selection.
Average height and spread
Five years
7 × 7 ft (2.1 × 2.1 m)
Ten years
12 × 12 ft (3.7 × 3.7 m)
Twenty years
14 × 14 ft (4.3 × 4.3 m)
Protrudes 3–4 ft (91 cm–1.2 m) from support.

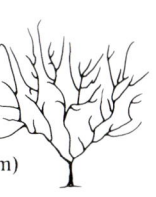

CHIMONANTHUS PRAECOX (*C. fragrans*)

FRAGRANT WINTERSWEET

Calycanthaceae   *Wall Shrub*
Deciduous
An interesting scented flowering shrub for winter display.

Origin From China.
Use As a freestanding wall shrub or for fan-training on sunny walls.
Description *Flower* Lemon yellow, waxy, hanging, bell-shaped flowers with purple anthers, frost-hardy, borne in late winter on mature branches three to four years old. *Foliage* Leaves light green to yellow green, medium-sized, elliptic, 3–7 in (7.5–18 cm) long, some yellow autumn colour. *Stem* Light green, strong, upright when young, ageing into brittle, twiggy branches. Medium growth rate, slowing with age. *Fruit* Insignificant.

Hardiness Tolerates a minimum winter temperature of 4°F (−15°C) but expect some die-back on growth produced in late autumn.
Soil requirements Does well on most soils, especially alkaline.
Sun/Shade aspect All but the most severe aspects. Full sun ripens the wood best for flowering. Tolerates light shade.
Pruning Normally not required.
Training Requires wires or individual anchor points to secure and encourage the fan-trained shape or can be allowed to grow freestanding.
Propagation and nursery production From semi-ripe cuttings taken in mid summer. Best planting height 15–30 in (38–76 cm). Purchase container grown from specialist nurseries.
Problems Can be slow to flower, especially in years when wood is slow to ripen. Looks unattractive in a container but grows rapidly once planted out.

Chimonanthus praecox in flower

Similar forms of interest *C. p.* **'Grandiflora'** Deeper yellow flowers with an interesting red stain in the throat. More difficult to find. *C. p.* **'Luteus'** Clear bright yellow flowers with even more waxy texture, more open than in the parent. Not easy to find but worth pursuing.
Average height and spread
Five years
6 × 6 ft (1.8 × 1.8 m)
Ten years
8 × 8 ft (2.4 × 2.4 m)
Twenty years
10 × 10 ft (3 × 3 m)
Protrudes up to 4 ft (1.2 m) from support if fan-trained, 10 ft (3 m) untrained.

CHOISYA TERNATA

MEXICAN ORANGE BLOSSOM

Rutaceae   *Wall Shrub*
Evergreen
A very attractive, scented, late spring to early summer flowering shrub.

Origin From Mexico.
Use As an evergreen shrub for summer flowering, but with training can become an interesting wall shrub.
Description *Flower* Fragrant, single, ½ in (1 cm) wide, white, orange-scented flowers, borne in flat-topped clusters, late spring to early summer. *Foliage* Leaves glossy, mid to dark green, trifoliate, 3–6 in (7.5–15 cm) long, which when crushed give off an aromatic scent. *Stem* Light to bright green, glossy, upright, becoming spreading and twiggy with age, forming a fan-shaped shrub. Fast to medium rate of growth when young or pruned back, slowing with age. *Fruit* Insignificant.
Hardiness Tolerates a minimum winter temperature of 14°F (−10°C). Leaf damage

235

can occur in lower temperatures or in severe wind chill. In some winters, may die back to ground level but can normally rejuvenate itself the following spring.
Soil requirements Does well on most, although very severe alkaline soils may lead to chlorosis.
Sun/Shade aspect Tolerates all but the most exposed aspects. Equally good in full sun or deep shade.
Pruning Two methods of pruning are advocated; cut back to within 12 in (30 cm) of ground level after three to four years so it can rejuvenate itself and repeat process every third or fourth year following. This keeps the foliage glossy and encourages flowering. Otherwise, on mature shrubs, remove one third of the oldest wood to ground level after flowering to encourage rejuvenation from centre and base and repeat annually.
Training Requires wires or individual anchor

Choisya ternata in flower

points to secure and encourage the fan-trained shape.
Propagation and nursery production From softwood cuttings taken in summer. Best planting height 1–2 ft (30–60 cm). Purchase container grown. Available from most garden centres and nurseries.
Problems If pruning is neglected, plant becomes old, woody and unproductive. Plants are always relatively small when purchased but quickly mature when planted out.
Similar forms of interest *C. t.* 'Sundance' Yellow/green foliage in spring, quickly becoming golden yellow which persists through winter. Slightly more tender. Two thirds average height and spread. Requires very light shade.
Average height and spread
Five years
5 × 5 ft (1.5 × 1.5 m)
Ten years
6 × 6 ft (1.8 × 1.8 m)
Twenty years
7 × 7 ft (2.1 × 2.1 m)
Protrudes up to 3 ft (91 cm) from support.

CISTUS

ROCK ROSE
Cistaceae   **Wall Shrub**
Evergreen
A tender shrub worth experimenting with for the beauty it can give in its summer display of flowers when fan-trained on a wall.

Origin From Southern Europe and North Africa.
Use As a fan-trained shrub for walls and fences.
Description *Flower* Single, 1½–2½ in (4–6 cm) wide, flat flowers, lasting only for one day but followed by new buds opening in rapid succession, early to late summer. White or white with brown or purple spots, through shades of pink to dark purple-pink. *Foliage* Leaves ovate or lanceolate, medium sized,

Cistus × purpureus in flower

1–2 in (2.5–5 cm) long. Light green or grey, glossy surfaced in green-leaved forms, grey-leaved having a grey down or bloom. Leaf buds and new growth are often sticky. *Stem* Light green or grey/green. Some forms grow upright, others are of spreading habit. Medium growth rate.
Hardiness Tolerates a minimum winter temperature of 23°F (−5°C).
Soil requirements Does well on all soil types, tolerating quite high alkalinity.
Sun/Shade aspect Requires a sheltered aspect. Best in full sun for free flowering.
Pruning Remove one third of oldest growth to ground level after flowering on plants established more than three years, or cut the plants hard back in spring to within 4 in (10 cm) from ground level every three to four years.
Training Requires wires or individual anchor points to secure and encourage the fan-trained shape.
Propagation and nursery production From softwood cuttings taken early to mid summer. Purchase container grown. A wide range of forms normally obtainable in mild areas, otherwise must be sought from specialist sources. Best planting height 8–30 in (20–76 cm).
Problems Container grown plants often look leggy and unsightly but are the best choice because the wood is harder. Plants looking soft and fleshy at purchase have been grown under glass and are less reliable, but for early to mid summer planting shrubs of any age are suitable.
Forms of interest *C. × aguilari* Large white flowers throughout early and mid summer. Grey/green foliage. *C. × a.* 'Maculatus' Large, flat flowers with a central ring of crimson blotches on each of the five petals. Dark green, glossy foliage 2–4 in (5–10 cm) long. Upright. *C. algarvensis* Yellow flowers

Cistus aguilari in flower

with brown central discs. Grey upright foliage. Tender. *C.* 'Anne Palmer' Pink flowers, dark green crinkled foliage. Attractive. *C.* 'Barnsley Pink' Pale pink flowers, green foliage. Dense habit. Spreading in nature. *C. × corbariensis* Small, 1 in (2.5 cm) wide, pure white flowers, opening from crimson-tinted buds. Small, light green to grey/green leaves. Hardier than most. *C. × cyprius* Large white flower with crimson blotches on each of the five petals. Shiny green foliage, 4–6 in (10–15 cm) long. Upright and tall-growing. Said to be one of the hardiest. *C. × c.* 'Albiflora' Pure white flowers, otherwise similar to parent plant. *C.* 'Greyswood Pink' A profusion of pink flowers. Downy grey to grey/green foliage. *C. ladanifer* (Gum Cistus) White flowers up to 4 in (10 cm) in width with chocolate stains at base of each crimped petal. Leaves narrow, lanceolate, 3–4 in (7.5–10 cm) long, bright to mid green with glossy surface. Strong-growing. *C. l.* 'Albiflorus' A pure white form of the above. *C. laurifolius* Flat, white flower with yellow marking in centre. Very leathery, dark blue/green leaves, 4–5 in (10–12 cm) long. A hardier variety. *C. × lusitanicus* Large, white flowers with crimson basal blotches, green leaves. Height and spread 12–24 in (30–60 cm). *C. × l.* 'Decumbens' Covered in small, white flowers, again with crimson basal blotches. *C.* 'Peggy Sannons' Flowers a soft shade of pink contrasting well with grey/green foliage with a downy texture, which is repeated on the upright stems. *C. populifolius* White flowers with yellow staining at the base of each petal. Rounded, poplar-shaped, hairy leaves produced on an erect, upright shrub. One of the hardiest. *C. × pulverulentus* (syn. *C. albidus × crispus*, *C.* 'Warley Rose') Cerise flowers, foliage sage green and waxy-textured. A low spreading shrub. *C. × purpureus* Flowers rosy crimson, each with a chocolate basal blotch. Narrow, green to grey/green foliage on strong, upright stems. *C. salvifolius* White flowers with a yellow basal stain to each petal, sage green leaves forming a foil. Height 12–24 in (30–60 cm); spread 24 in (60 cm). *C.* 'Silver Pink' Silver-pink flowers borne on long clusters, very attractive grey foliage. *C. × skanbergii* Clear pink flowers and small grey leaves make this one of the most attractive forms. *C.* 'Sunset' Deep cerise-pink flowers, grey foliage. Upright. *C.* 'Alan Fradd' White flowers with purple blotches. Good grower.
Average height and spread
Five years
3 × 3 ft (91 × 91 cm)
Ten years
4 × 4 ft (1.2 × 1.2 m)
Twenty years
5 × 5 ft (1.5 × 1.5 m)
Protrudes 1½–2 ft (45–60 cm) from support.

CLEMATIS ALPINA

ALPINE CLEMATIS, COLUMBINE

Ranunculaceae   *Woody Climber*
Deciduous

With its good range of colours and delicate flowers, this group of clematis enhances any spring garden.

Origin From northern Europe, northern Asia and of garden origin.
Use For walls, fences, pillars, trellis and pergolas and to ramble through large shrubs. Can be used as a ground cover with or without support.

Clematis alpina in flower

Description *Flower* Pendent, nodding flowers in shades from violet blue to purple, pink and white, carried singly on stalks up to 4 in (10 cm) long. No petals but four coloured sepals up to 1½ in (4 cm) long. Some double-flowered varieties. Produced from early to late spring with spasmodic later flowering in early summer, but not reliable. *Foliage* Leaves comprise nine leaflets, each ovate to lanceolate, up to 2 in (5 cm) long, forming a total length of 6 in (15 cm) with a width of 3 in (7.5 cm). Light green with yellow/brown autumn colour. *Stem* Twining, leaf stalks acting as supporting tendrils. Light green, ageing to grey/brown. Medium growth rate. *Fruit* Attractive, silky white seedheads up to 1½ in (4 cm) long produced after flowering and maintained well into summer/early autumn.

Clematis alpina 'Ruby' in flower

Clematis alpina 'White Moth' in flower

Hardiness Tolerates a minimum winter temperature of 0°F (−18°C).
Soil requirements Does well on all soil types with no particular preference, but must have adequate moisture throughout the year to induce good growth.
Sun/Shade aspect Tolerates all but the most exposed aspects. Requires its roots to be shaded and prefers its growth in the sun, but will tolerate very light shade.
Pruning Cut all current season's growth to within 6–12 in (15–30 cm) of its origin after main flowering period to produce new flowering growth for next spring throughout the summer; alternatively, lightly trim and tidy after flowering. The latter method will suffice for one to five years, but the plant will become old and woody and will need hard cutting back every five to six years.
Training Requires wires or a trellis on walls and fences. Allow to ramble freely through shrubs and trees.
Propagation and nursery production From semi-ripe internodule cuttings taken late spring/early summer. Always purchase container grown, best planting height 1½–3 ft (45–91 cm). Most varieties readily available from good garden centres and specialist nurseries, although some may need searching for.
Problems If not pruned correctly can become very woody and poor in growth and flowers.

Similar forms of interest *C. a.* 'Columbine' Pale blue flowers. One third more than average height and spread. *C. a.* 'Francis Rivis' Mid blue with white centres to the lantern-shaped flowers, which are larger than other varieties. *C. a.* 'Pamela Jackson' Mid blue, attractive, lantern-shaped flowers. *C. a.* 'Ruby' Purple/pink flowers with white centres. Generally reliable second late summer flowering. *C. a.* 'Siberica' (syn. *C. a.* 'Alba') White flowers, flushed pink/mauve at base. Good flowering performance. Scarce. *C. a.* 'White Moth' Pure white, double, nodding flowers. Most attractive.
Average height and spread
Five years
8 × 8 ft (2.4 × 2.4 m)
Ten years
8 × 8 ft (2.4 × 2.4 m)
Twenty years
8 × 8 ft (2.4 × 2.4 m)
Protrudes up to 2 ft (60 cm) from support.

CLEMATIS ARMANDII

EVERGREEN CLEMATIS

Ranunculaceae   *Woody Climber*
Evergreen to semi-evergreen

Given enough space, this evergreen climber produces an abundant display of pure white flowers in mid to late spring.

Origin From China.
Use As a climber for large walls or fences, to cover pergolas and roofs on small buildings. Can be allowed to grow through large shrubs or small trees.
Description *Flower* Saucer-shaped flowers up to 2 in (5 cm) wide, produced profusely in clusters at the leaf joints on wood two years old or more. Four to six narrow, oblong sepals, each ½ in (1 cm) wide, normally pure white, with some pale pink varieties, in mid to late spring. *Foliage* Trifoliate leaves with ovate to lanceolate leaflets, each 3–5 in (7.5–12 cm) long. Dark, glossy green with blue/silver undersides and prominent veins. Normally evergreen, but in severe winters may become semi-evergreen. *Stem* Light green, ageing to dark green. Limited branching; gains height very quickly and is reluctant to produce lower shoots. Twining, leaf tendrils giving some additional support. Medium to fast rate of growth. *Fruit* May produce small white fluffy seedheads.
Hardiness Requires a minimum winter

Clematis armandii in flower

temperature of 4°F (−15°C), although may be damaged by severe wind chill at this temperature.

Soil requirements Does well on all soil types, tolerating both alkaline and acid conditions, but must be moisture-retentive yet well drained.

Sun/Shade aspect Requires a sheltered aspect with protection from severe cold winds. Needs stem, leaves and flowers in full sun with roots in the shade.

Pruning Normally requires no pruning, but old shoots can occasionally be reduced to encourage rejuvenation.

Training Tie to wires or trellis, or allow to ramble along the top of a wall or through large shrubs and small trees.

Propagation and nursery production From semi-ripe cuttings or by grafting; the former method is difficult. Always purchase container grown; will have to be sought from specialist nurseries and named varieties may be very scarce in production and sometimes difficult to find. Best planting height 9 in–3 ft (23 cm–91 cm).

Problems Foliage may be completely defoliated by severe, cold winds. Size and production of flower may vary according to individual plants. Named varieties extremely scarce.

Similar forms of interest. *C. a.* 'Apple Blossom' Flowers suffused with pale mother-of-pearl pink. *C. a.* 'Everest' A good, strong-growing variety with pure white flowers. *C. a.* 'Snowdrift' Profusely borne snow-white flowers.

Average height and spread
Five years
10 × 10 ft (3 × 3 m)
Ten years
20 × 20 ft (6 × 6 m)
Twenty years
25 × 25 ft (7.6 × 7.6 m)
Protrudes up to 2 ft (60 cm) from support.

CLEMATIS CAMPANIFLORA

CAMPANULA FLOWERING CLEMATIS

Ranunculaceae   **Woody Climber**
Deciduous

A clematis with charming growth and foliage, bearing flowers which are among the most delicate found in the genus.

Origin From Portugal.
Use To ramble through wall shrubs and other climbers.

Clematis campaniflora **in flower**

238

Clematis cirrhosa balearica **in flower**

Description *Flower* Small, nodding, bell-shaped flowers with wide open mouths, carried singly on stalks up to 3 in (7.5 cm) long. White with delicate violet/blue-tinged sepals. Flowering from mid to late summer. *Foliage* Pinnate, with five to nine divisions to each of the three ovate to lanceolate leaflets. Each leaflet up to 3 in (7.5 cm) long. Very wide range of leaf shapes produced on a single plant. Clinging leaf tendrils. *Stem* Wispy, ranging, of thin constitution. Moderate rate of growth. *Fruit* No fruit of interest.

Hardiness Tolerates a minimum winter temperature of 14°F (−10°C).

Soil requirements Tolerates a wide range of soil types, only requiring adequate moisture and food retention.

Sun/Shade aspect Requires a sheltered aspect. Prefers light shade but will tolerate full sun as long as roots are shaded.

Pruning Requires only a light trim in early spring.

Training Allow to ramble through wall shrubs or other climbers or through a framework of wires. Rarely needs tying as is self-clinging by leaf tendrils.

Propagation and nursery production From semi-ripe internodal cuttings taken early to mid summer. Always purchase container grown, best planting height 6 in–3 ft (15–91 cm). Will have to be sought from specialist nurseries.

Problems Not a strong grower, although not delicate in constitution. Can be difficult to arrange and often grows into positions not intended, although not invasive.

Similar forms of interest None.

Average height and spread
Five years
6 × 6 ft (1.8 × 1.8 m)
Ten years
12 × 12 ft (3.7 × 3.7 m)
Twenty years
18 × 18 ft (5.5 × 5.5 m)
Protrudes up to 2 ft (60 cm) from support.

CLEMATIS CIRRHOSA (*C. calycina*)

WINTER-FLOWERING EVERGREEN CLEMATIS

Ranunculaceae   **Woody Climber**
Evergreen

With both attractively shaped evergreen foliage and winter flowers, this clematis is rightly gaining recognition as a good garden plant.

Origin From southern Europe.
Use For walls, fences, pillars, or to ramble

---

through large shrubs. Can be used as a ground cover with or without support.

Description *Flower* Up to 1 in (2.5 cm) wide, pendent, cup-shaped flowers, produced singly at leaf joints on stalks up to 1 in (2.5 cm) long. Yellow/white with red spots. Produced from early to late winter, whenever the weather is mild. *Foliage* Up to 3 in (7.5 cm) long and wide. Five individual leaflets, deeply lobed or toothed along the edges, giving a fern-like appearance. Bright green in spring, turning purple/bronze in winter. Shiny, evergreen surface. *Stem* Very twining, forming an interlocking mat. Light green when young, turning brown/grey. Medium rate of growth. *Fruit* May produce small, feathery seedheads after flowering, but they are of no real merit.

Hardiness Tolerates a minimum winter temperature of 14°F (−10°C).

Soil requirements Tolerates both acid and alkaline, but requires adequate moisture content to induce good growth.

Sun/Shade aspect Sheltered aspect for best winter flowering results. Resents strong, frost-laden winds, which cause foliage and flower damage. Requires its roots shaded and prefers its growth in light shade, although will tolerate full sun.

Pruning Trim the plants lightly after flowering each year for up to five years. Every fifth year prune hard back into the main structure to encourage a new head of growth. This may slightly reduce the flowering effect in the winter following such pruning, but it generally has the effect of encouraging better growth throughout.

Training Requires wire or trellis on walls or fences, or can be allowed to ramble through medium to large shrubs.

Propagation and nursery production From semi-ripe internodal cuttings taken late spring/early summer. Always purchase container grown; best planting height 1–3 ft (30–91 cm). May have to be sought from specialist nurseries, but not normally hard to find from such sources.

Problems Always crowns out on the top of a wall or shrub; rarely produces any low stem growth.

Similar forms of interest *C. c. balearica* (Fern-leaved clematis) Possibly slightly more tender. Pale yellow flowers with brown specks. Foliage slightly more cut. *C. c.* 'Wisley' Lighter green foliage, otherwise identical to its parent.

Average height and spread
Five years
6 × 6 ft (1.8 × 1.8 m)
Ten years
8 × 8 ft (2.5 × 2.5 m)
Twenty years
12 × 12 ft (3.7 × 3.7 m)
Protrudes up to 3 ft (91 cm) from support.

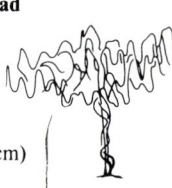

CLEMATIS
(Double and Semi-Double)

DOUBLE CLEMATIS

Ranunculaceae   *Woody Climber*
Deciduous

Attractive clematis with semi-double or double flowers in a wide range of colours, adding an extra dimension to this beautiful garden plant.

Origin Of garden or nursery origin.
Use As climbers for fences, walls, pillars, to ramble through medium-sized shrubs or over brushwood or pea-sticks. Can be used in large tubs and containers.
Description *Flower* Numerous sepals in a wide range of colours and flowering times, depending on variety, make up either semi-double or double flowers. Some varieties produce double flowers for the first crop and later flowering may well be single. *Foliage* Sometimes single ovate leaves, sometimes three to five lanceolate to ovate leaflets, each up to 4 in (10 cm) long, make up a single pinnate leaf. Mid green to grey/green. Tendril-type leaf stalks aid support. *Stem* Light green when young, ageing to green/brown and finally brown. Twining, with additional support from leaf tendrils. Medium to fast growth rate. *Fruit* Some varieties may produce white, tufted seedheads of some winter attraction.
Hardiness Tolerates a minimum temperature of 4°F (−15°C).
Soil requirements Tolerates all soil conditions except for extremely alkaline, but will produce average results even in the latter given a moisture-retentive soil with adequate organic material. For best results a higher degree of organic compost is advised than the normal planting recommendations for climbers and wall shrubs. When grown in tubs and containers use a soil-based potting compost, allow for adequate drainage and choose a container not less than 21 in (53 cm) wide and 18 in (45 cm) deep.
Sun/Shade aspect Tolerates all aspects, except extremely exposed. Full sun to light shade with full sun for preference. Roots should be shaded.
Pruning There are three different pruning methods, each particular to an individual variety. These are as follows:
1 Simply tidy and thin end shoots back to main framework after flowering.
2 Reduce by one third the growth made in the previous season in late winter, early spring.
3 Prune to within 12 in (30 cm) all growth produced in previous year during late winter, early spring.
It is important that the recommended pruning methods are carried out on an annual

Clematis **'Vyvyan Pennell'** in flower

basis, otherwise flowering will decrease and the overall well-being of the plant will suffer. The pruning method for each variety is given in 'Forms of Interest'. If the variety is not known use Method 2 as a safe option.
Training Allow to ramble through wires, trellis, branches of large shrubs or trees or any other support where the leaf tendrils can wrap around to anchor.
Propagation and nursery production From internodal cuttings taken in early to late summer, inserted into a sand/peat mixture, with some bottom heat to encourage rooting. Best planting height 6 in–3 ft (15–91 cm). Clematis is often available in a choice of 3–4 in (7.5–10 cm) or 1 litre (5–7 in/12–18 cm) pots; both sizes are equally good as long as the plants have been well hardened off and are not purchased before mid spring. Plant from mid spring through until late summer. Plants propagated in the spring are ripe enough for planting from mid autumn through until mid spring; these are normally offered in 3–4 in (7.5–10 cm) or 1 litre (5–7 in/12–18 cm) pots. Many varieties are standards and will be readily available, others will need searching for from specialist nurseries.
Problems Clematis wilt is an airborne fungus disease for which there is no prevention. The entire plant, both leaves and stem, turns black in early to late summer. Once an attack is seen, cut all growth to ground level, wash walls, fences or trellis with a mild disinfectant and protect any other plants growing below with plastic sheeting. There is then a 60 per

cent chance that new, uninfected shoots will develop from below ground level from the root system. Replanting with a new clematis in the same position is not a risk, but the area should be washed with mild disinfectant first.

The main reason for the failure of clematis plants is that they are often placed too close to the wall (see 'Planting Climbers and Wall Shrubs on p.168) and that not enough organic material such as garden compost, well rotted farmyard manure or mushroom compost is added to the soil. They will also fail to thrive if the roots are subject to direct summer sunshine. Blackfly can also be a problem and should be treated with a proprietary control. Where certain varieties are grown in association with other climbers such as roses pruning can be difficult and this should be taken into account before planting.
Forms of interest *C.* **'Beauty of Worcester'** Double, rich, deep mauve/blue flowers with bold creamy-white anthers from early to mid summer, followed by single flowers from late summer to early autumn. Two thirds average height and spread. Tolerates all but the most shady of aspects. Pruning Code 2. Readily available. *C.* **'Countess of Lovelace'** Large, violet-blue, rosette-shaped flowers, double in early summer, becoming single in late summer. Central white stamens. Tolerates all aspects. No pruning required. Readily available. *C.* **'Glynderek'** Double, deep blue flowers with central dark stamens late spring or early summer, late summer and early autumn flowers becoming single. Tolerates all but the most shady of aspects. No pruning required. Will have to be sought from specialist nurseries. *C.* **'Jackmanii Alba'** First flowering large, off-white, double in early summer, second flowering single with pronounced brown anthers from late summer to early autumn. Reaches one third more than average height and spread. Tolerates all aspects. Pruning Code 2. Normally readily available. *C.* **'Jackmanii Rubra'** Double crimson flowers carried on old wood in early summer, with attractive central creamy anthers. Late summer flowers become single. Pruning Code 2 or 3, although the latter will deter some of the double flowers. Will have to be sought from specialist nurseries. *C.* **'Jim Hollis'** Early spring flowers double, lavender-blue, followed in mid to late summer by single lavender-blue flowers. Tolerates all but the most shady of aspects. Pruning Code 2. Will have to be sought from specialist nurseries. *C.* **'Kathleen Dunford'** Large, rich rosy-purple flowers. Semi-double. Late spring to early summer, becoming single in late summer/early autumn. Does not always flower continuously. Pruning Code 2. Will have to be sought from specialist nurseries. *C.* **'Louise Rowe'** Pale mauve flowers with golden stamens, single, semi-double and

Clematis **'Beauty of Worcester'** in flower

double all at the same time from early to mid summer. Late summer and early autumn flowers are all single. Tolerates all but the most shady of aspects. No pruning required. Will have to be sought from specialist nurseries. *C.* **'Lady Caroline Nevill'** Mauve flowers, large, semi-double, early to mid summer, becoming single in late summer. Tolerates all but the most shady of aspects. Pruning Code 2. Normally readily available. *C.* **'Mrs George Jackman'** Large creamy-white flowers with beige anthers. Semi-double in early spring to early summer, single in late summer and early autumn. Tolerates all but the most shady of aspects. No pruning required. Will have to be sought from specialist nurseries. *C.* **'Mrs Spencer Castle'** Good, medium-sized double flowers, heliotrope pink with golden stamens, in late spring, early summer, becoming single in late summer and early autumn. Tolerates all aspects. No pruning required. Will have to be sought from specialist nurseries. *C.* **'Proteus'** Mauve-pink double flowers, large, with green anthers, in late spring, early summer, becoming single in late summer. Dislikes deep shade. Pruning Code 2. Will have to be sought from specialist nurseries. *C.* **'Sylvia Denny'** Medium-sized, semi-double pure white flowers with pink anthers from late spring to early summer, becoming single in late summer. Good, tight growth habit. Tolerates all but the most shady of aspects. Pruning Code 2. Will have to be sought from specialist nurseries. *C.* × **'Vyvyan Pennell'** Large, truly double flowers, deep violet-blue with tinges of red, in late spring early summer. Late summer flowers single lavender-blue. Tolerates all but the most shady of aspects. Pruning Code 2. Readily available. *C.* **'Walter Pennell'** Deep pink flowers with mauve tinge, large, double, in late spring early summer; single flowers in late summer and early autumn. Tolerates all aspects. Readily available.

Average height and spread
Five years
5 × 5 ft (1.5 × 1.5 m)
Ten years
7 × 7 ft (2.1 × 2.1 m)
Twenty years
9 × 9 ft (2.7 × 2.7 m)
Protrudes up to 18 in (45 cm) from support.

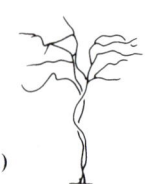

CLEMATIS × DURANDII

KNOWN BY BOTANICAL NAME

Ranunculaceae   *Perennial Climber*
Deciduous

An interesting cross between the perennial clematis *C. integrifolia* and the woody clematis *C.* × *jackmanii*.

Origin Of garden origin.
Use As a climber for walls and fences and to ramble through large shrubs, up poles and pillars, or as ground cover, with or without supports.
Description *Flower* Four to six oval dark-blue/violet sepals with yellow stamens from early summer to mid autumn. Flowers up to 4½ in (11 cm) across. *Foliage* Attractive dark green, oval and pointed leaves, up to 6 in (15 cm) long. *Stem* Light grey/green, angled, predominantly upright in habit without support. *Fruit* No fruit of any interest.
Hardiness Tolerates a minimum winter temperature of up to 4°F (−15°C), lower if given adequate root protection.
Soil requirements Tolerates a wide range of soil conditions as long as adequate organic material and plant nutrients are available. Dislikes waterlogging or very dry conditions.
Sun/Shade aspect Tolerates a wide range of aspects but not the most exposed. Full sun to medium shade with light shade for preference.
Pruning If any growth has survived the winter cut it back to ground level in early spring.
Training Tie to wires or individual anchor

Clematis × *durandii* in flower

points or allow to ramble through netting or large shrubs.
Propagation and nursery production From semi-ripe cuttings taken in early summer. Should always be purchased container grown; may have to be sought from specialist nursery. Best size to purchase; root-clumps to 2½ ft (76 cm).
Problems Suffers from attacks of its own specific mildew which is difficult to control but responds to perseverance.
Similar forms of interest None.
Average height and spread
One year
8 × 8 ft (2.4 × 2.4 m)
Protrudes up to 2 ft (60 cm) from support.

CLEMATIS × ERIOSTEMON (*C.* 'Hendersonii', *C. intermedia*)

KNOWN BY BOTANICAL NAME

Ranunculaceae   *Perennial Climber*
Deciduous

Although a perennial, in a good growing year it can reach a height of 6–10 ft (1.8–3 m), therefore it is worthy of consideration.

Origin Of garden origin.
Use As an attractive climber for small areas on walls and fences or to ramble through small to medium shrubs or other climbing plants.
Description *Flower* Deep blue, up to 1¼ in (3.5 cm) long, fragrant, nodding flowers carried on stalks up to 4 in (10 cm) long, giving a most attractive and delicate display in mid summer to early autumn. *Foliage* Single or pinnate leaves, each leaf oval and up to 2 in (5 cm) long, sparsely produced. Mid green with a slightly purple hue. *Stem* Wispy, ranging, forming no solid structure. Medium annual growth rate. *Fruit* Attractive small silver tufted seedheads in late summer/early autumn.
Hardiness Dies to ground level, therefore tolerates a minimum winter temperature of 4°F (−15°C), as long as roots are protected by a layer of organic material.
Soil requirements Tolerates all soils, except extremely waterlogged, but requires moisture retention and good organic material.
Sun/Shade aspect Requires a sheltered aspect. Best in light shade, although will tolerate full sun as long as roots are shaded.
Pruning Dies to ground level in winter and

therefore pruning is not required, but if shoots should be sustained over winter it is advisable to cut these hard back in spring.
Training Allow to ramble through trees or over wires. Very open in habit and therefore difficult to train.
Propagation and nursery production From internodal cuttings taken in early summer. Should always be purchased container grown; will have to be sought from specialist nurseries. Best planting height 6 in–3 ft (15–91 cm).
Problems None.
Similar forms of interest None.
Average height and spread
Five years
6 × 6 ft (1.8 × 1.8 m)
Ten years
6 × 6 ft (1.8 × 1.8 m)
Twenty years
6 × 6 ft (1.8 × 1.8 m)
Protrudes up to 2 ft (60 cm) from support.

Clematis × *eriostemon* in flower

CLEMATIS FLAMMULA

FRAGRANT VIRGIN'S BOWER

Ranunculaceae   *Woody Climber*
Deciduous

A charming, late summer/early autumn flowering clematis worthy of wider planting for its delicate foliage and flowers.

Origin From southern Europe.
Use As a late summer/early autumn flowering climber for walls, fences, archways and pillars. Can be allowed to grow through large shrubs and small trees to good effect.

Description *Flower* Very fragrant, up to 1 in (2.5 cm) wide, in large, loose bunches, each bunch up to 12 in (30 cm) long. Pure white in colour, varieties with red/purple markings, produced mid summer to mid autumn. *Foliage* Leaves bright green when young, becoming more grey/green with age; up to 1½ in (4 cm) long, composed of three to five lanceolate to almost round leaflets. Some bronze then yellow autumn colour. *Stem* Grey/green, ageing to grey/brown, finally grey. Graceful, twining, leaf tendrils assisting support. Fast rate of growth. *Fruit* Small, white, tufted seedheads follow flowers, providing interest for a limited time.

Hardiness Tolerates a minimum winter temperature of 4°F (−15°C).

Soil requirements Tolerates all soil types, as long as they are moisture-retentive.

Sun/Shade aspect Requires a sheltered aspect. Best in full sun, but will tolerate very light shade; roots must be in the shade.

Pruning Can either be left unpruned to attain more size and reduced every five years by very hard cutting back, or can be trimmed in early spring.

Training Will require wires or trellis or individual anchor points to train over any given area.

Propagation and nursery production From semi-ripe, internodal cuttings taken in early summer or from seed. Always purchase container grown; best planting height 1–2½ ft (30–76 cm). Normally available from good garden centres and specialist nurseries.

Problems May seem a little weak when purchased and often takes two or three years to gain any real stature.

Clematis flammula **'Rubra' in flower**

Similar forms of interest *C. f.* **'Rubra'** (syn. *C. × triternata* **'Rubromarginata'**) White flowers tipped with red to violet. Fragrant and extremely attractive but less vigorous. More difficult to find. *C. apiifolia* Small white flowers in abundance from mid to late summer. One third additional height and spread. Tolerates all areas but possibly best in a wild garden. *C. fargesii souliei* Medium-sized white flowers, strong-growing and free-flowering. Tolerates all aspects. *C. paniculata* (syn. *C. maximowicziana*) Large numbers of very small, white, scented flowers from late summer to early autumn. Prefers a sheltered, sunny aspect.

Average height and spread
Five years
5 × 5 ft (1.5 × 1.5 m)
Ten years
10 × 10 ft (3 × 3 m)
Twenty years
15 × 15 ft (4.6 × 4.6 m)
Protrudes up to 18 in (45 cm) from support.

Clematis flammula **in flower**

Clematis fargesii souliei **in flower**

CLEMATIS FLORIDA 'BICOLOR' (*C. f.* 'Sieboldii')

PASSION FLOWER CLEMATIS

Ranunculaceae   *Woody Climber*
Deciduous to semi-evergreen

A most beautiful and interesting flower, but a plant often of a weak, insipid constitution which necessitates some pampering.

Origin From China.

Use To ramble through wires, medium sized shrubs or other climbers and wall shrubs. It benefits from being grown under the protection of a conservatory or greenhouse.

Description *Flower* Up to 3 in (7.5 cm) wide with two leaf bracts in the centre and four to six oval sepals surrounding. Creamy-white with a green stripe and purple stamens. Carried singly on stalks up to 4 in (10 cm) from May to June. *Foliage* Leaves consisting of three segments of oval to lanceolate shape up to 2 in (5 cm) long, with toothed or lobed edges and downy undersides. Light green. Attractive, but sparsely produced. *Stem* Light green, ageing to green/brown, finally grey/brown. Ranging, wispy, of normally weak constitution. Medium growth rate. *Fruit* No fruit of any interest.

Hardiness Tolerates a minimum winter temperature of 14°F (−10°C).

Soil requirements Tolerates all but the most alkaline of soils, but does need a moisture retentive soil with additional organic material incorporated to encourage root growth. Suffers in drought conditions.

Sun/Shade aspect Requires a sheltered aspect in full sun to light shade. Roots should be in the shade.

Pruning Cut hard back in late winter or early spring to induce new flowering growth.

Training Allow to ramble over walls, through shrubs or other climbers, or tie to wires or individual anchorage points.

Propagation and nursery production From semi-ripe internodal cuttings taken in early summer. Extremely difficult to propagate. May be scarce in supply and will have to be sought from specialist nurseries. Always purchase container grown; best planting height 1–3 ft (30–91 cm).

Clematis florida **'Bicolor' in flower**

Problems Its weak constitution often causes it to die before becoming established, although additional organic material will go some way towards preventing this. Commands a premium price.

Similar forms of interest *C. f.* **'Alba'** Pure white flowers. Requires the protection of a conservatory or greenhouse in all but the most favoured areas.

Average height and spread
Five years
5 × 5 ft (1.5 × 1.5 m)
Ten years
10 × 10 ft (3 × 3 m)
Twenty years
12 × 12 ft (3.7 × 3.7 m)
Protrudes up to 18 in (45 cm) from support.

CLEMATIS × JOUINIANA PRAECOX

KNOWN BY BOTANICAL NAME
Ranunculaceae   *Perennial Climber*
Deciduous
A vigorous perennial clematis with attractive autumn flowers.

Origin Of garden origin.
Use As a fast-growing perennial climber for walls and fences, to cover large uninteresting shrubs, tree stumps or small sheds.
Description *Flower* Yellow/white, four-sepal flowers up to $\frac{3}{4}$ in (2 cm) wide, carried in panicles up to 2 ft (60 cm) long from late summer to mid autumn. In full sun flowers will be white/yellow, in light to medium shade flowers will be pale blue. *Foliage* Large leaves composed of three to five leaflets, oval in shape, each up to 4 in (10 cm) long with very coarsely toothed edges. Mid green with attractive light silver reverse. *Stem* Light green, ageing to grey/green. Rigid, upright, becoming more spreading with age. Fast growth rate. *Fruit* May produce small clusters of white, fluffy seedheads of some limited interest in autumn and retain into early winter.

Clematis × jouiniana praecox in flower

Hardiness Tolerates a minimum winter temperature of 4°F (−15°C). Dies to ground level in winter to be regenerated in following spring.
Soil requirements Tolerates all soils provided they are well-fed and moisture retentive.
Sun/Shade aspect Tolerates all aspects in full sun to medium shade. Roots should be in the shade.
Pruning Normally dies to ground level in winter. If it does not, in a mild winter, remove top growth to ground level to induce new, vigorous summer growth and autumn flowers.
Training Allow to ramble over large shrubs or tie to wires or individual anchorage points. When used through shrubs it may, in time, kill its host.
Propagation and nursery production From semi-ripe cuttings taken early to mid summer. Should always be purchased container grown; may have to be sought from specialist nurseries, although it is becoming more widely available. Best planting height 6 in–3 ft (15–91 cm).
Problems It dies to ground level in winter when in pot which can deter potential purchasers.
Similar forms of interest *C.* × 'Mrs Robert Brydon' Dark blue, very attractive flowers and dark green foliage.
Average height and spread
Five years
8 × 8 ft (2.4 × 2.4 m)
Ten years
12 × 12 ft (3.7 × 3.7 m)
Twenty years
12 × 12 ft (3.7 × 3.7 m)
Protrudes up to 3 ft (91 cm) from support.

CLEMATIS (Large-flowered Hybrids)

KNOWN BY BOTANICAL NAME
Ranunculacee   *Woody Climber*
Deciduous
The clematis is one of the jewels of all summer flowering climbers and rightly takes its place as the most well-known of all climbing plants.

Origin Of garden or nursery origin.
Use As a flowering climbing plant for walls, fences, trellis, pergolas, bowers or similar lattice-type constructions, or to be part of walkways, in particular laburnum. If provided with a low wire frame, can be used as ground cover. Some varieties can be encouraged to grow through trees or large shrubs. Can be grown in large tubs or containers.
Description *Flower* Depending on variety, four to eight oval sepals make up a flat, plate-shaped flower 3–8 in (7.5–20 cm) wide, many with attractive stamens. Colours range through white, pink, purple, red, blue, mauve and yellow, some varieties with a pronounced coloured bar through the centre of each sepal. The flowers are borne singly on flowering shoots. *Foliage* Three to five lanceolate to ovate leaflets, each up to 4 in (10 cm) long, make up a leaf, mid green to grey/green, with tendril-type leaf stalks which aid support. *Stem* Light green when young, ageing to green/brown, finally brown. Twining. Medium to fast growth rate. *Fruit* Some varieties may produce white tufted seedheads of some winter attraction.
Hardiness Tolerates a minimum temperature of 4°F (−15°C).
Soil requirements Tolerates all soil conditions, except for extremely alkaline, but even in the latter, with adequate organic material added to the soil and moisture retention being achieved, average results can be expected. Overall, the more preparation and the more organic material added – up to 50 per cent of the volume – the better the results. When grown in tubs and containers, use a soil-based potting compost, allow for adequate drainage and choose a container not less than 21 in (53 cm) wide and 18 in (45 cm) deep.
Sun/Shade aspect Tolerates all aspects, except extremely exposed. Full sun to light shade with full sun for preference. Roots should be in the shade.

Pruning There are three different pruning methods, each particular to an individual variety. The three methods are as follows:
1 Simply tidy and thin end shoots back to main framework after flowering.
2 Reduce by one third the growth made in the previous season in late winter/early spring.
3 Remove all last season's growth to within 12 in (30 cm) of its origin during late winter/early spring.

It is important the recommended pruning methods are carried out on an annual basis, otherwise flowering will decrease and the overall well-being of the plant will suffer. The individual pruning for each variety is given in 'Forms of Interest'. If the variety is not known use method 2 as a safe option.

Clematis 'Dr Ruppel' in flower

Training Allow to ramble through wires, trellis, branches of large shrubs or trees or any other support where the leaf tendrils can wrap round to anchor.
Propagation and nursery production From internodal cuttings taken in early to late summer, inserted into a sand/peat mixture, with some bottom heat to encourage rooting. Best planting height 6 in–3 ft (15–91 cm). Clematis is often sold in a choice of 3–4 in (7.5–10 cm) or 1 litre (5–7 in/12–18 cm) pots; both sizes are equally good as long as the plants have been well hardened off and are not purchased before mid spring. Plant from mid spring through until late summer. Plants propagated

Clematis 'Comtesse de Bouchard' in flower

Clematis 'Edith' in flower

in the spring are ripe enough for planting from mid autumn through until mid spring; these are normally offered in 3–4 in (7.5–10 cm) or 1 litre (5–7 in/12–18 cm) pots. Many varieties will be readily available, others will need searching for from specialist nurseries.

Problems Clematis wilt is an airborne fungus disease for which there is no prevention. The entire plant, both leaves and stem, turns black in early to late summer. Once an attack is seen, cut all growth to ground level, wash walls, fences or trellis with a mild disinfectant and protect any other plants growing below with plastic sheeting. There is then a 60 per cent chance that new, uninfected shoots will develop from below ground from the root system. Planting a new clematis in the same position is not a risk, but the area should be washed with mild disinfectant first.

The main reason for the failure of clematis hybrids is that they are often planted too close to the wall (see 'Planting Climbers and Wall Shrubs' on p.168) and that not enough organic material such as garden compost, well-rotted farmyard manure or mushroom compost is added to the soil. Problems may also be experienced if roots are subject to direct summer sunshine. Blackfly is a common pest but can be treated with a proprietary control. Where certain varieties are grown in association with other climbers such as roses, the requisite pruning can be difficult and this should be taken into account when planting is planned.

Forms of interest *C.* **'Alice Fisk'** Large, pointed sepals with wavy edges, mid to pale blue. Dark brown stamens. Late spring to early summer, with repeat flowering in late summer, early autumn. Any aspect. Reaches two thirds average height and spread. No pruning required. Will have to be sought from specialist nurseries. *C.* **'Allanah'** Large, ruby-red flowers with well-spaced sepals and dark brown anthers produced from early summer to early autumn. Tolerates all but the most shady of aspects. Two thirds average height and spread. Pruning code 3. Will have to be sought from specialist nurseries. *C.* **'Annabel'** Large blue flowers, white stamens. Will have to be sought from a specialist nursery. Pruning code 2. *C.* **'Asao'** Large red/pink sepals with white bar. Free-flowering from late spring to early summer, with repeat flowering from late summer to early autumn. Tolerates any aspect. No pruning required. Will have to be sought from specialist nurseries. *C.* **'Ascotiensis'** Large sepals, long and pointed, bright blue with central green stamens, produced continously from mid summer to early autumn. Tolerates any aspect. Pruning code 3. Will have to be sought from specialist nurseries. *C.* **'Barbara Dibley'** Large, deep red flowers with purple/red central anthers from late spring to

early summer. Good on all aspects. Two thirds average height and spread. Pruning code 2. Readily available. *C.* **'Barbara Jackman'** Large blue flowers with deep pink bar through each sepal and creamy-yellow anthers from late spring to early summer. Tolerates all aspects, although shade will enhance the coloured bar. Two thirds average height and spread. Pruning code 2. Readily available. *C.* **'Beauty of Richmond'** Large, lavender-blue flowers with chocolate-coloured central anthers from early to late summer. A variety for any aspect. Pruning code 2. Will have to be sought from specialist nurseries. *C.* **'Bees Jubilee'** Central brown anthers contrast well with deep pink sepals, each having a deep rose-pink bar. Tolerates all aspects, but the flower contrast colour is best in light shade. Pruning code 2. Readily available. *C.* **'Blue Gem'** Purple anthers contrast with large lavender-blue flowers in late spring to early autumn. Good on any aspect. Pruning code 2. Will have to be sought from specialist nurseries. *C.* **'Belle Nantaise'** Lavender-blue flowers, very large, with central white stamens, from early summer to early autumn. Tolerates any aspect. Pruning code 2 or 3. Will have to be sought from specialist nurseries. *C.* **'Bracebridge Star'** Lavender-blue sepals with deep crimson bar

Clematis 'Ernest Markham' in flower

and red stamens. Large flowers from late spring to early summer. Pruning code 2. Will have to be sought from specialist nurseries. *C.* **'Captain Thuilleaux'** Large, crimson sepals with strawberry-pink bar, contrasting with brown anthers, from late spring to early summer. Tolerates any aspect. Pruning code 2. Readily available. *C.* **'Cardinal Wyszynski'** Large, crimson flowers with brown stamens from early summer to early autumn. Tolerates all but the most shady of aspects. Pruning code 3. Will have to be sought from specialist nurseries. *C.* **'Carnaby'** Attractive crimped edges to raspberry-pink flowers and a central deep pink bar. Flowers from late spring to early summer. Two thirds average height and spread. A variety with a neat growth habit, making it ideal for confined spaces. Pruning code 2. Readily available. *C.* **'Charissima'** Cerise-pink pointed sepals make up a large flower, each sepal with a deep pink bar and delicate but pronounced vein markings. Flowers from early summer to early autumn. Tolerates all aspects. No pruning required. Will have to be sought from specialist nurseries. *C.* **'Comtesse de Bouchard'** One of the most attractive medium-sized flowering clematis. Mauve-pink in colour with cream anthers, very free-flowering throughout summer. Tolerates any aspect. Two thirds average height and spread. Pruning code 3. Readily available. *C.* **'Corona'** Sepals with a velvety texture, purple pink in colour with deep red anthers, from late spring to early summer. Extremely free-flowering.

Clematis 'Jackmanii Rubra' in flower

Good compact habit. Tolerates any aspect. Pruning code 2. Will have to be sought from specialist nurseries. *C.* **'Crimson King'** Large crimson sepals with brown stamens from early summer to early autumn. Tolerates all but the most shady of aspects. Pruning code 2 or 3. Will have to be sought from specialist nurseries. *C.* **'C.W. Dowman'** Lavender flowers shaded pink from centre, carmine bar. Pruning code 2. Will have to be sought from specialist nurseries. *C.* **'Dawn'** Very large pearly-pink flowers from late spring to early summer. Close growing, tidy plant. Avoid full sun, otherwise any aspect. Pruning code 2. Will have to be sought from specialist nurseries. *C.* **'Dr Ruppel'** Carmine bar boldly presented on rose-pink sepals. Gold anthers in centre. Flowers from late spring to early summer, repeating in late summer, early autumn. Good in any aspect. Pruning code 2. Readily available. *C.* **'Duchess of Sutherland'** A lighter red bar down a wine-red sepal makes a medium-sized flower with golden stamens, from early summer to early autumn. Tolerates all but the most shady of aspects. Pruning code 2 or 3. Will have to be sought from specialist nurseries. *C.* **'Edith'** Red/brown anthers show off the large white sepals. Good flowering performance from late spring to early autumn. Tolerates any aspect. Pruning code 2. Will have to be sought from specialist nurseries. *C.* **'Edouard de Fosse'** Red/purple anthers show off large, deep mauve/purple sepals in late spring to early summer. Tolerates any aspect. Pruning code 2. Will have to be sought from specialist nurseries. C. 'Elizabeth Foster' Delicate pink flowers, deep carmine bar, maroon centre. Pruning code 2. Will have to be sought from specialist nurseries. C. 'Elsa Spath' Large, deep violet-blue sepals surround red/purple anthers from early summer to early autumn. Tolerates any aspect. Pruning code 2. Will have to be sought from specialist nurseries. C. 'Ernest Markham' Glowing red flowers of medium size with golden anthers from mid summer to mid autumn. Dislikes exposed aspects. Pruning code 3. Readily available. C. 'Etoile de Paris' Large, pointed, mauve/blue sepals with red central bar and deep red anthers in centre from late spring to early summer. Good in all aspects. Pruning code 2. Will have to be sought from specialist nurseries. C. 'Fairy Queen' Large flesh-pink flowers with deep rose bar from early to late summer. Dislikes full sun. Pruning code 2. Will have to be sought from specialist nurseries. C. 'Fargesii Soulei' Medium-sized flower, pure white. Free-flowering and very strong growing, reaching 40 per cent more than average height and spread, making it ideal for growing through trees or large shrubs. Tolerates any aspect. Pruning code 3. Will have to be sought from specialist nurseries. C. 'Four Star' Lavender-coloured sepals with deeper blue bar make up a medium to large flower from late spring to early summer with repeat flowering late summer, early autumn. Tolerates all aspects. Pruning not normally required. Will have to be sought from specialist nurseries. C. 'General Sikorski' Crinkled edges to mid blue sepals with central golden anthers from early summer through until early autumn. Prefers a sheltered aspect. Pruning code 3. Will have to be sought from specialist nurseries. C. 'Gipsy Queen' Red/purple anthers surrounded by medium-sized rich violet/purple sepals. Very free-flowering from mid summer through to mid autumn. Tolerates any aspect. Pruning code 3. Readily available. C. 'Guiding Star' Blue/purple sepals making a large flower with brown anthers from late spring to early summer. Young foliage bronze in colour. Good compact habit for confined spaces. Tolerates any aspect. Pruning code 2. Will have to be sought from specialist nurseries. C. 'Hagley Hybrid' Medium-sized flowers, cup-shaped, shell-pink with brown anthers from early to late summer. Dislikes full sun which will fade the colour. Compact and good for restricted areas. Pruning code 3. Readily available. C. 'Haku Ookan' Very attractive deep violet-blue flowers with white central anthers from late spring to early summer. Tolerates any aspect. Pruning code 2. Will have to be sought from specialist nurseries. C. 'Henryii'

Clematis 'Nellie Moser' in flower

Medium-sized creamy-white flowers with coffee-coloured central anthers from early to late summer. A variety grown for many years. Pruning code 2. Will have to be sought from specialist nurseries. C. 'Herbert Johnson' Very large red/mauve flowers, maroon centre. Pruning code 2. Will have to be sought from specialist nurseries. C. 'H. F. Young' Wedgwood blue sepals make up large flowers with cream-coloured anthers. Free-flowering late spring to early summer, repeating in late summer, early autumn. Tolerates any aspect. Pruning code 2. Readily available. C. 'Horn of Plenty' Cup-shaped, rich rosy-purple flowers, large, with red anthers, produced from late spring to early summer with repeat flowering in late summer, early autumn. Reaches two thirds average height and spread. Good on any aspect. Pruning code 2. Will have to be sought from specialist nurseries. C. 'Hybrida Sieboldii Ramona' Overlapping sepals make a very round flower, lavender-blue, large, with dark coloured anthers, from early summer to early autumn. Good in any aspect. Pruning code 2. Readily available. C. 'Ishobel' Wavy edges to large white sepals with dark stamens from late spring to early summer, repeated in late summer to early autumn. Tolerates all but the most shady of aspects. Pruning not normally required. Will have to be sought from specialist nurseries. C. 'Jackmanii' Divided, deep purple blue sepals from mid summer to early autumn. Tolerates any aspect. Pruning code 3. Readily available. C. 'Jackmanii Superba' An improved form of the well known C. 'Jackmanii'. Larger sepals, more rounded,

dark purple, from mid summer to early autumn. Tolerates any aspect. Pruning code 3. Readily available. C. 'Jim Hollis' Vigorous, double lavender flowers. Pruning code 2. Will have to be sought from specialist nurseries. C. 'Joan Picton' Large flowers the colour of wild lilac with a white bar down each sepal and brown stamens produced from mid spring to early summer with repeat flowering in autumn. Tolerates all aspects. Pruning not normally required. Will have to be sought from specialist nurseries. C. 'John Paul II' Large, extremely attractive sepals, creamy white with pink shading which becomes deeper as summer passes and forms a pink bar. Sepals have creased edges and overlap. Flowers late spring, early summer with repeat flowering late summer, early autumn. Toler-

Clematis 'Niobe' in flower

ates all aspects. Pruning not normally required. Will have to be sought from specialist nurseries. C. 'John Warren' Extremely large flowers, consisting of pale pink sepals with dark pink bar surrounding central cluster of brown anthers; from early to late summer. Dislikes strong sunlight, otherwise good on all aspects, particularly shady. Pruning code 2. Readily available. C. 'Kaoper' Large, deep violet, crinkly-edged sepals with violet stamens from late spring to early summer, repeated in late summer, early autumn. Tolerates all aspects. Pruning not normally required. Will have to be sought from specialist nurseries. C. 'Kathleen Wheeler' Very attractive large plum/mauve-coloured flowers with pronounced golden anthers, early summer. Tolerates any aspect. Pruning code 2. Will have to be sought from specialist nurseries. C. 'Ken Donson' Large, deep blue flowers, golden anthers. Pruning code 2. Will have to be sought from specialist nurseries. C. 'King Edward VII' Large, violet/puce sepals with

Clematis 'Mrs Cholmondeley' in flower

crimson bar and brown anthers from early to late summer. Pruning code 2. Will have to be sought from specialist nurseries. *C.* **'King George V'** Large, flesh-pink sepals with bright pink bar surrounding brown anthers from mid to late summer. Tolerates any aspect. Pruning code 2. Will have to be sought from specialist nurseries. *C.* **'Lady Betty Balfour'** Royal purple sepals make up a large flower with a central cluster of yellow anthers from late summer to early autumn. Very strong growing, 25 per cent more than average height and spread. Requires full sun to encourage flowering, dislikes shady aspects. Pruning code 3. Readily available. *C.* **'Lady Londesborough'** Red anthers surrounded by medium-sized sepals, pale mauve/blue in colour, from late spring to early summer. Good compact growth habit; two thirds average height and spread. Tolerates all aspects, except extremely exposed. Pruning code 2. Readily available. *C.* **'Lady Northcliffe'** Medium-sized sepals of Wedgwood blue, surrounding yellow anthers, from early to late summer. Good compact growth, only reaching two thirds average height and spread. Tolerates any aspect. Pruning code 2. Readily available. *C.* **'Lasurstern'** Creamy-white anthers surrounded by large, pale mauve sepals from early spring to early summer with repeat flowering in late summer to early autumn. Tolerates any aspect. Pruning code 2. Readily available. *C.* **'Lawsoniana'** Sky-blue sepals flushed with mauve and pale brown anthers make up a large flower from late spring to early summer, with repeat flowering in late summer. Tolerates any aspect. Pruning code 2. Readily available. *C.* **'Lilacina Floribunda'** Large, very rich purple flowers from mid summer through until early autumn. Tolerates any aspect. Pruning code 3. Will have to be sought from specialist nurseries. *C.* **'Lincoln Star'** Raspberry-pink sepals make up a large flower with deep red anthers from late spring to mid summer. The repeat flowers in late summer are paler pink but the bar is of a deeper colour. Two thirds average height and spread. Tolerates all aspects, except those in strong sunlight. Pruning code 2. Readily available. *C.* **'Lord Neville'** Red anthers surrounded by a ring of dark blue sepals, making a large flower from early to late summer. Tolerates any aspect. Pruning code 2. Readily available. *C.* **'Madame Baron Veillard'** Medium-sized flowers consisting of rosy/lilac sepals from late summer through to early autumn. Tolerates all aspects except extremely exposed or shady. Pruning code 3. Readily available. *C.* **'Madame Edouard André'** Dusky red sepals of medium size with attractive velvety texture, creamy anthers. Early to late summer, good flowering performance. Tolerates any aspect. Pruning code 3. Readily available. *C.* **'Madame Grange'** Sepals have silvery underside with dusky purple top,

Clematis **'Rouge Cardinale'** in flower

making up a medium-sized flower with dark anthers produced from late summer to early autumn. Tolerates any aspect. Pruning code 3. Will have to be sought from specialist nurseries. *C.* **'Margaret Hunt'** Good-sized, dusky pink flowers carried from early to late summer. Strong growing, reaching one third more than average height and spread. Tolerates any aspect. Pruning code 3. Will have to be sought from specialist nurseries. *C.* **'Marcel Moser'** Large, delicate mauve sepals with crimson bar and red/purple stamens. Flowers from late spring to early summer with repeat flowering late summer to early autumn. Tolerates all aspects. Pruning not normally required. Will have to be sought from specialist nurseries. *C.* **'Marie Boisselot'** (syn. *C.* **'Madame le Coultre'**) Very large white flowers with beige-coloured anthers from early to late summer. Very free-flowering. Tolerates all aspects. Pruning code 3. Readily available. *C.* **'Maureen'** Light brown anthers surrounded by red/purple sepals, making a large flower from early to late summer. Tolerates any aspect. Pruning code 2. Will have to be sought from specialist nurseries. *C.* **'Miss Bateman'** Very pretty medium-sized creamy-

white flowers with chocolate-coloured anthers from mid spring to early summer. Two thirds average height and spread. Tolerates all but the most exposed or shady aspects. Pruning code 2. Readily available. *C.* **'Moonlight'** (syn. *C.* **'Yellow Queen'**) Primrose-yellow sepals with yellow anthers from late spring to early summer. Glossy green foliage of interesting shape. Dislikes full sun or deep shade. Pruning code 2. Readily available. *C.* **'Mrs Bush'** Very large, deep lavender sepals with light chocolate anthers from late spring to early summer. Tolerates all aspects. Pruning code 2. Will have to be sought from specialist nurseries. *C.* **'Mrs Cholmondeley'** Long, narrow, lavender-blue sepals make up a large flower with brown central anthers. Long flowering period from late spring to early autumn, not necessarily continuous. Tolerates any aspect. Pruning code 2. Readily available. *C.* **'Mrs N. Thompson'** Large blue sepals with a red bar and red anthers from late spring to early summer. Some repeat flowering in late summer. Tolerates any aspect. Pruning code 2. Readily available. *C.* **'Mrs Spencer Castle'** Single and double flowers, pale mauve/pink, from June to October. Pruning code 2. Will have to be sought from specialist nurseries. *C.* **'Mrs P. B. Truax'** Mid to deep blue medium-sized flowers with yellow anthers from late spring to early summer. Tolerates all but the most shady of aspects. Makes a compact plant, ideal for small gardens. Pruning code 2. Will have to be sought from specialist nurseries. *C.* **'Myojo'** Large, velvety red sepals with deep red bar and attractive cream-coloured stamens from late spring to early summer, repeated in late summer, early autumn. Tolerates all aspects. Pruning not normally required. Will have to be sought from specialist nurseries. *C.* **'Nellie Moser'** Large, pale pink/mauve sepals with a crimson-coloured bar and red/brown anthers from late spring to early summer. Repeat flowering late summer, early autumn. A very popular variety, suspected of not being easy to establish, but this is normally attributable to bad planting (see 'Planting Climbers and Wall Shrubs' on p.168). Good on exposed, shady aspects. Pruning code 2. Readily available. *C.* **'Niobe'** Large, velvety textured deep ruby red sepals with yellow anthers from early to late summer. Tolerates all aspects. Plant against a light background to show off flowers to full extent. Pruning code 2. Readily available. *C.* **'Patens Yellow'** Creamy yellow se-

Clematis **'Perle d'Azur'** in flower

pals and yellow stamens. Pruning code 2. Will have to be sought from specialist nurseries. C. 'Perle d'Azur' Delicate sky-blue sepals make up a medium-sized, semi-pendent flower, each sepal edge slightly corrugated, with green/yellow anthers. Good grower in any aspect. Flowers from mid summer to early autumn. Pruning code 3. Normally readily available. C. 'Pennell's Purity' White flowers with firm, crimped sepals, golden centre. Pruning code 2. Will have to be sought from specialist nurseries. C. 'Peveril Pearl' Violet-tipped creamy stamens, ringed by lilac sepals, each with a pink bar, making a large flower from late spring to early summer with repeat flowering early autumn. Strong growing. Tolerates any aspect. Pruning code 2. Will have to be sought from specialist nurseries. C. 'Pink Champagne' An interesting new large-flowering variety from Japan with strong pink petals. Pruning code 2. Readily available. C. 'Pink Fantasy' Large, shell pink sepals with deep pink bar and wavy edges; brown stamens. Flowers from early summer to early autumn. Tolerates all but the most shady of aspects. Pruning code 3. Will have to

Clematis 'The President' in flower

be sought from specialist nurseries. C. 'Prince Charles' Mauve-blue sepals make up medium-sized flowers which cover the entire plant from early summer to early autumn, although not necessarily continuously. Tolerates all but the most shady of aspects. Two thirds average height and spread. Pruning code 2. Will have to be sought from specialist nurseries. C. 'Princess of Wales' Satin mauve flowers from early summer to early autumn, not necessarily continuous. Tolerates all aspects. Strong growing. Named after Queen Alexandra when she was Princess of Wales in the 1800s. Pruning not normally required. Will have to be sought from specialist nurseries. C. 'Proteus' Rosy lilac, semi-double flowers with yellow stamens. Pruning code 2 on established plants. Will have to be sought from specialist nurseries. C. 'Richard Pennell' Pronounced red and cream anthers show off the large rosy/purple sepals from late spring to early summer. Repeat flowering late summer. Tolerates any aspect. Pruning code 2. Will have to be sought from specialist nurseries. C. 'Rouge Cardinale' Large, rich-textured, crimson square-tipped sepals around yellow anthers. Long flowering period from early summer to early autumn. Two thirds average height and spread. Tolerates all aspects. Pruning code 2. Readily available. C. 'Sally Cadge' Deep crimson bar runs down centre of large, mid blue sepals produced from late spring to early summer, with repeat flowering late summer, early autumn. Tolerates all aspects. Pruning not required. Will have to be sought from specialist nurseries. C. 'Saturn' Lavender blue sepals with maroon bar. White stamens. Pruning code 2. Will have to be sought from specialist nurseries. C. 'Scartho Gem' Attractive, large, deep pink sepals with deeper pink bar surrounding golden stamens. Very free flowering from early summer to

early autumn. Tolerates all aspects. Pruning not required. Normally readily available. C. 'Scotiensis' Attractive purple/blue flowers with yellow/green central anthers in mid to late summer. Tolerates any aspect. Pruning code 2. Will have to be sought from specialist nurseries. C. 'Sealand Gem' A deep mauve bar down the centre of rosy/mauve sepals makes up an attractive medium-sized flower with brown central stamens. Flowers early summer to early autumn, but not necessarily continuously. Tolerates any aspect. Pruning code 2 or 3. Normally readily available. C. 'Seranata' Deep purple flowers with bright yellow anthers from early summer to early autumn. Tolerates any aspect. Pruning code 3. Will have to be sought from specialist nurseries. C. 'Sir Garnet Wolseley' Mauve/blue sepals with a pale purple bar make up a medium-sized flower with red anthers produced from late spring to early summer. Good tight growth habit. Tolerates all aspects. Pruning code 2. Will have to be sought from specialist nurseries. C. 'Silver Moon' Large pale lilac flowers from early summer to early autumn. Good in exposed aspects but will flower in all conditions. Pruning code 2. Will have to be sought from specialist nurseries. C. 'Snow Queen' Medium- to large-sized flowers, pure white with blue-tinted edges, from early to mid summer. Tolerates all but the most shady of aspects. Pruning not required. Will have to be sought from specialist nurseries. C. 'Star of India' Deep purple sepals with a red bar, making up a medium-sized flower from early summer to early autumn. Tolerates any aspect. Pruning code 3. Will have to be sought from specialist nurseries. C. 'Susan Allsop' Large, rosy-purple flowers with red bar down each sepal and magenta-red centre shading, surrounding golden yellow anthers from early summer to early autumn. Tolerates any aspect. Pruning code 2. Will have to be sought from specialist nurseries. C. 'The President' Medium sized blue/purple flowers with a paler stripe and velvety texture, saucer-shaped in formation. Anthers red/purple. Flowers from late spring to early autumn, not necessarily continuously. Tolerates all aspects. Grown for many years and well established. Pruning code 2. Readily available. C. 'Twilight' Large, dark mauve flowers with yellow stamens from mid summer to early autumn, often continuous. Tolerates all but the most shady of aspects. Pruning code 3. Will have to be sought from specialist nurseries. C. 'Veronica's Choice' Large, semi-double lavender flowers with crimped edges. Pruning code 2. Will have to be sought from specialist nurseries. C. 'Victoria' Buff-coloured anthers surrounded by rosy/purple sepals, making up a large flower from early summer to early autumn. Strong growing, tolerates any aspect. Pruning code 3. Will have to be sought from specialist nurseries. C.

Clematis 'Ville de Lyon' in flower

Clematis 'William Kennet'

'Ville de Lyon' Carmine-pink sepals with crimson shading around their edges surround golden anthers to make up a medium-sized flower from early summer to mid autumn, often flowering continuously in good conditions. Dislikes full sun, otherwise tolerates a wide range of aspects. A well proven form. Pruning code 3. Readily available. C. 'Violet Charm' Long, rich violet/blue sepals with pointed ends and wavy edges with beige-coloured stamens from early summer to early autumn, not necessarily continuous. Tolerates all but the most shady of aspects. Two thirds average height and spread. Pruning code 2 or 3. Will have to be sought from specialist nurseries. C. 'Voluceau' Medium-sized red flowers with yellow anthers from early summer to early autumn, often continuous. Good strong grower, tolerating any aspect. Pruning code 3. Will have to be sought from specialist nurseries. C. 'Wadas Primrose' Pale primrose-yellow flowers with deeper central bar through each sepal from late spring to early summer with repeat flowers in early autumn. Strong growing. Prefers a shady position. Pruning code 2. Will have to be sought from specialist nurseries. C. 'Warsaw Nike' Good, glowing purple with central golden stamens on large flowers. Free-flowering from early summer to early autumn, not necessarily continuous. Tolerates all but the most shady of aspects. Pruning code 3. Readily available. C. 'W. E. Gladstone' Purple stamens surrounded by lilac blue sepals. Very large flowers from early summer to early autumn, not necessarily continuous. Tolerates all but the most shady of aspects. Pruning code 2 or 3. Readily available. C. 'Wilhelmina Tull' A broad crimson bar down the centre of large, deep violet blue sepals with central golden stamens from late spring to early summer, repeated late summer to early autumn. Two thirds average height and spread. Tolerates all but the most shady of aspects. Said to be an improved form of C. 'Mrs N. Thompson'. Pruning not normally required. Will have to be sought from specialist nurseries. C. 'Will Goodwin' Wavy edges to lavender blue sepals make up a large flower with golden central stamens from early summer to early autumn, not necessarily continuous. Tolerates all but the most shady of aspects. Pruning code 2 or 3. Will have to be sought from specialist nurseries. C. 'William Kennett' Crimped margins to the large, lavender blue flowers with dark blue anthers from early to late summer. Tolerates any aspect. Pruning code 2. Readily available.

Average height and spread
Five years
8 × 8 ft (2.4 × 2.4 m)
Ten years
12 × 12 ft (3.7 × 3.7 m)
Twenty years
15 × 15 ft (4.6 × 4.6 m)
Protrudes up to 18 in (45 cm) from support.

CLEMATIS MACROPETALA

KNOWN BY BOTANICAL NAME
Ranunculaceae   *Woody Climber*
Deciduous
One of the stars of the spring flowering clematis with a wide range of colours and extremely attractive flowers.

Origin From China and Siberia and of garden origin.
Use As a climber for walls, fences, pillars, pergolas and trellis, or to ramble through large shrubs. Can be used as ground cover with or without support.
Description *Flower* Nodding flowers carried singly on stalks 2 in (5 cm) long. Each has four blue sepals. *Foliage* Nine coarsely toothed, ovate to lanceolate leaflets, each $\frac{1}{2}$–$1\frac{1}{2}$ in (1–4 cm) long, making up a total size of 4 in (10 cm) long and 2 in (5 cm) wide. Light to mid green. Yellow/brown autumn colour. *Stem* Light green when young, ageing to grey/brown. Twining. Leaf tendrils act as additional anchorage. Medium rate of growth. *Fruit* May, in hot summers and mild locations, produce small silver/white fluffy seedheads.
Hardiness Tolerates a minimum winter temperature of 4°F (−15°C).
Soil requirements Does well on all soil types with no particular preference, but must have adequate moisture throughout the year to induce good growth.

Clematis macropetala **in flower**

Sun/Shade aspect Tolerates all but the most exposed of aspects. Requires its roots in the shade and prefers its top in the sun but will tolerate very light shade.
Pruning Cut all current season's growth to within 6–12 in (15–30 cm) of its origin after main flowering period so that through the summer new flowering growth will be produced for next spring, or lightly trim and tidy after flowering. The latter method is normally suitable over one to five years, but the plant will become old and woody and will need hard cutting back every five to six years.
Training Requires wires or trellis on walls and fences. Allow to ramble freely through large shrubs. Good as ground cover, with or without support.
Propagation and nursery production From semi-ripe internodal cuttings taken late spring/early summer. Always purchase container grown; best planting height 1½–3 ft (45–91 cm). Most varieties readily available from good garden centres and specialist nurseries, although some varieties may need searching for.
Problems If not pruned correctly can become very woody and uninteresting in its growth.

Clematis montana **'Elizabeth' in flower**

Similar forms of interest *C. m.* **'Blue Bird'** Large flowers, up to 4 in (10 cm) wide, lavender blue in colour. Strong growing, reaching one third more than average height and spread. *C. m.* **'Maidwell Hall'** Deep blue flowers. A decided improvement on the parent plant. *C. m.* **'Markham's Pink'** (syn. *C. m.* **'Markhamii'**) Pink flowers 3 in (7.5 cm) wide, pendent and nodding. Can produce a later summer display of flowers. *C. m.* **'Rosie O'Grady'** Pink flowers up to 3 in (7.5 cm) wide in spring with an occasional late summer flowering. *C. m.* **'Snowbird'** Pure white flowers 3 in (7.5 cm) across. *C. m.* **'White Lady'** Masses of star-shaped white flowers. Will have to be sought from specialist nurseries. *C. m.* **'White Swan'** Double white pendent flowers up to 2 in (5 cm) across. Extremely attractive.
Average height and spread
Five years
10 × 10 ft (3 × 3 m)
Ten years
10 × 10 ft (3 × 3 m)
Twenty years
10 × 10 ft (3 × 3 m)
Protrudes up to 18 in (45 cm) from support.

CLEMATIS MONTANA

KNOWN BY BOTANICAL NAME
Ranunculaceae   *Woody Climber*
Deciduous
A well known and very widely planted clematis which offers great value as a flowering climbing plant.

Origin From the Himalayas.
Use As a free-flowering, rambling climber for large walls or fences, or to grow over large shrubs and small trees. Ideal for pergolas, trellis, archways, in fact wherever a climber is required. It is also good as a widespreading ground cover with or without support.
Description *Flower* White, borne singly on stalks up to 5 in (12 cm) long, with up to five stalks at each leaf joint. Flowers 2–2½ in (5–6 cm) across with four sepals in a star formation. *Foliage* Three ovate to lanceolate leaflets on a stalk up to 4 in (10 cm) long. Each leaflet up to 4 in (10 cm) long, making a total leaf size of 4 × 5 in (10 × 12 cm). Bright mid green on white flowering forms or purple/green to quite deep purple on pink or purple flowering forms. *Stem* Light green ageing to purple/green, finally purple/brown. Twining with leaf petioles giving extra anchorage. Fast growth rate. *Fruit* May produce small, uninteresting seedheads following flowering.
Hardiness Tolerates a minimum winter temperature of 4°F (−15°C).

Soil requirements Does well on all soil types with no particular preference, but must have adequate moisture throughout the year to induce good growth.
Sun/Shade aspect Prefers its roots in the shade but its top growth will tolerate from full sun through to medium shade, although in medium shade the plant may become more open and lax and flower less.
Pruning Normally requires no pruning until five years after planting and then can be heavily reduced after flowering. Quickly rejuvenates itself, giving a better display of foliage in the following season and good flowering performance in the following spring. This treatment should be repeated every four to five years for best results.
Training Requires wires or trellis on walls and fences. Allow to ramble freely through shrubs and trees.
Propagation and nursery production From semi-ripe internodal cuttings taken in late spring/early summer. Always purchase container grown, best planting height 1½–3 ft (45–91 cm). Most varieties readily available from good garden centres and specialist nurseries, although some varieties may need searching for.
Problems Often outgrows the area intended for it. An extremely vigorous climber, needing care and attention regarding its ultimate size. May self-seed in the garden but resulting seedling will not be true to parent.
Similar forms of interest *C. m.* **'Alexandria'** Creamy white flowers 3 in (7.5 cm) wide with attractive scent and yellow stamens. Light green foliage. *C. m.* **'Elizabeth'** Scented, soft pale pink flowers up to 3 in (7.5 cm) wide. Extremely attractive. Light bronze foliage. *C. m.* **'Freda'** Cherry pink flowers with each sepal having a crimson red edge. Bronze foliage. *C. m. grandiflora* Flowers up to 4 in (10 cm) wide, white with yellow centres. Strong growing. *C. m.* **'Lilacine'** Purple/green foliage and blue/lilac flowers. *C. m.* **'Marjorie'** Semi-double, creamy pink flowers with orange tint. A most unusual variety. Two thirds average height and spread. *C. m.* **'Mayleen'** Deep pink flowers up to 3 in (7.5 cm) across with golden stamens. Bronze foliage. *C. m.* **'Odorata'** Possibly the original form of *C. montana* with soft pink flowers and some scent, produced in late spring/early summer. Tolerates all aspects. *C. m.* **'Picton's Variety'** Strawberry-pink flowers with golden centres. Purple to purple/green foliage. Reaches only half average height and spread. *C. m.* **'Pink Perfection'** Smaller flowers than most, but attractive pink colour with round-ended sepals. Two thirds average height and spread. Purple/green foliage. *C. m.* **'Rubens'** Purple/green foliage. Pale mauve pink flowers. Most reliable. *C. m.* **'Tetrarose'** Bronze foliage. Medium deep rose/mauve

247

Clematis montana in flower

flowers of good size. Half average height and spread. *C. m.* **'Vera'** Foliage dark green with large, pink, fragrant flowers. *C. m.* **'Wilsonii'** Creamy white flowers with yellow anthers. Later flowering than most varieties, not producing its flowers until early to mid summer. Two thirds average height and spread. *C. chrysocoma* Strong-growing attractive foliage with downy surface. Small soft pink flowers in late spring/early summer. Tolerates all aspects. *C.c. sericea* (syn. *C.c. spooneri*) Pure white flowers up to 3 in (7.5 cm) in diameter with bold yellow anthers, produced from late spring to early summer. Tolerates all aspects. *C. vedrariensis* Medium-sized bold rosy/mauve flowers borne in profusion from late spring to early summer. Tolerates all aspects. *C. v.* **'Hidcote'** Small, deep pink flowers from late spring to early summer. Tolerates all positions. Wide-ranging, rambling habit. *C. v.* **'Highdown'** Large pink flowers very similar to others in its group. Tolerates all aspects.

Average height and spread
Five years
15 × 15 ft (4.6 × 4.6 m)
Ten years
30 × 30 ft (9 × 9 m)
Twenty years
40 × 40 ft (12 × 12 m)
Protrudes up to 3 ft (91cm) from support. These
heights and spreads are only achieved when the vine is unpruned, but over ten years or more this can lead to deterioration in flower and foliage size and colour.

CLEMATIS ORIENTALIS
(*C. graveolens*)

ORIENTAL CLEMATIS

Ranunculaceae   *Woody Climber*
Deciduous

An attractive, late summer flowering clematis with a charm all of its own. Excellent for rambling over walls, fences and buildings but also a good ground cover plant.

Origin From Northern Asia.
Use For walls, fences, pillars, pergolas, roofs of small buildings or to ramble through large shrubs or small trees. Can be used as ground cover with or without support.
Description *Flower* Four oval, pointed, downy covered sepals carried on up to 4 in (10 cm) long flower stalks. Each flower 11–12 in (2.5–5 cm) wide, yellow in colour. Flowers late summer, early autumn. Attractive when in bud, forming a small balloon shape, opening to a more reflexed shape. *Foliage* Light green, dissected, ferny leaves offer-

ing some yellow autumn colour until the first hard frost. *Stem* Light green, becoming grey/green, finally grey/brown. Twining. More fragile than most. Additional support supplied by leaf tendrils. Medium rate of growth. *Fruit* Small, white, tufted seedheads follow flowers.
Hardiness Tolerates a minimum winter temperature of 14°F (−10°C).
Soil requirements Tolerates both alkaline and acid conditions, but must have a moist soil high in organic material. Resents drought.
Sun/Shade aspect Tolerates all but the most severe of aspects. Full sun to very light shade. Roots must be in the shade.
Pruning Cut hard back by at least two thirds or more in early to mid spring. Rapid regrowth follows pruning and the plant quickly gains its full height. It is sometimes suggested that no pruning is necessary unless it becomes too large, but this can lead to a very woody structure.
Training Will require wires on walls or fences, or can be allowed to ramble through large shrubs and small trees.
Propagation and nursery production From seed or, for named varieties, semi-ripe internodal cuttings taken early to mid summer. Should always be purchased container grown, best planting height 1–3 ft (30–91cm). Named varieties will need searching for from specialist nurseries.
Problems If not pruned as suggested can become old and woody, with resulting decrease in foliage and flower production. Stems in early spring often look dead.

Similar forms of interest *C. o.* **'Bill McKenzie'** The largest flowering of all the varieties and possibly the most spectacular. Yellow flowers followed by very attractive silver/white seedheads. *C. o.* **'Burford Variety'** One of the strongest-growing varieties with good-sized flowers of deep lemon yellow. *C. o.* **'L. & S. 13342'** Orange flowers with thick sepals in a nodding formation. *C. o.* **'Orange Peel'** Grey/green foliage, finely cut. Nodding, cup-shaped flowers with yellow, waxy sepals. Derives its name from the fact that the sepals are the same thickness as orange peel. L. & S. 13342 and 'Orange Peel' may well be the same plant masquerading under two different names. *C. o.* **'Sherriffii'** Very attractive grey/green foliage with deep yellow, cowslip-scented flowers in mid summer. *C. tangutica* Lemon yellow nodding flowers with silver/white stamens in mid summer to late autumn.
Average height and spread
Five years
7 × 7 ft (2.1 × 2.1 m)
Ten years
12 × 12 ft (3.7 × 3.7 m)
Twenty years
18 × 18 ft (4.5 × 4.5 m)
Protrudes up to 2 ft (60cm) from support.

CLEMATIS REHDERIANA
(*C. nutans thyrsoidea*)

KNOWN BY BOTANICAL NAME

Ranunculaceae   *Woody Climber*
Deciduous

One of the choicest of all clematis and other flowering climbing plants, not without its problems but worthy of perseverance.

Origin From China.
Use As a large spreading climber for walls, fences or pergolas and to ramble through medium-sized uninteresting shrubs and small trees.
Description *Flower* Bell-shaped, nodding, fragrant flowers 1½–2½ in (4–6 cm) wide carried in upright panicles. Each flower panicle up to 9 in (23 cm) long and 5 in (12 cm) wide. Sepals pale yellow, ¾ in (2 cm) long with recurved tips. Flowers in late summer to mid autumn. *Foliage* Pinnate leaves 6–9 in (15–23 cm) long, each with seven or nine wide, ovate, veined, light green leaflets up to 3 in (7.5 cm) long, deeply lobed with toothed edges and downy texture. *Stem* Light grey/green, becoming grey/ brown. Vigorous, upright habit. Limited twining. *Fruit* No fruit of interest.

Clematis orientalis **'Bill McKenzie'** – flowers and seedheads

Hardiness Tolerates a minimum winter temperature of 14°F (−10°C).
Soil requirements Does well on all soil types with no particular preference other than a high degree of good organic material and moisture retention.
Sun/Shade aspect Sheltered aspect. Full sun to very light shade. Roots must be in the shade.
Pruning Lightly trim in spring. Every five to ten years prune hard to induce new growth.
Training Will require tying to wires on walls or fences. Allow to ramble through large shrubs and small trees.
Propagation and nursery production From seed sown soon after fruiting or semi-ripe internodal cuttings taken in early to mid summer. Should always be purchased container grown, best planting height 9 in–3 ft (23–91 cm) Will have to be sought from specialist nurseries.

Clematis rehderiana

Problems Can take up to three years to establish. Unless adequate organic material is available in the soil, may often give poor results.
Similar forms of interest None.
Average height and spread
Five years
8 × 8 ft (2.4 × 2.4 m)
Ten years
12 × 12 ft (3.7 × 3.7 m)
Twenty years
20 × 20 ft (6 × 6 m)
Protrudes up to 3 ft (91 cm) from support.

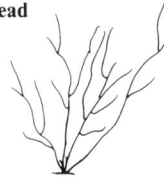

CLEMATIS TEXENSIS

KNOWN BY BOTANICAL NAME
Ranunculaceae   *Woody Climber*
Deciduous

A gem of a clematis but lacking the vigour and robustness of many of its stronger growing relations and needing care with siting.

Origin From Texas, USA.
Use Best allowed to ramble through medium to large shrubs but can also be used for walls and fences or small pillars.
Description *Flower* Single, urn-shaped flowers carried on slender, graceful stalks up to 6 in (15 cm) long, each flower up to 1¼ in (3.5 cm) long and 1 in (2.5 cm) wide in shades of red. Sepals thick and reflexed at tips. *Foliage* Leaves composed of four to eight broadly ovate leaflets up to 3 in (7.5 cm) long with two or three lobes. *Stem* Light grey/green to grey/brown. Fragile, sparse, medium to fast growth rate. Twining. *Fruit* Small tufts of silver/white seedheads follow flowers.

Hardiness Tolerates a minimum winter temperature of 14°F (−10°C).
Soil requirements Tolerates all soil types including alkaline but requires a good moisture-retentive soil with high organic content.
Sun/Shade aspect Needs a sheltered aspect. Stems, leaves and flowers in full sun and roots in shade.

Clematis texensis 'Gravetye Beauty' in flower

Pruning Cut hard back to within 6 in (15 cm) of previous season's point of origin in early to mid spring.
Training Allow to ramble through large shrubs or, if used on walls and fences, tie to wires.
Propagation and nursery production From semi-ripe cuttings. Difficult to propagate. Always purchase container grown; best planting height 9 in–3 ft (23–91 cm). Will have to be sought from specialist nurseries.
Problems Its sparse, wispy growth pattern is not always understood. Should not be compared with other forms of clematis, in particular the *montana* or large-flowering forms.
Similar forms of interest *C. t.* 'Duchess of Albany' Flowers deep pink with red stripe through each sepal. Can be semi-herbaceous, often dying to ground level in winter. *C. t.* 'Etoile Rose' Cherry pink flowers with silver margins. Semi-herbaceous. *C. t.* 'Gravetye Beauty' Small ruby-red flowers. Ends of flowers open wide. Semi-herbaceous. *C. t.* 'Pagoda' Early pink flowers. Semi-herbaceous. *C. t.* 'Princess of Wales' Deep pink, attractive flowers.
Average height and spread
Five years
7 × 7 ft (2.1 × 2.1 m)
Ten years
10 × 10 ft (3 × 3 m)
Twenty years
12 × 12 ft (3.7 × 3.7 m)
Protrudes up to 2 ft (60 cm) from support.

Clematis texensis 'Princess of Wales' in flower

CLEMATIS VITALBA

TRAVELLER'S JOY, OLD MAN'S BEARD
Ranunculaceae   *Woody Climber*
Deciduous

Although this is considered in many regions to be a wild plant, the beauty of its seedheads should allow it a wider use within gardens.

Origin From Europe and North Africa.
Use To ramble through large shrubs and small trees, as large-scale ground cover or to climb large walls, fences and pergolas.
Description *Flower* Off-white, ¾ in (2 cm) wide, carried at the ends and from the side shoots of stems. Almond scented, flowering from mid summer to early autumn. *Foliage* Pinnate, between 3–10 in (7.5–25 cm) long, depending on geographical location. Five ovate to lanceolate leaflets, each up to 4 in (10 cm) long, with coarsely toothed edges and downy grey/green appearance. *Stem* Grey/green, becoming grey/brown. Free-ranging and vigorous, making up to 16 ft (4.9 m) of growth per year. *Fruit* Attractive silky-white, ageing to grey/white, finally grey seedheads of very feathery texture, carried en masse. Often retained well into winter.

Clematis vitalba in flower

Hardiness Tolerates a minimum winter temperature of 0°F (−18°C).
Soil requirements Does extremely well on soils with high alkaline content but will tolerate all conditions. Requires a moist soil high in organic material for best results.
Sun/Shade aspect All aspects with foliage and flowers in light shade to full sun and roots in the shade.
Pruning Normally requires no pruning, but can be drastically reduced in size if required on a five or ten year basis. Pruning is best carried out in early spring.
Training Allow to ramble through large shrubs and trees
Propagation and nursery production From seed or from semi-ripe cuttings taken early summer. Always purchase container grown; although common, it may be difficult to find and will have to be sought from specialist nurseries. Best planting height 9 in–3 ft (23–91 cm).
Problems Does require a very large area to reach its full potential.
Similar forms of interest None.
Average height and spread
Five years
8 × 8 ft (2.4 × 2.4 m)
Ten years
12 × 12 ft (3.7 × 3.7 m)
Twenty years
20 × 20 ft (6 × 6 m)
Protrudes up to 3 ft (91 cm) from support.

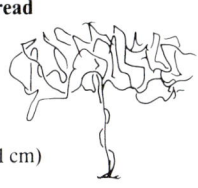

CLEMATIS VITICELLA

VINE BOWER

Ranunculaceae *Woody Climber*
Deciduous

A species with a large number of varieties covering a whole range of colours, but all retaining the delicate charm of the flowers.

Origin From south-east Europe.
Use For walls, fences, pillars and pergolas or to ramble through medium to large shrubs. Good for ground cover with support or through small, low-growing shrubs such as heathers.
Description *Flower* Broadly bell-shaped violet or purple/red flowers, 1½–2½ in (4–6 cm) across, presented singly on 3–4 in (7.5–10 cm) stems from mid summer to early autumn. *Foliage* Up to 5 in (12 cm) long dark green leaves composed of numerous ovate leaflets. *Stem* Thin, wispy, limited number of shoots, twining. Leaves offer support by leaf tendrils. Light green when young, ageing to grey/brown. Slow to medium rate of growth. *Fruit* May produce small silver seedheads of limited attraction.
Hardiness Tolerates a minimum winter temperature of 4°F (−15°C). Although some stem die-back may be suffered at this temperature, spring rejuvenation normally occurs.

Clematis viticella **'Alba Luxuriens' in flower**

Soil requirements Tolerates all types but must have substantial organic content. Resents drying out.
Sun/Shade aspect Tolerates all but the most exposed of situations. Foliage and flowers should be in full sun to very light shade and roots in shade.
Pruning Either allow to grow unpruned for one to ten years and then cut hard back every subsequent ten years and only lightly trim in interim years, or cut hard back each spring to induce new production of flowering growth. The latter pruning will reduce the overall size and possibly the architectural features of a mature climber.
Training Will require wires on walls and fences.
Propagation and nursery production From semi-ripe internodal cuttings taken in early summer. Should always be purchased container grown; best planting height 1–3 ft (30–91 cm). Many varieties will require searching for from specialist nurseries.
Problems Its sparse growth must be appreciated. May suffer attacks of mildew in late summer/early autumn but a proprietary control will normally eradicate this. Availability of some varieties may be extremely scarce. Difficult to propagate.
Similar forms of interest *C. v.* 'Abundance' Extremely attractive, light red to purple flowers. Responds well to hard pruning in

Clematis viticella **'Abundance' in flower**

spring. *C. v.* **'Alba Luxuriens'** One of the most interesting clematis with twisted, creamy white sepals with green bands of colour through each. Colour variegations will occur according to season. *C. v.* **'Albiflora'** White flowers, medium size. May be scarce. *C. v.* **'Caerulea'** Deep blue/violet flowers borne in profusion. Scarce. *C. v.* **'Elvan'** Purple flowers with cream edging. Extremely good when allowed to ramble through shrubs. Scarce. *C. v.* **'Etoile Violet'** Violet flowers of good size with pale yellow anthers. Very free-flowering throughout late summer and early autumn. Good for ground cover. *C. v.* **'Flora Plena'** Double, rose-coloured flowers. A very old variety scarce in production. *C. v.* **'Huldine'** Most attractive white flowers, suffused with mother-of-pearl pink, carried in large clusters. Well recommended. *C. v.* **'Kermesina'** Wine-red flowers up to 3 in (7.5 cm) across. Well recommended. *C. v.* **'Little Nell'** Small mauve flowers with cream central bar down each sepal. *C. v.* **'Madame Julia Correvon'** Wine-red flowers up to 2½ in (6 cm) across. Up to six twisted sepals with central golden anthers. Less vigorous than most. Flowers from early summer to early autumn. *C. v.* **'Margot Koster'** Mauve-pink flowers of medium size with pure white anthers. Sepals well spaced. Two thirds average height and

spread. *C.v.* **'Minuet'** White flowers, small, with each sepal having pronounced mauve edging. Two thirds average height and spread. Extremely attractive. *C. v.* **'Purpurea Plena Elegance'** Violet-purple, double flowers resembling a small chrysanthemum. Two thirds average height and spread. Grow through medium to large shrubs for best display. *C. v.* **'Royal Velours'** Royal purple, velvety textured, medium-sized flowers with deep black anthers. Position carefully to show off flowers to best advantage. *C. v.* **'Rubra'** Wine-red flowers, small, with contrasting black anthers. Good for ground cover. *C. v.* **'Venosa Violacea'** Sepals boat-shaped, medium-sized, purple with bold white and purple veined centres.
Average height and spread
Five years
8 × 8 ft (2.4 × 2.4 m)
Ten years
12 × 12 ft (3.7 × 3.7 m)
Twenty years
12 × 12 ft (3.7 × 3.7 m)
Protrudes up to 2 ft (60 cm) from support.

Clematis viticella **'Minuet' in flower**

Clematis viticella **'Etoile Violet' in flower**

Cleyera fortunei

CLEYERA FORTUNEI

KNOWN BY BOTANICAL NAME

Theaceae   *Wall Shrub*
Evergreen

An attractive foliage shrub responding well to fan-training or planting in association with walls or fences where it derives some protection.

Origin from Japan.
Use As a fan-trained foliage shrub for sheltered walls and fences both in favourable open areas and under the protection of greenhouses or conservatories.
Description *Flower* Flowers up to $\frac{1}{2}$ in (1 cm) across, pale yellow with red/brown calyxes, presented in clusters or singly in late summer/early autumn. *Foliage* Ovate to lanceolate, up to 4 in (10 cm) long, bright green with golden yellow and scarlet variegation around the edges. *Stems* Mid to dark green, semirigid, adapting well to fan-training. Slow to medium growth rate. *Fruit* No fruit of interest.
Hardiness Tolerates a minimum temperature of 14°F (−10°C) with the protection of a wall or fence.
Soil requirements Neutral to acid, although its full range of tolerance of alkalinity has not yet been fully tested. Requires a soil high in organic material and plant nutrients for best growth and foliage.
Sun/Shade aspect Sheltered aspect. Light shade for preference but will tolerate full sun to medium shade.
Pruning Not normally required other than for training.
Training Fan-train on to wires or individual anchor points or allow to grow freestanding.
Propagation and nursery production From semi-ripe cuttings. Should always be purchased container grown; will have to be sought from specialist nurseries. Best planting height 9 in-2 ft (23-60 cm).
Problems Its range of hardiness, soil conditions and growth have not yet been fully tested outside its native environment.
Similar forms of interest. None.
Average height and spread
Five years
4 × 4 ft (1.2 × 1.2 m)
Ten years
8 × 8 ft (2.4 × 2.4 m)
Twenty years
12 × 12 ft (3.7 × 3.7 m)
Protrudes up to 2½ ft (76 cm) from support if fan-trained, 5 ft (1.5 m) untrained.

CLIANTHUS PUNICEUS

PARROT'S BILL, LOBSTER CLAW, KAKA BEAK

Leguminosae   *Tender Wall Shrub*
Evergreen

An attractive and interesting flowering wall shrub requiring a very mild planting location to achieve any results.

Origin From New Zealand.
Use As a tender wall shrub for very warm sheltered walls or fences or under the protection of a conservatory or greenhouse.
Decription *Flower* 4 in (10 cm) long clusters of six or more brilliant scarlet flowers from early summer onwards. *Foliage* Pinnate, up to 6 in (15 cm) long, with 12–24 oblong leaflets each up to 1¼ in (3.5 cm) long. Mid green. *Stem* Mid green, semi-rigid, forming fan shape. Not self-clinging. Medium growth rate. *Fruit* Produces no fruit of interest.

Clianthus puniceus **in flower**

Hardiness Tolerates a minimum winter temperature of up to 23°F (−5°C) but only in very sheltered locations.
Soil requirements Tolerates all soil types but resents very dry conditions.
Sun/Shade aspect Requires a very warm, sheltered aspect in full sun to very light shade.
Pruning Not normally required but individual branches can be removed if necessary.
Training Fan-train on to wires or individual anchorage points.
Propagation and nursery production Seed or semi-ripe cuttings inserted into sand with some additional bottom heat. Always purchase container grown; will have to be sought from a specialist nursery or from florists, where it is sometimes sold as a house plant. Best planting height 6 in-2 ft (15-60 cm).
Problems Its hardiness is always suspect and even in the most sheltered of areas it is at risk in winter. Availability scarce.
Similar forms of interest *C. p.* **'Albus'** Pure white, less hardy. Very scarce. *C. p.* **'Flamingo'** Deep rose pink flowers. Less hardy. *C. p.* **'Red Cardinal'** Bright scarlet. Less hardy. *C. p.* **'White Heron'** Ivory white flowers with green shading. Less hardy.
Average height and spread
Five years
4 × 4 ft (1.2 × 1.2 m)
Ten years
8 × 8 ft (2.4 × 2.4 m)
Twenty years
12 × 12 ft (3.7 × 3.7 m)
Protrudes up to 2 ft (60 cm) from support.

COBAEA SCANDENS

CLIMBING CATHEDRAL BELLS

Polemoniaceae   *Annual Climber*
Deciduous

An attractive annual climber producing fascinating flowers when a hot summer prevails.

Origin From Central South America, including Mexico.
Use As an annual climber for walls and fences, to cascade through medium to large shrubs, or as a greenhouse or conservatory climber in large containers.
Description *Flower* Large, bell-shaped flowers, up to 2-3 in (5-7.5 cm) long, carried singly on leaf stalks up to 8 in (20 cm) long from May to October. Green inner colouring towards the base, violet/purple on outer side. Rounded, lobed ends. *Foliage* Pinnate leaves with three pairs of leaflets, up to 4 in (10 cm) long and 2 in (15 cm) wide. Grey/green with purple hue. Attractive. *Stem* Self-clinging by twining and tendrils. Light green. Fast annual growth produced from seedlings. *Fruit* No fruit of interest.
Hardiness Tolerates a minimum winter temperature of 32°F (0°C).
Soil requirements Any average garden soil with adequate moisture retention and nutrients. For plants grown in containers in conservatories and greenhouses, a good quality soil-based potting compost should be used.
Sun/Shade aspect Requires a sheltered summer planting area to prevent wind and rain damage to flowers. Full sun, dislikes shade in which it will be very shy to flower.
Pruning Not required. Of an annual nature and dies in autumn.
Training Allow to ramble through large shrubs, up pillars and posts. May need wires or strings where adequate training material is unavailable.
Propagation and nursery production From seed raised under protection and not planted out until all danger of spring frosts has passed. Purchase seed from specialist seed merchants. If plants are available in spring from nurseries, garden centres or other sources, best planting height is 2-6 in (5-15 cm).
Problems The requirement to grow it annually can be a little frustrating, but it is well worth the effort.

Cobaea scandens

Similar forms of interest *C. s.* 'Alba' Pure white form. Seed extremely scarce. *C. s.* 'Variegata' Variegated foliage. Seed extremely scarce.

Average height and spread
One year
15 × 15 ft
(4.6 × 4.6 m)
Protrudes up to
12 in (30 cm)
from support.

COLLETIA CRUCIATA

ANCHOR PLANT

Rhamnaceae   *Wall Shrub*
Evergreen

An unusual shrub without foliage but viciously armed with cactus-like spikes so must be carefully sited and handled.

Colletia cruciata

Origin From Brazil and Uruguay.
Use As a curiosity shrub for growing against walls and fences where it benefits from the protection.
Description *Flower* Small, white, fragrant, pitcher-shaped, borne in profusion on mature shrubs mid summer to early autumn. *Foliage* None. *Stem* Light grey/green stems, surmounted by branches which take on a flat, triangular shape, each triangulation topped by a very sharp spike. Slow-growing. *Fruit* Insignificant.
Hardiness Tolerates a minimum winter temperature of 14°F (−10°C), but susceptible to frost when young and needs protection in early stages.

Soil requirements Any soil, including dry to very dry.
Sun/Shade aspect Sheltered aspect. Full sun to light shade.
Pruning None required.
Training No training necessary, simply grow against a wall.
Propagation and nursery production From seed. Purchase container grown. Not easy to find. Best planting height 8–15 in (20–40 cm).
Problems Its spines are vicious, so handle with due care.
Similar forms of interest *C. armata* Vanilla-scented flowers, borne profusely late summer to early autumn. Shoots form rounded spines, each tipped with a single thorn. From Chile. *C. a.* 'Rosea' A stronger-growing, very scarce form with flowers pink in bud, opening to pink-tinged white.
Average height and spread
Five years
3 × 3 ft (91 × 91 cm)
Ten years
4 × 4 ft (1.2 × 1.2 m)
Twenty years
8 × 8 ft (2.4 × 2.4 m)
Protrudes up to 5 ft (1.5 m) from wall or fence.

CORNUS FLORIDA
(*Benthamidia florida*)

NORTH AMERICAN FLOWERING DOGWOOD

Cornaceae   *Wall Shrub*
Deciduous

A large shrub not always recognized as being suitable for a wall, but in such situations, particularly if lightly shaded, makes a handsome specimen.

Origin From eastern USA.
Use As a large, fan-trained flowering and autumn foliage shrub against walls and fences.
Description *Flower* A 3 in (7.5 cm) wide head with four medium-sized white bracts resembling petals, late spring to early summer, bracts being slightly twisted and curled. Some pink flowering forms. *Foliage* Leaves oblong, 2–3 in (5–7.5 cm) long, purple-green, giving good red/orange autumn colour. Some white, yellow and pink variegated forms. *Stem* Upright when young, becoming branching with age. Medium to fast growth rate when young, slowing with age. *Fruit* Insignificant.
Hardiness Tolerates a minimum winter temperature of 14°F (−10°C).
Soil requirements Prefers neutral to acid soil, dislikes any alkalinity.
Sun/Shade aspect Tolerates all but the most severe aspect. As a wall shrub best in light shade.
Pruning None, apart from that required for training.

Training Requires wires or individual anchor points to secure and encourage a fan-trained shape.
Propagation and nursery production From softwood cuttings taken in mid summer or from layers. Purchase container grown. Readily available in its native environment, scarce elsewhere. Best planting height 2–2½ ft (60–76 cm).
Problems None, given suitable conditions.
Similar forms of interest *C. f.* 'Apple Blossom' Apple blossom pink bracts. Good autumn colour. *C. f.* 'Cherokee Chief' Flower bracts deep rose red. Good autumn colour. *C. f.* 'Cherokee Princess' Large, round, white flower bracts. Good autumn colour. *C. f.* 'Pendula' White flower bracts. Good autumn colour. An interesting weeping form, of less than average height but with more spread. *C. f.* 'Rainbow' White flower bracts. Golden-margined dark green leaves, in autumn turning plum/purple. Slightly less hardy and less than average height and

Cornus florida in flower

spread. *C. f.* 'Rubra' One of the most sought-after varieties, with rosy-pink flower bracts in spring. Leaves red when young. Slightly less hardy and less than average height and spread. *C. f.* 'Spring Song' Deep rose-red flower bracts. *C. f.* 'Tricolor' White flower bracts in spring. Foliage green with white irregular margin, flushed rose-pink, turning purple in winter with rose-red edges. Less than average hardiness, height and spread. *C. f.* 'White Cloud' Large white flower bracts freely borne. Foliage bronze/green.
Average height and spread
Five years
7 × 7 ft (2.1 × 2.1 m)
Ten years
15 × 15 ft (4.6 × 4.6 m)
Twenty years
18 × 18 ft (5.5 × 5.5 m)
Protrudes up to 4 ft (1.2 m) from support.

Cornus florida 'Rubra' in flower

CORNUS KOUSA

CHINESE DOGWOOD, JAPANESE DOGWOOD, KOUSA DOGWOOD
Cornaceae   *Wall Shrub*
Deciduous
A truly magnificent late spring flowering shrub with good autumn colour, adapting well as a large fan-trained specimen.

Origin From Japan and Korea.
Use As a large, spreading, fan-shaped wall or fence shrub.
Description *Flower* Four 2 in (5 cm) wide, large creamy-white bracts surround purple-green flower clusters in late spring, early summer. *Foliage* Leaves elliptic, slightly curled, 2–3 in (5–7.5 cm) long, olive-green with some purple shading, giving exceptional orange/red autumn colour. Foliage in autumn is retained well and is not usually damaged or destroyed by wind. *Stem* Upright, light green to grey/green, becoming grey/brown and branching with age. With training forms a wide, spreading, fan-shaped shrub. Slow growth rate at first, then medium, finally again becoming less vigorous. *Fruit* Attractive dull red, strawberry-like fruits on mature shrubs.
Hardiness Tolerates a minimum winter temperature of 4°F (−15°C).
Soil requirements Neutral to acid, in moist conditions and light shade, however, may be happy on slightly alkaline soils.
Sun/Shade aspect Tolerates all aspects, except extremely severe. Prefers light shade, tolerates medium shade to full sun.
Pruning None required, other than for training.
Training Requires wires or individual anchor points to secure and encourage the fan-trained shape.
Propagation and nursery production From layers or softwood cuttings taken in mid summer. Purchase container grown. Moderately easy to find. Best planting height 2–3 ft (60–91 cm).

Cornus kousa **in flower**

Problems Slow to establish, taking three to four years to flower really well.
Similar forms of interest *C. k. chinensis* A Chinese form which produces larger flowering bracts than its parent. *C. k.* **'Gold Spot'** A variety with white flower bracts and with the green foliage mottled with golden variegation. Hard to find. *C. k.* **'Milky Way'** An American variety of garden origin with larger white flower bracts. *C. k.* **'Norman Haddon'** Light grey/green foliage with good autumn colour. Pink bracts in late spring. Fruits insignificant.
Average height and spread
Five years
5 × 5 ft (1.5 × 1.5 m)
Ten years
10 × 10 ft (3 × 3 m)
Twenty years
15 × 15 ft + (4.6 × 4.6 m +)
Protrudes up to 3 ft (91 cm) from support.

COROKIA COTONEASTER

WIRE-NETTING BUSH
Cornaceae   *Wall Shrub*
Evergreen
An interesting shrub that fan-trains well against a wall and often benefits from the protection.

Origin From New Zealand.
Use As a low, spreading, fan-trained shrub with interesting stems, foliage and flowers for walls and fences.
Description *Flower* Very small, single, bright yellow, star-like flowers, late spring to early summer. *Foliage* Leaves very small, ovate, ½–¾ in (1–2 cm) long, dark green to purple/green. Some yellow autumn colour. *Stem* Contorted, very twiggy branchlets, forming an intricate tracery. When trained, forms a fan-shape. Stems purple/green. Very slow rate of growth. *Fruit* In hot summers, produces small, round, orange fruits.

Corokia cotoneaster **in flower**

Hardiness Tolerates a minimum winter temperature of 14°F (−10°C).
Soil requirements Alkaline to acid. Prefers a well-drained soil.
Sun/Shade aspect Sheltered aspect. Best in full sun or very light shade.
Pruning None, other than that required for training.
Training Allow to grow against a wall. Normally needs no fixing.
Propagation and nursery production From softwood cuttings taken in early summer. Purchase container grown. Not easy to find. Best planting height 4–12 in (10–30 cm).
Problems Extremely slow-growing. Young plants often look weak and insipid before being planted out, but gradually acquire an interesting, contorted form.
Similar forms of interest *C. c.* **'Little Prince'** Black bark, green leaves with silver sheen. Small star-shaped yellow flowers. *C. buddleioides* Upright lanceolate grey leaves with silver reverse. Clusters of bright yellow star-shaped flowers. *C.* **'Coppershine'** foliage turns purple in sun. Yellow flowers.
Average height and spread
Five years
2½ × 2½ ft (76 × 76 cm)
Ten years
3½ × 3½ ft (1 × 1 m)
Twenty years
4 × 4 ft (1.2 × 1.2 m)
Protrudes up to 2 ft (60 cm) from support.

COROKIA × VIRGATA

WIRE NETTING PLANT
Cornaceae   *Wall Shrub*
Evergreen
A shrub benefiting from the protection of a wall or fence where it can show off its small yellow flowers to the best advantage.

Origin From New Zealand.
Use As a freestanding or fan-trained shrub for walls and fences. Good as a container-grown plant for greenhouses and conservatories in colder areas.
Description *Flower* Small, star-shaped yellow flowers up to ½ in (1 cm) wide, normally presented in threes at the ends of branches in mid to late spring. *Foliage* Lanceolate, up to 1¾ in (4.5 cm) long and ½ in (1 cm) wide. Grey/green shiny upper surface, white downy undersides. Attractive. *Stem* Grey/green, becoming grey/brown. Rigid, branching. Attractive. Slow to medium growth rate. *Fruit* In hot summers may produce oval, ¼ in (5 mm) long, orange/yellow fruits.
Hardiness Tolerates a minimum winter temperature of up to 23°F (−5°C) but needs shelter from cold winds.
Soil requirements Tolerates all soil types except extremely waterlogged and very dry.
Sun/Shade aspect Requires a sheltered aspect in full sun to light shade, with full sun for preference.
Pruning Normally not required but individual branches can be removed if necessary.
Training Allow to grow freestanding or fan-train on to wires or individual anchorage points.
Propagation and nursery production From semi-ripe cuttings taken in mid to late summer. Should always be purchased container grown; may have to be sought from specialist nurseries. Best planting height 6 in–2 ft (15–60 cm).

Corokia × virgata **in flower**

Problems Its hardiness can be suspect. Availability is scarce and the plants offered are normally small.
Similar forms of interest *C. × v.* **'Bronze King'** Bronze lanceolate leaves and small yellow star-shaped flowers. *C. × v.* **'Red Wonder'** Bronze/green leaves, yellow star-shaped flowers. May produce deep red berries. *C. × v.* **'Yellow Wonder'** Green leaves. Yellow fruits follow yellow star-shaped flowers.
Average height and spread
Five years
3 × 4 ft (91 cm × 1.2 m)
Ten years
8 × 8 ft (2.4 × 2.4 m)
Twenty years
10 × 12 ft (3 × 3.7 m)
Protrudes up to 18 in (45 cm) from support if fan-trained, 5 ft (1.5 m) untrained.

253

CORONILLA GLAUCA

KNOWN BY BOTANICAL NAME

Leguminosae　　　　　　*Wall Shrub*
Evergreen

Although long cultivated, certainly in Europe, too rarely used as a wall shrub today.

Origin From southern Europe.
Use As a freestanding shrub against walls and fences or up pillars.
Description *Flower* Yellow pea-flowers, ½ in (1 cm) wide and long, very freely produced, mid to late spring, with further intermittent flowering through summer and early autumn. *Foliage* Leaves round to ovate, 1–1½ in (2.5–4 cm) long, grey/green, giving some yellow autumn colour on older leaves. *Stem* Grey/green. Upright, becoming branching, forming a flat-backed, arching shrub. Can be fan-trained if required. *Fruit* Small, grey/green pea pods.
Hardiness Tolerates a minimum winter temperature of 23°F (−5°C).
Soil requirements Any soil, including very alkaline.
Sun/Shade aspect Sheltered aspect in full sun to very light shade.
Pruning Remove some three to four year old branches to ground level occasionally to help annual rejuvenation.
Training May need tying to anchor points or wires to prevent it falling forward.
Propagation and nursery production From softwood cuttings taken mid summer. Purchase container grown. Often looks weak when purchased, but quickly establishes itself. Best planting height 2–2½ ft (60–76 cm). Available from specialist nurseries and good garden centres.

Coronilla glauca in flower

Problems May suffer from blackfly but not a major problem.
Similar forms of interest *C.g.* **'Variegata'** Grey/green leaves with white variegation when old. Yellow flowers. Slightly more tender. *C. emerus* Dark yellow pea-flowers produced mainly in spring, but continuing intermittently into summer and early autumn. A slightly more hardy form. *C. valentina* Bright golden yellow fragrant flowers. Grey/green foliage. Slightly more tender.
Average height and spread
Five years
6 × 6 ft (1.8 × 1.8 m)
Ten years
7 × 7 ft (2.1 × 2.1 m)
Twenty years
9 × 9 ft (2.7 × 2.7 m)
Protrudes 2½–3 ft (76–91 cm) from wall or fence.

CORYNABUTILON VITIFOLIUM
(*Abutilon vitifolium*)

FLOWERING MAPLE, VINE-LEAVED ABUTILON

Malvaceae　　　　　　*Wall Shrub*
Deciduous

A tall, impressive plant but susceptible to winter cold. In very mild areas it may be considered a semi-evergreen. Recently reclassified as *Corynabutilon vitifolium*, this shrub is often still available under the name *Abutilon vitifolium*.

Origin From Chile.
Use As a tall flowering shrub for mild areas grown on a wall, or in large conservatories for colder areas.
Description *Flower* Large, saucer-shaped, deep to pale mauve or white flowers, depending on variety, borne freely late spring through to early summer. *Foliage* Vine-shaped leaves, 4–6 in (10–15 cm) long with a grey, downy covering. Some yellow autumn colour. *Stem* Upright, grey/green with a grey downy covering. Fast to medium rate of growth. *Fruit* Insignificant.
Hardiness Tolerates a minimum winter temperature of 23°F (−5°C).
Soil requirements Most soils, although thin alkaline soils will cause chlorosis.

Corynabutilon vitifolium 'Album' in flower

Sun/Shade aspect Sheltered aspect in full sun. Resents any shade.
Pruning None required, although individual branches or limbs may be removed.
Training Tie upright shoots to individual anchor points.
Propagation and nursery production The basic form from seed, named varieties may be produced from softwood cuttings taken mid to late summer. Purchase container grown, often difficult to find. Best planting height 14 in–3 ft (40–91 cm).
Problems Stems at risk to severe temperature changes in winter and to wind chill factor. If stems die off, regeneration from the base is possible.
Similar forms of interest *C. v.* **'Album'** A white flowering form. *C. v.* **'Veronica Tennant'** Produces masses of large mauve flowers, deeper in colour than those of the parent. *C. v.* **'Suntense'** Slightly later-flowering, with mauve flowers. More tender than parent.
Average height and spread
Five years
8 × 6 ft (2.4 × 1.8 m)
Ten years
16 × 12 ft (5 × 3.7 m)
Twenty years
24 × 18 ft (7.3 × 5.5 m)
Protrudes up to 4 ft (1.2 m) from support.

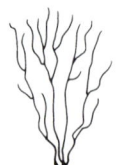

COTINUS COGGYGRIA

SMOKE TREE, SMOKE BUSH, BURNING BUSH, CHITAM WOOD, VENETIAN SUMACH

Anacardiaceae　　　　　　*Wall Shrub*
Deciduous

A shrub for summer and autumn display, producing fine foliage colours, profuse feathery flowers and good structural shape.

Origin From central and southern Europe.
Use As a foliage and flowering shrub fan-trained against a wall or fence.
Description *Flower* Large, open, pale pink inflorescences resembling plumes 6–8 in (15–20 cm) long, borne profusely on all wood three years old or more in summer, persisting into early and late autumn, turning smoky grey. *Foliage* Leaves ovate to oblong, 1½–2 in (4–5 cm) long, grey/green when young, opening to mid green, turning vivid orange/yellow in autumn; purple-leaved varieties turning scarlet-red. *Stem* Light green, becoming streaked with orange or red shading, finally grey/brown. Fast growing and upright when

Corynabutilon vitifolium 'Suntense' in flower

254

young, becoming slower and very branching and twiggy, forming a fan-shaped shrub when trained. *Fruit* Feathery flowers change to seedheads.
Hardiness Tolerates a minimum winter temperature of 4°F (−15°C). Some winter die-back may occur at tips of new growth.
Soil requirements Prefers rich, deep soil, but tolerates most.
Sun/Shade aspect Tolerates all but the most severe of aspects. Green-leaved varieties tolerate very light shade to full sun; purple-leaved varieties must have full sun, otherwise they turn green.
Pruning Remove some mature shoots one to three years old each spring, so inducing some foliage rejuvenation and improved flowering.
Training Requires wires or individual anchor points to secure and encourage the fan-trained shape.
Propagation and nursery production From layers. Purchase container grown. Relatively easy to find in nursery production, some varieties will be found in garden centres. Best planting height 15 in–2½ ft (38–76 cm).
Problems Some purple-leaved varieties susceptible to mildew. Slow to establish, taking two to three years after planting to gain full stature.
Similar forms of interest *C. c.* 'Flame' One of the best varieties for autumn colour. Pink flowers and bright orange-red foliage in autumn. Hard to find and often confused with the parent plant. *C. c.* 'Foliis Purpureis' Pink inflorescence, young foliage rich plum/purple, ageing to lighter red to purple/red, late summer. Good autumn colours. *C. c.* 'Notcutt's Variety' Pink to purple-pink inflorescence with good red/purple autumn colours. Very deep purple leaves, slightly larger than those of its parent. Of slightly less height and spread. *C. c.* 'Royal Purple'

Cotinus coggygria 'Royal Purple' in leaf

Purple/pink inflorescence, purple/wine foliage, almost translucent when seen with sunlight behind it, becoming duller purple towards autumn, finally red. Slightly smaller than *C. coggygria*. Of garden origin. *C. c.* 'Rubrifolius' Pink inflorescence, deep wine red leaves when young, translucent in sunlight, becoming red/green towards autumn. Good autumn colour. *C. obovatus* (syn. *C. o. americanus*) (Chitam Wood). Light pink inflorescence. Round, ovate, light to mid green foliage with some orange shading, brilliant orange-red in autumn. Leaves much larger than *C. coggygria*. A large, spreading shrub from south-eastern USA.
Average height and spread
Five years
7 × 7 ft (2.1 × 2.1 m)
Ten years
12 × 12 ft (3.7 × 3.7 m)
Twenty years
20 × 20 ft (6 × 6 m)
Protrudes up to 4 ft (1.2 m) from support.

Cotoneaster horizontalis in fruit

COTONEASTER HORIZONTALIS

FISHBONE COTONEASTER, HERRINGBONE COTONEASTER
Rosaceae　　　　　　　　　*Wall Shrub*
Deciduous
An excellent foliage, flowering and fruiting shrub ideal for fan-training against exposed walls and fences where height is limited.

Origin Possibly a native of Western China or of garden origin.
Use A deciduous shrub with attractive autumn foliage and fruit. Ideal for fan-training on low exposed walls or fences.
Description *Flower* Masses of small, white, four-petalled, cup-shaped flowers, borne singly at leaf joints, each with a red calyx, in late spring to early summer. *Foliage* Leaves ovate, ½–1 in (1–2.5 cm) long and wide, dark green to grey/green, with white edge variegation, depending on variety. Good autumn colours. *Stem* Upright when young, becoming branching, twiggy and spreading with age. Soft grey/green when young ageing to grey/brown. Medium growth rate. *Fruit* Small, red, round, glossy fruits in autumn.
Hardiness Tolerates a minimum winter temperature of 4°F (−15°C).
Soil requirements Tolerates all soil types.
Sun/Shade aspect Good in exposed situations. Full sun to medium shade.
Pruning None required but can be reduced in size in early spring if necessary.
Training Normally requires individual anchor points or wires to secure main branches. Side branches form a fish-bone pattern without assistance.
Propagation and nursery production From semi ripe cuttings taken in mid summer or seed sown in spring. Purchase container grown for best results. Best planting height 12–18 in (30–45 cm).
Problems Relatively slow growing.
Similar forms of interest *C. h.* 'Major' Very similar in growth and habit to its parent but slightly more vigorous, with larger, rounder leaves and larger but fewer fruit. *C. h.* 'Robusta' A larger, faster-growing variety in which the herringbone pattern is more widely spaced. Leaves larger, rounder, more cup-shaped. Good red, round autumn fruits. One third more average height and spread. Not always easy to find. *C. h.* 'Variegatus' Herringbone growth pattern. Light grey/green foliage, round, with creamy white margins and lined with red round the outer edge. White flowers and red fruit, less glossy and less freely produced. Two thirds average height and spread. Of garden origin. *C.*

Cotoneaster 'Coral Beauty' in flower

'Coral Beauty' A similar variety with small, round, grey/green, ovate leaves giving good orange autumn colour. White flowers followed by coral red fruits in autumn. Slightly less than average height and spread. Of garden origin.
Average height and spread
Five years
3 × 4 ft (91 cm × 1.2 m)
Ten years
5 × 6 ft (1.5 × 1.8 m)
Twenty years
8 × 6 ft (2.4 × 1.8 m)
Protrudes up to 18 in (45 cm) from support.

COTONEASTER × HYBRIDUS PENDULUS

WEEPING COTONEASTER
Rosaceae　　　　　　　　　*Wall Shrub*
Semi-evergreen to evergreen
A fast-growing shrub that when grown as a fan-trained specimen provides a brilliant display of berries against walls and fences.

Origin Of garden origin.
Use As a fan-trained shrub for large walls and fences where its rapid growth provides quick cover.
Description *Flower* Numerous small white flowers produced in clusters in early summer. *Foliage* Ovate, 2–3 in (5–7.5 cm) long, dark green, with reddish veins on silver undersides. Often retained well into winter. *Stem* Light green ageing to purple/brown, finally grey/

brown. Fast growing, slowing with age. *Fruit* Clusters of red fruits in autumn carried along full length of mature stems.

Hardiness Tolerates a minimum winter temperature of 0°F (−18°C).

Soil requirements Any good soil, but shows some resistance on severe alkaline types by being slower growing.

Sun/Shade aspect Tolerates a wide range of aspects. Full sun to light shade, preferring full sun.

Pruning None required other than that for fan-training.

Training Tie to wires or individual anchor points in a fan shape.

Propagation and nursery production From cuttings or by grafting or budding on to *C. bullatus*. Best purchased container-grown; tree-trained plants of 6 ft (1.8 m) or more are available from garden centres and general nurseries. Both plants grown for ground cover and tree-trained can be used for fan-training.

Cotoneaster × hybridus pendulus in fruit

Problems Susceptible to stem canker, which can kill entire branches. On poor soils may well lose its vigour. Also susceptible to fire blight; if this disease is confirmed the tree must be completely destroyed by burning and its occurrence should be reported to the appropriate government agency.

Similar forms of interest None.

Average height and spread
Five years
6 × 6 ft (1.8 × 1.8 m)
Ten years
9 × 9 ft (2.7 × 2.7 m)
Twenty years
15 × 15 ft (4.6 × 4.6 m)
Protrudes up to 3 ft (91cm) from support.

COTONEASTER (Low-growing, Spreading, Evergreen Forms for Walls)

KNOWN BY BOTANICAL NAME
Rosaceae *Wall Shrub*
Evergreen
Useful flowering and fruiting evergreens for growing as wall shrubs.

Origin Mainly native to China, but many varieties now of garden or nursery origin.

Use As low, spreading, fan-trained shrubs for walls and fences.

Description *Flower* Small, white, four-petalled flowers with red calyxes, borne singly or in clusters, depending on variety, late spring to early summer. *Foliage* Leaves ½–¾ in (1–2 cm) long, lanceolate to ovate and in some varieties round. Dark shiny green with grey undersides, veins and stalks often red-shaded. *Stem* Green to dark brown, spreading, forming a fan-shape. Slow to medium growth rate. *Fruit* Round to ovate, glossy fruits, normally red, some purple/red depending on variety. **Hardiness** Tolerates a minimum winter temperature of 4°F (−14°C).

Cotoneaster congestus in fruit

Soil requirements Tolerant of any soil.

Sun/Shade aspect Tolerates all aspects. Good in full sun to medium shade.

Pruning None required other than keeping within bounds.

Training Requires wires or individual anchor points to secure and encourage the fan-trained shape.

Propagation and nursery production From semi-ripe cuttings taken early to mid summer. Purchase container grown. Best planting height 8–15 in (20–38 cm). There is a wide range of varieties, most easy to find.

Problems Relatively small when purchased and rather slow-growing.

Forms of interest *C. adpressus* 'Praecox' Small, round, semi-evergreen, ovate leaves, giving some contrasting autumn red and orange colour from old leaves which are discarded. Orange/red, good-sized fruits in autumn. Also known sometimes as *C.* 'Nanshan'. From western China. *C. buxifolius* A very dense dwarf evergreen variety with white flowers and grey to dull green round foliage. Oval to round red fruits in autumn. From south-western India. *C. congestus* A very thick, creeping form with white flowers and blue/green foliage. Good red round fruits. From the Himalayas. *C. dammeri* Wide-spreading, with long ranging shoots. White flowers, bright red fruits in autumn. Leaves ovate, dark green above with grey/green underside, some having slight purple shading on veins and outer edges. Quite free in habit and has self-rooting tips. From China. *C. d.* 'Radicans' A vigorous variety with ovate leaves and flowers borne closely together, often in pairs. Red fruits in autumn. Slightly less than average height and spread.

Average height and spread
Five years
4 × 4 ft (1.2 × 1.2 m)
Ten years
6 × 6 ft (1.8 × 1.8 m)
Twenty years
7 × 7 ft (2.1 × 2.1 m)
Protrudes up to 4 ft (1.2 m) from support.

COTONEASTER (Medium Height, Spreading Evergreen and Semi-Evergreen Forms for Walls)

KNOWN BY BOTANICAL NAME
Rosaceae *Wall Shrub*
Evergreen or semi-evergreen
Useful, medium height flowering and fruiting shrubs to cover low exposed areas.

Origin Mostly native to China or of garden origin.

Use As medium size, fan-trained shrubs, often useful in difficult positions.

Description *Flower* Small ¼ in (5 mm) wide, white, four-petalled flowers with small red calyxes, borne singly or in clusters, late spring to early summer, depending on variety. *Foliage* Leaves small, 1–1½ in (2.5–4 cm) long, round to broad ovate, dark green or grey/green, depending on variety. *Stem* Upright when young, becoming spreading with age forming a fan-shape. Medium growth rate. *Fruit* Round to ovate, glossy, normally red, but some purple/red varieties.

Hardiness Tolerates a minimum winter temperature of 0°F (−18°C).

Soil requirements Any soil conditions.

Sun/Shade aspect Tolerates all aspects. Full sun to medium shade.

Pruning None required but may be cut back to keep within bounds.

Training Requires wires or individual anchor points to secure and encourage the fan-trained shape.

Cotoneaster microphyllus in fruit

Propagation and nursery production From semi-ripe cuttings taken in mid summer. Purchase container grown. Most varieties relatively easy to find. Best planting height 1–2 ft (30–60 cm).

Problems May suffer from fire blight, for which there is no control or prevention. The plant should be burnt and the relevant government agency should be informed, since this is a notifiable disease.

Forms of interest *C. conspicuus* Arching branches with small, ovate, grey to grey/green foliage, covered in small white flowers, late spring to early summer. Bright red fruits freely borne. From south-eastern Tibet. *C. c.* 'Decorus' Equal to the parent in flowers and fruit, lower growing, height up to 4 ft (1.2 m) at maturity, spread 8 ft (2.4 m). *C. salicifolius* 'Autumn Fire' Narrow, ovate, dark green leaves, white flowers followed by orange-red fruit borne in clusters. Some foliage dies in winter, turning scarlet and enhancing the

Cotoneaster 'Skogholm' in flower

overall effect of fruit and colour. C. 'Donards Gem' Small, round, ovate leaves, semi-evergreen. Good red autumn fruits. C. microphyllus Round, dark, glossy grey/green leaves, ½ in (1 cm) long, very thickly produced along very twiggy branches. Round, dark scarlet fruits in autumn. Very hardy. Slower growth rate than average. From the Himalayas and south-western China. C. m. 'Cochleatus' Slower than its parent and lower growing. Dull grey/green, ovate leaves and dull grey/red, round fruits in autumn. Of less height and vigour but of equal hardiness. From western China, south-eastern Tibet through eastern Nepal. C. m. 'Thymifolius' Very small, narrow, ovate, shiny, dark green leaves produced very thickly and close to stems and branches. Purple/red berries in autumn. Not easy to find. C. salicifolius 'Gnome' Small, elliptic to round, dark grey/green leaves. Small red fruits. Of garden origin. C. s. 'Parkteppich' (Park Carpet) Leaves evergreen, narrow to medium width, lanceolate, dark green upper surface with grey underside, purple marking on veins. Round, red fruit, borne singly and in clusters. Of garden origin. C. × 'Skogholm' Leaves small, lanceolate to oval, mid green to dark green. Large, oval to round, coral red fruits in autumn. Attractive weeping habit. From western China.

Average height and spread
Five years
4 × 4 ft (1.2 × 1.2 m)
Ten years
8 × 8 ft (2.4 × 2.4 m)
Twenty years
12 × 12 ft (3.7 × 3.7 m)
Protrudes up to 4 ft (1.2 m) from support depending on variety.

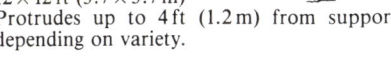

COTONEASTER
(Tall Deciduous Forms)

Origin Mostly from China and the Himalayas, or of garden origin.
Use As a fanned or spreading, fruiting shrub for large walls and fences.
Description *Flower* White, four-petalled flowers with prominent red calyxes, borne either singly or in clusters, 3 in (7.5 cm) across, in late spring to early summer. *Foliage* Basically elliptic, 1–1½ in (2.5–4 cm) long, normally green to grey/green with good autumn colours. *Stem* Strong, quick-growing, upright, becoming spreading with age. *Fruit* Red fruits, borne singly or in clusters, depending on variety.
Hardiness Tolerates a minimum winter temperature of 4°F (−15°C).
Soil requirements Any soil, except very dry.
Sun/Shade aspect Tolerates exposed aspects. Full sun to medium shade.
Pruning None required, but can be reduced in size.
Training Requires wires or individual anchor points to secure and encourage a fan-trained shape.
Propagation and nursery production From softwood cuttings, grafting or seed, depending on variety. Purchase container grown. Best planting height 2–4 ft (60 cm–1.2 m).
Problems The variety C. × 'Firebird' is normally grafted on to understock of C. bullatus which can sucker and, given time, kill the variety. Keep a careful watch for suckering growth and remove from the plant immediately when seen.
Forms of interest C. bullatus Foliage large, round to oblong, dark green with corrugated surfaces, giving good autumn colour. 2 in (5 cm) clusters of white flowers give way to dark red fruit, enhancing autumn colour of foliage. Often used as grafting understock for

Cotoneaster simonsii in fruit

many of the large-leaved, evergreen forms. From western China. C. distichus Foliage very distinct, round, ¾–1 in (2–2.5 cm) long and wide, glossy green with red autumn colour. Bright red elliptic-shaped fruits borne singly in each leaf axil, being held well into early spring. Two thirds average height and spread. Often found in commercial production under the incorrect name of C. rotundifolius. From the Himalayas and south-western China. C. divaricatus Very good orange-red autumn foliage, contrasting with some later held green leaves. Many fruits, dark red and glossy. From western China. C. × 'Firebird' Foliage dark grey/green, 3–4 in (7.5–10 cm) long, oval to oblong, giving good autumn colour. Large, round, orange/red fruits borne in thick clusters in autumn. Possibly more spreading than average and very hardy. Not easy to find but obtainable. A hybrid of C. bullatus, probably being a cross between this and C. franchetii. C. simonsii Semi-evergreen, round to elliptic leaves, borne on upright stems, covered in late spring with single or small clusters of white flowers in each leaf axil, culminating in red fruits. Adapts well to training. Very good for propagation from seed. From Assam.

Average height and spread
Five years
8 × 8 ft (2.4 × 2.4 m)
Ten years
12 × 12 ft (3.7 × 3.7 m)
Twenty years
15 × 15 ft (4.6 × 4.6 m)
Protrudes up to 4 ft (1.2 m) from support.

COTONEASTER
(Tall Evergreen Forms)

Origin Mostly native to western China or of garden origin.
Use As wide-spreading, evergreen shrubs for flower, foliage and fruit effect on large walls and fences.
Description *Flower* White flowers with red calyxes borne either in clusters 3 in (7.5 cm) across, or singly, depending on variety, in early summer. *Foliage* Ovate, lanceolate to round, 2–4 in (5–10 cm) long, dark to mid or grey/green, depending on variety. *Stem* Strong, upright, green to green/red or grey/green, becoming branching with age. Fast rate of growth. *Fruit* Yellow, red or

orange fruits, produced in clusters in early autumn. Not normally attacked by birds until late in the winter.
Hardiness Tolerates a minimum winter temperature of 4°F (−15°C)
Soil requirements Most soil conditions, but may show distress on very thin alkaline types.
Sun/Shade aspect Tolerates exposed aspects. Full sun to medium shade.
Training Tie to wires or to individual anchor points in a spreading fan shape.
Pruning None required but may be reduced in size in early spring, if necessary.
Propagation and nursery production From semi-ripe cuttings taken in early summer or by grafting in winter. Purchase container grown. Most varieties fairly easy to find in nurseries and in some garden centres. Best planting height 2–4 ft (60 cm–1.2 m).
Problems C. 'Cornubia', C. 'Exburiensis', C. 'Inchmery' and C. 'Rothschildianus' may be grafted on to understocks of C. bullatus which can sucker and, given time, kill the variety. If suckers are seen they should be removed as soon as possible. Certain varieties are susceptible to the fungus disease fire blight, while others are liable to stem canker. In areas where the fire blight problem cannot be controlled, some named varieties are likely to be withdrawn from commercial production. Stem canker can be dealt with by cutting back affected shoots and treating with pruning compound.

Cotoneaster lacteus in fruit

Forms of interest *C.* **'Cornubia'** Broad, lanceolate leaves. Large clusters of white flowers followed by bright red fruits. Of garden origin. *C.* **'Exburiensis'** Light to mid green, broad, lanceolate leaves. White flowers followed by clusters of bright yellow fruits. Two thirds average height and spread. Raised in Exbury Gardens, Hampshire, England. *C. franchetii sternianus* (syn. *C. wardii*) Round, silver/grey to grey/green foliage. White flowers followed by good orange fruits. Two thirds average height and spread. From southern Tibet and northern Burma. *C. henryanus* Leaves long, lanceolate, dark green, slightly corrugated, with duller grey undersides. Clusters of white flowers borne on brown/red, strong, upright shoots, followed by red fruits. *C.* **'Inchmery'** Foliage light green, slightly broader, lanceolate. Clusters of white flowers, followed by yellow fruit with coral shading in autumn. Two thirds average height and spread. Raised in the Inchmery Gardens, Hampshire, England. *C. lacteus* Elliptic, dark green foliage, clusters of white flowers followed by dark red rounded fruits borne profusely and held sometimes until they rot. Very good on large walls. From China. *C.* **'Rothschildianus'** Large clusters of yellow fruit set off against light green, broad, lanceolate leaves, 4–6 in (10–15 cm) long, spectacular in autumn. Raised at Exbury Gardens, Hampshire, England. *C. salicifolius floccosus* Long, graceful, pendulous branches with narrow, elliptic, shiny-surfaced leaves, green with white undersides, giving a drooping, fan-like appearance. Clusters of white flowers give way to small, dull, red fruits. Very good for using on an exposed, shady wall. Two thirds average height and spread. From China.

Average height and spread
Five years
8 × 8 ft (2.4 × 2.4 m)
Ten years
12 × 12 ft (3.7 × 3.7 m)
Twenty years
15 × 15 ft (4.6 × 4.6 m)
Protrudes up to 3 ft (91 cm) from support.

CRATAEGUS (Autumn Foliage Forms)

ORNAMENTAL THORN

Rosaceae   **Hardy Tree**
Deciduous

Not normally thought of for fan-training but their ability to make regrowth allows them to be used for this purpose with good results, particularly in exposed areas.

Origin Throughout the northern hemisphere.
Use As a large fan-trained tree for walls, particularly in exposed locations.
Description *Flower* Small white florets, ½ in (1 cm) across, produced in clusters 2–3 in (5–7.5 cm) wide in early summer, often with a musty scent enjoyed by bees. *Foliage* Ovate, 2–3 in (5–7.5 cm) long, light or mid green, sometimes glossy, depending on variety. Good yellow or orange autumn colour. *Stem* Medium rate of growth, forming a large fan-shaped tree with training. Very closely branched. Stems grey/green and attractive in winter. Most varieties have large thorns, normally curved and mahogany brown. *Fruit* Clusters up to 6 in (15 cm) in diameter of round, orange or crimson fruits produced in autumn; in some varieties very late to ripen.
Hardiness Tolerates a minimum winter temperature of 0°F (−18°C).
Soil requirements Most soil conditions, except very dry.
Sun/Shade aspect Tolerates all aspects. Full sun to medium shade, preferring light shade.
Pruning Prune young trees hard in the spring following planting. Select and train resulting five to seven shoots and tie into a fan-trained shape. In subsequent years, remove all side growths back to two points from their origin

Crataegus crus-galli in fruit

and maintain original main branches in fan shape.
Training Will require tying to wires or individual anchor points.
Propagation and nursery production From budding or grafting on to understock of *C. monogyna*. Purchase container grown or bare-rooted; most varieties relatively easy to find from general nurseries. Select trees of 6 ft (1.8 m), ensuring that they have been either bottom budded or grafted.
Problems Slow to establish and may need two full springs to recover from transplanting. Sharp thorns can be a hazard in close garden planting.
Forms of interest *C.* **'Autumn Glory'** Excellent yellow and orange autumn colour; white flowers and red berries. Of garden origin. *C. crus-galli* (Cockspur thorn) A flat-topped tree of more spreading habit and slightly less height. Foliage ovate to narrowly ovate, up to 3 in (7.5 cm) long, with toothed edges. Good orange/yellow autumn colours. White flowers in May followed by large clusters of long-lasting red fruits. Large thorns up to 3 in (7.5 cm) long. From North America. *C. durobrivensis* Leaves ovate with good autumn colours. White flowers followed by shining crimson berries maintained well into winter. Two thirds average height and spread and may also be grown as a large shrub. Limited in commercial production outside its native environment of North America. *C. × grignonensis* Leaves ovate, up to 2½ in (6 cm) long, very glossy upper surface, downy grey underside. Large clusters of white flowers followed by oval to globe-shaped red fruits in autumn. Two thirds average height and spread. Originating in France. Not readily available. *C. × lavallei* (syn. *C. carrierei*) Ovate, dark glossy green foliage with downy, paler undersides. Good autumn colour. White flowers in clusters with dominant anthers of red and yellow in June. Fruits orange/red, globe-shaped, ripening in September/October and maintained well into winter. Two thirds average height and spread. *C. mollis* Ovate leaves up to 4 in (10 cm) long with double-toothed edges. Light downy grey/green at first, ageing to light green. Good autumn colours. Flowers white with yellow anthers followed by large, globe-shaped red fruits, with downy texture. From North America. *C. pedicellata* (syn. *C. coccinea*) (Scarlet haw). Foliage light to mid green, slightly glossy, with toothed edges. Good yellow/orange autumn colour, ageing to red. Numerous short thorns. White flowers in early to mid spring, followed in autumn by bunches of scarlet fruits. Two thirds average height and spread. From north-eastern USA. *C. pinnatifida* Interesting crimson fruits with small, dark, red/brown to black dots over their surface. Leaves light to mid green, slightly glossy, with deeply cut lobes. Good orange/red autumn colour. Few thorns, in some cases none at all. Two thirds average

height and spread. From northern China. Not readily available and must be sought from specialist nurseries. *C. prunifolia* Dark green, glossy foliage, round to ovate, with slightly downy undersides. Round clusters of white flowers up to 2½ in (6 cm) across, on downy stalks, followed by rounded crimson fruits which are rarely maintained far into winter. Two thirds average height and spread. Thought to be a cross between *C. macrantha* and *C. crus-galli*. *C. submollis* Similar to *C. mollis*, but not reaching such a height. Plants may be intermixed in nursery production and difficult to differentiate; *C. submollis* has 10 stamens to the flower and *C. mollis* has 20.

Average height and spread
Five years
13 × 13 ft (4 × 4 m)
Ten years
20 × 20 ft (6 × 6 m)
Twenty years
26 × 26 ft (8 × 8 m)
Protrudes up to 5 ft (1.5 m) from support.

CRATAEGUS OXYACANTHA (C. laevigata)

THORN, MAY, HAWTHORN

Rosaceae   **Hardy Tree**
Deciduous

A tree responding well to hard pruning and forming a close fan-trained wall specimen.

Origin From Europe.
Use As a large, flowering, fan-trained specimen for large, exposed walls.
Description *Flower* Clusters up to 2 in (5 cm) across of white flowers produced in late spring. Musty scent attractive to bees. *Foliage* Basically ovate, 2 in (5 cm) long, very deeply lobed with three or five indentations. Grey/green with some yellow autumn tints. *Stem* Light grey/green, becoming grey/brown. Strong, upright when young, quickly branching. Armed with small, extremely sharp spines up to ½ in (1 cm) long. Medium rate of growth. *Fruit* Clusters 2–4 in (5–10 cm) wide of small, dull red, round to oval fruits containing two stone seeds, produced in autumn.
Hardiness Tolerates a minimum winter temperature of 0°F (−18°C).
Soil requirements Any soil conditions, but shows signs of distress on extremely dry areas, where growth may be stunted.
Sun/Shade aspect Tolerates all aspects. Full sun to medium shade, preferring light shade.
Pruning Prune young trees hard in the spring following planting. Select and train resulting five to seven shoots and tie into a fan-trained

shape. In subsequent years remove all side growths back to two points from their origin and maintain original main branches in fan shape.
Training Will require tying to wires or individual anchor points.
Propagation and nursery production From seed from the parent; all varieties budded or grafted. Purchase bare-rooted or container grown as young trees 3–6 ft (91 cm–1.8 m) tall, ensuring that they are grafted or budded at the base.
Problems The sharp spines can make cultivation difficult. Suckers of understock may appear and must be removed.

Crataegus oxyacantha **'Coccinea Plena' in flower**

Similar forms of interest *C. o.* 'Alba Plena' Double white flowers in mid spring, followed by limited numbers of red berries in autumn. *C. o.* **'Crimson Cloud'** Single, dark pink to red flowers with yellow eyes profusely borne in late spring. Good yellow/bronze/orange autumn colour. *C. o.* **'Gireoudii'** New foliage on new growth mottled pink and white, ageing to green. New foliage on old wood green. Two thirds average height and spread. Shy to flower, but can produce white, musty-scented blooms. Limited fruit production. Difficult to find and must be sought from specialist nurseries. *C. o.* **'Paul's Double Scarlet Thorn'** (syn. *C. o.* 'Coccinea Plena') Dark pink to red double flowers produced in late spring, early summer. Limited red berries in autumn. *C. o.* **'Rosa Plena'** Double pink flowers, produced late spring, early summer. Some limited berrying. *C. laciniata* (syn. *C. orientalis*) Attractive, dark grey/green, cut-leaved foliage, light grey undersides. White

flowers and large dull orange/red fruits. Two thirds average height and spread. Somewhat scarce in production. From the Orient. *C. monogyna* (Hedgerow thorn, singleseed hawthorn) Single white flowers. Red, round fruits containing one stone. Foliage dark green. Rarely offered as a tree, normally used as a hedgerow plant. From Europe. *C.* × *mordenensis* **'Toba'** Double, creamy white flowers in the spring, followed by red berries in autumn.

Average height and spread
Five years
13 × 4 ft (4 × 1.2 m)
Ten years
20 × 10 ft (6 × 3 m)
Twenty years
20 × 20 ft (6 × 6 m)
Protrudes up to 3 ft 91 cm from support.

CRINODENDRON HOOKERIANUM
(*Tricuspidaria lanceolata*)

LANTERN TREE
Elaeocarpaceae   *Wall Shrub*
Evergreen
A beautiful shrub responding well when grown against walls and fences where it will derive some additional protection.

Origin From Chile.
Use As a freestanding or fan-trained shrub for walls and fences on acid soils.
Description *Flower* Crimson, lantern-shaped flowers, 1–1¼ in (2.5–3.5 cm) long, hanging from stalks along underside of branches, on mature wood only, late spring to early summer. *Foliage* Leaves dark green, lanceolate, 1½–2 in (4–5 cm) long, with silver undersides. *Stem* Dark green, upright. Slow to medium growth rate. *Fruit* Rarely fruits, but occasionally produces brown, leathery capsules.
Hardiness Tolerates a minimum winter temperature of 14°F (−10°C), but some leaf scorch may be suffered in severe wind chill conditions.
Soil requirements Acid soil; dislikes any alkalinity.
Sun/Shade aspect Sheltered to moderately sheltered aspect. Full sun to light shade.
Pruning Not normally required.
Training Grow freestanding or tie to wires or individual anchor points for fan-training.
Propagation and nursery production From softwood cuttings taken in early summer. Purchase container grown; may be rather hard to find. Best planting height 15 in–2½ ft (38 – 76 cm).

Crinodendron hookerianum **in flower**

Problems Slow to establish and achieve full beauty and may take some years to become fan-trained.
Similar forms of interest *C. patagua* (syn. *Tricuspidaria dependens*) White bell-shaped flowers, late summer. Attractive round purple/green foliage. A stronger growing shrub, reaching greater height and spread. More tender. Very hard to find.
Average height and spread
Five years
4 × 3 ft (1.2 × 91 cm)
Ten years
7 × 4 ft (2.1 × 1.2 m)
Twenty years
12 × 7 ft (3.7 × 2.1 m)
Protrudes up to 2 ft (60 cm) from support if fan-trained, 5 ft (1.5 m) untrained.

CYDONIA OBLONGA

QUINCE
Rosaceae   *Hardy Fruit Tree*
Deciduous
Beautiful spring-flowering trees, with autumn fruits both edible and attractive, adapting well to fan-training.

Origin Parent from northern Iran and Turkestan; many varieties of garden origin, particularly from France.
Use As fan-trained edible fruit-producing trees for large walls and fences.
Description *Flower* Saucer-shaped flowers coloured delicate mother-of-pearl to light rose-pink, up to 2 in (5 cm) wide, produced in good numbers in mid to late spring. Slightly scented. *Foliage* Leaves ovate, mid to dark green with silver undersides, 3–6 in (7.5–15 cm) long and 1½–2½ in (4–6 cm) wide. Good yellow autumn colour. *Stem* Attractive

Cydonia oblonga **'Vranja' in flower**

Crataegus oxyacantha **'Rosa Plena' in flower**

Cydonia oblonga 'Vranja' in fruit

growth formation even when fan-trained. Branches dark brown to purple/ brown. Medium to fast rate of growth. *Fruit* Medium-sized, round or pear-shaped yellow fruits, 2–5 in (5–12 cm) long and 2–3 in (5–7.5 cm) wide, depending on variety, abundantly produced in late summer, early autumn. Used to make quince jelly.
Hardiness Tolerates a minimum winter temperature of 4°F (−15°C).
Soil requirements Most soil conditions, only dislikes extremely dry or very waterlogged areas.
Sun/Shade aspect Tolerates all but the most exposed of aspects. Best in full sun to allow the fruit to ripen.
Pruning Prune young trees hard in spring following planting. Select and train resulting five to seven shoots and tie into a fan-trained shape. In subsequent years remove all side growths back to two points from their origin and maintain original main branches in fan shape.
Training Will require tying to wires or individual anchor points.
Propagation and nursery production *C. oblonga* from seed; all named varieties by budding or grafting. Purchase bare-rooted or container grown from specialist nurseries. Best planting height 3–5 ft (91 cm–1.5 m). Always select young trees that have been grafted or budded at the base.
Problems Fruiting can be a little erratic, especially after hard, late spring frosts at flowering time. Trees tend to look misshapen and irregular when young.
Similar forms of interest *C. o.* **'Meech's Prolific'** Round, squat, pear-shaped fruits produced in good quantities. Flowers not the best feature. *C. o.* **'Portugal'** Small, pear-shaped fruits in profusion; a good culinary variety. Generous flower production. *C. o.*

'Vranja' Large, pear-shaped, yellow fruits. Large flowers, somewhat sparsely produced, but the tree is a fine ornamental form.
Average height and spread
Five years
10 × 10 ft (3 × 3 m)
Ten years
16 × 16 ft (4.9 × 4.9 m)
Twenty years
23 × 23 ft (7 × 7 m)
Protrudes up to 4 ft (1.2 m) from support.

CYTISUS BATTANDIERI

MOROCCAN BROOM, PINEAPPLE BROOM
Leguminosae   *Wall Shrub*
Deciduous
An elegant, summer-flowering shrub, which in mild areas may be semi-evergreen. In exposed gardens prefers a wall position.

Origin From Morocco.
Use As a fan or spreading large shrub for high walls and fences.
Description *Flower* Upright racemes 4 in (10 cm) long, 2 in (5 cm) wide, of bright yellow, pineapple-scented flowers, early to mid summer. *Foliage* Trifoliate silvery green leaves, up to 4 in (10 cm) long and wide, comprising three leaflets 1½ in (4 cm) long. *Stem* Silvery grey when young, ageing to grey/ green to grey/brown. Strong, upright, becoming branching and spreading with age. Fast-growing when young, slowing with age. *Fruit* Insignificant.
Hardiness Tolerates a minimum winter temperature of 14°F (−10°C).
Soil requirements Does well on moist soils but unhappy on very thin chalk types.
Sun/Shade aspect Sheltered, warm aspect in full sun.
Pruning None required, best planted where it can attain full maturity without pruning. If size must be restricted, cut young wood after flowering. Cutting into old wood may lead to die-back.
Training Requires wires or individual anchor points to secure and encourage the fan-trained shape.
Propagation and nursery production From seed, grafting or semi-ripe cuttings taken in mid summer. Available from most good garden centres. Best planting height 2–4 ft (60 cm–1.2 m).
Problems None, if given space to attain its full height and spread.
Similar forms of interest None.
Average height and spread
Five years
12 × 12 ft (3.7 × 3.7 m)
Ten years
16 × 16 ft (5 × 5 m)
Twenty years
25 × 25 ft (7.6 × 7.6 m)
Protrudes up to 4 ft (1.2 m) from support.

DECAISNEA FARGESII

KNOWN BY BOTANICAL NAME
Lardizabalaceae   *Wall Shrub*
Deciduous
Interesting green flowers which produce turquoise blue seed pods after a good summer.

Origin From China.
Use A tall, unusual, upright shrub for growing as a specimen against a wall, producing better flowers and fruit in this situation than when grown in the open.

Decaisnea fargesii in fruit

Description *Flower* Open panicles, up to 12 in (30 cm) long, of lime-green to yellow-green tubular flowers, produced at ends of upright shoots, late spring to early summer. *Foliage* Large leaves, 2–3 ft (60–91 cm) long, with six to 12 pairs of opposite leaflets each 3–6 in (7.5–15 cm) long. Light grey/green with yellow autumn colour. *Stem* Upright, becoming leaning with age, forming an upright central clump with some spreading side growths. Medium to fast growth rate. *Fruit* Metallic blue bean pods up to 15 in (38 cm) long in autumn following dry, hot summers.
Hardiness Tolerates a minimum winter temperature of 14°F (−10°C).
Soil requirements Does well on most soils but very moist, well-drained soil produces better growth and therefore more flowers and, in favourable years, fruit.
Sun/Shade aspect Requires some protection. Prefers light shade, but tolerates full sun provided adequate moisture is available.
Pruning None required. Weak, unattractive stems may be removed.
Training Normally grown in front of the wall and not necessarily against it, therefore needs no training or fixing.
Propagation and nursery production From seed. Purchase container grown plants or root-balled (balled-and-burlapped). Available from specialist nurseries. Best planting height 15 in–3 ft (38–91 cm).
Problems Can take several years to produce fruit.
Similar forms of interest None.
Average height and spread
Five years
6 × 6 ft (1.8 × 1.8 m)
Ten years
10 × 13 ft (3 × 4 m)
Twenty years
13 × 13 ft (4 × 4 m).
Protrudes up to 6 ft (1.8 m) from wall or fence.

Cytisus battandieri in flower

DECUMARIA BARBARA

KNOWN BY BOTANICAL NAME

Saxifragaceae   ***Woody Climber***
Deciduous to semi-evergreen

An interesting climbing plant related both to the climbing hydrangea and schizophragma but possibly a little less hardy.

Origin From USA.
Use As a self-clinging climber for lightly shady walls and fences.
Description *Flower* 2–3 in (5–7.5 cm) wide clusters of white, sterile flowers up to $\frac{1}{4}$ in (5 mm) across, carried at the ends of each branch in mid to late summer. *Foliage* Oval, pointed, up to 5 in (12 cm) long. Initially grey in colour, quickly becoming bright green ageing to dark green. Yellow autumn colour. *Stem* Light grey/green, becoming mahogany/brown, finally brown/grey. Short-branching, clinging by aerial roots. Slow to medium growth rate. *Fruit* Fruiting not reliable but may produce white $\frac{1}{4}$ in (5 mm) long fruits in early autumn.
Hardiness Tolerates a minimum winter temperature of 14°F (−10°C) although will require additional protection from cold winds.
Soil requirements Tolerates all types of soil including dry, once established.
Sun/Shade aspect Light to medium shade, dislikes full sun.
Pruning Requires no pruning but can be reduced in size if required and will slowly rejuvenate.
Training Normally self-clinging but may need additional tying to individual anchor points or wires, particularly when young.
Propagation and nursery production From layers or cuttings, rooted into sand under the protection of glass. Always purchase container grown; will have to be sought from a specialist nursery. Best planting height 6 in–2 ft (15–60 cm).
Problems Slow to establish and availability may be scarce.
Similar forms of interest *D. sinensis* A very similar variety except that it reaches 50 per cent less height and spread. Not readily available.
Average height and spread
Five years
5 × 5 ft (1.5 × 1.5 m)
Ten years
10 × 10 ft (3 × 3 m)
Twenty years
20 × 20 ft (6 × 6 m)
Protrudes up to 12 in (30 cm) from support.

Decumaria barbara in flower

Desfontainea spinosa in flower

DESFONTAINEA SPINOSA

KNOWN BY BOTANICAL NAME

Potaliaceae   ***Wall Shrub***
Evergreen

A moderately hardy shrub which will nevertheless benefit from the protection of a wall or fence.

Origin From Chile and Peru.
Use As a freestanding or fan-trained shrub for walls and fences.
Description *Flower* Scarlet, $1\frac{1}{2}$ in (4 cm) long, tubular flowers, yellow shaded in mouth, mid summer. *Foliage* Leaves oval, 1–2 in (2.5–5 cm) long, dark green, glossy, holly-like with silver undersides. *Stem* Moderately stout, dark green, forming a clump of upright flowering shoots. Slow growth rate. *Fruit* Insignificant.
Hardiness Tolerates a minimum winter temperature of 14°F (−10°C) but may suffer foliage damage in severe wind chill conditions.
Soil requirements Acid soil, dislikes any alkalinity or waterlogging. Thrives on soil high in organic content such as in woodland.
Sun/Shade aspect Requires a sheltered aspect. Prefers light to medium shade, dislikes full sun.
Pruning None required.
Training Allow to grow freestanding or fan-train by tying to individual anchor points.
Propagation and nursery production From semi-ripe cuttings taken mid summer or from rooted suckers from outer edges of mature clumps. Purchase container-grown. Hard to find; may be available from specialist nurseries. Best planting height 8–15 in (20–38 cm).
Problems Often takes a number of years to settle down. Slow to form a fan-trained habit but worthy of the effort once established.
Similar forms of interest *D. s.* 'Harold Comber' A variety collected from the wild, with varying flower colour, from vermillion to orient red. Very scarce.
Average height and spread
Five years
2 × 2$\frac{1}{2}$ ft (60–80 cm)
Ten years
3 × 4 ft (1 × 1.2 m)
Twenty years
7 × 5 ft (2.2 × 1.5 m)
Protrudes up to 2 ft (60 cm) from support if fan-trained, 4 ft (1.2 m) untrained.

DRIMYS WINTERI

WINTER'S BARK

Winteraceae   ***Wall Shrub***
Evergreen

An evergreen which often prefers the protection of a wall, making a tall, interesting shrub.

Origin From South America.
Use As a freestanding, large shrub for walls and fences.
Description *Flower* White to ivory-white loose clusters of jasmine-scented flowers, mid spring. *Foliage* Leaves large, 5–10 in (12–25 cm) long, ovate, grey/green with glaucous undersides and leathery texture. *Stem* Upright, light to bright green to grey/green, becoming branching and spreading with age. Medium growth rate. *Fruit* Insignificant.

Drimys winteri in flower

Hardiness Tolerates a minimum winter temperature of 23°F (−5°C). Foliage resents severe wind chill.
Soil requirements Acid, preferably open, woodland, leafy soil, dislikes any alkalinity.
Sun/Shade aspect Requires a sheltered aspect. Best in light or dappled shade.
Pruning None required, but any arching, spreading branches may be removed to confine the shrub to the planting area.
Training Main stems will require fixing to individual anchor points for support.

Propagation and nursery production From layers or semi-ripe cuttings taken late summer. Purchase container grown or root-balled (balled-and-burlapped). Rather hard to find and must be sought from specialist nurseries. Best planting height 2–2½ ft (60–76 cm).

Problems Wind chill may cause extensive defoliation and some die-back.

Similar forms of interest None.

Average height and spread
Five years
6 × 6 ft (1.8 × 1.8 m)
Ten years
10 × 8 ft (3 × 2.4 m)
Twenty years
15 × 10 ft (4.6 × 3 m)
Protrudes up to 6 ft (1.8 m) from support.

ECCREMOCARPUS SCABER

CHILEAN GLORY FLOWER

Bignoniaceae Annual/Perennial Climber
Evergreen

A somewhat tender climber worthy of effort but may need to be propagated annually from seed.

Origin From Chile.

Use As an annual climber for sunny walls and fences or for rambling over an uninteresting shrub. Good for training up pillars.

Eccremocarpus scaber **'Rubra' in flower**

Description *Flower* Racemes of narrow tubular flowers, deep orange/red in colour, up to 1 in (2.5 cm) long. Yellow and deep red flowering varieties also available. *Foliage* 1½ in (4 cm) long light green pinnate leaves; may be evergreen in very mild areas, although in less favourable conditions may be damaged and replaced in following spring by foliage on new basal growth. *Stem* Angular with ribbed edges, light green becoming mid green, not readily forming a woody structure except in the mildest areas. Rambling, not self-clinging. Medium to fast growing. *Fruit* None of interest, although seeds are required for annual propagation.

Hardiness Tolerates a minimum winter temperature of 23°F (−5°C). Not winter hardy except in the most favourable areas and best repropagated on an annual basis.

Soil requirements Alkaline to acid well drained soil with good plant food and adequate moisture retention to achieve good rate of annual growth.

Sun/Shade aspect Very sheltered aspect if overwintering is required. Full sun to light shade.

Pruning Normally killed over winter, therefore no pruning required other than cleaning away damaged material. If not killed by winter weather, a light trim in spring will secure new growth and good flowering in the following summer.

Eccremocarpus scaber **in flower**

Training Allow to ramble through a framework of wires, trellis or netting stretched over areas to be covered. Can be allowed to ramble over an uninteresting shrub to good effect.

Propagation and nursery production From seed collected after flowering in late summer/early autumn, sown in trays with a good quality potting compost under cover in late winter/early spring and then planted out on an annual basis once danger of frost has passed. Always purchase container grown, best planting height 3 in–2 ft (7.5–60 cm). Available from specialist nurseries and good garden centres.

Problems The possible need to produce this plant annually can be a disadvantage but its flowering display is sufficiently interesting to make the effort well worthwhile.

Similar forms of interest *E. s.* 'Aureus' Golden yellow flowers. Not readily available. *E. s.* 'Rubra' Orange/red flowers. Not readily available.

Average height and spread
One year
6 × 6 ft (1.8 × 1.8 m)
Protrudes up to 2 ft (60 cm) from support.

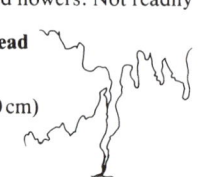

ELAEAGNUS COMMUTATA
(E. argentea)

SILVER BERRY

Elaeagnaceae   Wall Shrub
Deciduous

Interesting, useful, silver-foliage shrub attractive when fan-trained on a large wall.

Origin From North America.

Use As an attractive foliage shrub for training on large walls and fences.

Elaeagnus commutata **in leaf**

Description *Flower* Small, inconspicuous, scented, pale yellow flowers in mid spring. *Foliage* Ovate, deep silver green leaves. *Stem* Grey/green, upright, becoming arching. Medium to fast growing. *Fruit* Small, oval, ½ in (1 cm) long, silver fruits in early to mid summer, often retained into autumn.

Hardiness Tolerates a minimum winter temperature of 4°F (−15°C).

Soil requirements Does well on all soil types, tolerating quite dry areas.

Sun/Shade aspect Tolerates all but the most exposed aspects. Prefers full sun.

Pruning None required, but can be cut back and will rejuvenate.

Training Requires wires or individual anchor points to secure and encourage a fan-trained shape.

Propagation and nursery production From softwood cuttings taken in early summer. Purchase container grown. Easy to find. Best planting height 1½–4 ft (45 cm–1.2 m).

Problems Root disturbance may lead to die-back with foliage dropping and stems dying back to ground level, but often rejuvenates again.

Forms of interest See further entries.

Average height and spread
Five years
5 × 5 ft (1.5 × 1.5 m)
Ten years
10 × 10 ft (3 × 3 m)
Twenty years
15 × 15 ft (4.6 × 4.6 m)
Protrudes up to 5 ft (1.5 m) from support.

ELAEAGNUS × EBBINGEI

KNOWN BY BOTANICAL NAME

Elaeagnaceae   Wall Shrub
Evergreen

Very useful, quick-growing evergreen with hidden scented flowers, adapting well to use on walls.

Origin Of garden origin; a hybrid of *E. macrophylla* and *E. pungens*, both from Japan.

Use As a fan-trained evergreen to give winter foliage colour on exposed walls and fences.

Elaeagnus × ebbingei **in flower**

Description *Flower* Sulphur yellow petals with silver calyx in late summer to late autumn. *Foliage* Leaves ovate, 3–6 in (7.5–15 cm) long, grey/green with some gold-variegated varieties. *Stem* Upright when young, becoming twiggy and dense with age. Can be trained into a fan or spreading shape. Normally grey/green when young, ageing to grey/brown. Medium to fast rate of growth. *Fruit* Small, egg-shaped red/orange fruits on mature shrubs in hot summers.

Hardiness Tolerates a minimum winter temperature of 4°F (−15°C).

Soil requirements Tolerates most soils, although unhappy on extremely alkaline or dry soils.

Sun/Shade aspect Tolerates all aspects. Full sun to deep shade, although shrub may be more lax and loose in shape in deep shade.
Pruning Trim hard back by two thirds in spring to encourage bushy habit once main structure has been trained.
Training Requires wires or individual anchor points to secure and encourage a fan-trained shape.
Propagation and nursery production From semi-ripe cuttings taken in early summer or by grafting. Purchase container grown; most forms fairly easy to find. Best planting height 1½–4 ft (45–1.2 m).
Problems Roots very susceptible to any dramatic change in soil conditions such as drought or waterlogging; whole areas of root may die, resulting in death of limb or even whole side of shrub directly related to damaged root area. Cultivation damage can also lead to this demise. If gold-variegated foliage varieties start to revert, producing all green foliage, remove this once seen.
Similar forms of interest *E.* × *e.* 'Limelight' Centres of leaves irregularly splashed gold to pale gold, variegation best in spring and summer, declining slightly in autumn. Upright habit. *E.* × *e.* 'Salcombe Seedling' Fragrant white flowers produced in good numbers in autumn. Foliage dark green with silver reverse. *E. macrophylla* Slightly broader, larger, grey/green leaves with silver undersides. Good flower production. Less hardy than *E.* × *ebbingei* and rather hard to find in colder areas.
Average height and spread
Five years
6 × 6 ft (1.8 × 1.8 m)
Ten years
8 × 8 ft (2.4 × 2.4 m)
Twenty years
12 × 12 ft (3.7 × 3.7 m)
Protrudes up to 4 ft (1.2 m) from support.

Origin From Japan. Most variegated forms of garden origin.
Use As a green or golden variegated fan-trained shrub for walls and fences.
Description *Flower* Small, inconspicuous, fragrant sulphur yellow or silvery-white flowers, mid to late autumn. *Foliage* Leaves 2–3 in (5–7.5 cm) long, ovate with pointed ends, dark olive-green with glossy upper surfaces and duller undersides. The young leaves of golden variegated forms, grey/brown in spring, do not become variegated until early summer. *Stem* Grey/green, ageing to brown/green. Upright when young, quickly becoming spreading, branching and twiggy, forming a wide, spreading shrub. Medium rate of growth. *Fruit* Insignificant.
Hardiness Tolerates a minimum winter temperature of 0°F (−18°C). Foliage may be damaged in severe wind chill conditions.
Soil requirements Tolerates most soils but unhappy on extremely alkaline or dry conditions.
Sun/Shade aspect Tolerates all but the most exposed aspects in full sun to deep shade, but may be lax and poorly shaped in deep shade.
Pruning Prune and train new shoots into fan or spreading shape on wall or fence. Cut back all forward-growing growths in spring to encourage a thick covering of branches and foliage.
Training Will require the support of wires, trellis or individual anchor points to form a fan-trained shape.
Propagation and nursery production By grafting in winter or semi-ripe cuttings taken late spring/early summer. Purchase container grown or root-balled (balled-and-burlapped). Best planting height 1½–4 ft (45 cm –1.2 m).

Elaeagnus pungens 'Maculata'

Problems May suffer from aphid attack. Shoots of variegated forms showing reversion to green should be cut out once seen. On grafted shrubs, rip out any suckers of understock.
Similar forms of interest *E. p.* 'Dicksonii' Leaves narrow, almost holly-like, with bright gold margins. Two thirds average height and spread. Slightly less hardy than *E. pungens*. Of garden origin. *E. p.* 'Fredericia' Light green, cream-splashed, narrow, very pointed leaves. Very branching habit. Slightly less than average height and spread. *E. p.* 'Gold Rim' A European variety, having round to ovate leaves with extensive gold variegation.

Elaeagnus pungens 'Gold Rim'

Two thirds average height and spread. *E. p.* 'Maculata' Dark leaves with central gold patches. Reaching just less than average height and spread. Very susceptible to green reversion, remove any reverted shoots once seen or golden colouring will be destroyed. Of garden origin and very widely grown. *E. p.* 'Variegata' Foliage margined creamy white. Some sulphur yellow fragrant flowers on very old plants. Prune back old limbs occasionally to preserve foliage size and quality. Two thirds average height and spread. Of garden origin. *E.* × *ebbingei* 'Gilt Edge' Attractive gold-margined leaves. Severe wind chill and extreme waterlogging may damage foliage. Slightly tender – minimum winter temperature 23°F (−5°C). Two thirds average height and spread. Of garden origin.
Average height and spread
Five years
5 × 5 ft (1.5 × 1.5 m)
Ten years
8 × 8 ft (2.4 × 2.4 m)
Twenty years
14 × 14 ft (4.3 × 4.3 m)
Protrudes up to 4 ft (1.2 m) from support.

Origin From China.
Use As a freestanding, large, multi-stemmed shrub or tree for sheltered large walls, or in very large conservatories and greenhouses.
Description *Flower* Terminal panicles up to 6 in (15 cm) long consisting of yellow/white, ¾ in (2 cm) wide, fragrant, closely packed flowers in late summer, early autumn, rarely flowers outside its native environment. *Foliage* Leathery, lanceolate leaves, up to 12 in (30 cm) long and 6 in (15 cm) wide on young growth, smaller on older shoots. Dark glossy green above, with pronounced veins and silver/white, woolly undersides. Attractive. *Stem* Light green when young, becoming darker green with age. Stiff,

Eriobotrya japonica

upright, glossy surface, forming a round, upright pillar. Medium growth rate. *Fruit* Pear-shaped, yellow, edible fruits up to 1½ in (4 cm) long, but rarely fruits outside its native environment.
Hardiness Tolerates a minimum winter temperature of 14°F (−10°C) in sheltered wall locations.
Soil requirements Tolerates all types of soil as long as it is well drained.

Sun/Shade aspect Requires a very sheltered aspect. Best in a sunny, sheltered corner.
Pruning Normally requires no pruning for its own well-being but individual limbs can be removed if becoming an obstruction.
Training Allow to grow freestanding in front of walls. May need securing to individual anchor points.
Propagation and nursery production From seed sown in spring or autumn individually in pots, or by semi-ripe cuttings taken in late summer, in both cases under protection. Always purchase container grown; may be difficult to find, except from specialist nurseries. Best planting height 6–18 in (15–45 cm).
Problems Often outgrows the area allowed for it. Its availability may be scarce and its hardiness is suspect.
Average height and spread
Five years
5 × 3 ft (1.5 × 91 cm)
Ten years
15 × 6 ft (4.6 × 1.8 m)
Twenty years
20 × 12 ft (6 × 3.7 m)
Protrudes up to 12 ft (3.7 m) from wall or fence.

ERYTHRINA CRISTA-GALLI

CORAL TREE, COXCOMB
Leguminosae　　　　　*Wall Shrub*
Deciduous
An interesting shrub for a sun-facing wall.

Origin From Brazil.
Use As a medium-sized, summer-flowering sub-shrub for sunny walls and fences.
Description *Flower* Racemes of deep scarlet florets, waxy-textured, borne in large clusters at ends of shoots, mid to late summer. *Foliage* Leaves medium-sized, trifoliate, 4–5 in (10–12 cm) long, leaflets ovate. Grey/green, giving some yellow autumn colour. *Stem* Strong, upright, becoming arching with weight of flowers, grey to grey/green. Small sharp spines at each leaf axil. Dies to ground level each winter, regenerating following spring. Fast-growing, annually produced stems. *Fruit* Insignificant.
Hardiness Tolerates a minimum winter temperature of 23°F (−5°C).
Soil requirements Good, light, open soil.
Sun/Shade aspect Plant facing full sun, in a protected position against a wall or fence.
Pruning None required, but in early spring remove any old shoots retained through winter.
Training Requires no training, freestanding.
Propagation and nursery production Produced by root division and from seed. Purchase container-grown without stems while dormant, if available, or as plants up to

12–18 in (30–45 cm). Extremely scarce, should be sought from specialist nurseries.
Problems Difficult to establish, taking some years to flower.
Similar forms of interest None.
Average height and spread
Five years
5 × 5 ft (1.5 × 1.5 m)
Ten years
6 × 8 ft (1.8 × 2.4 m)
Twenty years
6 × 10 ft (1.8 × 3 m)
Protrudes up to 3 ft (91 cm) from wall or fence.

ESCALLONIA

KNOWN BY BOTANICAL NAME
Escalloniaceae　　　　*Wall Shrub*
Evergreen
Handsome early summer flowering shrubs with an attractive range of flower colours.

Origin From South America. Nearly all varieties offered are of garden or nursery origin.
Use As a loose, fan-trained or spreading shrub for walls or fences. Can be grown as a pillar subject.

Escallonia 'Crimson Spire' in flower

Description *Flower* Short racemes of bell-shaped flowers in various shades of pink to pink/red. Some white forms. Late spring through to early summer, with intermittent flowering through late summer and early autumn. *Foliage* Leaves ovate, 1–1½ in (2.5–4 cm) long, with indented edges. Light to dark green glossy upper surfaces, grey undersides. Size of foliage varies according to form. *Stem* Upright when young, becoming arching according to species and variety, and branching with age. Grey/green. Fast rate of growth when young, slowing with age. *Fruit* Insignificant.

Hardiness Tolerates a minimum winter temperature of 4°F (−15°C).
Soil requirements Tolerates most soils but liable to chlorosis on extremely alkaline types.
Sun/Shade aspect Tolerates all aspects, except extremely exposed. Full sun through to medium shade.
Pruning Remove one-third of old flowering

Escallonia 'Iveyi' in flower

wood after main flowering period. If cut to ground level will rejuvenate after second or third spring.
Training Wires, trellis or individual anchor points will be required for training.
Propagation and nursery production From softwood cuttings taken mid summer. Purchase container-grown; most varieties comparatively easy to find. Best planting height 2–2½ ft (60–76 cm).
Problems Susceptible to severe wind chill which may kill leaves and stems completely. Normally rejuvenates from ground level but may take two to three years to reach previous height.
Forms of interest E. 'Apple Blossom' Flowers apple blossom pink, large and normally borne singly. Slightly upright. E. 'C. F. Ball' Rich red flowers and medium-sized foliage. Slightly arching. E. 'Crimson Spire' Crimson flowers and medium-sized foliage. Upright. Slightly less than average spread. E. 'Donard Beauty' Large rose-carmine flowers and large green foliage. Slightly arching. E. 'Donard Brilliance' Large rose-red flowers and large foliage. Upright, becoming spreading with age. E. 'Donard Radiance' Medium-sized, rich pink flowers and large foliage. Slightly arching. E. 'Donard Seedling' Medium-sized flowers, pink in bud, opening to white tinted rose. Large leaves. Slightly arching. E. 'Edinensis' Medium-sized, carmine-pink flowers. Large foliage. Arching. E. 'Gwendolyn Anley' Small flowers, pink in bud, opening to paler pink. Small foliage. Arching, spreading and very twiggy. Two-thirds average height and spread. Best planting height 12–15 in (30–40 cm). E. 'Ingramii' Rose-pink flowers, medium size. Large foliage. Arching. E. 'Iveyi' Large pure white flowers. Large, dark green, shiny foliage. Upright, becoming slightly spreading with age. E. 'Langleyensis' Small, bright carmine-rose flowers in profusion along arching branches. Small-leaved. E. 'Peach Blossom' Good-sized clear pink flowers. Large foliage. Arching. E. 'Slieve Donard' Small pale pink flowers in profusion, borne on long, arching branches. Slightly less than average height and spread. E. macrantha Medium-sized, rose-carmine flowers, large, scented. Dark green foliage. Upright and branching. Good in coastal areas. Less than average hardiness. Height and spread slightly larger than average in favourable areas.
Average height and spread
Five years
8 × 8 ft (2.4 × 2.4 m)
Ten years
14 × 14 ft (4.3 × 4.3 m)
Twenty years
14 × 14 ft (4.3 × 4.3 m)
Protrudes up to 3 ft (91 cm) from support.

Erythrina crista-galli **in flower**

Eucalyptus gunnii in flower (juvenile foliage)

EUCALYPTUS

EUCALYPTUS, GUM TREE

Myrtaceae   *Semi-hardy Tree*
Evergreen

Attractive, blue-leaved and ornamental-stemmed trees good for growing in association with walls in mild locations.

Origin From Australia and Tasmania.
Use As a freestanding or fan-trained ornamental foliage tree for growing against large walls, where it can gain protection.
Description *Flower* White or cream tufted clusters 1–3 in (2.5–7.5 cm) long in autumn, early winter. *Foliage* The main attraction. Ovate to broadly linear, depending on variety. Stiff leathery texture, blue/green to glaucous blue. Adult foliage often different from juvenile in shape and form. *Stem* Grey/green to blue/green, often with peeling bark, revealing primrose-yellow underskin. Winter stems attractive. Often very fast-growing. *Fruit* Small, spinning-top-shaped fruit, glaucous blue ageing to brown, but rarely fruits outside its native environment.
Hardiness Tolerates a minimum winter temperature of 4°F (−15°C). Wind chill and very hard frosts can cause severe damage.
Soil requirements Light, well-drained soil for best results; tolerates a wide range of conditions.
Sun/Shade aspect Tolerates all but the most exposed of aspects. Full sun to light shade, preferring full sun.
Pruning Prune young trees hard in spring following planting. Select and train resulting five to seven shoots and tie into a fan-trained shape. In subsequent years, remove all side growths back to two points from their origin and maintain original main branches in fan shape. Alternatively, cut hard back in spring to induce a multi-stemmed large shrub effect.
Training Requires either tying to wires or individual anchor points or can be grown freestanding.
Propagation and nursery production From seed. Purchase container-grown, best planting height between 15 in (38 cm) and 4 ft (1.2 m). Transplanting larger trees is not recommended. Hardiest varieties generally available; outside favourable locations, more tender varieties must be sought from specialist nurseries.
Problems May look weak when young in nursery or garden centre, but develops rapidly once planted out. Reacts badly very quickly to poor planting; requires good preparation for speedy results.
Forms of interest *E. coccifera* Juvenile foliage glaucous blue, round to oval, up to 1½ in (4 cm) long. Adult foliage grey/green, narrow, oblong to lanceolate, up to 4 in (10 cm) long. Yellow flower umbels, produced in clusters late autumn, early winter. May produce conical fruits. One-third average height and spread. From Tasmania. *E. dalrympleana* Attractive foliage and patchwork bark, light brown with grey/white undercolour progressing in area as tree matures. Young foliage light green; mature grey/green, broadly lanceolate, up to 5 in (12 cm) long. *E. globulus* (Blue gum) Juvenile foliage glaucous white, up to 6 in (15 cm) long and 2⅓ in (6 cm) wide. Adult foliage is dark, shiny green, lanceolate, up to 12 in (30 cm) long and 2 in (5 cm) wide. Tufted clusters of three individual white florets produced late autumn, early winter. From Tasmania. In its native environment can exceed 100 ft (30 m) but usually achieves around 48 ft (15 m) in height. Tender; for use in non-mild areas only as annual foliage bedding plant. *E. gunnii* (Cider gum) Juvenile foliage orbicular, up to 2 cm (5 in) across, glaucous white to blue/green, produced on short stalks. Adult foliage green, lanceolate, up to 4 in (10 cm) long. White flowers in threes produced late autumn to early winter. May reach more than average height and spread in mild areas. From Tasmania and southern Australia. One of the hardiest of all eucalyptus. *E. niphophila* (Snow gum) Slow-growing with large, leathery, grey/green, ovate to lanceolate leaves 8 in (20 cm) long. Attractive trunk patched with grey/green, red/brown and cream. Relatively hardy. From Australia. *E. parvifolia* Narrow, lanceolate, blue/green leaves, up to 6 in (15 cm) long. Tolerates more alkalinity than most. Moderately hardy. *E. pauciflora* (syn. *E. coriacea*) (Cabbage gum) Juvenile foliage round to broadly lanceolate, glaucous green, up to 8 in (20 cm) long and 6 in (15 cm) wide. Adult foliage produced on long stalks up to 8 in (20 cm) long and 1 in (2.5 cm) wide. White flowers in tufted clusters of five to twelve. From Australia and Tasmania. Relatively hardy. *E. perriniana* Juvenile foliage small, silver/blue, ovate to round. Adult foliage lanceolate, glaucous blue. Stems white with dark purple/brown blotches. Tender. *E. pulverulenta* Attractive peeling bark. Juvenile foliage ovate to round, sometimes kidney-shaped, up to 2½ in (6 cm) long. Adult foliage similar, slightly larger. Tufted clusters of three white florets produced late autumn, early winter. Two thirds average height and spread. From New South Wales. Tender.

There are a large number of available eucalyptus forms. Those listed are among the most hardy, but even so, careful selection is required to suit the plant to any particular location.

Average height and spread
Five years
16 × 16 ft (4.9 × 4.9 m)
Ten years
25 × 25 ft (7.6 × 7.6 m)
Twenty years
35 × 35 ft (10.5 × 10.5 m)
Protrudes up to 5 ft (1.5 m) from support if fan-trained, 13 ft (4 m) untrained.

EUCRYPHIA

BRUSH BUSH

Eucryphiaceae   *Wall Shrub*
Evergreen or deciduous

A magnificent specimen shrub if a suitable site can be found with enough height and breadth.

Origin Native to South America, Australia and Tasmania but many forms of garden origin and garden crosses.
Use As a freestanding specimen shrub to be grown against a wall for protection, producing late summer flowers of interest, and ultimately reaching the proportions of a small tree.
Description *Flower* White, single, saucer-shaped flowers with pronounced stamens, freely borne. Size depends on variety, from numerous small flowers to flowers up to 2 in (5 cm) across, late summer to early autumn. *Foliage* Mostly evergreen. Ovate, 1½–3 in (4–7.5 cm) long, some with indented outer edges. Glossy upper surfaces, grey undersides. *Stem* Upright, becoming branching with age, but maintaining upright effect. Grey/green ageing to green/brown. Medium growth rate. *Fruit* Insignificant.
Hardiness Tolerates a minimum winter temperature of 4°F (−15°C).
Soil requirements Neutral to acid soil, tolerating low alkalinity provided soil is rich, deep and moist.
Sun/Shade aspect Prefers a moderately sheltered, shady location but will tolerate full sun.

Eucryphia × nymansensis 'Nymansay'

Pruning None required. Individual limbs can be cut back in late spring, encouraging new growth.
Training Normally freestanding but may be secured to the wall by individual anchor points.
Propagation and nursery production From semi-ripe cuttings taken in late summer. Purchase container grown; some varieties may have to be sought from specialist nurseries. Best planting height 15 in–3 ft (38–91 cm). May be available taller but normally very slow to establish.
Problems In severe cold, especially with high wind chill, foliage can be damaged and branches denuded. New growth normally appears in late spring.
Forms of interest *E. cordifolia* (syn. *E. ulmo*) White flowers and heart-shaped evergreen leaves. Possibly tolerates more alkalinity than most. Slightly less hardy than average and slightly less in height and spread. *E. glutinosa* White flowers 2½–2¾ in (6–7 cm) across, mid to late summer. Foliage deciduous, pinnate, grey/green, giving good autumn colour. A very upright form. From Chile. *E. × intermedia* 'Rostrevor' Produces a very good display of fragrant white flowers from 1–2 in

(2.5–5 cm) across, with yellow centres. Branches more pendulous and graceful than in most varieties. Fast growth rate. *E. lucida* Fragrant, white, hanging flowers 2 in (5 cm) across, mid to late summer. Very thick, dark, grey/green evergreen foliage, oblong with glaucous underside. Slightly less hardy. From Tasmania. *E. milliganii* Flowers 1–1½ in (2.5–4 cm) across, white and cup-shaped, often produced freely on young shrub. Small dark grey/green leaves open from sticky buds. Half average height and spread. From Tasmania. *E.* × *nymansensis* 'Nymansay' The most popular of all the varieties. Pure white flowers, 2½ in (6 cm) across, late summer to early autumn. Interesting dark green evergreen foliage. Raised in the Nymans Gardens, Sussex, England.

Average height and spread
Five years
8 × 4 ft (2.4 × 1.2 m)
Ten years
15 × 8 ft (4.6 × 2.4 m)
Twenty years
30 × 15 ft (9 × 4.6 m)
Protrudes 6 ft (1.8 m) or more from wall or fence.

EUONYMUS FORTUNEI
(E. radicans)

WINTERCREEPER EUONYMUS
Celastraceae   *Wall Shrub*
Evergreen
A very useful climbing foliage shrub for shady, exposed sites.

Origin From China. Most modern varieties of garden origin.
Use As an evergreen climbing shrub for exposed and shady positions. Tolerates dry conditions once established.
Description *Flower* Small, rather inconspicuous, green flowerheads. *Foliage* Leaves small, ovate, ¾–1 in (2–2.5 cm) long, ranging from green through to golden and silver variegated depending on variety, some changing to pink in winter. *Stem* Close-growing, spreading, grey/green when young ageing to grey/brown. Medium rate of growth. *Fruit* Sometimes produces small, round, red fruits, but rarely seen in variegated forms.
Hardiness Tolerates a minimum winter temperature of 0°F (–18°C). May suffer foliage loss in extremely cold conditions.
Soil requirements Accepts a wide range of soil conditions. Once established, will tolerate quite dry areas.
Sun/Shade aspect Tolerates all but the most exposed of aspects. Full sun to fairly deep shade, although golden varieties prefer full sun to medium shade.
Pruning None required, although old shoots may be removed occasionally to help rejuvenation.
Training Normally self-supporting by forming a bushy habit, but when reaching considerable height may need some individual anchor points.

Euonymus fortunei 'Emerald 'n' Gold'

Propagation and nursery production From softwood cuttings. Purchase container grown; most varieties easy to find. Best planting height 8–18 in (20–45 cm).
Problems Container grown shrubs sometimes appear weak when purchased, but once planted grow quickly.
Similar forms of interest *E. f.* 'Coloratus' Green leaves, turning purple in winter, becoming green again in spring. Best effect is achieved if grown on dry or starved soils. One of the quickest growing varieties. *E. f.* 'Emerald Gaiety' Round grey/green leaves with white margins, turning pink in very cold winter weather. *E. f.* 'Emerald 'n' Gold' Round to ovate, dark grey/green leaves with gold margins. Variegation turns pink/red in very cold conditions. One of the most popular varieties. *E. f.* 'Gold Spot' Dark green foliage with bright gold splashes. Branches more upright and stronger than most. Very attractive in winter and early spring. *E. f.* 'Silver Queen' Elliptic leaves, creamy yellow in spring, becoming green with creamy white margin through summer. Two thirds average height and spread. *E. f.* 'Sunshine' Bold gold edges, grey/green leaves. Slightly quicker growth rate than most variegated forms. *E. f.* 'Sunspot' Foliage splashed gold on dark green. Attractive. *E. f.* 'Variegatus' A small-leaved variety with grey/green ovate foliage, margined white, tinged pink in cold winter weather. *E. f.* 'Vegetus' Leaves ovate to round, very thick, dull green; bears profuse round, red fruits in autumn.

Average height and spread
Five years
3 × 4 ft (91 cm × 1.2 m)
Ten years
4 × 6 ft (1.2 × 1.8 m)
Twenty years
6 × 10 ft (1.8 × 3 m)
Protrudes 1–2 ft (30–60 cm) from wall or fence.

EUONYMUS JAPONICUS

JAPANESE EUONYMUS
Celastraceae   *Wall Shrub*
Evergreen
A useful range of evergreens including green and variegated forms, most of which should be treated as slightly tender and therefore benefiting from a wall or fence for protection.

Origin From Japan, with many forms of garden origin.
Use As an evergreen shrub to grow against walls and fences where it will benefit from protection in colder areas.
Description *Flower* Clusters of green/yellow flowers, late spring and early summer. *Foliage*

Leaves medium-sized, elliptic, up to 2½ in (6 cm) long, shiny upper surfaces, dull, matt undersides. All-green, or some silver or gold variegation, depending on variety. *Stem* Upright, becoming branching with age; becoming spreading after ten years or more. Light to dark green. Slow growth rate in cold areas, faster in warmer regions. *Fruit* Pink capsules with orange, hanging seeds if the shrub is planted in an ideal situation.
Hardiness Tolerates a minimum winter temperature of 14°F (–10°C). In many cold areas the protection of a wall or fence is imperative, particularly with the variegated forms.
Soil requirements Any soil conditions, including sandy, coastal locations.

Euonymus japonicus 'Albomarginatus'

Sun/Shade aspect Requires a sheltered aspect. Full sun to moderately deep shade; variegated forms may resent very deep shade and lose some variegation.
Pruning None required, but may be clipped and trimmed.
Training Normally requires no training, growing as a freestanding shrub in front of a wall or fence.
Propagation and nursery production From softwood cuttings taken in early summer. Purchase container grown; most forms easy to find. Best planting height 15 in–3 ft (38–91 cm).
Problems Very severe wind chill may damage foliage, but new foliage is normally regenerated the following spring. Some variegated forms, especially golden varieties, may revert to green, but this can be controlled by pruning.

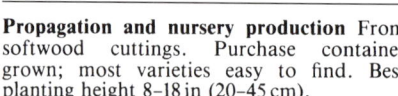

Euonymus fortunei 'Silver Queen'

Similar forms of interest *E. j.* 'Albomargin-atus' Leaves 2–2½ in (5–6 cm) long, mid to dark green with white outer margins. Slightly less than average height and spread. *E. j.* 'Aureopictus' Dark green foliage, with bold gold centre. *E. j.* 'Robusta' A green-leaved, strong growing variety with good fruiting. *E. kiautschovicus* (syn. *E. patens*) (Spreading euonymus) Light green foliage, small yellow/green flowers, followed by pink autumn fruits.

Average height and spread
Five years
6 × 6 ft (1.8 × 1.8 m)
Ten years
10 × 10 ft (3 × 3 m)
Twenty years
15 ft × 15 ft (4.6 × 4.6 m)
Protrudes up to 6 ft (1.8 m) from wall or fence.

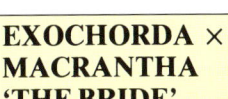

EXOCHORDA × MACRANTHA 'THE BRIDE'

KNOWN BY BOTANICAL NAME
Rosaceae   *Wall Shrub*
Deciduous
A late spring flowering shrub which is not always thought of as a fan-trained wall shrub.

Origin Of garden origin.
Use For fan-training on walls and fences.
Description *Flower* 4 in (10 cm) long by 2 in (5 cm) wide racemes of bold, single, white, saucer-shaped flowers ¾ in (2 cm) across, arching outwards and downwards from main stem in late spring. *Foliage* Leaves medium-sized, lance-shaped, 1½–3 in (4–7.5 cm) long, grey/green, offering some yellow autumn colour. *Stem* Forms an arching, cascading wall shrub. Grey/green. Medium rate of growth. *Fruit* Sometimes produces small, red, round fruits.

Exochorda × macrantha 'The Bride' in flower

Hardiness Tolerates a minimum winter temperature of 4°F (−15°C).
Soil requirements Does well on most soil conditions, but extreme alkalinity will lead to chlorosis which, if not fatal to the shrub, will certainly curtail its performance.
Sun/Shade aspect Tolerates all but the most exposed aspects. Prefers full sun to light shade. May become leggy and spreading in medium to deep shade.
Pruning Remove one third of old flowering wood after flowering to encourage new growth for flowering in subsequent seasons.
Training Requires wires or individual anchor points to encourage a fan-trained shape.

Propagation and nursery production From layers or softwood cuttings taken in early summer. Purchase container grown or root-balled (balled-and-burlapped); comparatively easy to find.
Problems None, apart from its dislike of extremely alkaline soils.
Similar forms of interest None.
Average height and spread
Five years
7 × 7 ft (2.1 × 2.1 m)
Ten years
9 × 9 ft (2.7 × 2.7 m)
Twenty years
12 × 12 ft (3.7 × 3.7 m)
Protrudes up to 3 ft (91 cm) from support.

FABIANA IMBRICATA

KNOWN BY BOTANICAL NAME
Solanaceae   *Wall Shrub*
Evergreen
An attractive shrub benefiting greatly from the protection of a wall or fence.

Origin From Chile.
Use As a freestanding or fan-trained shrub against walls and fences in areas where acid soils are available.
Description *Flower* White tubular flowers covering the branches so thickly when in full bloom the entire branch appears to be a flower spike; late spring. *Foliage* Lanceolate leaves ½ in (1 cm) long, bright green to grey/green, bunched closely on stems. *Stem* Upright at first, quickly becoming arching, forming a wide spreading shrub or fan-shape if trained. Bright green to light grey/green. Slow growth rate. *Fruit* Insignificant.
Hardiness Tolerates a minimum winter temperature of 14°F (−10°C).
Soil requirements Must have an acid soil, dislikes any alkalinity.
Sun/Shade aspect Requires a moderately sheltered aspect. Prefers full sun or very light shade. Resents deeper shade.

Fabiana imbricata **in flower**

Pruning None required other than occasional cutting back of any overlong branches.
Training Allow to grow freestanding or tie to wires or individual anchor points.
Propagation and nursery production From softwood cuttings taken in early summer. Purchase container grown; rather difficult to find. Best planting height 4 in–2 ft (10–60 cm).
Problems Its need for a sheltered aspect and acid soil restrict its possible locations.
Similar forms of interest None.
Average height and spread
Five years
3 × 6 ft (91 cm × 1.8 m)
Ten years
5 × 9 ft (1.5 × 2.7 m)
Twenty years
5 × 12 ft (1.5 × 3.7 m)
Protrudes up to 18 in (45 cm) from support if fan-trained, 4 ft (1.2 m) untrained.

× FATSHEDERA LIZEI

ARALIA IVY
Araliaceae   *Wall Shrub*
Evergreen
Very good for large-scale fence, wall or stump covering in shade, as long as adequate moisture is available.

Origin Of garden origin. Said to be a cross between *Hedera helix* 'Hibernica' and *Fatsia japonica* 'Moseri'.
Use As a spreading, rambling evergreen for shady walls and fences. Can also be fan-trained.

× *Fatshedera lizei* 'Aureovariegate'

Description *Flower* Clusters of round, green flowers on 2–2½ in (5–6 cm) stalks, borne almost year-round, either as buds, flowers or dead flowers, only produced on mature branches three to four years old. *Foliage* Leaves large, palmate, 4–10 in (10–25 cm) wide, dark green, leathery textured with shiny upper surfaces and duller undersides. *Stem* Allow to ramble. Light to dark green. Fast rate of growth in favourable conditions. *Fruit* Round clusters of black fruits follow flowers.
Hardiness Tolerates a minimum winter temperature of 14°F (−10°C).
Soil requirements Prefers rich, moist soils but does well on all types.
Sun/Shade aspect Requires some protection from exposed aspects. Very good in deep shade, but tolerates medium to light shade. Unhappy in full sun, unless adequate moisture is available.
Pruning None required, although very old shoots can be removed.
Training Tie to wires or individual anchor points or allow to ramble over shrubs or stumps.
Propagation and nursery production From softwood cuttings taken in late spring to early summer. Purchase container grown; normally easy to find. Best planting height 15 in–2½ ft (38–76 cm).

× *Fatshedera lizei*

Problems When purchased in container may look weak and floppy but grows quickly once planted out.
Similar forms of interest ×*F. l.* 'Variegata' Grey/green leaves with creamy white margins. May be hard to find. More tender than the parent. ×*F. l.* 'Aureovariegata' Grey/green leaves with dull gold edges. May be hard to find. More tender than its parent.
Average height and spread
Five years
6×6 ft (1.8×1.8 m)
Ten years
13×13 ft (4×4 m)
Twenty years
20×12 ft (6×3.7 m)
Protrudes up to 4 ft (1.2 m) from support.

FATSIA JAPONICA

CASTOR OIL PLANT, JAPANESE FATSIA

Araliaceae   *Wall Shrub*
Evergreen

One of the best of all large, shade-loving wall shrubs for quick effect.

Origin From Japan.
Use As a freestanding evergreen shrub against walls and fences, requiring space.
Description *Flower* 2 in (5 cm) wide clusters of silver-green opening to milk white flowers; produced in spring from round clusters of buds developed in previous autumn and maintained over a long period. *Foliage* Leaves large to very large, dark to mid-green, 2½–6½ in (6–16 cm) wide and 9–18 in (23–45 cm) long, palmate, with seven, nine, or 11 leaflets. Glossy upper surface and paler underside. *Stem* Light to mid green, strong, stout, upright, forming a tall, rigid structure. Fast growth rate in the right conditions. *Fruit* Large clusters of black fruits follow flowers. Removal will increase leaf size.
Hardiness Tolerates a minimum winter temperature of 4°F (−15°C).
Soil requirements Does well on most types, but rich, deep, moist soil produces largest leaves.
Sun/Shade aspect Requires a partially protected aspect. Best in deep to medium shade, dislikes full sun.
Pruning None required but can be reduced in size and will rejuvenate itself from below pruning cuts.
Training Normally requires no training and is freestanding.
Propagation and nursery production From semi-ripe cuttings taken in early summer. Purchase container grown; easy to find. Best planting height 1½–3 ft (45–91 cm).

Fatsia japonica 'Variegata'

Problems When purchased may look weak and sickly but once planted out it soon flourishes.
Similar forms of interest *F. j.* 'Variegata' White to creamy white variegation to lobes and tips of foliage. Slightly less than average height and spread. More tender but wall planting will give added protection.
Average height and spread
Five years
8×8 ft (2.4×2.4 m)
Ten years
12×12 ft (3.7×3.7 m)
Twenty years
15×15 ft (4.6×4.6 m)
Protrudes up to 8 ft (2.4 m) from wall or fence.

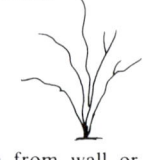

FEIJOA SELLOWIANA

GUAVA, PINEAPPLE GUAVA

Myrtaceae   *Wall Shrub*
Evergreen

An interesting, attractive, foliage evergreen, even without its summer flowers, worthy of a sheltered wall or fence.

Origin From Brazil and Uruguay.
Use As a freestanding or fan-trained wall shrub for sheltered walls and fences or for use in a conservatory or greenhouse.
Description *Flower* Fleshy, edible, crimson and white petals with central bunch of long

crimson stamens, but shy to flower in most areas. *Foliage* Leaves round to ovate, 1½–3 in (4–7.5 cm) long, grey/green with white felted undersides and grey leaf stalks. Some older leaves die in autumn, giving contrasting colour. *Stem* Grey/green when young, ageing to grey/brown, forming a loose fan-trained shrub on a wall or fence. Slow to medium rate of growth. *Fruit* Good-sized, egg-shaped, yellow fruits, edible, with aromatic flavour. Rarely fruits outside its native environment.
Hardiness Tolerates a minimum winter temperature of 23°F (−5°C).
Soil requirements Does well on most soils, disliking only extreme alkalinity.

Feijoa sellowiana

Sun/Shade aspect Requires a very sheltered aspect. Prefers light shade, tolerates full sun.
Pruning None required.
Training Requires wires or individual anchor points for fan-training or allow it to grow freestanding.
Propagation and nursery production From softwood cuttings. Purchase container grown; difficult to find.
Problems None.
Similar forms of interest *F. s.* 'David' A large-foliage variety from New Zealand. *F. s.* 'Variegata' Foliage variegated creamy white. Two thirds average height and spread; more tender than the parent.
Average height and spread
Five years
5×5 ft (1.5×1.5 m)
Ten years
9×9 ft (2.7×2.7 m)
Twenty years
15×15 ft (4.6×4.6m)
Protrudes up to 3 ft (91 cm) from support if fan-trained, 13 ft (4 m) untrained.

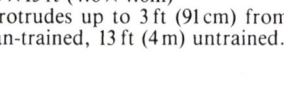

FICUS CARICA

COMMON FIG, FIG

Moraceae   *Wall Shrub*
Deciduous

A very decorative fruiting shrub, and there is nothing better on a late summer or early autumn afternoon than picking the ripe figs from your own plants.

Origin From western Asia.
Use As a wall or fence grown specimen to protect and encourage its edible fruits to ripen. Train horizontally or in a fan shape. Can be grown in large containers against walls and fences with less height and spread.
Description *Flower* Pale green, insignificant. *Foliage* Leaves large, palmate, 8–10 in (20–25 cm) long and wide, mid green to grey/green, abundantly produced, giving good yellow autumn colour. Attractive. *Stem* Light green when young, becoming spreading and arching with time, needing support. Fast

Fatsia japonica in flower

268

Ficus carica 'Brown Turkey' in fruit

growing. *Fruit* Large, green-skinned, becoming edible after two years. Fruits become wrinkled and purple/black when ripe in late summer to early autumn. Leave on shrub and use as required.

Hardiness Tolerates a minimum winter temperature of 4°F (−15°C). In cold winters or cold areas, stems should be protected with bracken or other open material.

Soil requirements Does well on most soil types. For best fruiting, roots must be restricted within a 6 ft (1.8 m) box sunk in soil, or the shrub will grow too vigorously and not fruit adequately. This treatment is imperative when it is grown against a wall or fence.

Sun/Shade aspect Requires a warm, protected aspect. Full sun to light shade needed to ripen fruit.

Pruning Reduce thin, weak growth either in winter or summer. Train fan or horizontal but avoid hard cutting back as this will only lead to stronger, unfruiting growth.

Training Requires wires or individual anchor points for training into a fan shape.

Propagation and nursery production From softwood cuttings taken in late summer or from layering, suckers, or grafting. Purchase container grown; easy to find. Best planting height 1½–2½ ft (45–76 cm).

Problems Can become invasive unless root development is restrained.

Similar forms of interest *F. c.* 'Brown Turkey' Large, brown to purple-skinned pear-shaped fruits. Flesh creamy white with red tinge. *F. c.* 'Brunswick' Pear-shaped, yellow-skinned fruits, flesh white flushed red. Hard to find, but not impossible. *F. c.* 'White Marseilles' Round, white-skinned fruits, flesh almost translucent. Very sweet and juicy. Very scarce.

Average height and spread
Five years
10 × 10 ft (3 × 3 m)
Ten years
15 × 15 ft (4.6 × 4.6 m)
Twenty years
20 × 20 ft (6 × 6 m)
Protrudes up to 8 ft (2.4 m) from support.

FORSYTHIA SUSPENSA

CLIMBING FORSYTHIA, WEEPING FORSYTHIA
Oleaceae   ***Wall Shrub***
Deciduous

A long, arching, ranging shrub requiring space to mature to its full potential.

Origin From China, but many forms of garden origin.

Use As a rambling, spreading, permanent wall or fence shrub, requiring space.

Description *Flower* Clear yellow to golden yellow, hanging, bell-shaped flowers covering all mature stems in early to mid spring. Last season's shoots do not flower. *Foliage* Leaves ovate, 2–4 in (5–10 cm) long, tooth-edged. Light to mid-green. Yellow autumn colour. *Stem* Green to grey/green to grey/brown. Long, arching, ranging branches. Medium to fast growing. *Fruit* Insignificant.

Hardiness Tolerates a minimum winter temperature of 4°F (−15°C).

Soil requirements Does well on all soil types.

Sun/Shade aspect Tolerates all aspects. Full sun to medium shade. In deep shade becomes very open, lax and shy to flower.

Pruning Sometimes very difficult to prune because it forms a very woody structure, but if possible remove one third of oldest flowering wood to ground level, or to lowest young, strong shoot, after flowering.

Training Will require tying to wires, trellis or individual anchor points.

Propagation and nursery production From semi-ripe cuttings taken mid summer or hardwood cuttings taken in winter. Purchase bare-rooted or container grown. Most varieties easy to find, some must be sought from specialist nurseries. Best planting height 1–3 ft (30–91 cm).

Problems Unlike other forsythias it may look loose and weak in constitution when young, but quickly generates new growth once planted out.

Similar forms of interest *F. s. atrocaulis* Pale lemon yellow flowers produced on black/purple stems, arching in habit in early spring. One of the earliest to flower. More open than most varieties in its habit. *F. s.* 'Nymans' Flowers primrose yellow, of good size and borne on an open arching shrub. Early flowering. From Nymans Gardens, England.

Average height and spread
Five years
8 × 8 ft (2.4 × 2.4 m)
Ten years
12 × 12 ft (3.7 × 3.7 m)
Twenty years
18 × 18 ft (5.5 × 5.5 m)
Protrudes up to 3 ft (91 cm) from support.

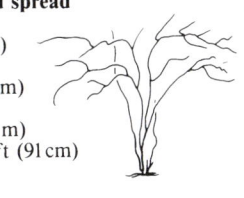

FREMONTODENDRON CALIFORNICUM

FREMONTIA
Sterculiaceae   ***Wall Shrub***
Evergreen

One of the most spectacular shrubs for planting against a high sunny wall.

Origin From California and Arizona, USA.

Use As a fan-trained wall shrub where it will reach a height of at least 26–30 ft (8–9 m) relatively quickly, flowering from late spring until well into late autumn.

Description *Flower* No actual petals, large, saucer-shaped yellow calyxes form flowers up to 3 in (7.5 cm) wide with golden yellow stigma and stamens protruding from centre, profusely borne in late spring through to late autumn. *Foliage* Trifoliate leaves with heart-shaped leaflets, three- to seven-lobed, 2–4 in (5–10 cm) long, grey/green. *Stem* Upright when young, becoming more spreading to form a fan-shaped shrub. Grey/green, down-covered. Fast growth rate. *Fruit* Insignificant.

Hardiness Tolerates a minimum winter temperature of 4°F (−15°C), although often wrongly considered more tender.

Soil requirements Any soil type, tolerating high alkalinity.

Sun/Shade aspect Tolerates all but the most exposed of aspects. Prefers full sun, tolerates very light shade.

Pruning Prune current season's growth back by two thirds in mid to late summer. This will generate new growth and more flowers. The height of the plant cannot be easily controlled by pruning.

Fremontodendron californicum 'California Glory' in flower

Training Requires wires or individual anchor points to achieve a fan-trained shape.

Propagation and nursery production From softwood cuttings taken in early spring, or from seed. Purchase container grown. May not be easy to find outside its native environment. Best planting height 15 in–3 ft (38–91 cm).

Problems When young, susceptible to severe winter weather, but after reaching 5 ft (1.5 m) rarely affected by cold. Often outgrows the area allowed for it. The dusty down from the stems can be very painful if it gets into the eyes or on the skin.

Forsythia suspensa in flower

Similar forms of interest *F. c.* 'California Glory' The best variety to try, with good large yellow flowers. *F. mexicanum* Orange/yellow flowers, more tender, requiring a very sheltered site.
Average height and spread
Five years
10 × 6 ft (3 × 1.8 m)
Ten years
20 × 10 ft (6 × 3 m)
Twenty years
30 × 15 ft (9 × 4.6 m)
Protrudes up to 3 ft (91 cm) from support.

GARRYA ELLIPTICA

TASSEL BUSH, SILK TASSEL TREE

Garryaceae   *Wall Shrub*
Evergreen

It is hard to imagine a finer sight than *Garrya elliptica* fully clothed with its catkins in late spring.

Origin From California and Oregon, USA.
Use A shrub for shady walls and fences. Produces evergreen foliage and attractive mid to late winter catkins; good in maritime areas.
Description *Flower* Male plants have long, grey/green, hanging catkins, up to 5 in (12 cm) long, made up of a series of bell-shaped flowers. Female plants have insignificant flowers in short catkins so are not normally considered as attractive. *Foliage* Leaves broadly ovate, 1½–3 in (4–7.5 cm) long, dark green, glossy upper surfaces, glaucous undersides and of leathery texture. *Stem* Upright, becoming branching and spreading with age. Grey/green to green/brown. Medium growth rate. *Fruit* Small, round, purple/brown fruits in long clusters produced by female plants when there is a male plant nearby.
Hardiness Tolerates a minimum temperature of 4°F (−15°C).
Soil requirements Prefers fertile soil, but does well on all types.

Garrya elliptica in flower

Sun/Shade aspect Requires a partially sheltered aspect in full sun to medium shade. In deep shade becomes long, open and straggly and catkins diminish in number.
Pruning None required but can be reduced by removing major branches in late winter or early spring and will rejuvenate itself in following spring.
Training Requires a limited number of individual anchor points to assist training and give support.
Propagation and nursery production From semi-ripe cuttings taken in early to mid summer. Purchase container grown; available from good garden centres and nurseries. Best planting height 1½–3 ft (45–91 cm).

Ginkgo biloba in autumn

Problems Rather untidy habit of growth. Black spots often appear on old foliage just before it drops in late winter and early spring. In cold springs there may be a delay between fall of old evergreen foliage and growth of new leaves, leaving the plant temporarily defoliated. This can also sometimes happen after planting.
Similar forms of interest *G. e.* 'James Roof' Good, bright, leathery leaves, with catkins twice the length of *G. elliptica* and thicker in texture. Strong-growing. Worth looking for. *G. fremontii* Similar to its parent but with more ovate, twisted leaves and short catkins. From western USA.
Average height and spread
Five years
8 × 8 ft (2.4 × 2.4 m)
Ten years
12 × 12 ft (3.7 × 3.7 m)
Twenty years
15 × 15 ft (4.6 × 4.6 m)
Protrudes up to 4 ft (1.2 m) from support.

GINKGO BILOBA

MAIDENHAIR TREE

Ginkgoaceae   *Conifer*
Deciduous

A deciduous conifer that adapts itself well to training horizontally or in a fan shape, producing a very attractive display of foliage.

Origin From northern China.
Use As a wide spreading, horizontal or fan-trained wall shrub for large walls and fences.
Description *Flower* Male flowers are short yellow catkins, appearing at the same time as the leaves; female flowers grow in pairs after hot summers. Flowering is rare except in the most favourable sunny positions or at great maturity. *Foliage* Fan-shaped, 2 in (5 cm) wide and long, with a cleft in the centre and irregular, notched, indented edges. Almost leathery in texture. Extremely attractive. Grey/green with beautiful yellow autumn colour. *Stem* Light grey/green when young, becoming grey/brown. Attractive both in summer and in winter. Rigid. Medium growth rate. *Fruit* Rarely fruits outside its native environment, but may produce plum-shaped yellow fruits which are edible and sweet tasting.
Hardiness Tolerates a minimum winter temperature of 0°F (−18°C).
Soil requirements Tolerates all soil conditions with no particular preference, although a high degree of organic material incorporated at planting time will encourage the root system to grow quickly and to support good top growth.

Sun/Shade aspect Tolerates all aspects with no particular preference, although may be more likely to flower and fruit in sunny positions.
Pruning Cut newly purchased plants in spring to within 12 in (30 cm) of ground level and the resulting two or three shoots can be trained horizontally or in a fan shape. Additional side branches as they arise can also be fan-trained. In subsequent springs, shorten back surplus side branches not needed for framework to within 2 in (5 cm) of their origin. The shortening back of side branches then becomes an annual process to encourage new growth and attractive foliage.
Training Tie to wires or individual anchor points.
Propagation and nursery production From seed. Should always be purchased container grown or root-balled (balled-and-burlapped). Young plants of trainable age will normally be found in garden centres but may have to be sought from general and specialist nurseries. Best planting height not more than 3 ft (91 cm) and ideally not more than three years old.
Problems Will take up to five years to produce a good fan-trained or horizontal shape and from then on will increase in size to cover a large area which must be allowed for.
Similar forms of interest *G. b.* 'Autumn Gold' Golden-yellow foliage in autumn. Neat habit. *G. b.* 'Saratoga' Very hardy. Good yellow autumn tints.
Average height and spread
Five years
4 × 4 ft (1.2 × 1.2 m)
Ten years
8 × 8 ft (2.4 × 2.4 m)
Twenty years
16 × 16 ft (4.9 × 4.9 m)
Protrudes up to 2 ft (60 cm) from support.

GLEDITSIA TRIACANTHOS 'SUNBURST'

GOLDEN-LEAVED HONEYLOCUST

Leguminosae   *Hardy Tree*
Deciduous

An attractive foliage tree responding well to fan-training.

Origin From North America.
Use As a fan-trained tree for large walls and fences.
Description *Flower* Green/white male flowers in pendent racemes 2 in (5 cm) long. Female flowers limited and inconspicuous. Borne in mid summer but may be shy to flower. *Foliage* Pinnate or bipinnate, up to 8 in (20 cm) long with up to 32 leaflets 1 in (2.5 cm) long. Golden-yellow becoming lime-green then deep

yellow in autumn. *Stem* Light grey/green, somewhat fragile. May suffer from wind damage. Mature bark grey and deeply channelled, of architectural interest. Medium rate of growth. *Fruit* Small, pea-shaped, grey/green pods, sword-shaped and twisted, often retained well into winter, but may be shy to fruit in all but the most favourable areas. Poisonous.
Hardiness Tolerates a minimum winter temperature of 4°F (−15°C).
Soil requirements Best results on well-drained, deep, rich soil. Tolerates limited amount of alkalinity through to fully acid types.
Sun/Shade aspect Tolerates all but the most exposed of aspects. Full sun to very light shade.

Gleditsia triacanthos 'Sunburst' in leaf

Pruning Prune young trees hard in spring following planting. Select and train resulting five to seven shoots and tie into a fan-trained shape. In subsequent years remove all side growths back to two points from their origin and maintain original main branches in fan shape.
Training Will require fixing to wires or individual anchor points.
Propagation and nursery production Grafted on to *G. triacanthos*. Best purchased container grown although root-balled (balled-and-burlapped) or bare-rooted plants can be successful if planted between late autumn and early spring. Should be sought from general or specialist nurseries and garden centres. Best planting height 3–5 ft (91 cm–1.5 m). Always choose young trees for preference.
Problems Normally slow growing and young plants may often look irregular and weak in constitution in nurseries and garden centres.

Similar forms of interest *G. t.* 'Elegantissima' Compact, only reaching one third ultimate height and spread; can be considered as large shrub. Interesting light green, fern-like foliage. Yellow autumn colour. *G. t.* 'Ruby Lace' Foliage purple to purple/green in spring; new growth red/purple. Some yellow/bronze autumn colour. White flowers, but rarely occurring. Reaches only one third average height and spread.
Average height and spread
Five years
7 × 7 ft (2.1 × 2.1 m)
Ten years
14 × 14 ft (4.3 × 4.3 m)
Twenty years
25 × 25 ft (7.6 × 7.6 m)
Protrudes up to 3 ft (91 cm) from support.

HALESIA

MOUNTAIN SILVERBELL, MOUNTAIN SNOWBALL TREE, CAROLINA SILVERBELL
Styracaceae   *Wall Shrub*
Deciduous
A very attractive large shrub that adapts well to being grown fan-trained on a wall or fence, showing off its flowers to the best effect.

Origin From south-eastern USA.
Use As a shrub for fan-training on walls and fences where the protection helps its performance.
Description *Flower* Small to medium-sized, hanging, nodding, bell-shaped flowers borne in 3 in (7.5 cm) long clusters of three or five along the underside of each branch in mid to late spring. *Foliage* Broad, ovate, 2–5 in (5–12 cm) long, light grey/green, giving good yellow autumn colour. *Stem* Upright when young, becoming branching and twiggy with age. Grey to grey/green. Medium growth rate. *Fruit* Small, green, winged fruits in autumn.
Hardiness Tolerates a minimum winter temperature of 4°F (−15°C).
Soil requirements Does well on most types but best on well-drained soil.
Sun/Shade aspect A sheltered aspect. Prefers full sun to light shade.
Pruning None required.
Training Requires individual anchorage points or wires to secure plant to wall or fence to achieve a fan shape.
Propagation and nursery production From layers or semi-ripe cuttings taken in early summer. Purchase container grown; must be sought from specialist nurseries. Best planting height 1½–2½ ft (45–76 m).

Problems Often looks weak when purchased; begins to achieve full potential only after three to five years.
Forms of interest *H. carolina* (syn. *H. tetraptera*) (Carolina Silverbell) White, bell-shaped nodding flowers, grouped along branches, early to mid spring. Fruits are four-winged and narrowly pear-shaped. *H. monticola* (Mountain Silverbell) A variety with larger flowers and fruit, fruit clusters being up to 2 in (5 cm) in length. From the mountain regions of south-eastern USA. *H. m.* 'Vestita' A large-flowering form with flower clusters up to 1–1½ in (2.5–4 cm) across, sometimes pink-tinged. When mature, the leaves are downy with a glabrous covering.
Average height and spread
Five years
6 × 6 ft (1.8 × 1.8 m)
Ten years
13 × 13 ft (4 × 4 m)
Twenty years
20 × 20 ft (6 × 6 m)
Protrudes up to 3 ft (91 cm) from support.

HAMAMELIS

WITCH HAZEL, CHINESE WITCH HAZEL
Hamamelidaceae   *Wall Shrub*
Deciduous
A shrub not normally considered for fan-training on a wall but when used in this form it is very attractive in winter.

Origin Most forms originating from Japan, with many varieties now of garden origin.
Use As a large, fan-trained shrub for walls and fences.

Hamamelis × intermedia 'Diane' in autumn

Description *Flower* Clusters of strap-like 1–1½ in (2.5–4 cm) long petals ranging from lemon-yellow through gold, brown, orange to dark red, according to variety, borne in winter through early spring, withstanding harsh weather conditions undamaged. Some varieties fragrant. Exact flowering time depends on winter temperatures; the first mild spell after the shortest day of the year often triggers the first flower bud to open. *Foliage* Leaves ovate, 3–5 in (7.5–12 cm) long, with good orange/yellow autumn colours. *Stem* Upright when young, becoming spreading with age when it can be adapted to become fan-shaped. Grey to grey/green. Medium growth rate.
Hardiness Tolerates a minimum winter temperature of 4°F (−15°C).
Soil requirements Neutral to acid; moderate to heavy alkalinity will rapidly cause signs of chlorosis. Mulch annually with peat or garden compost to a depth of 2 in (5 cm) over 1–2 sq yd (1–2 sq m) to maintain health and encourage flowering.
Sun/Shade aspect Tolerates all aspects. Prefers full sun to light shade; too much shade will spoil shape and reduce flowers.

Halesia carolina in flower

Best planted where winter sunlight can enhance flowering effect.

Pruning None required.

Training Requires wires or individual anchor points to secure and encourage a fan-trained shape.

Propagation and nursery production H. virginiana, grown from seed, is the understock on to which all other varieties are grafted. Purchase container grown or root-balled (balled-and-burlapped). Best planting height 2–4 ft (60 cm–1.2 m).

Hamamelis mollis in flower

Problems Young grafted plants often fail so it is advisable to purchase plants four years old or more, although these are expensive.

Forms of interest H. × intermedia 'Diane' Flowers 1–1½ in (2.5–4 cm) across, rich copper-red. Good autumn foliage colours. Slightly less than average height and spread. H. × i. 'Jelena' Flowers 1½ in (4 cm) across, bright copper-orange. H. × i. 'Ruby Glow' Flowers 1 in (2.5 cm) across, copper-red. Narrow foliage with good autumn colour. H. × i. 'Westersteide' Flowers 1 in (2.5 cm) across, clear yellow and freely borne, produced later than most. Small to medium sized foliage. Good autumn colour. H. japonica 'Zuccariniana' (Japanese witch hazel) Grey, curled flower buds open to release pale yellow, lemon-scented flowers in early spring. Flowers less than 1 in (2.5 cm) across but borne profusely. Growth habit makes it extremely good for fan-training. H. mollis (Chinese witch hazel) Pure golden yellow, very fragrant flowers 1½ in (4 cm) across, late winter. Large, oval to round leaves with good autumn colour. H. m. 'Brevipetala'

Scented, bronze-yellow flowers, 1½ in (4 cm) across, borne on strong upright branches, more vigorous than most. Broad ovate leaves with good autumn colour. Spreading habit. One of the most beautiful forms. **H. virginiana** (Common witch hazel) Slightly scented, golden yellow flowers, 1 in (2.5 cm) across. Good golden yellow autumn colour. Apart from its use as understock, this is the basic form from which the essence known as witch hazel is distilled. Its growth makes it extremely good for fan-training.

Average height and spread
Five years
8 × 8 ft (2.4 × 2.4 m)
Ten years
15 × 15 ft (4.6 × 4.6 m)
Twenty years
20 × 20 ft (6 × 6 m)
Protrudes up to 3 ft (91 cm) from support.

HEDERA CANARIENSIS

CANARY ISLAND IVY

Araliaceae   *Woody Climber*
Evergreen

Amongst one of the most attractive of all evergreen climbing plants.

Origin From the Canary Islands and North Africa.

Use As a free-growing climber for all aspects, growing up walls and fences, through trees and for covering large areas where required. Can be adapted to creep along the ground as an evergreen ground-covering carpet.

Description *Flower* Small clusters of green flowers in late winter on very mature plants. *Foliage* Up to 8 in (20 cm) wide with five to seven lobes and of leathery texture, matt dark green in colour with silver reverse. *Stem* Light green becoming dark green and glossy, twining and twisting, partially self-clinging. Medium to fast growing. *Fruit* Mature plants may carry clusters of dark blue/black fruits in early to late autumn.

Hardiness Tolerates a minimum winter temperature of 4°F (−15°C). Requires protection from north and east winds to stop leaf scorch.

Soil requirements All soil types, both alkaline and acid, only showing signs of distress in extremely dry conditions.

Sun/Shade aspect Full sun to deep shade, although in deep shade may be more lax and open in habit. Good in all aspects except for those exposed to severe winds in winter.

Pruning No pruning for its own well-being but can be contained in a specific area if required by cutting back.

Hedera canariensis

Training Allow to ramble over trellis and wires. Weight of leaf and stems will need support when used on walls and fences.

Propagation and nursery production From semi-ripe cuttings or rooted natural layers. Purchase container grown; readily available from good garden centres and general nurseries. Best planting height 1–4 ft (30 cm–1.2 m).

Problems It is susceptible to very cold winds and extreme conditions but if it is damaged it normally recovers quite quickly.

Similar forms of interest H. c. 'Azorica' Good grower, attractive formation, matt green. H. c. 'Ravensholst' Tender, needing protection, but strong growing. H. c. 'Variegata' (syn. 'Gloire de Marengo') Oval, leathery, dark green leaves with silvery reverse and bold creamy white margins. Smaller than the parent.

Average height and spread
Five years
7 × 7 ft (2.1 × 2.1 m)
Ten years
12 × 12 ft (3.5 × 3.5 m)
Twenty years
30 × 30 ft (9 × 9 m)
Protrudes up to 3 ft (91 cm) from support.

HEDERA COLCHICA

PERSIAN IVY

Araliaceae   *Woody Climber*
Evergreen

Amongst the most attractive of all the green-leaved ivies. Not startling in its display but able to present a bold appearance.

Origin From Persia.

Use As a dark green climber for covering all types of construction to give an evergreen attraction. Good as ground cover.

Hedera colchica

Hedera canariensis 'Variegata'

Hedera colchica 'Sulphur Heart'

Description *Flower* Clusters of dark green buds open to petal-less flowers throughout autumn and winter. Produced mainly on very mature plants and of some limited attraction. *Foliage* Oval to elliptic in shape, leathery texture, dark matt green in colour. *Stem* Vigorous red/brown stems becoming grey/brown with age. Self-clinging but some initial

Hedera colchica 'Dentata Variegata'

support needed in exposed positions or following heavy snow or rain. Medium to fast growing. *Fruit* Produces poisonous dark blue to black fruits through winter and early spring.
Hardiness Tolerates a minimum winter temperature of 14°F (−10°C). Hardy but may show signs of foliage damage in extremely severe winters.
Soil requirements Does well on all soil types.
Sun/Shade aspect Prefers light shade but will tolerate full sun to deep shade.
Pruning No pruning for its own well-being but can be maintained within bounds by removing offending shoots as and when required.
Training Partially self-clinging but will need some support to climb walls or fences.
Propagation and nursery production From semi-ripe cuttings or rooted natural layers. Always purchase container grown; although relatively common it may be difficult to find. Best planting height 1–4 ft (30 cm–1.2 m).

Problems Often becomes larger than the area intended, although slow to establish – will take up to three years before vigorous growth becomes apparent.
Similar forms of interest *H. c.* 'Amur River' Dark green, leathery texture. May be less hardy. *H. c.* 'Blair Castle' Dark green with indentations on outer edges of leaves. *H. c.* 'Dentata' Foliage larger, good strong grower. *H. c.* 'Dentata Variegata' Mid green foliage with yellow/cream variegation. *H. c* 'Sulphur Heart' (syn. 'Paddy's Pride') Central yellow splash on pale to mid green leaves. Tolerates a minimum winter temperature of 4°F (−15°C).
Average height and spread
Five years
6 × 6 ft (1.8 × 1.8 m)
Ten years
12 × 12 ft (3.7 × 3.7 m)
Twenty years
25 × 25 ft (7.5 × 7.5 m)
Protrudes up to 2 ft (60 cm) from support.

HEDERA HELIX

ENGLISH IVY, COMMON IVY

Araliaceae   *Woody Climber*
Evergreen

Although common, this still has a place for planting where an evergreen climber is needed in the most inhospitable areas, as it tolerates both shade and exposure to the extreme.

Origin From Europe, Asia Minor and into Persia.
Use A climber for all aspects, up walls, fences or banks. Can be trained both up or down a wall. Is also useful as a spreading ground cover.
Description *Flower* Round, ball-shaped clusters of dark green flower buds open to lighter green in late winter/spring, attracting many bees and other insects. *Foliage* Young leaves are three- to five-lobed, dark green with white undersides and some white veining. Adult foliage is oval to oval-triangular in shape. Some forms have interesting leaf shapes and green colour variations. *Stem* Light green turning dark green, finally grey-brown, self-clinging by suckering aerial roots on undersides of stems. Medium rate of growth. *Fruit* Clusters of poisonous black fruits produced in spring and summer on mature stems.
Hardiness Tolerates a minimum winter temperature of 0°F (−18°C).

Soil requirements Does well on all soil types, although it dislikes dry conditions until established.
Sun/Shade aspect Tolerates all aspects. Deep shade to full sun.
Pruning No pruning for its own well-being but can be cut back in spring to contain, as required.
Training Allow to form a self-clinging, area-defining climber. Normally requires no additional support.
Propagation and nursery production From semi-ripe cuttings taken in early summer or from self-rooted layers. May also be grown from seed. Always purchase container-grown, available all year round; may have to be sought from a specialist nursery. Those purchased as house plants will make useful climbers, once hardened off. Best planting height 1–3 ft (30–91 cm).
Problems Can become invasive. In Europe it is commonly believed to damage walls and trees, but this is only the case with very poor brickwork or trees that are relatively unhealthy, although climbing of trees should be discouraged.
Similar forms of interest *H. h.* 'Atropurpurea' Dark green leaves changing to purple/black in winter. Sparse open habit. *H. h.* 'Brokamp' Good dark green narrow leaves. Slow growing. *H. h.* 'Chicago' Strong grower. Soft green foliage of good shape. *H. h.* 'Deltoides' Interesting blunt leaf shape, dark green turning purple in winter. Growth thick and stiff. *H. h.* 'Digitata' Good large leaf shape, dark green. *H. h.* 'Glymii' (syn. *H. h.* 'Tortuosa') Leathery, average-sized, dark glossy green juvenile foliage, shaped more than when older. Turns purple to purple/green in winter. *H. h.* 'Gracilis' Mid green leaves supported on red stems, open habit, turning wine red in winter. *H. h.* 'Green Ripple' Dark green leathery texture, deeply lobed edges, strong grower. Copper colouring in winter. *H. h.* 'Hamilton' Strong growth, large, interestingly shaped, mid green leathery leaves. Very good overall shape. *H. h.* 'Hibernica' (Irish ivy) Average-sized, glossy dark green, five-lobed leaves. *H. h.* 'Ivalace' Small, dark green, leathery, curly-edged, five-lobed leaves, turning copper/orange in cold winters. *H. h.* 'Manda's Crested' (syn. *H. h.* 'Curly Locks') Pale green, soft-textured, five-pointed average-sized leaves, becoming blood red in cold winters. *H. h.* 'Marmorata Minor' (syns. *H. h.* 'Discolor', *H. h.* 'Minor Marmorata') Leaves mid green with darker green blotches. *H. h.* 'Meagheri' Open, trailing nature, presenting three-lobed, small, mid green leaves over a wide area. Trim in spring to encourage a lighter stem formation. *H. h.* 'Minima' (syn. *H. h.* 'Spetchley') Very small,

Hedera helix 'Green Ripple'

Hedera helix

Hedera helix angularis 'Aurea' in summer

three- or five-lobed, dark green foliage turning bronze to dull orange in winter. *H. h.* 'Nymans' The centre lobe is long compared with the other four. Bright green in colour, good white veins in summer. Slow to attain any size. *H. h.* 'Neilsonii' (syn. *H. h.* 'Neilson') Coppery red in winter, bright green in summer. Five-lobed interesting shape. Shrubby habit. Slower growing than its parent. *H. h.* 'Palmata' Brown/red winter colour, mid green palm-shaped leaves of average size. *H. h.* 'Pittsburgh' Medium-sized, five-lobed, bright green leaves, attaining a copper shading in winter. Strong growing. *H. h.* 'Poetica' (syn. *H. h.* 'Chrysocarpa') Stiff upright grower with bright green, average-sized leaves. *H. h.* 'Russelliana' (syn. *H. h.* 'Erecta') Very stiff, upright grower, covered with leathery dark green leaves of average size. *H. h.* 'Très Coupé' Slow growing. Narrow, mid green, five-lobed leaves with pronounced points to ends of lobes.
Average height and spread
Five years
6×6 ft (1.8×1.8 m)
Ten years
12×12 ft (3.5×3.5 m)
Twenty years
25×25 ft (7.6×7.6 m)
Protrudes up to 18 in (45 cm) from support.

HEDERA HELIX ANGULARIS 'AUREA'

KNOWN BY BOTANICAL NAME

Araliaceae   *Woody Climber*
Evergreen
One of the most attractive of all the golden-foliaged ivies with an irregular yet interesting growth pattern.

Origin From Europe.
Use As a self-clinging climber for walls and fences, useful for growing over old stumps and other similar features. Can be used as a spreading ground cover.
Description *Flower* Small round clusters up to 1 in (2.5 cm) across of mid to dark green round flower heads produced through autumn and winter on mature growth. *Foliage* Three- to five-fingered, triangular, curly, wavy leaves up to 1½ in (4 cm) wide by 2 in (5 cm) long, green ageing to bright yellow, turning to an attractive chocolate brown in cold winters. *Stem* Light green/gold aging to

dark green, finally brown. Self-clinging, often more irregular in habit than many small-leaved ivies. Small, suckering, hanging aerial roots produced from underside of branches. Slow to medium growing. *Fruit* Round clusters of black poisonous fruit may be produced in spring on mature plants.
Hardiness Tolerates a minimum winter temperature of 4°F (−15°C). Foliage may be damaged by severe wind chill in winter in exposed positions, although it normally rejuvenates the following spring.
Soil requirements Both alkaline or acid but with a good moisture content to assist production of new golden foliage.
Sun/Shade aspect All but the most exposed aspects. Must be in full sun, deep shade will turn foliage green.
Pruning Not normally required other than for retaining within bounds which is best carried out in spring.
Training Self-clinging on both stone and brick walls, tying-in may be required on timber fences and posts.
Propagation and nursery production From semi-ripe cuttings taken in early summer. Self-rooted layers may be available at ground level on established plants. Always purchase

container grown in early autumn to early summer, may have to be sought from specialist nurseries. Best planting height 6 in–3 ft (15–91 cm).
Problems Its requirement for full sun to prevent fading. Can be irregular in its growth habit, particularly in the early years.
Similar forms of interest None.
Average height and spread
Five years
12×5 ft (3.7×1.5 m)
Ten years
15×10 ft (4.6×3 m)
Twenty years
25×20 ft (7.6×6 m)
Protrudes up to 2 ft (60 cm) from support.

HEDERA HELIX 'BUTTERCUP'

BUTTERCUP IVY

Araliaceae   *Woody Climber*
Evergreen
One of the few shrubs to improve its foliage colour in winter.

Origin Of garden origin.
Use As a climber for walls, fences, pillars and large banks.
Description *Flower* On mature plants, round ball-shaped clusters of dark green flower buds open to lighter green in late winter/spring. *Foliage* Three- to five-lobed leaves, bright golden-yellow in autumn, winter and spring, may turn green/yellow in summer but regain their full yellow colouring in following

Hedera helix angularis 'Aurea' in winter

autumn/winter. *Stem* Grey/green ageing to yellow/green, finally green/brown, twining. Self-clinging. Medium rate of growth. *Fruit* Shy to fruit, only very mature shrubs may produce the round, dark red fruits typical of ivy in clusters during late winter.

Hardiness Tolerates a minimum winter temperature of 4°F (−15°C).

Soil requirements Tolerates both acid and alkaline soil with no particular preference. Requires moisture to establish and then will tolerate drier soil conditions.

Hedera helix 'Buttercup'

Sun/Shade aspect All but the most exposed aspects. Full sun to medium shade, although sun improves colour. In early to late summer, some shade protection should be given during midday sun to prevent scorching.

Pruning Normally requires no pruning for its own well-being but can be cut back in late winter/early spring if required.

Training Self-clinging pads attach to bricks. On fences will require wires or individual anchor points to assist training.

Propagation and nursery production From semi-ripe cuttings taken in early summer. Should always be purchased container grown; best planting height 1–3 ft (30–91 cm). Normally available from good garden centres and nurseries.

Problems May scorch in direct, strong, midday sunlight, but nevertheless will require sunny conditions to maintain good golden colouring. Although these two elements may seem contradictory, they are normally possible with a little attention to detail.

Similar forms of interest *H. h.* 'Clotted Cream' Soft cream-coloured foliage all year round. *H. h.* 'Light Fingers' (syn. 'Tampa Gold') All golden yellow foliaged leaves, may be susceptible to scorching in strong sunlight.

Average height and spread
Five years
8 × 8 ft (2.4 × 2.4 m)
Ten years
16 × 16 ft (4.9 × 4.9 m)
Twenty years
20 × 20 ft (6 × 6 m)
Protrudes up to 12 in (30 cm) from support.

HEDERA HELIX 'CRISTATA'

CRESTED IVY, PARSLEY CRESTED IVY

Araliaceae   *Woody Climber*
Evergreen

Leaves uncharacteristic for its species, offering an attractive alternative to the more familiar ivies.

Origin Sport of common English ivy.
Use As an attractive winter-foliaged shrub for walls, fences, pillars and pergolas.

Hedera helix 'Cristata'

Description *Flower* Round clusters of dark green flower buds open to lighter green in late winter/spring, attracting bees and other insects. *Foliage* Uncharacteristically round, with curled edges; 2–2½ in (5–6 cm) in diameter, shiny, pale-green, turning red on both sides, with more pronounced red blotching on upper surface in cold winters. *Stem* Twining. Brown/green to red/brown when young, becoming grey/green, finally green/brown. Medium rate of growth. *Fruit* Rarely fruits but may produce the round dark red fruits typical of ivy on mature shrubs.

Hardiness Tolerates a minimum winter temperature of 0°F (−18°C).

Soil requirements No particular preference, tolerating both acid and alkaline conditions. Will tolerate dry types, once established.

Sun/Shade aspect Tolerates all but the most exposed aspects. Prefers light shade but will tolerate full sun to deep shade, although in deep shade may become more lax in habit.

Pruning Requires no pruning for its own well-being but can be reduced in size if required, normally in spring.

Training Partially self-clinging. Allow to ramble over wires and trellis, secure individual branches to anchor points.

Propagation and nursery production From semi-ripe cuttings taken in early summer. Purchase container grown all the year round or plant those purchased as house plants, once hardened off. Will have to be sought from specialist nurseries and garden centres. Best planting height 1–3 ft (30–91 cm).

Problems Can become invasive. May be slightly difficult to obtain.

Similar forms of interest None

Average height and spread
Five years
8 × 8 ft (2.4 × 2.4 m)
Ten years
12 × 12 ft (3.7 × 3.7 m)
Twenty years
16 × 16 ft (4.9 × 4.9 m)
Protrudes up to 2 ft (60 cm) from support.

HEDERA HELIX 'GLACIER'

KNOWN BY BOTANICAL NAME

Araliaceae   *Woody Climber*
Evergreen

A useful variegated climber with attractive winter colour brightening the dullest of corners.

Origin Of garden origin.
Use A variegated, close-growing, self-clinging climber for most situations including exposed. Can be used as a spreading ground cover.

Description *Flower* Small clusters, 1½ in (4 cm) across, of green-yellow flower heads produced on very mature plants from autumn through until spring, not normally of great attraction. *Foliage* Three- to five-fingered triangular leaves up to 1½ in (4 cm) across and 1¾ in (4.5 cm) long; silver/grey/green with silvery white edges. *Stem* Self-clinging, tight formation, attractive red colouring on new growth, older wood becoming grey/green to grey/brown. Upright to fan-shaped formation on wall. Medium rate of growth. *Fruits* Clusters of poisonous black to blue/black fruits in spring on mature plants.

Hedera helix 'Glacier'

Hardiness Tolerates a minimum winter temperature of 4°F (−15°C). Hardy, although some juvenile and adult foliage may be damaged in severe wind chill conditions; rejuvenation is normal the following spring.

Soil requirements Does well on all soil types, only disliking extremely dry soil conditions.

Sun/Shade aspect All but the most exposed aspect. Medium shade to full sun.

Pruning Requires no pruning for its own well-being but can be reduced in size if required, in early to mid spring.

Training Self-clinging to most types of support; may need some tying in when young.

Propagation and nursery production From semi-ripe cuttings taken in early summer or from self-rooted layers. Always purchase container grown, in early autumn to early summer; readily available from most good garden centres and nurseries. Best planting height 9 in–3 ft (23–91 cm).

Problems In extremely dry summers may show some signs of distress by foliage drying; will normally rejuvenate the following spring.
Similar forms of interest *H. h.* **'Adam'** Silver variegated with the edges of leaves turning pink in extremely cold weather. *H. h.* **'Cavendishii'** (syns. *H. h.* **'Tricolor'**, *H. h.* **'Silver Sheen'**) White to silver variegation, some pink colouring in winter. *H. h.* **'Colebri'** Foliage mottled silver, attractive. *H. h.* **'Heise'** Silver variegation, more bushy habit.
Average height and spread
Five years
5 × 5 ft (1.5 × 1.5 m)
Ten years
10 × 10 ft (3 × 3 m)
Twenty years
20 × 20 ft (6 × 6 m)
Protrudes up to 12 in (30 cm) from support.

HEDERA HELIX 'GOLD HEART' (*H. h.* 'Jubilee')

GOLD HEART IVY
Araliaceae   *Woody Climber*
Evergreen

Rightly amongst the most popular of all the small-leaved golden variegated ivies.

Origin Possibly of Italian garden origin.
Use As a self-clinging climber for walls, will cling to timber fences and timber posts but will need some support. When young can be used as a spreading ground-cover.
Description *Flower* Small clusters, 1½ in (4 cm) across, of green-yellow flower heads produced on mature shrubs from autumn through until spring. *Foliage* Juvenile foliage three- or five-pointed, very regular in shape, dark green with golden splashes of yellow in centre; adult foliage often has no indentations and variegation is more irregular. *Stem* Smooth light green ageing to dark green, finally to grey/brown. Very close formation with little or no forward growth except on very mature plants. Small, suckering climbing aerial roots produced from underside of branches. Medium rate of growth. *Fruit*

Mature plants may produce clusters of poisonous black to blue/black round fruits in spring.
Hardiness Tolerates a minimum winter temperature of 4°F (−15°C). Hardy, although some juvenile and adult foliage may be damaged in severe wind chill conditions; rejuvenation is normal the following spring.
Soil requirements Does well on both alkaline and acid soils, requires adequate moisture for establishment and good growth.
Sun/Shade aspect All aspects. Medium shade to full sun with no particular preference.
Pruning Remove any very mature forward-facing shoots and any green reverting shoots. Otherwise prune to maintain within desired area in early to late spring.
Training It normally requires no support on stone and brick walls, but on timber fences and posts it may need some assistance by tying in when young.
Propagation and nursery production From semi-ripe cuttings taken in early to mid summer or by self-propagated layers at ground level. Always purchase container grown, in early autumn to early summer; readily obtainable from all garden centres and nurseries. Best planting height 6 in–3 ft (15–91 cm).
Problems May, in some situations, become slightly invasive and allowance must be made for its ultimate size. On acid soils there will be a greater degree of reversion and production of green shoots, which must be removed.
Similar forms of interest *H. h.* **'Chicago Variegata'** Leaves green with cream variegation below; average-sized leaves, three- to five-lobed, soft textured on strong-growing climbers. *H. h.* **'Eva'** Interesting, small, pointed-lobed, green, cream-edged leaves. Bushy habit. *H. h.* **'Gold Child'** Foliage golden-edged, may have to be sought from specialist nurseries. *H. h.* **'Herald'** Shiny yellow/green variegated leaves of average size. Yellow ageing to cream.
Average height and spread
Five years
12 × 5 ft (3.7 × 1.5 m)
Ten years
15 × 10 ft (4.6 × 3 m)
Twenty years
25 × 20 ft (7.6 × 6 m)
Protrudes up to 12 in (30 cm) from support.

Hedera helix 'Gold Heart'

HEDERA HELIX 'LUZII'

GOLDEN SPECKLED IVY
Araliaceae   *Woody Climber*
Evergreen

An interesting colour variation within the variegated ivies, often attractive all year round but of particular interest in winter.

Origin Of garden origin.
Use As an evergreen variegated climber to cover walls, fences, pillars and banks.
Description *Flower* Round ball-shaped clusters of dark green flower buds, opening to light green in late winter/spring. *Foliage* Leaves 1½–1¾ in (4–4.5 cm) wide and long, three- to five-lobed. Light gold to light yellow leaves with green blotching and marbled effect. Very attractive in winter. *Stem* Grey/green ageing to grey/brown. Twining, self-clinging, but may require additional support. Medium rate of growth. *Fruit* Mature shrubs may, in hot summers, produce small clusters of round black fruits but not reliable.
Hardiness Tolerates a minimum winter temperature of 14°F (−10°C).
Soil requirements Does well on all soil conditions but requires good moisture to establish.
Sun/Shade aspect All but the most exposed aspects. Full sun to very light shade; deep shade will spoil coloration.
Pruning None required but can be cut back to contain in early spring, if required.
Training Partially self-clinging, may need support by tying in when young or heavy at maturity.

Hedera helix 'Luzii'

Propagation and nursery production From semi-ripe cuttings taken in early summer. Always purchase container grown; available from good garden centres and nurseries. Plants purchased for the house can be planted in the garden between late spring and late summer and will usually establish. Best planting height 1–3 ft (30–91 cm).
Problems May suffer some summer scorching if adequate moisture is not available.
Similar forms of interest *H. h.* **'Bodil'** White and yellow leaves blotched with green; three pointed lobes. *H. h.* **'Hibernica Pallida'** Cream mottling on shiny, five-lobed, blunt-ended, dark green leaves. *H. h.* **'Masquerade'** Yellow blotched variegation. Small leaved. *H. h.* **'Sicilia'** Attractive crinkly edged leaves with creamy yellow variegation ageing to white.
Average height and spread
Five years
5 × 5 ft (1.5 × 1.5 m)
Ten years
10 × 10 ft (3 × 3 m)
Twenty years
15 × 15 ft (4.6 × 4.6 m)
Protrudes up to 18 in (45 cm) from support.

Hedera helix sagittifolia

HEDERA HELIX SAGITTIFOLIA

CROW'S FOOT IVY, ARROWHEAD IVY

Araliaceae   *Woody Climber*
Evergreen

A compact, fan-shaped, evergreen climber with attractive and interesting foliage.

Origin Possibly of Italian garden origin.
Use As a close-clinging, spreading climber for all walls, including exposed situations.
Description *Flower* Only very mature plants may occasionally produce typical ivy flowers. *Foliage* Five-fingered, arrow-shaped leaves 3 in (7.5 cm) long by 2½ in (6 cm) wide, forming an overall triangular shape. Mid to dark green with some light purple or silver veining, depending on season. **Stem** Very close, self-clinging, purple-green when young becoming dark green and finally brown/green. Upright at first, becoming spreading with age. Slow to medium growing. **Fruits** Rarely fruits.
Hardiness Tolerates a minimum winter temperature of 4°F (−15°C).
Soil requirements Does well on all soil types, only resenting extremely dry conditions until established.
Sun/Shade aspect All but the most exposed positions. Full sun to deep shade.
Pruning Normally requires no pruning but can be cut back if required.
Training Allow to spread by self-clinging stems to the area it is required to cover.

Hedera helix sagittifolia 'Variegata'

Propagation and nursery production From semi-ripe cuttings taken in early summer or from self-rooted layers. Always purchase container grown, in early autumn to early summer; will have to be sought from specialist nurseries and good garden centres. Best planting height 1–3 ft (30–91 cm).
Problems Slow to establish, can take up to three years to form a good-sized covering plant.
Similar forms of interest *H. h. s.* 'Variegata' (syn. *H. h. s.* 'Konigers Variegated') Cream and green variegated form. Very attractive. *H. h. pedata* 'Heron' Somewhat sparse in its green foliage presentation. *H. h. caenwoodiana* (syn. *H. h.* 'Grey Arrow') Grey/green leaves, attractive white veins down centre, upright growth. *H. h.* 'Silver Diamond' Attractive distinctive, silver/white-edged foliage.

Average height and spread
Five years
4 × 2 ft (1.2 m × 60 cm)
Ten years
8 × 4 ft (2.4 × 1.2 m)
Twenty years
16 × 8 ft (4.9 × 2.4 m)
Protrudes up to 6 in (15 cm) from support.

HEDERA (Shrubby Forms)

SHRUBBY IVIES

Araliaceae   *Wall Shrub*
Evergreen

Attractive and useful upright evergreen shrubs when used against walls and fences.

Origin From various areas of the northern hemisphere.
Use As slow-growing, upright, shrub-forming evergreens; suitable for growing against walls and fences. Ideal for under windows.
Description *Flower* 2 in (5 cm) wide heads of green to lime-green flowers, borne mainly in early to late spring but either in bud or open throughout year. *Foliage* Leaves small to large, diamond-shaped, 1½–5 in (4–12 cm) long, with glossy upper surfaces, duller underside, ranging from dark through mid green to golden or silver variegated forms. **Stem** Densely foliaged, upright stems. Some varieties with slightly looser habit. Slow growth rate. **Fruit** Round clusters of black poisonous fruits, 1½ in (4 cm) wide, in autumn and winter.
Hardiness Tolerates a minimum winter temperature of 4°F (−15°C).
Soil requirements Any soil, including very alkaline and very acid. Tolerates both dry and wet areas, although forms more growth on moister soils. In very dry areas should be watered to help it establish itself.

Sun/Shade aspect Tolerates all but the most exposed aspects. Prefers medium shade but good in deep shade through to full sun.
Pruning None required, but can be cut to contain or train. Best done in early spring.
Training Self-clinging. Normally needs no additional support but the larger leaved varieties may require individual anchor points to prevent damage from snow or heavy rain.
Propagation and nursery production From semi-ripe cuttings or rooted natural layers. Purchase container grown; best planting height 4–15 in (10–38 cm). May be difficult to find.
Problems Can be slow to attain any substantial size.

Hedera colchica arborescens

Similar forms of interest *H. colchica arborescens* (Shrubby Persian ivy) Foliage light green, oblong to ovate, up to 3 in (7.5 cm) in width and length, forming a dense, upright wall shrub. Good for flower and fruit production. *H. c. a.* 'Variegata' Attractive, white-edged variegated foliage. *H. helix* 'Arborescens' (Shrubby common ivy) Dark green, broad, spreading, tightly growing in fan shape. *H. h.* 'Conglomerata' (Shrubby common ivy) Small foliage, no more than 1–1½ in (2.5–4 cm) in width, lobed with wavy edges. Dark green upper surface, silvery underside. Forms a tight, fan-trained shape or column to be grown adjacent to a wall or fence. Very slow growth rate.

Average height and spread
Five years
3 × 3 ft (91 × 91 cm)
Ten years
5 × 5 ft (1.5 × 1.5 m)
Twenty years
7 × 7 ft (2.1 × 2.1 m)
Protrudes up to 2 ft (60 cm) from wall or fence.

HEDYSARUM MULTIJUGUM

KNOWN BY BOTANICAL NAME

Leguminosae   *Wall Shrub*
Deciduous

An attractive shrub in which rose-purple flowers are enhanced by soft green foliage.

Origin From Mongolia.
Use As an interesting fan-trained summer flowering shrub on walls and fences where it can gain protection in more hostile winter areas.
Description *Flowers* Racemes 4 in (10 cm) long by 2 in (5 cm) wide of openly spaced, small, rose-purple pea-flowers, borne mainly during mid to late summer but produced intermittently into early autumn. **Foliage** Leaves pinnate, 4–6 in (10–15 cm) long, with narrow leaflets, grey/green to sea-green, some yellow autumn colour. **Stem** Upright at first, becoming arching with age. Grey/green

Hedysarum multijugum in flower

to green. Growth rate fast when new, slowing with age. **Fruit** Insignificant.
Hardiness Tolerates a minimum winter temperature of 14°F (−10°C) but stems may die back in winter. Rejuvenation from base usually occurs the following spring.
Soil requirements Prefers light, open soil, dislikes extremely waterlogged, but tolerates both alkalinity and acidity.
Sun/Shade aspect Requires a sheltered aspect in full sun to very light shade.
Pruning None required but one third or more of oldest wood may be removed in early spring to encourage new shoots for flowering in late summer.
Training Secure by individual anchor points or wires.
Propagation and nursery production From softwood cuttings taken in early summer. Purchase container grown from specialist nurseries; best planting height 1–2 ft (30–60 cm).
Problems Rather slow to establish. In areas where its hardiness may be suspect, best not planted before late spring.
Similar forms of interest None.
Average height and spread
Five years
7 × 7 ft (2.1 × 2.1 m)
Ten years
10 × 10 ft (3 × 3 m)
Twenty years
12 × 12 ft (3.7 × 3.7 m)
Protrudes up to 2 ft (60 cm) from support.

HOHERIA

KNOWN BY BOTANICAL NAME

Malvaceae *Wall Shrub*
Deciduous

An interesting, very free-flowering, handsome shrub.

Origin From New Zealand.
Use As a tall, mid summer flowering shrub benefiting in winter from the protection of a wall or fence.
Description *Flower* Clusters of round, white, saucer-shaped, fragrant flowers in mid summer. *Foliage* Leaves ovate, 2–4½ in (5–11 cm) long, sometimes toothed and lobed, especially when young. Grey to grey/green, giving some yellow autumn colour; some variegated and purple-leaved forms. *Stem* Upright, becoming branching and spreading with age, forming a tall, upright wall shrub. Grey/green to grey/brown. Medium growth rate. *Fruit* Insignificant.

Hardiness Tolerates a minimum winter temperature of 4°F (−15°C).
Soil requirements Prefers deep, rich, light soil, acid to neutral but tolerates some alkalinity.
Sun/Shade aspect Requires a sheltered aspect. Prefers light shade but will tolerate full sun.
Pruning None required other than for training or removing any over-sized branches.
Training Allow to stand free or secure to wall or fence by individual anchor points or wires.
Propagation and nursery production From softwood cuttings or layers. Purchase container grown; all varieties must be sought from specialist nurseries. Best planting height 15 in–3 ft (38–91 cm).
Problems Appears quite tough, so its tenderness is not always appreciated. Stems can be damaged quite severely in winter.
Forms of interest *H. lyallii* (syn. *Gaya lyallii*, *G. lyallii ribifolia*, *Plagianthus lyallii*) Clusters of white flowers borne in profusion in mid summer. Foliage more glabrous grey than most. Somewhat variable in growth, forming a large fan-trained shrub. *H. populnea* Pure white flowers 1 in (2.5 cm) across, in large clusters, flowering later than most varieties. Large, broad, ovate leaves, very similar to poplar leaves. *H. p.* 'Foliis

Purpureis' White flowers. Leaves plum-coloured on undersides. Very scarce. *H. p.* **'Variegata'** White flowers. Leaves yellow/green, ageing to white with deep green margins. Benefits greatly from being grown against a wall or fence. Very scarce.
Average height and spread
Five years
6 × 6 ft (1.8 × 1.8 m)
Ten years
9 × 9 ft (2.7 × 2.7 m)
Twenty years
15 × 15 ft (4.6 × 4.6 m)
Protrudes up to 4 ft (1.2 cm) from support.

HOYA CARNOSA

WAX FLOWER
Asclepiadaceae *Tender Greenhouse Climber*
Evergreen

An attractive greenhouse or conservatory climber which is not winter hardy and always needs protection.

Origin From Queensland, Australia.
Use As a climbing vine for heated greenhouses and conservatories.
Description *Flower* Round clusters 3 in (7.5 cm) wide of waxy, very fragrant pink/white flowers throughout summer. *Foliage* Ovate, pointed, 3 in (7.5 cm) long and 1½ in (4 cm) wide, leathery, thick leaves, grey/green with slight purple hue. *Stem* Round, brown/green, trailing, limited branching, supporting by twisting. Slow to medium rate of growth. *Fruit* No fruit of interest.
Hardiness Not winter hardy. Tolerates a temperature of 32°F (0°C) but under protection.
Soil requirements If grown in large containers, a good quality potting compost is required. If grown in soil in greenhouse beds, add liberal quantities of sharp sand and sedge peat to lighten.
Sun/Shade aspect Must be under protection of greenhouse or conservatory. Light shade to full sun.
Pruning Remove overcrowded shoots in spring.
Training Allow to twine around wire or timber supports.
Propagation and nursery production From cuttings of previous year's growth taken in early to late spring and encouraged to root in a protecting frame with some bottom heat, or by layering from the parent plant into a pot containing good-quality potting compost; the latter method normally takes two to three years for rooting to occur. Always purchase container grown; will normally have to be

Hoheria lyalli in flower

Hoya carnosa in flower

sought from specialist nurseries or flower shops selling house plants. Best planting height 6 in–3 ft (15–91 cm).

Problems Its availability may be scarce. The correct temperature balance must be achieved to encourage good flowering; it does not necessarily need heat, but it must not be allowed to be frosted.

Similar forms of interest *H. carnosa* 'Variegata' A form with variegated leaves. Extremely scarce, but worth looking for.

Average height and spread
Five years
7 × 7 ft (2.1 × 2.1 m)
Ten years
14 × 14 ft (4.3 × 4.3 m)
Twenty years
20 × 20 ft (6 × 6 m)
Protrudes up to 6 in (15 cm) from support.

HUMULUS LUPULUS 'AUREUS' (*H. l. 'Luteus'*)

GOLDEN HOP
Urticaceae   *Perennial Climber*
Deciduous

Amongst one of the most attractive of all fast-growing, hardy perennials with attractive golden foliage and yellow/green fruits.

Origin Most of Europe, Asia and North America.
Use As a fast growing, golden foliaged perennial useful for growing over walls, buildings and trees and up poles. Good when used to cover a 10–12 ft (3–3.7 m) high tripod of poles. The young shoots can be blanched and used as a culinary herb.
Description *Flower* Green-yellow male flowers are in small panicles and the female flowers are in spikes; neither are of any real attraction. *Foliage* Leaves trifoliate, each leaflet round to oval, deeply veined with toothed edges, 5 in (12 cm) wide and 6 in (15 cm) long. *Stem* Light grey/green, twining, self-supporting, ageing to grey/brown. Dies back in winter. Very fast growing. *Fruit* Green panicles of round fruits each consisting of many overlapping scales, up to 10–12 in (25–30 cm) in size, ageing to yellow/brown. Can be used for brewing of beer.
Hardiness Root clumps hardy, all growth above ground level dies in winter to be replaced in following spring.
Soil requirements Does well on acid, neutral and moderately alkaline soils, may show some distress on extremely alkaline types. Requires a deep, moist, well-fed soil to achieve best results.
Sun/Shade aspect All but the most exposed aspects. Full sun to very light shade to maintain golden foliage.

Pruning All top growth should be removed to within 9 in (23 cm) of ground level in autumn to afford winter protection, remaining 9 in (23 cm) removed to ground level the following spring.
Training Requires some form of easily accessible wire, pole or branch to twine on.
Propagation and nursery production By division of root-stools in spring. Purchase bare-rooted in early spring or container grown in spring and early summer, the latter for preference. Will have to be sought from specialist nurseries.

Humulus lupulus 'Aureus' in leaf

Problems Can often over-exceed the intended area allocated for it. In all but light shade, the leaves will turn green. The need to cut down a large amount of growth each year could be considered a disadvantage.
Similar forms of interest *Humulus lupulus* The green leaf brewer's hop with attractive yellow autumn fruits, not readily available outside hop growing areas.
Average height and spread
Five years
12 × 12 ft (3.7 × 3.7 m)
Ten years
20 × 20 ft (6 × 6 m)
Twenty years
25 × 25 ft (7.6 × 7.6 m)
Protrudes up to 2½ ft (76 cm) from support.

HYDRANGEA (Large-leaved Forms)

LARGE-LEAVED HYDRANGEA
Hydrangeaceae   *Wall Shrub*
Deciduous

The aristocrats of the hydrangea group and a truly delightful sight when in full flower in late summer.

Origin From China, Taiwan and the Himalayas.
Use As a wall or fence fan-trained shrub where flowers show to best advantage.
Description *Flower* Large, porcelain blue clusters with ring of lilac, pink or white florets on outer edges in late summer, maintained into early autumn. *Foliage* Leaves large to very large, 4–9 in (10–23 cm) long and 3–6 in (7.5–15 cm) wide, ovate, purple/green, down-covered with silver undersides. *Stem* Cream and brown to cream and purple/brown stems. Upright when young, producing attractive coloured bark and becoming branching, gnarled and spreading, forming a fan-shape with training. Stem formation and colour provides interest in winter. Medium growth rate. *Fruit* None, but may retain old flower-heads with ray florets turning brown.
Hardiness Tolerates a minimum winter temperature of 14°F (−10°C).
Soil requirements Neutral to acid soil, tolerates some alkalinity.
Sun/Shade aspect Requires a sheltered aspect in medium to light shade. Dislikes full sun and deep shade may spoil shape.
Pruning None required and may resent it.
Training Train branches when young into a fan-shaped shrub, tying to individual anchor points or to wires.
Propagation and nursery production From semi-ripe cuttings taken in early summer. Purchase container grown; rather hard to find, must be sought from specialist nurseries. Best planting height 15 in–2½ ft (38–76 cm).
Problems Often looks uninteresting when purchased but grows rapidly once planted out.
Forms of interest *H. aspera* Purple/blue florets around central light blue clusters. Foliage purple/green, down-covered. *H. sargentiana* (Sargent's hydrangea) Foliage up to 6–10 in (15–25 cm) long. Flat flower up to 8 in (20 cm) wide. Blue central fertile petals, outer white ray florets, slightly pink on alkaline soils. Mid summer.
Average height and spread
Five years
5 × 5 ft (1.5 × 1.5 m)
Ten years
8 × 8 ft (2.4 × 2.4 m)
Twenty years
12 × 12 ft (3.7 × 3.7 m)
Protrudes up to 3 ft (91 cm) from support.

Hydrangea aspera in flower

HYDRANGEA PANICULATA

PANICK HYDRANGEA

Hydrangeaceae   *Wall Shrub*
Deciduous

Spectacular, late summer flowering shrub, requiring specific pruning to do well.

Origin From Japan, China and Taiwan.
Use As a fan-trained shrub for walls and fences.
Description *Flower* Large panicles of sterile white bracts, mid to late summer, fading to pink, finally to brown. *Foliage* Leaves broad, ovate, 3–6 in (7.5–15 cm) long and 1½–3 in (4–7.5 cm) wide, light green, giving yellow autumn colour. *Stem* Upright and brittle. Light green to green/brown, becoming mahogany-brown with age. Fast growth rate. *Fruit* None, but dead flowerheads maintained into early winter.
Hardiness Tolerates a minimum winter temperature of 4°F (−15°C).

Hydrangea paniculata 'Tardiva' in flower

Soil requirements Prefers rich, deep soil, tolerates a wide range including alkaline or acid, but dislikes extreme alkalinity.
Sun/Shade aspect Tolerates all but the most exposed aspects. Prefers light shade but will tolerate full sun. Deep shade spoils shape.
Pruning Once a fan framework of branches has been formed, ensure new growth and improved flowering by cutting back hard each spring, reducing previous year's wood back to two buds from point of origin, otherwise the shrub becomes old, woody and very brittle, with small flowers.
Training As new shoots emerge tie them to wires or individual anchor points in fan-trained formation. The reduction of the previous season's growth by pruning will entail retying each year.
Propagation and nursery production From softwood cuttings taken in early summer. Purchase container grown or root-balled (balled-and-burlapped). Easy to find, especially when in flower. Best planting height 15 in–3 ft (38–91 cm).
Problems As the shrub is very brittle, take care to avoid damage.
Similar forms of interest *H. p.* **'Grandiflora'** (Pee gee hydrangea) Very large white panicles of flowers, fading to pink. The most common form. *H. p.* **'Kyushu'** Large white flowers. Said to be an improvement on *H. p.* 'Grandiflora'. *H. p.* **'Praecox'** Earlier, white flowers, although slightly less profuse, larger foliage and stronger stems. *H. p.* **'Tardiva'** A later-flowering form, with slightly more open flowers and a wider base to the panicles.
Average height and spread
Five years
6 × 6 ft (1.8 × 1.8 m)
Ten years
10 × 10 ft (3 × 3 m)
Twenty years
13 × 13 ft (4 × 4 m)
Protrudes up to 2 ft (60 cm) from support.

HYDRANGEA PETIOLARIS
(*H. p. scandens*)

CLIMBING HYDRANGEA

Hydrangeaceae   *Wall Shrub*
Deciduous

Rightly deserving its widespread use as a vigorous wall shrub but its full potential is not always appreciated.

Origin From Japan.
Use Ideal as a self-clinging climber for exposed, shady walls, for adorning the trunks of trees, or as a free-growing pillar. Will spread to form a ground-cover.
Description *Flower* White, round florets, up to ¾ in (2 cm) wide, form a flat lace-cap flower up to 10 in (25 cm) wide with central clusters of tufted, creamy white, fertile flowers, each consisting of up to 20 stamens in each cluster; profusely borne from June onwards, ageing to pink/brown in late autumn, finally turning brown in winter and retained in good order until late winter. *Foliage* Oval, tapering, slightly curled shape, up to 5 in (12 cm) long and 1½ in (4 cm) wide, light green turning a good yellow in autumn. *Stem* Light green ageing to light brown, becoming mahogany brown in winter with peeling bark; very attractive. Clinging aerial roots produced on undersides of shoots and branches. Slow growing at first, becoming faster with time. *Fruit* No fruits of interest, but brown flowerheads have winter attraction.
Hardiness Tolerates a minimum winter temperature of 4°F (−15°C).
Soil requirements Does well on both alkaline or acid soils but requires adequate moisture for establishment when young.
Sun/Shade aspect All aspects. Medium shade to full sun with no particular preference.
Pruning No pruning required although can be cut back in spring to keep within bounds as required.
Training Self-clinging but young plants will appreciate being tied to some form of cane framework to assist their initial establishment.
Propagation and nursery production From layers. Purchase container grown or root-balled (balled-and-burlapped) from late summer through to early spring; generally available from most good garden centres and nurseries. Best planting height 6 in–3 ft (15–91 cm).
Problems Slow to establish, often taking up to three years to produce any amount of new growth, but from then on quickly establishes. Can become larger than anticipated.
Similer forms of interest None.
Average height and spread
Five years
6 × 6 ft (1.8 × 1.8 m)
Ten years
20 × 20 ft (6 × 6 m)
Twenty years
40 × 40 ft (12 × 12 m)
Protrudes up to 2½ ft (76 cm) from support.

HYDRANGEA QUERCIFOLIA

OAK LEAF HYDRANGEA

Hydrangeaceae   *Wall Shrub*
Deciduous

A very good autumn-flowering shrub with interestingly shaped foliage giving good autumn colours.

Origin From south-eastern USA.
Use Good as a fan-trained climber for shady walls and fences.
Description *Flower* Panicles of white florets produced from mid summer, maintaining their white texture well into early autumn when they begin to turn pink and finally, in winter, brown. *Foliage* Leaves round to ovate, 4–8 in (10–20 cm) long and 2–4 in (5–10 cm) wide, with large pointed lobes simi-

Hydrangea quercifolia in flower

Hydrangea petiolaris in flower

Hydrangea villosa in flower

lar in shape to oak leaves. Dark green, deeply veined, turning vivid orange/red in autumn. *Stem* Upright when young, becoming lax and ranging. Brown, peeling bark. Very brittle in constitution. Medium growth rate. *Fruit* None, but dead flowerheads retained into winter.
Hardiness Tolerates a minimum winter temperature of 14°F (−10°C).
Soil requirements Does well on most soils, both alkaline and acid.
Sun/Shade aspect Tolerates all but the most exposed aspects in full sun to deep shade.
Pruning None, other than that required for training.
Training Secure to individual anchor points or wires to form a fan-trained formation.
Propagation and nursery production From semi-ripe cuttings taken in mid summer. Purchase container grown; easy to find in nurseries, less common in garden centres. Best planting height 15–20 in (38–60 cm).
Problems The wood is brittle. Young plants when purchased look extremely misshapen but, once trained, quickly form an attractive all-round shape.
Similar forms of interest None.
Average height and spread
Five years
6 × 6 ft (1.8 × 1.8 m)
Ten years
8 × 8 ft (2.4 × 2.4 m)
Twenty years
10 × 12 ft (3 × 3.7 m)
Protrudes up to 2 ft (60 cm) from support.

HYDRANGEA VILLOSA
(*H. rosthornii*)

KNOWN BY BOTANICAL NAME

Hydrangeaceae   *Wall Shrub*
Deciduous

Has always been one of the most interesting of hydrangeas but its use as a fan-trained specimen is often overlooked although it is a form in which it does extremely well.

Origin From western China.
Use As a fan-trained shrub for walls and fences.
Description *Flower* Round, flat, lilac-blue clusters with toothed outer marginal sepals up to 6 in (15 cm) wide in late summer to early autumn. May be more pink on alkaline soils. *Foliage* Leaves narrow, elliptic, 4–9 in (10–23 cm) long, purple/green, velvet textured. Some autumn yellow colours. *Stem* Numerous upright stems, surmounted by slight branching, each branch producing a terminal flower. Grey/green when young, ageing to creamy/brown, attractive in winter.

Fast growth rate when young, becoming slower with age. *Fruit* None. Flowers retained into winter as brown flowerheads.
Hardiness Tolerates a minimum winter temperature of 14°F (−10°C). Foliage may be destroyed by late spring frosts, although a wall or fence will offer some protection; recovery is normal.
Soil requirements Prefers acid soil, but tolerates considerable alkalinity. Does best on rich, deep, moist woodland leaf mould.
Sun/Shade aspect Preferably an aspect which gives protection from late spring frosts to prevent foliage damage. Light dappled shade; dislikes full sun. Very useful in exposed, shady positions if wind protection is provided.
Pruning Remove one third of oldest stems on established plants in early spring to encourage good basal rejuvenation and maintain foliage and flower size.
Training Tie to individual anchor points or to wires to achieve a fan shape.
Propagation and nursery production From softwood cuttings taken in early summer or from micropropagated nursery stock. Purchase container grown or root-balled (balled-and-burlapped). Best planting height 15 in–2 ft (38–60 cm).
Problems Stock often appears short and deformed when purchased.
Similar forms of interest None.
Average height and spread
Five years
4 × 4 ft (1.2 × 1.2 m)
Ten years
8 × 8 ft (2.4 × 2.4 m)
Twenty years
12 × 12 ft (3.7 × 3.7 m)
Protrudes up to 3 ft (91 cm) from support.

ILEX × ALTACLAERENSIS

ALTACLAER HOLLY, HIGHCLERE HOLLY

Aquifoliaceae   *Wall Shrub*
Evergreen

A beautiful range of evergreen shrubs with female plants producing interesting berries and foliage.

Origin Basic form raised in Highclere, Berkshire, England as a cross between *I. aquifolium* and *I. perado*; from this came all the varieties.
Use As an attractive shrub for growing against walls and fences where it can be allowed to grow free or can be trimmed and clipped.
Description *Flower* Small clusters of small white flowers with prominent stamens in late spring and early summer. *Foliage* Leaves ovate, 2–4 in (5–10 cm) long and 2–3 in

(5–7.5 cm) wide, some spines at outer lobed edges. Glossy, waxy upper surfaces, dull undersides. Mainly green, but with silver and golden variegated forms. *Stem* With pruning, forms a tight, upright wall shrub for all year round attraction. Light to mid green, glossy. Medium rate of growth. *Fruit* Round berries, green at first, becoming bright glossy red, borne on female plants in autumn and maintained into winter. Male shrubs do not fruit. Male and female specimens must be planted together to achieve fruits on the female.
Hardiness Tolerates a minimum winter temperature of 4°F (−18°C).
Soil requirements Prefers rich, moist, open soil but does well on most types.
Sun/Shade aspect Tolerates all aspects in full sun to medium shade. In deep shade it will lose its compact shape and become sprawling and less likely to fruit.
Pruning None required, but can be reduced in size, or trimmed to shape in early spring.
Training Normally freestanding but may need securing to individual anchor points.
Propagation and nursery production From semi-ripe cuttings taken in early summer. Purchase container grown or root-balled (balled-and-burlapped). Availability varies. Best planting height 1½–4 ft (45 cm–1.2 m).
Problems Although less spiny than most hollies, handling can be uncomfortable. When purchased, young plants often look uninteresting, but shape improves once planted out. Newly planted shrubs may lose their leaves entirely in early spring following planting.

Ilex × altaclaerensis 'Lawsoniana'

Similar forms of interest *I. × a.* 'Atkinsonii' Large, dark green, glossy foliage with spiny edges. A male form often used as a pollinator for other forms. *I. × a.* 'Camelliifolia' Large, camellia-like foliage, red/purple when young, becoming dark green and shiny with age. A female form with large clusters of dark red fruits and purple stems. *I. × a.* 'Golden King' Almost round, shiny, spineless leaves, green with bright yellow to gold margins. A female form producing red to orange/red fruit. One of the finest of all golden variegated hollies. Slightly less than average height. *I. × a.* 'Hodginsii' Dark green, oval leaves with some spines produced irregularly. A male form useful as a pollinator. *I. × a.* 'Lawsoniana' Spineless leaves, dark green with bright yellow centres. A female form with orange/red fruit. Reversion may be a problem and any completely green shoots should be removed at once. *I. × a.* 'Silver Sentinel' Creamy white to yellow margins around pale green to grey/green, narrow, elliptic leaves. Very few spines. A female form with orange/red fruit. Quick-growing and upright in habit. Slightly less than average height and spread.
Average height and spread
Five years
6 × 6 ft (1.8 × 1.8 m)
Ten years
13 × 10 ft (4 × 3 m)
Twenty years
20 × 13 ft (6 × 4 m)
Protrudes up to 13 ft (4 m) from wall or fence.

ILEX AQUIFOLIUM

COMMON HOLLY, ENGLISH HOLLY

Aquifoliaceae   *Wall Shrub*
Evergreen

An extremely interesting and useful range of very hardy evergreen shrubs.

Origin From Europe and western Asia.
Use As an evergreen shrub planted to grow up against a wall or fence, either trimmed and shaped or left free-growing.
Description *Flower* Small clusters of small white flowers with prominent stamens in late spring to early summer. *Foliage* Leaves ovate, lobed, 1–3 in (2.5–7.5 cm) long and 1–2 in (2.5–5 cm) wide, with very spiny edges. Glossy green, golden or silver variegated, depending on variety. *Stem* Upright, becoming very spreading, forming an individual upright shrub against a wall or fence. Mid to dark green, glossy. Fast to medium growth rate when young, slowing with age. *Fruit* Female forms, if pollinated by a male form nearby, produce round, red fruits in autumn, maintaining them into winter.
Hardiness Tolerates a minimum winter temperature of 4°F (−15°C).
Soil requirements Does well on any soil except extremely dry.
Sun/Shade aspect Tolerates all aspects in full sun to medium shade. In deep shade will lose its compact shape, becoming sprawling and less likely to fruit.
Pruning None required, but can be reduced in height and spread, or closely trimmed and shaped, in early spring.
Training Normally freestanding but may need individual anchor points.
Propagation and nursery production From semi-ripe cuttings taken in early summer. Purchase container-grown or root-balled (balled-and-burlapped). Most varieties easy to find in nurseries and some offered through garden centres. Best planting height 1½–4 ft (45 cm–1.2 m).
Problems Newly planted shrubs may lose leaves entirely in early spring following transplanting. Young plants when purchased often look thin and sparse but shape quickly improves once planted out.
Similar forms of interest *I. a.* **'Argenteo-marginata'** (Broad-leaved silver holly) White margins on dark green or grey/green spiny leaves. Normally offered in female form, but can also be found as a male plant so care should be taken to ascertain sex before purchasing. Fruit orange to orange/red in female form, retained into mid winter, but fruit production requires a nearby male shrub for pollination. *I. a.* **'Bacciflava'** (syn. *I. a.* **'Fructu-luteo'**) (Yellow-fruited holly) Female form with dark green to mid green spiny leaves and good bright yellow fruits maintained into winter. *I. a.* **'Golden Milkboy'** (syn. *I. a.* **'Aurea Mediopicta Latifolia'**) Leaves flat in presentation, green with large golden centre variegation and spiny edges. May revert and all-green shoots should be removed when seen. Male form, no fruits. Very attractive foliage. *I. a.* **'Golden Queen'** (syn. *I. a.* **'Aurea Regina'**) Very dark green leaves with broad, bright golden yellow margins. Male form so no fruits. Slightly less than average height and spread. *I. a.* **'Golden van Tol'** Foliage less spiny than most forms, more ovate to lanceolate, and with very good golden margins. Female form producing red fruits, maintained into winter. *I. a.* **'Handsworth New Silver'** Narrow, ovate leaves with creamy white margins around deep green to grey/green centres. Female, producing orange to orange/red fruits. *I. a.* **'J.C. van Tol'** Narrow, ovate, dark green foliage, almost spineless. Female, producing large regular crops of red fruits. *I. a.* **'Madame Briot'** Glossy, dark yellow to gold margins around dark green leaves, presented on purple stems. Female, producing orange/red fruits. *I. a.* **'Myrtifolia Aureomaculata'** Smaller foliage than most, narrow and ovate, very dark green with central golden variegation. Male form, no fruit. Slightly less than average height and spread. *I. a.* **'Silver Milkboy'** Flattish, spiny leaves, dark green with central creamy white variegation. Some reversion may occur and all-green shoots should be removed when seen. Male form, no fruit. Slightly less than average height and spread. *I. a.* **'Silver Queen'** Young shoots purple to black in colour. Creamy white margins to dark green leaves. Male form, no fruit.
Average height and spread
Five years
6 × 6 ft (1.8 × 1.8 m)
Ten years
14 × 14 ft (4.3 × 4.3 m)
Twenty years
20 × 20 ft (6 × 6 m)
Protrudes up to 20 ft (6 m) from wall or fence.

Indigofera heterantha in flower

INDIGOFERA HETERANTHA
(*I. gerardiana*)

INDIGO BUSH

Leguminosae   *Wall Shrub*
Deciduous

An attractive, mid to late summer flowering shrub useful for its ability to adapt to dry conditions.

Origin From the north-western Himalayas.
Use As a mid summer, early autumn flowering fan-trained or freestanding wall or fence shrub, especially suitable for colder areas.
Description *Flower* Moderately open racemes of purple/pink pea-flowers in mid summer through to early autumn. *Foliage* Leaves pinnate, 2–4 in (5–10 cm) long with 13–21 leaflets up to ½ in (1 cm) long. Grey/green, giving some good yellow autumn colour. *Stem* Long, arching, becoming twiggy in second year in milder areas. Grey/green. Fast growth rate when young, slowing with age. *Fruit* May produce small, grey/green pea-pods of some winter interest.
Hardiness Tolerates a minimum winter temperature of 14°F (−10°C). Stems may die back to ground level in severe winters, but normally rejuvenate in the spring.
Soil requirements Does well on all soil conditions, especially dry areas, once established.
Sun/Shade aspect Requires a sheltered aspect in full sun to very light shade.
Pruning If they are not destroyed by winter cold, reduce long arching stems by two thirds or more in spring to encourage new growth.
Training Allow to stand free or secure to wall or fence by individual anchor points or wires.
Propagation and nursery production From softwood cuttings taken in late spring/early summer. Purchase container grown; may need to be sought from specialist nurseries. Best planting height 15 in–2½ ft (38–76 cm).
Problems Often very late to produce new leaves and growth, which may not appear until early summer. May look weak and insipid when purchased.
Similar forms of interest *I. potaninii* Longer racemes of pink flowers, up to 4–5 in (10–12 cm) long, late summer to early autumn. Slightly less than average height and spread. From north-western China.
Average height and spread
Five years
6 × 6 ft (1.8 × 1.8 m)
Ten years
8 × 8 ft (2.4 × 2.4 m)
Twenty years
12 × 12 ft (3.7 × 3.7 m)
Protrudes up to 4 ft (1.2 m) from support if fan-trained, 11 ft (3.4 m) untrained.

Ilex aquifolium 'Golden Queen'

IPOMOEA HEDERACEA

MORNING GLORY

Convolvulacaea   *Annual Climber*
Deciduous

One of the most spectacular of all the annual climbing plants, needing some care for good results, but worth the effort.

Origin From northern Australia.
Use As an annual climber for walls and fences, to ramble over large shrubs and ideal for growing up the south side of conifers or in containers or greenhouses.
Description *Flower* Five intense blue petals make up a 2½ in (6 cm) wide, reflex-mouthed trumpet with white central eye. *Foliage* Broad, ovate, up to 5 in (12 cm) long and 1 in (2.5 cm) wide. Light to mid green, showing off the flowers to good advantage. *Stem* Light green, twining, self-supporting. Fast rate of growth. *Fruit* No fruit of any interest.
Hardiness Not winter hardy. Must be planted when all danger of spring frosts has passed. Will be killed by autumn frosts. Seeds do not overwinter in soil.
Soil requirements Tolerates all soil conditions, but an addition of a good quantity of organic material will aid root run and therefore good growth above ground.
Sun/Shade aspect Requires a sheltered aspect with protection from wind and heavy rain if possible. Can be grown in greenhouses or conservatories. Best in full sun. Will tolerate light shade, although flowering will be reduced the greater the amount of shade.

Ipomoea hederacea **in flower**

Pruning Not required. Is killed in autumn by frosts.
Training Allow to twine over wires, trellis or through the branches of medium to large shrubs. If used for growing up the face of conifers a cane or pole may be needed against the tree to give the vine additional support.
Propagation and nursery production From seeds planted under protection with bottom heat in early spring; do not plant out until all danger of frosts has passed, normally early summer. Always purchase container grown; often found in garden centres in late spring, early summer with the bedding plants. Best planting height 2–4 in (5–10 cm).
Problems Its need for annual planting does make it less appealing than some other climbers, but it is well worth the effort. Can be difficult to establish unless adequate organic material is introduced into the soil prior to planting.
Similar forms of interest None.
Average height and spread
One year
12 × 12 ft (3.7 × 3.7 m)
Protrudes up to 12 in (30 cm) from support.

Itea ilicifolia **in flower**

ITEA ILICIFOLIA

KNOWN BY BOTANICAL NAME

Iteaceae   *Wall Shrub*
Evergreen

One of the most spectacular late summer flowering shrubs to benefit from the protection of a wall or fence in cold conditions.

Origin From central China.
Use As a fan-trained shrub for large walls and fences in sheltered positions.
Description *Flower* Racemes up to 15 in (38 cm) long of fragrant green to green/white flowers, ageing to yellow/green, reminiscent of large catkins, late summer. *Foliage* Leaves lobed, dark green, glossy, ovate, 2–5 in (5–12 cm) long, similar to those of holly, with purple hue underlying base colour and silver undersides. *Stem* Strong and upright when young, becoming weeping with age, forming a cascading effect as a weeping shrub when grown against a wall or fence. Light green to purple/green. Fast growth rate when young, slowing with age. *Fruit* Insignificant.
Hardiness Tolerates a minimum winter temperature of 14°F (−10°C). Leaves easily damaged by severe wind chill.
Soil requirements Does well on all soil types.
Sun/Shade aspect Requires a sheltered aspect. Prefers light shade, tolerates full sun to medium shade.
Pruning None required.
Training Requires wires or individual anchor points to secure and encourage a fan-trained shape.
Propagation and nursery production From semi-ripe cuttings taken in early summer. Purchase container grown from garden centres and specialist nurseries. Best planting height 15 in–2 ft (38–60 cm).
Problems Gives the appearance of poor rooting and irregular shape when young but soon grows away once planted.
Average height and spread
Five years
3 × 3 ft (91 × 91 cm)
Ten years
6 × 6 ft (1.8 × 1.8 m)
Twenty years
12 × 12 ft (3.7 × 3.7 m)
Protrudes up to 3 ft (91 cm) from support.

JASMINUM FRUTICANS

KNOWN BY BOTANICAL NAME

Oleaceae   *Wall Shrub*
Evergreen to semi-evergreen

An interesting, shrubby jasmine, presenting itself more attractively when used against a wall or fence than when free-growing.

Origin From the Mediterranean region.
Use As a freestanding shrub or fan-trained against walls or fences.
Description *Flower* Clusters of small yellow flowers carried in early to mid summer. Little individual attraction but effective en masse. *Foliage* Narrowly lanceolate up to ¾ in (2 cm) long, with rounded ends. Deep green, blue/green towards edges, sparse in number, some yellow autumn colour. *Stem* Light green when young, becoming mid green and finally grey/green. Arching, ranging, semi-rigid. Medium to fast growth. *Fruit* Black, globe-shaped fruits up to ⅛ in (5 mm) wide carried in late summer/early winter.
Hardiness Tolerates a minimum winter temperature of up to 0°F (−18°C).

Jasminum fruticans **in flower**

Soil requirements Tolerates a wide range of soils with no particular preference, except it dislikes very dry or waterlogged conditions.
Sun/Shade aspect Requries a sheltered aspect. Full sun to medium shade, full sun for preference.
Pruning If freestanding, remove one third of oldest growth to ground level after flowering. If fan-trained, do the same as far as is practical.
Training Tie to wires or individual anchor points or allow to grow freestanding.
Propagation and nursery production From semi-ripe cuttings taken in early summer. Should always be purchased container-grown; may have to be sought from specialist nurseries. Best planting height 6 in–3 ft (15–91 cm).
Problems Its open habit does not appeal to all. Flowers and foliage are limited but are of interest.
Similar forms of interest None.
Average height and spread
Five years
4 × 4 ft (1.2 × 1.2 m)
Ten years
8 × 8 ft (2.4 × 2.4 m)
Twenty years
8 × 8 ft (2.4 × 2.4 m)
Protrudes up to 18 in (45 cm) from support if fan-trained, 4 ft (1.2 m) untrained.

JASMINUM HUMILE REVOLUTUM

SHRUBBY JASMINE
Oleaceae   *Wall Shrub*
Evergreen

An attractive, late-summer flowering wall shrub often overlooked and not planted as widely as it deserves.

Origin From the Himalayas.
Use As a freestanding wall shrub for walls and fences where it benefits from the protection offered.

Jasminum humile revolutum **in flower**

Description *Flower* Small, five-petalled, ½ in (1 cm) wide bright yellow flowers are intermittently produced in small clusters through late summer and early autumn. *Foliage* Attractive foliage, bright green, glossy, 1¼ in (3.5 cm) long and ½ in (1 cm) wide. *Stem* Light, bright green. Upright, becoming branching. Produces underground suckers which in some circumstances become invasive. *Fruit* No fruits of any significance.
Hardiness Tolerates a minimum winter temperature of 4°F (−15°C).
Soil requirements Tolerates all soil conditions, with no particular preference, only disliking extremely dry soil when becoming established.

Sun/Shade aspect Moderately sheltered aspect in light shade to full sun.
Pruning Remove one third of oldest growth to ground level in early to mid spring to encourage rejuvenation.
Training Allow to stand free, growing against a wall or fence.
Propagation and nursery production From semi-ripe cuttings taken in early to mid summer. Purchase container grown from good garden centres and specialist nurseries. Best planting height 1–2½ ft (30–76 cm).
Problems Can outgrow the area allowed for it, particularly in a forward direction.
Similar forms of interest None.
Average height and spread
Five years
4 × 3 ft (1.2 m × 91 cm)
Ten years
6 × 6 ft (1.8 × 1.8 m)
Twenty years
8 × 6 ft (2.4 × 1.8 m)
Protrudes up to 6 ft (1.8 m) from wall or fence.

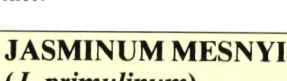

JASMINUM MESNYI
(*J. primulinum*)

PRIMROSE JASMINE
Oleaceae   *Tender Woody Climber*
Deciduous to semi-evergreen

An attractive climber but requires protection in all but the mildest areas.

Origin From China.
Use As a rambling climber for very sunny, hot walls but best grown under the protection of a greenhouse or conservatory.
Description *Flower* Small, solitary flowers are up to 1¾ in (4.5 cm) long, bright yellow, with seven to eight petals carried at each leaf joint during mid to late summer, earlier under protection. *Foliage* Trifoliate, leaflets 1–3 in (2.5–7.5 cm) long and ½–¾ in (1–2 cm) wide. The degree of evergreen will depend on its location. *Stem* Shoots four-angled, bright green, loosely twisting. Very ranging and spreading in habit. Fast growing. *Fruit* Does not normally fruit outside its native environment.
Hardiness Tolerates a minimum winter temperature of 32°F (0°C).
Soil requirements Any average soil as long as adequate moisture is available. If grown in tubs they must be at least 21 in (53 cm) in diameter and filled with a good soil-based potting compost.
Sun/Shade aspect Protected aspect in full sun to light shade.
Pruning Not normally required, but can be reduced in size if necessary.
Training Semi-self-twining, but will need tying to train over wires or similar support system.

Jasminum mesnyi **in flower**

Propagation and nursery production From semi-ripe cuttings taken in early spring to mid summer. Best planting height 1–3 ft (30–91 cm). Always purchase container grown; will have to be sought from specialist nurseries or may be found in florists.
Problems Its tenderness must always be allowed for, and its overall size, particularly when grown in confined spaces under protection.
Similar forms of interest None.
Average height and spread
Five years
6 × 6 ft (1.8 × 1.8 m)
Ten years
12 × 12 ft (3.7 × 3.7 m)
Twenty years
18 × 18 ft (5.5 × 5.5 m)
Protrudes up to 3 ft (91 cm) from support.

JASMINUM NUDIFLORUM

WINTER JASMINE
Oleaceae   *Wall Shrub*
Deciduous

One of the most widely planted of all wall shrubs, well deserving its rightful place as a true gem for its winter flowers.

Origin From western China.
Use As a winter flowering shrub for walls, fences and pillars.
Description *Flower* Five-petalled, ¾ in (2 cm) wide, butter-yellow in colour, carried at the leaf joints in profusion from early winter to early spring whenever mild weather persists. *Foliage* ¾ in (2 cm) long and ¼ in (5 mm) wide, dark bright green, carried in groups of three.

Jasminum nudiflorum **in flower**

Stem Angled, arching, possibly lax in habit, ribbed. Bright green in winter. Medium rate of growth, slowing with age. *Fruit* Insignificant.
Hardiness Tolerates a minimum winter temperature of 4°F (−15°C).
Soil requirements Any soil, often tolerating extremely poor conditions, both acid and alkaline.
Sun/Shade aspect Tolerates all but the most exposed of aspects in full sun through to medium shade and even deep shade in some circumstances.
Pruning Remove one third of old flowering shoots to ground level on shrubs established more than three years as an annual operation to encourage regrowth.
Training Allow to ramble through wires and trellis. May need some initial support by tying in.
Propagation and nursery production From semi-ripe cuttings taken in early summer. Purchase container grown; most forms easy to find. Best planting height 15 in–2½ ft (38–76 cm).
Problems Can often outgrow the desired area because it is assumed to be smaller than it truly is. Some winter die-back may be experienced and flowers may be damaged in severe cold spells, but new buds normally open once warmer conditions arrive.
Similar forms of interest None.
Average height and spread
Five years
5 × 5 ft (1.5 × 1.5 m)
Ten years
10 × 10 ft (3 × 3 m)
Twenty years
10 × 15 ft (3 × 4.6 m)

Protrudes up to 3 ft (91 cm) from support.

JASMINUM OFFICINALE

COMMON JASMINE

Oleaceae　　　　　　*Woody Climber*
Deciduous

A vigorous, sweetly-scented, summer flowering climber to cover large areas.

Origin From Persia, North India and China.
Use As a very free-growing climber, quickly spreading in all directions both sidewards and forward through large shrubs, small trees, over fences, walls and buildings.
Description *Flower* Small clusters of three to eight small white flowers at the ends of branches, each flower made up of four to five ¾ in (2 cm) wide petals, flowering from mid summer to early autumn. Very fragrant. *Foliage* Pinnate leaves 4–6 in (10–15 cm) long and 3 in (7.5 cm) wide with five, seven or nine leaflets, oval in shape and from 1½–2½ in (4–6 cm) long and ½ in (1 cm) wide; end leaflet on short stalk. Mid to grey-green, some good yellow autumn colour. *Stem* Grey-green when young, becoming darker with age, angled, partially twining in habit. Very vigorous, able to produce in excess of 10 ft (3 m) of growth in one year. *Fruit* Produces no fruit of interest.
Hardiness Tolerates a minimum winter temperature of 14°F (−10°C). Tips of new growth often killed in winter but replaced in following spring.
Soil requirements Does well on all soil types but requires adequate moisture to sustain vigorous growth.
Sun/Shade aspect Good in all but the most severe, exposed aspects. Sull sun to very light shade.
Pruning Prune hard in late winter and early spring, cutting back growth as much as is required for training; will reflower on current season's new growth.
Training Allow to ramble through shrubs and trees, provide wires or other suitable support for twining over walls and fences. Long shoots may need some tying in, especially in windy situations.
Propagation and nursery production From semi-ripe cuttings taken in early summer. Always purchase container-grown from early

Jasminum officinale **in flower**

to late summer; best planting height 1½–4 ft (45 cm–1.2 m). Normally available from good garden centres and nurseries.
Problems Its ultimate size and rate of growth are often underestimated. Requires a large area to show itself off to best advantage.
Forms of interest See further entries.
Average height and spread
Five years
12 × 12 ft (3.7 × 3.7 m)
Ten years
24 × 24 ft (7.3 × 7.3 m)
Twenty years
36 × 36 ft (11 × 11 m)

Protrudes up to 4 ft (1.2 m) from support.

JASMINUM OFFICINALE (Variegated Forms)

VARIEGATED SUMMER-FLOWERING JASMINE

Oleaceae　　　　　　*Woody Climber*
Deciduous

Given some protection, these are amongst the finest of both ornamental foliage and summer-flowering climbers.

Origin From north India and China.
Use As a climber for walls, fences, pillars and pergolas where protection can be given. Can be planted in large conservatories and greenhouses.
Description *Flower* White, five-petalled, small flowers up to ½ in (1 cm) across and long, carried in small open clusters. Very frag-

Jasminum officinale **'Aureum' in leaf**

rant. *Foliage* Pinnate with up to five leaflets, each leaflet grey/green, splashed boldly with yellow and golden or silver variegation. More intense on new foliage than old. *Stem* Light grey/green to mid green, ridged. Partially twining. Medium to fast growing. *Fruit* Insignificant.
Hardiness Tolerates a minimum winter temperature of 23°F (−5°C).
Soil requirements Alkaline to acid. Needs a rich, moist soil type.
Sun/Shade aspect Requires a sheltered aspect. Full sun to very light shade.
Pruning Remove one third of oldest growth to ground level in early spring on plants more than three years old. Repeat annually.
Training Semi-twining. Will need some additional support by tying into wires, trellis or individual anchor points when young and when becoming heavy at maturity.
Propagation and nursery production From semi-ripe cuttings taken in early to mid summer. Best planting height 1–3 ft (30–91 cm). Always purchase container grown; will have to be sought from specialist nurseries.
Problems Can be slow to establish, taking up to three years to promote any new growth. Occasionally branches may revert to all-green and should be removed at point of origin as soon as seen. Some frost die-back may be experienced but rejuvenation normally occurs the following spring.
Forms of interest *J. o.* **'Argentea'** White variegated foliage. White, very fragrant flowers. *J. o.* **'Aureum'** Golden variegated foliage. White, very fragrant flowers.
Average height and spread
Five years
5 × 5 ft (1.5 × 1.5 m)
Ten years
10 × 10 ft (3 × 3 m)
Twenty years
18 × 18 ft (5.5 × 5.5 m)
Protrudes up to 2 ft (60 cm) from support.

JASMINUM POLYANTHUM

FLORIST'S JASMINE

Oleaceae　　　　*Tender Woody Climber*
Deciduous, semi-evergreen or evergreen depending on location and cultivation

Although this beautiful climber requires an extremely sheltered position or, ideally, the protection of a conservatory or greenhouse, it is well worth consideration.

Origin From China.
Use As a climber for very sheltered walls in very mild areas, or to adorn the roof of a conservatory or large greenhouse.

Description *Flower* Small, five-petalled flowers up to $\frac{3}{4}$ in (2 cm) long and wide with reflexed petals, white on the inside and rosy-white on the outside. Produced in panicles at the leaf joints, each panicle up to 2–4 in (5–10 cm) long. Very fragrant. Flowers from late spring to late summer, but may be earlier depending on location. *Foliage* Light green, pinnate, with five to seven curling leaflets, each 2 in (5 cm) long. Overall size of leaf 5 in (12 cm) long by 4 in (10 cm) wide. *Stem* Bright green when young ageing to mid-green. Semi-twining. Fast growing. *Fruit* Insignificant.

Jasminum polyanthum in flower

Hardiness Tolerates a minimum winter temperature of 32°F (0°C).
Soil requirements Moderately alkaline to acid, with a good degree of organic material and moisture retention. If grown in pots in conservatories, requires a pot with a diameter of not less than 21 in (53 cm), filled with a good quality soil-based potting compost.
Sun/Shade aspect Must be in a very warm, sheltered position. Full sun to light shade.
Pruning Plants establishd for three years should have one third of the oldest growth removed to ground level in late winter when under protection or in early spring when outside to encourage rejuvenation. Repeat annually.
Training Will twine around wires and timber structures. When young may need some additional support by tying in. May also need tying in for training in a specific direction.
Propagation and nursery production From semi-ripe cuttings taken in late spring and early summer. Always purchase container grown; will have to be sought from specialist nurseries, good garden centres or florists. Can be planted out in the greenhouse or con-

servatory once the plant has finished flowering and it will grow away well. Best planting height 1–3 ft (30–91 cm).
Problems Its tenderness and its ability to outgrow a conservatory or greenhouse in a very short time.
Similar forms of interest None.
Average height and spread
Five years
8 × 8 ft (2.4 × 2.4 m)
Ten years
16 × 16 ft (4.9 × 4.9 m)
Twenty years
30 × 30 ft (9 × 9 m)
Protrudes up to 2 ft (60 cm) from support.

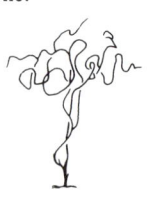

JASMINUM × STEPHANENSE

PINK FLOWERING SUMMER JASMINE
Oleaceae　　　　　*Woody Climber*
Deciduous
An attractive, summer-flowering climber with strongly scented blooms; often sadly overlooked.

Origin Of garden origin.
Use For climbing walls, fences and pillars and for rambling through medium-sized shrubs.
Description *Flower* Five-petalled, pale pink flowers up to $\frac{1}{2}$ in (1 cm) across and long; carried in small open clusters of three to five flowers. Very fragrant. *Foliage* Pinnate, with up to five leaflets 1$\frac{1}{2}$–2 in (4–5 cm) long. Dull green to grey/green, with downy undersides. When young, each leaf has an attractive yellow/orange margin around the edge. *Stem* Light green when young, becoming dark green and eventually green/brown. Smooth, ridged, semi-twining. Medium growth rate. *Fruit* None.
Hardiness Tolerates a minimum winter temperature of 14°F (−10°C). Below this temperature some damage may be caused in winter but rejuvenation is normal in spring.
Soil requirements Alkaline to acid. Moist, rich, well-fed soil for best results.
Sun/Shade aspect Tolerates all but the most exposed aspect. Best in full sun to light shade.
Pruning Remove one third of oldest growth to ground level in early spring to induce rejuvenation on plants more than three years old and continue on annual basis.
Training Will require the support of wires or trellis through which it will twine.
Propagation and nursery production From semi-ripe cuttings taken in early to mid summer. Always purchase container grown; normally available from good garden centres and nurseries. Best planting height 15 in–3 ft (38–91 cm).

Problems Can be slow to establish unless soil conditions are satisfactory.
Similar forms of interest *J. beesianum* A darker red/pink variety.
Average height and spread
Five years
6 × 6 ft (1.8 × 1.8 m)
Ten years
10 × 10 ft (3 × 3 m)
Twenty years
15 × 15 ft (4.6 × 4.6 m)
Protrudes up to 18 in (45 cm) from support.

KERRIA JAPONICA

JAPANESE KERRIA, JEW'S MALLOW, BACHELOR'S BUTTON, SAILOR'S BUTTON
Rosaceae　　　　　*Wall Shrub*
Deciduous
A shrub of attractive flowers, foliage and stems, rather invasive because of its suckering habit but useful for dense cover against walls and fences on poor soils.

Origin From China and Japan.
Use As a spring-flowering shrub that can be grown in close association with walls and fences.
Description *Flower* Yellow buttercup-shaped flowers, single or double, dependent on variety, mid to late spring. *Foliage* Leaves oval with narrow pointed ends, 1$\frac{1}{2}$–3 in (4–7.5 cm) long, bright green, giving some yellow autumn colour. *Stem* Glossy green, strong, upright, spreading. Lateral underground shoots are produced which emerge some distance from parent. Good winter stem attraction if pruned as suggested. Medium to fast rate of growth. *Fruit* Insignificant.
Hardiness Tolerates a minimum winter temperature of 0°F (−18°C).
Soil requirements Any soil type and condition, often tolerating extremely poor areas.

Kerria japonica 'Pleniflora' in flower

Sun/Shade aspect Tolerates all aspects in full sun to medium shade.
Pruning Cut back one third to half of oldest flowering wood to ground level, after flowering.
Training Allow to stand free or secure to wall or fence by individual anchor points or tying to wires.
Propagation and nursery production From rooted underground suckers or from hardwood cuttings taken in winter or semi-ripe cuttings taken in early summer. Purchase container grown; all forms easy to find. Best planting height 2–3 ft (60–91 cm).
Problems Needs space to perform to best advantage.
Similar forms of interest *K. j.* 'Pleniflora' (syn. *K. j.* 'Flore Pleno') Double, golden

Jasminum stephanense in flower

yellow flowers on an upright, vigorous shrub. Good green stems in winter. *K. j.* **'Splendens'** A single-flowered form with buttercup yellow flowers larger than most varieties.

Average height and spread
Five years
10 × 5 ft (3 × 1.5 m)
Ten years
10 × 8 ft (3 × 2.4 m)
Twenty years
10 × 12 ft (3 × 3.7 m)
Protrudes up to 4 ft (1.2 m) from support.

KOLKWITZIA AMABILIS

BEAUTY BUSH

Caprifoliaceae   *Wall Shrub*
Deciduous

An often overlooked shrub for late spring to early summer flowering, but a real beauty, as its common name implies. Adapts well, if given space, as a fan-trained wall shrub.

Origin From western China.
Use As a medium- to large-sized shrub for walls and fences where adequate space can be allowed for it to be fan-trained.
Description *Flower* Bell shaped, ¾ in (2 cm) long and ½ in (1 cm) wide, soft pink with yellow throat, hanging in small clusters 3 in (7.5 cm) wide and long along wood three years old, late spring, early summer and mid summer. *Foliage* Leaves medium to small, ovate, 1–1½ in (2.5–4 cm) long, slightly tooth-edged, light olive-green to grey/green with red shading and silver undersides. Yellow autumn colour. *Stem* Young shoots light green to green/brown, strong and upright, produced mainly from ground level. Growth more than two years old becomes slightly arched, spreading and twiggy. Fast to medium rate of growth. *Fruit* Small, grey/brown, slightly translucent seeds.
Hardiness Tolerates a minimum winter temperature of 4°F (−15°C).
Soil requirements Any soil, no preferences.
Sun/Shade aspect Tolerates all aspects in full sun to light shade.
Pruning Remove one third of older flowering wood by cutting to ground level after the flowering period and retraining in its fan shape.
Training Allow to stand free or secure to wall or fence by individual anchor points and wires.
Propagation and nursery production From soft to semi-ripe cuttings taken mid summer or from hardwood cuttings taken in winter. Plant bare-rooted or container grown. Best planting height 1½–2½ ft (45–76 cm).

Kolkwitzia amabilis **in flower**

Laburnocytisus × *adamii* **in flower**

Problems When purchased, container grown plants present a very fragile appearance but rapidly become robust once planted out.
Similar forms of interest *K. a.* 'Pink Cloud' A cultivar of garden origin with large, strong pink flowers, possibly better for garden planting than its parent.

Average height and spread
Five years
5 × 7 ft (1.5 × 2.1 m)
Ten years
8 × 10 ft (2.4 × 3 m)
Twenty years
10 × 12 ft (3 × 3.7 m)
Protrudes up to 5 ft (1.5 m) from support if fan-trained, 10 ft (3 m) untrained.

LABURNOCYTISUS × ADAMII

PINK LABURNUM

Leguminosae   *Hardy Tree*
Deciduous

A very attractive tree to give interest and variety, adapting well to fan-training.

Origin A graft hybrid of *Laburnum anagyroides* and *Cytisus purpureus* originated in the early 1800s in France.
Use As a fan-trained tree for large walls.
Description *Flower* Individual limbs bear either racemes of yellow *Laburnum anagyroides* flowers 9–12 in (23–30 cm) long or shorter racemes of the pink *Cytisus purpureus* flowers 4–6 in (10–15 cm) long. Some limbs may even present both flowers, one type interspersed with the other. Small areas of true *Cytisus purpureus* may occur in small clusters at the end of older branches. *Foliage* Light grey/green trifoliate leaves, rather untidy. On mature trees tend to be yellow and sickly. Some yellow autumn colour. *Stem* Grey/green, rubbery texture. Slow to medium rate of growth. *Fruit* Small, poisonous pea-pods sometimes produced.
Hardiness Tolerates a minimum winter temperature of 0°F (−18°C).
Soil requirements Does well on most soil conditions; tolerates heavy alkalinity or acid types. Dislikes wet conditions, when root damage may cause poor anchorage.
Sun/Shade aspect Tolerates all aspects. Best in full sun; tolerates light shade.
Pruning Resents pruning, therefore plant only young or bush trees.
Training Tie branches as they appear into a fan-trained shape on wires or individual anchor points.
Propagation and nursery production By grafting or budding on to *Laburnum vulgaris*

understocks. Purchase bare-rooted or container grown. Best planting height: one- or two-year-old grafted trees 3 ft (91 cm) high or bush trees with multiple stems two years old and up to 5 ft (1.5 m) high.
Problems Root systems often feeble at time of purchase.
Similar forms of interest None.

Average height and spread
Five years
7 × 7 ft (2.1 × 2.1 m)
Ten years
14 × 14 ft (4.3 × 4.3 m)
Twenty years
28 × 28 ft (8.5 × 8.5 m)
Protrudes up to 3 ft (91 cm) from support.

LABURNUM × WATERERI 'VOSSII'

GOLDEN CHAIN TREE, WATERER LABURNUM

Leguminosae   *Hardy Tree*
Deciduous

A well-known flowering tree adapting well to fan-training, particularly useful for exposed aspects, but care must be taken with its poisonous fruits.

Origin Of garden origin.
Use As a large fan- or upright-trained flowering tree for walls and fences.
Description *Flower* Long, pendent racemes up to 12 in (30 cm) long of numerous deep yellow to golden-yellow pea-flowers produced

Laburnum × *watereri* **'Vossii' in flower**

287

in late spring. *Foliage* Leaves trifoliate, each leaflet up to 3 in (7.5 cm) long. Grey/green to dark green, glossy upper surfaces and lighter, often hairy undersides. *Stem* Dark glossy green with a slight grey sheen. Strong, upright when young, branching and twiggy with age; finally a spreading tree of medium vigour. Medium rate of growth. *Fruit* Grey/green, pendent pods containing poisonous, black, pea-shaped fruits. Produces fewer seed pods than other laburnum.

Hardiness Tolerates a minimum winter temperature of 0°F (−18°C).

Soil requirements Any soil conditions; tolerates high alkalinity.

Sun/Shade aspect Tolerates all aspects. Full sun to light shade, preferring full sun.

Pruning Prune young trees hard in spring following planting. Select and train resulting five to seven shoots and tie into a fan-trained shape. In subsequent years remove all side growths back to two points from their origin and maintain original main branches in fan shape.

Training Will require tying to wires or individual anchor points.

Propagation and nursery production By grafting. Plant bare-rooted or container grown; stocked by most garden centres and nurseries. Choose bush or young trees of one to two years old.

Problems All laburnums have poor root systems and require permanent staking. Relatively short-lived trees, showing signs of distress after 40 or more years. Poisonous fruits can be dangerous to children.

Similar forms of interest None.

L. alpina (Scotch laburnum) Dark glossy foliage, short racemes of yellow flowers. Adapts well to training.

Average height and spread
Five years
10 × 10 ft (3 × 3 m)
Ten years
20 × 20 ft (6 × 6 m)
Twenty years
28 × 28 ft (8.5 × 8.5 m)
Protrudes up to 3 ft (91 cm) from support.

LAGERSTROEMIA INDICA

CRAPE MYRTLE

Lythraceae **Tender Tree**
Deciduous

An attractive tender tree, benefiting from the protection of a wall both in very warm sheltered areas and when grown under protection of a greenhouse or conservatory.

Origin From China.

Use As a large fan-trained wall shrub for very sheltered positions or under glass.

Description *Flower* Six crescent-shaped crinkle-edged stalked petals make up a round flower, petunia-pink to deep red, carried in panicles 7–8 in (17–20 cm) long and 4–5 in (10–12 cm) wide in summer/early autumn. *Foliage* Lanceolate leaves up to 2½ in (6 cm) long and 1 in (2.5 cm) wide, presented in threes on short stalks. Mid-green, with pronounced veining. Yellow autumn colour. *Stem* Grey/green, rigid, branching, interesting in winter. Medium rate of growth. *Fruit* No fruit of interest.

Hardiness Tolerates a minimum winter temperature of 23°F (−5°C) but requires shelter from cold winds.

Soil requirements Tolerates most soil conditions with no particular preference, except dislikes extreme alkaline types.

Sun/Shade aspect A sheltered aspect in full sun.

Pruning Flowering shoots should be shortened back to within two buds of their origin in early spring to induce new flowering growth.

Training Fan-train on walls by fixing wires or individual anchor points.

Propagation and nursery production From seed or semi-ripe cuttings. Always purchase

---

Lagerstroemia indica in flower

container grown; best planting height 1–3 ft (30–91 cm). Not readily available, will have to be sought from specialist nurseries.

Problems Its tenderness must not be underestimated and it does require a very sheltered position for any degree of success. May be hard to find.

Similar forms of interest *L. i.* 'Alba' Pure white flowers, more tender and even scarcer in commercial production.

Average height and spread
Five years
4 × 4 ft (1.2 × 1.2 m)
Ten years
6 × 6 ft (1.8 × 1.8 m)
Twenty years
10 × 10 ft (3 × 3 m)
Protrudes up to 2 ft (60 cm) from support.

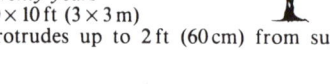

LAPAGERIA ROSEA

KNOWN BY BOTANICAL NAME

Philesiaceae **Tender Woody Climber**
Evergreen

Of all tender climbing plants, has among the most spectacular flowers.

Origin From Chile.

Use As a climbing plant for very sheltered gardens or for use in conservatories and greenhouses where it is possibly more at home.

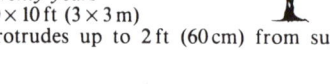

Lapageria rosea 'Nashcourt' in flower

Description *Flower* Large, 3 in (7.5 cm) long, bell-shaped, fleshy, pendent pink flowers from late summer to early autumn. *Foliage* Broad, ovate, pointed leaves, leathery, grey/green, up to 3 in (7.5 cm) long, borne either singly or in twos and threes. *Stem* Grey/green, loosely twining, not self-clinging or supporting. Medium rate of growth. *Fruit* May produce oblong to oval berries containing many seeds in late summer/early autumn.

Hardiness Tolerates a minimum winter temperature of 32°F (0°C).

Soil requirements If grown in containers use a good-quality potting compost. If grown in soil add additional sharp sand and peat to provide well-drained soil with moisture retention.

Sun/Shade aspect Very sheltered aspect or under the protection of a conservatory or greenhouse. Light shade, dislikes full sun. Also requires a moist atmosphere.

Pruning Not normally required but can be cut back if necessary in early spring.

Training Encourage to twine around wires. Ideal when allowed to climb above walkways, where the pendent flowers can be seen to full advantage.

Propagation and nursery production From seed, from cuttings taken from semi-ripe wood in mid summer, or layered. Always purchase container grown; will have to be

Lapageria rosea in flower

sought from specialist nurseries. Best planting height 6 in–3 ft (15–91 cm).

Problems Can be attacked by greenfly, mealy bug, scale insects and thrips. Eradicate with a proprietary control. In some circumstances can become invasive by underground shoots appearing some considerable distance from parent, although this is normally only seen in greenhouse or conservatory beds.

Similar forms of interest *L. r.* 'Albiflora' Pure white flowers. *L. r.* 'Flesh Pink' Mother-of-pearl to very pale pink flowers. *L. r.* 'Nashcourt' Dark pink to red flowers.

Average height and spread
Five years
7 × 7 ft (2.1 × 2.1 m)
Ten years
14 × 14 ft (4.3 × 4.3 m)
Twenty years
20 × 20 ft (6.6 × 6.6 m)
Protrudes up to 6 in (15 cm) from support.

Lathyrus latifolius in flower

LATHYRUS LATIFOLIUS

EVERLASTING PEA

Leguminosae   *Perennial Climber*
Deciduous

When allowed to grow free and wild, one of the most spectacular of flowering climbers.

Origin From Europe.
Use As a climbing perennial for walls, fences, pillars and pergolas but at its best when allowed to ramble through large shrubs. Useful for ground cover with or without support.
Description *Flower* A large back shield, with two wings either side and two forward-facing petals, forming a typical sweet-pea flower, 1½ in (4 cm) wide and deep. Rose-pink in colour and scented, produced from June to September. *Foliage* Grey/green, elliptic, 1½–2 in (4–5 cm) long, normally in pairs, with clinging tendrils from the stalk. Some scent. *Stem* Heavily angled stems, almost square, with ribs. Grey/green to mid green. Ranging. Fast rate of growth. *Fruit* Small, grey/green pea pods containing poisonous black fruits.
Hardiness Roots tolerate a minimum winter temperature of 14°F (−10°C). Top growth dies to ground level in winter.
Soil requirements Tolerates both alkaline and acid conditions. Requires moisture retention and a high degree of organic material for best results.
Sun/Shade aspect Tolerates all aspects. Full sun to light shade for preference.
Pruning Dies to ground level in winter.
Training Tie to wires or individual anchor points or allow to ramble through wires, trellis or large shrubs.
Propagation and nursery production From seed sown in early spring under protection, or from division of root clump. Should always be purchased container grown; relatively easy to find in garden centres and specialist nurseries. Best planting height: either pot-grown clumps in early spring with no top growth, or up to 18–20 in (45–55 cm) tall in later spring, early summer.
Problems Often outgrows the position allocated for it. If happy can be invasive.
Similar forms of interest *L. l. albus* Pure white flowers. Scarce. *L. l. a.* **'Snow Queen'** A very good form. Extremely scarce. *L. l. roseus* Becoming more plentiful. Attractive bright flowers. *L. l. splendens* A form sometimes seen, with large flowers.
Average height and spread
One year
6–10 ft × 5–6 ft
(1.8–3 m × 1.5–1.8 m)
Protrudes up to
2 ft (60 cm) from
support.

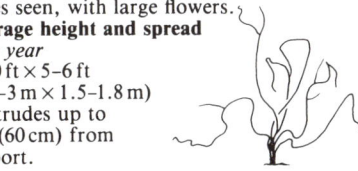

Lathyrus latifolius albus in flower

LATHYRUS ODORATUS

SWEET PEA

Leguminosae   *Annual Climber*
Deciduous

Rightly well loved by every gardener for its summer display of delicately coloured, fragrant flowers.

Origin From Sicily.
Use As a climber for walls, fences, pillars and pergolas, supported by brushwood or twigs, or grown up netting. Allow to ramble through large shrubs. Grown individually for flower size against canes. Use for ground cover where wire supports are provided.
Description *Flower* Two wings stand either side of a large back flower up to 1½ in (4 cm) wide and deep in a range of colours from white, blue, mauve, purple, pink, red and yellow from mid summer to early autumn. Highly fragrant, carried in clusters at the end of long flower stalks. *Foliage* Grey/green, ovate leaves, normally in pairs, the stalk ending in a clinging tendril. *Stem* Angular, grey/green, loosely twining. Fast rate of growth. *Fruit* Small, grey/green pea pods of little attraction containing black, poisonous fruits.
Hardiness Seeds do not overwinter in soil.
Soil requirements Requires a deep, well prepared soil with liberal quantities of organic material, moisture retaining yet well drained. Both alkaline and acid.
Sun/Shade aspect Tolerates most aspects but requires protection from wind and driving rain which will damage the flowers. Full sun to light shade.
Pruning Allow to ramble or reduce to a central cordon stem on which the flowers will be larger and on long flowering stalks. Plants will die with the first winter frosts.
Training Allow to grow up nets, up strings, poles or wires, or to ramble through brushwood. Clings by leaf tendrils, as well as loosely twining habit.
Propagation and nursery production From seed grown under protection and planted out in mid to late spring. Best grown in pots, either singly or in threes using good-quality soil-based potting compost. Seeds may also be sown in late autumn or early winter and overwintered under protection or sown directly into the soil in early spring in their flowering position, although the flowers may not be as large when grown this way. Container grown plants can be purchased from garden centres and nurseries, best planting size 2–4 in (5–10 cm).
Problems May suffer attacks of blackfly, greenfly or mildew. May well run out of energy unless adequate moisture and plant food are available.
Forms of interest SPENCER HYBRIDS **'Aerospace'** Pure white flowers. **'Air Warden'** Cerise-scarlet flowers. **'Alan Williams'** Large, mid blue flowers. **'Apricot Queen'** Orange/pink flowers. **'Beaujolais'** Burgundy flowers. **'Blue Danube'** Deep blue, frilled flowers. **'Blue Ripple'** White flowers, flushed lavender. **'Carlotta'** Brilliant carmine flowers. **'Corinne'** Rose-pink and carmine flowers with white blotch at base. **'Cream Beauty'** Frilled cream flowers. **'Dorothy Sutton'** Rose-pink on cream ground. **'Elizabeth Taylor'** Clear mauve, strongly fragrant flowers. **'Garden Party'** Wavy flowers, bright red with orange tinge. **'Geranium Pink'** Deep rose-pink flowers. **'Grace of Monaco'** Soft pink on white ground. Slightly waved petals. **'Hunter's Moon'** Large, creamy yellow, frilled blooms. **'Lady Diana'** Pale violet-blue flowers. **'Lady Fairbairn'** Lavender/rose flowers. **'Larkspur'** Soft blue to lavender flowers. **'Leamington'** Deep lavender flowers. **'Lilac Ripple'** Pure white flowers splashed with purple. **'Lillie Langtry'** Deep cream, wavy flowers. **'Maggie May'** Sky-blue flowers flushed white. **'Marietta'** Rose/mauve flowers on strong stems. **'Mrs Bernard Jones'** Large, frilled, cerise on white flowers. **'Mrs C. Kay'** Lavender flowers. **'Mrs R. Bolton'** Almond-

Lathyrus odoratus in flower

blossom pink. **'Noel Sutton'** Large, rich blue flowers. **'North Shore'** Navy blue flowers with pale violet wings. **'Pageantry'** Large, red/purple flowers. **'Pennine Floss'** Large, wavy, red/purple flowers. **'Princess Elizabeth'** Salmon-pink on cream flowers. **'Radar'** Salmon-pink flowers. **'Red Ensign'** Red flowers on long stems. **'Rosy Frills'** Large, frilled white flowers with deep rose edge. **'Royal Flush'** Salmon-pink on cream flowers. **'Royal Wedding'** Large, snow-white flowers on long stems. **'Snowdonia Park'** Pure white, frilled flowers. **'Superstar'** Deep rose-pink flowers with white base. **'Swan Lake'** White flowers. **'Welcome'** Deep scarlet flowers. **'Wiltshire Ripple'** White flowers flushed claret red. **'Winston Churchill'** Rich crimson flowers.

GRANDIFLORA VARIETIES (old-fashioned scented types) Very fragrant, flowers smaller than average, with a good range of colours. **'Jet Set Mixed'** Large flowers on good stems in a range of colours including scarlet, crimson, blue, salmon, cerise and mauve. **'Old Spice Mixed'** Old variety in a wide range of colours. Very strongly scented flowers, small compared with modern varieties. **'Romance'** Bicoloured flowers, each delicately frilled around outer edges, with very pleasant perfume. **'Royals Mixed'** Good mixture for cutting. Strong plants with very attractive blooms on long stems. **'Ruffle-edged Mixed'** Sweetly scented. Ruffled edges to petals in pinks, white and creams. **'Summer Breeze Mixed'** Soft pastel colours. Some deep red and blue bi-colours. Good scent. Wavy petals with frilled edges.

OLD 18th CENTURY VARIETY **'Fragrant Beauty'** Royal blue and rich violet purple. Fewer flowers. Highly scented.

MIXED SEED COLLECTIONS **'Candyman'** A range of colours, with some striped flowers. Large flowers with good fragrance. **'Early-bird Mammoth Mixed'** Early flowering, fragrant variety, with a wide range of colours. Good for greenhouses or outdoor display. **'Early Multiflora Gigantea Mixed'** Good range of colours. Strong growing, flowers produced over a long period. **'Floriana Mixed'** Similar to Spencer Hybrids. Good number of blooms on individual stems. Good range of colours. **'Galaxy Mixed'** Very vigorous and prolific in flower. Frilled petals. Very fragrant and good stems.

Average height and spread
One year
9 × 3 ft (2.7 m × 91 cm)
Protrudes up to 2 ft (60 cm) from support.

LAURUS NOBILIS

BAY, SWEET BAY, BAY LAUREL, POET'S LAUREL
Lauraceae   *Wall Shrub*
Evergreen
A useful evergreen shrub with protection gained from winter cold by wall planting.

Origin From the Mediterranean region.
Use A large, freestanding evergreen for mild locations, planted in front of walls and fences.
Description *Flower* Small ½ in (1 cm) wide tufts of almost petalless flowers with very prominent stamens, yellow/green. Flowers are either male or female, the latter being more commonly produced. *Foliage* Elliptic leaves 2–4 in (5–10 cm) long, dark green, glossy upper surfaces with dull grey/green undersides. *Stem* Upright, becoming branching. Mid green sometimes tinged purple. Medium to fast growth rate, slowing with age. *Fruit* Small, shiny, black fruits sometimes produced on mature shrubs growing in mild areas. Fruit not always an asset as it can mean that the shrub is maturing, when the leaves become small and uninteresting, or even in ill health.

Laurus nobilis in flower

Hardiness Tolerates a minimum winter temperature of 14°F (−10°C). Susceptible to leaf scorch in cold winter winds.
Soil requirements Prefers light open sc∷, but tolerates most types, including considerable degrees of alkalinity and acidity.
Sun/Shade aspect Requires a moderately protected aspect in full sun to light shade but will tolerate deep shade.
Pruning Can be clipped hard, as required. If reduced to ground level it rejuvenates but takes some years to reach former height.
Training Allow to stand free in front of walls and fences.
Propagation and nursery production From semi-ripe cuttings taken mid summer. Purchase container grown; if not found in garden centres and nurseries, may be sought from suppliers of culinary herbs. Best planting height 1½–6 ft (45 cm–1.8 m).
Problems Often outgrows its intended space; may be damaged in winter.
Similar forms of interest *L. n.* 'Aurea' An attractive variety with golden yellow evergreen leaves. Slightly less than average height and spread and slightly more tender.
Average height and spread
Five years
7 × 4 ft (2.1 × 1.2 m)
Ten years
10 × 7 ft (3 × 2.1 m)
Twenty years
20 × 15 ft (6 × 4.6 m)
Protrudes up to 7 ft (2.1 m) from wall or fence.

LAVATERA OLBIA

TREE MALLOW, TREE LAVATERA
Malvaceae   *Wall Shrub*
Deciduous
A tall-growing, rather tender shrub which adapts well to being grown against a wall where some protection is afforded against wind damage to flowers.

Lavatera olbia in flower

Origin From southern France.
Use As a freestanding wall shrub.
Description *Flower* Single, 3 in (7.5 cm) wide, saucer-shaped flowers, pink to silvery/pink, produced from early summer through until autumn frosts. Flowers open, progressively from midway down flowering shoot, last for a short period and then perish as the next two or three flowers open, continuing in this way to the end of flowering spike. *Foliage* Leaves broad, elliptic, five-lobed, 3–4 in (7.5–10 cm) long and wide, light green to grey/green. *Stem* Strong, upright, grey/green shoots produced from a central crown in spring. Very fast annual growth of new shoots. *Fruit* Insignificant.

Hardiness Tolerates a minimum winter temperature of 14°F (−10°C).
Soil requirements Light, open, dry soil. Requires adequate feeding to encourage production of long flower shoots.
Sun/Shade aspect Requires a sheltered aspect in full sun. Tolerates very light shade.
Pruning All previous year's shoots should be cut hard back annually in early spring. Hard pruning encourages flowering.
Training Allow to stand free in front of a wall or fence.
Propagation and nursery production From hardwood cuttings taken in winter or from semi-ripe cuttings taken in early summer. Purchase container grown; normally found in specialist nurseries. Best planting height 15 in–2 ft (40–60 m).
Problems Can be late to break into growth and may show little life until early summer, but grows rapidly once dormancy is broken. Plants in containers tend to look unsightly and sickly. Flower colour and size is dependent on maximum sunlight – the duller the summer, the less showy the flowers.
Similar forms of interest *L. o.* 'Barnsley' Attractive light purple/pink flowers in abundance. Slightly less height and spread. May be more tender. *L. o.* 'Variegata' Bright pink flowers. Foliage golden variegated. Requires protection.
Average height and spread
Five years
8 × 6 ft (2.4 × 1.8 m)
Ten years
10 × 10 ft (3 × 3 m)
Twenty years
10 × 10 ft (3 × 3 m)
Protrudes up to 7 ft (2.1 m) from wall or fence.

LIGUSTRUM LUCIDUM

GLOSSY PRIVET, CHINESE PRIVET, WAXLEAF PRIVET

Oleaceae   *Wall Shrub*
Evergreen

A useful group of evergreen shrubs with some striking variegated varieties, not so widely planted as they deserve. Adapts well to wall-training.

Origin From China.
Use As an upright evergreen shrub for walls and fences, suitable for trimming and shaping if required.
Description *Flower* Good-sized, 6 in (15 cm) long and 4 in (10 cm) wide, triangular-shaped panicles of musty scented flowers on mature wood in late summer. *Foliage* Pointed ovate leaves, 3–6 in (7.5–15 cm) long, glossy green upper surfaces, duller undersides, some

Ligustrum lucidum 'Tricolor'

white, silver and gold variegated forms. *Stem* Strong, vigorous, upright, becoming branching and spreading with age. Light green to green/purple. Medium to fast rate of growth. *Fruit* Clusters of dull blue to blue/black fruits in late summer, early autumn.
Hardiness Tolerates a minimum winter temperature of 4°F (−15°C).
Soil requirements Any soil, but dislikes very dry, alkaline conditions or extremely water-logged areas.
Sun/Shade aspect Tolerates all aspects. Full sun to medium shade. Tolerates deep shade but may become slightly deformed.
Pruning None required. Can be cut hard back and will quickly rejuvenate, even from very old stems.
Training Allow to stand free or secure to wall or fence by individual anchor points or tying to wires.
Propagation and nursery production From semi-ripe cuttings taken in mid to late summer, or hardwood cuttings taken in late autumn to early winter. Purchase container grown or root-balled (balled-and-burlapped); quite easy to find. Best planting height 1½–2½ ft (45–76 cm).
Problems The green form is vigorous and may outgrow the area allowed for it. Golden-leaved varieties may suffer sun scorch in strong sunlight.
Similar forms of interest *L. l.* 'Aureum' Bright yellow to golden spring foliage, this colouring declining as summer progresses. Two thirds average height and spread and less hardy than the parent. May be hard to find. *L. l.* 'Excelsum Superbum' Foliage margined and mottled with deep yellow or creamy

white. Two thirds average height and spread. May be hard to find. *L. l.* 'Tricolor' White to cream variegation, leaf margins tinged pink when young, making a very pleasant colour combination. Two thirds average height and spread. Slightly less hardy than the parent. May be hard to find.
Average height and spread
Five years
8 × 8 ft (2.4 × 2.4 m)
Ten years
15 × 15 ft (4.6 × 4.6 m)
Twenty years
26 × 26 ft (8 × 8 m)
Protrudes up to 5 ft (1.5 m) from support if fan-trained, 20 ft (6 m) untrained.

LIGUSTRUM OVALIFOLIUM

OVAL-LEAVED PRIVET, HEDGING PRIVET, CALIFORNIA PRIVET

Oleaceae   *Wall Shrub*
Deciduous to semi-evergreen

In its gold and silver form can make an interesting fan-trained shrub. The green form will grow against walls but is of little attraction, other than for clipping to a specific shape.

Origin From Japan.
Use As a freestanding or fan-trained shrub for walls and fences.

Ligustrum ovalifolium 'Aureum' in leaf

Description *Flower* 3 in (7.5 cm) long and 1½ in (4 cm) wide panicles of off-white flowers with a musty scent, borne only on untrimmed shrubs in mid summer. *Foliage* Leaves ovate, pointed, 1–1½ in (2.5–4 cm) long, mid to dark green with glossy upper surfaces and lighter, duller green undersides. Often maintained to mid winter in mild conditions, but cannot be relied upon as a full evergreen. *Stem* Strong, grey/green to green/brown, upright, vigorous, becoming very branching with age. Fast growth rate. *Fruit* Small, 2 in (5 cm) long and 1½ in (4 cm) wide, clusters of dull black berries produced only on mature, untrimmed shrubs.
Hardiness Tolerates a minimum winter temperature of 4°F (−15°C).
Soil requirements Most soils, but distressed by extremely dry or very alkaline types.
Sun/Shade aspect Tolerates any aspect in full sun to deep shade, but the latter leads to very open and lax habit.
Pruning May be reduced to ground level and will rejuvenate quite quickly, or may be trimmed back hard. Both operations decrease flowering on mature plants.
Training Allow to stand free or secure to wall or fence by individual anchor points or tying to wires.
Propagation and nursery production From semi-ripe cuttings taken in late spring or early summer, or hardwood cuttings taken in

Ligustrum lucidum in flower

winter. Plant bare-rooted or container grown. Availability varies. Best planting height 15 in–3 ft (38–91 cm).

Problems Privet roots are extremely invasive and draw all plant nutrients out of surrounding soil up to 6–10 ft (1.8–3 m) from base of shrub. Consequently it is difficult for other plants to grow in its immediate vicinity. Golden-leaved varieties may suffer sun scorch in strong sunlight.

Similar forms of interest *L. o.* **'Argenteum'** Leaves grey/green with creamy white margins. Mature, unpruned shrubs may produce off-white flowers not particularly visible against foliage. Current season's growth should be cut by half or more in early spring to encourage prolific regeneration of creamy white foliage. Two thirds average height and spread. *L. o.* **'Aureum'** (Golden privet) Rich golden yellow leaves with green centres. If required for foliage effect, current season's growth should be cut back by half or more in early spring to encourage regeneration of golden foliage. Two thirds average height and spread. *L.* × **'Vicaryi'** Attractive all-golden foliage. May be semi-evergreen in some areas. White flowers.

Average height and spread
Five years
10 × 10 ft (3 × 3 m)
Ten years
13 × 13 ft (4 × 4 m)
Twenty years
20 × 20 ft (6 × 6 m)

Protrudes up to 6 ft (1.8 m) from support if fan-trained, 20 ft (6 m) untrained.

LIGUSTRUM QUIHOUI

KNOWN BY BOTANICAL NAME

Oleaceae　　　　　　　**Wall Shrub**
Deciduous

An attractive, graceful privet adapting well to being fan-trained on walls and fences.

Origin From China.
Use As a late summer flowering shrub for large walls and fences where it can be grown freestanding or trained in a fan shape.
Description *Flower* Large, 10–12 in (25–30 cm) long, white to creamy white flower panicles produced profusely, mid to late summer. *Foliage* Leaves narrow, ovate, 1–1½ in (2.5–4 cm) long, somewhat sparsely produced on branches. Grey to olive-green giving some yellow autumn colour. *Stem* Upright at first, quickly becoming ranging and arching, producing an informal, open and somewhat lax effect. Light grey/green when young, becoming dark green to green/brown with age. Medium growth rate, slowing with age. *Fruit* Clusters of blue to blue/black fruits, early to late summer.

Ligustrum quihoui in flower

Ligustrum sinense in flower

Hardiness Tolerates a minimum winter temperature of 4°F (−15°C).
Soil requirements Most soils, but may be distressed by very alkaline or waterlogged conditions.
Sun/Shade aspect Tolerates all but the most exposed aspects in full sun to medium shade.
Pruning None required. May be cut back hard but will rejuvenate.
Training Allow to stand free or secure to wall or fence by individual anchor points or wires.
Propagation and nursery production From semi-ripe cuttings taken in mid summer or hardwood cuttings taken in early winter. Purchase container grown; rather hard to find, should be sought from specialist nurseries. Best planting height 2–2½ ft (60–76 cm).
Problems The young plant gives no indication of its true flowering potential.

Average height and spread
Five years
5 × 5 ft (1.5 × 1.5 m)
Ten years
7 × 7 ft (2.1 × 2.1 m)
Twenty years
15 × 15 ft (4.5 × 4.5 m)
Protrudes up to 4 ft (1.2 m) from support if fan-trained, 15 ft (4.6 m) untrained.

LIGUSTRUM SINENSE

CHINESE PRIVET

Oleaceae　　　　　　　**Wall Shrub**
Deciduous

A handsome shrub, with an attractive variegated form providing useful material for flower arranging. Adapts well to fan-training.

Origin From China.
Use As a tall, freestanding shrub or as a fan-trained specimen against walls and fences.
Description *Flower* 3–4 in (7.5–10 cm) long panicles of white flowers freely produced in mid summer. *Foliage* Leaves narrow, pointed, ovate, 1–3 in (2.5–7.5 cm) long, mid green, giving some limited autumn colour. *Stem* Strong and upright when young, becoming branching with age, forming a high, wide-spreading clump or fan-trained specimen. Medium growth rate. *Fruit* Clusters of black to blue/black fruits follow flowers in early autumn and are often retained into winter.
Hardiness Tolerates a minimum winter temperature of 4°F (−15°C).
Soil requirements Most soils but dislikes extremely dry, very alkaline or waterlogged conditions.
Sun/Shade aspect Tolerates all aspects in full sun to medium shade. In deep shade may become open, lax and shy to flower.

Pruning None required other than for training. May be trimmed very hard and will regenerate rapidly; can even be cut to ground level.
Training Allow to stand free or secure to wall or fence by individual anchor points or wires.
Propagation and nursery production From semi-ripe cuttings taken in mid summer or hardwood cuttings taken in late winter. Purchase container grown or bare-rooted; variegated forms easy to find, green form less often seen. Best planting height 2–2½ ft (60–76 cm).
Problems The true beauty of this privet is not always appreciated, because it is often kept trimmed and not allowed to reach full stature.
Similar forms of interest *L. s.* **'Variegatum'** Narrow, ovate, grey/green leaves, with white to creamy white margins, white flowers. Slightly less than average height and spread. Current season's growth may be cut back by half in early spring to encourage new, prolific growth of variegated leaves.

Average height and spread
Five years
10 × 10 ft (3 × 3 m)
Ten years
20 × 20 ft (6 × 6 m)
Twenty years
20 × 26 ft (6 × 8 m)
Protrudes up to 3 ft (91 cm) from support if fan-trained, 8 ft (2.4 m) untrained.

LIPPIA CITRIODORA
(*Aloysia triphylla*)

LEMON VERBENA, SHRUBBY VERBENA, LEMON PLANT

Verbenaceae　　　　**Tender Wall Shrub**
Deciduous

A very attractive shrub with lemon-scented foliage and stems, benefiting from the protection of a wall or fence in many areas.

Origin From Chile.
Use As a fan-trained shrub on a sunny wall or fence. In cold areas can be grown in conservatories or greenhouses. Can be used as a culinary herb.
Description *Flower* 6 in (15 cm) long and 3 in (7.5 cm) wide, open terminal panicles of small pale blue/mauve florets in mid summer. *Foliage* Leaves lanceolate, 3–4 in (7.5–10 cm) long, grey/green to sea/green, giving off lemony, aromatic scent when crushed. *Stem* In cooler areas dies back completely in winter, in milder areas is maintained as a loose, open shrub. Grey/green, lemon-scented. *Fruit* Insignificant.
Hardiness Tolerates a minimum winter temperature of 23°F (−5°C).

Lippia citriodora **in flower**

Soil requirements Open, well-drained, warm soil.

Sun/Shade aspect Requires a sheltered aspect in full sun to very light shade.

Pruning Either cut to ground level or remove one third of oldest wood each spring to encourage new strong shoots with good aromatic foliage.

Training Allow to stand free or secure to wall or fence by individual anchor points or wires.

Propagation and nursery production From semi-ripe cuttings taken in mid summer for overwintering under cover and planting out in spring, or earlier if planting area is completely frost-free. Purchase container grown. Best planting height 15 in–2 ft (38–60 cm).

Problems Often a poor-looking specimen when purchased container grown but grows rapidly once planted.

Average height and spread

Five years
4 × 4 ft (1.2 × 1.2 m)
Ten years
5 × 5 ft (1.5 × 1.5 m)
Twenty years
7 × 7 ft (2.1 × 2.1 m)
Protrudes up to 4 ft (1.2 m) from support.

LOMATIA MYRICOIDES
(*L. longifolia*)

KNOWN BY BOTANICAL NAME

Proteaceae　　　　　　　　*Wall Shrub*
Evergreen

A shrub which tolerates lower temperatures and a larger geographical range when given the protection of a wall or fence.

Origin From Chile.

Use As a large, fan-trained or freestanding foliage shrub for walls or fences, requiring an acid soil.

Description *Flower* Long, strap-like white petals, very fragrant, borne in late summer. *Foliage* Long, narrow, light grey/green leaves up to 8 in (20 cm) long and ½ in (1 cm) wide. *Stem* Upright, erect stems with red/brown velvety texture. Medium growth rate. *Fruit* Insignificant.

Hardiness Tolerates a minimum winter temperature of 14°F (−10°C), although at this temperature there may be some wind chill scorch to foliage.

Soil requirements Acid soil, dislikes any alkalinity. Must have good drainage.

Sun/Shade aspect Sheltered aspect in full sun to light shade.

Pruning None required, other than that for fan-training.

Training Allow to grow freestanding or tie to wires or individual anchorage points in a fan shape.

Propagation and nursery production From softwood cuttings taken in late summer, or from seed. Purchase container grown; rather hard to find. Best planting height 1½–2 ft (45–60 cm).

Problems Often looks very poor and weak in containers but on correct soil quickly presents itself attractively.

Similar forms of interest *L. ferruginea* Attractive foliage and flowers, possibly slightly stronger growing. *L. tinctoria* A variety with pinnate to double pinnate leaves with very narrow, long leaflets, light to mid green. Flowers sulphur yellow in bud, becoming creamy white, produced in long racemes at the terminals of each shoot. Half average height and spread. From Tasmania.

Average height and spread

Five years
7 × 7 ft (2.1 × 2.1 m)
Ten years
12 × 12 ft (3.7 × 3.7 m)
Twenty years
15 × 15 ft (4.6 × 4.6 m)
In favourable conditions may eventually reach 26 ft (8 m) height and spread. Protrudes up to 3 ft (91 cm) from support if fan-trained, 8 ft (2.4 m) untrained.

Lomatia myricoides **in flower**

HONEYSUCKLE

Caprifoliaceae　　　　　　*Woody Climber*
Deciduous

The attractive flower colouring and strong scent make this a notable honeysuckle worthy of a place in any garden.

Origin From south-east Europe.

Use As a mid summer to early autumn flowering, strong-growing climber for trellis, wires, walls, fences, poles, large shrubs and small trees. Can be grown as a weeping standard.

Description *Flower* Clusters of funnel-shaped trumpets up to 2 in (5 cm) long, carried in panicles up to 12 in (30 cm) long and 8 in (20 cm) wide at the ends of the branches. Yellow with purple shading. The tubular trumpets have two distinct lips at their ends up to 1–1½ in (2.5–4 cm) in diameter. Flowering is from early to mid summer with some later flowers in late summer and early autumn. *Foliage* Oval, up to 3 in (7.5 cm) long and 1½ in (4 cm) wide, carried in pairs along the stems. Light green with some purple shading. *Stem* Light yellow/green when young, ageing to yellow/brown, finally grey/brown. Twining and twisting, making it self-supporting. Medium to fast growing. *Fruit* Small, red, almost translucent fruits are produced in hot summers.

Lonicera americana **in flower**

Hardiness Tolerates a minimum winter temperature of 4°F (−15°C). Young foliage may be damaged by spring frosts but normally rejuvenates.

Soil requirements Does well on all soil types, although may show distress on very severe alkaline or dry conditions.

Sun/Shade aspect Tolerates all aspects. Performance is enhanced in light shade but will tolerate full sun to medium shade with varying degrees of success. Requires its roots to be shaded.

Pruning Once plants are established more than three years, remove one third of the oldest growth to ground level in early spring.

Training Allow to ramble and twine over wire trellis or branch supports. Normally needs no tying in except when young.

Propagation and nursery production From semi-ripe cuttings taken in mid summer or hardwood cuttings taken in mid to late winter. Always purchase container grown; best planting height 1½–3½ ft (45 cm–1.1 m) from late summer through to early autumn. Normally available from good garden centres and specialist nurseries.

Problems May suffer attacks of blackfly and mildew. Protect with a proprietary control.

Similar forms of interest None.

Average height and spread

Five years
12 × 12 ft (3.7 × 3.7 m)
Ten years
20 × 20 ft (6 × 6 m)
Twenty years
25 × 25 ft (7.6 × 7.6 m)
Protrudes up to 2 ft (60 cm) from support.

LONICERA × BROWNII

SCARLET TRUMPET HONEYSUCKLE

Caprifoliaceae   **Woody Climber**
Deciduous

A most attractive group of honeysuckles, with unusual orange-red flowers.

Origin From North America.

Use Attractive climbers with all the characteristics of honeysuckle but with an interestingly different flower colour. Can be grown as a weeping standard.

Description *Flower* Red to scarlet-red, tubular trumpets up to 1–1½ in (2.5–4 cm) long are produced together at the ends of flowering shoots from early summer to early autumn. Moderately scented. *Foliage* Oval, up to 3½ in (9 cm) long and 1–1½ in (2.5–4 cm) wide. Downy, blue undersides, light fresh green upper. Carried in pairs along the shoots. *Stem* Grey/green when young, ageing to yellow/brown, finally grey/brown. Twining. Self-supporting. Medium to fast growing. *Fruit* May produce small, round, red fruits ⅛ in (2 mm) in diameter following hot summers.

Hardiness Tolerates a minimum winter temperature of 14°F (−10°C). May suffer foliage damage from late spring frosts in excess of 23°F (−5°C).

Soil requirements Requires a moist, deep, rich soil to produce adequate growth and flowers. Tolerates moderately alkaline to acid soil.

Sun/Shade aspect Best on a slightly protected aspect in light shade, although will tolerate full sun to medium shade, as long as roots have adequate moisture and shade.

Pruning Remove one third of oldest growth to ground level in early spring on plants established more than three years. To restrict growth, prune offending shoots in early spring.

Training Self-twining over trellis, wires and branches and normally requires no other support, except tying in when young.

Propagation and nursery production From semi-ripe cuttings in early summer or hardwood cuttings in winter. Always purchase container grown from late spring through to early summer; normally available from good garden centres and specialist nurseries. Best planting height 1½–3½ ft (45 cm–1.1 m).

Problems May suffer from attacks of blackfly and mildew. Protect with a proprietary control. May be slow to establish. Make sure adequate organic material is available to retain moisture to produce best growth results.

Similar forms of interest *L. × b.* **'Dropmore Scarlet'** Bright scarlet-red tubular flowers from mid summer to early autumn. The main and best form offered by garden centres and nurseries. *L. × b.* **'Fuchsoides'** A variety with larger flowers. May be slightly less hardy but not to any marked degree. Scented. May have to be sought from specialist nurseries. *L. × b.* **'Plantierensis'** Coral-red flowers. Will have to be sought from specialist nurseries.

Average height and spread

Five years
7 × 7 ft (2.1 × 2.1 m)
Ten years
14 × 14 ft (4.3 × 4.3 m)
Twenty years
21 × 21 ft (6.4 × 6.4 m)
Protrudes up to 2½ ft (76 cm) from support.

LONICERA CAPRIFOLIUM
(L. 'Early Cream')

PERFOLIATE HONEYSUCKLE

Caprifoliaceae   **Woody Climber**
Deciduous

Amongst the most pleasant of all the scented honeysuckles, but it may be a little difficult to find the true plant.

Origin From Europe.

Use As a useful scented climber for wires, trellis, walls and fences or for rambling through large shrubs. Can be grown as a weeping standard.

Description *Flower* One to two clusters carried at the ends of flowering shoots, made up of a number of tubular trumpet-shaped florets, each 1½–2 in (4–5 cm) long. Yellow/white with pink tinge. Two lips at the end of each floret up to 1 in (2.5 cm) across. Very fragrant. *Foliage* Oval, up to 4 in (10 cm) long and 1½ in (4 cm) wide. Grey/green. End pairs of shoots may be united, forming an attractive saucer-shaped leaf. *Stem* Light grey/green, ageing to yellow/brown, finally grey/brown. Twining. Medium to fast in growth. *Fruit* In autumn following hot summers may produce red fruits up to ¼ in (5 mm) across, in clusters.

Hardiness Tolerates a minimum winter temperature of 4°F (−15°C).

Soil requirements Does well on all soil types, but prefers a moist soil, high in organic material. Tolerates both alkalinity and acidity, only showing distress on extremely alkaline soil types.

Sun/Shade aspect Light shade for preference, but will tolerate full sun to medium shade as long as roots are shaded. Does well on all aspects.

Pruning Remove one third of oldest growth to ground level in early spring on plants established more than three years.

Training Self-twining over trellis, wires and branches and requires no other support except tying in when young.

Lonicera caprifolium **in flower**

Propagation and nursery production From semi-ripe cuttings in early summer or hardwood cuttings in winter. Always purchase container grown; best planting height 1½–3½ ft (45 cm–1.1 m). Normally available from garden centres and specialist nurseries.

Problems May suffer from attacks of blackfly and mildew. Protect with a proprietary control.

Similar forms of interest *L. c.* **'Alba'** (syn. *L. praecox*) All white flowers. Not readily available and will have to be sought from specialist nurseries. *L. c.* **'Pauciflora'** Rose-tinged outer colouring to flowers.

Average height and spread

Five years
6 × 6 ft (1.8 × 1.8 m)
Ten years
12 × 12 ft (3.5 × 3.5 m)
Twenty years
18 × 18 ft (5.5 × 5.5 m)
Protrudes up to 2½ ft (45 cm) from support.

LONICERA ETRUSCA

ETRUSCAN HONEYSUCKLE

Caprifoliaceae   **Woody Climber**
Deciduous to semi-evergreen

The grey/green foliage of this very strong-growing honeysuckle, combined with the fragrant, creamy-yellow flowers, makes it a must for the larger gardens.

Origin From the Mediterranean region.

Use As a strong-growing climber for larger walls, fences, trellis, poles, pillars or for growing over large shrubs and medium-sized trees.

Description *Flower* Clusters of very fragrant tubular florets up to 1¾ in (4.5 cm) long. Creamy-yellow in colour with purple/red shading, produced from late summer to early autumn. *Foliage* Up to 3½ in (9 cm) long, blue/grey with downy undersides and light grey/green upper. End pairs of leaves join at the base. Some yellow autumn colouring.

Lonicera × *brownii* **'Fuchsoides' in flower**

Lonicera etrusca 'Superba' in flower

Stem Grey/green when young, becoming yellow/brown, finally grey/brown. Twining. Medium to fast growing. **Fruit** May in very hot summers produce small red fruits in clusters.
Hardiness Tolerates a minimum winter temperature of 4°F (−15°C).
Soil requirements Requires a moist, deep soil high in organic material. Tolerates moderately alkaline to acid.
Sun/Shade aspect Tolerates all aspects. Best in light shade, although will tolerate full sun to medium shade as long as adequate shade and moisture are provided for the roots.
Pruning Remove one third of oldest growth to ground level in spring once plants are established more than three years.
Training Allow to ramble over wire, trellis or branches. Self-supporting and only requires tying in when young.
Propagation and nursery production From semi-ripe cuttings in early summer or hardwood cuttings in winter. Always purchase container grown; best planting height 1½–3½ ft (45 cm–1.1 m). Normally available from garden centres and specialist nurseries.
Problems May suffer from attacks of blackfly and mildew. Protect with a proprietary control. Often underestimated for its ultimate height and spread.
Similar forms of interest *L. e.* 'Pubescens' Downy surface to leaves. Yellow flowers. *L. e.* 'Superba' Creamy-yellow flowers turning orange. More tender than its parent. The variety most frequently offered in garden centres and nurseries.
Average height and spread
Five years
6 × 6 ft (1.8 × 1.8 m)
Ten years
12 × 12 ft (3.5 × 3.5 m)
Twenty years
18 × 18 ft (5.5 × 5.5 m)
Protrudes up to 2½ ft (76 cm) from support.

LONICERA FRAGRANTISSIMA

HONEYSUCKLE, WINTER HONEYSUCKLE, SHRUBBY HONEYSUCKLE

Caprifoliaceae   **Wall Shrub**
Semi-evergreen
A delightful, scented winter shrub, adapting well to training against walls and fences.

Origin From China.
Use As a shrub for growing against a wall or fence to show off its highly scented winter flowers.
Description *Flower* Small, ¼ in (5 mm) long and wide, sweetly scented, creamy white flowers, produced on almost leafless branches in mild spells, late autumn through to mid spring, standing up well to light to moderate frosts. *Foliage* Leaves ovate, 1–2½ in (2.5–6 cm) long, dark green tinged purple with lighter undersides. Some autumn colour. *Stem* Upright when young, becoming branching and spreading with age. Light grey/green when young, becoming darker green with some purple veining. Medium growth rate when young, becoming slower with age. *Fruit* May produce bunches of red fruits in early to mid spring.
Hardiness Tolerates a minimum winter temperature of 0°F (−18°C).
Soil requirements Any soil, tolerating quite dry areas.
Sun/Shade aspect Tolerates all but the most exposed aspects, full sun to medium shade.
Pruning None required, but may be reduced in size or trained.
Training Allow to stand free or secure to wall or fence by individual anchor points or wires, or fan-train.
Propagation and nursery production From semi-ripe cuttings taken in early summer. Purchase container grown; quite easy to find. Best planting height 15 in–2½ ft (38–76 cm).
Problems Young plants give no indication of their true potential.
Similar forms of interest *L.* × *purpusii* Fragrant white flowers, early winter to early spring. Slightly more vigorous than *L. fragrantissima* and larger leaves but fewer flowers. *L.* × *p.* 'Spring Purple' White

Lonicera fragrantissima in flower

fragrant flowers in winter. New foliage purple/green and stems purple when young. *L. standishii* Fragrant white flowers in winter, produced on stems covered in fine downy hair. Red fruits in early summer.
Average height and spread
Five years
6 × 6 ft (1.8 × 1.8 m)
Ten years
8 × 10 ft (2.4 × 3 m)
Twenty years
10 × 13 ft (3 × 4 m)
Protrudes up to 3 ft (91 cm) from support if fan-trained, 4 ft (1.2 m) untrained.

LONICERA × HECKROTII (*L.* × *heckrotii* 'Gold Flame')

SHRUBBY HONEYSUCKLE, EVER-FLOWERING HONEYSUCKLE

Caprifoliaceae   **Woody Climber**
Deciduous
A shrubby, pillar-forming honeysuckle without the characteristic twining branches, but with attractive flowers.

Origin Not known.
Use As a shrubby climber for pillars and against walls and fences.
Description *Flower* Clusters of slender, tubular flowers up to 2 in (5 cm) long, produced on long stalks. Orange/yellow in colour with purple/crimson shading on the outside. Flowering in early to mid summer. *Foliage* Oblong to oval, up to 2½ in (6 cm) long. Grey/blue underside. End pairs of leaves join at the base. *Stem* Grey/green to grey/brown, twining. Medium to fast growth rate. *Fruit* May produce small red clusters of fruits in very hot summers but not reliable.
Hardiness Tolerates a minimum winter temperature of 4°F (−15°C).
Soil requirements Does well on all soil conditions except very dry.
Sun/Shade aspect All but the most exposed of aspects. Light shade for preference but will tolerate full sun with root shading.

Lonicera × *heckrotii* in flower

Pruning Not normally required, but to assist rejuvenation of foliage and flowers remove one third of oldest growth on three-year-old plants and repeat every three to five years.
Training Will need tying to an upright support. Normally not twining.
Propagation and nursery production From semi-ripe cuttings taken in early summer. Should always be purchased container grown; best planting height 1½–3 ft (45–91 cm). Available from good garden centres and specialist nurseries.
Problems May suffer from attacks of blackfly and mildew. Protect with a proprietary control. Its non-climbing habit is not always understood.

Similar forms of interest Although it should truly be called *Lonicera × heckrotii*, it is often offered in nurseries and garden centres as *L. × h.* 'Gold Flame'.
Average height and spread
Five years
5 × 3 ft (1.5 × 91 cm)
Ten years
8 × 5 ft (2.4 × 1.5 m)
Twenty years
10 × 6 ft (3 × 1.8 m)
Protrudes up to 2½ ft (76 cm) from support.

LONICERA JAPONICA 'AUREORETICULATA'

GOLDEN VARIEGATED JAPANESE HONEYSUCKLE

Caprifoliaceae *Woody Climber*
Semi-evergreen

Given the right conditions an attractive climber for all aspects except those extremely exposed.

Origin From Japan, Korea and China.
Use As an ornamental foliage climber for walls, fences, trellis, wires, pergolas, stumps of old trees and for rambling through medium to large shrubs.
Description *Flower* Somewhat inconspicuous, fragrant, bright yellow flowers, up to 1¼ in (3 cm) long in clusters. Purple in bud and ageing to yellow in early to mid summer, but often not noticeable due to the golden variegated foliage. *Foliage* Oval to round. Green to blue/green, splashed yellow to gold. Normally semi-evergreen, but in favourable

Lonicera japonica 'Aureoreticulata' in leaf

conditions can be evergreen. *Stem* Light grey/green ageing to grey/brown. Twining, partially self-supporting. Medium rate of growth. *Fruit* Clusters of small black fruits mainly produced following hot summers.
Hardiness Tolerates a minimum winter temperature of 14°F (−10°C). Severe wind chill in winter may cause damage to the foliage. Rejuvenation is normal the following spring.
Soil requirements Alkaline to acid. Must always be moist and rich in organic material.
Sun/Shade aspect Does well in all aspects except extremely exposed. Full sun to light shade.
Pruning When plants are established for more than three years, remove one third of oldest growth to ground level in early spring to encourage rejuvenation of new, good-coloured foliage and feed with a liquid fertilizer in mid summer. Every five to ten years it may be advantageous to cut plants back to within 2 ft (60 cm) of ground level to induce new regrowth.
Training Self-twining over wires and trellis but may need some initial tying in when young.

Propagation and nursery production From semi-ripe cuttings taken in early summer. May require a little bottom heat for good root development. Always purchase container grown; normally available from garden centres and nurseries. Best planting height 1½–3 ft (45–91 cm).
Problems Can be a little shy in its growth in early years. Growth without pruning can be old and woody.
Similar forms of interest None.
Average height and spread
Five years
5 × 5 ft (1.5 × 1.5 m)
Ten years
10 × 10 ft (3 × 3 m)
Twenty years
15 × 15 ft (4.6 × 4.6 m)
Protrudes up to 2 ft (60 cm) from support.

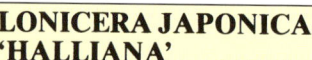

LONICERA JAPONICA 'HALLIANA'

EVERGREEN HONEYSUCKLE

Caprifoliaceae *Woody Climber*
Evergreen

Whenever a quick-growing, flowering evergreen climber is required, this vigorous honeysuckle is ideal.

Origin From Japan.
Use As an evergreen climber for fences, trellises and pillars. Can be allowed to ramble over large shrubs and through small trees. Useful, with support, as an evergreen climbing screen.
Description *Flower* Borne in pairs at the leaf joints; five to seven tubular florets ½ in (1 cm) long. White, sometimes tinged purple, ageing to yellow, from early to mid summer. *Foliage* Oval to oblong, pointed. Up to 3½ in (9 cm) long and 1¾ in (4.5 cm) wide. Light spring green to light grey/green with a downy covering on both sides. *Stem* Light green ageing to dark green, finally green/brown. Twining. Fast growing. *Fruit* Clusters of small, black, round fruits produced after hot summers.
Hardiness Tolerates a minimum winter temperature of 4°F (−15°C). Foliage may be damaged in severe wind chill conditions, but is normally replaced in the following spring. Some stem damage may also be caused under these conditions.
Soil requirements Does well on both moderately alkaline and acid. Requires a moist, rich, organic type for best results.
Sun/Shade aspect Does well on all aspects. Light shade for preference, although will tolerate full sun to medium shade as long as roots are in a shady, moist position.
Pruning Allow to grow for four to five years and then shear off all forward growing shoots

to a height of 6 ft (1.8 m) or more after flowering. Rejuvenation is rapid. This process should be repeated every four to six years.
Training Allow to twine through wires, trellis, and branches of large shrubs and small trees. May need tying in when young.
Propagation and nursery production From semi-ripe cuttings taken in early to mid summer. Always purchase container grown; best planting height 2–4 ft (60 cm–1.2 m). Normally available from all garden centres and nurseries.
Problems Foliage can look a little damaged after winter, making the plant appear untidy, but is quickly rejuvenated in spring with fresh, soft, light green growth.
Similar forms of interest None.
Average height and spread
Five years
8 × 8 ft (2.4 × 2.4 m)
Ten years
16 × 16 ft (4.9 × 4.9 m)
Twenty years
30 × 30 ft (9 × 9 m)
Protrudes up to 3 ft (91 cm) from support.

LONICERA JAPONICA HENRYII

EVERGREEN HONEYSUCKLE

Caprifoliaceae *Woody Climber*
Evergreen

Possibly the most attractive of the evergreen honeysuckles with good foliage and interesting flowers.

Origin From China.
Use As an evergreen climber for trellis, wires, fences, walls, pergolas and pillars, for growing over large shrubs and small trees and for a screen which is attractive all year round.
Description *Flower* Clusters up to 3 in (7.5 cm) wide of small, tubular, purple/red to yellow/red flowers, ¾ in (2 cm) long with reflexed lips at outer edges, normally in pairs at ends of flower shoots. Produced in early to mid summer. Scented. *Foliage* Ovate to lanceolate, light green ageing to mid green. Some yellow autumn colour. *Stem* Grey/green to green, finally grey/brown. Twining. Fast growing. *Fruit* In hot summers small clusters of blue/black fruits are produced.
Hardiness Tolerates a minimum winter temperature of 4°F (−15°C). In extremely cold wind chill conditions the evergreen foliage will be damaged, but is normally replaced in the following spring. Some stem damage also may be suffered in such conditions.
Soil requirements Does well on both moderately alkaline and acid soils where a moist, high organic content is found.
Sun/Shade aspect Does well on all aspects

Lonicera japonica 'Halliana' in flower

except extremely exposed. Light shade for preference, but will tolerate from full sun to medium shade as long as roots are adequately shaded.

Pruning None required until five years after established, then it can be heavily pruned, with the removal of all branches up to 4 ft (1.2 m) from ground level in early spring. It will quickly rejuvenate and cover the same space from which it was cut. Repeat every five years.

Training Twining in habit but may need some initial support when young by tying in. It can also develop a very heavy head of evergreen foliage which may need tying in to prevent damage in heavy snow.

Lonicera japonica henryii in flower

Propagation and nursery production From semi-ripe cuttings taken in early to mid summer. Should always be purchased container grown; normally available from garden centres and nurseries. Best planting height 1½–3 ft (45–91 cm).

Similar forms of interest *L. similis delavayi* (*L. delavayi*) Fragrant white flowers, ageing to pale yellow, produced in late summer, early autumn. From western China. Scarce and will have to be sought from specialist nurseries.

Average height and spread
Five years
10 × 10 ft (3 × 3 m)
Ten years
20 × 20 ft (6 × 6 m)
Twenty years
25 × 25 ft (7.6 × 7.6 m)
Protrudes up to 3 ft (91 cm) from support.

LONICERA JAPONICA REPENS (*L. flexuosa*)

JAPANESE HONEYSUCKLE

Caprifoliaceae   **Woody Climber**
Deciduous to semi-evergreen

One of the most attractive of the Japanese flowering honeysuckles, with good flowers and attractive purple-tinged foliage.

Origin From Japan.
Use For covering trellis, walls, fences, pillars and for growing through large shrubs. Can be trained into a small weeping standard if required.
Description *Flower* Borne in pairs at leaf joints, up to 2 in (5 cm) long. Five to six tubular florets, heavily scented; purple/red and pink/white. *Foliage* Oval, up to 2½ in (6 cm) long and 1¼ in (3.5 cm) wide. Olive green to purple/olive green. Leaves have a distinct purple veining, contrasting well with the flowers. *Stem* Red/purple when young, ageing to purple/green, finally grey/green. Twining. Fast growing. *Fruit* May produce small clusters of black fruits following hot summers, but not reliable.
Hardiness Tolerates a minimum winter tem-

Lonicera japonica repens in flower

perature of 0°F (−18°C). Although its semi-evergreen habit may be turned fully deciduous by severe winters, spring rejuvenation normally follows.
Soil requirements Does well on moderately alkaline to acid soils. Requires a moist, rich, organic type.
Sun/Shade aspect Happy in all aspects. Light shade for preference but will tolerate full sun to medium shade as long as roots are adequately shaded.
Pruning Allow to grow free for up to five years and then prune heavily in early spring. This process should be repeated every five years.
Training Normally self-supporting by twining around wires and trellis but may need some tying in when young.
Propagation and nursery production From semi-ripe cuttings taken in early summer or hardwood cuttings taken in winter. Always purchase container grown; best planting height 2–4 ft (60 cm-1.2 m). Normally available from good garden centres and specialist nurseries.
Problems May suffer from attacks of blackfly and mildew. Protect with a proprietary control.
Similar forms of interest *L. j. r.* 'Halls Prolific' Heavily scented, creamy/white flowers produced even on young plants. *L. j. chinensis* Attractive rose/pink flowers. Not entirely hardy. Is scarce and will have to be sought from specialist nurseries. *L. reflexa* Flowers red and white in mid summer, foliage dark red/purple. Fragrant.
Average height and spread
Five years
8 × 8 ft (2.4 × 2.4 m)
Ten years
16 × 16 ft (4.9 × 4.9 m)
Twenty years
28 × 28 ft (8.5 × 8.5 m)
Protrudes up to 3 ft (91 cm) from support.

LONICERA NITIDA

BOXLEAF HONEYSUCKLE, POOR MAN'S BOX

Caprifoliaceae   **Wall Shrub**
Evergreen

A useful evergreen shrub that can be trimmed and shaped into pillars adjacent to walls and fences. Both the green and golden-leaved varieties are attractive in winter.

Origin From western China.
Use As a trimmed or untrimmed freestanding specimen adjacent to walls or fences.
Description *Flower* Inconspicuous, small, scented, sulphur-yellow flowers. *Foliage* Leaves small, ovate, ¼–½ in (5 mm–1 cm) long, mid green, shiny, with silver undersides, freely produced along branches and shoots. Slight purple tinge when young. *Stem* Upright

and branching when young, becoming very full and twiggy with age. Purple/green when young, becoming dark green with purple shading. Fast growth rate. *Fruit* Insignificant.
Hardiness Tolerates a minimum winter temperature of 4°F (−15°C).
Soil requirements Tolerates most soils but dislikes extreme dryness or waterlogging.
Sun/Shade aspect Tolerates all aspects except very dry. Full sun to deep shade.
Pruning Remove one third of old wood each spring to rejuvenate. May be cut to ground level and will regrow. Trim as required to almost any shape.
Training Normally requires no support other than possibly securing with individual anchor points when plants become mature.
Propagation and nursery production From semi-ripe cuttings taken in late summer or hardwood cuttings in mid winter. Purchase bare-rooted or container grown, best planting height 1–2 ft (30–60 cm).
Problems Becomes woody and uninteresting as it ages, although hard pruning corrects this. Mature plants may become lax and suffer damage from heavy snow.

Lonicera nitida 'Baggesen's Gold'

Similar forms of interest *L. n.* 'Baggesen's Gold' Yellow foliage, often turning more golden in winter when in full sun. In light shade turns yellow/green in autumn. Dislikes dry or waterlogged soils and shows ill effects immediately by losing foliage. Two thirds average height and spread. *L. n.* 'Yunnan' (syn. *L. yunnanensis*) A green-leaved variety with large foliage, more freely flowering and fruiting. Upright, typically offered by nurseries for hedging.
Average height and spread
Five years
6 × 6 ft (1.8 × 1.8 m)
Ten years
8 × 12 ft (2.4 × 3.7 m)
Twenty years
9 × 16 ft (2.7 × 4.9 m)
Protrudes up to 8 ft (2.4 m) from wall or fence.

Lonicera periclymen 'Belgica' in flower

LONICERA PERICLYMENUM (Hybrids)

HONEYSUCKLE

Caprifoliaceae　　　　　*Woody Climber*
Deciduous

The hybrid varieties are a decided improvement on the wild parent, offering large, fragrant flowers.

Origin Of nursery origin.
Use As a twining climber for trellis, walls, fences and pergolas and to grow over medium to large shrubs or small trees. Can be grown as a small standard or used as loose, flowing ground cover.
Description *Flower* Tubular florets, each up to 2 in (5 cm) long, make up a cluster up to 2½–3 in (6–7.5 cm) across in May and June. See 'Forms of interest' for colour. Depending on location, very fragrant. *Foliage* Oval to oblong. Up to 2½ in (6 cm) long and 1¾ in (4.5 cm) wide. Blue/grey undersides, light to mid green upper with a purple hue. *Stem* Light grey/green, ageing to grey/brown. Twining, fast growing. *Fruit* Small clusters of red fruits may sometimes be formed following hot summers.
Hardiness Tolerates a minimum winter temperature of 4°F (−15°C).
Soil requirements Moderately alkaline to acid. Must have a deep, moist, organic soil for best results.
Sun/Shade aspect Tolerates all but the most exposed aspects. Best in light shade but will

Lonicera periclymen 'Serotina' in flower

tolerate full sun to medium shade as long as roots are shaded.
Pruning Remove one third of oldest growth to ground level in early spring on all shrubs established more than three years and repeat annually.
Training Allow to twine through wires and trellis. May need some support in early years by tying in.
Propagation and nursery production From semi-ripe cuttings taken in early summer. Always purchase container grown, normally readily available from all garden centres and nurseries. Best planting height 2–4 ft (60 cm–1.2 m).
Problems May suffer from attacks of blackfly and from mildew. Protect with a proprietary control.
Forms of interest *L. p.* 'Belgica' (Early Dutch honeysuckle) flowers tubular, up to 2 in (5 cm) long in May and June, deep purple/red, fading to yellow/red. Fragrant. *L. p.* 'Graham Thomas' Large, yellow, scented flowers from mid summer through to early autumn. *L. p.* 'Red Gables' Attractive, large, red/yellow flowers. Strong grower. *L. p.* 'Serotina' (Late Dutch honeysuckle). Red/purple flowers, creamy white inside, from July to October. *L. p.* 'Serotina Winchester' Flowers red/purple outside, pale cream inside. From USA.
Average height and spread
Five years
6 × 6 ft (1.8 × 1.8 m)
Ten years
12 × 12 ft (3.7 × 3.7 m)
Twenty years
20 × 20 ft (6 × 6 m)
Protrudes up to 2 ft (60 cm) from support.

LONICERA SPLENDIDA

SPANISH HONEYSUCKLE

Caprifoliaceae　　　　　*Woody Climber*
Evergreen

An interesting honeysuckle but only useful in mild locations or under protection.

Origin From Spain.
Use As an evergreen climber for favoured positions against walls.
Description *Flower* Semi-open, yellow, tubular flowers, 1½–2 in (4–5 cm) long, with two red lips at the ends up to 1 in (2.5 cm) across, produced from mid summer to early autumn. *Foliage* Oval to oblong, grey/blue to green/blue. Somewhat sparsely produced. *Stem* Blue/grey, ageing to grey/green. Loosely twining, not normally self-supporting. Downy texture. Slow to medium rate of growth. *Fruit* Small grey/blue fruits may be produced, but not reliable.

Hardiness Tolerates a minimum winter temperature of 23°F (−5°C).
Soil requirements Moderately alkaline to acid. Must have a moist, deep, rich soil.
Sun/Shade aspect Will require protection in all aspects, except those which are very warm. Light shade or full sun, provided roots are shaded.
Pruning Does not normally require pruning.
Training Not completely self-clinging and will require tying to wires or trellis.
Propagation and nursery production From semi-ripe cuttings taken in early summer. Very difficult to propagate. Always purchase container grown; will have to be sought from specialist nurseries. Best planting height 1–3 ft (30–91 cm).

Lonicera splendida in flower

Problems Not completely hardy and requires specific degrees of shade and good soil to get good results, but worth an attempt for collectors. Scarce in production.
Similar forms of interest None.
Average height and spread
Five years
5 × 5 ft (1.5 × 1.5 m)
Ten years
10 × 10 ft (3 × 3 m)
Twenty years
15 × 15 ft (4.6 × 4.6 m)
Protrudes up to 2 ft (60 cm) from support.

LONICERA TATARICA

TATARIAN HONEYSUCKLE

Caprifoliaceae　　　　　*Wall Shrub*
Deciduous

An interesting mid spring, early summer flowering shrub, provided mildew attacks are controlled. Not often thought of as a wall shrub.

Origin From the USSR.
Use As a freestanding or fan-trained shrub for large walls and fences.
Description *Flower* Small, ⅓ in (1 cm) long, tubular, pink flowers profusely borne in mid spring to early summer. White and red flowering varieties also available. *Foliage* Elliptic, grey/green, somewhat sparsely produced. *Stem* Upright, becoming branching with age, grey/green. Medium growth rate when young, becoming slower with age. *Fruit* Small, red fruits, late summer, early autumn.
Hardiness Tolerates a minimum winter temperature of 0°F (−18°C).
Soil requirements Any soil type except very dry.
Sun/Shade aspect Tolerates all aspects. Prefers light shade, but will tolerate full sun to medium shade.
Pruning Remove one third of old wood after flowering period to induce new growth.
Training Will require wires or individual anchor points for fan-training.
Propagation and nursery production From semi-ripe cuttings taken in early summer. Purchase container grown; quite easy to find in nurseries and some garden centres, especi-

Lonicera tatarica in flower

LONICERA TRAGOPHYLLA

CHINESE WOODBINE

Caprifoliaceae   *Woody Climber*
Deciduous

A most spectacular, tall, climbing plant when given the correct environment.

Origin From western China.
Use As a tall, vigorous climber to grow through large shrubs and trees.
Description *Flower* Terminal whorls of 10–20 bright yellow, tubular flowers up to 3½ in (9 cm) long in June and July. Slender in shape with a 1–1½ in (2.5–4 cm) lobed mouth. Not fragrant. *Foliage* Ovate to lanceolate, up to 5 in (12 cm) long and 1¼ in (3.5 cm) wide. Light grey/green. Some yellow autumn colour. *Stem* Light green to yellow/brown, smooth textured. Loosely twining, fast growing, up to 9 ft (2.7 m) per year. *Fruit* Small clusters of bright red fruits up to ¼ in (5 mm) across may be produced following hot summers.
Hardiness Tolerates a minimum winter temperature of 14°F (−10°C).
Soil requirements Alkaline to acid; a rich, moist soil for best results.
Sun/Shade aspect Requires protection from exposed aspects and light shade.
Pruning Normally impractical to prune due to

ally when in flower. Best planting height 15 in–2½ ft (38–76 cm).
Problems Suffers badly from attacks of mildew which can lead to complete defoliation. Use a proprietary control.
Similar forms of interest *L. t.* 'Alba' White flowers and red fruits. *L. t.* 'Hack's Red' Deep rose pink flowers and red fruits.
Average height and spread
Five years
10 × 10 ft (3 × 3 m)
Ten years
10 × 13 ft (3 × 4 m)
Twenty years
10 × 20 ft (3 × 6 m)
Protrudes up to 2 ft (60 cm) from support if fan-trained, 6 ft (1.8 m) untrained.

LONICERA × TELLMANNIANA

KNOWN BY BOTANICAL NAME

Caprifoliaceae   *Woody Climber*
Deciduous

Its unusual colouring makes it possibly one of the most interesting of flowering honeysuckles.

Origin From Hungary.
Use As a summer-flowering climber for walls, fences and pillars.
Description *Flower* 2 in (5 cm) long, coppery yellow in colour, when in bud flushed red. Produced in good-sized clusters from early to mid summer. *Foliage* Oval, produced in pairs opposite each other at source, forming an interesting double shield effect. Grey/green in colour. *Stem* Light grey/green to grey/brown in colour. Not self-supporting but does cling by twisting growth. Medium to fast growing. *Fruit* Small red fruits may be produced in hot summers.
Hardiness Tolerates a minimum winter temperature of 14°F (−10°C).
Soil requirements Alkaline to acid. Requires a well drained but good, moisture retentive soil. The inclusion of large quantities of organic material will help growth and keep roots in the cool.
Sun/Shade aspect Best on a south-west wall. Flowers in full sun to light shade, roots in deep shade.

Pruning Not normally required, but can be reduced in size if necessary with no ill effects. The occasional removal of lower growth will encourage new shoots.
Training Allow to twine through wires and trellis. Normally needs no assistance other than when young or heavy at maturity.
Propagation and nursery production From semi-ripe cuttings taken in mid summer. Should always be purchased container grown; best planting height 1½–2½ ft (45–76 cm). Available from most garden centres and nurseries.
Problems Can be attacked by mildew in warm, humid summers. Control with a proprietary control.
Similar forms of interest None.
Average height and spread
Five years
5 × 5 ft (1.5 × 1.5 m)
Ten years
12 × 12 ft (3.7 × 3.7 m)
Twenty years
15 × 15 ft (4.6 × 4.6 m)
Protrudes up to 3 ft (91 cm) from support.

Lonicera × *tellmanniana* in flower

Lonicera tragophylla in flower

its height and rambling nature.
Training Allow to ramble and twine through large shrubs and trees.
Propagation and nursery production From semi-ripe cuttings. May be difficult to propagate without specialist equipment. Always purchase container grown; will have to be sought from specialist nurseries. Best planting height 1½–4 ft (45 cm–1.2 m).
Problems Its requirement for a shady woodland situation must not be overlooked, or its performance will be disappointing.
Similar forms of interest *L. hildebrandiana* (Giant honeysuckle) Tender, requiring very favourable conditions. Up to 80 ft (25 m) with fragrant, creamy-white flowers, which deepen to orange with age. Flowering from mid summer to early autumn. From Burma.
Average height and spread
Five years
15 × 10 ft (4.6 × 3 m)
Ten years
30 × 20 ft (9 × 6 m)
Twenty years
40 × 30 ft + (12 × 9 m +)
Protrudes up to 2 ft (60 cm) from support.

LYCIUM BARBARUM

DUKE OF ARGYLL'S TEA TREE

Solanaceae   *Wall Shrub*
Deciduous

A very useful shrub for coastal areas, standing up to salt-laden winds and adapting well to fan-training.

Origin From China, but naturalized throughout Europe and western Asia.
Use As a fan-trained shrub for walls and fences.
Description *Flower* Small, ½ in (1 cm) long, purple, trumpet-shaped flowers in clusters 1¾–3 in (4.5–7.5 cm) in diameter, produced from each leaf axil from early to mid summer. *Foliage* Leaves narrow, 1–4 in (2.5–10 cm) long, sea green, giving some yellow autumn colour. *Stem* Upright when young, quickly forming a spreading, arching wall shrub. Grey to grey/green with sparsely distributed small spines. Medium to fast growth rate. *Fruit* Small, egg-shaped, orange/red fruits in autumn.
Hardiness Tolerates a minimum winter temperature of 4°F (−15°C).
Soil requirements Light, open, sandy soil. Dislikes waterlogging.
Sun/Shade aspect Tolerates all but the most severe of aspects, including salt-laden winds. Full sun, dislikes shade.
Pruning Reduce one third of oldest shoots on mature shrubs to ground level in early spring to encourage growth of new shoots for following year.

Lycium chilense in flower

Training Requires wires or individual anchor points to secure and encourage a fan-trained shape.
Propagation and nursery production From seed or semi-ripe cuttings taken in early summer. Purchase container grown; availability varies but may be found in coastal areas. Best planting height 15 in–2 ft (38–60 cm).
Problems Can become a little invasive if happily situated, but if not, may be difficult to establish.
Similar forms of interest *L. chilense* Spreading branches covered with yellow, white and purple funnel-shaped flowers, mid to late summer, followed by red, egg-shaped fruits. A spineless variety more open and of slightly less than average height and spread. Less hardy than *L. barbarum*. From Chile.
Average height and spread
Five years
8 × 8 ft (2.4 × 2.4 m)
Ten years
12 × 12 ft (3.7 × 3.7 m)
Twenty years
16 × 16 ft (4.9 × 4.9 m)
Protrudes up to 4 ft (1.2 m) from support.

Magnolia grandiflora in flower

MAGNOLIA GRANDIFLORA

EVERGREEN MAGNOLIA, LAUREL MAGNOLIA, SOUTHERN MAGNOLIA, BULL BAY

Magnoliaeceae   *Wall Shrub*
Evergreen

When grown in the correct position, given space and time to mature, this is a truly spectacular evergreen flowering shrub.

Origin From south-eastern USA.
Use As a freestanding shrub or even small tree, grown adjacent to large walls or fences, where it benefits from the protection in colder areas.
Description *Flower* Up to 10 in (25 cm) across, creamy white, very fragrant flowers, produced in late spring, summer and early autumn. Late flowers susceptible to early autumn frost damage. *Foliage* Leaves elliptic, 8–10 in (20–25 cm) long, attractive bright green when young, maturing to duller green, with brown, felted undersides. *Stem* Strong, upright, becoming branching with age, forming a tall shrub. Mid to dark green. Medium rate of growth. *Fruit* Green fruit pods 4 in (10 cm) long, following flowering.
Hardiness Tolerates a minimum winter temperature of 14°F (−10°C).
Soil requirements Does well on most soils including alkaline as long as there is 6 ft (1.8 m) of good topsoil above underlying alkaline soil.
Sun/Shade aspect Requires a sheltered aspect in full sun to mid shade.
Pruning None required but can be trimmed to control size.

Magnolia grandiflora 'Exmouth' in flower

Training Individual anchor points will be required to secure main stems as they develop.
Propagation and nursery production From semi-ripe cuttings taken in early summer. Purchase container grown; one or other of the varieties likely to be found in most general nurseries and some good garden centres. Best planting height 2–4 ft (60 cm–1.2 m).
Problems May take five to six years to come into flower.
Similar forms of interest *M. g.* 'Exmouth' Richly scented, large, creamy white flowers, produced at a relatively early age in its life span. Dark green, polished, narrow foliage with red or brown felted undersides. A useful variety for colder areas. *M. g.* 'Ferruginea' Scented, large, white flowers. Leaves elliptic to ovate, dark shiny green, brown and heavily felted undersides. Upright and of average height but slightly less spread. More tender than other varieties and should be planted only in very mild areas. *M. g.* 'Goliath' Scented, white, globe-shaped flowers, produced three to five years after planting. Foliage elliptic, slightly concave with rounded ends; dark glossy green to light green, grey/green undersides. Slightly less than average spread. Not suited to temperatures below 23°F (−5°C). *M. g.* 'Maryland' Large, fragrant white flowers of more open shape, produced two years after planting. Foliage elliptic to ovate, mid green, shiny upper surfaces and grey/brown undersides. Slightly less than average height and spread. *M. g.* 'Russet' Russet brown buds to leaves. Creamy white flowers. Less hardy. *M. g.* 'Samuel Sommer' Good flowering variety but difficult to obtain. Less hardy.
Average height and spread
Five years
6 × 6 ft (1.8 × 1.8 m)
Ten years
12 × 12 ft (3.7 × 3.7 m)
Twenty years
26 × 26 ft (8 × 8 m)
Protrudes up to 12 ft (3.7 m) from wall or fence.

MAGNOLIA (Large-growing, Star-flowered Forms)

STAR MAGNOLIA

Magnoliaceae   *Wall Shrub*
Deciduous

Large, early spring flowering shrubs for a very attractive flower display. Even more spectacular when fan-trained.

Origin From Japan
Use As a freestanding shrub grown adjacent to large walls and fences or for fan-training.
Description *Flower* Multi-petalled, star-shaped, white or pink depending on variety,

fragrant flowers produced in small numbers for 15–20 years after planting, after which flowering increases to give a glorious display in mid to late spring. *Foliage* Leaves elliptic, 2½–4in (6–10cm) long, light to mid green. Some yellow autumn colour. *Stem* Strong, upright, becoming branching with age, eventually forming a very dense, twiggy framework. Dark green to green/brown. Medium growth rate. *Fruit* Small green fruit capsules in late summer.

Hardiness Tolerates a minimum winter temperature of 0°F (−18°C).

Soil requirements Most soil types, tolerates alkaline conditions provided 18in (45cm) of topsoil is provided.

Sun/Shade aspect Tolerates all but the most severe of aspects. Requires protection from early morning sun. Prefers full sun, tolerates light shade.

Pruning None required.

Training Requires individual anchor points or wires to achieve a fan-trained effect or allow to grow freestanding.

Propagation and nursery production From layers or semi-ripe cuttings taken in early summer. Purchase container grown or root-balled (balled-and-burlapped); best planting height 1½–3 ft (60–91 cm). Generally available from garden centres and nurseries.

Problems Some varieties, such as *M. kobus*, are very slow to come into flower and can take as long as 15 years to produce a full display of blooms.

Forms of interest *M. kobus* (Northern Japanese magnolia, Kobus magnolia) Fragrant white flowers produced only after 10–15 years from date of planting. *M. × loebneri* A cross between *M. kobus* and *M. stellata*, which from an early age produces a profusion of multi-petalled, fragrant white flowers in early to mid spring. Does well on all soil types, including alkaline. Reaches two thirds average height and spread. Of garden origin. *M. × l.* 'Leonard Messel' Fragrant, multi-petalled flowers are deep pink in bud, opening to lilac/pink. Said to be a cross between *M. kobus* and *M. stellata* 'Rosea'. Of garden origin. *M. × l.* 'Merrill' Large, fragrant, star-shaped flowers produced from an early age on a shrub two thirds average height and spread. From USA. *M. salicifolia* Fragrant, white, star-shaped flowers with six narrow petals in mid spring. Slightly more than average height but slightly less spread. Leaves, bark and wood are lemon-scented if bruised. From Japan.

Average height and spread
Five years
8 × 8 ft (2.4 × 2.4 m)
Ten years
15 × 15 ft (4.6 × 4.6 m)
Twenty years
26 × 26 ft (8 × 8 m)
Protrudes 3 ft (91 cm) from support if fan-trained, 15 ft (4.6 m) or more untrained.

Magnolia × soulangiana **in flower**

Origin From France.

Use As a fan-trained or freestanding shrub grown adjacent to large walls or fences.

Description *Flower* Light pink with purple shading in centre and at base of each petal. Flowers produced before leaves in early spring; buds large with hairy outer coat. Some secondary flowering in early summer. *Foliage* Leaves elliptic to ovate, 3–6in (7.5–15 cm) long, light green to grey/green. Some yellow autumn colour. *Stem* Upright, strong when young and light grey/green. In maturity branches become very short, twiggy and almost rubbery in texture. Medium rate of growth. *Fruit* May produce long, orange/red, pod-shaped fruits in hot summers.

Hardiness Tolerates a minimum winter temperature of 4°F (−15°C).

Soil requirements Does well on heavy clay soils and most other types, except extremely alkaline areas which will lead to chlorosis.

Sun/Shade aspect Requires a moderately sheltered aspect and must be planted away from early morning spring sun, otherwise flowers frozen by late spring frosts thaw too quickly and cell damage causes browning.

Pruning None required, but remove any small crossing branches in winter to prevent rubbing.

Training Will require individual anchor points or wires for tying in when grown fan-trained.

Propagation and nursery production From layers or semi-ripe cuttings taken in early summer. Purchase container grown or root-balled (balled-and-burlapped). *M. × soulangiana* easy to find but some varieties must be sought from specialist nurseries. Best planting height 2–4 ft (60 cm–1.2 m)

Problems Can take up to five years or more to flower well.

Magnolia × soulangiana 'Alba Superba' **in flower**

Similar forms of interest *M. s.* × 'Alba Superba' (syn. *M. × s.* 'Alba') Large, scented, pure white, erect, tulip-shaped flowers, flushed purple at base. Growth upright and strong, but forms slightly less spread than the parent. *M. × s.* 'Alexandrina' Large, upright, white flowers with purple-flushed bases. A good, vigorous, upright, free-flowering variety, sometimes difficult to obtain. *M. × s.* 'Amabilis' Ivory white, tulip-shaped flowers, flushed light purple inside at base of petals. Upright habit. May have to be obtained from specialist nurseries. *M. × s.* 'Brozzonii' Longer than average white flowers with purple shading at the base. May have to be sought from specialist nurseries. *M. × s.* 'Lennei' Flowers goblet-shaped with fleshy petals rose purple outside, creamy white stained purple on inside, in mid to late spring. Sometimes limited repeat flowering in autumn. Broad, ovate leaves, up to 10–12 in (25–30 cm) long. *M. × s.* 'Lennei Alba' Ivory white, extremely beautiful goblet-shaped flowers, presenting themselves upright along branches. May need to be obtained from specialist nurseries. Slightly more than average spread. *M. × s.* 'Picture' Purple outer colouring to petals, white inside. Flowers

Magnolia × loebneri 'Merrill' **in flower**

Magnolia × soulangiana **'Lennei' in flower**

borne erect, often appearing early in the shrub's lifespan. Leaves up to 10 in (25 cm) long. Somewhat upright branches, reaching less than average spread. Best sought from specialist nurseries. *M. × s.* **'Speciosa'** White flowers with very little purple shading, leaves smaller than average. Slightly less than average height and spread. Best sought from specialist nurseries. *M. liliiflora* **'Nigra'** Buds resemble slender tulips, gradually opening to a reflexed star shape. Flowers deep purple outside, creamy white stained purple inside, late spring through to early summer. From Japan.

Average height and spread
Five years
8 × 8 ft (2.4 × 2.4 m)
Ten years
16 × 16 ft (4.9 × 4.9 m)
Twenty years
26 × 26 ft (8 × 8 m)
Protrudes up to 3 ft (91 cm) from support if fan-trained, 20 ft (6 m) untrained.

MAGNOLIA STELLATA

STAR MAGNOLIA, STAR-FLOWERED MAGNOLIA

Magnoliaceae   **Wall Shrub**
Deciduous

A well-loved, exceptionally beautiful early spring-flowering garden shrub which, when grown on a wall or fence, gains an added dimension.

Origin From Japan.
Use As a freestanding or fan-trained wall or fence shrub where it gains extra protection.
Description *Flower* Slightly scented, white, multi-petalled, star-shaped flowers 2–2½ in (5–6 cm) wide, borne in early spring before leaves. Usually flowers within two years of planting. *Foliage* Leaves medium-sized, elliptic, 2–4 in (5–10 cm) long, light green, giving some yellow autumn colouring. *Stem* Upright when young, quickly branching and spreading. Can be trained into a fan shape or close-branched network adjacent to walls and fences. Grey/green. Slow growth rate.
Fruit Insignificant.
Hardiness Tolerates a minimum winter temperature of 4°F (−15C°).
Soil requirements Any soil type except extremely alkaline.
Sun/Shade aspect Plant in a position where the shrub will not get early morning sun in early spring, so allowing frozen flowers to thaw out slowly and incur less tissue damage and browning. Prefers full sun to light shade.
Pruning None required.
Propagation and nursery production From layers or cuttings taken in early summer. Purchase container grown or root-balled (balled-and-burlapped); normally available from

Magnolia stellata **in flower**

garden centres and nurseries, especially in spring when in flower. Best planting height 15 in–2½ ft (38–76 cm).
Training Will require individual anchor points or wires for tying in when grown fan-trained.
Problems Young plants when purchased may look small and misshapen due to their slow growth rate, and are expensive, but the investment is well worthwhile.
Similar forms of interest *M. s.* **'King Rose'** A variety with good pink flowers. *M. s.* **'Rosea'** Star-shaped flowers deep pink in bud, opening to flushed pink. *M. s.* **'Royal Star'** Slightly larger white flowers with numerous petals making a very full star shape. *M. s.* **'Rubra'** Flowers multi-petalled and purple/pink, deeper colouring while in bud. Rather scarce in production. *M. s.* **'Water Lily'** Larger flowers with more petals. Extremely attractive but usually a little weaker in constitution than the parent.

Average height and spread
Five years
4 × 4 ft (1.2 × 1.2 m)
Ten years
7 × 7 ft (2.1 × 2.1 m)
Twenty years
15 × 15 ft (4.6 × 4.6 m)
Protrudes up to 3 ft (91 cm) from support if fan-trained, 10 ft (3 m) untrained.

MAGNOLIA
(Summer-flowering Forms)

KNOWN BY BOTANICAL NAME

Magnoliaceae   **Wall Shrub**
Deciduous

Summer-flowering magnolias are among the most magnificent of summer shrubs and adapt well to fan-training on walls and fences.

Origin From western China.
Use As a fan-trained or freestanding shrub adjacent to large walls and fences.
Description *Flower* Hanging, egg-shaped buds, opening into outward-facing, 3–4 in (7.5–10 cm) wide, white, fragrant, cup-shaped flowers. Flowers mainly in late spring and early summer, then intermittently until late summer. *Foliage* Leaves elliptic, 4–5 in (10–12 cm) long, light grey/green giving some autumn colours. Liable to be blackened by hard early autumn frosts, making them unsightly. *Stem* Strong, forming a goblet-shaped shrub. Grey/green. Medium growth rate. *Fruit* Crimson-red fruit clusters in late summer, early autumn.
Hardiness Tolerates a minimum winter temperature of 4°F (−15°C).

Soil requirements Does well on most soils, only disliking extreme alkalinity.
Sun/Shade aspect Requires some protection in exposed aspects. Best in full sun through to medium shade.
Pruning None required other than that needed for training. May be reduced in size with care; best done in late autumn or early winter.
Training Will require individual anchor points or wires for tying when grown fan-trained.

Magnolia × watsonii

Propagation and nursery production From layers. Purchase container grown or root-balled (balled-and-burlapped); fairly easy to find in nurseries, less often stocked by garden centres. Best planting height 1½–4 ft (45 cm –1.2 m).
Problems Often planted in areas where it is unable to reach its full size and potential.
Forms of interest *M. denudata* (syn. *M. conspicua*) Fragrant, cup-shaped, pure white flowers, with broad, thick, fleshy petals, produced mid spring to early summer. Foliage ovate with rounded ends, 3–6 in (7.5–15 cm) long, mid to grey/green. Upright habit, becoming rounded with time. Benefits well from a wall or fence. *M. × highdownensis* Fragrant, white, hanging flowers with purple central cone 4 in (10 cm) across. Good on very chalky soils. Of slightly less than average height and spread. Rather hard to find except in specialist nurseries. Possibly a clone of *M. wilsonii*. *M. hypoleuca* (syn. *M. obovata*) Fragrant, white to creamy white flowers up to 6 in (15 cm) across with central crimson stamens, early summer, followed by interesting fruit clusters in late summer, early autumn. Obovate leaves up to 6 in (15 cm) long, grey/green. Dislikes alkaline soils, best on neutral to acid types. Benefits from the protection of a wall or large fence. From

Japan. *M. sieboldii* (syn. *M. parviflora*) Fragrant, white, cup-shaped flowers 3–3½ in (7.5–9 cm) across, with rose pink to crimson central stamens. Stems upright when young, spreading with age. Foliage grey/green with some yellow autumn colour. Two thirds average height and spread. *M. sinensis* Lemon-scented, white, hanging flowers with red central cones, 4–6 in (10–15 cm) across, early summer. Foliage grey/green, 5 in (12 cm) long, obovate. Slightly less than average height and slightly more than average spread. From western China. *M. virginiana* (Swamp bay, sweetbay magnolia) Fragrant creamy white flowers 2 in (5 cm) across, globular and slightly hanging, early to late summer. Ovate to elliptic, semi-evergreen leaves with glossy upper surfaces and blue/white undersides. Tolerates some alkalinity as long as 18 in (45 cm) of good topsoil is available. Rather hard to find. From eastern USA. *M. × watsonii* Strongly scented, creamy white, saucer-shaped flowers facing upwards, with rosy crimson anthers and pink sepals, 5½–6 in (13–15 cm) across, early to mid summer. Grey/green, oval, leathery leaves. Will tolerate some alkalinity provided 18 in (45 cm) of good topsoil is available. Extremely difficult to find. Benefits well from wall or fence protection. *M. wilsonii* Slightly scented, hanging, white, saucer-shaped flowers with crimson stamens, in late spring to early summer. Foliage elliptic to lanceolate, 4–5 in (10–12 cm) long, grey/green. A wide, spreading shrub. Will tolerate a limited amount of alkalinity, provided 18 in (45 cm) of topsoil is available. Scarce in production. From western China.

Average height and spread
Five years
7 × 7 ft (2.1 × 2.1 m)
Ten years
15 × 15 ft (4.6 × 4.6 m)
Twenty years
26 × 26 ft (8 × 8 m)
Protrudes up to 3 ft (91 cm) from support if fan-trained, 13 ft (4 m) untrained.

MAHONIA × 'CHARITY'

KNOWN BY BOTANICAL NAME
Berberidaceae   *Wall Shrub*
Evergreen

Among the aristocrats of tall-growing, winter-flowering shrubs, particularly adaptable for growing against walls or fences where a tall evergreen is required.

Origin Of garden origin.
Use As a freestanding or fan-trained tall-growing shrub for large walls and fences.

Mahonia × 'Charity' in flower

Description *Flower* Circular groups of bright yellow racemes, up to 8–10 in (20–25 cm) long, in late autumn through to mid winter. Flowers borne at terminals of upright shoots, upright at first, becoming more arching with age. *Foliage* Eight to 16 spined leaflets, each 1½–2 in (4–5 cm) long, make up a pinnate leaf in excess of 12 in (30 cm) in length and 4 in (10 cm) in width. Light grey/green in colour. *Stem* Rigid, upright, light grey/green becoming grey/brown. Medium growth rate. *Fruit* Racemes of blue/black fruits with white downy covering follow flowers in mid to late winter.
Hardiness Tolerates a minimum winter temperature of 4°F (−15°C). Young foliage may suffer leaf scorch in severe winters, but normally rejuvenates the following spring. Some damage may be caused to new foliage by late spring frosts.
Soil requirements Prefers open, leafy, moist, peaty soil. Tolerates wide range of types but may show distress on very dry, alkaline soils.
Sun/Shade aspect Tolerates all aspects. Full sun to very light shade. In deeper shade will become more open and lax in habit.
Pruning It is very important that young plants have terminal clusters of foliage removed after flowering to encourage branching, otherwise growth becomes very tall and upright, often placing the flowers at a height where they cannot be appreciated to the full.
Training Normally requires no support and is freestanding but may also be fan-trained.
Propagation and nursery production From semi-ripe cuttings taken in early summer. Very slow to root. Purchase container grown or root-balled (balled-and-burlapped). Easy to find from good garden centres and nurseries. Best planting height 12–15 in (30–38 cm).
Problems None if pruning advice is followed.
Similar forms of interest *M.* 'Buckland' Good, thick racemes of yellow flowers in late autumn to early winter, followed by blue/black fruits. Flowers are initially upright in habit but become lax with age. Foliage pinnate, mid green and 10 in (25 cm) long, but with smaller leaflets than *M. ×* 'Charity'. Upright form, slightly more than average spread. Obtainable from specialist nurseries. *M.* 'Lionel Fortescue' Upright racemes of yellow scented flowers in clusters of up to 15. Good large foliage, up to 12 in (30 cm) long and pinnate. One third more than average spread. Obtainable from specialist nurseries. *M. lomariifolia* Flowers deep yellow, produced from terminal clusters of up to 20 flower spikes, racemes retaining an upright habit, 6–10 in (15–25 cm) high. Leaves pinnate, dark to olive-green, up to 2 ft (30 cm) long, with moderately small leaflets. Minimum winter temperature 23°F (−5°C). Benefits from the protection of a wall or

fence. Reaches one third more than average height. *M.* 'Winter Sun' Upright racemes of scented yellow flowers in winter. Dark green, pinnate leaves. Slightly less than average height and spread. Usually found in specialist nurseries.
Average height and spread
Five years
7 × 4 ft (2.1 × 1.2 m)
Ten years
12 × 6 ft (3.7 × 1.8 m)
Twenty years
16 × 10 ft (4.9 × 3 m)
Protrudes up to 3 ft (91 cm) from support if fan-trained, 10 ft (3 m) untrained.

MAHONIA FREMONTII

KNOWN BY BOTANICAL NAME
Berberidaceae   *Tender Wall Shrub*
Evergreen

Given the right conditions, this attractively-foliaged mahonia offers a great deal of all-year-round interest.

Origin From south-west USA.
Use As a wall shrub for sheltered walls and fences.
Description *Flower* Racemes up to 2 in (5 cm) long of yellow flowers in mid to late spring. *Foliage* Up to 4 in (10 cm) long, consisting of two to three pairs of leaflets, curled when juvenile, often with coarsely toothed edges. Grey/green with white undersides. *Stem*

Mahonia fremontii

Grey/green becoming grey/brown. Predominantly upright. Slow growth rate. *Fruit* Blue/black, oval-shaped berries follow flowers.
Hardiness Tolerates a minimum winter temperature of 14°F (−10°C) as long as there is adequate shelter from wind.
Soil requirements Tolerates all soil conditions except extremely dry.
Sun/Shade aspect Must have a sheltered aspect with protection from cold winds. Performs best in light shade but will tolerate full sun.
Pruning Not normally required but individual branches can be removed if necessary.
Training Tie to individual anchorage points or wires.
Propagation and nursery production From semi-ripe cuttings taken in early to mid summer. Should always be purchased container grown; will have to be sought from specialist nurseries. Best planting height 6 in–2 ft (15–60 cm).
Problems Slow to establish and scarce in production.
Similar forms of interest None.
Average height and spread
Five years
3 × 3 ft (91 × 91 cm)
Ten years
6 × 4 ft (1.8 × 1.2 m)
Twenty years
8 × 6 ft (2.4 × 1.8 m)
Protrudes up to 18 in (45 cm) from support if fan-trained, 4 ft (1.2 m) untrained.

MALUS
(Fruiting Forms)

FRUITING CRAB, CRAB APPLE

Rosaceae   **Hardy Tree**
Deciduous

A group of trees not often enough used for fan-training even though they adapt well for the purpose.

Origin Some varieties of natural origin, others of garden or nursery extraction.
Use As large, fan-trained, fruiting trees for walls and fences.
Description *Flower* White, pink or wine-red, depending on variety. Flowers 1–1½ in (2.5–4 cm) across, produced singly or in multiple heads of five to seven flowers in mid spring. *Foliage* Ovate, tooth-edged, 2 in (5 cm) long. Green or wine-red, depending on variety. Some yellow autumn colour. *Stem* Light green to grey/green when young, becoming green/brown. Moderately upright, becoming spreading and branching. Moderate rate of growth. *Fruit* Colours from yellow, orange/red, through to purple/red. Shaped like miniature apples, 1–2 in (2.5–5 cm) wide. Produced in late summer and early autumn, all edible and used for making jelly.
Hardiness Tolerates a minimum winter temperature of 4°F (−15°C).
Soil requirements Tolerates most soil conditions; dislikes extreme waterlogging.
Sun/Shade aspect Tolerates all but the most severe of aspects. Full sun to light shade.
Pruning Prune young trees hard in spring following planting. Select and train resulting five to seven shoots and tie into a fan-trained shape. In subsequent years, remove all side growths back to two points from their origin and maintain original main branches in fan shape.
Training Requires tying to wires or individual anchorage points.
Propagation and nursery production Mainly grafted or budded on to wild apple understock. Purchase bare-rooted or container grown; most varieties readily available from general or specialist nurseries, also sometimes from garden centres. Choose one- to two-year-old trees 3–6 ft (91 cm–1.8 m) in height, ensuring that they are grafted at the base.
Problems Liable to fungus diseases such as apple scab and apple mildew, damaging both foliage and fruit. Some stem canker may occur; remove by pruning and treat cuts with pruning compound.

Forms of interest *M.* **'Dartmouth'** White flowers followed by sizeable red/purple fruits. Green foliage. *M.* **'Dolgo'** A white-flowering form with yellow fruits held well into autumn. Foliage mid green, often with light purplish hue. Used as a universal pollinator for garden or orchard trees. *M.* **'Golden Hornet'** White flowers, followed by a good crop of bright yellow fruits which may remain on the tree well into winter. Green foliage. Can be used as a universal pollinator for garden or orchard trees. *M.* **'John Downie'** Pear-shaped fruits, 1 in (2.5 cm) long, bright orange shaded scarlet. One of the best for making jelly. Green foliage. More susceptible to apple scab and mildew than most varieties. *M.* **'Professor Sprenger'** Flowers pink in bud, opening to white. Good crop of amber fruits retained until mid winter. Green foliage. Half average height and spread. *M.* **'Red Sentinel'** White flowers followed by deep red fruits maintained beyond mid winter. Green foliage. *M.* × *robusta* (Siberian crab) Two forms available, both with pink-tinged white flowers: *M.* × *r.* **'Red Siberian'** Large crop of red fruits, green foliage; *M.* × *r.* **'Yellow Siberian'** Yellow fruits, green foliage; *M. sylvestris* (Common crab apple) Flowers white shaded with pink. Fruits yellow/green, sometimes flushed red, 1–1½ in (3–4 cm) wide. Light green foliage. Not readily available; must be sought from specialist nurseries. The parent of many ornamental crabs and the garden apple. *M.* **'Wintergold'** White flowers, pink in bud. Good crop of yellow fruit retained into winter. Green foliage with autumn colour. *M.* **'Wisley'** Strong-growing tree with limited purple/red to bronze/red flowers, with reddish shading and slight fragrance. Large purple/red fruits in autumn which, although sparse, are attractive for their size. Purple to purple/green foliage.

Average height and spread
Five years
12 × 12 ft (3.7 × 3.7 m)
Ten years
24 × 24 ft (7.3 × 7.3 m)
Twenty years
30 × 30 ft (9 × 9 m)
Protrudes up to 3 ft (91 cm) from support.

Malus floribunda **in flower**

Malus **'John Downie' in fruit**

MALUS
(Green-leaved Flowering Forms)

FLOWERING CRAB, CRAB APPLE

Rosaceae   **Hardy Tree**
Deciduous

A group of trees responding well to fan-training, giving a good flowering display in a wide range of aspects.

Origin Mostly of garden origin; a few direct species.
Use As a fan-trained tree for walls and fences.
Description *Flower* White, pink or bi-coloured pink and white flowers 1½ in (4 cm) across, singly or in clusters of five to seven flowers, producing a mass display. *Foliage* Green, ovate, 2 in (5 cm) long, tooth-edged, giving some yellow autumn colour. *Stem* Purple/red to purple/green. Upright when young, but very trainable. Medium to fast rate of growth. *Fruit* Normally green to yellow/green and of little attraction.
Hardiness Tolerates a minimum winter temperature of 4°F (−15°C).
Soil requirements Tolerates most soil conditions; dislikes waterlogging.
Sun/Shade aspect Tolerates all aspects and full sun to light shade, preferring full sun.

Pruning Prune young trees hard in spring following planting. Select and train resulting five to seven shoots and tie into a fan-trained shape. In subsequent years, remove all side growths back to two buds from their origin and maintain main branches in fan shape.
Training Secure to wires or individual anchor points.
Propagation and nursery production From budding or grafting on to wild apple under-stock. Best planting height for fan-training 3–5 ft (91 cm–1.5 m). Purchase bare-rooted or container grown one- to two-year-old trees, ensuring they are grafted at the base; most varieties readily available from garden centres and general nurseries.
Problems Can suffer from severe attacks of apple mildew and lesser attacks of apple scab.
Forms of interest *M. baccata* White flowers up to 1½ in (4 cm) across in mid spring, followed by bright red, globe-shaped fruits. One third more than average height and spread. From eastern Asia and north China. Normally sold in the form *M. baccata mandshurica*, which has slightly larger fruits. *M. floribunda* A pendulous variety, branches on mature trees reaching to the ground. Flowers rose-red in bud, opening to pink, finally fading to white, produced in mid to late spring in great profusion. Foliage smaller than most, ovate and deeply toothed. *M. hupehensis* Fragrant flowers soft pink in bud, opening to white. Fruits yellow with red tints. Two thirds average height and spread. Somewhat upright in habit. From China and Japan. *M. 'Katherine'* Semi-double flowers, pink in bud, finally white. Bright red fruits with yellow flushing. Two thirds average height and spread. Not readily found in production, but worth searching for. *M. 'Lady Northcliffe'* Carmine-red buds, opening to white with blush shading. Fruits small, yellow and round. Two thirds average height and spread. Not always available. *M. 'Magdeburgensis'* A tree similar to cultivated apple. Flowers deep red in bud, opening to blush-pink, finally becoming white. In-significant fruits, light green to green/yellow. Two thirds average height and spread. Not readily available. *M. sargetii* Foliage oblong with three lobes up to 2¼ in (6 cm) long. Some yellow autumn colour. Flowers pure white with greenish centres in clusters of five and six; petals overlap. Fruits bright red. Very floriferous. One third average height and spread. From Japan. *M. spectabilis* Grey/green foliage susceptible to apple scab. Flowers rosy red in bud, opening to pale blush pink, up to 2 in (5 cm) across and borne in clusters of six to eight in early spring. Fruits yellow and globe-shaped. From China. *M. 'Strathmore'* Light green foliage and a pro-fusion of pale pink flowers. *M. toringoides* Foliage ovate to lanceolate, up to 3 in (7.5 cm) long, deeply lobed when new; that produced on older wood is less indented. Pastel autumn colours. Flowers light pink in bud opening to creamy-white, produced in clusters of six to eight. Fruit globe-shaped, yellow with scarlet flushing. Two thirds average height and spread. From China. *M. transitoria* Small-lobed foliage, small pink/white flowers and rounded yellow fruits. Excellent autumn colour. Two thirds average height and spread. From north-west China. *M. trilobata* Leaves maple-shaped, deeply lobed, mid to dark green with good autumn colour. White flowers, followed by infrequently produced yellow fruits. Two thirds average height and spread. Scarce in production and will have to be sought from specialist nurseries. From eastern Mediterranean and north-eastern Greece. *M. 'Van Eseltine'* Flowers rose/scar-let in bud, opening to shell-pink, semi-double. Small yellow fruits. Two thirds average height and spread.

Average height and spread
Five years
12 × 12 ft (3.7 × 3.7 m)
Ten years
18 × 18 ft (5.5 × 5.5 m)
Twenty years
25 × 25 ft (7.6 × 7.6 m)
Protrudes up to 3 ft (91 cm) from support.

MALUS
(Purple-leaved Forms)

PURPLE-LEAVED CRAB APPLE
Rosaceae　　　　　　　　*Hardy Tree*
Deciduous

These attractive foliage trees respond well to fan-training where pruning encourages even better foliage display but may lessen flowering.

Origin Of garden or nursery origin.
Use As fan-trained trees for large walls and fences.
Description *Flower* Wine-red to purple/red flowers up to 1 in (2.5 cm) across in clusters of five to seven, produced in great profusion in mid spring. *Foliage* Ovate, sometimes tooth-ed, purple/red to purple/bronze. *Stem* Purple/red to purple/green. Upright when young, quickly becoming spreading and branching, particularly with training. Moder-ate rate of growth. *Fruit* Small, wine-red fruits in early autumn, sometimes inconspic-uous against the purple foliage.
Hardiness Tolerates a minimum winter tem-perature of 4°F (−15°C).
Soil requirements Does well on most soils; dislikes very poor or waterlogged types.
Sun/Shade aspect Tolerates all aspects. Full sun to very light shade. Deeper shade spoils foliage colour and shape of tree.
Pruning Prune young trees hard in spring following planting. Select and train resulting five to seven shoots and tie into a fan-trained shape. In subsequent years, remove all side growths back to two buds from their origin and maintain original main braches in fan shape.

Malus 'Royalty' in flower

Training Requires tying to wires or individual anchor points.
Propagation and nursery production From budding or grafting on to wild apple under-stocks. For fan-training best purchased 3–5 ft (91 cm–1.5 m) high as one- to two-year-old trees, ensuring that they are grafted at the base. Available from garden centres and nur-series.
Problems Can suffer severe attacks of apple mildew, lesser attacks of apple scab.
Forms of interest *M.* × *eleyi* Large red/purple flowers, followed by conical, purple/red fruits. Foliage red/purple, up to 4 in (10 cm) long. Initial growth somewhat weak. *M.* × *lemoinei* An early nursery cross of merit. Purple foliage, crimson/purple flowers, bronze/purple fruits. *M. 'Liset'* Modern hybrid with good foliage and flowers,

adequate dark red fruit. *M. 'Neville Cope-man'* Foliage dull wine red, flowers pink/purple, fruits purple. Not as intensely col-oured as some, but worth consideration. Not easy to find in commercial production. *M. 'Profusion'* An early nursery cross. Good purple/wine flowers, purple/red fruits and copper/crimson spring foliage. Originally ro-bust but recently appears to be losing its over-all vigour. *M. 'Red Glow'* Wine red flowers, large leaves and fruit. A good introduction, but not readily available. *M. 'Royalty'* Pos-sibly the best purple/red leaf form. Large, di-sease-resistant, wine-coloured foliage and wine-red flowers followed by large purple/red fruits. Becoming more widely available.

Average height and spread
Five years
12 × 12 ft (3.7 × 3.7 m)
Ten years
24 × 24 ft (7.3 × 7.3 m)
Twenty years
30 × 30 ft (9 × 9 m)
Protrudes up to 3 ft (91 cm) from support.

MALUS PUMILA
(Apple Hybrids)

APPLE
Rosaceae　　　　　　　　*Fruit Tree*
Deciduous

Trained fruit trees have for a long time been recognized as attractive wall specimens, not only for their fruit but also their flowers.

Origin Of orchard, garden and nursery ori-gin.
Use As fan-trained, horizontal-trained or cordon-shaped trees for walls and fences.
Description *Flower* Five oval to round petals make up a saucer-shaped flower ½–¾ in (1–2 cm) wide carried in clusters on two-year-old growth or more, from white through shades of pink, profusely borne in mid spring. Specific varieties cross-pollinate each other and in 'Forms of Interest' (UK) each variety is given a pollination number. For best results varieties of the same number should be planted together, but in many cases numbers to either side will also produce good pollin-ation. *Malus* 'Golden Hornet' in close prox-imity to any variety will also encourage fruit-ing as it is a universal pollinator. *Foliage* Oval, up to 4 in (10 cm) long, some with toothed edges. Mid green with yellow autumn colour. *Stem* Grey/green, becoming brown to grey/brown. Stiff, branching. Medium to fast growing. *Fruit* Round edible apples in a wide range of varieties, each with its own texture, colour, flavour and usage for dessert or culi-

Malus pumila 'Discovery'

nary purposes. Green, yellow or shaded orange, red or striped depending on variety.

Hardiness Tolerates a minimum winter temperature of 0°F (−18°C), although late spring frosts may damage flowers and consequently fruiting.

Soil requirements Tolerates a wide range of soil conditions. Some varieties may show distress on extremely wet soils.

Sun/Shade aspect Tolerates all aspects. Best in full sun to very light shade to aid ripening of fruit. Will grow in deeper shade but flowering and fruiting will be decreased.

Pruning For all three shapes of fan, horizontal (espalier) and cordon: shoots produced in late spring and early summer are reduced to two buds from their point of origin in late summer, early autumn.

Malus pumila 'Ellison's Orange'

Training Tie to wires. Purchase pre-trained trees or one-year-old trees for training. For fan-training, cut back to within 12 in (30 cm) of their graft union to induce numerous side shoots which are then trained into a fan shape. For horizontal training, cut back to 18 in (45 cm) from their graft union. This will induce three shoots, two of which are trained horizontally either side of the main stem and the third upwards. After one year the upward shoot is again cut back to 18 in (45 cm) from the lower tier. This again produces three new shoots and the process continues until up to five or six tiers are formed. Once a tier is formed, all growth made between early and late summer is cut back in early autumn to two buds from its origin, so forming a fruiting spur. For cordon fruit trees, all side growth is cut back to two buds from its origin each year in late summer/early autumn to form fruiting spurs.

Propagation and nursery production Should always be grafted or budded on to an understock with a known characteristic. There are many combinations, but in the main, fan, horizontal and cordon fruit trees are propagated on to semi-dwarfing root stocks. Can be planted bare-rooted from early autumn to early spring or container grown as available, with autumn and early spring planting for preference. Pre-trained trees in each of the shapes or young trees can be purchased at a height of 3–5 ft (91 cm–1.5 m) from a garden centre or from a specialist nursery.

Problems May suffer from attacks of apple mildew and apple scab, but a proprietary spray will normally control these diseases. Greenfly and blackfly may be a problem in some circumstances, but again proprietary sprays can be used. Codling moth may attack fruit; prevention is by grease bands applied to the trees in early to mid summer. Apple trees are also susceptible to stem and bud canker; remove affected wood immediately.

Forms of interest
UK
The numbers in brackets denote the pollination group.

Malus pumila 'James Grieve'

LATE SUMMER DESSERT APPLES Harvest in late summer and eat within ten days. **'Beauty of Bath'** (2) Fruit mottled orange, scarlet or yellow. Sharp, crisp, sweet, juicy apple. **'Discovery'** (2) Scarlet flushed. Juicy fruit with good flavour. Resistant to apple scab. **'George Cave'** (2) Crisp, good flavoured fruit, green with orange flush. Gives a good crop. **'Greensleeves'** (2) Green apples with sweet flavour. Can be small in some years. Cross between 'James Grieve' and 'Golden Delicious'. **'Scarlet Pimpernel'** (syn. **Stark Earliest'**) (1) An American variety. Pale green or white with scarlet shading, crisp and juicy. Scab resistant.

EARLY AUTUMN DESSERT APPLES Ripening early autumn and keeping until mid autumn. Eaten immediately after picking, although late varieties will keep for up to 14 days. **'Ellison's Orange'** (4) Very similar to 'Cox's Orange Pippin'. Strong-scented apple with pleasant flavour. Attractive orange/yellow speckling to fruit. **'Fortune'** (3) Very good flavour. Very heavy crop of fruit, needs thinning for best results. **'James Grieve'** (3) Juicy and sharp flavoured. Handsome round fruits with orange stripes. Dislikes waterlogged conditions. Susceptible to canker. **'Laxton's Fortune'** (3) Crisp, well-flavoured fruit with good orange/red striped colouring. May fruit only every other year. Attractively shaped tree resistant to apple scab. **'Merton Worcester'** (3) A 'Cox's Orange Pippin' cross with the colour of Cox and the flavour of 'Worcester Pearmain'. **'Tydeman's Early Worcester'** (3)

Juicy fruit of good colour and a tree of good garden merit. **'Worcester Pearmain'** (3) Crisp, juicy fruit, strawberry flavoured. Allow to ripen on tree and eat immediately.

LATE AUTUMN DESSERT APPLES Ripening in mid autumn and keeping through late autumn into early winter. Pick the fruit when ripe and store until required. **'Charles Ross'** (3) Large, round fruit, shiny, juicy, pleasant flavour. Does not keep well. Scab resistant. Extremely good on alkaline soils. **'Egremont Russet'** (2) Golden russet-skinned with good flavour. Will keep until mid winter in good conditions. Upright growing. Resistant to scab. **'Lord Lambourne'** (2) Bright coloured fruit, juicy and tender with good mellow flavour. Good cropping. **'Spartan'** (3) Dark crimson, almost black fruit when ripe. Crisp, juicy with good flavour. Scab resistant. A variety from Canada. **'Sunset'** (3) Considered by many to be better than Cox, crisp with a rich Cox flavour. Resistant to scab and tolerates a wider soil range than Cox, its parent.

MIDWINTER DESSERT APPLES Ripening late autumn to late winter with good storage. Pick in mid autumn and store until required. **'Blenheim Orange'** (3) Crisp, acid-flavoured orange/yellow fruit. Makes a large, spreading tree. **'Cox's Orange Pippin'** (3) Flavour and texture are unsurpassed, but it is susceptible to all fungus, bacteria and soil deficiencies and can be a most disappointing and difficult tree to grow. Consider 'Sunset' as an alternative. Using as a fan-trained or horizontal tree can improve its growth performance.

Malus pumila 'Worcester Pearmain'

'**Crimson**' (syn. '**Mutsu**') (2) Green/yellow and shiny, juicy and hard with good flavour. Strong growing. From Japan. '**Jupiter**' (3) A Cox cross. Very similar to its parent but larger fruit. Strong grower. '**Kidd's Orange Red**' (3) A derivative of Cox with bright-coloured, good-flavoured fruit. Resistant to both scab and mildew. '**Merton Russet**' (3) Russet-flavoured. Good russet foliage and crimson pink flowers. Small growing and upright in habit. '**Queen Cox**' (3) An improved form of 'Cox's Orange Pippin' with brightly coloured fruit. '**Ribston Pippin**' (2) Strong flavour, crisp, juicy and aromatic. Suffers from canker on poor soils. Strong growing, does well as a cordon. '**Suntan**' (3) Cox flavour. Very similar to 'Cox's Orange Pippin'.

Malus pumila 'Lord Lambourne'

Malus pumila 'Egremont Russet'

NEW YEAR DESSERT APPLES Harvest in late autumn and store through the winter until required. '**Midwinter**' (3) Often keeping until early spring, with good storage, but may become somewhat shrivelled towards the end. '**Claygate Pearmain**' (3) Rich flavour. Small apple, greenish, carried in profusion. Upright growth and resistant to scab. '**Laxton's Pearmain**' (4) Similar to a red-coloured Cox. Crisp, with a pleasant flavour. '**Laxton's Superb**' (4) Good apple which has a dual purpose, both culinary and dessert, with the flavour and colouring of Cox, but a larger fruit. Well recommended. '**Orleans Reinette**' (4) Golden yellow with brown red shading. Good fruiting but susceptible to scab in poor soil conditions. '**Rosemary Russet**' (3) Russet flavour. Keeps well. Somewhat light in cropping performance. '**Winston**' (5) An improved form of 'Laxton's Superb'. Fruit small, highly coloured but flavour not always of the best. Resistant to scab. Upright growth

SPRING DESSERT APPLES Harvest in late autumn/early winter and keep in store until required. '**Granny Smith**' (2) An Australian variety with bright green fruit. Benefits well from the extra warmth afforded by a wall or fence. '**Sturmer Pippin**' (3) Thrives on the extra warmth provided by a wall or fence. Pale green, russet-flavoured fruit. Harvest as late as possible in the year. '**Tydeman's Late Orange**' (3) A form of Cox. Bright orange/red with russety markings. Good Cox flavour, crisp. Stores well. '**Wagener**' (3) An American variety of some standing. Crisp and juicy, with good flavour. Keeps well until early spring. Golden yellow in colour with carmine shading when ripe. Free fruiting and free from scab.

SUMMER AND AUTUMN COOKING APPLES Use as required from the tree. '**Arthur Turner**' (3) Late summer to mid autumn. Green/yellow fruit, turning to orange/brown on sun side. Good cropper. '**Early Victoria**' (syn. '**Emneth Early**') (3) Mid summer to early autumn. Pale yellow fruit of average size. Cooks well. '**George Neal**' (2) Early to mid autumn. Pale golden fruit with rosy red shading on sun side. Good for cooking, becoming golden in colour. Good cropping and disease free. '**Grenadier**' (3) Late summer to mid autumn. Large green fruit. Good cropping and disease free. '**Lord Derby**' (5) Mid autumn to early winter. Green fruit, turning red with cooking. Heavy cropper. Disease free. '**Peasgood's Nonesuch**' (3) Early to late autumn. Very large golden fruit with attract-

ive orange/red stripe. Sweet. '**Rev. W. Wilks**' (2) Early to late autumn. Creamy white in colour, becoming pale yellow with cooking. Good cropper.

LATE KEEPING COOKING APPLES Harvest in late autumn/early winter and store until required. '**Annie Elizabeth**' (4) Mid winter to early summer, with good storage. Large fruit, golden yellow in colour with red shading. Cooks well with crisp flavour. '**Bramley Seedling**' (3) Must be one of the best known of all cooking apples. Late autumn to early spring. Large fruit, cooking exceedingly well and very high in vitamin C. Will require a large wall or fence for results. '**Crawley Beauty**' (8) Mid winter to early spring. Fruit pale yellow/green with red striping. Cooks well, with good flavour. Self-fertile. Scab re-

Malus pumila 'Cox's Orange Pippin'

sistant. '**Edward VII**' (6) Mid winter to mid spring. Pale yellow, turning red with cooking. '**Howgate Wonder**' (4) Mid autumn to late spring. Fruit creamy yellow with red shading. Heavy cropping. Cooks well. Scab resistant. '**Newton Wonder**' (5) Late autumn to early spring. Golden yellow with red stripes and shading. Strong grower.

USA

All varieties are self-fertile unless otherwise stated; however, pollination is improved by planting any two varieties together.

EARLY SEASON DESSERT APPLES Use from the tree in mid to late summer. '**Akane**' A crisp, red-skinned, juicy apple with white flesh. Good flavour. Ideal variety for areas where the fungus disease fire blight is a problem. From Japan. '**Anna**' A variety for Florida and Southern California. Green skin with red

Malus pumila 'Spartan'

blush. Ripens in July but may produce a second crop of flowers giving late fruits. Requires pollination from 'Dorset Golden' or 'Ein Shemer'. A low-chill variety. From Israel. **'Dorset Golden'** Large fruit, reminiscent of 'Golden Delicious'. Cannot tolerate any degree of frost or winter chill. Good for eating or cooking. Use 'Anna' or 'Ein Shemer' as pollinators. A low-chill variety. From Bahamas. **'Ein Shemer'** A derivative of 'Golden Delicious', does best in the deep South, Texas and Southern California. Early fruiting. Use 'Dorset Golden' as a pollinator.

Malus pumila 'Jupiter'

A low-chill variety. From Israel. **'Jerseymac'** Moderately firm, juicy, red-skinned fruit of good quality ripening in August. From New Jersey. **'Liberty'** An apple of medium size, sweet and juicy but can be coarse-grained in its texture. Skin shaded red, often over entire area. Resistant to rust, scab, mildew and attacks of fire blight. From New York. **'Tydeman's Early Worcester'** Juicy fruit of good colour and a tree of good garden merit. From the UK.

EARLY TO MID SEASON DESSERT APPLES Use from the tree from mid summer to early fall. **'Gravenstein'** Large, light green fruit with red shading. Yellow/green flesh. Good texture. Firm, crisp and juicy. Strong-growing, spreading habit. From Germany. **'Jonamac'** A 'McIntosh' type apple with very good flavour and texture. From New York. **'McIntosh'** One of the best of all the apple varieties. Medium to large in size, with sweet, tender, juicy white flesh. The skin is yellow, flushed red. Good as a dessert and as a cooking apple. From Ontario.

MID SEASON DESSERT APPLES Harvest in mid fall to early winter. May store for a short period. **'Empire'** Medium-sized fruit, red striped skin. Flesh white to cream. Good texture, crisp and juicy. Trees moderately vigorous and spreading. From New York.

Malus pumila 'Golden Delicious'

Malus pumila 'Laxton's Superb'

'Jonathon' A widely grown variety, both by the amateur and in commercial orchards. Fruit of medium size, skin shaded red and pale yellow. Crisp, juicy, firm fruit with good flavour. Good for dessert or cooking. Very heavy crop. From New York. **'Winter Banana'** Large, attractive fruit. Pale skin colour of waxy texture with a pink blush which spreads as it ripens. Tender flesh. Very attractive perfume and interesting flavour. Use 'Red Astrachan' as a pollinator. A low-chill variety. From Indiana. **'Winter Pearmain'** Large green-skinned apples with moderately firm flesh of very good quality. For use in southern California. A low-chill variety. Origin unknown.

Malus pumila 'Grenadier'

MID TO LATE SEASON DESSERT APPLES Harvest in mid fall to early winter. Will keep in store until required. **'Golden Delicious'** Fruit medium to large in size. Yellow/green skin, sometimes with a bright pink blush. Crisp, juicy and sweet. From West Virginia. **'Honeygold'** Good flavoured fruit of medium to large size, similar to 'Golden Delicious'. Yellow to golden yellow skin. Yellow flesh which is crisp, smooth and juicy. Extra hardy. From Minnesota. **'Jonagold'** This cross between 'Jonathon' and 'Golden Delicious' produces large yellow/green fruit shaded red, with crisp, juicy, cream-coloured flesh of good flavour. Can be used as a dessert and cooking apple. From New York. **'Red Delicious'** A popular variety with fruit of medium to large size, either striped or solid red skin colour. Flesh moderately firm, sweet and juicy. From Iowa.

LATE SEASON DESSERT APPLES Harvest in late fall to late winter and keep in store until

required. **'Fuji'** Can be left on the tree to ripen for at least 200 days, can also be ripened in store. Medium to large fruit, fresh, tart flavour. Can be used for dessert or cooking. From Australia. **'Idared'** This cross between 'Jonathon' and 'Wagner' produces large, bright red apples with smooth, glossy skin. Good flavoured sweet white flesh. Stores well. From Idaho. **'Regent'** A medium-sized variety with red skin. Crisp, juicy, creamy-white flesh. Can be used for dessert and cooking. Maintains its quality well into winter. Extra hardy. From Minnesota. **'Stayman'** Very late to ripen, a good keeper. Fruit juicy with tart flavour and bright red skin. Can be used for dessert and cooking. From Kansas.

COOKING APPLES Use from the tree in late fall or store until required. **'Beverley Hills'** An early-cropping variety with small to medium sized fruit. Skin striped or splashed with red on a yellow background. Tender, juicy with a tart flavour. Best production achieved in coastal areas of California and other southern states. A low-chill variety. From California.

Malus pumila 'Bramley Seedling'

'Courtland' Said by many growers to be even an improvement on McIntosh, both as a dessert and cooking apple. Skin red striped, flesh white and of good flavour. Mid season fruiting. From New York. **'Paulard'** Fruit red, flushed bright yellow, ripening late summer/early fall. White flesh, sometimes cream. Possible tart flavour. Can be used for cooking or dessert. From Michigan. **'Yellow Newton'** Medium-sized fruit with green to yellow skin for dessert or cooking. Firm, crisp flesh of good flavour. Strong growth, vigorous habit. Mid season to late fruiting. From New York.

Average height and spread
Fan-trained
Five years
8 × 8 ft   (2.4 × 2.4 m)
Ten years
10 × 12 ft   (3 × 3.7 m)
Twenty years
12 × 16 ft   (3.7 × 4.9 m)
Protrudes up to 3 ft (91 cm) from support.

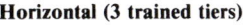

Horizontal (3 trained tiers)
Five years
6 × 8 ft   (1.8 × 2.4)
Ten years
8 × 12 ft   (2.4 × 3.7 m)
Twenty years
10 × 20 ft   (3 × 6 m)
Protrudes up to 3 ft (91 cm) from support.

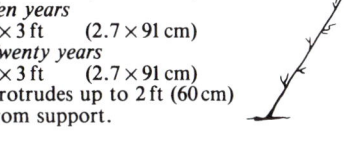

Cordon
Five years
8 × 2 ft   (2.4 × 60 cm)
Ten years
9 × 3 ft   (2.7 × 91 cm)
Twenty years
9 × 3 ft   (2.7 × 91 cm)
Protrudes up to 2 ft (60 cm) from support.

MANDEVILLA SPLENDENS

ACHILLEAN JASMINE

Apocynaceae   *Tender Greenhouse Climber*

Evergreen

Possibly one of the most charming of tender flowering climbers but its need for protection must be appreciated.

Mandevilla 'Alice du Pont'

Origin From Argentina.
Use As a climber for very sheltered favourable conditions or for preference under protection of a greenhouse or conservatory.
Description *Flower* Five bold petunia-pink or white flowers 2 in (5 cm) across make a tubular, trumpet-shaped flower arrangement carried in clusters. Flowers through summer to early autumn or, if grown under protection, over a wider timespan. Fragrant. *Foliage* Oblong, up to 3½ in (9 cm) long, with pointed ends. Dark green, glossy, attractive, borne well apart on short stalks along the stems. *Stem* Twisting, twining, self-supporting to suitable wires or string. Light green to mid green. Medium to fast rate of growth, depending on planting position *Fruit* Slender seeds, often up to 15 in (38 cm) long, presented in pairs in ideal growing conditions.

Melianthus major

Hardiness Tolerates a minimum winter temperature of 32°F (0°C).
Soil conditions Tolerates a wide range of soil conditions except extremely alkaline. Dislikes drying out.
Sun/Shade aspect Very sheltered, warm aspect or under protection. Light shade for preference, will tolerate full sun but may show signs of scorching in very open positions.
Pruning Requires no pruning other than that to keep it within bounds.
Training Allow to ramble up wire supports by twining its branches around the support.
Propagation and nursery production From cuttings of well-ripened side-shoots taken approximately 2–3 in (5–7.5 cm) long and rooted into a sand mixture in a propagating case. Always purchase container grown; best planting height 6 in–2 ft (15–60 cm). Will have to be sought from specialist nurseries.
Problems Rarely succeeds for any great length of time in a container, so if planted in a conservatory or greenhouse should be planted into soil beds rather than pots or tubs.
Similar forms of interest *M.* 'Alice Du Pont' Petunia-pink, scented, large flowers. Good climbing habit. *M. suaveolens* White, scented flowers. The original plant in commercial production but today often superseded in interest by the pink-flowering forms.
Average height and spread
Five years
10 × 10 ft (3 × 3 m)
Ten years
20 × 20 ft (6 × 6 m)
Twenty years
25 × 25 ft (7.6 × 7.6 m)
Protrudes up to 2 ft (60 cm) from support.

MELIANTHUS MAJOR

HONEY BUSH

Melianthaceae   *Tender Wall Shrub*
Evergreen

A useful evergreen shrub suitable for only very mild areas and benefiting from the protection of walls and fences.

Origin From South Africa.
Use As a wall shrub for warm, sunny, sheltered walls and fences. In colder areas, for conservatories and greenhouses.
Description *Flower* Crimson to tawny/crimson, tubular flowers in erect terminal racemes up to 6 in (15 cm) long, borne on mature shrubs throughout summer, but only in very favourable areas. *Foliage* Leaves pinnate, up to 12–18 in (30–45 cm) long, with nine to 11 leaflets 3–4 in (7.5–10 cm) long. Deeply

toothed edges. Sea-green to glaucous blue. *Stem* Hollow, upright stems, grey/green. In cold areas may die back to ground level in winter but can show a fast rate of growth in following spring. *Fruit* Insignificant.
Hardiness Tolerates a minimum winter temperature of 23°F (−5°C) when given the protection of a wall or fence.
Soil requirements Rich, moist, deep soil to produce good foliage growth.
Sun/Shade aspect Requires a very sheltered aspect in full sun to very light shade, otherwise leaf colour and shrub shape are spoiled.
Pruning None required.
Training Normally freestanding, but may need individual anchor points for securing.
Propagation and nursery production From root division or from seed. Purchase container grown; very hard to find outside mild areas and even in these locations available only from specialist nurseries. Best planting height 4 in–2 ft (10–60 cm).
Problems Only suitable for very mild areas.
Average height and spread
Five years
5 × 5 ft (1.5 × 1.5 m)
Ten years
8 × 8 ft (2.4 × 2.4 m)
Twenty years
12 × 12 ft (3.7 × 3.7 m)
Protrudes up to 6 ft (1.8 m) from wall or fence.

MORUS

MULBERRY

Moraceae   *Hardy Tree*
Deciduous

Although not often seen as a fan-trained wall tree, it should be considered more widely for this use.

Origin From China, North America and western Asia, depending on variety.
Use As a large fan-trained tree for large walls.
Description *Flower* Very short, male or female catkins of little interest; produced in early spring. *Foliage* Ovate, lobed or unlobed, grey/green leaves. Individual branches may carry both leaf shapes. Good yellow autumn colour. The lighter green leaves of *Morus alba* are used to feed silkworms. *Stem* Grey/green, corky, almost rubbery texture. Slow to medium rate of growth. *Fruit* Blackberry-shaped, dark red to black clusters of fruit. Edible and very juicy.
Hardiness Tolerates a minimum winter temperature of 4°F (−15°C).
Soil requirements Best results on rich, moist, deep soil. Must be well drained.
Sun/Shade aspect Tolerates all but the most

Morus nigra **in fruit**

severe of aspects. Best in full sun to allow ripening of fruit, but tolerates light shade.
Pruning Purchase bush trees to encourage branches to become fan-trained. Remove forward growing branches after fruiting to two buds from their point of origin. After 10–20 years thin out some of the major branches.
Training Secure to wires or individual anchor points.
Propagation and nursery production From hardwood cuttings taken in early winter. Purchase container grown bush or young trees not more than 3 ft (91 cm) in height from garden centres and nurseries.
Problems Roots very fleshy and often poorly anchored.
Forms of interest *M. alba* (White mulberry) Fast-growing. Light grey/green stems with large, ovate, light green foliage. Black fruits in autumn. One third more than average height and spread and slightly more tender.

Morus nigra **in autumn**

From China. *M. a.* **'Laciniata'** Leaves deeply cut and toothed. Good autumn colour. Fruits as parent. *M. nigra* (Common or black mulberry) Good fruiting ability in all areas. Of reliable hardiness. From western Asia. *M. rubra* (Red mulberry) Fruits red to dark purple and sweet. May exceed average height and spread by one third or more. From North America and not always successful as a garden tree elsewhere.
Average height and spread
Five years
8 × 8 ft (2.4 × 2.4 m)
Ten years
16 × 16 ft (4.9 × 4.9 m)
Twenty years
24 × 24 ft (7.3 × 7.3 m)
Protrudes up to 3 ft (91 cm) from support.

MYRTUS COMMUNIS

COMMON MYRTLE
Myrtaceae *Wall Shrub*
Evergreen
An attractive flowering evergreen shrub for mild locations which benefits from the protection of a wall or fence.

Origin From southern Europe and the Mediterranean region.
Use As a freestanding evergreen shrub for planting in front of walls and fences.
Description *Flower* White, round, tufted flowers, single or double depending on variety, mid to late summer. *Foliage* Leaves round to ovate, 1–2 in (2.5–5 cm) long, dark green with duller undersides, aromatic and very profusely borne. *Stem* Upright, becoming branching with age. Purple green to dark green. Fast growth rate when young, slowing with age. *Fruit* Small purple/black fruits in autumn.
Hardiness Tolerates a minimum winter temperature of 14°F (−10°C).
Soil requirements Does best on well-drained soil, tolerating both alkalinity and acidity.
Sun/Shade aspect Requires a sheltered aspect in full sun to mid shade.
Pruning None required but can be cut back hard or trimmed.
Training Allow to grow freestanding; normally needs no additional support.
Propagation and nursery production From semi-ripe cuttings taken in early summer. Purchase container grown; easy to find but varieties must be sought from specialist nurseries. Best planting height 15 in–2 ft (38–60 cm).
Problems Sometimes thought to be hardier than it actually is, but a wall will give it protection.
Similar forms of interest *M. c.* **'Variegata'** A form with broad, creamy white margins to the foliage. A tender variety which will not withstand temperatures below freezing. *M. c. tarentina* White tufted flowers in summer with white berries in autumn. Leaves narrower than those of *M. communis*. From the Mediterranean region. *M. c. t.* **'Variegata'** Flowers white, fragrant, produced in good numbers. Fruits black, foliage edged creamy white. *M. apiculata* (syn. *M. luma*) Tufted white flowers borne singly at each leaf axil cover the entire shrub during late summer and early autumn, followed by red to black fruits which are edible and sweet. Foliage oval, dark, dull green, ending in an abrupt point. Bark cinnamon-coloured and peels off in patches, showing creamy under surface. In

mild sheltered locations reaches size of small tree. From Chile.
Average height and spread
Five years
4 × 4 ft (1.2 × 1.2 m)
Ten years
8 × 8 ft (2.4 × 2.4 m)
Twenty years
15 × 10 ft (4.6 × 3 m)
Protrudes up to 8 ft (2.4 m) from wall or fence.

NERIUM OLEANDER

OLEANDER
Apocynaceae *Tender Wall Shrub*
Evergreen
Although tender, this shrub is worthy of mention, especially for wall-trained use in conservatories and large heated greenhouses.

Nerium oleander **in flower**

Origin From the Mediterranean region.
Use As a wall shrub under protection in greenhouses or conservatories.
Description *Flower* Medium-sized, flat, five-petalled flowers ranging from white through yellow, buff and various shades of pink in mid summer to mid autumn. *Foliage* Leaves lanceolate, 3–6 in (7.5–15 cm) long, borne in clusters. Grey/green to sea-green with some silver and gold variegated forms. *Stem* Upright stems, becoming branching at extremities, forming a narrow-based, wide-headed shrub. Grey/green to sea-green. *Fruit* Insignificant.

Myrtus communis **in flower**

Nerium oleander in flower

Hardiness Tolerates a minimum winter temperature of 32°F (0°C).

Soil requirements Requires a light, open soil, tolerates both alkalinity and acidity.

Sun/Shade aspect Requires the protection of a conservatory or greenhouse or may be planted outside in areas where the minimum temperature can be guaranteed year long. Full sun to light shade.

Pruning None required. Individual branches may be shortened and will rebranch.

Training Can either be allowed to grow freestanding or trained into a fan shape using individual anchor points or wires.

Propagation and nursery production From softwood cuttings taken in early spring. Scarce, normally found in the house-plant sections of larger garden centres. Purchase container grown; best planting height 15 in–2 ft (38–60 cm).

Problems Scarce in production.

Similar forms of interest None.

Average height and spread
Five years
3 × 3 ft (91 × 91 cm)
Ten years
6 × 6 ft (1.8 × 1.8 m)
Twenty years
13 × 13 ft (4 × 4 m)
Protrudes up to 3 ft (91 cm) from support.

OSMANTHUS

KNOWN BY BOTANICAL NAME

Oleaceae   *Wall Shrub*
Evergreen
A useful group of evergreen shrubs with scented flowers adapting well to wall training.

Origin Most varieties from western China or Japan.

Use As a freestanding shrub to grow in front of walls or fences, or to train in a fan shape.

Description *Flower* Small, white or cream, trumpet-shaped flowers, sweetly scented, late autumn or early spring, depending on variety. *Foliage* Leaves elliptic, ovate or oblong, ½–1½ in (1–4 cm) long, some larger according to variety, some with toothed, spiny edges. White and gold variegation in some forms. *Stem* Upright, becoming branching. Dark green to grey/green. Slow to medium growth rate. *Fruit* Broad seedheads may be formed after flowering.

Hardiness Tolerates a minimum winter temperature of 14°F (−10°C).

Soil requirements Tolerates most soils, but resents waterlogging.

Sun/Shade aspect Tolerates all but the most exposed aspects. Prefers light dappled shade but tolerates full sun through to medium shade.

Pruning Some pruning will be required to achieve a fan-trained shrub. May be cut and trimmed after flowering to keep within bounds.

Training Allow to stand free or secure to walls or fences by individual anchor points or wires.

Propagation and nursery production From semi-ripe cuttings taken in early summer. Purchase container grown; most varieties fairly easy to find. Best planting height 15 in–2½ ft (38–76 cm).

Problems Liable to leaf scorch in cold winters.

Forms of interest *O. americanus* (Devil wood) A variety reaching more than one third average height and spread. White fragrant flowers borne in short panicles in axils of each leaf joint in spring. Dark blue fruits, late summer. Slightly tender. Outside its native environment needs a warm, sheltered position, or can be grown in a conservatory. From south-eastern USA. *O. armatus* White, sweetly scented flowers in autumn. Elliptic, oblong or lanceolate leaves, thick and rigid in texture, from 3–8 in (7.5–20 cm) long. Edges of leaves have a series of hooked spiny teeth. From western China. *O. × burkwoodii* Round to ovate leaves, 1 in (2.5 cm) long. Dark, shiny upper surface, leathery textured, silvery undersides with toothed edges. Clusters of tubular, white, very fragrant flowers, early to mid spring. Will tolerate shallow chalk soils. A nursery hybrid formerly known as *Osmarea burkwoodii*. *O. delavayi* Dark grey/green foliage. Reaches two thirds average height and spread. Covered with a good display of very sweetly scented white flowers, early to mid spring. Foliage may be damaged by severe wind chill. *O. heterophyllus* (syns. *O. aquifolium, O. ilicifolius*) (Holly osmanthus, false holly) Small to medium holly-shaped leaves, dark green, shiny upper surface, paler undersides, spined and toothed edges, 1½ in (4 cm) long. White, scented, tubular flowers in small clusters in autumn. From Japan. *O.h.* **'Aureomarginatus'** Holly-like leaves with golden yellow to deep yellow variegated margins. White flowers. Two thirds average height and spread. *O.h.* **'Gulftide'** Leaves twisted and lobed with strong spines. White scented flowers. Very dense habit of growth. Slightly less than average height and spread. *O.h.* **'Purpureus'** Young growth holly-shaped, purple, later becoming very dark green tinged purple. White flowers in autumn. Two thirds average height and spread. *O.h.* **'Rotundifolius'** Almost round, tooth-edged, leathery, dark green foliage on dark, green/black stems. White trumpet flowers, scented, freely borne in spring. Very scarce but not impossible to find. *O.h.* **'Variegatus'** Holly-like leaves, grey/green with creamy white bold bordered edges. Shy to flower. Grows slowly to two thirds average height and spread.

Average height and spread
Five years
6 × 6 ft (2 × 2 m) freestanding
8 × 8 ft (2.5 × 2.5 m) fan-trained
Ten years
10 × 10 ft (3 × 3 m) freestanding
12 × 12 ft (3.5 × 3.5 m) fan-trained
Twenty years
13 × 13 ft (4 × 4 m) freestanding
15 × 15 ft (4.5 × 4.5 m) fan-trained
Protrudes up to 2 ft (60 cm) from support if fan-trained, 7 ft (2.1 m) untrained.

PALIURUS SPINA-CHRISTI

CHRIST'S THORN, JERUSALEM THORN

Rhamnaceae   *Wall Shrub*
Deciduous
An interesting shrub for specimen planting, traditionally reputed to have been used in Christ's crown of thorns.

Paliurus spina-christi in flower

Origin From southern Europe to the Himalayas and through to northern China.

Use As a fan-trained or freestanding wall shrub where it will benefit from the protection afforded.

Description *Flower* Small, open racemes of yellow to green/yellow flowers, late summer. *Foliage* Leaves ovate, ¾–1½ in (2–4 cm) long, slightly tooth-edged, light yellow/green, giving good yellow autumn colour. *Stem* Upright when young, quickly becoming

Osmanthus delavayi in flower

311

branching, twiggy and spreading. Mid green to green/yellow. Armed with numerous pairs of thorns. Slow to medium growth rate. *Fruit* Rounded, curiously shaped, green/yellow fruits in autumn and early winter.
Hardiness Tolerates a minimum winter temperature of 14°F (−10°C).
Soil requirements Most soils, provided adequate feeding is supplied.
Sun/Shade aspect Tolerates all but the most exposed aspects in full sun to very light shade. Dislikes deeper shade.
Pruning None required, but may be reduced in size or fan-trained if necessary.
Training Allow to stand free or secure to wall or fence with individual anchor points or wires.
Propagation and nursery production From semi-ripe cuttings taken in late spring/early summer, or from layers. Difficult to propagate. Purchase container grown; hard to find. Best planting height 8–18 in (20–45 cm).
Problems A little slow to develop into an interesting plant.
Similar forms of interest None.
Average height and spread
Five years
3 × 3 ft (91 × 91 cm) freestanding
5 × 5 ft (1.5 × 1.5 m) fan-trained
Ten years
6 × 6 ft (1.8 × 1.8 m) freestanding
8 × 8 ft (2.4 × 2.4 m) fan-trained
Twenty years
13 × 13 ft (4 × 4 m) freestanding
16 × 16 ft (5 × 5 m) fan-trained
Protrudes up to 2 ft (60 cm) from support if fan-trained, 10 ft (3 m) untrained.

Origin From central China.
Use As a free-growing climber for walls, fences, trellises, pergolas and to cover large shrubs and small trees. Can be used as a loose-flowing ground cover.
Description *Flower* Insignificant, minutely petalled, green to green/yellow flowers carried in clusters through spring and summer. *Foliage* Three to five leaflets, each up to 5 in (12 cm) long and 1½ in (4 cm) wide, form a hand-shaped leaf. Leaflets oval to narrowly oval, with tapered ends and toothed edges. Dark green when young, with distinct silver-white veining, turning more purple/red during summer, finally vivid orange/scarlet in autumn. *Stem* Purple/green with red/purple leaf stalks. Becoming brown/grey. Twining. Fast growing. *Fruit* May produce small

clusters of round blue/black fruits in late summer and early autumn.
Hardiness Tolerates a minimum temperature of 0°F (−18°C).
Soil requirements Tolerates both alkaline and acid conditions. Requires a rich, moisture-retaining soil.
Sun/Shade aspect Tolerates all aspects with no particular preference. Best in light to medium shade; will tolerate full sun but the variegation will be reduced.
Pruning No pruning required for its own wellbeing but previous season's growth can be cut back to within two buds of its origin to allow the build-up of a fan-trained shape. Pruning is normally carried out in early spring and should be commenced on an annual basis as early as possible to achieve a good fan-trained shape. Alternatively can be allowed to ramble with no pruning although foliage may decrease in size.
Training Produces tendrils which make it self-clinging, but may need some additional support when young or very heavy with age.
Propagation and nursery production From cuttings taken in mid summer. Always purchase container grown; best planting height 1½–4 ft (45 cm–1.2 m). Generally available from good garden centres and nurseries.
Problems None.
Similar forms of interest *P. thomsonii* (syn. *Vitis thomsonii*) Five oval leaflets make up a hand-shaped leaf with a length and width of up to 4½ in (11 cm), with teeth indents along the outer edges. Glossy dark green. Leaf shoots purple when young, becoming a duller purple with age. Very good red/purple autumn colours, black fruits. From the Himalayas and China. Not readily available.
Average height and spread
Five years
5 × 5 ft (1.5 × 1.5 m)
Ten years
10 × 10 ft (3 × 3 m)
Twenty years
20 × 20 ft (6 × 6 m)
Protrudes 9–12 in (23–30 cm) from support.

Origin From eastern North America.
Use For covering large walls, fences and hedges, will ramble through large shrubs and trees; can be allowed to fall down banks to make a spreading ground cover.

Description *Flower* Insignificant green to green/yellow flowers, minutely petalled, carried in clusters through spring and summer. *Foliage* Five leaflets form a hand shape, with each leaf up to 4 in (10 cm) long and 6 in (15 cm) wide. Oval in shape with slender points and closely toothed edges. Dull to mid-green when young with grey/blue undersides, turning vivid orange and crimson in autumn. *Stem* Light brown/red when young, becoming deep brown to grey/brown with age. Small sucker pads make it self-clinging. Fast growth rate. *Fruit* In hot seasons may produce small blue/black fruits.
Hardiness Tolerates a minimum winter temperature of 0°F (−18°C).
Soil requirements Does well on all soil types with no particular preference.
Sun/Shade aspect Tolerates all aspects. Best in full sun but will tolerate light shade.

Parthenocissus quinquefolia in autumn

Pruning Allow to cover a desired area and then reduce its extremities by up to 2 ft (60 cm) to expose windows, doors or roofs; will rejuvenate in following spring/summer. Will also respond to cutting down to ground level and over the next two to three years will rejuvenate itself.
Training Self-clinging with tendrils and pads. May need some assistance by tying in when young or when excessively heavy with age.
Propagation and nursery production From cuttings taken in early to mid summer, or self-rooted layers from undersides of established plants. Always purchase container grown; best planting height 1–3 ft (30–91 cm). Normally available from all good garden centres and nurseries.
Problems Often outgrows the area allowed for it. Can take up to two years to establish before any real growth is seen.
Similar forms of interest *P. q.* 'Engelmannii' Smaller leaved. Good autumn colour. *P. q.* 'Murorum' Very similar. Found in central and southern European nurseries and garden centres and occasionally in the UK.
Average height and spread
Five years
8 × 8 ft (2.4 × 2.4 m)
Ten years
16 × 16 ft (5 × 5 m)
Twenty years
32 × 32 ft (10 × 10 m)
Protrudes up to 12 in (30 cm) from support.

Origin From Japan, Korea, China and also Taiwan.
Use As a fast-growing, close, self-clinging climber for all large walls and fences.

Parthenocissus henryana in leaf

Parthenocissus tricuspidata 'Veitchii' in autumn

Description *Flower* Insignificant, minutely petalled, green to green/yellow flowers produced only on mature growth. *Foliage* Leaves vary according to age; on young plants ovate and toothed, sometimes carried in threes, becoming three-lobed with pointed fingers on older plants. Glossy upper surface, downy underside. Mid to dark green in spring and summer, turning vivid crimson and scarlet in autumn. *Stem* Light green, becoming green/purple, particularly on outer sides, finally brown/grey. Covered with self-clinging pads. Very tight formation, hand-shaped effect achieved naturally. Medium to fast growth rate. *Fruit* Round, blue-black with downy covering, produced on mature wood in warm seasons. Often hidden by foliage.
Hardiness Tolerates a minimum winter temperature of 0°F (−18°C).
Soil requirements Does well on all soil types with no particular preference, except it must have adequate moisture and organic material at the roots to produce good growth.
Sun/Shade aspect Tolerates all aspects. Full sun to light shade.
Pruning Normally requires no pruning, but can be reduced at its extremities by up to 2 ft (60 cm) with no harm and will rejuvenate very quickly. Can be cut down to within 3 ft (91 cm) of the base and will regain its original size over three to four years.
Training Self-clinging, requires no support.

Parthenocissus tricuspidata 'Beverley Brook' in summer

Propagation and nursery production From cuttings taken in mid summer. Best planting height 1–3 ft (30–91 cm). Always purchase container grown; normally available from most garden centres and nurseries.
Problems Slow to establish, often taking up to two years, but then very rapid in its further development.
Similar forms of interest *P. t.* 'Lowii' Three- to seven-lobed leaves when young, becoming three-lobed with age. Very good dark foliage both in summer and autumn. *P. t.* 'Beverley Brook' Almost purple foliage in summer with very good orange/scarlet autumn tints. A decided improvement on the parent plant. *P. t.* 'Green Spring' Smaller-growing with small leaves. Good autumn colour. Not readily available outside southern Europe. *P. t.* 'Purpurea' Green leaves in summer with purple/red autumn tints. *P. t.* 'Boskoop' Purple/green leaves in summer, turning red in autumn. A central European variety but sometimes found outside this region.
Average height and spread
Five years
6 × 6 ft (1.8 × 1.8 m)
Ten years
15 × 15 ft (4.6 × 4.6 m)
Twenty years
30 × 30 ft (9 × 9 m)
Protrudes up to 6 in (15 cm) from support.

PASSIFLORA CAERULEA

PASSION FLOWER, BLUE PASSION FLOWER

Passifloraceae   *Perennial Climber/*
                               Woody Climber

Deciduous

This woody climber is technically a perennial plant, but because of its woody nature is now considered a woody climber. May revert back to the perennial form in severe winters.

Origin From Brazil.
Use As a climber for walls, fences, to cover archways or pillars and to grow over pergolas.

Passiflora caerulea 'Constance Elliott' in flower

Description *Flower* Borne singly, up to 4 in (10 cm) wide. Sepals and petals white or pink-white. Central area 2 in (5 cm) across with blue spiky filaments, white in the middle and purple at the base. Fragrant. Flowers produced from mid to late summer. *Foliage* Mid green, palmate with five to seven lobed leaflets, each oblong to lance-shaped, 4–7 in (10–18 cm) long and 2 in (5 cm) wide. Yellow autumn colouring. *Stem* Twining, light green when young ageing to dark green, may turn light grey in winter. Fast growing. *Fruit* Oval, up to 1¼ in (3.5 cm) long. Turning yellow to orange/yellow from mid summer through until autumn. Edible, but not necessarily palatable.
Hardiness Tolerates a minimum winter temperature of 23°F (−5°C).
Soil requirements No particular soil preference, only disliking extremely alkaline and dry types.
Sun/Shade aspect Requires a very sheltered, warm aspect in full sun to very light shade.
Pruning The entire shrub can be cut to within

Passiflora caerulea in flower

3 ft (91 cm) of ground level every five to six years and will quickly rejuvenate itself with young, fresher foliage. Best carried out in early spring.
Training Allow to ramble through wires, over trellis or similar support. Can be tied to individual anchor points if required, but this is not normally the best method.
Propagation and nursery production From seed sown in autumn or semi-ripe cuttings taken in early summer. Always purchase container grown; normally available from good garden centres and nurseries. Best planting height 1–3 ft (30–91 cm).
Problems May die back in severe winters but rejuvenates from ground level the following spring.
Similar forms of interest *P. c.* 'Constance Elliott' White flowers with pale blue spiky filaments. Produced mid summer through until autumn. Slight fragrance.
Average height and spread
Five years
8 × 8 ft (2.4 × 2.4 m)
Ten years
16 × 16 ft (4.9 × 4.9 m)
Twenty years
20 × 20 ft (6 × 6 m)
Protrudes up to 12 in (30 cm) from support.

PELARGONIUM

ZONAL AND IVY-LEAF GERANIUMS

Geraniaceae   *Tender Greenhouse*
                          Climber

Deciduous

Many zonal and ivy-leaf geraniums adapt well to fan-training in very favourable sheltered areas or under protection.

Origin Of garden origin.
Use As fan-trained plants for sheltered warm walls and fences or in conservatories or greenhouses.

Pelargonium 'Gustave Emich' in flower

Description *Flower* Five or more petals, ends indented, make up either a single or double flower up to 1–1½ in (2.5–4 cm) across, carried in clusters on the ends of 4 in (10 cm) long flower stalks, normally from mid summer to mid autumn, often longer, particularly when under protection. Colours range through white, pink, red, and purple with some bi-colours, particularly in ivy-leaf varieties. *Foliage* Three-lobed, each lobe heart-shaped, grey/green to red/green on zonal varieties, glossy green on ivy-leaf varieties. Some variegated varieties in both forms. Leaves 1½–4 in (4–10 cm) in width and length, depending on variety. *Stem* Light green becoming darker green, finally grey/green. Vigorous growth, rigid formation; can be easily broken. *Fruit* No fruit of interest.
Hardiness Tolerates a minimum winter temperature of 32°F (0°C) but may survive lower winter temperatures in very sheltered areas.
Soil requirements Tolerates a wide range of soil conditions as long as adequate plant nutrients and organic material are available on an annual basis. Dislikes waterlogging.
Sun/Shade aspect Very sheltered aspect or under protection. Light shade for preference but will tolerate full sun to medium shade.
Pruning Remove all forward-growing shoots in early winter, only retaining those required to form a fan-shaped framework.
Training Train into a fan-shaped form, securing to wires or individual anchor points. Ivy-leaf and thin-stemmed varieties may need the assistance of small canes between wires or anchorage points.
Propagation and nursery production From semi-ripe cuttings taken in early summer or in late autumn. Should always be purchased container grown; an extremely wide range of varieties is normally available from garden centres and nurseries. Best planting height 2 in–2 ft (5–60 cm).
Problems Under protection of greenhouse or conservatory may suffer from attacks of whitefly. Can be slow to form fan-shaped formation and will need patience.
Forms of interest Almost any variety of zonal or ivy-leaf geraniums can be fan-trained on a wall but the following would be my first choices:
ZONAL *P.* 'Gustave Emich' Double bright red flowers. *P.* 'Mrs Lawrence' Double pink

flowers. *P.* 'Paul Crampel' Single red flowers; one of the best varieties. *P.* 'Queen of the Whites' Single white flowers.
IVY-LEAF *P. peltatum* 'Abel Carrière' Magenta flowers. *P. p.* 'L'Elegante' Pink flowers; white-edged foliage. *P. p.* 'La France' Mauve flowers. *P. p.* 'Galilee' Pink flowers.
Average height and spread
Five years
4 × 4 ft (1.2 × 1.2 m)
Ten years
8 × 8 ft (2.4 × 2.4 m)
Twenty years
12 × 12 ft (3.7 × 3.7 m)
Protrudes up to 2 ft (60 cm) from support.

PHASEOLUS COCCINEUS

RUNNER BEAN, SCARLET RUNNER

Leguminosae   *Annual Climber*
Deciduous

Normally considered as a vegetable but equally attractive when grown as a flowering climber with the added bonus of its edible bean pods.

Origin From North America.
Use As an annual climber against walls and fences, up a pyramidal support or can be trained into walkways and arches. Produces edible runner beans.
Description *Flower* Racemes up to 12 in (30 cm) long of pea-shaped flowers each ⅜ in (1 cm) wide and long, bright scarlet with some white and pink varieties. *Foliage* Three oval to round leaflets each 3–4 in (8–10 cm) long and wide make up a leaf up to 6–8 in (15–20 cm) long and wide, on long leaf stalks. Larger size depends on available moisture. Light green with yellow autumn colour. *Stem* Light grey/green, fast-growing, self-climbing by twining habit. *Fruit* Long green bean pods up to 18 in (45 cm) long and 1 in (2.5 cm) wide, toxic when raw but edible cooked.
Hardiness Tolerates a minimum winter temperature of 32°F (0°C). Seeds are not winter hardy in soil.
Soil requirements Does well on all soil types with no particular preferences although adequate moisture and plant food must be

available for maximum growth.
Sun/Shade aspect Tolerates all aspects. Full sun to light shade with full sun for preference.
Pruning Requires no pruning. Growth dies back totally in winter.
Training Provide supports on wires, canes, poles or other solid structures to form a framework.
Propagation and nursery production From seed planted in early to mid spring under protection and transplanted into final position after all danger of frost has passed, or planted in situ as seed in mid spring and protected if frosts are likely. Purchase container grown plants from garden centres and nurseries; best planting height 4–5 in (10–12 cm).
Problems Suffers attacks from blackfly. Dislikes drought and poor soils; supply extra feed and irrigation in such conditions.
Similar forms of interest *P. c.* 'Achievement' Red flowers. Good flavoured variety, good for freezing. *P. c.* 'Best of All' Good, heavy crop of bean pods. Orange/red flowers. *P. c.* 'Butter' Stringless pods of medium length. Red flowers. *P. c.* 'Bokkie' Early cropping in large amounts. Red flowers. *P. c.* 'Crusader' Red flowers. Long pods. *P. c.* 'Enoma' Very heavy cropper, beans have good shape and colour. Red flowers. *P. c.* 'Erecta' White flowers. Tolerant of wide range of growing conditions. *P. c.* 'Goliath' (Prize Taker) Bright orange/red flowers. Very long beans and heavy crop. *P. c.* 'Kelvedon Marvel' Red flowers, good as ground cover. *P. c.* 'Gower Emporer' Good early variety with good pod production. Red flowers. *P. c.* 'Mergoles' Beans are stringless and of good length. Red flowers. *P. c.* 'Painted Lady' Attractive flowers, both red and white. *P. c.* 'Polestar' Stringless, fleshy pods. Good cropper. Red flowers. *P. c.* 'Prizewinner' Well tested variety with good pod production. Red flowers. *P. c.* 'Purple Podded' Red flowers. Pods attractive

Phaseolus coccineus in flower

deep blue to purple colour. *P. c.* 'Red Knight' Stringless pods. Red flowers. *P. c.* 'Scarlet Emperor' Can be used up supports or as large-scale ground cover. Red flowers. *P. c.* 'Stringline' Smooth podded variety. Very strong growing. Red flowers. *P. c.* 'Sunset' Pale pink flowers. Early cropping, *P. c.* 'White Achievement' White flowers with good pod production
Average height and spread
One year
12 × 3 ft (3.7 m × 91 cm)
Protrudes up to 18 in (45 cm) from support.

PHILADELPHUS
(Medium-sized Forms)

MOCK ORANGE, SYRINGA

Philadelphaceae   ***Wall Shrub***
Deciduous

A group of plants not often thought of as wall shrubs, but if fan-trained or allowed to grow freestanding adjacent to a wall or fence will make a very elegant display.

Origin Most varieties of garden origin.
Use To be planted freestanding in front of a wall or fence or fan-trained.

Philadelphus 'Belle Etoile' in flower

Description *Flower* White, often fragrant flowers, $\frac{3}{4}$–1$\frac{1}{2}$ in (2–4 cm) wide, single or double depending on variety, some with purple throat markings, borne in mid summer. *Foliage* Leaves ovate, 1$\frac{1}{2}$–4 in (4–10 cm) long, slightly tooth-edged, light to mid green with good yellow autumn colour. Some varieties variegated. *Stem* Forming an upright, round-topped shrub if freestanding or can be fan-trained. Grey/green. Medium to fast growth rate. *Fruit* Insignificant.
Hardiness Tolerates a minimum winter temperature of 4°F (−15°C).
Soil requirements Most soils, including alkaline and acid types.
Sun/Shade aspect Tolerates all aspects and full sun through to medium shade.
Pruning On established plants three years old or more, remove one third of older flowering shoots to ground level after flowering each year.
Training Will require tying to wires or individual anchor points if fan-trained.
Propagation and nursery production From semi-ripe cuttings taken in early summer, or from hardwood cuttings taken in winter. Purchase container grown or bare-rooted from garden centres and nurseries; best planting height 15 in–2$\frac{1}{2}$ ft (38–76 cm).
Problems Harbours blackfly, which can transfer to other plants.

Forms of interest *P.* 'Avalanche' Small, pure white, single, scented flowers borne in profusion. An upright-growing shrub with smaller than average foliage. *P.* 'Belle Etoile' White, single, fragrant flowers with maroon central blotches. *P. coronarius* 'Variegatus' Attractive white-margined green/grey foliage and scented white flowers. Must have light shade and protection from strong midday sun to avoid scorching. Round bushy habit if freestanding but can also be fan-trained. *P.* 'Erectus' Pure white, single, scented flowers, very free-flowering. Distinct upright branches. *P.* 'Galahad' Single, scented, white flowers. Stems mahogany-brown in winter. Rather hard to find. *P.* 'Innocence' Single, white, fragrant flowers. Foliage sometimes has creamy white variegation. *P.* 'Sybille' Almost square, purple-stained, single, white flowers, orange-scented, on arching branches.
Average height and spread
Five years
3 × 3 ft (91 × 91 cm) freestanding
5 × 5 ft (1.5 × 1.5 m) fan-trained
Ten years
5 × 5 ft (1.5 × 1.5 m) freestanding
7 × 7 ft (2.1 × 2.1 m) fan-trained
Twenty years
5 × 6 ft (1.5 × 1.8 m) freestanding
9 × 9 ft (2.7 × 2.7 m) fan-trained
Protrudes up to 3 ft (91 cm) from support if fan-trained, 6 ft (1.8 m) untrained.

PHILADELPHUS
(Tall-growing Forms)

MOCK ORANGE, SYRINGA

Philadelphaceae   ***Wall Shrub***
Deciduous

These fragrant, summer-flowering shrubs make a spectacular display on walls and fences.

Origin Most varieties of garden origin.
Use As large, freestanding shrubs to be planted adjacent to walls or fences or for fan-training.
Description *Flower* Single or double, depending on variety, very fragrant, $\frac{3}{4}$–1$\frac{1}{2}$ in (2–4 cm) wide, white with purple throat markings, borne in clusters in early to mid summer. *Foliage* Leaves ovate, 1$\frac{1}{2}$–4 in (4–10 cm) long, slightly tooth-edged, light to mid green, good yellow autumn colour. Some varieties with golden foliage. *Stem* Forms an upright, round-topped shrub or large fan-trained specimen. Grey/green. Fast growth rate. *Fruit* Insignificant.
Hardiness Tolerates a minimum winter temperature of 4°F (−15°C).
Soil requirements Most soils, including alkaline and acid types.
Sun/Shade aspect Tolerates all aspects in full sun through to medium shade.
Pruning Remove one third of old flowering

wood to ground level annually after flowering on plants established three years or more.
Training Will require tying to wires or individual anchor points if grown fan-trained.
Propagation and nursery production From semi-ripe cuttings taken in early summer or hardwood cuttings in winter. Plant bare-rooted or container grown from garden centres and nurseries; best planting height 15 in–3 ft (38–91 cm).
Problems Harbours blackfly which can transfer to other plants.
Forms of interest *P.* 'Beauclerk' Large, somewhat square, fragrant, single, white flowers with pink centres. *P.* 'Burfordensis' Large, fragrant, semi-double, white flowers with yellow stamens. More upright in growth than most. *P.* 'Burkwoodii' Very fragrant, single, cup-shaped white flowers with pink-stained centres. *P. coronarius* Single, white to creamy white, scented flowers. The form from which many garden hybrids derived. *P.* 'Aureus' Single, white, scented flowers, somewhat obscured by lime green to golden-yellow foliage, which is susceptible to late frost damage and strong summer sun scorching. Plant in light dappled shade – in deeper shade foliage turns green. *P.* 'Minnesota Snowflake' Large, double, fragrant, white flowers, weighing down branches. *P.* 'Norma' Single, large, white flowers, less fragrant than most, on long arching branches. *P.* 'Virginal' Fragrant, pure white, double flowers up to 1$\frac{1}{2}$ in (4 cm) across, covering upright branches in mid summer. The most popular variety. It is worth looking out for a very rare variegated form, *P.* 'Virginal Variegata'. *P.* 'Voie Lactée' Strong-growing, broad-petalled, pure white, single, scented flowers up to 2 in (5 cm) across.
Average height and spread
Five years
5 × 3 ft (1.5 m × 91 cm)
freestanding
5 × 5 ft (1.5 × 1.5 m)
fan-trained
Ten years
8 × 6 ft (2.4 × 1.8 m)
freestanding
8 × 8 ft (2.4 × 2.4 m)
fan-trained
Twenty years
13 × 13 ft (4 × 4 m) freestanding
15 × 15 ft (4.6 × 4.6 m) fan-trained
Protrudes up to 4 ft (1.2 m) from support if fan-trained, 12 ft (3.7 m) untrained.

PHILLYREA

JASMINE BOX

Oleaceae   ***Wall Shrub***
Evergreen

A group of evergreen shrubs with small but fragrant spring flowers adapting well to fan training which shows off their elegant evergreen foliage to the full.

Origin From Africa, southern Europe and the Mediterranean regions.
Use As a large, fan-trained or freestanding shrub for large walls and fences.
Description *Flower* Small, tubular, 1 in (2.5 cm) fragrant flowers, borne in clusters 5 in (12 cm) long on mature shrubs in late spring/early summer. *Foliage* Leaves ovate to lanceolate, small or medium-sized, 2–5 in (5–12 cm) long and $\frac{1}{2}$–$\frac{3}{4}$ (1–2 cm) wide, leathery textured. Dark glossy green upper surfaces, duller glabrous undersides. *Stem* Branching from an early age to form a dome-shaped or fan-trained specimen. Light olive-green to green/brown. Medium growth rate. *Fruit* Small, oval, purple to purple/black fruits in autumn.
Hardiness Tolerates a minimum winter temperature of 4°F (−15°C)
Soil requirements Tolerates most soils but distressed by extremely alkaline or very wet conditions.
Sun/Shade aspect Tolerates all but the most severe of aspects. Prefers light shade, tolerates full sun to medium shade.

Philadelphus 'Virginal' in flower

Pruning None required but may be clipped and shaped.

Training Will require tying to wires or individual anchor points if grown fan-trained.

Propagation and nursery production From semi-ripe cuttings taken in early summer. Purchase container-grown; normally available from specialist nurseries but becoming more widely stocked. Best planting height 15 in–2½ ft (38–76 cm).

Problems Young plants when purchased often look misshapen and require time to develop.

Forms of interest *P. angustifolia* Flowers creamy yellow, fragrant, produced in clusters at leaf axils in late spring, early summer on mature shrubs. Foliage narrow ovate, 1½–2 in (4–5 cm) long, mid grey/green, lighter green when young. Very good for coastal planting. *P. decora* (syn. *P. medwediewii*) Small clusters of fragrant white flowers in mid to late spring. Foliage large, ovate to lanceolate, leathery texture, dark glossy green upper surfaces and paler glabrous undersides. The best variety for general garden use. Now classified as

Phillyrea decora

Osmarea decora by some authorities. *P. latifolia* Flowers white, some fragrance, but less interesting than most, borne in late spring, sometimes followed by small, blue/black fruits. Small, glossy green leaves presented opposite each other on arching stems. Slightly less than average height.

Average height and spread
Five years
4 × 4 ft (1.2 × 1.2 m) freestanding
6 × 6 ft (1.8 × 1.8 m) fan-trained
Ten years
6 × 6 ft (1.8 × 1.8 m) freestanding
8 × 8 ft (2.4 × 2.4 m) fan-trained
Twenty years
10 × 10 ft (3 × 3 m) freestanding
12 × 12 ft (3.7 × 3.7 m) fan-trained
Protrudes up to 3 ft (91 cm) from support if fan-trained, 10 ft (3 m) untrained.

PHLOMIS FRUTICOSA

JERUSALEM SAGE

Labiatae   *Wall Shrub*
Evergreen

An attractive, silver-leaved, summer flowering shrub which benefits from the protection of a wall or fence.

Origin From the Mediterranean region.
Use As a loose, open, summer flowering shrub using a wall or fence to give added height and protection.
Description *Flower* Yellow, interestingly shaped flowers up to 2 in (5 cm) wide and 3 in (7.5 cm) long in June and July. *Foliage* Leaves oval, pointed, 3 in (7.5 cm) long and 2 in (5 cm) wide, with grey/green downy surface, giving some yellow autumn colour. *Stem* Upright when young, grey/green and downy, becoming open and straggly with age, forming a loose fan shape. Old growth of less interest. Fast rate of growth in spring, slowing with season. *Fruit* Insignificant.

Phlomis fruticosa in flower

Hardiness Tolerates a minimum winter temperature of 14°F (–10°C).
Soil requirements Dry, open soil, dislikes any waterlogging.
Sun/Shade aspect Requires a moderately protected aspect in full sun. Dislikes any degree of shade.
Pruning As the best foliage and flowers are borne on new wood, trim by one-third in late spring to induce new growth.
Training Requires tying to wires on individual anchor points if grown fan-trained or allow to grow freestanding.
Propagation and nursery production From softwood cuttings taken in mid to late summer. Purchase container grown; normally stocked by garden centres and general nurseries. Best planting height 8–18 in (20–45 cm).
Problems May die off in a severe winter so rooted cuttings should be over-wintered under protection to ensure replacements for following spring.

Average height and spread
Five years
3 × 3 ft (91 × 91 cm)
Ten years
3 × 4 ft (91 cm × 1.2 m)
Twenty years
3 × 4 ft (91 cm × 1.2 m)
Protrudes up to 4 ft (1.2 m) from support if fan-trained, 5 ft (1.5 m) untrained.

PHOTINIA × FRASERI 'RED ROBIN'
(*P. glabra* 'Red Robin')

KNOWN BY BOTANICAL NAME

Rosaceae   *Wall Shrub*
Evergreen

An attractive winter foliage shrub not normally considered for its attractive wall-training ability.

Origin From New Zealand and Australia.
Use As a short or tall fan-trained shrub for walls and fences, adapting well to hard pruning to maintain it within a metre of the ground, but exceeding this if required.
Description *Flower* Clusters of 3 in (7.5 cm) long, white flowers in late spring but outside its native environment rarely flowers except in very warm areas. *Foliage* Leaves large, ovate to lanceolate, 4–8 in (10–20 cm) long and 1½–3½ in (4–9 cm) wide, often tooth-edged. Dark glossy green when mature. Young growth, produced from late autumn, starts as dark or brilliant red and increases in intensity through winter into early spring, becoming bronze during mid spring to early summer. *Stem* Upright, becoming branching with age,

forming a close-foliaged, fan-trained wall shrub. Dark green tinged red. Medium growth rate. *Fruit* Rarely fruits outside its native environment or very hot climates, where it produces clusters of red or orange/red fruits.
Hardiness Tolerates a minimum winter temperature of 4°F (–15°C), although young foliage may become slightly misshapen or split in severe cold.
Soil requirements Does well on most soils, only disliking extreme alkalinity.
Sun/Shade aspect Tolerates all but the most severe aspects in full sun to light shade.

Photinia × fraseri 'Red Robin'
(juvenile foliage)

Pruning Can be cut hard to maintain within a required area through late spring and early summer. Responds by producing large amounts of new red foliage growth in late summer, autumn and winter.
Training Will require tying to wires or individual anchor points if grown fan-trained.
Propagation and nursery production From semi-ripe cuttings taken in early summer. Purchase container-grown; normally stocked by good garden centres and nurseries. Best planting height 15 in–3 ft (38–91 cm).
Problems Often on sale as a single shoot plant but soon grows thicker.
Similar forms of interest *P. × f.* 'Birmingham' Very dark coppery red to dark red new growth. Slightly broader foliage. Raised in North America. *P. × f.* 'Pink Lady' Foliage pink variegated. An attractive plant for sheltered aspects. *P. glabra* 'Rubens' Foliage

shorter but broader than many of the hybrid types. Can be more reliable in its production of clusters of white flowers followed by red fruits. Young growth and foliage brilliant red in winter and spring. Slightly more tender. From Japan. *P. serrulata* (Chinese Hawthorn) Upright coppery red foliage from mid winter through spring and maintained over a long period. Sometimes bears clusters of white flowers in mid to late spring, followed by red fruits in early autumn.

Average height and spread
Five years
7 × 7 ft (2.1 × 2.1 m)
Ten years
9 × 9 ft (2.7 × 2.7 m)
Twenty years
15 × 15 ft (4.6 × 4.6 m)
Protrudes up to 3 ft (91 cm) from support.

PHOTINIA STRANVAESIA

CHINESE STRANVAESIA

Rosaceae   *Wall Shrub*
Semi-evergreen

An interesting, semi-evergreen, fruiting shrub not always thought of as a potential fan-trained wall shrub.

Origin From western China.
Use As a wall-trained specimen for large walls and fences, where it can display its flowers and berries to the fullest effect.
Description *Flower* Clusters of 3 in (7.5 cm) wide white to off-white flowers borne on wood two years old or more, in early summer. *Foliage* Leaves lanceolate to broad lanceolate, 2–4 in (5–10 cm) long, leathery-textured, mid green with duller, lighter undersides. Some bright red autumn colour, contrasting with remaining semi-evergreen leaves, extent of evergreen depending upon severity of winter. Better foliage on young shoots, decreasing in interest and colour intensity on older wood. *Stem* Upright, vigorous stems, dark green tinged red, becoming spreading and branching with age and will form a good fan-shape with training. Fast growth rate when young, slowing with age. *Fruit* Dull red/orange fruits, resistant to birds, produced in clusters in late summer through to early autumn.
Hardiness Tolerates a minimum winter temperature of 4°F (−15°C) but some stem damage may be suffered.
Soil requirements Does well on most, but performs less well on very poor soils.
Sun/Shade aspect Tolerates most aspects in full sun to medium shade.
Pruning Cut as required for training and from time to time reduce young shoots. Prune shoots more than four years old to ground level in early spring, to induce new, upright,

Phygelius capensis 'Coccineus' in flower

very leafy shoots, improve overall appearance and still maintain older fruiting branches.
Training Will require wires or fixing to individual anchor points.
Propagation and nursery production From semi-ripe cuttings taken in early summer or by layering. Purchase container-grown. Some nurseries offer plants from seed, but resulting shrubs can be extremely variable, so check the method of propagation with nursery staff. Availability varies. Best planting height 2–3 ft (60–91 cm).
Problems This shrub and its forms are susceptible to the fungus disease fire blight. Its future is uncertain, since production may be discontinued on this account. Affected shrubs should be destroyed immediately by burning and the appropriate government agency should be notified. Also liable to stem canker fungus, especially older stems, which are strangled by the collar-like effect of the canker.
Similar forms of interest *P. s.* 'Fructu-luteo' Yellow fruits in autumn but somewhat sparser than the parent. Two thirds average height and spread. *P. s.* 'Palette' Attractive foliage splashed creamy white or pink. Fruits red. *P. s. var. undulata* Slightly later flowering than the species, but otherwise very similar. Two-thirds average height and equal spread. *P. × stranvinia* 'Redstart' Foliage red to dark red. Flowers white. Fruits red and yellow.

Average height and spread
Five years
7 × 7 ft (2.1 × 2.1 m)
Ten years
10 × 10 ft (3 × 3 m)
Twenty years
20 × 20 ft (6 × 6 m)
Protrudes up to 4 ft (1.2 m) from support.

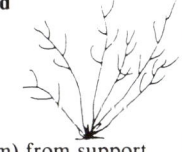

PHYGELIUS

CAPE FIGWORT

Scrophulariaceae   *Wall Shrub*
Deciduous

Interesting, late summer flowering sub-shrubs that adapt well to wall conditions.

Origin From South Africa.
Use As a fan-trained or trailing wall shrub for walls, fences and pillars or through shrubs.
Description *Flower* Yellow or orange tubular flowers, up to 1½ in (4 cm) long, hanging from upright panicles, late summer/early autumn. *Foliage* Leaves ovate to lanceolate, 2–3 in (5–7.5 cm) long, tooth-edged, dark glossy green upper surfaces, grey undersides. Some yellow autumn colour. *Stem* Upright shoots may be annually produced after die-back in winter but against a wall or fence may become permanent. Dark green. Fast annual spring and summer growth. *Fruit* Insignificant.
Hardiness Tolerates a minimum winter temperature of 14°F (−10°C).
Soil requirements Does well on all soil types except very dry.
Sun/Shade aspect Requires a sheltered aspect. Prefers full sun, tolerates light shade.
Pruning Should be lightly trimmed in early spring and new growths tied in as they appear.
Training Will require tying to wires or individual anchor points if fan-trained.
Propagation and nursery production From seed or semi-ripe cuttings taken in mid summer. In cold areas young plants may be overwintered under cover in case replacements are needed. Purchase container-grown. Fairly easy to find; yellow-flowering forms less often available. If purchased when in flower keep under cover until next spring. Best planting height 8–15 in (20–38 cm).
Problems As sub-shrubs may die back in winter or become very woody; if so, cut hard back in spring to encourage rejuvenation.
Forms of interest *P. aequalis* 'Yellow Trumpet' Large, yellow, tubular flowers, up to 1½ in (4 cm) long, hanging from upright panicles, 12–18 in (30–45 cm) in mid summer to early autumn. Light green foliage. May not withstand any frost. *P. capensis* 'Coccineus' Crimson-scarlet, hanging, tubular flowers, 1 in (2.5 cm) long, produced from terminal panicles 12 in (30 cm) long in late summer, early autumn. *P. c.* 'Indian Chief' Good-sized red tubular flowers with yellow throats.

Average height and spread
Five years
4 × 4 ft (1.2 × 1.2 m)
Ten years
7 × 7 ft (2.1 × 2.1 m)
Twenty years
8 × 8 ft (2.4 × 2.4 m)
Protrudes up to 2½ ft (76 cm) from support.

Photinia stranvaesia 'Palette'

PILEOSTEGIA VIBURNOIDES

CLIMBING VIBURNUM

Hydrangeaceae   *Woody Climber*
Evergreen

An interesting climbing evergreen if given adequate protection.

Origin From China, Taiwan and India.
Use For covering large walls and fences where it can be trained in a fan shape.
Description *Flower* Many small flowers, creamy-white in bud, opening to white, with prominent stamens in the centre. Carried in panicles up to 4–6 in (10–15 cm) long at ends of shoots during late summer to early autumn. *Foliage* Narrow oblong or lanceolate leaves 2½–6 in (6–15 cm) long. Leathery in texture, dark green above with pitted silver underside and prominent veins. *Stem* Green ageing to brown. Slow to medium rate of growth. Aerial roots on underside. *Fruit* No fruits of interest.
Hardiness Tolerates a minimum winter temperature of 14°F (−10°C). Foliage may be damaged by severe winter chill.
Soil requirements Does well on all soil types with no particular preference, except needing adequate moisture.
Sun/Shade aspect Does well on all aspects, only disliking extreme degrees of wind chill and draught. Full sun to light shade with no particular preference.

Pileostegia viburnoides in flower

Pruning Normally does not require pruning, but can be retained within an area by cutting back the offending shoots in early spring.
Training Best fan-trained, using individual pins or wires to train branches or canes. Clings by aerial roots which need additional support when young and when plants become heavy with age, particularly during snowfalls.
Propagation and nursery production From semi-ripe cuttings taken in early to mid summer or self-produced layers. Always purchase container grown; will have to be sought from specialist nurseries. Best planting height 9–18 in (23–45 cm).
Problems Some foliage may be damaged by severe winters. This is normally replaced the following spring. Slow to establish, taking up to five years to become a plant of any substance.
Similar forms of interest. None.
Average height and spread
Five years
3 × 3 ft (91 × 91 cm)
Ten years
6 × 6 ft (1.8 × 1.8 m)
Twenty years
18 × 18 ft (5.5 × 5.5 m)
Protrudes up to 2 ft (60 cm) from support.

Piptanthus laburnifolius in flower

PIPTANTHUS LABURNIFOLIUS (*P. nepalensis*)

EVERGREEN LABURNUM

Leguminosae   *Wall Shrub*
Evergreen

An interesting evergreen shrub, particularly useful as a wall specimen.

Origin From the Himalayas.
Use As a fan-trained or spreading specimen for walls and fences.
Description *Flower* Bright yellow pea-flowers in 4 in (10 cm) long, 2 in (5 cm) wide racemes, in mid to late spring. *Foliage* Leaves trifoliate, 3–6 in (7.5–15 cm) long, with dark glossy green upper surfaces and duller undersides. Old leaves drop in autumn giving some good autumn colour. *Stem* Upright with some branching, forming a narrow-based, wide-headed, fan-shaped shrub. Dark glossy green. Medium growth rate. *Fruit* Medium to long pea-pods, grey to grey/brown in autumn and winter.
Hardiness Tolerates a minimum winter temperature of 14°F (−10°C) but may be defoliated in severe winters, with some stem die-back.
Soil requirements Prefers a well-drained soil but accepts most, tolerating high alkalinity.
Sun/Shade aspect Tolerates all but the most exposed aspects in full sun or shade, but may be slightly taller in shade.
Pruning Each year shorten back one or two stems on established shrubs after flowering to encourage new wood.
Training Will require tying to wires or individual anchor points if grown fan-trained or allow to grow freestanding.
Propagation and nursery production From seed. Purchase container grown; easy to find in some areas, otherwise obtainable from specialist nurseries. Best planting height 15 in–2½ ft (38–76 cm).
Problems Apart from possible winter dieback, occasional shoots can die from drought, waterlogging, or mechanical damage to roots.
Average height and spread
Five years
6 × 5 ft (1.8 × 1.5 m)
Ten years
10 × 8 ft (3 × 2.4 m)
Twenty years
13 × 12 ft (4 × 3.7 m)
Protrudes up to 2½ ft (76 cm) or more from support if fan-trained, 6 ft (1.8 m) untrained.

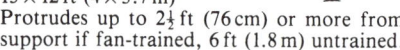

PITTOSPORUM

PITTOSPORUM, KOHUHA

Pittosporaceae   *Wall Shrub*
Evergreen

Attractive foliage shrubs, in most areas benefiting from the protection of a wall or fence.

Origin From New Zealand.
Use As freestanding shrubs to be grown adjacent to walls and fences, both in the open and in the protection of conservatories and greenhouses. Foliage useful for flower arranging.
Description *Flower* Small, ¼–½ in (5 mm–1 cm) wide, unusual, chocolate brown to purple, honey-scented flowers in late spring, only on very mature shrubs in favourable areas. *Foliage* Leaves ovate to oblong, 1–2½ in (3–6 cm) long, glossy, thick textured, in varying shades of olive-green, purple, silver, gold or variegated. *Stem* Upright stems, becoming branching and twiggy with age, forming an upright, pyramidal shrub. Dark brown to almost black. Medium growth rate. *Fruit* Insignificant.
Hardiness Tolerates a minimum winter temperature of 23°F (−5°C).
Soil requirements Well-drained soil.

Pittosporum tenuifolium

318

Sun/Shade aspect Requires a sheltered, protected aspect in full sun to very light shade.
Pruning None required but can be trimmed quite harshly in early spring. If cut to ground level will rejuvenate.
Training Normally freestanding and requires no additional support.
Propagation and nursery production From semi-ripe cuttings taken in early summer. Purchase container grown; easy to find, especially in mild coastal areas. Best planting height 1–3 ft (30–91 cm).
Problems Often purchased in mild coastal areas and taken inland for planting in the open, where it rarely succeeds.

Pittosporum tenuifolium **'Wendle Channon'**

Forms of interest *P.* **'Margaret Turnbull'** Mid green foliage edged with yellow. *P. tenuifolium* Attractive grey/green foliage, leaves slightly twisted with good black twigs. Flowers well in mild areas. One of the hardiest forms. *P. t.* **'Garnettii'** Round, white-variegated leaves, flushed pink in winter. One of the hardiest of variegated forms. Slightly less than average height and

Pittosporum tenuifolium **'Purpureum'**

spread. Of garden origin. *P. t.* **'Irene Paterson'** Marbled, creamy white foliage. Interesting leaf display, forming a slightly rounder shrub than most. Half average height and spread. *P. t.* **'James Stirling'** Small, silver/green, rounded or oval leaves, very densely presented on black/purple twigs. Tender and will not withstand frost. Slightly less than average height and spread. *P. t.* **'Purpureum'** Young pale green leaves turning to deep bronze/purple, ageing to purple/green, giving an outer canopy of good purple foli-

age. Two thirds average height and spread. *P. t.* **'Saundersii'** Foliage golden variegated. *P. t.* **'Silver Queen'** Narrow, ovate leaves, pointed and tightly bunched on black stems, forming a neat, conical shape. Foliage variegated silver/grey, black twigs. Two thirds average height and spread. *P. t.* **'Tresederi'** Amber leaves, mottled gold when young. *P. t.* **'Warnham Gold'** Yellow to yellow/green young foliage, ageing to golden yellow. Half average height and spread. Of garden origin. *P. t.* **'Wendle Channon'** Attractive mid to dark green foliage with yellow edging ageing to off-white, sometimes gold. *P. tobira* (Japanese pittosporum, tobiri pittosporum) Large, obovate, bright glossy green leaves, presented in whorls, bearing orange-blossom scented, creamy-coloured flowers in mid summer. Good in mild coastal areas; in less favourable conditions use as a conservatory specimen. Two thirds average height and spread. From China, Taiwan and Japan, now frequently seen in southern Europe. *P. t.* **'Variegatum'** Variegation irregular, with creamy white margins and central patches. Needs a favourable, mild location. In colder areas use as a conservatory shrub.

Average height and spread
Five years
5 × 3 ft (1.5 m × 91 cm)
Ten years
10 × 6 ft (3 × 1.8 m)
Twenty years
20 × 10 ft (6 × 3 m)
Protrudes up to 8 ft (2.4 m) from wall or fence.

PLUMBAGO CAPENSIS

BLUE PLUMBAGO, PLUMBAGO, CAPE LEADWORT

Plumbaginaeceae   *Tender Greenhouse Climber*

Deciduous

A well distributed climber with attractive flowers but requiring greenhouse or conservatory protection in winter.

Origin From South Africa.
Use As a large climbing shrub for conservatories and greenhouses, either in large containers or planted into greenhouse beds.
Description *Flower* Large panicles up to 5 in (12 cm) across, consisting of numerous $\frac{1}{2}$ in (1 cm) wide, five-petalled flowers of a delicate

pale blue from April to November. Clusters supported on flower shoots up to 4–6 in (10–15 cm) long. *Foliage* Light grey/green, elliptic, pointed, soft in texture. Some yellow autumn colour. *Stem* Light green to grey/green. Upright, branching, twiggy. Not self-clinging or supporting after 3 ft (91 cm) in height. Medium to fast rate of growth. *Fruit* No fruit of interest.
Hardiness Tolerates a minimum winter temperature of 32°F (0°C). Requires the protection of conservatory or greenhouse in areas experiencing frost.
Soil requirements No particular preference. If planted in greenhouse beds, additional grit, sand and sedge peat should be added in liberal quantities. If grown in containers, a good-quality proprietary soil-based potting compost should be used.
Sun/Shade aspect Needs the protection of a conservatory or greenhouse in all but the most favourable of areas. Can be grown in containers, given winter protection and moved out in spring or summer. Prefers light shade but will tolerate a wide range of sun/shade conditions, except for deep shade.
Pruning Shorten back flowering shoots after flowering to encourage new production for following year's blooms. Thin main branches once area required has been covered. Can be cut hard back if required and will rejuvenate.
Training Secure to wires or individual anchor points.
Propagation and nursery production From semi-ripe cuttings taken late spring/early summer and rooted by providing some bottom heat in a propagating frame. Best purchased container grown at a height of 6 in–3 ft (15–91 cm). Will have to be sought from specialist nurseries or from florists where it is sold as a house plant when in flower in late summer.
Problems Requires a large root run or large container. Top growth can quickly outgrow the space it is intended for; pruning will control it to a certain extent, but it can become invasive. Can be attacked by whitefly.
Similar forms of interest *P. c.* **'Alba'** Pure white flowers. Scarce. More tender.
Average height and spread
Five years
5 × 5 ft (1.5 × 1.5 m)
Ten years
8 × 8 ft (2.4 × 2.4 m)
Twenty years
16 × 16 ft (4.9 × 4.9 m)
Protrudes up to 2 ft (60 cm) from support.

Plumbago capensis **in flower**

POLYGONUM BALDSCHUANICUM

RUSSIAN VINE, MILE A MINUTE
Polygonaceae   *Woody Climber*
Deciduous
Possibly one of the quickest growing of all climbing plants, even to the extent of becoming invasive.

Origin From south-eastern USSR.
Use As a fast-growing climber for walls, fences and to grow through trees or large shrubs. Ideal for covering an ugly building.
Description *Flower* Small white flowers in panicles 10–18 in (25–45 cm) long and 3 in (7.5 cm) wide entirely covering the whole framework of branches from late summer to early autumn and then ageing to pink and finally brown by mid winter. *Foliage* Oval-shaped, up to 2 in (5 cm) long and 1 in (2.5 cm) wide. Light green, ageing to a yellow/green with veins and edges turning to orange/brown, then good autumn colour of yellow and bronze. *Stem* Smooth, light grey/green when young, ageing to light brown and finally dark grey/brown. Loosely twining and very fast growing – can make up to 15 ft (4.6 m) of growth in one season. *Fruit* The ageing flowers, as they turn brown, are retained as seed heads and have some attraction in certain winter light.
Hardiness Tolerates a minimum winter temperature of 0°F (–18°C).
Soil requirements Does well on all soil types, only disliking extremely dry types when first planted.
Sun/Shade aspect Does well in all aspects, including very exposed. Full sun to medium shade, although will flower best in a sunny position.
Pruning Requires no pruning for its own well being but often needs containing within an allotted area. Prune back in early spring.
Training Not self-clinging but its loose, twining habit secures it to wires, trellis, open fences or through the branches of trees and large shrubs. Normally needs no other form of support.
Propagation and nursery production From seed or semi-ripe cuttings taken in summer. Can also be produced from hardwood cuttings taken in winter. Always purchase container grown, normally available from good garden centres and nurseries. Best planting height 2–4 ft (60 cm–1.2 m).
Problems Almost always underestimated for its rate and spread of growth. This variety is often confused in commercial horticulture with *P. aubertii*.
Similar forms of interest *P. aubertii* (Silver lace or fleece vine) A variety flowering in mid summer, otherwise indistinct.
Average height and spread
Five years
20 × 20 ft (6 × 6 m)
Ten years
40 × 40 ft (12.2 × 12.2 m)
Twenty years
60 × 60 ft (18.3 × 18.3 m)
Protrudes up to 2 ft (60 cm) from support.

Poncirus trifoliata in flower

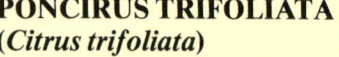

PONCIRUS TRIFOLIATA
(*Citrus trifoliata*)

HARDY ORANGE, JAPANESE BITTER ORANGE, TRIFOLIATE ORANGE
Rutaceae   *Wall Shrub*
Deciduous
An interesting shrub forming a round, rigid, architectural shape with attractive scented flowers and yellow autumn fruit. Now referred to by some authorities as *Aegle sepiara*.

Origin From northern China.
Use As a freestanding or fan-trained shrub of interest for walls and fences.
Description *Flower* Very sweetly scented, white, orange-blossom type flowers 1½–2 in (4–5 cm) across in mid to late spring. *Foliage* Leaves somewhat sparsely presented, obovate, 1–2½ in (2.5–6 cm) long, light, bright green, yellow autumn colour. *Stem* Branching, stout, armed with spines 1 in (2.5 cm) long, grey to bright green. Forms a ball-shaped shrub or can be fan-trained. Slow growth rate. *Fruit* Green, ripening to yellow, globular, miniature orange fruits 1½–1¾ in (4–4.5 cm) across.
Hardiness Tolerates a minimum winter temperature of 14°F (–10°C).
Soil requirements Well-drained, open, light soil, tolerates alkaline and acid types.
Sun/Shade aspect Tolerates all but the most severe of aspects in full sun to light shade.
Pruning None required.
Training Free-standing or, if fan-trained, will require wires or individual anchor points.
Propagation and nursery production From seed. Purchase container grown from specialist nurseries. Best planting height 8–18 in (20–45 cm).
Problems Relatively short-lived, can deteriorate after 25 years. Dislikes soil cultivation close to roots.
Average height and spread
Five years
3 × 3 ft (91 × 91 cm)
Ten years
6 × 6 ft (1.8 × 1.8 m)
Twenty years
10 × 10 ft (3 × 3 m)
Protrudes up to 2 ft (60 cm) from support if fan-trained, 6 ft (1.8 m) untrained.

Polygonum baldshuanicum in flower

PRUNUS ARMENIACA (Apricot Hybrids)

APRICOT

Rosaceae　　　　　　　　*Fruit Tree*
Deciduous

A fruiting tree that benefits greatly from being grown against a warm wall or fence, possibly the only location in which it will produce any fruit of use, other than under protection.

Origin From China.
Use As a fan-trained tree for large walls and fences, both in the open and under the protection of conservatory or greenhouse.
Description *Flower* Five-petalled, 1 in (2.5 cm) wide, saucer-shaped pink flowers produced in clusters on short spurs on the naked branches in early spring. Self fertile. *Foliage* Oval, 3 in (7.5 cm) long by 2 in (5 cm) wide, dark green with some yellow autumn colour. *Stem* Mid green when young, quickly ageing to mahogany brown. Shiny, attractive, rigid. Medium rate of growth. *Fruit* Edible, round, yellow flushed, sweet-flavoured, with a stone.
Hardiness Tolerates a minimum winter temperature of 4°F (−15°C).
Soil requirements Tolerates all types, but prefers an alkaline soil. Requires large amounts of organic compost applied prior to planting and an annual mulch to achieve best results.
Sun/Shade aspect Requires a sheltered aspect. Full sun or fruits will not ripen.
Pruning Remove old fruiting shoots in autumn and avoid pruning at other times.
Training Purchase already fan-trained and tie to wires and individual anchor points.
Propagation and nursery production From grafting or budding of chosen variety on to a seedling understock. Seedling understocks do not control vigour or ultimate size. Best size to purchase: minimum of five pre-trained shoots, overall size 3–5 ft (91 cm–1.5 m). Can be planted bare-rooted from mid autumn to early spring or container grown as available, with autumn and winter for preference.
Problems May suffer attacks of silver leaf disease, so take care not to prune other than in autumn. Severe attacks of blackfly may occur; control with a proprietary spray. Mildew may be another problem. Some bleeding on established plants may be found; normally this is a sign of extreme age and impending demise.
Forms of interest
UK
'Alfred' A canadian variety with mid summer fruiting. **'Farmindale'** A Canadian variety which is gaining in popularity. Mid summer fruiting. **'Moorpark'** Possibly the best variety to attempt. Mid to late summer.
USA
All varieties are self-fertile unless otherwise stated but pollination is improved by planting two varieties. **'Blenheim'** ('**Royal**') A very good variety for eating or drying. Used extensively in California for canning. Medium sized, dull orange fruit. May produce green shoulders to the fruit which is detrimental. Can only tolerate temperatures up to 90°F at harvesting stage. From the UK. **'Chinese'** ('**Morman**') A good variety for the West Coast and cooler regions. Late flowering allows blossom to escape the worst of spring frosts. Heavy crops of small, sweet, juicy fruit. From Utah. **'Flora Gold'** A dwarf-growing variety. Medium-sized, good quality fruit. Used as a dessert and also for canning. Heavy cropping. Requires relatively cool conditions when at harvest stage. From California. **'Goldcot'** A hardy variety which will survive temperatures down to −20°F. Late flowering and late cropping. Good in Eastern and Midwestern states. Produce medium to large-sized, good flavoured fruit with tough skins. Good for canning or for dessert. From Michigan. **'Harcot'** A variety for cool conditions, late flowering but earlier ripening than most. Heavy crops of large to medium good flavoured fruits, on good-shaped, compact

trees. Resistant to bacterial spot and brown rot. From Ontario, Canada. **'Harogem'** Fruits shaded bright red over an orange background. Small to medium-sized with firm flesh. Mid-season ripening and good long-lasting qualities, once picked. The tree is resistant to perennial canker and brown spot. From Ontario, Canada. **'Moorpark'** Large orange fruit with deep blush shading. Orange flesh with good flavour and scent. Thought by many to be the best variety and is the most reliable for all areas. From the UK. **'Perfection'** ('**Goldbeck**') Very large, light orange-yellow, oval fruits. Flesh bright orange and of moderate quality. Strong-growing tree, flowering early and therefore possibly frost-prone. Requires another variety for pollination to be grown with it. Crops possibly light. Good for the South and West. From Washington. **'Rival'** Large, firm, mild-flavoured fruit, heavily blushed. Good variety for canning. Tree large and spreading. Use 'Perfection' as a pollinating variety. Good in Northwestern States. From Washington. **'Royal Rosa'** Bright yellow, firm-fleshed fruit with strong aromatic scent and tart to sweet flavour. Best eaten fresh from the tree. Large crops. Medium-sized tree, early in its fruiting. Good in Southern States. From California. **'Scout'** Medium to large bronze-skinned fruit with deep yellow flesh. Good as dessert, for canning or as jam. Trees strong-growing, upright in habit and hardy. Midsummer fruiting, good in the Midwest. From Manitoba, Canada. **'Sungold'** Round, medium-sized, tender fruit with a golden skin, blushed orange. Mild, sweet flavour. Good as dessert or for jam. Strong-growing, upright habit, medium-sized tree. Good in all areas and zones. From Minnesota. **'Tilton'** Yellow-orange fruit, can tolerate high temperatures. It has a high-chill requirement of over 1,000 hours below 45°F. From California. **'Wenatchee'** Large, oval fruit with orange-yellow skin and flesh of good flavour. Trees have a lifespan of up to 30 years. Good on the Pacific Northwest coast and in the West. From Washington.
Average height and spread
Five years
7 × 7 ft (2.1 × 2.1 m)
Ten years
15 × 15 ft (4.6 × 4.6 m)
Twenty years
15 × 20 ft (4.6 × 6 m)
Protrudes up to 2½ ft (76 cm) from support.

Prunus armeniaca 'Moorpark'

PRUNUS AVIUM, P. CERASUS, P. MAHALEB (Fruiting Cherry Hybrids)

FRUITING CHERRIES

Rosaceae　　　　　　　*Fruit Trees*
Deciduous

The fan-training of cherries on walls will aid the ripening of the fruit and will help protect it from the ravages of birds.

Origin Of nursery origin.
Use As fan-trained trees with edible fruits for large walls and fences.
Description *Flower* Five-petalled, ½–¾ in (1–2 cm) wide, saucer-shaped white flowers in mid spring. Carried in clusters at leaf joints on mature wood. *Foliage* Oval, 4 in (10 cm) long and 2 in (5 cm) wide, mid to dark green. Some yellow autumn colour. *Stem* Brown when young, ageing to brown/grey, finally grey. Strong, unbranching. Medium to fast growth rate. *Fruit* Clusters of round, white, red or black edible stoned fruits in mid to late summer depending on variety.
Hardiness Tolerates a minimum winter temperature of 4°F (−15°C).
Soil requirements Tolerates all soil conditions, except very waterlogged. Ideal for alkaline soils.
Sun/Shade aspect Tolerates all aspects, but fruit ripens best in full sun.
Pruning Normally not required other than the removal of dead or crossing branches. This is best carried out in late summer/early autumn to avoid attacks of silver leaf.
Training Purchase fan-trained trees and tie to wire or individual anchor points.
Propagation and nursery production From grafting or budding of a main variety on to a seedling understock. Seedling understocks do not control vigour or ultimate size but some dwarfing rootstocks which do are now available. Best size to purchase: minimum of five pre-trained shoots, overall size 3–5 ft (91 cm–1.5 m). Can be planted bare-rooted from mid autumn to early spring, or container grown as available, with autumn and winter for preference.
Problems Susceptible to silver leaf. Fruit often ravaged by birds and once ripe requires netting to protect.

321

Forms of interest
UK
More than one variety normally has to be used for pollination. The letters A–N indicate suitable pollinating varieties.

DESSERT CHERRIES A 'Early Rivers' Pollinator J. Heart-shaped black fruits with rich flavour and tender flesh. B 'Florence' Pollinators E, L. Heart-shaped white fruits with sweet flavour. Late summer. C 'Frogmore' Pollinators D, I, J, L, M. Round white fruits with good flavour. A well established variety. Mid summer. D 'Governor Wood' Pollinators C, E, F, G, I, L, M, H, K. Round white fruits of good flavour. Mid summer. E. 'Kent Bigarreau' Pollinators F, G, D, B, I. Round white fruits of good flavour. Mid to late summer. F 'Merton Bigareau' Pollinators D, E, I, J, L, M, K. Round black fruits of good colour. Mid summer. G 'Merton Bounty' Pollinators D, E, I, J, L, M, K. Black fruits. Mid summer. Can be used as a universal pollinator. H 'Merton Crane' Pollinators D, I, K. Round black fruits with good flavour. Mid summer. I 'Merton Glory' Pollinators F, G, D, C, E, J, L, M. White fruits. Mid summer. Can be used as a universal pollinator. J 'Merton Heart' Pollinators A, C, F, G, I, M. Heart-shaped black fruit with good flavour. Early to mid summer. K 'Merton Late' Pollinators D, F, G, H. White fruits of good flavour. Mid to late summer. L 'Napoleon Bigarreau' Pollinators B, C, D, F, G, I. White fruits. Late summer. A well established variety. M 'Roundel' Pollinators C, D, F, I, J, L. Black fruits. Mid summer.

SELF-FERTILE VARIETIES 'Stella' Large red fruits with firm flesh. Mid summer. Can be used as a universal pollinator. 'Van' Large red fruits with sweet flavour. Mid summer. Fruits at only three years old and is very hardy. 'Morello' Fruits with an acid flavour, ideal for culinary purposes. A universal pollinator except for 'Early Rivers' and 'Merton Heart'.

USA
EARLY SEASON VARIETIES 'Black Tartarian' Medium-sized, sweet black cherry. Early to fruit. A useful pollinator for many other varieties and easily pollinated itself. Good for all areas. From California. 'May Duke' Dark red fruit of medium size with very good flavour. Best used for cooking or preserving. Most other varieties of cherry will pollinate it and it is a good pollinator for other varieties. Good for the Western states. From France. 'Northstar' A morello-type, sour cherry of dwarf constitution. Red fruit and flesh, rarely cracking, presented on a small yet vigorous and hardy tree. Resists diseases such as brown rot. Self fertile. Good for all areas. From Minnesota. 'Sam' Medium to large sweet, black fruits which are juicy and firm. Good dessert variety. Use varieties 'Bing', 'Lam-

Prunus **'Morello Cherry' in autumn**

bert' or 'Van' as pollinators. Very vigorous and heavy cropping. Good for the North and West. From British Columbia, Canada.

MID SEASON VARIETIES 'Bing' Heavy crops of sweet, juicy, black cherries. May suffer from bacterial leaf spot in humid climates. Use varieties 'Sam', 'Van' or 'Black Tartarian' as pollinators. Good for the West. From Oregon. 'Chinook' Large, sweet, heart-shaped fruit with mahogany skin and deep red flesh. Produces good crops. Use varieties 'Bing', 'Sam' or 'Van' as pollinators. Good in Western states. From Washington. 'Corum' A sweet cherry particularly good for pollinating the variety 'Royal Ann'. Fruit yellow, sweet and firm. Use 'Royal Ann', 'Sam' or 'Van' as pollinators. Extremely good in Western areas. From Oregon. 'Emperor Francis' Large, yellow blushed cherry red. Firm, sweet flesh. Heavy cropping and hardy stature. Use 'Rainier' or 'Hedelfingen' as pollinators. Particularly good in the North. From Europe. 'Garden Bing' Sweet, dark red fruit. A dwarf form of the variety 'Bing', not reaching more than 8 ft (2.4 m) in height. Self-fertile. Good in Western areas. From California. 'Kansas Sweet' ('Hansen Sweet') A sweet form of the pie cherry. Fruit red with a firm flesh. Best used for pies, where it has a fresh flavour. Both blossom and tree are hardy in Kansas. Self-fertile. From Kansas. 'Meteor' A dwarf amarelle sour cherry, only

reaches up to 10 ft (3 m). Large, bright red fruit with yellow flesh. Good for pies. Very hardy but tolerating milder conditions well. Self-fertile. Good in all areas. From Minnesota. 'Montmorency' A sour cherry for both commercial and home use. Brilliant red fruits with yellow, firm flesh. Self-fertile. Good in all areas. From France. 'Rainier' Fruit blushed yellow with firm, sweet, juicy flesh. Use 'Bing', 'Sam' or 'Van' as pollinators. Ideal for the South and West. From Washington. 'Royal Ann' ('Napoleon') Blushed yellow fruits with firm juicy flesh. Mostly used commercially for candies and maraschino cherries. Not entirely hardy. Use 'Corum', 'Windsor' or 'Hedelfingen' as pollinators. Good in all areas. From France. 'Schmidt' Large, mahogany coloured fruit with thick skin and wine-red sweet flesh. Fruit buds may suffer from spring frosts. Use 'Bing', 'Lambert' or 'Royal Ann' as pollinators. Good for the North and South. From Germany. 'Stella' Large red fruits with firm flesh. A strong-growing tree. Self-fertile and can be used as a universal pollinator. From the UK. 'Van' Large red fruits with sweet flavour. Fruits at only three years old and is very hardy. From the UK.

LATE SEASON VARIETIES 'Angela' Large, sweet, dark red fruits. The tree is vigorous and produces large crops. Use 'Emperor Francis' or 'Lambert' as pollinators. From Utah. 'Black Republican' ('Black Oregon') Very dark, sweet, firm fruits. Heavy cropping. Any other variety of cherry will act as a pollinator. From Oregon. 'Hedelfingen' Medium sized, sweet dark fruits with firm flesh. Very heavy cropping. Any sweet cherry variety will act as pollinator. Good for North and South. From Germany. 'Lambert' Large, dark, sweet fruits. Use 'Van' or 'Rainier' as pollinators. Good in all areas. From British Columbia, Canada. 'Late Duke' Large, light red fruits. Best for cooking or preserving. Normally self-fertile but in colder climates requires any other sour variety to pollinate. Good for the West. From France. 'Morello' Black fruits with acid flavour, ideal for culinary purposes. Good in cold locations. From the UK. 'Windsor' Small, sweet, dark red fruits. Often succeeds where other varieties may fail. Any sweet variety will pollinate with the exception of 'Van' and 'Emperor Francis'. Ideal for easterly areas as well as North and South. Origin unknown.

Average height and spread
Five years
7 × 7 ft (2.1 × 2.1 m)
Ten years
15 × 12 ft (4.6 × 3.7 m)
Twenty years
15 × 18 ft (4.6 × 5.5 m)
Protrudes up to 3 ft (91 cm) from support.

Prunus **'Morello Cherry' in fruit**

PRUNUS DOMESTICA
(Damson Hybrids)

DAMSON

Rosaceae   **Fruit Tree**
Deciduous

As long as adequate space can be given, damsons make extremely fine fan-trained specimens, the ripening of the fruit being enhanced by the warmth afforded by the wall.

Origin Most hybrids of garden origin.

Use As a large fan-trained tree for walls and fences.

Description *Flower* Five-petalled, 1 in (2.5 cm) wide white flowers in profusion in mid spring. Plums can be used to pollinate and corresponding numbers of both plums or damsons can be considered. *Foliage* Oval, 1½–2 in (4–5 cm) and ¾ (2 cm) wide, mid to dark grey/green with some yellow autumn colour of attraction. *Stem* Twiggy, branching. Green/brown when young, becoming mahogany brown with age. Medium growth rate. *Fruit* Round to oval, blue skinned, yellow flesh with stones. Sweet and juicy when ripe.

Hardiness Tolerates a minimum winter temperature of 0°F (−18°C) but flowers may be damaged by spring frosts.

Soil requirements Tolerates all soil conditions, responding particularly well to alkaline types.

Sun/Shade aspect Tolerates all aspects but benefits from some protection from wind to encourage pollinating insects. Full sun to enhance ripening, but will tolerate light shade.

Pruning Rarely requires pruning other than removal of dead, damaged or crossing branches in late summer/early autumn.

Training One-year-old trees should be cut to within 12 in (30 cm) of graft union and will make up to three to five new shoots. These can then be trained into a fan shape. Trees can be purchased pre-trained, which is possibly the best solution. Will require wires or individual anchor points for support.

Propagation and nursery production From grafting or budding on to 'St Julien A' rootstock. Can be planted bare-rooted from mid autumn to early spring or container grown as available. Purchase one-year-old trees of 2½–4 ft (76 cm–1.2 m) or pre-trained trees with not less than five branches up to 3–4 ft (91 cm–1.2 m). Not readily available either as one-year-old or pre-trained trees and will have to be sought from specialist nurseries.

Prunus domestica **(greengage) in flower**

Problems Suffers from attacks of silver leaf. May take up to five to eight years to come into full fruit production.

Forms of interest 'Farleigh Damson' (2) Mid autumn. When in full crop, large numbers of small blue-black fruits with very good flavour. Can be used either as a dessert or cooking damson. Most other varieties of plum, if in flower at the correct time, will pollinate this variety. **'Merryweather'** (3) Early to mid autumn. The main variety worthy of consideration and often the only variety available. Large, blue/black fruits with good damson flavour. Sweet and juicy. Self-fertile but benefits from additional pollination from other plum varieties. **'Shropshire Prune'** (Prune Damson) (5) Oval, blue/black fruits. Very good flavour, particularly when cooked. Reliable cropper.

Average height and spread

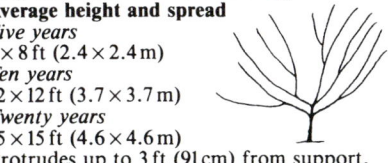

Five years
8 × 8 ft (2.4 × 2.4 m)
Ten years
12 × 12 ft (3.7 × 3.7 m)
Twenty years
15 × 15 ft (4.6 × 4.6 m)
Protrudes up to 3 ft (91 cm) from support.

PRUNUS DOMESTICA
(Gage Hybrids)

GAGES, GREENGAGES

Rosaceae   **Fruit Tree**
Deciduous

Attractive flowering and fruiting trees, adapting well to fan-training on walls, where the warmth aids their ripening.

Origin Most varieties of garden or nursery origin.

Use As a large fan-trained tree for walls and fences.

Description *Flower* Small, five-petalled white flowers, 1 in (2.5 cm) wide, carried in profusion in early to mid spring. Consideration has to be given to the selection of varieties for cross-pollination. *Foliage* 1–1½ in (2.5–4 cm) long and ½–¾ in (1–2 cm) wide, light green, often with notched edges. Good yellow autumn colour. *Stem* Twiggy, light green, ageing to brown/green. Medium growth rate. *Fruit* Round, 1 in (2.5 cm) across, normally light green to yellow. Edible and sweet, with stones.

Hardiness Will tolerate a minimum winter temperature of 0°F (−18°C) but flowers may be damaged by spring frost.

Soil requirements Tolerates all soil conditions, particularly benefiting from alkaline types. Dislikes waterlogging.

Sun/Shade aspect Tolerates all aspects, but pollination may be better in sheltered positions, where strong wind does not deter pollinating insects. Best in full sun to aid the ripening of the fruit.

Pruning Dead, crossing and congested branches may be removed in late summer/early autumn. No other pruning except that required for training.

Training Purchase fan-trained trees or train one-year-old plants by cutting to within 12 in (30 cm) of graft union and tying in subsequent growth in following season. Will require wires or individual anchor points for support.

Propagation and nursery production From budding or grafting on to an understock, normally 'St Julien A'. Some dwarfing stocks are becoming available but are not recommended for fan-training. Purchase bare-rooted from mid autumn to early spring or container grown as available. Fan-trained trees should not have less than five shoots and should be 3–6 ft (91 cm–1.8 m) high. Young trees for training should be no more than one

Prunus domestica **'Merryweather' in fruit**

year old. Will have to be sought from specialist nurseries and young trees may be difficult to find.

Problems Suffers from attacks of silver leaf. Can be slow to fruit. Poor availability.

Forms of interest Two trees, of appropriate varieties, will be needed for pollination. Choose varieties with the same number or numbers to either side except in the case of self-fertile forms.

LATE SUMMER VARIETIES **'Denniston's Superb'** (2) Very reliable, tough variety with true gage fruit of good flavour. Self-fertile. **'Early Transparent Gage'** (4) Very sweet and good flavoured fruit, apricot yellow with white shading. Self-fertile. **'Cambridge Gage'** (4)

Prunus domestica (Greengage) **in fruit**

Good cropper. Yellow/green fruits with sweet and good flavour. **'Old Greengage'** (5) The original variety of greengage. Small green fruit with green/yellow flesh. Attractive scent and very rich, sweet flavour. May be shy to fruit. **'Oullins Gage'** (4) Neither a true plum nor a gage but, due to its shape, its golden gage colour and sweet flavour, listed here.

MID AUTUMN VARIETIES **'Jefferson'** (1) Considered by many to be a plum, but listed here because of its strong greengage flavour. Oval, green, plum-shaped fruit. A very tough dessert variety. **'Reine Claude De Bavay'** (2) Large, yellow/green fruit. Strong gage flavour. Self-fertile.

Average height and spread
Five years
7 × 7 ft (2.1 × 2.1 m)
Ten years
10 × 10 ft (3 × 3 m)
Twenty years
12 × 16 ft (3.7 × 4.9 m)
Protrudes up to 3 ft (91 cm) from support.

PRUNUS DOMESTICA (Plum Hybrids)

PLUM

Rosaceae   *Fruit Tree*
Deciduous

Flowering and the ripening of the plums are enhanced by growing the trees as fan-trained specimens

---

Origin Of garden and nursery origin.
Use As fan-trained trees for walls and fences where adequate space can be allowed.
Description *Flower* Small, 1 in (2.5 cm) wide, five-petalled white flowers carried along branches in early to mid spring. Consideration has to be given to the selection of varieties for cross pollination. *Foliage* Oval, pointed, grey/green, up to 2½ in (6 cm) long

Prunus domestica **'Victoria' in fruit**

and 1½ in (4 cm) wide. Yellow autumn colour. *Stem* Green to green/brown when young, ageing to brown. Shiny. Medium rate of growth. *Fruit* Oval, 2 in (5 cm) long, thin skinned, yellow or purple, with stones. Juicy, sweet, edible.
Hardiness Tolerates a minimum winter temperature of 0°F (−18°C) but flowers may be damaged by spring frosts.
Soil requirements Tolerates most soil conditions, particularly benefiting from alkaline types. Dislikes waterlogging.
Sun/Shade aspect Tolerates all aspects but pollination may be better in sheltered positions where strong wind does not deter pollinating insects. Best in full sun to aid ripening of fruit.
Pruning Dead, crossing and congested branches may be removed in late summer/early autumn. No other pruning except that required for training.
Training Purchase fan-trained trees or train one-year-old plants by cutting to within 12 in (30 cm) of graft union and tying in subsequent growth in following season. Will require wires or individual anchor points for support.
Propagation and nursery production From budding or grafting on to an understock,

Prunus domestica **'Coe's Golden Drop'**

normally 'St Julien A'. Some dwarfing stocks are available but not recommended for fan-training. Purchase bare-rooted from mid autumn to early spring or container grown as available; one-year-old trees for training, fan-trained trees with not less than five shoots. 'Victoria' may be available from good garden centres and nurseries. Other varieties will have to be sought from specialist nurseries. Young trees may be hard to find.
Problems In the UK the Ministry of Agriculture annually destroy young plum trees due to their extreme susceptibility to many diseases, including plum pox, silver leaf and plum canker. This can, in certain seasons, make their availability extremely scarce. These diseases are also suffered by mature trees and in most cases there are no controls. Plum gall mite and red plum maggot may also attack, but to a certain extent these can be deterred by proprietary controls.
Forms of interest
UK
Consideration has to be given to selecting cross-pollinating trees. For best results varieties of the same number should be planted together but in many cases numbers to either side will also give good pollination. The numbers are indicated alongside each variety.
MID TO EARLY SUMMER RIPENING VARIETIES **'Czar'** (3) Very heavy cropper. Purple/black fruits. Only suitable for cooking. **'Early Laxton'** (3) Small golden yellow fruits, round with rosy pink shading to skin. Good flavour, juicy. **'River's Early Prolific'** (3) A cooking plum with violet/purple fruits. Can be used as a dessert variety and is sweet when ripe.
LATE SUMMER RIPENING VARIETIES **'Belle De Louvain'** (5) Purple fruit for cooking. **'Goldfinch'** (3) Golden yellow dessert plum with greengage flavour. **'Victoria'** (3) Possibly the most reliable of all the varieties. Oval, red plums of very good flavour. Sweet and juicy when ripe. Can be cooked. Self-pollinating therefore the best tree to choose if there is room for only one.
EARLY AUTUMN RIPENING VARIETIES **'Kirke's Blue'** (3) Sweet, juicy, violet/red fruit. Ideal for wall use. **'Warwickshire Drooper'** (2) Yellow fruit. Best variety for preserving.
MID AUTUMN RIPENING VARIETIES **'Coe's Golden Drop'** (2) Amber-yellow fruits with apricot texture. Sweet when ripe. Only does well when grown as a wall-trained tree. Does not cross-pollinate with 'Jefferson' gage (see *P. domestica* Gage Hybrids). **'Marjorie's Seedling'** (5) Black cooking plum often retained on tree well into autumn. Can be used when ripe as a dessert variety. **'Pond's Seedling'** (4) Red, dry flavoured cooking plum. Not recommended for dessert. **'Severn Cross'** (3) Good dessert variety. Sweet flavour. Oval with golden skin.

USA

EARLY SEASON VARIETIES **'Bruce'** Large, red-skinned fruit, red-fleshed with good flavour. Ripens in early summer. A Japanese plum. Use 'Santa Rosa' as a pollinator. Good for the North and South. From Texas. **'Earliblue'** Blue skinned with tender green/yellow flesh. Medium production. Ideal for home garden use. Hardy. Ripens mid summer. Good for the North. Self-fertile. Origin unknown. **'Early Golden'** Round yellow fruit of medium quality. Can be biennial bearing. A Japanese plum. Use 'Shiro' or 'Burbank' as pollinators. Fruit ripening in Michigan area in mid summer. Good for Northern areas. From Canada. **'Mariposa'** Large, heart-shaped, yellow-skinned fruit with red mottling. Sweet red flesh. Good for dessert and cooking. Ripening mid summer. Requires winter cold to aid pollination. A Japanese plum. Use 'Late Santa Rosa', 'Santa Rosa' or 'Wickson' as pollinators. From California. **'Methley'** Medium-sized, red/purple fruit with red flesh of very good flavour. A Japanese plum. Use 'Shiro' or 'Burbank' as pollinators. Mid summer ripening but needs several pickings. Good for the North. From South Africa. **'Santa Rosa'** Large fruit with deep crimson skin and purple flesh, streaked yellow towards the centre. Good for dessert or canning. An early or mid season plum will pollinate. Early to mid summer ripening. A Japanese plum. Good for all areas. From California.

EARLY MID SEASON VARIETIES **'Abundance'** Purple/red skin and yellow flesh. Use for dessert or cooking. Bears fruit biennially from an early age. Use 'Methley' or 'Shiro' as pollinators. Ripens mid summer. A Japanese plum. Good for the North. From California. **'Satsuma'** Small to medium fruit with dull, dark red skin and red flesh. Use either as a dessert or for preserves. Use 'Santa Rosa' or 'Wickson' as pollinators. A Japanese plum. Good in all areas. From California. **'Shiro'** Medium to large round yellow fruit with very good flavour. Use as a dessert or for cooking. Very heavy crops. Use 'Early Golden', 'Methley', or 'Santa Rosa' as pollinators. A Japanese plum. Mid to late summer ripening, depending on area. Good for all areas. From California.

MID SEASON VARIETIES **'Burbank'** Large, red-skinned fruit with amber flesh with very good flavour. Use for dessert or canning. Use 'Early Golden' or 'Santa Rosa' as pollinators. A Japanese plum. Late summer ripening. Good for all areas. From California. **'Ozark Premier'** Very large, red fruit with yellow flesh. Very productive. Disease resistant and hardy. Ripening late summer. A Japanese plum. Good for the North, Midwest, and South. Self-fertile. From Missouri. **'Queen Ann'** Large purple fruit with golden orange flesh. Very juicy and sweet with good flavour. Ripens in mid summer. Use 'Santa Rosa' as a pollinator. A Japanese plum. Less vigorous than many. From California. **'Stanley'** Large, dark blue fruit with yellow, good flavoured flesh. Heavy cropping. Hardy. Self-fertile. A European plum. Ripening late summer/early autumn. Good for the North but also very popular in the East, Midwest and South. From New York. **'Sugar'** Dark blue, medium-sized, sweet fruit in mid to late summer. Good for drying and canning. Self-pollinating. Biennial bearing with some limited cropping in intermediate years. A European plum. Good in all areas. From California. **'Yellow Egg'** Thick golden-yellow skin and yellow flesh. Self-fertile. A European plum. Late summer/early autumn ripening. Good for the North and West. Origin unknown.

LATE SEASON VARIETIES **'Bluefre'** Large blue-skinned fruit with yellow flesh in early fall. Self-fertile. A European plum. May suffer from attacks of brown rot. Good for the North. From Missouri. **'French Prune'** Small, red to purple/black fruit with very sweet flesh and mild flavour. A European plum. Ripening late summer/early autumn. Good for the South and West. From France. **'Italian Prune'** (**'Fellenberg'**) Dark blue fruit with sweet yellow flesh in late summer/early fall. Use for dessert, canning or drying. A European plum. Self-fertile. Good for the South and West. From Germany. **'President'** Dark blue fruit with amber flesh in early fall. Flavour somewhat indifferent, but good for cooking and canning. Use any late European plum as pollinator. A European plum. Good for the North. From the UK.

HARDY PLUMS Varieties suitable for the coldest Northern areas. **'Pipestone'** Large, well-flavoured red fruit with yellow flesh. Use 'Toka' or 'Superior' as pollinators. From Minnesota. **'Superior'** Large red fruit with russet dots. Firm yellow flesh with very good flavour. Use 'Toka' as a pollinator. From Minnesota. **'Toka'** Large red pointed fruit with red/orange skin. Firm yellow flesh with a good flavour. Very heavy crops. Use 'Superior' as a pollinator. From Minnesota. **'Underwood'** Very large red fruit with golden yellow flesh. Very good for dessert. Fruits become available over a long ripening period from mid-summer onwards. Use 'Superior' as a pollinator. From Minnesota.

Average height and spread
Five years
8 × 8 ft (2.4 × 2.4 m)
Ten years
12 × 12 ft (3.7 × 3.7 m)
Twenty years
15 × 15 ft (4.6 × 4.6 m)
Protrudes up to 3 ft (91 cm) from support.

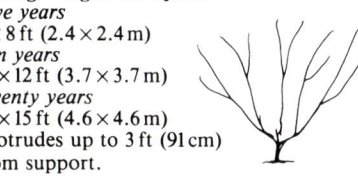

Prunus domestica **'Victoria' in flower**

PRUNUS PERSICA
(Nectarine Hybrids)

NECTARINE

Rosaceae   *Fruit Tree*
Deciduous

Not always reliable for their fruiting achievements out of doors and will benefit from the protection of a greenhouse or large conservatory in all but the most southerly of areas.

Origin Of garden origin
Use As fan-trained trees for large, sheltered walls and fences or under protection.
Description *Flower* Five-petalled, $\frac{3}{4}$ in (2 cm) wide, saucer-shaped flowers in early spring, sugar pink in colour. Self-fertile. *Foliage* Lanceolate, 2–4 in (5–10 cm), mid-green with silvery reverse. Some red hue towards autumn. *Stem* Green when young with orange/red shading towards the sun,

Prunus persica **'Lord Napier'**

becoming purple/red. Rigid. Medium to fast growth rate. *Fruit* Smooth, sometimes glossy, yellow stone fruit with bold patches of dark red shading. Edible. Mid to late summer. May be hard and unripe in all but the sunniest of positions.
Hardiness Tolerates a minimum winter temperature of 4°F (−15°C).
Soil requirements Tolerates all soil conditions, preferring alkaline types.
Sun/Shade aspect Requires a sheltered aspect. Needs a position in full sun for fruit to ripen.
Pruning Remove all old fruiting or flowering shoots after fruiting.
Training Tie to wire on walls or fences. Plant pre-trained trees for preference. Alternatively, one-year-old trees 3–4 ft (91 cm–1.2 m) tall should be cut hard back to within 12 in (30 cm) of their graft union. This will induce numerous side branches, which can then be tied into a fan-trained shape. Tie in annually produced growths as they become large enough.
Propagation and nursery production From budding or grafting. Can be planted bare-rooted from late autumn to early spring or container grown as available. Best size to purchase: fan-trained trees with five or more branches, one-year-old trees 3–5 ft (91 cm–1.5 m) high. Pre-trained trees normally available from garden centres. Younger trees may have to be sought from specialist nurseries.
Problems Suffers from peach leaf curl, which is extremely difficult to control; possibly only a physical barrier, such as a polythene sheet stretched over the tree in early spring, will prevent, and even this is not infallible. Also susceptible to silver leaf disease.

Forms of interest 'Early Rivers' Mid summer fruiting. Reliable. **'Humbolt'** Late summer fruiting. Scarce but worth consideration. **'Lord Napier'** Late summer fruiting. Good variety. **'Pineapple'** Late summer, early autumn fruiting. Interesting flavour. Scarce.

Average height and spread
Five years
7 × 7 ft (2.1 × 2.1 m)
Ten years
12 × 15 ft (3.7 × 4.6 m)
Twenty years
15 × 18 ft (4.6 × 5.5 m)
Protrudes up to 2 ft (60 cm) from support.

PRUNUS PERSICA (Peach Hybrids)

PEACH

Rosaceae　　　　　　　　*Fruit Tree*
Deciduous

In many areas peaches demand a wall or fence for best fruiting performance.

Origin Of garden origin.
Use As fan-trained trees for large walls and fences, both in the open and under protection of greenhouse or conservatory.
Description *Fruit* Five-petalled, ½ in (1 cm) wide, saucer-shaped flowers, attractive sugar-pink in colour in early spring. Produced close to the stems on two-year-old shoots. Self-fertile. *Foliage* Lanceolate, 2–4 in (5–10 cm) long, mid green with silvery reverse. Some red hue towards autumn. *Stem* Green when young with orange/red shading towards the sun, becoming purple/red. Rigid. Medium to fast growth rate. *Fruit* Round, orange/ red to red fruits with stone, edible and juicy. Fruiting early autumn.
Hardiness Tolerates a minimum winter temperature of 4°F (−15°C).
Soil requirements Tolerates all soil, except heavily waterlogged.
Sun/Shade aspect Requires a sheltered aspect. Must be in full sun or fruits will not ripen.
Pruning Remove all old fruiting or flowering shoots after training.
Training Tie to wires on walls or fences. Buy fan-trained trees for preference. Alternatively, one-year-old trees can be cut hard back to within 12 in (30 cm) of their graft union. This will induce numerous side branches, which can then be tied into a fan-trained shape. Tie in annually produced growths as they become large enough to become fruiting shoots for following year.

Prunus persica **'Peregrine' in flower**

Propagation and nursery production From budding or grafting. Can be planted bare-rooted from late autumn to early spring or container grown as available. Best size to purchase: fan-trained trees with five or more shoots, one-year-old trees for training 3–5 ft (91 cm–1.5 m) high. Pre-trained trees normally available from garden centres. One-year-old trees may have to be sought from specialist nurseries.
Problems Suffers from peach leaf curl. Possibly only a physical barrier, such as a polythene sheet stretched over the tree in early spring, will prevent, and even this is not infallible. Also susceptible to silver leaf disease. Avoid planting dwarf patio peaches for fan-training on walls as they are too small.
Forms of interest
UK
'Duke of York' Large yellow fruits, flushed crimson. Excellent flavour. Mid summer. **'Peregrine'** The best of all varieties and most reliable. Good colour and flavour. Late summer. **'Royal George'** A white-fleshed variety with good colour. Ripens late summer. Very scarce in production.

USA
EARLY SEASON VARIETIES **'Fairhaven'** A large peach with bright yellow skin with red sheen. Yellow flesh with a red centre. Good for Western states. From Michigan. **'Flavorcrest'** Large yellow fruit with good flavour. Skin has a red blush. From California. **'Garnet Beauty'** Medium to large firm fruit with yellow, red-streaked flesh. Heavy cropping but susceptible to bacterial leaf spot. Good in Northern states. From Ontario, Canada. **'Golden Jubilee'** Medium to large fruits with yellow, mottled red skin. Yellow flesh. Large cropper. Good in all areas. From New Jersey. **'Redhaven'** Medium-sized fruit, deep red skin on a yellow background. Yellow, firm flesh. Resistant to bacterial leaf spot. Good in all areas. From Michigan. **'Redtop'** Pinkish blush on yellow skin, flesh firm and yellow. May be susceptible to bacterial leaf spot. Good in Western states. From California. **'Reliance'** Large fruit with dark red skin over a yellow background. Flesh bright yellow. Winter hardy. Good for Northern and Western states. From New Hampshire. **'Springcrest'** Medium-sized, good flavoured fruits with yellow flesh. Good for Western states. From California. **'Sunhaven'** Large to medium-sized fruits, bright red skin over a golden background. Flesh yellow flecked with red. Good in all areas. From Michigan. **'Ventura'** Yellow skin with red blush. Firm, yellow, good flavoured flesh. Hardy. From California. **'Veteran'** Medium to large fruit with yellow skin splashed red. Soft yellow flesh. Heavy cropping. Good on Pacific coast. From Ontario, Canada. **'Babcock'** Small to medium-sized fruit, light pink with blush-red shading. Flesh white with red towards the centre. Juicy. Thin fruits for better performance. Ideal for Western states, particularly Southern California. From California. **'Early Elberta'** (**'Gleason Strain'**) Large fruits with yellow flesh of good flavour. Can be used for canning and freezing. Good for the South and West. From Utah. **'J. H. Hale'** Very large yellow fruit with deep crimson shading. Golden yellow, firm flesh. Use a second variety for cross-pollination. Good in all areas. From Connecticut. **'July Elberta'** (**'Kim Elberta'**) Medium-sized fruit with green/yellow skin and pink to dark red blush. Yellow, good quality flesh. Reliable large crops. May be susceptible to bacterial leaf spot. Good for the West. From California. **'Loring'** Medium-sized fruit, blush red over a yellow background. Firm yellow flesh of good texture. Said to be resistant to bacterial leaf

Prunus persica **'Peregrine'**

spot. Good for the North and South. From Missouri. **'Suncrest'** Large firm fruit with a red blush covering to the yellow skin. Susceptible to bacterial leaf spot and for this reason should only be grown in the West and other areas where the disease is not prevalent. Hardy enough for Northern areas. From California.

LATE SEASON VARIETIES **'Belle of Georgia'** (**'Georgia Belle'**) A white peach with red blush. Firm white flesh with very good flavour. Use for dessert and freezing but not suitable for canning. Susceptible to brown rot. Good for North and South. From Georgia. **'Blake'** Red skin and firm yellow flesh. Use mainly for freezing and canning. Susceptible to bacterial canker. Good for North and South. From New Jersey. **'Cresthaven'** Medium to large golden fruit with bright red shading. Yellow flesh. Hardy. Good for North and South. From Michigan. **'Elberta'** Deep golden skin with red blush. Fruits may drop from the tree at maturity.

Prunus persica in flower

Said to be resistant to brown rot. Good in all areas. From Georgia. **'Fay Elberta'** In its own state of California rates amongst one of the most reliable and widely-planted varieties. Skin yellow with red blush. Useful for all purposes of eating, cooking, canning and freezing. Not entirely winter-hardy. **'Jefferson'** Orange skin with bright red blush. Yellow, firm flesh. Good for canning and freezing. Said to be resistant to brown rot. Good in areas where late spring frosts are experienced and in Northern and Southern states. From Virginia. **'Madison'** Medium-sized orange/

yellow fruit with bright red blush. Flesh orange/yellow with good texture and flavour. Good in the North and South. From Virginia. **'Raritan Rose'** Red skin with white flesh. Very good flavour. Strong growing and hardy. Best used in the East and North. From New Jersey. **'Redskin'** Red-skinned fruit with good yellow flesh. Ideal for freezing, canning and dessert. Well-established in the East and North. From Maryland. **'Rio Oso Gem'** Large fruit with yellow skin and red shading. Yellow flesh. Can be used for dessert and freezing. Best in the South and West. From California. **'Sunhigh'** Medium to large fruit with bright red blush over a yellow skin. Yellow flesh. Said to be very susceptible to bacterial leaf spot and must have protection throughout the summer. Good in North and South. From New Jersey.

Average height and spread
Five years
7 × 7 ft (2.1 × 2.1 m)
Ten years
12 × 12 ft (3.7 × 3.7 m)
Twenty years
15 × 15 ft (4.6 × 4.6 m)
Protrudes up to 2 ft (60 cm) from support.

PYRACANTHA

PYRANTHA, FIRETHORN
Rosaceae   *Wall Shrub*
Evergreen
This attractive fruiting shrub is most often seen against walls and fences where it is generally found as a fan-trained shrub.

Origin Most forms originate in China; many varieties of garden origin.
Use As a large, fan-trained or lattice formation, flowering and fruiting shrub for walls and fences; can also be grown up pillars. Although it is well-known as a wall shrub, its other possibilities as a freestanding, small, mop-headed shrub or even as a hedging plant should not be overlooked.
Description *Flower* Good-sized clusters of white, hawthorn-like flowers up to 5 in (12 cm) across with a musty scent in early summer. Loved by bees. *Foliage* Leaves ovate, 1–2 in (2.5–5 cm) long and ½–¾ in (1–2 cm) wide. Some grey and variegated leaved varieties. *Stem* ½–1 in (1–2.5 cm) long, sharp spines at most leaf axils. Red/brown, ageing to dark green to brown. Forms a flat, fan-shaped shrub on walls and fences. Some attractive pendulous stem forms. Fast growth rate. *Fruit* Clusters of round, good-sized fruits, yellow, orange or red depending on variety,

early autumn, maintained well into winter.
Hardiness Tolerates a minimum winter temperature of 4°F (−15°C) although severe wind chill may damage some leaves and tips of younger growth. Damage is usually corrected the following spring.
Soil requirements Most soils, but distressed on extremely alkaline types.
Sun/Shade aspect Tolerates all aspects, only needing protection from very exposed winter wind conditions. Full sun to deep shade, but produces fewer fruit in deeper shade, and may also become more open.
Pruning None required other than that needed for fan-training. Can be drastically reduced in size, although following year's flowering and fruiting will be poorer. Pruning is best carried out in late winter to early spring to avoid attacks of fungus disease. Can be trimmed and shaped as desired.
Training Will require tying to wires or individual anchor points to achieve a fan-trained shape.
Propagation and nursery production From semi-ripe cuttings taken in early summer. Propagates readily from seed, but seed-raised plants are variable in fruit quality and colour. Purchase container grown; easy to find a good range of named varieties. Best planting height 15 in–3 ft (38–91 cm).

Pyracantha atalantoides in flower

Problems Subject to the airborne fungus disease fire blight in summer months which completely destroys all foliage. There is no cure or prevention. All affected or suspect shrubs should be burnt and the appropriate government agency informed of its occurrence.
Forms of interest *P.* **'Alexander Pendula'** A variety with interesting, long, weeping branches, forming a somewhat unruly shrub. Fruit coral red. *P. angustifolia* Grey/green, long, narrow, oblong leaves with grey, felted undersides. Fruit orange/yellow, may be maintained into winter. *P. atalantioides* (syn. *P. gibbsii*) (Gibbs firethorn) Strong, fast-growing variety with good-sized, oval, dark, glossy green leaves. Scarlet fruits produced late in autumn. Good for difficult shady walls, but may be withdrawn from production due to fire blight. *P. a.* **'Aurea'** (syn. *P. a.* **'Flava'**) A yellow-fruited form of the above, which may also be withdrawn for the same reason as its parent. *P. coccinea* **'Lalandei'** A somewhat upright but wide form with small, obovate leaves. Orange/red fruits in autumn, may be maintained into winter. *P. c.* **'Sparklers'** Rather small, narrow, ovate foliage variegated with white margins. Orange/red fruits. Reaches two thirds average height and spread. *P.* **'Golden Charmer'** Good dark foliage. Large clusters of golden-yellow fruits. *P.* **'Golden Sun'** Large golden-yellow fruits. Dark green foliage. *P.* **'Harlequin'** Grey/green foliage with silver/white edges, sometimes pink-tinged, especially in cold

Pyracantha **'Orange Charmer'** in fruit

Pyracantha rogersiana 'Flava' in fruit

weather. Small, orange/red fruits. Two thirds average height and spread. *P.* 'Mojave' Good, large, mid to dark green foliage. Large orange/red fruits in autumn, may be maintained into winter. Very winter hardy and resistant to wind scorch. *P.* 'Orange Charmer' Dark green foliage. Fruits large, deep orange in colour. *P.* 'Orange Glow' Interesting dark green foliage with dark purple/black stems. Good clusters of orange/red fruits very freely produced in autumn. Smaller than average. *P.* 'Red Column' Large red fruits. Upright habit. Good foliage. May have to be sought from specialist nursery. *P.* 'Red Cushion' Light green foliage, ageing to grey/green. White flowers, followed by red fruits in autumn. Low-growing, reaching only 3 ft (91 cm) in height and 6 ft (1.8 m) spread. *P. rogersiana* Very dark, oblong to lanceolate leaves. Especially recommended for shaded walls, giving a display of orange/red fruits in autumn. *P. r.* 'Flava' Small foliage, clusters of bright yellow fruits. Good for shaded walls. Somewhat weeping habit. *P.* 'Shawnee'

Pyracantha 'Orange Glow'

Dense foliage and numerous spines. Fruit yellow to light orange, colouring early in season and maintained well into winter. Said to be resistant to fire blight and scab fungus. Raised in USA. *P.* 'Soleil d'Or' Good mid to light green foliage. Very large clusters of deep yellow fruits. *P.* 'Teton' Good, small, dark green foliage. Clusters of orange/yellow fruits. Upright growth.
Average height and spread
Five years
6 × 4 ft (1.8 × 1.2 m)
Ten years
12 × 6 ft (3.7 × 1.8 m)
Twenty years
13 × 10 ft (4 × 3 m)
Protrudes up to 3 ft (91 cm) from support.

Origin From Europe through to Asia and Japan.
Use As medium to large fan-trained trees for walls and fences.
Description *Flower* Clusters of up to 3 in (7.5 cm) across composed of five to seven white, single, cup-shaped florets, each up to 1½ in (4 cm) across, produced in mid spring. *Foliage* Ovate to lanceolate. Either green or grey, depending on variety. Good autumn colours. *Stem* Grey/green to grey/brown. Upright, spreading or pendulous, depending on variety, and all easy to train. Medium rate of growth. *Fruit* Small, oval or rounded pears up to 2 in (5 cm) long, in autumn. Fruits not edible.
Hardiness Tolerates a minimum winter temperature of 4°F (−15°C).
Soil requirements Any soil type, but shows distress in extremely poor soils.
Sun/Shade aspect Tolerates all aspects. Green forms tolerate moderate shade, silver-leaved varieties prefer full sun.
Pruning Prune young trees hard in spring following planting. Select and train resulting five to seven shoots and tie into a fan-trained shape. In subsequent years remove all side growths back to two buds from their origin and maintain original main branches in fan shape.

Pyrus salicifolia

Training Secure to wires or individual anchor points.
Propagation and nursery production From grafting or budding on to understocks of quince or *P. communis*. Available bare-rooted or container grown. For fan-training choose trees of not more than two years old and 3–6 ft (91 cm–1.8 m) tall. Most forms easily found in garden centres and general nurseries; some must be sought from specialist nurseries.
Problems All varieties are poor-rooted, especially when young. Extra peat or other organic material should be added to the planting hole to encourage establishment of young roots. Also needs staking and extra watering in spring.
Forms of interest *P. amygdaliformis* Foliage grey/green and ovate, white undersides. Rarely available. *P. calleryana* 'Chanticleer' Dark glossy green leaves maintained into late autumn and early winter, then turning orange/red. Narrow, columnar habit. *P. communis* (Wild pear) Attractive white flowers in spring. Light green, ovate foliage with good autumn colour. Limited in commercial production, must be sought from specialist nurseries. *P. c.* 'Beech Hill' Dark green, round leaves; interesting leathery texture, glossy surface and wavy edges. Good autumn colours. *P. elaeagrifolia* Silver grey, ovate, sometimes lanceolate foliage. Interesting white flowers. Slightly more tender than most but above average height. *P. nivalis* Attractive, ovate, white to silver/grey foliage. Graceful white flowers. Small, globed, yellow/green fruits. Two thirds average height and spread. Must be sought from specialist nurseries. *P. salicifolia* 'Pendula' (Weeping willow-leaved pear) Narrow, lanceolate leaves up to 2 in (5 cm) long. Not usually supplied as large as other forms.
Average height and spread
Five years
12 × 12 ft (3.7 × 3.7 m)
Ten years
18 × 18 ft (5.5 × 5.5 m)
Twenty years
22 × 22 ft (6.7 × 6.7 m)
Protrudes up to 3 ft (91 cm) from support.

Origin Of garden origin.
Use As fan-trained, horizontal-trained (espalier) or cordon trees for walls and fences.
Description *Flower* Five-petalled, 1 in (2.5 cm) wide, pink in bud opening to white in early to mid spring. Consideration must be given to the selection of varieties for cross-pollination. *Foliage* Round to oval, mid green glossy foliage up to 2½ in (6 cm) long and 1½ in (4 cm) wide. Yellow autumn colour. *Stem* Light grey/green when young, becoming grey/brown, finally grey. Upright. Rigid. Medium growth rate. *Fruit* Oval, pointed, up to 5 in (12 cm) long and 2½ in (6 cm) wide, grey/green to green/brown, often with orange and red shading. Fruiting late summer to early winter depending on variety.
Hardiness Tolerates a minimum winter temperature of 0°F (−18°C) but spring frosts may damage flowers.
Soil requirements Tolerates all soil conditions with no particular preference.
Sun/Shade aspect Tolerates all aspects. Best in full sun to ripen fruit, but will tolerate very light shade.
Pruning Once trained, shorten all annual growth back to within 2 in (5 cm) of its origin in early autumn, without reducing the main framework.

Training Buy fan-trained, horizontal-trained (espalier) or cordon trees and tie to wires or individual anchor points.

Propagation and nursery production Should always be grafted or budded with understock of known varieties. There are many combinations, but in the main, fan, horizontal and cordon fruit trees are propagated on to a semi-dwarf rootstock. Can be planted bare-rooted from early autumn to early spring or container grown as available, with autumn and spring planting for preference. Pre-trained trees in each of the forms measure 3–5 ft (91 cm–1.5 m) overall and will normally have to be obtained from a specialist supplier, although some varieties are available from garden centres and general nurseries.

Problems May suffer from attacks of mildew and blackfly. Treat with a proprietary control. Susceptible to pear scab – again, treat with a proprietary control.

Pyrus communis 'Conference' in flower

Pyrus communis 'Gorham'

Forms of interest
UK
The number in brackets denotes the pollination group. Each variety should be grown in proximity with one of a similar number for successful pollination.

EARLY TO LATE AUTUMN RIPENING PEARS These varieties should be picked from the tree while still green and stored in a cool dark room for seven to 14 days to ripen. '**Dr Jules Guyot**' (3) Pale golden fruit, often flushed scarlet. Interesting flavour. Mid autumn. '**Fertility Improved**' (3) Yellow with russet skin. Good flavour, sweet and juicy. Good red autumn foliage. Disease resistant. Mid to late autumn. '**Glow Red William**' (3) A form of 'Williams' Bon Chrétien'. Very juicy fruit with a tart flavour and bright crimson skin. Resistant to scab. Mid autumn. '**Gorham**' (4) Pale yellow with russet shading. Long and narrow fruit shape. Mid to late autumn. A variety from America bred from 'Williams' Bon Chrétien', and inheriting its good flavour. '**Laxton's Foremost**' (4) Good-sized fruit with good flavour. Disease resistant. Mid to late autumn. '**Williams' Bon Chrétien**' (3) Pale yellow fruit. Very juicy and with a very good flavour. Will tolerate exposed positions. Susceptible to scab, but worthy of garden merit. Mid autumn.

LATE AUTUMN/EARLY WINTER RIPENING PEARS Harvest unripe in mid autumn, store to ripen, use as required. '**Beurre Hardy**' (4) Coppery colour with red shading. Good grower. Scab resistant. Attractive scarlet foliage in autumn. '**Beurre Superfin**' (3) Fruit long and yellow with russet patches on skin. Susceptible to scab. Late autumn. '**Conference**' (3) Long, pale green, bulbous-based fruit with russet skin. Very free fruiting. Self-fertile, but fruiting enhanced if another pollinating variety is used in association. Resistant to scab. '**Doyenne du Comice**' (4) Large golden fruit with some russeting. In good summers has a red shading. Fruiting can be unreliable and it is susceptible to scab unless regular control is applied. Early winter. '**Louise Bonne of Jersey**' (2) Fruits long, golden yellow with brown/red shading. Late autumn. Should not be considered for pollinating 'Williams' Bon Chrétien'. '**Packham's**

Triumph' (3) 'Williams' Bon Chrétien' type with golden yellow fruit. Early winter.

MID TO LATE WINTER RIPENING PEARS Harvest in mid to late autumn, store to ripen, use as required. '**Glou Morceau**' (4) Medium-sized green pears, becoming yellow when ripe. Benefits from training on a sunny wall or fence in a sheltered position. Mid to late winter. '**Joséphine de Malines**' (3) Green, becoming yellow when ripe. Benefits from use on a wall or fence. Will take some time to ripen in store. Early to mid winter. '**Winter Nellis**' (4) Green/yellow, small with some russeting to skin. Late autumn to mid winter.

Pyrus communis 'Williams' Bon Chrétien'

LATE COOKING PEARS Harvest in late autumn and allow to ripen in store. '**Pitmaston Duchess**' (4) Fruit pale yellow with a light brown russeting. Not of dessert flavour but good as a cooking pear. In wet, moist areas very susceptible to scab. Good red autumn colours. Late autumn to early winter.

USA
Most American varieties will pollinate each other, although the varieties 'Bartlett', 'Magness' and 'Seckel' are poor pollinators and should not be considered for such use.

EARLY SEASON VARIETIES Eat from the tree when ripe. '**Clapp's Favorite**' Large yellow fruit with red shading. Sweet soft flesh for dessert or canning. Attractively shaped tree very susceptible to fire blight. Hardy for all areas but particularly good for North and West. From Massachusetts. '**Gorham**' See UK varieties. '**Moonglow**' Large, soft, juicy fruit with attractive flavour. Use for dessert

Pyrus communis 'Fertility Improved'

Pyrus communis 'Pitmaston Duchess'

and canning. Heavy cropping from an early age. Resistant to fire blight. Good for all areas. From Maryland. **'Orient'** Almost round fruit with firm flesh. Good for canning but not to all tastes as a dessert. Moderate cropping. Resistant to fire blight. Good for Southen areas. From California. **'Red Clapp'** (**'Starkrimson'**) Good quality red-skinned fruit. Susceptible to fire blight. Good for the West and North. From Michigan.

MID SEASON VARIETIES Eat from the tree when ripe. **'Bartlett'** Yellow, medium to large fruit with thin skin. Very sweet and juicy, good for dessert and canning, and is commercially used in large quantities for this purpose. Subject to fire blight. Does not resent hot summers, assuming that there are cold winters. All varieties pollinate with the exception of 'Seckel' or 'Magness'. From the UK. **'Doyenne du Comice'** See UK varieties. **'Lincoln'** Large fruits carried in large numbers. Tree hardy and resistant to fire blight. Very reliable, particularly in the Midwest, but good for the North and South too. Origin unknown, possibly the Midwest. **'Magness'** Oval, medium sized fruit with russet skin and very attractive perfume. Fruit variable in its quality. Highly resistant to fire blight. Good for Southern and Western areas. From Maryland. **'Maxine'** (**'Starking Delicious'**) Large, good-looking fruit with firm, juicy, sweet white flesh. Moderately resistant to fire blight. Good in North and South. From Ohio. **'Parker'** Medium to large fruit, yellow-skinned with red blush. White, juicy, sweet flesh. Moderately hardy. Susceptible to fire

Pyrus communis 'Beurre Superfin'

blight. Good for northern areas. From Minnesota. **'Sensation Red Bartlett'** (**'Sensation'**) Yellow skins with heavy red blushing. White juicy flesh. Susceptible to fire blight. Good for the West. From Australia.

LATE SEASON VARIETIES Harvest in mid to late fall and allow to ripen in store. Can be used from the tree from mid fall onwards for cooking. **'Anjou'** Large green fruit with mild flavoured, firm flesh, somewhat on the dry side. Very good for storing. Can be eaten fresh or from store or used for canning. Susceptible to fire blight. Good for North and West. Variety 'Red Anjou' may also be worth consideration. From France. **'Bosc'** Long, narrow fruit with russet colouring. Crisp, firm flesh with good scent. Considered by many to be the best of all pears. Can be used for dessert, fresh from the tree, or canned. Cooks well. Susceptible to fire blight. Good for the North and West. From France.

Pyrus communis 'Doyenne du Comice'

'Duchess' Green/yellow, extremely large fruits with good-flavoured flesh. Good for Northern areas. From France. **'Keiffer'** Large yellow fruit, somewhat gritty texture to the flesh not universally well-liked. Good for keeping, cooks well and is ideal for canning. High resistance to fire blight. Needs some winter cold to encourage flowering and pollination, but will accept summer heat. Good for the East, North, South and Midwest. From Pennsylvania. **'Mericourt'** Green to yellow/green, sometimes blushed deep red. Creamy white flesh. Good for dessert or canning. Hardy, resistant to both fire blight and leaf spot. Good for the South. From Tennessee. **'Patten'** Large, juicy fruits. Good for dessert. Hardy, good for Northern areas. From Louisiana. **'Seckel'** Small yellow/brown fruits. Very good flavour and scent and an ideal pear for gardens. Either use for dessert directly from the tree or for preserving. Very

heavy crops. Resistant to fire blight. Good for all areas. From New York. **'Winter Nellis'** See UK varieties.

Average height and spread
Fan-trained
Five years
8 × 8 ft (2.4 × 2.4 m)
Ten years
10 × 12 ft (3 × 3.7 m)
Twenty years
12 × 16 ft (3.7 × 4.9 m)
Protrudes up to 3 ft (91 cm) from support

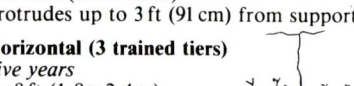

Horizontal (3 trained tiers)
Five years
6 × 8 ft (1.8 × 2.4 m)
Ten years
8 × 12 ft (2.4 × 3.7 m)
Twenty years
10 × 20 ft (3 × 6 m)
Protrudes up to 3 ft (91 cm) from support
Cordon
Five years
8 × 2 ft (2.4 m × 60 cm)
Ten years
29 × 3 ft (2.7 m × 91 cm)
Twenty years
9 × 3 ft (2.7 m × 91 cm)
Protrudes up to 2 ft (60 cm) from support.

RAPHIOLEPSIS UMBELLATA

KNOWN BY BOTANICAL NAME
Roseaceae   **Wall Shrub**
Evergreen
An attractive shrub which needs the protection of walls or fences against wind damage which can scorch and defoliate in winter.

Origin From Japan and Korea.
Use As a freestanding or fan-trained shrub for sheltered walls and fences.
Description *Flower* White suffused with pink, fragrant, $\frac{3}{4}$ in (2 cm) across, carried in terminal clusters 3–4 in (7.5–10 cm) across in late spring/summer. *Foliage* Oval leaves up to 3 in (7.5 cm) long. Grey/green with some red shading. *Stem* Short-branched, grey/green becoming green/ brown. Slow to medium growth rate. *Fruit* Pear-shaped, blue/black fruits in late summer/early autumn.
Hardiness Tolerates a minimum winter temperature of up to 14°F (−10°C) against a wall.
Soil requirements Requires a neutral to acid soil with adequate organic material and plant nutrient. Dislikes waterlogging or extremely dry conditions.
Sun/Shade aspect Requires a sheltered aspect. Tolerates full sun to medium shade, light shade for preference.
Pruning Normally not required other than that for training.
Training Allow to grow freestanding or fan-trained against walls or fences.

Raphiolepsis umbellata **in flower**

Rhamnus alaterna 'Argenteo-variegata'

Propagation and nursery production From semi-ripe cuttings taken in early summer and from seed. Should always be purchased container-grown; will have to be sought from specialist nurseries. Best planting height 3 in–2 ft (7.5–60 cm).
Problems Slow-growing and will not thrive in poor or alkaline soils. Full range of hardiness not yet explored.
Similar forms of interest *R.u.* 'Enchantress' Rose pink flowers, earlier than species and carried over a long period. Possibly less hardy than its parent.
Average height and spread
Five years
3 × 3 ft (91 × 91 cm)
Ten years
5 × 5 ft (1.5 × 1.5 m)
Twenty years
8 × 8 ft (2.4 × 2.4 m)
Protrudes up to 18 in (45 cm) from support if fan-trained, 3 ft (91 cm) untrained.

RHAMNUS ALATERNA 'ARGENTEO-VARIEGATA'

BUCKTHORN
Rhamnaceae   *Wall Shrub*
Evergreen
The variegated leaves of this moderately hardy shrub can add interest to any winter garden.

Origin From Europe and the Mediterranean region.
Use As a large, freestanding or fan-trained foliage shrub for large walls and fences.
Description *Flower* Inconspicuous, small, pale cream flowers in early summer. *Foliage* Leaves ovate, 1–2½ in (2.5–6 cm) long. Olive-green with white variegation. *Stem* Upright, becoming very branching with age. Dark purple to purple/brown. Fast growth rate. *Fruit* Produces small red fruits in autumn.
Hardiness Tolerates a minimum winter temperature of 4°F (−15°C).
Soil requirements Most soil types, but dislikes extremely wet conditions.
Sun/Shade aspect Requires a sheltered aspect in medium to full shade.
Pruning None required and is best left free-growing but may be trimmed or trained as necessary.
Propagation and nursery production From semi-ripe cuttings taken in summer. Purchase container grown; normally available from good garden centres and nurseries, particularly those in warm coastal areas. Best planting height 1–3 ft (30–91 cm).
Training Allow to stand free or secure to wires or individual anchor points to encourage a fan-trained shape.

Problems Its hardiness is suspect in cold winds and it often requires a larger space than anticipated.
Similar forms of interest None.
Average height and spread
Five years
6 × 3 ft (1.8 m × 91 cm)
Ten years
12 × 6 ft (3.7 × 1.8 m)
Twenty years
16 × 13 ft (4.9 × 4 m)
Protrudes up to 3 ft (91 cm) from support if fan-trained, 6 ft (1.8 m) untrained.

RHODOCHITON ASTROSANGUINEUM

PURPLE BELLS
Scrophulariaceae   *Climbing Perennial*
Deciduous
A tender, interesting, flowering climber for sheltered gardens, greenhouses and conservatories.

Rhodochiton atrosanguineum **in flower**

Origin From Mexico.
Use As a short climber for very warm, sunny positions in favourable locations, or as a greenhouse climber where the temperature is not allowed to fall below freezing.
Description *Flower* Blood-red to dark blood-red, ageing to purple-red. Five petals form a hanging bell-shaped flower arranged along the stems in early to mid summer. *Foliage* Narrow-pointed, with some small teeth and a light downy covering. Mid green with purple hue around veins. *Stem* Dark green, dull, partially twining. Medium rate of growth. *Fruit* None of interest.
Hardiness Tolerates a minimum winter temperature of 32°F (0°C).
Soil requirements Moderately alkaline to acid with no particular preference. If grown in containers in conservatories, a container of at least 21 in (53 cm) in diameter, filled with a good quality soil-based potting compost, is required.
Sun/Shade aspect Requires a very sheltered, warm, sunny position with no frost. Best in full sun, but will tolerate light shade.
Pruning Normally requires no pruning.
Training Allow to ramble over wires or other similar constructions. May require some tying in, particularly when young. Can be worked around a wire framework in a circle when used in greenhouses and conservatories.
Propagation and nursery production From seed, which is often self-set on existing plants and germinates well when sown with a little bottom heat in early spring. Always purchase container grown; will have to be sought from specialist nurseries or as seed from a specialist seed merchant. Best planting height 5–18 in (12–45 cm).
Problems Its dislike of frost must always be considered, while its open habit of growth can make it difficult to accommodate in some greenhouses and conservatories.
Similar forms of interest None.
Average height and spread
Five years
4 × 4 ft (1.2 × 1.2 m)
Ten years
8 × 8 ft (2.4 × 2.4 m)
Twenty years
10 × 10 ft (3 × 3 m)
Protrudes up to 6 in (15 cm) from support.

RIBES GROSSULARIA (Gooseberry Hybrids)

GOOSEBERRY
Grossulariaceae   *Fruiting Shrub*
Deciduous
A fruiting bush not always considered for wall training but responding well, particularly in ripening, when used in such situations.

Origin The basic form is from the temperate regions of Europe, but most varieties today are hybrids of garden or nursery origin.
Use As single, double or treble cordons, fan-trained or upright bushes against a wall or fence.
Description *Flower* Small, pendent, silver/yellow, tubular flowers in mid to late spring.

Ribes grossularia 'Careless'

Ribes grossularia 'Whinham's Industry'

Not attractive. Self-fertile. *Foliage* Hand-shaped with lobed edges, $\frac{1}{2}$–$\frac{3}{4}$ in (1–2 cm) across, light green with yellow autumn colour. *Stem* Light grey/green, becoming grey to grey/brown. Thorned. Medium growth rate. Branching habit. *Fruit* Oval, $\frac{3}{4}$–1 in (2–2.5 cm) long, pendent, often covered with hairs. Green in colour, ripening either to yellow/green or red/green. Normally sweet and edible when ripe. Fruits mid to late summer on wood two years old or more. Good for freezing.
Hardiness Tolerates a minimum winter temperature of 0°F (−18°C).
Soil requirements Tolerates most soils, except very alkaline, dry or waterlogged conditions.
Sun/Shade aspect Tolerates all aspects. Full sun to light shade, with full sun for preference to ripen fruit to best advantage.
Pruning Once structure is trained, shorten all last season's growth back to two buds after fruiting to encourage formation of fruit spurs.
Training Purchase one-year-old rooted cuttings and train central upright stems for cordons. If double and treble cordons are required, stop initial shoot at 6 in (15 cm) from its origin; this will induce one, two or three shoots which can be trained. Continue to train, shortening back side growths. For fan-training, cut back central shoot to 18 in (45 cm) from ground level then train ensuing side shoots into a fan-trained shape. Cut back any side shoots not required for training to within two buds. Will require wires or individual anchor points for support.
Propagation and nursery production From hardwood cuttings taken in winter. Can be planted bare-rooted from mid autumn to early spring or container grown as available. One-year-old rooted cuttings are best for training. On very rare occasions, pre-trained cordons can be found; fan-trained shrubs are never available.
Problems Thorns can make cultivation uncomfortable. May suffer from American gooseberry mildew; a proprietary control will normally treat it, but do not apply during flowering period. There are a number of other mildews and rust fungus diseases which attack gooseberries and all can be treated as for American gooseberry mildew. Gooseberry sawfly and gooseberry red spider mite also can be problems requiring treatment with a proprietary control. The need to grow shrubs in situ for training may also be a drawback, but they are nevertheless worth persistence.
Forms of interest 'Careless' The most useful variety where only a few can be accommodated. Milky green skin when ripe in mid summer. Strong, upright growth. **'Keepsake'** Pale green when ripe in mid summer. **'Lancer'** Green when ripe. Mid to late summer fruiting. **'Leveller'** Yellow when ripe. Late summer fruiting. **'Whinhams Industry'** Dark red when ripe. Mid summer fruiting. **'Whitesmith'** Pale green when ripe. Mid summer fruiting.
Average height and spread
Fan-trained
Five years
5 × 5 ft (1.5 × 1.5 m)
Ten years
7 × 7 ft (2.1 × 2.1 m)
Twenty years
9 × 9 ft (2.7 × 2.7 m)
Protrudes up to 18 in (45 cm) from support
Cordon
Five years
5 × 1 ft (1.5 m × 30 cm)
Ten years
7 × 3 ft (2.1 × 91 cm)
Twenty years
7 × 3 ft (2.1 m × 91 cm)
Protrudes up to 12 in (30 cm) from support.

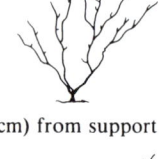

RIBES LAURIFOLIUM

EVERGREEN WINTER FLOWERING CURRANT
Grossulariaceae   *Wall Shrub*
Evergreen
A unique evergreen, winter flowering variety deserving more attention.

Origin From western China.
Use As a winter-flowering and evergreen shrub for sheltered walls and fences.
Description *Flower* Small, hanging racemes of male flowers, 3 in (7.5 cm) long and 1$\frac{1}{2}$ in (4 cm) wide, yellow to light green, are produced in mid to late winter at each leaf joint on mature shrubs. Female flowers are borne on shorter, erect racemes. *Foliage* Oval, 4 in (10 cm) long and 1$\frac{1}{2}$ in (4 cm) wide, pointed with slightly notched edges. Dull grey/green, lighter grey underside, contrasting well with flowers. *Stem* Light red/brown when young, ageing to darker brown, finally grey/brown with age. Branching. A little weak in constitution but this is overcome by fan-training. Slow to medium rate of growth. *Fruit* Rarely fruits, but may produce small, round blackcurrants in autumn and early winter.
Hardiness Tolerates a minimum winter temperature of 4°F (−15°C).
Soil requirements Most soils, only disliking extremely wet or dry conditions.
Sun/Shade aspect Requires a sheltered aspect in full sun to light shade.
Pruning Normally needs no pruning, other than that required to achieve a fan shape on walls or fences.
Training Will require tying to wires or individual anchor points to achieve a fan-trained shape.
Propagation and nursery production From semi-ripe cuttings taken in early summer or hardwood cuttings in winter. Always purchase container grown; will have to be sought from specialist nurseries. Best planting height 9 in–2 ft (23–60 cm).
Problems Slow to establish, often taking two years to produce new growth, but once established relatively fast growing.
Similar forms of interest None.
Average height and spread
Five years
5 × 3 ft (1.5 m × 91 cm)
Ten years
6 × 6 ft (1.8 × 1.8 m)
Twenty years
8 × 10 ft (2.4 × 3 m)
Protrudes up to 12 in (30 cm) from support.

RIBES SATIVUM (Red and White Currant Hybrids)

RED AND WHITE CURRANTS
Grossulariaceae   *Fruiting shrub*
Deciduous
Fruiting bushes which adapt very well to cultivation against a wall or fence, where they benefit from the protection and warmth.

Origin Of European origin. Possibly crosses between *Ribes sativum* and *R. rubrum* (Northern red currant). Most varieties of nursery or garden origin.
Use As fan-trained shrubs or single, double or treble cordons, upright or at an angle of 35°.

Ribes laurifolium in flower

Ribes sativum 'Laxton's No. 1'

Description *Flower* Small, inconspicuous sulphur-yellow flowers carried in small clusters at leaf axils in late spring. *Foliage* Hand-shaped, five-fingered, 3 in (7.5 cm) across and 3½ in (8.5 cm) long. Grey/green with silver reverse and yellow autumn colour. Currant-scented when crushed. *Stem* Grey/green, becoming grey, finally grey/brown. Upright. Limited branching. Currant-scented when crushed. *Fruit* Hanging clusters of red or white fruits, depending on variety, in mid summer. Sweet, edible, for dessert or culinary uses.

Hardiness Tolerates a minimum winter temperature of 0°F (−18°C).

Soil requirements Tolerates all soil conditions, except extremely waterlogged and extremely dry.

Sun/Shade aspect Tolerates all aspects but benefits from full sun to aid ripening of fruit.

Pruning Once trained shape has been achieved, remove all side growth back to within 2 in (5 cm) of its origin annually.

Training Purchase one-year-old plants or propagate as hardwood cuttings in winter. The second year, cut central shoot back to within 12 in (30 cm) of ground level in early to mid spring and fan-train the resulting side growths. If training as a single cordon, remove all but the central shoot. If required for two and three cordons, train two shoots horizontally and turn upwards once they have reached approximately 12–15 in (30–38 cm) from the central stem. Tie to wires or individual anchor points.

Propagation and nursery production From hardwood cuttings taken in mid to early spring. For preference purchase plants no more than one year old if required for training; these will normally be 12–18 in (30–45 cm) in height, bare-rooted, although some may be found container grown. Older plants can be used for training, but will require hard cutting back the first year after planting so that ensuing shoots can be trained and these may be slow to appear. Young plants may have to be sought from specialist growers but are sometimes found in garden centres.

Problems Apart from the need for careful training they have few problems except for mildew, which can be treated with a proprietary control.

Forms of interest
UK
RED CURRANTS 'Laxton's No. 1' Still one of the best varieties. Large fruits. Reliable. 'Red Lake' Very dark red fruits. Extremely good variety. 'Red Start' Heavy cropping. Late season fruit of medium size. 'Rondon' Cropping can be inconsistent but of good size. Late fruiting.
WHITE CURRANTS 'White Dutch' White to yellow/white fruits. 'White Versailles' Scarce but worth seeking out. Good sprays of fruits.

USA
RED CURRANTS 'Perfection' Medium-sized fruits in open clusters. Vigorous growing and highly productive. Good for Washington and Oregon areas. From New York. 'Red Lake' Very dark red fruits with good flavour. Good over a wide area. From the UK. 'Stephen's No. 9' Medium-sized red berries in medium clusters. Spreading habit but highly productive. From Ontario, Canada. 'Wilder' Large, firm, tender red berries with tart flavour. Plants vigorous, hardy and very long-lived. From Indiana.
WHITE CURRANTS 'Jumbo' Large, pale green, sweet fruits. Strong-growing, upright habit. Origin unknown. 'White Grape' Good quality and very widely used. From Europe.

Average height and spread
Five years
5 × 5 ft (1.5 × 1.5 m)
Ten years
7 × 7 ft (2.1 × 2.1 m)
Twenty years
9 × 9 ft (2.7 × 2.7 m)

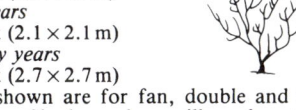

Sizes shown are for fan, double and treble cordons. Single cordons will produce 60 per cent less spread. Protrudes up to 12 in (30 cm) from support prior to annual pruning.

RIBES SPECIOSUM

FUCHSIA-FLOWERING CURRANT

Grossulariaceae   *Wall Shrub*
Deciduous to semi-evergreen

This flowering gooseberry is an attractive shrub with interesting flowers and adapts well to fan-training.

Origin From California.

Use As a flowering fan-trained shrub of medium size for walls and fences.

Description *Flower* Small, tubular, 1–1½ in (2.5–4 cm) long, pendent, fuchsia-shaped flowers, dark red, carried very regimentally along branches in mid to late spring. Some later flowering. *Foliage* Three or five lobed, mid-green ovate to obovate leaves, 1½ in (4 cm) wide, but sometimes smaller on older wood. Some yellow autumn colour. Leaves may be retained through winter in favoured localities. *Stem* Grey/green, becoming grey/brown. Upright, becoming spreading. Limited branching. Armed with soft thorns. Good fan shape can be achieved with time. *Fruit* Bright red fruit with bristly surface, up to ½ in (1 cm) long. Not reliable and not generally of interest.

Hardiness Tolerates a minimum winter temperature of 4°F (−15°C).

Soil requirements Tolerates all soil types, except extremely dry and extremely waterlogged.

Sun/Shade aspect Tolerates any aspect. Full sun to medium shade with medium shade for preference.

Pruning Not normally required for this plant, but can be cut back if necessary after flowering.

Training Fan-train on to wires or individual anchor points.

Propagation and nursery production From semi-ripe cuttings. Should be purchased container grown, best planting height 6–18 in (15–45 cm). Normally available from garden centres and nurseries.

Ribes speciosum

Problem May be slow to establish.
Similar forms of interest None.
Average height and spread
Five years
4 × 4 ft (1.2 × 1.2 m)
Ten years
5 × 7 ft (1.5 × 2.1 m)
Twenty years
7 × 10 ft (2.1 × 3 m)
Protrudes up to 12 in (30 cm) from support.

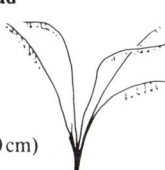

ROBINIA
(Pink-flowering Forms)

FALSE ACACIA

Leguminosae   *Hardy Tree*
Deciduous

A group of trees rapidly becoming appreciated for their adaptability to fan-training.

Origin From south-western USA, some named varieties from France.

Use As medium to large fan-trained trees for walls and fences.

Description *Flower* Clusters of pea-flowers, up to 3 in (75 cm) long, on wood two years old or more in early summer. *Foliage* Pinnate leaves, up to 6 in (15 cm) long, with nine to 11 oblong or ovate leaflets each 2 in (5 cm) long. Light grey/green with yellow autumn colour. *Stem* Light grey/green to grey/brown with small prickles. Upright when young, branching with age, easy to fan-train. Branches and twigs appear dead in winter, but produce leaves from apparently budless stems in late spring. Medium growth rate. *Fruit* Small, grey/green, bristly pea-pods up to 4 in (10 cm) long, in late summer and early autumn.

Hardiness Tolerates a minimum winter temperature of 4°F (−15°C). Stems may suffer some tip damage in severe winters.

Soil requirements Thrives in most soils; particularly tolerant of alkaline types. Resents waterlogging.

Sun/Shade aspect Tolerates all but the most

exposed of aspects. Full sun to very light shade.

Pruning Prune very young trees back to 9–12 in (23–30 cm) from the graft point in early spring following planting. Select and train resulting five to seven shoots and tie into a fan-trained shape. In subsequent years, remove all side growths back to two points from their origin and maintain original main branches in fan shape.

Training Will require securing to wires or individual anchor points.

Propagation and nursery production From seed or grafting. Purchase container grown. Choose either bush grown or young trees not more than two years old and between 3–5 ft (91 cm–1.5 m) tall. Must be sought from general or specialist nurseries; most varieties not offered by garden centres.

Problems Notorious for poor establishment; container grown trees provide best results. Branches may be damaged by high winds and need shelter.

Robinia kelseyi in flower

Forms of interest *R.* × *ambigua* Light pink flowers. Pinnate leaves with 13 to 21 light grey/green leaflets. Must be sought from specialist nurseries. *R.* 'Casque Rouge' Rose-pink to pink-red flowers. An interesting variety from France. Difficult to find. *R. fertilis* 'Monument' Rosy-red flowers. Half average height and spread. from south-eastern USA. *R.* × *hillieri* Slightly fragrant lilac-pink flowers. Originally raised in UK. *R. hispida* 'Macrophylla' Large flowers similar to those of wisteria. The variety most often planted. *R. kelseyi* Flowers white with purple/pink shading. Attractive pale grey/green foliage with nine to 11 leaflets. From south-eastern

USA. *R. luxurians* Rose-pink flowers. Leaves up to 12 in (30 cm) long, pinnate and with 15 to 25 ovate, bright green leaflets. Slightly more than average height and spread. Not readily available; must be sought from specialist nurseries. From south-western USA.

Average height and spread
Five years
12 × 12 ft (3.7 × 3.7 m)
Ten years
18 × 18 ft (5.5 × 5.5 m)
Twenty years
24 × 24 ft (7.3 × 7.3 m)
Protrudes up to 2 ft (60 cm) from support.

ROBINIA PSEUDOACACIA

ACACIA, BLACK LOCUST, FALSE ACACIA

Leguminosae　　　　　　*Hardy Tree*
Deciduous

This tree makes an elegant fan-trained shape, as long as sufficient space can be provided.

Origin From the USA.
Use As a large fan-trained tree for walls.
Description *Flower* Racemes 4–7 in (10–18 cm) long of fragrant white flowers in early summer; florets have blotched yellow bases inside. Size varies with age of tree and location. *Foliage* Pinnate, up to 10 in (25 cm) long, with 11 to 23 ovate leaflets. Light grey/green with good yellow autumn colour. *Stem* Grey/green to grey/brown and sparsely covered in thorns. Fast-growing when young, slowing and branching with maturity. Appears completely dead in winter, breaks leaf from almost budless stems. *Fruit* Small, grey/green pea-pods up to 4 in (10 cm) long in autumn.
Hardiness Tolerates a minimum winter temperature of 4°F (−15°C). Some stem tip damage may occur in severe winters.
Soil requirements Tolerates most soil conditions but dislikes waterlogging. Good on dry, sandy soils.
Sun/Shade aspect Tolerates a moderately exposed aspect but may need wind protection. Full sun to medium shade, preferring full sun.
Pruning Prune young trees back to 9–12 in (23–30 cm) from base or graft point in early spring following planting. Select and train resulting five to seven shoots and tie into a fan-trained shape. In subsequent years, remove all growths back to two points from their origin and maintain original main branches in fan shape.
Training Secure to wires or individual anchor points.
Propagation and nursery production From seed. Named varieties from grafting on to

understock of *R. pseudoacacia*. Best purchased container grown or root-balled (balled-and-burlapped); difficult to establish bare-rooted. Plants for fan-training should not be more than two years old, either bush or single stem, between 3–5 ft (91 cm–1.5 m) tall. Stocked by general nurseries and specialist outlets.
Problems Subject to wind damage. Sometimes difficult to establish; extra organic or peat composts should be added to soil, and adequate watering supplied in first spring. Produces suckers at ground level, often far from the central stem.
Similar forms of interest *R. p.* 'Bessoniana' White flowers in early summer. Two thirds average height and spread; the best white-flowering form for small gardens. Possibly the best variety for fan-training.

Average height and spread
Five years
10 × 10 ft (3 × 3 m)
Ten years
20 × 20 ft (6 × 6 m)
Twenty years
30 × 30 ft (9 × 9 m)
Protrudes up to 4 ft (1.2 m) from support.

ROBINIA PSEUDOACACIA 'FRISIA'

GOLDEN ACACIA

Leguminosae　　　　　　*Hardy Tree*
Deciduous

One of the most spectacular of all golden foliage wall specimens.

Robinia pseudoacacia 'Frisia' in leaf

Origin Of garden origin, from Europe.
Use As a medium to large fan-trained tree for walls and tall fences.
Description *Flower* Short racemes of white pea-flowers, produced only on very mature trees in mid summer. *Foliage* Pinnate, with seven to nine leaflets 6 in (15 cm) long. Bright yellow to yellow/green in spring, lightening in early summer. Turns deeper yellow in late summer to early autumn. *Stem* Brown to grey/brown. Strong shoots with definite red prickles on new growth. Wood may appear dead in winter but quickly grows away in late spring or early summer from apparently budless stems. Medium to fast growth rate. *Fruit* Insignificant.
Hardiness Tolerates a minimum winter temperature of 4°F (−15°C). Some stem-tip damage may occur in severe winters.
Soil requirements Most soil conditions; growth is limited on very alkaline, permanently wet or poor soils. Grows best on moist, rich, loamy soil.
Sun/Shade aspect Tolerates all but the most severe of aspects. Full sun to very light shade.
Pruning Prune young trees back to 9–12 in (23–30 cm) from the graft point in early spring following planting. Select and train resulting

Robinia pseudoacacia in flower

five to seven shoots and tie into a fan-trained shape. In subsequent years remove all side growths back to two points from their origin and maintain original main branches in fan shape.

Training Secure to wires or individual anchor points.

Propagation and nursery production From grafting on to *R. pseudoacacia*. Purchase container grown; best planting height for fan-trained trees between 2–5 ft (60 cm–1.5 m), using trees of not more than two years old. Normally available from general nurseries and garden centres.

Problems Branches may be brittle and easily damaged by severe weather conditions. Late to break leaf, often bare until early summer, but grows rapidly once started. Notoriously difficult to establish unless container grown.

Average height and spread

Five years
12 × 12 ft (3.7 × 3.7 m)
Ten years
18 × 18 ft (5.5 × 5.5 m)
Twenty years
24 × 24 ft (7.3 × 7.3 m)
Protrudes up to 4 ft (1.2 m) from support.

ROSA
(Climbing Musk Roses and Similar Forms)

KNOWN BY VARIETY NAME

Rosaceae   **Rose**
Deciduous

There are many beautiful white and off-white musk roses but many can, in all but the largest of spaces, be invasive, so alternative, less vigorous varieties have also been included in this entry.

Origin From the Himalayas and of garden origin.

Use Ideal for covering very large walls, fences, hedgerows, trellises and pergolas, or to climb through trees to give a cascading effect. Careful selection regarding size should be made as many are extemely vigorous.

Description *Flower* Single, semi-double or double, normally carried in clusters, each flower up to 1 in (2.5 cm) across, white to pale pink and flowering from mid summer to early autumn with attractive musk scent. *Foliage* Five, seven or nine small round leaflets make up a pinnate leaf, 5–7 in (12–18 cm) long. Normally light to mid green. Some good yellow autumn colour depending on variety. *Stem* Strong, ranging, arching, heavily thorned, thorns often attractive. Medium to

Rosa **'Francis E. Lester'**

fast growth rate. *Fruit* Small, round, orange/red hips carried in clusters in many varieties offering winter attraction.

Hardiness Tolerates a minimum winter temperature of 0°F (−18°C).

Soil requirements Will grow on all soil types, including alkaline and acid, but needs a deep root run to achieve full growth potential.

Sun/Shade aspect Tolerates all aspects in full sun to medium shade.

Pruning Not normally required, but the occasional removal of an old branch almost to ground level in early spring will induce new growth and keep the rose in good order.

Training Tie to wires or individual anchor points. Allow to ramble through branches of large trees or hedgerows.

Propagation and nursery production Commercial production is by budding, which entails the insertion of a single bud into a predetermined rootstock. Plants are normally sold 18 months on from this process. This group of roses can also be grown from hardwood cuttings. Purchase bare-rooted or root-wrapped from early autum to mid spring; may be available container grown from late spring to early autumn and beyond, but try to ensure the plant has not been potted for more than one year. Best planting height 1–3 ft (30–91 cm) depending on variety. Some varieties are available from garden centres and nurseries, but most may have to be sought from specialist growers.

Problems Often heavily thorned, many thorns with barbs. May suffer from mildew, although not normally to any great extent. Greenfly and blackfly may also be a problem, but normally pests and diseases are simply controlled by the great vigour of the plant, which can, in many cases, end in it growing out of its intended area.

Forms of interest 'Autumnalis' (syn. *Rosa moschata autumnalis*, *Rosa* × *noisettiana*, **'Champney's Rose'**) White flowers, single, flushed pink and red with central yellow stamens, carried in round clusters in mid summer. Needs some protection. Introduced prior to 1812. Average height and spread 8 × 8 ft (2.4 × 2.4 m). Protrudes up to 3 ft (91 cm) from support. **'Bleu Magenta'** Multiflora hybrid. Deep purple flowers in large clusters and bright green leaves. Rarely has thorns and is scentless. Ideal for growing up walls and trellis and through small trees or shrubs. Introduced 1899. Average height and spread 6 × 6 ft (1.8 × 1.8 m). Protrudes up to 2 ft (60 cm) from support. **'Bobbie James'** Creamy-white flowers, single, in large, pendent clusters in mid summer. Extremely vigorous. Introduced 1961. Average height and spread 30 × 20 ft (9 × 6 m). Protrudes up to 6 ft (1.8 m) from support. **'Brenda Colvin'** Soft pink flowers with yellow stamens, often appearing apricot overall. Very strong-growing. Ideal for growing in trees or for covering old buildings. Date of introduction

Rosa **'Paul's Himalayan Musk'**

unknown. Average height and spread 30 × 30 ft (9 × 9 m). Protrudes up to 5 ft (1.5 m) from support. **'Francis E. Lester'** Flowers single, some white, most pink. Good hip production in autumn. Introduced 1946. Average height and spread 15 × 15 ft (4.6 × 4.6 m). Protrudes up to 5 ft (1.5 m) from support. **'Paul's Himalayan Musk'** Double, soft pink flowers, ageing to white, fragrant. Flowers in mid summer. Ideal for ground cover. Introduced in the late 19th century. Average height and spread 20 × 15 ft (6 × 4.6 m). Protrudes up to 5 ft (1.5 m) from

Rosa **'Rambling Rector'**

support. **'Rambling Rector'** (syn. **'Shakespeare's Musk'**) Small, semi-double white flowers in clusters in mid summer. Origin is undetermined and has been in cultivation for many years. Average height and spread 20 × 20 ft (6 × 6 m). Protrudes up to 7 ft (2.1 m) from support. *R. brunonii* (syn. *R. b.* **'Himalayan Musk Rose'**, *R. moschata nepalensis*) Single white flowers in large clusters, larger than most, with golden yellow stamens, flowering mid summer. Long, ranging branches. Good hip production in autumn. Introduced 1822. Average height and spread 25 × 20 ft (7.6 × 6 m). Protrudes up to 7 ft (2.1 m) from support. *R. filipes* **'Kiftsgate'** Creamy-white flowers in huge trusses in mid summer. Extremely vigorous, flowering well in shade and producing good fruit. Tolerates full sun to almost full shade. Ideal for trees or hedgerows, but requires considerable space. Can also be allowed to grow as widespreading ground cover. Ideal for covering difficult banks, again where space can be allowed.

Introduced 1954. Average height and spread 30×30 ft (9×9 m). Protrudes up to 8 ft (2.4 m) from support. **R. helenae** The grey/green foliage shows off clusters of creamy-white single flowers. Introduced 1907. Average height and spread 20×18 ft (6×5.5 m). Protrudes up to 5 ft (1.5 m) from support. **R. moschata** Off-white flowers in trusses in mid summer to late summer. Strong-growing but not tall. Recorded as early as the 16th century. Average height and spread 12×12 ft (3.7×3.7 m). Protrudes up to 6 ft (1.8 m) from support. **R. mulliganii** (formerly known as *R. longicuspis*) Flowers white, single, banana-scented. Foliage coppery-tinted when young. Shoots and foliage glossy green, can be evergreen in some situa-

Rosa brunonii

tions. Good fruit production maintained into winter. Can be used as large-scale ground cover. Introduced 1917. Average height and spread 15×10 ft (4.6×3 m). Protrudes up to 5 ft (1.5 m) from support. **R. multiflora** Creamy-white single flowers with central yellow stamens, produced in large trusses in mid summer. Branches almost thornless. Bright, shiny mid green foliage. Date of introduction not known, but extremely old. Average height and spread 15×15 ft (4.6×4.6 m). Protrudes up to 5 ft (1.5 m) from support. **R. m. carnea** Flowers lilac/pink in hanging clusters, double and globe-shaped. Thin, ranging branches. Dark green foliage. Introduced 1804. Average height and spread 20×15 ft (6×4.6 m). Protrudes up to 6 ft (1.8 m) from support. **R. m. cathayensis** Single, flat, pink flowers, produced in mid summer. Introduced 1907. Average height and spread 15×15 ft (4.6×4.6 m). Protrudes up to 15 ft (4.6 m) from support. **R. m. platyphylla** (Seven Sisters Rose) Flowers deep lilac/pink to white, sweetly scented and flowering in mid summer. Thought to have been introduced from China prior to 1816. Average height and spread 15×15 ft (4.6×4.6 m). Protrudes up to 8 ft (2.4 m) from support. **R. m. watsoniana** Small, single

Rosa multiflora

flowers carried in large panicles in spring and again in mid summer, followed by small red hips. Very thin wispy growth. Leaves have wavy edges and stems are very thorny. Requires a sheltered aspect. Introduced 1870. Average height and spread 6×6 ft (1.8×1.8 m). Protrudes up to 3 ft (91 cm) from support. **R. sinowilsonii** Single white flowers, shown off well against glossy green foliage. Flowers mid summer. Strong-growing but needs a sheltered aspect. In existence for a great number of years. Average height and spread 12×10 ft (3.7×3 m). Protrudes up to 4 ft (1.2 m) from support. **'Seagull'** Truly a rambler, but closely resembles *R. brunonii* 'Himalayan Musk Rose'. Clusters of single, pure white flowers with bright golden-yellow stamens on vigorous-shooted plant. Introduced 1907. Average height and spread 25×20 ft (7.6×6 m). Protrudes up to 7 ft (2.1 m) from support. **'The Garland'** Semi-double, creamy-white, narrow petals, with age becoming tinged pink. Mid summer flowering. Good fragrance. Introduced 1835 and possibly the best Himalayan Musk type for a small garden. Average height and spread 15×12 ft (4.6×3.7 m). Protrudes up to 4 ft (1.2 m) from support. **'Treasure Trove'** Clusters of creamy-apricot flowers in mid summer, against good mid green foliage. Long red hips in autumn. Can be used for ground cover or woodland planting. Introduced 1979. Average height and spread 30×30 ft (9×9 m). Protrudes up to 8 ft (2.4 m) from support. **R. longicuspis 'Wedding Day'** Single white flowers with lemon and pink tinge, carried in clusters in mid summer. Introduced 1950. Average height and spread 30×20 ft (9×6 m). Protrudes up to 7 ft (2.1 m) from support.

Average height and spread Growth at ten years given for each variety.

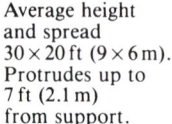

Rosa felipes **'Kiftsgate'**

ROSA
(Climbing Species Roses)

KNOWN BY BOTANICAL NAME
Rosaceae   Rose
Deciduous
These varieties are gathered together under one heading and described in some detail. Some have specific requirements which are covered within 'Forms of interest', but all can be fan-trained as climbers.

Origin As per individual variety.
Use For large walls and fences, to cover pergolas and trellis. Some varieties suitable for cold greenhouse and conservatory use.
Description *Flower* Yellow or pink, 1–3 in (2.5–7.5 cm) across, early to mid summer, depending on variety. Some with fragrance. *Foliage* Five or seven leaflets, some light green, some darker, depending on variety, up to 5–7 in (12–18 cm) long. *Stem* Light green, ageing to brown, normally strong and upright. Medium to fast growth rate. *Fruit* Not normally produced.

Rosa chinensis mutabilis

Hardiness Tolerates a minimum winter temperature of 4°F (−15°C).
Soil requirements Tolerates all soil types, except for extremely alkaline, which may cause distress. Requires good moisture and plant food retention.
Sun/Shade aspect Sheltered aspect in full sun to light shade.
Pruning Not normally required, but after five to six years some old branches can be removed to within 2 ft (60 cm) of ground level. This will usually induce new shoots.
Training Tie to wires or individual anchor points, or to trellis.
Propagation and nursery production Always purchase container-grown with the exception of *R. chinensis* **'Climbing Cécile Brunner'**, which is also sold bare-rooted. Best planting height 1–3 ft (30–91 cm). Not always stocked by garden centres and will often have to be sought from specialist nurseries.
Problems May suffer from attacks of greenfly and blackfly; treat with a proprietary control. Mildew and blackspot may be a problem. Correct pruning, cultivation, feeding and treatment with a proprietary control will normally keep the disease to a minimum. Thorns can make handling uncomfortable and plants which are not pruned as suggested can become woody. Suckers from rootstocks may appear and are difficult to identify as they closely resemble the climbing variety. The indication is that if they develop from below ground level they are normally suckers.

Rosa banksiae lutea

Forms of interest *R. chinensis* **'Climbing Cécile Brunner'** Miniature hybrid tea flowers, mother-of-pearl pink, presented in large flowing sprays in mid summer. Extremely attractive. Tolerates all soil conditions and aspects in full sun to medium shade. Can be trained through trees. Introduced prior to 1904. Average height and spread 25 × 20 ft (7.6 × 6 m). Protrudes up to 5 ft (1.5 m) from support. *R. bracteata* **'Mermaid'** Single, large flowers, up to 4 in (10 cm) across, buff yellow with apricot shading and central russet-red stamens. Recurrent flowering, with main flowering period early to mid summer. Dark, almost purple/green foliage with glossy surface. Tolerates all aspects, in full sun to medium shade, but some die-back may be seen in very severe winters. Thorns are bright purple/red on purple/red foliage. Its natural habit is to climb high towards the sun and produce little or no basal growth. Dislikes severe alkalinity. Introduced 1917. Average height and spread 30 × 30 ft (9 × 9 m). Protrudes up to 5 ft (1.5 m) from support. *R. banksiae lutea* Double, almost carnation-like canary-yellow flowers 1 in (2.5 cm) across in large trusses. Flowering late spring, early summer, some later intermittent flowering. Attractive light green foliage and bright green stems with some winter attraction. Remove one-third of oldest flowering growth to ground level after flowering. Strong-growing. Requires a sheltered position; tolerates a minimum winter temperature of 14°F (−10°C). Introduced prior to 1824. Average height and spread 20 × 20 ft (6 × 6 m). Protrudes up to 4 ft (1.2 m) from support. *R. chinensis mutabilis* Flowers single, honey yellow, ageing to orange and finally red. Produced continuously from late spring to early or even late autumn, making an interesting contrast against dark wine-red, suffused dark green foliage. Flowers carried on attractive shoots, giving a flowing texture. Tolerates exposed positions and any soil in full sun to light shade. Introduced in the 18th century. Average height and spread 7 × 7 ft (2.1 × 2.1 m). Protrudes up to 2 ft (60 cm) from support. **Average height and spread** Growth at ten years given for each variety.

Origin Varied, but most of nursery origination.
Use For walls, fences, pillars, pergolas and trelliswork, for wire or timber frames in the open or to adorn the front of buildings. Can be grown in large containers.

Description *Flower* Either single or semi-double, with round-ended, triangular-shaped petals held open, closed or in cupped clusters. Many varieties with strong scent. Colours range from red, yellow, white, pink, blue and shades in between. Some bi-coloured and multi-coloured varieties. Flowers from early summer to late autumn, not continuously but repeating. *Foliage* Five ovate to round leaflets make up a pinnate leaf, dark green to mid green, some varieties with purple hue or purple backing. Ribs on leaves may have small thorns. *Stem* Light to dark green, becoming green/brown. Basically upright and branching. Frequent thorns, large, red to red/brown in colour. Medium to fast growth rate. *Fruit* Large orange to orange/red rose hips produced in autumn and often retained well into winter.
Hardiness Tolerates a minimum winter temperature of 4°F (−15°C).
Soil requirements Does well on all soils but may show some signs of chlorosis on severe alkaline types. Requires an annual mulch with spent mushroom compost, garden compost or farmyard manure.
Sun/Shade aspect Tolerates all aspects. Full sun to medium shade, but full sun to light shade for best performance. Some varieties will grow in a shady planting position, but may be shy to flower.

Rosa 'Schoolgirl'

Pruning Do not prune plants prior to planting as this may cause reversion to bush forms; just remove any very weak shoots or any dead tips. Apart from deadheading, the removal of any dead wood during the winter period and shaping, this group of roses needs little pruning for the first three to five years after planting. From then on it is a good practice to remove up to one-third of the oldest shoots to ground level or to near ground level in March, to encourage regrowth from the base. This can be carried out every two to three years thereafter to prevent the rose from becoming old and woody.
Training Tie to wires or individual anchor points, either upright or in a fan shape, depending on situation. Some varieties respond well to having their branches trained horizontally from the fan shape. This practice reduces sap flow and encourages flower buds to form.
Propagation and nursery production Commercial production is from budding, entailing the insertion of one single bud into a predetermined rootstock. Plants are normally sold 18 months on from this process. Can be grown from hardwood cuttings, but lifespan may be shortened due to lack of root vigour. Purchase bare-rooted from early autumn and beyond; if buying container grown plants try to ensure they have not been potted for more than one year. Best planting height 1–3 ft (30–91 cm), depending on variety. Many varieties readily available from garden centres

Rosa chinensis **'Climbing Cécile Brunner'**

Rosa 'Superstar, Climbing'

and nurseries, others may have to be sought from specialist growers.

Problems May suffer from attacks of greenfly and blackfly. Treat with a proprietary control. Mildew and blackspot may be a problem; correct pruning, cultivation and feeding and treatment with a proprietary control will normally keep to a minimum. Thorns can make handling uncomfortable and plants which are not pruned as suggested can become very woody. Newly planted modern hybrid climbing roses may be slow to produce strong shoots during the building up of a good root system.

Suckers from rootstocks may appear and are difficult to identify as they closely resemble the actual climbing variety. The indication is that if they develop from below ground they are normally suckers. Careful observation should, however, make it possible to differentiate between the two foliage patterns.

Many climbing roses in this group are sports of hybrid tea or floribunda roses and therefore care must be taken when purchasing to ensure that they are climbing and not bush forms.

Rosa 'White Cockade'

Forms of interest 'Alchemist' Double yellow flowers with interspersed egg-yolk yellow petals in late spring and mid to late summer. Good scent. Attractive light green to mid green foliage. Strong growth. Good on all soils. Introduced in 1956. **'Alec's Red, Climbing'** Fragrant cherry red flowers, not strong growing but worthy of space. Good foliage. Ideal as a pillar rose. Introduced 1975. **'Allen Chandler'** Flowers up to 4 in (10 cm) across, single to semi-double, with central golden stamens setting off the vivid red petals. Repeat flowering. Requires a shel-

tered position with good, well prepared, moisture-retentive soil. Needs plenty of space. Introduced in 1923. **'Allgold, Climbing'** Double, golden yellow flowers carried in clusters from early to late summer, not necessarily continously. Slightly scented. Good growth. Mid green foliage. Good for all soil types. Introduced 1961. **'Aloha'** Very full, double, rich pink flowers with darker backs to petals. Very attractive. Flowering from late spring then almost continuously throughout summer. Leathery dark green foliage. Disease resistant. Tolerates all soil types. May be found in some garden centres in the shrub rose section. **'Altissimo'** Single, slightly scented, blood-red flowers, $3\frac{1}{2}$–4 in (9–10 cm) across, carried in large trusses. Spectacular. Almost continuous flowering throughout early summer to mid autumn. Good dark green, disease resistant foliage. Tolerates all soil types. Introduced 1966. **'Ash Wednesday'** Unusual double flowers, off-white to lilac/blue. Repeat flowering. Strong growing with good foliage. Can be grown in exposed aspects from full sun to medium shade. Introduced 1955. **'Bantry Bay'** Bold pink semi-

Rosa 'Aloha'

double to double flowers presented in clusters. Repeat flowering. Tolerates all soils with adequate preparation and exposed aspects from full sun to medium shade. Introduced 1967. **'Bettina, Climbing'** Coppery-orange, fragrant, hybrid tea-shaped flowers. Repeat flowering from mid to late summer. Dark green leaves carried on strong, dark, upright stems. Needs a sheltered position for best results. Train branches horizontally to aid flowering. Introduced 1958. **'Blessings, Climbing'** Strongly fragrant, hybrid tea shaped-flowers from mid to late summer, salmon to bright pink in bud, opening to a full double clear pink. Dark green foliage on good, dark, upright stems. Introduced 1975. **'Blue Moon, Climbing'** Large, well-formed flowers, lilac to mid blue. Attractive in bud, opening to loose formation. Very heavily scented. Foliage susceptible to blackspot. Recurrent flowering. Requires full sun and good soil conditions. Introduced 1967. **'Breath of Life'** (syn. **'Harquanne'**) Large, fragrant, hybrid tea flowers of apricot/pink presented both singly and in sprays. Good foliage. Introduced 1982. **'Casino'** (syn. **'Gerbe d'Or'**) Very good yellow, double, scented flowers carried on strong stems. Repeat flowering. Good golden-green foliage. Tolerates a wide range of soils and exposed aspects in full sun to medium shade. Introduced 1963. **'Chaplin's Pink'** Well established variety with semi-double to single flowers, bright pink in colour, produced en masse from mid to late summer and more sparsely in early autumn. Good growth and foliage. Ideal for exposed aspects and for growing in trees. Introduced 1928. **'Chateau-de-Clos-Vougeot, Climbing'** Dark rich velvet crimson blooms, opening to

form a flat and wide flower. Very strongly scented. Possibly repeat flowering. Strong growing for trellis and walls. Introduced 1920. **'City of York'** Creamy-white flowers in clusters. Short flowering season. Clean foliage. Introduced 1945. **'Compassion'** Apricot to copper, fragrant flowers with yellow and pink highlights and shading. Repeat flowering. Good glossy foliage and strong

Rosa 'Altissimo'

growth. Tolerates all soil conditions. Introduced 1974. **'Comtesse Vandal'** Orange in bud and opening to salmon. Reverse of petals orange/pink. Flowers from mid summer to early autumn, not necessarily continuously. Some scent. Introduced 1936. **'Coral Dawn'** Double, wide, balloon-shaped coral-pink flowers. Repeat flowering from mid summer to early autumn. Good dark green foliage and strong growth. Tolerates all soil conditions and aspects. Introduced 1952. **'Crimson Descant'** Fragrant, double, crimson flowers in good numbers. Good dark green foliage. No recorded date of introduction. **'Crimson Glory'** Possibly the most fragrant of all red roses. Double, attractively shaped flowers with a velvety texture from early to late summer. Foliage not of the best, subject to blackspot and mildew but worth perseverance. Introduced 1946. **'Danse de Feu'** (syn. **'Spectacular'**) Brick-red flowers, globe-shaped, opening to a looser formation. Almost continuous flowering from early summer to early autumn. Foliage dark green. Subject to attacks of blackspot and mildew but worth perseverance for its use on exposed and shady positions. Introduced 1952. **'Dream Girl'** Soft coral-pink, very attractively shaped flowers, rosette formation. Very strong characteristic rose scent. For pillars and trellis. Introduced

Rosa 'Compassion'

Rosa 'Ena Harkness, Climbing'

1944. **'Dreaming Spires'** Bright golden yellow flowers with good scent. Dark green foliage and strong, upright growth. Introduced 1973. **'Dublin Bay'** Scented, bright crimson flowers in clusters from mid to early autumn, almost continuous. Dark green foliage. Two thirds average height and spread. Makes a good pillar rose for all soil types. Introduced 1976. **'Elegance'** Double, shapely flowers, mid yellow, ageing to lemon-yellow. Repeat flowering once established. Mid green, glossy foliage, strong growing. Tolerates all soil types. Introduced 1937. **'Elizabeth Harkness, Climbing'** Ivory to white flowers on upright growth. Ideal for pillars in all aspects and soils. Introduced 1975. **'Elizabeth Heather Grierson'** (syn. **'Mattnot'**) Fragrant, soft

Rosa 'Golden Showers'

pink hybrid tea flowers. Good growth and foliage. No recorded date of introduction. **'Ena Harkness, Climbing'** Double, velvety red, very highly scented, hybrid tea flowers carried on weak flower stalks which give the flowers a nodding effect. Flowers from early to late summer. Foliage may suffer from attacks of mildew and blackspot but worth perseverance. Tolerates all soil types. Ideal for exposed or shady walls. Introduced 1954. **'Fashion, Climbing'** A floribunda type rose. Large, fragrant, coral/peach flowers carried in clusters. Free-flowering from late summer to early autumn. Introduced 1951. **'Fragrant Cloud, Climbing'** Well-shaped coral-red flowers, very highly scented. Repeat flowering. Ideal for pillars in all aspects and soils. Introduced 1966. **'General MacArthur'** Flowers scented, loose in formation, rosy-red in colour, borne from mid summer to early autumn. Strong growing. Tolerates all soil types. Introduced 1923. **'Golden Showers'** Deep golden-yellow ageing to cream, loosely

floppy flowers with wavy-edged petals. Attractive and continuous flowering. Foliage dark green. Stems strong, upright, mid green with very few thorns. Ideal for exposed aspects from full sun to medium shade. Introduced 1956. **'Grand Hotel'** Bright red double flowers shaded scarlet. Medium-sized hybrid tea. Unfading good foliage and strong growth. Introduced 1973. **'Grand'mère Jenny, Climbing'** Hybrid tea flowers, creamy yellow with splashes of pink and red both through petals and on outer edges. Scented, flowering from mid summer to late summer. Leaves dark green. Strong growing, one third more than average height and spread. Introduced 1958. **'Guinée'** Very heavily scented, attractively shaped, deep wine-red petals opening to form a saucer-shaped flower. Repeat flowering. Suffers attacks of mildew which is difficult to control, but worth growing for its scent. Tolerates wide range of soil conditions as long as adequate preparation has been made. Introduced 1938. **'Handel'** Most attractive semi-double large flowers, white to silvery white with pink or red markings around each petal edge. Scented. Continuous flowering. Introduced 1956. **'Highfield'** Attractive buds open to medium-sized bright yellow flowers, often with peach tints. Good foliage and growth in all aspects. Introduced 1981. **'Iceberg, Climbing'** (syn. **'Schneewittchen'**) Floribunda type flowers in large sprays. Pure white semi-double, flowering from mid summer to early autumn. In wet conditions flower buds may spot pink, but still worthy of merit. Ideal for exposed aspects in full sun to medium shade. Introduced 1968. **'Joseph's Coat'** Flowers change

Rosa 'Handel'

from yellow through orange to red. Continuous flowering over a long period. Good on all aspects and soil types. Introduced 1954. **'Josephine Bruce, Climbing'** Extremely good, fragrant, deep velvet red flowers up to 4 in (10 cm) across. Flowers from mid summer to early autumn, not necessarily continuously. Good growth of dark green foliage but very thorny. Introduced 1954. **'Korona, Climbing'** Attractive buds open to semi-double orange/red flowers from mid summer to early autumn. Growth strong and upright in habit. Foliage dark green. Disease resistant. Tolerates exposed aspects in full sun to medium shade. Introduced 1957. **'Lady Sylvia, Climbing'** Hybrid tea flowers, rich pink with attractive perfume from early summer to early autumn, not necessarily continuous. Strong growing. Good foliage. Introduced 1933. **'La Rêve'** Mainly single, sometimes semi-double pale yellow flowers with attractive scent. Good glossy green foliage. Strong growing for trellis and walls. Introduced 1923. **'Lavinia'** Deep pink flowers, some scent. Strong growth and good foliage. Good on all soil types and aspects. Introduced 1983. **'Leaping Salmon'** Salmon-pink, good shaped, very fragrant flowers. Perpetual flowering. Good for trellis and pillars. Introduced 1986. **'Leverkusen'** Large sprays of semi-double, pale yellow flowers. Continuous flowering. Good light green foliage. Strong growing, tolerating exposed aspects in full sun to medium shade. Introduced 1954. **'Leys Perpetual'** Yellow and cream, flat, double flowers, scented and repeat flowering. Growth may be suspect and requires good soil conditions. Introduced 1937. **'Maigold'** Copper/yellow flowers, semi-double, somewhat open in habit, flowering from early summer to early autumn, not

Rosa 'Leverkusen'

necessarily continuously. Foliage mid green, glossy. Good growth. Tolerates a wide range of soil conditions in full sun to medium shade. Good for exposed aspects or for growing through trees. May be found in some garden centres in the shrub rose section. Introduced 1953. **'Masquerade, Climbing'** Clusters of semi-double saucer-shaped floribunda type flowers, first opening yellow, ageing to pink, finally red. Flowers from early to late summer. Dark green foliage, disease resistant. Strong growth. Tolerates all soil conditions. Introduced 1958. **'Meg'** Single buff-yellow to apricot flowers with central russet-red stamens. Scented. Repeat flowering from early summer to early autumn. Good dark green foliage, disease resistant. Tolerates a wide range of aspects. Introduced 1954. **'Mme Butterfly, Climbing'** Hybrid tea flowers in different shades of pink through to blush with lemon centres. Fragrant. Repeat flowering from early to late summer. Strong growth. Attractive foliage. Tolerates full sun to medium shade and most soil conditions. May benefit from horizontal training to induce more flowers. Introduced 1926. **'Mme**

Edouard Herriot, Climbing' Hybrid tea coral flowers with yellow shading. Semi-double, flowering from mid summer to early autumn. Good foliage. Tolerates all aspects and soils. Introduced 1921. **'Mme Grégoire Staechelin'** Large, open, double, pale pink flowers with deeper pink reverse and attractive veining. Flowers in early summer and intermittently thereafter. Good foliage. Tolerates all soil types in exposed locations in full sun to medium shade. Introduced 1927. **'Mme Henri Guillot'** Loose, semi-double, burnt orange flowers from mid to late summer. Some fragrance. Introduced 1942. **'Mrs Aaron Ward'** Strong, fragrant, bright yellow flowers, splashed with salmon pink. Weather will vary colour intensity. Flowers from mid to late summer. Introduced 1922. **'Mrs G. A. Van Rossem'** Orange/apricot, globe-shaped flowers with golden background, the reverse of petals deeper coloured. Repeat flowering. Tolerates exposed aspects in full sun to medium shade. Introduced 1937. **'Mrs Herbert Stevens'** Pure white, scented flowers. Very old variety still deserving consideration. Repeat flowering. Tolerates exposed aspects in full sun to medium shade. Introduced 1922. **'Mrs**

Rosa 'Parade'

Rosa 'Maigold'

Sam McGredy, Climbing' Copper/orange scented flowers presented against a coppery-red foliage, making an attractive combination. Flowers from early to late summer. Strong growing, ideal for growing in trees. Tolerates all soil conditions. Introduced 1937. **'Malaga'** Repeat flowering, deep rose pink blooms with an interesting apple fragrance. Good growth and foliage. Good on all soil types. Introduced 1971. **'Morning Jewel'** Bright pink flowers of medium size produced in good numbers. Very good first flowering and some later flowers. Clean foliage. Good for walls, fences, trellis and pillars. Introduced 1968. **'New Dawn, Climbing'** Numerous repeat to almost continuous production of pale pink, perfumed flowers. Almost rambler-like growth. Tolerates exposed positions but dislikes shade. Introduced 1930. **'Night Light'** (syn **'Poullight'**) Fragrant, deep yellow flowers which age to orange/red. Strong growth and good foliage. Ideal for all aspects and soils. Introduced 1987. **'Norwich Pink'** Semi-double flowers, bright cerise, opening flat. Strong-growing for walls and trellis. Good clean foliage. Introduced 1962. **'Norwich Salmon'** Pale salmon-pink flowers in clusters. Dark green foliage. Strong growing, disease resistant. Good for all aspects on pillars, walls and trellis. Introduced 1962. **'Ophelia, Climbing'** Most beautifully shaped hybrid tea flowers in flesh pink with deeper shading, often with lemon tints towards the centre. Very fragrant. Repeat flowering. Good foliage. Growth may benefit from horizontal training to encourage flowering. Introduced 1920. **'Parade'** Deep rose pink flowers, sweetly scented and carried over a

long period. Good on a wide range of aspects, including very exposed. Strong growing with good foliage. Introduced 1953. **'Parkdirektor Riggers'** Single, deep red flowers in large clusters. Continuous flowering. Strong growing with good foliage. Tolerates exposed aspects in full sun to medium shade. Introduced 1957. **'Paul's Scarlet'** One of the first of all the modern climbers and still worthy of merit. Repeat flowering with double scarlet flowers borne in clusters. Tolerates all soil conditions in exposed positions in full sun to medium shade. Introduced 1931. **'Picture, Climbing'** Fragrant hybrid tea flowers, pink with deeper shading and lemon at base, borne from mid to late summer. Strong growth which may benefit from being horizontally trained to encourage flowering. Good dark green foliage. Tolerates exposed positions in full sun to medium shade. Introduced 1942. **'Pink Perpetue'** Flowers in clusters, double, fragrant, deep pink in colour, borne in abundance in June and July and again in September. Good growth and dark green foliage. Tolerates all soil types and exposed aspects in full sun to medium shade. Introduced 1965. **'Queen Elizabeth, Climbing'** Extremely vigorous and often unruly by its sheer height. Double, silver/pink flowers of large proportions, carried in singles or up to five in a cluster from mid to late summer. Tolerates all soil types. Certainly needs horizontal training to encourage flowering, but will reach at least 20 ft (6 m) in width. Introduced 1956. **'Réveil Dijonnais'** Hybrid tea-shaped flowers, semi-double, crimson with yellow reverse. May be repeat flowering. Can be susceptible to blackspot. Introduced 1931. **'Ritter von Bramstedt'** Large, dark rose-pink flowers on strong, vigorous, upright growth. Repeat flowering.

Rosa 'Paul's Scarlet'

Requires careful placing due to its vigour. Date of introduction unknown. **'Rosy Mantle'** Double to semi-double mid pink flowers. Repeat flowering. Fragrant. Good dark glossy green foliage. Tolerates exposed aspects in full sun to medium shade and all soil types. Introduced 1968. **'Royal Gold'** Golden yellow flowers, scented. Continuous flowering. Good dark green foliage. Needs horizontal training to encourage maximum flower production. Requires good soil and dislikes extreme exposure. Introduced 1957. **'Schoolgirl'** Flowers copper/orange, double, good scent. Continuous flowering. Leaves and growth can be sparse. Requires good soil preparation and annual feeding for best results. Introduced 1964. **'Senateur Amic'** Strongly fragrant, semi-double to single, bright carmine flowers. Repeat flowering. Strong growing. Responds well to horizontal training to increase flowering. Introduced 1924. **'Shot Silk, Climbing'** Hybrid tea flowers of cherry cerise with golden yellow base. Flowers from early to late summer, not

Rosa 'Masquerade, Climbing'

necessarily continuously. Good growth responding to horizontal training to increase flower production. Disease resistant. Bright glossy green foliage. Tolerates exposed positions in full sun to medium shade. Introduced 1931. **'Souvenir de Claudius Denoyel'** Strongly fragrant double flowers, cup-shaped, rich red to scarlet. Repeat flowering. Strong growth but a little erratic and becoming somewhat ragged in formation. Introduced 1920. **'Speke's Yellow, Climbing'** Bright golden-yellow flowers, open in stature. Good flower production. Repeat flowering. Good for pillars and trellis on all aspects and soils. Date of introduction unknown. **'Sterling Silver, Climbing'** Hybrid tea flowers becoming loose when open, silver/lavender in colour. Mid to late summer flowering, highly fragrant. Foliage subject to blackspot and mildew and growth might be suspect without adequate preparation and good annual feeding. Nevertheless, worth consideration. Introduced 1963. **'Summer Wine'** (syn. **'Korizont'**) Semi-double coral pink flowers with bold red stamens. Good foliage and growth. Tolerates all

Rosa **'Pink Perpetue'**

aspects and soils. Introduced 1984. **'Super Star, Climbing'** Bright vermillion, good shaped, fragrant flowers. Repeat flowering. For pillars and trellis. Dislikes shade, requires a good well-fed soil. Introduced 1965. **'Swan Lake'** Pure white with pale powdery pink shading on older flowers. Double. Continuous flowering. May spot pink in wet weather or in heavy dews or fog. Dark green foliage, disease resistant. Two thirds average height and spread. Introduced 1968. **'Sutters Gold, Climbing'** Fragrant yellow hybrid tea flowers with pink veining. Repeat flowering. Makes a good cut flower. Upright growth and good foliage. Introduced 1950. **'Sympathie'** Bright scarlet, fragrant, hybrid tea flowers on strong growths with good foliage. Tolerates a wide range of soils and aspects. Introduced 1964. **'Vicomtesse Pierre de Fou'** Rosette-shaped

Rosa **'Royal Gold'**

coppery-pink flowers hanging down from branches. Good rose scent. Repeat flowering. Very strong growing for walls. Introduced 1923. **'White Cockade'** Pure white double flowers carried on upright thorny stems. Flowers continuously. Good dark foliage. Two thirds average height and spread. Introduced 1969.

Average height and spread
Five years
9 × 9 ft (2.7 × 2.7 m)
Ten years
15 × 15 ft (4.6 × 4.6 m)
Twenty years
15 × 15 ft (4.6 × 4.6 m)
Protrudes up to 3 ft (91 cm) from support.

ROSA
(Old Climbing Roses Prior to 1920)

KNOWN BY VARIETY NAME

Rosaceae   Rose
Deciduous

A group of climbing roses that have stood the test of time and still merit consideration for planting today.

Origin Mostly of garden or nursery origin.
Use As climbing roses for fences, walls, pergolas, trellis and archways.
Description *Flower* Single, semi-double and double in a range of colours through white, pink, red and apricot in sizes from 2–4 in (5–10 cm) across, many with fragrance, and carried from early summer to early autumn, depending on variety. *Foliage* Five round leaflets make up a pinnate leaf, up to 5–7 in (12–18 cm) long, normally dark to mid green. *Stem* Normally upright, becoming spreading. Mid green to dark green and covered with thorns. Medium to fast growth rate. *Fruit* May produce large, round rose hips, orange in colour and of some winter attraction.
Hardiness Tolerates a minimum winter temperature of 4°F (−15°C).
Soil requirements Tolerates all soil types, with no preference unless otherwise stated.
Sun/Shade aspect Tolerates all aspects and full sun to medium shade unless otherwise stated.
Pruning Requires no pruning when first planted. Remove one third of oldest wood to ground level after roses are established for more than three to four years.
Training Tie to wires or individual anchor points.
Propagation and nursery production Commercial production is from budding which entails the insertion of one single bud into a predetermined rootstock. Plants are normally sold 18 months on from this process. Can be grown from hardwood cuttings, but lifespan may be shortened due to lack of root vigour. Can be purchased bare-rooted or root-wrapped from early autumn onwards, but ensure that plants are one year old. May also be found container grown in spring, summer and autumn as two-year-old plant. Best planting height 1–3 ft (30–91 cm), depending on variety. Many varieties available from garden centres and nurseries, but some may have to be sought from specialist nurseries.
Problems Mildew and blackspot may be a problem, but correct pruning, cultivation, feeding and treatment with a proprietary control will normally keep the diseases to a minimum. May also suffer from attacks of greenfly and blackfly which should be treated

with a proprietary control. Thorns can make handling uncomfortable and plants which are not pruned as suggested can become very woody. Suckers may appear and are difficult to identify as they closely resemble the actual climbing variety. The indication is that if they develop from below ground level they are indeed suckers. Careful observation should, however, make it possible to differentiate between the two foliage patterns.
Forms of interest 'Adam' Peach/pink double flowers in clusters of three, sometimes produced singly. Good scent. Strong-flowering and recurrent. Introduced 1833. Average height and spread 7 × 7 ft (2.1 × 2.1 m). Protrudes up to 2 ft (60 cm) from support. **'Aimée Vibert'** (Noisette) Graceful sprays of scented double white flowers, showing yellow stamens when open. Repeat flowering. Good glossy foliage. For large trellises and walls. Introduced in 1823. Average height and spread 12 × 15 ft (3.7 × 4.6 m). Protrudes up to 5 ft (1.5 m) from support. **'Anemone Rose'** Flowers almost papery, single, silver/pink, sometimes with mauve shading. Recurrent flowering. Introduced 1895. Average height and spread 10 × 10 ft (3 × 3 m). Protrudes up to 2 ft (60 cm) from support. **'Ards Rover'** Crimson flowers with very strong scent. Normally flowers once in early summer, but may give later flowers. Introduced 1898. Average height and spread 10 × 8 ft (3 × 2.4 m). Protrudes up to 3 ft (91 cm) from support. **'Belle Portugaise'** (Hybrid) Semi-double, pale pink flowers in spring. Needs a sheltered position. Introduced 1900. Average height and spread 15 × 12 ft (4.6 × 3.7 m). Protrudes up to 3 ft (91 cm) from support. **'Blairi No.1'** (Bourbon) Large fluffy open flowers, soft pink and scented. Recurrent flowering. Introduced prior to 1845. Average height and spread 12 × 8 ft (3.7 × 2.4 m). Protrudes up to 3 ft (91 cm) from support. **'Blairi No.2'** (Bourbon) Pale pink, flat, almost saucer-shaped flowers with deeper coloured pink centres. Double and fragrant. Recurrent and profuse flowering. Introduced 1845. Average height and spread 18 × 10 ft (5.5 × 3 m). Protrudes up to 3 ft (91 cm) from support. **'Blush Boursault'** Double, blush-pink flowers, somewhat irregular in petal formation. Spring and mid summer flowering. Thornless stems, strong growing. Introduced 1848. Average height and spread 15 × 12 ft (4.6 × 3.7 m). Protrudes up to 3 ft (91 cm) from support. **'Captain Christy'** (Hybrid Tea) Soft pink with deeper pink centres, globe-shaped and good fragrance. Repeat flowering. Introduced 1881. Average height and spread 12 × 10 ft (3.7 × 3 m). Protrudes up to 3 ft (91 cm) from support. **'Céline Forestier'** (Noisette) Pale pink, silky textured, cabbage-like flowers. Good perfume. Repeat flowering. Requires a warm wall and regular feeding for good results. For pillars and trel-

Rosa **'Gloire de Dijon'**

Rosa 'Anemone Rose'

lis. Introduced 1842. Average height and spread 9×9 ft (2.7×2.7 m). Protrudes up to 3 ft (91 cm) from support. **'Champney's Pink Cluster'** A hybrid of *R. chinensis* and *R. moschata*. Pink double flowers in large clusters in mid summer. Requires a sheltered position. Introduced 1802. Average height and spread 15×10 ft (4.6×3 m). Protrudes up to 3 ft (91 cm) from support. **'Claire Jacquier'** (Noisette) Yolk-yellow flowers, medium sized and fragrant. Recurrent flowering. Strong-growing for walls and large trellis. Introduced 1888. Average height and spread 15×15 ft (4.6×4.6 m). Protrudes up to 5 ft (1.5 cm) from support. **'Cooper's Burmese'** (*R. laevigata*) (Species) Creamy-white flowers, single, unscented, in mid summer. Good dark green foliage. Strong growing but requires a sheltered aspect. Introduced 1920. Average height and spread 20×20 ft (6×6 m). Protrudes up to 3 ft (91 cm) from support. **'Cupid'** (Hybrid Tea) Large single flowers, sometimes semi-double, up to 5 in (12 cm) across. Flesh-pink with apricot shading. Petals crinkly-edged. Strong growing. Not repeat flowering but produces good hips. Ideal for trees and large walls. Introduced 1915. Average height and spread 15×15 ft (4.6×4.6 m). Protrudes up to 5 ft (1.5 cm) from support. **'Desprez à Fleur Jaune'** (Noisette) Strongly fragrant, many petalled, silky-textured flowers of warm yellow, shaded with peach and apricot. Good perfume. Strong growing. Repeat flowering. Needs a warm situation. For walls and fences and for growing through trees. Introduced 1826. Average height and spread 18×18 ft (5.5×5.5 m). Protrudes up to 6 ft (1.8 m) from support. **'Devoniensis, Climbing'** (Tea) Large creamy-white flowers with apricot flushing. Strong tea scent. Perpetual flowering. Must have a warm wall. Strong growing for walls and large trellis. Introduced 1858. Average height and spread 15×15 ft (4.6×4.6 m). Protrudes up to 5 ft (1.5 cm) from support. **'Etoile de Hollande'** Rich velvety crimson, highly scented flowers from early to late summer, carried on strong, well-foliaged shoots. Introduced 1919. Average height and spread 8×8 ft (2.4×2.4 m). Protrudes up to 2½ ft (76 cm) from support. **'General Schablikine'** (China) Pendent coppery carmine-pink flowers with some fragrance. For trellis and pillars with a warm aspect. Introduced 1878. Average height and spread 9×9 ft (2.7×2.7 m). Protrudes up to 3 ft (91 cm) from support. **'Gloire de Dijon'** (Noisette) Fragrant buff to peach flowers, tight in bud, opening to a more frothy nature. Repeat flowering. Introduced 1853. Average height and spread 12×10 ft (3.7×3 m). Protrudes up to 2 ft (60 cm) from support. **'Grüss an Teplitz'** An old fashioned rose with large, rich crimson flowers and very strong fragrance. Repeat flowering. For pillars and trellis. Introduced 1897. Average height and spread 8×8 ft (2.4×2.4 m). Protrudes up to 2 ft (60 cm) from support. **'Kathleen Harrop'**

(Bourbon) Soft pink flowers, scented, thornless. Spring flowering with recurrent blooms after. Tolerates exposed aspects and all soil types. Introduced 1919. Average height and spread 10×8 ft (3×2.4 m). Protrudes up to 3 ft (91 cm) from support. **'Lady Hillingdon'** (Tea Rose) Very popular. Apricot-yellow flowers carried on purple-coloured shoots with grey/green leaves. Fragrant and recurrent flowering. Introduced 1917. Average height and spread 10×10 ft (3×3 m). Protrudes up to 3 ft (91 cm) from support. **'Lady Waterlow'** (Hybrid Tea) Soft pink flowers to salmon semi-double flowers with deeper pink edges. Recurrent flowering. Strong climber with good foliage. Tolerates all aspects. Introduced 1903. Average height and spread 12×10 ft (3.7×3 m). Protrudes up to 3 ft (91 cm) from support. **'La France, Climbing'** Pale pink cupped flowers. Moderately strong-growing for walls and trellis. Introduced 1893. Average height and spread 10×10 ft (3×3 m). Protrudes 3 ft (91 cm) from support. **'Lawrence Johnston'** Yellow, semi-double with some tints of buff. Mid summer flowering. Tolerates all aspects. Vigorous. Introduced prior to 1900. Average height and spread 25×25 ft (7.6×7.6 m). Protrudes up to 4 ft (1.2 m) from support. **'Leys Perpetual'** (Noisette) Globe-shaped flowers, lemon-yellow in colour and fragrant. Strong-growing for trellis and walls. Date of introduction unknown. Average height and spread 12×12 ft (3.7×3.7 m). Protrudes 4 ft (1.2 m) from support. **'Maréchal Niel'** (Noisette) Bright golden yellow flowers, repeating. Fragrant. Attractive when in bud. Needs good protection, even to the extent of a greenhouse or conservatory. Introduced 1864. Average height and spread 15×10 ft (4.7×3 m). Protrudes up to 3 ft (91 cm) from support. **'Mme Abel Chatenay'** (Hybrid Tea) Soft silver/pink flowers plump when in bud, opening to a good shaped flower. Mid summer flowering. Less vigorous than some with weak foliage. Introduced 1917. Average height and spread 8×8 ft (2.4×2.4 m). Protrudes up to 2 ft (60 cm) from support. **'Mme Alfred Carrière'** (Noisette) Double, globe-shaped flowers, flesh pink/white, profusely borne. Continuous flowering. May need training horizontally to encourage best flowering performance. Strong growing, tolerating all soils and aspects. Can be grown through trees. Introduced 1879. Average height and spread 10×10 ft (3×3 m). Protrudes up to 3 ft (91 cm) from support. **'Mme Caroline Testout'** (Hybrid Tea) Satin-pink with deeper pink centre. Very large blooms carried on strong and upright stems with good foliage. Repeat flowering. Tolerates all aspects and soil types. Introduced 1890. Average height and spread 15×10 ft (4.6×3 m). Protrudes up to 3 ft (91 cm) from support. **'Mme Jules Gravereux'** Yellow/buff double flowers with some peach and pink shading. Repeat flowering. Some scent. Good dark green foliage. Introduced 1901. Average

height and spread 8×8 ft (2.4×2.4 m). Protrudes up to 3 ft (91 cm) from support. **'Paul Lédé'** (Tea Rose) Yellowish-buff flowers with carmine shading at the centre. Attractive sweet scent. Very free flowering from early to late summer. Introduced 1913. Average height and spread 6×6 ft (1.8×1.8 m). Protrudes up to 3 ft (91 cm) from support. **'Paul's Lemon Pillar'** (Hybrid Tea) Creamy lemon blooms tinged with green at the centre. Strongly fragrant. Flowering mid summer. Vigorous climber with strong upright branches and very good foliage. Tolerates all soil types and aspects. Introduced 1915. Average height and spread 15×12 ft (4.6×3.7 m). Protrudes up to 18 in (45 cm) from support. **'Pompon De Paris, Climbing'** A climbing form of the miniature rose, twiggy growth. Rose-pink, pompon-shaped flowers carried in good profusion in early to mid summer. Not repeat flowering. Foliage small and grey/green. For growing through shrubs, up pillars or wherever a small climber is required. Date of introduction unknown. Average height and spread 9×9 ft (2.7×2.7 m). Protrudes 3 ft (91 cm) from support. **'Richmond, Climbing'** (Hybrid Tea). Narrow, elegant, bright scarlet, ageing to crimson. Moderately strong-growing for trellis and walls. Introduced 1912. Average height and spread 12×12 ft (3.7×3.7 m). Protrudes 4 ft (1.2 m) from support. **'Sombreuil'** (Tea Rose) Pure white in humid conditions becoming flushed with pink. Flat, quartered flowers with delightful tea fragrance. Repeat flowering. Introduced 1850. Average height and spread 8×6 ft (2.4×1.8 m). Protrudes up to 3 ft (91 cm) from support. **'Souvenir de la Malmaison,**

Rosa 'Zéphirine Drouhin'

Climbing' (Bourbon) Blush-pink, globe-shaped flowers. Repeat flowering. Good foliage. Strong-growing for trellis and walls. Introduced 1893. Average height and spread 12×12 ft (3.7×3.7 m). Protrudes up to 3 ft (91 cm) from support. **'Souvenir de Madame Léonie Viennot, Climbing'** (Tea) Tea rose flowers, pale yellow, shaded to coppery-pink. Fragrant. Free-flowering. Tolerant of most aspects. For trellis and large walls. Introduced 1898. Average height and spread 12×12 ft (3.7×3.7 m). Protrudes up to 3 ft (91 cm) from support. **'Tea Rambler'** Salmon-pink, double, small, fragrant flowers in mid summer and carried in profusion on vigorous, strong-growing shrub. Tolerates all soils and aspects. Can be used to grow through trees. Introduced 1904. Average height and spread 12×10 ft (3.5×3 m). Protrudes up to 3 ft (91 cm) from support. **'Zéphirine Drouhin'** (Bourbon) Cerise-pink, semi-double, very fragrant flowers. Some spring flowers, after which continuous. Thornless. Tolerates all aspects and soil conditions. Introduced 1868. Average height and spread 10×10 ft (3×3 m). Protrudes up to 18 in (45 cm) from support.

Average height and spread Growth at ten years given for each variety.

ROSA (Rambler Roses)

KNOWN BY VARIETY NAME

Rosaceae   **Rose**
Deciduous

Rambler roses have a pedigree that stretches back often more than 100 years. They had their heyday in the late 1800s to early 1900s followed by a period of unpopularity, but are now becoming rightly appreciated for their flower display.

Origin Mostly derivatives of *R. wichuraiana* with various crosses with other forms. Often of garden or nursery descent.

Use As large, rambling climbers for fences, walls, pergolas and archways, to ramble through trees or to be used for ground cover with or without support.

Description *Flower* Single or semi-double, carried in clusters, either single or repeat flowering. Individual flowers between $\frac{1}{2}$–$1\frac{1}{2}$ in (1–4 cm) in diameter. Normally flowering mid to late summer in a range of colours through white, yellow, pink, red, mauve, all depending on variety. *Foliage* Five or seven round leaflets make up a pinnate leaf. Light green to mid green depending on variety. *Stem* Light green, rambling, branching, with thorns, degree according to variety. Fast to medium growth rate. *Fruit* Some varieties may produce clusters of small, red rose hips which have winter attraction.

Hardiness Tolerates a minimum winter temperature of 4°F (−15°C).

Soil requirements Tolerates all soil types, both alkaline and acid, but on severe alkaline soils may show signs of chlorosis. Good preparation and annual feeding and mulching will normally control this.

Sun/Shade aspect Tolerates all aspects. Best in full sun to light shade, but some varieties may tolerate deeper degrees of shade.

Pruning Prior to planting young plants the top growth should be reduced to 2–3 ft (60–91 cm), and any very weak shoots removed. Roses established more than 12 months should always be pruned in early to late autumn and all old flowering shoots cut to ground level. Tie in new shoots to supports to flower in following year. If production of new growth is limited, retain a proportion of older shoots, choosing the healthiest, but cut back all their lateral shoots to within 3 in (7.5 cm) of their point of origin.

Training Allow to ramble through trees or over wire or timber supports. On walls and fences tie to wires or individual anchor points. Up pillars and trelliswork some limited tying may be required.

Rosa 'Albertine'

Propagation and nursery production Commercial production is from budding, entailing the insertion of one single bud into a predetermined root stock. Plants are normally sold 18 months on from this process. Can be grown from hardwood cuttings, but lifespan may be shortened due to possible lack of root vigour. Can be purchased bare-rooted or root-wrapped from early autumn to mid spring. May be available container grown from late spring to early autumn and beyond, but ensure that the plant has not been potted for more than one year. Best planting height 1–3 ft (30–91 cm), depending on variety. Many varieties are readily available from garden centres and nurseries, but some have to be sought from specialist rose growers.

Problems Many varieties suffer from black spot and mildew. In the case of mildew, this is often specific to the particular plant and is hard to control, but worth the effort. Suckers from root stools may appear and are difficult to identify as they closely resemble the rambler variety. The indication is that if they develop from below ground they may be suckers and they should be ripped away from the root stool. Careful observation should, however, make it possible to differentiate between the two foliage patterns.

Forms of interest '**Abbandonata**' (Sempervirens) Mid-pink flowers, strong-growing. Ideal for large walls and for growing through trees. Date of introduction unknown. '**Adeläide d'Orléans**' (Sempervirens) Creamy-pink, semi-double, small, pendent flowers with attractive foliage. Primrose scent.

Moderately strong-growing in trees and on large walls. Introduced 1826. '**Albéric Barbier**' Double, very fragrant flowers, creamy-white flushed yellow. Somewhat floppy when open. Flowering mid to late summer. Good dark glossy green foliage, in some years almost evergreen. Ideal for trees. Tolerates exposed aspects in full sun to medium shade, but may be shy to flower in shade. Introduced 1900. '**Albertine**' The exquisitely fragrant rambler of cottage doors. Copper/orange in bud and opening to pink with gold shading. Flowering early to mid summer. Strong growing. Tolerates all soil conditions. Ideal for growing in trees. Introduced 1921. '**Alexandre Girault**' Deep rose pink and copper double flowers with a raspberry fragrance in mid summer. Useful as ground cover. Tolerates full sun to medium shade. Two thirds average height and spread. Introduced 1909. '**Alister Stella Gray**' (syn. '**Golden Rambler**') Yellow with deeper yellow centre, ageing to creamy-white. Good scent over a long flowering period. Repeat flowering. Ideal for trees and can be grown in full sun to medium shade. Introduced 1894. '**American Pillar**' Single flowers, pink with central white eye. Good dark green foliage. Strong and upright stems. Ideal for growing in trees. Will tolerate a wide range of conditions. Introduced 1909. '**Apple Blossom**' Apple-blossom pink flowers, each petal with a crinkled edge, from mid to late summer. Tolerates all soil conditions in full sun to medium shade. Two thirds average height and spread.

Rosa 'American Pillar'

Unusual in being a rambler suitable for pillars. Introduced 1932. '**Auguste Gervais**' Semi-double, fragrant, salmon-pink flowers with a coppery tinge, ageing to soft pink. Flowers mid summer. Mid green shiny foliage on good shoots. Tolerates all soil conditions and exposed positions, but dislikes shade. Ideal for growing up trees. Introduced 1918. '**Blush Rambler**' Pink flowers in very large clusters, reminiscent of apple blossom, in mid summer. For walls, fences, trellis and trees. Introduced 1903. '**Breeze Hill**' Clear pink flowers flushed tawny orange. Double to semi-double, cup-shaped. Mid summer flowering. Strong growing even to the extent of being gangly and gaunt. Tolerates all soil conditions and aspects. Full sun to medium shade. Ideal for growing in trees. Introduced 1926. '**Chaplin's Pink Climber**' Attractive single or semi-double mid to deep pink flowers, in profusion mid summer. Tolerates all soil types. Good for use on fences, walls, pergolas and arches. Introduced 1928. '**Coralle**' Coral-pink flowers in clusters in mid summer. Good green foliage. Two thirds average height and spread. Introduced 1919 and should be more widely grown. '**Crimson Conquest**' Semi-double flowers, bright crimson, carried in clusters in mid summer.

Rosa 'Albéric Barbier'

Good foliage. Tolerates all aspects in full sun to medium shade and all soil conditions. Introduced 1932. **'Crimson Shower'** Semi-double crimson flowers in bold clusters from mid to late summer. Some later autumn flowers. Small glossy foliage. Introduced 1951. **'Debutante'** Fragrant, rose-pink flowers carried in clusters. Mid summer flowering. Foliage dark green, showing off the flowers to good advantage. Tolerates all soil types. Ideal for extensive ground cover or for growing through trees. Introduced 1902. **'Dorothy Perkins'** Clear pink double flowers in clusters. Mid summer flowering. Light green foliage on wispy stems. Widely planted but prone to its own specific strain of mildew which is difficult to control. Introduced 1902. **'Dr Van Fleet'** Double, fragrant, flesh-pink flowers, fading to white, with attractive texture to petals. Mid summer flowering. Strong growing with good foliage. Tolerates full sun to medium shade. Ideal for growing in trees. Introduced 1910. **'Easlea's Golden Rambler'** Rich yellow, double flowers in clusters on long, ranging flower stems. Mid summer flowering. Good mid green foliage and good growth. Ideal on all soil types in full sun to medium shade. Ideal for growing in trees. Introduced 1932.

Rosa 'François Juranville'

Rosa 'Dorothy Perkins'

'Emily Gray' Large, fragrant, golden yellow flowers in mid summer showing off well against the glossy, deep green foliage. Tolerates all soil types in full sun to medium shade. Ideal for growing in trees. Introduced 1918. **'Ethel'** Mauve/pink flowers in clusters mid summer set against good foliage. Tolerates all soil types and exposed positions in full sun to medium shade. Ideal for growing in trees. Introduced 1912. **'Evangeline'** Creamy-white flowers diffused pink, flowering in late summer. Good, dark green, leathery foliage. Ideal for exposed positions in full sun to medium shade. Tolerates all soil conditions. Ideal for growing up trees. Introduced 1906. **'Excelsa'** Double, light crimson flowers carried in large trusses on thin wispy shoots. Flowers in mid summer. Strong growing. Tolerates exposed aspects in full sun to medium shade on all soil types. Ideal for ground cover or for climbing through trees. Introduced 1909. **'Félicité et Perpétue'** Fragrant, creamy-white flowers, small in size and globe-shaped, from mid to late summer. Light green foliage, almost evergreen. Tolerates all soil types. Ideal for climbing in trees. Introduced 1827. **'Flora'** Double flowers, off-white with lilac shading. Cup-shaped, scented, flowering mid to late summer. Good dark green foliage. Tolerates exposed positions in full sun to medium shade. Ideal for growing in trees. Introduced 1929. **'François Juranville'** Mid pink, double flowers in mid to late summer. Tolerates exposed conditions in full sun to medium shade and almost any soil type. Ideal for

growing in trees or for ground cover. Introduced 1906. **'Goldfinch'** Large golden-yellow and primrose-yellow petals with golden central anthers. Double, fragrant flowers freely borne from late spring to mid summer. Good foliage. Tolerates full sun to medium shade but dislikes exposed positions. Less robust than some varieties. Two thirds average height and spread. Introduced 1907. **'Hiawatha'** Each single crimson flower has a white eye and pronounced golden anthers in the centre. Flowers produced in clusters from mid to late summer. Tolerates all soil conditions in full sun to medium shade. Ideal for climbing up trees. Introduced 1904. **'Kew Rambler'** Soft pink single flowers with deeper pink margins to each petal. Flowers from mid to late summer. Foliage grey to grey/green. Tolerates all aspects in full sun to medium shade. Ideal for growing in trees. Introduced 1912. **'Lykkefund'** Creamy-yellow, semi-double, fragrant flowers in large tight clusters from mid to late summer. Stems thornless, supporting glossy dark green foliage. Requires some protection but will tolerate full sun to medium shade on a wide range of soil conditions. Very good for woodland planting and for use in trees. Introduced 1930. **'Léontine Gervais'** Double flowers, clear pink

Rosa 'Easlea's Golden Rambler'

with copper shading. Fragrant. Foliage clean and attractive. Very strong growing; ideal for covering buildings or large walls or for growing through trees. Introduced 1903. **'May Queen'** Lilac/pink flowers, ageing to white, semi-double, produced in mid to late summer in an abundance of large clusters. Tolerates all aspects in full sun to medium shade. Ideal for growing through trees or for ground cover. Introduced 1898. **'Minnehaha'** Pale pink flowers ageing to white in large clusters from mid to late summer. Strong growing. Large quantity of small, dark green foliage. Tolerates full sun to medium shade in exposed positions on all soil types, ideal for growing through trees or for ground cover. Introduced 1905. **'Mme Alice Garnier'** Bright orange/pink flowers with orange centres and attractive scent from mid to late summer. Foliage dark green on slender, arching branches. Ideal for ground cover. Introduced 1906. **'Mme d'Arblay'** Blush-pink to white fragrant, cup-shaped flowers from mid to late summer. Strong growing and vigorous. Ideal for exposed positions. Can be grown in trees. One third more than average height and spread. Introduced 1835. **'Paul Transon'** Medium-sized double flowers, rich salmon with coppery shading and yellow base. Repeat flowering from mid summer to mid autumn. Foliage coppery when young, shiny, becoming light green, making a good contrast. Requires some protection. Tolerates full sun to medium shade and a wide range of soil types. Introduced 1901. **'Phyllis Bide'** Semi-double pink, salmon and gold flowers, continuous flowering. Some fragrance. Medium growth rate, reaching two thirds average height and spread. Introduced 1923. **'René André'** Semi-double, fragrant flowers with saffron and carmine tinges, paling with age. Flowers from mid to late summer. Tolerates exposed situations in full sun to medium shade on all soil types. Ideal for training in trees. Introduced 1901. **'Russelliana'** (syns. **'Old Spanish Rose'**, **'Russell's Cottage'**, **'Scarlet Grevillea'**) Was once called 'Souvenir de la Bataille de Marengo'. Magenta/crimson flat flowers borne in clusters in mid to late summer. Tolerates exposed situations in full sun to light shade on all soil types. Two thirds average height and spread. Ideal for growing in trees. Introduced 1840. **'Sanders White'** Distinct pure white rosette-shaped flowers, carried in cascading clusters. Scented and flowering from mid to late summer. Foliage bright green. Tolerates most soil conditions in full sun to light shade. Can be used to climb in trees. Introduced 1912. **'Silver Moon'**

Rosa 'Phyllis Bide'

Strongly fragrant, creamy-white single flowers with golden anthers. Flowers mid summer. Date of introduction unknown. **'Thelma'** Coral-pink and red large semi-double flowers, from mid to late summer. Stems have very few thorns. Tolerates full sun to light shade on all soil types. Ideal for training in trees. Introduced 1927. **'Veilchenblau'** (syn. **'Violet Blue'**) Semi-double, violet/purple flowers with white towards centre, fading to blue/lilac and lilac/grey with age. Good scent. Best in light to medium shade for flower colour, although will tolerate full sun if required, but may fade. Tolerates exposed aspects and all soil types. Ideal for training in trees. Introduced 1909. **'Violette'** Violet/purple very full double

Rosa 'Veilchenblau'

flowers carried in large trusses from mid to late summer. Strong growing. Tolerates full sun to medium shade on most soil types. Ideal for training in trees. Introduced 1921. **'White Flight'** Semi-double pure white flowers from mid to late summer. Light green foliage. Tolerates full sun to light shade in exposed positions and most soils. Introduced 1900.

Average height and spread
Five years
8 × 8 ft (2.4 × 2.4 m)
Ten years
15 × 15 ft (4.6 × 4.6 m)
Twenty years
20 × 20 ft (6 × 6 m).
Protrudes up to 3 ft (91 cm) from support.

ROSA
(Shrub and Species Roses for Fan-training)

KNOWN BY VARIETY NAME
Rosaceae   *Rose*
Deciduous

Apart from climbing and rambler roses there are many shrub and species roses which adapt extremely well to fan-training, often showing off their flowers to better advantage.

Origin Various.
Use As fan-trained shrubs for walls and fences.
Description *Flower* Either single, semi-double or double in a wide range of colours through pink, white, purple, red, flowering from spring through summer, depending on variety. Many are fragrant. *Foliage* Five or seven leaflets make up a pinnate leaf, 5–7 in (12–18 cm) long. Some yellow autumn colour. *Stem* Bushy habit, normally green thorny stems. Some may have winter attraction. Medium to fast growth rate. *Fruit* Some varieties may produce orange/red or scarlet hips in autumn.

Rosa 'Alba Maxima'

Hardiness Tolerates a minimum winter temperature of 4°F (−15°C).
Soil requirements Tolerates all soil conditions with no particular preference except that it must be moisture- and nutrient-retentive.
Sun/Shade aspect Tolerates all aspects in full sun to light shade.
Pruning Remove one third of oldest growth to ground level on roses established more than three years.
Training Tie to wires or individual anchor points.
Propagation and nursery production Commercial production is from budding, entailing the insertion of one single bud into a predetermined rootstock. Plants are normally sold 18 months on from this process. Some varieties can be grown successfully from hardwood cuttings. Purchase bare-rooted or root-wrapped from early autumn to mid spring; may be available container grown from late spring to early autumn and beyond. Best planting height 1–3 ft (30–91 cm). Many varieties are readily available from garden centres and nurseries, but some may have to be sought from specialist nurseries.
Problems May suffer from greenfly and blackfly. Treat with a proprietary control. Mildew and blackspot may be a problem, but correct pruning, cultivation, feeding and treatment with a proprietary control will normally keep the disease to a minimum. Thorns can make handling uncomfortable and plants which are not pruned as suggested can become very woody. Suckers from rootstocks

may appear and are difficult to identify as they closely resemble the actual variety. The indication is that if they develop from below ground they are normally suckers. Careful observation should, however, make it possible to differentiate between the two foliage patterns.
Forms of interest The genetic group which each variety fits into is shown in brackets: **'Adam Messerich'** (Bourbon) Very bright pink, semi-double flowers in bold sprays. Continuous flowers, strong growing. Introduced 1920. Average height and spread 6 × 6 ft (1.8 × 1.8 m). Protrudes up to 2 ft (60 cm) from support. **'Alba Maxima'** (Alba) ('Jacobite Rose', 'White Rose of York') Flat double flowers, slightly incurved, white on outer edges, more creamy-white in centre. Early to late summer flowering. Leaves dark, almost grey/green. Flowers followed by a good crop of orange/red autumn fruits. Introduced prior to the 16th century. Average height and spread 6 × 6 ft (1.8 × 1.8 m). Protrudes up to 2 ft (60 cm) from support. **'Ballerina'** (Modern Shrub) Small, single flowers, up to $\frac{1}{2}$ in (1 cm) across, pink with white centres, carried in sprays. Continuous flowering. Introduced 1937. Average height and spread 5 × 5 ft (1.5 × 1.5 m). Protrudes up to 18 in (45 cm) from support. Of garden origin. **'Belle de Crécy'** (Gallica) Flowers up to 3 in (7.5 cm) across, bright pink and mauve with good fragrance. Early to mid summer flowering, tolerating full sun to light shade. Stems almost thornless. Introduced mid 19th century. Average height and spread 4 × 4 ft (1.2 × 1.2 m). Protrudes up to 18 in (45 cm) from support. Of garden origin. **'Bloomfield Abundance'** (China Rose) Small, attractive,

Rosa 'Ballerina'

mid pink hybrid tea blooms, carried on large sprays. Continuous flowering. Introduced 1920. Average height and spread 6 × 6 ft (1.8 × 1.8 m). Protrudes up to 2 ft (60 cm) from support. Of garden origin. **'Blush Damask'** (Damask) Strongly fragrant flowers, pale pink with deeper pink centre. Early to mid summer flowering. An extremely old variety. Average height and spread 4 × 4 ft (1.2 × 1.2 m). Protrudes up to 12 in (30 cm) from support. Of garden origin. **'Bonn'** (Hybrid Musk) Large, double flowers up to 4 in (10 cm) across, orange/scarlet, fragrant, carried in clusters. Continuous flowering. Introduced 1915. Average height and spread 5 × 5 ft (1.5 × 1.5 m). Protrudes up to 18 ft (45 cm) from support. Of nursery origin. **'Buff Beauty'** (Hybrid Musk) Fragrant apricot/yellow flowers, ageing to buff-yellow, carried in large trusses, each flower up to 3 in (7.5 cm) across and semi-double. Continuous flowering. Tolerates full sun to medium shade. Introduced 1939. Average height and spread 7 × 7 ft (2.1 × 2.1 m). Protrudes up to 2 ft (60 cm) from support. Of garden origin.

'**Capitaine John Ingram**' (Moss) Dark crimson, but can be purple in some weather conditions. Very strongly scented. Flowers from early to mid summer. Introduced 1856. Average height and spread 7×7 ft (2.1×2.1 m). Protrudes up to 18 in (45 cm) from support. Of garden origin. '**Cardinal de Richelieu**' (Gallica) Strongly fragrant purple flowers with a velvety texture in mid summer. Good foliage. Introduced 1840. Average height and spread 5×5 ft (1.5×1.5 m). Protrudes up to 18 ft (45 cm) from support. Of garden origin. '**Celestial**' (syn. '**Celeste**') (Alba) Semi-double, soft pink flowers from late spring to early summer. Foliage grey/green and attractive after flowering. Average height and spread 7×7 ft (2.1×2.1 m). Protrudes up to 12 ft (30 cm) from support. Origin not known, but a very old variety. '**Cerise Bouquet**' (Modern Shrub) Flowers deep crimson pink, double and scented, borne from early to late summer with

Rosa 'Bloomfield Abundance'

main flowering period mid summer. Foliage grey to grey/green with indented edges to leaves. Tolerates full sun to medium shade. Often underestimated for its height and spread potential. Introduced 1958. Average height and spread 15×15 ft (4.6×4.6 m). Protrudes up to 6 ft (1.8 m) from support. '**Chapeau de Napoleon**' (syns. '**Crested Moss**', '**Cristata**') (Centifolia) Large, double, lettuce-like flowers, silvery pink, very fragrant, from mid to late summer. Flower buds before opening have attractive moss covering. Introduced 1826. Average height and spread 7×7 ft (2.1×2.1 m). Protrudes up to 2 ft (60 cm)

Rosa 'Celestial'

Rosa 'Fruhlingsgold'

from support. Of garden origin. '**Charles de Mills**' (Gallica) Flowers bi-coloured, purple and deep red. Mid summer flowering. Foliage attractive, mid-green, indented edges. Origin unknown. Average height and spread 5×5 ft (1.5×1.5 m). Protrudes up to 12 in (30 cm) from support. '**Constance Spry**' (Modern Shrub) Flowers up to 5 in (12 cm) across, pink and very heavily scented. Mid summer flowering. Full sun to medium shade. Does best against a wall. Introduced 1960. Average height and spread 20×12 ft (6×3.7 m). Protrudes up to 2 ft (60 cm) from support. '**Copenhagen**' (Shrub) Often considered as a climbing rose and stocked by garden centres and nurseries in that area, but technically is a shrub rose with its scarlet flowers borne in clusters. Continuous flowering. A truly handsome red rose for climbing or for pillars. Introduced 1964. Average height and spread 8×6 ft (2.4×1.8 m). Protrudes up to 2 ft (60 cm) from support. '**Cornelia**' (Hybrid Musk) Apricot flowers with pink shading continuously from mid summer to early autumn. Foliage bronze when young. Flowers in full sun to medium shade. Introduced 1925. Average height and spread 7×7 ft (2.1×2.1 m). Protrudes up to 2 ft (60 cm) from support. '**Dortmund**' (Hybrid Pimpinellifolia) Fragrant red flowers with white eye, opening from long, pointed, attractive buds and carried in clusters. often considered as a climber. Repeat flowering. Foliage glossy dark green, strong growing. Tolerates exposed aspects in full sun to medium shade. Introduced 1955. Average height and spread 12×12 ft (3.7×3.7 m). Protrudes up to 3 ft (91 cm) from support. '**Elmshorn**' (Modern Shrub) Vivid pink blooms, repeat flowering from mid summer to early autumn. Tolerates full sun to medium shade. Introduced 1951. Average height and spread 7×7 ft (2.1×2.1 m). Protrudes up to 2 ft (60 cm) from support. '**Erfut**' (Hybrid Musk) Pink and white flowers with central dull yellow stamens. Continuous flowering. Tolerates full sun to medium shade. Introduced 1939. Average height and spread 7×7 ft (2.1×2.1 m). Protrudes up to 2 ft (60 cm) from support. '**Fantin-Latour**' (Centifolia) Pale pink scented flowers from early to mid summer. Introduced prior to 1890. Average height and spread 7×7 ft (2.1×2.1 m). Protrudes up to 2 ft (60 cm) from support. '**Felicia**' (Hybrid Musk) Silver/pink fading to salmon-pink. Continuous flowering. One of the best. Introduced 1928. Average height and spread 6×6 ft (1.8×1.8 m). Protrudes up to 2 ft (60 cm) from support. '**Félicité Parmentier**' (Alba) Soft pink petals, reflexed, forming a double flower. Scented. Early to mid summer flowering. Foliage grey/green, attractive. Full sun to medium shade. Intro-

duced 1834. Average height and spread 5×5 ft (1.5×1.5 m). Protrudes up to 18 in (45 cm) from support. '**Ferdinand Pichard**' (Hybrid Perpetual) Double, deep pink flowers with lighter pink candy stripes, profusely borne. Recurrent flowering. Good foliage. Introduced 1921. Average height and spread 7×7 ft (2.1×2.1 m). Protrudes up to 18 in (45 cm) from support. '**Francesca**' (Hybrid Musk) Single to semi-double apricot-coloured flowers produced in large sprays. Good scent. Continuous flowering. Strong growth with attractive foliage. Introduced 1921. Average height and spread 5×5 ft (1.5×1.5 m). Protrudes up to 2 ft (60 cm) from support. '**Fred**

Rosa 'Dortmund'

Loads' (Modern Shrub) Orange to vermillion/orange, semi-double, scented, continuous flowering. Strong upright growth. Introduced 1968. Average height and spread 7×7 ft (2.1×2.1 m). Protrudes up to 2 ft (60 cm) from support. '**Fritz Nobis**' (Modern Shrub) Pale pink flowers, semi-double with clove scent, from mid to late summer. Good hips in autumn. Introduced 1940. Average height and spread 7×7 ft (2.1×2.1 m). Protrudes up to 2 ft (60 cm) from support. '**Frühlingsanfang**' (Hybrid Pimpinellifolia) Ivory-white, single flowers 4 in (10 cm) across with attractive golden anthers. Scented. Flowers late spring to early summer. Strong growing. Introduced 1950. Average height and spread 10×10 ft (3×3 m). Protrudes up to 4 ft (1.2 m) from support. '**Frühlingsgold**' (Hybrid Pimpinellifolia) Single pale yellow flowers, 4 in (10 cm) across, from mid to late spring. May produce purple tomato-shaped fruits. Strong growing, with numerous thorns. Full sun to medium shade. Introduced 1951. Average height and spread 9 × 9 ft (2.7 × 2.7 m). Protrudes up to 2 ft (60 cm) from support. **'Frühlingsmorgen'** (Hybrid Pimpinellifolia) Very attractive single cherry-pink flowers with primrose-coloured centres. Flowers up to 4 in (10 cm) across pro-

Rosa 'Ferdinand Pichard'

duced in mid to late spring. Second flowering late summer to early autumn. Normally a good display of tomato-shaped purple hips. Introduced 1942. Average height and spread 8 × 8 ft (2.4 × 2.4 m). Protrudes up to 2 ft (60 cm) from support. **'Frühlingsschnee'** (Hybrid Pimpinellifolia) Double pure white flowers in mid summer in full sun to light shade. Good foliage, thorny stems. Introduced 1954. Average height and spread 8 × 8 ft (2.4 × 2.4 m). Protrudes up to 2 ft (60 cm) from support. **'Frühlingszauber'** (Hybrid Pimpinellifolia) Semi-double, silver/pink, scented flowers from late spring to early summer, shown off well against dark green foliage. Introduced 1942. Average height and spread 9 × 9 ft (2.7 × 2.7 m). Protrudes up to 2 ft (60 cm) from support. **'Gipsy Boy'** (syn. **'Zigeunerknabe'**) (Bourbon) Double, crimson to deep crimson fading to purple, with primrose yellow centres. Flowers mid summer. Full sun to light shade. Introduced 1909. Average height and spread 8 × 8 ft (2.4 × 2.4 m). Protrudes up to 2 ft (60 cm) from support. **'Golden Chersonese'** (*R. ecae* hybrid) Single flowers, bright yellow, 1½ in

Rosa 'Fred Loads'

Rosa 'Fritz Nobis'

(4 cm) across, carried singly at each leaf joint in mid to late spring. Foliage fern-like, attractive. Full sun to medium shade. Introduced 1963. Average height and spread 6 × 6 ft (1.8 × 1.8 m). Protrudes up to 18 in (45 cm) from support. **'Golden Wings'** (Modern Shrub) Single flowers, up to 4 in (10 cm) across, yellow with red central anthers. Continuous flowering. Can be prone to black-

Rosa 'Gipsy Boy'

spot, but worth consideration. Introduced 1953. Average height and spread 7 × 7 ft (2.1 × 2.1 m). Protrudes up to 3 ft (91 cm) from support. **'Helen Knight'** (*R. ecae* hybrid) 1 in (2.5 cm) wide, saucer-shaped flowers of deep, clear yellow, carried en masse close to stems in late spring to early summer in full sun to medium shade. Introduced 1953. Average height and spread 9 × 4 ft (2.7 × 1.2 m). Protrudes up to 12 in (30 cm) from support. **'Ispahan'** (Damask) Pink, semi-double flowers up to 2½ in (6 cm) across from mid to late summer. Attractive foliage. Introduced prior to 1842. Average height and spread 7 × 7 ft (2.1 × 2.1 m). Protrudes up to 2 ft (60 cm) from support. **'Kassel'** (Modern Shrub) Clusters of orange/scarlet, semi-double flowers contrasting well with dark green foliage. Continuous flowering from mid summer to early autumn. Introduced 1957. Average height and spread 7 × 7 ft (2.1 × 2.1 m). Protrudes up to 2 ft (60 cm) from support. **'La Noblesse'** (Centifolia) Soft silver/pink flowers, semi-double, with good scent. Flowering from mid to late summer, although not continuously. Introduced 1856.

Average height and spread 6 × 6 ft (1.8 × 1.8 m). Protrudes up to 2 ft (60 cm) from support. **'La Reine Victoria'** (Bourbon) Warm rose pink, cup-shaped flowers. Perpetual flowering. Attractive soft green foliage. Introduced 1872. Average height and spread 6 × 3 ft (1.8 × 91 cm). Protrudes up to 18 in (45 cm) from support. **'Maiden's Blush'** (syn. **'Cuisse de Nymphe'**) (Alba) Double, somewhat floppy, blush-pink, sweet-scented flowers from mid to late summer. Foliage grey/green. Tolerates full sun to medium shade. Introduced in the late 18th century. Average height and spread 12 × 12 ft

Rosa 'Buff Beauty'

(3.7 × 3.7 m). Protrudes up to 2 ft (60 cm) from support. **'Marguerite Hilling'** (Modern Shrub) Mid-pink, semi-double flowers. Recurrent flowering. Without good soil preparation and adequate organic material and additional feed, can be a little insipid in growth. Introduced 1959. Average height and spread 7 × 7 ft (2.1 × 2.1 m). Protrudes up to 2 ft (60 cm) from support. **'Mme Isaac Pereire'** (Bourbon) Large, purple/crimson rosette flowers. Continuous flowering in full sun to medium shade. Introduced in 1881. Average height and spread 8 × 8 ft (2.4 × 2.4 m). Protrudes up to 2 ft (60 cm) from support. **'Mme Lauriol de Barny'** (Bourbon) Flowers quartered, flat, with unusual fragrance. Silver/pink in colour. Recurrent flowering in full sun to medium shade. Introduced 1868. Average height and spread 8 × 8 ft (2.4 × 2.4 m). Protrudes up to

Rosa 'Helen Knight'

Rosa 'Marguerite Hilling'

2 ft (60 cm) from support. **'Mme Pierre Oger'** (Bourbon) Cup-shaped flowers, pale silver/pink with good scent. Continuous flowering. Dislikes shade. Introduced 1878. Average height and spread 6 × 6 ft (1.8 × 1.8 m). Protrudes up to 18 in (45 cm) from support. **'Moonlight'** (Hybrid Musk) Flowers single, lemon to white on attractive long stems. Continuous flowering in full sun to medium shade. Introduced 1913. Average height and spread 7 × 7 ft (2.1 × 2.1 m). Protrudes up to 18 ft (45 cm) from support. **'Mrs John Laing'** (Hybrid Perpetual) Good shaped, double, soft pink flowers. Recurrent flowering. Requires good soil preparation and additional organic material and food for best

Rosa 'Nevada'

results. Introduced 1887. Average height and spread 3 × 2 ft (91 × 60 cm). Protrudes up to 3 ft (91 cm) from support. **'Nevada'** (Modern Shrub) Large creamy-white blooms up to 3 in (7.5 cm) across with wavy edged petals. Repeat flowering through late spring to early summer. Introduced 1927. Average height and spread 10 × 10 ft (3 × 3 m). Protrudes up to 2 ft (60 cm) from support. **'Nozomi'** (Modern Shrub) Pearly-pink flowers, ageing to white, carried in profusion on long cascading sprays. Normally considered for ground cover, but with care can be persuaded to be an attractive low climbing wall shrub in full sun to medium shade. Introduced 1968. Average height and spread 7 × 6 ft (2.1 × 1.8 m). Protrudes up to 12 ft (30 cm) from support. **'Penelope'** (Hybrid Musk) Creamy-pink, semi-double flowers with good scent. Recurrent flowering

from early summer to early autumn. Introduced 1924. Average height and spread 7 × 7 ft (2.1 × 2.1 m). Protrudes up to 18 in (45 cm) from support. **'Pink Prosperity'** (Hybrid Musk) Double pink flowers carried in small sprays. Continuous flowering. Introduced 1931. Average height and spread 7 × 7 ft (2.1 × 2.1 m). Protrudes up to 2 ft (60 cm) from support. **'Pomifera Duplex'** ('Wolly-Dod's Rose') (Pomifera) Soft clear pink flowers, semi-double, from mid to late summer with light grey/green foliage. Attractive purple autumn hips. Ideal in full sun to medium shade. Tol-

Rosa complicata

erates exposed aspects. Introduced 1900. Average height and spread 7 × 7 ft (2.1 × 2.1 m). Protrudes up to 2 ft (60 cm) from support. **'Prince Charles'** (Bourbon) Maroon to lilac, scented, double flowers from mid to late summer but not necessarily continuous. Full sun to medium shade. Origin unknown. Average height and spread 7 × 7 ft (2.1 × 2.1 m). Protrudes up to 18 in (45 cm) from support. **'Prosperity'** (Hybrid Musk) Trusses of double, creamy-white flowers with strong scent. Recurrent flowering from summer to early autumn in both sun and medium shade. Introduced 1919. Average height and spread 7 × 7 ft (2.1 × 2.1 m). Protrudes up to 2 ft (60 cm) from support. *R. californica plena* (Species) Deep pink, small, semi-double flowers profusely borne from mid summer to early autumn, not necessarily continuous. Introduced 1894. Average height and spread 10 × 10 ft (3 × 3 m). Protrudes up

Rosa moyesii 'Geranium'

to 2 ft (60 cm) from support. *R. complicata* (Of Gallica origin) Possibly one of the most attractive of all species roses with large, pink, single flowers with a paler centre and bold golden stamens in early to mid summer. Strongly fragrant. Stems and foliage green to grey/green. Will flower in full sun to medium shade. Origin unknown. Average height and spread 11 × 11 ft (3.4 × 3.4 m). Protrudes up to 12 in (30 cm) from support. *R. ecae* (Species) Deep yellow flowers, up to 1 in (2.5 cm) across, from mid to late spring. Small, dark green foliage. Stems red/brown with attractive red/brown thorns. Introduced 1880. Average height and spread 6 × 6 ft (1.8 × 1.8 m). Protrudes up to 18 in (45 cm) from support. *R. farreri persetosa* ('Three-penny Bit Rose') (Species) Small lilac/pink flowers carried on graceful, heavily soft-thorned shoots with attractive fern-like foliage. Flowers from mid to late summer, not necessarily continuously. Small orange/red hips in autumn, making an all-round attraction. Tolerates full sun to medium shade. Introduced 1914. Average height and spread 7 × 7 ft (2.1 × 2.1 m). Protrudes up to 12 in (30 cm) from support. *R. fedtschenkoana* (Species) Flowers single, white, but main attraction is the grey/green to silver green foliage. Red hips in autumn. Flowers recurrent. Introduced 1880. Average height and spread 10 × 10 ft (3 × 3 m). Protrudes up to 5 ft (1.5 cm) from support. *R. foetida* (syn. **'Yellow Austrian Briar'**) (Species) Attractive single yellow flowers from mid to late spring. Dark glossy green foliage and brown stems. The protection of a

Rosa bicolor

wall adds to its overall vigour. Introduced earlier than the 16th century. Average height and spread 8 × 8 ft (2.4 × 2.4 m). Protrudes up to 18 in (45 cm) from support. **R. f. bicolor** (syn. **'Austrian Copper'**) (Species) Single, copper/scarlet with rich yellow reverse to petals in mid to late spring. Can be prone to diseases, such as mildew and blackspot; spraying can normally counteract this. Fan-training against a wall often enhances its growth. Introduced 1590. Average height and spread 7 × 7 ft (2.1 × 2.1 m). Protrudes up to 18 in (45 cm) from support. **R. f. persiana** (syn. **'Persian Yellow'**) (Species) Globe-shaped, bright yellow double flowers against dark, glossy green foliage and brown stems with brown thorns. Flowers from mid spring to mid summer. Fan-training against a wall enhances its growth. Introduced 1837.

Rosa × *paulii*

Average height and spread 7 × 7 ft (2.1 × 2.1 m). Protrudes up to 2 ft (60 cm) from support. **R.** × **highdownensis** (Hybrid) Light crimson flowers, followed in autumn by attractive bottle-shaped hips. Mid spring to early summer flowering. Strong growing. Introduced 1903. Average height and spread 9 × 9 ft (2.7 × 2.7 m). Protrudes up to 2 ft (60 cm) from support. **R. moyesii** (Species) Pink flowers in mid summer and good hips in autumn. Red, thorny branches and dark green foliage. Tolerates full sun to medium shade. Origin unknown. Average height and spread 10 × 10 ft (3 × 3 m). Protrudes up to 2 ft (60 cm) from support. **R. m. 'Geranium'** (Species) Very bright scarlet flowers with good orange/red hip production. Flowers mid summer in full sun to medium shade. A good form. Introduced 1938. Average height and spread 9 × 9 ft (2.7 × 2.7 m). Protrudes up to 18 in (45 cm) from support. **R. m. 'Sealing Wax'** Deep pink flowers in mid summer. Good autumn hips. Introduced 1938. Average height and spread 9 × 9 ft (2.7 × 2.7 m). Protrudes up to 18 in (45 cm) from support. **R. omeiensis pteracantha** (syn. **R. sericea pteracantha**) (Species) Attractive bright red stems that age to brown in winter. Large bright red thorns of interesting shape. Single, small white flowers in late spring to early summer, followed by red hips in autumn. Tolerates full sun to medium shade. Will require an annual reduction of growth by one third, selecting oldest shoots and removing to ground level. This will entail an annual retying in to wires or individual anchorage points. Introduced 1890. Average height and spread 12 × 12 ft (3.7 × 3.7 m). Protrudes up to 18 in (45 cm) from support. **R. o. 'Red Wing'** (Species) Very similar to *R. o. pteracantha* but thorns are stronger red with yellow flowers in early summer. Full sun to medium shade. Origin unknown. Average height and spread 7 × 7 ft (2.1 × 2.1 m). Protrudes up to 18 in (45 cm) from support. **R.** × **paulii** (Hybrid) Single white flowers, 3 in (7.5 cm) across with bold

Rosa woodsii fendleri in fruit

yellow stamens, on very thorny branches. Flowers mid summer. Somewhat ranging in habit. Normally considered for ground cover, but very useful for covering large walls in difficult situations. Introduced prior to 1903. Average height and spread 15 × 15 ft (4.6 × 4.6 m). Protrudes up to 6 ft (1.8 m) from support. **R.** × **p. 'Rosea'** (Hybrid) Similar in all respects to *R.* × *p.* but with pink flowers. **R. rubrifolia** (syn. **R. glauca**) (Species) Attractive purple stems and purple foliage, covered in mid summer by numerous single, pink flowers 1 in (2.5 cm) across with white eyes. Some red hip production. Tolerates full sun to medium shade in exposed positions. Introduced prior to 1830. Average height and spread 8 × 8 ft (2.4 × 2.4 m). Protrudes up to 3 ft (91 cm) from support. **R. willmottiae** (Species) The fern-shaped foliage of this rose is its main attraction, but it does produce lilac/pink flowers ½ in (1 cm) across in mid summer. Introduced 1904. Average height and spread 8 × 8 ft (2.4 × 2.4 m). Protrudes up to 18 in (45 cm) from support. **R. woodsii fendleri** (Species) Lilac/pink flowers followed by good production of round, small hips. Mid summer flowering. Tolerates full sun to medium shade. Introduced prior to 1888. Average height and spread 7 × 7 ft (2.1 × 2.1 m). Protrudes up to 18 in (45 cm) from support. **'Schariachglut'** (syn. **'Scarlet Fire'**) (Hybrid Gallica) Bright scarlet/crimson, single flowers 3 in (7.5 cm) across with scarlet golden stamens in early summer to late autumn, but not necessarily continuous. Flowers followed by urn-shaped hips which

Rosa 'Wilhelm'

are retained well into winter. Full sun to medium shade. Introduced 1952. Average height and spread 10 × 10 ft (3 × 3 m). Protrudes up to 3 ft (91 cm) from support. **'Souvenir de la Malmaison'** (Bourbon) Fragrant, pink/white mother-of-pearl flowers up to 3 in (7.5 cm) across with deeper shading. Continuous flowering through summer. Difficult in wet weather. Introduced 1843. Average height and spread 5 × 5 ft (1.5 × 1.5 m). Protrudes up to 18 in (45 cm) from support. **'Swany'** (Procumbens) Cup-shaped white flowers up to ½ in (1 cm) across, double. Continuous flowering. This variety is often considered for ground cover but makes an attractive climbing rose for small areas. Introduced 1978. Average height and spread 6 × 6 ft (1.8 × 1.8 m). Protrudes up to 12 in (30 cm) from support. **'Tuscany Superb'** (Gallica) Flowers deep crimson/purple, semi-double with central golden stamens, from mid summer to late summer. Introduced prior to 1848. Average height and spread 5 × 5 ft (1.5 × 1.5 m). Protrudes up to 18 in (45 cm) from support. **'Vanity'** (Hybrid Musk) Fragrant, semi-double, rose-pink flowers carried in large sprays. Recurrent flowering from mid summer to early autumn. Introduced 1920. Average height and spread 7 × 7 ft (2.1 × 2.1 m). Protrudes up to 2 ft (60 cm) from support. **'White Wings'** (Hybrid Tea) Beautiful large single white flowers with papery texture and chocolate-coloured anthers. Foliage thick, dark green. Repeat flowering. Tolerates all aspects, except extremely shady. May be susceptible to blackspot. Introduced 1947. Average height and spread 6 × 5 ft (1.8 × 1.5 m). Protrudes up to 3 ft (91 cm) from support. **'Wilhelm'** (syn. **'Skyrocket'**) (Hybrid Musk) Flowers semi-double, dark red and carried in large clusters. Repeat flowering with good autumn displays. Good foliage, although it does congregate towards the top ends of the branches. Tolerates all aspects. Introduced 1944. Average height and spread 7 × 6 ft (2.1 × 1.8 m). Protrudes up to 3 ft (91 cm) from support. **'William Lobb'** (syn. **'Old Velvet Moss'**) (Moss) Strong growing, needs a large wall. Buds have heavy moss effect. Flowers purple magenta, scented. Introduced 1855. Average height and spread 9 × 9 ft (2.7 × 2.7 m). Protrudes up to 4 ft (1.2 m) from support. **'Yesterday'** (Modern Shrub) Semi-double fragrant flowers, rose-pink, carried in sprays. Good as a cut flower. Continuous flowering. Introduced 1973. Average height and spread 5 × 5 ft (1.5 × 1.5 m). Protrudes up to 2 ft (60 cm) from support.

Average height and spread
Growth at ten years given for each variety.

ROSMARINUS

ROSEMARY
Labiatae   *Wall Shrub*
Evergreen
Although these shrubs are well-known as aromatic herbs, their potential for fan-training is not often appreciated.

Origin From southern Europe and Asia Minor.
Use As medium height, fan-shaped, scented foliage shrubs with attractive flowers for walls and fences or up pillars.
Description *Flower* Small mauve/blue flowers of varying shades produced in leaf axils, often in small clusters, on branches two years old or more. Pink and white flowering varieties are less hardy. Flowers late spring to early summer, and intermittently later in summer and sometimes in autumn. *Foliage* Leaves linear $\frac{1}{4}$–$1\frac{1}{4}$ in (5 mm–3.5 cm) long, grey to grey/green with white undersides. Some golden variegated forms. Aromatic when crushed. *Stem* Upright, becoming arching and spreading with age, grey/green. Medium growth rate when young, slowing with age. *Fruit* Insignificant.
Hardiness Tolerates a minimum winter temperature of 14°F (−10°C).
Soil requirements Light open soils, dislikes waterlogging and liable to chlorosis on extreme alkalinity.
Sun/Shade aspect Requires protection from the most severe weather. Full sun, but tolerates light shade.
Pruning Remove one third of old wood each spring to ground level plus any pruning required for training.
Training Requires individual wires or anchor points to achieve a fan shape.
Propagation and nursery production From semi-ripe cuttings taken in late spring to early summer. Purchase container grown, most varieties easy to find. Best planting height 8 in–2 ft (20–60 cm).
Problems Can become very woody, but pruning should control this.
Forms of interest *R. angustifolia* (Narrow-leaved rosemary) Very narrow foliage and mauve/blue flowers. Not easy to find. *R. a.* 'Corsican Blue' Narrow foliage, bright blue flowers. Difficult to find. *R. officinalis* (Common rosemary) Grey/green, aromatic foliage with white undersides. Pale to mid blue flowers produced in axillary clusters along wood two years old or more. One of the best forms for culinary use. *R. o.* 'Albus' A variety with white flowers. More tender than most. Not easy to find. *R. o.* 'Aurea Variegata' Foliage intermittently splashed pale

Rosmarinus angustifolia **in flower**

gold. Light blue flowers. *R. o.* 'Benenden Blue' Bright blue flowers and dark green, narrow foliage. Reaches two thirds average height and spread. *R. o.* 'Fastigiatus' (syn. *R. o.* 'Pyramidalis') Mid-blue flowers. An upright growing variety, forming an upright pillar. Often sold as 'Miss Jessop's Upright'. *R. o.* 'Roseus' Lilac-pink flowers, more tender than average. Reaches two thirds average height and spread. *R. o.* 'Severn Sea' Brilliant blue flowers. A low-growing variety, reaching one third average height and spread. *R. o.* 'Tuscan Blue' Good bright blue flowers, broader leaves than most. One third average height and spread.
Average height and spread
Five years
2½ × 3 ft (76 × 91 cm)
Ten years
4 × 5 ft (1.2 × 1.5 m)
Twenty years
4 × 5 ft (1.2 × 1.5 m)
Protrudes up to 4 ft (1.2 m) from support.

RUBUS FRUTICOSUS
(Blackberry Hybrids)

BLACKBERRY, BRAMBLE
Rosaceae   *Fruiting Canes*
Deciduous, Semi-evergreen or Evergreen
The wild blackberry and the many hybrids offer ornamental foliage, flowers and edible fruit.

Origin From Europe. Named hybrids of nursery and garden extraction.

Use As fan-trained canes for walls and fences and for the open garden when supported by posts and wires.
Description *Flower* Five-petalled, saucer-shaped flowers $\frac{3}{4}$ in (2 cm) wide in varying colours, from white to pink, carried in large, open clusters 4 in (10 cm) across during mid summer. *Foliage* Three- to five-lobed, 4 in (10 cm) wide and long, light to dark green, often with silver reverse. Some yellow autumn colour. *Stem* Either thorned or thornless, ranging, strong-growing. Light grey/green when young, quickly becoming red, particularly on sun side. Canes die after fruiting. *Fruit* Berries black to red/black in a range of sizes according to variety. Sweet, edible, fruiting in late summer to early autum. Good for freezing.
Hardiness Tolerates a minimum winter temperature of 0°F (−18°C).
Soil requirements Tolerates all soil conditions with no particular preference, often succeeding on the poorest of environments.
Sun/Shade aspect Tolerates all aspects. Full sun; will grow in light to medium shade, but fruiting may be diminished.
Pruning In first year of establishment, cut plants to ground level in spring. In subsequent years, remove all fruiting canes to ground level after fruiting, retaining new growth for following year's fruit.
Training Tie to wires or individual anchor points. Must be planted at least 10 ft (3 m) apart for training against a wall and if planted in rows in the open garden the rows must be at least 8 ft (2.4 m) apart.
Propagation and nursery production Layer tips of stems into garden soil in early summer and remove in autumn for planting in final position. Purchase bare-rooted from mid autumn to early spring or container grown as available, with autumn, winter or spring planting for preference. Named varieties normally available from good garden centres and general nurseries, but wild form may be very difficult to purchase.
Problems The thorns can make cultivation difficult. Its habit of tip layering itself and spreading can become invasive. Removing old fruiting canes each year can be laborious, but is worthwhile for fruit production.
Forms of interest
UK
'Bedfordshire Giant' Large fruit with good flavour. 'Himalayan Giant' Very large fruit. Possibly the most popular fruiting variety. One third more height and spread. 'Merton Thornless' Thornless stems and semi-evergreen foliage. Good fruit with good flavour. 'Oregon Thornless' ('Cut-leaved Bramble') Semi-evergreen to evergreen foliage, dark green and attractive. Good production of medium-sized fruit with good flavour. Two thirds average height and spread. **Wild varieties** are also good for fruiting; flowers, foliage and fruit are smaller.
USA
ERECT BLACKBERRIES Winter protection of canes, straw and burlap is required in Northern regions.
'Alfred' Large, firm, dark red berries. Good for the North. From Michigan. 'Bailey' Large, good quality fruit. Good productivity. Good for the North and parts of the Pacific Northwest. From New York. 'Black Satin' Thornless, semi-erect. Large crops of dark berries, good for eating directly from the cane or for cooking. Good for the South. Origin Maryland. 'Brainerd' Large fruit of good quality. Very productive, strong-growing canes. Hardy. Good for the South. From Georgia. 'Brazos' Large fruit carried over a long fruiting period. Strong-growing, resistant to disease. Good for the South. From Texas. 'Cherokee' Upright canes, moderately thorny. Large crops of good quality fruit of medium size. Good for the South. From Arkansas. 'Comanche' Large fruits, good for eating directly from the canes but can be cooked. Good for the South. From Arkansas. 'Darrow' Large, somewhat irregular in shape, ripening over a long period. Good flesh. Hardy but dislikes the coldest areas. From New York. 'Ebony King' Large, glossy,

Rosmarinus officinalis **in flower**

Rubus fruticosus 'Oregon Thornless'

BOYSENBERRY, JOHN INNES BERRY,
TAYBERRY, WORCESTER BERRY

Rosaceae   ***Fruit Canes***

Deciduous to semi-evergreen

A group of fruit canes where the parentage is very diverse, listed together here due to their shared general characteristics.

---

black, sweet and tangy fruit. Early ripening. Resistant to orange rust. Good for the South, North and Pacific Northwest. From Michigan. **'Eldorado'** Extremely hardy. Good production of good dark red fruits. Resistant to orange rust. Good for the South and North. From Ohio. **'Flint'** Requiring only moderate winter chill. Large fruit in clusters of up to 15. Resistant to leaf spot and anthracnose. Good for the South. From Georgia. **'Hendrick'** Large fruit, medium firm, with a sharp flavour. Very productive. Good for the North. From New York. **'Humble'** Soft fruit and a limited number of thorns. Good for the South. From Texas. **'Jerseyblack'** Large, good flavoured fruits. Strong-growing. Rust resistant. Good for the South. From New Jersey. **'Ranger'** Large, firm fruit, best eaten when really ripe. Good in Virginia. From

Rubus fruticosus 'Himalayan Giant'

Maryland. **'Raven'** Large fruit of very good quality. Very productive. Tender. From Maryland. **'Smoothstem'** Soft, late ripening fruits. Good production in large clusters. Thornless. Hardy. Best from Maryland southwards. From Maryland. **'Thornfree'** Good, medium to large, sharp flavoured fruit. Canes semi-erect up to 10 ft (3 m). Very large cropping. Tender. From Maryland. **'Williams'** Medium-sized fruit, ripening early summer. Good flavour. Strong growing, semi-erect habit with thorns. Resistant to most diseases. Good for the South. From North Carolina.

TRAILING BLACKBERRIES **'Aurora'** Large, firm, early fruit of good flavour. Canes reach up to 6 ft (1.8 m). From Oregon. **'Boysen'** (**'Nectar'**) Large, good flavoured, scented fruit produced over a long season. Strong-growing. Some thorns. In some Southern states can have a first crop in late spring/early summer and a second crop in late summer. Good for the South and Pacific Northwest. From California. **'Carolina'** A good strong-growing cane with large fruits. A dewberry. Resistant to leaf spot diseases. Good for the South. Origin North Carolina. **'Cascade'** Very good flavour. Good production but canes are tender. Good in milder parts of the Pacific Northwest. From Oregon. **'Early June'** Large round fruits of very good flavour. Good for jam, jelly and pies. A dewberry ripening in early summer. Limited number of thorns. Partially resistant to anthracnose and leaf spot. Good for the South. From Georgia. **'Floragrand'** Large, soft, rather tart fruit. Good for cooking and preserving. Early fruiting. Canes remain over winter. A dewberry. Requires 'Oklawaha' for pollination. Good for the South. From Florida. **'Himalayan Berry'** Very large fruits. One third more height and spread than average. Good over a wide area. From the UK. **'Lavaca'** Hardy, resistant to disease. Good fruits with acid flavour. Good for the South. Origin unknown. **'Lucretia'** Hardy, strong-growing, well-established variety. Heavily productive with early ripening, large, soft fruits. Will require protection in winter in the North. A dewberry. From North Carolina. **'Marion'** Medium to large, very good flavoured fruit. Mid season variety. Limited number of canes but extremely long, often up to 20 ft (6 m) or more. Very thorny. Requires mild areas for best results, particularly good in Pacific Northwest. From Oregon. **'Oklawaha'** Normally used as a pollination variety for 'Floragrand' but has a good fruit in its own right. Good for the South. **'Olallie'** Large, shiny black fruit of very high quality, sweet and good flavoured. The canes have thorns and are very productive. Resists verticillium wilt and mildew. Good for Southern California. From Oregon. **'Thornless Evergreen'** See 'Oregon Thornless' in UK varieties for full description. **'Young'** Large purple/black fruits of good flavour. A black dewberry making limited long canes. Anthracnose is a serious problem. Good for the South. From Louisiana.

Average height and spread

One year

8 × 8 ft (2.4 × 2.4 m)

Protrudes up to 5 ft (1.5 m) from support prior to training, 2 ft (60 cm) after training.

Origin Various crosses between recognized soft fruit canes.

Use Fan-trained for walls and fences, or for growing over archways and similar constructions.

Description *Flower* Attractive white or pink flowers, up to $\frac{1}{2}$ in (1 cm) across, saucer-shaped, carried in short clusters in mid to late spring. *Foliage* Hand-shaped, dark green with silver reverse, 4 in (10 cm) wide and 5 in (12 cm) long. Some good yellow autumn colour. Deciduous or semi-evergreen according to variety. *Stem* Light green, becoming shaded red on sun side, finally red/brown. Mostly armed with thorns, but some varieties thornless. *Fruit* Either round or oblong fruits, normally dark red to purple red in colour. Sweet and edible.

Hardiness Tolerates a minimum winter temperature of 4°F (−15°C).

Rubus 'Tayberry'

Soil requirements Tolerates all soils, except for extremely waterlogged or dry conditions.

Sun/Shade aspect Tolerates all aspects, but best fruiting in sheltered, sunny positions where pollination and ripening can benefit.

Pruning In first spring following planting, cut all existing canes to ground level. This will induce a vigorous root development and build up a root stool from which will develop strong canes for the following year's fruiting. In subsequent years remove old fruiting canes to ground level after fruiting and tie in new canes to replace them.

Training Tie in new canes to a fan-shape formation as they develop through spring and summer.

Propagation and nursery production From stem top layers. Can be purchased bare-rooted from mid autumn to early spring or container grown as available. Some forms may have to be sought from specialist nurseries. Look for good root formation; stem size is irrelevant as it has to be removed in spring after planting.

Problems Thorny stems can make cultivation difficult. The need to cut hard back and lose the fruit in the first year is a drawback.

Forms of interest *Boysenberry* Large, dark red fruits, ageing to black. Good sharp distinct flavour. *Thornless boysenberry* A thornless variety with lighter stems. Possibly less fruit production. *John Innes berry* Similar to a blackberry, sweetly flavoured. *Tayberry* A hybrid between a blackberry and a raspberry with large, deep purple, good flavoured fruit. Good for freezing. *Worcester berry* A cross between a gooseberry and a blackcurrant. Berries red/purple, sweet flavoured. Resistant to most diseases, such as mildew. Two thirds average height and spread.

Average height and spread
One year
10 × 10 ft (3 × 3 m)
New canes protrude
up to 3 ft (91 cm) from
wall or fence, 12 in
(30 cm) once tied in.

RUBUS HENRYI BAMBUSARUM
(R. bambusarum)

KNOWN BY BOTANICAL NAME
Rosaceae   *Wall Shrub*
Evergreen
A somewhat unruly bramble but worthy of interest for its attractive foliage.

Origin From China.
Use As a rambling climber through other large wall shrubs. Normally not seen trained in its own right.
Description *Flower* Racemes 4 in (10 cm) long consisting of six to ten ¾ in (2 cm) wide pink sepals resembling long, thin tails, early summer. *Foliage* Attractive, three-lobed, up to 6 in (15 cm) wide and long. Light green when young, becoming mid to dark green. White felted undersides. *Stem* Light green becoming dark green, finally green/brown.

Rubus henryii bambusarum

Ranging and arching in habit. Stems die after flowering. Medium to fast annual growth rate. *Fruit* Round, black and shiny, of some limited interest.
Hardiness Tolerates a minimum winter temperature of 14°F (−10°C).
Soil requirements Requires a soil high in organic material and nutrients. Neutral to acid, although as long as adequate moisture is available, may tolerate degrees of alkalinity.
Sun/Shade aspect Requires a sheltered aspect. Full sun to medium shade with light to medium shade for preference.
Pruning Remove all flowering stems to ground level in winter.
Training Normally allowed to ramble through freestanding shrubs or through other wall specimens, although can be tied to wires or individual anchor points.
Propagation and nursery production From tip cuttings or semi-ripe cuttings taken in early summer. Should always be purchased container grown; will have to be sought from specialist nurseries. Best planting height: from root clumps to 3 ft (91 cm).

Problems Rarely looks attractive as a young plant and requires time to establish. Can become invasive in ideal situations.
Forms of interest None.
Average height and spread
Five years
6 × 6 ft (1.8 × 1.8 m)
Ten years
10 × 10 ft (3 × 3 m)
Twenty years
10 × 10 ft (3 × 3 m)
Protrudes up to 4 ft (1.2 m) from support.

RUBUS LOGANOBACCUS
(Loganberry Hybrids)

LOGANBERRY
Rosaceae   *Fruiting Canes*
Deciduous
A cane fruit that responds well to fan-training, but its robustness of growth must be taken into account.

Origin Raised in the USA in the late 1800s.
Use As a large, ranging, loosely fan-trained cane fruit for walls and fences. Ideal for growing in the open garden with the support of stakes and wires.
Description *Flower* White, ½ in (1 cm) wide, five-petalled flowers carried in clusters in early summer; each cluster up to 5 in (12 cm) in width. *Foliage* Hand-shaped, attractive, light grey/green with silvery reverse. Often up to 8 in (20 cm) in width and length. Some yellow autumn colouring. *Stem* Both thorned and thornless varieties. Stems grey/green when young, quickly becoming shaded red, ageing to red/green, finally grey/brown. Upright at first, quickly becoming spreading, ranging, vigorous. After fruiting, canes die and must be removed. Fast rate of growth. *Fruit* Cone-shaped berries, red with grey sheen. Sweet, edible. Produced late summer/early autumn. Good for freezing.
Hardiness Tolerates a minimum winter temperature of 0°F (−18°C).
Soil requirements Tolerates all but the most alkaline. Dislikes excess waterlogging. Must contain a high degree of organic material and adequate plant nutrients both when planting and as an annual addition.
Sun/Shade aspect Tolerates all aspects, but adequate windbreaks should be provided so that pollinating insects are able to reach flowers. Full sun for fruit, but will tolerate very light shade although with decrease in fruiting.
Pruning In year of planting reduce all shoots to ground level in spring to induce the formation of root stool and strong shoots for fruiting in following year. Once established,

remove all fruiting canes to ground level after fruiting, tying in new canes into fan-trained shape.
Training Requires wires or individual anchor points. Tie in annually produced canes alternately to the left and right where they will flower and fruit the following year. Alternatively, gather all new canes loosely in the centre then retie to form the fan-trained shape when the old fruiting canes have been removed. Must be planted at least 12 ft (3.7 m) apart on any wall, fence or wire structure, and if more than one row is planted on wires rows must be at least 8 ft (2.4 m) apart.
Propagation and nursery production From self-rooted tips by inserting into the soil in mid summer and removing in following year for replanting in new positions. Purchase bare-rooted from mid autumn to early spring, container grown as available, with autumn, winter and spring for preference. Normally available from garden centres and general nurseries. Size is irrelevant and a good root system is the predominant factor.
Problems Suffers attacks of raspberry cane spot and mildew, otherwise is one of the most reliable amongst all of the fruiting canes.
Forms of interest 'LY59' The best of the thorned varieties, giving the biggest crop and best flavour. 'Loganberry Thornless' Canes without thorns. Fruiting good but crop may be slightly decreased in quantity.
Average height and spread
After one year
8 × 8 ft (2.4 × 2.4 m)
Protrudes up to
6 ft (1.8 m) from
support prior to training,
2 ft (60 cm) after training.

RUBUS PHOENICOLASIUS

JAPANESE WINEBERRY
Rosaceae   *Woody Fruiting Climber*
Deciduous
An edible, fruiting shrub which can make an attractive display in the garden.

Origin From Japan.
Use As a fan-trained wall or fence shrub to show off its flowers, fruit, foliage and stems, all of which are of attraction. Can also be grown freestanding.
Description *Flower* Clusters up to 3 in (7.5 cm) across and 4 in (10 cm) long of single pink flowers produced in early summer. *Foliage* Trifoliate leaves, 6 in (15 cm) long and 4 in (10 cm) wide, with five leaflets, each 1¾ in (4.5 cm) long and 1 in (2.5 cm) wide. Mid green with good yellow autumn colour. Felted

Rubus loganobaccus 'LY59'

Rubus phoenicolasius

texture with silver undersides. *Stem* Long, arching, orange/red in colour. Extremely attractive in winter. Downy, with soft thorns. Fast rate of growth. *Fruit* Small clusters, up to 6 in (15 cm) long and 4 in (10 cm) wide, of round, bright red, sweet, edible fruits in late summer, early autumn.
Hardiness Tolerates a minimum winter temperature of 0°F (−18°C). At this temperature some stem damage may be caused but is not terminal.
Soil requirements Most soils suitable.
Sun/Shade aspect Requires only light protection. Prefers light shade, tolerates full sun to deep shade.
Pruning Cut all old fruit canes to ground level after fruiting. Retain new shoots for next year's display and fruit. Young plants should be cut severely prior to planting to induce a good formation of root stool. This normally leads to no production of fruit in the subsequent year.
Training Tie to wires or individual anchor points.
Propagation and nursery production From semi-ripe cuttings taken in early summer. Purchase container grown from good nurseries and some garden centres. Best planting height 15 in–2½ ft (38–76 cm).
Problems Often requires more room than is anticipated. May be difficult to find.
Similar forms of interest None.
Average height and spread
Five years
3 × 3 ft (91 × 91 cm)
Ten years
6 × 6 ft (1.8 × 1.8 m)
Twenty years
10 × 10 ft (3 × 3 m)
Protrudes up to 2 ft (60 cm) from support if fan-trained, 10 ft (3 m) untrained.

RUBUS TRICOLOR

CREEPING BRAMBLE

Rosaceae　　　　　　　　*Wall Shrub*
Evergreen but may become deciduous in severe winters

A low-growing, carpeting shrub that can be easily adapted to a low climber for more inhospitable areas.

Origin From China.
Use As a low climber for walls, fences or pillars, or to ramble through medium-sized shrubs where many other climbers will not succeed.
Description *Flower* Single white flowers, ½ in (1 cm) wide, produced in early summer. *Foliage* Leaves oval, up to 4 in (10 cm) long

and 1½ in (4 cm) wide, with pronounced veins. Dark green with a silver sheen and silver reverse. *Stem* Arching, orange/red to orange/brown, covered with soft brown to grey/brown hairy thorns. Tips and buds root on contact with soil. Slow to medium growth rate. *Fruit* Round berries, up to ¼ in (5 mm) in diameter, bright red, edible, but with a bland taste.

Rubus tricolor

Hardiness Tolerates a minimum winter temperature of 0°F (−18°C).
Soil requirements Does well on all soil types with no particular preference, tolerating poor conditions if adequate moisture is available.
Sun/Shade aspect All aspects in full sun to deep shade, although deep shade may make it more lax in its habit.
Pruning Remove old fruiting wood after fruiting to encourage new growth.
Training Allow to ramble through shrubs or over wires.
Propagation and nursery production From layers or from semi-ripe cuttings taken in early summer. Purchase container grown; may have to be sought from specialist ground cover nurseries or from good garden centres. Best planting height 6–18 in (15–45 cm).
Problems May be shy to flower, particularly when in deep shade.
Similar forms of interest *R. idaeus* 'Aureus' A golden form of the raspberry, annually producing new golden shoots with white flowers. Red and yellow edible fruits. *R. microphyllus* 'Variegatus' Green foliage splashed cream and pink. Flowers magenta-rose in colour. Fragrant. Edible orange fruits.
Average height and spread
Five years
4 × 4 ft (1.2 × 1.2 m)
Ten years
6 × 6 ft (1.8 × 1.8 m)
Twenty years
6 × 6 ft (1.8 × 1.8 m)
Protrudes up to 18 in (45 m) from support.

RUBUS ULMIFOLIUS 'BELLIDIFLORUS' (Rusticanus bellidiflorus)

DOUBLE FLOWERING BLACKBERRY

Rosaceae　　　　　　　　*Wall Shrub*
Deciduous

A gem of a climber, tolerating the worst degrees of cultivation, yet still producing a flowering display.

Origin From Europe.
Use As a climber for fences and walls and for growing through large shrubs. Can be allowed to grow untrained, but requires some support.
Description *Flower* Attractive, mauve/pink, cylinder-shaped panicles, up to 12 in (30 cm) long and 5 in (12 cm) wide, consisting of numerous button-like, lacerated petals. Produced in mid to late summer. *Foliage* Three to five leaflets make up a pinnate leaf up to 3½ in (9 cm) long and 2 in (5 cm) wide, with white felted undersides and dark green upper. *Stem* Dark green, angular, thorny, ranging in habit, although not invasive. Flowering in second year. *Fruit* Small, black, inedible round berries produced in late summer to early autumn.
Hardiness Tolerates a minimum winter temperature of 0°F (−18°C).
Soil requirements Does well on all soil types with no particular preference, tolerating quite poor conditions as long as adequate moisture is available.
Sun/Shade aspect Tolerates all aspects with no particular preference. Full sun to deep shade, but in deep shade may be more lax in habit.
Pruning Remove all flowering shoots to ground level once flowering is finished, encouraging new shoots for following year.
Training Allow to grow freestanding against a wall or fan-train to wires or individual anchor points.

Rubus ulmifolius 'Bellidiflorus' in flower

Propagation and nursery production From layers or softwood cuttings taken in early summer. Purchase container grown; will have to be sought from specialist nurseries. Best planting height 6 in–3 ft (15–91 cm).
Problems Apart from its limited availability it has no problems.
Similar forms of interest None.
Average height and spread
Five years
5 × 5 ft (1.5 × 1.5 m)
Ten years
7 × 7 ft (2.1 × 2.1 m)
Twenty years
12 × 12 ft (3.7 × 3.7 m)
Protrudes up to 2 ft (60 cm) from support if fan-trained, 10 ft (3 m) untrained.

Schisandra grandiflora rubrifolia **in flower**

SCHISANDRA GRANDIFLORA RUBRIFOLIA (S. rubrifolia)

KNOWN BY BOTANICAL NAME

Magnoliaceae, syn.   *Woody Climber*
Schisandraceae
Deciduous
An attractive climber which is not planted as often as it deserves.

Origin From China.
Use As a climber for walls, fences, poles and pergolas for its interesting flowers.
Description *Flower* Borne singly at the leaf joints. Pendulous, hanging on stalks 1–2 in (2.5–5 cm) long. Seven petals, deep crimson in colour. Mid to late spring. *Foliage* Ovate to lanceolate, 5½ in (14 cm) long and 2 in (5 cm) wide. Deep green with attractive autumn colour. *Stem* Green/brown, ageing to grey/brown, finally dark brown. Moderately twining. Slow to medium growth rate. *Fruit* Flower spikes give way to fruit spikes, with globe-shaped, round, red fruits borne in late summer/early autumn, although not always reliable.
Hardiness Tolerates a minimum winter temperature of 14°F (–10°C).
Soil requirements Does equally well on both moderately alkaline and acid soil, as long as adequate moisture is available to achieve good growth.
Sun/Shade aspect Tolerates all but the most exposed aspects. Best growth and flowers in light shade, but will tolerate full sun as long as roots have adequate shade and moisture.
Pruning Normally requires no pruning, but can be reduced in size in early spring if necessary.
Training Shoots semi-twining over wires, trellis or similar framework, but may need additional support by tying in when young or heavy with age.
Propagation and nursery production From seed sown in spring or semi-ripe cuttings taken in early to mid summer. In both instances propagation is not easy. Always purchase container grown; will have to be sought from specialist nurseries. Best planting height 1–3 ft (30–91 cm).
Problems Limited availability. Takes up to three to five years to reach any size and to give a good flowering display.
Similar forms of interest None.
Average height and spread
Five years
5 × 5 ft (1.5 × 1.5 m)
Ten years
10 × 10 ft (3 × 3 m)
Twenty years
18 × 18 ft (5.5 × 5.5 m)
Protrudes up to 3 ft (91 cm) from support.

SCHIZOPHRAGMA HYDRANGEOIDES

KNOWN BY BOTANICAL NAME

Hydrangeaceae   *Woody Climber*
Deciduous
This attractive climber that is closely related to *Hydrangea petiolaris* (climbing hydrangea) is worthy of inclusion in the garden, but its availability and the time it takes to mature must be considered.

Origin From Japan.
Use As an attractive climber for large walls and fences or to climb through large trees, over stumps or medium to large shrubs.
Description *Flower* Small yellow-white flowers ringed by 1½ in (4 cm) long white bracts make up a flat, lace-capped flower up to 10–12 in (25–30 cm) in diameter in late summer. Bracts age to brown and are retained until Christmas. *Foliage* Broad, oval-shaped leaves with coarsely toothed edges, up to 3–5 in (7.5–12 cm) long, supported on stalks 1–2 in (2.5–5 cm) long. Deep green above, glaucous pale green beneath. *Stem* Green/brown ageing to grey/brown, branching. Slow growing when young, faster with age. *Fruit* No fruit of interest.
Hardiness Tolerates a minimum winter temperature of 4°F (–15°C).
Soil requirements Does well on all soil types, only disliking extremely alkaline conditions.
Sun/Shade aspect Tolerates all but the most severe of aspects, benefiting from a small amount of protection. Best in full sun, but will tolerate very light shade.
Pruning Normally needs no pruning, but can be cut back if required in early spring.
Training Tie to wires, trellis or individual fixings to form a broad, fan-shaped formation.
Propagation and nursery production From semi-ripe cuttings taken in late summer. Can be layered if parent plant has young enough shoots. Propagation at times can be difficult, making it scarce. Should always be purchased container grown; may have to be sought from specialist nurseries. Best planting height 6 in–2 ft (15–60 cm).
Problems Slow to establish and not readily available but, once established, a true gem of a plant.
Similar forms of interest *S. h.* 'Roseum' A variety with pink bracts. Very scarce in commercial production. *S. integrifolia* Larger leaves and flower heads than *S. hydrangeoides*. Scarce in production. From China.
Average height and spread
Five years
5 × 5 ft (1.5 × 1.5 m)
Ten years
10 × 10 ft (3 × 3 m)
Twenty years
20 × 20 ft (6 × 6 m)
Protrudes up to 2 ft (60 cm) from support.

SOLANUM CRISPUM

CLIMBING POTATO, CHILEAN POTATO TREE
Solanaceae   *Wall Shrub*
Deciduous, semi-evergreen
One of the most attractive of all wall shrubs, requiring space to develop to its full potential but, if allowed, making a splendid display.

Origin From Chile.
Use As a freestanding wall shrub for walls and fences.
Description *Flower* Five purple/blue, fleshy petals, surmounted in the centre by an attractive bright yellow bunch of anthers, make up star-shaped flowers 1 in (2.5 cm) across, which are borne in 3–6 in (7.5–15 cm) wide clusters from June to September. *Foliage* Narrowly ovate leaves dark green above, paler beneath. Some yellow autumn colour. *Stem* Bright glossy green, upright, becoming branching. Medium to fast growing. *Fruit* Yellow/white, pea-sized, poisonous fruits.

Schizophragma hydrangeoides **in flower**

Hardiness Tolerates a minimum winter temperature of 14°F (−10°C).
Soil requirements Does well on most soils, except in very dry or poor conditions. Dislikes high degrees of alkalinity.
Sun/Shade aspect Requires a sheltered aspect in full sun to light shade.
Pruning Remove one third of oldest growth to ground level in early spring.
Training Tie to wires or trellis against walls or fences to obtain a fan-trained shape.

Solanum crispum in flower

Propagation and nursery production From semi-ripe cuttings taken in early summer. Always purchase container grown; best planting height 12 in–2 ft (30–60 cm). Relatively easy to find in garden centres and nurseries, particularly when in flower.
Similar forms of interest *S. c.* 'Glasnevin' Larger flower clusters of slightly darker blue.
Average height and spread
Five years
4 × 4 ft (1.2 × 1.2 m)
Ten years
6 × 6 ft (1.8 × 1.8 m)
Twenty years
12 × 12 ft (3.7 × 3.7 m)
Protrudes up to 4 ft (1.2 m) from support.

SOLANUM DULCAMARA 'VARIEGATUM'

VARIEGATED BITTERSWEET, VARIEGATED FELLONWOOD, VARIEGATED WOODY NIGHTSHADE
Solanaceae   *Perennial Climber*
Deciduous

An attractive variegated perennial climber which does not form a woody structure. Produces poisonous fruits.

Origin From Europe, Asia and Africa.
Use As an attractive foliage climber for covering walls, fences and pillars. Can be grown through large, uninteresting shrubs.
Description *Flower* Five-lobed petals, purple to purple/blue, each flower ½ in (1 cm) across, carried in clusters up to 4 in (10 cm) wide and long. *Foliage* Oval with three to five leaflets, each up to 3 in (7.5 cm) long, with narrow points. Olive/green, splashed with creamy-white variegation. *Stem* Mid green, semi-twining, spreading and lax in habit. Medium growth rate. *Fruit* Small, round, red, sometimes yellow, up to ½ in (1 cm) across, poisonous, produced in late summer/early autumn.
Hardiness Tolerates a minimum winter temperature of 4°F (−15°C). All top growth is killed annually to ground level and rejuvenates from root stool.
Soil requirements Does well on all soil conditions, with no particular preference.
Sun/Shade aspect Tolerates all aspects in full sun to light shade.

Solanum dulcamara 'Variegatum' in flower

Pruning Requires no pruning other than removal of dead growth above ground level in autumn or early spring.
Training Allow to ramble through wires, netting, trelliswork or over large uninteresting shrubs.
Propagation and nursery production From seed or semi-ripe cuttings taken in early summer. Always purchase container grown; best planting height 6 in–3 ft (15–91 cm). Normally available from good garden centres and nurseries.
Problems Fruits are poisonous, making it an inadvisable plant where there are young children.
Similar forms of interest None.
Average height and spread
Five years
8 × 8 ft (2.4 × 2.4 m)
Ten years
8 × 8 ft (2.4 × 2.4 m)
Twenty years
8 × 8 ft (2.4 × 2.4 m)
Protrudes up to 4 ft (1.2 m) from support.

SOLANUM JASMINOIDES

POTATO VINE, JASMINE NIGHTSHADE
Solanaceae   *Woody Climber*
Deciduous to semi-evergreen

Possibly too tender for most gardens, but if a sheltered position can be found it is well worth consideration as it is one of the most fragrant of all climbers.

Origin From South America.
Use As a climber for walls, fences or pillars in sheltered environments, or for greenhouses or conservatories.

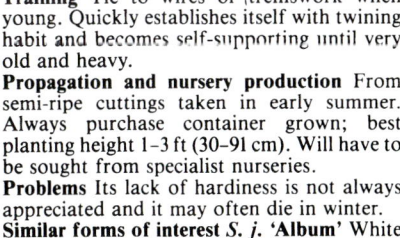

Solanum jasminoides 'Album' in flower

Description *Flower* Blue/white to pale blue, each flower up to ¾ in (2 cm) across, consisting of five thick, fleshy petals with golden-yellow anthers. Fragrant. Flowers produced in branched clusters. *Foliage* Three leaflets, each oval, terminal leaflet up to 2 in (5 cm) long. Pair of lower lateral leaflets 1½ in (4 cm) long and wide. Grey/green, supported on short leaf shoot. *Stem* Dark grey/green, twining, not self-clinging. Medium rate of growth. *Fruit* No fruit of interest.
Hardiness Tolerates a minimum winter temperature of 23°F (−5°C).
Soil requirements Tolerates all soil conditions, except extremely dry.
Sun/Shade aspect Must have a sheltered aspect. Fully sun to light shade.

Solanum jasminoides in flower

Pruning Remove one third of oldest growth to ground level in early spring to induce rejuvenation.
Training Tie to wires or trelliswork when young. Quickly establishes itself with twining habit and becomes self-supporting until very old and heavy.
Propagation and nursery production From semi-ripe cuttings taken in early summer. Always purchase container grown; best planting height 1–3 ft (30–91 cm). Will have to be sought from specialist nurseries.
Problems Its lack of hardiness is not always appreciated and it may often die in winter.
Similar forms of interest *S. j.* 'Album' White flowers, more fragrant than the parent plant. Slightly smaller in height and spread.
Average height and spread
Five years
6 × 6 ft (1.8 × 1.8 m)
Ten years
12 × 12 ft (3.7 × 3.7 m)
Twenty years
18 × 18 ft (5.5 × 5.5 m)
Protrudes up to 2 ft (60 cm) from support.

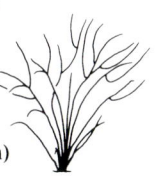

SOLLYA FUSIFORMIS

AUSTRALIAN BLUEBELL CREEPER

Pittosporaceae   *Tender Woody Climber*
Evergreen

An attractive climbing plant requiring protection in all but the most sheltered of areas.

Origin From Australia.
Use As a self-clinging climber for very sheltered walls and fences or for growing under the protection of a conservatory or greenhouse.
Description *Flower* Pendent clusters of four to 12 blue flowers each approximately $\frac{1}{2}$ in (1 cm) across, consisting of five small pointed petals. *Foliage* Ovate to lanceolate, pointed, up to $2\frac{1}{2}$ in (6 cm) long and $\frac{3}{4}$ in (2 cm) wide. Light bright green when young, becoming darker with age. *Stem* Light green becoming dark green. Glossy, self-supporting by twining habit. Medium growth rate. *Fruit* No fruit of interest.

Sollya fusiformis in flower

Hardiness Requires a very sheltered sunny position or the protection of a greenhouse or conservatory. Tolerates a minimum winter temperature of 23°F (−5°C) but only in very sheltered, favourable areas.
Soil requirements Requires a soil high in organic material. Resents drying out.
Sun/Shade aspect Needs a sheltered aspect. Semi-shade for best results but will tolerate full sun.
Pruning Normally requires no pruning but can be reduced in size if necessary and will quickly rejuvenate.
Training Allow to twine through trellis or over a wire support.
Propagation and nursery production From semi-ripe cuttings, rooted into sharp sand, under the protection of a greenhouse. Should always be purchased container grown; may have to be sought from specialist nurseries, although sometimes available in good garden centres. Best size to purchase 6 in–3 ft (15–91 cm).
Problems Its lack of hardiness is often underestimated and attempts are made to grow it in areas where it cannot tolerate the low winter temperatures.
Similar forms of interest None.
Average height and spread
Five years
6 × 6 ft (1.8 × 1.8 m)
Ten years
9 × 9 ft (2.7 × 2.7 m)
Twenty years
15 × 15 ft (4.6 × 4.6 m)
Protrudes up to 2 ft (60 cm) from support.

Sophora tetraptera in flower

SOPHORA TETRAPTERA

NEW ZEALAND LABURNUM, KOWHAI

Leguminosae   *Wall Shrub*
Evergreen

An interesting large shrub or small tree for foliage and flowers in mild locations provided the protection of a wall or fence is available.

Origin From New Zealand.
Use As an attractive foliage wall shrub for walls and fences in very mild areas, or for growing against walls in conservatories or large greenhouses.
Description *Flower* Small, yellow tubular flowers $1\frac{1}{2}$–2 in (4–5 cm) long in pendent clusters in late spring. *Foliage* Pinnate leaves consisting of numerous pairs of small, ovate to oblong leaflets, light green to green/brown. *Stem* Upright, becoming attractively branched, slightly drooping. Light green/brown to brown. Slow to medium growth rate. *Fruit* Unusual seed pods looking like beads, each with four broad wings, late summer to early autumn.
Hardiness Tolerates a minimum winter temperature of 23°F (−5°C).
Soil requirements Light, open, well-drained soil, alkaline or acid.
Sun/Shade aspect Requires a very sheltered, warm aspect in full sun.
Pruning None required.
Training Requires wires or individual anchor points to achieve a good trained shape.
Propagation and nursery production From seed or semi-ripe cuttings taken in early summer. Purchase container grown; may need to be obtained from specialist nurseries. Best planting height 15 in–$2\frac{1}{2}$ ft (38–76 cm).
Problems Often purchased in mild areas for transplanting to colder regions where it rarely succeeds.
Similar forms of interest *S. t.* ‘Early Gold’ Fern-like foliage and lemon-yellow flowers. *S. microphylla* A very similar form to *S. tetraptera* but with smaller leaflets and flowers. Juvenile foliage may be dense and wiry. *S. m.* ‘Early Gold’ Yellow flowers and rich green leaves. Upright habit.
Average height and spread
Five years
3 × 3 ft (91 × 91 cm)
Ten years
6 × 6 ft (1.8 × 1.8 m)
Twenty years
10 × 10 ft (3 × 3 m)
Protrudes up to 3 ft (91 cm) from support.

SORBUS ARIA

WHITEBEAM

Rosaceae   *Hardy Tree*
Deciduous

A tree which is easy to grow in most soils and situations and offers attractive berries and autumn foliage.

Sorbus aria ‘Lutescens’ in leaf

Origin From Europe.
Use As a fan-trained tree on large walls or on open wire fences.
Description *Flower* Clusters 4 in (10 cm) across of white fluffy flowers in late spring. *Foliage* Ovate, toothed leaves, up to 4 in (10 cm) long with downy white undersides and dark green to grey/green upper surfaces. Good russet and gold autumn colour. *Stem* Grey/green to grey/brown with downy covering. Strong, upright and fast-growing when young. Responds well to fan-training. Medium to fast growth rate. *Fruit* Clusters of scarlet-red, oval to globe-shaped fruits 4 in (10 cm) across in autumn.
Hardiness Tolerates a minimum winter temperature of 4°F (−15°C).
Soil requirements Any soil conditions; particularly tolerant of alkalinity.

Sun/Shade aspect Tolerates all aspects. Best in full sun to maintain colouring.
Pruning Prune young trees hard in the spring following planting. Select and train resulting five to seven shoots and tie into a fan-trained shape. In subsequent years, remove all side growths back to two points from their origin and maintain original main branches in fan shape.
Training Secure to wires or individual anchor points.
Propagation and nursery production Parent plant from seed; named varieties grafted or budded on to understocks of *S. aria*. Available bare-rooted or container grown. Select young trees of not more than two years old and 3–6 ft (91 cm–1.8 m) in height for fan-training. Usually available from specialist nurseries and garden centres.
Problems Young trees suitable for fan-training may be hard to find.
Forms of interest *S. a.* **'Chrysophylla'** Leaves predominantly primrose yellow through summer with some silver undersides, turning buttercup yellow in autumn. White flowers. Orange/red fruits. *S. a.* **'Lutescens'** Best of all silver whitebeams. Pale green foliage with silver down, white flowers and orange berries. Most adaptable variety for fan-training. *S. a.* **'Majestica'** Large, round to ovate foliage 6 in (15 cm) long, grey upper surface with white, downy undersides. Orange fruits in autumn. *S. bristoliensis* Small, ovate, silver leaves. Little clusters of white flowers and orange/red berries. Not readily available but worth researching for a planting collection. Two thirds average height and spread. Originating in the UK. *S. folgneri* Slender growths with lanceolate to narrowly ovate, finely tapering leaves up to 4 in (10 cm) long. White felted undersides and dark grey upper surfaces. Good yellow autumn colour. Clusters of white flowers 4 in (10 cm) across, followed by red, oval fruits. Half average height and spread. Not easy to find and will have to be sought from specialist nurseries. From China. *S. hybrida* **'Gibbsii'** Dark grey leaves with silver undersides, lobed with three to five indentations. Large clusters of white flowers up to 5 in (12 cm) across, followed by orange/red berries in autumn. Two thirds average height and spread. Not readily available and must be sought from specialist nurseries. *S. intermedia* (Swedish whitebeam) Lobed leaves up to 4 in (10 cm) long, with steel-grey upper surfaces and white undersides. White flowers and orange/red fruits. *S. latifolia* (Service tree of Fontainbleau) Large, silver/grey foliage with white, downy undersides, round to ovate, 4 in (10 cm) long or more, toothed edges. White flower clusters up to 3 in (7.5 cm) across, followed by globe-shaped, brown/red fruits. From Europe. *S.*

'Mitchellii' Among the largest of whitebeam foliage forms. Leaves up to 8 in (20 cm) long and 6 in (15 cm) wide, grey/green to silver/grey upper surfaces with white, downy, felted undersides. Good yellow autumn colour. White flower clusters up to 4 in (10 cm) across, followed by red fruits in autumn, more sparse than in most varieties. Stems brown, solid, stout. Two thirds average height and spread. Not readily available but found in specialist nurseries.
Average height and spread
Five years
10 × 10 ft (3 × 3 m)
Ten years
20 × 20 ft (6 × 6 m)
Twenty years
30 × 30 ft (9 × 9 m)
Protrudes up to 4 ft (1.2 m) from support.

STACHYURUS PRAECOX

KNOWN BY BOTANICAL NAME
Stachyuraceae   *Wall Shrub*
Deciduous
A winter-flowering shrub which deserves to be more widely planted and is not always considered for its use as a wall shrub.

Origin From Japan.
Use As a fan-trained wall shrub for its late winter and early spring flowers.
Description *Flower* Numerous pendent racemes, 1½–4 in (4–10 cm) long, consisting of over 20 pale yellow, cup-shaped flowers, borne along branches from late winter through to early spring. Exact flowering time is dependent on winter temperatures. Flowers produced on wood two years old or more. Frost resistant. *Foliage* Leaves large, ovate, pointed, 3–6 in (7.5–15 cm) long and 2–3 in (5–7.5 cm) wide. Dark green tinged purple, with purple veining. Some yellow autumn colour. *Stem* Strong, upright, arching when young, becoming more branching in subsequent years. Purple/green. When fan-trained forms a large, interesting shrub. Medium to fast growth rate. *Fruit* Insignificant.
Hardiness Tolerates a minimum winter temperature of 0°F (−18°C).
Soil requirements Does well in most soils.
Sun/Shade aspect Requires some protection from cold winter winds to maintain flowers over a long period. Full sun to mid shade.
Pruning On established shrubs five years old or more, remove one third of old flowering wood to ground level following flowering to encourage rejuvenation.
Training Will require wires or individual

anchor points to achieve a fan-trained shape.
Propagation and nursery production From layers or semi-ripe cuttings taken early to mid summer. Purchase container grown; fairly easily found in nurseries. Best planting height 2–4ft (60 cm–1.2 m).
Problems Young plants often appear lopsided, with one or more strong shoots, but grow quickly once planted.
Similar forms of interest *S. chinensis* Foliage slightly narrower and possibly duller green than that of *S. praecox*. Racemes shorter but comprising more numerous small florets. Flowers dependent on weather, but can be slightly earlier. *S. c.* **'Magpie'** White to creamy white margins on leaves. Often splashed with pale green mottling and tinged rose pink. Yellow flowers. Extremely scarce.
Average height and spread
Five years
8 × 8 ft (2.4 × 2.4 m)
Ten years
12 × 12 ft (3.7 × 3.7 m)
Twenty years
15 × 15 ft (4.6 × 4.6 m)
Protrudes up to 3 ft (91 cm) from support.

STAPHYLEA COLCHICA

BLADDERNUT, COLCHIS BLADDERNUT
Staphyleaceae   *Wall Shrub*
Deciduous
One of the aristocrats of flowering shrubs but not always thought of for its use as a large wall shrub, where it can show off its flowers and interesting fruits to the best advantage.

Staphylea colchica **in flower**

Origin From southern Caucasus.
Use As a very large, fan-trained wall shrub for its flowers and fruit.
Description *Flower* Small, white, fragrant flowers ½ in (1 cm) wide with five reflexed outer petals and five central petals forming a loose trumpet shape. Borne in panicles over 5 in (12 cm) long in mid spring. *Foliage* Leaves ovate to oblong, composed of three to five leaflets up to 4 in (10 cm) long. Dull grey/green upper surfaces, glossy undersides. Some yellow autumn colour. *Stem* Upright with limited branching. Light green to grey/green, becoming grey/brown. Fast growth rate when young, slowing with age. *Fruit* Conspicuous, translucent, bladder-shaped capsules, 3–4 in (7.5–10 cm) long, in late summer/early autumn.
Hardiness Tolerates a minimum winter temperature of 4°F (−15°C).
Soil requirements Best in neutral to acid fertile soil, but tolerates moderate alkalinity.

Stachyurus praecox **in flower**

Sun/Shade aspect Requires a sheltered aspect in full sun or light shade.

Pruning None required for its well-being but for fan-training the previous season's growth can be cut back by two-thirds or more. Older shoots of four years or more should be cut hard to ground level to induce new foliage shoots which can be trained. This method can also be used to enhance the foliage production on varieties not fan-trained.

Training Will require wires or fixing to individual anchor points.

Propagation and nursery production From seed or layers. Purchase container grown; rather hard to find. Best planting height 2–2½ ft (60–76 cm).

Problems Susceptibility to severe winters can be a problem if the shrub is to establish and show its full potential. Must be provided with the right type of soil.

Similar forms of interest *S. holocarpa* Flowers pink in bud, opening to white, in short pendent panicles in mid to late spring. Foliage trifoliate, oblong to lanceolate leaflets, light grey/green. From central China. Difficult to find. *S. h.* 'Rosea' Flowers pink to soft pink, in pendent panicles mid to late spring. Young foliage trifoliate and bronze in colour, ageing to mid green. Yellow autumn colour. Difficult to find. *S. pinnata* (Antoney nut) White flowers in thin, pendent panicles in late spring to early summer. Foliage pinnate, grey/green, five to seven leaflets. Easy to find. From central Europe.

Average height and spread
Five years
6 × 6 ft (1.8 × 1.8 m)
Ten years
9 × 9 ft (2.7 × 2.7 m)
Twenty years
16 × 16 ft (4.9 × 4.9 m)
Protrudes up to 3 ft (91 cm) from support.

STAUNTONIA HEXAPHYLLA

KNOWN BY BOTANICAL NAME

Lardizabalaceae   *Woody Climber*
Evergreen

An interesting evergreen climber which requires a protected aspect in very favourable areas, but is well worth inclusion if the right conditions can be provided.

Origin From Japan and Korea.

Use As a self-twining climber for warm walls and fences or for growing in conservatories and large greenhouses.

Description *Flower* Male and female flowers each have between three and seven slender, white to violet tinged, fragrant petals. Flowers produced in short racemes in mid spring. *Foliage* Pinnate leaves with oval leaflets each up to 4 in (10 cm) long. Leathery texture. Light green to grey/green. Leaves presented in bunches of three to seven at each leaf joint. *Stem* Wiry, dark olive green, twining. Medium rate of growth. *Fruit* Egg-shaped, tinged with purple, up to 2 in (5 cm) long. Dark green, ageing to yellow. Fruit juice is sweet to taste. Only produces fruit in hot, dry summers.

Hardiness Tolerates a minimum winter temperature of 23°F (−5°C).

Soil requirements Does well on all soil conditions with no particular preference.

Sun/Shade aspect Requires a sunny, sheltered, warm wall in full sun to light shade.

Pruning Shorten back trailing tendrils in winter to induce new growth the following spring.

Training Allow to twine through wires on fences, walls, pillars and pergolas. Will also ramble through large shrubs or small trees.

Propagation and nursery production From semi-ripe cuttings taken in early to mid summer. Should always be purchased container grown; will have to be sought from specialist nurseries. Best planting height 1–3 ft (30–91 cm).

Stauntonia hexaphylla in flower

Problems Not entirely hardy and may be better in a large conservatory or greenhouse unless very mild, warm conditions can be guaranteed all year round.

Similar forms of interest None.

Average height and spread
Five years
6 × 6 ft (1.8 × 1.8 m)
Ten years
12 × 12 ft (3.7 × 3.7 m)
Twenty years
24 × 24 ft (7.3 × 7.3 m)
Protrudes up to 9 in (23 cm) from support.

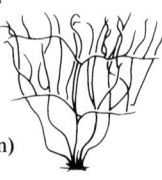

SYRINGA MICROPHYLLA 'SUPERBA'

DAPHNE LILAC, LITTLELEAF LILAC

Oleaceae   *Wall Shrub*
Deciduous

A lilac giving two flowering displays, one in spring and one in autumn, adapting well as a fan-trained wall specimen.

Origin Originally from northern and western China, but the form 'Superba' is of garden origin.

Use As a fan-trained flowering shrub for walls, fences or pillars.

Description *Flower* Panicles 3 in (7.5 cm) long and 1½ in (4 cm) wide of fragrant, rosy pink

Syringa microphylla 'Superba' in flower

flowers in mid to late spring, often with some autumn display. *Foliage* Leaves small, ovate, pointed, up to 2 in (5 cm) long. *Stem* Thin, wispy, light brown to grey/brown, forming a widespreading fan shape. Medium to slow growth rate. *Fruit* Insignificant.

Hardiness Tolerates a minimum winter temperature of 4°F (−15°C).

Soil requirements Does well on all but severely alkaline soils.

Sun/Shade aspect Tolerates all aspects in full sun to light shade.

Pruning Fan-train but once the plant is more than three years old remove one third of its branches to ground level in spring every five years to induce regrowth.

Training Requires tying to wires or individual anchor points.

Propagation and nursery production From semi-ripe cuttings taken in early summer, or from budding or grafting on to *Ligustrum vulgaris*. Plant bare-rooted or container grown. Availability varies. Best planting height 1–2 ft (30–60 cm).

Problems Young plants often look small and insipid and require a year or more to develop.

Average height and spread
Five years
3 × 3 ft (91 × 91 cm)
Ten years
5 × 5 ft (1.5 × 1.5 m)
Twenty years
7 × 7 ft (2.1 × 2.1 m)
Protrudes up to 2 ft (60 cm) from support.

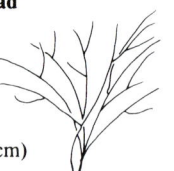

SYRINGA × PERSICA

PERSIAN LILAC

Oleaceae   *Wall Shrub*
Deciduous

A small-flowering, very fragrant lilac whose growth adapts well to fan-training.

Syringa × *persica* 'Alba' in flower

Origin From Iran to China.

Use As a fan-trained shrub or freestanding against walls or fences. Can also be pillar-trained.

Description *Flower* Fragrant panicles, 4 in (10 cm) long and 2 in (5 cm) wide, of mauve or off-white flowers produced in late spring/early summer. *Foliage* Leaves lanceolate, 1¼–2½ in (3.5–6 cm) long, dark green, giving some autumn colour. *Stem* Upright when young, becoming more spreading with age, forming a wide, fan-shaped shrub with training. Dark green tinged purple. Medium growth rate. *Fruit* Insignificant.

Hardiness Tolerates a minimum winter temperature of 4°F (−15°C).

358

Syringa × *persica* in flower

Soil requirements Most soils, although extremely alkaline types may cause chlorosis.
Sun/Shade aspect Tolerates all but the most severe of aspects. Full sun to very light shade.
Pruning None required.
Training Allow to stand free or secure to walls and fences by wires or individual anchor points.
Propagation and nursery production From budding, layers or semi-ripe cuttings taken in early summer. Purchase container grown; relatively easy to find. Best planting height 15 in–2½ ft (38–76 cm).
Problems Young plants may be shy to flower and can take two or three years to establish.
Similar forms of interest *S.* × *p.* **'Alba'** A variety of off-white, scented flower panicles. Interesting but not showy. *S.* × *p.* **'Lanciniata'** (syn. *S.* × *p. afghanica*) Very attractive, finely divided leaves unlike those of any other syringa, not more than 1¼ in (3.5 cm) long. Flowers slightly fragrant, lilac-pink but often sparsely produced. Benefits well from the protection of a wall or fence. Two thirds average height and spread.
Average height and spread
Five years
5 × 5 ft (1.5 × 1.5 m)
Ten years
7 × 7 ft (2.1 × 2.1 m)
Twenty years
9 × 9 ft (2.7 × 2.7 m)
Protrudes up to 3 ft (91 cm) from support if fan-trained, 10 ft (3 m) untrained.

SYRINGA × PRESTONIAE

CANADIAN LILAC
Oleaceae   ***Wall Shrub***
Deciduous
An elegant, large-flowering shrub which can be fan-trained to good effect but needs a large space.

Origin A range of hybrids produced in Canada by crossing *S. reflexa* and *S. villosa*.
Use As a shrub for large walls and fences, either fan-trained or freestanding.
Description *Flower* Graceful, 6–12 in (15–30 cm) long and 3 in (7.5 m) wide panicles of scented pink to red/pink or purple/pink flowers, depending on the variety, produced in late spring/early summer. All varieties single-flowered. *Foliage* Leaves long, lanceolate, 2–6 in (5–15 cm) long and 2–3 in (5–7.5 cm) wide, dark green with prominent veins. *Stem* Upright when young, strong-growing, forming a narrow-based, wide-topped, large shrub or can be fan-trained. Grey/green to grey/brown. Fast growth rate when young, becoming slower with age. *Fruit* Grey/brown seedheads may be retained into winter.
Hardiness Tolerates a minimum winter temperature of 0°F (−18°C).
Soil requirements Does well on most but alkaline types may produce narrow, yellow,

chlorotic leaf margins. On very alkaline soils this may become acute and cause distress.
Sun/Shade aspect Tolerates all aspects in full sun to medium shade.
Pruning Remove one third of oldest flowering wood to ground level after flowering on established plants more than three years old to encourage rejuvenation from base, otherwise the shrub can become very large and woody and will produce fewer flowers.
Training Requires wires or individual anchor points if fan-trained.
Propagation and nursery production From semi-ripe cuttings taken in early summer, from budding, grafting or from layers. Purchase container grown for best results, but may also be planted bare-rooted or root-balled (balled-and-burlapped). Some varieties rather scarce. Best planting height 2–3 ft (60–91 cm).

Syringa × *prestoniae* 'Elinor' in flower

Problems Often underestimated for its overall size and in too small an area may become dominating.
Similar forms of interest *S.* × *p.* **'Audrey'** Deep pink flowers, early summer. May not be easy to find. *S.* × *p.* **'Desdemona'** Pink panicles of flowers, late spring. Not easy to find. *S.* × *p.* **'Elinor'** Pale pink/lavender flowers purple/red in bud, borne in upright panicles, late spring, early summer. The most widely available variety. *S.* × *p.* **'Isabella'** Red/pink flowers in erect panicles in late spring/early summer. May be hard to obtain.
Average height and spread
Five years
9 × 9 ft (2.7 × 2.7 m)
Ten years
15 × 15 ft (4.6 × 4.6 m)
Twenty years
20 × 20 ft (6 × 6 m)
Protrudes up to 4 ft (1.2 m) from support if fan-trained, 20 ft (6 m) untrained.

TEUCRIUM FRUTICANS

SHRUBBY GERMANDER
Labiatae   ***Wall Shrub***
Evergreen
A useful silver grey/blue flowering shrub that benefits from winter protection from walls and fences.

Origin From southern Europe and North Africa.
Use As a fan-trained shrub or freestanding specimen in front of walls and fences. Can be trained up a pillar if required.
Description *Flower* Light blue flowers, ½ in (1 cm) long and wide, in 3 in (7.5 cm) long terminal racemes, throughout summer.

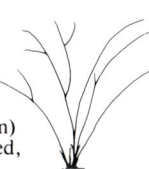

Syringa × *prestoniae* 'Isabella' in flower

Teucrium fruticans **in flower**

Foliage Leaves ovate, ½–1½ in (1–4 cm) long, light grey/green upper surfaces, white undersides. Some limited yellow autumn colour. **Stem** Upright stems, becoming branching, forming a dome-shaped shrub if free-standing. Grey to silver-grey. Fast growth rate in spring, slowing with age. **Fruit** Insignificant.
Hardiness Tolerates a minimum winter temperature of 23°F (−5°C).
Soil requirements Well-drained soil, dislikes any waterlogging.
Sun/Shade aspect Requires a sheltered aspect in full sun.
Pruning Previous season's growth should be shortened by two thirds or three quarters in mid spring to encourage new foliage and flowering.
Training Requires wires or individual anchor points to fan-train.
Propagation and nursery production From semi-ripe cuttings. Always purchase container grown. Generally available in mild districts, less easy to find elsewhere. Best planting height 6–12 in (15–30 cm).
Problems Its susceptibility to cold and wet conditions is not always appreciated.
Similar forms of interest *T. chamaedrys* (Wall germander) Deep green foliage and dark, red/mauve flowers in mid to late summer. Upright habit, reaching only one quarter average height and spread.
Average height and spread
Five years
5 × 5 ft (1.5 × 1.5 m)
Ten years
7 × 7 ft (2.1 × 2.1 m)
Twenty years
9 × 9 ft (2.7 × 2.7 m)
Protrudes up to 2 ft (60 cm) from support if fan-trained, 7 ft (2.1 m) untrained.

Teucrium chamaedrys **in flower**

THUNBERGIA ALATA

BLACK-EYED SUSAN

Acanthaceae   *Annual Climber*
Deciduous

An extremely colourful, attractive annual climber, needing some attention to cultivation to achieve good results.

Origin From South Africa.
Use As an annual climber for walls, fences, archways and trelliswork, or to ramble through medium to large shrubs, brushwood or pea sticks. Ideal for container growing on patios or in conservatories and greenhouses and for hanging baskets.

Thunbergia alata **in flower**

Description *Flower* Bell-shaped tubes consisting of twelve calyxes opening to form a reflexed mouth 1 in (2.5 cm) across. Bright yellow with purple base and a distinct black centre. Mid summer to early autumn. *Foliage* Oval, up to 1½ in (4 cm) wide and 3 in (7.5 cm) long. Light to mid green with slight grey overtones. *Stem* Light to mid green. Upright. Not self-supporting except for limited twining. Fast rate of growth. *Fruit* No fruits of interest.
Hardiness Not winter hardy. Seeds do not survive in soil over winter.
Soil requirements Requires a well prepared, highly organic, moisture-retaining soil to encourage good growth. If grown in containers, use a soil-based potting compost and a large container.

Sun/Shade aspect Requires a sheltered aspect out of strong winds to prevent damage to foliage and flowers. Full sun to medium shade with light shade for preference.
Pruning Not required. Dies in winter.
Training Tie to individual anchor points and wires. Allow to ramble freely through medium-sized shrubs or over pea sticks or brushwood.
Propagation and nursery production From seed sown under protection with bottom heat in early to mid spring. Do not plant out until all danger of spring frosts has passed. Always purchase container grown; will have to be sought from specialist nurseries or garden centres, in particular those growing a wide range of bedding plants. Best size to purchase 3–15 in (7.5–38 cm).
Problems The need to produce this plant annually can be a drawback, but it is well worth the effort to achieve good summer colour.
Similar forms of interest *T. a.* 'Alba' Pure white flowers with black centres. Not readily available and will have to be sought from specialist nurseries. *T. a. lutea* Pale yellow with white centre. Scarce.
Seed will have to be sought from specialist seedsmen.
Average height and spread
One year
6 × 6 ft (1.8 × 1.8 m)
Protrudes up to 12 in (30 cm) from support.

TRACHELOSPERMUM ASIATICUM
(*T. divaricatum,*
T. crocostemon)

KNOWN BY BOTANICAL NAME

Apocynaceae   *Woody Climber*

One of the most fragrant of all climbing plants, but requiring a little attention to its needs.

Origin From Japan and Korea.
Use As a close-growing evergreen climber for walls, fences and pillars or for conservatories and greenhouses.

Trachelospermum asiaticum **in flower**

Description *Flower* Open clusters up to 2½ in (6 cm) across at the ends of short shoots consisting of small, creamy-white, five-petalled flowers ¾ in (2 cm) wide. Flowers from mid to late summer. *Foliage* Leaves dark glossy green, narrowly ovate, pointed and up to 2 in (5 cm) long and ¾ in (2 cm) wide. *Stem* Dull, dark green, predominantly upright in close formation. Slow to medium growth rate. *Fruit* No fruit of interest.
Hardiness Tolerates a minimum winter temperature of 14°F (−10°C) although severe wind chill conditions may damage evergreen foliage and stems. Established plants will rejuvenate in spring, but young plants may be killed outright.

Soil requirements Moderately alkaline to acid. Resents drying out. If grown in a conservatory or greenhouse, the container must have a diameter of not less than 21 in (53 cm) and a good quality soil-based potting compost must be used.

Sun/Shade aspect Sheltered aspect. Full sun to light shade.

Pruning Normally requires no pruning.

Training Will require tying to wires, trellis or individual anchor points.

Propagation and nursery production From semi-ripe cuttings taken early to mid summer. Always purchase container grown; may have to be sought from specialist nurseries. Best planting height 6 in–3 ft (15–91 cm).

Problems Slow to establish. Young plants particularly susceptible to cold winds in early years. Can often become stunted and lacking in vigour; a liquid fertilizer given annually in mid summer normally corrects this.

Similar forms of interest None.

Average height and spread

Five years
3 × 3 ft (91 × 91 cm)
Ten years
6 × 6 ft (1.8 × 1.8 m)
Twenty years
12 × 12 ft (3.7 × 3.7 m)

Will continue to grow, probably reaching 20 × 20 ft (6 × 6 m) in 30 years. Protrudes up to 12 in (30 cm) from support.

TRACHELOSPERMUM JASMINOIDES

KNOWN BY BOTANICAL NAME

Apocynaceae　　　　　*Woody Climber*
Evergreen

An attractive climber requiring a protected aspect in mild areas.

Origin From China.

Use As an evergreen climber for very sheltered, warm walls or for growing in conservatories or large greenhouses.

Description *Flower* White, five-petalled, fragrant, 1 in (2.5 cm) wide and carried in terminal racemes up to 4 in (10 cm) long. Flowers produced in mid summer. *Foliage* Ovate to lanceolate, up to 4½ in (11 cm) long and 1½ in (4 cm) wide. Grey/green, variegated forms with white marginal irregular variegation, turning pink towards winter and regaining full variegation in following spring. *Stem* Dull, dark green. Moderately close formation. Upright, becoming spreading with age. Slow to medium rate of growth. *Fruit* No fruit of interest.

Hardiness Tolerates a minimum winter temperature of 14°F (−10°C). Cannot be fully trusted except in the warmest areas unless grown under protection.

Trachelospermum jasminoides **in flower**

Trachelospermum jasminoides 'Variegatum' **in autumn**

Soil requirements Moderately alkaline to acid. Dislikes extremely dry conditions.

Sun/Shade aspect Requires a very sheltered aspect. Full sun to light shade.

Pruning Not normally required, but can be restricted in size to a certain extent by cutting back shoots after flowering.

Training Will require tying to wires, trellis or individual anchor points.

Propagation and nursery production From semi-ripe cuttings taken in early to mid summer. Always purchase container grown; will have to be sought from specialist nurseries. Best planting height 6 in–3 ft (15–91 cm).

Problems Slow to establish, often dormant for two to three years after planting.

Similar forms of interest *T. j.* 'Variegatum' Variegated form, less hardy. Two-thirds average height and spread. Scarce.

Average height and spread

Five years
5 × 5 ft (1.5 × 1.5 m)
Ten years
10 × 10 ft (3 × 3 m)
Twenty years
15 × 15 ft (4.6 × 4.6 m)

Growth rate and size may well be exceeded under protection of conservatories and greenhouses. Protrudes up to 12 in (30 cm) from support.

TROPAEOLUM MAJUS

NASTURTIUM, INDIAN CREST

Tropaeolaceae　　　　*Annual Climber*
Deciduous

Some people may think nasturtiums mundane, but used in the right location they can make a very spectacular climber, particularly useful for infilling while other more permanent plantings establish.

Origin From Peru.

Use For walls, fences and pillars, to ramble through medium to large shrubs or over brushwood, pea sticks or pyramids of canes. Ideal for ground cover with or without support. Can be used in hanging baskets and tubs. Best out of doors but can be grown under cover as an early flowering plant if required.

Description *Flower* Various range of colours, including orange, red, pink, yellow. Up to 2 in (5 cm) across, consisting of either single or double, dependent on variety. Flowering from early to late summer, possibly into autumn. *Foliage* Up to 3–4 in (7.5–10 cm) across, light green with some variegated forms. Orb-like in shape with heart-type indentations at base. Yellow autumn colour. *Stem* Light green, trailing, not self-clinging. Medium annual growth rate. *Fruit* Small,

round fruits of limited interest other than for collecting for following season's plants.

Hardiness Not winter hardy. Plant killed by first frosts. Seeds can stay viable in soil throughout winter to regerminate in following spring. Plants raised under protection can be planted out from mid to late spring when danger of frost has passed.

Soil requirements Tolerates all types, including extremely poor. Will tolerate dry conditions as long as adequate moisture is provided for establishment.

Sun/Shade aspect Tolerates all aspects from full sun to medium shade, with light to medium shade for preference.

Pruning Not required, although can be cut back when young to encourage branching and to keep within bounds.

Training Allow to ramble over wires, trellis, brushwood or pea sticks, or through branches of medium to large shrubs.

Tropaeolum majus 'Alaska Mixed' **in flower**

Propagation and nursery production From seed, either saved from previous season's crop or purchased from seed merchants or garden centres, all of which carry a good range of varieties. Young plants can be purchased from garden centres in spring. Best planting height 2–6 in (5–15 cm).

Problems Can suffer attacks of blackfly which are difficult to control. Can become invasive in ideal conditions. Seed naturally sown in autumn can sometimes become a nuisance in following year.

Similar forms of interest *T. m.* 'Alaska Mixed' A good range of colours. Flowers double to semi-double. Foliage very attractive, white and cream splashed variegation on pale green. Not vigorous, only reaching 50 per cent of average height and spread. *T. m.* 'Climbing Mixed' A general mixture. Single flowers

Tropaeolum majus 'Jewel Mixed'

basically yellow and red. Strong growing. *T. m.* **'Double Gleam Mixed'** Semi-trailing variety. Large double flowers. Reaches only 50 per cent of average height and spread. *T. m.* **'Dwarf Cherry Rose'** Semi-double, brilliant cerise-red flowers held well above foliage. Reaches only 20 per cent of average height and spread. *T. m.* **'Empress of India'** Very dark foliage. Crimson-scarlet flowers. Only reaches 20 per cent average height and spread. *T. m.* **'Fiery Festival'** Scented, deep scarlet flowers carried in good amounts. Reaches only 20 per cent of average height and spread. *T. m.* **'Golden Gleam'** All yellow, semi-double. Reaches 40 per cent of average height and spread. *T. m.* **'Jewel Mixed'** Semi-double. Very bright colours. Only reaching 20 per cent of average height and spread. *T. m.* **'Orange Gleam'** Deep orange/red flowers. Reaches 40 per cent of average height and spread. *T. m.* **'Peach Melba'** Light yellow with scarlet blotches. Only reaching 20 per cent of average height and spread. *T. m.* **'Red Roulette'** Semi-double. Fiery red blooms carried well above foliage. *T. m.* **'Scarlet Gleam'** Orange/scarlet, semi-double flowers. Only reaching 40 per cent of average height and spread. *T. m.* **'Whirlybird Mixed'** Semi-double. Good range of colours through cherry, rose, gold, mahogany, orange, tangerine and cream. Flowers face upwards. *T. lobbianum* **'Spitfire'** Deep yellow to orange and red blooms. Strong growing. Ideal as a climber.
Average height and spread
One year
5 × 5 ft (1.5 × 1.5 m)
Protrudes up to 2 ft (60 cm) from support.

TROPAEOLUM PEREGRINUM

CANARY CREEPER
Tropaeolaceae   *Perennial/Annual Climber*
Deciduous
Canary creeper does not produce its flowers until mid to late summer, but they do make an extremely attractive display.

Origin From Peru.
Use As a fast-growing perennial or annual climber for walls, fences and arches, or to cover brushwood or pea sticks. Ideal for trailing through medium to large shrubs and for intermingling amongst other climbers. Attractive when grown with a dark background, such as conifers.

Description *Flower* Three large, lower spade-shaped petals with two smaller upper petals with fringed edges; ⅓–¾ in (1–2 cm) across from mid to late summer and early autumn. Bright yellow. *Foliage* Five ovate, lobed leaflets make up a hand-shaped leaf, light to medium green with a grey sheen, 2–4 in (5–10 cm) across. Some yellow autumn colours. *Stem* Light green, becoming yellow/green. Twining, but not entirely self-supporting. Weak in constitution and easily broken during cultivation. Fast annual rate of growth. *Fruit* No fruit of any significance.
Hardiness Not winter hardy above ground but seeds may survive in soil over winter.

Tropaeolum peregrinum

Root stool will tolerate a minimum winter temperature of 14°F (−10°C), particularly if some additional mulch or covering of organic material is given to the area in which it is growing to act as insulation. Young spring-grown plants from seed should not be planted out until all danger of spring frost has passed.
Soil requirements Tolerates all soil. Does best in a moist, well fed type high in organic material where the roots can spread fast and produce good growth at the top.
Sun/Shade aspect Tolerates all aspects. Full sun to light shade, with full sun for preference, although roots should be in the shade.
Pruning Not required. Dies to ground level in winter.

Training Allow to ramble over wires, trellis, brushwood or pea sticks, or through medium to large shrubs.
Propagation and nursery production From seed sown in spring under protection. Always purchase container grown from 2–12 in (5–30 cm) tall. Not readily available and will have to be sought from specialist nurseries and garden centres. Seed will have to be sought from specialist seed merchants.
Problems Can often be forgotten, as is relatively late flowering, and is unobtrusive until the flowers burst into bloom.
Similar forms of interest None.
Average height and spread
One year
8 × 4 ft (2.4 × 1.1 m)
Protrudes up to 6 in (15 cm) from support.

TROPAEOLUM SPECIOSUM

FLAME NASTURTIUM, FLAME CREEPER
Tropaeolaceae   *Perennial Climber*
Deciduous
One of the most spectacular of all perennial climbers, but requiring specific cultivation to achieve results.

Origin From Chile.
Use As a perennial creeper, best used when grown over medium to large shrubs or against a background of *Taxus baccata* (English yew) or other evergreens.
Description *Flower* Five reflexed petals, the lower three larger, make up a flat flower, presented in small groups in mid to late summer. Flowers ⅓–¾ in (1–2 cm) across, carried on short shoots 1–1¼ in (2.5–3.5 cm) long. *Foliage* Five, sometimes six, lobes make up a hand-shaped leaf, 1½ in (4 cm) across with a hairy covering. *Stem* Light green, trailing but not twining. Supported by intermingling. Medium growth rate. *Fruit* Round, turquoise/blue seeds often appear in attractive clusters with late flowers.
Hardiness Dies to ground level in winter. Roots are able to survive a temperature of 14°F (−10°C).
Soil requirements Alkaline to acid. Requires an extremely well prepared soil with over 50 per cent of organic material added to a depth of not less than 18 in (45 cm). Material such as sedge peat, spent mushroom compost, garden compost or very well rotted manure is ideal and 1 sq yd (1 sq m) should be prepared for any chance of good results.

Tropaeolum speciosum **in flower**

Tropaeolum speciosum in fruit

Sun/Shade aspect Flowers in full sun to light shade. Roots should be in shade.
Pruning Not required. Dies to ground level in winter.
Training Allow to ramble over medium to large shrubs, including shrub roses. If grown against a hedge of *Taxus baccata* (yew) or other conifer it often needs some additional support with canes.
Propagation and nursery production From root cuttings, selecting underground rhizomes and rooting into pots containing a soil-based potting compost. Propagation is difficult and it is therefore only obtainable at a high price from specialist nurseries. Best height is as clumps in pots before growth matures in spring or later with top growth of not more than 2–3 in (5–7.5 cm).
Problems Without adequate soil preparation rhizomes normally die. A number of attempts may be needed to establish. Grows better in wet, humid areas.
Similar forms of interest
None.
Average height and spread
One year
6 × 4 ft (1.8 × 1.2 m)
Protrudes up to 12 in (30 cm) from support.

TROPAEOLUM TUBEROSUM

KNOWN BY BOTANICAL NAME
Tropaeolaceae   *Perennial Climber*
Deciduous
A climber which is rarely seen but deserves a wider planting.

Origin From Peru and Bolivia.
Use Grow over medium to large shrubs or against a background of *Taxus baccata* (English yew) or other evergreens.
Description *Flower* Extremely attractive red and yellow tubular flowers with a basal spur, carried on long, graceful stalks. Up to ¾ in (2 cm) long, produced in mid to late summer. *Foliage* Five-lobed leaves on long stalks. Light to mid green with some purplish hue, particularly towards autumn. Attractive white veining. *Stem* Light green to grey/green. Trailing, not truly self-clinging but intertwining. May need additional support. Medium growth rate. *Fruit* No fruits of interest.
Hardiness Root stool tolerates a minimum winter temperature of 14°F (−10°C) but will require a good surface mulch of not less than 3 in (7.5 cm) of organic material over at least 1 sq yd (1 sq m) to protect it in most locations.
Soil requirements Tolerates all soil types, but requires extremely good preparation with up to 50 per cent of organic material added to a depth of not less than 18 in (45 cm). Material such as sedge peat, spent mushroom compost, garden compost or very well rotted manure is ideal and 1 sq yd (1 sq m) should be prepared for good results. Tubers should be planted at least 6 in (15 cm) deep, and in light soils up to 8 in (20 cm) is not unacceptable.
Sun/Shade aspect Flowers should be in full sun to light shade and roots in shade.
Pruning Not required. Dies down in winter.
Training Allow to ramble over medium to large shrubs, including shrub roses. If grown against a hedge of *Taxus baccata* (English yew) or other evergreens, often needs some additional support from canes.
Propagation and nursery production By removal of large root rhizomes and planting into individual pots containing a soil-based compost in late winter/early spring. Propagation generally reliable by this method. Purchase as root rhizomes or as pot-grown plants up to 2½ ft (76 cm) in height. Will have to be sought from specialist nurseries.
Problems Care must be taken with soil preparation and some winter protection given by mulching.
Similar forms of interest
T. t. 'Ken Aslet' Good-sized orange flowers in summer.
Average height and spread
One year
4 × 4 ft (1.2 × 1.2 m)
Protrudes up to 12 in (30 cm) from support.

VIBURNUM (Best Fruiting Forms)

KNOWN BY BOTANICAL NAME
Caprifoliaceae   *Wall Shrub*
Deciduous
Useful large shrubs that can adapt well to fan-training to show off their fruit to full effect.

Origin From China.
Use For growing freestanding against large walls or, for better results, fan-trained.
Description *Flower* Clusters of small, white flowers, 3 in (7.5 cm) wide, in mid spring to early summer. *Foliage* Leaves ovate, 2–4 in (5–10 cm) long, mid to dark green, often with good golden autumn colour. *Stem* Dark green or purple/green, upright, strong, becoming branching and spreading with age, forming a large, wide, narrow-based shrub if left untrained. Fast growth rate when young, slowing with age. *Fruit* In bunches, either red or black, depending on variety, produced profusely in autumn.
Hardiness Tolerates a minimum winter temperature of 4°F (−15°C).
Soil requirements Any soil.
Sun/Shade aspect Tolerates all but the most severe of aspects. Prefers light shade, tolerates full sun to medium shade.

Viburnum lantana

Pruning On established plants more than five years old, remove one third of oldest wood to ground level every two to three years to encourage new basal growth and keep the shrub healthy.
Training Requires wires or individual anchor points to achieve a fan-trained shape.
Propagation and nursery production From semi-ripe cuttings taken in early summer or from seed. Always purchase container grown, most forms will need to be obtained from specialist nurseries. Best planting height 1–3 ft (30–91 cm).
Problems Takes five years or more to reach maturity and produce good fruiting displays.
Forms of interest *V. betulifolium* Clusters of white flowers in early summer, set off by ovate to rhomboidal, coarsely toothed, dark to olive-green foliage. Red fruits hanging in large bunches from long, arching branches, maintained well into winter. It is often recommended that several shrubs should be planted nearby to encourage fertilization, but in fact single plants will fruit well at maturity. May have to be obtained from specialist nurseries. From central China. *V. hupehense* White clusters of flowers in late spring/early summer give way to interesting, egg-shaped, orange/yellow fruits in autumn, ageing to red in early winter. Ovate, coarsely toothed leaves, light to mid green, giving good autumn colour. Obtain from specialist

Tropaeolum tuberosum in flower

nurseries. From central China. *V. lantana* (Wayfaring tree) Creamy white clusters of flowers 3–4 in (7.5–10 cm) across in late spring/early summer, giving way to clusters of oblong fruits, red at first, ageing to black. Grey/green, felted, upright leaves, broad and ovate, 6 in (15 cm) long, velvety-textured on undersides, turning dark crimson in autumn. Fairly easy to find. Used as understock for all viburnums that are grafted. From central and southern Europe, spreading through northern Asia to Asia Minor and North Africa. *V. l.* **'Variegatum'** (syn. *V. l.* **'Auratum'**) Young foliage golden-yellow to yellow/green. Colour not retained well and turns light green in mid to late summer, although fruiting is very good. *V. lentago* (Sheepberry, nannyberry) Upright growth, strong in stature. Creamy white flowers produced in terminal clusters in late spring, early summer, giving way to clusters of blue/black, bloom-covered damson-shaped fruits. Ovate leaves, dark shiny green tinged purple or red. Good rich autumn colour. From eastern North America. *V. l.* **'Pink Beauty'** A variety with pink flowers, otherwise identical to its parent. *V. prunifolium* (Black haw) Upright stems with horizontal side branches. Clusters of white flowers give way to large blue/black, bloom-covered, edible fruits. Foliage bright green and shiny-textured, ovate to obovate, giving good autumn colour. Hard to find outside its native environment of eastern North America. *V. sargentii* White flowers with purple anthers in early summer give way to large clusters of translucent red fruits lasting well into winter. Strong, grey/green stems support maple-shaped leaves with good autumn colour. Stems develop a cork-like bark. From north-eastern Asia.

Average height and spread
Five years
5 × 5 ft (1.5 × 1.5 m)
Ten years
9 × 9 ft (2.7 × 2.7 m)
Twenty years
15 × 15 ft (4.6 × 4.6 m)
Protrudes up to 3 ft (91 cm) from support if fan-trained, 13 ft (4 m) untrained.

Origin Forms originating from China, many of garden origin.
Use As fan-trained or upright, freestanding shrubs in front of walls and fences.
Description *Flower* Clusters of pink or white, 1½ in (4 cm) wide, tubular, fragrant flowers produced on bare stems in early to mid spring. Also some intermittent flowering from late autumn through winter during periods of mild weather. *Foliage* Leaves lanceolate to ovate, ½–4 in (1–10 cm) long, dark green with red tinge, intensifying towards late summer/early autumn. Some autumn colour. *Stem* Forms an upright shrub of stout habit, or a widespreading, fan-trained shrub with training. Purple/green when young, becoming mahogany-brown with age. Medium growth rate. *Fruit* Insignificant.
Hardiness Tolerates a minimum winter temperature of 4°F (−15°C), but any flowers produced in winter may be damaged by sudden temperature changes.
Soil requirements Any soil, including alkaline.
Sun/Shade aspect Tolerates all aspects, although some protection will prolong the flowers in very severe winters. Full sun to medium shade.
Pruning Four or five years after planting, re-

Viburnum × bodnantense in flower

move one third of oldest flowering growth to ground level in early to mid spring, thereafter pruning every two to three years, depending on shrub's vigour.
Training Requires tying to wires or individual anchor points if fan-trained.
Propagation and nursery production From semi-ripe cuttings taken in early summer. Purchase container grown, bare-rooted or root-balled (balled-and-burlapped). Availability varies. Best planting height 2–3 ft (60–91 cm).
Problems Grows larger than is generally foreseen.
Forms of interest *V. × bodnantense* A cross between *V. farreri* and *V. grandiflorum.* Strong, upright habit, covered with clusters of scented, rose-tinted flowers from December to February. *V. × b.* **'Dawn'** An improved variety, with vigorous habit and larger foliage. Large, pink scented flowers. *V. × b.* **'Deben'** Flowers pink in bud, opening to white. Sweetly scented. Flowers may be damaged by severe frosts or excessive wetness. *V. farreri candidissimum* Slightly less freely flowering, producing terminal and lateral clusters of white, tubular, scented flowers. Foliage light green. *V. foetens* (syn. *V. koreana* Gaunt, stout stems, mahogany-brown tinged purple, forming large, goblet-shaped shrub. Clusters of good-sized, tubular, white, very fragrant flowers, susceptible to frost damage. Large, ovate, dark green leaves, tinged purple. Scarce, must be obtained from specialist nurseries. Very slow-growing, eventually reaching up to 6 ft (1.8 m) in height and spread.
Average height and spread
Five years
5 × 5 ft (1.5 × 1.5 m)
Ten years
9 × 9 ft (2.7 × 2.7 m)
Twenty years
15 × 15 ft (4.6 × 4.6 m)
Protrudes up to 3 ft (91 cm) from support if fan-trained, 13 ft (4 m) untrained.

Origin From China.
Use As large fan-trained or freestanding shrubs for walls and fences.
Description *Flower* White to dull white clus-

ters up to 4–6 in (10–15 cm) across of short, tubular flowers produced mid spring through early summer, according to variety. *Foliage* Leaves ovate, 2–6 in (5–15 cm) long, dark to mid green upper surface. Some smooth, some glossy, others felted, often with grey undersides. *Stem* Normally strong, upright, becoming branching, forming a large, spreading fan-trained shrub or round-topped if untrained. Smooth or felty-textured, depending on variety. *Fruit* Blue or black clusters of round or oval fruits in autumn.
Hardiness Tolerates a minimum winter temperature of 4°F (−15°C).
Soil requirements Any soil.
Sun/Shade aspect Tolerates all aspects. Prefers light shade, tolerates full sun to medium shade.

Viburnum henryi in flower

Pruning May be left unpruned. Alternatively, remove one third of oldest wood on shrubs five years old or more to encourage rejuvenation.
Training Requires tying to wires or individual anchor points if fan-trained.
Propagation and nursery production From semi-ripe cuttings taken in early summer. Purchase container grown or root-balled (balled-and-burlapped). All forms fairly easy to find. Best planting height 1–3 ft (30–91 cm).
Problems Space must be allowed for eventual size of mature shrubs.
Forms of interest *V. buddleifolium* An almost evergreen shrub which occasionally defoliates in hard winters. White clusters of flowers up to 2¾ in (6.5 cm) across in early summer, fol-

lowed by clusters of red fruits which turn black. Oblong to lanceolate leaves, pale green, velvety-textured and with grey felted undersides. Two thirds average height and spread. From central China. *V. cinnamomifolium* Small clusters of dull white flowers up to 4–6 in (10–15 cm) across in early summer, dark, glossy, leathery-textured leaves. Clusters of egg-shaped, blue/black, shiny fruits in autumn. Good in light to medium shade. May be slightly less hardy and more difficult to find. Two thirds average height and spread. From China. *V. henryi* A variety of open, upright habit. Pyramid-shaped panicles of yellow/white flowers, early to late summer. Elliptic, glossy green, leathery-textured leaves on stiff, upright red/green branches. Bright red fruits, later turning black. Obtain from specialist nurseries. From central China. *V.* × *hillieri* (syn. *V.* × *hillieri* 'Winton') Semi-evergreen, depending on winter severity. Creamy white panicles of flowers in early summer, red fruits, later turning black. Foliage narrow, ovate, dark to mid green tinged copper, turning bronze/red in winter. From China. *V. japonicum* (Japanese viburnum) White fragrant flowers in rounded trusses produced in early summer, only on mature shrubs, followed by red fruits. Leathery-textured foliage, up to 6 in (15 cm) long and 4 in (10 cm) wide, glossy dark green with paler undersides. Stout, orange/green leaf stalks. Two thirds average height and spread. From Japan. *V.* 'Pragense' Clusters of pink buds opening to creamy white, produced at terminals of branches in late spring. Corrugated, elliptic leaves 2–3 in (5–7.5 in) long, dark green with white felted undersides. Extremely hardy. Obtain from specialist nurseries. Two thirds average height and spread. Raised in Prague, Czechoslovakia. *V.* × *rhytidophylloides* (Lantanaphyllum viburnum) Clusters of yellow/white flower buds produced in autumn and carried through winter before opening in late spring. Foliage elliptic to ovate or oblong to ovate, rough-textured, often hanging limp in cold weather, giving drooping, almost shabby effect. Some orange/brown autumn colour. Of garden origin. *V. rhytidophyllum* (Leatherleaf viburnum) A strong-growing variety, with clusters of creamy yellow flowers in late spring, formed in previous autumn and carried through winter as felted closed buds. Leaves large, up to 8 in (20 cm) long, elliptic or oblong, with corrugated surface, dark glossy green, undersides grey and covered with matted hairs. Oval red fruits, finally turning black, produced in autumn and maintained well into early winter. For good fruiting effect a number of shrubs should be planted fairly close together. Relatively easy to find. From central western China. *V. r.* 'Roseum' Flowers tinted rose-pink, otherwise identical to its parent. *V. r.* 'Variegatum' Rarely seen variety

with white-splashed leaves, otherwise identical to its parent. *V. utile* (Service viburnum) Clusters of white, scented flowers in late spring followed by blue/black fruits in autumn. Sparsely produced branches. Ovate to oblong, glossy dark green foliage with white undersides. Obtain from specialist nurseries. Two thirds average height and spread. From central China.

Average height and spread
Five years
7 × 7 ft (2.1 × 2.1 m)
Ten years
12 × 12 ft (3.7 × 3.7 m)
Twenty years
18 × 18 ft (5.5 × 5.5 m)
Protrudes up to 3 ft (91 cm) from support if fan-trained, 16 ft (4.9 m) untrained.

VIBURNUM OPULUS

GUELDER ROSE, WATER ELDER, EUROPEAN CRANBERRY BUSH VIBURNUM

Caprifoliaceae   *Wall Shrub*
Deciduous

Fruiting shrubs with flat lacecap or snowball-shaped flowers and an attractive autumn display, adapting well to fan-training on walls and fences.

Origin From Europe, northern and western Asia and north Africa.
Use As fan-trained shrubs on walls or fences where they will afford an attractive display in autumn.

Description *Flower* Either white, flat, 4 in (10 cm) wide lacecap flowers with central small flowers surrounded by a ring of ray florets or a globular 3 in (7.5 cm) cluster of florets. The lacecap forms are fertile whereas the globular forms are not. *Foliage* Leaves light green, 2–5 in (5–12 cm) long, often with orange tinge or shading in late summer. Five-lobed, maple-shaped, giving good autumn colour. *Stem* Upright when young, becoming spreading and adapting well to fan-training. Grey to grey/green. *Fruit* Lacecap forms produce clusters of red and yellow round fruits in autumn, depending on variety. Globular varieties do not fruit.
Hardiness Tolerates a minimum winter temperature of 0°F (−18°C).
Soil requirements Any soil, tolerating alkalinity and acidity, dry and even boggy or water-logged areas.
Sun/Shade aspect Tolerates all aspects. Prefers full sun, but will tolerate medium shade.
Pruning Can be fan-trained or alternatively, on shrubs more than five years old, remove one third of oldest wood to ground level in spring. Repeat every two to three years to induce new growth from ground level and to prevent shrub from becoming woody.
Training Requires tying to wires or individual anchor points.
Propagation and nursery production From semi-ripe cuttings taken in early summer or hardwood cuttings taken in winter. Purchase container grown, bare-rooted or root-balled (balled-and-burlapped). Most forms fairly easy to find. Best planting height 2–3 ft (60–91 cm).
Problems Can outgrow its desired area unless adequate space is allowed.
Similar forms of interest *V. o.* 'Aureum' White, lacecap flowers with some red fruits in autumn. New spring growth bright yellow, ageing to yellow/green. Unfortunately will scorch in full sun and therefore requires very light shade protection. Not readily available. One third average height and spread. *V. o.* 'Fructu-luteo' White, lacecap flowers in early summer. Lemon-yellow fruits, tinged pink and ageing to chrome yellow, retaining faint shadings of pink. Obtain from specialist nurseries. *V. o.* 'Notcutt's Variety' A very good form with large white lacecap flowers, followed by bunches of succulent red fruits in autumn. *V. o.* 'Xanthocarpum' White lacecap flowers, pure golden yellow fruit, becoming darker and attaining translucent appearance when ripe. *V. o.* 'Sterile' (syn. *V. o.* 'Roseum' (Snowball shrub) Globular, creamy white, snowball-type flowerheads. Non-fruiting.
Average height and spread
Five years
7 × 7 ft (2.1 × 2.1 m)
Ten years
12 × 12 ft (3.7 × 3.7 m)
Twenty years
18 × 18 ft (5.5 × 5.5 m)
Protrudes up to 3 ft (91 cm) from support.

Viburnum opulus 'Notcutt's Variety' in flower

Viburnum rhytidophyllum 'Roseum' in flower

VIBURNUM PLICATUM

DOUBLEFILE VIBURNUM, JAPANESE
SNOWBALL, LACECAP VIBURNUM

Caprifoliaeceae   *Wall Shrub*
Deciduous

Of all the viburnums, the *plicatum* forms adapt themselves best for use against walls and fences.

Viburnum macrocephalum in flower

Origin From Japan and China.

Use As a medium to large fan-trained or freestanding shrub, depending on variety, for growing against walls and fences.

Description *Flower* Either globular heads of white florets, 4–5 in (10–12 cm) wide, or heads of flat, lacecap flowers up to 6 in (15 cm) across with central fertile small flowers, surrounded by white ray florets, ½ in (1 cm) wide. *Foliage* Leaves 2–4 in (5–10 cm) long, ovate, with pleated effect and pronounced channelling along veins, light to mid green. Some good autumn colour, deep orange to orange/red, particularly after a dry, hot summer. *Stem* Upright when young, quickly becoming spreading, forming a dome-shape if freestanding. Light green to grey/green. The tiered effect which is seen on some varieties can be maintained when fan-trained. Medium growth rate. *Fruit* Clusters of oval, red to red/orange fruits in autumn on shrubs more than five years old on relatively dry to average soil. Fruiting can be erratic.

Hardiness Tolerates a minimum winter temperature of 4°F (−15°C).

Soil requirements Tolerates a wide range of soils but care must be taken not to allow root system to dry out or become waterlogged or its fine-textured root system will be damaged.

Sun/Shade aspect Tolerates all aspects. Prefers light shade but tolerates full sun to medium shade.

Pruning None, other than that required for training as and when necessary.

Training Tie to wires or individual anchor points to achieve a fan-trained shape or allow to grow freestanding.

Propagation and nursery production From semi-ripe cuttings taken in early summer or from layers. Purchase container grown or root-balled (balled-and-burlapped); most varieties fairly easy to find. Best planting height 15 in–3 ft (38–91 cm).

Problems Very susceptible to root damage from cultivation such as hoeing and from drought or waterlogging. If roots are damaged a section of top growth will die back. In severe cases this can be terminal.

Similar forms of interest *V. p.* 'Cascade' Large, white, flat, lacecap flowers. Fertile inner small tufted florets, surrounded by a ring of bold white ray florets. Red fruits after hot summer. Large, ovate, pointed foliage, giving good autumn colour, on branches arching to give semi-weeping effect. Less likely to die back than other varieties. Two thirds average height and spread. *V. p.* 'Grandiflorum' (Japanese snowball) Sterile flowerheads in spring, large, round to globular, attractive green, ageing to white, taking on pink margin then flushing overall pink before becoming purple and eventually brown. Good autumn colour. Must have good moist soil and light dappled shade to do well. Resents full sun. Two thirds average height and spread. *V. p.* 'Lanarth' White, large, flat lacecap flowers produced in defined tiers along horizontal branches. Large, ovate, mid to dark green foliage, giving good autumn colour. Sometimes listed as shorter growing shrub; when this variety was first catalogued it was confused with *V. p.* 'Mariesii', 'Lanarth' being described as the lower of the two varieties, but the reverse is the case. *V. p.* 'Mariesii' White lacecap flowers borne on very horizontal, tiered branches. Good fruiting with fine autumn colour. Reaches half average height and spread. *V. p.* 'Pink Beauty' Ray florets of white lacecap flowers,

ageing to attractive pink. Small to medium, ovate, dark to olive/green foliage. Good autumn colour. Two thirds average height and spread. *V. p.* 'Rowallane' White lacecap flowers in late spring. Small, ovate, tooth-edged foliage, mid to dark green, giving some autumn tints. Closely tiered branch effect. Extremely reliable fruiting form. May be more difficult to find than most. Two thirds average height and spread. *V. p. tomentosum* Lacecap flowers 2–4 in (5–10 cm) across, creamy white and surrounded by white ray florets, giving way to red, oval fruits, which age eventually to black. Bright green, pleated, ovate foliage. Good autumn colour. Not easy to find. From Japan, China and Taiwan. *V. p.* 'Watanabe' (syn. *V. semperflorens*) Very compact, producing good-sized white lacecap flowers in early summer through to mid autumn. Some orange/red fruits, late summer through to autumn. Reaching only two thirds average height and spread. *V. macrocephalum* Semi-evergreen. White, round, large, globular heads of sterile flowers, 4–6 in (5–10 cm) long. across, late spring. Medium-sized, ovate foliage, light green, up to 2–4 in (5–10 cm) long. Less hardy than most, requiring a favourable sunny position where temperature does not fall below 23°F (−5°C) Difficult to obtain.

Average height and spread

Five years
8 × 6 ft (2.4 × 1.8 m)
Ten years
12 × 10 ft (3.7 × 3 m)
Twenty years
16 × 16 ft (4.9 × 4.9 m)
Protrudes up to 4 ft (1.2 m) from support if fan-trained, 16 ft (4.9 m) untrained.

Viburnum plicatum 'Grandiflorum' in flower

Viburnum plicatum 'Pink Beauty' in flower

VIBURNUM (Spring Flowering, Scented Forms)

KNOWN BY BOTANICAL NAME

Caprifoliaceae   *Wall Shrub*
Deciduous and semi-evergreen

Some of the most highly scented of all spring flowering shrubs, adapting extremely well to fan-training.

Origin Basic forms from Korea, with many garden hybrids.
Use As large, fan-trained shrubs for walls and fences.
Description *Flower* Round, dense clusters, 3–4 in (7.5–10 cm) across, consisting of many tubular, very fragrant flowers in varying shades of pink to white in early to late spring. *Foliage* Leaves ovate, medium-sized, 2–4 in (5–10 cm) long, grey/green, some yellow autumn display. *Stem* Upright but can be trained into a narrow fan-shape. Covered with grey scale when young. Medium growth rate. *Fruit* May produce blue/black fruits in autumn.

Viburnum carlesii in flower

Viburnum × burkwoodii in flower

Hardiness Tolerates a minimum winter temperature of 4°F (−15°C).
Soil requirements Most soils, disliking only very dry or very wet types.
Sun/Shade aspect Tolerates all aspects, but will benefit from some protection when in flower. Prefers light shade, accepts full sun to medium shade.
Pruning None, other than that required for training which should be done in early spring, but remove any suckering growths appearing below graft or soil level.
Training Requires wires or individual anchor points to create a fan shape.
Propagation and nursery production Normally from grafting on to an understock of *V. lantana*. Some varieties from semi-ripe cuttings taken in early summer. Purchase container grown or root-balled (balled-and-burlapped); best planting height 2–3 ft (60–91 cm). Most varieties fairly easy to find, especially when in flower.
Problems All forms, particularly *V. carlesii* and its varieties, suffer from aphid attack. Root systems of all forms are very fibrous and surface-rooting and react badly, sometimes succumbing completely, to damage caused by cultivation, drought or waterlogging.
Forms of interest *V. bitchiuense* (Bitchiu viburnum) Clusters of pink, scented flowers, mid to late spring. Foliage ovate to elliptic, dark metallic green. Open habit. From Japan. *V. × burkwoodii* (Burkwood viburnum) Clusters of pink buds open into fragrant, white, tubular flowers, early to mid spring, followed by clusters of blue/black fruits. Semi-evergreen ovate foliage with dark green shiny surface. As leaves die in autumn they turn scarlet, red and orange, contrasting with remaining dark green foliage. Reaches one third more than average height and spread. *V. × b.* '**Anne Russell**' Semi-evergreen. Large clusters of pale pink, fragrant flowers in mid spring, dark pink in bud. *V. × b.* '**Chenaultii**' Semi-evergreen. Flowers similar to parent, but does not reach same overall proportions.

Two thirds average height and spread. Not easy to find. *V. × b.* '**Fulbrook**' Large white flowers, pink in bud and sweetly scented. *V. × b.* '**Park Farm Hybrid**' A form with larger, more vigorous habit of growth. Flowers, mid spring, slightly larger than the form. Good glossy green foliage. *V. × carlcephalum* A deciduous hybrid producing large, white, tubular florets, pink in bud, very fragrant, borne in extremely attractive clusters 4–5 in (10–12 cm) across. Large, ovate to round, grey/green foliage, may produce good autumn colours. *V. carlesii* (Koreanspice viburnum) Clusters of highly fragrant, pure white, tubular flowers, opening from pink buds in mid to late spring. Ovate to round, downy, grey to grey/green leaves with grey felted undersides, producing good red/orange autumn colouring. Some forms of *V. carlesii* are weak in constitution and named varieties may be more successful. *V. c.* '**Aurora**' Red flower buds, opening to fragrant pink tubular flowers produced in clusters, mid to late spring. Good ovate grey/green foliage. Good constitution. *V. c.* '**Charis**' Good, vigorous growth. Flowers red in bud, opening to pink and finally fading to white. Very good scent. Foliage disease free and grey/green. May be difficult to find. *V. c.* '**Diana**' A good variety of compact habit. Flowers pink, red in bud. Good fragrance. May be difficult to find. *V. × juddii* (Judd viburnum) Clusters of scented, pink-tinted tubular flowers, produced at terminals of branching stems in mid to late spring. Grey/green ovate foliage with some autumn colour. Open in habit when young, becoming denser with age. Two thirds average height and spread.

Average height and spread
Five years
5 × 5 ft (1.5 × 1.5 m)
Ten years
8 × 8 ft (2.4 × 2.4 m)
Twenty years
12 × 12 ft (3.7 × 3.7 m).
Protrudes up to 3ft (91 cm) from support.

VIBURNUM TINUS

LAURUSTINUS

Caprifoliaceae   *Wall Shrub*
Evergreen

An attractive evergreen winter flowering shrub, benefiting from the protection of a wall or fence.

Origin From the Mediterranean regions of eastern Europe.
Use As a loose, fan-trained or freestanding shrub for planting in front of walls or fences.

Viburnum × carlcephalum in flower

Description *Flower* Small, 3 in (7.5 cm) wide clusters of white tubular flowers often pink in bud, in late autumn through until late spring, with peak performance in mild spells. Resistant to frost. *Foliage* Leaves broadly ovate, 1½–4 in (4–10 cm) long, evergreen, dark green with lighter silver undersides. *Stem* Upright when young, becoming spreading and branching. Can be loosely fan-trained or left to form a round-topped, broad-based shrub. Dark green to green/brown. Fast growth rates when young, slowing with age. *Fruit* Clusters of oval black fruits in autumn.
Hardiness Tolerates a minimum winter temperature of 14°F (−10°C).

Viburnum tinus in flower

Soil requirements Most soils. Tolerates both alkalinity and acidity but resents extremely dry or waterlogged conditions.
Sun/Shade aspect Requires a sheltered aspect. Prefers light shade, good from full sun to medium shade.
Pruning Can be reduced drastically and will rejuvenate quickly, flowering from early stage. Alternatively, on shrubs more than five years old, remove one third of oldest wood to ground level in early spring to encourage rejuvenation from base.
Training Allow to stand free or secure to wires or individual anchor points for fan-training.
Propagation and nursery production From semi-ripe cuttings taken in early summer. Purchase container grown; best planting height 12 in–2½ ft (30–76 cm). Easy to find in good garden centres and nurseries.
Problems Suffers from severe winter cold, especially wind chill. Can appear old and straggly if pruning is neglected.

Similar forms of interest *V. t.* 'Eve Price' Flowers carmine-red in bud, opening to white with pink shading. Foliage smaller than parent. Two thirds average height and spread. *V. t.* 'French White' A good, strong-growing variety, producing large white flowers. *V. t.* 'Gwenllian' Flowers deep pink in bud, opening to white with pink tinge. Small leaves, compact habit, reaching only two thirds average height and spread. *V. t. lucidum* Flowers large, white, in early to late spring. Good, vigorous form with larger, glossier leaves than parent. Slightly tender and benefits from the protection of a wall or fence. *V. t.* 'Purpureum' White flowers. New growth tinged purple, older foliage very dark green. Benefits from the protection of a wall or fence. Two thirds average height and spread. *V. t.* 'Variegatum' White flowers. Attractive white to creamy white variegated foliage. Two thirds height and spread. Benefits from the protection of a wall or fence.

Average height and spread
Five years
3 × 3 ft (91 × 91 cm) freestanding
5 × 5 ft (1.5 × 1.5 m) fan-trained
Ten years
6 × 6 ft (1.8 × 1.8 m) freestanding
9 × 9 ft (2.7 × 2.7 m) fan-trained
Twenty years
12 × 12 ft (3.7 × 3.7 m) freestanding
15 × 15 ft (4.6 × 4.6 m) fan-trained
Protrudes up to 4 ft (1.2 m) from support if fan-trained, 10 ft (3 m) untrained.

VINCA MAJOR

GREATER PERIWINKLE, LARGE PERIWINKLE
Apocynaceae   *Perennial Climber*
Evergreen
A creeping ground-cover plant rarely considered as a climbing specimen but can be adapted to such and, in its variegated form, extremely attractive when used in this manner.

Origin From Europe.
Use As a low climber for walls and fences or to creep through other established shrubs. Can also be used to cover a wall by cascading down from the top.
Description *Flower* Blue to purple/blue, five-petalled, 1 in (2.5 cm) wide, forming a single saucer-shaped flower in mid to late spring. *Foliage* Ovate, pointed, 1–1½ in (2.5–4 cm) long. Light green when young, quickly becoming glossy dark green. Some yellow autumn colour. *Stem* Light green, becoming

darker with age. Glossy texture, rambling habit. *Fruit* No fruit of interest.
Hardiness Tolerates a minimum winter temperature of 0°F (−18°C).
Soil requirements Tolerates all soil conditions with no particular preference, although for good climbing results adequate plant nutrient must be available.
Sun/Shade aspect Tolerates all aspects and deep shade but growth will be thicker and more lush in full sun to light shade.
Pruning Reduce all previous season's growth to ground level in early spring to encourage new annual formation of climbing structure.
Training Allow to ramble through other shrubs or wires.
Propagation From rooted tips, from layers or from semi-ripe cuttings. Can be purchased bare-rooted from late autumn to early spring or container grown as available. Normally stocked by most garden centres. Best planting height: root clumps to 18 in (45 cm).
Problems Plants often look insipid when young, quickly establishing once planted. Can become invasive in ideal situations.
Similar forms of interest *V. m.* 'Variegata' Attractive white to yellow/white variegated foliage. *V. m.* 'Maculata' A rarely seen variety with a central splash of green/yellow on each young leaf. Splash deteriorates as leaf ages and is produced only in open, sunny positions.

Average height and spread
One year
3 × 3 ft (91 × 91 cm)
Protrudes up to 12 in (30 cm) from support.

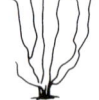

VITEX AGNUS-CASTUS

CHASTE TREE
Verbenaceae   *Wall Shrub*
Deciduous
An attractive, uncommon shrub useful for its autumn flowers and benefiting from the protection of a large wall or fence.

Origin From the Mediterranean area and central Asia.
Use As a fan-trained shrub for walls and fences.

Vitex agnus-castus in flower

Description *Flower* Fragrant, violet, small tubular flowers in slender racemes, 3–7 in (7.5–18 cm) long and 3 in (7.5 cm) wide, produced at ends of current season's growth, early to mid autumn. *Foliage* Compound leaves consisting of five to seven ovate to lanceolate leaflets, 2–6 in (5–15 cm) long, on short stalks borne in pairs along stems and shoots. *Stem* Graceful, grey, downy shoots, when trained forming a rigid fan shape. Medium growth rate. *Fruit* Small grey/brown seedheads of little interest.
Hardiness Tolerates a minimum winter temperature of 14°F (−10°C).

Vinca major in flower

Soil requirements Well-drained soil, dislikes any degree of waterlogging.
Sun/Shade aspect Requires a sheltered aspect in full sun.
Pruning Once fan framework is formed shorten all previous season's growth back to two or three buds.
Training Tie to anchor points or to wires.
Propagation and nursery production From seed or semi-ripe cuttings taken in early summer. Purchase container grown from specialist nurseries. Best planting height 15 in–2 ft (38–60 cm).
Problems Likely to succumb in wet, cold winters, hardier in dry, cold conditions.
Average height and spread
Five years
5 × 5 ft (1.5 × 1.5 m)
Ten years
7 × 7 ft (2.1 × 2.1 m)
Twenty years
12 × 12 ft (3.7 × 3.7 m)
Protrudes up to 3 ft (91 cm) from support.

Vitis coignetiae in autumn

VITIS 'BRANT' (*V. vinifera* 'Brandt' *V. v.* 'Brant')

ORNAMENTAL GRAPE VINE
Vitaceae   *Fruiting Vine*
Deciduous
Good covering foliage in summer but coming into its own in autumn with flame-orange foliage.

Origin From Asia Minor and the Caucasus.
Use To cover walls, fences, wires, ideal for growing through large shrubs and trees. Will cover buildings, pergolas and other similar structures.
Description *Flower* Small, creamy green clusters 3 in (7.5 cm) long of inconspicuous flowers, often hidden by foliage, are produced in early to mid summer. *Foliage* Five-fingered maple-shaped leaves, up to 5 in (12 cm) wide and long. Light green when young becoming slightly duller and yellower in summer, in autumn turning startling copper/orange/red, often in a defined pattern on the upper surface of the leaf. *Stem* Light grey/green when young, ageing to yellow/brown, finally grey/brown when mature. Vigorous, able to make in excess of 9 ft (2.7 m) of growth in one season. Clinging tendrils produced from leaf axils make it partially self-supporting. *Fruit* Small clusters up to 4–5 in (10–12 cm) long containing numerous small grapes ¼–½ in (5 mm–1 cm) in diameter, purple blue with a silver bloom. Edible in early to late autumn.
Hardiness Tolerates a minimum winter temperature of 4°F (−15°C).
Soil requirements Does well on all soil types but requires adequate moisture to aid the production of the vigorous vine growth.
Sun/Shade aspect Tolerates all aspects in light shade to full sun.
Pruning Can be left to ramble or treated as a fruiting vine. All current season's shoots not required for covering can be removed in early to mid spring, cutting back flush with the main stem. This will lead to larger foliage and possibly increased fruit production.
Training Wires or some form of framework will be required. The vine is semi self-clinging by the use of leaf tendrils, although young and heavy branches may need tying in.
Propagation and nursery production Semi-ripe cuttings taken in early to mid summer. Always purchase container grown, best planting height 1½–3 ft (45–76 cm). Available from good garden centres and nurseries.
Problems Birds may take fruit. Autumn colour may be poor following wet summers.
Similar forms of interest None.
Average height and spread
Five years
10 × 10 ft (3 × 3 m)
Ten years
20 × 20 ft (6 × 6 m)
Twenty years
30 × 30 ft (9 × 9 m)
Protrudes up to 3 ft (91 cm) from support.

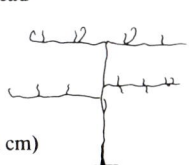

VITIS COIGNETIAE (*V. kaempferi*)

JAPANESE CRIMSON GLORY VINE
Vitaceae   *Woody Climber*
Deciduous
A most attractive autumn foliage climber with particularly interesting leaves.

Origin From Japan.
Use As a fast-growing climber for walls, fences, trelliswork, pergolas, gazebos and small buildings. Attractive when grown up large conifers, large shrubs and other trees.
Description *Flower* Small clusters of round, light green flower buds of little overall attraction. *Foliage* Up to 12 in (30 cm) across, heart-shaped, coarsely toothed, with three to five lobes. Glabrous above with downy, orange underside. Particularly brilliant autumn colours of crimson and scarlet. *Stem* Light grey/green, becoming green/brown, finally grey/brown. Fast growing. *Fruit* Small, ½ in (1 cm) wide, black, grape fruits in short clusters. Not always reliable in their production and not normally edible.
Hardiness Tolerates a minimum winter temperature of 4°F (−15°C).
Soil requirements Tolerates all soil types, except extremely alkaline. On very wet soils autumn colour may be decreased.
Sun/Shade aspect Tolerates all aspects. Best in light shade but will tolerate full sun to medium shade.
Pruning Select and train a vine system of shoots with individual laterals no closer than 18 in (45 cm). Prune back all other laterals flush with main vine system in early spring. This will induce larger foliage display.
Training Allow to ramble through wires, trellis or the branches of large shrubs and trees. Not self-clinging, not twining, but normally interlaces itself giving support. May need tying to wires and individual anchor points when young but can be left free-growing without pruning.
Propagation and nursery production From semi-ripe cuttings taken in mid summer. Always purchase container grown, best planting height 1½–4 ft (45 cm–1.2 m). Readily available from garden centres and nurseries.
Problems Often slow to establish, taking up to two years to make any amount of growth, but, once established, very fast-growing and can outgrow the area allocated for it.
Similar forms of interest None.
Average height and spread
Five years
10 × 10 ft (3 × 3 m)
Ten years
20 × 20 ft (6 × 6 m)
Twenty years
40 × 40 ft (12 × 12 m)
Protrudes up to 3 ft (91 cm) from support.

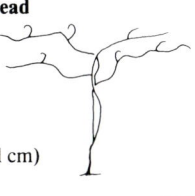

Vitis 'Brant' in fruit

VITIS VINIFERA
(Grape Vine Hybrids)

GRAPE VINE
Vitaceae
Deciduous
Fruiting Vine

Attractive climbers, displaying both pleasing foliage and edible fruits.

Origin Most hybrids of garden or nursery origin.
Use As a climber for walls and fences, to cover pergolas and similar structures. Can be used in the open garden on a post and wire framework or can be allowed to grow through trees. Many varieties are suitable for greenhouses and conservatories. Fruit can be eaten or used for making wine.
Description *Flower* Very small, soft green flowers of little interest, produced in mid to late spring. *Foliage* Three- or five-lobed, up to 6 in (15 cm) long and wide. Normally light green to grey/green. Good yellow autumn colour. *Stem* Light grey/green, quickly becoming grey/brown, finally grey. Not self-supporting. Attractive in winter. *Fruit* Hanging bunches up to 12 in (30 cm) long of round fruits, either purple or light green to soft yellow, with grey bloom. Edible, sweet flavoured. Some varieties suitable for wine-making.
Hardiness Tolerates a minimum winter temperature of 0°F (−18°C), except for those varieties marked for growing under protection.
Soil requirements Tolerates all soil conditions, except extremely dry and extremely wet. Varieties used for growing in greenhouses or conservatories should be planted outside and laid in through an appropriate opening to allow their roots adequate space.
Sun/Shade aspect Tolerates a wide range of aspects but the more sheltered and sunny the aspect, then the better the fruit. Some varieties must be grown under protection, not only to ripen fruit but to induce flowering.
Pruning Once established remove all annual side growth back as close to stem as possible in early spring to encourage fruiting.
Training Young plants should be cut to within 12 in (30 cm) of soil level in early spring. Train ensuing shoot or shoots upwards and horizontally. In second year again reduce vertical shoots to within 12 in (30 cm) of origin and reduce side shoots to one bud. Tie in ensuing growths vertically and horizontally and

Vitis vinifera 'Black Hamburg'

repeat this process until adequate height and area is covered. Tie to wires or individual anchor points.
Propagation and nursery production From vine eyes (buds) by taking one single bud and a small area of stem growth and inserting shallowly in a grit and sand rooting medium with assisted heat, or from hardwood cuttings approximately 4–6 in (10–15 cm) long taken in winter. Can also be layered by taking the tip of an existing plant and burying it in a sand/soil mixture. Once rooted, replant in final position. Purchase bare-rooted from mid autumn to early spring or container grown as available. Some varieties can be found in garden centres and good nurseries, others may have to be sought from specialist growers. Best planting height 10 in–4 ft (25 cm–1.2 m).
Problems The training requirements can be daunting. May suffer from mildew and vine weevils may attack. Proprietary controls will normally eradicate. Needs a large root run area to achieve good results.

Forms of interest
UK
OUTDOOR VARIETIES - WHITE 'Chardonnay' White fruit. A variety good for the cooler areas. A good wine grape. 'Nimrod' Seedless golden fruit for wine-making. Must have a sunny position. 'Madeleine Angevine 7972' Early autumn fruiting. Pale green fruit can be used for both dessert and for wine-making. Good in less favourable areas. 'Madeleine Sylvaner 2851' Good flavoured fruit used for both dessert and wine-making. Tolerates cool conditions. 'Mueller-Thurgau' Mid autumn fruiting. Possibly the best wine-making variety for the amateur. Golden brown fruits. Good in cool areas. 'Muscat de Saumur' A very old variety for wine-making with good muscat flavour. Must have a sunny position. 'Pinot Blanc' Very good wine-making grape which does well in cool English summers. 'Précoce de Malingre' Early fruiting. Good flavoured small grapes for dessert or wine-making. Must have a warm, sunny position. 'Seyval' Mid autumn fruiting. Large grapes with good flavour, best for wine-making. Prefers a warm, sheltered aspect. 'Seyve-Villard 5/276' Early fruiting. Golden fruit can be used for both dessert and wine-making. Good on alkaline soils. Must have a sheltered, warm position. 'Siegerrebe' Late fruiting. Good-flavoured, golden yellow fruit for dessert or wine-making. Heavy cropper. Dislikes severely alkaline soils. 'Traminer' Can be late fruiting. Rosy coloured grapes best for wine-making.
OUTDOOR VARIETIES - BLACK 'Baco 1' A wine-making variety. Strong growing, needs space. Heavy cropper. 'Black Hamburg' Often said to be best under protection but in the author's observation grows well outdoors in sheltered, sunny positions. For dessert and wine-making. 'Brant' Small fruits with a tart flavour for dessert and wine-making. Good ornamental foliage. 'Léon Millot' Good wine-making variety. Large crop. Strong growing, needs space. 'Marshall Joffre' Good wine-making variety. Strong growing, needs space. 'Millers Burgundy' ('Pinot Meunier') Small fruits, good for wine-making. Foliage attractive with a white woolly down when young. 'Pirovano 14' Good flavour. Can be used for both dessert or wine-making. Must have a sheltered, sunny position. 'Seibel 13053' Vigorous, needs space. Good for red or rosé wine-making. Must have a sunny position. 'Schuyler' Good flavour for wine-making. Strong-growing. Must have a sunny, warm wall. 'Strawberry Grape' Small dessert fruit with musky flavour. Of medium crop and vigour. Must have a warm, sunny position. 'Triomphe d'Alsace' Strong growing. Must have a warm, sheltered position. Good for red wine production.
VARIETIES FOR GREENHOUSES - WHITE 'Buckland Sweetwater' Very popular variety for dessert and wine-making. Sweet, juicy, amber-coloured fruit in good numbers. Best in unheated greenhouses. 'Chasselas d'Or' Late fruiting. Good-flavoured, golden yellow fruit for both dessert or wine-making. Must have heated greenhouse for success. 'Foster's Seedling' Early to mid season. Large crop of amber-coloured fruits of good flavour, sweet and juicy. For dessert or wine-making. Can be grown in both cold or heated greenhouses. 'Lady Hutt' White, sweet, good-flavoured fruits, borne late. For wine-making. Medium to heavy cropper. Best in unheated greenhouses. 'Mireille' Early fruiting. Large white grapes with good muscat flavour for wine. Best in unheated greenhouse. 'Mrs Pearson' White/yellow grapes with good muscat flavour for wine. Vigorous. Needs heat to induce flowering and ripening. 'Muscat of Alexandria' Possibly the best white variety under glass. Amber-coloured fruits, sweet and good-flavoured, for dessert or wine. May need additional heat to ripen fruits in poor summers. 'Syrian' White grapes, borne late. For wine-making. Strong growing, needs space. Heat is required for ripening and to induce good flavour. 'Trebbiano' Late, large crop of white fruits, good for wine. Very

Vitis vinifera 'Mueller-Thurgau'

Vitis vinifera 'Muscat of Alexandria'

strong growing, needs space. Must have heat to induce flowering.

VARIETIES FOR GREENHOUSES - BLACK **'Alicante'** Becoming very popular for wine-making. Black fruits with good flavour, borne late. May be shy to flower without additional heat in spring. **'Black Hamburg'** Performs well in greenhouses, both heated and cool. May also succeed outdoors in favourable sunny positions. For dessert and wine-making. Early to mid season fruiting. **'Frontignan'** Good muscat-flavoured black fruits for wine-making. Must have heat in the greenhouse. **'Gros Colmar'** Late, large crop of black grapes for wine-making. Very strong growing. Must have heat to induce flowering. **'Lady Downe's Seedling'** Black grapes. An old wine-making variety of medium vigour. Must have heat to induce flowering. **'Mrs Pince'** Old muscat variety with black fruits of good flavour for wine-making. Strong growing. Must have heat for flowering and ripening. **'Muscat of Hamburg'** Late, heavy cropping. Red to purple, muscat-flavoured fruit for dessert or wine-making. Needs heat for flowering and ripening.

USA

These varieties are best in the areas shown but may be suitable for wider areas depending on their hardiness.

Grapes for the Northeast and Midwest

AMERICAN VARIETIES **'Buffalo'** Mid season ripening. Medium to large clusters of red/black grapes, good for wine or juice. Good also in the Pacific Northwest. From New York. **'Catawba'** A red grape, very popular commercially for wine or juice. Ripens over a long season. Also good in southerly areas. From North Carolina. **'Cayuga White'** White grapes in tight clusters. Good for dessert. From New York. **'Concord'** Dark blue grapes with a rich flavour which is retained in wine. Late fruiting. From Massachusetts. **'Delaware'** A green grape used for wine-making and juice but also good as a dessert. Vines subject to mildew. From New Jersey. **'Edelweiss'** Medium-sized grapes good for dessert use. Hardy. From Minnesota. **'Fredonia'** Black fruits of good flavour for wine-making. Hardy. May sometimes be difficult to pollinate. From New York. **'Nimrod'** White seedless grape for wine-making. Vines can be brittle and need careful handling. Moderately hardy. Good also for the Northern states. From New York. **'Interlaken Seedless'** Medium-sized clusters of small, sweet, seedless, green/white grapes for wine-making. Moderately hardy. Also ideal for the Pacific Northwest. From New York. **'New York**

Muscat'** Red/black berries in medium clusters. Rich, fruity muscat scent. Good for wine and juice. Not entirely hardy. From New York. **'Niagara'** A very popular white grape for wine-making. Strong-growing and moderately hardy. From New York. **'Ontario'** White fruits in large open clusters. Good for wine. Strong-growing and moderately hardy. Does well on heavy soils. Also good for the Pacific Northwest. From Ontario, Canada. **'Schuyler'** Soft and juicy fruit with a tough skin. Good for wine. Moderately hardy. Good disease resistance. Ideal also for the Northwest. From New York. **'Seneca'** Somewhat small golden-skinned fruits which have a sweet, aromatic flavour. Good for wine. Hardy. Also good in the Pacific Northwest. From New York. **'Swenson Red'** Medium to large red fruits with good flavour for wine-making. Hardy. From Minnesota. **'Veesport'** Black fruits in medium-sized clusters. Good for wine and juice but also can be eaten as dessert. Strong-growing. From Ontario, Canada.

FRENCH HYBRIDS **'Aurore'** (**'Seibel 5279'**) Early fruiting soft white grape with a good flavour for wine. Strong-growing. Dislikes heavy soils. From France.

Grapes for the West 'Baco 1' Black grapes for wine-making. Heavy cropper. Strong-growing, needs space. Good over a wide area. From France. **'Cabernet Sauvignon'** Black fruits. Used to make red Bordeaux wine in France. From France. **'Cardinal'** Medium-sized clusters of large, dark red dessert fruits with green flesh. Early ripening. Ideal for training purposes. Good in coastal and valley areas. From California. **'Chardonnay'** White fruits for wine-making. Good in cooler areas. From France. **'Chenin Blanc'** White, strong-growing, medium-sized fruits for wine-making. Ideal for coastal areas. From France. **'Delight'** Early-ripening green/yellow dessert fruits with good firm flesh and muscat flavour. Best in coastal valley conditions. From California. **'Emperor'** Late-ripening large red dessert fruits with firm, crisp flesh. Stores well. Origin unknown. **'Flame Seedless'** Red to light red, seedless, sweet dessert fruits with a crisp texture. Early-ripening, medium-sized, loose clusters. Requires heat to ripen but best colour is developed with cool nights. From California. **'French Colombard'** Medium-sized white fruits with high acid content for wine-making. Good in coastal valleys and also the Central Valley of California. From France. **'Muscat of Alexandria'** Amber-coloured dessert grapes, sweet and good-flavoured. Best in Southern states or under protection in the North. From

North Africa. **'Niabell'** Large black dessert fruit of good flavour. Performs well in coastal valleys but will tolerate hot inland areas. Mid season ripening. Strong-growing. Resistant to powdery mildew. From California. **'Pierce'** Black dessert fruit requiring a hot summer to ripen. Strong-growing. Best in central California. From New York. **'Pinot Noir'** Small black grapes, used to make the French Burgundy wines. From France. **'Ribier'** Large, jet black fruits. Use for dessert. Best in hot inland areas. Fruits do not store well. From France. **'Thompson Seedless'** Green, mild-flavoured dessert fruits in good-shaped clusters, ripening early mid season. Used for raisin production. Will only grow well in hot climates. From Asia Minor. **'Tokay'** Large clusters of large, firm, red grapes with limited flavour for wine or dessert. Ripens late mid season. Does well in cool valley climates. From Algeria. **'Zinfandel'** Used both for red and white wines. Very reliable and a good variety to start with. Origin unknown.

Grapes for the Southwest

MUSCADINE VARIETIES Attention must be given to providing pollinators for some of these varieties. **'Hunt'** Dull black fruits of good quality. Ideal for wine-making and juice. Strong-growing and very productive. Use any other variety as a pollinator. From Georgia. **'Jumbo'** As its name would imply, very large black fruits. Good for wine. Ideal for garden use. Vines disease resistant. Use any other variety as a pollinator. From southern USA. **'Magoon'** Red/purple berries of medium size with attractive aromatic flavour for wine-making. Vines vigorous and heavy-cropping. Self-fertile. From Mississippi. **'Scuppernong'** Green to red/bronze fruit, depending on the amount of sun. Late ripening, sweet, juicy aromatic flavour. Good for dessert or wine-making. Use any other variety as a pollinator. From North Carolina. **'Southland'** Very large, purple, dull-skinned fruits with good flavour and high sugar content. Good for wine. Moderately vigorous and productive. Good for the central and southern Gulf Coast states. Self-fertile. From Mississippi. **'Thomas'** Small to medium red/black, very sweet fruits for wine-making. Excellent for fresh juice. Use any other variety as a pollinator. From southern USA. **'Topsail'** Sweet, green fruit, splotched with bronze, for winemaking. Good muscadine flavour. Limited production of fruit. Not hardy but disease resistant. Use any other variety as a pollinator. From North Carolina. **'Yuga'** Sweet red/bronze fruits of good quality for wine. Late ripening but somewhat irregular. Ideal for gardens. From Georgia.

Average height and spread
Five years
9 × 9 ft (2.7 × 2.7 m)
Ten years
18 × 18 ft (5.5 × 5.5 m)
Twenty years
30 × 30 ft (9 × 9 m)
Protrudes up to 2 ft (60 cm)
from support prior to spring pruning.

VITIS VINIFERA 'PURPUREA'

PURPLE-LEAVED COMMON GRAPE VINE, TEINTURIER GRAPE

Vitaceae   *Woody Climber*
Deciduous

A vine with attractive foliage and good ornamental fruit.

Origin From the Caucasus and Asia Minor.
Use As a rambling vine for walls, fences, pillars and pergolas and for covering roofs of gazebos and small buildings. Looks extremely fine when used in association with silver-leaved shrubs and trees.
Description *Flower* Small, oval, green, ageing to black, often with a blue bloom. *Foliage* Up to 6 in (15 cm) long and wide, hand-shaped, with three or five lobes. Claret red, downy

Vitis vinifera 'Purpurea' in leaf

upper surface, purple/blue undersides. Good purple autumn colour. *Stem* Purple becoming purple/green, eventually purple/brown. Loosely twining, medium to fast growing. *Fruit* Round grapes, $\frac{1}{4}$ in (5 mm) across, purple with blue bloom, carried in bunches.
Hardiness Tolerates a minimum winter temperature of 4°F (−15°C).
Soil requirements Tolerates all soil conditions with no particular preferences, only disliking extremely dry soils.
Sun/Shade aspect Tolerates all aspects, except extremely exposed. Best in full sun but will tolerate light shade.
Pruning Train into a vine system with each vine 15–18 in (38–45 cm) apart. Attempt to remove all other surplus vines in early spring. This will encourage better foliage, shape, size and colour. However, can be left free-growing without pruning.
Training Not self-clinging, but semi-twining and will climb by twisting around wires, trellis or other suitable thin supports. May need tying in if trained or when young.
Propagation and nursery production Semi-ripe cuttings taken in early summer. Always purchase container grown, best planting height 1$\frac{1}{2}$–3 ft (45–91 cm). Readily available from good garden centres and nurseries.
Problems Can suffer from a form of mildew which seems to be specific to this variety. Can be cured by a proprietary controlled fungicide and is not normally terminal or disfiguring.
Similar forms of interest None.
Average height and spread
Five years
10 × 10 ft (3 × 3 m)
Ten years
15 × 15 ft (4.6 × 4.6 m)
Twenty years
20 × 20 ft (6 × 6 m)
Protrudes up to 2 ft (60 cm) from support.

WATTAKA-KA SINENSIS (*Dregea sinensis*)

WATTAKAKA
Asclepiadaceae   **Tender Woody Climber**
Deciduous

An interesting climber but requiring a very sheltered growing location or the protection of a greenhouse or conservatory.

Origin From China.
Use As a small flowering climber for very sheltered, sunny walls.
Description *Flower* Up to 25 long-stalked, downy florets make up a flower 3 in (7.5 cm) across, with a central trumpet $\frac{1}{2}$ in (1 cm) wide with five lobes, red with small white dots. Produced in mid summer. *Foliage* Ovate, broad, pointed leaves up to 4 in (10 cm) long

and 3 in (7.5 cm) wide, produced on stalks up to 1$\frac{1}{2}$ in (4 cm) long. Light grey/green, with downy, velvet undersides. Some yellow autumn colour. *Stem* Mid green, sparse, loosely twining. Medium growth rate. *Fruit* In extremely favourable conditions and under the protection of a greenhouse or conservatory may produce interesting incurving fruits, consisting of two parts, down-covered, up to 2 in (5 cm) long and $\frac{1}{2}$ in (1 cm) wide and tapering in shape.
Hardiness Tolerates a minimum winter temperature of 25°F (−5°C).

Wattaka-ka sinensis in flower

Soil requirements Prefers a neutral to acid soil, but will tolerate a certain amount of alkalinity.
Sun/Shade aspect Requires a very sheltered aspect or the protection of a greenhouse or conservatory. Full sun to light shade.
Pruning Not normally required.
Training Allow to twine through wires or trellis.
Propagation and nursery production From seed grown under protection. Always purchase container grown; very scarce in commercial production and will have to be sought from specialist growers. Best planting height 6 in–2 ft (15–60 cm).
Problems Difficult to find and not fully hardy.
Similar forms of interest None.
Average height and spread
Five years
4 × 4 ft (1.2 × 1.2 m)
Ten years
8 × 8 ft (2.4 × 2.4 m)
Twenty years
12 × 12 ft (3.7 × 3.7 m)
Protrudes up to 2 ft (60 cm) from support.

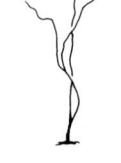

WEIGELA

KNOWN BY BOTANICAL NAME
Caprifoliaceae   **Wall Shrub**
Deciduous

Early summer flowering shrubs adapting well to fan-training on walls and fences.

Origin Most forms from Japan, Korea, north China and Manchuria but many cultivars and hybrid varieties are of garden origin.
Use As a freestanding shrub grown in front of walls or fences, or can be fan-trained to good effect.
Description *Flower* Funnel-shaped, 1 in (2.5 cm) long and $\frac{1}{2}$ in (1 cm) wide, in varying shades from yellow to white, pink and red. Flowers produced on wood two years old or more, late spring through early summer, possibly with intermittent flowering through late summer and early autumn. *Foliage* Leaves ovate, 1$\frac{1}{2}$–5 in (4–12 cm) long, dark to mid green with some light green varieties and golden and silver variegated. Yellow autumn colour. *Stem* Upright, becoming spreading with age. Grey/green to grey/brown. Medium growth rate.
Fruit Seedheads dark to mid brown, of some attraction in winter.
Hardiness Tolerates a minimum winter temperature of 0°F (−8°C).
Soil requirements Any soil.
Sun/Shade aspect Good in exposed situations. Prefers full sun, tolerates light to medium shade.
Pruning From two years after planting remove one third of old flowering wood annually to ground level, after flowering.
Training Allow to stand free or secure to the wall or fence by wires or by individual anchor points.
Propagation and nursery production From semi-ripe cuttings taken in early summer or hardwood cuttings in winter. Purchase container grown or bare-rooted. Most varieties easy to find but some may have to be obtained from specialist nurseries. Best planting height 15 in–2$\frac{1}{2}$ ft (38–76 cm).
Problems If unpruned can become too woody and flowers will diminish in size and number. Large, established shrubs can be cut to ground level and will regenerate, but will take two years to come into flower. *Weigela* was once classified with the closely related *Diervilla* but in recent years these shrubs have been classified separately.
Forms of interest *W.* 'Abel Carrière' Large trumpet-shaped, rose-carmine flowers with gold markings in throat, opening from purple/carmine buds. Good, bold, green foliage. *W.* 'Avalanche' A good, strong-growing, white flowering variety. May have to be obtained from specialist nurseries. *W.* 'Boskoop Glory' Large, trumpet-shaped

Weigela 'Abel Carrière' in flower

Weigela 'Aureovariegata' in leaf

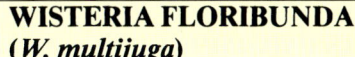

WISTERIA FLORIBUNDA
(*W. multijuga*)

JAPANESE WISTERIA

Leguminosae   **Woody Climber**
Deciduous

An attractive group of climbers not always receiving the attention they deserve.

Origin From Japan.

Use As a large climber for sunny walls, fences, pergolas or growing through large shrubs or trees. Can be trained into a small weeping standard.

Description *Flower* Racemes 10–12 in (25–30 cm) long of small pea-shaped flowers up to $\frac{1}{2}$–$\frac{3}{4}$ in (1–2 cm) long and wide. Violet/blue, pink or white, depending on variety. *Foliage* Light grey/green to olive green. Up to 15 in (38 cm) long, pinnate, consisting of up to 19 leaflets each 1 in (2.5 cm) wide and 3 in (7.5 cm) long. Oval in shape. Good yellow autumn colour. *Stem* Grey/green, twining,

flowers, rose pink ageing to salmon pink. A beautiful form. Two thirds average height and spread. *W.* **'Bristol Ruby'** Possibly the most popular of all flowering forms. Ruby red flowers profusely borne on upright, strong shrub in late spring/early summer. *W.* **'Candida'** Pure white flowers with slightly green shading. Light green foliage on arching stems. Two thirds average height and spread. *W.* **'Carnival'** Flowers pink and white on the same shrub. *W.* **'Eve Rathke'** Bright crimson-red flowers with yellow anthers, produced over a long period from late spring to late summer. *W. florida* **'Albovariegata'** Attractive creamy white edges to ovate leaves.

Weigela 'Boskoop Glory' in flower

Pale to mid pink flowers produced profusely on stems two or three years old. Two thirds average height and spread. *W. f.* **'Aureovariegata'** A variety with yellow variegation, often producing pink to red tinged leaves, particularly during autumn. Pink flowers profusely produced in late spring to early summer. *W.* **'Looymansii Aurea'** Flowers pale pink, contrasting with foliage which is light golden-yellow in spring, ageing to lime-yellow in autumn. Must be in light shade or it scorches. Obtain from specialist nurseries. Often looks weak when young. One third average height and spread. *W.* **'Lucifer'** Very large red flowers carried later than most. *W. middendorffiana* Arching branches with attractive grey/green winter wood. Flowers bell-shaped, sulphur yellow with dark orange markings on lower lobes, mid to late spring.

Ovate, light green foliage with some yellow autumn colour. Prefers light shade, although not fussy. An all-round attractive variety, reaching two thirds average height and spread. From Japan, northern China and Manchuria. *W.* **'Mont Blanc'** Fragrant white flowers, strong-growing. Obtain from specialist nurseries. *W.* **'Newport Red'** Good, dark red flowers. Two thirds average height and spread. *W. praecox* **'Variegata'** A variety with ovate to obovate, creamy white variegated foliage, rigid and deeply veined. Flowers honey-scented, rose pink with yellow markings in throat in late spring/early summer. Obtain from specialist nurseries. From Japan, Korea and Manchuria. *W.* **'Rubidor'** Golden yellow variegated foliage which may be susceptible to scorching in strong sunlight. Dark pink flowers. *W.* **'Stelzneri'** Good mid pink flowers borne in profusion. Interesting upright growth. Not widely available. *W.* **'Styriaca'** Carmine-red flowers produced in good quantities in late spring/early summer. Strong, old-fashioned variety.

Average height and spread
Five years
4 × 4 ft (1.2 × 1.2 m) freestanding
5 × 5 ft (1.5 × 1.5 m) fan-trained
Ten years
5$\frac{1}{2}$ × 5$\frac{1}{2}$ ft (1.6 × 1.6 m) freestanding
7 × 7 ft (2.1 × 2.1 m) fan-trained
Twenty years
7 × 7 ft (2.1 × 2.1 m) freestanding
9 × 9 ft (2.7 × 2.7 m) fan-trained
Protrudes up to 3 ft (91 cm) from support if fan-trained, 6 ft (1.8 m) untrained.

Wisteria floribunda in flower

ageing to grey/brown. Fast growing. *Fruit* May produce, in warm summers, pods $\frac{1}{3}$ in (1 cm) wide and 6 in (15 cm) long, velvety, grey/green in colour. Poisonous.

Hardiness Tolerates a minimum winter temperature of 14°F (−10°C).

Soil requirements Does well on all soil conditions, only disliking extremely alkaline types. Must have adequate root-run to allow it to spread and produce the maximum amount of top growth possible.

Wisteria floribunda 'Alba' in flower

373

Sun/Shade aspect Warm, sheltered aspect. Full sun to very light shade.

Pruning Prune tendrils in late summer and autumn to form a framework, then remove all current season's tendrils back to two buds from the point of origin.

Training Train to a vine system with one vine every 18 in (45 cm) laterally from a central upright stem.

Propagation and nursery production Grafted plants should always be chosen from a known parent source. Always purchased container grown, best planting height 2–4 ft (60 cm–1.2 m). May have to be sought from specialist nurseries.

Problems Often planted in areas where it cannot reach its full potential. Some varieties may be hard to find.

Similar forms of interest *W. f.* 'Alba' Attractive white flowers in racemes up to 2 ft (60 cm) long. *W. f.* 'Rosea' Pale rose-pink flowers. Scarce. *W. f.* 'Violacea Plena' Double, violet-purple flowers.

Average height and spread
Five years
10 × 10 ft (3 × 3 m)
Ten years
20 × 20 ft (6 × 6 m)
Twenty years
30 × 30 ft (9 × 9 m)
Protrudes up to 4 ft (1.2 m) from support.

WISTERIA FLORIBUNDA 'MACROBOTRYS'

JAPANESE WISTERIA

Leguminosae   **Woody Climber**
Deciduous

Possibly one of the most spectacular of all wisterias, with its flowers exceeding any other variety in length.

Origin From Japan.
Use As a free-growing climber for walls, fences, pergolas, through trees and over large shrubs. Can also be grown as a small weeping standard tree.
Description *Flower* Racemes of dark violet/purple, pea-shaped flowers up to 21 in (53 cm) long and 5 in (12 cm) wide. Racemes produced very uniformly along the branches and giving an effect of a waterfall. *Foliage* Pinnate leaves up to 18 in (45 cm) long and 8 in (20 cm) wide with each leaflet up to 4 in (10 cm) long. Light green, giving good yellow autumn colour. *Stem* Twining, smooth, light grey/green when young, ageing to yellow/brown and finally dark/grey brown. Fast growing. *Fruit* May produce inedible, green pea pods in very hot summers but not reliable or attractive.
Hardiness Tolerates a minimum winter temperature of 14°F (−10°C). In very cold wind chill conditions stems may be killed in winter but rejuvenation from the base normally occurs the following spring.

Wisteria floribunda **'Macrobotrys' in flower**

Wisteria sinensis **in flower**

Soil requirements Does well on all soil types but to achieve full potential, needs a root run of 50–64 ft (15–20 m). Very alkaline soils may induce chlorosis, identified by a yellowing of the leaves in mid summer. This will restrict flowering but normally will not kill the plant.
Sun/Shade aspect Must be in full sun to very light shade, otherwise will not flower.
Pruning Cut back all current season's growth to within two buds of point of origin in mid to late autumn. This pruning is carried out once the framework has been achieved. Over shrubs and large trees it is not normally possible to prune and the climber must mature before flowering to its full potential.
Training It is not a good policy to retain all of the season's growth. Only use tendrils produced in current season to form a framework of shoots, ideally one or two major upright shoots with side branches encouraged to grow 18 in (45 cm) apart. Use wires on walls and fences to achieve the framework training.
Propagation and nursery production Must always be grafted, will not come true from seed. Always purchase container-grown plants, may have to be sought from specialist nurseries. Best planting height 2–6 ft (60 cm–1.8 m).
Problems May be difficult to find. May be slow to establish, taking three years before major growth is forthcoming. Takes up to six or eight years to flower and then possibly another four years to come into full flowering potential but truly worth the wait.
Similar forms of interest None.
Average height and spread
Five years
18 × 18 ft (5.5 × 5.5 m)
Ten years
36 × 36 ft (11 × 11 m)
Twenty years
50 × 50 ft + (15 × 15 m +)
Will continue to spread beyond this for up to 50 years. Protrudes up to 3 ft (91 cm) from support.

WISTERIA SINENSIS

CHINESE WISTERIA

Leguminosae   **Woody Climber**
Deciduous

Of all the climbing plants from China this surely must be one of the most spectacular with its long flowing racemes of blue flowers.

Origin From China.
Use As a fast-growing climber for sunny walls and fences, for growing over pergolas, buildings and through trees. Can be trained as a small weeping tree.

Description *Flower* Racemes of mid blue pea-like flowers, 6–9 in (15–23 cm) long and 3–4 in (8–10 cm) wide, are produced in profusion from mid to late spring. *Foliage* Pinnate leaves up to 12–18 in (30–45 cm) long and 4–6 in (10–15 cm) wide with seven to nine leaflets. Each leaflet 3 in (7.5 cm) long by ½ in (1 cm) wide. Light green with good yellow autumn colour. *Stem* Twining, smooth, light grey/green when young, ageing to yellow/brown and finally dark/grey brown. Fast growing. *Fruit* Long grey/green, bean-like pods, 6–8 in (15–20 cm) long, ¼–½ in (5 mm–1 cm) wide, produced in hot summers. Poisonous.

Wisteria sinensis **'Alba' in flower**

Hardiness Tolerates a minimum winter temperature of 14°F (−10°C). In very cold wind chill conditions stems may be killed in winter but rejuvenation from the base normally occurs the following spring.
Soil requirements Does well on all soil types but must have adequate root run – needs up to 20–30 ft (6–9 m) to perform to its full potential. On very alkaline soils may show signs of chlorosis in the form of a yellowing of the leaves in mid summer; although this will restrict flowering, it is not normally terminal.
Sun/Shade aspect Must be in full sun to very light shade, otherwise will not flower.
Pruning Cut back all current season's wood in mid to late autumn to within two buds from point of origin. This procedure is carried out once basic framework is achieved and the more that can be removed, the sooner it will flower. Over shrubs and large trees this

is not normally possible and the climber must mature to achieve full flowering.

Training Use the long tendrils produced in current season to form a frame-work of shoots, ideally with one to two upright shoots with side branches encouraged to grow 18 in (45 cm) apart. It is not a good policy to attempt to retain all of the season's growth. Wires will be required on walls and fences to achieve the framework system.

Propagation and nursery production Can be grown from seed but resulting plants may vary greatly in their final flowering performance, flowers being between a light grey to a good blue and racemes from 3–9 in (7.5–23 cm) long. Best results are obtained from plants grafted on to rootstocks of *W. sinensis* and the graft taken from a known, good source. Can be planted bare-rooted from mid autumn to mid spring or container grown any time except mid summer, with early summer for preference. Best planting height 2–6 ft (60 cm–1.8 m). Readily available from garden centres and nurseries.

Problems Its requirement for a sunny wall and the area that it covers are often underestimated. Because of its propagation method, can be expensive.

Similar forms of interest *W. s.* **'Alba'** A white-flowering variety. *W. s.* **'Plena'** Rare double blue flowering variety. *W. s.* **'Prematura'** Flowers earlier in its life than its parent. Mauve blooms. *W. s.* **'Prematura Alba'** An earlier, white-flowering variety. *W. s.* **'Variegata'** Golden variegated foliage, violet blue flowers. Poor grower.

Average height and spread
Five years
18 × 18 ft (5.5 × 5.5 m)
Ten years
36 × 36 ft (11 × 11 m)
Twenty years
50 × 50 ft+ (15 × 15 m+)
Will continue to grow for
up to 50 years. Protrudes up   4 ft (1.2 m)
from support.

WISTERIA SINENSIS (Modern Hybrids)

HYBRID CHINESE WISTERIA

Leguminosae   *Woody Climber*
Deciduous

A range of varieties with blooms of differing colours and all reliable for their flowering performance.

Origin Of nursery and garden origin.
Use As fast-growing climbers for walls, fences, pergolas, buildings and through trees and large shrubs. Can be trained as a small weeping standard.
Description *Flower* Racemes of pea-like flowers, 6–9 in (15–23 cm) long, 3–4 in (8–10 cm) wide are produced in profusion from mid to late spring. Colour will depend on variety, ranging through pink, white, purple and blue with some bi-colours. *Foliage* Light green pinnate leaves up to 12–18 in (30–45 cm) long, with seven to nine leaflets. Each leaflet 3 in (7.5 cm) long by ½ in (1 cm) wide. Good yellow autumn colour. *Stem* Twisting, smooth light grey/green when young, ageing to yellow/brown and finally dark grey/brown. Fast growing. *Fruit* In very hot summers may produce inedible, green bean pods but not reliable.
Hardiness Tolerates a minimum winter temperature of 14°F (−10°C). Stems may be killed by very cold wind chill conditions but rejuvenation normally occurs the following spring.
Soil requirements Does well on all soil types but requires root run of at least 16 ft (4.9 m). On very alkaline soils may show signs of chlorosis in the form of a yellowing of the leaves in mid summer.
Sun/Shade aspect Plant in a sheltered aspect in full sun to very light shade otherwise it will not flower.
Pruning Once the framework of climber is

Wisteria sinensis **'Pink Ice'** in flower

established, cut back all current season's wood in mid to late autumn to within two buds of point of origin.
Training Use the long tendrils with, ideally, one or two major upright shoots and side branches encouraged to grow 18 in (45 cm) apart. Wires will be required on walls and fences to support the trained growth. Over shrubs and large trees training is not normally possible and the climber must mature to achieve full flowering.
Propagation and nursery production Will not come true from seed, must always be grafted. Always purchase container-grown, may have to be sought from specialist nurseries. Best planting height 2–6 ft (60 cm–1.8 m).
Problems May be in limited supply. Can take up to five to eight years to come into full flowering. May be expensive due to its method of propagation.
Forms of interest *W. s.* **'Black Dragon'** Double, dark purple flower racemes up to 30 in (76 cm) long. *W. s.* **'Caerulea'** White flowers. *W. s.* **'Caroline'** Deep blue, scented flowers, free-flowering. *W. s.* **'Domino'** (syn. *W. s.* **'Issai Fuji'**) Lilac blue flower racemes 20 in (50 cm) long well before the foliage. *W. s. formosa* **'Issai'** Lilac blue flowers from an early age. *W. s.* **'Peaches and Cream'** (syn. *W. s.* **'Kuchibeni'**) Rose pink in bud opening to off-white flowers in racemes 40 in (1 m) long. *W. s.* **'Pink Ice'** (syn. *W. s.* **'Hond Beni'**) Rosy-pink flower racemes up to 40 in (1 m) long. *W. s.* **'Purple Patches'** (syn. *W. s.* **'Murasaki Naga Fuji'**) Racemes of violet-purple flowers 30 in (76 cm) long. *W. s.* **'Snow Showers'** (*W. s.* **'Shiro Naga Fuji'**) Racemes up to 50 in (1.3 m) long of pure white flowers.
Average height and spread
Five years
18 × 18 ft (5.5 × 5.5 m)
Ten years
36 × 36 ft (11.5 × 11.5 m)
Twenty years
50 × 50 ft+ (15 × 15 m+)
Will continue to grow for 50 years. Protrudes up to 4 ft (1.2 m) from support.

XANTHOCERAS SORBIFOLIUM

YELLOWHORN

Sapindaceae   *Wall Shrub*
Deciduous

An interesting, rare, spring flowering shrub with attractive foliage, adapting well for planting against a tall wall or fence.

Origin From northern China.
Use As a tall, upright, trained wall specimen.

Description *Flower* White flowers with carmine eyes, over 1 in (2.5 cm) wide, presented in upright panicles 4 in (10 cm) wide and 5 in (12 cm) long on previous year's wood in late spring. *Foliage* Leaves pinnate, consisting of nine to 17 lanceolate, sharply toothed leaflets 8–10 in (20–25 cm) long, giving some yellow autumn colour. *Stem* Upright, can be trained flat against a wall to good effect. Grey/green. *Fruit* Shaped like a child's top, three-valved, walnut-like seed pods, containing large numbers of chestnut-like seeds.
Hardiness Tolerates a minimum winter temperature of 14°F (−10°C).
Soil requirements Tolerates most soils, including alkaline.

Xanthoceras sorbifolium in flower

Sun/Shade aspect Requires some protection In full sun to light shade.
Pruning None required except for training.
Training Tie to wires or individual anchor points to form an upright, rather than fan-trained, shape.
Propagation and nursery production From seed. Purchase container grown; very hard to find, must be sought from specialist nurseries. Best planting height 1–5 ft (30 cm–1.5 m).
Problems Slightly slow to develop and plants are usually small when purchased.
Average height and spread
Five years
5 × 7 ft (1.5 × 2.1 m)
Ten years
7 × 9 ft (2.1 × 2.7 m)
Twenty years
9 × 15 ft (2.7 × 4.6 m)
Protrudes up to 2 ft (60 cm)
from support.

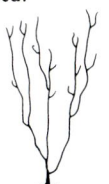

PLANTING TREES, SHRUBS AND CLIMBING PLANTS

Successful establishment of a tree, shrub or climbing plant begins with the preparation of a correctly sized planting hole. The planting process may seem somewhat laborious, but it is worthwhile providing the best conditions in which the plant can grow and thrive rather than merely survive, as a tree, shrub or plant may be the focal point of the garden and can give years of pleasure if allowed to establish itself properly.

Any perennial weed roots, such as couch grass, dock or thistles, must be cleared from the site, otherwise they become almost impossible to remove and grow strongly in competition. If the planting area is grass-covered, turves should be removed before preparation of the planting hole. If free from weeds, they may later be replaced upside down at least 9in (23cm) deep in the planting hole to conserve moisture in the turned-over soil after planting.

If a broad area is to be planted, such as a shrub border, it can be prepared by double-digging the whole area and incorporating compost or well-rotted manure. Holes can then be dug for the shrubs, which need only accommodate the depth and width of the roots, since the ground has been prepared.

Preparing the planting hole

The diameter and depth of the planting area depends upon the size of a tree, shrub or climbing plant as follows: for trees up to 9ft (2.8m) when purchased, the planting hole should be 3ft (1m) in diameter. For larger trees, up to 16ft (5m), the required size of planting hole is 4½ft (1.4m) or more.

The depth of preparation is the same for trees of all heights—18in (50cm). The soil is worked over in two stages: planting depth corresponds to the original depth of soil around the tree or shrub, whether it is supplied bare-rooted or root-balled (balled-and-burlapped) or has been grown in a container, but a similar depth of soil is broken up and prepared below the actual planting depth.

Remove the topsoil to a depth of 9in (25cm). Store the soil on a flat board beside the planting hole and add to it half a bucket of compost or well-rotted farmyard manure. Fork over and break up a further 9in (25cm) of subsoil. Remove any weed roots from the soil as it is turned. Dig in half a bucket of compost

or well-rotted farmyard manure.

For shrubs reaching an ultimate height of not more than 18in (50cm), the minimum diameter of the planting hole should be 2ft (60cm). For shrubs which will ultimately exceed this height, the required size of planting hole is 3ft (1m). The depth of the prepared planting area is the same in both cases, 18in (50cm); soil preparation is in two stages as described above for trees.

The planting of shrubs and climbing plants raises a number of problems, the most important of which is that the plants will dry out if in close proximity to walls and fences. It is no less vital than with any other plant to prepare the soil thoroughly to an adequate depth to allow for good plant growth and to provide a hole of a diameter not less than 3ft (91cm).

It is often the case when planting shrubs and climbing plants that the planting hole is in close proximity to a path or other obstruction, so a diameter of 3ft (91cm) may not always be practicable. In such cases prepare the hole to the same overall volume but in a different shape - for example, an oblong of 4 × 2ft (1.2m × 60cm). Occasionally even this is not possible and if the open surface area cannot be achieved then excavation under the surrounding obstruction must be considered.

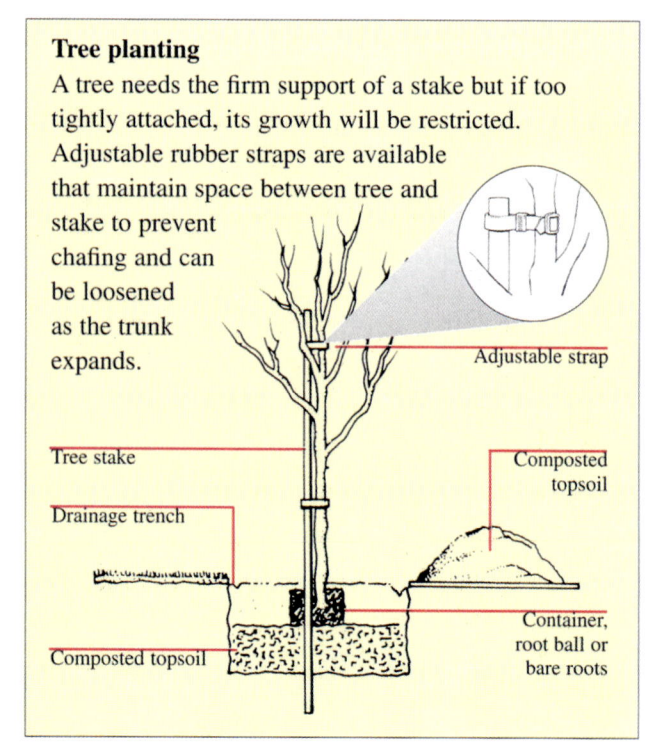

Tree planting
A tree needs the firm support of a stake but if too tightly attached, its growth will be restricted. Adjustable rubber straps are available that maintain space between tree and stake to prevent chafing and can be loosened as the trunk expands.

Adjustable strap

Tree stake

Drainage trench

Composted topsoil

Composted topsoil

Container, root ball or bare roots

Remove the top 9in (23cm) of topsoil and fork over the lower 9in (23cm). Add a good quantity of organic material to the latter and at least 25 per cent by volume into the former. Organic material holds moisture and plant foods and allows for rapid root growth and, therefore, fast establishment of the shrub or climbing plant.

Planting bare-rooted trees, shrubs or climbing plants

The basic planting method is the same for bare-rooted or root-balled (balled-and-burlapped) plants as for container-grown, but it is even more important that the roots should not be allowed to dry out. When returning the topsoil to the planting hole, take care to work it well in around the roots of the tree, shrub or climbing plant, leaving no air pockets in the soil.

Planting container-grown trees, shrubs or climbing plants

A container-grown tree, shrub or plant should be well-watered before it is planted, ideally at least one hour beforehand.

Place the plant, still in its container, in the planting hole and adjust the depth so the rim of the container is just below the surrounding soil level. If the tree, shrub or plant is in a rigid or flexible plastic container, you can now remove this, taking care not to disturb the soil ball around the roots. (A young tree will need staking at this point—see below.) Peat composition or treated paper pots can be left in place as the material is decomposable, but if the surrounding soil is dry or the planting takes place in midsummer, it is best to remove the pot.

Never lift the plant by its trunk or stems, as this can tear and damage the roots. Handle the whole rootball carefully and, once the container had been removed, take care that small exposed roots do not dry out.

Replace the prepared topsoil around the root ball of the plant to the level of the soil around the planting hole. Tread the soil gently all around the plant to compress it evenly. Unless the soil is very wet, pour a bucket of water into the depressed area.

Fill the area with more prepared topsoil; bringing the level up just above that of the surrounding soil. When planting a tree or shrub, dig a small V-shaped trench, 3in (8cm) deep and wide, around the planting area to allow drainage. Any surplus topsoil should be used elsewhere in the garden, not heaped around the newly planted tree, shrub or plant.

Staking a tree

A young tree should be staked as soon as it is placed in the planting hole. The stake should be at least 1-1½in (3-4cm) thick, round or square-sectioned with a pointed tip, and treated to resist rotting. Select a suitable length to support the height and weight of the tree—the top of the stake should extend well into the upper stems or branches, and the point should go into the subsoil to a depth of 18in (50cm).

When the tree is in place in the planting hole, push the stake through the soil ball and into the prepared subsoil below. If it meets a definite obstruction, remove the stake and try in a different spot, but not more than 2in (5cm) from the tree stem. When you find a point where the stake pushes easily, drive it into the subsoil to the required depth.

To support the tree, fasten two adjustable ties, with small spacing blocks to hold the tree clear of the stake, one on the stem among the branches as high as is practical, the other halfway up the main stem or trunk. This encourages straight growth and supports the tree in high wind or heavy snow. From time to time as the tree grows, check the ties and loosen them as necessary so they do not restrict the trunk. Ties of plastic or rubber are available, fastened with a buckle so they are easily loosened as the tree grows. Most trees need staking for up to five years before they are sufficiently strong to support themselves.

Planting climbers to ramble through trees

Climbing plants that are intended to ramble through trees should have holes prepared as for those planted against walls. Dig the planting hole on the outer edges of the tree canopy and lead a string, chain or wire from ground level to a branch, taking care not to damage the latter. A black flexible tree strap around the branch will protect it. The plant can then be tied to the wire and left to climb up into the tree branches. This process is slow and it may often be several years before the full effect is achieved. It should also be borne in mind that the climbing plant may eventually damage the tree.

Planting position for climbing plants

Plant the shrub or climbing plant at least 15-18in (38-45cm) away form its support, laid back towards it, using either the cane it was supplied with or providing one if not. Planting this distance away from the wall will ensure that the plant always has an adequate moisture supply.

Growing trees, shrubs and climbing plants in pots and tubs

As a general rule it is not advisable to attempt to grow shrubs and climbing plants in containers because the lack of soil will render the plant unable to make growth. Even with a large container filled with a good potting compost and given regular feeding, only 20-30 per cent of the potential growth of any shrub or climbing plant will be achieved and this growth will always be at risk from drying out in summer thus causing damage, often of lasting effect. There are certain exceptions to this rule, and they are the trees *Acacia* (Hardy Forms); *Cotoneaster* (all entries); *Cytisus × praecox* and *scoparius; Ilex.* Exceptions among shrubs include: *Acer japonicum* and *palmatum; Aucuba japonica; Buxus sempervirens; Callistemon; Camellia japonica; Cotoneaster* (all entires); *Cytisus* (all entries bar *battandieri*); *Euonymus* (evergreen varieties); *Fatsia japonica; Ficus carica; Fuchsia* (Hardy Forms); *Griselinia littoralis; Hedera* (all entries); *Hydrangea* (all entries); *Ilex* (all entries); *Laurus nobilis; Ligustrum* (all entries); *Nerium oleander; Olearia; Phormium; Pittisporum; Prunus laurocerasus* and *lusitanicus; Rhododendron* (all entries); *Rosmarinus; Salvia officinalis; Santolina; Skimmia; Trachycarpus fortunei; Yucca.*

Watering

The danger period for loss of any plant is in the spring and summer following planting. In dry conditions, water the tree, shrub or plant well at least three times a week. Stems and foliage benefit from an all-over fine spray of water at the same time; if new leaves and wood become dehydrated the plant may die.

Feeding

If the tree, shrub or plant is planted in spring, early or late in the season, apply one gloved handful of bonemeal per sq. yd (sq.m). If planting is carried out in any other season, apply the bonemeal early in the spring following planting, and repeat annually thereafter. A general purpose liquid fertilizer can also be given annually in midsummer.

Planting times—UK and Europe

Bare-rooted or root-balled trees, shrubs or climbing plants can be planted at any time from late autumn to early spring, except in the harshest of winter conditions. Do not plant when the ground is frozen or waterlogged. Container-grown trees, shrubs or climbing plants can be planted at any time of year, but with autumn and spring for preference, unless weather conditions are extreme. Do not plant when the ground is frozen, dried hard, or water-logged.

Planting times—USA

The best time to plant bare-rooted trees, shrubs or climbing plants is in late winter or early spring, just before bud-break. Bare-rooted plants lose most of their root surface—and water-absorbing capacity—during transplanting. New roots will not develop until spring, so if you plant in fall, there is the risk that buds and twigs will dry out over winter.

Autumn is the best time to plant balled-and-burlapped and container-grown trees, shrubs and climbing plants, because it gives them a long season of cool air and warm soil for strong root growth. Root put on most of their year's growth after leaf-fall. Trees, shrubs and climbing plants planted as early as possible after their leaves have dropped will be able to establish a powerful root system before the soil temperature drops, and will therefore require less watering in the following season.

Planting times—Australia and New Zealand

Follow the same rules as for **Planting times—UK and Europe.** It is advisable to avoid planting container-grown trees and shrubs in midsummer when conditions are extremely hot and dry.

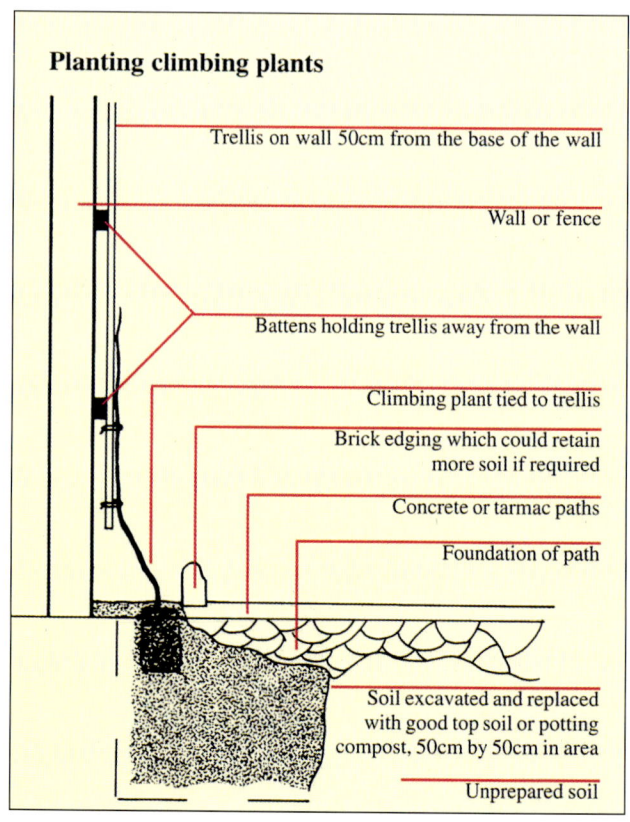

Planting climbing plants

Trellis on wall 50cm from the base of the wall

Wall or fence

Battens holding trellis away from the wall

Climbing plant tied to trellis

Brick edging which could retain more soil if required

Concrete or tarmac paths

Foundation of path

Soil excavated and replaced with good top soil or potting compost, 50cm by 50cm in area

Unprepared soil

PRUNING

Correct pruning of a tree or shrub can improve and increase the plant's flowering and fruiting, the size and colour of foliage, and the appearance of attractive stems and bark. Most importantly for the gardener, it controls and shapes the growth of the tree or shrub, so that it remains an attractive garden feature suited to its location.

Specific pruning instructions for individual trees or shrubs are given in the dictionary entries. In some cases it is not necessary to prune, but all trees and shrubs should be inspected every spring for broken twigs and branches and other signs of damaged wood, which is not only unsightly but also vulnerable to disease. Soft or woody growth which has suffered winter die-back should also be cut back in spring. Trees frequently develop crossing branches which may rub together and cause a lesion that may become the site of various diseases; this also occurs in shrubs, but less commonly. The weaker of the two branches should be removed in winter, or while the plant is dormant, and this is also the best time to check for signs of fungus disease, as fungus spores are inactive in winter and if the damaged area is removed there is little chance that the disease will spread.

Methods of pruning

There are some slow-growing trees and shrubs which require no pruning, while others actively resent pruning and if cut will tend to die back. It may also be inadvisable to prune if this will remove flower buds at the terminals of twigs or branches. Otherwise, a plant can be lightly trimmed or pruned back hard, or one-third of the growth may be selectively removed to encourage new growth production.

Many shrubs gain improved flowering performance in the coming season if all growth is cut back to ground level in mid to late spring. The shrub will rapidly rejuvenate and there is no advantage to pruning less severely where cutting right back is recommended. This type of pruning also improves the production of winter stems in plants which show good winter shape or colour, and will also encourage larger foliage in some varieties. Hard pruning is slightly less drastic, consisting of cutting back all growth to new growth points on woody stems near the plant base. Some young shrubs needing time to become established will gain a more compact shape and finer foliage if the previous year's growth is cut back by half in early spring.

Another method of encouraging better flowering is to remove one-third of the oldest growth or old flowering wood as soon as the flowering period has ceased. If this is not carried out on shrubs which can benefit from this type of pruning, the stems grow taller and less strong and the flower display becomes less attractive. This type of pruning is also used to encourage rejuvenation of shrubs which, if pruned back hard, would lose all flowering shoots and subsequently fail to flower for up to three years following pruning. One-third of old growth is removed in spring of the first year; in the following spring, another one-third of the mature wood, and again in the third year when the last of the old growth is removed.

Pollarding is a system of pruning applied to some trees, in which all growth produced in the previous year is cut back in late winter or early spring to form a crown of old growth from which the new spring

Pruning cuts

It is important to cut a stem at the correct angle when pruning, at a point slightly above a bud with the cut sloping away from it at a gentle angle. Sharply angled or horizontal cuts are incorrect as is a cut sloping the wrong way which can cause damage by directing excessive moisture towards a bud.

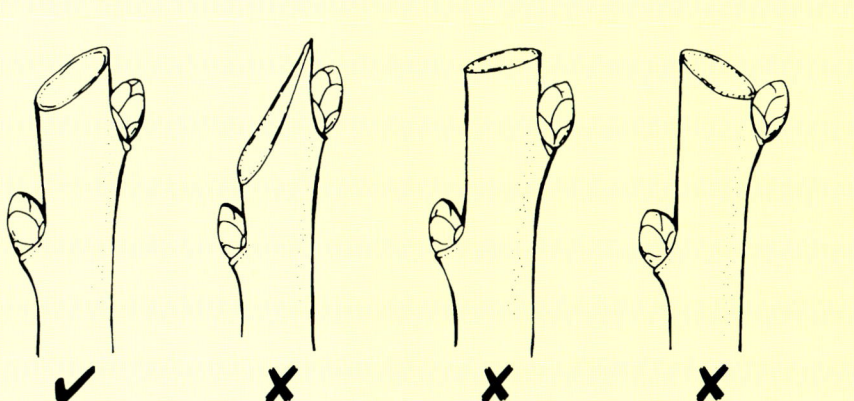

display emerges vigorously. It is also possible to cease pruning for two to three years to allow the tree to increase in size, resuming the pollarding process again when the tree becomes too large or its features are losing some of their interest.

You can distinguish between one- or two-year-old wood and old or mature wood by colour and texture.

One-year-old wood is new growth produced between the spring and autumn of the same year, light in colour and relatively flexible with a soft texture. Two-year-old wood is the previous season's growth, usually darker and stiffer with the beginnings of a bark. Old or mature wood is shown by thick stems, dark colour and noticeably tougher bark texture.

Pruning by one-third

Removing one-third of old growth, or of wood which has flowered in the previous season, encourages vigorous flowering and rejuvenation of the shrub. It is also an opportunity to remove any damaged or crossing branches. Forsythia is an example of a shrub which benefits from this treatment.

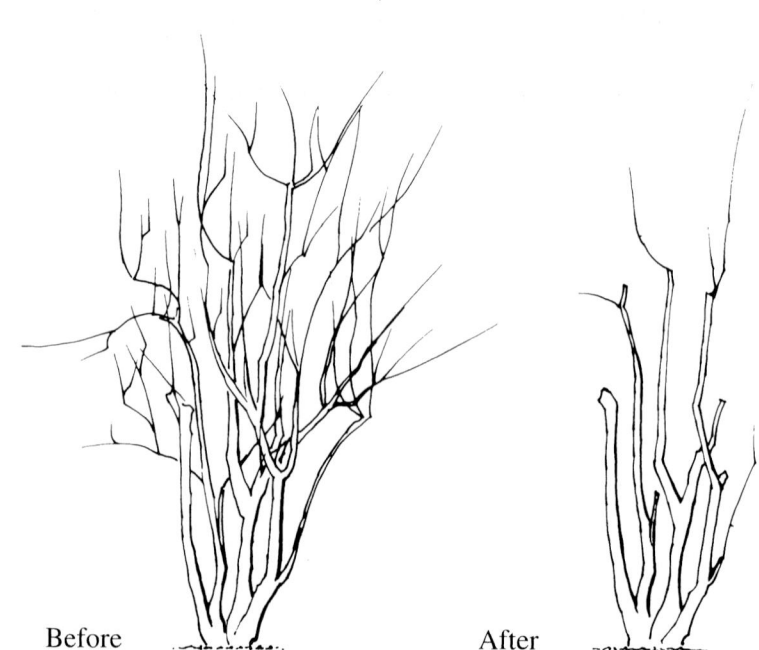

Before   After

Hard pruning

Hard pruning consists of cutting back all growth down to new growing points at the woody base of the shrub. Many shrubs, including Buddleia, respond very well to this measure, with improved shape and performance in the next growing season.

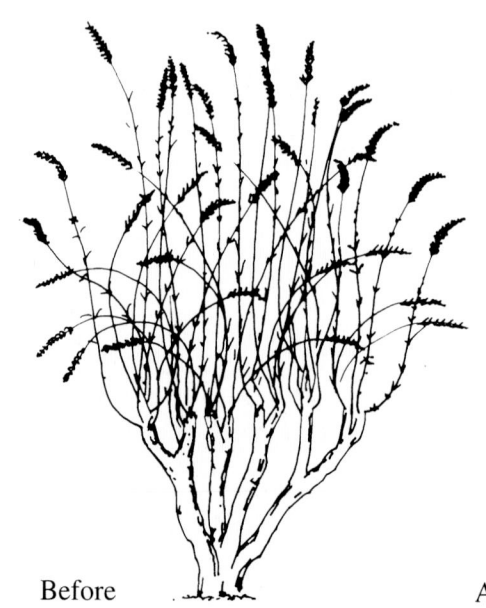

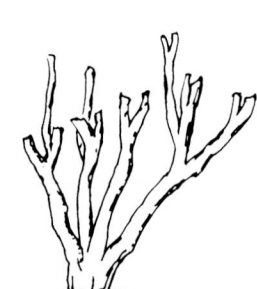

Before   After

380

SUPPORTS FOR SHRUBS AND CLIMBING PLANTS

Once you have chosen the shrub or climbing plant you wish to plant you must consider what kind of support it will need to assist it in covering the required area.

The support can take a number of different forms but no matter what kind you use you must bear in mind the following points:

1. The support must be able to accommodate the weight of the fully developed plant — a weight which may be considerable, especially when the plant is subjected to rain, snow and wind.

2. The support should not be unsightly, especially in the years while the plant is attaining its full coverage.

3. A clear air space of at least 2in (5cm) must be left between the support and the fence or wall. This allows free passage of air, keeping attacks of fungus disease such as mildew to a minimum.

Individual anchor points

Many climbing plants and wall shrubs and trees in particular can be adequately supported by individual anchor points consisting of some type of DIY masonry nail. Adjustable tree straps, secured to the wall or fence by masonry nails, are worth considering, particularly for heavier, freestanding plants.

Vine eyes and wires

Vine eyes come in two forms, normally 4-6in (10-15cm) long. The first has a thread, so that it may be screwed into a rawlplug inserted into a pre-drilled hole in the mortar. The other kind is wedge-shaped, and again a predrilled hole is made in the mortar course and the vine eye carefully driven in, making sure that it is horizontal, with the holes facing latitudinally along the wall. On wooden fences, screw-thread vine eyes can be simply screwed into wooden support posts.

The spacing of vine eyes is important. Place them 6ft (1.8m) apart in width and at 18in (45cm) intervals up the wall from ground level to the total area required for covering. Stretch a strong wire, either galvanized or plastic coated, between the vine eyes. On a long stretch of wall a system of tension bolts may be necessary to form a secure climbing framework.

Plant the climber or shrub at the base of the support, with a 3-5ft (91cm-1.5m) cane inserted near the base of the plant and tied to the first one or two wires. Encourage the plant to climb this cane until it reaches the wires, where it will probably grow up of its own accord, although it may require some help to climb to the next wire, simply by tying in once it is long enough.

This form of support is one of the simplest and, if done correctly and neatly, can be one of the tidiest. It has an additional advantage in that if the wall has to be remortared or the fence painted or creosoted, the wire can simply be cut and detached and the plant laid

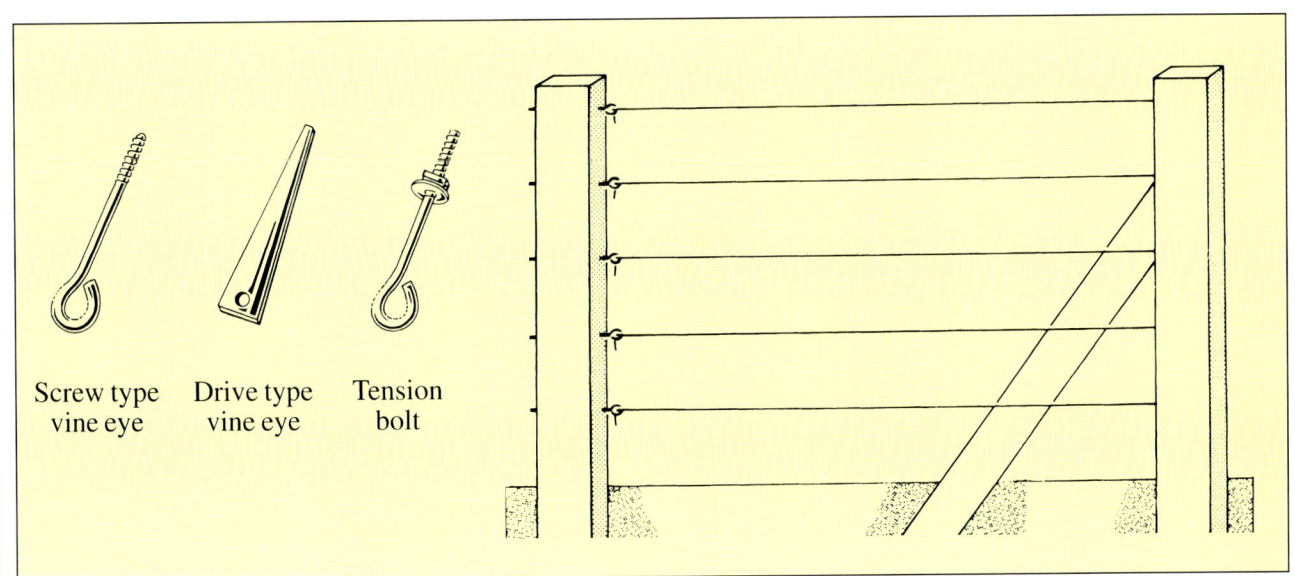

Screw type vine eye   Drive type vine eye   Tension bolt

down away from the wall. New wires can then be supplied, the old wires cut out and the plant reinstated in its former climbing position.

Plastic or wire netting

Galvanized or plastic covered wire netting, with a minimum of 1in (2.5cm) wide mesh, can be secured to the face of the wall, either by vine eyes or by screwing 2in (5cm) thick battens to the wall and then attaching the netting to them. Whichever method is used the fixing must be secure as it will take the full weight of the plant. Vine eyes and 2in (5cm) battens both allow air space between the plant and the wall. Extra netting can be added as required, although it is a good precaution to buy enough in the first instance to cover the entire surface and then store the surplus until it is needed as it is often difficult to find identical netting in subsequent years. Netting has the advantage of reducing the attention given to training as the plants find their own way between the relatively small squares in the wire mesh.

Extruded plastic netting is available in green, white and brown, providing a good match for the wall or fence. It is often sold in convenient lengths designed to accommodate average plant growth.

Right: Heavy square trellis

Below: Expanding trellis

Below right: Fan trellis

Timber trellis

A number of timber trellises are available with either a square ar diamond pattern. Some are rigid, others work on a concertina basis, and they are supplied either in a natural wood colour or painted white. The trellis is attached to the wall on a 2in (5cm) thick support batten, which again allows for the free passage of air behind and supports the weight of the plant.

This is one of the most attractive forms of support but it has two disadvantages: the trellis will only last for 10-15 years and maintenance of the wall and support is more problematical.

Wires and posts in open garden

Fruit trees, shrubs and climbing plants (particularly roses) can make very good open garden divisions and screens. To support these, drive posts securely into the ground, leaving a height of 5-8ft (1.5-2.4m). The posts should not be more than 8ft (2.4m) apart and the end posts will require bracing to allow wires to be stretched tightly between them. Attach the wires to the posts by wire staples, with the first wire 18in (45cm) above the ground and the remaining wires every 18in (45cm) thereafter.

Securing climbing plants to supports

It is best to avoid all forms of wire, plastic (except adjustable types) or combinations of wire and paper; any tie that does not naturally decompose within 18 months is not advisable as severe restriction is caused to stems and branches and in many cases damage is terminal. Soft string or raffia is the best material to use.

PRACTICAL GLOSSARY

Acid The term applied to soils with a pH of 6.5 or below, and usually containing no free lime.

Alkaline The term applied to soils with a pH of 7.4 or above, commonly but not exclusively associated with chalk or limestone soils.

Annual climber A plant normally grown from seed under protection in the spring, producing its display in summer and autumn then dying completely in winter. Seeds are not viable when overwintered in the soil and have to be harvested, dried and stored until the following spring.

Annual/perennial climber There are a few plants that con be treated as either annual or perennial. Normally the local climate will dictate the best growth pattern to adopt.

Aphids The general term for a number of small sap-sucking insects which cause damage to leaves, stems and new plant tissues. *See* Blackfly, Greenfly.

Avenue planting Group planting of trees, normally of the same species or variety, evenly spaced in single line on either side of a road or driveway.

Balled-and-burlapped The description of a tree, shrub or climbing plant which has been dug from the soil on which it was grown with a ball of soil surrounding the roots, kept intact by a wrapping of coarse cloth or net. This root ball should not be disturbed. *See* root-balled.

Bare-rooted
The description of a tree, shrub or climbing plant which has been dug from the ground and is sold without soil around its roots, or the roots are contained in a polythene bag with a small amount of peat.

Blackfly A small sucking insect which attacks the young parts of plants, damaging the tissue.

Bottom heat Heat applied to rooting compost and rooting mediums to encourage the production of roots on cuttings.

Brushwood Small pieces of branching shoots used as support for certain climbers.

Budding
A propagation method in which a single bud from a selected tree, shrub or climbing plant 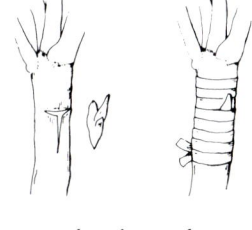 is grafted on to a rooted understock to produce a new specimen with the characteristics of the selected parent.

Bush A term applied to a shrub or small tree when main branches arise from soil level or on a small stem up to 2ft (60cm) high.

Canker An airborne fungus disease which enters damaged stem tissues and gradually surrounds the stem. If left untreated it will cut off the growth system of the plant with fatal result.

Chlorosis Yellowing of the leaves of a plant caused by lack of iron and nitrogen, seen particularly in plants grown on alkaline soils such as chalk or limestone which lock up these necessary elements.

Clone A single selection of a plant chosen from one or more parent plants and propagated by cuttings, grafting etc., but not from seed, thereby retaining the characteristics of the first selection.

Collar rot Damage to bark and underlying growth tissue at the base of a tree, shrub or climbing plant caused by excessive build-up of moisture in the affected area. This problem is commonly associated with heavy clay soils; good soil preparation and staking of the plant help to prevent it.

Conifer A cone-bearing tree which adapts well to fan-training and is normally hardy under all conditions.

Container-grown
The description of a tree, shrub or climbing plant grown in, or potted on into, a container of rigid or flexible plastic, treated paper, peat composition, or earthenware.

Containerized The description of a tree, shrub or climbing plant permanently grown in a container, e.g. as a patio or conservatory plant.

Coppice A plantation of trees closely spaced to induce multi-stemmed growth.

Coppicing A practice applied to certain shrubs, and some trees when grown as bushes, of cutting back all branches to ground level at five or ten year intervals to encourage healthy new growth.

Coral spot A fungus disease which first attacks dead wood and then spreads into healthy growth. It is visible as small, coral pink spots.

Cordon
Many wall shrubs, in particular fruiting forms, are 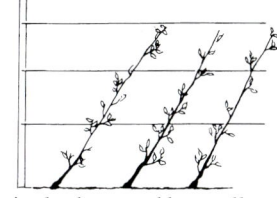 trained as a single shoot and have all side growths reduced in late winter/early spring to within 2in (5cm) of the main shoot to encourage flowering. They are often planted at an angle of 35° for the same reason.

Cross pollination (as in fruit) The requirement of certain groups of fruiting plants to have pollen taken from one plant to another to achieve fertilization.

Crown The base of a shrub, or the part of a tree from which the branching head grows.

Cultivar A plant originated under cultivation as a variant of an existing plant, with some characteristic distinctively different from the parent. The term is also used of a variant

383

occurring in the wild but kept under garden cultivation in the form of clones.

Deciduous (of a tree, shrub or climbing plant) Losing leaves seasonally, i.e. at the end of the growing season, to regenerate new foliage at the start of the following season.

Die-back The condition of a plant in which twigs, stems and sometimes whole branches dry out or become very brittle and turn dark brown or black. All damaged shoots must be cut back to healthy tissue and cuts of any size treated with pruning compound. Die-back usually occurs due to winter frost, summer drought, or mechanical root damage.

Division A method of propagating shrubs by digging up the root ball and dividing the clump at the roots to replant the sections separately.

Double-digging Preparing ground for planting by digging out the soil to a spade's depth, then turning over the same depth within the planting area before returning the topsoil.

Ericaceous The description of plants belonging to the family Ericaceae, including Calluna, Erica, Rhododendron and Azalea, which typically require acid soil conditions for successful growth.

Espalier *See* **Horizontal-trained**.

Evergreen (of a tree, shrub or climbing plant) Retaining foliage throughout the year, except for a small proportion which is unobtrusively shed.

Fan-training A method of training a tree, shrub or climbing plant against a wall so the branches radiate from the base or trunk, forming a fan shape. Some form of support is required to train the growth.

Field planting The practice of planting large specimen trees in parkland or large estates where the full growth potential can be accommodated.

Fire blight A fungus disease which becomes visible when a plant or section of a plant appears as if burned. It cannot be controlled and the plant must be destroyed, but it is also a notifiable disease and the occurrence should be reported to the appropriate authority.

Fruiting canes Annually produced canes which produce edible fruit and benefit from training as wall specimens.

Fruiting shrubs Shrubs (bushes) that produce edible fruit and adapt to training as wall shrubs.

Fruiting vines Vines which bear fruit and will train against walls, fences or similar constructions, where they benefit from the protection given.

Garden or nursery origin Many plants originate in the wild, but others are hybridized by nurserymen and gardeners and then grown on as individual varieties in their own right.

Gazebo A construction made out of metal, wood or stone in the shape of a building but with open lattice-work to the sky, over which various climbing plants can be grown.

Grafting A propagation method in which a section of wood from a selected plant is grafted on to a rooted understock and later develops into a new plant with the characteristics of the selected parent.

Greenfly A small sucking insect which attacks young parts of a plant, damaging the tissues.

Ground cover A planting of low shrubs or climbing plants relatively closely spaced, that spread to cover a particular area. Apart from the good display effect, this assists in soil retention and keeps down weeds.

Half-standard The term given to a tree with a bare stem or trunk of 4-4½ ft (1.2-1.4m) below the head of branches.

Hardwood cutting A section of stem taken from firm or hard wood of the

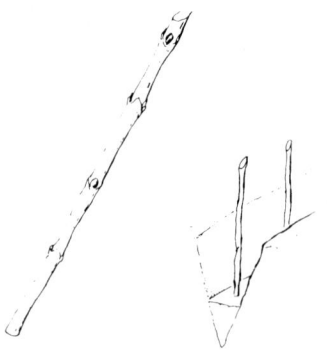

previous season's growth, placed in soil to root to propagate a new plant.

Hardy The description of a tree, shrub or climbing plant able to withstand winter frost.

Hedging A planting of shrubs in a single, or occasionally double, row, which gradually spread together to form a hedge. Growth may be shaped or controlled by pruning.

Heel A small fragment of the previous season's wood left at the end of a cutting for propagation purposes.

Honey fungus (Bootlace fungus) A disease affecting trees and shrubs caused by two types of fungus living together on the plant. The first sends out fine black underground shoots, hence the alternative name Bootlace fungus, which enter the root system of the host plant and kill the roots. The second fungus feeds on the decomposing material. A sweet honey scent from the dying host plant is one of the signs of the disease.

Horizontal-trained (espalier) Many wall-trained trees, shrubs, and climbers are trained into horizontal tiered shape to show off their beauty to the full and also to aid ripening of fruit. Support is required to maintain this shaping.

Humus Fully or partly decomposed organic material of animal or vegetable origin. This may occur naturally within a soil due to decaying plant matter, or can be added to soil to improve its texture and moisture retention and supply additional plant nutrients.

Hybrid A plant derived from a cross between two species, to combine or improve upon the characteristics of one or both of the parent plants.

Internodal cutting 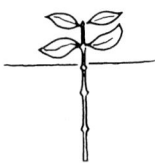 A section of stem taken between leaf joints (nodes) for propagation.

Layering A method of propagating a new shrub from existing stock by making a small cut below a bud on one side of a healthy young shoot and bending the shoot downwards to bury the cut section in the soil. After one year the buried bud has produced roots and new shoots and can be detached from the parent plant.

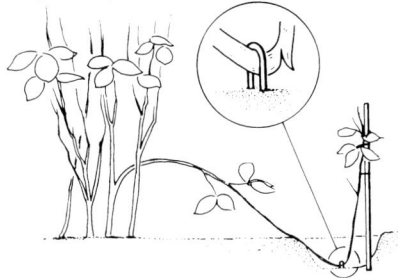

Leaf mold A dark-coloured powdery or flaky material resulting from the decay of fallen leaves.

Leaf scorch Discoloration and shrivelling of foliage due to strong wind, sun or heavy frost.

Mass planting A group planting of numbers of the same tree, shrub or climbing plant to form a spectacular display of foliage, flowers or fruit.

Micropropagation A commercial method of propagation using single cells from a parent plant to develop new specimens.

Mildew An airborne fungus disease which produces a white, downy coating over leaves, stems and fruits. It is encouraged by high humidity and warmth and commonly appears in late summer or early autumn.

Mulch A dressing applied to the soil surface around a tree or shrub consisting of decayed leaves, grass cuttings, manure etc., which retains moisture in the soil, provides natural foods for the plant and keeps down weeds.

New wood A stem, branch or twig borne in the current growing season.

Old wood A stem, branch or twig grown in the previous year's growing season or earlier.

Organic material Rotted and decomposed organic vegetable material such as garden compost and farmyard manure which, when added to garden soil, forms moisture- and nutrient-holding material. Also opens and lightens clay, heavy or difficult soil.

Patio planting The positioning of a tree, shrub or climbing plant in a small area of soil associated with a patio site, or in a container. Growth may be limited by the planting area, or should be controlled by pruning.

Peat Organic matter in a condition of partial decay, forming naturally in waterlogged areas and used to improve soil texture and supply plant nutrients in cultivated areas.

Perennial climber A hardy plant normally dying to ground level during the winter and rejuvenating the following spring. May, under certain circumstances in mild areas, produce a permanent woody structure but this can often be to the detriment of future foliage and flower production.

Pergola A timber, metal or stone construction, built in such a way as to make an open or lattice-type roof over which plants can grow. Sometimes used to form tunnels and semi-enclosed walkways.

pH The measurement of acidity and alkalinity in the soil which in practical terms can be ascertained using a soil-testing kit. The numeric scale used defines soils with pH of 6.5 or under as acid, 6.6 to 7.3 neutral, 7.4 or over alkaline. Commonly chalk or limestone soils tend to be alkaline, while lime-free soils tend to be acid, but the determining factor is not the actual soil composition but the origin of the water supply underlying the soil.

Pollarding A system of pruning in which all growth produced in the previous year is cut back in early spring to form a crown of old growth from which the new season's growth develops.

Pollinator A tree, shrub or climbing plant planted for the purpose of supplying pollen to another plant to help formation of fruit.

Pleaching A form of pruning or training in which a tree's growth is encouraged sideways rather than upwards. Lateral growths are tied in to a framework of wire, canes or stakes and all other growths shortened back until a block shape is achieved, maintained in subsequent seasons by pruning.

Rejuvenate (of a tree, shrub or climbing plant) To generate new growth after severe pruning, damage or die-back.

Rich soil A soil high in organic material such as garden compost, leaf-mould or well-rotted farmyard manure, used to encourage root growth and subsequent stem and foliage production.

Root-balled The description of a tree, shrub or climbing plant dug from the soil on which it was grown with a ball of soil surrounding the roots, kept intact by a covering of coarse cloth or net. The root ball should not be disturbed. Also called balled-and-burlapped.

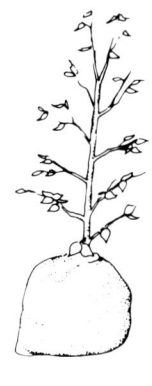

Root cutting A method of propagation in which a section of fleshy root is cut from a plant and set in a tray or pot filled with a mixture of peat and sand. New roots and shoots are formed from the root section.

Root rot A fungus disease which attacks plants with fibrous root systems, visible when the plant's leaves wilt for no apparent reason. The plant eventually dies. Good soil preparation lessens the likelihood of this occurring.

Root stooling A propagation method applied to trees. The tree is pruned back to a stump and soil is heaped around the stump, encouraging new roots and shoots to grow which can be separated from the parent.

Rust An airborne fungus producing a red, rust-like coating on the plant.

Scab An airborne fungus disease which produces grey-brown lesions on leaves and fruit, particularly seen in Malus varieties.

Semi-evergreen (of a tree, shrub or climbing plant) Normally evergreen, but likely to shed some or all leaves in unusually cold conditions to which the plant is not acclimatized.

Semi-ripe cutting A section of stem taken from wood one year old in early or midsummer, placed in soil to root to propagate a new plant.

Shrub A woody plant with a growth pattern of stems branching from or near to the base, not with a distinct single main stem or trunk.

Silver-leaf fungus A fungus disease apparent as a silver sheen on leaves in late spring or early summer, generally persisting until midsummer of the following year when the foliage appears burned and drops, at which stage there is also some stem damage. There is no effective treatment and the tree should be destroyed.

Single planting, solo planting Planting of an individual tree, shrub or climbing plant in an area which will accommodate its full ultimate size and allow a distinctive display of its overall shape and special features.

Softwood cutting A section of soft stem taken from current season's growth in spring, placed in soil to root to propagate a new plant.

Specimen plant A tree, shrub or climbing plant planted singly for display in an area able to accommodate its full growth potential.

Sport A shoot occurring by chance that differs in some characteristic from the parent plant; propagated by cuttings, grafting etc., to maintain the distinctive characteristics.

Standard A term applied to a tree with a bare stem or trunk of 5-6ft (1.5-1.8m) below the head of branches.

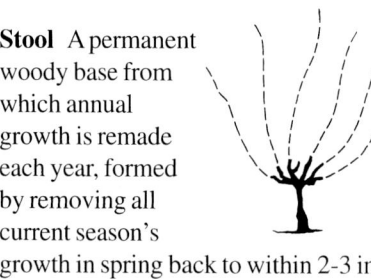

Stool A permanent woody base from which annual growth is remade each year, formed by removing all current season's growth in spring back to within 2-3 in (5-7.5 cm) of its origin.

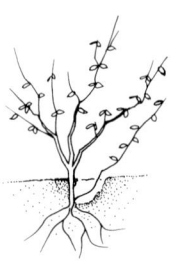

Subsoil The layer of soil directly below the uppermost layer (topsoil), generally having a poor composition and containing fewer nutrients.

Sucker New growth on a plant, generally arising from an underground root, which may appear some distance from the parent. In some plants, this is a natural method of spreading new growth, but suckers from undergrowth of grafted plants are usually undesirable and should be ripped out to remove all dormant buds in the new growth.

Suckering The production of suckers.

Sun scorch Damage to leaves in the form of browing, mottling and shrivelling, caused by exposure to strong sunlight.

Tender The description of a tree or shrub liable to frost damage.

Tender greenhouse climber A climber maintaining a woody structure but unable to tolerate any winter frost and requiring a protected growing environment such as a greenhouse or conservatory except in very mild areas.

Tender tree A tree that is not fully hardy but which under some conditions may be grown against walls and fences where it is afforded extra protection.

Tender wall shrub A shrub not normally considered fully hardy but adapting well to fan-training on walls and fences or growing free-standing in

front of them, where it gains added protection. Also of use grown under the extra protection or greenhouses and conservatories.

Tender woody climber A climber not fully hardy under most winter conditions, but benefitting from the protection of a wall, fence, greenhouse or conservatory. Normally maintains a woody structure.

Thicket A dense growth of trees or shrubs with upright stems.

Topsoil The uppermost layer of soil having a workable texture and containing a high proportion of plant nutrients.

Topworking A method of grafting used for weeping trees in which the understock is grown to a height of 6-7ft (2-2.3m) and the graft applied at this height.

Tree A woody plant typically producing a single stem (trunk) below a head of branches.

Understock A selected or specially grown young plant used as the rooted base for creating new plants by the propagation techniques of budding and grafting. The physical characteristics of the new plant are gained from the grafted material, not the understock, but this may dictate overall size and growth rate.

Variety Strictly, a term specifying a naturally occurring variant of a plant, but also used loosely to describe any variant closely related to a particular form.

Vegetative propagation Methods of raising new plants from existing specimens, e.g. by budding, cuttings, division, grafting, rather than from seed. These methods ensure that the new stock is identical to the parent.

Vine eyes Wedge-shaped metal supports that carry wires for training plants (see page 381).

Wall shrub A hardy shrub that adapts well to training on walls or fences, normally in a fan-trained shape. Can also be planted untrained and will benefit from the protection of a wall or fence.

Weeping standard Some climbing plants will allow themselves to be trained with one or two single upright stems from which top growth emerges and forms a weeping habit. These plants are not normally sold in this form and have to be trained in the garden.

Wild garden A cultivated garden area allowed to retain natural growth patterns and features, such as grass growing among fowering plants and around trees or shrubs, unpruned shrubby growth and trees developing naturally as in woodland.

Wind chill The condition of frost accelerated by gale force winds, which substantially lower the overall temperature.

Wind scorch Damage to leaves caused by wind when water vapour is too rapidly lost from the leaves.

Woody climber Normally a hardy climbing plant maintaining a rigid or semi-rigid structure of growth throughout the year. Can be evergreen or deciduous depending on species or form.

BOTANICAL GLOSSARY

Alternate (of leaves) Growing one above the other on a stem from separate leaf nodes; not opposite.

Anther The usually coloured, pollen-bearing part of the male organ of a flower.

Ascendant, ascending (of a growth pattern)

Axil The angle between a stem and leaf stalk, or between two stems or branches.

Berry A succulent, normally several-seeded fruit that does not split or burst.

Bipinnate (of leaves) Divided into segments that are themselves divided; doubly pinnate.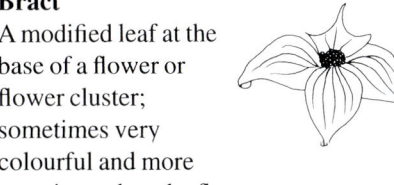

Bisexual (of a flower) Incorporating male and female organs.

Bloom A fine, powdery deposit, as on the surface of a fruit.

Bract A modified leaf at the base of a flower or flower cluster; sometimes very colourful and more prominent than the flowers.

Bud The early stage of a new shoot or flower. The term refers not only to an obvious growth bud, but also to an incipient swelling.

Calyx The circlet or whorl of sepals that encloses the petals of a flower before it opens.

Capsule A dry seed pod of several cells.

Catkin A flower spike or spike-like raceme consisting of very small stemless flowers covered with scale-like bracts. The term is also applied to a similar arrangement of tiny fruits.

Clusters Flowers carried in moderately condensed groups.

Compound (of a leaf) Divided into several distinct parts, or leaflets.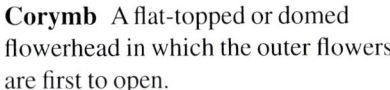

Corymb A flat-topped or domed flowerhead in which the outer flowers are first to open.

Cyme A flat-topped or domed flowerhead in which the inner flowers are first to open.

Dioecious (of a plant species) Bearing either male or female flowers on a particular plant. *See also* Monoecious.

Dissected (of a leaf) Deeply cut into a number of narrow segments.

Double (of a flower) Having more than the usual number of petals. Sometimes having petals in place of style and stamens.

Downy Lightly coated with hairs.

Elliptic (of a leaf) Narrowing at both ends, with the widest point at or near the middle of the leaf.

Embryo A rudimentary plant.

Fastigiate (of a tree or shrub) Having a growth habit in which branches and stems are dense and erect.

Fibrous (of roots) Thin, but densely growing.

Floret A single, small flower in a flowerhead or clustered inflorescence.

Floriferous Flowering, flower-bearing.

Flowerhead A short, dense inflorescence; a flower cluster or spike.

Fruit buds Buds laid down in previous seasons, as in the case of apples, pears, etc.

Glabrous (of leaf or stem) Smooth and hairless.

Glaucous (of leaf or stem) Covered with a blue/white or blue/grey bloom.

Hermaphrodite A bisexual flower, incorporating male and female parts.

Inflorescence The flower-bearing part of a plant. *See also* Corymb, Cymbe, Panicle, Raceme, Spike, Umbel.

Internode A section of a stem between two leaves joints, or nodes.

Lacerated (of a leaf) Irregularly cut at the margins.

Lacinate (of a leaf) Fringed; with the margin cut into narrow pointed lobes.

Lanceolate (of a leaf) Spear-shaped; swelling above the base and tapering to a point.

Leaflet An individual section of a compound leaf

Linear (of a leaf) Long and narrow, with almost parallel margins tapering briefly at the tip.

Lobe A protruding part of a leaf, distinct but not separated from the other segments.

Midrib The vein or rib along the centre of a leaf.

Monoecious (of a plant species) Bearing both male and female flowers on the same individual plant.

Monotypic (of a plant genus) Having only one species.

Node The point on a stem where the leaf is attached; the leaf joint.

Nut A hard, one-seeded fruit which does not split or open.

Oblong (of a leaf) Longer than it is wide, with long edges almost parallel.

Obovate (of a leaf) Broadening towards the tip; the reverse of ovate.

Opposite (of a leaves) Borne from the same node and opposite each other on the stem.

Orbicular (of a leaf or petal) Disc-shaped; almost circular.

Ovate (of a leaf) Swelling from the base and tapering towards the tip.

Palmate (of a leaf) Lobed or divided, resembling fingers on a hand. A palmate leaf is usually five- or seven-lobed.

Panicle A flower spike or raceme with several small branches, themselves also branching.

Pea-flower A flower having the typical shape of a sweet-pea blossom.

Pendulous Hanging or weeping.

Perennials (herbaceous) Plants with root systems that become dormant or semi-dormant in winter, with all top growth dying to ground level in winter and becoming regenerated the following spring.

Petal A modified leaf, usually brightly coloured, which forms part of a flower.

Petiole A leaf stalk.

Pinnate (of leaves) Divided and composed of leaflets arranged on either side of a central stalk.

Propagation The act of increasing a particular plant under controlled conditions by softwood cutting, grafting, etc.

Prostrate (of a stem or growth habit) Lying along the ground.

Raceme An elongated flowerhead composed of stalked flowers on a central stem.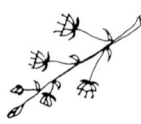

Recurved Curving down or back.

Reniform Kidney-shaped.

Reticulate Networked, as in the veins of a leaf or petal.

Rhomboidal Diamond-shaped.

Scale A small leaf or bract; also a small flat growth on the surface of a flower, leaf or fruit.

Self-clinging A plant that supports itself by stem, root or sucker pads, or by leaf tendrils or twining habit, and that does not normally require further assistance.

Semi-double (of a flower) More than one layer of petals, but not double.

Sepal Green shields that protect the petals before they open, and in some cases provide added interest to the flower. In some plants, such as clematis, they replace the petals.

Serrated (of a leaf) Having saw-toothed edges, with the teeth pointing towards the leaf tip.

Single (of a flower) Having only one ring or layer of petals.

Spike An elongated flowerhead composed of stalkless flowers.

Stamen The male organ of a flower, consisting of a pollen-bearing anther supported by a filament.

Stigma The female organ of a flower; the part that receives pollen for reproduction.

Stolon A shoot running from the plant at or below ground level, and that gives rise to a new plant at its tip.

Style Part of the female organ of a flower, a stalk linking the stigma to the reproductive organ.

Tendril A fine, modified stem or leaf that twines to provide support for a plant.

Trifoliate (of a leaf) Having three separate leaflets.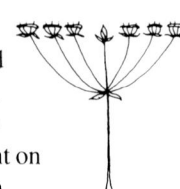

Twining (as in stems) Twisting in an anti-clockwise or clockwise motion around suitable supports, encouraging semi or full self-support.

Umbel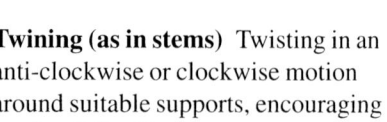
A usually flat-topped flowerhead in which stalked flowers arise from a common point on the main flower stem.

Universal pollinator A variety of fruit tree that is able to pollinate itself and a wide range of other varieties. Some ornamental trees can also pollinate dessert or culinary fruits, for example Malus 'Golden Hornet', which is a universal pollinator for most apples.

Vine eyes A name given to the buds of Vitus vinifera (vine).

✕ Preceding a species or variety, this indicates a botanical hybrid cross.

CHOOSING A TREE, SHRUB OR CLIMBING PLANT

When a tree, shrub or plant is first seen as a young specimen in the garden centre, features such as vivid autumn colour, curious winter stems or an unusual habit of growth may not be apparent. You may wish to select a shrub with flowers that are fragrant as well as beautiful, shrubs for ground cover, a tree that will not outgrow a small garden, varieties that can flourish in a shady location or on acid soil. The following lists are intended to help you in the overall planning of your garden by pinpointing these individual features.

EVERGREEN TREES

ACACIA

ARBUTUS *andrachne*
   × *andrachnoides*
   menziesii
   unedo

COTONEASTER 'Cornubia'
   'Rothschildianus'
   salicifolius var. flocosus

EUODIA *hupenhensis*

ILEX × *altaclarensis*
   aquifolium
   × *kochneana* 'Chestnut Leaf'

LAURUS *nobilis*

LIGUSTRUM *lucidum*

MAGNOLIA *delavayi*
   grandiflora

NOTHOFAGUS *dombeyi*
   fusca
   menziesii

QUERCUS*ilex*

× KEWENSIS
   suber
   × *turneri*

TREES WITH ORNAMENTAL BARK AND STEMS

ACER *campestre*
   capillipes
   cappadocicum
   crataegifolium
   davidii
   grosseri
   griseum
   hersii
   negundo
   pennsylvanicum
   pennsylvanicum 'Erythrocladum'
   rufinerve

ALNUS *incana* 'Aurea'

ARBUTUS

BETULA

CORYLUS COLURNA

EUCALYPTUS

FRAXINUS *excelsior* 'Aurea Pendula'
   excelsior 'Jaspidea'

ILEX × *altaclarensis* 'Camelliifolia'
   × *altaclarensis* 'Hodginsii'
   aquifolium 'Handsworth'
   aquifolium 'New Silver'
   aquifolium 'Mme. Briot'
   aquifolium 'Silver Queen'

JUGLANS *regia*

LIQUIDAMBAR *styraciflua*

PHELLODENDRON *amurense*

PLATANUS × *hispanica*

POPULUS *alba*
   nigra

PRUNUS *cerasifera*
   'Cistena'
   padus
   serrula

RHUS *typhina*

SALIX *acutifolia* 'Blue Streak'
   acutifolia 'Pendulifolia'
   alba 'Britzensis'
   alba 'Vitellina'
   × *chrysocoma*
   daphnoides
   × *erythroflexuosa*
   matsudana 'Tortuosa'
   purpurea 'Pendula'
   × *smithiana*

***Eucalyptus niphophila* – branches**

SOPHORA *japonica*

SORBUS *aucuparia* 'Beissneri'
   cashmiriana
   hupehensis

STUARTIA

TILIA *platyphyllos* 'Rubra'

TREES FOR SMALL GARDENS

ACER *griseum*
   grosseri
   pseudoplatanus 'Brilliantissimum'
   pseudoplatanus 'Prinz Handjery'
   rufinerve

ALNUS *incana* 'Aurea'

ARBUTUS *menziesii*
   unedo

BETULA *pendula* 'Purpurea'
   pendula 'Youngii'

CERCIS *siliquastrum*

CORNUS *nuttallii*

COTONEASTER

CRATAEGUS 'Autumn Glory'
   crus-galli
   laciniata
   × *lavallei*
   × *mordenensis* 'Toba'
   oxyacantha
   prunifolia

EUCALYPTUS *coccifera*

ILEX × *altaclarensis*
   aquifolium

KALOPANAX *pictus*

LABURNOCYTISUS *adamii*

LABURNUM × *waterei* 'Vossii'

LIGUSTRUM *lucidum*

MAGNOLIA × *loebneri*
   soulangiana

MALUS 'Golden Hornet'
   hupehensis
   'Lizet'
   'Red Jade'
   'Red Glow'
   'Profusion'
   'Royalty'
   'Strathmore'
   toringoides
   transitoria
   'Van Eseltine'
   'Wintergold'

MESPILUS *germanica*

OXYDENDRUM *arboreum*

PHOTINIA *villosa*

PRUNUS × *amygdalo-persica*
   'Asano'
   × *blireana*
   cerasifera 'Rosea'
   dulcis
   'Fudanzakura'
   × *hillieri* 'Spire'
   incisa
   'Hally Jolivette'
   'Kiku-Shidare Sakura'
   'Kursar'
   'Okame'
   padus 'Albertii'
   'Pandora'
   serrula
   serrulata 'Amanogawa'

PYRUS *calleryana* 'Chanticleer'
   nivalis
   salicifolia 'Pendula'

RHUS *typhina*

SALIX *caprea* 'Pendula'
   purpurea 'Pendula'

SORBUS *americana*
   aucuparia 'Beissneri'
   aucuparia 'Fastigiata'
   cashmiriana
   commixta
   'Embley'
   hupehensis
   'Joseph Rock'
   sargentiana
   vilmorinii

STUARTIA *malacodendron*

STYRAX *japonica*

QUICK-GROWING TREES

ACACIA

ACER *negundo*
   platanoides
   pseudoplatanus
   saccharinum

AESCULUS *hippocastanum*

AILANTHUS *altissima*

ALNUS

AMELANCHIER

CASTANEA *sativa*

COTONEASTER 'Cornubia'
   'Rothschildianus'

EUCALYPTUS

FRAXINUS *americana*
   excelsior

GLEDITSIA *triacanthos*

KALOPANAX *pictus*

JUGLANS *nigra*
   regia

LIGUSTRUM *lucidum*

LIRIODENDRON *tulipifera*

NOTHOFAGUS *obliqua*
   procera

PHELLODENDRON *amurense*

POPULUS

PRUNUS *avium*

PTEROCARYA (on moist soils)

RHUS

ROBINIA

SALIX

SORBUS

WEEPING TREES

ALNUS INCANA 'PENDULA'

BETULA *pendula* 'Dalecarlica'
   pendula 'Tristis'
   pendula 'Youngii'

CARAGANA *arborescens* 'Pendula'
   arborescens 'Walker'

CARPINUS *betulus* 'Pendula'

COTONEASTER 'Hybridus Pendulus'

FAGUS *sylvatica* 'Aurea Pendula'
   sylvatica 'Pendula'
   sylvatica 'Purpurea Pendula'

FRAXINUS *excelsior* 'Pendula'

ILEX *aquifolium* 'Argentea pendula'

LABURNUM *alpinum* 'Pendulum'
   anagyroides 'Pendulum'

MALUS 'Echtermeyer'
   'Red Jade'

MORUS *alba* 'Pendula'

PARROTIA *persica* 'Pendula'

POPULUS *tremula* 'Pendula'

PRUNUS *avium* 'Pendula'
   'Hilling's Weeping'
   'Kiku-shidare Sakura'
   persica 'Crimson Cascade'
   subhirtella 'Pendula Rosea'
   subhirtella 'Pendula Rubra'
   × *yedoensis* 'Shidare Yoshino'

PYRUS *salicifolia* 'Pendula'

QUERCUS *robur* 'Pendula'

SALIX *babylonica*
   caprea 'Pendula'
   × *chrysocoma*
   × *erythroflexuosa*
   matsudana 'Pendula'
   purpurea 'Pendula'

SOPHORA *japonica* 'Pendula'

SORBUS *aucuparia* 'Pendula'

TILIA *petiolaris*

ULMUS *glabra* 'Camperdownii'
   glabra 'Pendula'

TREES WITH FRAGRANT FLOWERS

ACACIA

CLADRASTIS *lutea*

FRAXINUS *ornus*

LABURNUM *alpinum*

LIGUSTRUM *lucidum*

Malus hupehensis in flower

MAGNOLIA *delavayi*
   grandiflora
   kobus
   × *loebneri*
   salicifolia 'Superba'
   × *soulangiana* 'Alba'

MALUS *coronaria* 'Charlottae'
   hupehesis
   'Profusion'
   × *robusta*
   toringoides
   transitoria

PRUNUS *padus*
   serrulata 'Amanogawa'
   × *yedoensis*

ROBINIA *pseudoacacia*

TILIA × *euchlora*
   × *europaea*
   petiolaris

TREES WITH MULTICOLORED FOLIAGE

ACER *negundo* 'Flamingo'
   pseudoplatanus 'Brilliantissimum'
   pseudoplatanus 'Leopoldii'
   pseudoplatanus 'Prinz Handjery'
   pseudoplatanus 'Simon-Louis Freres'

CRATAEGUS *oxyacantha* 'Gireoudii'

FAGUS *sylvatica* 'Roseomarginata'

POPULUS × *candicans* 'Aurora'

TREES FOR ALKALINE SOIL

This is a selection of trees that thrive in alkaline conditions. Others that are tolerant of alkalinity will be found listed in the dictionary of trees.

ACER *platanoides*
   pseudoplatanus

CARAGANA *arborescens*

EUODIA *hupehensis*

LABURNUM *alpinum*

LIQUIDAMBAR *styraciflua*

SORBUS *aria*
   aucuparia

TREES FOR ACID SOIL

This is a selection of trees that specifically require acid soil, or that thrive on acid conditions. Others that are tolerant of acidity will be found listed in the dictionary of trees.

ACER *rubrum*

CERCIDIPHYLLUM *japonicum*

CERCIS *siliquastrum*

CYTISUS × *praecox*
   scoparius

GYMNOCLADUS *dioicus*

IDESIA *polycarpa*

MAGNOLIA *campbellii*

NYSSA *sylvatica*

OXYDENDRON *arboreum*

PARROTIA *persica*

PHOTINIA *villosa*

QUERCUS (Autumn foliage varieties)

SASSAFRAS *albidum*

STUARTIA *pseudocamellia*

STYRAX *japonica*

TREES FOR DRY SITES IN SUN OR SHADE

ACER *platanoides*

BETULA

CARAGANA

CASTANEA

COTONEASTER

EUCALYPTUS

FAGUS *sylvatica*

GLEDITISIA *triacanthos*

ILEX

KOELREUTERIA *paniculata*

QUERCUS *robur*

RHUS

TREES FOR MOIST SITES

ALNUS *cordata*
   glutinosa
   incana

CRATAEGUS *oxyacantha*

FRAXINUS *excelsior*

LIQUIDAMBAR *styraciflua*

MESPILUS *germanica*

NYSSA *sylvatica*

POPULUS

PTEROCARYA

QUERCUS (Autumn foliage varieties)

SALIX

SORBUS *aucuparia*

TREES WITH GOOD AUTUMN FOLIAGE

ACER *campestre*
   capillipes
   griseum
   nikoense
   pennsylvanicum
   platanoides
   rubrum
   rufinerve
   saccharinum

AESCULUS *flava*

AMELANCHIER

BETULA

CARPINUS *betulus*

CERCIDIPHYLLUM

CRATAEGUS 'Autumn Glory'
   crus-galli
   durobrivensis
   × *grignonensis*
   × *lavallei*
   mollis

phaenopyrum
pedicellata
pinnatifida major
FRAXINUS *excelsior* 'Jaspidea'
   oxycarpa 'Flame'
   oxycarpa 'Raywood'
KOELREUTERIA *paniculata*
LIQUIDAMBAR
LIRIODENDRON
MALUS *toringoides*
   transitoria
   trilobata
   tschonoskii
MESPILUS *germanica*
NOTHOFAGUS
NYSSA *sylvatica*
OXYDENDRUM *arboreum*
PARROTIA *persica*
PHELLODENDRON *amurense*
PHOTINIA *villosa*
PRUNUS *avium*
   × *hillieri* 'Spire'
   'Kursar'
   sargentii
   serrulata 'Amanogawa'
   'Shosar'
   Japanese Cherries
PTELEA *trifoliata*
PTEROCARYA
PYRUS *calleryana* 'Chanticleer'
QUERCUS *coccinea*
   palustris
   petraea
   rubra
RHUS *typhina*
SASSAFRAS *albidum*

Acer platanoides in autumn

SORBUS *aria commixta*
   'Embley'
   esserteauana
   folgneri
   hupehense
   'Joseph Rock'
   matsumurana
   'Mitchellii'
   'November Pink'
   sargentiana
   scalaris
   vilmorinii
STUARTIA

ACER *negundo*
ALBIZIA *julibrissin*
COTONEASTER (Small-leaved standards)
EUCALYPTUS
FORSYTHIA
ILEX
LABURNUM *alpinum* 'Pendulum'
LAURUS *nobilis*
VIBURNUM *alpinum*
WEIGELA 'Bristol Ruby'

SHRUBS AND CLIMBING PLANTS WITH GOOD AUTUMN COLOR

ACER *ginnala*
   japonicum
   japonicum 'Aureum'
   negundo
   palmatum
   palmatum 'Dissectum'
   palmatum 'Linearilobum'
   palmatum 'Senkaki'
ACTINIDIA *arguta*
   chinensis
   kolomikta
AESCULUS *parviflora*
AKEBIA *quinata*
AMELANCHIER *lamarkii*
AMPELOPSIS *brevipendiculata*
ARISTOLOCHIA *macrophylla*
ARONIA *arbutifolia*
BERBERIS *aggregata*
   'Buccaneer'
   calliantha
   jamesiana
   koreana
   'Pirate King'
   rubrostilla
   sieboldii
   thunbergii
   thunbergii 'Erecta'
   thunbergii 'Kobold'
   wilsoniae
BIGONIA *capreolata*
BUDDLEIA *alternifolia*
CALLICARPA *bodinieri*
CAMPSIS *grandiflora*
   radicans
CARAGANA *alborescens* 'Lorbergii'
CATALPA *bignoides*
CEANOTHUS (Deciduous forms)
CELASTRUS *orbiculatus*
CERCIS *siliquestrum*
CERATOSTIGMA *willmottianum*
CHAENOMELES
CHIMONANTHUS *praecox*
CLEMATIS *alpina*
   flammula
   macropetala
   orientalis
CORNUS *alba*
   controversa
   florida
   kousa
   mas
   nuttallii
   sanguinea
COROKIA *cotoneaster*
CORONILLA *glauca*

CORYNABUTILON *vitifolium*
COTINUS *coggygria*
   coggygria 'Flame'
   obovatus
COTONEASTER *divaricatus*
   horizontalis
   simonsii
CRATAEGUS (Autumn foliage forms)
   oxyacantha
CYDONIA *oblonga*
DECAISNEA *fargesii*
DECUMARIA *barbara*
DIERVILLA *sessifolia*
DISANTHUS *cercidifolius*
ENKIANTHUS
ERYTHRINA *crista-galli*
EUCRYPHIA *glutinosa*
EUONYMUS *alatus*
   europaeus
   phellomanus
   sachalinensis
EXOCHORDA × *macrantha* 'The Bride'

Bignonia capreolata in flower

FICUS *carica*
FORSYTHIA *suspensa*
FOTHERGILLA
GINKO *biloba*
GLEDITSIA *triacanthos* 'Sunburst'
HALESIA
HAMAMELIS
HEDYSARUM *multijugum*
HOHERIA
HYDRANGEA *paniculata*
   petiolaris
   quercifolia
   serrata 'Preziosa'
INDIGOFERA *heterantha*
ITEA *virginica*
JASMINUM *fruticans*
   officianale
KERRIA *japonica*
KOLKWITZIA *amabilis*
LABURNOCYTISUS × *adamii*
LAGERSTROEMIA *indica*
LONICERA *fragrantissima*
   tragophylla
LYCIUM *barbarum*
MAGNOLIA (Large-growing, star-flowered forms)

×*soulangiana*
stellata

MAHONIA *aquifolium*

MALUS (Green-leaved flowering forms)
  pumila (Apple hybrid)

MORUS

NANDINA *domestica*

PALIURUS *spina-christi*

PARTHENOCISSUS *henryana*
  quinquefolia
  tricupidata 'Veitchii'

PASSIFLORA *caerulea*

PHASEOLUS *coccineus*

PHILADELPHUS (Medium-sized and tall-growing forms)

PHYGELIUS

PIPTANTHUS *laburnifolius*

PLUMBAGO *capensis*

POLYGONUM *baldschuanicum*

PONCIRUS *trifoliata*

PRUNUS *armeniaca*
  avium, P. cerasus, P. mahaleb
  domestica (Damson, gage and plum hybrids)
  persica (Nectarine and peach hybrids)

PYRUS *communis*

RHODODENDRON *luteum*

RIBUS *glossularia*
  ordoratum
  sativum
  speciosum

ROBINIA (Pink-flowering forms)
  pseudoacacia
  pseudoacacia 'Frisia'

ROSA
(Climbing musk roses and similar forms;
Modern hybrid climbing; Old climbing roses
prior to 1920; Rambler roses; Shrub and species
roses for fan-training)

RUBUS *fruticosus*
  loganobaccus
  phoenicolasius

SOLANUM *crispum*

SORBUS *aria*
  reducta

SPIREA *prunifolia* 'Plena'

STACHYURUS *praecox*

STAPHYLLEA *colchica*

STEPHANDRA *incisa*

STRANVAESIA *davidiana*

SYMPLOCOS *paniculata*

TEUCRIUM *fruticans*

TROPAEOLUM *majus*
  peregrinum
  tuberosum

VIBURNUM (Best fruiting; early flowering, and
spring flowering, scented forms)
  acerifolium
  carlesii
  hupehense
  opulus
  plicatum
  prunifolium
  sargentii

VINCA *major*

VITIS 'Brant'
  coignetiae
  vinifera
  vinifera 'Purpurea'

WATTAKA-KA *sinensis*

WEIGELA

WISTERIA *floribunda*
  floribunda 'Macrobotrys'
  sinensis
  sinensis (Modern hybrids)

XANTHOCERAS *sorbifolium*

ABELIA *grandiflora*
  triflora

ACER *negundo*
  palmatum

ACTINIDIA *arguta*
  chinensis
  kolomikta

ARBUTUS

BERBERIS *dictyophylla*
  temolaica

COLLETIA *armata*
  cruciata

CORNUS *alba*
  alba 'Elegantissima'
  alba 'Kesselringii'
  alba 'Sibirica'
  alba 'Spaethii'
  controversa
  sanguinea
  stolonifera 'Flaviramea'

Abelia chinensis in flower

COROKIA *cotoneaster*
  ×*virgata*

CORYLUS *avellana* 'Contorta'

COTONEASTER *horizontalis*

DECAISNEA

ELAEAGNUS *commutata*

EUONYMUS *alatus*
  phellomanus
  sachalinensis

FICUS *carica*

FORSYTHIA *suspensa*

GENISTA

GINKO *biloba*

HIPPOPHAE *rhamnoides*

HYDRANGEA (Large-leaved forms)
  petiolaris
  villosa

ILEX ×*altaclarensis* 'Camellifolia'
  ×*altaclarensis* 'Hodginsii'
  aquifolium 'Handsworth New Silver'
  aquifolium 'Mme. Briot'
  aquifolium 'Silver Queen'

JASMINUM *fruticans*
  mesnyi
  nudiflorum
  officinale
  officinale (Variegated form)
  ×*stephanense*

KERRIA *japonica*

KOLKWITZIA *amabalis*

LABURNUM ×*wateri* 'Vossii'

LEYCESTERIA *formosa*

LIPPIA *citriodora*

LONICERA *fragrantissima*

PEROVSKIA *atriplicifolia*

PHILADELPHUS 'Galahad'

PITTISPORUM *tenufolium*

PONCIRUS *trifoliate*

PSEUDOSASA *japonica*

RUBUS *biflorus*
  cockburnianus
  phoenicolasius
  thibetanus

SALIX *acutifolia* 'Blue Streak'
  acutifolia 'Pendulifolia'
  alba 'Britzensis'
  alba 'Vitellina'
  daphnoides
  fargesii
  irrorata
  phylicifolia
  purpurea
  sachalinensis 'Sekka'
  ×*smithiana*

SCHISANDRA *grandiflora rubrifolia*

SCHIZOPHRAGMA *hydrangeoides*

SOLANUM *crispum*
  jasminoides

SORBARIA

SORBUS *aria*

SPARTIUM *junceum*

STACHYURUS *praecox*

STEPHANANDRA *tanakae*

VIBURNUM *opulus*

VITIS 'Brant'
  coignetiae
  vinifera 'Purpurea'

ABELIA 'Edward Goucher'
  floribunda
  ×*grandiflora*

ANDROMEDA *polifolia*

ARBUTUS

ARCTOSTAPHYLOS

ATRIPLEX *halimus*

AUCUBA *japonica*

AZALEA 'Addy Wery'
  'Blaauw's Pink'
  'Blue Danube'
  'Hatsugiri'
  'Hinomayo'
  Japanese azaleas
  'John Cairns'
  'Kure-no-yuki'
  'Leo'
  'Mother's Day'
  'Naomi'
  'Orange Beauty'
  'Palestrina'
  'Rosebud'
  'Vuyk's Rosy Red'
  'Vuyk's Scarlet'

AZARA *dentata*
  lanceolata
  microphylla
  serrata

BALLOTA *pseudodictamnus*

BERBERIS *atrocarpa*
  ×*bristolensis*
  buxifolia 'Nana'
  calliantha
  candidula
  'Chenaultii'

darwinii
gagnepainii
julianae
knightii
linearifolia
× *logogensis*
panlanensis
'Parkjuweel'
sargentiana
× *stenophylla*
verruculosa
'Walitch Purple'

BERBERIDOPSIS *corollina*

BUPLEURUM *fruticosum*

BUXUX *sempervirens*

CALLISTEMON *citrinus*

CALLUNA *vulgaris*

CAMELLIA *japonica*

CARPENTERIA *californica*

CASSINIA *fulvida*

CEANOTHUS *arboreus*
'A. T. Johnson'
'Autumnal Blue'
'Burkwoodii'
'Cascade'
'Delight'
dentatus
'Dignity'
'Edinburgh'
(Evergreen forms)
impressus
'Italian Skies'
× *lobbianus* 'Russellianus'
prostratus
rigidus
'Southmead'
thrysiflorus
'Topaz'
× *veitchianus*
'Yankee Point'

CESTRUM *elegans*

Itea illicifolia in flower

CHOISYA *ternata*

CISTUS

CLEMATIS *armandii*
cirrhosa

COLLETIA *armata*
cruciata

CONVOLVULUS *cneorum*

CORDYLINE *australis*

COROKIA *cotoneaster*
× *virgata*

CORONILLA *glauca*
emerus

CORTADDERIA *selloana*

COTONEASTER *buxifolia*
congestus
conspicuus
'Cornubia'
dammeri
'Donard's Gem'
'Exburiensis'
henryanus
'Hybridus Pendulus'
'Inchmery'
lacteus
microphyllus

Ceanothus 'Topaz' in flower

pannosus
'Rothschildianus'
'Skogholme'
'St. Monica'
salicifolius
× *watereri*

CRINODENDRON *hookeranum*
patagua

DABOECIA *cantabrica*

DANAE *racemosa*

DAPHNE *cneorum*
collina
laureleola
odora
pontica
retusa
tangutica

DESFONTAINEA *spinosa*

DISTYLIUM *racemosum*

DRIMYS *winteri*

ELAEAGNUS × *ebbingei*
macrophylla
pungens

EMBOTHRIUM *coccineum*

ERIBOTRYA *japonica*

ERICA

ESCALLONIA

EUCRYPHIA *cordifolia*
cordifolia × *lucida*
lucida
× *nymansensis* 'Nymansay'

EUONYMUS *fortunei*
japonicus
kiautschovicus

EUCALYPTUS

FABIANA *imbricata*

× FATSHEDRA *lizei*

FATSIA *japonica*

FEIJOA *sellowiana*

FREMONTODENDRON *californicum*

GARRYA *elliptica*

× GAULNETTYA

GAULTHERIA

GRISELINIA *littoralis*

× HALIMIOCISTUS

HALIMIUM

HEBE

HEDERA *canariensis*
colchica
helix
helix angularis 'Aurea'
helix 'Buttercup'
helix 'Cristata'
helix 'Glacier'
helix 'Gold Heart'
helix 'Luzii'
helix sagittifolia
(Shrubby forms)

HELIANTHEMUM

HELICHRYSUM

HOYA *carnosa*

HYPERICUM *calycinum*
'Hidcote'
'Rowallane'

ILEX × *altaclarensis*
aquifolium
crenata
× *meservae*
pernyi

ITEA *illicifolia*

JASMINUM *fruticans*
humile 'Revolutum'

KALMIA

LAPAGERIA *rosea*

LAURUS *nobilis*

LAVANDULA

LEPTOSPERMUM

LEUCOTHOE *fontanesiana*

LIGUSTRUM *japonicum* 'Rotundifolium'
lucidum
ovalifolium

LOMATIA *myricoides*

LONICERA *fragrantissima*
japonica 'Aureoreticulata'
japonica 'Halliana'
japonica henryii
japonica repens
nitida
pileata
× *purpusii*

MAGNOLIA *grandiflora*

MAHORBERBERIS *aquisargentii*

MAHONIA 'Charity'
fremontii

Photinia × *fraseri* 'Red Robin'

MANDEVILLA *splendens*
MELIANTHUS *major*
MYRICA *cerifera*
  californica
MYRTUS *communis*
NANDINA *domestica*
NERIUM *oleander*
OLEARIA
OSMANTHUS
OZOTHAMNUS *rosmarinfolius*
PACHYSANDRA
PARAHEBE *catarractae*
PERNETTYA *mucronata*
PHILLYREA
PHLOMIS *fruticosa*
PHORMIUM
PHOTINIA × *fraseri* 'Red Robin'
  stranvaesia
PIERIS
PILEOSTEGIA *viburnoides*
PIPTANTHUS *laburnifolius*
PITTISPORUM
PRUNUS *laurocerasus*
  lusitanica
PSEUDOSASA
PYROCANTHA
RAPHIOLEPSIS *umbellata*
RHAMNUS *alaterna* 'Argenteovariegata'
RHODODENDRON
RIBES *laurifolium*
  viburnifolium
ROSMARINUS
RUBUS *calycinoides*
  henryi bambusarum
  microphyllus
  tricolor
RUSCUS *aculeatus*
  hypoglossum
RUTA *gravens*
SALVIA *officinalis*
SANTOLINA
SARCOCCA
SENECIO
SKIMMIA
SOLANUM *crispum*
  jasminoides
SOLLYA *fusiformis*
SOPHORA *tetraptera*
STANVAESIA *davidiana*
STAUNTONIA *hexaphylla*
TEUCRIUM *fruticans*
TRACHELOSPERMUM *asiaticum*
  jasminoides
TRACHYCARPUS *fortunei*
VACCINEUM *ovatum*
  vitis-idaea
VIBURNUM (Large-leaved varieties)
  buddleifolium
  × *burkwoodii*
  cinnamimifolium
  cylindricum
  davidii
  henryi
  × *hillieri*
  japonicum
  'Pragense'
  rhytidophyllum
  tinus
  utile
VINCA *major*

394

SHRUBS AND CLIMBING PLANTS FLOWERING IN WINTER AND EARLY SPRING

ABELIOPHYLLUM *distichum*
AZARA *microphylla*
CHAENOMELES
CHIMONANTHUS *praecox*
CLEMATIS *cirrhosa*
CORNUS *mas*
CORYLOPSIS *spicata*
CORYLUS *avellana*
DAPHNE *mezereum*
  odora
ELAEAGNUS × *ebbingei*
× FATSHEDRA *lizei*
ERICA *herbacea*
  × *darleyensis*
FATSIA *japonica*
FORSYTHIA *suspensa*
GARRYA *elliptica*
HAMAMELIS
HEDERA *canariensis*
  colchica
  helix (all forms)
JASMINUM *nudiflorum*
LONICERA *fragrantissima*
  japonica 'Halliana'
  × *purpusii*
  standishii

***Daphne mezereum* in flower**

MAHONIA 'Buckland'
  'Charity'
  fremontii
  japonica
  'Lionel Fortescue'
  lomarifolia
  pinnata
  'Winter Sun'
RIBES *laurifolium*
SALIX
SARCOCOCCA
STACHYURUS *praecox*
VIBURNUM × *bodnantense* 'Dawn'
  × *bodnantense* 'Deben'
  farreri
  foetens
  tinus

SHRUBS AND CLIMBING PLANTS WITH INTERESTING SUMMER COLOR OR SHAPE OF FOLIAGE

ABELIA *chinensis*
  'Edward Goucher'
  grandiflora
  schumannii
ABUTILON
ACACIA (Hardy forms)
ACER *negundo* (Variegated forms)
ACTINIDIA *chinensis*
  kolomikta
AKEBIA *quinata*
ALBIZA *julibrissin*
AMORPHA *canescens*
AMPELOPSIS *brevipendunculata* 'Elegans'
ARALIA *elata*
ARISTOLOCHIA *macrophylla*
AUCUBA *japonica*
AZARA *microphylla*
BUDDLEIA *alternifolia*
  crispa
  davidii
  fallowiana 'Alba'
  'Lochinch'
  × *weyerana*
CARYOPTERIS *incana*
CATALPA *bignonioides*
CEANOTHUS 'A. T. Johnson'
  'Autumnal Blue'
  'Burkwoodii'
  'Gloire de Versailles'
  'Henri Desfosse'
  impressus
CERATOSTIGMA *willmottiianum*
CHOISYA *ternata*
CISTUS (Grey-leaved forms)
CLERODENDRUM *trichotomum*
CLEYERA *fortunei*
COLUTEA *arboresdens*
CORTADERIA *selloana*
CORYNABUTILON *vitifolium*
COTINUS *coggygria* (Purple-leaved forms)
COTONEASTER *horizontalis* (Variegated forms)
CYTISUS *battandieri*
DABOECIA *cantabrica*
DECAISNEA *fargesii*
DECUMARIA *barbara*
DESFONTAINEA *spinosa*
DORYCNIUM *hirsutum*
DRIMYS *winteri*
ELAEAGNUS *commutata*
  × *ebbingei*
  pungens
ERIOBOTRYA *japonica*
ESCALLONIA
EUONYMUS *fortunei* (Variegated forms)
  japonicus (Variegated forms)
× FATSHEDRA *lizei*
FATSIA *japonica*
FEIJOA *sellowiana*
FICUS *carica*
FREMONTODENDRON *californicum*
FUCHSIA
GARRYA *elliptica*
GINKO *biloba*

GLEDITSIA *triacanthos* 'Sunburst'

HEBE (Low-growing varieties)

HEDERA *canariensis* (Variegated forms)
  colchica (Variegated forms)
  helix (all forms)

HEDYSARUM *multijugum*

HIBISCUS *syriacus*

HUMULUS *lupulus* 'Aureus'

HYDRANGEA *aspera*
  involucrata
  paniculata
  quercifolia
  villosa

HYPERICUM

ILEX × *altaclaerensis* (Variegated forms)
  aquifolium (Variegated forms)

INDIGOFERA *heterantha*

ITEA *illicifolia*

JASMINUM *officianale* (Variegated forms)

LAVANDULA

LAVATERA *olbia*

LESPEDEZA *thunbergii*

LEYCESTERIA *formosa*

LIGUSTRUM *lucidum* (Variegated forms)
  ovalifolium (Variegated forms)
  sinense (Variegated forms)

LONICERA *japonica* 'Aureoreticulata'
  japonica repens
  splendida

MAGNOLIA *grandiflora*

Vitis **'Brant' in fruit**

MAHONIA × Charity
  fremontii

MALUS (Purple-leaved forms)

MELIANTHUS *major*

MYRTUS *communis*

PARTHENOCISSUS *henryana*
  quinquefolia
  tricuspidata
  tricuspidata 'Veitchii'

PEROVSKIA *atriplicifolia*

PHILADELPHUS

PHLOMIS *fruticosa*

PHOTINIA × *fraseri* 'Red Robin'
  stranvaesia

PHYGELIUS *capensis*

PILEOSTEGIA *viburnoides*

PIPTANTHUS *laburnifolius*

Magnolia sinensis **in flower**

PITTOSPORUM

POTENTILLA

PYRACANTHA (Variegated forms)

PYRUS (Silver-leaved forms)

RHAMNUS *alaterna* 'Argenteovariegata'

ROBINIA *pseudoacacia* 'Frisia'

RUBUS *henryi bambusarum*

SCHIZOPHRAGMA *hydrangeoides*

SOLANUM *dulcamara* 'Variegatum'

SOPHORA *tetraptera*

SORBUS *aria*

STAUNTONIA *hexaphylla*

TAMARIX *pentandra*

TEUCRIUM *fruticans*

TRACHELOSPERMUM

  jasminoides 'Variegatum'

VIBURNUM *plicatum* 'Watanabe'
  tinus (Purple and variegated forms)

VINCA *major* (Variegated forms)
  minor

VITEX *agnus-castus*

VITIS 'Brant'
  coignetiae
  vinifera 'Purpurea'

WEIGELA

XANTHOCERAS *sorbifolium*

SHRUBS AND CLIMBING PLANTS WITH FRAGRANT FLOWERS

ABELIA *chinensis*
  triflora

ABELIOPHYLLUM *distichum*

AZARA *dentata*
  microphylla

BERBERIS × *stenophylla*

BUDDLEIA

CEANOTHUS 'Gloire de Versailles'

CHIMONANTHUS *praecox*

CHIONANTHUS *virginicus*

CHOISYA *ternata*

CLEMATIS *montana* (some forms)

CLERODENDRUM

CLETHRA

COLLETIA *armata*
  cruciata

CORYLOPSIS *spicata*

pauciflora
  willmottiae

CRATAEGUS *monogyna* 'Compacta'

CYDONIA *oblonga*

CYTISUS *battandieri*
  × *praecox*

DAPHNE

DEUTZIA × *elegantissima*

DRIMYS *winteri*

ELAEAGNUS *augustifolia*
  commutata
  × *ebbingei*
  macrophylla
  pungens

ERICA × *darleyensis*

FOTHERGILLA

GENISTA *aetnensis*
  cinerea
  tenera 'Golden Showers'

HAMAMELIS

HOHERIA *glabrata*
  lyalli

HOYA *carnosa*

ITEA *illicifolia*

LIGUSTRUM *lucidum*
  ovalifolium

LOMATIA *myricoides*

LONICERA × *brownii*
  caprifolium
  etrusca
  fragrantissima
  japonica 'Aureoreticulata'
  japonica 'Halliana'
  japonica henryii
  japonica repens
  nitida
  periclymenum (Hybrids)
  × *purpusii*
  standishii
  syringgantha
  tragophylla

MAGNOLIA *denudata*
  hypoleuca
  grandiflora
  kobus
  × *highdownensis*
  × *loebneri*
  salicifolia
  sieboldi
  sinensis
  × *soulangiana*
  stellata
  (Summer flowering forms)
  × *watsoni*
  wilsonii

MAHONIA

MANDEVILLA *splendens*

MYRTUS *communis*

OLEARIA *haastii*
  macrodonta

OSMANTHUS

OSMARONIA *cerasiformis*

PASSIFLORA *caerulea*

PHILADELPHUS (Medium-sized and tall-growing forms)

PHILLYREA

PITTISFORUM *tobira*

PONCIRUS trifoliata

PRUNUS *lusitanica*
  mume

PYRACANTHA

RAPHIOLEPSIS *umbrellata*

RHODODENDRON *luteum*

RIBES *odoratum*

ROMNEYA *coulteri*

ROSA (Climbing musk roses and similar forms; Climbing and species roses; Modern hybrid climbing; Old climbing roses prior to 1920; Rambler roses; Shrub and species roses for fan-training)

SAMBUCUS *nigra*

SARCOCCA

SKIMMIA

SPARTIUM *junceum*

STAUNTONIA *hexahylla*

SYRINGA *microphylla* 'Superba'
   × *persica*
   × *prestoniae*

TRACHELOSPERMUM *asiaticum*
   jasminoides

ULEX

VIBURNUM (Early flowering and spring flowering scented forms)
   bitchiuense
   × *bodnantense*
   × *burkwoodii*
   × *carlcephalum*
   carlesii
   farreri
   × *juddii*

VITEX *agnus-castus*

SHRUBS WITH MULTICOLORED FOLIAGE

ACER *palmatum* 'Roseomarginatum'

BERBERIS × *stenophylla* 'Pink Pearl'
   thunbergii 'Harlequin'
   thunbergii 'Rose Glow'

CORNUS *mas* 'Elegantissima'

EUONYMUS *fortunei* 'Emerald Gaiety'

HYDRANGEA *macrophylla* 'Tricolor'

LEUCOTHOE *fontanesiana* 'Rainbow'

LIGUSTRUM *lucidum* 'Excelsum Superbum'
   lucidum 'Tricolor'

PHORMIUM *tenax* 'Dazzler'
   tenax 'Sundowner'
   tenax 'Tricolor'

PITTISPORUM *tenuifolium* 'Garnetti'

PYRACANTHA *coccinea* 'Harlequin'
   coccinea 'Sparkers'

RUBUS *microphyllus* 'Variegatus'

SALIX *cinerea* 'Tricolor'
   'Harlequin'
   integra 'Albomaculata'

SALVIA *officianalis* 'Tricolor'

Phormium tenax **'Sundowner'**

SHRUBS AND CLIMBING PLANTS TO GROW IN LARGE TUBS AND CONTAINERS

Many other shrubs and climbing plants may perform well with this type of cultivation, but those listed are the most reliable.

In all cases, the container should be no less that 18in (45cm) in depth, and 21in (53cm) in diameter. The larger the container, the more success will be achieved with growth, flower and fruit.

The container should have one or more drainage holes in the base, and should have adequate drainage material. It should be filled with good quality potting compost, and be regularly watered throughout spring, summer and autumn. Feed at least once a week with a liquid fertilizer throughout late spring and early and mid-summer.

ABUTILON (Large-leaved and flowering forms)
   megapotamicum

ACACIA (Hardy forms)

ACER *negundo*
   palmatum 'Dissectum'

ALBIZIA *julibrissin*

AMPELOPSIS *brevipenduculata* 'Elegans'

AZARA *dentata*
   microphylla

BALLOTA *pseudodoctamnus*

BETULA *nana*

BOUGAINVILLEA *spectabilis*

BUXUS *sempervirens*

CALLISTEMON *citrinus*

CAMELLIA *japonica*

CARAGANA *arborescens* 'Lorbergii'
   pygmaea

CESTRUM *elegans*

CHOISYA *ternata*

CLEMATIS (Double and semi-double)
   flammula
   (Large-flowered hybrids)
   montana
   orientalis

CLEYERA *fortunei*

CONVOLVULUS *cneorum*

CORDYLINE *australis*

COROKIA *cotoneaster*
   × *virgata*

CORONILLA *glauca*

COTONEASTER (Low-spreading and medium-height spreading evergreens)

CYTISUS × *beanii*
   × *kewensis*
   × *praecox*
   purpureus

DAPHNE *collina*
   tangutica

DESFONTAINE *spinosa*

DRIMYS *winteri*

ERICA

EUONYMUS *fortunei*
   japonicus

FABIANA *imbricata*

× FATSHEDRA *lizei*

FATSIA *japonica*

FEIJOA *sellowiana*

FICUS *carica*

FUCHSIA

GENISTA *lydia*

HEBE

HEDERA *canariensis*
   colchica
   helix (all forms)

HELIANTHEMUM

HELICHRYSUM

HOYA *carnosa*

HYDRANGEA *petiolaris*
   quercifolia

ILEX × *altaclaerensis*
   aquifolium

JASMINUM *humile* 'Revoltum'
   nudiflorum

KALMIA *latifolia*

LAGERSTROEMIA *indica*

LAURUS *nobilis*

LEPTOSPERMUM

LIGUSTRUM *japonicum* 'Rotundifolium'
   lucidum
   ovalifolium
   quihoui
   sinense

LIPPIA *citriodora*

LONICERA *fragrantissima*

MELIANTHUS *major*

MYRTUS *communis*

NERIUM *oleander*

OSMANTHUS

PELARGONIUM

PHORMIUM

PHOTINIA × *fraseri* 'Red Robin'
   stranvaesia

PIERIS

PITTOSPORUM

PLUMBAGO *capensis*

Clematis montana **in flower**

PONCIRUS *trifoliata*

PRUNUS × *cistena*

PYRACANTHA

RAPHIOLEPSIS *umbellata*

RHAMNUS *alaterna* 'Argenteo-Variegata'

RHODODENDRON and AZALEA

ROSMARINUS

RUTA *graveolens*

SKIMMIA

SOPHORA *tetraptera*

STEPHANDRA *incisa*

SYRINGA *meyeri var. palibin*

TRACHYCARPUS *fortunei*

VIBURNUM (Spring-flowering, scented forms)

YUCCA

SHRUBS FOR DRY SITES

ABELIA
ARTEMISIA *pseudodictamnus*
BALLOTA
BERBERIS
BUDDLEIA
CARAGANA *pygmaea*
CARYOPTERIS *incana*
CEANOTHUS
CISTUS
COLUTEA *arborescens*
CONVOLVULUS *cneorum*
CORNUS *canadensis*
COROKIA *cotoneaster*
COTINUS *coggygria*
COTONEASTER
CYTISUS
DORYCNIUM *hirsutum*
ELAEAGNUS
ELSHOLTZIA *stauntonii*
FREMONTODENDRON *californicum*
GENISTA
HEBE
HEDERA
HELIANTHEMUM
HELICHRYSUM
HIPPOPHAE *rhamnoides*
HYPERICUM
INDIGOFERA *heterantha*
KERRIA *japonica*
LAVANDULA
LAVATERA *olbia*
LESPEDEZA *thunbergii*
LONICERA
NITIDA *pileata*
MAHONIA *aquifolium*
OLEARIA
PEROVSKIA
PHILADELPHUS
PHLOMIS
PHORMIUM
POTENTILLA
RIBES
ROMNEYA *coulteri*
ROSMARINUS
RUBUS
RUSCUS *aculeatus*
SALVIA *officinalis*
SANTOLINA
SENECIO
SPARTIUM *junceum*
TAMARIX
TEUCRIUM *fruticans*
ULEX
VINCA
YUCCA

SHADE-TOLERANT SHRUBS

ACER *japonicum*
AUCUBA *japonica*
BERBERIS

BUXUS *sempervirens*
CAMELLIA *japonica*
  × *williamsii* 'Donation'
CHOISYA *ternata*
COLUTEA *arborescens*
CORNUS *alba*
  canadensis
COTONEASTER
CRINODENDRON *hookeranum*
DANAE *racemosa*
DAPHNE *blagayana*
  cneorum
  laureola
  mezereum
  odora
  pontica
  tangutica
ELAEAGNUS × *ebbingei*
ENKIANTHUS
EUONYMUS *europaeus*
  fortunei
× FATSHEDERA *lizei*
FATSIA *japonica*
FOTHERGILLA
GARRYA *elliptica*
× GARULNETTYA
GAULTHERIA
HEDERA
HYDRANGEA *sargentiana*
  serrata 'Grayswood'
  villosa
HYPERICUM *androsaemum*
  calycinum
ILEX × *altaclarensis*
  aquifolium
LEUCOTHOE *fontanesiana*
LEYCESTERIA *formosa*

***Danae racemosa* in leaf**

LIGUSTRUM
LONICERA *nitida*
  pileata
MAHONIA
NIELLIA *thibetica*
OSMANTHUS *armatus*
  burkwoodii
  delavayi
  heterophyllus
PACHYSANDRA
PERNETTYA *mucronata*
PIERIS

PRUNUS *laurocerasus*
  lusitanica
PSEUDOSASA *japonica*
PYRACANTHA
RHODODENDRON and AZALEA
RIBES *alpinum*
ROSMARINUS
RUBUS
RUSCUS *aculeatus*
SAMBUCUS
SARCOCOCCA
SKIMMIA
SPIRAEA *prunifolia*
  thunbergii
STACHYURUS *praecox*
SYMPHORICARPOS
VACCINEUM
VIBURNUM × *bodnantense*
  davidii
  farreri
  foetens
  × *burkwoodii*
  plicatum
  tinus
VINCA

SHRUBS FOR ACID SOILS

This is a selection of shrubs that specifically require acid soil or thrive on acid conditions. Others that are also tolerant of acidity will be found listed in the dictionary.

ACER *japonicum*
  palmatum
  palmatum 'Dissectum'
  palmatum var. heptalobum
ANDROMEDA *polifolia*
ARCTOSTAPHYLOS *uva-ursi*
ARONIA ARBUTIFOLIA
CALLUNA *vulgaris*
CAMELLIA *japonica*
CASSINIA *fulvida*
CLETHRA *alnifolia*
CORNUS *florida*
  kousa
  nuttallii
CORYLOPSIS *spicata*
CRINODENDRON *hookeranum*
DABOECIA *cantabrica*
DAPHNE *blagayana*
DESFONTAINEA *spinosa*
DISANTHUS *cercidifolius*
DRIMYS *winteri*
ENKIANTHUS *campanulatus*
ERICA *arborea*
  cillarts
  cinerea
  × *darleyensis*
  erigena
  herbacea
  tetralix
  vagans
FOTHERGILLA
× GAULNETTYA
GAULTHERIA
HYDRANGEA *serrata*
ILEX *verticillata*
KALMIA
LEPTOSPERMUM
LEUCOTHOE *fontanesiana*
LOMATIA *ferruginea*

MAGNOLIA *liliiflora*

MYRICA

PACHYSANDRA

PERNETTYA *mucronata*

PIERIS

RHODODENDRON and AZELEA

SKIMMIA

SYMPLOCOS *paniculata*

VACCINIUM

ZENOBIA *pulvurulenta*

SHRUBS FOR ALKALINE SOILS

This is a selection of shrubs that specifically require alkaline soil or thrive on alkaline conditions. Others that are also tolerant of alkalinity will be found listed in the dictionary.

ACER *ginnala*

CARAGANA *pygmaea*

CARYOPTERIS *incana × mongolica* 'Arthur Simmonds'

CHIMONANTHUS *praecox*

CISTUS

COLUTEA *arborescens*

CORONILLA *glauca*

DAPHNE *mezereum*
   odora

DEUTZIA

FREMONTODENDRON *californicum*

SPARTIUM *junceum*

XANTHOCERAS *sorbifolium*

Caryopteris incana × mongolica 'Arthur Simmonds'

AROMATIC SHRUBS

CALYCANTHUS

CARYOPTERIS *incana*

CHIMONANTHUS *praecox*

CHOISYA *ternata*

CLERODENDRUM *bungei*
   trichotomum

ELSHOLTZIA *stauntonii*

ESCALLONIA

HELICHRYSUM *plicatum*
   serotinum

LAURUS *noblis*

LAVANDULA

LIPPIA *citriodora*

MAGNOLIA *salicifolia*

MYRICA

MYRTUS *communis*

OLEARIA *mollis*

PEROVSKIA *atriplicifolia*

RIBES *viburnifolium*

ROSMARINUS

RUTA *graveolens*

SALVIA *officinalis*

SANTOLINA

SHRUBS FOR GOOD GROUND COVER
*Suitable for dry, poor soils

ARCTOSTAPHYLOS *uva-ursi*

BERBERIS × *bristolensis*
   calliantha
   candidula
   verruculosa
   wilsonae

CALLUNA *vulgaris*

CEANOTHUS *thyrsiflorus* 'Repens'

CISTUS × *corbariensis*
   × *lusitanicus* 'Decumbens'

CORNUS *canadensis*

COTONEASTER *adpressus var. praecox*
   congestus
   conspicuus 'Decorus'
   'Coral Beauty'
   dammeri
   'Donard's Gem'
   horizontalis
   'Hybridus Pendulus'
   microphyllus
   salicifolius 'Parkteppich'
   salicifolius 'Repens'
   'Skogholm'

CYTISUS × *beanii**
   × *kewensis**
   × *praecox**

ERICA *carnea**
   × *darleyensis**
   *vagans**

EUONYMUS *forunei**

× FATSHEDERA *lizei*

× GAULNETTYA *wisleyensis* 'Wisley Pearl'

GAULTHERIA

GENISTA *hispanica**
   *lydia**
   *pilosa**

HALIMIUM*

HEBE *albicans*
   'Carl Teschner'
   pinguifolia 'Pagei'
   pinguifolia 'Quicksilver'
   rakaiensis

HEDERA *helix* 'Buttercup'

HELIANTHEMUM*

HYDRANGEA *petiolaris*

HYPERICUM *calycinum*
   × *moseranum*

MAHONIA *aquifolium*

PACHYSANDRA

PERNETTYA *mucronata*

POTENTILLA *dahurica* 'Abbotswood'
   fruticosa 'Beesii'*
   fruticosa 'Elizabeth'*
   'Longacre'*
   'Red Ace'*
   'Tangerine'*
   'Tilford Cream'*

Erica × darleyensis 'Silberschmelze' in flower

PRUNUS *laurocerasus* 'Otto Luyken'
   laurocerasus 'Zabeliana'

PSEUDOSASA *japonica*

PYRACANTHA 'Alexander Pendula'
   'Soleil D'or'

RUBUS *calycinoides*
   tricolor

SALIX *repens* 'Argentea'

SANTOLINA*

SARCOCOCCA *humilis*
   ruscifolia

STEPHANANDRA *incisa* 'Crispa'

STRANVAESIA *davidiana* 'Prostrata'

SYMPHORICARPOS 'Hancock'

VACCINEUM *vitis-idaea*

VIBURNUM *davidii*

VINCA

SHRUBS FOR SCREENING AND WINDBREAK

BERBERIS *julianae*

BUDDLEIA *globosa*

CORYLUS *avellana*

ELAEAGNUS × *ebbingei*

FORSYTHIA

HIPPOPHAE *rhamnoides*

LIGUSTRUM *ovalifolium*
   sinense
   vulgare

LYCIUM *barbarum*

PRUNUS *lusitanica*

PSEUDOSASA *japonica*

RHAMNUS

SALIX (Medium-height and tall-growing varieties)

STRANVAESIA *davidiana*

SYRINGA × *hyacinthiflora*
   × *prestoniae*
   vulgaris

TAMARIX

VIBURNUM (Large-leaved varieties)
   tinus

SHRUBS FOR MOIST SITES

AMELANCHIER *lamarckii*

ARONIA

CLETHRA *alnifolia*

CORNUS *alba*
   stolonifera 'Flaviramea'

DANAE *racemosa*

MYRICA *gale*

NEILLIA *thibetica*

PHYSOCARPUS *opulifolius* 'Luteus'

PSEUDOSASA *japonica*

RHAMNUS *frangula*

SALIX

SAMBUCUS

SORBARIA

SPIRAEA

SYMPHORICARPOS

VACCINIUM

VIBURNUM *lentago*
   opulus

SHRUBS AND CLIMBING PLANTS WITH LEAST PROTRUDING GROWTH WHEN FAN-TRAINED

ABELIOPHYLLUM *distichum*

ACTINIDIA *kolomikta*

AKEBIA *quinata*

AMPELOPSIS *brevipendunculata*
   brevipendunculata 'Elegans'

BERBERIDOPSIS *corallina*

BILLARDIER'A *longiflora*

BUPLEURUM *fruticosum*

CHAENOMELES

CLEMATIS *alpina*
   armandii
   campaniflora
   cirrhosa
   (Double and semi-double forms)
   × *durandii*
   × *eriostemon*
   flammula
   florida 'Bicolor'
   × *jouinana praecox*
   (Large-flowered hybrids)
   macropetala
   rehderiana
   texensis
   viticella

COBAEA *scandens*

COTONEASTER *horizontalis*

DECUMARIA *barbara*

ECCREMOCARPUS *scaber*

EUONYMUS *fortunei*

HEDERA *canariensis*
   colchica
   helix
   helix angularis 'Aurea'
   helix 'Buttercup'
   helix 'Cristata'
   helix 'Glacier'
   helix 'Gold Heart'
   helix 'Luzii'
   helix sagittifolia

JASMINUM × *stephanense*

MAHONIA *fremontii*

PARTHENOCISSUS *henryana*
   quinquefolia
   tricuspidata 'Veitchii'

RIBES *laurifolium*
   speciosum

RUBUS *tricolor*

SCHIZOPHRAGMA *hydrangeoides*

THUNBERGIA *alata*

TROPAEOLUM *majus*
   peregrinum
   speciosum
   tuberosum

TRACHELOSPERMUM *asiaticum*
   jasminoides

WATTAKA-KA *sinensis*

SHRUBS AND CLIMBING PLANTS FOR DIFFICULT AND EXPOSED CONDITIONS

ACER *negundo*

AMPELOPSIS *brevipendunculata*

AUCUBA *japonica*

CARAGANA *arborescens* 'Lorbergii'

CELASTRUS *orbiculatus*

CHAENOMELES

CLEMATIS *montana*
   vitalba

COTINUS *coggygria*

COTONEASTER × *horizontalis*
   hybridus pendulus
   (Low-growing, spreading evergreen forms for walls; Medium height, spreading evergreen and semi-evergreen forms for walls; Tall deciduous and tall evergreen forms)

CRATAEGUS (Autumn foliage forms)
   oxyacantha

CYDONIA *oblonga*

ELAEAGNUS × *ebbingei*
   pungens

EUONYMUS *fortunei*

FORSYTHIA *suspensa*

GINKGO *biloba*

HEDERA *helix*
   helix angularis 'Aurea'
   helix 'Buttercup'
   helix 'Cristata'
   helix 'Glacier'
   helix 'GoldHeart'
   helix 'Luzii'
   helix sagittifolia

HYDRANGEA *petiolaris*

ILEX × *altaclaerensis*
   aquifolium

JASMINUM *nudiflorum*

KERRIA *japonica*

Kolkwitizia amabilis in flower

KOLKWITZIA *amabilis*

LABURNOCYTISUS × *adamii*

LABURNUM × *watereri* 'Vossii'

LIGUSTRUM *lucidum*
   ovalifolium
   quihoui
   sinense

***Pharthenocissus tricuspidata* 'Veitchii' in autumn**

LONICERA *japonica* 'Halliana'
   japonica repens
   nitida
   periclymenum (Hybrids)
   tatarica

LYCIUM *barbarum*

MAHONIA × 'Charity'

MALUS (Fruiting forms; Green-leaved flowering forms; Purple-leaved forms)

PARTHENOCISSUS *henryana*
   quinquefolia
   tricuspidata 'Veitchii'

PHILADELPHUS (Medium-sized and tall-growing forms)

PHILLYREA

POLYGONUM *baldschuanicum*

PYRACANTHA

PYRUS

ROBINIA *pseudoacacia*

ROSA (Climbing musk roses and similar forms; Modern hybrid climbing) The following varieties:
   Alchemist
   Aloha
   Compassion
   Dance du Feu
   Dreaming Spires
   Golden Showers
   Leverkusen
   Maigold
   Masquerade, Climbing
   Mme Edouard Herriot
   Mme Grégoire Staechelin
   Mme Henri Guillot
   Morning Jewel
   New Dawn, Climbing
   Paul's Scarlet
   Pink Perpetue
   Queen Elizabeth, Climbing
   Rosy Mantle
   (Rambler roses; Shrub and species roses for fan-training)

RUBUS *phoenicolasius*
   tricolor
   ulmifolius 'Bellidiflorus'

SORBUS *aria*

VIBURNUM (Best fruiting forms; Early-flowering forms; Large-leaved forms)
   opulus
   plicatum
   (Spring flowering, scented forms)

VINCA *major*

VITIS 'Brant'
  coignetiae

WEIGELA

SHRUBS AND CLIMBING PLANTS WITH ORNAMENTAL FRUIT

ACTINIDA *arguta*
  chinensis
  kolomikta

AKEBIA *quinata*

AMPELOPSIS *brevipendunculata*
  brevipendunculata 'Elegans'

ARBUTUS

AUCUBA *japonica*

BILLARDIER'A *longiflora*

BUPLEURUM *fruticosum*

CALLISTEMON *citrinus*

CELASTRUS *orbiculatus*

CLEMATIS *alpina* (seed heads)
  armandii (seed heads)
  campaniflora (seed heads)
  cirrhosa (seed heads)
  (Double and semi-double) (seed heads)
  flammula (seed heads)
  × *jouiniana praecox* (seed heads)
  (Large-flowered hybrids) (some seed heads)
  macropetala (seed heads)
  orientalis (seed heads)
  vitalba (seed heads)
  viticella (some seed heads)

COTINUS *coggygria*

Cotoneaster simonsii in fruit

COTONEASTER × *horizontalis*
  hybridus pendulus
  (Low-growing, spreading evergreen forms for walls; Medium height, spreading evergreen and semi-evergreen forms for walls; Tall deciduous and tall evergreen forms)

CRATAEGUS (Autumn foliage forms)

DECAISNEA *fargesii*

EUONYMUS *fortunei*

× FATSHEDERA

LIZEI

FATSIA *japonica*

FEIJOA *sellowiana*

HEDERA *canariensis*
  colchica
  helix (all forms)
  (Shrubby forms)

HUMULUS *lupulus* 'Aureus'

HYDRANGEA *aspera* (sterile bracts)
  paniculata (sterile bracts)
  petiolaris (sterile bracts)
  quercifolia (sterile bracts)

villosa (sterile bracts)

ILEX × *altaclaerensis*
  aquifolium

LIGUSTRUM *lucidum*
  ovalifolium
  quihoui
  sinense

LONICERA × *americana*
  × *brownii*
  caprifolium
  etrusca
  fragrantissima
  × *heckrottii*
  japonica 'Aureoreticulata'
  japonica 'Halliana'
  japonica henryii
  japonica repens
  nitida
  periclymenum (Hybrids)
  splendida
  tatarica
  × *tellmanniana*
  tragophylla

LYCIUM *barbarum*

MALUS (Fruiting forms; Green-leaved flowering forms; Purple-leaved forms)

MORUS *alba*

MYRTUS *communis*

PASSIFLORA *caerulea* (following hot summers)

PHASEOLUS *coccineus*

PHOTINIA *stranvaesia*

PIPTANTHUS *laburnifolius*

PITTOSPORUM (following hot summers)

PONCIRUS *trifoliata*

PYRACANTHA

PYRUS

RAPHIOLEPIS *umbellata*

RHAMNUS *alaterna* 'Argenteo-Variegata'

ROSA (Climbing musk roses; Climbing species roses; Modern hybrid climbing; Old climbing roses prior to 1920; Rambler roses; Shrub and species roses for fan-training)

RUBUS *tricolor*

SCHISANDRA *grandiflora rubrifoloia*

SOLANUM *crispum*
  dulcamara 'Variegatum'

SORBUS *aria*

STAPHYLEA *colchica*

STAUNTONIA *hexaphylla*

SYMPLOCOS *paniculata*

TROPAEOLUM *speciosum*

VIBURNUM (Best fruiting forms; Large-leaved forms)
  opulus
  plicatum
  tinus

VITIS 'Brant' (edible)
  vinifera 'Purpurea' (edible)

WATTAKA-KA *sinensis*

XANTHOCERAS *sorbifolium*

SHRUBS AND CLIMBING PLANTS WITH EDIBLE FRUIT

ACTINIDIA *chinensis*

CYDONIA *oblonga*

FEIJOA *sellowiana*

FICUS *carica*

MALUS (Fruiting forms)
  pumila (Apple hybrids)

MORUS

PASSIFLORA *caerulea*

PHASEOLUS *coccineus*

PRUNUS *armeniaca*
  avium, P. cerasus, P. mahaleb
  domestica (Damson, gage and plum hybrids)
  persica (Nectarine and peach hybrids)

PYRUS *communis*

RIBES *grossularia*
  sativum

RUBUS *fruticosus*
  (Hybrid forms)
  loganobaccus
  phoenicolasius

VITIS 'Brant'
  vinifera (Grape vine hybrids)

SHRUBS AND CLIMBING PLANTS NEEDING OR RESPONDING WELL TO CONSERVATORY OR GREENHOUSE

ABUTILON *megapotamicum*
  megapotamicum (Large-leaved and flowering forms)

ACACIA (Hardy forms)

ALBIZIA *julibrissin*

AMPELOPSIS *brevipendunculata* 'Elegans'

AZARA *dentata*

BERBERIDOPSIS *corallina*

BIGNONIA

CAPREOLATA

BILLADIER'A *longiflora*

BOUGAINVILLEA *spectabilis*

BUDDLEIA (Tender forms)

CALLISTEMON *citrinus*

CAMELLIA *japonica*

CEANOTHUS (Evergreen forms)

CESTRUM *elegans*

CLEMATIS *florida* 'Bicolor'

CLIANTHUS *puniceus*

COROKIA × *virgata*

CORONILLA *glauca*

DECUMARIA *barbara*

DRIMYS *winteri*

ECCREMOCARPUS *scaber*

ERIOBOTRYA *japonica*

× FATSHEDERA *lizei*

Prunus persica 'Lord Napier' in fruit

FATSIA *japonica*

FEIJOA *sellowiana*

HOYA *carnosa*

IPOMOEA *hederacea*

JASMINUM *mesnyi*
  officinale (Variegated forms)
  polyanthum

LAGERSTROEMIA *indica*

LAPAGERIA *rosea*

LOMATIA *myricoides*

LONICERA *splendida*

MANDEVILLA *splendens*

MELIANTHUS *major*

MYRTUS *communis*

PASSIFLORA *caerulea*

PITTOSPORUM

PLUMBAGO *capensis*

RHODOCHITON *atrosanguineum*

SOLANUM *jasminoides*

SOLLYA *fusiformis*

SOPHORA *tetraptera*

STAUNTONIA *hexaphylla*

TEUCRIUM *fruticans*

THUNBERGIA *alata*

TRACHELOSPERMUM *asiaticum*
  jasminoides

WATTAKA-KA *sinensis*

QUICK-GROWING SHRUBS AND CLIMBING PLANTS

ACACIA (Hardy forms)

ACER *negundo*

ACTINIDIA *arguta*
  chinensis

AMPELOPSIS *brevipendunculata*

ARISTOLOCHIA *macrophylla*

ATRIPLEX *halimus*

AZARA *dentata*

BUDDLEIA *alternifolia*
  davidii
  (Tender forms)

CAMPSIS *grandiflora*
  radicans

CATALPA *bignoniodes*

CEANOTHUS (Evergreen and deciduous forms)

CELASTRUS *oriculaus*

CLEMATIS *alpina*
  armandii
  (Double and semi-double)
  flammula
  × *jouiniana praecox*
  (Large-flowered hybrids)
  montana
  orientalis
  vitalba

COLUTEA *arborescens*

CORNUS *alba*
  stononifera 'Flaviramea'

CORYNABUTILON *suntense*
  vitifolium

COTONEASTER (Medium height, spreading evergreen and semi-evergreen forms for walls; Tall deciduous forms; Tall evergreen forms)
  × *hybridus pendulus*
  Exburiensis
  henryanus
  lacteus

CRATAEGUS (Autumn foliage forms)
  oxycantha

CYDONIA *oblonga*

CYTISIS *battandieri*
  multiflorus
  scoparius

DEUTZIA

ECCREMOCARPUS *scaber*

ELAEGNUS *angustifolia*
  × *ebbingei*

ESCALLONIA

EUCALYPTUS

× FATSHEDRA *lizei*

FATSIA *japonica*

FICUS *carica*

FREMONTODENDRON *californica*

GENISTA *aetnensis*
  tenera 'Golden Shower'

HEDERA *canariensis*
  colchica

HUMULUS *lupulus* 'Aureus'

HYDRANGEA *macrophylla*
  paniculata
  villosa

HYPERICUM 'Hidcote'
  × *inodorum*
  × *moseranum*

INDIGOFERA *heterantha*

JASMINUM *officianale*

KOLKWITZIA *amabilis*

LATHYRUS *latifolius*
  odoratus

LAVANDULA

LAVATERA *olbia*

LESPEDEZA *thunbergii*

LIGUSTRUM *ovalifolium*

LONICERA × *americana*
  caprifolium
  etrusca
  involucrata
  japonica 'Halliana'
  japonica henryii
  japonica repens
  nitida
  periclymenus (Hybrids)
  pileata
  syringantha
  tragophylla

MAGNOLIA *grandiflora*
  (Large-growing, star-flowered forms)
  × *soulangiana*
  (Summer flowering forms)

MAHONIA 'Charity'

MALUS (Fruiting forms; Green-leaved flowering forms; Purple-leaved forms)

PARTHENOCISSUS *henryana*
  quinquefolia
  tricuspidata 'Veitchii'

PASSIFLORA *caerulea*

PHASEOLUS *coccineus*

PHILADELPHUS (Tall-growing forms)

PIPTANTHUS *laburnifollus*

PLUMBAGO *capensis*

POLYGONUM *baldschuanicum*

PYRACANTHA

PYRUS

ROBINIA (Pink-flowering forms)
  pseudoacacia
  pseudoacacia 'Frisia'

ROMNEYA *coulteri*

ROSA (Climbing musk roses and similar forms; Rambler roses)

RUBUS *fruticosus*
  loganobaccus

SALIX (Medium-height and tall-growing varieties)

SALVIA *officinalis*

SAMBUCUS

SOLANUM *crispum*
  dulcamara 'Variegatum'

SORBARIA

SORBUS *aria*

SPARTIUM *junceum*

SPIRAEA

STRANVAESIA *davidiana*

SYMPHORICARPOS

TAMARIX

TROPAEOLUM *majus*

VIBURNUM × *bodnantense*
  opulus
  rhytidophyllum

VITIS 'Brant'
  coignetiae
  vinifera (Grape vine hybrids)

WEIGELA

WISTERIA *floribunda*
  floribunda 'Macrobotyrs'
  sinensis

PRONUNCIATION GLOSSARY

Accurate identification of trees and shrubs can only be guaranteed by using the universally accepted botanical (Latin) name. This glossary provides a simple syllabic guide to pronunciation. Stressed syllables are indicated by italic type. The basic elements of pronunciation are as follows:

a	as in ago, cat
ah	as in calm
ay	as in say
e	as in let
ee	as in treat
ew	as in few
ewr	as in pure
g	as in get
i	as in pin
I	as in pie
j	as in jam
k	as in can
o	as in cot
oh	as in note
oo	as in clue
ow	as in now
oy	as in noise
th	as in thin
u	as in cup
v	as in vat
zh	as in vision

A

abelia	a-*bee*-lee-a
abeliophyllum	a-bee-lee-oh-*fil*-um
abrotanum	a-broh-*tay*-num
abutilon	a-*bew*-ti-lon
acacia	a-*kay*-shee-a
acanthopanax	a-kan-thoh-*pan*-aks
acer	*ay*-sir
acerifolia	ay-sir-i-*foh*-lee-a
aconitifolium	a-ko-ni-tee-*foh*-lee-um
aculeatus	a-kew-lee-*ah*-tus
acuminata	a-kew-mi-*nah*-ta
acutifolia	a-kew-ti-*foh*-lee-a
adamii	a-*dam*-ee-I
adenophylla	ad-en-oh-*fil*-a
adpressus	ad-*pres*-us
aegyptiaca	I-jip-tee-*ah*-ka
aequalis	I-*kwah*-lis
aesculus	*es*-kew-lus
aetnensis	et-*nen*-sis
afghanica	af-*gan*-ik-a
aggregata	ag-re-*gah*-ta
aglaia	ag-*ll*-a
agnus-castus	*ag*-nus-*kas*-tus
aguilari	a-gwi-*lah*-ree
ailanthus	I-*lan*-thus
aitchsonii	aych-i-*son*-ee-I
alabamensis	al-a-bam-*en*-sis
alaterna	a-la-*tern*-a
alatus	a-*lay*-tus
alba	*al*-ba
albertii	al-*ber*-tee-I
albescens	al-*bes*-enz
albicans	*al*-bi-kanz
albicaulis	al-bi-*kaw*-lis
albidum	*al*-bi-dum
albidus	*al*-bi-dus
albiflorus	al-bi-*flor*-us
albiplena	al-bi-*plen*-a
albizia	al-*bitz*-ee-a
albomarginata	al-boh-mar-ji-*nah*-ta
albomarginatus	al-boh-mar-ji-*nah*-tus
albovariegata	al-bo-ver-ee-e-*gay*-ta, al-bo-var-ee-e-*gah*-ta
album	*al*-bum
albus	*al*-bus
algarvense	al-gar-*ven*-see
alnifolia	al-ni-*foh*-lee-ah
alnus	*al*-nus
aloysia	al-o-*ee*-she-a
alpestre	al-*pes*-tree
alpina	al-*pI*-na
alpinum	al-*pI*-num
altaclarensis	al-ta-kla-*ren*-sis

alternifolia	al-ter-ni-*foh*-lee-a
altissima	al-*tis*-i-ma
amabilis	a-*mah*-bil-is
amanogawa	a-*man*-oh-gow-a
amara	a-*mah*-ra
ambigua	am-*big*-ew-a
amelanchier	am-el-*ang*-kee-er, am-el-*an*-chee-er
americana	a-mer-i-*kah*-na
americanus	am-er-i-*kah*-nus
amoenum	a-*mee*-num
amorpha	a-*mor*-fa
amurense	a-mur-*en*-se
amygdaliformis	a-mig-day-li-*for*-mis
amygdalus	a-mig-*day*-lus
anagyroides	an-a-gi-*roy*-deez
andersonii	an-der-son-ee-I
andrachne	an-*drak*-nee
andrachnoides	an-drak-*noy*-deez
andreanus	an-dree-*ah*-nus
andromeda	an-*drom*-e-da
androsaemum	an-dro-*say*-mum
angustifolia	ang-gus-ti-*foh*-lee-a
anomala	a-*nom*-a-la
antarctica	ant-*ark*-ti-ka
apiculata	a-pik-ew-*lah*-ta
aquatica	a-*kwat*-i-ka
aquifolium	a-kwi-*foh*-lee-um
aquisargentiae	a-kwi-sar-*jen*-ti-ee
aralia	a-*ray*-lee-a, a-*ral*-ee-a
arborea	ar-bor-*ee*-a
arborescens	ar-bor-*es*-enz
arboreum	ar-bor-*ee*-um
arboreus	ar-bor-*ee*-us
arbuscula	ar-*bus*-kew-la
arbutifolia	ar-bew-ti-*foh*-lee-a
arbutus	ar-*bew*-tus
arctostaphylos	ark-toh-*sta*-fil-os
arcuata	ar-kew-*ah*-ta
ardens	ar-*denz*
argentea	ar-*jen*-tee-a
argenteovariegatum	ar-*jen*-tee-oh-ver-ee-e-tum, ar-*jen*-tee-oh-var-ee-e-*gah*-tum
argenteum	ar-*jen*-tee-um
arguta	ar-*gew*-ta
aria	*ah*-ree-a
armata	ar-*mah*-ta
armatus	ar-*mah*-tus
armeniaca	ar-men-ee-*ah*-ka
armstrongii	arm-*strong*-ee-I
aronia	a-*roh*-nee-a
artemisia	ar-ti-*me*-see-a
arundinaria	a-run-di-*ner*-ee-a
asano	as-*say*-no
aspera	*as*-pe-ra
asplenifolia	as-ple-ni-*foh*-lee-a
atalantioides	at-a-lan-tee-*oy*-deez
atkinsonii	at-kin-*son*-ee-I
atriplex	*a*-tri-pleks
atriplicifolia	a-tre-pli-ki-*foh*-lee-a
atrocarpa	a-tro-*kar*-pa
atrocaulis	a-troh-*kaw*-lis
atropurpurea	a-troh-pur-*pewr*-ee-a
atropurpureum	a-troh-pur-*pewr*-ee-um
atrorubens	a-troh-*roo*-benz
aucuba	*aw*-kew-ba
aucuparia	aw-kew-*par*-ee-a
auratum	aw-*rah*-tum
aurea	aw-*ree*-a
aureomaculata	aw-ree-oh-mak-ew-*lah*-ta
aureomarginata	aw-ree-oh-mar-ji-*nah*-ta
aureopictus	aw-ree-oh-*pik*-tus
aureovariegata	aw-ree-oh-ver-ee-e-*gay*-ta, aw-ree-oh-var-ee-e-*gah*-ta
aureovariegatum	aw-ree-oh-ver-ee-e-*gay*-tum, aw-ree-oh-var-ee-e-*gah*-tum
aureum	aw-*ree*-um
auricoma	aw-ree-*koh*-ma
aurora	aw-*ror*-a
australis	aw-*stra*-lis
autumnalis	aw-tum-*na*-lis
avellana	a-ve-*lah*-na
avium	*ay*-vee-um
azalea	a-*zay*-lee-a
azara	a-*zah*-ra
azorica	a-*zor*-i-ka

B

babylonica	bab-i-*lon*-i-ka
baccata	bak-*ka*-ta
bacciflava	bak-ee-*flah*-va
baileyana	bay-lee-*ah*-na
ballota	ba-*lo*-ta
balsamifera	bawl-sa-*mi*-fe-ra
barbarum	*bar*-ba-rum
battandieri	ba-tan-dee-*er*-I
baumannii	bow-*man*-ee-I
bealei	*beel*-ee-I
beanii	*been*-ee-I
beesii	*beez*-ee-I
beissneri	*bes*-ner-ee
beni-shi-don	ben-i-*shi*-don
benthamia	ben-*thay*-mee-a
benthamidia	ben-thay-mi-*dee*-a
berberis	*ber*-ber-is
bessoniana	bes-on-ee-*ah*-na
betula	*bet*-ew-la
betulifolia	bet-ew-li-*foh*-lee-a
betulifolium	bet-ew-li-*foh*-lee-um
betuloides	bet-ew-*loy*-deez
betulus	*bet*-ew-lus
bholua	*boh*-loo-a
bignonioides	big-non-ee-*oy*-deez
billiardii	bi-lee-*ard*-ee-I
bioritsensis	bi-or-it-*zen*-sis
bitchiuense	bit-choo-*zen*-se
blagayana	bla-gay-*ah*-na
blireana	blir-ee-*ah*-na
bockii	*bok*-ee-I
bodinieri	bo-din-ee-*er*-ee
bodnantense	bod-nan-*ten*-see
borealis	bor-ee-*ah*-lis, bor-ee-*ay*-lis
boydii	*boy*-dee-I
brachysiphon	brak-*ee*-sI-fon
brevipetala	brev-i-*pe*-ta-la
brilliantissimum	bril-ee-an-*tis*-ee-mum
briotii	bree-*ot*-ee-I
bristolensis	bris-tol-*en*-sis
bristoliensis	bris-tol-ee-*en*-sis
britzensis	brit-*zen*-sis
brocklebankii	brok-il-*bank*-ee-I
bronxensis	bronks-*en*-sis
brozzonii	broz-*on*-ee-I
buddleia	*bud*-lee-a
buddleifolium	bud-lee-i-*foh*-lee-um
bujoti	boo-*jot*-i
bullatus	bul-*ay*-tus, bul-*ah*-tus
bumalda	bum-*al*-da
bungei	*bun*-jee-I
bupleurum	boo-*plur*-um
burfordensis	bur-for-*den*-sis
burkwoodii	burk-*wud*-ee-I
buxifolia	buks-i-*foh*-lee-a
buxifolius	buks-i-*foh*-lee-us
buxus	*buk*-sus

C

caerulea	kI-*ru*-lee-a
caeruleus	kI-*ru*-lee-us
californica	kal-i-*for*-ni-ka
californicum	kal-i-*for*-ni-kum
calleryana	kal-er-ee-*ah*-na
calliantha	kal-ee-*anth*-a
callicarpa	kal-i-*kar*-pa
callistemon	kal-*is*-tee-mon
calluna	kal-*oo*-na
calycanthus	kal-i-*kanth*-us
calycinum	kal-i-*sI*-num
camellia	ka-*meel*-ee-a
camelliifolia	ka-meel-ee-i-*foh*-lee-a
campanulatus	kam-pan-ew-*lah*-tus
campbellii	kam-*bel*-lee-I
camperdownii	kam-per-*down*-ee-I
campestre	kam-*pes*-tree
canadensis	kan-a-*den*-sis
canariensis	kan-ayr-ee-*en*-sis
candicans	*kan*-di-kanz
candidissima	kan-did-*is*-i-ma
candidissimum	kan-did-*is*-i-mum
candidula	kan-*did*-ew-la
canescens	kan-*es*-enz
cantabrica	kan-*tab*-ri-ka
capensis	kap-*en*-sis
capillipes	ka-*pil*-lee-pees, ka-*pil*-li-pees
capitata	ka-pi-*tah*-ta
cappadocicum	kap-pah-*doh*-see-kum

402

caprea	*kap*-ree-a
caragana	ka-ra-*gah*-na
cardinalis	kar-di-*nah*-lis
carica	*kar*-i-ka
carlcephalum	karl-*sef*-a-lum
carlesii	kar-*leez*-ee-I
carminea	kar-*min*-ee-a
carnea	kar-*nee*-a
carneum	kar-*nee*-um
carolina	ka-roh-*ll*-na
caroliniana	ka-roh-lin-ee-*ah*-na
carpenteria	kar-pen-ter-*ee*-a
carpinifolia	kar-pi-ni-*foh*-lee-a
carpinifolium	kar-pi-ni-*foh*-lee-um
carpinus	kar-*pI*-nus
carya	*ka*-ree-a
caryopteris	ka-ree-*op*-ter-is
cashmiriana	kash-mi-ree-*ah*-na
cassinia	kas-*in*-ee-a
castanea	kas-*tan*-ee-a
castaneifolia	kas-tan-ee-i-*foh*-lee-a
catalpa	ka-*tal*-pa
catesbaei	ka-tes-*bI*-ee
cathartica	ka-*thar*-ti-ka
ceanothus	see-a-*no*-thus
cerasifera	ser-as-i-*fee*-ra
cerasiformis	ser-as-i-*for*-mis
cerasus	ser-*a*-sus
ceratostigma	ser-at-oh-*stig*-ma
cercidifolius	ser-si-di-*foh*-lee-us
cercidiphyllum	ser-si-di-*fil*-um
cercis	*ser*-sis
cerifera	ser-*i*-fe-ra
cernuus	*ser*-new-us
cerris	*ser*-is
chaenomeles	kee-*noh*-me-leez
chamaecyparissus	kam-ee-si-*pa*-ri-sus
chapmanii	chap-*man*-ee-I
chenaultii	che-*naul*-tee-I
chermesina	cher-*me*-si-na
chilense	chi-*len*-see
chimonanthus	kI-mo-*nanth*-us
chinensis	chi-*nen*-sis
chionanthus	kI-oh-*nanth*-us
chitoseyama	chi-toh-*see*-ah-ma
choisya	*choy*-zee-a
chosan	cho-san
chrysantha	kris-*anth*-a
chrysocoma	krI-soh-*koh*-ma
chrysophylla	krI-soh-*fil*-a
chrysophyllus	krI-so-*fil*-us
chunii	*chun*-ee-I
ciliaris	sil-ee-*ar*-is
cineraria	sin-er-*ayr*-ee-a
cinerea	sin-er-*ay*-a
cinnamomifolium	sin-na-mom-i-*foh*-lee-um
cistena	sis-*teen*-a
cistus	*sis*-tus
citrinus	*sit*-rin-us
citriodora	sit-ree-oh-*dor*-a
cladrastis	klad-*ras*-tis
clandonensis	klan-don-*en*-sis
clerodendrum	kler-oh-*den*-drum
clethra	*kleth*-ra
cneorum	nee-*or*-um
coccifera	kok-*si*-fer-a
coccinea	kok-*sin*-ee-a
coccineum	kok-*sin*-ee-um
coccineus	kok-*sin*-ee-us
cochleatus	kok-lee-*ah*-tus
coggygria	kog-*ig*-ree-a
colchica	*kol*-chi-ka
colensoi	kol-en-soy
colletia	kol-*ee*-tee-a
collina	kol-*ee*-na
colorata	kol-o-*rah*-ta
coloratus	kol-o-*rah*-tus
columnare	kol-um-*nar*-ee
columnaris	kol-um-*nar*-ris
colurna	kol-*urna*-a
colutea	kol-*oo*-tee-a
commixta	kom-*iks*-ta
communis	kom-*ew*-nis
commutata	kom-ew-*tah*-ta
compacta	kom-*pak*-ta
compactus	kom-*pak*-tus
concordia	kon-*kor*-dee-a
confusa	kon-*few*-sa
congestus	kon-*jes*-tus
conglomerata	kon-glom-er-*ah*-ta
conspicua	kon-*spik*-ew-a
conspicuus	kon-*spik*-ew-us
contorta	kon-*tor*-ta
controversa	kon-troh-*ver*-sa

convexa	kon-*veks*-a
convolvulus	con-*vol*-vew-lus
cookianum	kook-ee-*ah*-num
corallina	kor-a-*lee*-na
corbariensis	kor-bar-ee-*en*-sis
cordata	kor-*dah*-ta
cordifolia	kor-di-*foh*-lee-a
cordiformis	kor-di-*form*-is
cordyline	kor-di-*llne*
coreana	kor-ee-*ah*-na
coriacea	kor-i-a-*see*-a
coriaceum	kor-i-a-*see*-um
coris	*kor*-is
cornubia	kor-*new*-bee-a
cornubiensis	kor-new-bee-*en*-sis
cornus	*kor*-nus
corokia	ko-*roh*-kee-a
coronaria	kor-on-*ayr*-ee-a
coronarius	kor-on-*ayr*-ee-us
coronilla	kor-on-*il*-a
corsica	*kor*-si-ka
cortaderia	kor-ta-*der*-ee-a
corylopsis	kor-il-*op*-sis
corylus	*kor*-il-us
corymbosa	kor-im-*boh*-sa
corymbosum	kor-im-*boh*-sum
costata	kos-*tah*-ta
cotinus	ko-*tI*-nus
cotoneaster	ko-ton-ee-*as*-ter
coulteri	kool-ter-*I*
crataegifolium	kra-tee-gi-*foh*-lee-um
crataegus	kra-*tee*-gus
crenata	kren-*ah*-ta
crinodendron	krI-noh-*den*-dron
crispa	*kris*-pa
crispus	*kris*-pus
crista-galli	*kris*-ta-*gal*-ee
cristata	kris-*tah*-ta
crotonifolia	kroh-ton-i-*foh*-lee-a
cruciata	kroo-see-*ah*-ta
crus-galli	kroos-*gal*-lee
cunninghamii	cun-ing-*ham*-ee-I
cuprea	*kup*-ree-a
cupressoides	kup-res-*oy*-deez
cydonia	sI-*doh*-nee-a
cylindricum	sil-*in*-dri-kum
cyprius	*sip*-ree-us
cytisus	sI-*tis*-us

D

daboecia	da-*boh*-she-a
dahurica	da-*hur*-i-ka
dalecarlica	dal-e-*kar*-li-ka
dalrympleana	dal-rim-plee-*ah*-na
dammeri	*dam*-mer-I
danae	*dan*-ay
daniellii	dan-ee-*el*-ee-I
daphne	*daf*-nee
daphnoides	*daf*-noy-deez
darleyensis	dar-lee-*en*-sis
darwinii	dar-*win*-ee-I
davidia	da-*vid*-ee-a
davidiana	da-vid-ee-*ah* na
davidii	da-*vid*-ee-I
daviesii	da-*vee*-see-I
dawyckii	doy-*ik*-ee-I
dealbata	dee-al-*bah*-ta
decaisnea	dee-*kayz*-nee-a
decaisneana	dee-kayz-nee-*ah*-na
decora	de-*kor*-a
decorus	de-*kor*-us
decumbens	dee-*kum*-benz
delavayi	del-a-*vay*-ee
dentata	den-*tah*-ta
dentatus	den-*tah*-tus
denudata	de-nu-*dah*-ta
desfontainea	des-fon-*tay*-nee-a
deutzia	*doyt*-zee-a, *doot*-zee-a
dichotoma	dI-koh-*toh*-ma
dicksonii	dik-*son*-ee-I
dictyophylla	dik-ti-oh-*fil*-a
diervilla	dee-*er*-vil-la
difformis	di-*for*-mis
digitata	di-ji-*tah*-ta
digyna	*di*-ji-na
dioicus	dee-*oh*-i-kus
dipelta	di-*pel*-ta
diplopappus	di-ploh-*pap*-us
disanthus	dI-*santh*-us
dissectum	di-*sek*-tum
distichum	dis-*tik*-um

distichus	dis-*tik*-us
distylium	di-*stIl*-ee-um
divaricatus	di-ver-i-*kah*-tus
diversifolia	di-verz-si-*foh*-lee-a
dombeyi	*dom*-bee-I
domestica	do-*mes*-ti-ka
dorrenbosii	dor-en-*bohz*-ee-I
dorycnium	do-*rik*-nee-um
douglasii	dug-*las*-ee-I
drimys	*drim*-is
drummondii	drum-*mon*-dee-I
dulcis	*dul*-sis
durobrivensis	*doo*-roh-bri-*ven*-sis

E

ebbingei	e-*bin*-gee-I
echtermeyer	*ek*-ter-may-er
edinensis	ed-i-*nen*-sis
elaeagnos	e-lee-*ag*-nos
elaeagnus	e-lee-*ag*-nus
elaeagrifolia	e-lee-ag-ri-*foh*-lee-a
elata	e-*lay*-ta
elegans	*el*-e-gans
elegantissima	el-e-gan-*tis*-i-ma
elegantissimum	el-e-gan-*tis*-ee-mum
eleyi	ee-*lee*-I
elliptica	e-*lip*-ti-ka
elsholtzia	el-*sholtz*-ee-a
embothrium	em-*both*-ree-um
emerus	*em*-er-us
englerana	en-gler-*ah*-na
enkianthus	en-kee-*anth*-us
erecta	e-*rek*-ta
erectum	e-*rek*-tum
erica	e-*rik*-a
erigena	e-ri-*jee*-na
ermanii	er-*man*-ee-I
erythrina	e-*rith*-reen-a
erythrocladum	e-rith-roh-*klad*-um
escallonia	es-ka-*loh*-nee-a
esserteauana	e-ser-tee-ew-*ah*-na
eucalyptus	ew-kal-*ip*-tus
euchlora	ew-*klor*-a
eucryphia	ew-*krif*-ee-a
eugenei	ew-*jen*-ee-I, ew-*jeen*-ee-I
euodia	ew-*oh*-dee-a
euonymus	ew-*on*-i-mus
europaea	ew-roh-*pee*-a
europaeus	ew-roh-*pee*-us
exburiensis	eks-bur-ee-*en*-sis
excelsior	eks-*sel*-see-or
excelsum	eks-*sel*-sum
exigua	eks-*ig*-ew-a
eximia	eks-*im*-ee-a
exochorda	eks-oh-*kor*-da

F

fabiana	fab-ee-*ah*-na
fagus	*fay*-gus
fallowiana	fal-oh-wee-*ah*-na
fargesii	far-*jee*-see-I
farreri	*fa*-rer-I
fascicularis	fas-ik-ew-*lah*-ris
fastigiata	fas-ti-jee-*ah*-ta
fastuosum	fas-tu-*oh*-sum
fatshedera	fats-*hed*-e-ra
fatsia	*fat*-see-a
feijoa	fee-*joh*-a
ferox	*fer*-oks
ferruginea	fer-u-*jin*-ee-a
fertilis	fer-*til*-is
ficus	*fI*-kus
filamentosa	fil-a-men-*toh*-sa
filicifolium	fi-li-see-*foh*-lee-um
fisheri	*fish*-er-I
flaccida	*fla*-si-da
flava	*flah*-vah
flavescens	flah-*ves*-enz
flaviramea	flah-vi-*rah*-me-a
floccosus	flo-*koh*-sus
flore	*flo*-re
floribunda	flor-i-*bun*-da
florida	*flor*-i-da
floridus	*flor*-i-dus
foetens	*foh*-tenz
folgneri	*folg*-ner-I
foliis	*foh*-lee-is
fontanesiana	fon-tan-ez-ee-*ah*-na

foremanii for-*man*-ee-I
formosa for-*moh*-sa
formosana for-moh-*sah*-na
formosus for-*moh*-sus
forrestii fo-*rest*-ee-I
forsythia for-*sIth*-ee-a, for-*sith*-ee-a
fortunei for-*tew*-nee-I
fothergilla foth-er-*gil*-a
fragifera fra-*gi*-fe-ra
fragans *fray*-granz
fragrantissima fray-gran-*tis*-i-ma
frainetto fray-*net*-oh
franchettii fran-*shet*-ee-I
franciscana fran-sis-*kah*-na
frangula *fran*-gew-la
fraseri *frayz*-er-I
fraxinifolia fraks-i-ni-*foh*-lee-a
fraxinus *fraks*-in-us
fremontii free-*mon*-tee-I
fremontodendron free-mon-toh-*den*-dron
frigidus *fri*-ji-dus
frisia *fris*-ee-a
fructu-albo *fruk*-too-*al*-boh
fructuluteo fruk-too-*loo*-tee-oh
frutex *froo*-tex
fruticans *froo*-ti-kanz
fruticosa froo-ti-*koh*-sa
fruticosum froo-ti-*koh*-sum
fuchsia *few*-sha
fudanzakura foo-dan-za-ku-ra
fuiri-koriyangi fi-ew-ri-koo-re-an-ji
fukubana foo-koo-bah-na
fulgens *ful*-jenz
fulvida *ful*-vi-da
fusca *fus*-ka

G

gangepainii gan-ya-*pan*-ee-I
gardenii gar-*den*-ee-I
garnettii gar-*net*-ee-I
garrya *ga*-ree-a
gaulnettya gawl-*net*-ee-a
gaultheria gawl-*ther*-ee-a
gaya *gay*-a
genista je-*nis*-ta
geraldiniae jer-al-*din*-ee-I
gerardiana jer-*ar*-dee-ay-na
germanica jer-*man*-i-ka
gibbsii *gibbs*-ee-I
ginme *gin*-me
ginnala ji-*na*-la
giraldiana jir-*al*-dee-*ah*-na
giraldii jir-*al*-dee-I
glabosa gla-*boh*-sa
glabra *gla*-bra
glabrata glab-*rah*-ta
glabrescens gla-*bres*-enz
glandulosa glan-dew-*loh*-sa
glauca *glaw*-ka
glaucophylla glaw-koh-*fil*-a
gleditsia gled-*it*-see-a
globosa gloh-*boh*-sa
globosum gloh-*boh*-sum
globulus *glob*-ew-lus
gloriosa glo-*ree*-oh-sa
glutinosa gloo-ti-*noh*-sa
gordonianum gor-don-ee-*ah*-num
gracilis *gra*-sil-is
gracilistyla gra-sil-is-*til*-a
grandiflora gran-di-*flor*-a
grandiflorus gran-di-*flor*-us
grandifolia gran-di-*foh*-lee-a
graveolens gra-*vee*-oh-lenz
griffithii gri-*fith*-ee-I
grignonensis grig-non-*en*-sis
griselinia gri-se-*lin*-ee-a
griseum *gris*-ee-um
grosseri *gro*-ser-I
gunniana gun-ee-*ay*-na
gunnii *gun*-ee-I
gymnocladus jim-no-*kla*-dus

H

haastii *has*-tee-I
hakuro hak-*ew*-ro
halesia hal-*ree*-zee-a
halimiocistus ha-lim-ee-oh-*sis*-tus
halimium ha-*lim*-ee-um

halimodendron ha-lim-oh-*den*-dron
halimus *ha*-lim-us
hamamelis *ham*-a-mel-is
handsworthiensis hands-wor-thee-*en*-sis
hastata has-*ta*-ta
hebe *hee*-bee
hedera *hed*-e-ra
hedysarum hed-ee-*sar*-um
helianthemum hel-ee-*anth*-e-mum
helichrysum hel-i-*krIs*-um
helix *hel*-iks
helleri *hel*-er-I
helvetica hel-*vet*-i-ka
hemsleyana hems-le-*ah*-na
henryana hen-ree-*ah*-na
henryanus hen-ree-*ah*-nus
henryi *hen*-ree-I
heptalobum hep-ta-*loh*-bum
hersii *her*-see-I
heterantha he-te-*ranth*-a
heterophylla he-te-roh-*fil*-a
heterophyllus he-te-roh-*fil*-us
hibernica hI-*bern*-i-ka
hibiscus hI-*bis*-kus
highdownensis hI-down-*en*-sis
hillieri *hil*-ee-er-I
hippocastanum hip-oh-kas-*tah*-num
hippophae *hip*-oh-fay
hirsutum hir-*soo*-tum
hisakura hI-sa-*kew*-ra
hispanica hi-*span*-i-ka
hispida *his*-pi-da
hodginsii hod-*jinz*-ee-i
hoheria hoh-*her*-ee-a
hokusai hok-*ew*-sI
hollandia ho-*lan*-dee-a
hollandica ho-*lan*-di-ka
holocarpa ho-lo-*kar*-pa
holodiscus ho-loh-*dis*-kus
hookeranum hook-er-*ah*-num
hookeriana hook-er-ree-*ah*-na
hookerianum hook-e-ree-*ah*-num
horizontalis hor-i-zon-*tah*-lis
hortensia hor-*ten*-see-a
hortensis hor-*ten*-sis
hulkeana hul-kee-*ah*-na
humile *hew*-mi-le
himilis *hew*-mi-lis
hupehense hew-pe-*hen*-se
hupehensis hew-pe-*hen*-sis
hyacinthiflora hI-a-sinth-i-*flor*-a
hybrida *hI*-bri-da
hybrididus *hI*-bri-dus
hydrangea hI-*drayn*-jee-a
hypericum hI-*per*-i-kum
hypoglossum hI-poh-*glos*-um
hypoleuca hI-poh-*loo*-ka

I

ichiyo i-*chee*-oh
icterinus ik-ter-*ee*-nus
idesia I-*de*-zee-a
ilex *I*-leks
ilicifolia i-lis-i-*foh*-lee-a
ilicifolius i-lis-i-*foh*-lee-us
illinoinensis il-li-noy-*nen*-sis
imbricata im-bri-*kah*-ta
imose im-*o*-say
impeditum im-*pe*-di-tum
imperalis im-peer-ee-*ah*-lis
impressus im-*pres*-us
inabashidare in-a-bash-i-dah-re
incana in-*kah*-na
incisa in-*sIz*-a
indica *in*-di-ka
indigofera in-di-*gof*-e-ra
inodorum in-oh-*dor*-um
inermis i-*ner*-mis
ingramii *in*-gram-ee-I
ingwersenii ing-wer-*sen*-ee-I
integra in-*teg*-ra
intermedia in-ter-*mee*-dee-a
involucrata in-vol-ew-*kra*-ta
irrorata ir-roh-*rah*-ta
irwinii ir-*win*-ee-I
italica i-*tal*-i-ka
itea *I*-tee-a
ivensii I-*vens*-ee-I
iveyi *I*-vee-I

J

jacquemontii jahk-*mon*-tee-I
jamesiana jaym-see-*ah*-na
japonica ja-*pon*-i-ka
japonicum ja-*pon*-i-kum
japonicus ja-*pon*-i-kus
jasminum jas-*min*-um
jaspidea jas-*pI*-da
josiflexa jo-si-*fleks*-a
josikaea jo-si-*kee*-a
juddii *jud*-ee-I
juglans *jug*-lanz
julianae ju-lee-*ah*-nee
julibrissin joo-lee-*bris*-sin
junceum *jun*-see-um

K

kalmia *kal*-mee-a
kalmiiflora kal-mee-i-*flor*-a
kalopanax kal-oh-*pan*-aks
kanzan *kan*-zan
kelleriis *kel*-er-is
kelseyi *kel*-see-I
kerria *ker*-ee-a
kewensis kew-*en*-sis
kiautschovicus koit-shoi-vik-us
kiku-shidare sakura kI-koo-shi-dah-ra sak-ew-ra
kluis *kloo*-is
knightii *nI*-tee-I
kobus *koh*-bus
koelreuteria kol-roo-*teer*-ee-a
kolkwitzia kolk-*witz*-ee-a
koreana kor-ee-*ah*-na
kousa *koo*-sa
kouytchense koo-eet-*chen*-se
kurome koo-*roh*-me

L

laburnifolius la-bur-ni-*foh*-lee-us
laburnocytisus la-bur-noh-sI-*tis*-us
laburnum la-*bur*-num
laciniata la-sin-ee-*ah*-ta
laciniatum la-sin-ee-*ah*-tum
lacteus *lak*-tee-us
ladanifer lad-an-i-*fer*
laevis *lay*-vis
lalandei lay-lan-*dee*-I
lamarckii lam-*ark*-ee-I
lanata la-*nah*-ta
lanceolata lan-see-oh-*lah*-ta
lanceolatum lan-see-oh-*lah*-tum
langleyensis lang-lee-*en*-sis
lantana lan-*tah*-na
lapponum lap-*oh*-num
lasianthum las-ee-*anth*-um
lasiocarpa las-ee-oh-*kar*-pa
latifolia la-ti-*foh*-lee-a
laureola law-*ree*-oh-la
lauriflorum law-ri-*flor*-um
laurifolius law-ri-*foh*-lee-us
laurocerasus law-roh-sir-*ah*-sus
laurus *law*-rus
lavallei la-*val*-ee-I
lavandula la-*van*-dew-la
lavandulaceus la-van-dew-*lay*-see-us
lavatera la-va-*tee*-ra
lawsoniana law-*soh*-nee-ah-na
laxifolius laks-i-*foh*-lee-us
ledebourii led-e-*boor*-ee-I
lemoinei le-*moy*-nee-I
lennei *len*-ee-I
lenta *len*-ta
lentago len-*tah*-goh
leopoldii lee-oh-*pold*-ee-I
leptospermum lep-to-*sperm*-um
lespedeza les-pe-*dez*-a
leucocarpa loo-koh-*kar*-pa
leucostachys loo-koh-*sta*-kis
leucothoe loo-ko-*thoh*-ee
leycesteria lI-ces-*teer*-ee-a
liempde *leemp*-de
ligustrinum lig-us-*tree*-num
ligustrum lig-*us*-trum
liliiflora lil-ee-i-*flor*-a
linearifolia lin-ee-er-i-*foh*-lee-a

linearilobum	lin-ee-er-i-*loh*-bum	moerloosii	more-*loo*-see-I	ostrya	*os*-tree-a
lippia	*lip*-ee-a	mollicomata	mol-lee-ko-*mah*-ta	ottawensis	ot-a-*wen*-sis
liquidambar	lik-wid-*am*-bar	mollis	*mol*-lis	ovalifolium	oh-va-li-*foh*-lee-um
liriodendron	lir-ee-oh-*den*-dron	monbeigii	mon-*bay*-gee-I	ovalis	oh-*vah*-lis
littoralis	li-to-*rah*-lis	mongolica	mon-*gol*-i-ka	ovata	oh-*vah*-ta
lizei	*liz*-ay	monogyna	mon-oh-*jI*-na	ovatus	oh-*vah*-tus
lobbianus	lob-ee-*ah*-nus	monophylla	mon-oh-*fil*-a	oxycarpa	oks-ee-*kar*-pa
loebneri	*lurb*-ner-I	monroi	mon-*roh*-ee	oxycoccos	oks-ee-*kok*-kos
lologensis	lo-lo-*gen*-sis	monticola	mon-tik-*oh*-la	oxydendrum	oks-ee-*den*-drum
lomariifolia	lom-ayr-ee-i-*foh*-lee-a	morus	*maw*-rus	ozothamnus	o-zoh-*tham*-nus
lomatia	lo-*mah*-tee-a	moserianum	moh-ser-ree-*ay*-num		
longifolia	long-i-*foh*-lee-a	moseri	*mo*-ser-I		
longifolium	long-i-*foh*-lee-um	moupinense	moo-pin-*en*-see		
longiracemosa	long-i-ras-ee-*moh*-sa	moupinensis	moo-pin-*en*-sis		
lonicera	lon-i-*ser*-a	mucronata	mew-kro-*nah*-ta		
lorbergii	lor-*berg*-ee-I	multiflora	mul-ti-*flor*-a		
lucida	*loo*-sid-a	multiflorus	mul-ti-*flor*-us		
lucidum	*loo*-sid-um	multijugum	mul-ti-*joo*-gum		
ludlowii	lud-*loh*-ee-I	multiplex	*mul*-ti-pleks		
lunulatum	loon-ew-*lah*-tum	mume	*mu*-mee		
lusitanica	loo-si-*tan*-i-ka	murieliae	*mew*-ree-el-ee-I		
lusitanicus	loo-si-*tan*-i-kus	myrica	*mir*-i-ka		
lutea	*loo*-tee-a	myrsinites	mir-sin-*I*-tees		
luteovariegata	loo-tee-oh-ver-ee-e-*gay*-ta,	myrtifolia	mir-tee-*foh*-lee-ah		
	loo-tee-oh-var-ee-e-*gah*-ta	myrtillus	mir-*til*-us		
lutescens	loo-*tes*-enz	myrtus	*mir*-tus		
luteus	*loo*-tee-us	mytifolius	mi-ti-*foh*-li-us		
luxurians	luk-*zhewr*-ee-ans				
lyallii	*lI*-al-ee-I				
lycium	*lis*-ee-um				
lydia	*li*-dee-a				

P

pachysandra	pa-kis-*an*-dra
padus	*pay*-dus
paeonia	*pee*-oh-nee-a
pagei	*pay*-jee-I
palibin	*pal*-i-bin
palibiniana	pal-i-bin-ee-*ah*-na
paliurus	pa-lee-*ew*-rus
pallida	*pa*-le-da
palmatum	pal-*may*-tum
palustris	pa-*lus*-tris
paniculata	pan-ik-ew-*lah*-ta
panlanensis	pan-lan-*en*-sis
pannosus	pa-*noh*-sus
papyrifera	pa-pi-*ri*-fe-ra
parkeri	*park*-er-I
parkjuweel	*park*-joo-wel
parrotia	pa-*ro*-tee-a
parrotiopsis	pa-ro-tee-*op*-sis
parviflora	par-vi-*flor*-a
parvifolia	par-vi-*foh*-lee-a
patagua	pat-ag-*ew*-a
patens	*pay*-tenz
patulum	*pat*-ew-lum
pauciflora	*paw*-si-flor-a
paulownia	paw-*loh*-nee-a
pavia	*pah*-vee-a
pemakoense	pe-ma-koh-*en*-see
pendula	*pen*-dew-la
pendiflorum	pen-dew-li-*flor*-um
pendulifolia	pen-dew-li-*foh*-lee-a
pendulum	*pen*-dew-lum
pendulus	*pen*-dew-lus
pennsylvanica	pen-sil-*van*-i-ka
pensylvanicum	pen-sil-*van*-i-kum
pentandra	pen-*tan*-dra
pentaphyllus	pen-ta-*fil*-us
pernettya	per-*net*-ee-a
pernyi	*per*-nee-I
perovskia	pe-*rof*-skee-a
perriniana	pe-rin-ee-*ah*-na
persica	*per*-si-ka
persistens	per-*sis*-tenz
perulatus	pe-ru-*lah*-tus
petiolaris	pet-ee-oh-*lah*-ris
petraea	pe-*tree*-a
phaenopyrum	fee-noh-*pI*-rum
pellodendron	fel-oh-*den*-dron
phellos	*fel*-os
phellomanus	fel-oh-*man*-us
philadelphus	fil-a-*del*-fus
philippii	fil-*ip*-ee-I
phillyrea	fil-i-*ree*-*u*
phlogopappa	flog-oh-*pa*-pa
phlomis	*floh*-mis
phormium	*for*-mee-um
photinia	foh-*tin*-ee-a
phygelius	fI-*jee*-lee-us
phylicifolia	fil-i-si-*foh*-lee-a
physocarpus	fi-soh-*kar*-pus
pia	*pee*-a
picta	*pik*-ta
picturata	pik-tewr-*ah*-ta
pictus	*pik*-tus
pieris	pee-*e*-ris
pileata	pi-lee-*ah*-ta
pilosa	pi-*loh*-sa
pinguifolia	ping-gwi-*foh*-lee-a
pinnata	pin-*nah*-ta
piptanthus	pip-*tanth*-us
pissardii	pi-*sard*-ee-I
pittosporum	pit-toh-*spor*-um
plagianthus	play-jee-*an*-thus
planipes	*plan*-i-pees
platanoides	pla-ta-*noy*-deez
platanus	*plat*-an-us
platyphylla	pla-ti-*fil*-a
platyphyllos	pla-ti-*fil*-os
plena	*plee*-na
pleno	*plee*-no

M

maackii	*mak*-ee-I
maanelys	*man*-e-leez
macrantha	ma-*kranth*-a
macrocephalum	mak-roh-*sef*-a-lum
macrodonta	mak-roh-*don*-ta
macrophylla	mak-roh-*fil*-a
maculata	ma-kew-*lah*-ta
maculatus	ma-kew-*la*-tus
magdeburgensis	mag-de-bur-*gen*-sis
magellanica	ma-je-*lan*-i-ka
magnifica	mag-*ni*-fi-ka
magnificum	mag-*ni*-fi-kum
magnolia	mag-*nol*-ee-a
magnoliifolia	mag-nol-ee-i-*foh*-lee-a
mahoberberis	ma-hoh-*ber*-ber-is
mahonia	ma-*hon*-ee-a
majestica	ma-*jes*-ti-ka
major	*may*-jor
malacodendron	ma-la-koh-*den*-dron
malus	*may*-lus
mandshurica	man-*shoo*-ri-ca
marginata	mar-ji-*nah*-ta
mariesii	ma-*reez*-ee-I
mas	mas
mastacanthus	mas-ta-*kanth*-us
mathotiana	ma-*thoh*-tee-ah-na
matsudana	mat-soo-*dah*-na
matsumurana	mat-soo-mur-*ah*-na
maweana	ma-we-*ah*-na
maxima	*maks*-ee-ma
maximowicziana	maks-i-moh-vits-ee-*ah*-na
maximowiczii	maks-i-moh-*vits*-ee-I
medemii	me-*dem*-ee-I
media	*mee*-dee-a
mediopicta	med-i-oh-*pik*-ta
mediterranea	med-i-ter-*ay*-nee-a
medwediewii	med-we-dee-*ev*-ee-I
megapotamicum	meg-a-pot-*am*-i-kum
melanocarpa	me-la-noh-*kar*-pa
melanostachys	me-la-noh-*sta*-kis
melianthus	mel-ee-*anth*-us
menziesia	men-zeez-ee-a
menziesii	men-zeez-ee-I
meserveae	me-*ser*-vee-I
mespilus	*mes*-pil-us
mexicanum	meks-i-*kah*-num
meyeri	*may*-er-I
mezereum	me-*ze*-ree-um
micranthum	mi-*kran*-thum
microglossa	mI-kroh-*glos*-a
mecrophylla	mI-kroh-*fil*-a
microphyllus	mI-kroh-*fil*-us
middendorffiana	mid-en-dor-fee-*ah*-na
milleri	*mil*-er-I
milliganii	mil-li-*gan*-ee-I
minor	*mI*-nor
mitchellii	*mit*-chel-ee-I
moreheimii	more-*hem*-ee-I

N

nagasaki	na-ga-*sa*-ki
nana	*nah*-na
nandina	nan-*dee*-na
napaulensis	*na*-paw-len-sis
narcissiflorum	nar-sis-i-*flor*-um
neapolitana	nee-a-pol-i-*tah*-na
negundo	ne-*gun*-doh
neillia	*neel*-ee-a
nepalensis	ne-pa-*len*-sis
nerium	*nay*-ree-um
nichollsii	ni-kolz-ee-I
nigra	*nI*-gra
nigricans	*nI*-gri-kanz
nigrum	*nI*-grum
nikoense	ni-koh-*en*-see
niphophila	nI-fo-*fil*-a
nipponica	ni-*pon*-i-ka
nishikii	ni-shi-*kee*-I
nitida	*ni*-ti-da
nivalis	ni-*val*-is
nobilis	no-*bil*-is
nothofagus	no-tho-*fay*-gus
nudiflorum	new-di-*flor*-um
nummulariifoilia	num-ew-layr-i-*foh*-lee-a
nummularium	num-ew-*layr*-ee-um
nuttallii	nut-*al*-ee-I
nuttallis	nut-*al*-is
nymansensis	nI-manz-*en*-sis
nyssa	*nis*-sa

O

o-moi-no-wac	oh-*moy*-noh-wak
obassia	oh-*ba*-see-a
obliqua	ob-*lee*-kwa
oblonga	ob-*lon*-ga
obovata	ob-oh-*vah*-ta
obtusum	ob-*tew*-sum
occidentalis	osk-i-den-*tal*-is
ochracea	ok-ra-*see*-a
ocymoides	ok-*im*-oy-deez
odora	oh-*dor*-a
odoratum	oh-do-*rah*-tum
officinalis	o-fik-in-*ah*-lis
ojochin	*oh*-joh-chin
okame	oh-*kah*-me
olbia	*ol*-bee-a
oleander	o-lee-*an*-der
olearia	o-*ler*-ee-a
opalus	*op*-ah-lus
opulifolius	op-ew-lee-*foh*-lee-us
opulus	*op*-ew-lus
orbiculatis	or-bik-ew-*lah*-tis
orientalis	or-ee-en-*tah*-lis
ornatum	or-*nah*-tum
ornus	*or*-nus
osa kazuki	oh-sa *ka*-zoo-ki
osmanthus	os-*manth*-us
osmaria	os-*may*-ree-a
osmaronia	os-ma-*roh*-nee-a

plenus	*plee*-nus
plicatum	pli-*kay*-tum
plumosa	ploo-*moh*-sa
pohuashanensis	poh-ew-ash-a-*nen*-sis
polifolia	pol-i-*foh*-lee-a
pollardii	po-*lard*-ee-I
polycarpa	po-lee-*kar*-pa
pontica	*pon*-ti-ka
ponticum	*pon*-ti-kum
populifolia	pop-ew-li-*foh*-lee-a
populifolius	pop-ew-li-*foh*-lee-us
populnea	po-*pul*-nee-a
populus	*pop*-ew-lus
porcina	por-*sI*-na
potaninii	pot-an-*in*-ee-I
potentilla	poh-ten-*til*-a
poteriifolia	pot-er-ee-*foh*-lee-a
praecox	*pray*-coks, *prI*-coks
praegerae	*pray*-ge-rI
pragense	pra-*gen*-see
prattii	pra-*tee*-I
prestoniae	pres-*ton*-ee-I
preziosa	pre-zee-*oh*-sa
procera	pro-*ser*-a
procumbens	proh-*kum*-benz
prolificum	proh-*li*-fi-kum
prostrata	pro-*strah*-ta
prostratus	pro-*strah*-tus
prunifolia	proo-ni-*foh*-lee-a
prunus	*proo*-nus
pseudoacacia	soo-doh-a-*kay*-see-a
pseudocamellia	soo-doh-ka-*mee*-lee-a
pseudodictamnus	soo-doh-dik-*tam*-nus
pseudoplatanus	soo-doh-*plat*-an-us
pseudosasa	soo-doh-*sas*-a
ptelea	*tee*-lee-a
pterocarya	ter-oh-*kar*-ee-a
pubescens	pew-*bes*-senz
pulchellus	pul-*chel*-us
pulchra	*pul*-kra
pulverulenta	pul-ver-ew-*len*-ta
pulverulentus	pul-ver-ew-*len*-tus
pumila	pew-*mi*-la
pungens	*pun*-jenz
puniceus	pew-*ni*-see-us
purpurea	pur-pewr-*ee*-a
purpureis	pur-pewr-*ee*-is
purpureum	pur-pewr-*ee*-um
purpusii	pur-*pew*-see-I
pygmaea	pig-*mee*-a
pyracantha	pI-ra-*kanth*-a
pyramidale	pi-ra-mi-*dah*-lee
pyramidalis	pr-ra-mi-*dah*-lis
pyrus	*pI*-rus

Q

quercifolia	kwerk-see-*foh*-lee-a
quecus	*kwer*-kus
quihoui	kee-*hoo*-ee
quinquepeta	kwin-kwe-*pet*-a

R

racemosa	ras-em-*oh*-sa
racemosum	ras-em-*oh*-sum
radicans	*ra*-di-kanz
rakaiensis	rak-I-*en*-sis
ramosissima	rah-moh-*sis*-i-ma
recurvifolia	re-kur-vee-*foh*-lee-a
reducta	re-*duk*-ta
reflexa	re-*fleks*-a
regia	*ree*-ja
rehderiana	re-der-ee-*ah*-na
rendatieri	ren-da-tee-*er*-ee
repens	*re*-penz
reticulatum	re-ti-kew-*lah*-tum
retusa	re-*tew*-sa
retusus	re-tew-sus
revolutum	rev-oh-*lu*-tum
rhamnoides	ram-*noy*-deez
rhamnus	*ram*-nus
rhexii	*reks*-ee-I
rhodanthe	roh-*dan*-thee
rhododendron	roh-doh-*den*-dron
rhodotypos	roh-doh-*tI*-pos
rhus	roos
rhtidophylloides	rI-ti-doh-fil-*oy*-deez
rhytidophyllum	rI-ti-doh-*fil*-um
ribes	*rI*-beez

ribifolia	rI-be-*foh*-lee-a
riccartonii	ri-kar-*ton*-ee-I
richardii	ri-*char*-dee-I
ricinifolius	ri-sin-i-*foh*-lee-us
rigidus	*ri*-ji-dus
riversica	ri-*verz*-i-ka
riversii	ri-*verz*-ee-I
rivularis	riv-ew-*lah*-ris
robinia	ro-*bin*-ee-a
robur	*roh*-bur
robusta	roh-*bus*-ta
rogersiana	ro-jer-zee-*ah*-na
rogersii	ro-*jer*-zee-I
rohanii	roh-*hahn*-ee-I
rollissonii	rol-lis-*on*-ee-I
romneya	rom-*nay*-a
rosalba	roh-*zal*-ba
rosea	roh-*zee*-a
rosemarginata	roh-zee-oh-mar-ji-*nah*-ta
rosmarinifolius	ros-ma-rin-i-*foh*-lee-us
rosmarinus	ros-*ma*-rin-us, ros-ma-*rI*-nus
rothschildianus	wroths-chil-dee-*ah*-nus
rotundifolia	roh-tun-di-*foh*-lee-a
rotundifolium	roh-tun-di-*foh*-lee-um
rotundifolius	roh-tun-di-*foh*-lee-us
rubella	roo-*bel*-a
rubra	*rou*-bra
rubrifolium	rou-bri-*foh*-lee-um
rubrifolius	rou-bri-*foh*-lee-us
rubrostilla	rou-broh-*sti*-la
rubrum	*rou*-brum
rufinerve	roof-i-*ner*-vee
ruscifolia	rus-ki-*foh*-lee-a
ruscus	*rus*-kus
russellianus	ru-sel-ee-*ah*-nus
rustica	*rus*-ti-ka
ruta	*roo*-ta

S

saccharum	sak-*ar*-rum
sachalinensis	sak-a-lin-*en*-sis
sahucii	*sook*-ee-I
salicifolia	sal-is-i-*foh*-lee-a
salicifolius	sal-is-i-*foh*-lee-us
salix	*say*-liks
saluenense	sal-ew-en-en-see
salvia	*sal*-vee-a
salvifolius	sal-vi-*foh*-lee-us
sambucus	sam-*bew*-kus
sanguineum	san-*gwin*-ee-um
santolina	san-toh-*leen*-a, san-toh-*lI*-na
sarcococca	sar-ko-*ko*-ka
sargentiana	sar-jen-tee-*ah*-na·
sargentii	sar-*jen*-tee-I
sassafras	*sas*-sa-frass
sativa	sat-*Iv*-a
saugeana	saw-jee-*ah*-na
scabra	*ska*-bra
scalaris	skal-*ar*-is
scandens	*skan*-denz
scheideckeri	*shI*-dek-er-I
schipkaensis	ship-ka-*en*-sis
schumannii	*shoo*-mahn-ee-I
schwedleri	*shwed*-ler-I
scilloniensis	si-lon-ee-*en*-sis
scoparium	skoh-*pah*-ree-um
scoparius	skoh-*pah*-ree-us
sekka	*sek*-ka
selloana	sel-oh-*ah*-na
sellowiana	se-*loh*-ee-ah
semperflorens	sem-per-*flor*-enz
sempervirens	sem-per-*vir*-enz
senecio	se-*nee*-see-oh, se-*ne*-kee-oh
senkaki	sen-*ka*-kee
septentrionalis	sep-ten-tri-oh-*nal*-is
sericea	ser-*ee*-see-a, ser-i-*see*-a
serotina	se-ro-*tee*-na
serotinum	se-ro-tee-num
serpyllifolium	ser-pil-i-*foh*-lee-um
serrata	se-*rah*-ta
serrula	se-*roo*-la
serrulata	se-roo-*lah*-ta
sessilifolia	ses-i-li-*foh*-lee-a
setchuanensis	se-chu-wahn-*en*-sis
setsuka	set-*soo*-ka
sexstylosa	seks-sti-*loh*-sa
shidare yoshino	*shI*-dah-re *yosh*-in-oh
shimidsu sakura	shi-*mid*-soo sa-ku-ra
shinonome	shI-no-*no*-me
shirobana	shI-roh-*bah*-na
shirofugen	*shi*-roh-fu-gen

shirotae	*shI*-roh-tay
shosar	*shoh*-sar
sieboldii	see-*bold*-ee-I
sieboldianus	see-*bold*-ee-ah-nus
silberschmelze	*sil*-ber-shmel-ze
siliquastrum	si-li-*kwas*-trum
simonii	sI-*mon*-ee-I
sinense	si-*nen*-se
sinensis	si-*nen*-sis
skanbergii	skan-*ber*-jee-I
skimmia	*skim*-ee-a
smithiana	smi-thee-*ah*-na
sophora	*so*-fo-ra
sorbaria	sor-*bay*-ree-a
sorbifolia	sor-bi-*foh*-lee-a
sorbifolium	sor-bi-*foh*-lee-um
sorbus	*sor*-bus
soulangiana	soo-lan-jee-*ah*-na
spartium	*spar*-tee-um
speciosa	spee-she-*oh*-sa, *spek*-i-oh-sa
speciosus	spee-she-*oh*-sus, *spek*-i-oh-sus
spectabilis	spek-*ta*-bi-lis
spica	*spi*-ka
spicata	spi-*kah*-ta
spina-christi	*spI*-na-kris-tee
spinosa	spin-*oh*-sa
spiraea	spI-*ree*-a
splendens	*splen*-denz
splendidum	*splen*-di-dum
stachyrus	sta-chee-*ew*-rus
standishii	stan-*dish*-ee-I
staphylea	sta-fil-*ee*-a
stauntonii	stawn-*ton*-ee-I
stellata	ste-*lah*-ta
stellulata	ste-lew-*lah*-ta
stenophylla	sten-oh-*fil*-a
stenoptera	sten-*op*-te-ra
stephanandra	ste-fan-an-dra
sternianus	ster-nee-*ah*-nus
stoechas	*stoh*-chas, *stoh*-kas
stolonifera	stoh-lon-*if*-er-a
stranvaesia	stran-*veez*-ee-a
stricta	*strik*-ta
stuartia	stew-*art*-ee-a
styraciflua	stI-ra-chi-*flu*-a
styrax	*stI*-raks
subalpina	sub-al-*pI*-na
suber	*soo*-ber
subhirtella	sub-hir-*tel*-la
submollis	sub-*mol*-lis
subrepanda	sub-re-*pan*-da
suffruticosa	suf-roo-ti-*koh*-sa
sulphurea	sul-*few*-ree-a
suntense	sun-*ten*-se
superba	soo-*per*-ba
superbum	soo-*per*-bum
suspensa	sus-*pen*-sa
sweginzowii	sway-gin-*zoh*-wee-I
sylvatica	sil-*vat*-i-ka
sylvestris	sil-*ves*-tris
symphoricarpos	sim-for-i-*kar*-pos
symplocos	sim-*ploh*-kos
syriacus	si-*ree*-ah-kus
syringa	si-*ring*-ga
syringantha	si-ring-*ganth*-a

T

tai haku	*tI*-hak-u
taiwanensis	tI-wahn-*en*-sis
tamarix	*tam*-a-riks
tanakae	*tan*-a-kay
tangutica	tan-*gew*-ti-ka
taoyama zakura	ta-oh-*ya*-ma za-*ku*-ra
tardiva	*tar*-di-va
tarentina	tar-en-*tee*-na
tatarica	ta-*ta*-ri-ka
temolaica	tem-oh-*ll*-ka
tenax	*ten*-aks
tenella	ten-*el*-a
tenera	*ten*-e-ra
tenuifolium	ten-ew-i-*foh*-lee-um
terminalis	ter-mi-*nay*-lis
ternata	ter-*nah*-ta
tetralix	te-*tray*-liks
tetrandra	te-*tran*-dra
tetraptera	te-*trap*-te-ra
teucrium	*too*-kree-um
thibetica	ti-*bet*-i-ka
thunbergii	thun-*ber*-jee-I

thuringiaca	thu-ring-ee-*ah*-ka
thymifolius	tI-mi-*foh*-lee-us
thyrsiflorus	thir-si-*flor*-us
tilia	*til*-ee-a
tinctoria	tink-*tor*-ree-a
tinus	*tI*-nus
toba	*toh*-ba
tobira	*toh*-bi-ra
tomentosa	toh-men-*toh*-sa
toringoides	tor-in-*goy*-deez
torminalis	tor-min-*ah*-lis
tortuosa	tor-tew-*oh*-sa
tosaensis	tos-a-*en*-sis
trachycarpus	tra-kee-*kar*-pus
transitoria	tran-si-*tor*-ee-a
traversii	tra-*ver*-see-I
tremula	*trem*-ew-la
tresederi	tres-e-*der*-I
triacanthos	tri-a-*kanth*-os
trichocarpa	tri-koh-*kar*-pa
trichotomum	trI-koh-*toh*-mum
tricuspidaria	trI-*kus*-pi-da-ree-a
triflora	trI-*flor*-a
triflorum	trI-*flor*-um
trifoliata	trI-foh-lee-*ah*-ta
triloba	trI-*loh*-ba
triphylla	trI-*fil*-a
tristis	*tris*-tis
triumphans	trI-*um*-fanz
tschonoskii	chon-*os*-kee-I
tsugaluensis	soo-gal-ew-*en*-sis
tulipifera	tew-lip-*i*-fe-ra
turneri	*tur*-ner-I
tutsam	*tut*-sam
typhina	tI-*fee*-na

U

ukon	*ew*-kon
ulex	*ew*-leks
uliginosum	ew-lij-in-*oh*-sum
ulmus	*ul*-mus
umbellata	um-bel-*ah*-ta
umbellatum	um-bel-*ah*-tum
umbilicata	um-bi-li-*kah*-ta
umineko	ew-min-ek-oh
undulata	un-dew-*lah*-ta
unedo	ew-*nee*-doh
utile	*ew*-til-ee
utilis	*ew*-til-is
uva-ursi	oo-va-*ur*-see

V

vaccinium	vak-*sin*-ee-um
vagans	*vay*-ganz
vanhouttei	van-*hoot*-ee-I
variegata	ver-ee-e-*gay*-ta, var-ee-e-*gah*-ta
variegatum	ver-ee-e-*gay*-tum, var-ee-e-*gah*-tum
variegatus	ver-ee-e-*gay*-tus, var-ee-e-*gah*-tus
vegetus	*vej*-e-tus
veitchiana	vee-chi-*ah*-na
veitchianus	vee-chi-*ah*-nus
veitchii	*vee*-chee-I
velutina	vel-ew-*tee*-na
ventricosa	ven-tri-*koh*-sa
verrucosa	ver-roo-*koh*-sa
verruculosa	ve-roo-kew-*loh*-sa
versicolor	vers-e-*kol*-or
vestita	*ves*-ti-ta
viburnifolium	vI-bur-ni-*foh*-lee-um
viburnum	vI-*bur*-num
vicaryi	vI-*kar*-ee-I
villosa	vi-*loh*-sa
vilmoriniana	vil-mor-in-ee-*ah*-na
vinca	*vin*-ka
violacea	vI-oh-*lay*-see-a
violaceum	vI-oh-*lay*-see-um
virens	*vi*-renz
virgata	vir-*gah*-ta
vigatum	vir-*gah*-tum
virginiana	vir-jin-ee-*ah*-na
virginica	vir-*jin*-i-ka
virginicus	vir-*jin*-i-kus
viridis	vi-*ri*-dis
viridissima	vir-i-*dis*-i-ma
virisdistriata	vi-ri-di-stree-*ah*-ta

vitellina	vi-te-*leen*-a
vitex	*vI*-teks
vitifolium	vI-ti-*foh*-lee-um
vitilifolium	vI-ti-li-*foh*-lee-um
vitus-idaea	vI-tis-I-*dee*-a
vivellii	vI-vel-ee-I
vossii	*vos*-see-I
vulgarc	vul-*gah*-ree
vulgaris	vul-*gah*-ris

W

walitch	*wol*-itch
watanbe	wat-an-*ah*-bee
watereri	waw-te-*rer*-ee
watsonii	wat-*son*-ee-I
wehrhahnii	vayr-*hahn*-ee-I
weigela	wI-*jee*-la
westersteide	*wes*-ter-stead
weyeriana	way-er-ee-*ah*-na
whipplei	*wip*-lee-ee
williamsii	wil-ee-*am*-see-I
willmottiae	wil-*mot*-ee-I
willmottianum	wil-mot-ee-*ah*-num
wilsoniae	wil-*son*-ee-I
wilsonii	wil-*son*-ee-I
winteri	*win*-ter-I
wintonensis	win-ton-*en*-sis
wisleyensis	wiz-lee-*en*-sis
wisteria	wis-*ter*-ee-a
worleei	*wor*-lee-I
wredei	*ree*-dee-I

X

xanthocarpa	zanth-oh-*kar*-pa
xanthocarpum	zanth-oh-*kar*-pum
xanthoceras	zanth-o-*ser*-ras

Y

yakushimanum	ya-koo-shee-*mah*-num
yedoensis	yed-oh-*en*-sis
youngii	*youn*-gee-I
ysella	I-*sel*-a
yucca	*yuk*-ah
yunnan	*yun*-an
yunnanensis	yun-an-*en*-sis

Z

zabeliana	za-bel-ee-*ah*-na
zeelandia	zee-*lan*-dee-a
zelkova	zel-*koh*-va
zenobia	zen-*oh*-bee-a
zlatia	*zlat*-ee-a
zuccariniana	zoo-kar-in-ee-*ah*-na

INDEX OF LATIN AND COMMON NAMES

This is an index of all plant names listed as main entries or under Varieties/Forms of Interest in the Dictionaries. Varieties that can be grown as either trees and wall shrubs, or shrubs and climbing plants are given separate entries in both Dictionaries. Italic text signifies a Latin name; common names are in plain text.

C

M

Q

426

S